INORGANIC CHEMISTRY

The Elements

Element	Symbol	Atomic number	Molar mass/g mol^{-1}
Actinium	Ac	89	227.03
Aluminum	Al	13	26.98
Americium	Am	95	241.06
Antimony	Sb	51	121.75
Argon	Ar	18	39.95
Arsenic	As	33	74.92
Astatine	At	85	210
Barium	Ba	56	137.34
Berkelium	Bk	97	249.08
Beryllium	Be	4	9.01
Bismuth	Bi	83	208.98
Boron	B	5	10.81
Bromine	Br	35	79.91
Cadmium	Cd	48	112.40
Calcium	Ca	20	40.08
Californium	Cf	98	251.08
Carbon	C	6	12.01
Cerium	Ce	58	140.12
Cesium	Cs	55	132.91
Chlorine	Cl	17	35.45
Chromium	Cr	24	52.01
Cobalt	Co	27	58.93
Copper	Cu	29	63.54
Curium	Cm	96	247.07
Dysprosium	Dy	66	162.50
Einsteinium	Es	99	254.09
Erbium	Er	68	167.26
Europium	Eu	63	151.96
Fermium	Fm	100	257.10
Fluorine	F	9	19.00
Francium	Fr	87	223
Gadolinium	Gd	64	157.25
Gallium	Ga	31	69.72
Germanium	Ge	32	72.59
Gold	Au	79	196.97
Hafnium	Hf	72	178.49
Helium	He	2	4.00
Holmium	Ho	67	164.93
Hydrogen	H	1	1.008
Indium	In	49	114.82
Iodine	I	53	126.90
Iridium	Ir	77	192.2
Iron	Fe	26	55.85
Krypton	Kr	36	83.80
Lanthanum	La	57	138.91
Lawrencium	Lr	103	257
Lead	Pb	82	207.19
Lithium	Li	3	6.94
Lutetium	Lu	71	174.97
Magnesium	Mg	12	24.31
Manganese	Mn	25	54.94
Mendelevium	Md	101	258.10
Mercury	Hg	80	200.59
Molybdenum	Mo	42	95.94
Neodymium	Nd	60	144.24
Neon	Ne	10	20.18
Neptunium	Np	93	237.05
Nickel	Ni	28	58.71
Niobium	Nb	41	92.91
Nitrogen	N	7	14.01
Nobelium	No	102	255
Osmium	Os	76	190.2
Oxygen	O	8	16.00
Palladium	Pd	46	106.4
Phosphorus	P	15	30.97
Platinum	Pt	78	195.09
Plutonium	Pu	94	239.05
Polonium	Po	84	210
Potassium	K	19	39.10
Praseodymium	Pr	59	140.91
Promethium	Pm	61	146.92
Protactinium	Pa	91	231.04
Radium	Ra	88	226.03
Radon	Rn	86	222
Rhenium	Re	75	186.2
Rhodium	Rh	45	102.91
Rubidium	Rb	37	85.47
Ruthenium	Ru	44	101.07
Samarium	Sm	62	150.35
Scandium	Sc	21	44.96
Selenium	Se	34	78.96
Silicon	Si	14	28.09
Silver	Ag	47	107.87
Sodium	Na	11	22.99
Strontium	Sr	38	87.62
Sulfur	S	16	32.06
Tantalum	Ta	73	180.95
Technetium	Tc	43	98.91
Tellurium	Te	52	127.60
Terbium	Tb	65	158.92
Thallium	Tl	81	204.37
Thorium	Th	90	232.04
Thulium	Tm	69	168.93
Tin	Sn	50	118.69
Titanium	Ti	22	47.90
Tungsten	W	74	183.85
Uranium	U	92	238.03
Vanadium	V	23	50.94
Xenon	Xe	54	131.30
Ytterbium	Yb	70	173.04
Yttrium	Y	39	88.91
Zinc	Zn	30	65.37
Zirconium	Zr	40	91.22

INORGANIC CHEMISTRY

Second Edition

Duward F. Shriver
NORTHWESTERN UNIVERSITY

Peter Atkins
OXFORD UNIVERSITY

Cooper H. Langford
CONCORDIA UNIVERSITY

W. H. FREEMAN AND COMPANY
NEW YORK

Cover illustration: The structure of $Ba_2Cu_3O_4Cl_2$ showing a planar Cu-O net with red Ba^{2+} and yellow Cl^-, from H. Müller-Buschbaum, Agnew. Chem. International Edn., 30, 723-744 (1991). We thank Prof. Müller-Buschbaum for supplying a photograph of this computer-generated image.

Library of Congress Cataloging-in-Publication Data

Shriver, D. F. (Duward F.), 1934-
      Inorganic chemistry / Duward E. Shriver, Peter Atkins,
  Cooper H. Langford. –2nd ed.
              p.   cm.
      Includes index.
      ISBN O-7167-2398-0
      1. Chemistry, Inorganic. I. Atkins, P. W. (Peter William, 1940-
      II. Langford, Cooper Harold, 1934- . III. Title
  QD151.5.S57   1994
  546–dc20

Printed in the United States of America

1 2 3 4 5 6 7 8 9 0 VB 9 9 8 7 6 5 4 3

Preface

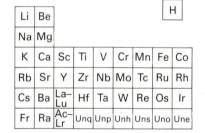

To the student

This text introduces you to the thriving discipline of inorganic chemistry. Inorganic chemistry deals with the properties of over a hundred elements, ranging from highly reactive metals such as sodium to noble metals such as gold; it deals with aggressive nonmetals such as fluorine and unreactive gases such as helium. This variety is one of the great attractions of the subject, but we have to admit that it causes some difficulty to the newcomer. To master and appreciate the broad sweep of the subject you should focus on trends in reactivity, structure, and properties of the elements and their compounds in relation to their position in the periodic table. These general periodic trends provide a foundation for an initial understanding. As you go deeper into the subject it will be apparent that the elements also have their characteristic chemical personalities, which again may be associated with their position in the periodic table.

Inorganic compounds vary from ionic solids, which can be described by simple applications of classical electrostatics, to covalent compounds and metals, which are best described by models that have their origin in quantum mechanics. For the rationalization and interpretation of most inorganic properties we use qualitative models that are based on quantum mechanics, such as the properties of atomic orbitals and their use to form molecular orbitals. Similar models may already be familiar to you from your study of organic chemistry. Theory has contributed greatly to our understanding of inorganic chemistry, and qualitative models of bonding and reactivity clarify and systematize the subject. Nevertheless, it should never be forgotten that inorganic chemistry, like organic chemistry and biochemistry, is essentially an experimental subject. The ultimate authority consists of observations and measurements, such as the identities of products of a reaction, structures,

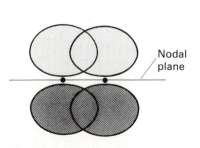

2.9 Two *p* orbitals can overlap to form a π

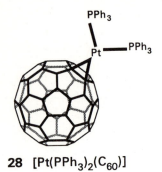

28 [Pt(PPh₃)₂(C₆₀)]

*Much of the interpretation of i
models of bonding and molecul
augmented by noting the cor
strengths of bonds, spectrosco
patterns of chemical reactiv
widely used and very helpf
principal types of models th*

Example 8.1: *Explaining trends in*
Use simple bonding models
but beryllium does not.

Answer. Large anions are g
tion 4.8). Therefore, bariu
beryllium peroxide. In fac
eously on exposure of bari

thermodynamic properties, spectroscopic signatures, and measurements of reaction rates.

Large areas of inorganic chemistry remain unexplored, so new and often unusual inorganic compounds are constantly being synthesized in laboratories. Such exploratory inorganic syntheses continue to enrich the field with compounds that give us new perspectives on structure, bonding, and reactivity. This exciting aspect of the subject is conveyed in Parts 2 and 3 of the text. You should gain sufficient command of inorganic chemistry in the early chapters to savor these newer discoveries.

In addition to its intellectual attractions, you will learn that inorganic chemistry has considerable practical impact and touches on all other branches of science. If you enter the chemical industry professionally or pursue a chemically related discipline, it will be very useful for you to have a clear sense of these applications. Eight of the top ten industrial chemicals, by weight, are inorganic. Inorganic chemistry is also essential to the formulation and improvement of modern materials such as catalysts, semiconductors, light guides, nonlinear optical devices, superconductors, and advanced ceramic materials.

The environmental impact of inorganic chemistry is also huge. In this connection, the extensive role of metal ions in plants and animals led to the thriving area of bioinorganic chemistry. These current topics are mentioned throughout the book and developed more thoroughly in Part 3. If your formal course in inorganic chemistry is too short to cover the final chapters, you may find it interesting and informative to browse through them at your leisure. They will bring you to the brink of modern research.

In writing this book we gave considerable thought to clarity of expression, logical organization, visual representation, and learning aids. The brief *Introductions* to the chapters are not filled with facts and principles that might be directly tested but we urge you to pay attention to them and to the accompanying table of contents for each chapter, because they set the stage for the material that follows. Read them initially, and read them again after completion of the chapter to appreciate the context of the material and thereby gain a better grasp and working knowledge of the subject.

We are all too well aware that the process of reading a text can quickly become a passive undertaking, without significant comprehension or retention. To turn it into a more active learning experience, we have provided *Examples* coupled with self-test *Exercises* at appropriate intervals throughout each chapter. The *Exercises* are not so difficult as to distract you from the immediate subject matter; they are designed to engage you actively in thinking about the material. You will find that although it might seem like more work at the time, it really does facilitate the learning process.

The *Key points* at the end of each chapter provide broad summaries of some of the main topics. These are intended to be read after completion of the chapter. You can refresh your memory and gauge

KEY POINTS

1. *Complexes*

Assemblies of units (ligands) are called complexes; they

EXERCISES

3.1 What shapes would you expe (b) SO_3^{2-}, (c) IF_5?

PROBLEMS

3.1 Consider a molecule IF_3O_2 (w How many isomers are possible?

your mastery of the subject by the extent to which you can fill out these broad organizing statements with principles and facts.

The *Exercises* at the each of the chapter will further consolidate your understanding of the material. To give rapid feedback on whether you are on the right track, we provide (at the end of the book) brief answers to each exercise. The *Exercises* are worked out in detail and thoroughly explained in S. H. Strauss's accompanying *Guide to solutions for inorganic chemistry*. Students who used the first edition of Strauss's guide were enthusiastic about its clear and helpful perspective.

Some or all of the *Problems* at the end of the chapter may be assigned by your instructor. The *Problems* are intentionally more demanding than the *Exercises*, and are designed to extend your skills and knowledge beyond the confines of the text. They encourage you to look into the professional literature, which is by far the best way of achieving a sense of the current vitality of the subject.

The planning, writing, and refinement of this text (always with the advice of students and instructors) has been a stimulating undertaking for us. The true test of the enterprise, however, is not our pleasure but must be judged by seeing how effectively it interests you in the subject and helps you to learn it. Write to us if you spot points where we might improve the next edition.

To the instructor

One aim in this revision was to build on the success of the first edition by refining the presentation throughout and broadening the development of the systematic descriptive chemistry of the elements. You will see that a considerable amount of reorganization of the material has taken place, and that the book is now divided into three parts.

In Part 1, *Foundations*, the discussion of symmetry is covered with improved clarity and efficiency in a single chapter and the development of bonding models is extended and reorganized. Economies in space and improved clarity have been achieved by the combination of Brønsted and Lewis acids and bases into a single chapter.

Part 2, *Systematic chemistry of the elements*, now includes a major new chapter, *The metals*, which provides a survey of the trends in properties of this important class of elements. The remaining five chapters on the main-group elements have been extensively revised to improve balance and clarity. The combination of descriptive material and principles in the first two parts of the book provide a solid foundation in inorganic chemistry. Each of the descriptive chapters in Part 2 presents chemistry within the framework of the principles covered in Part 1. This type of rationalization is necessary to bring the mass of observations into a coherent context. However, we emphasize throughout the text that bonding models will continue to evolve, that new experimental data may lead to changes in the proposed mechanisms for chemical reactions, and that even some thermodynamic data will be superseded through more accurate and extensive measurements.

In Part 3, *Advanced topics*, we present material that we consider too advanced for inclusion in the two preceding parts. Much of this material is in a state of flux, for it is at the center of attention in laboratories around the world. We have had to be highly selective, and have aimed only to give a sense of the activity that is currently characteristic of the individual subjects. This material helps to convey the vigor of inorganic chemistry and to point the way for further study, both in the literature and the laboratory.

Comments from instructors using the first edition confirmed our original perception of the wide divergence of thought on the appropriate topics and detailed content of inorganic chemistry courses. In this connection, Part 3 has been used in a variety of ways. Its chapters are frequently employed in part or in their entirety for undergraduate courses. They sometimes form the backbone of graduate-level courses in North America. Elsewhere, they are often utilized toward the end of a 3- or 4-year course in inorganic chemistry. This section of the text has been augmented, refined, and extensively updated.

We continue to aim for a style that is clear and interesting to students and to supplement information on classical inorganic chemistry with recent developments in the subject. Among the special aids to the student we have expanded the number of in-chapter worked *Examples* coupled with *Exercises*. Comments from both students and instructors have reinforced our belief that this device aids in the assimilation of the material. The favorable reception to our extensive use of three-dimensional structural drawings has prompted us to expand their number so as to convey this important aspect of inorganic chemistry. You will now see that many reactions are expressed structurally, rather than in simple formulas.

A new aid to learning is the introduction of a set of *Key points* at the end of every chapter. These brief statements provide the context for recalling the material in the chapter. We have also added more *Exercises* and *Problems* at the end of each chapter. The *Exercises* are intended to facilitate review and extend comprehension of the material without putting excessive demands on the student. Short answers are provided in the back of the book, and more extensive answers with explanations are available in a new edition of S. H. Strauss's well-received *Guide to solutions for inorganic chemistry*.

The *Problems* provide a useful vehicle for deepening the students' understanding of inorganic chemistry. Typically, they are assigned sparingly in short and fast-paced courses. Many of them include citations to the literature that are necessary for their solution. A few require the use of computer programs, and the sources of suitable programs are indicated. We have found for example that it is very useful for students to run Extended Hückel calculations on diatomic and triatomic molecules and to interpret the output. The point is not to make each student an expert in molecular orbital theory or the programs, but rather to emphasize the nature of the input information, the ordering of energy levels, and the form of molecular orbitals for simple molecules. Our students have found this approach to be very instructive.

We continue to adhere to IUPAC recommendations and SI units almost everywhere. However, we continue to follow much of the current literature in using ångströms for bond lengths. We also continue to use the 1–18 labeling for groups in the periodic table as recommended by IUPAC. This numbering is supplemented with the older roman numeral designator for the groups in the *p* block. In this new edition we moved to the IUPAC recommended -ane suffix for common hydrides, such as phosphane for PH_3 and its derivatives, because it is increasingly adopted in the literature.

In another departure from the first edition, *Further information* sections have been moved from the chapters to the back of the book. This permits some expansion of the material, for example on NMR, without intruding on a chapter. The new arrangement also acknowledges the generality of this material, which may have been lost in the original edition by association with a specific chapter. The *Appendices* now include a section on nomenclature as well as data on atoms and standard potentials, pictorial representations of symmetry adapted orbitals, and Tanabe–Sugano diagrams. At the suggestion of many instructors, we have included more references to the literature throughout the text. To the extent possible, we refer to the leading and most widely available journals.

Some useful general reference books are the two single-volume compendia of inorganic chemistry: *Chemistry of the elements*, by N. N. Greenwood and A. Earnshaw (Pergamon Press, Oxford, 1982) and *Advanced inorganic chemistry*, by F. A. Cotton and G. Wilkinson (Wiley, New York, 1988). Three useful multi-volume compendia are *Comprehensive inorganic chemistry*, edited by J. C. Bailar, H. J. Emeleus, R. Nyholm, and A. F. Trotman-Dickenson (Pergamon Press, Oxford, 1973), *Comprehensive organometallic chemistry*, edited by G. Wilkinson, F. G. A. Stone, and E. W. Abel (Pergamon Press, Oxford, 1987; a second edition, now in production, will provide recent advances), and *Comprehensive coordination chemistry*, edited by G. Wilkinson, R. D. Gillard, and J. McCleverty (Pergamon Press, Oxford, 1987). A mine of specific information is available in the *Gmelin handbook of inorganic chemistry* (Springer-Verlag, Berlin). Inorganic industrial processes are nicely summarized in *Industrial inorganic chemistry*, by W. Buchner, R. Schliebs, G. Winter, and K. H. Buchel (VCH, Deerfield Beach, 1989). Two outstanding multi-volume compendia are *Kirk–Othmer Encyclopedia of chemical technology* (Wiley-Interscience, 1978–1984; volumes of the fourth edition are now appearing), and *Ullmann's Encyclopedia of industrial chemistry* (VCH, Deerfield Beach). Of the two multi-volume references, Ullmann's generally provides more satisfying chemical detail.

We have taken particular care to ensure that the text is free of errors. This is difficult in a rapidly changing field, where today's knowledge is soon replaced by tomorrow's. We acknowledge below all those readers who so unstintingly gave their time and expertise to a careful reading of a variety of draft chapters. We are particularly grateful to Dr. C. S. G. Phillips, who read the entire galley proofs, and

Dr. W. Levason, who read the entire page proofs and trapped a number of points at a late stage.

Yet again, we wish to thank our publishers for the understanding and unstinting assistance they have provided at all stages of the intricate and time consuming task of producing a book of such structural complexity as this. We owe a considerable debt to their patience, wisdom, and understanding.

Evanston　　　　　　　　　　　　　　　　　　　　　　D. F. S.
Oxford　　　　　　　　　　　　　　　　　　　　　　　P. W. A.
Calgary　　　　　　　　　　　　　　　　　　　　　　C. H. L.

Acknowledgements

As with the first edition we have benefitted from the suggestions and expertise of others. Many instructors and students gave us useful suggestions for this revision, as did several of the foreign language translators and our colleagues. In particular we acknowledge the help of

Dr H C Aspinall, University of Liverpool
Professor B Averill, University of Virginia, Charlottesville, Virginia
Professor F Basolo, Northwestern University, Evanston, Illinois
Professor P Bianconi, Pennsylvania State University, University Park, Pennsylvania
Dr P Brain, University of Edinburgh
Dr J Burgess, University of Leicester
Dr S K Chapman, University of Edinburgh
Professor M Clarke, Boston College, Chestnut Hill, Massachusetts
Dr P I Clemenson, University of Huddersfield
Professor D J Cole-Hamilton, University of St Andrews
Dr E C Constable, University of Cambridge
Dr S R Cooper, Inorganic Chemistry Laboratory, Oxford
Professor F DiSalvo, Cornell University, Ithaca, New York
Professor P Dorhout, Colorado State University, Fort Collins, Colorado
Dr J Emsley, Imperial College, London
Professor D Farrar, University of Toronto, Ontario
Professor T Fehlner, University of Notre Dame, Indianapolis, Indiana
Professor D E Fenton, University of Sheffield
Dr P W Fowler, University of Exeter
Professor M Geselbracht, Reed College, Portland, Oregon
Dr J B Gill, University of Leeds
Dr R O Gould, University of Edinburgh
Dr R Greatrex, University of Leeds
Professor M L H Green, Inorganic Chemistry Laboratory, Oxford
Professor D T Haworth, Marquette University, Milwaukee, Wisconsin
Professor B Henry, Mississippi State University, Mississippi State, Mississippi

Professor R H Holm, Harvard University, Cambridge, Masachusetts
Professor B F G Johnson, University of Edinburgh
Dr T P Kee, University of Leeds
Professor C Kelley, US Air Force Academy, Colorado Springs, Colorado
Dr K Kite, University of Exeter
Dr W Levason, University of Southampton
Professor C Lock, McMaster University, Hamilton, Ontario
Dr G R Moore, University of East Anglia, Norwich
Professor T V O'Halloran, Northwestern University, Evanston, Illinois
Professor P Paetzold, Reinisch-Westfälische Technische Hochschule, Aachen, Germany
Dr C S G Phillips, Inorganic Chemistry Laboratory, Oxford
Professor D Phillips, Wabash College, Crawfordsville, Indianapolis
Dr C Pulham, University of Edinburgh
Professor D W H Rankin, University of Edinburgh
Professor E Rosenberg, California State University, Northridge, California
Dr M Schröder, University of Edinburgh
Professor R Shepherd, University of Pittsburgh, Pittsburgh, Pennsylvania
Professor E Sinn, University of Hull
Professor A Sleight, Oregon State University, Corvallis, Oregon
Dr A K Smith, University of Liverpool
Professor S H Strauss, Colorado State University, Fort Collins, Colorado
Professor D Talham, University of Florida, Gainesville, Florida
Professor R Trautman, San Francisco State University, San Francisco, California
Professor R J P Williams, Inorganic Chemistry Laboratory, Oxford
Dr R E P Winpenny, University of Edinburgh
Dr M J Winter, University of Sheffield
Professor M Zeldin, Indiana University/Purdue University at Indianapolis, Indiana
Professor M Zisk, Otterbein College, Westerville, Ohio

Summary of contents

Contents

xiv

Contents

PART ONE

Foundations

The seven chapters of this part lay the foundations of inorganic chemistry. The first four chapters develop an understanding of the theory as it relates to structural, spectroscopic, and some chemical properties of atoms, molecules, solids. Because all models of bonding are based on atomic properties, atomic structure is described in Chapter 1. The following two chapters show how the ideas developed for atoms may be applied to molecules, first in terms of their electronic structures and then in terms of their geometrical shapes. We shall see that the symmetries of molecules can be defined precisely and used to rationalize a number of their properties. The ideas developed on the electronic and geometrical structures of molecules can be extended to solids, which is the subject of Chapter 4.

The following chapter (Chapter 5) develops more fundamental principles by examining the reactions that take place by the transfer of protons or by the sharing of electron pairs. As we shall see, many reactions can be classified into one type or the other, and the introduction of the classes helps to systematize inorganic chemistry. Indeed, it also helps to rationalize the structures and properties of a certain class of compound, the coordination compounds, that are so typical of the transition elements, and Chapter 6 explores some elementary aspects of their structures and reactions. Chapter 7 rounds off the introduction of the foundations of chemical reactions by introducing another great class of reactions, those proceeding by oxidation and reduction.

1

Atomic structure

This chapter describes the origin, abundance, and classification of the elements. It also introduces the atomic properties of the elements and shows how they can be rationalized in terms of the behavior of electrons in atoms. To do this rationalization to an adequate level, we need to introduce some of the concepts of quantum mechanics. These concepts are introduced qualitatively with emphasis on pictorial representations rather than mathematical rigor. In the course of the chapter we meet some of the parameters that characterize the properties of atoms, and which are used to systematize the chemical properties of the elements and to help organize inorganic chemistry. These parameters include measures of atomic and ionic size (which help to govern the manner in which atoms pack together), the energetics of electron removal and addition to atoms, and one of the systematizing parameters of inorganic chemistry, the electronegativity of an element.

The observation that the universe is expanding has led to the view that about 15 billion years ago all observable matter was concentrated into a pointlike region which exploded in the event called the **Big Bang**. With initial temperatures immediately after the Big Bang thought to be about 10^9 K, the fundamental particles produced in the explosion had too much kinetic energy to bind together in the forms we know today. However, the universe cooled as it expanded, the particles moved more slowly, and they soon began to adhere together under the influence of a variety of forces. In particular, the **strong force**—a short-range but powerful attractive force between protons, neutrons, and each other—bound particles together into nuclei. As the temperature fell still further, the **electromagnetic force**—a relatively weak but long-range force between electric charges—bound electrons to nuclei to form atoms. The properties of the only subatomic

1 Atomic structure

Table 1.1 Subatomic particles of relevance to chemistry

Particle	Symbol	Mass/u*	Mass number	Charge/$e^\dagger$	Spin
Electron	e^-	5.486×10^{-4}	0	-1	$\frac{1}{2}$
Proton	p	1.0073	1	$+1$	$\frac{1}{2}$
Neutron	n	1.0087	1	0	$\frac{1}{2}$
Photon	γ	0	0	0	1
Neutrino	ν	$c.\ 0$	0	0	$\frac{1}{2}$
Positron	e^+	5.486×10^{-4}	0	$+1$	$\frac{1}{2}$
α particle	α	[$^4_2$He^{2+} nucleus]	4	$+2$	0
β particle	β	[e^- ejected from nucleus]	0	-1	$\frac{1}{2}$
γ photon	γ	[electromagnetic radiation from nucleus]	0	0	1

*Masses are expressed in atomic mass units, u, with $1\ u = 1.6605 \times 10^{-27}$ kg.
$\dagger$The elementary charge e is 1.602×10^{-19} C.

particles that we need to consider in chemistry are summarized in Table 1.1. The 100 or so known elements that are formed from these subatomic particles are distinguished by their **atomic number**, Z, the number of protons in the nucleus of an atom of the element. The **isotopes** of elements, which are atoms with the same atomic number but different atomic masses, are distinguished by the **mass number**, A, the total number of protons and neutrons in the nucleus.

THE ORIGIN OF THE ELEMENTS

If current views are correct, by about 2 h after the start of the universe the temperature had fallen so much that most of the matter was in the form of H atoms (89 percent) and He atoms (11 percent). In one sense, not much has happened since then, for as Fig. 1.1 shows, hydrogen and helium remain the most abundant elements in the universe. However, nuclear events have formed a wide assortment of other elements and have immeasurably enriched the variety of matter in the universe.

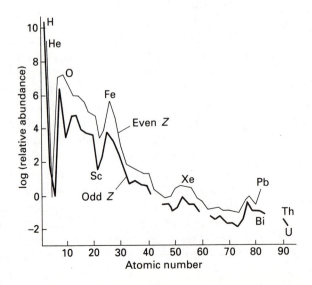

1.1 The abundances of the elements in the universe. Elements with odd Z are less stable than their neighbors with even Z. The abundances refer to the number of atoms of each element relative to Si taken as 10^6.

1.1 The nucleosynthesis of light elements

The condensation of clouds of H and He atoms is thought to have led to the formation of the earliest stars. The collapse of these stars under the influence of gravity gave rise to high temperatures and densities within them, and fusion reactions began as nuclei merged together. The earliest nuclear reactions are closely related to those now being studied in connection with the development of controlled nuclear fusion.

Energy is released when light nuclei fuse together to give elements of higher atomic number. For example, the nuclear reaction in which an α particle (a ^4_2He nucleus, consisting of two protons and two neutrons) fuses with a carbon-12 nucleus to give an oxygen-16 nucleus and a γ-ray photon (γ) is

$$^{12}_6\text{C} + {}^4_2\alpha \rightarrow {}^{16}_8\text{O} + \gamma$$

and releases 7.2 MeV.[1] In this nuclear equation, the **nuclide**, a nucleus of specific atomic number Z and mass number A, is designated ^A_ZE, where E is the chemical symbol of the element. Note that in a balanced nuclear equation, the mass numbers of the reactants sum to the same value as the mass numbers of the products ($12 + 4 = 16$). The atomic numbers sum likewise ($6 + 2 = 8$) so long as an electron, e^-, when it appears as a β-particle, is ascribed an atomic number of -1 and a positron,[2] e^+, is ascribed $Z = +1$.

Elements of atomic number up to 26 were formed inside stars. Such elements are the products of the nuclear fusion events referred to as 'nuclear burning'. The burning reactions (which should not be confused with chemical combustion) involved H and He nuclei and a complicated fusion cycle catalyzed by C nuclei. (The stars that formed in the earliest stages of the evolution of the universe lacked C nuclei and used noncatalyzed H burning reactions.) Some of the most important reactions in the cycle are

Proton (p) capture by carbon-12:	$^{12}_6\text{C} + {}^1_1\text{p} \rightarrow {}^{13}_7\text{N} + \gamma$
Positron decay accompanied by neutrino (ν) emission	$^{13}_7\text{N} \rightarrow {}^{13}_6\text{C} + \text{e}^+ + \nu$
Proton capture by carbon-13:	$^{13}_6\text{C} + {}^1_1\text{p} \rightarrow {}^{14}_7\text{N} + \gamma$
Proton capture by nitrogen-14:	$^{14}_7\text{N} + {}^1_1\text{p} \rightarrow {}^{15}_8\text{O} + \gamma$
Positron decay, accompanied by neutrino emission	$^{15}_8\text{O} \rightarrow {}^{15}_7\text{N} + \text{e}^+ + \nu$
Proton capture by nitrogen-15:	$^{15}_7\text{N} + {}^1_1\text{p} \rightarrow {}^{12}_6\text{C} + {}^4_2\alpha$

---

[1] An electronvolt (1 eV) is the energy required to move an electron through a potential difference of 1 V. It follows that $1\text{ eV} = 1.602 \times 10^{-19}$ J; $1\text{ MeV} = 10^6$ eV.

[2] A positron is a positively charged version of an electron: it has zero mass number and a single positive charge. When it is emitted, the mass number of the nuclide is unchanged but the atomic number decreases by 1 because the nucleus has lost one positive charge. Its emission is equivalent to the conversion of a proton in the nucleus into a neutron:

$$^1_1\text{p} \rightarrow {}^1_0\text{n} + \text{e}^+ + \nu$$

A neutrino, ν, resembles an uncharged electron: it is electrically neutral and has a very small (possibly zero) mass.

The net result of this sequence is the conversion of four protons (four ^{1}H nuclei) into an α particle (a ^{4}He nucleus):

$$4{}^1_1\text{p} \rightarrow {}^4_2\alpha + 2\text{e}^+ + 2\nu + 3\gamma$$

The nuclear reactions in the sequence are rapid at temperatures between 5 and 10 MK (where 1 MK $= 10^6$ K).

Heavier elements begin to be produced in significant quantities when hydrogen burning is complete and the collapse of the star's core raises the density there to 10^8 kg m^{-3} (about 10^5 times the density of water) and the temperature to 100 MK. Under these extreme conditions, **helium burning** becomes viable. The low abundance of lithium, beryllium, and boron is consistent with the events taking place during this phase in the life of a star, because any $^8_4$Be formed by collisions between α particles goes on to react with more α particles:

$$^8_4\text{Be} + {}^4_2\alpha \rightarrow {}^{12}_6\text{C} + \gamma$$

Thus, the helium-burning stage of stellar evolution does not result in the formation of beryllium as a stable end product; nor, for similar reasons, does it result in lithium and boron. The origin of these three elements is still uncertain, but they may result from events in which C, N, and O nuclei undergo **spallation**, or fragmentation by collisions with high-energy particles. Elements can also be produced by nuclear reactions such as neutron (n) capture accompanied by proton emission:

$$^{14}_7\text{N} + {}^1_0\text{n} \rightarrow {}^{14}_6\text{C} + {}^1_1\text{p}$$

This reaction still continues in our atmosphere as a result of the impact of cosmic rays and contributes to the steady-state concentration of radioactive carbon-14 on earth.

The high abundance of iron in the universe is consistent with its having the most stable of all nuclei. This stability can be assessed from its **binding energy**, the difference in energy between the nucleus itself and the same numbers of individual protons and neutrons. This binding energy is often expressed in terms of a difference in mass between the nucleus and its individual protons and neutrons, for according to Einstein's theory of relativity, mass and energy are related by $E = mc^2$, where c is the speed of light. Therefore, if the mass of a nucleus differs from the total mass of its components by Δm, then its binding energy is

$$\Delta E = \Delta m \times c^2$$

The binding energy of ^{56}Fe, for example, is the difference in energy between the ^{56}Fe nucleus and 26 protons and 30 neutrons. Figure 1.2 shows the binding energy (expressed as an energy per nucleon, a nuclear particle) for all the elements, and we see that iron occurs at the maximum of the curve, showing that its nucleons are bound more strongly than in any other nuclide.

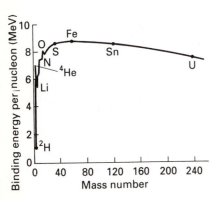

1.2 Nuclear binding energies. The greater the binding energy, the more stable the nucleus. The most stable nuclide is $^{56}_{26}$Fe.

1.2 The nucleosynthesis of heavy elements

Elements heavier than iron are produced by a variety of energy-consuming processes. These include the capture of free neutrons, which are not present in the earliest stages of stellar evolution but are produced later in reactions such as

$$^{23}_{10}\text{Ne} + ^{4}_{2}\alpha \rightarrow ^{26}_{12}\text{Mg} + ^{1}_{0}\text{n}$$

Under conditions of intense neutron flux, as in a supernova (the explosion of a star), a given nucleus may capture a succession of neutrons and become a progressively heavier isotope. However, there comes a point at which it will eject an electron from the nucleus as a β particle (a high-velocity electron, e^-). Because β-decay leaves the mass number of the nuclide unchanged but increases its atomic number by 1 (the nuclear charge increases by 1 unit when an electron is ejected), a new element is formed. An example is

Neutron capture: $^{98}_{42}\text{Mo} + ^{1}_{0}\text{n} \rightarrow ^{99}_{42}\text{Mo} + \gamma$

Followed by β-decay accompanied
by neutrino emission: $^{99}_{42}\text{Mo} \rightarrow ^{99}_{43}\text{Tc} + e^- + \nu$

The **daughter nuclide**, the product of a nuclear reaction ($^{99}_{43}\text{Tc}$, an isotope of technetium, in this example), can absorb another neutron, and the process can continue, gradually building up the heavier elements.

Example 1.1: *Balancing equations for nuclear reactions*

Synthesis of heavy elements occurs in the neutron capture reactions believed to take place in the interior of cool 'red giant' stars. One such reaction is the conversion of $^{68}_{30}\text{Zn}$ to $^{69}_{31}\text{Ga}$ by neutron capture to form $^{69}_{30}\text{Zn}$, which then undergoes β-decay. Write balanced nuclear equations for this process.

Answer. Neutron capture increases the mass number of a nuclide by 1 but leaves the atomic number (and hence the identity of the element) unchanged:

$$^{68}_{30}\text{Zn} + ^{1}_{0}\text{n} \rightarrow ^{69}_{30}\text{Zn} + \gamma$$

The excess energy is carried away as a photon. β-decay, the loss of an electron from the nucleus, leaves the mass number unchanged but increases the atomic number by 1. Because zinc has atomic number 30, the daughter nuclide has $Z = 31$, corresponding to gallium. Therefore, the nuclear reaction is

$$^{69}_{30}\text{Zn} \rightarrow ^{69}_{31}\text{Ga} + e^-$$

(In fact, a neutrino is also emitted, but this cannot be inferred from the data as a neutrino is massless and electrically neutral.)

Exercise E1.1. Write the balanced nuclear equation for neutron capture by $^{80}_{35}\text{Br}$.

1.3 The classification of the elements

Some substances that we now recognize as chemical elements have been known since antiquity: they include carbon, sulfur, iron, copper, silver, gold, and mercury. The alchemists and their immediate successors the early chemists had added about another 18 elements by 1800. By that time, the precursor of the modern concept of an element had been formulated as a substance that consists of only one type of atom. (Now, of course, by 'type' of atom we mean an atom with a particular atomic number.) By then a variety of experimental techniques were available for breaking oxides and other compounds down into elements, and these techniques were considerably enhanced by the introduction of electrolysis. The list of elements grew rapidly in the late nineteenth century. This growth was in part a result of the development of atomic spectroscopy, in which thermally excited atoms are observed to emit electromagnetic radiation with a unique pattern of frequencies, so permitting a much easier identification of previously unknown elements.

The recognition of patterns

A useful broad division of elements is into **metals** and **nonmetals**. Metallic elements (such as iron and copper) are typically lustrous, malleable, ductile, electrically conducting solids at about room temperature. Nonmetals are often gases (oxygen), liquids (bromine), or solids that do not conduct electricity appreciably (sulfur). The unifying aspect of this classification will become increasingly clear. For example, metallic elements combine with nonmetallic elements to give compounds that are typically hard, nonvolatile solids (for example, sodium chloride). When combined with each other, the nonmetals often form volatile molecular compounds (such as phosphorus trichloride). When metals combine (or simply mix together) they produce alloys which have most of the physical characteristics of metals.

Further patterns that act as an aid to classification are found by noting the molecular formulas of the compounds each element forms. For example, the following formulas for the compounds of some elements with hydrogen suggest that the elements belong to two families:

$$CH_4 \ SiH_4 \ GeH_4 \ SnH_4$$

$$NH_3 \ PH_3 \ AsH_3 \ SbH_3$$

Other compounds of these elements show similar family relationships, as in the formulas CCl_4 and $SnCl_4$. However (as so often in inorganic chemistry), analogies are not perfect, and although $SnCl_2$ is known as a white solid that melts at $246\,°C$, the compound CCl_2 cannot be isolated. A positive outcome of the recognition of family relationships is that they point to the existence of new, interesting compounds. Thus, if we recognize that carbon and silicon are in the same family, as suggested by the formulas of their hydrogen compounds, then the existence of alkenes (R_2C=CR_2) suggests that R_2Si=SiR_2 ought to exist too. They do indeed exist, but it was not until 1981 that inorganic chemists succeeded in isolating a stable member of the family.

By the middle of the nineteenth century, approximately 60 elements had been recognized, the rudiments of their chemical properties were known, and a number of family relationships had been discerned. The great synthesis of this knowledge occurred in 1869, when the Russian chemist Dmitri Mendeleev arranged the known elements in order of increasing atomic weight (which we would now call molar mass) and found that the arrangement resulted in families of elements with similar chemical properties lying close to each other in his table. Mendeleev concentrated on the chemical properties of the elements, but at about the same time Lothar Meyer in Germany was investigating their physical properties, and found that similar values repeated periodically with increasing atomic weight. A classic example is shown in Fig. 1.3, where the molar volume of the element (its volume per mole) in its normal form is plotted against atomic number. The work of Mendeleev and Meyer led to the construction of the **periodic table** of the elements and the recognition of its spectacular usefulness for the prediction of unknown elements from gaps in the table. The same process of inference from periodic trends is used widely by inorganic chemists to rationalize properties and to suggest new avenues of chemical synthesis.

The modern periodic table

We presume that the general structure of the modern periodic table is familiar. The elements are listed in order of atomic number (not atomic weight), for atomic number is the more fundamental property. The

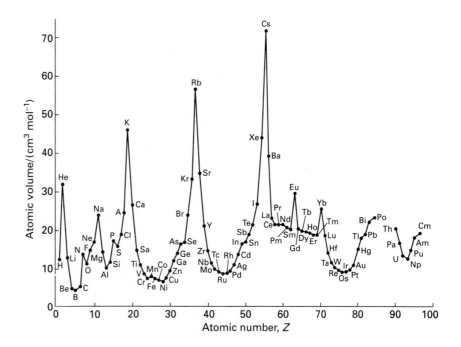

1.3 The periodicity of molar volume with increasing atomic number.

horizontal rows of the table are called **periods** and the vertical columns
are called **groups**. The table divides into four **blocks** which are
labeled as shown in Fig. 1.4. The members of the s and p blocks are
collectively called the **main-group elements**. The numbering system of the groups is still in contention. In the illustration we show
both the traditional numbering of the main groups (with the roman
numerals from I to VIII) and the current IUPAC[3] recommendations,
in which the groups are numbered from 1 through 18.

The essential key to atomic structure and the explanation of periodicity was unknown in Mendeleev's day. One crucial step in coming to
an understanding was the identification of the electron as a universal
constituent of matter. That discovery opened the way to the realization
that atomic spectra can be interpreted as the outcome of transitions of
electrons between precisely defined energy levels in the atom. The
puzzle as to why electrons appeared to occupy discrete energy levels
motivated the development of quantum mechanics in the twentieth
century. From these developments we now recognize that the periodic
table is a reflection of the electronic structure of atoms. To understand
the properties of an element and the compounds it forms requires us
to understand how electrons are arranged in atoms and how those
arrangements may be modified. The language now used to describe
the electronic structure of an atom is based on quantum mechanics,
but it depends as much on the information derived from the systematic
analysis of atomic spectra as it does on calculation from first principles.
We see the outlines of the results in this chapter. In Chapter 14, we
shall analyze somewhat more quantitatively the information available
from atomic spectra.

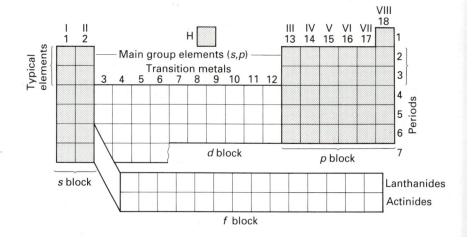

1.4 The general structure of the periodic table.
The tinted areas denote the main-group
elements. Compare this template with the
complete table inside the front cover for the
identities of the elements that belong to each
block.

[3] IUPAC denotes the International Union of Pure and Applied Chemistry. This international body establishes commissions that develop agreements on nomenclature, symbols,
units, and sign conventions. The collection of rules that relate specifically to inorganic
chemistry is published as *Nomenclature of inorganic chemistry*, Blackwell, Oxford (1990),
which is known colloquially as the 'Red Book'.

THE STRUCTURE OF HYDROGENIC ATOMS

A general feature of the structure of any atom is that:

> A neutral atom of atomic number Z has Z extranuclear electrons, and their total negative charge balances the positive charge of the nucleus.

The electrons are attracted electrostatically to the nucleus. However, a complicating feature of **many-electron atoms** (atoms that contain more than one electron) is that the electrons repel each other with a strength that is comparable to that of the electron–nuclear attraction. It therefore proves sensible to develop a description of atomic structure in two stages. Initially, we consider **hydrogenic atoms**, which (like hydrogen) have only one electron and so are free of the complicating effects of electron–electron repulsions. Hydrogenic atoms include ions such as He^+ and C^{5+} as well as hydrogen itself. Then we use the concepts these atoms introduce to build up an approximate description of the structures of many-electron atoms.

Because the electronic structures of atoms must be expressed in terms of quantum mechanics, we need to review some of the concepts of this theory of matter.

1.4 Some principles of quantum mechanics

A fundamental concept of quantum mechanics is that *matter has wave-like properties*. As we shall see, this character of matter implies that a particle, such as an electron in an atom, is described by a **wavefunction**, ψ, which is a mathematical function of the position coordinates x, y, z, and of the time t. A wavefunction describes the distribution of electrons in atoms, and hence is central to interpretations of the properties of atoms and the compounds they form. We need to spend some time describing its essential characteristics and understanding its significance.

The de Broglie relation and the kinetic energy

In 1924, the French physicist Louis de Broglie made the radical suggestion that a particle has associated with it a wave—later this wave was to become the wavefunction. He asserted that the wavelength λ of the wave (its peak-to-peak distance) is inversely proportional to the momentum p of the particle (its mass times its velocity), and proposed the **de Broglie relation**:

$$p = \frac{h}{\lambda}$$

The constant h is **Planck's constant**, a fundamental constant with the value 6.626×10^{-34} J s. This constant had already been introduced into the description of electromagnetic radiation by Max Planck.

The de Broglie relation shows that *the greater the momentum of a particle, the smaller the wavelength of its wavefunction* (Fig. 1.5). The

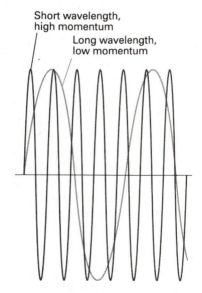

Short wavelength, high momentum

Long wavelength, low momentum

1.5 The wavefunction of a freely moving particle has a long wavelength if the particle's linear momentum is low but a short wavelength if the momentum is high.

relation has been confirmed by showing that rapidly moving electrons undergo diffraction, which is a typical wave property. The details of the patterns produced by the diffraction of electron beams are in agreement with the wavelength predicted by the de Broglie relation. This early realization that matter has wave-like properties has been applied in the technique of **electron diffraction**, in which a potential difference of about 40 kV is used to accelerate electrons to a speed at which they have a wavelength of about 0.05 Å (where $1 \text{ Å} = 10^{-10}$ m). The diffraction pattern that results when the beam passes through a gaseous sample or is reflected from a surface is analyzed to obtain bond lengths and bond angles.

The uncertainty principle

An implication of the wave character of matter (its possession of wave-like properties) is that *it is impossible to specify the exact position and momentum of a particle simultaneously.*[4] One of the important consequences of the uncertainty principle for chemistry is that it denies the possibility of ascribing electrons to precise orbits around nuclei. Such a planetary model had been proposed by Ernest Rutherford, who first discovered the atomic nucleus, and had been used by Niels Bohr as a basis for a quantitative model of the hydrogen atom. For an orbit to be a meaningful concept, it is necessary to specify the position and momentum of an electron at each instant. The uncertainty principle, however, tells us that this is impossible. The concept of orbit works for massive objects such as planets only because we do not try to determine their positions and momenta precisely enough for the uncertainty principle to be relevant.

The Schrödinger equation

De Broglie's revolutionary (and then unclear) concept of a particle having a wavelength led the Austrian physicist Erwin Schrödinger to formulate (in 1926) an equation which, when solved, gave the actual wavefunction. When the **Schrödinger equation** is solved for a free particle (such as an electron in a beam), it is found that the wavefunction has a wavelength given by the de Broglie relation, and that solutions of the equation exist for any wavelength. However, when the equation is solved for a particle that is confined to a small region of space or is bound to an attractive center (like an electron in an atom), it is found that acceptable solutions can be obtained only for certain energies. That is, the energy of such particles is **quantized**, or confined to discrete values. Later we shall see that certain other properties (for instance, angular momentum) are also quantized. This quantization of physical observables is of the most profound importance in

---

[4] This conclusion is expressed quantitatively by the **uncertainty principle**, which was proposed by Werner Heisenberg. This principle states that if the uncertainty in the position of a particle is Δx and the uncertainty in its momentum is Δp, then the product of these two uncertainties must satisfy the relation

$$\Delta p \Delta x \geq \tfrac{1}{2} \hbar$$

The fundamental constant $\hbar$ stands for $h/2\pi$: $\hbar = 1.052 \times 10^{-34}$ J s.

chemistry for it endows atoms and molecules with stability and governs the bonds they can form.

The Schrödinger equation is central to the discussion of electrons, atoms, and molecules. Although we shall not need to find its solutions explicitly, it is appropriate to see what the equation looks like. For the simple case of a particle of mass m moving in a one-dimensional region of space where its potential energy is V, the equation is:

$$-\frac{\hbar^2}{2m}\frac{d^2\psi}{dx^2} + V\psi = E\psi$$

where $\hbar = h/2\pi$. Although this equation is probably best regarded as a fundamental postulate, some motivation for its form comes from noting that the first term, the one proportional to $d^2\psi/dx^2$, is essentially the kinetic energy of the particle.[5] Therefore, the equation simply expresses in quantum mechanical terms the fact that the total energy, E, is the sum of the kinetic energy and the potential energy V.

Quantization

A simple illustration of how energy quantization emerges from the Schrödinger equation is obtained by considering a **particle in a box**, a particle confined to a one-dimensional region of constant potential energy between two impenetrable walls. An acceptable wavefunction of this system is very similar to the acceptable states of a vibrating violin string: the displacement must be zero at the walls, and between the walls consists of a standing wave with an integral number of half-wavelengths (Fig. 1.6). For an integral number of half-wavelengths to fit, the wavelength of the wavefunction must satisfy $n \times \frac{1}{2}\lambda = L$, where n is an integer and L is the length of the box. The acceptable solutions of the Schrödinger equation for the particle in a box are therefore waves of wavelength $\lambda = 2L/n$. Waves of this kind have the mathematical form

$$\psi = \sin\frac{2\pi x}{\lambda} = \sin\frac{n\pi x}{L}$$

with $n = 1, 2,...$ The number n is an example of a **quantum number**, an integer that labels the wavefunction. In a moment we shall see that n also determines the allowed value of certain properties of the system. For this one-dimensional system, we need only one quantum number to specify the wavefunction.

The wavefunction with $n = 1$ in Fig. 1.6 goes to zero at the walls of the box (at $x = 0$ and $x = L$) and is positive everywhere inside the box. The wavefunction with $n = 2$ is positive for $0 < x < \frac{1}{2}L$, passes through zero at the midpoint of the box, and is negative for $\frac{1}{2}L < x < L$. A point at which the wavefunction passes *through* zero (as distinct from merely

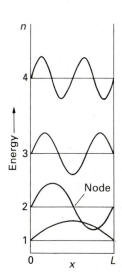

1.6 Energies (horizontal lines) of the lowest four levels of a particle in a box. Superimposed on the levels are the wavefunctions. The first four wavefunctions of a particle in a box have wavelengths $2L$, L, $\frac{2}{3}L$, and $\frac{1}{2}L$. In general, only waves of wavelength $2L/n$, with n an integer, fit between the walls.

---

[5] The identification of the first term with the kinetic energy can be justified by appealing to the de Broglie relation, and noting that $d^2\psi/dx^2$ is proportional to the curvature of the wavefunction ψ. Thus, as the wavelength of ψ decreases, the linear momentum of the particle increases, and so too therefore does its kinetic energy (which is proportional to p^2). However, as the wavelength shortens, the wave becomes more curved, and the value of $d^2\psi/dx^2$ increases. Thus, the kinetic energy is proportional to $d^2\psi/dx^2$.

approaching zero) is called a **node**, so the wavefunction with $n = 1$ has no nodes but that with $n = 2$ has one node. The illustration shows that for $n = 3$ there are two nodes and for $n = 4$ there are three nodes.

The permitted energies of the particle are given by the expression

$$E = \frac{n^2 h^2}{8mL^2}, \quad n = 1, 2, \dots$$

We now see that as well as labeling the wavefunction, the quantum number n determines the allowed energies of the system.

There are two features that are exhibited by a particle in a box that are echoed in the more complex atoms and molecules we consider shortly:

1. As the number of nodes increases, so too does the energy.

2. The larger the system, the smaller the separation of neighboring energy levels.

For example, as an illustration of the first point, the wavefunction ψ_n of a particle in a box has $n - 1$ nodes, and its energy is proportional to n^2. As an illustration of the second point, the size of a box is expressed by the length of the box L, and the separation in energy of the energy levels with quantum numbers $n + 1$ and n is

$$\Delta E = E_{n+1} - E_n = (2n + 1)\frac{h^2}{8mL^2}$$

We see that as L increases, the separation decreases (Fig. 1.7). Later we shall see that the energy levels of atoms and molecules behave similarly, for they come closer together as the dimensions of the system increase. A small helium atom, for instance, has widely spaced energy levels and is so unresponsive to its environment that it forms no compounds. A large xenon atom, on the other hand, has closely spaced energy levels and responds strongly enough to its environment to form several compounds.

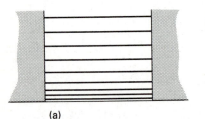

(a)

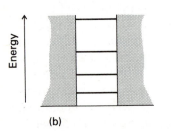

(b)

1.7 The separations of the energy levels of a particle in a square well increase as the length of the box is reduced from (a) to (b).

Transitions

An experimental confirmation of energy quantization is the observation that discrete frequencies of electromagnetic radiation are absorbed and emitted by atoms. This observation is explained on the grounds that electromagnetic radiation of frequency v consists of a stream of particles called **photons**, each one of which has an energy hv. Two points to note are:

1. The higher the frequency of the light, the more energetic each of its photons.

2. The greater the intensity of the light, the greater the number of photons in the beam.

The energy of an atom or molecule increases by hv when it absorbs a photon of frequency v. Similarly, if an atom or molecule emits a

Table 1.2 Color, frequency, and wavelength of electromagnetic radiation

Color	Frequency $v/10^{14}$ Hz	Wavelength λ/nm	Energy per photon $hv/(10^{-19}$ J$)$	E/eV
X-rays and γ-rays	10^3 and above	3 and below	660 and above	420 and above
Ultraviolet radiation	10	300	6.6	4.1
Visible light:				
Violet	7.1	420	4.7	2.9
Blue	6.4	470	4.2	2.6
Green	5.7	530	3.7	2.3
Yellow	5.2	580	3.4	2.1
Orange	4.8	620	3.2	2.0
Red	4.3	700	2.8	1.8
Infrared radiation	3.0	1000	2.0	1.3
Microwaves and radiowaves	3×10^{-3} and below	1 mm and above	2.0×10^{-3} and below	1.3×10^{-3} and below

photon of frequency v, then its energy decreases by hv. These remarks are summarized by the **Bohr frequency condition**, which states that if the change in energy of the atom or molecule is ΔE, then the frequency of the light absorbed or emitted must satisfy

$$\Delta E = hv$$

Because the energy of a bound system, like an atom or molecule, is quantized, only certain changes of energy ΔE are possible, so only certain values of v will occur in the light emitted or absorbed. Transitions between widely separated energy levels emit (or absorb) high frequency radiation; transitions between closely spaced energy levels emit (or absorb) low frequency radiation. Moreover, because the frequency v and wavelength λ of light are related by

$$\lambda = \frac{c}{v}$$

it follows that transitions between widely separated energy levels emit (or absorb) short wavelength radiation whereas transitions between closely spaced energy levels emit (or absorb) long wavelength radiation. If the wavelength of the radiation lies in the range 400 to 800 nm, we perceive it as visible light. The relation between color and wavelength is given with other information in Table 1.2.

Because the properties of light, particularly its frequency, are an important source of information on energy levels, it is common to express the energies of the levels themselves in terms of the quantities and units normally used for discussing the properties of radiation. Of these quantities the most common is the **wavenumber**, $\tilde{v}$:

$$\tilde{v} = \frac{v}{c} = \frac{1}{\lambda}$$

The dimensions of wavenumber are 1/length and the unit employed is normally reciprocal centimeters, cm^{-1}. The wavenumber in cm^{-1}

can be pictured as the number of wavelengths of the radiation that fit into 1 cm, so the shorter the wavelength (the higher the frequency) the higher the wavenumber. Infrared spectrometers for the study of molecular vibrations typically cover the wavenumber range 200 to 4000 cm^{-1}. Wavenumbers of visible light are typically about 2×10^4 cm^{-1}; those of ultraviolet radiation are closer to 10^5 cm^{-1}. In terms of the wavenumber, the Bohr frequency condition is

$$\Delta E = hc\tilde{\nu}$$

Example 1.2: *Interpreting the energies of characteristic atomic transitions*
Sodium street lamps give off a characteristic yellow light of wavelength 588 nm. What is the frequency of this light? What is the energy per mole (in kJ mol^{-1}) of these photons?

Answer. The frequency is

$$\nu = \frac{c}{\lambda} = \frac{3.00 \times 10^8 \text{ m s}^{-1}}{5.88 \times 10^{-7} \text{ m}} = 5.10 \times 10^{14} \text{ s}^{-1}$$

The energy of a single photon is

$$E = h\nu = (6.63 \times 10^{-34} \text{ J s}) \times (5.10 \times 10^{14} \text{ s}^{-1}) = 3.38 \times 10^{-19} \text{ J}$$

The energy per mole is the energy of a single photon multiplied by Avogadro's constant, N_A (the number of entities, such as photons, per mole):

$$E = (6.02 \times 10^{23} \text{ mol}^{-1}) \times (3.38 \times 10^{-19} \text{ J}) = 2.03 \times 10^5 \text{ J mol}^{-1}$$

or 203 kJ mol^{-1}.

Exercise E1.2. What is the energy per mole of the 514 nm photons emitted in the Ar$^+$ transition which is exploited to make green laser light?

1.8 The probability density for a particle in a box with $n = 1$ and 2. Note that there is zero probability of finding a particle at the midpoint of a box when it is in a state $n = 2$: this point corresponds to a node of ψ_2, for the wavefunction passes through zero at $x = \frac{1}{2}L$.

The Born interpretation

De Broglie did not really know what he meant by the 'wave' that is 'associated' with a particle. The interpretation was clarified by the German physicist Max Born, who proposed that the square ψ^2 of the wavefunction[6] is proportional to the *probability* of finding the particle in an infinitesimal region of space. According to the **Born interpretation**, there is a high chance of finding the particle where ψ^2 is large and the particle will not be found where ψ^2 is zero. The quantity ψ^2 is called the **probability density** of the particle.

Figure 1.8 shows the probability density for the lowest two states of a particle in a box. We see that the most probable region of finding the particle when it is in a state with $n = 1$ is in the center of the box,

---

[6] If the wavefunction is complex, in the sense of having real and imaginary parts, the probability is proportional to the square modulus, $\psi^*\psi$, where ψ^* is the complex conjugate of ψ. For simplicity, we shall usually assume that ψ is real and write all formulas accordingly.

but when $n = 2$ it is most likely to be found on either side of the center and there is zero probability of its being found in an infinitesimal region at the midpoint itself.

The precise interpretation of ψ is expressed in terms of the volume element $d\tau$, an infinitesimal region of space (such as a minute region of an atom). If we wish to know the probability that an electron will be found in a volume element $d\tau$ at a specified point, then we evaluate the probability density ψ^2 at that point and multiply it by the size of the volume element, obtaining $\psi^2 d\tau$ (this procedure is similar to calculating the mass of a region of a solid by multiplying the *mass* density of the solid by the volume of the region of interest). The total probability of finding the electron somewhere in the universe is the sum (integral) of the probabilities of finding it in all the volume elements into which the universe can be pictured as being divided. However, because we also know that the probability of being *somewhere* is 1 (that is, a particle is certainly somewhere), we can conclude that the wavefunction must satisfy the relation

$$\int \psi^2 d\tau = 1$$

Wavefunctions that satisfy this relation are said to be **normalized**. The Born interpretation is valid only for normalized wavefunctions.

The very important conclusion from this discussion is that, with its emphasis on the *probability* of finding particles in various regions rather than *precise predictions* of their locations, quantum mechanics does away with the classical concept of an orbit.

The sign of the wavefunction

Like other waves, wavefunctions may have regions of positive or negative amplitude. However, unlike waves in water, where a positive displacement of the wave is an accumulation of water and a negative displacement is a trough, these signs do not have a direct physical significance. That is, a positive region of a wavefunction does *not* mean that there is a greater chance of finding an electron there. Indeed, with its emphasis on the *square* of ψ, the Born interpretation means that the sign of the wavefunction has no direct physical significance. When we want to interpret a wavefunction, we should focus on its magnitude, not whether it happens to be positive or negative. For example, the wavefunction of a particle in a box with $n = 2$ is positive on the left of the box and negative on the right, but the probability density is the same on either side of the midpoint. We shall see that the wavefunctions of electrons in atoms also have positive and negative regions; in each case, though, the probability of finding the electron somewhere is independent of the sign of the wavefunction there.

The sign of the wavefunction, however, cannot be ignored. It is of crucial importance when carrying out calculations, and it is only when we seek a physical interpretation that we take the square of the wavefunction and lose the significance of its signs. The sign of a wavefunction is particularly important when two wavefunctions spread into

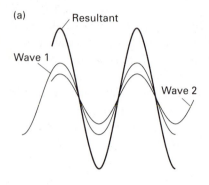

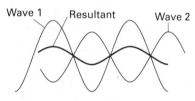

1.9 Wavefunctions interfere where they spread into the same region of space. (a) If they have the same sign in a region, they interfere constructively and the total wavefunction has an enhanced amplitude in the region. (b) If the wavefunctions have opposite signs, then they interfere destructively, and the resulting superposition has a reduced amplitude.

the same region of space, for then a positive region of one wavefunction may add to a positive region of the other wavefunction to give an enhanced wavefunction. This enhancement of amplitude (Fig. 1.9) is called **constructive interference**. It means that, where the two wavefunctions overlap, there may be a significantly enhanced probability of finding the particles described by the wavefunctions. As we shall see, this aspect of wavefunctions is of great importance in the explanation of chemical bonding. Alternatively, a positive region of one wavefunction may be canceled by a negative region of the second wavefunction. This **destructive interference** between wavefunctions will greatly lessen the probability that a particle will be found in that region.

The extent to which two wavefunctions ψ_1 and ψ_2 interfere with one another is measured by their **overlap integral** S, which is defined as

$$S = \int \psi_1^* \psi_2 \, d\tau$$

If two wavefunctions are widely separated in space (for example, if they describe the particles in two different boxes), then $S = 0$ because where one function is large the other is small, so their product is small everywhere (Fig. 1.10). If the two wavefunctions are very similar, and have the same signs in the regions of space they share, then S may approach 1 (the value it has when two identical wavefunctions are coincident). If the two wavefunctions interfere constructively in one region but interfere destructively to a similar extent in another region, then S will be close to zero. An example of this behavior is the overlap integral between the $n = 1$ and $n = 2$ wavefunctions of a particle in a box. Two wavefunctions that have identically zero overlap (such as these two wavefunctions) are said to be **orthogonal**. It is often helpful to think of the value of the overlap integral as indicating the *similarity* of two wavefunctions, with $S \approx 1$ indicating close similarity and $S \approx 0$ indicating that they are very dissimilar (orthogonal).

It is not necessary to know the absolute sign of a wavefunction for most purposes in inorganic chemistry, but it is often important to know the relative signs of two wavefunctions or of different regions of one wavefunction, particularly when assessing the degree of overlap. Therefore, we shall label regions of opposite sign with dark and light shading (sometimes white in the place of light shading).

1.5 Atomic orbitals

The wavefunctions of an electron in an atom are called **atomic orbitals**. Hydrogenic atomic orbitals are central to a large part of the interpretation of inorganic chemistry, and we shall spend some time describing their shapes and significance.

Hydrogenic energy levels

We begin by considering the Schrödinger equation for a hydrogenic atom consisting of a positive nucleus of atomic number Z and charge

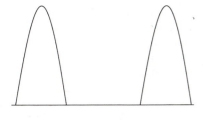

(a)

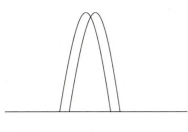

(b)

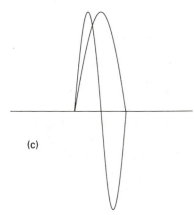

(c)

1.10 The overlap integral, S, is a measure of the similarity of two wavefunctions, as represented here for the wavefunctions of a particle in a box. (a) The boxes are so widely separated that the wavefunctions occupy different regions of space and $S = 0$. (b) The boxes are close together, the two wavefunctions resemble each other closely, and $S \approx 1$. (c) The boxes are coincident, but one wavefunction has a node at the center, and $S = 0$. These two wavefunctions are mutually orthogonal.

Ze and one electron (of charge $-e$) bound to it. The potential energy of an electron at a distance r from the nucleus is

$$V = -\frac{Ze^2}{4\pi\varepsilon_0 r}$$

where ε_0 is a fundamental constant, the vacuum permittivity, with the value 8.854×10^{-12} J^{-1} C^2 m^{-1}. This expression is called the **Coulomb potential energy**. The Schrödinger equation for this three-dimensional system is

$$-\frac{\hbar^2}{2m_e}\left(\frac{\partial^2}{\partial x^2} + \frac{\partial^2}{\partial y^2} + \frac{\partial^2}{\partial z^2}\right)\psi + V\psi = E\psi$$

where the potential energy V is given above. It is important to remember in what follows that the solutions we obtain are for the 'centro-symmetric' Coulomb potential.

Just as there are only certain acceptable solutions for the particle in a box, so there are only certain acceptable solutions of the Schrödinger equation for a hydrogenic atom. Because the atom consists of a single particle (the electron) in a three-dimensional region of space, the solutions need three quantum numbers for their complete specification. These quantum numbers are designated n, l, and m_l. The allowed energies of a hydrogenic atom are determined solely by the **principal quantum number** n, and are given by

$$E = -\frac{hc\Re}{n^2} \quad n = 1, 2,...$$

The zero of energy corresponds to the electron and nucleus widely separated and stationary; the constant $\Re$ is a collection of fundamental constants called the **Rydberg constant**:

$$\Re = \frac{m_e e^4}{8h^3 c\varepsilon_0^2}$$

Its numerical value is 1.097×10^5 cm^{-1}, corresponding to an energy of 13.6 eV. The $1/n^2$ dependence of the energy leads to a rapid convergence of energy levels at high (less negative) energies (Fig. 1.11). The zero of energy, which occurs when $n = \infty$, corresponds to an infinite separation of a stationary nucleus and electron, and therefore to ionization of the atom. Above this zero of energy, the electron is unbound and may travel with any energy.

In a hydrogenic atom, all orbitals with the same value of n correspond to the same energy and hence are said to be **degenerate**. The principal quantum number therefore defines a series of **shells** of the atom, or sets of orbitals with the same value of n and hence (in a hydrogenic atom) with the same energy.

Atomic quantum numbers

The orbitals belonging to each shell (that is, all orbitals of a given value of n) are classified into **subshells**. Each subshell of a shell is

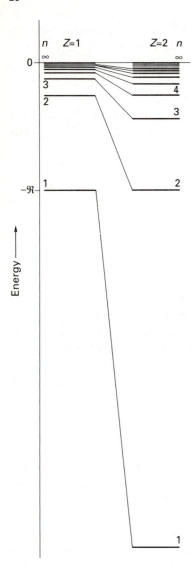

1.11 The quantized energy levels of an H atom ($Z = 1$) and an He$^+$ ion ($Z = 2$). The separation of the energy levels of a hydrogenic atom is proportional to Z^2.

distinguished by a quantum number l called the **orbital angular momentum quantum number**. For a given principal quantum number n, the quantum number l can have the values

$$l = 0, 1,..., n-1$$

giving n different values in all. Thus, the shell with $n = 2$ consists of two subshells of orbitals, one with $l = 0$ and the other with $l = 1$. It is common to refer to each subshell by a letter:

l: 0 1 2 3 4 ...

$\quad$ s $\ p$ $\ d$ $\ f$ $\ g$ $\ ...$

It follows that there is only one subshell (an s subshell) in the shell with $n = 1$, two subshells in the shell with $n = 2$ (the s and p subshells), three in the shell with $n = 3$ (the s, p, and d subshells), four when $n = 4$ (the s, p, d, and f subshells), and so on. For most purposes in chemistry we need consider only s, p, d, and f subshells.

A subshell with quantum number l consists of $2l + 1$ individual orbitals. These may be distinguished by the **magnetic quantum number** m_l which can take the $2l + 1$ values

$$m_l = l, l-1, l-2,..., -l$$

Thus, a d subshell of an atom consists of five individual atomic orbitals that are distinguished by the following values of m_l:

$$m_l = 2, 1, 0, -1, -2$$

The practical conclusion for chemistry from these remarks is that there is only one orbital in an s subshell ($l = 0$), the one with $m_l = 0$: this orbital is called an **s orbital**. There are three orbitals in a p subshell ($l = 1$), with quantum numbers $m_l = +1$, 0, and -1; they are called **p orbitals**. The five orbitals of a d subshell ($l = 2$) are called **d orbitals**, and so on.

Orbital angular momentum

In a centrosymmetric system (such as a hydrogenic atom) the **angular momentum**, a measure of the momentum of a particle travelling around the central nucleus, is quantized. Its magnitude is limited to the values

$$\text{orbital angular momentum} = \sqrt{\{l(l+1)\}} \times \hbar$$

This expression shows that an electron in an s orbital (with $l = 0$) has zero orbital angular momentum around the nucleus. As l increases on going from subshell to subshell, the orbital angular momentum rises too, and an electron in a d orbital has a higher orbital angular momentum than an electron in a p orbital. The angular momentum plays an important role in determining the shape of the orbital close to the nucleus, because an electron with angular momentum cannot approach the nucleus very closely on account of the strong centrifugal

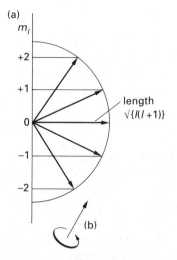

1.12 (a) The significance of the quantum numbers l and m_l for $l = 2$. The quantum number l gives the magnitude of the orbital angular momentum, as represented here by the length of a vector. The quantum number m_l specifies the component of momentum around the z axis as represented by the $2l + 1$ orientations of the vector. (b) The relation between the vector (for $m_l = +2$ in this case) and the direction of the motion.

effect of the angular momentum.[7] However, an s electron, with its zero orbital angular momentum, experiences no centrifugal force and can approach the nucleus closely. As we shall see, this difference between orbitals is largely responsible for the structure of the periodic table.

Solution of the Schrödinger equation also shows that angular momentum can occur with only certain values around a designated axis. If the magnitude of the angular momentum is represented by a vector of length $\sqrt{\{l(l+1)\}}$ units, then the only components that the vector can have on a selected axis (for instance, an axis defined by an applied magnetic field) are m_l units (Fig. 1.12). The restriction of the orientation of the angular momentum to certain values is yet another example of quantization and is often called **space quantization**.

Electron spin

In addition to the three quantum numbers required to specify the spatial distribution of an electron in a hydrogenic atom, two more quantum numbers are needed to define the state of an electron completely. These additional quantum numbers relate to the intrinsic angular momentum of an electron, its **spin**. This evocative name suggests that an electron can be considered as having an angular momentum arising from an intrinsic spinning motion, rather like the diurnal rotation of a planet about its polar axis as it travels in its annual orbit around the sun. However, spin is a purely quantum mechanical property and differs considerably from its classical namesake.

Although (like orbital angular momentum) the spin angular momentum is specified by a quantum number, in this case s, and has magnitude $\sqrt{\{s(s+1)\}}\hbar$, the value of s for an electron is *invariably* $\frac{1}{2}$. Electron spin is similar to electron charge and mass: it has a value characteristic of the particle and cannot be changed. Like orbital angular momentum, the spin of an electron can have only certain orientations relative to a chosen axis. However, for electron spin, only two orientations are allowed. They are distinguished by the quantum number m_s, which can take the values $+\frac{1}{2}$ and $-\frac{1}{2}$ but no others. These two spin states of the electron, which classically can be pictured as the rotation of an electron on its axis either clockwise or counterclockwise, are often represented by the two arrows ↑ ('spin-up') and ↓ ('spin-down').

Because the spin state of an electron must be specified if the state of the atom is to be specified fully, it is common to say that the state of an electron in a hydrogenic atom is characterized by four quantum numbers, namely n, l, m_l, and m_s (the fifth quantum number s is fixed at $\frac{1}{2}$).

---

[7] The repulsive centrifugal force on an electron is proportional to $l(l+1)/r^3$ whereas the attractive Coulombic force is proportional to $1/r^2$. Therefore, if $l \neq 0$, the centrifugal force will become stronger than the Coulombic force when r is sufficiently small. If $l = 0$, the centrifugal force is identically zero at all radii, and the electron experiences only the Coulombic attraction whatever its distance from the nucleus.

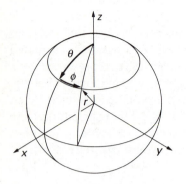

1.13 The spherical polar coordinates: r is the radius, θ the colatitude, and ϕ the azimuth.

The radial shapes of hydrogenic orbitals

The mathematical expressions for some of the hydrogenic orbitals are shown in Table 1.3. Because the Coulomb potential of the nucleus is spherically symmetric, the orbitals are best expressed in terms of the spherical polar coordinates r, θ, and ϕ defined in Fig. 1.13. In these coordinates, the orbitals all have the form

$$\psi_{n,l,m_l} = R_{n,l}(r)\,Y_{l,m_l}(\theta,\,\phi)$$

This formula (and the entries in the table) may look somewhat complicated, but it expresses the simple idea that a hydrogenic orbital can be written as the product of a function R of the radius and a function Y of the angular coordinates. The **radial wavefunction** R determines the variation of the orbital with distance from the nucleus. The **angular wavefunction** Y expresses the orbital's angular shape. Most of the time we shall use pictorial representations and not the expressions themselves.

The variations of the wavefunction with radius are shown in Figs 1.14 and 1.15. A 1s orbital, the wavefunction with $n = 1$, $l = 0$, and $m_l = 0$, decays exponentially with distance from the nucleus, but never passes through zero. All orbitals decay exponentially at sufficiently great distances from the nucleus, but some orbitals oscillate through zero close to the nucleus, and thus have one or more **radial nodes**, before beginning their final exponential decay. This oscillation is evident in the 2s orbital, the orbital with $n = 2$, $l = 0$, $m_l = 0$, which passes through zero once and hence has one radial node. A 3s orbital passes through zero twice and so has two radial nodes. A 2p orbital (one of the three orbitals with $n = 2$ and $l = 1$) has no radial nodes because its

1.14 The amplitudes of 1s, 2s, and 3s hydrogen orbitals as a function of distance from the nucleus. Note that the number of radial nodes is 0, 1, and 2, respectively. Each orbital has a nonzero amplitude at the nucleus (at $r = 0$). Although not clear on the scale of this diagram, the innermost nodes of 2s and 3s are not exactly coincident.

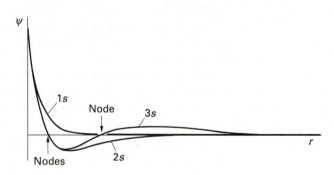

1.15 The amplitudes of 2p and 3p hydrogen orbitals as a function of distance from the nucleus. Note that the number of radial nodes is 0 and 1, respectively. Each orbital has zero amplitude at the nucleus (at $r = 0$).

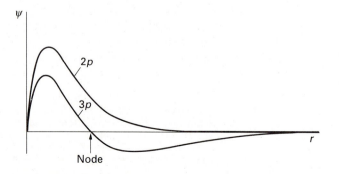

Table 1.3 Hydrogenic orbitals

(a) Radial wavefunctions
$$R_{nl}(r) = f(r)(Z/a_0)^{3/2}e^{-\rho/2}$$
where a_0 is the Bohr radius (0.53 Å) and
$\rho = 2Zr/na_0$

n	l	$f(r)$
1	0	2
2	0	$(1/2\sqrt{2})(2-\rho)$
2	1	$(1/2\sqrt{6})\rho$
3	0	$(1/9\sqrt{3})(6-6\rho+\rho^2)$
3	1	$(1/9\sqrt{6})(4-\rho)\rho$
3	2	$(1/9\sqrt{30})\rho^2$

(b) Angular wavefunctions
$$Y_{l,m}(\theta,\phi) = (1/4\pi)^{1/2}y(\theta,\phi)$$

l	m_l	$y(\theta,\phi)$
0	0	1
1	0	$3^{1/2}\cos\theta$
1	±1	$(3/2)^{1/2}\sin\theta\ e^{\pm i\phi}$
2	0	$(5/4)^{1/2}(3\cos^2\theta-1)$
2	±1	$(15/4)^{1/2}\cos\theta\sin\theta\ e^{\pm i\phi}$
2	±2	$(15/8)^{1/2}\sin^2\theta\ e^{\pm 2i\phi}$

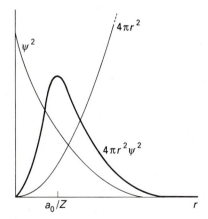

1.16 The radial distribution function of a hydrogenic 1s orbital. The product of $4\pi r^2$ (which increases as r increases) and ψ^2 (which decreases exponentially) passes through a maximum at $r = a_0/Z$.

radial wavefunction does not pass through zero anywhere. However, a 2p orbital, like *all* orbitals other than s orbitals, is zero at the nucleus. Although an electron in an s orbital may be found at the nucleus, an electron in any other type of orbital will not be found there. We shall soon see that this apparently minor detail, which is a consequence of the absence of orbital angular momentum when $l = 0$, is one of the key concepts for understanding the periodic table.

One of the reasons for displaying the mathematical expressions in Table 1.3 is to provide a more quantitative basis for the points we have been making. For example, because the 2s radial wavefunction ($R_{2,0}$) is proportional to the factor $2-\rho$, where ρ is proportional to r as specified in the table, we know at once that $\psi=0$ when $\rho=2$. That is, there is a radial node at $r = 2a_0/Z$, where a_0 is the **Bohr radius**:

$$a_0 = \frac{4\pi\varepsilon_0\hbar^2}{m_e e^2} = 5.29\times10^{-11}\ \text{m}\ (0.529\ \text{Å})$$

Similarly, the 3s orbital (with $n = 3$ and $l = 0$) has radial nodes at the two solutions of the quadratic equation $6-6\rho+\rho^2 = 0$. Table 1.3 also illustrates how (as a consequence of their lack of orbital angular momentum) s orbitals do not vanish at the nucleus (at $\rho=0$) but that all other orbitals do vanish there. For instance, the 3d orbital (with $n = 3$ and $l = 2$) is proportional to ρ^2, which is zero when $\rho=0$.

The radial distribution function

The force that binds the electron is centered on the nucleus, so it is often of interest to know the probability of finding an electron at a given distance from the nucleus regardless of its direction. This information enables us to judge how tightly the electron is bound to a nucleus. The total probability of finding the electron in a spherical shell of radius r and thickness dr is the integral of ψ^2 over all angles. For a spherical wavefunction (which is independent of angle), this integral can be evaluated easily:

$$\int_0^\pi\int_0^{2\pi}\psi^2\sin\theta\,d\theta\,d\phi\ r^2dr = 4\pi\psi^2r^2dr$$

The result is often written Pdr, where $P = 4\pi r^2\psi^2$ is called the **radial distribution function**. If we know the value of P at some radius r (which we can find once we know ψ), then we can state the probability of finding the electron somewhere in a shell of thickness dr at that radius simply by multiplying P by dr.

Because a 1s orbital decreases exponentially with distance from the nucleus, and r^2 increases, the radial distribution function of the orbital goes through a maximum (as shown in Fig. 1.16). Therefore, there is a distance at which the electron is most likely to be found. In general, this distance decreases as the nuclear charge increases (because the electron is attracted more strongly to the nucleus). It increases as n increases, because the higher the energy, the more likely it is that the electron will be found far from the nucleus. The most probable distance of an electron from the nucleus in the ground state of a hydrogenic atom is at the point where the radial distribution function goes through

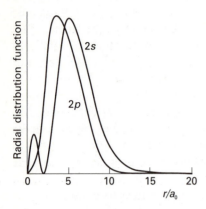

1.17 The radial distribution functions of hydrogen 2*s* and 2*p* orbitals. Although the 2*p* orbital is *on average* closer to the nucleus (note where its maximum lies), an electron in a 2*s* orbital has a higher probability of being close to the nucleus on account of the inner maximum.

a maximum. This maximum occurs at

$$r = \frac{a_0}{Z}$$

We see that the most probable radius decreases as the atomic number increases.

Example 1.3: *Interpreting radial distribution functions*

Figure 1.17 shows the radial distribution function for 2*s* and 2*p* wavefunctions. Which orbital gives the electron a greater probability of close approach to the nucleus?

Answer. The radial distribution function of a 2*p* orbital approaches zero near the nucleus faster than that of a 2*s* electron. This difference is a consequence of the fact that a 2*p* orbital has zero amplitude at the nucleus on account of its orbital angular momentum. Thus, the 2*s* electron has a greater probability of close approach to the nucleus.

Exercise E1.3. Which orbital, 3*p* or 3*d*, gives an electron a greater probability of being found close to the nucleus?

The angular shapes of atomic orbitals

An *s* orbital has the same amplitude at a given distance from the nucleus whatever the angular coordinates of the point of interest: that is, an *s* orbital is spherically symmetrical. It is normally represented by a sphere with the nucleus at its center. The sphere is called the **boundary surface** of the orbital, and defines the region of space within which there is a high (typically 90 percent) probability of finding the electron. All other types of orbitals (those with $l \neq 0$) have amplitudes that vary with the angle. Their boundary surfaces are more complicated in general, and are surfaces of constant ψ^2 that capture 90 percent (or whatever percentage is specified) of the probability of finding an electron.

In the most common graphical representation, the boundary surfaces of the three *p* orbitals of a given shell are identical (Fig. 1.18) apart from the fact that they lie parallel to each of the three different cartesian axes. This representation is the origin of the labels p_x, p_y, and p_z, which are alternatives to the m_l values as labels for individual orbitals. We shall refer to p_x, p_y, and p_z as the **real forms** of the *p* orbitals (because the wavefunctions are real functions). The orbitals labeled with m_l are the **complex forms** (because the wavefunctions are complex in the sense that their mathematical forms contain the imaginary[8] quantity

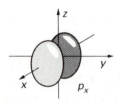

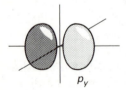

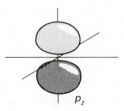

1.18 The boundary surfaces of *p* orbitals. Each orbital has one nodal plane running through the nucleus. For example, the nodal plane of the p_z orbital is the *xy* plane. The lightly shaded lobe has a positive amplitude, the darker shaded one is negative.

---

[8] For example, an orbital with $m_l = +1$ is proportional to $e^{i\phi}$ and one with $m_l = -1$ is proportional to $e^{-i\phi}$, and both are complex. The linear combinations

$$\frac{e^{i\phi} + e^{-i\phi}}{2} = \cos\phi \qquad \frac{e^{i\phi} - e^{-i\phi}}{2i} = \sin\phi$$

are both real and correspond to the p_x and p_y orbitals respectively. We see, therefore, that the complex forms correspond to definite orientations of the orbital angular momentum, but that the real forms are mixtures—superpositions—of these complex forms. It is an important feature of quantum theory that linear combinations of solutions of the Schrödinger equation with the same energy are equally valid solutions.

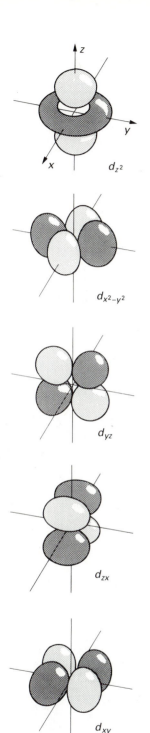

1.19 One representation of the five *d* orbitals of a shell. Four of the orbitals have two perpendicular nodal planes that intersect in a line passing through the nucleus. In the d_{z^2} orbital, the nodal surface forms two cones that meet at the nucleus.

$i = \sqrt{(-1)}$). The two forms differ in that the complex form of the orbitals with a definite value of m_l are traveling waves, and correspond to the electron circulating in a definite sense around the nucleus. The real forms are standing waves, and do not correspond to a definite sense of circulation. We need the complex forms when we are discussing a free atom, for the electrons are then free to circulate. We need the real forms when the electron is concentrated in a definite region of space, such as in a chemical bond. In most applications, we shall use only the real forms of the *p* orbitals, and represent them by diagrams like those in Fig. 1.18. As in that illustration, we shall show the positive and negative amplitudes of the wavefunction by different shading: positive amplitude will be shown by light gray and negative by dark gray. Each real orbital has a **nodal plane**, a plane on which there is zero probability that the electron will be found. The nodal planes cut through the nucleus and separate the regions of positive and negative sign.

The forms and labels we use for the *d* and *f* orbitals are shown in Figs 1.19 and 1.20, respectively. The boundary surfaces and phases shown there are the (real) standing wave versions of the orbitals; these are the only ones we need to consider explicitly.

MANY-ELECTRON ATOMS

As remarked at the start of the chapter, by a 'many-electron atom' is meant an atom with more than one electron; so even He, with two electrons, is a many-electron atom. The exact solution of the Schrödinger equation for an atom with N electrons would be a function of the $3N$ coordinates of all the electrons. There is no hope of finding exact formulas for such complicated functions, but as computing power has increased it has become possible to perform numerical computations that provide ever more precise energies and probability densities. However, the price of numerical precision is the loss of the ability to visualize the solutions. Chemists have therefore found it useful to make the **orbital approximation**, in which each electron is pictured as occupying (that is, having the distribution of) an atomic orbital that resembles those found in hydrogenic atoms.

1.6 The orbital approximation

In mathematical terms, the orbital approximation is expressed by writing the true wavefunction of an N-electron atom as a product of N one-electron wavefunctions:

$$\Psi = \psi(\mathbf{r}_1)\psi(\mathbf{r}_2)...\psi(\mathbf{r}_N)$$

This expression means that electron 1 is described by the wavefunction $\psi(\mathbf{r}_1)$, electron 2 by the wavefunction $\psi(\mathbf{r}_2)$, and so on.

The orbital approximation treats the repulsion between electrons in an approximate manner by supposing that the electronic charge is spherically distributed around the nucleus. Then each electron moves in the attractive field of the nucleus plus this average repulsive charge

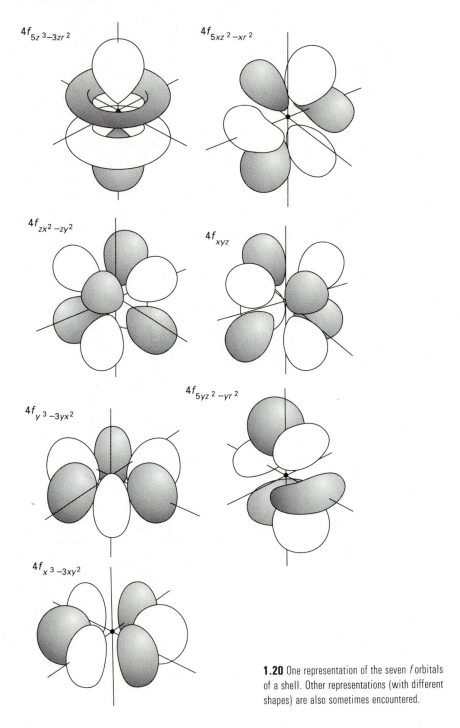

1.20 One representation of the seven *f* orbitals of a shell. Other representations (with different shapes) are also sometimes encountered.

distribution. According to classical electrostatics, the field that arises from a spherical distribution of charge is equivalent to the field generated by a point charge at the center of the distribution. The magnitude of the point charge is equal to the total charge within a sphere with a radius equal to the distance of the point of interest to the center of the distribution. In the orbital approximation, we assume that each elec-

Table 1.4 Effective nuclear charges, Z_{eff}

	H							He
Z	1							2
$1s$	1.00							1.69

	Li	Be	B	C	N	O	F	Ne
Z	3	4	5	6	7	8	9	10
$1s$	2.69	3.68	4.68	5.67	6.66	7.66	8.65	9.64
$2s$	1.28	1.91	2.58	3.22	3.85	4.49	5.13	5.76
$2p$			2.42	3.14	3.83	4.45	5.10	5.76

	Na	Mg	Al	Si	P	S	Cl	Ar
Z	11	12	13	14	15	16	17	18
$1s$	10.63	11.61	12.59	13.57	14.56	15.54	16.52	17.51
$2s$	6.57	7.39	8.21	9.02	9.82	10.63	11.43	12.23
$2p$	6.80	7.83	8.96	9.94	10.96	11.98	12.99	14.01
$3s$	2.51	3.31	4.12	4.90	5.64	6.37	7.07	7.76
$3p$			4.07	4.29	4.89	5.48	6.12	6.76

Source: E. Clementi and D. L. Raimondi, *Atomic screening constants from SCF functions*, IBM Research Note NJ-27, 1963

tron experiences its own characteristic **central field**. This central field is the sum of the field of the nucleus and the average field of all the electrons other than the one of interest. The average field (which depends on the value of *l* of the electron of interest, because electrons in different orbitals occupy different regions of space) is modeled by a point negative charge centered on the nucleus. This point negative charge reduces the nuclear charge from its true value Ze to an effective nuclear charge $Z_{eff}e$. This reduction is called **shielding**, and the **shielding parameter**, σ, is a correction to the true nuclear charge:

$$Z_{eff} = Z - \sigma$$

Because the average field is taken to be centrosymmetric, the *angular* components of the wavefunctions will be the same as those of the hydrogenic orbitals we have already discussed. However, because the effective nuclear charge varies with distance from the nucleus, the radial wavefunctions will be markedly different from those in a hydrogenic atom and their shapes and energies will respond in complex ways to changing electron populations.

Once we know the effective nuclear charge we can write approximate forms of the atomic orbitals and begin to make estimates of their extent and other properties. This was first done by J. C. Slater, who devised a set of rules for estimating the value of Z_{eff} for an electron in any atom, and using the value to write down an approximate atomic orbital. The rules have been superseded by more accurate calculated values, which are listed in Table 1.4, and we shall consider shortly the implications of these values. Some points to note from the values in Table 1.4 are that across each period the effective nuclear charge for the valence electrons increases in line with the atomic number. However, the effective nuclear charge for an electron in a valence *s* orbital is greater than that for the corresponding *p* orbital of the same atom. An additional point is that the effective nuclear charges for the valence electrons of Period 3 elements are only slightly greater than those of Period 2 elements even though the nuclear charge itself is considerably greater.

The Pauli principle

It is quite easy to account for the electronic structure of the helium atom in its **ground state**, its state of lowest energy. According to the orbital approximation, we suppose that both electrons occupy an atomic orbital that has the same spherical shape as a hydrogenic $1s$ orbital, but with a more compact radial form: the effective nuclear charge is $+1.69e$, and so the electrons are drawn in toward the nucleus more closely than is the one electron of an H atom. The ground-state **configuration** of an atom is the statement of the orbitals its electrons occupy in its ground state. For helium, with two electrons in the $1s$ orbital, the ground-state configuration is denoted $1s^2$.

As soon as we come to the next atom in the periodic table, lithium ($Z = 3$) we encounter several major new features. The ground-state configuration is *not* $1s^3$. This configuration is forbidden by a fundamental feature of nature known as the **Pauli exclusion principle**:

> No more than two electrons may occupy a single orbital and, if two do occupy a single orbital, then their spins must be paired.

By 'paired' we mean that one electron spin must be ↑ and the other ↓; the pair is denoted ↑↓. Another way of expressing the principle is to note that because an electron in an atom is described by four variable quantum numbers, *no two electrons can have the same four quantum numbers*. The Pauli principle was introduced originally to account for the absence of certain transitions in the spectrum of atomic helium.

It follows that whereas the ground-state configuration of He that we have proposed, namely $1s^2$, is feasible (with the two electrons paired), the configuration of Li cannot be $1s^3$, for that would require all three electrons to occupy the same orbital, which is forbidden. Therefore, the third electron must occupy an orbital of the next higher shell, a shell with $n = 2$.

The question that now arises is whether the third electron occupies a $2s$ orbital or one of the three $2p$ orbitals. To answer this question, we need to examine the energies of the two subshells, for although they have the same energy in a hydrogenic atom, that is not the case in a many-electron atom.

Penetration and shielding

Our task now is to see why a $2s$ electron has a different (in fact, lower) energy than a $2p$ electron. To do so, we need to establish how the effective nuclear charge reflects the radial distribution of the electrons and, in particular, their closeness of approach to the nucleus.

For a specific electron in the atom, the closer to the nucleus that an electron can penetrate, the higher the value of Z_{eff} that it experiences (Fig. 1.21). With this point in mind, consider a $2s$ electron outside an

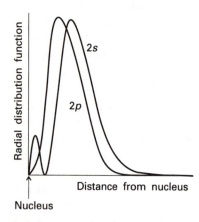

1.21 The penetration of a $2s$ electron through the inner core is greater than that of a $2p$ electron because the latter vanishes at the nucleus. Therefore, the $2s$ electrons are less shielded than the $2p$ electrons.

inner shell of two $1s$ electrons, as in the Li atom. If the $2s$ electron were entirely outside the two $1s$ electrons, it would experience a net centrosymmetric nuclear charge of $3e - 2e = +e$ (as in an H atom); that is, $Z_{eff} = 1$. If this description were accurate, then the energy of a $2s$ electron in Li would be exactly the same as the energy of a $2s$ electron in H, or $-\frac{1}{4}hc\Re$, corresponding to $-3.4\,\text{eV}$. However, the experimental energy of a $2s$ electron in a Li atom is $-5.4\,\text{eV}$. (As we see later, these energies are equal, with change of sign, to the ionization energy of the atom, and can be measured spectroscopically.) Therefore, although the calculation gives the right order of magnitude of the strength with which the electron is bound, it underestimates its value.

The error in the model on which the calculation is based is that the $2s$ electron of Li does not remain outside the $1s$ electrons. There is a nonzero probability that the $2s$ electron **penetrates** the $1s$ shell and experiences a nuclear charge greater than $+e$ and closer to the true value of $+3e$. Moreover, the lowering of energy arising from penetration is greater for a $2s$ electron than for a $2p$ electron: the latter, as we have seen, does not approach the nucleus so closely as a $2s$ electron. We can conclude that a $2s$ electron has a lower energy (is bound more tightly) than a $2p$ electron, and therefore that the ground-state electron configuration of Li is $1s^2 2s^1$.

Before going further, we shall introduce some nomenclature. The outermost occupied s and p orbitals of an atom are called its **valence orbitals**, and belong to the **valence shell**. Thus, the valence orbital of a Li atom is a $2s$ orbital and the valence shell is the shell with $n = 2$. The shells inside the valence shell (for Li, the $n = 1$ shell) constitute the **core** of the atom, and the orbitals are called **core orbitals**. In the case of Li, the core has the electron configuration $1s^2$. Because this configuration is the same as the configuration of He, it is sometimes denoted [He]. The electron configuration of Li is then denoted [He]$2s^1$, which emphasizes that it has a single $2s$ electron outside a helium-like core. When (as is usually the case) all the core electrons are paired, we speak of a **closed core**.

The pattern of energies ($2s$ lower than $2p$) is a general feature of many-electron atoms. This pattern can be seen from Table 1.4, which gives the effective nuclear charges for a number of valence-shell atomic orbitals in the ground-state electron configuration of atoms. (Remember that the greater the effective nuclear charge, the more strongly the electron is bound.) In general:

> The overall trend in effective atomic charge is an increase across a period, for in general the increase in nuclear charge in successive groups is not canceled by the additional electron.

Furthermore, a valence-shell ns electron is generally less shielded than an np electron. So, for example, $Z_{eff} = 5.13$ for a $2s$ electron in an F atom, whereas for a $2p$ electron $Z_{eff} = 5.10$, a lower value. Similarly, the value of Z_{eff} is larger for an electron in an np orbital than for an nd electron of the same shell. As a result of penetration and shielding,

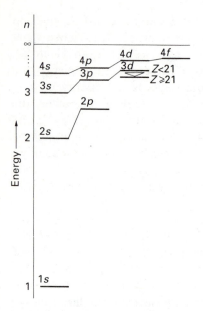

1.22 A schematic diagram of the energy levels of a many-electron atom with $Z < 21$ (as far as calcium). There is a change in order for $Z \geq 21$ (from scandium onward). This is the diagram that justifies the building-up principle, with up to two electrons being allowed to occupy each orbital.

the order of energies in many-electron atoms is

$$ns < np < nd < nf$$

because, in a given shell, s orbitals are the most penetrating and f orbitals are the least.

The overall effect of penetration and shielding is depicted in the energy-level diagram for a neutral atom shown in Fig. 1.22. As can be seen, the effects are quite subtle, and the order of the orbitals depends strongly on the numbers of electrons present in the atom. For example, the effects of penetration are very pronounced for $4s$ electrons in K and Ca, and in these atoms the $4s$ orbitals lie lower in energy than the $3d$ orbitals. However, from Sc through Zn, the $3d$ orbitals in the neutral atom lie close to but lower than the $4s$ orbitals.

The energies of the orbitals through the periodic table are shown in Fig. 1.23. In atoms from Ga ($Z = 31$) onward, the $3d$ orbital lies well below the $4s$ orbital in energy, and the outermost electrons are unambiguously those of the $4s$ and $4p$ subshells; in these elements, the $3d$ orbitals are no longer considered to be valence orbitals.

1.7 The building-up principle

The ground-state electron configurations of many-electron atoms are experimentally determined characteristics, and are deduced spectro-

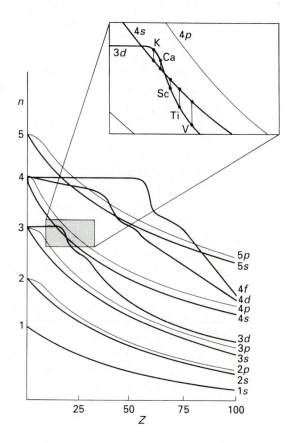

1.23 A more detailed portrayal of the energy levels of many-electron atoms in the periodic table. The inset shows a magnified view of the order near $Z = 20$, where the $3d$ elements begin.

scopically. The experimentally determined ground state electron configurations for atoms are given in Appendix 1. To account for them, we need to take into account both the effects of penetration and shielding on the energies of the orbitals and the role of the Pauli principle. The **building-up principle** (which is also called by its German name, the *Aufbau* principle, and is described below) is a procedure that leads to plausible ground-state configurations. It is not infallible, but it is an excellent starting point for the discussion. Moreover, as we shall see, it provides a theoretical framework for the periodic table.

Ground-state electron configurations

In the building-up principle, orbitals of the neutral atoms are treated as being occupied in the order determined in part by the principal quantum number and in part by penetration and shielding:

$1s\ 2s\ 2p\ 3s\ 3p\ 4s\ 3d\ 4p...$

Each orbital, according to the Pauli principle, is allowed to accommodate two electrons. Thus, a p subshell can accommodate up to six electrons and a d subshell can accommodate up to ten. The ground-state configurations of the first five elements, for instance, are

H	He	Li	Be	B
$1s^1$	$1s^2$	$1s^22s^1$	$1s^22s^2$	$1s^22s^22p^1$

When more than one orbital is available for occupation—such as when the $2p$ orbitals are starting to be filled in B and C—we adopt the following rule:

> **Hund's rule**: When more than one orbital has the same energy, electrons occupy separate orbitals and do so with parallel spins $(\uparrow, \uparrow)$.

The occupation of separate orbitals (such as a p_x orbital and a p_y orbital) can be understood in terms of the weaker repulsive interactions between electrons that occupy different regions of space than those that occupy the same orbital and are found in the same region of space. The requirement of parallel spins for electrons that do occupy different orbitals is a consequence of a quantum mechanical effect called **spin correlation**, the tendency of two electrons with parallel spins to stay apart from one another and hence to repel each other less.

Example 1.4: *Accounting for trends in effective nuclear charge*

Suggest a reason why the increase in Z_{eff} (Table 1.4) for a $2p$ electron is smaller between N and O than between C and N given that the configurations of the three atoms are

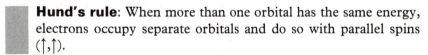

C: $[He]2s^22p^2$ N: $[He]2s^22p^3$ O: $[He]2s^22p^4$

Answer. On going from C to N, the additional electron occupies an empty $2p$ orbital. On going from N to O, the additional electron must occupy a $2p$ orbital that is already occupied by one electron. It therefore experiences a stronger repulsion and the increase in nuclear charge is more completely canceled than between C and N.

Exercise E1.4. Account for the larger increase in effective nuclear charge for a $2p$ electron on going from B to C compared with a $2s$ electron on going from Li to Be.

It follows from the building-up principle that the ground-state configuration of C is $1s^2 2s^2 2p_x^1 2p_y^1$ (it is arbitrary which of the p orbitals is occupied first: it is common to adopt the order p_x, p_y, p_z) or, more briefly, $1s^2 2s^2 2p^2$. If we recognize the helium-like core ($1s^2$), an even briefer notation is [He]$2s^2 2p^2$, and we can think of the electronic structure of the atom as consisting of two paired $2s$ electrons and two parallel $2p$ electrons surrounding a closed helium-like core. The electron configurations of the remaining elements in the period are similarly

$$\text{C [He]}2s^2 2p^2 \quad \text{N [He]}2s^2 2p^3 \quad \text{O [He]}2s^2 2p^4$$

$$\text{F [He]}2s^2 2p^5 \quad \text{Ne [He]}2s^2 2p^6$$

The $1s^2 2s^2 2p^6$ configuration of neon is another example of a closed shell, and is denoted [Ne].

The ground-state configuration of Na is obtained by adding one more electron to a neon-like core, and is [Ne]$3s^1$, showing that it consists of a single electron outside a closed shell. Now a similar sequence begins again, with the $3s$ and $3p$ orbitals complete at argon, with configuration [Ne]$3s^2 3p^6$, denoted [Ar]. Because the $3d$ orbitals are so much higher in energy, this configuration is effectively a closed shell. Moreover, the $4s$ orbital is next in line for occupation, so the configuration of K is analogous to that of Na, with a single electron outside a closed shell: specifically, it is [Ar]$4s^1$. The next electron, for Ca, also enters the $4s$ orbital, giving [Ar]$4s^2$, the analog of Mg. However, it is observed that the next element, scandium, accommodates the added electron into a $3d$ orbital, and the **_d_ block** of the periodic table begins.

In the d block, the d orbitals of the atoms are in the process of being occupied (according to the formal rules of the building-up principle). However, the energy levels in Figs 1.22 and 1.23 are for individual atomic orbitals and do not fully take into account interelectronic repulsions. A particularly important contribution to the energy arises when two electrons are present in the same d orbital and repel each other more strongly than do two electrons in an s orbital. For most of the d block, the spectroscopic determination of ground states (and detailed computation) shows that it is advantageous to occupy *higher* energy orbitals (the $4s$ orbitals). The explanation is that the occupation of orbitals of higher energy can result in a reduction in the repulsions between electrons that would occur if the lower-energy $3d$ orbitals were occupied. It is essential to consider all contributions to the energy of a configuration, not merely the one-electron orbital energies. Spectroscopic data show that the ground-state configurations of *d*-block atoms are of the form $3d^n4s^2$, with the $4s$ orbitals fully occupied despite individual $3d$ orbitals being lower in energy.

In some cases, a lower total energy may be obtained by forming a half-filled or filled *d* subshell at the expense of an *s* electron. Therefore, close to the center of the block the ground state configuration is likely to be d^5s^1 and not d^4s^2 (as for Cr), and close to the right of the block the configuration is likely to be $d^{10}s^1$ rather than d^9s^2 (as for Cu). A similar effect occurs in the *f* block, where *f* orbitals are being occupied. Thus the electron configuration of Gd is $[\text{Xe}]4f^75d^16s^2$.

The complication of the orbital energy not being a true guide to the total energy disappears late in Period 4 when the $3d$ orbital energies fall well below that of the $4s$ orbitals, for then the competition is less subtle. The same is true of the cations of the *d* block elements, where the removal of electrons reduces the complicating effects of electron-electron repulsions: all *d*-block cations have d^n configurations: the Fe^{2+} ion, for instance, has a d^6 configuration outside an argon-like closed shell. In later chapters (starting in Chapter 6), we shall see the great significance of the configurations of the *d*-metal ions, for the subtle modulations of their energies is the basis of important properties of their compounds. For the purposes of chemistry, the electron configurations of the *d*-block ions are more important than those of the neutral atoms.

Example 1.5: *Deriving an electron configuration*
Give the ground state electron configurations of the Ti atom and the Ti^{3+} ion.

Answer. For the atom, we add $Z = 22$ electrons in the order specified above, with no more than two electrons in any one orbital. This results in the configuration

$$\text{Ti } 1s^22s^22p^63s^23p^64s^23d^2, \text{ or } [\text{Ar}]4s^23d^2$$

with the two $3d$ electrons in different orbitals with parallel spins. However, because the $3d$ orbitals lie below the $4s$ orbitals once we are past Ca, it is appropriate to reverse the order in which they are written. The configuration is therefore reported as $[\text{Ar}]3d^24s^2$. The configuration of the cation is obtained by removing *first* the *s* electrons and *then* as many *d* electrons as required. We must remove three electrons in all, two *s* electrons and one *d* electron. The configuration of Ti^{3+} is therefore $[\text{Ar}]3d^1$.

Exercise E1.5. Give the ground state electron configurations of Ni and Ni^{2+}.

1.8 The format of the periodic table

Each period of the periodic table corresponds to the completion of the *s* and *p* subshells. The period number is the value of the principal quantum number *n* of the shell currently being occupied.

The group numbers are closely related to the number of electrons in the valence shell, but the precise relation depends on the group number G (and the numbering system adopted). In the '1-18' numbering system recommended by IUPAC:

Block	Number of valence electrons
s, d	G
p	$G - 10$

(For the purpose of this expression, the 'valence electrons' of a d-block element include its outermost d electrons.) The number of valence electrons for the p-block element selenium (Group 16) is therefore $16 - 10 = 6$. Alternatively, in the roman numeral system, the group number is equal to the number of s and p valence electrons for the s and p blocks. Thus, selenium belongs to Group VI; hence it has six valence (s and p) electrons. Thallium belongs to Group III, so it has three valence s and p electrons.

ATOMIC PARAMETERS

Certain characteristic properties of atoms, particularly their sizes and the energies associated with the removal and addition of electrons, show periodic variations with atomic number. These atomic properties are of considerable importance for accounting for the chemical properties of the elements. A knowledge of the variation enables chemists to rationalize observations and predict likely chemical and structural behavior without resort to tabulated data for each element.

1.9 Atomic and ionic radii

One of the most useful atomic properties of an element is the size of its atoms and ions. As we shall see in later chapters, geometrical considerations are central to the structures of many solids and individual molecules, and the distance of electrons from an atom's nucleus correlates well with the energy needed to remove them in the process of ion formation.

The quantum theory of the atom does not result in precise atomic or ionic radii because the radial distribution function falls off gradually with increasing distance from the nucleus. Despite this lack of a precise radius, we can expect atoms with numerous electrons to be larger, in some sense, than atoms that have only a few electrons. Such considerations have led chemists to determine a variety of measures of atomic radius on the basis of empirical considerations.

The **metallic radius** of a metallic element is defined as half the experimentally determined distance between the nuclei of nearest neighbor atoms in the solid (**1**, but see Section 4.3 for a refinement of this definition). The **covalent radius** of a nonmetallic element is similarly defined as half the internuclear separation of neighboring atoms of the same element in a molecule.[9] The periodic trends in radii

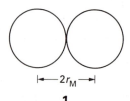

1

---

[9] We anticipate that atoms may be linked by single, double, and triple bonds, and that different bond orders result in different covalent radii. The covalent radii in Fig. 1.24 refer to single bonds between atoms. Multiple bonds are shorter than single bonds.

Table 1.5 Metallic radii (in angstroms)*

Li	Be													
1.57	1.12													

Na	Mg											Al		
1.91	1.60											1.43		

K	Ca	Sc	Ti	V	Cr	Mn	Fe	Co	Ni	Cu	Zn	Ga		
2.35	1.97	1.64	1.47	1.35	1.29	1.37	1.26	1.25	1.25	1.28	1.37	1.53		

Rb	Sr	Y	Zr	Nb	Mo	Tc	Ru	Rh	Pd	Ag	Cd	In	Sn	
2.50	2.15	1.82	1.60	1.47	1.40	1.35	1.34	1.34	1.37	1.44	1.52	1.67	1.58	

Cs	Ba	Lu	Hf	Ta	W	Re	Os	Ir	Pt	Au	Hg	Tl	Pb	Bi
2.72	2.24	1.72	1.59	1.47	1.41	1.37	1.35	1.36	1.39	1.44	1.55	1.71	1.75	1.82

*The values refer to coordination number 12 (see Section 4.3)
Source: A. F. Wells, *Structural inorganic chemistry*, 5th edn, Clarendon Press, Oxford (1984)

can be discerned in Table 1.5 and are illustrated in Fig. 1.24. We shall refer to metallic and covalent radii jointly as **atomic radii**. The **ionic radius** of an element is related to the distance between the nuclei of neighboring cations and anions. An arbitrary decision has to be taken on the apportionment, and in one common scheme the radius of the O^{2-} ion is taken to be 1.40 Å (**2**; see Section 4.6 for a refinement of this definition). For example, the ionic radius of Mg^{2+} is obtained by subtracting 1.40 Å from the internuclear separation of neighboring Mg^{2+} and O^{2-} ions in solid MgO. Some ionic radii are listed in Table 1.6.

The data in Table 1.5 show that *atomic radii increase down a group*, and within the s and p blocks *they decrease from left to right across a period*. These trends are readily interpreted in terms of the electronic structures of the atoms. On descending a group, the valence electrons are found in orbitals of successively higher principal quantum number, and hence occupy larger orbitals. Across a period, the valence electrons

$\longleftarrow r_+ + r_- \longrightarrow$

2

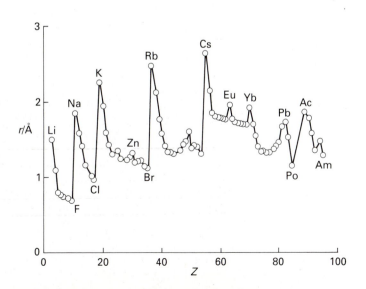

1.24 The variation of atomic radii through the periodic table. Note the contraction of radii following the lanthanides in Period 6. Metallic radii have been used for the metallic elements and covalent radii have been used for the nonmetallic elements.

enter orbitals of the same shell; however, the increase in effective nuclear charge across the period draws in the electrons and results in progressively more compact atoms (Table 1.5 and Fig. 1.24). The general increase in radius down a group and decrease across a period should be remembered as it correlates well with trends in many chemical properties.

Period 6 shows an interesting and important modification to these otherwise general trends. We see from Fig. 1.24 that the metallic radii in the third row of the *d* block are very similar to those in the second row, and not significantly larger as might be expected given their considerably larger numbers of electrons. For example, the radii of molybdenum and tungsten are 1.40 Å and 1.41 Å, respectively. The reduction of radius below what extrapolation might suggest is called the **lanthanide contraction**. The name points to the origin of the effect. The elements in the third row of the *d* block (Period 6) are preceded by the elements of the first row of the *f* block, the lanthanides, in which the 4*f* orbitals are being occupied. These orbitals have poor shielding properties,[10] and the repulsions between electrons being added on crossing the *f* block fail to compensate for the increasing nuclear charge, so Z_{eff} increases from left to right across a period. The dominating effect of the latter is to draw in all the electrons, and hence to result in a more compact atom. Another factor that contributes to the contraction is the relativistic effect of the increase in mass of electrons. This effect plays a significant role in the structures of elements of high atomic number, for their inner electrons move at very high velocities.

A similar contraction is found in the elements that follow the *d* block. For example, although there is a substantial increase in atomic radius between boron and aluminum (from 0.88 Å for B to 1.43 Å for Al), the atomic radius of gallium (1.53 Å) is only slightly greater than that of aluminum. As for the lanthanide contraction, this effect can be traced to the poor shielding characteristics of the elements earlier in the period. Because atomic radii have a central role to play in determining the chemical properties of the elements, these apparently small modifications of radii in fact have profound consequences. The consequences of trends in atomic radii will form a major unifying theme in the remainder of this text.

A general feature apparent from Table 1.6 is that:

> All anions are larger than their parent atoms and all cations are smaller (in some cases, markedly so).

The increase in radius of an atom on anion formation is a result of the greater electron–electron repulsions that occur in an anion compared to those in the neutral atom. The smaller radius of a cation compared with its parent atom is partly a consequence of the reduction in electron–electron repulsions that follows electron loss. It also arises

---

[10] A careful analysis (see D. R. Lloyd, *J. Chem. Educ.*, **53**, 502 (1986)) supports the view that the poor shielding abilities of *f* electrons stems from their radial distribution, not, as is sometimes expressed, from their highly angular shapes.

Table 1.6 Ionic radii (in angstroms)[*]

$Li^+(4)$ 0.59	$Be^{2+}(4)$ 0.27	$B^{3+}(4)$ 0.12	N^{3-} 1.71	$O^{2-}(6)$ 1.40	$F^-(6)$ 1.33
$Na^+(6)$ 1.02	$Mg^{2+}(6)$ 0.72	$Al^{3+}(6)$ 0.53	P^{3-} 2.12	$S^{2-}(6)$ 1.84	$Cl^-(6)$ 1.81
$K^+(6)$ 1.38	$Ca^{2+}(6)$ 1.00	$Ga^{3+}(6)$ 0.62	As^{3-} 2.22	$Se^{2-}(6)$ 1.98	$Br^-(6)$ 1.96
$Rb^+(6)$ 1.49	$Sr^{2+}(6)$ 1.16	$In^{3+}(6)$ 0.79		$Te^{2-}(6)$ 2.21	$I^-(6)$ 2.20
$Cs^+(6)$ 1.70	$Ba^{2+}(6)$ 1.36	$Tl^{3+}(6)$ 0.88			

[*]Numbers in parentheses are the coordination numbers of the ions. Values for ions without a coordination number stated are estimates.
Source: R. D. Shannon and C. T. Prewitt, *Acta Crystallogr.*, **A32**, 751 (1976).

because cation formation typically results in the loss of the valence electrons, often leaving only the much more compact closed-shell core of the atom. Once these gross differences are taken into account, the variation in ionic radii through the periodic table mirrors that of the atoms.

1.10 Ionization energy

The ease with which an electron can be removed from an atom is measured by its **ionization energy**, I, the minimum energy needed to remove an electron from a gas phase atom:

$$A(g) \rightarrow A^+(g) + e^-(g)$$

The **first ionization energy**, I_1, is the ionization energy of the least tightly bound electron of the neutral atom, the **second ionization energy**, I_2, is the ionization of the resulting cation, and so on. Ionization energies are conveniently expressed in **electronvolts** (eV), where 1 eV is the energy acquired by an electron when it falls through a potential difference of 1 V. Because this energy is equal to $e \times 1$ V, it is easy to deduce that 1 eV = 96.49 kJ mol^{-1}. The ionization energy of the H atom is 13.6 eV, so to remove an electron from an H atom is equivalent to dragging the electron through a potential difference of 13.6 V. In thermodynamic calculations it is often more convenient to use the **ionization enthalpy**, the standard enthalpy of the process written above. The molar ionization enthalpy is larger by RT than the ionization energy (because 2 mol of gas particles replace 1 mol when ionization occurs). However, because RT is only 2.5 kJ mol^{-1} (corresponding to 0.026 eV) at room temperature and ionization energies are of the order of 10^2 to 10^3 kJ mol^{-1} (1 to 10 eV), the difference between ionization energy and enthalpy can often be ignored. In this book we express ionization energies in electronvolts and ionization enthalpies in kilojoules per mole.

Table 1.7 First and second (and some higher) ionization energies of the elements (in electron-volts*)

H							He
13.60							24.58
							54.40
Li	**Be**	**B**	**C**	**N**	**O**	**F**	**Ne**
5.39	9.32	8.30	11.26	14.53	13.61	17.42	21.56
75.62	18.21	25.15					
	153.85	37.92					
		259.30					
Na	**Mg**	**Al**	**Si**	**P**	**S**	**Cl**	**Ar**
5.14	7.64	5.98	8.15	10.48	10.36	13.01	15.76
47.29	15.03	18.82					
	80.12	28.44					
		119.96					
K	**Ca**	**Ga**	**Ge**	**As**	**Se**	**Br**	**Kr**
4.34	6.11	6.00	8.13	9.81	9.75	11.84	14.00
31.81	11.87						
	51.21						
Rb	**Sr**	**In**	**Sn**	**Sb**	**Te**	**I**	**Xe**
4.18	5.69	5.79	7.34	8.64	9.01	10.45	12.13
27.5	11.03						
Cs	**Ba**	**Tl**	**Pb**	**Bi**	**Po**	**At**	**Rn**
3.89	5.21	6.11	7.42	7.29	8.43	10.75	10.74
25.1	10.00						
	Ra						
	5.28						
	10.15						

*To convert to kJ mol^{-1}, multiply by 96.485. See Appendix 1 for a longer list.
Source: C. E. Moore, *Atomic energy levels*, NBS Circular 467, (1949–1958).

To a large extent, the first ionization energy of an element is determined by the energy of the highest occupied orbital of its ground state atom.[11] The general trend in values can be summarized as follows:

First ionization energies vary systematically through the periodic table (Table 1.7 and Fig. 1.25), being smallest at the lower left (near cesium) and greatest near the upper right (near fluorine).

The variation follows the pattern of effective nuclear charge already mentioned in connection with the building-up principle, and (as Z_{eff} itself shows) there are some subtle modulations arising from the effect of electron–electron repulsions within the same subshell. Ionization energies also correlate strongly with atomic radii, for elements that

---

[11] This approximation is valid if the electrons remaining after ionization do not adjust their spatial distributions significantly. For the ionization of He, for example, the ionization energy is the difference in energy between He and He$^+$, and as the electron in He$^+$ is much more tightly bound than it is in He, the ionization energy cannot be ascribed solely to the one-electron orbital energy of the neutral atom.

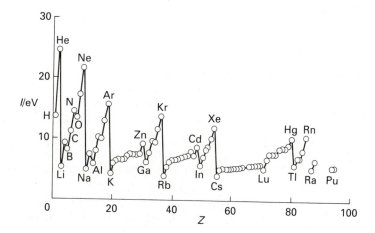

1.25 The variation of first ionization energies through the periodic table.

have small atomic radii generally have high ionization energies. The explanation of the correlation is that in a small atom an electron is close to the nucleus and experiences a strong Coulombic attraction.

Some differences in ionization energy can be explained quite readily. An example is the observation that the first ionization energy of boron is smaller than that of beryllium despite boron's higher nuclear charge. This anomaly is readily explained by noting that on going to boron, the outermost electron occupies a $2p$ orbital and hence is less strongly bound than if it had entered a $2s$ orbital. As a result, the value of I_1 falls back to a lower value. The decrease between nitrogen and oxygen has a slightly different explanation. The configurations of the two atoms are

$$\text{N } [\text{He}]2s^2 2p_x^1 2p_y^1 2p_z^1 \qquad \text{O } [\text{He}]2s^2 2p_x^2 2p_y^1 2p_z^1$$

(The choice of p_x, p_y, or p_z for double occupancy is arbitrary.) We see that in an O atom, two electrons are present in a single $2p$ orbital. They are so close together that they repel each other strongly, and this strong repulsion offsets the greater nuclear charge.

Example 1.6: *Accounting for a variation in ionization energy*
Account for the decrease in first ionization energy between phosphorus and sulfur.

Answer. The valence configurations of the two atoms are

$$\text{P } [\text{Ne}]3s^2 3p_x^1 3p_y^1 3p_z^1 \qquad \text{S } [\text{Ne}]3s^2 3p_x^2 3p_y^1 3p_z^1$$

As in the analogous case of N and O, in the ground configuration of S, two electrons are present in a $3p$ orbital. They are so close together that they repel each other strongly, and this increased repulsion offsets the effect of the greater nuclear charge of S compared with P.

Exercise E1.6. Account for the decrease in first ionization energy between fluorine and chlorine.

For fluorine and neon, the next electrons enter orbitals that are already half-full, and continue the trend from oxygen. The higher values of the ionization energies of these two elements reflects the high value of Z_{eff}. The value of I_1 falls back sharply from neon to sodium as the outermost electron occupies the next shell with an increased principal quantum number.

Another pattern in the ionization energies of the elements of considerable importance in inorganic chemistry is that <u>successive ionizations of a species require higher energies</u>. Thus, the second ionization energy of an element (the energy needed to remove an electron from the cation E^+) is higher than its first ionization energy, and its third ionization energy is higher still. The explanation is that a greater energy is required to remove an electron from a positively charged species. The difference in ionization energy is greatly magnified when the electron is removed from an inner shell of the atom (as is the case for the second ionization energy of lithium and any of its congeners) because the electron must then be extracted from a compact orbital in which it interacts strongly with the nucleus. The first ionization energy of lithium, for instance, is 5.3 eV, but its second ionization energy is 75.6 eV, more than ten times greater.

The pattern of successive ionization energies down a group is far from simple. Figure 1.26 shows the first, second, and third ionization energies of the boron group (Group 13/III) elements. Although they lie in the expected order $I_1 < I_2 < I_3$, the shape of each curve does not suggest any simple trend. The lesson to be drawn is that whenever an argument hangs on trends in small differences in ionization energies, it is always best to refer to actual numerical values rather than to guess a likely outcome.

Insofar as there is a pattern in the data shown in Fig. 1.26, it may reflect the presence of the d and f blocks earlier in a period. Figure 1.27 shows the variation in the *sum* of the first three ionization energies of

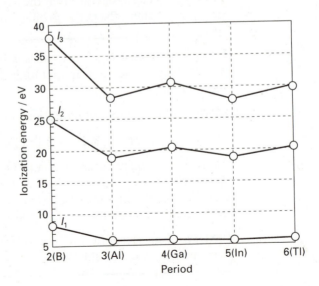

1.26 The first, second, and third ionization energies of the elements of Group 13/III. Successive ionization energies increase, but there is no clear pattern of ionization energies down the group.

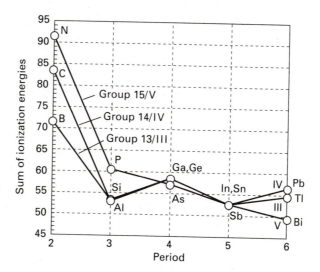

1.27 The sum of the first three ionization energies for the elements in Groups 13/III to 15/V.

the members of Groups 13/III to 15/V. The increase in value between Period 3 and 4 in Groups 13/III and 14/IV may reflect the shrinkage of the atoms that occurs as a result of the presence of the $3d$ series earlier in the row, and the corresponding increase between Periods 5 and 6 may reflect the similar effect of the $4f$ series. (However, the individual ionization energies, as we have seen, do not show such a simple correlation, so this correlation may not have a simple explanation either.)

It is conventional in inorganic chemistry to take note of the greater energy required to form the ions E^{3+} and E^{4+} in Period 6 compared with Period 5 (particularly for thallium and lead compared with indium and tin, respectively) and to call it the **inert-pair effect**, to signify the difficulty of removing s electrons and their reluctance to participate readily in compound formation. There is good experimental evidence for an effect of this kind, for many Tl^+ compounds are known, and Tl^{3+} compounds are easily reduced to Tl^+. Similarly, compounds that formally contain Pb^{4+} are easily reduced to compounds of Pb^{2+}. However, an experimental observation of a trend does not necessarily mean that the conventional explanation is correct. In this case, it should be borne in mind that the elements of Period 4 (gallium, germanium, and arsenic) actually have a *higher* total ionization energy than the elements of Period 6 for the formation of E^{3+} ions, yet they do not show the chemical characteristics ascribed to the inert pair effect. As we shall see in Chapter 2, it is always essential to consider *all* the contributions to the energy of bond formation, and the 'inert-pair effect' may be less a property of individual atoms and more a reflection of the influence of their radii on their abilities to form bonds with other atoms. That is, in the language introduced in Section 4.7, the true explanation of the inert-pair effect may be found in the lattice enthalpies of the compounds the elements form, not the electronic characteristics of the individual atoms.

Table 1.8 Electron affinities of the main-group elements (in electronvolts*)

H 0.754							**He** − 0.5
Li 0.618	**Be** − 0.5	**B** 0.277	**C** 1.263	**N** − 0.07	**O** 1.461 − 8.75	**F** 3.399	**Ne** − 1.2
Na 0.548	**Mg** − 0.4	**Al** 0.441	**Si** 1.385	**P** 0.747	**S** 2.077 − 5.51	**Cl** 3.617	**Ar** − 1.0
K 0.502	**Ca** − 0.3	**Ga** 0.30	**Ge** 1.2	**As** 0.81	**Se** 2.021	**Br** 3.365	**Kr** − 1.0
Rb 0.486	**Sr** − 0.3	**In** 0.3	**Sn** 1.2	**Sb** 1.07	**Te** 1.971	**I** 3.059	**Xe** − 0.8

*To convert to kJ mol^{-1}, multiply by 96.485.
The first values refer to the formation of the ion X^{-} from the neutral atom X; the second value to the formation of X^{2-} from X^{-}.
Source: H. Hotop and W. C. Lineberger, *J. Phys. Chem. Ref. Data,* **14**, 731 (1985).

1.11 Electron affinity

The **electron-gain enthalpy** ΔH_{eg} is the enthalpy change when a gas phase atom gains an electron:

$$A(g) + e^{-}(g) \rightarrow A^{-}(g)$$

Electron gain may be either exothermic or endothermic (Table 1.8). Although the electron-gain enthalpy is the thermodynamically appropriate term, much of inorganic chemistry is discussed in terms of a closely related property, the **electron affinity**, A_e, of an element, which is the negative of the energy change accompanying electron gain. If the electron affinity of an element A is higher than that of another element B, then electron gain is more exothermic for A than for B. As in the case of ionization energy, we can distinguish between the electron-gain enthalpy and the electron-gain energy; the former we report in kilojoules per mole and the latter in electronvolts.

The electron affinity of an element is determined in large part by the energy of the lowest unfilled (or half-filled) orbital of the ground state atom. This orbital is the second of the two **frontier orbitals** of an atom, its partner being the highest filled atomic orbital. The frontier orbitals are the site of many of the changes in electron distributions when bonds form, and we shall see more of their importance as the text progresses. An element has a high electron affinity if the additional electron can enter a shell where it experiences a strong effective nuclear charge. This is the case for elements toward the top right of the periodic table, as we have already explained. Therefore, elements close to fluorine (specifically oxygen, nitrogen, and chlorine) can be expected to have the highest electron affinities. The second electron-gain enthalpy, the enthalpy change for the attachment of a second electron to an initially neutral atom, is invariably positive because the electron repulsion outweighs the nuclear attraction.

Example 1.7: *Accounting for the variation in electron affinity*
Account for the large decrease in electron affinity between lithium and beryllium despite the increase in nuclear charge.

Answer. The electron configurations of the two atoms are $[He]2s^1$ and $[He]2s^2$. The additional electron enters the $2s$ orbital of lithium, but it must enter the $2p$ orbital of beryllium, and hence is much less tightly bound. In fact, the nuclear charge is so well shielded in beryllium that electron gain is endothermic.

Exercise E1.7. Account for the decrease in electron affinity between carbon and nitrogen.

1.12 Electronegativity

The **electronegativity** χ of an element is the power of an atom of the element to attract electrons to itself when it is part of a compound. If an atom has a strong tendency to acquire electrons, it is said to be highly **electronegative** (like the elements close to fluorine). If it has a tendency to lose electrons (like the alkali metals), it is said to be **electropositive**.

The electronegativity has been defined in many different ways and its precise interpretation is still the subject of debate.[12] Similarly, there have been a variety of attempts to formulate a quantitative scale of electronegativity. Pauling's original formulation (which results in the values called χ_P in Table 1.9 and illustrated in Fig. 1.28), draws on concepts relating to the energetics of bond formation which will be dealt with in more detail in Chapter 2. He argued that the excess

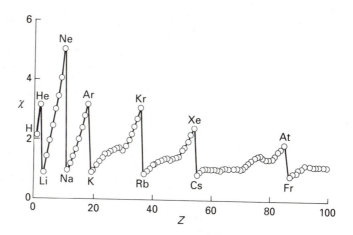

1.28 The variation of electronegativity in the periodic table.

[12] To get the flavor of this debate see: R.G. Pearson, *Acc. Chem. Res.*, **23**, 1 (1990) and L. C. Allen, *Acc. Chem. Res.*, **23**, 175 (1990).

Table 1.9 Pauling (*italics*) and Mulliken electronegativities

H							He
2.20							
3.06							

Li	Be	B	C	N	O	F	Ne
0.98	*1.57*	*2.04*	*2.55*	*3.04*	*3.44*	*3.98*	
1.28	1.99	1.83	2.67	3.08	3.22	4.43	4.60

Na	Mg	Al	Si	P	S	Cl	Ar
0.93	*1.31*	*1.61*	*1.90*	*2.19*	*2.58*	*3.16*	
1.21	1.63	1.37	2.03	2.39	2.65	3.54	3.36

K	Ca	Ga	Ge	As	Se	Br	Kr
0.82	*1.00*	*1.81*	*2.01*	*2.18*	*2.55*	*2.96*	*3.0*
1.03	1.30	1.34	1.95	2.26	2.51	3.24	2.98

Rb	Sr	In	Sn	Sb	Te	I	Xe
0.82	*0.95*	*1.78*	*1.96*	*2.05*	*2.10*	*2.66*	*2.6*
0.99	1.21	1.30	1.83	2.06	2.34	2.88	2.59

Cs	Ba	Tl	Pb	Bi			
0.79	*0.89*	*2.04*	*2.33*	*2.02*			

Source: Pauling values: A. L. Allred, *J. Inorg. Nucl. Chem.,* **17**, 215 (1961); L. C. Allen and J. E. Huheey, *ibid.,* **42**, 1523 (1980) Mulliken values: L. C. Allen, *J. Am. Chem. Soc.,* **111**, 9003 (1989). The Mulliken values have been scaled to the range of the Pauling values.

energy Δ of an A—B bond over the average energy of A—A and B—B bonds can be attributed to the presence of an ionic contributions to the covalent bonding. He defined the difference in electronegativity as

$$|\chi_A - \chi_B| = 0.102 \times \sqrt{(\Delta/\text{kJ mol}^{-1})}$$

where

$$\Delta = E(A—B) - \tfrac{1}{2}\{E(A—A) + E(B—B)\}$$

with $E(X—Y)$ the X—Y bond energy (the energy of the molecule relative to that of its dissociation products). Thus, if the A—B bond energy differs markedly from the average of the nonpolar A—A and B—B bonds, it is presumed that there is a substantial ionic contribution to the wavefunction, and hence a large difference in electronegativity between the two atoms.

There are complications with this definition, because Pauling electronegativities increase with increasing oxidation state of the element. (The values in Table 1.9 are for the maximum oxidation state of the element concerned.) Pauling electronegativities are useful for estimating bond enthalpies between elements of different electronegativity and for a qualitative assessment of the polarities of bonds.

Various alternative definitions of electronegativity can be understood in terms of the intrinsic properties of individual atoms rather than their behavior on bond formation. A widely used scale proposed by

A. L. Allred and Eugene Rochow, is based on the view that electronegativity is determined by the electric field at the surface of an atom. As we have seen, an electron in an atom experiences an effective nuclear charge Z_{eff}. The Coulombic potential at the surface of such an atom is proportional to Z_{eff}/r, and the electric field there is proportional to Z_{eff}/r^2. In the **Allred–Rochow definition** of electronegativity, χ_{AR} is assumed to be proportional this field, with r taken to be the covalent radius of the atom:

$$\chi_{AR} = 0.744 + \frac{0.3590 Z_{eff}}{(r/\text{Å})^2}$$

The parameters have been chosen so that values comparable to Pauling electronegativities are obtained. According to the Allred–Rochow definition, elements with high electronegativity are those with high effective nuclear charge and small covalent radius: these elements lie close to fluorine. The Allred–Rochow values parallel closely those of the Pauling electronegativities and are useful for discussing the electron distributions in compounds.

Robert Mulliken proposed a definition of electronegativity using data from atomic spectra. He observed that a simple view of electron redistribution on compound formation is one in which an atom becomes either a cation (by electron loss) or an anion (by electron gain). If an atom has a high ionization energy and a high electron affinity, then it will be likely to acquire electrons when it is part of a compound, and hence be classified as highly electronegative. Conversely, if its ionization energy and electron affinity are both low, then the atom will tend to lose electrons, and hence be classified as electropositive. These observations motivate the definition of the **Mulliken electronegativity** χ_M as the average value of the ionization energy and the electron affinity of the element:

$$\chi_M = \tfrac{1}{2}(I + A_e)$$

If both I and A_e are high, then the electronegativity is high; if both are low, then the electronegativity is low. Because the electron affinity and ionization energy increase rapidly for successively more positive ions, the Mulliken electronegativity increases greatly with the extent of ionization.

There is a complication in the definition of the Mulliken electronegativity, in as much as the ionization energy and electron affinity in the definition relate to a specified state of an atom, which is called its **valence state**. The valence state is the configuration that the atom is supposed to have when it is part of a molecule (in the terms we shall introduce in Chapter 2, the extent to which s and p orbitals are hybridized when bond formation occurs). Hence, some calculation is involved because the ionization energy and electron affinity to be used in calculating χ_M are mixtures of their values for the contributing spectroscopic states. We need not go into the calculation, but the resulting values given in Table 1.9 may be compared with the Pauling values. The two scales are approximately in line. One reasonably

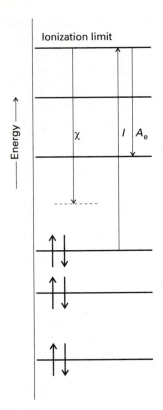

1.29 The interpretation of the electronegativity and the hardness of an element in terms of the energies of the frontier orbitals (the highest filled and lowest unfilled atomic orbitals).

reliable conversion between the two is

$$\chi_P = 1.35 \sqrt{\chi_M} - 1.37$$

Because the elements near fluorine are the ones with both high ionization energies and appreciable electron affinities, these elements have the highest Mulliken electronegativities. Because χ_M depends on atomic energy levels—and in particular the location of the frontier orbitals (Fig. 1.29)—the electronegativity of an element is high if the two frontier orbitals of its atoms are low in energy.

So-called **spectroscopic electronegativities**, χ_S, have also been defined by L. C. Allen. The definition is based on the average energy of the *s* and *p* electrons of an atom:

$$\chi_S = \frac{N_s \varepsilon_s + N_p \varepsilon_p}{N_s + N_p}$$

The energies ε_s and ε_p are obtained from averaged spectroscopic data. Because spectroscopic data are available, it is possible to calculate the electronegativities of the noble gases, for which there is no thermochemical data for the calculation of χ_P.

Electronegativities have numerous applications, and will be used in arguments throughout this text. These applications include a description of bond energies, the predictions of polarities of bonds and molecules, and the rationalization of the types of reactions that substances undergo.

1.13 Hardness and softness

We have seen that the mean of the energies of the frontier orbitals, half the sum of the ionization energy and the electron affinity, is one definition of the electronegativity of an element. The *separation* between the frontier orbitals, the difference of the ionization energy of the neutral atom and its anion is a measure of the chemical **hardness** η of the element:

$$\eta = \tfrac{1}{2}(I - A_e)$$

The hardness is large when the two levels are widely separated in energy. The hardness is low (the atom is 'soft') when the frontier orbitals are close together.[13] The concept of hardness was first introduced from empirical chemical considerations, and we shall describe this approach in Section 5.14.

According to the definition of hardness just given, the hardest atoms and ions are those with high ionization energies and low electron affinities. If the ionization energy is much larger than the electron affinity, as is often the case, hardness is correlated with high ionization energy. Hence, the hardest atoms and ions are the small atoms and ions near fluorine. The softest atoms and ions are the ones with low ionization energies and low electron affinities. These are the atoms and

---

[13] An instructive review of hardness and tables of values are given in R. G. Pearson, *Inorg. Chem.*, **27**, 734 (1988) and in an introductory article in *Chem. Brit.*, 441 (1991).

ions of the heavier alkali metals and the heavier halogens. The light atoms of a group are generally hard and the heavier atoms soft. The hardness of an atom is generally complementary to the electronegativity in determining the outcome of reactions, and it is important to acquire a sense of when one influence is likely to dominate the other.

The general significance of hardness is that *it indicates the responsiveness of the atom to the presence of electric fields*, particularly those arising from neighboring atoms and ions. Hardness is closely related to the concept of **polarizability**, α, the ease with which an atom or ion can be distorted by an electric field, and hard atoms and ions are species with low polarizabilities. One of the most important polarizing influences on an atom in chemistry is that arising from the presence of another atom nearby. The changes in wavefunction and energies that then result are responsible for the adhesion of one atom to another and result in chemical bonding, the topic we consider next.

FURTHER READING

P. A. Cox, *The elements: their origin, abundance, and distribution.* Oxford University Press (1989). A concise and readable account.

R. J. Puddephatt and P. K. Monaghan, *The periodic table of the elements.* Oxford University Press (1986). An elementary survey of the structure of the periodic table and the trends in properties of elements.

P. W. Atkins, *Physical chemistry.* Oxford University Press and W. H. Freeman & Co., New York (1994). Chapters 11 and 12 give an introduction to the principles of atomic structure.

P. W. Atkins, *Quanta: a handbook of concepts.* Oxford University Press (1991). A collection of nonmathematical accounts of the concepts of quantum chemistry arranged like an encyclopedia.

P. Pyykkö, Relativistic effects in structural chemistry. *Chem. Rev.,* **88**, 563 (1988). An accessible account of the role that relativistic considerations play in the chemical and physical properties of the elements.

J. Emsley, *The elements.* Oxford University Press (1991). A very useful collection of data and information in a handy format.

KEY POINTS

1. Hydrogen and helium formation
Hydrogen and helium formed during the earliest stages of the evolution of the universe, and they remain the most abundant elements.

2. Nuclear fusion
Light elements were synthesized in stars. Nuclear fusion cycles led to the formation of elements with atomic numbers up to 26 (iron). The heavier elements are not stable with respect to smaller fragments and were synthesized by more complex pathways.

3. The periodic table
The most comprehensive summary of relationships among the elements is the periodic table. It is divided into vertical groups and horizontal periods. The groups belong to four blocks. The *s* and *p* blocks constitute the main groups of the periodic table; the *d* block consists of the transition elements, and the *f* block consists of the lanthanides and actinides.

4. Quantum mechanics
The principal concepts of quantum mechanics in

chemistry are:
(a) the wavelike character of matter;
(b) the uncertainty principle;
(c) the quantization of certain properties, particularly energy; and
(d) the Pauli principle.

5. Quantum numbers

Solutions of the Schrödinger equation for hydrogenic atoms exist for specific values of the quantum numbers n, l, and m_l. The corresponding wavefunctions are called atomic orbitals, and describe the three-dimensional distribution of the electron around the nucleus. The square of the wavefunction represents the probability density for the distribution of an electron.

6. Electron spin

An additional quantum number m_s, which can have only the values $+\frac{1}{2}$ or $-\frac{1}{2}$, is needed to specify the spin state of the electron in an orbital.

7. Electron distributions

Electron distributions in many-electron atoms may be approximated by the occupation of atomic orbitals. The energies of orbitals are affected by penetration and shielding, with typically an ns orbital lying lower in energy than an np orbital.

8. The building-up principle

According to the building-up principle, electrons occupy orbitals in the order $1s$, $2s$, $2p$, $3s$, $3p$, $4s$, $3d$, with no more than two electrons in any one orbital. If more than one orbital is available for occupation, Hund's rule applies.

9. Trends in radii

Metallic, covalent, and ionic radii generally increase down a group and decrease across a period. The lanthanide contraction reduces the expected sizes of the atoms that follow the f block.

10. Ionization energy and electron affinity

The ionization energy and electron affinity are measures of the energy needed to remove an electron from and attach an electron to an atom, respectively. The periodic variation follows the variation in atomic radii and effective nuclear charge.

11. Electronegativity

The electronegativity of an element is the power of an atom of the element to attract electrons to itself when it is part of a compound. Electronegativities generally decrease down a group and increase from left to right across a period.

12. Hardness

The hardness of an element is a measure of the responsiveness of an atom or ion to an electric field, including the field arising from a neighboring atom in a chemical bond.

EXERCISES

1.1 Write balanced equations for the following nuclear reactions (show emission of excess energy as a photon of electromagnetic radiation, γ): (a) $^{14}N + {}^4He$ to produce ^{17}O; (b) $^{12}C + p$ to produce ^{13}N; (c) $^{14}N + n$ to produce ^3H and ^{12}C. (The last reaction produces a steady state concentration of radioactive ^3H in the upper atmosphere.)

1.2 One possible source of neutrons for the neutron-capture processes mentioned in the text is the reaction of ^{22}Ne with α particles to produce ^{25}Mg and neutrons. Write the balanced equation for the nuclear reaction.

1.3 Without consulting reference material, draw the form of the periodic table with the numbers of the groups and the periods and identify the s, p, and d blocks. Identify as many elements as you can. (As you progress through your study of inorganic chemistry, you should learn the positions of all the s, p, and d block elements and associate their positions in the periodic table with their chemical properties.)

1.4 What is the ratio of the energy of a ground state He^+ ion to that of a Be^{3+} ion?

1.5 The ionization energy of H is 13.6 eV. What is the difference in energy between the $n = 1$ and $n = 6$ levels?

1.6 What is the relation of the possible angular momentum quantum numbers to the principal quantum number?

1.7 How many orbitals are there in a shell of principal quantum number n? (Hint: begin with $n = 1$, 2, and 3 and see if you can recognize the pattern.)

1.8 Using sketches of 2*s* and 2*p* orbitals, distinguish between (a) the radial wavefunction, (b) the radial distribution function, and (c) the angular wavefunction.

1.9 Compare the first ionization energy of calcium to that of zinc. Explain the difference in terms of the balance between shielding with increasing numbers of *d* electrons and the effect of increasing nuclear charge.

1.10 Compare the first ionization energies of strontium, barium, and radium. Relate the irregularity to the lanthanide contraction.

1.11 The second ionization energies of some Period 4 elements are

Ca	Sc	Ti	V	Cr	Mn
11.87	12.80	13.58	14.15	16.50	15.64 eV

Identify the orbital from which ionization occurs and account for the trend in values.

1.12 Give the ground state electron configurations of (a) C, (b) F, (c) Ca, (d) Ga^{3+}, (e) Bi, (f) Pb^{2+}.

1.13 Give the ground state electron configurations of (a) Sc, (b) V^{3+}, (c) Mn^{2+}, (d) Cr^{2+}, (e) Co^{3+}, (f) Cr^{6+}, (g) Cu, and (h) Gd^{3+}.

1.14 Give the ground state electron configurations of (a) W, (b) Rh^{3+}, (c) Eu^{3+}, (d) Eu^{2+}, (e) V^{5+}, (f) Mo^{4+}.

1.15 Account for the trends, element by element, across Period 3 in (a) ionization energy, (b) electron affinity, and (c) electronegativity.

1.16 Account for the fact that the two Group 5 elements niobium (Period 5) and tantalum (Period 6) have the same metallic radii.

1.17 Discuss the trend in electronegativities across Period 2 from lithium to fluorine. Can you account for the difference in details from the trend in ionization energy.

1.18 Identify the frontier orbitals of a Be atom.

1.19 Compare the broad trends in ionization energy, atomic radius, and electronegativity. Account for the parallels.

PROBLEMS

1.1 Show that an atom with the configuration ns^2np^6 is spherically symmetrical. Is the same true of an atom with the configuration ns^2np^3?

1.2 According to the Born interpretation, the probability of finding an electron in a volume element dτ is proportional to $\psi^*\psi d\tau$. (a) What is the most probable location of an electron in an H atom in its ground state? (b) What is its most probable distance from the nucleus, and why is this different? (c) What is the most probable distance of a 2*s* electron from the nucleus?

1.3 The ionization energies of rubidium and silver are respectively 4.18 eV and 7.57 eV. Calculate the ionization energies of an H atom with its electron in the same orbitals as in these two atoms and account for the differences in values.

1.4 When 58.4 nm radiation from a helium discharge lamp is directed on a sample of krypton, electrons are ejected with a velocity of 1.59×10^6 m s^{-1}. The same radiation ejects electrons from rubidium atoms with a velocity of 2.45×10^6 m s^{-1}. What are the ionization energies (in eV) of the two elements?

1.5 Survey the early and modern proposals for the construction of the periodic table. You should consider attempts to arrange the elements on helices and cones as well as the more practical two-dimensional surfaces. What, in your judgement, are the advantages and disadvantages of the various arrangements?

1.6 The decision about which elements should be identified as belonging to the *f* block has been a matter of some controversy. A view has been expressed by W. B Jensen (*J. Chem. Educ.*, **59**, 635 (1982)). Summarize the controversy and Jensen's arguments.

2

Molecular structure

Much of the interpretation of inorganic chemistry is expressed in terms of models of bonding and molecular structure. This type of interpretation is augmented by noting the correlation of molecular shapes, trends in the strengths of bonds, spectroscopic and magnetic properties of molecules, and patterns of chemical reactivity. Pictorial representations of bonding are widely used and very helpful, and in this chapter we shall describe the principal types of models that are used. However, pictorial models only go so far, and their reliability must be established by reference to empirical observations and, where possible, to more detailed calculations. We shall see a little of this interplay between qualitative models, experiment, and calculation in this chapter and throughout the remainder of the text.

It is possible to correlate a range of experimental observations in inorganic chemistry with a number of qualitative concepts that are supported by approximate theories, and in this way to rationalize the rich diversity of facts that is so characteristic of the subject. Structural formulas showing the arrangement of chemical bonds still play a central role in such an approach, and it will be useful to establish the concepts involved. The simplest applications of bonding theory are found in the structures of molecules formed by *p*-block elements, and initially we concentrate on these species.

ELECTRON-PAIR BONDS

The earliest successful model of a chemical bond in terms of electrons was proposed by G. N. Lewis in the opening decades of this century. His approach gave a very direct interpretation of a bond in terms of electrons, and even in modern discussions it is often quite helpful to make use of Lewis's model. By 1916 he had identified a chemical bond in a molecule as a shared electron pair, but he had no idea why a *pair* and not some other number of electrons was responsible for bond formation.[1] In Section 2.1 we review Lewis's elementary (but powerful) theory of the electron-pair bond, the main features of which we assume to be familiar. This review will be followed by an exploration of certain aspects of the chemical properties of molecules that can be expressed in terms of the existence of definite electron pair bonds that are transferable from one species of molecule to another. In Section 2.3, we examine a simple quantum mechanical version of bonding theory and see an explanation of some of the concepts that Lewis adopted.

2.1 Lewis structures: a review

Lewis supposed that each pair of electrons involved in bonding lies between and is shared by two neighboring atoms. This **covalent bond**, a shared electron pair, is denoted A—B. Double (A=B) and triple (A≡B) bonds consist of two and three shared pairs of electrons, respectively. Unshared pairs of valence electrons on atoms are called **lone pairs**. Although lone pairs do not contribute directly to the bonding, they do influence the shape of the molecule and its chemical properties.

The octet rule

Lewis found that he could account for the existence of a wide range of molecules by proposing the **octet rule**: *each atom acquires shares in electrons until its valence shell achieves eight electrons*. As we have seen, a closed-shell, noble-gas configuration is achieved when eight electrons occupy the s and p subshells of the valence shell. A hydrogen atom completes a 'duplet', the $1s^2$ configuration of its neighbor helium.

The octet rule provides a simple way of constructing **Lewis structures**, which are diagrams that show the pattern of bonds in a molecule. In most cases we can construct a Lewis structure in three steps.

1. Decide on the number of electrons that are to be included in the structure by adding together the numbers of all the valence electrons provided by the atoms.

Each atom provides all its valence electrons (thus, H provides one electron and O, with the configuration $[He]2s^2 2p^4$, provides six). Each negative charge on an ion corresponds to an additional electron; each positive charge corresponds to one electron less.

2. Write the chemical symbols of the atoms in the arrangement showing which are bonded together.

---

[1] The full quantum mechanical basis of the electron-pair bond, which includes the role of the lone pairs, and the characteristic bond properties that can be transferred from one molecule to another, is only now becoming clear using results of the best numerical solutions of the Schrödinger equation for molecules. See R. F. W. Bader, *Atoms in molecules*. Oxford University Press (1990).

In most cases we know the arrangement or can make an informed guess. The less electronegative element is usually the central atom of a molecule, as in CO_2, SO_4^{2-}, and PCl_5.

> 3. Distribute the electrons in pairs so that there is one pair of electrons between each pair of atoms bonded together, and then supply electron pairs (to form lone pairs or multiple bonds) until each atom has an octet.

All bonding pairs are then represented by a single line. If we are dealing with a polyatomic ion, the net charge is assumed to be possessed by the ion as a whole, not by a particular individual atom.

Example 2.1: *Writing a Lewis structure*
Write a Lewis structure for the BF_4^- ion.

Answer. The atoms supply $3+(4\times7)=31$ valence electrons; the single negative charge of the ion reflects the presence of an additional electron. We must therefore accommodate 32 electrons in 16 pairs around the 5 atoms. One solution is

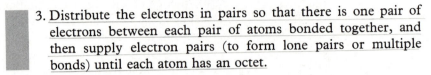

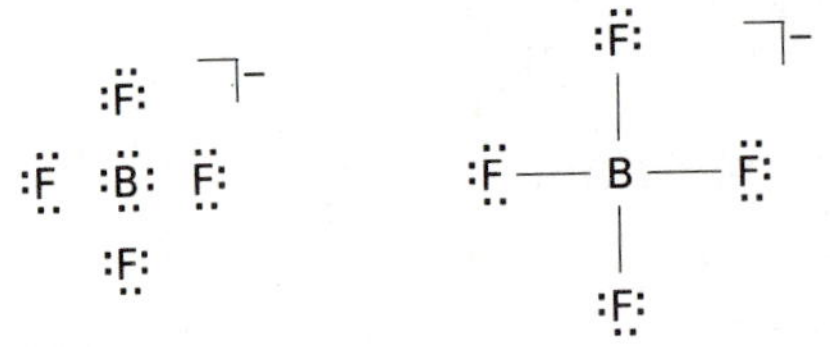

The negative charge is ascribed to the ion as a whole, not to an individual atom.

Exercise E2.1. Write a Lewis octet structure for the PCl_3 molecule.

Table 2.1 gives examples of Lewis structures of some common molecules and molecular ions. It is important to remember that Lewis structures do not in general convey the shape of the molecule, but only the pattern and numbers of bonds. The BF_4^- ion, for example, is a regular tetrahedron, not the planar structure depicted in Example 2.1. However, in simple cases (particularly for planar molecules), an attempt is often made to show the relative locations of the atoms, and the Lewis structures of the linear CO_2 molecule and the angular O_3 molecule are usually written

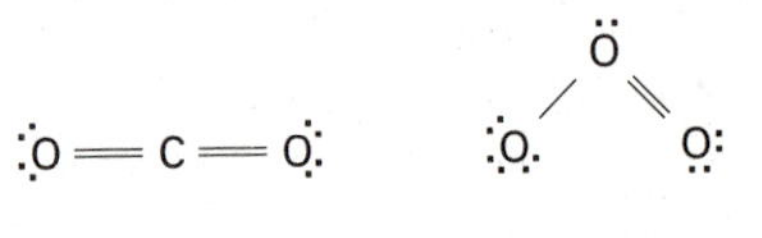

$$:\!\overset{..}{O}\!=\!C\!=\!\overset{..}{O}\!:\qquad\qquad :\!\overset{..}{O}\diagdown\!\!\!\overset{\textstyle\overset{..}{O}}{}\!\!\!\diagup\overset{..}{O}\!:$$

Formal charge and oxidation number

Although the overall charge of a species is possessed by the species as a whole, for some applications it is convenient to ascribe a **formal**

Table 2.1 Lewis structures of some common molecules

Molecule	Lewis structures*
H_2	H — H
N_2, CO	:N≡N: :C≡O:
O_3, SO_2, NO_2^-	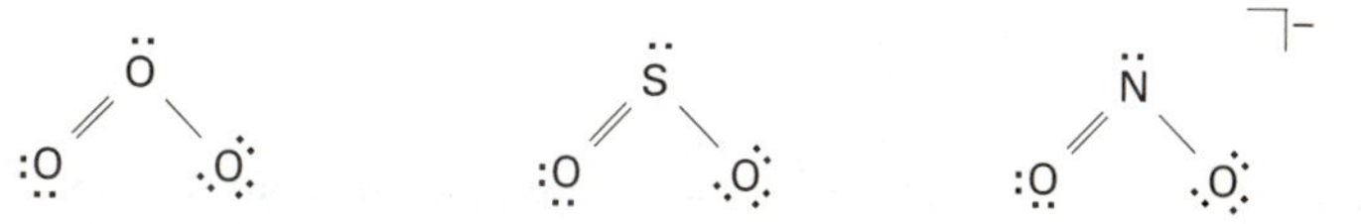 (resonance structures shown)
NH_3, SO_3^{2-}	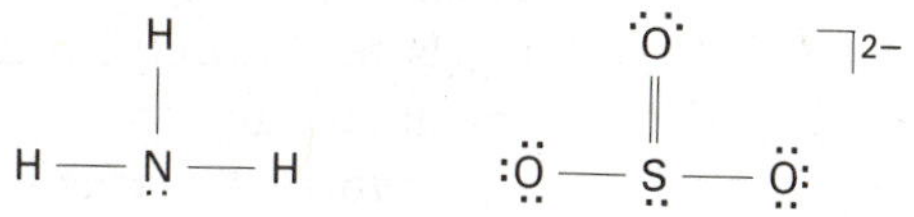 (structures shown)
PO_4^{3-}, SO_4^{2-}, ClO_4^-	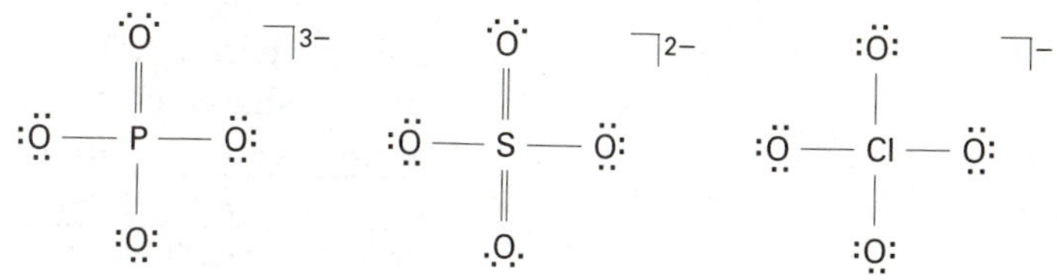 (structures shown)

*Only representative resonance structures are given. Shapes are indicated only for diatomic and triatomic molecules. Octet expansion is treated on p. 57.

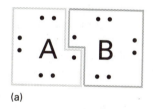

(a)

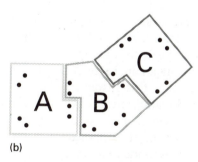

(b)

2.1 A pictorial representation of the calculation of formal charge. The lines show how the bonding electrons and lone pair electrons re-apportion to each atom (a) in a diatomic molecule with Lewis structure A—B and (b) in a triatomic molecule with Lewis structure A=B—C. The formal charge on each atom is the difference between the number of electrons obtained in this way and the number in the free, neutral atom.

charge (F.C.) to each atom using the definition

> F.C. = Number of valence electrons on the parent atom
> − Number of lone pair electrons
> − $\frac{1}{2}$ × Number of shared electrons

The formal charge is the difference between the number of valence electrons in the free atom and the number that the atom has in the molecule supposing it to own one electron of each shared pair and both electrons of any lone pair (Fig. 2.1). The formal charge represents (in some idealized sense) the number of electrons that an atom gains or loses when it enters into perfect covalent bonding with other atoms.

Formal charge is helpful for choosing between more than one possible Lewis structure, for it is generally the case that *the lowest energy structure is the one with the smallest formal charges on the atoms (and which usually lies between − 1 and +1)*. Thus, of the two structures

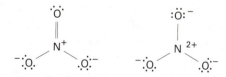

we select the first and reject the second. As in these diagrams, formal charges are ascribed to individual atoms, and their sum over all the atoms in the species is equal to the overall charge. The lowest energy structure is often the one in which the more electronegative element is ascribed a formal negative charge and the less electronegative element is ascribed a formal positive charge. Thus, we can suspect that of the two structures for BF_3,

we should adopt the former as the better first approximation to its actual structure, for although the B atom has an incomplete octet, in the latter structure the electronegative F atoms have a positive formal charge.

The formal charge is a parameter based on a 'pure covalent' view of the bonding, in which electron pairs are shared equally by neighbors. An alternative 'ideal' model of the bonding is one in which the partial drift of shared electrons toward the more electronegative of two bonded atoms is exaggerated to the point that the latter is regarded as possessing *both* electrons of a pair. For instance, in the NO_3^- ion depicted on p. 53, each O atom is ascribed a complete octet, so making its effective ionic charge -2 (whatever the multiplicity of the bonds joining it to the central N atom in the Lewis structure); the N atom is left with no valence electrons, and so it is ascribed an effective ionic charge of $+5$.

The effective ionic charge obtained by exaggerating electron drifts is called the **oxidation number** (O.N.) of the element, and we say that the oxidation number of oxygen in NO_3^- is -2 whereas that of nitrogen is $+5$. Positive oxidation numbers are often reported as a roman numeral, as in N(V) for the oxidation number of nitrogen in the nitrate ion. The element itself is said to be in a particular **oxidation state** when it has a specific oxidation number.

In practice, oxidation numbers are ascribed by applying a set of rules (Table 2.2) that are based on the effects of electronegativity but which formalize the procedure into a simple scheme. The name 'oxidation number' is inspired by the fact that, in all its compounds other than with fluorine or with itself, an O atom is ascribed an oxidation number of -2, and so the addition of oxygen to a compound (as in the oxidation of NO_2^- to NO_3^-) results in an increase in oxidation number of the element (from $+3$ to $+5$ in this instance). That is, oxidation (in both the classical sense of the addition of oxygen and in the modern sense of the loss of electrons, see Chapter 7) corresponds to an increase in oxidation number. Reduction, the opposite of oxidation (the gain of electrons, Chapter 7) corresponds to a decrease in the oxidation number of an element.

The oxidation number is a guide to the type of reactions that a substance can undergo, because an element in its highest oxidation state may only be reduced whereas one in an intermediate oxidation state may be either oxidized or reduced. For monatomic ions, such as Na^+ or Cl^-, the oxidation number is equal to the number of electrons that the atom has lost or gained. The modern definition of oxidation number is based on concepts relating to oxidation and reduction, as explained in Chapter 7.

Table 2.2 The determination of oxidation number*

	Oxidation number
1. The sum of the oxidation numbers of all the atoms in the species is equal to its total charge	
2. For atoms in their elemental form	0
3. For atoms of Group 1 For atoms of Group 2 For atoms of Group 13/III (except B) For atoms of Group 14/IV (except C, Si)	$+1$ $+2$ $+3$ (EX_3), $+1$ (EX) $+4$ (EX_4), $+2$ (EX_2)
4. For hydrogen	$+1$ in combination with nonmetals -1 in combination with metals
5. For fluorine	-1 in all its compounds
6. For oxygen	-2 unless combined with F -1 in peroxides (O_2^{2-}) $-\frac{1}{2}$ in superoxides (O_2^-) $-\frac{1}{3}$ in ozonides (O_3^-)
7. Halogens	-1 in most compounds, unless the other elements include oxygen or more electronegative halogens

*To determine an oxidation number, work through the following rules in the order given. Stop as soon as the oxidation number has been assigned.
These rules are not exhaustive, but they are applicable to a wide range of common compounds.

Example 2.2: *Determining the oxidation number of an element*
What is the oxidation number of (a) nitrogen in the azide ion, N_3^-, and (b) manganese in the permanganate ion, MnO_4^-.

Answer. We work through the steps set out in Table 2.2, in the order given. (a) The charge of the species is -1; all three atoms are of the same element; therefore their oxidation numbers are the same and $3 \times O.N.(N) = -1$. Therefore, $O.N.(N) = -\frac{1}{3}$. (b) The sum of the oxidation numbers is equal to -1, so $O.N.(Mn) + 4 \times O.N.(O) = -1$. Because oxygen is ascribed an oxidation number of -2, it follows that $O.N.(Mn) = -1 - 4 \times (-2) = +7$. That is, MnO_4^- is a compound of manganese(VII).

Exercise E2.2. What is the oxidation number of (a) oxygen in O_2^+, and (b) phosphorus in PO_4^{3-}.

Resonance

In many cases, a single Lewis structure is an inadequate description of the molecule. The Lewis structure of O_3 given earlier, for instance, suggests incorrectly that one OO bond is different from the other, whereas in fact they have identical lengths (1.28 Å) intermediate between those of typical single O—O and double O=O bonds (1.48

Å and 1.21 Å, respectively). This deficiency is repaired by introducing the concept of **resonance**, in which the actual structure of the molecule is taken to be a blend of all the feasible Lewis structures corresponding to a given atomic arrangement.

Resonance is indicated by a double-headed arrow:

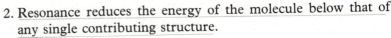

Resonance should be pictured as a *blending* of structures, not a flickering alternation between them. In quantum mechanical terms, the electron distribution of each structure is represented by a wavefunction, and the actual wavefunction ψ of the molecule is the superposition of the individual wavefunctions for each structure:

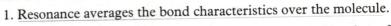

The first of the contributing wavefunctions, for instance, represents an electron distribution in which there is a double bond between the right-hand pair of atoms in O_3. The wavefunction is written as a superposition with equal contributions from each structure because the two structures have identical energies.

Resonance has two main effects:

1. Resonance averages the bond characteristics over the molecule.

2. Resonance reduces the energy of the molecule below that of any single contributing structure.

The energy of the O_3 resonance hybrid, for instance, is lower than that of either individual structure alone. Resonance is most important when there are several structures of identical energy that can be written to describe the molecule, as for O_3. In such cases, all the structures of the same energy contribute equally to the overall structure.

Structures with different energies may also contribute to an overall resonance hybrid, but in general the higher the energy of a Lewis structure, the smaller the contribution it makes to the overall structure. The BF_3 molecule, for instance, is a resonance hybrid of the structures

as well as several others, but the first structure dominates (see the discussion of formal charge, above). Consequently, BF_3 is regarded *primarily* as having single B—F bonds with a small admixture of double bond character. In contrast, of the analogous structures of the NO_3^- ion

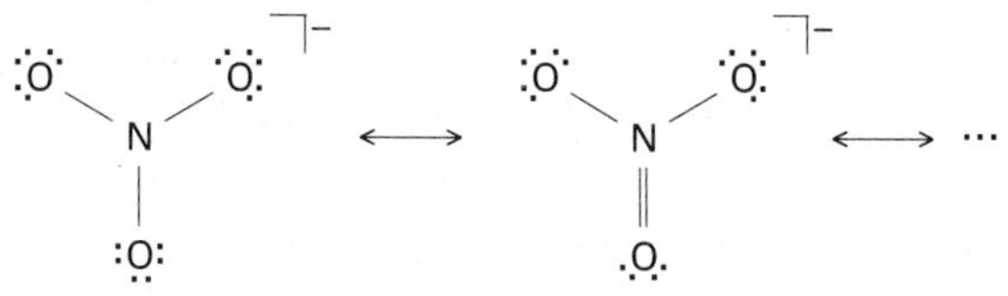

the three doubly-bonded structures dominate (as we saw earlier, they have smaller formal charges than the first structure), and we treat the ion as having partial double-bonded character.

Example 2.3: *Writing resonance structures*

Write resonance structures for the NO_2F molecule and identify the dominant structures.

Answer. Four Lewis structures and their formal charges are

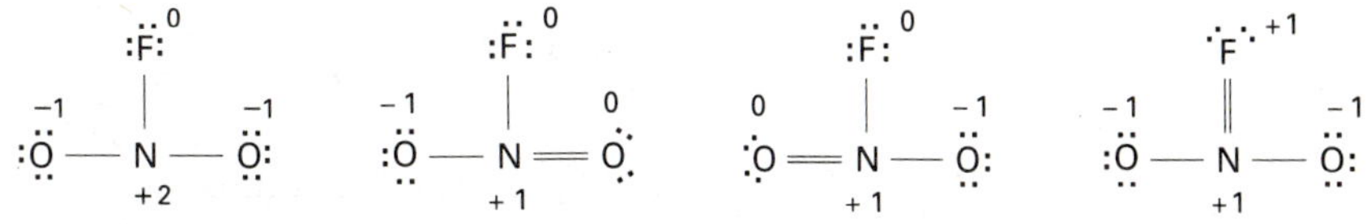

It is very unlikely that a low energy will be achieved with a positive charge on an F atom or with a high positive charge on the N atom, so the two structures with N=O bonds are most likely to dominate in the resonance.

Exercise E2.3 Write resonance structures for the NO_2^- ion.

Hypervalence

Although Period 2 elements obey the octet rule quite well, Period 3 and subsequent elements show deviations from it. For example, the existence of PCl_5 can be expressed in Lewis terms only if the P atom has ten electrons in its valence shell, one pair for each P—Cl bond. Similarly SF_6, a very stable molecule, must have 12 electrons around the S atom if each F atom is to be bound to the central S atom by an electron pair. Compounds of this kind, for which the Lewis structures demand the presence of more than an octet of electrons around at least one atom, are called **hypervalent compounds**.

The occurrence of hypervalence is widespread for the elements of Periods 3 to 6: for instance, it includes the structures SO_3^{2-}, PO_4^{3-}, SO_4^{2-}, and ClO_4^- in Table 2.1. The traditional explanation of hypervalence invokes the availability of low-lying unfilled *d* orbitals, which can accommodate the additional electrons. According to this explanation, a P atom can accommodate more than eight electrons if it uses its vacant 3*d* orbitals. In PCl_5, for instance, at least one 3*d* orbital must

Table 2.3 Bond lengths

	$R_e/\text{Å}$
H_2^+	1.06
H_2	0.74
HF	0.92
HCl	1.27
HBr	1.41
HI	1.60
N_2	1.09
O_2	1.21
F_2	1.44
Cl_2	1.99
I_2	2.67

Source: G. Herzberg, *Spectra of diatomic molecules.* Van Nostrand, Princeton (1950).

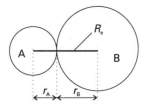

1 Covalent radii

Table 2.4 Covalent radii* ($r_{cov}/\text{Å}$)

H			
0.37			

C	N	O	F
0.71 (1)	0.74 (1)	0.66 (1)	0.64
0.67 (2)	0.65 (2)	0.57 (2)	
0.60 (3)			

	P	S	Cl
	1.10	1.04 (1)	0.99
		0.95 (2)	

	As	Se	Br
	1.21	1.04	1.14

	Sb	Te	I
	1.41	1.37	1.33

*Values are for single bonds except where otherwise stated (in parentheses).

be used. The rarity of hypervalence in Period 2 is then ascribed to the absence of $2d$ orbitals in these elements.[2] However, a more compelling reason for the rarity of hypervalence in Period 2 may be the geometrical difficulty of packing more than four atoms around a single small central atom, and may have little to do with the availability of d orbitals. Recent computations (of the kind described later) suggest that the traditional explanation overemphasizes the role of d orbitals in hypervalent compounds (for example, SF_6). In Section 2.8 we shall see how to account for hypervalent compounds without making use of d orbitals.

2.2 Bond properties

Certain properties of bonds are approximately the same in different compounds of the elements. Thus, if we know the strength of an O—H bond in H_2O, then with some confidence we can use the same value if we are interested in the strength of the O—H bond in CH_3OH. We shall confine our attention to two of the most important characteristics of a bond: its length and its strength.

Bond length

The **bond length** of a molecule is the equilibrium internuclear separation of the two bonded atoms. Bond lengths may be measured spectroscopically (for instance, from the rotational structure of gas-phase spectra) or by X-ray diffraction of solids and electron diffraction of gases. Some typical bond lengths (which are of the order of 1 to 2 Å) are shown in Table 2.3.

To a reasonable first approximation, bond lengths can be partitioned into characteristic contributions from each atom of the bonded pair. The contribution of an atom to a covalent bond is called the **covalent radius (1)** of the element; some values are given in Table 2.4. From such data we can predict, for example, that the length of a P—N bond is 1.10 Å + 0.74 Å = 1.84 Å (experimentally it is close to 1.8 Å in a number of compounds). Covalent radii vary through the periodic table in much the same way as metallic and ionic radii (Section 1.9), for the same reasons, and are smallest close to F.

Covalent radii are approximately equal to the separation of nuclei when the cores of the two atoms are in contact: the valence electrons draw the two atoms together until the repulsion between the cores starts to dominate (Fig. 2.2). The covalent radius expresses the closeness of approach of *bonded* atoms; the closeness of approach of *nonbonded* atoms in neighboring molecules that are in contact is expressed in terms of the **van der Waals radius** of the element, which is the internuclear separation when the *valence* shells of the two nonbonded atoms are in contact.

[2] Five- and six-coordinate B and C are known in carboranes and metal carbides (Chapter 11), metal clusters (see Chapter 16), and in other molecules.

Table 2.5 Mean bond enthalpies* (B/kJ mol^{-1})

	H	C	N	O	F	Cl	Br	I	S	P	Si
H	436										
C	412	348 (1) 612 (2) 518 (a)									
N	388	305 (1) 613 (2) 890 (3)	163 (1) 409 (2) 946 (3)								
O	463	360 (1) 743 (2)	157	146 (1) 497 (2)							
F	565	484	270	185	155						
Cl	431	338	200	203	254	242					
Br	366	276				219	193				
I	299	238				210	178	151			
S	338	259			496	250	212		264		
P	322									201	
Si	318			466							226

*Values are for single bonds except where otherwise stated (in parentheses); (a) denotes aromatic.

Bond strength

A convenient thermodynamic measure of the strength of a bond is the **bond dissociation enthalpy**, the enthalpy change for the dissociation

$$A-B(g) \rightarrow A(g) + B(g) \quad \Delta H^{\ominus}$$

The bond dissociation enthalpy of a polyatomic molecule is derived from the complete atomization of the molecule:

$$H_2O(g) \rightarrow 2H(g) + O(g) \quad \Delta H^{\ominus} = +920 \text{ kJ}$$

and the **mean bond enthalpy** B is this quantity divided by the number of bonds dissociated. In this example, the mean O—H bond enthalpy is one-half the atomization enthalpy, or 460 kJ mol^{-1}, because two O—H bonds are dissociated.

Mean bond enthalpies can be used to estimate reaction enthalpies. However, data on actual species should be used whenever they are available in preference to mean bond enthalpies: bond enthalpies between even the same two elements show such a wide variation that their mean value can be quite misleading. For instance, the Si—Si bond enthalpy ranges from 226 kJ mol^{-1} in Si_2H_6 to 322 kJ mol^{-1} in $Si_2(CH_3)_6$. The values in Table 2.5 are very much data of last resort: at best, they can be used to make rough estimates of reaction enthalpies when enthalpies of formation are unavailable.

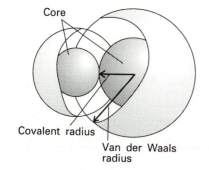

Core

Covalent radius

Van der Waals radius

2.2 The covalent radius can be interpreted as the radius of an atom's core, which in a covalent bond is (broadly speaking) in contact with the core of an adjacent atom. The van der Waals radius represents the overall size of the atom including its valence shells.

Example 2.4: *Making estimates using mean bond enthalpies*
Estimate the reaction enthalpy for the production of $SF_6(g)$ from $SF_4(g)$ given that the mean bond enthalpies of F_2, SF_4, and SF_6 are $+155$, $+343$, and $+327$ kJ mol^{-1}, respectively, at 25 °C.

Answer. The reaction is

$$SF_4(g) + F_2(g) \rightarrow SF_6(g)$$

One mole of F—F bonds and 4 mol S—F bonds (in SF_4) must be broken, corresponding to an enthalpy change of $155 + (4 \times 343) = +1527$ kJ. On the other hand, 6 mol S—F bonds (in SF_6) must be formed, corresponding to an enthalpy change of $6 \times (-327) = -1962$ kJ. The net enthalpy change is therefore

$$\Delta H = +1527 \text{ kJ} - 1962 \text{ kJ} = -429 \text{ kJ}$$

Hence, the reaction is strongly exothermic. The experimental value for the reaction is -434 kJ.

Exercise E2.4. Estimate the enthalpy of formation of H_2S from S_8 (a cyclic molecule) and H_2.

Bond enthalpy trends in the p block

The trend in bond enthalpies in the p block can be summarized as follows:

> For an element E that has no lone pairs, the E—X bond enthalpy decreases down the group.

For example:

	C—C	Si—C	Ge—C
B/kJ mol^{-1}	347	301	242

Another general trend is as follows:

> For an element that has lone pairs, the bond enthalpy decreases down the group, but the value for an element at the head of a group is typically anomalous and is smaller than the bond enthalpy of an element in Period 3.

Two examples are

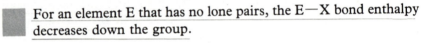

	N—O	<	P—O	>	As—O		
B/kJ mol^{-1}	157		368		330		
	C—Cl	<	Si—Cl	>	Ge—Cl	>	Sn—Cl
B/kJ mol^{-1}	338		401		339		314

The relative weakness of single bonds between Period 2 elements that have lone pairs is often ascribed to the closeness of the lone pairs on neighboring atoms and the consequent repulsion between them. An alternative explanation is the importance of antibonding interactions (a concept we explain later) between compact orbitals that are so close that they overlap appreciably.

A number of features of the p block can be interpreted (and remembered) with the aid of bond enthalpy arguments. For example, the bond dissociation enthalpy of gaseous BO is 788 kJ mol^{-1} whereas the

B—O single bond enthalpy is 523 kJ mol^{-1}. It follows that, even though the octet rule cannot be satisfied, the boron–oxygen bond in BO must be at least double and perhaps triple.

The process of **catenation**, the formation of chains of atoms of the same element, is common in Group 14/IV: the existence of alkanes (C_nH_{2n+2}) is one of the most striking manifestations of catenation. Although acyclic alkanes with less than four C atoms are thermodynamically stable with respect to decomposition into their elements, their silicon analogs the silanes (Si_nH_{2n+2}) are unstable with respect to Si(s) and H_2(g). This difference is a consequence of the weakness of the Si—H bond compared to that of the H—H bond. Because the Cl—Cl bond of Cl_2 is weaker than the H—H bond of H_2, and an Si—Cl bond is stronger than an Si—H bond, compounds of formula Si_nCl_{2n+2} can be expected to exist, and indeed are known for n as high as 10.

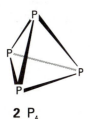

2 P_4

The trend in bond strengths for atoms with lone pairs is illustrated by the strength of the P—P bond, for which $B = 201$ kJ mol^{-1}, compared with that of the N—N bond, for which $B = 163$ kJ mol^{-1}. As is usual, the Period 2 element forms much stronger multiple bonds than its heavier congeners. This difference can account for the difference in the normally occurring forms of the elements, P_4 **(2)** and N_2 (:N≡N:). Moreover, the great difference in single and triple bond enthalpies can account for the rarity of catenation in compounds of nitrogen: hydrazine, H_2N—NH_2 is quite strongly endoergic (its standard free energy of formation is $+149$ kJ mol^{-1}) and analogs of higher alkanes are unknown.

The bond enthalpies of O_2 and S_2 are 498 kJ mol^{-1} and 431 kJ mol^{-1}, respectively. As the lone-pair trend suggests, the S—S single bond enthalpy (263 kJ mol^{-1}) is greater than that of the O—O single bond (142 kJ mol^{-1}). Thus, elemental sulfur forms rings or chains with S—S single bonds, whereas oxygen exists primarily as O=O. Similarly, sulfur catenation leads to polysulfides of formula $[S—S—S]^{2-}$ and $[S—S—S—S]^{2-}$, but oxygen's power of catenation is exhausted at O_4F_2.

One enlightening application of bond-enthalpy arguments concerns **subvalent compounds**, or molecules in which fewer bonds are formed than valence rules allow, such as PH_2. This subvalent compound is thermodynamically stable with respect to dissociation into the constituent atoms. Its instability lies in its tendency to undergo **disproportionation**, that is, to form compounds in which the oxidation number of an element both increases and decreases in a single reaction:

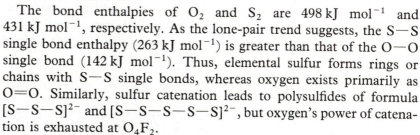

	$3PH_2$(g)	→ $2PH_3$(g)	+ P(s)
Oxidation number of P:	+2	+3	0

The origin of the exothermicity of this reaction is the strength of the P—P bonds in solid phosphorus (P_4). There are the same number (six) of P—H bonds among the reactants as there are among the products, but the reactants have no P—P bonds.

MOLECULAR ORBITALS OF DIATOMIC MOLECULES

We shall now generalize the *atomic* orbital description of atoms in a very natural way to **molecular orbitals**. These orbitals describe how electrons spread over all the atoms in a molecule and bind them together. In the spirit of this chapter, we shall continue to treat the concepts qualitatively and to give a sense of how inorganic chemists discuss the electronic structures of molecules.

2.3 An introduction to the theory

We begin by considering **homonuclear diatomic molecules**, which are molecules consisting of two identical atoms, such as N_2, O_2, and F_2. The same ideas apply to diatomic ions, such as the superoxide ion, O_2^-, and the mercury(I) ion, Hg_2^{2+}. The concepts these diatomics introduce are readily extended to **heteronuclear diatomic molecules**, or molecules built from two different atoms. They are also easily extended, as we see later, to polyatomic molecules and solids composed of huge numbers of atoms and ions. In parts of this section we shall include molecular fragments in the discussion, such as the SF diatomic group in the SF_6 molecule or the OO diatomic group in H_2O_2, for similar concepts also apply to pairs of atoms bound together as parts of larger molecules.

The approximations of the theory

As in the description of the electronic structures of atoms, we set out by making the **orbital approximation**. To do so, we assume that a reasonable first approximation is that the wavefunction Ψ of the N electrons in the molecule can be written as a product of N one-electron wavefunctions ψ:

$$\Psi(\boldsymbol{r}_1, \boldsymbol{r}_2, \ldots, \boldsymbol{r}_N) = \psi(\boldsymbol{r}_1)\psi(\boldsymbol{r}_2)\ldots\psi(\boldsymbol{r}_N)$$

The interpretation of this expression is that electron 1 is described by the wavefunction $\psi(\boldsymbol{r}_1)$, electron 2 by the wavefunction $\psi(\boldsymbol{r}_2)$, and so on. These one-electron wavefunctions are the **molecular orbitals** of the theory. As for atoms, the square of a one-electron wavefunction gives the probability distribution for that electron over the whole molecule: an electron in a molecular orbital is likely to be found where the orbital has a large amplitude, and will not be found at all at any of its nodes.

The next approximation is motivated by noticing that when an electron is close to the nucleus of one atom, its wavefunction closely resembles an atomic orbital of that atom. For instance, when an electron is close to the nucleus of an H atom in a molecule, its wavefunction is like a $1s$ orbital of that atom (we denote such an orbital H$1s$). Therefore we may suspect that we can construct a reasonable first approximation to the molecular orbital by superimposing—in effect, adding together—atomic orbitals contributed by each atom. This modeling of a molecular orbital in terms of contributing atomic orbitals is called the **linear combination of atomic orbitals** (LCAO) approximation. A 'linear combination' is a sum with various weighting coefficients.

In the elementary form of molecular orbital theory, only the valence shell atomic orbitals are used to form molecular orbitals. Thus the molecular orbitals of H_2 are approximated by adding together two hydrogen $1s$ orbitals, one from each atom:

$$\psi = c_A \phi_{1s}(A) + c_B \phi_{1s}(B)$$

In this case the **basis set**, the atomic orbitals ϕ from which the molecular orbital is built, consists of two hydrogen $1s$ orbitals, one on atom A and the other on atom B. The coefficients c in the linear combination show the extent to which each atomic orbital contributes to the molecular orbital: the greater the value of c^2, the greater the contribution of that orbital to the molecular orbital.

The linear combination that models the lowest energy exact solution of the Schrödinger equation for the H_2 molecule has equal contributions from each $1s$ orbital ($c_A^2 = c_B^2$). As a result, electrons in this orbital are equally likely to be found near each nucleus. Specifically, the coefficients are $c_A = c_B = 1$, and

$$\psi_+ = \phi_{1s}(A) + \phi_{1s}(B)$$

(We disregard normalization throughout the chapter.[3]) The combination that models the next higher energy orbital also has equal contributions from each $1s$ orbital ($c_A^2 = c_B^2$), but the coefficients have opposite signs ($c_A = +1$, $c_B = -1$):

$$\psi_- = \phi_{1s}(A) - \phi_{1s}(B)$$

The relative signs of the coefficients in molecular orbitals play a very important role in determining the energies of the orbitals. As we shall explain, the relative signs determine whether atomic orbitals interfere constructively or destructively in different regions of the molecule and hence lead to an accumulation or reduction of electron density in those regions.

Two more preliminary points should be noted. We see from this discussion that *two* molecular orbitals may be constructed from *two* atomic orbitals. In due course, we shall see the importance of the general point that:

> N molecular orbitals can be constructed from a basis set of N atomic orbitals.

For example, if we use all four valence orbitals on each O atom in O_2, then from the total of eight atomic orbitals we can construct eight

[3] The wavefunction ψ_+ (like all wavefunctions) must have the property that

$$\int \psi_+^2 \, d\tau = 1$$

if the Born interpretation is to be applicable. This is achieved by multiplying ψ_+ by a constant N called the 'normalization constant'. For clarity in the nonmathematical presentation that we are adopting, we omit N, because the form of the orbitals is then much easier to see.

molecular orbitals. Secondly, as in atoms, the Pauli exclusion principle implies that each molecular orbital may be occupied by two electrons with paired spins. Thus, in a diatomic molecule constructed from two Period 2 atoms and in which there are eight molecular orbitals available for occupation, up to 16 electrons may be accommodated before all the molecular orbitals are full.

Bonding and antibonding orbitals

The general pattern of the energies of molecular orbitals is that:

> One molecular orbital lies below that of the parent atomic energy levels, one lies higher in energy than they do, and the remainder are distributed between these two extremes.

An indication of the size of the energy gap between the two molecular orbitals in H_2 that we have been considering is the observation of a spectroscopic absorption in H_2 at 11.4 eV (in the ultraviolet at 109 nm) which can be ascribed to the transition of an electron from the lower orbital to the upper orbital. In this section we shall see how molecular orbital theory accounts for the difference in energy of the two orbitals. In the following section we shall see how similar concepts account for the spectra of more complex molecules.

The orbital ψ_+ is a **bonding orbital**. It is so called because the energy of the molecule is lowered if it is occupied by electrons. In elementary discussions of the chemical bond, the bonding character of ψ_+ is ascribed to the constructive interference between the two atomic orbitals and the enhanced amplitude this causes between the two nuclei (Fig. 2.3(a)). An electron that occupies ψ_+ has an enhanced probability of being found in the internuclear region, and can interact strongly with both nuclei. Hence **orbital overlap**, the spreading of one orbital into the region occupied by another, leading to enhanced probability of electrons being found in the internuclear region, is taken to be the origin of the strengths of bonds.[4]

The orbital ψ_- is an **antibonding orbital**. It is so called because if it is occupied, then the energy of the molecule is higher than for the two separated atoms. The greater energy of an electron in this orbital arises from the destructive interference between the two atomic orbitals, which cancels their amplitudes and gives rise to a nodal plane between the two nuclei (Fig. 2.3(b)). Electrons that occupy ψ_- are largely excluded from the internuclear region and are forced to occupy less favorable locations. It is generally true that:

> The energy of a molecular orbital in a polyatomic molecule is higher the more internuclear nodes it has.

The increase in energy reflects an increasingly complete exclusion of electrons from the regions between nuclei. The energies of the two

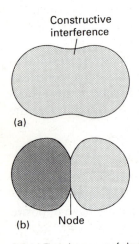

Constructive interference

(a)

(b)   Node

2.3 (a) The enhancement of electron density in the internuclear region arising from the constructive interference between the atomic orbitals on neighboring atoms. (b) The destructive interference that leads to a node when the overlapping orbitals have opposite phases.

---

[4] For a straightforward, readable discussion of conventional wisdom on the origin of the strengths of bonds according to molecular orbital theory, see R. L. DeKock and H. B. Gray, *Chemical structure and bonding*. Benjamin/Cummings, Menlo Park (1980).

orbitals are depicted in Fig. 2.4, which is an example of a **molecular orbital energy level diagram**.

The Pauli exclusion principle limits to two the number of electrons that can occupy any molecular orbital, and requires that those two electrons be paired ($\uparrow\downarrow$). The exclusion principle is the origin of the importance of the electron pair in bond formation: two electrons is the maximum number that can occupy an orbital that contributes to the stability of the molecule. The H_2 molecule, for example, has a lower energy than that of the separated atoms because two electrons can occupy the orbital ψ_+ and both can contribute to the lowering of its energy (as shown in Fig. 2.4). One electron is less effective but, nevertheless, H_2^+ is known as a transient gas-phase ion. Three electrons are also less effective than two electrons because the third electron occupies the antibonding orbital ψ_- and so destabilizes the molecule. With four electrons, when two can occupy ψ_+ and two can occupy ψ_-, the antibonding effect of the electrons in ψ_- completely overcomes the bonding effect of the pair in ψ_+. There is then no net bonding. Hence a four-electron molecule with only $1s$ orbitals available for bond formation, such as He_2, cannot be expected to be stable relative to dissociation into its atoms.

We can now appreciate that the 11.4 eV transition in H_2 corresponds to the removal of an electron from the bonding region in H_2 and its redistribution outside the internuclear region in the antibonding orbital.

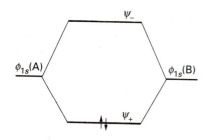

2.4 The molecular orbital energy level diagram for H_2 and analogous molecules.

2.4 Homonuclear diatomic molecules

Although ultraviolet absorption spectra have proved very useful for the analysis of the electronic structure of diatomic molecules more complex than H_2, a more direct portrayal of the molecular orbital energy levels can often be achieved by using **ultraviolet photoelectron spectroscopy** (UV-PES) in which electrons are ejected from the orbitals they occupy in molecules. In this technique, a sample is irradiated with 'hard' (high frequency) ultraviolet radiation (typically 21.2 eV radiation from excited He atoms) and the kinetic energies of the **photoelectrons**—the ejected electrons—are measured. Because a photon of frequency v has energy hv, if it expels an electron with ionization energy I from the molecule, the kinetic energy E_K of the photoelectron will be

$$E_K = hv - I$$

The lower in energy the electron lies initially (that is, the more tightly it is bound in the molecule), the greater its ionization energy and hence the lower its kinetic energy after it is ejected (Fig. 2.5). Because the peaks in a photoelectron spectrum correspond to the various kinetic energies of photoelectrons ejected from different energy levels of the molecule, the spectrum gives a vivid portrayal of the molecular orbital energy levels of a molecule.

The UV-PES spectrum of N_2 is shown in Fig. 2.6. We see that the

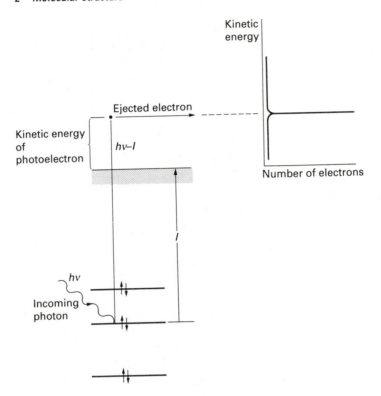

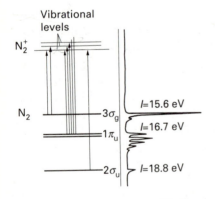

2.5 The photoelectron experiment. The incoming photon has energy $h\nu$; when it ejects an electron from an orbital of ionization energy I, the electron acquires a kinetic energy $h\nu - I$. The spectrometer detects the numbers of photoelectrons with different kinetic energies.

2.6 The UV photoelectron spectrum of N_2. The fine structure in the spectrum arises from the excitation of vibrations in the cation formed by photoejection.

photoelectrons have a series of discrete ionization energies close to 15.6 eV, 16.7 eV, and 18.8 eV. This pattern of energies strongly suggests a shell structure for the arrangement of electrons in the molecule. All the values are close to the ionization energy of the atom (14.5 eV). Because the ionization energy corresponds to the removal of a valence electron, the shell structure suggests that when a molecule forms, the valence electrons arrange themselves into the molecular analog of atomic shells. The electrons in these shells differ only slightly in the strengths with which they are bound to the molecule. The lines corresponding to the lowest ionization energy in N_2 (close to 15.6 eV) are photoelectrons ejected from the occupied molecular orbital with highest energy in the molecule (the orbital in which an electron is bound most weakly). The higher ionization energies of 16.7 eV and 18.8 eV must indicate the presence of molecular orbitals with successively lower energies (in which the electrons are bound more strongly), and indicate the existence of a ladder-like array of orbital energies in the molecule. There may also be more deeply lying orbitals, but the 21.2 eV ultraviolet photons cannot eject electrons from them, so they are not observed in this photoelectron spectrum.

The detailed structure of each group of lines in a photoelectron spectrum is a result of the ionized molecule being left in a vibrationally excited state after it is ionized. Some of the energy of the incoming photon is used to excite the vibration, so less energy is available for the kinetic energy of the ejected electron. Each quantum of vibration that is excited results in a corresponding reduction in kinetic energy of the photoelectron. As a result, the photoelectrons appear at a series of kinetic energies separated by the energy of each vibrational excitation that has occurred.

The vibrational structure in a photoelectron spectrum can be very helpful in assigning the origin of a spectral line. For example, extensive vibrational structure occurs when the ejected electron comes from an orbital in which it exerts a strong force on the nuclei, for the loss of the electron has a pronounced effect on the forces experienced by the nuclei. If the orbital is neutral in this sense, photoejection leaves the force field largely undisturbed and there is little vibrational structure.

Our task now it to see how molecular orbital theory can account for the features revealed by photoelectron spectroscopy (and the other techniques, principally absorption spectroscopy that are used to study diatomic molecules). As with H_2, the starting point in the theoretical discussion is the **minimal basis set**, the smallest set of atomic orbitals from which useful molecular orbitals can be built. In Period 2 diatomic molecules, the minimal basis set consists of the one valence s and three valence p orbitals on each atom, giving eight atomic orbitals in all. As remarked previously, N atomic orbitals can be used to construct N molecular orbitals. We shall now show how the minimal basis set of eight valence-shell atomic orbitals (four from each atom) is used to construct eight molecular orbitals. Then we shall use the Pauli principle to predict the ground-state electron configurations of the molecules.

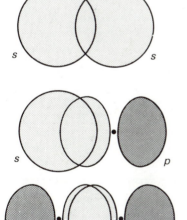

2.7 The molecular orbital energy level diagram for the later Period 2 homonuclear diatomic molecules. This diagram should be used for O_2 and F_2 molecules.

The orbitals

The energies of the atomic orbitals are shown on either side of the molecular orbital diagram in Fig. 2.7. We can form σ orbitals by allowing overlap between atomic orbitals that have cylindrical symmetry around the internuclear axis, which is conventionally labeled z. The notation σ is the analog of s for atomic orbitals and signifies that the orbital has cylindrical symmetry (and, when viewed along the internuclear axis, resembles an s orbital). Both orbitals in Fig. 2.3 are σ orbitals. Orbitals that can form σ orbitals include the $2s$ and $2p_z$ orbitals on the two atoms, as is shown in Fig. 2.8. From these four orbitals (the $2s$ and the $2p_z$ orbitals on atom A and the corresponding orbitals on atom B) with cylindrical symmetry we can construct four σ molecular orbitals. Two of the molecular orbitals will be bonding and two will be antibonding. Their energies will resemble those shown in Fig. 2.7, but it is difficult to predict the precise locations of the central two orbitals: the orbitals are labeled $1\sigma, 2\sigma,...$ starting with the orbital of lowest energy.

The remaining two $2p$ orbitals on each atom, which have a nodal plane including the z axis, overlap to give π orbitals (Fig. 2.9). These molecular orbitals are so called because, viewed along the internuclear axis of the molecule, they resemble p orbitals: more precisely, they have a single nodal plane that contains the internuclear axis (in the illustration, we see this nodal plane from the side). Bonding and

2.8 σ orbitals can be formed in various ways, which include (a) s,s overlap, (b) s,p overlap, and (c) p,p overlap, with the p orbitals directed along the internuclear axis in each case.

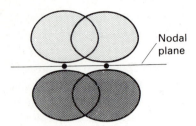

2.9 Two *p* orbitals can overlap to form a π orbital. The orbital has a nodal plane passing through the internuclear axis: this view shows the plane from the side.

antibonding π orbitals can be formed from the mutual overlap of the two $2p_x$ orbitals and, also, from the mutual overlap of the two $2p_y$ orbitals. This pattern of overlap gives rise to the two pairs of doubly degenerate energy levels shown in Fig. 2.7.

The procedure we have adopted for the description of diatomic molecules in terms of molecular orbitals can be summarized as follows:

1. From a basis set of *N* atomic orbitals, *N* molecular orbitals are constructed. In Period 2, $N = 8$.

2. The eight orbitals can be classified by symmetry into two sets: there are four σ orbitals and four π orbitals.

3. The four π orbitals form one doubly degenerate pair of bonding orbitals and one doubly degenerate pair of antibonding orbitals.

4. The four σ orbitals span a range of energies, one being strongly bonding and another strongly antibonding, with the remaining two σ orbitals lying between these extremes.

5. To establish the actual location of the energy levels, it is necessary to use electronic absorption spectroscopy, photoelectron spectroscopy, or detailed computation.

Photoelectron spectroscopy and detailed computation (the numerical solution of the Schrödinger equation for the molecules) let us build the orbital energy schemes shown in Fig. 2.10. As we see there, for B_2 through N_2 the arrangement of orbitals is that shown in Fig. 2.11 whereas for O_2 and F_2 the order of the 2σ and 1π orbitals is reversed and the array is that shown in Fig. 2.7.

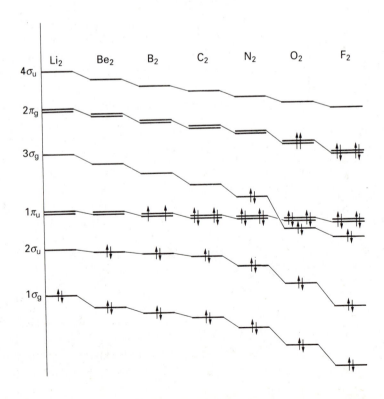

2.10 The variation in orbital energies for Period 2 homonuclear diatomic molecules as far as F_2.

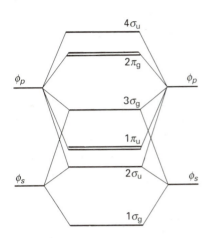

2.11 The alternative ordering of orbital energies found in homonuclear diatomic molecules from Li_2 to N_2.

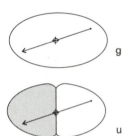

3 σ_g and σ_u orbitals

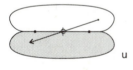

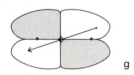

4 π_g and π_u orbitals

The reversal of order can be traced to the increasing separation of the 2s and 2p orbitals that occurs on traveling to the right across Period 2. It is a general principle of quantum mechanics that the mixing of wavefunctions is strongest if their energies are similar. Therefore, as the s and p energy separation increases, the molecular orbitals become more purely s like and p like. When the s,p energy separation is small, each molecular orbital is a more extensive mixture of s and p character on each atom.

If we were considering species containing two neighboring d-block atoms, as in $[Cl_4ReReCl_4]^{2-}$, we should also allow for the possibility of forming bonds from d orbitals. The d_{z^2} orbital has cylindrical symmetry with respect to the internuclear (z) axis, and hence can contribute to the σ orbitals that are formed from s and p_z orbitals. The d_{yz} and d_{zx} orbitals both look like p orbitals when viewed along the axis, and hence can contribute to the π orbitals formed from p_x and p_y. The new feature is the role of $d_{x^2-y^2}$ and d_{xy}, which have no counterpart in the orbitals discussed up to now. These two orbitals can overlap with matching orbitals on the other atom to give rise to doubly degenerate pairs of bonding and antibonding δ orbitals (Fig. 2.12; a δ orbital has two perpendicular nodal planes that intersect along the internuclear axis of the molecule). As we shall see in Section 8.9, δ orbitals are important for the discussion of bonds between metal atoms and lead to descriptions of some species in terms of quadruple bonds, as in $[Cl_4Re{\equiv}ReCl_4]^{2-}$.

For homonuclear diatomics, it is convenient (particularly for spectroscopic discussions) to signify the symmetry of the molecular orbitals with respect to their behavior under inversion through the center of the molecule. The operation of inversion consists of starting at an arbitrary point in the molecule, traveling in a straight line to the center of the molecule, and then continuing an equal distance out on the other side of the center. The orbital is designated g (for *gerade*, even) if it is identical under inversion, and u (for *ungerade*, odd) if it changes sign. Thus, a bonding σ orbital is g and an antibonding σ orbital is u (**3**). On the other hand, a bonding π orbital is u and an antibonding π orbital is g (**4**). These labels have been included in Figs 2.7, 2.10, and 2.11.

The building-up principle for molecules

We use the building-up principle in conjunction with the energy level diagram in the same way as for atoms. The order of occupation of the orbitals is the order of increasing energy as depicted in Figs 2.7 or 2.11. Each orbital can accommodate up to two paired electrons. If more than one orbital is available for occupation (because they happen to have identical energies, as in the case of pairs of π orbitals), then the orbitals are occupied separately. In that case, the electrons in the half-filled orbitals adopt parallel spins (↑↑), just as is required by Hund's rule for atoms.

With very few exceptions, these rules lead to the actual ground state

2.12 The formation of δ orbitals from d-orbital overlap. The orbital has two mutually perpendicular nodal planes that intersect along the internuclear axis.

configuration of the Period 2 diatomic molecules. For example, the electron configuration of N_2, with 10 valence electrons, is

$$N_2: 1\sigma_g^2 2\sigma_u^2 1\pi_u^4 3\sigma_g^2$$

Molecular orbital configurations are written like those for atoms: the orbitals are listed in order of increasing energy, and the number of electrons in each one is indicated by a superscript.

Example 2.5: *Writing electron configurations of diatomic molecules*
Give the ground-state electron configurations of the oxygen molecule, O_2, the superoxide ion, O_2^-, and the peroxide ion, O_2^{2-}.

Answer. The O_2 molecule has 12 valence electrons. The first 10 electrons recreate the N_2 configuration except for the reversal of the order of the $1\pi_u$ and $3\sigma_g$ orbitals (see Fig. 2.10). Next in line for occupation are the doubly degenerate $2\pi_g$ orbitals. The last two electrons enter these orbitals separately, and have parallel spins. The configuration is therefore

$$O_2 \quad 1\sigma_g^2 2\sigma_u^2 3\sigma_g^2 1\pi_u^4 2\pi_g^2$$

The O_2 molecule is interesting because the lowest energy configuration has two unpaired electrons in different π orbitals. Hence, O_2 is paramagnetic (tends to move into a magnetic field). The next two electrons can be accommodated in the $2\pi_g$ orbitals, giving

$$O_2^- \quad 1\sigma_g^2 2\sigma_u^2 3\sigma_g^2 1\pi_u^4 2\pi_g^3$$

$$O_2^{2-} \quad 1\sigma_g^2 2\sigma_u^2 3\sigma_g^2 1\pi_u^4 2\pi_g^4$$

(We are assuming that the orbital order does not change; this might not be the case.)

Exercise E2.5. Write the valence electron configuration for S_2^{2-} and Cl_2^-.

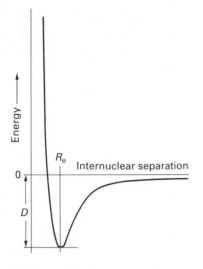

2.13 A molecular potential energy curve showing how the total energy of a molecule varies as the internuclear separation changes.

The **highest occupied molecular orbital** (HOMO) is the molecular orbital that, according to the building-up principle, is occupied last. The **lowest unoccupied molecular orbital** (LUMO) is the next higher molecular orbital. In Fig. 2.10, the HOMO of F_2 is $2\pi_g$ and its LUMO is $4\sigma_u$; for N_2 the HOMO is $3\sigma_g$ and the LUMO is $2\pi_g$. We shall increasingly see that these **frontier orbitals** play special roles in structural and kinetic studies.

A final fundamental feature of molecular structure is the **molecular potential energy curve**, the plot of the energy of the molecule against internuclear separation (Fig. 2.13). The curve shown in the illustration is the typical shape for the ground state of a diatomic molecule. The energy of the molecule—the combined effect of all the electrons in the molecule—decreases as the two atoms approach and bonds begin to form. However, below a certain separation the energy begins to rise again as the bonding electrons cannot accumulate sufficiently in the internuclear region to overcome the internuclear repulsion. The minimum of the curve is the **equilibrium bond length** R_e of the molecule (which, previously, we have called simply the 'bond length'). The depth of the minimum is denoted D_e. The deeper the minimum, the more strongly the atoms are bonded together. The steepness of the well shows how rapidly the energy of the molecule rises as the bond is stretched or compressed. The steepness therefore governs the force constant of the bond and (in combination with the masses of the atoms) the vibrational frequency of the molecule. The more confining the potential well, the greater the force constant and (for given masses) the higher the vibrational frequency.

2.5 Heteronuclear diatomic molecules

The molecular orbitals of heteronuclear diatomic molecules differ from those of homonuclear diatomic molecules by having unequal contributions from each atomic orbital. Each molecular orbital has the form

$$\psi = c_A \phi(A) + c_B \phi(B) + \ldots$$

as in homonuclear molecules. The unwritten orbitals include all the other orbitals of the correct symmetry for forming σ or π bonds, but which typically make a smaller contribution than the two valence shell orbitals we are considering. In contrast to orbitals for homonuclear species, the coefficients c_A and c_B are not necessarily equal in magnitude. If c_A^2 is larger than c_B^2, then the orbital is composed principally from $\phi(A)$ and an electron that occupies the molecular orbital is more likely to be found near atom A than atom B. The opposite is true for a molecular orbital in which c_B^2 is larger than c_A^2.

Molecular orbitals built from atoms of different elements

It is generally the case that

> The greater contribution to a bonding molecular orbital comes from the more electronegative atom.

The reason for this behavior is that the bonding electrons are then likely to be found close to that atom and hence be in an energetically favorable location. The extreme case of a **polar covalent bond**, a covalent bond formed by an electron pair that is unequally shared by the two atoms, is an **ionic bond** in which one atom gains complete control over the electron pair.

The contribution of atomic orbitals to antibonding molecular orbitals is different:

> The less electronegative atom contributes more to an antibonding orbital (Fig. 2.14).

That is, antibonding electrons are more likely to be found in an energetically unfavorable location, in an orbital close to the less electronegative atom.

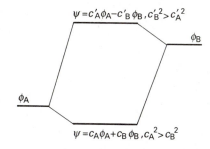

$\psi = c'_A \phi_A - c'_B \phi_B, \; c_B'^2 > c_A'^2$

ϕ_B

ϕ_A

$\psi = c_A \phi_A + c_B \phi_B, \; c_A^2 > c_B^2$

2.14 When two atomic orbitals with different energies overlap, the lower molecular orbital is primarily composed of the lower atomic orbital, and vice versa. Moreover, the shift in energies of the two levels is less than if they had had the same energy in the atoms.

A second difference from homonuclear diatomic molecules stems from the energy mismatch between the two sets of orbitals on the two different atoms. We have already remarked that two wavefunctions interact less strongly as their energies diverge. This dependence on energy separation implies the following general principle:

> The energy lowering as a result of the overlap of atomic orbitals on different atoms in a heteronuclear molecule is less pronounced than in a homonuclear molecule, when the orbitals have the same energies.

However, we cannot necessarily conclude that AB bonds are weaker than AA bonds, because other factors (which include orbital size and closeness of approach) are also important. The CO molecule, for example, which is isoelectronic with N_2, has an even higher bond enthalpy (1070 kJ mol^{-1}) than N_2 (946 kJ mol^{-1}).

Hydrogen fluoride

As an illustration of these points, consider a simple heteronuclear diatomic molecule, HF. The valence orbitals available for molecular orbital formation are the $1s$ orbital of H and the $2s$ and $2p$ orbitals of F; there are $1 + 7 = 8$ valence electrons to accommodate in the molecular orbitals.

The σ orbitals for HF can be constructed by allowing an H$1s$ orbital to overlap the $2s$ and $2p_z$ orbitals of F (z being the internuclear axis). These three atomic orbitals combine to give three σ molecular orbitals of the form

$$\psi = c_1 \phi_{1s}(\text{H}) + c_2 \phi_{2s}(\text{F}) + c_3 \phi_{2p_z}(\text{F})$$

This procedure leaves the F$2p_x$ and F$2p_y$ orbitals unaffected as they have π symmetry and there are no valence H orbitals of that symmetry. The π orbitals are therefore **nonbonding orbitals**, or orbitals that are neither bonding nor antibonding in character and which (in a diatomic molecule) are confined to a single atom.

The resulting energy level diagram is shown in Fig. 2.15. The 1σ bonding orbital is predominantly F$2s$ in character (in accord with the high electronegativity of fluorine) because that orbital lies so low in energy and plays the major role in the bonding orbitals it forms. The 2σ orbital is largely nonbonding and confined mainly to the F atom. The bulk of its density lies on the opposite side of the F atom to the H atom, so it barely participates in bonding. The 3σ orbital is antibonding, and principally H$1s$ in character: the $1s$ orbital has a relatively high energy (compared with the fluorine orbitals) and hence contributes predominantly to the high energy antibonding molecular orbital.

Two of the eight electrons enter the 1σ orbital, forming a bond between the two atoms. Six more enter the 2σ and 1π orbitals; these two sets of orbitals are largely nonbonding and confined mainly to the F atom. All the electrons are now accommodated, so the configuration of the molecule is $1\sigma^2 2\sigma^2 1\pi^4$. One important feature to note is that all

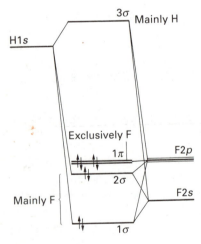

2.15 The molecular orbital energy level diagram for HF. The relative positions of the atomic orbitals reflects the ionization energies of the atoms.

the electrons occupy orbitals that are predominantly on the F atom. It follows that we can expect the HF molecule to be polar, with a partial negative charge on the F atom, which is found experimentally. The observed dipole moment is 1.91 D.[5] Unfortunately, simple molecular orbital theory is not very successful when used to calculate molecular dipole moments, and even the qualitative analysis of the contributions to dipole moments is fraught with difficulty.

Carbon monoxide

The molecular orbital energy level diagram for carbon monoxide (and the isoelectronic CN⁻ ion) is a somewhat more complicated example than HF because both atoms have $2s$ and $2p$ orbitals that can participate in the formation of σ and π orbitals. The energy level diagram is shown in Fig. 2.16. The ground configuration is

$$\text{CO} \quad 1\sigma^2 2\sigma^2 1\pi^4 3\sigma^2$$

The HOMO in CO is 3σ, which is a largely nonbonding lone pair on the C atom. The LUMO is the doubly degenerate pair of antibonding π orbitals, with mainly $C2p$ orbital character. This combination of frontier orbitals is very significant, and we shall see that it is one reason why metal carbonyls are such a characteristic feature of the chemical properties of the d elements (Chapter 16). In metal carbonyls, as we shall see in Chapter 16, the HOMO lone pair orbital of CO participates in the formation of a σ bond, and the LUMO antibonding π orbital participates in the formation of π bonds to the metal atom.

Although the difference in electronegativity between C and O is large, the experimental value of the electric dipole moment of the CO molecule (0.1 D) is small. Moreover, the negative end of the dipole is on the C atom despite that being the less electronegative atom. This odd situation stems from the fact that the lone pairs and bonding pairs have a complex distribution. It is wrong to conclude that because the bonding electrons are mainly on the O atom that O is the negative end of the dipole, for this ignores the balancing effect of the lone pair on the C atom. The inference of polarity from electronegativity is particularly unreliable when antibonding orbitals are occupied.

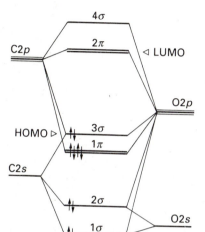

2.16 The molecular orbital energy level diagram of CO.

Example 2.6: *Accounting for the structure of a heteronuclear diatomic molecule*
The halogens form compounds among themselves. One of these 'interhalogens' is iodine monochloride, ICl. What is its ground-state electron configuration?

---

[5] The SI unit for reporting dipole moments, the product of charge in coulombs and distance in meters, is C m. However, it proves more convenient to adopt the non-SI Debye unit D, which was originally defined in terms of electrostatic units. All we need is the conversion

$$1 \text{ D} = 3.336 \times 10^{-30} \text{ C m}$$

A dipole consisting of charges e and $-e$ separated by 1 Å has a dipole moment of 4.8 D. In most cases, dipole moments are somewhat smaller since they arise from partial charges; typical values are close to 1 D.

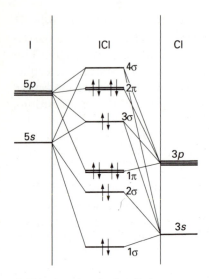

2.17 A schematic molecular orbital energy level diagram for the interhalogen molecule ICl.

Answer. First, we identify the atomic orbitals that are to be used to construct molecular orbitals: these are the $3s$ and $3p$ valence-shell orbitals of Cl and the $5s$ and $5p$ valence-shell orbitals of I. As for Period 2 elements, an array of σ and π orbitals can be constructed, and are shown in Fig. 2.17. The bonding orbitals are predominantly Cl in character (because Cl is the more electronegative element) and the antibonding orbitals are predominantly I in character. There are $7 + 7 = 14$ valence electrons to accommodate, which results in the ground-state electron configuration

$$1\sigma^2 2\sigma^2 1\pi^4 3\sigma^2 2\pi^4$$

Exercise E2.6. Describe the structure of the hypochlorite ion, ClO^-, in terms of molecular orbitals.

2.6 Bond properties

We now bring the discussion full circle, and show that although seemingly very different, molecular orbital theory does elucidate many features of the Lewis description of molecules. We have already seen the origin of the importance of the electron pair: two electrons is the maximum number that can occupy a bonding orbital and hence contribute to a chemical bond. We now extend this concept by introducing the concept of 'bond order'.

Bond order

The **bond order** (B.O.) is an attempt to judge the number of bonds between pairs of atoms in a molecule in the framework of molecular orbital theory and so to provide a link with the Lewis description. The bond order in effect identifies a shared electron pair as counting as a 'bond' and an electron pair in an antibonding orbital as an 'antibond' between two atoms. More precisely, the bond order is defined as

$$\text{B.O.} = \tfrac{1}{2} \times (\text{Number of electrons in bonding orbitals} - \text{Number of electrons in antibonding orbitals})$$

For example, for N_2, $1\sigma_g^2 2\sigma_u^2 1\pi_u^4 3\sigma_g^2$, because σ_g and π_u orbitals are bonding and σ_u are antibonding,

$$\text{B.O.} = \tfrac{1}{2} \times (2 + 4 + 2 - 2) = 3$$

The bond order of 3 corresponds to a triply bonded molecule, which is in line with the Lewis structure $:N \equiv N:$. The high bond order is reflected in the high enthalpy change accompanying dissociation of the molecule, $+946 \text{ kJ mol}^{-1}$, one of the highest for any diatomic molecule. The bond order of the isoelectronic CO molecule is also 3, in accord with the Lewis structure $:C \equiv O:$.

Electron loss from N_2 leads to the formation of a molecular cation N_2^+, in which the bond order is reduced from 3 to 2.5. This reduction in bond order is accompanied by a corresponding decrease in bond

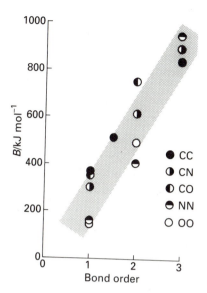

2.18 Bond order/bond strength correlation.

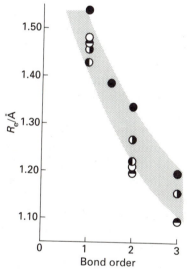

2.19 Bond order/bond length correlation. The key for the points is the same as in Fig. 2.18.

strength (to 855 kJ mol^{-1}) and increase of the bond length from 1.09 Å for N_2 to 1.12 Å for N_2^+. The bond order of F_2 is 1, which is consistent with the Lewis structure F—F and a description of the molecule as having a single bond.

The definition of bond order allows for the possibility that an orbital is only singly occupied. The bond order in O_2^-, for example, is 1.5, because three electrons occupy the $2\pi_g$ antibonding orbitals. Isoelectronic molecules and ions have the same bond order, so F_2 and O_2^{2-} both have bond order 1, and N_2, CO, and NO^+ all have bond order 3.

Bond correlations

The strengths and lengths of bonds correlate quite well with each other and with the bond order:

> Bond enthalpy for a given pair of atoms increases as bond order increases (Fig. 2.18).

> Bond length decreases as bond order increases (Fig. 2.19).

The strength of the dependence varies with the elements. In Period 2 it is relatively weak in CC bonds, with the result that a C=C double bond is less than twice as strong as a C—C single bond. This difference has profound consequences in organic chemistry, particularly for the reactions of unsaturated compounds. It implies, for example, that it is energetically favorable (but kinetically slow in the absence of a catalyst) for ethylene and acetylene to polymerize: in this process, C—C single bonds form at the expense of the appropriate numbers of multiple bonds.

Familiarity with carbon's properties must not be extrapolated without caution to the bonds between other elements. An N=N double bond (409 kJ mol^{-1}) is more than twice as strong as an N—N single bond (163 kJ mol^{-1}), and an N≡N triple bond (946 kJ mol^{-1}) is more than five times as strong. It is on account of this trend that NN multiply bonded compounds are stable relative to polymers or three-dimensional compounds having only single bonds. The same is not true of phosphorus, where the P—P, P=P, and P≡P bond enthalpies are 200 kJ mol^{-1}, 310 kJ mol^{-1}, and 490 kJ mol^{-1}, respectively. For this element, single bonds are stable relative to the matching number of multiple bonds. Thus, phosphorus exists in a variety of solid forms in which P—P single bonds are present, including the tetrahedral P_4 molecules of white phosphorus, and not as P≡P molecules.

The two correlations with bond order taken together imply that:

> For a given pair of elements, the bond enthalpy increases as bond length decreases (Fig. 2.20).

This correlation is a useful feature to bear in mind when considering the stabilities of molecules, because bond lengths may be readily available from independent sources.

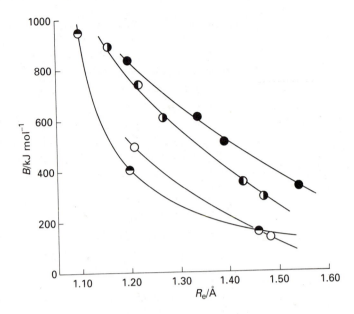

2.20 Bond strength/bond length correlation. The key for the points is the same as in Fig. 2.18.

MOLECULAR ORBITALS OF POLYATOMIC MOLECULES

Molecular orbital theory can be used to discuss in a uniform manner the electronic structures of triatomic molecules, finite groups of atoms, and solids. In each case the molecular orbitals resemble those of diatomic molecules, the only important difference being that the orbitals are built from a more extensive basis set of atomic orbitals. As remarked earlier, a key point to bear in mind is that from N atomic orbitals it is possible to construct N molecular orbitals.

We saw in Section 2.4 that the general structure of molecular orbital energy level diagrams can be derived by grouping the orbitals into different sets, the σ and π orbitals, according to their shapes. The same procedure is used in the discussion of the molecular orbitals of polyatomic molecules. However, as their shapes are more complex than diatomic molecules, we need a more powerful approach. The discussion of polyatomic molecules will therefore be carried out in two stages. In this chapter we use intuitive ideas about molecular shape to construct molecular orbitals. In the next chapter we discuss the shapes of molecules and the use of their symmetry characteristics in the construction of molecular orbitals (and for the discussion of other properties) systematically. That chapter will explain some of the procedures that are presented here.

2.7 The construction of molecular orbitals

We shall use the simplest polyatomic species H_3^+ and H_3 to identify some of the central features of all polyatomic molecules. Although H_3^+ and H_3 might seem far removed from real inorganic chemistry, the orbitals we are about to derive occur widely, as in NH_3 and, in a more disguised form, in other molecules, such as BF_3. The transitory gas phase molecular ion H_3^+ has been detected spectroscopically and tracer experiments have provided evidence that H_3^+ may also occur as a reaction intermediate in solution. Thus, D_2 gas undergoes an exchange reaction with mixtures of HF and SbF_5 (a mixture that we shall come to know as an example of a 'superacid'), presumably through the triangular intermediate D_2H^+, which can split up to give HD or D_2:

$$D_2 + HF/SbF_5 \rightarrow [D_2H]^+[SbF_6]^- \rightarrow HD + D^+ \text{ or } D_2 + H^+$$

A major question in connection with polyatomic molecules is what controls their shapes. We shall therefore begin by considering two possibilities for H_3 and H_3^+, namely linear and triangular, and look for reasons why one shape may have a lower energy than the other.

The linear H₃ molecule

The basis set we shall use to discuss the H_3 molecule are the three $1s$ orbitals of the H atoms, which we shall denote $1s(A)$, $1s(B)$, and $1s(C)$. As always, N molecular orbitals can be formed from N atomic orbitals, and in this case $N = 3$. The three molecular orbitals of the linear H_3 molecule are combinations of the three atomic orbitals, and are shown in Fig. 2.21: we can expect one combination to be strongly bonding, one strongly antibonding, and the third in between these two extremes. We shall continue to omit normalization constants because the form of the orbitals is then much clearer; where we judge it appropriate, we give the normalized forms in footnotes.

A specific calculation (which we shall not describe in detail) shows that the most strongly bonding of the three combinations is

$$1\sigma = \phi_{1s}(A) + \sqrt{2}\phi_{1s}(B) + \phi_{1s}(C)$$

This orbital has a low energy because it is bonding between H_A and H_B and between H_B and H_C. (It is also bonding between H_A and H_C, but as these are so far apart, this contribution is less important.) The 1σ orbital is designated σ because it has cylindrical symmetry around the molecular axis. The next higher orbital is also a σ orbital:

$$2\sigma = \phi_{1s}(A) - \phi_{1s}(C)$$

This molecular orbital has no contribution from the central atom, and since the outer two orbitals are so far apart there is negligible interaction between them. It is therefore called a **nonbonding orbital**, and neither lowers nor raises the energy of the molecule when it is occupied. The third molecular orbital that can be formed from the basis set is another σ orbital:

$$3\sigma = \phi_{1s}(A) - \sqrt{2}\phi_{1s}(B) + \phi_{1s}(C)$$

This orbital is antibonding between both neighboring pairs of atoms and therefore has the highest energy of the three σ orbitals.[6]

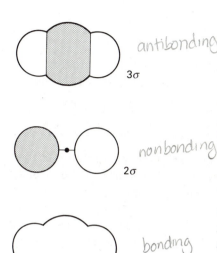

antibonding

3σ

nonbonding

2σ

bonding

1σ

2.21 The three molecular orbitals of the linear H_3 molecule formed from overlap of the H$1s$ orbitals. Throughout this chapter, white and gray denote opposite signs of the wavefunction.

---

[6] The normalized forms of these three orbitals, if we ignore overlap, are

$$1\sigma = \tfrac{1}{2}(\phi_A + \sqrt{2}\phi_B + \phi_C)$$

$$2\sigma = \frac{1}{\sqrt{2}}(\phi_A - \phi_C)$$

$$3\sigma = \tfrac{1}{2}(\phi_A - \sqrt{2}\phi_B + \phi_C)$$

The three σ orbitals introduce one feature that should always be kept in mind:

> The energies of molecular orbitals generally increase as the number of nodes between neighboring atoms increases.

The physical reason for this variation is that electrons are progressively excluded from the internuclear regions as the number of nodes increases.

The triangular H_3 molecule

Now suppose that H_3 is an equilateral triangle. Its three molecular orbitals have the same form as in the linear molecule, but the orbitals previously called 2σ and 3σ now have the same energy. It is not immediately obvious that 2σ and 3σ are degenerate, but symmetry arguments of the kind to be introduced in Chapter 3 can be used to show that they do in fact have exactly the same energy. That their energies are the same can be seen at least to be plausible by the following argument. In the triangular molecule, A and C are next to each other (Fig. 2.22), so 2σ is antibonding between them and hence higher in energy than in the linear molecule. On the other hand, the 3σ orbital is bonding between A and C (but still antibonding between A and B and between C and B), so its energy is lower than before. The two orbitals therefore converge in energy as the linear molecule is converted into the triangular molecule, and when all three H—H separations are the same the energies are found to be equal.

It is no longer strictly appropriate to use the label σ in the triangular molecule because that label applies to a linear molecule. However, it is often convenient to continue to use the notation σ (and in other molecules π too) when we are concentrating on the *local* form of an orbital, its shape relative to the internuclear axis between two neighboring atoms.

The correct procedure for labeling orbitals in polyatomic molecules according to their symmetry is described in Chapter 3. For our present purposes all we need know is the following:

> a, b denote a nondegenerate orbital
>
> e denotes a doubly degenerate orbital
>
> t denotes a triply degenerate orbital

Subscripts and superscripts are sometimes added to these letters, as in a_1, b'', e_g, and t_2 because it is sometimes necessary to distinguish different a, b, e, and t orbitals according to a more detailed analysis of their symmetries. These details are explained in Chapter 3.

The orbital we have called 1σ is nondegenerate and is labeled a'. The two orbitals 2σ and 3σ are doubly degenerate and so must be labeled e'. Therefore, the orbitals are

$$a' = \phi_{1s}(A) + \phi_{1s}(B) + \phi_{1s}(C)$$

$$e' = \begin{cases} \phi_{1s}(A) - \phi_{1s}(C) \\ \phi_{1s}(A) - 2\phi_{1s}(B) + \phi_{1s}(C) \end{cases}$$

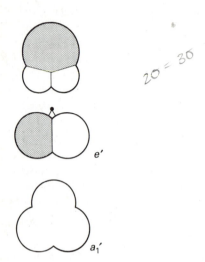

2.22 The three molecular orbitals of the equilateral triangular H_3 molecule formed from overlap of the H1s orbitals.

$2\sigma = 3\sigma$

e'

a_1'

$1\sigma = a'$ (non deg)

$2\sigma, 3\sigma = e'$ (double deg)

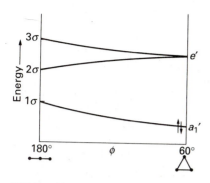

2.23 The orbital correlation diagram for the H_3 molecule shows how the energies of the orbitals change as the linear molecule is converted into the equilateral triangular molecule.

It is possible to compute the energies of these three orbitals as the bond angle is changed from 180° (linear) to 60° equilateral triangular) and the σ orbitals of the linear molecule evolve into the a and e orbitals of the equilateral triangular molecule. The resulting diagram (Fig. 2.23) is called a **correlation diagram**. We shall see that such diagrams play an important role in understanding the shapes, spectra, and reactions of polyatomic molecules.

Electron configurations in H_3^+

Now that the form of the molecular orbitals is available and their relative energies are known, it is possible to deduce the electron configuration of the two-electron molecule ion H_3^+ for any bond angle.

The two electrons occupy the orbital with the lowest energy. The resulting configuration for linear H_3^+ will be $1\sigma^2$; for equilateral triangular H_3^+ it is a^2. It is possible to decide which geometry has the lower energy only by detailed calculation. However, the fact that a is bonding between all three atoms whereas 1σ is bonding only between A and B and between B and C suggests that the lower energy is obtained for the equilateral triangular arrangement. This is in fact the case, and spectroscopic data and numerical computation indicate that H_3^+ is an equilateral triangular species with configuration a^2.

The important point, as can be seen from Fig. 2.22, is that the a orbital spreads equally over all the atoms. The same is true of the 1σ orbital in the linear molecule. Therefore, in each case *the two electrons bind the entire cluster together*. In other words, molecular orbitals are **delocalized orbitals**, in the sense that their bonding or antibonding influence is spread over several atoms and not localized between two atoms. The H_3^+ molecule ion is in fact the simplest example of a **three-center, two-electron bond** (a 3c,2e bond) in which three nuclei are bound by only two electrons. This type of bond is represented by the three-spoke pattern shown in structure (5). There should be nothing particularly puzzling about 3c,2e bonding, for it is a natural outcome of the formation of delocalized molecular orbitals.

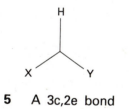

5   A 3c,2e bond

Molecular orbitals for chains and rings of atoms

It is quite easy to extend these ideas to build up the molecular orbitals for chains and rings of more than three atoms. To be as useful as possible for what is to follow, we shall consider the σ orbitals that may be formed by combining s orbitals on atoms in the chain. We shall see later that we can treat some large molecules as formed by combining the orbitals of a central metal atom with orbitals that resemble the orbitals we are about to construct, as in the compound bis(benzene)-chromium.

The molecular orbitals that can be formed from three, four, and five s orbitals are shown in Fig. 2.24. (The triatomic case is identical to the H_3 system that we have already treated.) The relative signs of the atomic orbitals in the linear combinations are represented by their different tints. The sizes of the circles indicate the approximate sizes of the coefficients in the molecular orbital: a small circle, for instance,

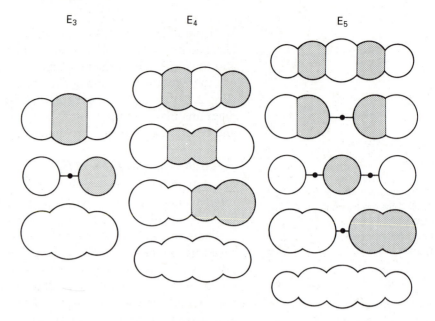

2.24 Nodal properties of orbitals in linear chains of atoms. Opposite signs of the orbitals are represented by the presence or absence of shading and the sizes of the atomic orbitals indicate the size of the coefficients. An open circle represents a positive sign, a shaded circle represents a negative sign. The diagram applies to s orbitals along the axis of a molecule or to the top view of p orbitals perpendicular to the plane of the page.

indicates an atomic orbital that makes only a small contribution to the molecular orbital. The molecular orbitals are arranged in order of increasing energy, with the lowest energy at the bottom. The regular pattern of nodal planes should be noted: it is a feature common to all linear chains that the molecular orbitals resemble the harmonics of a vibrating string. We have already seen that it is also a common feature of molecular orbitals that the energy increases as the number of nodes increases. The lowest energy orbital has no nodes between neighboring atoms and the most antibonding orbital has a node between each pair of neighboring atoms.

Example 2.7: *Constructing molecular orbitals for polyatomic molecules*
Draw the linear combinations of s orbitals that form molecular orbitals in a six-membered linear chain.

Answer. A simple approach is to note the resemblance of molecular orbitals to the waves of a vibrating string. Therefore, we draw the sine waves representing the first six vibrations of a string (Fig. 2.25(a)), and then replace the curves by a circle on each atom (Fig. 2.25(b)). The size of the circle is proportional to the amplitude of the curve at the atom in question, and the sign is that of the wave.

Exercise E2.7. Construct the σ molecular orbitals formed of an E_8 ring. We shall see below that π molecular orbitals can be treated similarly, with the circles representing the top view of the contributing p orbitals; with this interpretation this exercise is relevant to the structure of uranocene.

The power of molecular orbital theory can begin to be appreciated by noting that the π orbitals that can be formed by the overlap of p

orbitals of atoms have essentially the same form as those shown in Fig. 2.24, with a nodal plane in the plane of the page. That is, we can use the orbitals in Fig. 2.24 to discuss π orbitals so long as we interpret the illustration as a top view of the orbital and remember that there is a node in the plane of the page. Very similar diagrams can also be used to discuss zigzag chains of atoms. Thus, we can use the diagram for E_3 (where E denotes an element) to describe the π orbitals of the V-shaped allyl group ($CH_2=CH-CH_2 \leftrightarrow CH_2-CH=CH_2$) and the nitrite ion ($O-N=O^- \leftrightarrow O=N-O^-$).

Figure 2.26 shows the analogous diagrams for cyclic systems: the circles can be taken to represent either s orbitals or the top view of p orbitals, resulting in systems of σ and π orbitals, respectively. A new feature is that some of the molecular orbitals of cyclic systems are doubly degenerate (as in triangular H_3, the simplest cyclic system). The degeneracy of the orbitals in cyclic E_4 is probably easier to accept than in E_3 because the two degenerate orbitals are changed into each other by a simple 90° rotation of the molecule. As with chain systems, the greater the number of nodes, the greater the antibonding character and hence the higher the energy of the orbital.

2.8 Polyatomic molecules in general

We shall now introduce the molecular orbitals of some polyatomic molecules and generalize the discussion to include three-dimensional systems. First, we shall describe the result of a molecular orbital analysis of three important cases. The aim is to see what the theory has to say about the electronic structure of the molecules, and in particular the origin of the bonding in them.

(a)   (b)

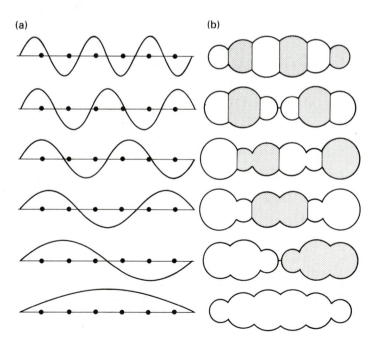

2.25 (a) The first six harmonics of a stretched string and (b) the molecular orbitals they suggest.

E_3 E_4 E_5 E_6

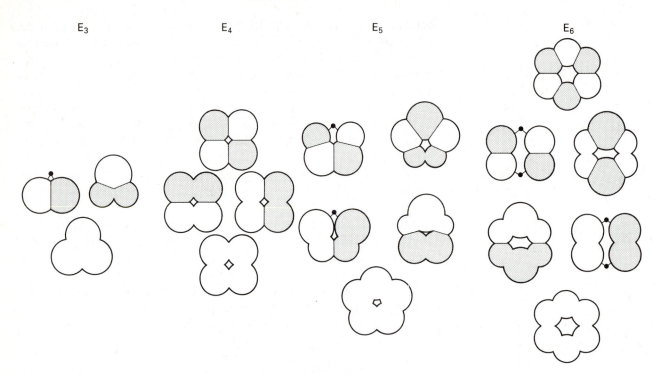

2.26 Nodal properties of π orbitals in cyclic molecules. The same conventions are used as in Fig. 2.24.

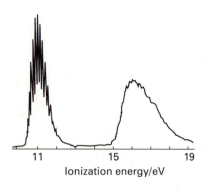

2.27 The UV photoelectron spectrum of NH_3 using helium 21 eV radiation.

The photoelectron spectrum of NH_3 (Fig. 2.27) indicates some of the features that a theory of the structure of polyatomic molecules must elucidate. The spectrum shows two bands. The one with the lower ionization energy (in the region of 11 eV) has considerable vibrational structure. This structure indicates that the orbital from which the electron is ejected plays a considerable role in the determination of the molecule's shape. The broad band in the region of 16 eV arises from electrons that are bound more tightly.

The formation of molecular orbitals

The features that have been introduced in connection with H_3 are present in all the molecules considered from now on. In each case, we can write the molecular orbital of a given symmetry (such as the σ orbitals of a linear molecule) as a sum of *all* the orbitals that can overlap to form orbitals of that symmetry:

$$\psi = \sum_i c_i \phi_i$$

In this linear combination the ϕ_i are atomic orbitals and the index i runs over all the orbitals of all the atoms in the molecule that have the appropriate symmetry. From N atomic orbitals we can construct N molecular orbitals. Then:

1. The greater the number of nodes in a molecular orbital, the greater the antibonding character and the higher the orbital energy.

2. Interactions between non-nearest neighbor atoms are weakly bonding (lower the energy slightly) if the orbital lobes on these

atoms have the same sign (and interfere constructively). They are weakly antibonding if the signs are opposite (and interfere destructively).

3. Orbitals constructed from lower energy atomic orbitals lie lower in energy.

(So atomic *s* orbitals produce lower energy molecular orbitals than atomic *p* orbitals of the same shell.)

For instance, to account for the features in the photoelectron spectrum of NH_3, we need to build molecular orbitals that will accommodate the eight valence electrons in the molecule. Each molecular orbital is the combination of seven atomic orbitals: the three H1*s* orbitals, the N2*s* orbital, and the three N2*p* orbitals. It is possible to construct seven molecular orbitals from these seven atomic orbitals. The form of the orbitals is shown in Fig. 2.28.

The construction of the orbitals is described in Chapter 3, but it is possible to obtain a sense of their origin by imagining viewing the NH_3 molecule along its threefold axis (designated *z*). The N2p_z and N2*s* orbitals both have cylindrical symmetry about that axis. If the three H1*s* orbitals are superimposed with the same sign relative to each other (that is, so that all have the same tint in the diagram), then they

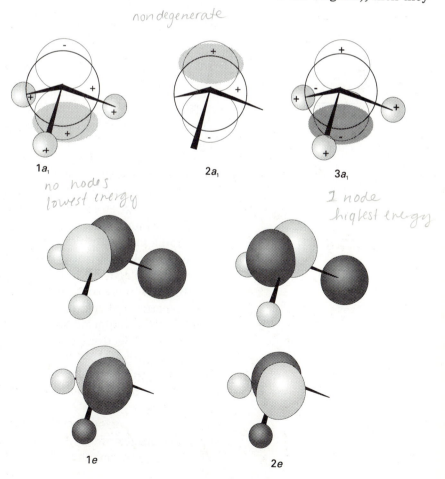

2.28 The composition of the molecular orbitals of NH_3. The signs are shown explicitly in the three *a* orbitals, and the light tinting there indicates the region of orbital amplitude arising from interference between *s* and *p* orbitals on the N atom. Note that the phase of this amplitude in 1*a* is opposite to that in 3*a*. The signs of the orbitals are represented by tinting in the *e* orbitals.

match this cylindrical symmetry. It follows that we can form molecular orbitals of the form

$$\psi = c_1\phi(N2s) + c_2\phi(N2p_z) + c_3\{\phi(H1s_A) + \phi(H1s_B) + \phi(H1s_C)\}$$

From these *three* orbitals (the specific combination of H1s orbitals counts as a single orbital), it is possible to construct three molecular orbitals (with different values of the coefficients c). The forms of these three orbitals are sketched in Fig. 2.28. The orbital with no nodes between the N and H atoms is the lowest in energy, that with a node between all the NH neighbors is the highest in energy, and the third orbital lies between the two. The three orbitals are nondegenerate and are labeled $1a_1$, $2a_1$, and $3a_1$ in order of increasing energy.

The $N2p_x$ and $N2p_y$ orbitals have π symmetry with respect to the z axis, and can be used to form orbitals with combinations of the H1s orbitals that have a matching symmetry. For example, one such superposition will have the form

$$\psi = c_1\phi(Np_x) + c_2\{\phi(H1s_A) - \phi(H1s_B)\}$$

As can be seen from the illustration, the signs of the H1s orbital combination match those of the $N2p_x$ orbital. The N2s orbital cannot contribute to this superposition, so only *two* combinations can be formed, one without a node between the N and H orbitals and the other with a node. The two orbitals differ in energy, the former being lower. A similar combination of orbitals can be formed with the $N2p_y$ orbital, and it turns out (by the symmetry arguments that we use in Chapter 3), that the two orbitals are degenerate with the two we have just described. The combinations are examples of e orbitals, and are labeled $1e$ and $2e$ in order of increasing energy.

The general form of the molecular orbital energy-level diagram is shown in Fig. 2.29. The actual location of the orbitals (particularly the relative positions of the a_1 set and the e_1 set, can only be found by detailed computation or by identifying the orbitals responsible for the photoelectron spectrum. We have indicated the probable assignment of the 11 eV and 16 eV peaks, which fixes the locations of two of the occupied orbitals. The third occupied orbital is out of range of the 21-eV radiation used to obtain the spectrum.

The photoelectron spectrum is consistent with the need to accommodate eight electrons in the orbitals. The electrons enter the molecular orbitals in increasing order of energy, starting with the orbital of lowest energy and taking note of the requirement of the exclusion principle that no more than two electrons can occupy any one orbital. The first two electrons enter $1a_1$ and fill it. The next four enter the doubly degenerate $1e$ orbitals, and fill them. The last two enter the $2a_1$ orbital, which calculations show is almost nonbonding and localized on the N atom. The resulting overall ground state electron configuration is therefore $1a_1^2 1e^4 2a_1^2$. No antibonding orbitals are occupied, so the molecule has a lower energy than the separated atoms. The conventional description of NH_3 as a molecule with a lone pair is also mirrored in the configuration: the HOMO is $2a_1$, which is largely confined to

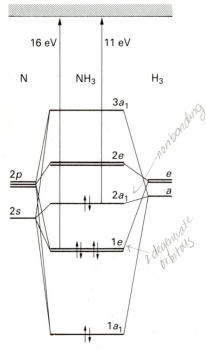

2.29 The molecular orbital energy level diagram of NH_3 when the molecule has its observed bond angle (107°) and bond lengths.

the N atom and makes only a small contribution to the bonding. We shall see in more detail in Chapter 3 that lone pair electrons play a considerable role in determining the shapes of molecules. The extensive vibrational structure in the 11 eV band of the photoelectron spectrum is consistent with this observation, for photoejection of a $2a_1$ electron removes the effectiveness of the lone pair and the shape of the ionized molecule is considerably different from that of the NH_3 molecule itself. Photoionization therefore results in extensive vibrational structure in the spectrum.

Hypervalence

A slightly more complicated example, but one that makes an important point, is the octahedral molecule SF_6. A simple basis set that is adequate to illustrate the point we want to make consists of the valence shell s and p orbitals of an S atom and one p orbital of each of the six F atoms and pointing toward the S atom. We use the $F2p$ orbitals rather than the $F2s$ orbitals because they match the S orbitals more closely in energy. From these 10 atomic orbitals it is possible to construct ten molecular orbitals. Calculations[7] indicate that four of the orbitals are bonding and four are antibonding; the two remaining orbitals are nonbonding (Fig. 2.30). The (abbreviated) labels of the orbitals[8] are

Antibonding:	$2t$	(triply degenerate)
	$2a$	(nondegenerate)
Nonbonding:	e	(doubly degenerate)
Bonding:	$1t$	(triply degenerate)
	$1a$	(nondegenerate)

There are 12 electrons to accommodate: the first two can enter $1a$ and the next six can enter $1t$. The remaining four complete the nonbonding pair of orbitals, resulting in $1a^2 1t^6 e^4$. As we see, none of the antibonding orbitals is occupied. The theory, therefore, accounts for the formation of SF_6: with four bonding orbitals and two nonbonding orbitals occupied, the average S—F bond order is $^2/_3$.

The important point is that the bonding in the hypervalent SF_6 molecule can be explained without using sulfur d orbitals to expand the octet. This does not mean that d orbitals cannot participate in the bonding, but it does show that they are not *necessary* for bonding six F atoms to the central S atom. Molecular orbital theory takes hypervalence into its stride by having available plenty of orbitals, not all of which are antibonding. Therefore the question of when hypervalence can occur appears to depend on factors other than d-orbital availability.

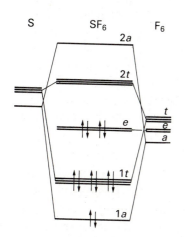

2.30 The molecular orbital energy level diagram of SF_6.

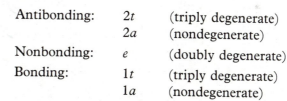

---

[7] For a detailed analysis of the electronic structure of sulfur hexafluoride, see A. E. Reed and F. Weinhold, *J. Am. Chem. Soc.*, **108**, 3586 (1986).
[8] For a broad idea of their shapes, refer to Appendix 4 and see the patterns of orbitals corresponding to each symmetry type.

Electron deficiency

The formation of molecular orbitals by combining several atomic orbitals accounts effortlessly for the existence of **electron-deficient compounds**, which are compounds for which there are insufficient electrons for a Lewis structure to be written. This point can be illustrated most easily with diborane (B_2H_6). In this molecule, the six atoms contribute a total of 14 valence orbitals for the construction of 14 molecular orbitals (four from each B atom, making eight, and one each from the six H atoms). About half these molecular orbitals will be bonding or nonbonding, which is more than enough to accommodate the 12 valence electrons provided by the atoms. There should therefore be nothing mysterious about the existence of this molecule.

A slightly different viewpoint is achieved by concentrating on the two BHB fragments and the three molecular orbitals that can be formed by combining the atomic orbitals contributed by the three atoms (Fig. 2.31).[9] The bonding orbital spanning these three atoms can accommodate two electrons and pull the molecule together, exactly as in H_3^+. Because the bridging bond is formed from two electrons in a molecular orbital built from orbitals on three atoms, it is called a **3-center, 2-electron bond** and is denoted (3c,2e). That two electrons can bind two pairs of atoms (BH and HB) in this way is not particularly remarkable, but it is an insurmountable difficulty for the Lewis approach. Electron deficiency is in fact well developed not only in boron (where it was first clearly recognized), but also in carbocations and a variety of other classes of compounds that we encounter later in the text. The electronic structure of metals may be regarded as an extreme example of its occurrence.

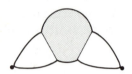

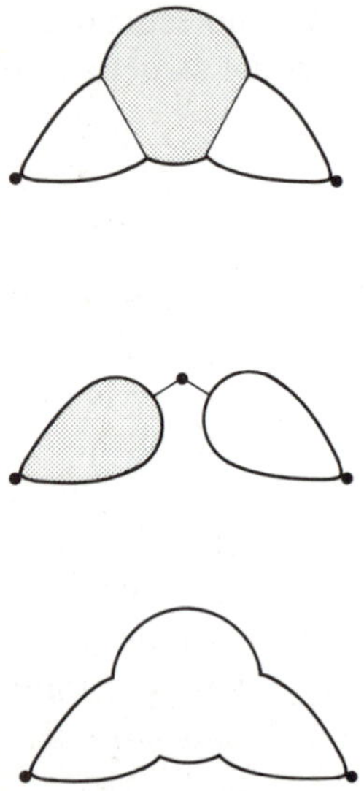

2.31 The molecular orbitals formed by two B orbitals and one H orbital on an atom lying between them, as in B_2H_6. Two electrons occupy the bonding combination and hold all three atoms together.

Localization

A striking feature of the Lewis approach to chemical bonding is its accord with chemical instinct, for it identifies something that can be called 'an A—B bond'. Both O—H bonds in H_2O, for instance, are treated as localized, equivalent structures, because each consists of an electron pair shared between O and H. This feature appears to be absent from molecular orbital theory, for molecular orbitals are delocalized, and the electrons that occupy them bind all the atoms together, not just a specific pair of neighboring atoms. The concept of an A—B bond as existing independently of other bonds in the molecule, and of being transferable from one molecule to another, appears to have been lost. However, we shall now show that the molecular orbital description is mathematically almost equivalent to a localized description of the overall electron distribution.

The demonstration hinges on the fact that quantum theory shows that we can take linear combinations of orbitals and still arrive at the same overall electron density distribution and energy of a molecule.

---

[9] This is the first of several examples where we construct molecular orbitals from the orbitals of molecular fragments. This approach is a half-way stage between the atomic orbitals and the final molecular orbitals, and is very useful for drawing analogies between molecules.

The same is true of molecular orbitals: combinations of occupied molecular orbitals can be formed that result in the same *overall* electron distribution, but the individual orbitals are distinctly different.

Consider the H_2O molecule. The two occupied bonding orbitals of the delocalized description, a_1 and b_2, are shown in Fig. 2.32. If we form the sum $a_1 + b_2$, the negative half of b_2 cancels half the a_1 orbital, leaving a localized orbital between O and the other H. Likewise, when we form the difference $a_1 - b_2$, the other half of the a_1 orbital is canceled, so leaving a localized orbital between the other pair of atoms. That is, by taking sums and differences of delocalized orbitals, localized orbitals are created (and vice versa). Because these are two equivalent ways of describing the same overall electron population, one description cannot be said to be better than the other. Hence it is reasonably well justified to use the localized description of a molecule when chemical evidence suggests that it is appropriate.

Table 2.6 suggests when it is appropriate to select a delocalized description or a localized description. In general:

> A delocalized description is needed for dealing with global properties of the entire molecule.

Such properties include electronic spectra (UV and visible transitions), photoionization spectra, ionization and electron attachment energies, and reduction potentials. In contrast,

> A localized description is most appropriate for dealing with properties of a fragment of a total molecule.

Such properties include bond strength, bond length, bond force constant, and some aspects of reactions (such as acid–base neutralizations). The localized description is more appropriate because it focuses attention on the distribution of electrons in and around a particular bond.

Localized bonds and hybridization

The localized molecular orbital description of bonding can be taken a stage further by introducing the concept of **hybridization**, the formation of mixtures of atomic orbitals on the same atom. Strictly speaking, hybridization belongs to the 'valence-bond theory' of the chemical

Table 2.6 A general indication of the properties for which localized and delocalized descriptions are appropriate

Localized appropriate	Delocalized appropriate
Bond strengths	Electronic spectra
Force constants	Photoionization
Bond lengths	Electron attachment
Brønsted acidity*	Magnetism
VSEPR description of molecular geometry†	Walsh description of molecular geometry
	Standard potentials‡

*Chapter 5.   †Chapter 3.   ‡Chapter 7.

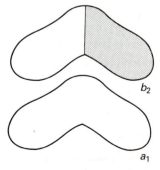

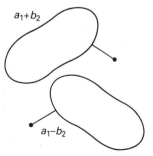

2.32 The two occupied $1a_1$ and $1b_2$ orbitals of the H_2O molecule and their sum $1a_1 + 1b_2$ and difference $1a_1 - 1b_2$. In each case we form a localized orbital between a pair of atoms.

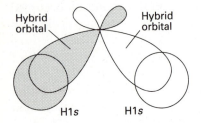

2.33 The formation of localized O—H orbitals in H_2O by the overlap of hybrid orbitals on the O atom and H1s orbitals. The hybrid orbitals are a close approximation to the sp^3-hybrids shown in Fig. 2.34.

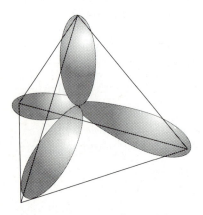

2.34 The four equivalent sp^3 tetrahedral hybrid orbitals. Each one points toward the corner of a regular tetrahedron.

bond, but it is commonly invoked in simple qualitative descriptions of molecular orbitals.

We have seen that in general a molecular orbital is constructed from atomic orbitals of the appropriate symmetry. However, instead of picturing an orbital as a linear combination of all possible atomic orbitals, it is sometimes convenient to form a mixture of orbitals on one atom (the O atom in H_2O, for instance), and then to use these **hybrid orbitals** to construct localized molecular orbitals. In H_2O, for instance, each O—H bond can be regarded as formed by the overlap of an H1s orbital and a hybrid orbital composed of O2s and O2p orbitals (Fig. 2.33).

The mixing of s and p orbitals on a given atom results in hybrid orbitals that have a definite direction in space. This directionality is a result of constructive and destructive interference between the contributing orbitals. For example, as a result of interference, the four equivalent **sp^3-hybrid orbitals**

$$h_1 = s + p_x + p_y + p_z \qquad h_2 = s - p_x - p_y + p_z$$

$$h_3 = s - p_x + p_y - p_z \qquad h_4 = s + p_x - p_y - p_z$$

are directed toward the corners of a regular tetrahedron (Fig. 2.34).

Once the hybrid orbitals have been selected, a localized molecular orbital description can be constructed. For example, four bonds in CF_4 can be formed by building bonding and antibonding localized orbitals by overlap of each hybrid and one F2p orbital directed toward it.

Hybrid orbitals can be used to match any molecular geometry. Because orbitals cannot simply be 'lost', N atomic orbitals combine to give N hybrid orbitals. To recreate a trigonal planar arrangement of orbitals, as in BF_3, we form three **sp^2-hybrids** from an s orbital and two p orbitals lying in the plane. Therefore, if we are seeking to describe the electron distribution of BF_3, we could consider each B—F σ orbital as formed by the overlap of an sp^2 hybrid with an F2p orbital.

To match the arrangements of bonds in molecules with more complicated shapes it is necessary to use d orbitals as well as s and p orbitals. To recreate five localized orbitals we must use one d orbital in addition to the four s and p orbitals. As Table 2.7 shows, sp^3d hybridization leads to five hybrids that point toward the corners of a trigonal bipyramid. If we insisted on a localized orbital description, a PCl_5 molecule would be pictured as having five P—Cl σ bonds formed by overlap of each of the five trigonal bipyramidal sp^3d hybrid orbitals with a 2p orbital of a Cl atom. Similarly, where we wanted to form six localized orbitals in a regular octahedral arrangement, we would need two d orbitals: the resulting six sp^3d^2 hybrids point in the required directions.

Isolobal analogies

A concept that is found throughout chemistry is the substitution of one fragment for another in a molecule. Thus, we may view $N(CH_3)_3$

Table 2.7 Some hybridization schemes

Coordination number	Arrangement	Composition
2	Linear	sp, pd, sd
	Angular	sd
3	Trigonal planar	sp^2, p^2d
	Unsymmetrical planar	spd
	Trigonal pyramidal	pd^2
4	Tetrahedral	sp^3, sd^3
	Irregular tetrahedral	spd^2, p^3d, pd^3
	Square planar	p^2d^2, sp^2d
5	Trigonal bipyramidal	sp^3d, spd^3
	Tetragonal pyramidal	sp^2d^2, sd^4, pd^4, p^3d^2
	Pentagonal planar	p^2d^3
6	Octahedral	sp^3d^2
	Trigonal prismatic	spd^4, pd^6
	Trigonal antiprismatic	p^3d^3

Source: H. Eyring, J. Walter, and G. E. Kimball, *Quantum chemistry.* Wiley, New York (1944).

as derived from NH_3 by substitution of a CH_3 fragment for each H atom. In current terminology, the structurally analogous fragments are said to be **isolobal**, and the relationship is expressed by the symbol ⟷. Two fragments are isolobal if their uppermost orbitals have the same symmetry (such as the σ symmetry of the H1s and a Csp^3 orbital), similar energies, and the same electron occupation (one in each case in H1s and Csp^3). The origin of the name is the lobe-like shape of an orbital in a molecular fragment. As will become clear, the concept of isolobality is rarely reliable for making predictions but it is a helpful device for rationalizing observations.

A simple localized orbital viewpoint can be adopted for many of the applications of isolobality. This viewpoint allows us to identify families of isolobal fragments, such as

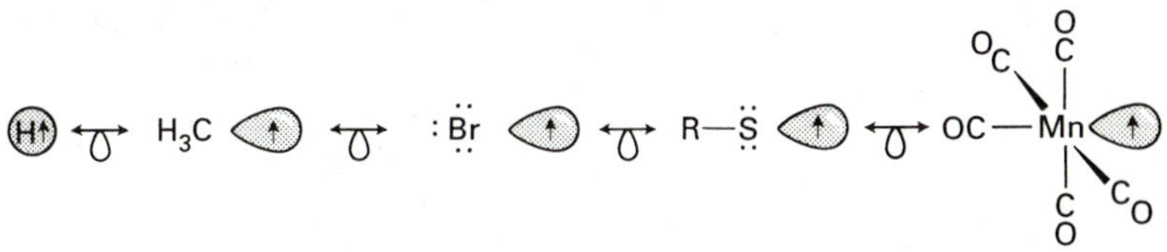

where, in this case, the arrow represents a single electron (other isolobal fragments may have two paired electrons in the corresponding orbitals). The recognition of this family permits us to anticipate by analogy with H—H that molecules such as H_3C—Br and $(OC)_5Mn$—CH_3 can be formed. The same delocalized orbital viewpoint can also be used to identify some isolobal fragments with two available singly occupied orbitals:

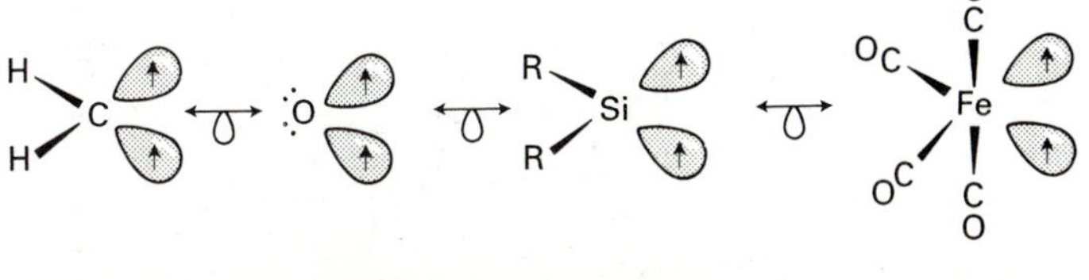

Some three-orbital isolobal fragments can also be identified:

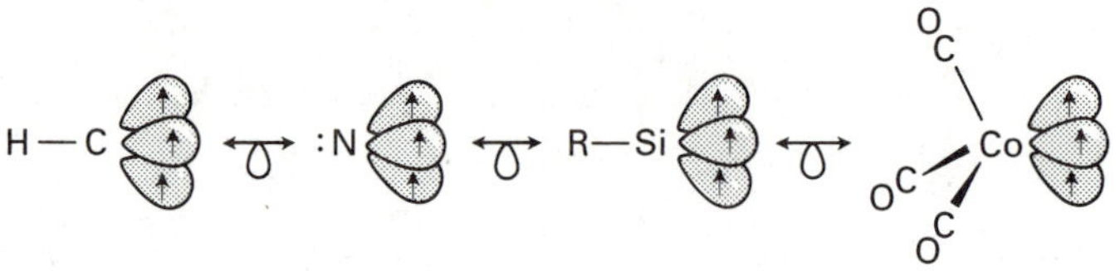

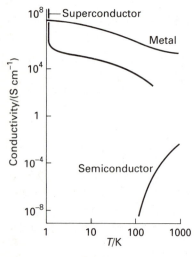

The existence of these families suggests that we should expect to encounter molecules such as *cyclo*-C_4H_8, $O(CH_3)_2$, and $N(CH_3)_3$, which may all be built from these and similar fragments. These species are all known. The two complexes $Co_4(CO)_{12}$ (**6**) and $Co_3(CO)_9CH$ (**7**) are two more examples. However, isolobal analogies must be used with care, for they may also tempt us to postulate the existence of $(OC)_5Mn—O—Mn(CO)_5$ and $(OC)_4Fe=Fe(CO)_4$, but both are unknown. Isolobal analogies—like molecular orbitals themselves—provide useful correlations and hints for the synthesis of new molecules, but they are no substitute for experimental facts.

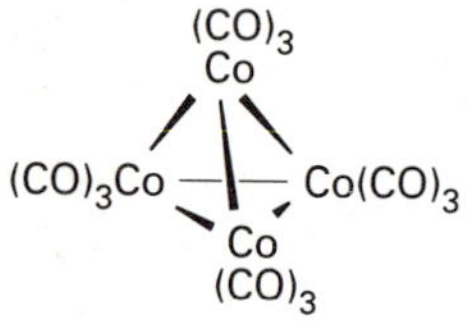

THE MOLECULAR ORBITAL THEORY OF SOLIDS

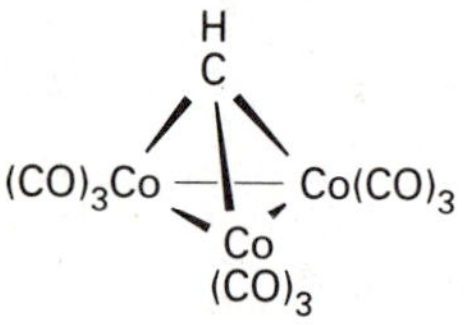

The molecular orbital theory of small molecules can be extended to account for the properties of solids, which are aggregations of a virtually infinite number of atoms. This approach is most strikingly successful for the description of metals, for it may be used to explain their characteristic luster, their good electrical and thermal conductivity, and their malleability. All these properties stem from the ability of the atoms to contribute electrons to a common sea. The luster and electrical conductivities stem from the mobility of these electrons, either in response to the oscillating electric field of an incident ray of light or to a potential difference. The high thermal conductivity is also a consequence of electron mobility, because an electron can collide with a vibrating atom, pick up its energy, and transfer it to another atom elsewhere in the solid. The ease with which metals can be mechanically deformed is another aspect of electron mobility, because the electron sea can quickly readjust to a deformation of the solid and continue to bind the atoms together.

Electronic conduction is also a characteristic of the solids known as **semiconductors**. The criterion for distinguishing between a metal and a semiconductor is the temperature dependence of the electric conductivity (Fig. 2.35):[10]

A metallic conductor is a substance with an electric conductivity that *decreases* with increasing temperature.

[10] The resistance R of a sample is measured in ohms, Ω. The inverse of the resistance is called the conductance G and is measured in siemens S, where $1\ S = 1\ \Omega^{-1}$. The resistance of a sample increases with its length l and decreases with its cross-sectional area A, and we write

$$R = \rho \frac{l}{A}$$

where ρ is the resistivity of the substance. The units of resistivity are Ω m. The conductivity σ is the reciprocal of the resistivity, and its units are $S\ m^{-1}$ (or $S\ cm^{-1}$).

6 $Co_4(CO)_{12}$

7 $Co_3(CO)_9CH$

2.35 The variation of the electrical conductivity of a substance with temperature is the basis of its classification as a metallic conductor, a semiconductor, or a superconductor.

A semiconductor is a substance with an electric conductivity that *increases* with increasing temperature.

It is also generally the case (but not the criterion for distinguishing them) that the conductivities of metals at room temperature are higher than those of semiconductors. An **insulator** is a substance with a very low electrical conductivity. However, when that conductivity can be measured, it is found to increase with temperature, like that of a semiconductor. For some purposes, therefore, it is possible to disregard the classification 'insulator' and to treat all solids as either metals or semiconductors. **Superconductors** are a special class of materials that have zero electrical resistance.

2.9 Molecular orbital bands

The central idea underlying the description of the electronic structure of solids is that the valence electrons donated by the atoms spread through the entire structure. This concept is expressed more formally by making a simple extension of molecular orbital theory in which the solid is treated like an indefinitely large molecule. (In solid state physics, this approach is called the 'tight-binding approximation'.) The description in terms of delocalized electrons is also valid for nonmetallic solids (such as ionic solids) but is virtually essential for the description of the properties of metals. We shall therefore begin by showing how metals are described in terms of molecular orbitals. Then we shall go on to show that the same principles can be applied, but with a different outcome, to ionic and molecular solids.

Band formation by orbital overlap

The overlap of a large number of atomic orbitals leads to molecular orbitals that are closely spaced in energy and so form a virtually continuous **band** that covers a range of energies (Fig. 2.36). On an energy diagram, bands are often separated by **band gaps**, which are values of the energy for which there are no orbitals.

The formation of bands can be understood by considering a line of atoms, and supposing that each atom has an *s* orbital that overlaps the *s* orbitals on its immediate neighbors (Figs 2.37 and 2.38). When the line consists of only two atoms, there is a bonding and an antibonding molecular orbital. When a third atom joins them, there are three orbitals. The central orbital of the set is nonbonding and the outer two are at low energy and high energy, respectively. As more atoms are added, each one contributes an atomic orbital, and hence one more molecular orbital is formed. When there are N atoms in the line, there are N molecular orbitals. The orbital of lowest energy has no nodes between neighboring atoms. The orbital of highest energy has a node between every pair of neighbors. The remaining orbitals have successively 1, 2,... internuclear nodes and a corresponding range of energies between the two extremes.

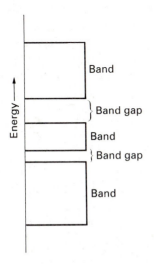

2.36 The electronic structure of a solid is characterized by a series of bands of orbitals which are separated by gaps in the energy for which there are no orbitals.

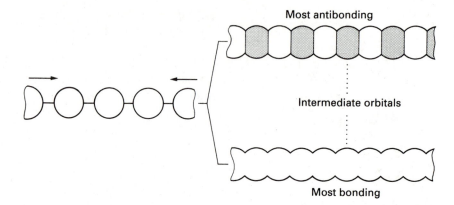

2.37 A band can be thought of as formed by bringing up atoms successively to form a line of atoms. N atomic orbitals give rise to N molecular orbitals.

The total width of the band, which remains finite even as N approaches infinity (as shown in Fig. 2.38), depends on the strength of the interaction between neighboring atoms. The greater the strength of interaction (in broad terms, the greater the degree of overlap between neighbors), the greater the energy separation of the no-node orbital and the all-node orbital. However, whatever the number of atomic orbitals used to form the molecular orbitals there is only a finite spread of orbital energies (as depicted in Fig. 2.38). It follows that the separation in energy between neighboring orbitals must approach zero as N approaches infinity, for otherwise the range of orbital energies could not be finite. That is:

> A band consists of a countable number but near continuum of energy levels.

The band just described is built from s orbitals and is called an **s band**. If there are p orbitals available, a **p band** can be constructed from their overlap as shown in Fig. 2.39. Because p orbitals lie higher in energy than s orbitals of the same valence shell, there is often an energy gap between the s band and the p band (Fig. 2.40). However, if the bands span a wide range of energy and the atomic s and p

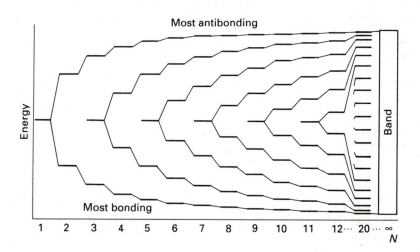

2.38 The energies of the orbitals that are formed when N atoms are brought up to form a line.

Most antibonding

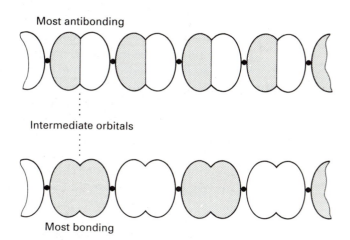

Intermediate orbitals

Most bonding

2.39 An example of a *p* band in a one-dimensional solid.

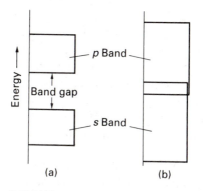

(a)   (b)

2.40 (a) The *s* and *p* bands of a solid and the gap between them. Whether there is in fact a gap depends on the separation of the *s* and *p* orbitals of the atoms and the strength of interaction between the atoms. (b) If the interaction is strong, the bands are wide and may overlap.

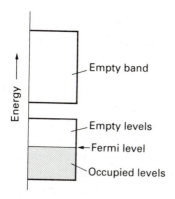

2.41 If each of the *N* atoms supplies one *s* electron, then at $T = 0$ the lowest $\frac{1}{2}N$ are occupied, and the Fermi surface lies near the center of the band.

energies are similar (as is often the case), then the two bands overlap. The **d band** is similarly constructed from the overlap of *d* orbitals.

The Fermi level

At $T = 0$, electrons occupy the individual molecular orbitals of the bands in accord with the building-up principle. If each atom supplies one *s* electron, then at $T = 0$ the lowest $\frac{1}{2}N$ are occupied. The highest occupied orbital at $T = 0$ is called the **Fermi level**; it lies near the center of the band (Fig. 2.41).

At temperatures above absolute zero, the population of higher levels is important because they lie very close to the occupied energy levels. The population P of the orbitals is given by the **Fermi–Dirac distribution**, which is a version of the Boltzmann distribution that takes into account the effect of thermal excitation and the requirement that no more than two electrons can occupy any level. The distribution has the form

$$P = \frac{1}{e^{(E - E_F)/kT} + 1}$$

E_F is the **Fermi energy**, the energy of the level for which $P = \frac{1}{2}$. The Fermi energy depends on the temperature, and at $T = 0$ is equal to the energy of the Fermi level. As the temperature is increased, the Fermi energy rises above the Fermi level because electrons start to occupy higher states and the level at which $P = \frac{1}{2}$ becomes higher in energy. The shape of the Fermi–Dirac distribution is shown in Fig. 2.42.

For energies well above the Fermi energy, the 1 in the denominator of the Fermi–Dirac distribution can be neglected, and the populations for $T > 0$ resemble a Boltzmann distribution in as much that they decline exponentially with increasing energy:

$$P \approx e^{-(E - E_F)/kT}$$

When the band is not completely full, the electrons close to the Fermi surface can easily be promoted to nearby empty levels. As a result,

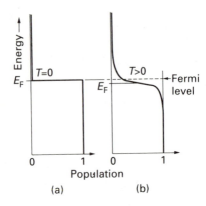

2.42 The shape of the Fermi distribution (a) at $T = 0$ and (b) at $T > 0$. The population decays exponentially at energies well above the Fermi level.

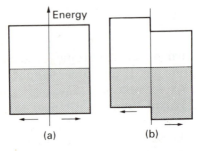

2.43 Another way of representing the bands is to draw the orbitals for motion to the right and motion to the left separately. In the absence of a potential difference applied across the metal (a), the corresponding orbitals are degenerate. However, when a field is applied (b) one 'half band' has a lower energy than the other. It is more heavily populated, and the net result is a flow of electrons.

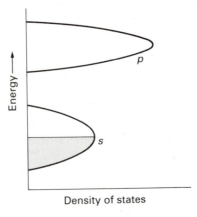

2.44 A typical density of states in a metal.

they are mobile, and can move relatively freely through the solid. The substance is an electronic conductor. One way of seeing that this is so is to think of the individual orbitals in a band as being standing waves. As we saw when discussing the real and complex representations of atomic p orbitals (Section 1.5), standing waves can be regarded as superpositions of traveling waves that correspond to motion in opposite directions. In the absence of a potential difference, the two directions of travel are degenerate and are equally populated up to the Fermi level (Fig. 2.43(a)). However, when a potential difference is applied, electrons traveling in one direction have a different energy from those traveling in the opposite direction, and the two sets of orbitals are no longer equally populated (Fig. 2.43(b)). Consequently, there are now more electrons traveling in one direction than in the other, and an electric current flows through the solid.

We have seen that the criterion of metallic conduction is the decrease of conductivity with increasing temperature. This behavior is the opposite of what we might expect if the conductivity were governed by a Boltzmann distribution of electrons. The competing effect can be identified once we recognize that the ability of an electron to travel smoothly through the solid in a conduction band depends on the uniformity of the arrangement of the atoms. An atom vibrating vigorously at a site is equivalent to an impurity that disrupts the orderliness of the orbitals. This decrease in uniformity reduces the ability of the electron to travel from one edge of the solid to the other, so the conductivity of the solid is less than at $T = 0$. If we think of the electron as described by a wave propagating through the solid, then we would say that it was 'scattered' by the impurity—the atomic vibration. This **carrier scattering** increases with increasing temperature as the lattice vibrations increase, and the increase accounts for the observed inverse temperature dependence of the conductivity of metals.

Densities of states

The number of energy levels per unit energy increment is called the **density of states**, ρ. The density of states is not uniform across a band because the energy levels are packed together more closely at some energies than at others. This variation is apparent even in one dimension, for—compared with its edges—the center of the band is relatively sparse in orbitals (as can be seen in Fig. 2.38). In three dimensions, the variation of density of states is more like that shown in Fig. 2.44, with the greatest density of states near the center of the band and the lowest density at the edges. The reason for this behavior can be traced to the number of ways of producing a particular linear combination of atomic orbitals. There is only one way of forming a fully bonding molecular orbital (the lower edge of the band) and only one way of forming a fully antibonding orbital (the upper edge). However, there are many ways (in a three-dimensional array of atoms) of forming a molecular orbital with an energy in the interior of the band.

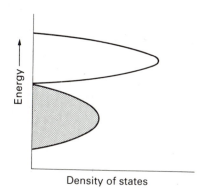

2.45 The density of states typical of a semimetal.

The density of states is zero in the band gap—there are no energy levels in the gap. In certain special cases, though, a full band and an empty band might coincide in energy,[11] but with a zero density of states at their conjunction (Fig. 2.45). Solids with this band structure are called **semimetals**. Because they have only a few electrons that can act as carriers, semimetals are characterized by a low metallic conductivity. One important example is graphite, which is a semimetal in directions parallel to the sheets of carbon atoms. The use of the term semimetal in this context should be carefully distinguished from another unrelated usage, in which it denotes a *chemical* character intermediate between that of metal and nonmetal; for this character we use the term **metalloid**.

Photoelectron and X-ray analysis of bands

Evidence for the existence of bands and a map of their densities of states can be obtained experimentally with photoelectron spectroscopy in much the same way as for discrete molecules. The densities of states of discrete molecules consist of a series of widely separated narrow peaks, each spike corresponding to the energy of a discrete molecular orbital. These peaks appear in the photoelectron spectrum as photoelectrons with discrete ionization energies.

Analogous information about solids can be obtained from **X-ray emission bands**. In this technique, electrons are ejected (by electron bombardment) from the inner closed shells of the atoms and X-rays are emitted as electrons from the valence bands fall into the resulting vacancies (Fig. 2.46). Because a valence electron can originate from any occupied level in the band, the X-ray emission that accompanies its transition covers a range of frequencies. The greatest emission intensities are obtained when there are many states in the valence band with similar energies, and the weakest intensities are obtained when there are only a few states with a particular energy. Hence, it follows that the shape of the X-ray emission band is an indication of the variation of the density of states across the band. The intensity profile does not match the band profile exactly, however, because allowance must be made for the differences between the ease with which the incoming photon may eject an electron from different types of orbital (more specifically, the differences in transition probabilities).

X-Ray emission bands give information about the densities of states of the occupied parts of bands. The analogous information for the unoccupied parts of bands comes from X-ray absorption data.

Special features of one-dimensional solids

In recent years, a series of solids have been investigated that have a partially filled band formed by a linear chain of metal atoms. An

[11] The coincident orbitals at the top and bottom of the two bands are not identical wavefunctions since they differ in wavelength, a feature not shown in the illustration. That is, the electrons that occupy orbitals at the top and bottom of the bands have the same energy but different linear momenta.

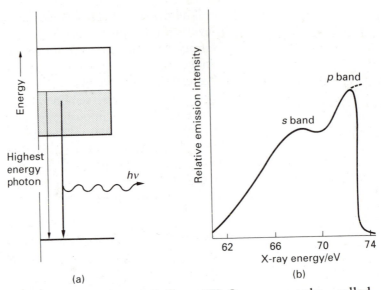

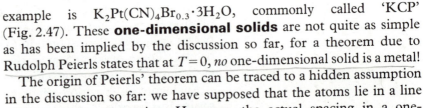

2.46 (a) The formation of an X-ray emission band and (b) a typical example (aluminum).

(a)

(b)

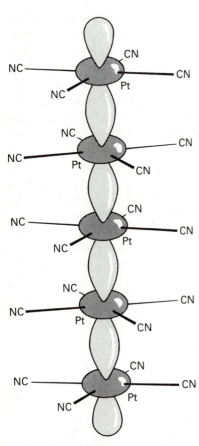

2.47 A representation of the infinite chain structure of KCP, $K_2Pt(CN)_4Br_{0.3} \cdot 3H_2O$.

example is $K_2Pt(CN)_4Br_{0.3} \cdot 3H_2O$, commonly called 'KCP' (Fig. 2.47). These **one-dimensional solids** are not quite as simple as has been implied by the discussion so far, for a theorem due to Rudolph Peierls states that at $T = 0$, *no* one-dimensional solid is a metal!

The origin of Peierls' theorem can be traced to a hidden assumption in the discussion so far: we have supposed that the atoms lie in a line with a regular separation. However, the actual spacing in a one-dimensional solid (and any solid) is determined by the distribution of the electrons, not vice versa, and there is no guarantee that the state of lowest energy is a solid with a regular lattice spacing. In fact, in a one-dimensional solid at $T = 0$, there always exists a distortion, a **Peierls distortion**, which leads to a lower energy than in the perfectly regular solid.

An idea of the origin and effect of a Peierls distortion can be obtained by considering a one-dimensional solid of N atoms and N valence electrons (Fig. 2.48). Such a line of atoms distorts to one that has long and short alternating bonds. Although the longer bond is energetically unfavorable, the strength of the short bond more than compensates for the weakness of the long bond, and the net effect is a lowering of energy below that of the regular solid. Now, instead of the electrons near the Fermi surface being free to move through the solid, they are trapped between the longer-bonded atoms (these electrons have anti-bonding character, and so are found outside the internuclear region between strongly bonded atoms). The Peierls distortion introduces a band gap in the center of the original conduction band, and the filled orbitals are separated from the empty orbitals. Hence, the distortion results in a semiconductor or insulator, not a metallic conductor.

The conduction band in KCP is a *d* band formed principally by overlap of platinum's $5d_{z^2}$ orbitals. The small proportion of bromine in the compound, which is present as Br^-, removes a small number of electrons from this otherwise full *d* band, so turning it into a conduction band. Indeed, at room temperature, doped KCP is a lustrous bronze color with its highest conductivity along the axis of the

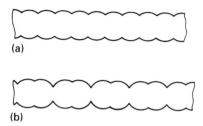

(a)

(b)

2.48 The formation of a Peierls distortion: the energy of the line of atoms with alternating bond lengths is lower than that of the uniformly spaced atoms.

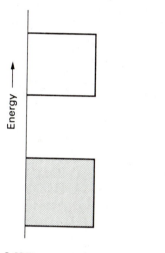

2.49 The structure of a typical insulator: there is a significant gap between the filled and empty bands.

Table 2.8 Some typical band gaps at 25°C

Material	E/eV
Carbon (diamond)	5.47
Silicon carbide	3.00
Silicon	1.12
Germanium	0.66
Gallium arsenide	1.42
Indium arsenide	0.36

Source: S. A. Schwartz, *Kirk–Othmer encyclopedia of chemical technology*, Wiley-Interscience, New York (1982). Vol. 20, p. 601.

Pt chain. However, below 150 K the conductivity drops sharply on account of the onset of a Peierls distortion. At higher temperatures, the motion of the atoms averages the distortion to zero, the separation is regular (on average), the gap is absent, and the solid is metallic.

Insulators

A solid is an insulator if enough electrons are present to fill a band completely and there is a considerable energy gap before an empty orbital becomes available (Fig. 2.49). In a sodium chloride crystal, for instance, the N Cl$^-$ ions are nearly in contact and their $3s$ and $3p$ valence orbitals overlap to form a narrow band of $4N$ levels. The Na$^+$ ions are also nearly in contact and also form a band. The electronegativity of chlorine is so much greater than that of sodium that the chlorine band lies well below the sodium band, and the band gap is about 7 eV. A total of $8N$ electrons are to be accommodated (seven from each chlorine atom, one from each sodium atom). These $8N$ electrons enter the lower chlorine band, fill it, and leave the sodium band empty. Because kT is equivalent to 0.03 eV at room temperature, very few electrons occupy the orbitals of the sodium band.

We normally think of an ionic or molecular solid as consisting of discrete ions or molecules. According to the picture we have just described, though, it appears that they should be regarded as having a band structure. The two pictures can be reconciled, because it is possible to show that a *full* band is equivalent to a sum of localized electron densities. In sodium chloride, for example, a full band built from Cl orbitals is equivalent to a collection of discrete Cl$^-$ ions. As with molecules, the delocalized band picture is needed for description of spectra where processes involve one electron at a time, such as photoelectron spectra and X-ray spectra.

2.10 Semiconduction

The characteristic physical property of a semiconductor is that its electrical conductivity increases with increasing temperature. At room temperature, the conductivities of semiconductors are typically intermediate between those of metals and insulators (in the region of 10^3 S cm^{-1}). The dividing line between insulators and semiconductors is a matter of the size of the band gap (Table 2.8); the conductivity itself is an unreliable criterion because, as the temperature is increased, a given substance may have a low, intermediate, or high conductivity. The values of the band gap and conductivity that are taken as indicating semiconduction rather than insulation depend on the application being considered.

Intrinsic semiconductors

In an **intrinsic semiconductor**, the band gap is so small that the Fermi–Dirac distribution results in some electrons populating the empty upper band (Fig. 2.50). This occupation of the conduction band

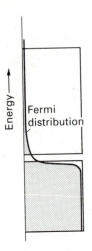

2.50 In an intrinsic semiconductor, the band gap is so small that the Fermi distribution results in some electrons populating the empty upper band.

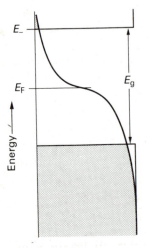

2.51 The relation between the Fermi distribution and the band gap.

introduces negative carriers into the upper level and positive holes into the lower, and as a result the solid is conducting. A semiconductor at room temperature generally has a much lower conductivity than a metallic conductor because only very few electrons and holes can act as charge carriers. The strong, increasing temperature dependence of the conductivity follows from the exponential Boltzmann-like temperature dependence of the electron population in the upper band.

It follows from the exponential form of the population of the conduction band that the conductivity of a semiconductor should show an Arrhenius-like temperature dependence of the form

$$\sigma = \sigma_0 e^{-E_a/kT}$$

The relation of the activation energy E_a to the band gap is established by deciding how $E - E_F$, which appears in the high-temperature form of the Fermi–Dirac distribution (p. 93) depends on E_g.

In a simple picture of the band structure, the Fermi energy (the energy at which $P = \frac{1}{2}$) is approximately half way between the upper and lower bands (Fig. 2.51), and the energy of the lowest level of the upper band E_- is related to E_F and E_g by

$$E_- - E_F \approx \tfrac{1}{2}E_g$$

It follows that the conductivity should follow the expression

$$\sigma = \sigma_0 e^{-E_g/2kT}$$

That is, the conductivity of a semiconductor can be expected to be Arrhenius-like with an activation energy equal to half the band gap, $E_a \approx \frac{1}{2}E_g$. This is found to be the case in practice.

Example 2.8: *Determining the band gap from the temperature dependence of the conductivity*

The conductance G of a sample of germanium varied with temperature as indicated below. Estimate the value of E_g.

T/K	312	354	420
G/S	0.0847	0.429	2.86

Answer. From the equation for σ we see that the analysis is similar to that used to obtain the activation energy of a chemical reaction. Because the conductance G is proportional to the conductivity σ, we can write

$$G = G_0 e^{-E_g/2kT}$$

Taking logarithms gives

$$\ln G = \ln G_0 - \frac{E_g}{2kT}$$

Therefore, a plot of $\ln G$ against $1/T$ should yield a straight line of slope $-E_g/2k$. The data give a slope of -4.26×10^3, and since $k = 8.614 \times 10^{-5}$ eV K^{-1}, we find $E_g = 0.73$ eV.

Exercise E2.8. What is the conductance of the sample at 370 K?

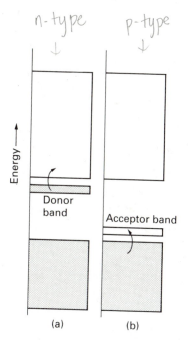

2.52 The band structure in (a) an n-type semiconductor and (b) a p-type semiconductor.

Extrinsic semiconductors

The number of electron carriers can be increased if atoms with more electrons than the parent element can be introduced by the process called **doping**. Remarkably low levels of dopant concentration are needed—only about 1 atom per 10^9 of the host material—so it is essential to achieve very high purity of the parent element initially.

If As atoms are introduced into a silicon crystal, then one additional electron will be available for each dopant atom that is substituted. Note that the doping is *substitutional* in the sense that the dopant atom takes the place of an Si atom. If the donor atoms, the As atoms, are far apart from each other, their electrons will be localized and the **donor band** will be very narrow (Fig. 2.52(a)). Moreover, the foreign atom levels will lie at higher energy than the valence electrons of the host lattice. The filled dopant band is commonly near the empty band of the lattice. For $T > 0$, some of its electrons will be thermally promoted into the empty conduction band. In other words, thermal excitation will lead to the transfer of an electron from an As atom into the empty orbitals on a neighboring Si atom. From there it will be able to migrate through the lattice in the molecular orbitals formed by Si–Si overlap. This process gives rise to **n-type semiconductivity**, the n indicating that the charge carriers are negative electrons.

An alternative substitutional procedure is to dope the silicon with atoms of an element with fewer valence electrons per atom, such as gallium. A dopant atom of this kind effectively introduces holes into the solid. More formally, the dopant atoms form a very narrow, empty **acceptor band** that lies above the full Si band (Fig. 2.52(b)). At $T = 0$ the acceptor band is empty, but at higher temperatures it can accept thermally excited electrons and hence withdraw them from the Si valence band. By doing so, it introduces holes into the latter and hence allows the remaining electrons in the band to be mobile. Because the charge carriers are now effectively positive holes in the lower band, this type of semiconductivity is called **p-type semiconductivity**.

Several d-metal oxides, including ZnO and Fe_2O_3, are n-type semiconductors. In this case, the property is due to nonstoichiometry and a small deficit of O atoms. The electrons that should occupy the localized O atomic orbitals (giving a very narrow oxide band, essentially localized individual O^{2-} ions) occupy a previously empty conduction band formed by the metal orbitals. The conductivity decreases when the solid is heated in oxygen because the deficit of O atoms is replaced, and as the atoms are added electrons are withdrawn from the conduction band.

p-Type semiconduction is observed for some low oxidation number d-metal chalcogenides and halides, including Cu_2O, FeO, FeS, and CuI. In these nonstoichiometric compounds, the loss of electrons is equivalent to the oxidation of some of the metal atoms, with the result that holes appear in the metal-ion band. The conductivity increases when these compounds are heated in oxygen because more holes are formed in the metal-ion band as oxidation progresses.

2.11 Superconduction

A superconductor is a substance that conducts electricity without resistance. Until 1987, the only known superconductors (which included metals, some oxides, and some halides) needed to be cooled to below about 20 K before they became superconducting. However, in 1987 the first **high-temperature superconductors** (HTSC) were discovered; their superconduction is well established at 120 K and spasmodic reports of superconduction at even higher temperatures have appeared. We will not consider these high-temperature materials at this stage (they are discussed in Chapter 18), but sketch the ideas behind the mechanism of low-temperature superconduction.

The central concept of low-temperature superconduction is the existence of a **Cooper pair**, a pair of electrons that exists on account of their indirect interaction with one another via their interaction with the nuclei of the atoms in the lattice. Thus, if one electron is in a particular region of a solid, the nuclei there move toward it to give a distorted local structure (Fig. 2.53). Because that local distortion is rich in positive charge, it is favorable for a second electron to join the first. Hence, there is a virtual attraction between the two electrons, and they move together as a pair. The local distortion can be easily disrupted by thermal motion of the ions, so the virtual attraction occurs only at very low temperatures.

A Cooper pair undergoes less scattering than an individual electron as it travels through the solid, because the distortion caused by one electron can attract back the other electron should it be scattered out of its path in a collision. This has been likened to the difference between the motion of a herd of cattle, with members of the herd that are deflected from their path by boulders in their way, and a team of cattle yoked together, which will travel forward largely regardless of obstacles. Because the Cooper pair is stable against scattering, it can carry charge freely through the solid, and hence give rise to superconduction.

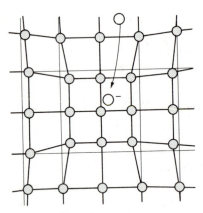

2.53 The formation of a Cooper pair. One electron distorts the crystal lattice, and the second electron has a lower energy if it goes to that region. This effectively binds the two electrons into a pair.

FURTHER READING

Bonding

The first three books are introductions to the treatment of molecules in terms of molecular orbitals:

R. L. DeKock and H. B. Gray, *Chemical structure and bonding*. Benjamin/Cummings, Menlo Park (1980).

T. A. Albright, J. K. Burdett, and M.-H. Wangbo, *Orbital interactions in chemistry*. Wiley, New York (1985). This text has a thorough discussion of isolobality.

J. N. Murrell, S. F. A. Kettle, and J. M. Tedder, *The chemical bond*. Wiley, New York (1978).

More thorough discussions are given in the following:

B. Webster, *Chemical bonding theory*. Blackwell Scientific, Oxford (1990).

B. M. Gimarc. *Molecular spectroscopy and bonding*. Academic Press, New York (1979).

R. McWeeney, *Coulson's valence*. Oxford University Press (1979).

For introductions to photoelectron spectroscopy and its molecular implications, see

R. E. Ballard. *Photoelectron spectroscopy*. Wiley, New York (1978).

J. H. D. Eland. *Photoelectron spectra*. Open University Press, Milton Keynes (1983).

Solids
The following three texts give further details on the solid state concepts that we have introduced, and do so at about the same level:

P. A. Cox, *Solid state chemistry*. Oxford University Press (1987).

M. F. C. Ladd, *Structure and bonding in solid state chemistry*. Wiley, New York (1979).

M. H. B. Stiddard, *The elementary language of solid state physics*. Academic Press, New York (1975).

KEY POINTS

1. Lewis structures
The earliest description of the formation of a covalent bond is that it consists of a shared electron pair. Lewis structures depict schematically how such pairs are shared and give a topological picture of bonding in a molecule.

2. Formal charge and oxidation number
The formal charge on an atom in a Lewis structure is an indication of the charge it would carry if it shared the electrons in a pair equally. The oxidation number of an element, which is independent of the Lewis structure proposed for a species, is an indication of the extent to which electrons transfer to or from an atom in a molecule.

3. Bond parameters
Certain properties of bonds, particularly their lengths and their strengths, are approximately transferrable between species.

4. Molecular orbitals
The principal theoretical description of chemical bonding is in terms of molecular orbitals, which are wavefunctions that spread over two or more atoms. Molecular orbitals are usually approximated by linear combinations of atomic orbitals.

5. Bonding and antibonding orbitals
A molecular orbital is built by superimposing all the atomic orbitals of the appropriate symmetry. From N atomic orbitals, N molecular orbitals can be formed. Approximately half will be bonding orbitals, about half will be antibonding orbitals, and the rest are nonbonding orbitals. The greatest bonding and antibonding effects are obtained by overlap of orbitals of similar energies.

6. Local symmetry classification of orbitals
Molecular orbitals are classified as σ, π, and δ orbitals depending on their symmetry with respect to rotation around the internuclear axis.

7. Electron configurations
The ground state electron configurations of molecules are predicted by applying the building-up principle to the molecular orbitals formed by overlap of atomic orbitals.

8. Electron distributions
In heteronuclear diatomic molecules, the greater contribution to bonding orbitals comes from the more electronegative atom; the less electronegative atom contributes more to the antibonding orbitals.

9. Bond correlations
Certain properties, particularly bond length and bond strength, correlate with the bond order (and with each other).

10. Electron delocalization
Molecular orbitals in polyatomic molecules are delocalized, and their bonding or antibonding influences are shared over all the atoms in the molecule. Thus, an electron pair can bind more than one pair of atoms together.

11. Localized descriptions

Although molecular orbitals are delocalized, it is possible to form mathematically equivalent localized descriptions. One way of modeling such localized orbitals is to build them from the overlap of hybrid orbitals on each atom. Such hybrid orbitals have definite geometrical arrangements that depend on their composition.

12. Isolobality

The concept of localized bonds and hybridization leads to the concept of isolobality, which allows certain bonding analogies to be predicted.

13. Band theory

The concept of molecular orbital formation can be extended to effectively infinite numbers of atoms in solids, where orbital overlap leads to the formation of bands of orbitals separated by energy gaps.

14. Electronic conduction

The occupation of bands and the existence of band gaps accounts for the classification of solids as metallic conductors and semiconductors.

EXERCISES

2.1 Write Lewis electron dot structures for (a) $GeCl_3^-$, (b) FCO_2^-, (c) CO_3^{2-} (d) $AlCl_4^-$ (e) NOF. Where more than one resonance structure is important, give examples of all major contributors.

2.2 Construct Lewis structures of typical resonance contributions of (a) ONC^- and (b) NCO^- and assign formal charges to each atom. Which resonance structure is likely to be the dominant contribution in each case?

2.3 (a) Write Lewis structures for the major resonance forms of NO_2^-. (b) Assign formal charges. (c) Assign oxidation numbers to the atoms. (d) Describe whether oxidation numbers or formal charges are appropriate for the following applications: (i) predominant resonance Lewis dot structure among several resonance forms, (ii) determining whether there is the possibility of nitrogen being oxidized or reduced, (iii) determining the physical charge on the nitrogen atom.

2.4 Write Lewis structures for (a) XeF_4, (b) PF_5, (c) BrF_3, (d) $TeCl_4$, (e) ICl_2^-.

2.5 Use the covalent radii in Table 2.4 to calculate the bond lengths in (a) CCl_4 (1.77 Å), (b) $SiCl_4$ (2.01 Å), (c) $GeCl_4$ (2.10 Å). (The values in parentheses are experimental bond lengths and are included for comparison.)

2.6 Given that $B(Si=O)$ is 640 kJ mol^{-1}, show that bond enthalpy considerations predict that silicon-oxygen compounds are likely to contain networks of tetrahedra with Si—O single bonds and not discrete molecules with Si=O double bonds.

2.7 The common forms of nitrogen and phosphorus are $N_2(g)$ and $P_4(s)$, respectively. Account for the difference in terms of the single and multiple bond enthalpies.

2.8 Use the data in Table 2.5 to calculate the standard enthalpy of the reaction $2H_2(g) + O_2(g) \rightarrow 2H_2O(g)$. The experimental value is -484 kJ. Account for the difference between the estimated and experimental values.

2.9 Predict the standard enthalpies of the reactions
(a) $S_2^{2-}(g) + \frac{1}{4}S_8(g) \rightarrow S_4^{2-}(g)$
(b) $O_2^{2-}(g) + O_2(g) \rightarrow O_4^{2-}(g)$
using mean bond enthalpy data. Assume that the unknown species O_4^{2-} is a singly bonded chain analog of S_4^{2-}.

2.10 Use molecular orbital diagrams to determine the number of unpaired electrons in (a) O_2^-, (b) O_2^+, (c) BN, and (d) NO^-.

2.11 Use Fig. 2.10 to write the electron configurations of (a) Be_2, (b) B_2, (c) C_2^-, and (d) F_2^+ and sketch the form of the HOMO in each case.

2.12 Determine the MO bond orders of (a) S_2, (b) Cl_2, and (c) $[NO]^-$ from their molecular orbital configurations and compare the values with the bond orders determined from Lewis structures. (NO has orbitals like O_2.)

2.13 What are the expected changes in bond order and bond distance that accompany the following ionization processes? (a) $O_2 \rightarrow O_2^+ + e^-$, (b) $N_2 + e^- \rightarrow N_2^-$, (c) $NO \rightarrow NO^+ + e^-$.

2.14 (a) How many independent linear combinations are possible for four $1s$ orbitals? (b) Draw pictures of the linear combinations of $H1s$ orbitals for a hypothetical linear H_4 molecule. (c) From a consideration of the number of nonbonding and antibonding interactions, arrange these molecular orbitals in order of increasing energy.

2.15 (a) Construct the form of each molecular orbital in linear $[HHeH]^{2+}$ using $1s$ basis atomic orbitals on each atom and considering successive nodal surfaces. (b) Arrange the MOs in increasing energy. (c) Indicate the electron population of the MOs. (d) Should $[HHeH]^{2+}$ be stable in isolation or in solution? Explain your reasoning.

2.16 Based on the MO discussion of NH_3 in the text, find the average NH bond order in NH_3 by calculating the net number of bonds and dividing by the number of NH groups.

2.17 From the relative atomic orbital and molecular orbital energies depicted in Fig. 2.30, describe the character as mainly F or mainly S for the frontier orbitals e (the HOMO) and $2t$ (the LUMO) in SF_6. Explain your reasoning.

2.18 Classify the hypothetical species (a) square H_4^{2+}, (b) bent O_3^{2-} as electron precise or electron deficient. Explain your answer and decide whether either of them are likely to exist.

2.19 Identify (a) the hydrogen–nitrogen molecule or molecular fragment that is isolobal with CH_3^-, (b) the hydrogen–boron molecule or molecular fragment that is isolobal with an O atom, (c) a nitrogen-containing species that is isolobal with $[Mn(CO)_5]^-$.

2.20 (a) Draw a simple band picture to distinguish a metallic conductor from a semiconductor. (b) Explain how the temperature dependence of the electrical conductivity can be used to distinguish a metallic conductor from a semiconductor. (c) Can the temperature dependence of the conductivity be used to distinguish an insulator from a semiconductor?

2.21 Decide whether the following systems are likely to be n-type or p-type semiconductors: (a) arsenic-doped germanium, (b) gallium-doped germanium, (c) silicon-doped germanium.

2.22 The promotion of an electron from the valence band into the conduction band in pure titanium(IV) oxide by light absorption requires a wavelength of less than 350 nm. Calculate the energy gap in electronvolts between the valence and conduction bands.

2.23 When titanium(IV) oxide is heated in hydrogen, a blue color develops, indicating light absorption in the red. Does reduction of Ti(IV) to Ti(III) correspond to n-doping or p-doping?

2.24 Gallium arsenide is a semiconductor that is widely used for the construction of the red-light-emitting displays and is under development for advanced central processor chips in supercomputers. If gallium arsenide (GaAs) is doped with selenium (on the As site), is it n-doped or p-doped?

2.25 Cadmium sulfide, CdS, is used as a photoconductor in light meters. The band gap is about 2.4 eV. What is the greatest wavelength of light that can promote an electron from the valence band to the conduction band in cadmium sulfide?

2.26 The band gap of silicon as determined from optical absorption spectra is about 1.12 eV. Calculate the ratio of the conductivities at 373 K and 273 K.

PROBLEMS

2.1 Use the concepts from Chapter 1, particularly the effects of penetration and shielding on the radial wavefunction, to account for the variation of single bond covalent radii with position in the periodic table.

2.2 Develop an argument based on bond enthalpies for the importance of SiO bonds in substances common in the Earth's crust in preference to SiSi or SiH bonds. How and why does the behavior of silicon differ from that of carbon?

2.3 When a He atom absorbs a photon to form the excited configuration $1s^1 2s^1$ (here called He*) a weak bond forms with another He atom to give the diatomic molecule HeHe*. Construct a molecular orbital description of the bonding in this species.

2.4 Construct an approximate molecular orbital energy diagram for a hypothetical planar form of NH_3. You may refer to Appendix 4 to determine the form of the appropriate orbitals on the central N atom and on the triangle of H_3 atoms. From a consideration of the atomic energy levels, place the N and H_3 orbitals on either side of a molecular-orbital energy-level diagram. Then use your judgement about the effect of bonding and antibonding interactions and energies of the parent orbitals to construct the molecular orbital energy levels in the center of your diagram and draw lines indicating the contributions of the atomic orbitals to each molecular orbital. Atomic orbital energy level data are $H1s = -13.6$ eV, $N2s = -26.0$ eV, $N2p = -13.4$ eV.

2.5 (a) Use an extended Hückel molecular orbital program[12] or input and output from such a program supplied by your instructor, to construct a molecular orbital energy level diagram to correlate the MO (from the output) and AO (from the input) energies and indicate the occupancy of the MOs (in the manner of Fig. 2.10) for one of the following molecules: HF (bond length 0.92 Å), HCl (1.27 Å), or CS (1.53 Å); (b) Use the output to sketch the form of the occupied orbitals, showing signs of the AO lobes by shading and their amplitudes by means of size of the orbital.

---

[12] Suitable programs are QCMP001, from QCPE, Chemistry Department, Indiana University, Bloomington, IN; CACAO, by C. Meali and D. M. Proserpio, *J. Chem. Educ.*, **67**, 3399 (1990); PLOT3D, by J. A. Bertrand and M. R. Johnson, School of Chemistry, Georgia Institute of Technology, Atlanta, GA.

2.6 Perform an extended Hückel MO calculation on H_3 by using the H energy given in Problem 2.4 and H—H distances from NH_3 (N—H length 1.02 Å, HNH bond angle 107°) and then carry out the same type of calculation for NH_3. Use energy data for N2s and N2p orbitals from Problem 2.4. From the output, plot the molecular orbital energy levels with proper symmetry labels and correlate them with the N orbitals and H_3 orbitals of the appropriate symmetries. Compare the results of this calculation with the qualitative description in Problem 2.4.

2.7 Assign the lines in the UV photoelectron spectrum of CO shown in Fig. 2.54. (p. 104) and predict the appearance of the UV photoelectron spectrum of the SO molecule.

2.8 The K-shell X-ray emission spectrum of aluminum oxide is illustrated in Fig. 2.46. It is so called because emission arises from valence band electrons falling into a vacancy in the K-shell (an alternative name for the $n=1$ shell) that has been produced by bombardment. The corresponding absorption arises from promotion of an electron from the K-shell into the conduction band. What is the band gap energy in Al_2O_3? Is alumina an insulator or a semiconductor? Are energy levels dense near the band edges or near the band centers? Which peak gives an indication of the distribution of levels derived mainly from O orbitals?

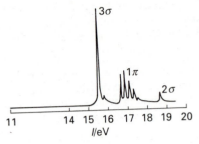

2.54 The ultraviolet photoelectron spectrum of CO using 21eV radiation.

2.9 The electrical conductivity of bismuth is 9.1×10^5 S m^{-1} at 273 K, 6.4×10^5 S m^{-1} at 373 K, and 7.8×10^5 S m^{-1} at 573 K. What type of material is bismuth? Bismuth melts at 271 °C.

2.10 The conductivity of VO increases sharply with increasing temperature up to 125 K, and reaches 1×10^{-4} S m^{-1}. At about 125 K the conductivity rises abruptly to 1×10^2 S m^{-1} and then declines slowly to about 5×10^1 S m^{-1} near 400 K. How would you classify (a) the low-temperature and (b) the high-temperature forms of VO?

3 Molecular shape and symmetry

The concept of molecular shape is of the greatest importance in inorganic chemistry, for not only does it affect the physical properties of the molecule, but it provides hints about how some reactions might occur. In this chapter we explore some of the consequences of molecular shape, and refine that concept into the powerful concept of molecular symmetry and the language of group theory. We shall see that considerations of symmetry allow us to make predictions about the physical and spectroscopic properties of molecules. Symmetry considerations can also be used to construct molecular orbitals, to discuss electronic structure, and to simplify the discussion of molecular vibrations.

The shape of a molecule can be reported, in simple cases, by stating whether it is linear, tetrahedral, and so on (Table 3.1) and giving the bond lengths. Individual bond angles are reported for molecules in which the angles are not implied by the shape itself. Thus, SF_6 (**1**) is octahedral and all its bond angles are necessarily 90°. The NH_3 molecule is trigonal pyramidal, and although its symmetry requires all three HNH angles to be the same, we need to specify that the angles are in fact 107°, for trigonal pyramidal molecules can occur with different bond angles.

Even when the shape of a molecule does not correspond to one of the symmetrical geometrical figures, one of these shapes may be a good starting point for the description of the molecule. The 'see-saw' shape of the SF_4 molecule (**2**), for example, is closely related to a trigonal bipyramid. If a molecule is less symmetrical than one of the shapes shown in Table 3.1, or if it is more complex, then to specify its shape it is necessary to give the individual coordinates of the atoms. We need to remember that the *name* of the shape of a molecule (such as linear or tetrahedral) is determined by the geometrical arrangement of the

Table 3.1 The description of molecular shapes

Description of shape	Shape	Examples
Linear	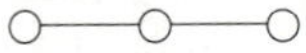	HCN, CO_2
Angular		H_2O, O_3, NO_2^-
Trigonal planar	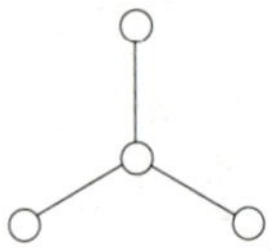	BF_3, SO_3, NO_3^-, CO_3^{2-}
Trigonal pyramidal		NH_3, SO_3^{2-}
Tetrahedral	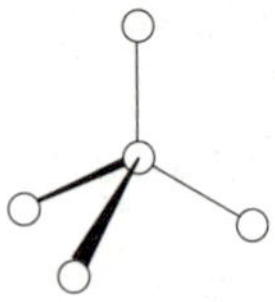	CH_4, SO_4^{2-}, NSF_3^*
Square planar	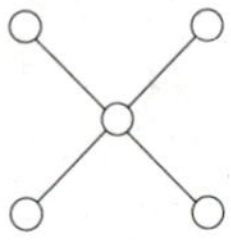	XeF_4
Square pyramidal	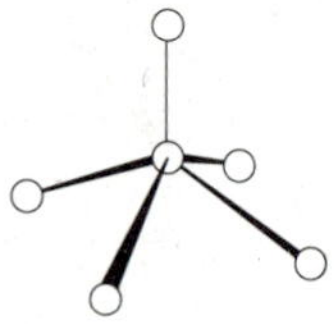	$Sb(Ph)_5$
Trigonal bipyramidal	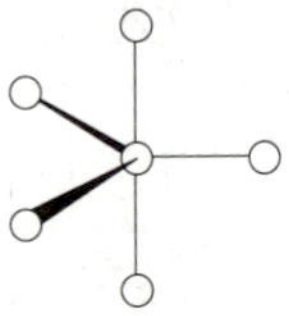	$PCl_5(g)$, SOF_4^*
Octahedral	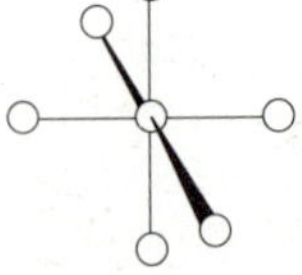	SF_6, PCl_6^-, $IO(OH)_5^*$

*Approximate shape.

atoms disregarding the location of any lone pairs. The NH_3 molecule, for example, is pyramidal, not tetrahedral; the NH_4^+ ion is tetrahedral.

THE ORIGIN OF MOLECULAR SHAPE

We shall describe two approaches to the explanation of molecular shape in terms of the electronic structures of molecules. The first

Table 3.2 The basic arrangement of electron pairs according to the VSEPR model

Number of electron pairs	Arrangement
2	Linear
3	Trigonal planar
4	Tetrahedral
5	Trigonal bipyramidal
6	Octahedral

approach explores to what extent repulsions between pairs of electrons localized in the valence shells of individual atoms determine the geometry the molecule adopts. The second expresses molecular shape in language that is more in the spirit of molecular orbital theory, with electrons in orbitals that are delocalized over an entire molecule.

3.1 The VSEPR model

The **valence shell electron pair repulsion model** (VSEPR model) of molecular shape is a simple extension of Lewis's ideas and is surprisingly successful for predicting the shapes of polyatomic molecules. The theory stems from suggestions made by Nevil Sidgwick and Herbert Powell in the years up to 1940 and later extended and put into a more modern context by Ronald Gillespie and Ronald Nyholm.[1]

The basic shapes

The primary assumption of the VSEPR model is that electron pairs repel each other and take up positions as far apart as possible. According to this model, three electron pairs on an atom will lie at the corners of an equilateral triangle, for that maximizes their separation. Similarly, four electron pairs will lie at the corners of a regular tetrahedron, five will lie at the corners of a trigonal bipyramid, and so on (Table 3.2). As an example, an SF_6 molecule, with six electron pairs around the central S atom, is predicted (and found) to be octahedral (**1**) and a PCl_5 molecule, with five pairs, is predicted (and found) to be trigonal bipyramidal (**3**). Apart from the fact that the VSEPR model is highly successful, however, there is no firm evidence for the assumptions about the distribution of electron density on which it is based.

Some basic shapes have electron pair repulsion energies that are not much lower than those in alternative arrangements, and a molecule may adopt one of these alternatives if there are other contributions to the energy that result in a lower energy overall. For example, a square pyramidal arrangement of electron pairs is only slightly higher in electron-repulsion energy than a trigonal bipyramidal arrangement, and there are several examples of the former (**4**).

The arrangement of electron pairs—both bonding and lone pairs—governs the shape of the molecule. However, as we have already remarked, the *name* of the shape is determined by the arrangement of *atoms*, not the arrangement of the electron pairs. For instance, the

1 SF_6

2 SF_4

3 PCl_5

4 $[InCl_5]^{2-}$

[1] For an excellent introduction to modern attitudes to VSEPR theory, see R. J. Gillespie and I. Hargittai, *The VSEPR model of molecular geometry*, Allyn and Bacon, Needham Heights (1991). An advanced discussion of the role of electron density distributions in molecules and their role in the determination of shape has been given by R. F. W. Bader, *Atoms in molecules*. Clarendon Press, Oxford (1990). See also R. F. W. Bader, R. J. Gillespie, and P. J. MacDougall, *J. Am. Chem. Soc.*, **110**, 7329 (1988).

NH_3 molecule has four electron pairs that are disposed tetrahedrally, but as one of them is a lone pair the molecule itself is classified as trigonal pyramidal. The apex of the pyramid is occupied by the lone pair. Similarly, H_2O has a tetrahedral arrangement of its electron pairs, but as two of the pairs are lone pairs, the molecule is reported as angular.

According to the VSEPR model, a multiple bond is treated as though it were a single electron pair, and its two or three electron pairs are treated as a single 'superpair'. A linear structure for $O=C=O$, for example, is predicted on the grounds that the C atom effectively has two superpairs that adopt a linear arrangement. This rule does away with any worry about which resonance structure to consider. Thus, the Lewis structure of SO_4^{2-} in which the S atom has an octet and all SO bonds are single and an expanded-octet structure with two SO double bonds and two SO single bonds are both predicted to be tetrahedral.

The basic shapes for seven electron pairs are less readily predicted than others, partly because so many different arrangements correspond to similar energies. Lone pairs are stereochemically less influential when they belong to heavy p-block elements. The $[SeF_6]^{2-}$ and $[TeCl_6]^{2-}$ ions, for instance, are octahedral despite the presence of a lone pair on the Se and Te atoms. Lone pairs that do not influence the molecular geometry are said to be **stereochemically inert**.

Modifications of the basic shape

Once the number of electron pairs has been used to identify the basic shape of a molecule, minor adjustments are made by taking into account the differences in electrostatic repulsion between bonding pairs and lone pairs. These repulsions are generally assumed to lie in the order

> lone pair/lone pair repulsions
> are stronger than
> lone pair/bonding pair repulsions
> are stronger than
> bonding pair/bonding pair repulsions

In elementary accounts, the greater repelling effect of a lone pair is explained by supposing that the lone pair is on average closer to the nucleus than a bonding pair and therefore repels other electron pairs more strongly. However, the true origin of the difference is obscure. An additional detail about this order of repulsions is that, given the choice between an axial and an equatorial site for a lone pair in a trigonal bipyramidal array, it occupies the equatorial site because it is then repelled less by the bonding pairs (Fig. 3.1).

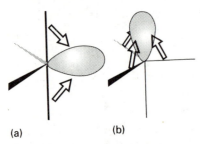

(a)   (b)

3.1 In the VSEPR model of molecular shape, a lone pair (a) in the equatorial position of a trigonal bipyramidal arrangement interacts strongly with two bonding pairs, but (b) in the axial position it interacts strongly with three bonding pairs. The former arrangement is generally of lower energy.

Example 3.1: *Using VSEPR theory to predict shapes*
Predict the shape of the SF_4 molecule.

Answer. First, we write the Lewis structure of SF_4:

This structure has five electron pairs around the central atom, and they adopt a trigonal bipyramidal arrangement (Table 3.2). Repulsion from the one lone pair is minimized if it occupies an equatorial site, for then it interacts with the two axial bonding pairs strongly, rather than an axial site, when it would interact strongly with three equatorial bonding pairs. The S—F bonds then bend away from the lone pair to give a molecular shape that resembles a see-saw, with the axial bonds forming the 'plank' of the see-saw and the equatorial bonds the 'pivot' (**5**).

Exercise E3.1. Predict the shape of an XeF_2 molecule.

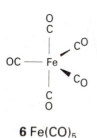

5 SF_4

The angle between the O—H bonds in H_2O decreases slightly from its tetrahedral value (109.5°) as the two lone pairs move apart. This decrease is in agreement with the observed HOH bond angle of 104.5°. A similar effect accounts for the HNH bond angle of 107° in NH_3. It also accords neatly with the regular tetrahedral shape of NH_4^+, in which all four bonds are equivalent.

3.2 The determination of molecular shape

Molecular shapes, including detailed bond lengths and angles, can be determined experimentally from X-ray diffraction studies of single crystals (Box 3.1) or electron diffraction and microwave spectroscopy on small gas-phase molecules. In most instances infrared and Raman spectroscopy give information on the symmetries of molecules but not on their detailed structures.

Occasionally, different techniques lead to apparently conflicting conclusions. For example, pentacarbonyliron ($Fe(CO)_5$; **6**) has infrared and Raman spectra consistent with a trigonal bipyramidal structure with distinct axial and equatorial CO groups. However, when studied by ^{13}C-NMR (see Further information 2) a single resonance line is observed, indicating that the five CO groups are identical.

The conflict is resolved by realizing that a molecule may be **fluxional**, or able to change from one conformation to another, and that each technique observes the molecule on a characteristic timescale. In the infrared and Raman experiments, the interaction of the incident photon with the molecule is almost instantaneous and the results are an almost instantaneous snapshot of its structure. The timescale of an NMR experiment is much longer (we shall say more precisely what

6 $Fe(CO)_5$

Box 3.1. X-Ray diffraction

X-Ray diffraction is the most widely used and least ambiguous method for the precise determination of the positions of atoms in molecules and solids. X-Ray structure determinations play a much more prominent role in inorganic chemistry than in organic chemistry because inorganic molecules and solids are structurally more diverse. Thus structural inferences from spectroscopic data frequently suffice for organic molecules, but spectroscopy is less successful for the unambiguous characterization of new inorganic compounds. Additionally, the bonding in inorganic molecules is more varied than in organic molecules, so inorganic chemists depend on bond distance and bond angle information to infer the nature of bonds.

A typical X-ray diffractometer (Fig. B3.1) consists of an X-ray source with a fixed wavelength, a mount for a single crystal of the compound being investigated, and an X-ray detector. The positions of the detector and the crystal, which is typically as small as 0.2 mm on a side, are controlled by a computer. For certain orientations of the crystal relative to the X-ray beam the crystal diffracts the X-rays at a fixed angle and the intensity is measured when the detector is placed in the direction of this diffracted beam. Under computer control, the detector is scanned through each reflection while the intensities are recorded and stored. It is common to collect data on the intensities and positions of over 1000 reflections, and to obtain more than 10 observed reflections for each structural parameter to be determined (positions of atoms and a range of locations associated with their thermal motion). A trial structure is chosen, either by means of a 'direct method' program or by hints from the diffraction data together with a knowledge of physically reasonable arrangements of atoms. This structural model is refined by systematic shifts in

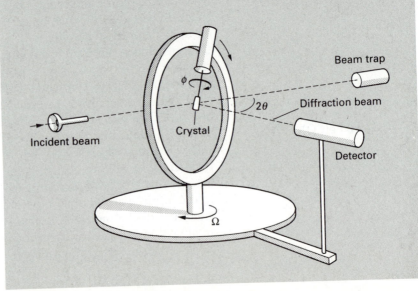

B3.1 A schematic diagram of an X-ray diffractometer.

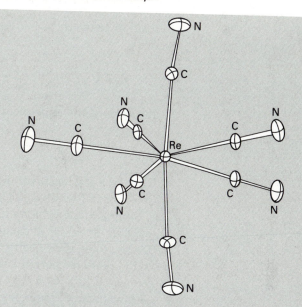

B3.2 An ORTEP diagram of $[Re(CN)_7]^{4-}$ in $K_4[Re(CN)_7] \cdot 2H_2O$.

the atom positions until satisfactory agreement between observed and calculated X-ray diffraction intensities is obtained.

The pictorial display of an X-ray structure often has the appearance of Fig. B3.2. This type of computer-generated drawing is called an **ORTEP diagram** (an acronym for Oak Ridge Thermal Ellipsoid Program). The ORTEP diagram depicts the bond distances and bond angles. In addition, the atoms are displayed as ellipsoids that indicate the amplitude of their thermal motion. Since the restoring force for bond angle bending is generally less than for bond stretching, the ellipsoids are generally elongated in directions perpendicular to the bonds, as clearly shown by the N atoms of the CN^- ligands in Fig. B3.2.

[References: A. K. Cheetham, Chapter 2 in *Solid state chemistry techniques* (ed. A. K. Cheetham and P. Day), Oxford University Press (1987); M. F. C. Ladd and R. A. Palmer, *Structure determination by X-ray crystallography*. Plenum, New York (1985); J. P. Glusker and K. N. Trueblood, *Crystal structure analysis: a primer*, Oxford University Press (1985).]

this means in a moment), and the absorption occurs at an average of the conformations of the molecule. Hence, in an NMR experiment, the axial and equatorial CO groups look identical.

The timescale of a spectroscopic technique is identified by considering the quantum mechanical effect called **lifetime broadening**. According to quantum mechanics, the energy of a state becomes less well defined as the lifetime of the state decreases. A state has a precisely determined energy only if it persists for ever, and the shorter the lifetime, the wider the range of energies of the state (Fig. 3.2). The

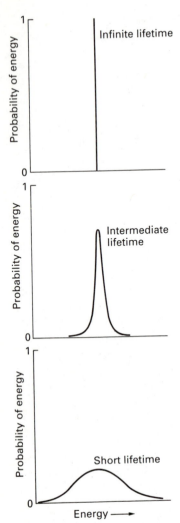

3.2 The relation between lifetime of a state and its energy. A state with an infinite lifetime has a precisely defined energy; a state that has a short lifetime has an energy that may be anywhere within a range.

relation between the lifetime τ of a state and the spread in energy ΔE is

$$\tau \Delta E \approx \frac{h}{2\pi}$$

In terms of the equivalent frequency spread (using $\Delta E = h\Delta \nu$),

$$\tau \Delta \nu \approx \frac{1}{2\pi} \qquad (1)$$

Thus, if a state lasts for only 1 ps, then its frequency may lie anywhere in a range of width 160 GHz (1 GHz $= 10^9$ Hz).[2]

Suppose that a static molecule has two inequivalent sites for an atom (such as the axial and equatorial sites for the C atoms in $Fe(CO)_5$). These atoms give rise to two distinct NMR signals at frequencies ν_1 and ν_2 that are characteristic of the two sites. However, if the molecule can adjust in such a way that the atoms in the sites are interchanged, each resonance frequency will span a range $\Delta \nu$ given by eqn (1), with τ the residence time of an atom in one of the sites. The distinction between the two sites is lost, and the molecule consequently appear to have higher symmetry, when the width of the absorption $\Delta \nu$ is approximately equal to $\nu_2 - \nu_1$ because the frequencies of the two transitions then become indistinguishable. By combining this condition with eqn (1), we see that the sites become indistinguishable when

$$\tau \approx \frac{1}{2\pi(\nu_2 - \nu_1)} \qquad (2)$$

The effect is illustrated in Fig. 3.3. If the distinct sites survive for a lifetime shorter than the τ given by eqn (2), then we cannot distinguish between the two absorptions and the experiment gives the average of the two. We can distinguish the two absorptions if the interchange is so slow that each site survives for longer than the τ given by eqn (2). The lifetime of a particular conformation of a molecule usually decreases with temperature because the interconversion of sites is an activated process (that is, there is an energy barrier to be overcome). Thus, the discrete absorption lines of some molecules may coalesce at higher temperatures as their rate of fluxional interchange increases.

It is generally the case that the higher the frequency of the technique, then the more precise the molecular coordinates it can be used to obtain. A high-frequency technique (like infrared spectroscopy) is generally accompanied by large frequency differences between different groups, and so fluxional lifetimes must be very short before they mar the resolution of the experiment. Table 3.3 summarizes the typical timescales of techniques commonly used in inorganic chemistry. If the difference in resonance frequency between sites for an NMR experiment is 100 Hz, then only lifetimes longer than 2 ms will allow the

---

[2] The lifetime broadening effect, and the equations that describe it, are reminiscent of the uncertainty principle for position and momentum (Section 1.4). However, they are derived in different ways and are best regarded as unrelated. See P. W. Atkins, *Molecular quantum mechanics*. Oxford University Press (1983), p.96.

(a)

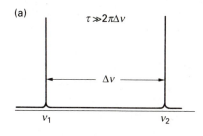

(b)

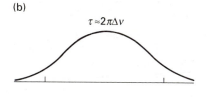

(c)

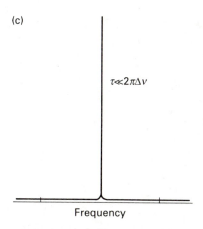

3.3 (a) When the lifetime τ of a conformation of a molecule is infinite, the spectrum shows discrete absorptions. (b) As the lifetimes shorten, the absorptions broaden and merge. (c) When the lifetime is very short, a single, sharp absorption is obtained at the mean of the two frequencies.

Table 3.3 Approximate timescales of common structural techniques*

Technique	Timescale/s
Ultraviolet spectroscopy	10^{-15}
Visible spectroscopy	10^{-14}
Infrared spectroscopy	10^{-13}
Vibrational Raman spectroscopy	10^{-13}
Mössbauer spectroscopy	10^{-7}
Electron spin resonance	10^{-4} to 10^{-8}
Nuclear magnetic resonance	10^{-1} to 10^{-9}
Chemical separation of isomers	10^{2} and longer

*In diffraction (X-ray, electron, and neutron) the image obtained is an envelope spanning all atomic locations, and the theoretical timescales are not particularly relevant.

conformations to be distinguished. In contrast, the difference in vibrational frequencies between sites might be 10^{12} Hz (corresponding to a difference in wavenumber of 30 cm^{-1}), so they can be distinguished even if they survive for little more than 0.2 ps. Hence, vibrational spectroscopy shows an almost instantaneous conformation of a fluxional molecule.

Example 3.2: *Judging the timescale of a technique*
In a certain fluxional molecule, two groups that are interchanged by the conversion have vibrational absorptions at 1650 cm^{-1} and 1655 cm^{-1}. Calculate the minimum lifetime of the states for which the separate absorptions could be distinguished.

Answer. The absorptions are indistinguishable if the lifetimes are less than $1/2\pi\Delta\nu$, or in terms of wavenumber, less than

$$\tau = \frac{1}{2\pi \times c\Delta\tilde{\nu}} = \frac{1}{2\pi \times (2.998 \times 10^{10}\ \text{cm s}^{-1}) \times (5\ \text{cm}^{-1})}$$

$$= 1 \times 10^{-12}\ \text{s, or 1 ps}$$

A lifetime of 1 ps is time for the molecule to undergo only about 10 oscillations in each conformation.

Exercise E3.2. Two groups in a molecule gave NMR absorptions with shifts separated by 550 Hz. What is the minimum lifetime for distinguishing them?

3.3 Molecular shape in terms of molecular orbitals

It will be recalled from Chapter 2 that in molecular obital theory the electrons responsible for bonding are delocalized over the entire molecule. A simple pictorial approach to the task of analyzing molecular shape in terms of delocalized molecular orbitals was devised by A. D. Walsh in a classic series of papers published in 1953. Walsh's approach relates molecular shape to the occupation of molecular orbitals and

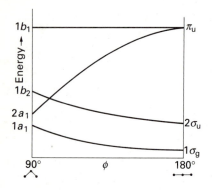

3.4 The Walsh diagram for XH$_2$ molecules. Only the bonding and nonbonding orbitals are shown.

does not consider electron pair repulsions explicitly. As in the VSEPR model, Walsh sought to identify a dominating influence on molecular shape, and the success of both approaches shows that it is possible to do so in more than one way. There is no simple theory of molecular shape that can take all the relevant factors into account at once.

Walsh's approach to the discussion of the shape of an H$_2$X triatomic molecule (such as BeH$_2$ and H$_2$O) is illustrated in Fig. 3.4. The illustration shows an example of a **Walsh diagram**, a graph of the dependence of orbital energy on molecular geometry. A Walsh diagram for an H$_2$X molecule is constructed by considering how the composition and energy of each molecular orbital changes as the bond angle changes from 90° to 180°. The diagram is in fact just a more elaborate version of the correlation diagram that we illustrated for H$_3^+$ in Section 2.7.

The molecular orbitals to consider in the angular molecule are

$$\psi_{a_1} = c_1 \phi_{2s} + c_2 \phi_{2p_z} + c_3 \phi_+$$

$$\psi_{b_1} = \phi_{2p_x}$$

$$\psi_{b_2} = c_4 \phi_{2p_y} + c_5 \phi_-$$

(We are continuing to use the letters a and b to label nondegenerate orbitals, and will explain their full significance later in the chapter.) The linear combinations ϕ_+ and ϕ_- are illustrated in Fig. 3.5. There are three a_1 orbitals and two b_2 orbitals; the lowest energy orbitals of each type (the only one occupied in H$_2$O) are shown on the left of Fig. 3.6. In the linear molecule, the molecular orbitals are

$$\psi_{\sigma_g} = c_1 \phi_{2s} + c_2 \phi_+$$

$$\psi_{\pi_u} = \phi_{2p_x} \text{ and } \phi_{2p_z}$$

$$\psi_{\sigma_u} = c_3 \phi_{2p_y} + c_4 \phi_-$$

The lowest energy molecular orbital in 90° H$_2$X is the one labeled $1a_1$, which is built from the overlap of the X2p_z orbital (the $2p_z$ orbital on X) with the ϕ_+ combination of H1s orbitals. As the bond angle changes to 180° the two H1s orbitals overlap less but the contribution of the X2s orbital increases. In the 180° molecule, X2s is the only contribution from the X atom to the $1a_1$ orbital (see Fig. 3.6). The replacement of X2p_z by X2s lowers the energy of the orbital. The energy of the $1b_2$ orbital is also lowered because the adverse H—H overlap decreases and the H1s orbitals move into a better position for overlap with the X2p_y orbital. The biggest change occurs for the $2a_1$ orbital. It has considerable X2s character in the 90° molecule, but correlates with a pure X2p_z orbital in the 180° molecule. Hence, it shows a steep rise in energy as the bond angle increases. The $1b_1$ orbital is a nonbonding X2p orbital perpendicular to the molecular plane in the 90° molecule and remains nonbonding in the linear molecule. Hence, its energy barely changes with angle.

The principal feature that determines whether or not the molecule is angular is whether the $2a_1$ orbital is occupied. This is the orbital

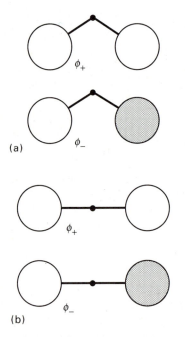

(a)

(b)

3.5 The combinations of H1s orbitals that are used to construct molecular orbitals in (a) angular and (b) linear XH$_2$ molecules.

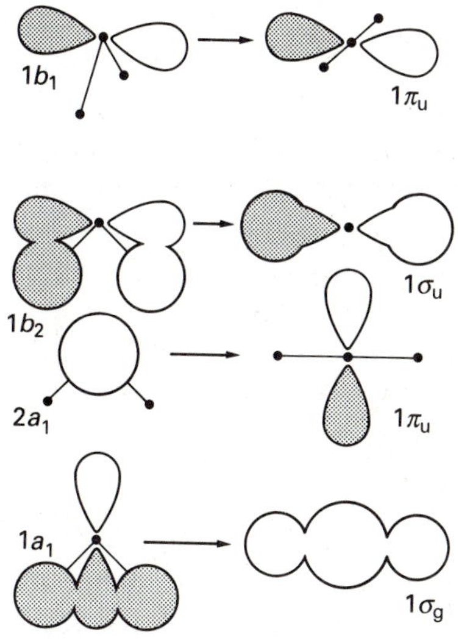

3.6 The composition of the molecular orbitals of an XH_2 molecule at the two extremes of the correlation diagram shown in Fig. 3.4.

that has considerable X2s character in the angular molecule but not in the linear molecule. Hence, a lower energy is achieved if, when it is occupied, the molecule is angular. The shape adopted by an H_2X molecule therefore depends on the number of electrons that occupy the orbitals.

The simplest XH_2 molecule in Period 2 is the transient gas-phase BeH_2 molecule (BeH_2 normally exists as a polymeric solid), in which there are four valence electrons. These four electrons occupy the lowest two molecular orbitals. If the lowest energy is achieved with the molecule angular, then that will be its shape. We can decide whether the molecule is likely to be angular by accommodating the electrons in the lowest two orbitals corresponding to an arbitrary bond angle in Fig. 3.4. We then note that the HOMO decreases in energy on going to the right of the diagram and that the lowest total energy is obtained when the molecule is linear. Hence, BeH_2 is predicted to be linear and to have the configuration $1\sigma_g^2 2\sigma_u^2$. In CH_2, which has two more electrons than BeH_2, three of the molecular orbitals must be occupied. In this case, the lowest energy is achieved if the molecule is angular and has configuration $1a_1^2 2a_1^2 1b_2^2$.

In general, any XH_2 molecule with from 5 to 8 valence electrons is predicted to be angular. The observed bond angles are

BeH_2	BH_2	CH_2	NH_2	OH_2
180	131	136	103	105

Whereas both the VSEPR model and the Walsh approach can account qualitatively for these results, neither provides a reliable way of estimating the angles quantitatively. Although these simple approaches cannot provide quantitative data, computational quantum chemistry is increasingly successful. It is now possible to carry out molecular orbital calculations with sufficient precision to determine reliable bond lengths and bond angles for small molecules (of up to about a dozen atoms).

Example 3.3: *Using a Walsh diagram to predict a shape*
Predict the shape of an H_2O molecule on the basis of a Walsh diagram for an XH_2 molecule.

Answer. We choose an intermediate bond angle along the horizontal axis of the XH_2 diagram in Fig. 3.4 and accommodate eight electrons. The resulting configuration is $1a_1^2 2a_1^2 1b_2^2 1b_1^2$. The $2a_1$ orbital is occupied, so we expect the nonlinear molecule to have a lower energy than the linear molecule.

Exercise E3.3. Is any XH_2 molecule, in which X denotes an atom of a Period 3 element, expected to be linear? If so, which?

Walsh applied his approach to molecules other than compounds of hydrogen, but the correlation diagrams soon become very complicated.

His approach represents a valuable complement to the VSEPR model because it traces the influences on molecular shapes of the occupation of orbitals spreading over the entire molecule and concentrates less on localized repulsions between pairs of electrons. Correlation diagrams like those introduced by Walsh are frequently encountered in contemporary discussions of the shapes of complex molecules, and we shall see a number of examples in later chapters. They illustrate how inorganic chemists can sometimes identify and weigh competing influences by considering two extreme cases (such as linear and 90° XH_2 molecules), and then rationalize the fact that the state of a molecule is a compromise intermediate between the two extremes.

MOLECULAR SYMMETRY

One aspect of the shape of a molecule is its 'symmetry' (we define the meaning of this term technically in a moment). The systematic treatment of symmetry is called **group theory**. Group theory is a rich and powerful subject, but we shall confine our use of it at this stage to the classification of molecules and to drawing some general conclusions about their properties.

3.4 An introduction to symmetry analysis

Our initial aim is to define the symmetries of molecules much more precisely than we have done so far, and to provide a scheme for specifying and reporting their symmetries. It will become clear in later chapters that symmetry analysis is one of the most pervasive techniques in inorganic chemistry.

Symmetry operations and symmetry elements

A fundamental concept of group theory is the **symmetry operation**, which is an action, such as a rotation through a certain angle, that leaves the molecule apparently unchanged. An example is the rotation of an H_2O molecule by 180° (but not any smaller angle) around the bisector of the HOH angle. Associated with each symmetry operation there is a **symmetry element**, a point, a line, or a plane with respect to which the symmetry operation is performed. The most important symmetry operations and their corresponding elements are listed in Table 3.4. All these operations leave at least one point of the molecule unmoved, just as a rotation of a sphere leaves its center unmoved, and hence they are the operations of **point-group symmetry**.

The **identity operation**, E, leaves the whole molecule unchanged. The rotation of an H_2O molecule by 180° around a line bisecting the HOH angle is a symmetry operation, so the H_2O molecule possesses a 'twofold' rotation axis C_2 (Fig. 3.7). In general, an n-fold rotation is a symmetry operation if the molecule appears unchanged after rotation by $360°/n$. The corresponding symmetry element is a line, the **n-fold rotation axis**, C_n, about which the rotation is performed. The trigonal pyramidal NH_3 molecule has a threefold rotation axis,

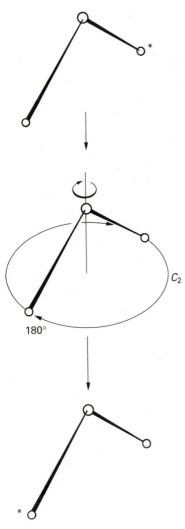

3.7 An H_2O molecule may be rotated through any angle about the bisector of the HOH bond angle, but only a rotation of 180°, C_2, leaves it apparently unchanged.

Table 3.4 Important symmetry operations and symmetry elements

Symmetry element	Symmetry operation	Symbol
	Identity*	E
n-Fold symmetry axis	Rotation by $2\pi/n$	C_n
Mirror plane	Reflection	σ
Center of inversion	Inversion	i
n-Fold axis of improper rotation†	Rotation by $2\pi/n$ followed by reflection perpendicular to rotation axis	S_n

*The symmetry element can be thought of as the molecule as a whole.
†Note the equivalences $S_1 = \sigma$ and $S_2 = i$.

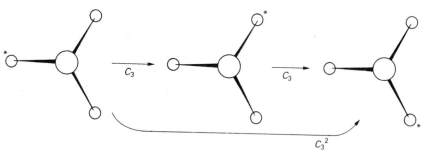

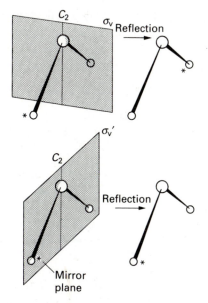

3.8 A threefold rotation and the corresponding C_3 axis in NH_3. There are two rotations associated with this axis, one through 120° (C_3) and the other through 240° (C_3^2).

3.9 The two vertical mirror planes σ_v and σ_v' in H_2O and the corresponding operations. Both planes cut through the C_2 axis.

denoted C_3, but there are two operations associated with it, one a rotation by 120° and the other a rotation through twice this angle (Fig. 3.8). The two operations are denoted C_3 and C_3^2. We do not need to consider C_3^3, a rotation through $3 \times 120° = 360°$, because it is equivalent to the identity E.

The **reflection** of an H_2O molecule in either of the two planes shown in Fig. 3.9 is a symmetry operation; the corresponding symmetry element is a **mirror plane**, σ. The H_2O molecule has two mirror planes that intersect at the bisector of the HOH angle. Because the planes are vertical (in the sense of being parallel to the rotational axis of the molecule), they are labeled σ_v and σ_v'. The C_6H_6 molecule has a mirror plane σ_h in the plane of the molecule. The subscript h signifies that the plane is 'horizontal' in the sense that the principal rotational axis of the molecule is perpendicular to it. This molecule also has two more sets of three mirror planes that intersect the sixfold axis (Fig. 3.10). In such cases, the members of one set are called 'vertical' and the members of the other are called 'dihedral'. The symmetry elements (and the associated operations) are denoted σ_v and σ_d, respectively. The planes designated σ_v pass through the C atoms of the ring whereas those designated σ_d bisect the angle between them.

To understand the **inversion** operation we need to imagine that each atom is projected in a straight line through a single point and then out to an equal distance on the other side (Fig. 3.11). If we consider CO_2, for instance, with the point at the center of the molecule (at the C nucleus), then the operation exchanges the two O atoms. In an octahedral molecule such as SF_6, with the point at the center of

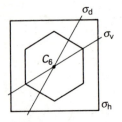

3.10 Some of the symmetry elements of the benzene ring. There is one horizontal reflection plane (σ_h) and two sets of vertical reflection planes (σ_v and σ_d); one example of each is shown.

the molecule, diametrically opposite pairs of atoms at the corners of the octahedron are interchanged. The symmetry element, the point through which the projections are made, is called the **center of inversion**, i: the center of inversion of CO_2 lies at the C nucleus; that of the SF_6 molecule lies at the nucleus of the S atom. There need not be an atom at the center of inversion: an N_2 molecule has a center of inversion midway between the two nitrogen nuclei. An H_2O molecule does not possess a center of inversion. No tetrahedral molecule has a center of inversion; for example, $Ni(CO)_4$ has no center of inversion. As Fig. 3.12 shows, although an inversion and a twofold rotation may sometimes achieve the same effect, that is not the case in general and the two operations should be clearly distinguished.

An **improper rotation** is a composite operation (and one of the most difficult to identify in a molecule). It consists of a rotation of the molecule through a certain angle around an axis, followed by a reflection in the plane perpendicular to that axis (Fig. 3.13). The illustration shows a fourfold improper rotation of a tetrahedral CH_4 molecule. In this case, the operation consists of a 90° rotation about an axis bisecting two HCH bond angles, followed by a reflection through a plane perpendicular to the rotation axis. Neither the operation C_4 nor the reflection σ_h alone is a symmetry operation for CH_4. In each case we can see that the molecule has been moved after either one of the operations has been applied, but their overall effect is a symmetry operation because we cannot tell that the molecule has been moved if the reflection is applied after the rotation. This fourfold improper rotation is denoted S_4. The symmetry element, the **improper-rotation axis** S_n (S_4 in the example), is the corresponding combination of an n-fold rotational axis and a perpendicular mirror plane.

An S_1 axis, a rotation through a full 360° followed by a reflection in the horizontal plane, is equivalent to a horizontal reflection alone, so S_1 and σ_h are the same; the symbol σ_h is generally used rather than S_1. Similarly, an S_2 axis, a rotation through 180° followed by a reflection in the horizontal plane, is equivalent to an inversion i (Fig. 3.14), and the symbol i is employed rather than S_2.

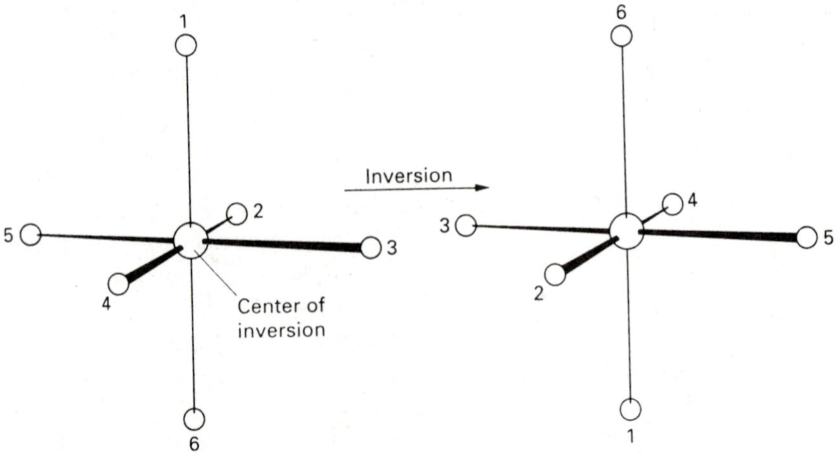

3.11 The inversion operation and the center of inversion i in SF_6.

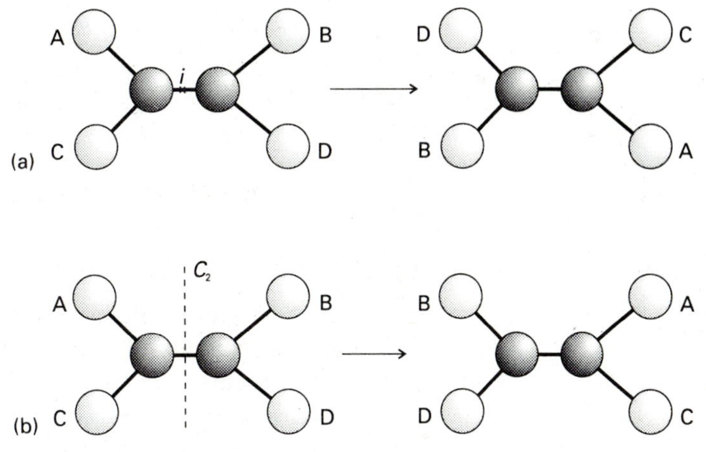

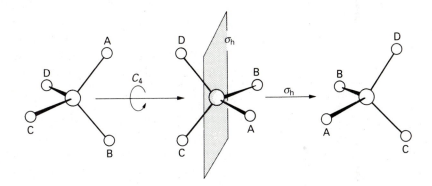

3.12 Care must be taken not to confuse (a) an inversion operation with (b) a twofold rotation. Although the two operations may sometimes appear to have the same effect, that is not the case in general.

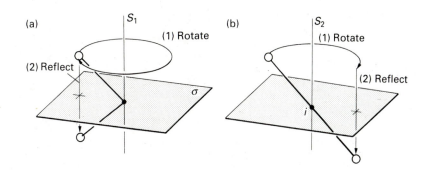

3.13 A fourfold axis of improper rotation S_4 in the CH_4 molecule.

3.14 (a) An S_1 axis is equivalent to a mirror plane and (b) an S_2 axis is equivalent to a center of inversion.

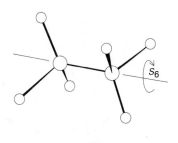

7

Example 3.4: *Identifying a symmetry element*

Which conformation of a CH_3CH_3 molecule has an S_6 axis?

Answer. We need to find a conformation that leaves the molecule looking the same after a 60° rotation followed by a reflection in a plane perpendicular to that axis. The conformation and axis are shown in (7); this 'staggered' conformation of the molecule also happens to be the conformation of lowest energy.

Exercise E3.4. Identify a C_3 axis of an NH_4^+ ion. How many of these axes are there in the ion?

---

8 CHBrClF

3.15 The decision tree for identifying a molecular point group. After passing through part (a), go to part (b) if necessary. The symbols at each decision point refer to the symmetry elements, not the symmetry operations.

The point groups of molecules

The symmetry elements possessed by a molecule determine the **point group** to which it belongs. The general starting point for a symmetry analysis of a molecule can be summarized as follows:

> To assign a molecule to a particular group, a list of symmetry elements it possesses is compiled and compared with the list that is characteristic of each point group.

For instance, if a molecule has only the identity element (CHBrClF is an example; **8**), then we list its elements as E alone and look for the group that has *only* this element. Because the group labeled C_1 has only the element E, the CHBrClF molecule belongs to that group. The molecule CH_2BrCl belongs to a slightly richer group: it has the elements E (all groups have that element) and a mirror plane. The group of elements (E, σ) is called C_s, so the CH_2BrCl molecule belongs to that group. This procedure can be continued, and molecules assigned to the group that matches the symmetry elements they possess. Some of the more common groups and their names are listed in Table 3.5. The assignment of a molecule to its group depends on listing the symmetry elements it possesses and then referring to the table: the shapes in the table give a very good clue to the identity of the group to which the molecule belongs, at least in simple cases. Figure 3.15 provides a chart that should be used to assign the point group systemat-

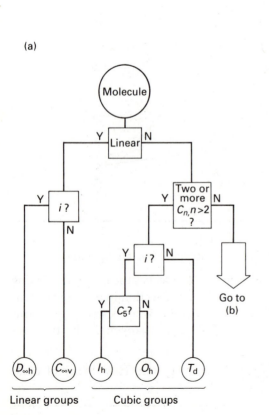

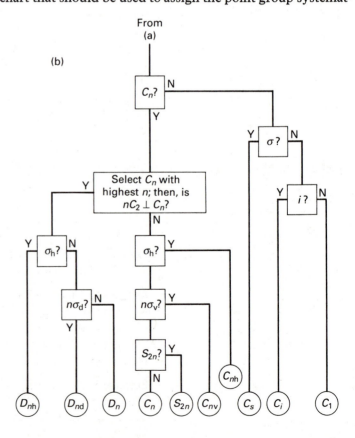

Table 3.5 The composition of some common groups*

Point group	Symmetry elements	Shape	Examples
C_1	E	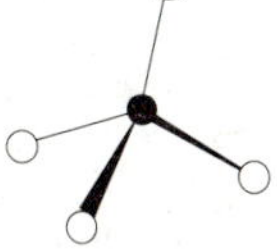	$SiBrClFI$
C_2	E, C_2		H_2O_2
C_s	E, σ		NHF_2
C_{2v}	$E, C_2, \sigma_v, \sigma_v$	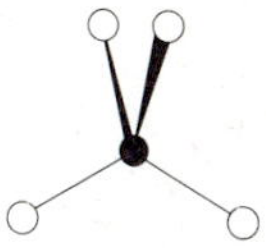	H_2O, SO_2Cl_2
C_{3v}	$E, C_3, 3\sigma_v$		NH_3, PCl_3, $POCl_3$
$C_{\infty v}$	$E, C_2, C_\infty, \ldots \infty \sigma_v$		CO, HCl, OCS
D_{2h}	$E, C_2 (x,y,z), \sigma (xy,yz,zx), i$	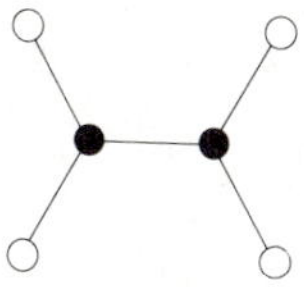	N_2O_4, B_2H_6
D_{3h}	$E, C_3, 3C_2, 3\sigma_v, \sigma_h, S_3$	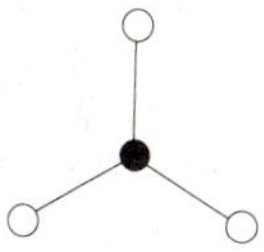	BF_3, PCl_5
D_{4h}	$E, C_4, C_2, 2C_2', 2C_2'', i, S_4, \sigma_h, 2\sigma_v, 2\sigma_d$	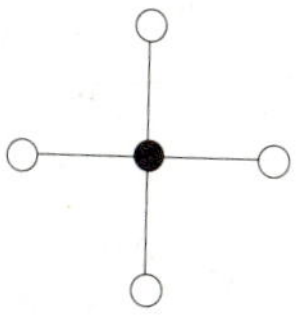	XeF_4, *trans*-MA_4B_2
$D_{\infty h}$	$E, C_\infty, \ldots, \infty \sigma_v, i, S_\infty, \ldots, \infty C_2$		H_2, CO_2, C_2H_2
T_d	$E, 3C_2, 4C_3, 6\sigma_d, 4S_4$		CH_4, $SiCl_4$
O_h	$E, 6C_2, 4C_3, 3C_4, 4S_6, 3S_4, i, 3\sigma_h, 3\sigma_v$	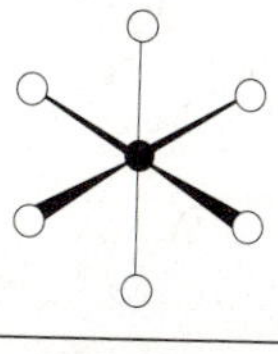	SF_6

*Not all the elements of each group are listed, but enough are listed for unambiguous assignments to be made.

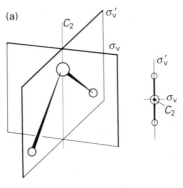

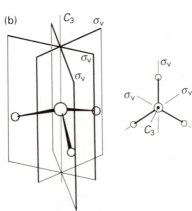

3.16 The symmetry elements of (a) H_2O and (b) NH_3. The diagrams on the right are views from above and summarize the diagrams to their left.

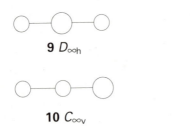

9 $D_{\infty h}$

10 $C_{\infty v}$

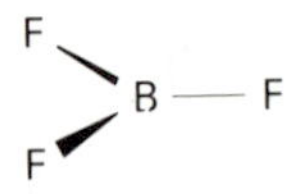

11 BF_3

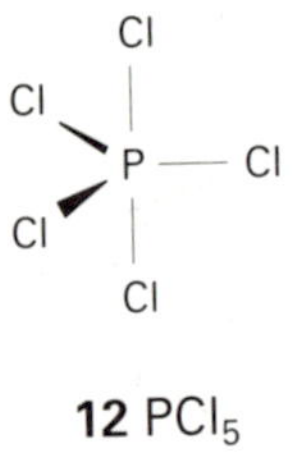

12 PCl_5

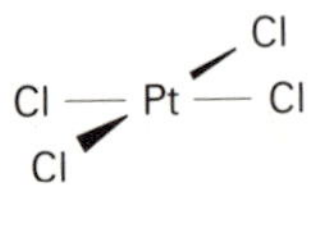

13 $PtCl_4$, D_{4h}

ically by answering the questions at each decision point on the tree. (Note that we need to be careful to distinguish the names of the groups, C_2 and so on, from the symbols for the symmetry elements, such as C_2, and the corresponding operations, also C_2. The context will always make it clear what interpretation is intended.)

Example 3.5: *Identifying the point group of a molecule*
To what point groups do H_2O and NH_3 belong?

Answer. Work through Fig. 3.15. The symmetry elements are shown in Fig. 3.16. (a) H_2O possesses the identity (E), a twofold rotation axis (C_2), and two vertical mirror planes (σ_v and σ_v'). The set of elements (E, C_2, σ_v, σ_v') corresponds to the group C_{2v}. (b) NH_3 possesses the identity (E), a threefold axis with which are associated two threefold rotations (C_3), and three vertical mirror planes ($3\sigma_v$). The set of elements (E, C_3, $3\sigma_v$) identifies the group as C_{3v}.

Exercise E3.5. Identify the point groups of (a) BF_3, a trigonal planar molecule, and (b) the tetrahedral SO_4^{2-} ion.

It is very useful to be able to recognize immediately the point groups of some common molecules. Linear molecules with a center of symmetry (H_2, CO_2, $HC\equiv CH$; **9**) belong to the point group $D_{\infty h}$. A molecule that is linear but has no center of symmetry (HCl, OCS, NNO; **10**) belongs to the point group $C_{\infty v}$. Tetrahedral (T_d) and octahedral (O_h) molecules (Fig. 3.17) have more than one non-collinear principal axis of symmetry: a tetrahedral CH_4 molecule, for instance, has four non-collinear C_3 axes, one along each C—H bond. A closely related group, the icosahedral group I_h characteristic of the icosahedron has non-collinear five-fold axes (see Fig. 3.17(c)). The icosahedral group is important for boron compounds and for the C_{60} fullerene molecule. A regular tetrahedron has four equilateral triangles for its faces, an octahedron has eight, and an icosahedron has 20.

The distribution of molecules over the various point groups is very uneven. Some of the most common groups for molecules are the low-symmetry groups C_1 and C_s, the groups for a number of polar molecules C_{2v} and C_{3v}, and the highly symmetrical tetrahedral and octahedral groups. There are many linear molecules, which belong to the groups $C_{\infty v}$ and $D_{\infty h}$, and a number of trigonal planar molecules, D_{3h} (such as BF_3; **11**), trigonal bipyramidal molecules (such as PCl_5; **12**), which are also D_{3h}, and square-planar molecules, D_{4h} (**13**). 'Octahedral' molecules with two substituents opposite each other, as in (**14**), are also D_{4h}. The last example shows that the point-group classification of a molecule is more precise than the casual use of the term 'octahedral' or 'tetrahedral'. For instance, a molecule may loosely be called octahedral even if it has six different groups attached to the central atom. However, it belongs to the octahedral point group O_h only if all six groups are identical (**15**).

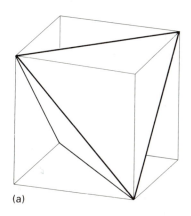

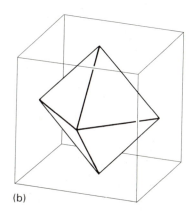

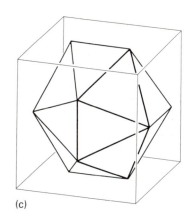

3.17 Shapes with the symmetries of the groups (a) T_d, the tetrahedron, (b) O_h, the octahedron, and (c) I_h, the icosahedron. They are all closely related to the symmetries of a cube.

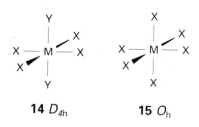

14 D_{4h} **15** O_h

3.5 Applications of symmetry

An important application of symmetry in this text will be to the construction and labeling of molecular orbitals (see Section 3.6). However, there are a number of simple applications of the group classification of molecules that we shall deal with at this point before getting down to the main business of the chapter. One is to use it to decide whether a compound is polar or chiral. In fact, in most cases we do not need to use the sledgehammer of group theory to decide whether a molecule has these characteristics, but these examples will give an impression of the approach that can be adopted when the result is less obvious. Symmetry arguments may also be applied to the vibrations of polyatomic molecules, and we shall also illustrate the role of group theory in vibrational spectroscopy.

Polar molecules

A **polar molecule** is a molecule with a permanent electric dipole moment. There are certain symmetry elements that rule out a dipole in molecules, or forbid it to lie in certain orientations in the molecule. First, a molecule cannot be polar if it has a center of inversion, for inversion implies that a molecule has matching charge distributions at *all* diametrically opposite points about a center. Although a molecule without a center of inversion may have a nonzero dipole moment, that dipole moment cannot lie perpendicular to any mirror plane or axis of rotation that the molecule may possess. For example, a dipole perpen-

3.18 (a) The presence of a mirror plane rules out a dipole in the direction shown here. (b) The presence of a symmetry axis rules out any dipole in the perpendicular plane. (c) If both a C_n symmetry axis and a perpendicular C_2 axis are present, a dipole is ruled out in all directions.

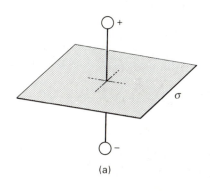

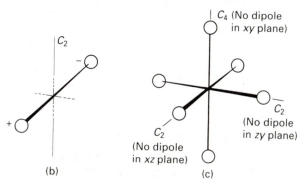

dicular to a mirror plane would require equal and opposite charges on either side of that plane (Fig. 3.18(a)), in which case the plane would not be a symmetry element. Similarly, a molecule with a dipole moment perpendicular to an axis of symmetry implies the presence of opposite charges on either side of the axis, which is inconsistent with the axis being a symmetry element (Fig. 3.18(b)). In summary:

1. A molecule cannot be polar if it has a center of inversion.
2. A molecule cannot have an electric dipole moment perpendicular to any mirror plane.
3. A molecule cannot have an electric dipole moment perpendicular to any axis of rotation.

Some molecules have a symmetry axis that rules out a dipole moment in one plane and another symmetry axis or mirror plane that rules it out in another direction. The two or more symmetry elements jointly forbid the presence of a dipole moment in *any* direction. For example, any molecule that has both a C_n axis and either a C_2 axis perpendicular to that C_n axis, or a σ_h plane perpendicular to the axis, cannot have a dipole moment in *any* direction (Fig. 3.18(c)). Any molecule belonging to a D point group is of this kind, so any such molecule must be nonpolar; a BF_3 molecule (D_{3h}) is therefore nonpolar. Likewise, molecules belonging to the tetrahedral, octahedral, and icosahedral groups have several perpendicular rotation axes that rule out dipoles in all three directions, so such molecules must be nonpolar; hence, SF_6 (O_h) and CCl_4 (T_d) are nonpolar. In summary, a molecule cannot be polar if it belongs to any of the following point groups:

1. Any group that includes a center of inversion.
2. Any of the D groups (including D_n, D_{nh}, and D_{nd}).
3. The cubic groups (T, O), the icosahedral group (I), and their modifications.

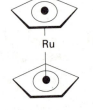

16 Ruthenocene

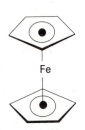

17 Ferrocene (excited)

Example 3.6: *Judging whether a molecule can be polar*
The ruthenocene molecule (**16**), is a pentagonal prism with the Ru atom sandwiched between two C_5H_5 rings. Is it polar?

Answer. We should decide whether the point group is D or cubic, because in neither case can it have a permanent electric dipole. Reference to Fig. 3.15 shows that a pentagonal prism belongs to the point group D_{5h}. Therefore, the molecule must be nonpolar.

Exercise E3.6. A conformation of the ferrocene molecule that lies $4\,kJ\,mol^{-1}$ above the lowest energy configuration is a pentagonal antiprism (**17**). Is it polar?

Chiral molecules

A **chiral molecule** (from the Greek word for hand) is a molecule that cannot be superimposed on its mirror image. A hand is chiral in the sense that the mirror image of a left hand is a right hand, and the two hands cannot be superimposed on each other. So long as they are sufficiently long-lived, chiral molecules are **optically active,** which means that they can rotate the plane of polarized light. A chiral molecule and its mirror image partner are called **enantiomers** (from the Greek word for 'both'). Enantiomeric pairs of molecules rotate the plane of polarization in equal but opposite directions.

The group-theoretical criterion of whether or not a molecule is chiral is that it should not have an improper rotation axis S_n. If it has such an axis, then it is not chiral. Groups in which S_n is present include D_{nh} (which include S_n), D_{nd}, and the cubic groups. Therefore, a tetrahedral molecule belonging to the group T_d is not chiral. That a 'tetrahedral' carbon atom leads to optical activity (as in CHClFBr) should serve as another reminder that group theory is much stricter in its terminology than casual conversation: a molecule such as CHClFBr belongs to the group C_1, not to the group T_d—the molecule is tetrahedral in a casual sense but it is not tetrahedral according to group theory.

It is important to be alert for improper rotation axes that are present in disguise. Thus, we have seen that a mirror plane alone is an S_1 axis and a center of inversion is equivalent to an S_2 axis. Therefore, molecules with either a mirror plane or a center of inversion have improper rotation axes and cannot be chiral. In summary, a molecule is not chiral if

1. It possesses an improper rotation axis.
2. It belongs to the groups D_{nh} or D_{nd} (but it may be chiral if it belongs to the groups D_n).
3. It belongs to a cubic group (T or O and their modifications).

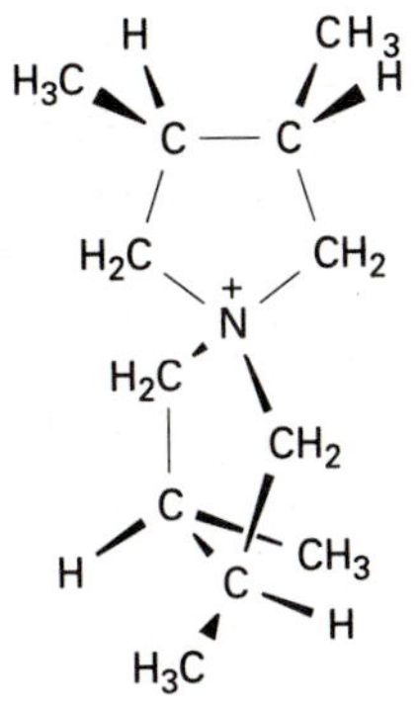

Molecules without a center of inversion and a mirror plane (and hence with neither S_1 nor S_2 axes) are usually chiral, but it is important to verify that a higher-order improper rotation axis is not present. For instance, the quaternary ammonium ion (**18**) has neither a mirror plane (S_1) nor an inversion center (S_2), but it does have an S_4 axis (it belongs to the group S_4, in fact) and so it is not chiral.

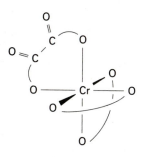

18

19 [Cr(C$_2$O$_4$)$_3$]$^{3-}$

20

Example 3.7: *Judging whether a molecule is chiral*

The complex ion [Cr(ox)$_3$]$^{3-}$, where ox denotes the oxalate ion [O$_2$CCO$_2$]$^{2-}$ has the structure shown as (**19**). Is it chiral?

Answer. We begin by identifying the point group. Working through the chart in Fig. 3.15 shows that the ion belongs to the point group D_3. This group consists of the elements (E, C_3, $3C_2$) and hence does not contain an improper-rotation axis (neither explicitly nor in a disguised form). The complex ion is chiral and hence, if a conformation is sufficiently long lived, optically active.

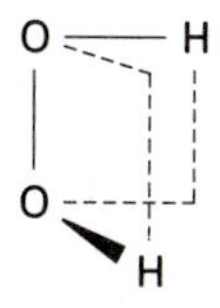

Exercise E3.7. Is the skew form of H$_2$O$_2$ (**20**) chiral?

3.6 The symmetries of orbitals

We shall now see in more detail the significance of the labels used for the orbitals, and gain more insight into the construction of molecular orbitals. The discussion will continue to be informal and pictorial, its aim being to give a flavor of group theory, but not the details of the specific calculations involved. The objective here is to show how to identify the symmetry label of an orbital from a drawing like those in Appendix 4 and, conversely, to appreciate what a symmetry label signifies. Many of the arguments later in the book are based on simply 'reading' molecular orbital diagrams qualitatively. If you are interested in the details of the calculations on which the illustrations are based, and in particular the construction of combinations of specific symmetry, then you will find further guidance in Further information 4.

Character tables and symmetry labels

Molecular orbitals of diatomic molecules (and linear polyatomic molecules) are labeled σ, π, and so on. These labels refer to the symmetries of the orbitals with respect to rotations around the principal symmetry axis of the molecule (or, when it is clear from the context, locally, with respect to the axis of a bond). Thus, a σ orbital does not change sign under a rotation through any angle, a π orbital changes sign when rotated by 180°, and so on (Fig. 3.19). This labeling of orbitals according to their behavior under rotations can be generalized and extended to nonlinear polyatomic molecules, where there may be reflections and inversions to take into account as well as rotations.

A related point is based on the fact that the labels σ and π can be assigned to individual *atomic* orbitals in a linear molecule. For example, we often speak of an individual p_z orbital as having σ symmetry. Thus, in a polyatomic molecule it is possible to ascribe symmetry labels to combinations of individual atomic orbitals on symmetry-related atoms (such as certain combinations of the three $1s$ orbitals of the three H atoms in NH_3). This ability to classify individual atomic orbitals is important, because, as we saw in Chapter 2, atomic orbitals that have different symmetry types cannot contribute to the same molecular

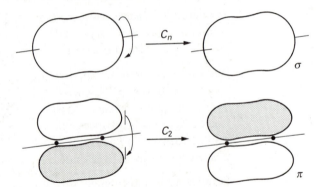

3.19 The σ and π classification of orbitals is based on their symmetries with respect to rotations about an axis. A σ orbital is unchanged by any rotation; a π orbital changes sign when rotated by 180°.

orbital. For example, an s orbital of σ symmetry on one atom and a p_x orbital of π symmetry on its neighbor cannot contribute to the same molecular orbital. In summary:

> Only atomic orbitals, or combinations of atomic orbitals, of the same symmetry type can contribute to a molecular orbital of a given symmetry type.

As this statement implies, symmetry labels may also be ascribed to linear combinations of atomic orbitals on symmetry-related atoms. Thus, a symmetry label can be ascribed to the combination $\phi_{1s}(A) + \phi_{1s}(B) + \phi_{1s}(C)$ in NH_3, and only the orbitals of the same symmetry type on the N atom will have a nonzero net overlap with this combination. The general point to bear in mind is that

> Through a symmetry analysis of orbitals and groups of orbitals, the general form of molecular orbitals can be found as combinations of units associated with fragments of the molecule.

The combinations of atomic orbitals from which molecular orbitals are built are called **symmetry-adapted linear combinations** of atomic orbitals.

The labels a, a_1, e, e_g, and so on, reflect the behavior of the orbitals under all the symmetry operations of the relevant molecular point group. The label is assigned by referring to the **character table** of the group, a table that characterizes the different symmetry types possible in a point group. Thus, when we assign the labels σ and π, we use

	C_2	(i.e. rotation by 180°)
σ	$+1$	(i.e. no change of sign)
π	-1	(i.e. change of sign)

This table is a fragment of a character table. The entry $+1$ shows that an orbital remains the same and the entry -1 shows that it changes sign under the operation C_2.

The entries in a complete character table are derived using the formal techniques of group theory and are called **characters**, χ. These numbers characterize the essential features of each symmetry type in a way that will be illustrated as we go on.[3] We shall these essential features by considering the C_{3v} character table (Table 3.6). The character tables for a selection of other groups are given in Appendix 3.

The columns in a character table are labeled with the symmetry operations of the group. If there are several operations of the same kind (technically, of the same class), then they are collected together into a single column. In the C_{3v} character table, for instance, there are

---

[3] For an account of the generation and use of character tables without too much mathematical background, see P. W. Atkins, *Physical chemistry*. Oxford University Press and W. H. Freeman and Co., New York (1994).

Table 3.6 The C_{3v} character table*

	E	$2C_3$	$3\sigma_v$			
A_1	1	1	1	z	$x^2 + y^2, z^2$	
A_2	1	1	-1			R_z
E	2	-1	0	(x,y)	$(x^2 - y^2, 2xy),\ (zx, yz)$	(R_x, R_y)

*(a,b) denotes a degenerate pair of orbitals; the characters in the table refer to the symmetry of the pair jointly. The symbol R_q denotes a rotation around the axis q.

two threefold rotations about a single axis, as indicated by the heading $2C_3$.

The rows, which summarize the characteristic symmetries of orbitals (and, as we shall see, other entities too), are labeled with the **symmetry type** (the analogs of σ and π). The symmetry types label the 'irreducible representations' of a group, which (in a technical sense that we shall not pursue) are the primitive types of symmetry that may occur in any molecule that belongs to the specific group. By convention, the symmetry types are given upper case roman letters (such as A_1 and E) but the orbitals to which they apply are labeled with the lower case italic equivalents (so an orbital of symmetry type A_1 is called an a_1 orbital). Note that care must be taken to distinguish the identity operation E (italic, a column heading) from the symmetry label E (roman, a row label).

The identities of the orbitals to which a row of characters refers are indicated by the letters (such as xy) on the right of the table. These letters indicate the angular variation of the orbitals. Thus, xy denotes the d_{xy} orbital. An s orbital appears in a slightly disguised form as $x^2 + y^2$, or some other similar expression. The reason for this notation is that the symmetry species apply to entities other than orbitals, and the letters capture the essential form of the entities of a given symmetry. The parentheses around pairs of functions indicate that they must be treated jointly as a pair.

The entry in the column headed by the identity operation E gives the degeneracy of the orbitals. Thus, in a C_{3v} molecule, any orbital with a symmetry label a_1 or a_2 (Fig. 3.20) must be nondegenerate and have a character of 1 in the column headed E. Conversely, if we know that we are dealing with a nondegenerate orbital in a C_{3v} molecule, then we also know that its symmetry type must be either A_1 or A_2 and the orbital will be labeled either a_1 or a_2. Similarly *any* doubly degenerate pair of orbitals in C_{3v} must be labeled e and have a character 2 in the column labeled E (Fig. 3.21). Because there are no characters with

3.20 (a) The combinations of H$1s$ orbitals that are used to build a_1 molecular orbitals in NH$_3$. An a_1 molecular orbital is formed by the overlap of the combination shown with N$2s$ and N$2p_z$ orbitals. None of the valence orbitals of nitrogen has the appropriate symmetry to form an a_2 orbital. There are no A_2 combinations in NH$_3$, but (b) this combination of F$2p$ orbitals in NF$_3$ contributes to a_2 orbitals.

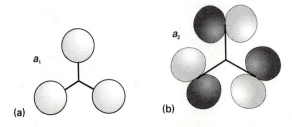

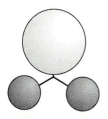

3.21 The combination of H1s orbitals that are used to form e orbitals in NH_3. They overlap the p_x and p_y orbitals on the N atom.

the value 3 in the column headed E, we know at a glance that there can be no triply degenerate orbitals in a C_{3v} molecule.

Example 3.8: *Using a character table to judge degeneracy*
Can a trigonal planar molecule such as BF_3 have triply degenerate orbitals?

Answer. Trigonal planar molecules belong to the point group D_{3h}. Reference to the character table for this group (Appendix 3) shows that the maximum degeneracy is 2, because no character exceeds 2 in the column headed E. Therefore, the orbitals cannot be triply degenerate.

Exercise E3.8. The SF_6 molecule is octahedral. What is the maximum possible degree of degeneracy of its orbitals?

A general point concerning the interpretation of a character table is as follows:

> The entries in a character table in the rows labeled A and B and in the columns headed by symmetry operations other than the identity E indicate the behavior of an orbital or a set of orbitals under the corresponding operations.

The specific interpretation is as follows:

Character	Significance
+1	the orbital is unchanged
−1	the orbital changes sign
0	the orbital undergoes a more complicated change

It follows that we can identify the symmetry label of the orbital by comparing the changes that occur to an orbital under each operation and then comparing the resulting +1 or −1 with the entries in a row of the character table for the point group concerned.

For the rows labeled E or T (which refer to sets of doubly and triply degenerate orbitals, respectively), the characters in a row of the table are the *sums* of the characters summarizing the behavior of the individual orbitals. Thus, if one member of a doubly degenerate E pair remains unchanged under a symmetry operation but the other changes sign, then the entry is reported as $\chi = +1 - 1 = 0$. We now see why the first column tells us the degeneracy: it is the sum of 1 for each orbital in the degenerate set (because each orbital remains unchanged under the identity operation), so a nondegenerate orbital gives $\chi = +1$, a doubly degenerate orbital gives $\chi = +1 + 1 = +2$, and so on.

As an example, consider a $2p_x$ orbital on the O atom of H_2O. Because H_2O belongs to the point group C_{2v}, we know by referring to the C_{2v} character table (Appendix 3) that the labels available for the orbitals are A_1, A_2, B_1, and B_2. We can decide the appropriate label for $2p_x$ by noting that under a 180° rotation (C_2) the orbital changes sign

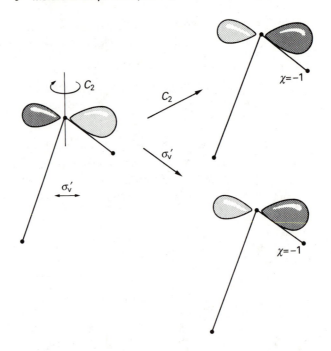

3.22 In a C_{2v} molecule such as H_2O, the $2p_x$ orbital on the central atom changes sign under a C_2 rotation, which signifies that it is a b orbital. That it also changes sign under the reflection σ_v' identifies it as b_1.

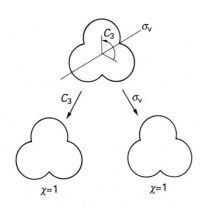

3.23 The combination $\phi_1 = \phi(A) + \phi(B) + \phi(C)$ of the three $H1s$ orbitals in the C_{3v} molecule NH_3 remains unchanged under a C_3 rotation and under any of the vertical reflections.

(Fig. 3.22), so it must be either B_1 or B_2 as only these two symmetry types have character -1 under C_2. The $2p_x$ orbital also changes sign under the reflection σ_v', which identifies it as B_1. As we shall see, any molecular orbital built from this atomic orbital will also be a b_1 orbital. Similarly, $2p_y$ changes sign under C_2 but not under σ_v' and so contributes to b_2 orbitals.

A slightly more complicated example is the symmetry classification of the combination $\phi_1 = \phi(A) + \phi(B) + \phi(C)$ of the three $H1s$ orbitals in the C_{3v} molecule NH_3 (Fig. 3.23). Because ϕ_1 is nondegenerate, it is either A_1 or A_2. It remains unchanged under a C_3 rotation and under any of the vertical reflections, so its characters are

E	$2C_3$	$3\sigma_v$
1	1	1

Comparison with the C_{3v} character table in Table 3.6 shows that ϕ_1 is of symmetry type A_1 and therefore that it contributes to a_1 molecular orbitals in NH_3. An extensive collection of orbitals is shown in Appendix 4, and it is usually a simple matter to identify the symmetry type of a combination of orbitals by comparing it with the diagrams provided there.

Example 3.9: *Identifying the symmetry type of orbitals*
Identify the symmetry type of the orbital

$$\psi = \phi - \phi'$$

in a C_{2v} NO_2 molecule, where ϕ is a $2p_x$ orbital on one O atom and ϕ' is a $2p_x$ orbital on the other atom.

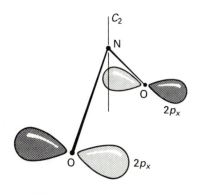

3.24 The combination of $O2p_x$ orbitals referred to in Example 3.9.

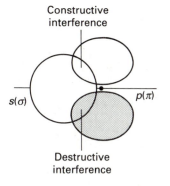

3.25 An s orbital (with σ symmetry) has zero net overlap with a p orbital (with π symmetry) because the constructive interference between the parts of the atomic orbitals with the same sign exactly matches the destructive interference.

Answer. The combination is shown in Fig. 3.24. Under a C_2 rotation, ψ changes into itself, implying a character of $+1$. Under the reflection σ_v' both orbitals change sign, so $\psi \to -\psi$, implying a character of -1. Under σ_v ψ also changes sign, so the character for this operation is also -1. The characters are therefore

E	C_2	σ_v	σ_v'
1	1	-1	-1

These values match the characters of the A_2 symmetry label, so ψ can contribute to an a_2 orbital.

Exercise E3.9. Identify the symmetry type of the combination $\phi_{1s}(A) - \phi_{1s}(B) + \phi_{1s}(C) - \phi_{1s}(D)$ in a square planar array of H atoms.

The construction of molecular orbitals from symmetry-adapted orbitals

We have already seen an illustration of the important conclusion deduced from group theory that

> Molecular orbitals are constructed from symmetry-adapted linear combinations of atomic orbitals of the same symmetry type.

Thus, in a linear molecule with the z axis as the internuclear axis, an s orbital and a p_z orbital both have σ symmetry and so may combine to form molecular orbitals. On the other hand, the s orbital and the p_x orbitals have different symmetries (σ and π, respectively) and hence cannot contribute to the same molecular orbital (Fig. 3.25). As can be seen from the illustration, the contribution from the region of constructive interference is canceled by the contribution from the region of destructive interference.

The analogous argument for a nonlinear molecule can be illustrated by considering NH_3 again. We have just seen that in NH_3 the combination ϕ_1 has A_1 symmetry (see Appendix 4). The N2s and N2p_z orbitals also have the same set of characters, as shown explicitly to the right of the character table for C_{3v}, and so these atomic orbitals also have A_1 symmetry. Because they have the same symmetry as the ϕ_1 symmetry-adapted combination, they can all contribute to a_1 molecular orbitals. All three of the resulting molecular orbitals have the form

$$\psi_{a_1} = c_1 \phi_{2s}(N) + c_2 \phi_{2p_z}(N) + c_3 \{\phi(A) + \phi(B) + \phi(C)\}$$

Only three such linear combinations are possible (because the H combination $\phi(A) + \phi(B) + \phi(C)$ counts as a single orbital), and they are labeled $1a_1$, $2a_1$, and $3a_1$ in order of increasing energy (the order of increasing number of internuclear nodes).

We have seen (and can confirm by referring to Appendix 4) that the symmetry-adapted combinations ϕ_2 and ϕ_3 have E symmetry in C_{3v}. The character table shows that the same is true of the N2p_x and N2p_y

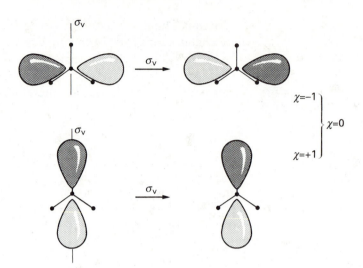

3.26 An $N2p_x$ orbital in NH_3 changes sign under a σ_v reflection but an $N2p_y$ orbital is left unchanged. Hence the degenerate pair jointly has character 0 for this operation. The plane of the paper is the xy plane.

orbitals, and this is confirmed by noting that jointly the two $2p$ orbitals behave exactly like ϕ_1 and ϕ_2 (Fig. 3.26). It follows that ϕ_2 and ϕ_3 can combine with these two $2p$ orbitals and form doubly degenerate bonding and antibonding e orbitals of the molecule. These orbitals have the form

$$\psi_e = \{c_1\phi_{2p_x}(N) + c_2\phi_2, c_3\phi_{2p_y}(N) + c_4\phi_3\}$$

The bonding pair is labeled $1e$ and the antibonding pair is labeled $2e$.

Example 3.10: *Identifying symmetry-adapted linear combinations*
The H1s orbitals in H_2O (point group C_{2v}) form two symmetry adapted linear combinations $\phi_+ = \phi(1) + \phi(2)$ and $\phi_- = \phi(1) - \phi(2)$ (**21**). What symmetry labels do they have? With what O orbitals will they overlap to form molecular orbitals?

Answer. Under C_2, ϕ_+ does not change sign but ϕ_- does; their characters are $+1$ and -1, respectively. Under the reflections, ϕ_+ does not change sign; ϕ_- changes sign under σ_v, so that its character is -1 for this operation. The characters are therefore

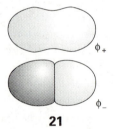

21

	E	C_2	σ_v	σ_v'
ϕ_+	1	1	1	1
ϕ_-	1	-1	-1	1

This table identifies their symmetry types as A_1 and B_2, respectively. The same conclusion could have been obtained more directly by referring to Appendix 4. According to the right of the character table, the O2s and O2p_z orbitals also have A_1 symmetry; O2p_y has B_2 symmetry. The linear combinations that can be formed are therefore

$$\psi_{a_1} = c_1\phi_{2s}(O) + c_2\phi_{2p_x}(O) + c_3\phi_+$$

$$\psi_{b_2} = c_4\phi_{2p_y}(O) + c_5\phi_-$$

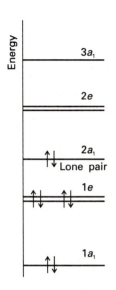

3.27 A schematic molecular orbital energy level diagram for NH_3 and an indication of its ground state electron configuration.

The a_1 orbitals may be bonding, intermediate, and antibonding in character according to the relative signs of the three coefficients. Similarly, depending on the relative signs of the two coefficients, one of the b_2 orbitals is bonding and the other is antibonding.

Exercise E3.10. What is the symmetry label of the symmetry-adapted linear combination $\phi = \phi(A) + \phi(B) + \phi(C) + \phi(D)$ in CH_4, where $\phi(J)$ is a $1s$ orbital on H atom J?

A symmetry analysis has nothing to say about the energies of orbitals other than to identify degeneracies. To calculate the energies it is necessary to resort to quantum mechanics; to assess them experimentally it is necessary to turn to techniques such as photoelectron spectroscopy. In simple cases, though, we can use the general rules set out previously (Sections 2.7 and 3.3) to judge the relative energies of the orbitals. For example, in NH_3, the $1a_1$ orbital, being composed of the low-lying N2s orbital, will lie lowest in energy and its antibonding partner, $3a_1$, will probably lie highest. The e bonding orbital is next after $1a_1$, and is followed by the $2a_1$ orbital, which is largely nonbonding. This analysis leads to the energy level scheme shown in Fig. 3.27.

The general procedure for constructing a qualitative molecular orbital scheme for a reasonably simple molecule may now be summarized as follows:

1. Assign a point group to the molecule.

2. Look up the shapes of the symmetry-adapted linear combinations (SALC) in Appendix 4.

3. Arrange the SALCs of each molecular fragment in increasing order of energy, first noting whether they stem from s, p, or d orbitals (and put them in the order $s < p < d$), and then their number of internuclear nodes.

4. Combine SALCs of the same symmetry type from the two fragments, and from N SALCs form N molecular orbitals.

5. Estimate the relative energies of the molecular orbitals from considerations of overlap and relative energies of the parent orbitals, and draw the levels on a molecular orbital energy level diagram (showing the origin of the orbitals, if desired).

THE SYMMETRIES OF MOLECULAR VIBRATIONS

Molecular vibrations are small periodic distortions from the equilibrium geometry of molecules. Their excitation by infrared radiation is the basis of infrared (IR) and Raman spectroscopy. The interpretation of IR and Raman spectra is greatly simplified by taking into account the symmetry of the molecule in question, and in this section we shall see a little of the reasoning involved.

3.7 Vibrating molecules: the modes of vibration

First, we review some general principles of molecular vibrations from the viewpoint of quantum theory.

Vibrational energy levels

The energy of a molecular vibration is quantized, and a vibrating molecule may possess only certain energies. We may think of a bond in a molecule as a spring, and stretching the spring through a distance x produces a restoring force. For small displacements from equilibrium the restoring force F of a spring is proportional to the displacement, and we write

$$F = -kx$$

The constant of proportionality k is the **force constant** of the bond: the stiffer the bond, the greater the force constant. A particle that experiences this type of restoring force is called a **harmonic oscillator**. The solutions of the Schrödinger equation for a harmonic oscillator of mass m shows that the allowed energies are

$$E_v = (v + \tfrac{1}{2})\hbar\omega \quad v = 0, 1, 2,...$$

These energy levels are illustrated in Fig. 3.28. The quantity ω is

$$\omega = \sqrt{\frac{k}{m}}$$

and has the dimensions of a frequency. Note that the frequency ω is high when the force constant is large (a stiff bond) and the mass of the oscillator is low (if only light atoms are moved during the molecular vibration). The mass m that should be used in the expression for ω is the **effective mass** of the vibrating molecule; for a diatomic molecule consisting of atoms of masses m_A and m_B it is given by

$$\frac{1}{m} = \frac{1}{m_A} + \frac{1}{m_B}$$

This expression is plausible, for if one atom is very heavy (in the sense $m_A \gg m_B$ so that $m \approx m_B$), then only the other moves appreciably during the vibration and the vibrational energy levels are determined largely by the mass of the lighter atom alone.

There are two important features to note about the energy levels of a harmonic oscillator. One is that there is a zero-point energy, the energy of the level with $v = 0$:

$$E_0 = \tfrac{1}{2}\hbar\omega$$

This minimum, irremovable energy is large if the bond is stiff (k large) and the masses of the vibrating atoms are small (m small). The second is that the separation of any two neighboring levels (for example, the separation of the levels with $v = 0$ and $v = 1$) is

$$\Delta E = \hbar\omega$$

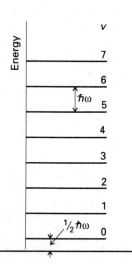

3.28 The energy levels of a harmonic oscillator. Note that all levels are equally spaced, and that the zero-point energy is half the spacing. The spacing increases as the force constant (stiffness) of the bond increases and as the effective mass of the vibrating molecule decreases.

Table 3.7 Vibrational wavenumbers and force constants of diatomic molecules

Molecule	$\tilde{v}/cm^{-1}$	$k/(N\ m^{-1})$
HCl	2885	4.8×10^2
Cl_2	557	3.2×10^2
Br_2	321	2.4×10^2
CO	2143	1.9×10^4
NO	1876	1.6×10^4

As for the zero-point energy, this separation is large if the bond is stiff and the masses of the vibrating atoms are small. Vibrational transition energies are normally expressed in terms of the wavenumber $\tilde{v}$, using $\Delta E = hc\tilde{v}$; then it follows from the preceding equation that

$$\tilde{v} = \frac{\omega}{2\pi c}$$

Typical values of $\tilde{v}$ lie in the range from 200 cm^{-1} to 3500 cm^{-1} and the correponding transitions lie in the infrared region of the electromagnetic spectrum. Table 3.7 shows some typical values of vibrational wavenumbers of diatomic molecules and the force constants derived from them.

Normal modes of vibration

The vibrations of polyatomic molecules present a much more complex problem than those of diatomic molecules. A diatomic molecule may vibrate in only one mode (its stretching mode) whereas a polyatomic molecule can vibrate in many modes.

The number of independent vibrational modes of a polyatomic molecule of N atoms is calculated by recognizing that the motion of each atom may be described in terms of displacements along three perpendicular directions in space. It follows that the complicated motion of the N atoms of a polyatomic molecule can be expressed in terms of $3N$ displacements. Three combinations of these displacements result in the motion of the whole molecule through space and correspond to translations of its center of mass (Fig. 3.29). If the molecule is nonlinear, another three combinations of the displacements are needed to specify the rotation of the whole molecule about its center of mass. That leaves $3N-6$ combinations of displacements that leave the center of mass and the orientation of the molecule unchanged, and which are distortions of the molecule. Linear molecules are a special case because there is no rotation about the axis of the molecule: only two combinations of displacements correspond to a change in orientation of the molecule, so there are $3N-5$ vibrational modes. In summary:

> A nonlinear molecule that consists of N atoms has $3N-6$ modes of vibration; if it is linear, it has $3N-5$ vibrational modes.

Thus, the linear triatomic molecule CO_2 has four modes of vibration ($3 \times 3 - 5 = 4$), the angular triatomic molecule H_2O has three modes ($3 \times 3 - 6 = 3$), and the octahedral SF_6 molecule has fifteen ($3 \times 7 - 6 = 15$).

The $3N-6$ modes of vibration of a nonlinear molecule (and the $3N-5$ modes of a linear molecule) can be described in a number of ways. One of the most useful descriptions is to identify the **normal modes** of the molecule by a combination of symmetry arguments and computation based on the force constants of the bonds. The normal modes of a molecule are particular collective motions of the atoms that are independent of each other (so long as the motion is harmonic) in

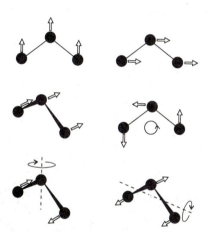

3.29 An illustration of the counting procedure for displacements of the atoms of a nonlinear molecule. There are nine atomic displacements in all. As shown here, three combinations of the displacements correspond to translation of the molecule through space, and three combinations correspond to rotations. Hence, three combinations (shown in Fig. 3.30) correspond to vibrations of the molecule that leave its center of mass and orientation unchanged.

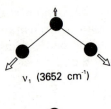

v_1 (3652 cm^{-1})

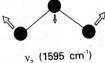

v_2 (1595 cm^{-1})

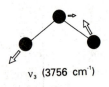

v_3 (3756 cm^{-1})

3.30 The three normal modes of H_2O and their wavenumbers.

Table 3.8 Group wavenumbers of some important groups often present in inorganic molecules

Group	Wavenumber/cm^{-1}
Terminal CN	2200–2000
Terminal CO	2150–1850
Bridging CO	1850–1700
Terminal MH of *d*-block hydride	1950–1750
Superoxo OO	1200–1100
Peroxo OO	920–750
MX in *d*-block Metal halides (X=Cl, Br, I)	450–150
Metal–metal bond	250–150*

*Metal–metal vibrations can occur at much lower wavenumbers if the metal atoms are heavy and the bonds are weak.

the sense that if any one is excited, it does not stimulate motion in another normal mode. A representation of the three independent modes of vibration of the angular triatomic molecule H_2O is shown in Fig. 3.30. Each diagram in the illustration represents an extreme point (in classical terms, a turning point) of the vibration. In diagrams of this kind, we shall show only the relative direction of displacement in only one phase of the vibration; we do not show the matching displacements as the atoms swing back to their original positions. Each mode has its own characteristic frequency, and the corresponding wavenumbers for H_2O are shown in the illustration.

Although normal modes are *collective* motions of all the atoms in a molecule, in many cases it is possible to identify the vibration as primarily bond-stretching or bond-bending. For example, two of the vibrations shown in Fig. 3.30 (v_1 and v_3) have similar wavenumbers because both modes mainly involve the stretching of O—H bonds. Whenever a vibration can be identified as primarily one type of displacement (such as a stretch or a bend) it is a useful approximation to refer to it as a particular type of vibration. Thus, we speak of the O—H stretching region of the spectrum, the C—O stretching region, or the H—O—H bending region.

Table 3.8 shows some of the characteristic wavenumbers of the absorption regions in the IR spectra of inorganic molecules. These characteristic wavenumbers are called **group wavenumbers** (more colloquially, 'group frequencies'). Their observation in a spectrum is a great help in the recognition of the presence of groups of atoms in a molecule and to its overall identification.

3.8 Symmetry considerations

Now we turn to a consideration of how the symmetry of a molecule can assist in the analysis of IR and Raman spectra. It is convenient to consider two aspects of symmetry. One is the information that can be obtained from (or about) the point group to which a molecule as a whole belongs. The other is the additional information that comes from a knowledge of the symmetry of each normal mode.

The general principles of IR and Raman spectroscopy are set out in Box 3.2. For the purpose of this discussion we need to note that absorption of infrared radiation can occur when a vibration results in a change in the electric dipole moment of a molecule. Similarly, a Raman transition can occur when the polarizability of a molecule changes during a vibration. Vibrations that can contribute to IR and Raman spectra are called **infrared active** or **Raman active**, respectively. Vibrations that cannot contribute to a spectrum are classified as **inactive**.

Information from the point group: the exclusion rule

It should be quite easy to see intuitively that all three displacements shown in Fig. 3.30 lead to a change in the molecular dipole moment

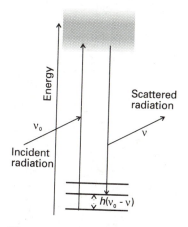

B3.3

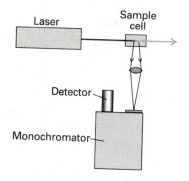

B3.4

Laser

Sample cell

Detector

Monochromator

B3.5

Box 3.2. Infrared and Raman spectroscopy

Infrared (IR) and Raman spectroscopy are used primarily to determine transition energies between vibrational energy levels in molecules and solids. Although they both measure the vibrational frequencies of atoms in molecules or solids, the principles of IR and Raman detection are quite different.

In an IR absorption process (Fig. B3.3), a photon of infrared radiation is completely absorbed and the molecule or solid is promoted to a higher vibrational energy state. For this absorption to occur, the energy of the photon must match the energy separation of vibrational states in the molecule or solid. By contrast, a Raman spectrometer generally uses visible light. The energy of a photon of visible light is far greater than the fundamental vibrational transitions in molecules or solids, so the visible photons are not absorbed. Instead, they undergo a scattering process in which a photon gives up some of its energy. The energy loss of the photon corresponds to the energy of a vibrational transition in the molecule. As a result of losing some of its energy, the scattered visible photon is shifted to lower frequency (Fig. B3.4).

Fourier transform infrared (FTIR) spectrometers are now widely used on account of their high sensitivity and their digital data handling capability. Such a spectrometer makes use of interference effects between two rays from a broad band source, and detects an interferogram, a signal at the detector that oscillates in time. The mathematical technique of Fourier transformation converts the interferogram to a function of frequency. For the spectroscopic determination, infrared radiation is directed through a sample in the form of a liquid, gas, or solid. Solid samples are commonly prepared as powdered suspensions in oil or in pressed potassium bromide disks. The spectrum produced by an infrared spectrometer is commonly displayed as percentage transmittance plotted against wavenumber.

In a Raman spectrometer, a monochromatic laser source of frequency v_0 is directed through the sample (Fig. B3.5). The scattered light of frequency v is collected at right angles to the incident laser beam, and it is directed through the monochromator. The latter scans frequencies from near the incident laser frequency to lower frequencies. The difference between the frequencies of the scattered and incident light is equal to the frequency of a vibrational transition in the sample.

The Raman effect is so feeble that only about one photon in every 10^{12} incident photons is scattered with a Raman shift. However the existence of powerful visible lasers and very efficient photomultiplier detectors for visible light make the technique quite practical.

The Raman experiment is generally more tedious and costly than IR spectroscopy, but chemists go to the trouble of collecting Raman data for many reasons, including:

1. The combination of IR and Raman data is more conclusive than either one alone for identifying the symmetries of simple molecules.
2. The Raman shifted light is polarized, and this additional information is useful for identifying the symmetry of molecular vibrations.
3. Raman spectroscopy on aqueous solutions and on single crystals is generally much more successful than IR spectroscopy.

[References: E. A. V. Ebsworth, D. W. H. Rankin, and S. Cradock, *Structural methods in inorganic chemistry*, Blackwell, Oxford (1991), Chapter 5; K. Nakamoto, *Infrared and Raman spectra of inorganic and coordination compounds*, Wiley, New York (1986).]

of H_2O. It follows that all three modes are IR active. It is more difficult to judge if a mode is Raman active, but symmetry analysis indicates whether a particular distortion of a molecule results in a change of polarizability. In fact, all three modes of H_2O are Raman active as well as IR active. The difficulty of deciding about the activity is partly overcome by the **exclusion rule**, which is sometimes very helpful:

If a molecule has a center of inversion, then none of its modes can be both IR and Raman active.

(A mode may be inactive in both.) There is no center of inversion in H_2O, so all three modes may be both IR and Raman active.

As an illustration of the usefulness of the exclusion rule, consider CO_2, a linear triatomic with the C atom at the center of inversion. Its four normal modes are shown in Fig. 3.31. One mode involves a symmetrical stretching motion of the two O atoms and is called the **symmetric stretch**. This mode of vibration leaves the electric dipole moment unchanged at zero and so it is IR inactive and may be Raman active (in fact, it is). In another mode, the **antisymmetric stretch**, the C atom moves in a direction opposite to that of the two O atoms. As a result, the electric dipole moment changes from zero in the course of the vibration and the mode is IR active. Because the CO_2 molecule has a center of inversion, it follows from the exclusion rule that this mode cannot be Raman active. The two **bending modes** cause a departure of the dipole moment from zero and are therefore IR active. It follows from the exclusion rule that the two bending modes are also Raman inactive.

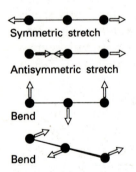

3.31 The four normal modes of CO_2. The two bending modes have the same frequency. The symmetric stretch leaves the electric dipole moment unchanged (at zero) but the antisymmetric stretch and the bending modes change the electric dipole moment.

Example 3.11: *Using the exclusion rule*

The symmetrical 'breathing' mode of the SF_6 is Raman active. Can it be detected in the IR spectrum?

Answer. The molecule belongs to the point group O_h and has a center of inversion (at the S atom). The exclusion rule applies, so the Raman active breathing mode cannot be IR active.

Exercise E3.11 The bending mode of N_2O is active in the IR. Can it be Raman active also?

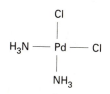

22 *cis*–[PdCl$_2$(NH$_3$)$_2$]

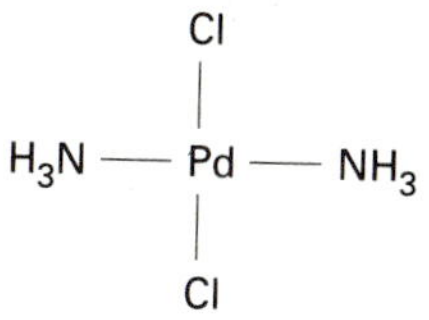

23 *trans*–[PdCl$_2$(NH$_3$)$_2$]

Information from the symmetries of normal modes

So far, we have remarked that it is often intuitively obvious whether a vibrational mode gives rise to a changing electric dipole and is therefore IR active. When intuition is unreliable (perhaps because the molecule is complex or the mode of vibration difficult to visualize), a symmetry analysis can be used instead. We shall illustrate the procedure by considering the two square-planar species (**22**) and (**23**).[4] The *cis* isomer has C_{2v} symmetry whereas the *trans* isomer is D_{2h}. Both species have bands in the Pd—Cl stretching region between 200 and 400 cm^{-1}. We know immediately from the exclusion rule that the two modes of the *trans* isomer may not be both active in IR and Raman. However, to decide which modes are IR active and which are Raman active, we may consider the characters of the modes themselves.

The symmetric and antisymmetric stretches of a Pd—Cl group are shown in Fig. 3.32 where the NH$_3$ group is treated as a single mass point. To decide on the activity of the modes we need to classify them according to their symmetry types in their respective point groups. The approach is similar to the symmetry analysis of molecular orbitals[5] in terms of SALCs.

The group C_{2v} has the operations E, C_2, σ_v, and σ_v'. Consider the symmetric stretch of the *cis* isomer labeled A$_1$ in Fig. 3.32. The point to notice is that the result of each symmetry operation of the group leaves the displacement vectors representing the vibration apparently unchanged. For example, the twofold rotation interchanges two equivalent displacement vectors. It follows that the character of each operation is $+1$:

	E	C_2	σ_v	σ_v'
χ	1	1	1	1

Comparison with the C_{2v} character table identifies the symmetry species of this mode to be A$_1$. Next consider the antisymmetric mode. The identity E leaves the displacement vectors unchanged; the same is true of σ_v', which lies in the plane containing the two Cl atoms. However, both C_2 and σ_v interchange the two oppositely directed displacement vectors, and so convert the overall displacement into -1 times itself:

---

[4] The Pt analogs of these species (and the distinction between them) are of considerable social and practical significance because the *cis* Pt isomer is used as a chemotherapeutic agent against certain types of cancer whereas the *trans* Pt isomer is therapeutically inactive.
[5] A helpful account of the analogies between symmetry adapted molecular orbitals and normal modes has been given by J. G. Verkade, *J. Chem. Educ.*, **64**, 411 (1987).

$$\begin{array}{ccccc} & E & C_2 & \sigma_v & \sigma_v' \\ \chi & 1 & -1 & -1 & 1 \end{array}$$

A comparison of this set of characters with the C_{2v} character table lets us identify the symmetry type of this mode as B_2. A similar analysis of the *trans* isomer, but using the D_{2h} group, results in the labels A_g and B_{2u} for the symmetric and antisymmetric Pt—Cl stretches, respectively.

Example 3.12: *Identifying the symmetry of vibrational displacements*
The *trans* isomer in Fig. 3.32 has D_{2h} symmetry. Verify that the anti-symmetric stretch is of the symmetry type B_{2u}.

Answer. The operations of D_{2h} are E, $C_2(x)$, $C_2(y)$, $C_2(z)$, i, $\sigma(xy)$, $\sigma(yz)$, $\sigma(zx)$. Of these, E, $C_2(y)$, $\sigma(xy)$, and $\sigma(yz)$ leave the displacement vectors unchanged and so have characters $+1$. The remaining operations reverse the directions of the vectors, so giving characters of -1:

$$\begin{array}{ccccccccc} & E & C_2(x) & C_2(y) & C_2(z) & i & \sigma(xy) & \sigma(yz) & \sigma(zx) \\ \chi & 1 & -1 & 1 & -1 & -1 & 1 & 1 & -1 \end{array}$$

Comparison of this set of characters with the D_{2h} character table shows that the symmetry type is B_{2u}.

Exercise E3.12. Confirm that the symmetric mode of the *trans* isomer is of symmetry A_g.

To decide whether a vibration produces a change in the dipole moment of the molecule we must consider the symmetry of the dipole vector:

A vibrational mode is IR active if it has the same symmetry as a component of the electric dipole vector.

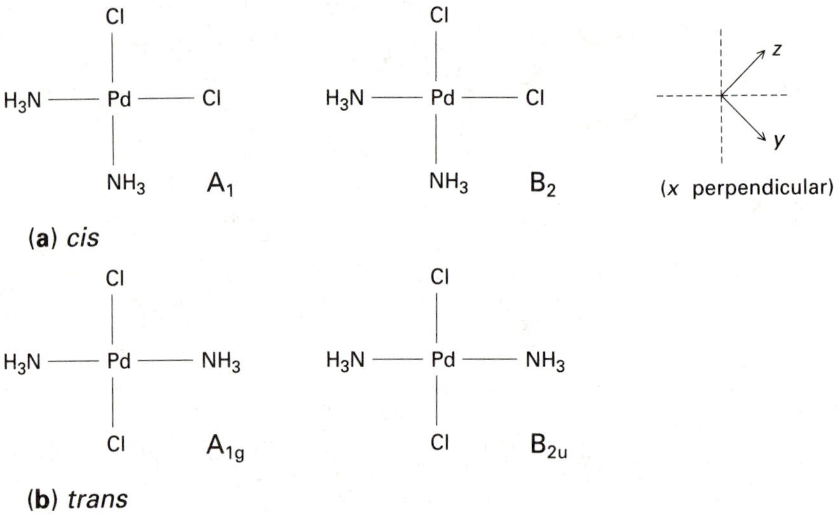

3.32 Some of the Pd—Cl stretching modes of a $[PdCl_2(NH_3)_2]$ square-planar complex. The motion of the Pd atom (which preserves the center of mass of the molecule) is not shown.

To use this rule, we need to know that the components of the dipole vector have the same symmetries as the functions x, y, and z, which are indicated on the right of the character table. For example, in C_{2v}, z is A_1 and y is B_2. Both A_1, and B_2 vibrations in C_{2v} are therefore IR active. In D_{2h}, x, y, and z are B_{3u}, B_{2u}, and B_{1u}, respectively, so only the B_{2u} vibration of the *trans* isomer can be IR active.

To judge whether a mode is Raman active we use a similar rule:

> A vibrational mode is Raman active if it has the same symmetry as a component of the molecular polarizability.

To use this rule, we need to know that the components of the polarizability have the same symmetries as the functions x^2, y^2, z^2, xy, yz, and zx. The symmetry species of these products may also be found by referring to the right of the character table for each group. It follows that in the group C_{2v}, modes of symmetry A_1, A_2, B_1, and B_2 are Raman active. In D_{2h}, on the other hand, only A_{1g}, B_{1g}, B_{2g}, and B_{3g} are Raman active. The experimental distinction between the *cis* and *trans* isomers now emerges. In the Pd—Cl stretching region, the *cis* (C_{2v}) isomer will have two bands in both the Raman and IR spectra. In contrast, the *trans* (D_{2h}) isomer will have one band at a different frequency in each spectrum. The IR spectra of the two isomers are shown in Fig. 3.33.

The assignment of molecular symmetry from vibrational spectra

One important application of vibrational spectra is to the identification of molecular symmetry and hence shape. An especially important example arises in metal carbonyls in which CO molecules are bound to a metal atom. Vibrational spectra are especially useful because the

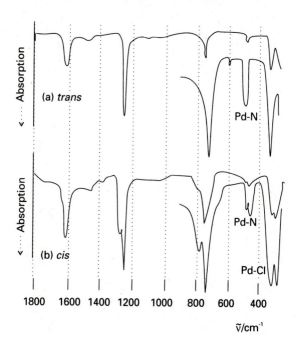

3.33 The IR spectra of *cis*- and *trans*-[PdCl$_2$(NH$_3$)$_2$]. (R. Layton, D. W. Sink, and J. R. Durig, *J. Inorg. Nucl. Chem.*, **28**, 1965 (1966).)

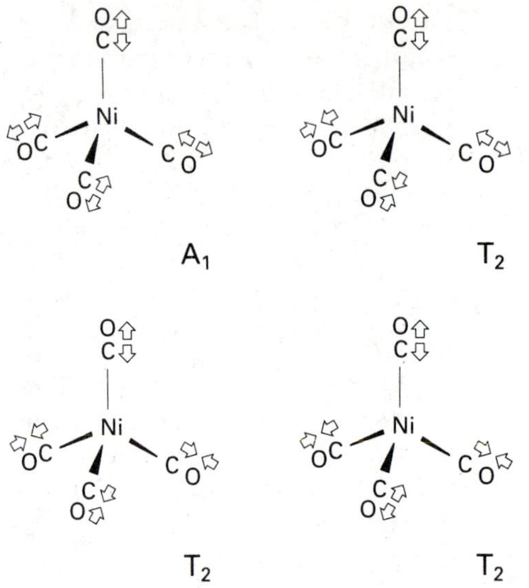

3.34 The modes of [Ni(CO)$_4$] that correspond to the stretching of the CO bonds. The A$_1$ mode is nondegenerate; the T$_2$ modes are triply degenerate. (Compare the relative phases of the displacements with the relative phases of the SALCs of the same symmetry in Appendix 4.)

CO stretch is responsible for strong characteristic absorptions between 1850 and 2100 cm^{-1}.

The first metal carbonyl to be characterized was Ni(CO)$_4$. This tetrahedral (T_d) molecule is shown in (Fig. 3.34) with vectors that represent stretches of the four CO groups. Because there are four CO groups, the vibrational modes of the molecule that arise from stretching motions of the CO groups will be four combinations of the four displacement vectors.

As suggested above, the problem of recognizing the appropriate linear combinations is similar to the problem of finding SALCs of atomic orbitals in the construction of molecular orbitals. Reference to Appendix 4 shows that that four atomic s orbitals give one A$_1$ SALC and three T$_2$ SALCs. The analogous linear combinations of the CO displacements are depicted in Fig. 3.34.

At this stage we consult the character table for T_d. We see that the combination labeled A$_1$ transforms like $x^2 + y^2 + z^2$ indicating that it will be Raman active but not IR active. In contrast, x, y, and z and the products xy, yz, and zx transform as T$_2$, and so the T$_2$ modes are both Raman and IR active. Consequently, a tetrahedral carbonyl is recognized by one IR band and two Raman bands in the CO stretching region.[6]

Example 3.13: *Predicting the IR bands of an octahedral molecule*
Consider an AB$_6$ molecule, such as SF$_6$. Sketch the form of linear combinations of A—B vibrations that have the symmetry types A$_1$ and T$_{1u}$ in the group O_h.

Answer. To proceed, we identify the SALCs that can be constructed from s orbitals in an octahedral arrangement (Appendix 4). These

---

[6] The number of IR bands expected for various symmetries of metal carbonyls are given in Table 16.5.

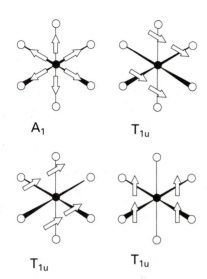

A_1 T_{1u}

T_{1u} T_{1u}

3.35 The A_1 and T_{1u} A—B stretching modes of an octahedral AB_6 molecule. The motion of the A atom, which preserves the center of mass of the molecule is not shown (it is stationary in the A_1 mode).

orbitals are the analogs of the displacements of the A—B bonds and the signs represent their relative phases. The resulting linear combinations are illustrated in Fig. 3.35.

Exercise E3.13. Does the single IR absorption band of SF_6 arise from the A_1 or the T_{1u} mode?

FURTHER READING
R. J. Gillespie, *Molecular geometry*. Van Nostrand-Reinhold, New York (1972). This book describes VSEPR theory in some detail.

R. J. Gillespie and I. Hargittai. *The VSEPR model of molecular geometry*. Allyn and Bacon, Needham Heights (1991). An introductory text, which includes an account of the model's quantum mechanical basis.

B. M. Gimarc, *Molecular structure and bonding*. Academic Press, New York (1979).

J. K. Burdett, *Molecular shapes: Theoretical models of inorganic stereochemistry*. Wiley, New York (1980).

Structural techniques
E. A. V. Ebsworth, D. W. H. Rankin, and S. Cradock, *Structural methods in inorganic chemistry*. Blackwell Scientific, Oxford (1991). A general book that covers the principles of spectroscopic techniques such as NMR and vibrational spectroscopy as well as X-ray diffraction.

K. Nakamoto, *Infrared and Raman spectra of inorganic and coordination compounds*. Wiley, New York (1986). A more specialized text which introduces vibrational spectroscopy and provides a very useful collection of data.

R. S. Drago, *Physical methods for chemists*. Saunders, Philadelphia (1992). A good general reference.

Point groups
Four reasonably elementary accounts of the use of point groups and character tables in chemistry are:

S. F. A. Kettle, *Symmetry and structure*. Wiley, New York (1985).

B. E. Douglas and C. A. Hollingsworth, *Symmetry in bonding and spectra*. Academic Press, New York (1985).

D. C. Harris and M. D. Bertolucci, *Symmetry and spectroscopy*. Oxford University Press (1978).

F. A. Cotton, *Chemical applications of group theory*. Wiley, New York (1990).

KEY POINTS

1. The VSEPR model
In the valence shell electron pair repulsion (VSEPR) model of molecular shape, it is supposed that electron pairs repel each other and take up positions as far apart as possible. The basic shapes of the theory are modified by allowing for the larger repulsions from lone pairs of electrons.

2. Fluxionality
When different experimental techniques suggest that a molecule has different shapes, the molecule may be fluxional on the timescale of the experiment. The resolution of experimental techniques is governed by the lifetimes of the conformations.

3. Molecular orbitals and molecular shape
The Walsh approach to the explanation of molecular shape is an attempt to identify the origin of shape within the context of delocalized molecular orbitals. In the approach, correlation diagrams for the orbitals are constructed, and the shape that results in the lowest energy is inferred from the variation of energy with bond angle.

4. Point group assignment
A molecule is assigned to a point group by identifying the symmetry elements it possesses and working through Fig. 3.15.

5. Polarity and chirality
Some properties of molecules, particularly their polarity and chirality, can be inferred from the identity of the point group alone. Other properties (particularly the compositions of orbitals and normal modes and selection rules) require a more detailed analysis in terms of character tables.

6. Symmetry-adapted linear combinations
Character tables are used to construct symmetry-adapted linear combinations of atomic orbitals as an initial step in the construction of molecular orbitals: only SALCs of the same symmetry type have nonzero overlap. Appendix 4 is a pictorial summary of a number of SALCs.

7. Vibrational modes
Molecular vibrations are conveniently expressed as normal modes. A nonlinear molecule that consists of N atoms has $3N-6$ modes of vibration; if it is linear, then it has $3N-5$ vibrational modes.

8. Infrared and Raman activity
A normal mode is infrared active if it corresponds to a change in electric dipole moment of the molecule; it is Raman active if the polarizability changes during a vibration. If the molecule has a center of inversion, a mode cannot be both infrared and Raman active.

EXERCISES

3.1 What shapes would you expect for the species (a) SO_3, (b) SO_3^{2-}, (c) IF_5?

3.2 Solid phosphorus pentachloride is an ionic solid composed of PCl_4^+ cations and PCl_6^- anions, but the vapor is molecular. What are the shapes of the ions in the solid?

3.3 Arrange the following techniques in order of increasing sensitivity to lifetime broadening: X-ray photoelectron spectroscopy, infrared vibrational spectroscopy, NMR spectroscopy, visible and UV electronic spectroscopy.

3.4 A molecule jumps between two conformations that have infrared absorptions differing by (a) $10\ cm^{-1}$, (b) 250 Hz. What is the minimum lifetime of the conformations for the two absorptions to be resolvable in each case?

3.5 A fluxional molecule shows two NMR proton signals separated by $\Delta\delta = 2.00$ at 90 MHz. What is the maximum rate at which the two can be interconverting?

3.6 Draw sketches to identify the following symmetry elements: (a) a C_3 axis and a σ_v plane in the NH_3 molecule; (b) a C_4 axis and a σ_h plane in the square planar $[PtCl_4]^{2-}$ ion.

3.7 Which of the following molecules and ions has (1) a center of inversion, (2) an S_4 axis: (a) CO_2, (b) C_2H_2, (c) BF_3, (d) SO_4^{2-}?

3.8 Determine the symmetry elements and assign the point group of (a) NH_2Cl, (b) CO_3^{2-}, (c) SiF_4, (d) HCN, (e) SiFClBrI, (f) BF_4^-.

3.9 Determine the symmetry elements of (a) an s orbital, (b) a p orbital, (c) a d_{xy} orbital, and (d) a d_{z^2} orbital.

3.10 (a) State the symmetry elements that imply that a molecule is nonpolar. (b) Use symmetry criteria to determine whether or not each species in Exercise 3.8 is polar.

3.11 (a) State the symmetry criteria for chirality. (b) Determine whether any of the species in Exercise 3.8 can be optically active.

3.12 (a) Determine the symmetry group appropriate to the SO_3^{2-} ion. (b) What is the maximum degeneracy of a molecular orbital in this ion? (c) If the sulfur orbitals are $2s$ and $2p$, which of them can contribute to molecular orbitals of this maximum degeneracy?

3.13 (a) Determine the point group of the PF_5 molecule. (Use VSEPR, if necessary, to assign geometry.) (b) What is the maximum degeneracy of its molecular orbitals? (c) Which phosphorus $3p$ orbitals contribute to a molecular orbital of this degeneracy?

3.14 In a metal complex $[MH_4L_2]$ (where L is a ligand), four H atoms form a square planar array around the central atom. (a) Draw the symmetry-adapted combinations of H1s orbitals (it may be helpful to refer to Appendix 4). (b) Determine the point group for the complex and assign the symmetry label of each of these symmetry-adapted orbitals. (c) Which metal d orbitals have the correct symmetry to form molecular orbitals with these H linear combinations?

3.15 (a) How many vibrational modes does an SO_3 molecule have in the plane of the nuclei? (b) How many vibrational modes does it have perpendicular to the molecular plane?

3.16 What are the symmetries of the vibrations of (a) SF_6, (b) BF_3 that are *both* IR and Raman active?

3.17 What are the symmetries of the vibrational modes of a C_{6v} molecule that are neither IR nor Raman active?

PROBLEMS

3.1 Consider a molecule IF_3O_2 (with I as the central atom). How many isomers are possible? Which is likely to be most stable? Assign point group designations to each isomer.

3.2 Group theory is often used by chemists as an aid in the interpretation of infrared spectra. For NH_4^+, four stretching modes are possible. There is the possibility that several vibrational modes are degenerate. A quick glance at the character table will tell if degeneracy is possible. (a) In the case of the tetrahedral NH_4^+ ion, is it necessary to consider the possibility of degeneracies? (b) Are degeneracies possible in any of the vibrational frequencies of $NH_2D_2^+$?

3.3 Figure 3.36 shows the energy levels for CH_3^+. What is the point group used for this illustration? What H1s orbital linear combination participates in a_1'? What C orbitals contribute to a_1'? What H linear combinations participate in the bonding e' pair? What C orbitals are e'? Which C orbitals are a_2''? Is any H linear combination a_2''? Now add two H1s orbitals on the z axis (above and below the plane), modify the linear combinations of each symmetry type accordingly, and construct a new a_2'' linear combination. Are there bonding and nonbonding (or only weakly antibonding orbitals) that can accommodate 10 electrons and allow carbon to become hypervalent? (*Hint*: Refer to Appendices 3 and 4.)

3.4 Construct a Walsh diagram correlating the orbitals of a square-planar H_4 (D_{4h}) molecule with those of a linear H_4 ($D_{\infty h}$) molecule.

3.5 Construct a Walsh diagram correlating the orbitals of a trigonal planar XH_3 (D_{3h}) molecule with those of a trigonal pyramidal XH_3 (C_{3v}) molecule.

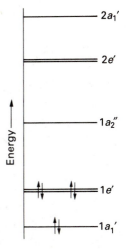

3.36 A schematic molecular orbital energy level diagram for planar CH_3^+.

4

The structures of solids

This chapter surveys the patterns adopted by atoms and ions in simple solids and explores the reasons why one arrangement may be preferred to another. We begin by seeing that the simplest model, in which atoms are represented by spheres and the structure is the outcome of stacking them together densely, is a good description of many metals and a very useful starting point for the discussion of alloys and ionic solids. It is then shown that the structures of many ionic solids can be expressed in terms of a few commonly occurring structures, which in a number of cases can also be related to the packing of spheres. We shall see that some of these structures are more likely when the bonding in the solid has a partially covalent character.

The final part of the chapter develops atomic parameters that help to rationalize why one structure is obtained rather than another. One parameter is the ionic radius, which makes use of the material introduced in Chapter 1. Then, by considering the energetics of crystal formation, it is possible to identify another parameter, the electrostatic parameter, which is relevant when ionic interactions are dominant. In later chapters we shall see that this parameter plays a role whenever charge–charge interactions are important. In the closing sections of this chapter, we see that the parameters we have introduced—the ionic radii and the electrostatic parameter—are properties that have implications for the formation and decomposition of solids.

The thermodynamically most stable stacking patterns adopted by atoms and ions in solids are the arrangements that correspond to minimum Gibbs free energy at the temperature and pressure in question. The Gibbs free energy is difficult to assess in general, but for ionic solids it is possible to analyze the contributions in terms of the coulombic interactions between the ions. A part of the difficulty with the analysis is that the forces that favor one structure over another are often finely balanced. As a result, many crystalline solids are **polymorphic**, or exist with different crystal forms. Such compounds undergo phase transitions to different structures as the temperature and pressure are changed. Polymorphism is a property of all kinds of solids, not only ionic compounds. Examples of polymorphism include the black and white phases of elemental phosphorus and the calcite and aragonite phases of calcium carbonate.

CRYSTAL STRUCTURE

The first task is to develop the concepts that are needed for the description of the structures of crystals. We also need to develop the methods that are used to generate simple models of the stacking arrangements.

4.1 Crystal lattices

The structures of crystalline solids are best discussed in terms of the unit cell. A **unit cell** is a component of the crystal that reproduces the whole crystal when it is stacked together repeatedly (Fig. 4.1), the stacked cells being related to one another by pure translations. That is, all the cells in the crystal are related to one another by displacements without rotation, reflection, or inversion. There is a wide range of choices in the selection of a unit cell, as the two-dimensional example in the illustration shows, but it is generally preferable to choose a cell that exhibits the full symmetry of the arrangement of the atoms. Thus, the unit cell in Fig. 4.1(a), which shows the fourfold rotation axis of the unit cell and its several mirror planes, is preferred to that shown in Fig. 4.1(b), which possesses only a single mirror plane.

The pattern of atoms, ions, or molecules in a crystal is depicted by an array of points called the **lattice**. The lattice points do not necessarily lie at the centers of atoms, but denote some common position of an **asymmetric unit**, the atom, ion, molecule, or group of ions or molecules from which the actual crystal is built. Each point in the

4.1 A two-dimensional solid and two choices of unit cell. The entire crystal is reproduced by translational displacements of either unit cell, but (a) is generally preferred because it displays the maximum symmetry of the structure, whereas (b) does not.

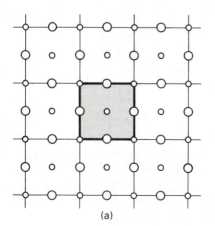

(a)

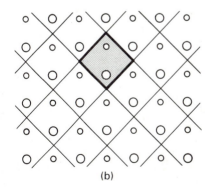

(b)

Asymmetric unit
Associated lattice point

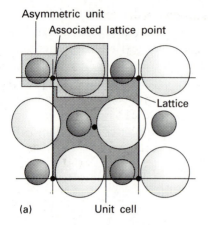

Lattice

(a)   Unit cell

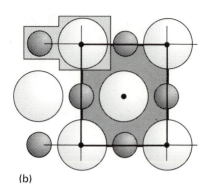

(b)

4.2 Each lattice point represents the location of an asymmetric unit, in this case an Na^+Cl^- formula unit. The relation of the point to the asymmetric unit is arbitrary (but, once chosen, fixed throughout the crystal). (a) The lattice point is chosen to lie between the Na^+ and Cl^- ions. (b) An equally valid choice places the lattice point on a Cl^- ion.

lattice shown in Fig. 4.2 indicates the location of an M^+ and X^- pair of ions (the asymmetric unit). However, the point may be located on the cation, on the anion, or at an arbitrary point relative to either of them. The relation of the point to the asymmetric unit is arbitrary, but once chosen it is the same throughout the crystal.

The unit cell is formed by joining the lattice points with straight lines. This may also be done in an arbitrary manner, so long as pure translational repetition of the unit cell reconstructs the entire lattice. However, in practice conventions have been adopted that systematize this choice.

4.2 The packing of spheres

The structures of many solids can be described in terms of the stacking of spheres that represent the atoms or ions. Metals are particularly simple in this respect because (for elemental metals such as sodium and iron) all the atoms are identical and the solid can be treated as though it consists of spheres of identical size. In many cases, the atoms are free to pack together as closely as geometry allows: this close packing occurs if there are no specific bonding forces that favor particular local arrangements. That is, metals are often **close-packed structures**, structures having least waste of space and with each sphere having its geometrically maximal number of neighbors. Close-packed structures are also useful starting points for the discussion of substances other than metals, and we shall introduce them in a general way.

The **coordination number** (C.N.) of an atom is the number of nearest neighbors it has in the lattice. The coordination number is often large (typically 8 or 12) for metals, intermediate for ionic solids (typically 4 to 8), and low for molecular solids (typically 1 to 6). This variation is reflected to some extent in the densities of the three types of solid, with the densest materials (metals) being most closely packed and having high coordination numbers. It is believed (but has not yet been shown experimentally) that all elements become metals when subjected to such high pressures that their atoms are forced into close-packed arrangements.

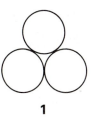

1

The close-packing of spheres

Close-packed structures of identical spheres are often pictured as formed by laying close-packed layers on top of each other. The structure is started by placing a sphere in the indentation between two touching spheres, so making a triangle (**1**). The layer is then formed by continuing this process of laying spheres together in the indentations between those already in place. A complete close-packed layer consists of spheres in contact (Fig. 4.3) with each sphere having six nearest neighbors in the plane. This arrangement is shown by the white spheres in the illustration.

The second layer is formed by placing spheres in the dips of the first layer. The third layer can be laid in either of two ways and can give rise to either of two **polytypes**, which are structures that are the same in two dimensions (in this case, in the planes) but different in the third. The coordination number is 12 for each polytype. (Later we shall see that many different polytypes can be formed; those described here are two very important special cases.)

In one polytype, the spheres of the third layer lie directly above the

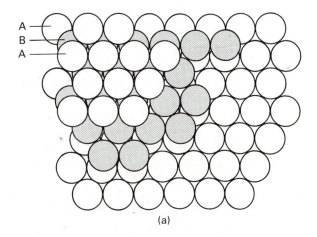

(a)

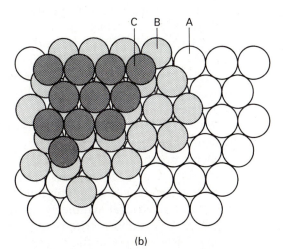

(b)

4.3 The formation of two close-packed polytypes. (a) The third layer reproduces the first, giving an ABA structure. (b) The third layer lies above the gaps in the first layer, giving an ABC structure. The different shadings denote the different layers of identical spheres.

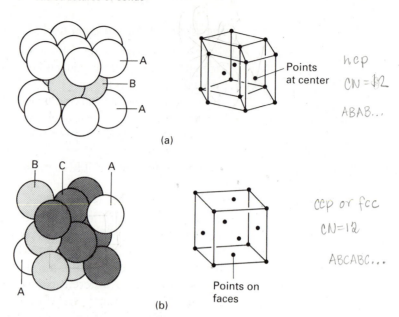

(a)

(b)

4.4 (a) The hexagonal (hcp) unit of the ABAB... close-packed solid and (b) the cubic (fcc) unit of the ABCABC... polytype. The tints of the spheres correspond to the layers shown in Fig. 4.3.

spheres of the first. This ABAB... pattern of layers gives a lattice with a hexagonal unit cell, and hence is said to be **hexagonal close-packed** (hcp, Fig. 4.4(a)). In the other polytype, the spheres of the third layer are placed above the *gaps* in the first layer. Thus the second layer covers half the holes in the first layer and the third layer lies above the remaining holes. This arrangement results in an ABCABC... pattern, and corresponds to a lattice with a face-centered cubic unit cell (Fig. 4.4(b)). Hence the crystal structure is **cubic close-packed** (ccp) or, more specifically, **face-centered cubic** (fcc; the origin of this name will become clear later).

Holes in close-packed structures

A feature of a close-packed structure that is modeled by a collection of hard spheres is the existence of two types of **hole**, an unoccupied space. (The space represented by the holes is not empty in a real solid because electron density does not end as abruptly as the hard-sphere model suggests.) The type and distribution of holes are important because many structures, including those of some alloys and many ionic compounds, can be regarded as formed from an expanded close-packed arrangement in which additional atoms or ions occupy some of the holes.

One type of hole is an **octahedral hole** (shaded in Fig. 4.5(a)). Such a hole lies between two oppositely directed planar triangles of spheres in adjoining layers. If there are N atoms in the crystal, then there are N octahedral holes. These holes are distributed in an fcc lattice as shown in Fig. 4.6(a); this illustration also shows that the hole has local octahedral symmetry (in the sense that it is surrounded by six nearest neighbor lattice points arranged octahedrally). If each hard sphere has radius r, then each octahedral hole can accommodate

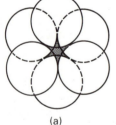

(a)

(b)

4.5 (a) An octahedral hole in the cleft between six spheres. (b) A tetrahedral hole in the cleft between four spheres in a close-packed lattice.

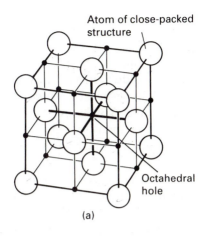

Atom of close-packed structure

Octahedral hole

(a)

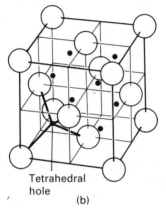

Tetrahedral hole

(b)

4.6 The locations of (a) octahedral holes and (b) tetrahedral holes relative to the atoms in an fcc structure. The holes take names from the disposition of lattice points around them.

another hard-sphere atom with a radius no larger than $0.414r$ (that is, $\sqrt{2} - 1$ times r).

A **tetrahedral hole** (T, shaded in Fig. 4.5(b)) is formed by a planar triangle of touching spheres that is capped by a single sphere lying in the dip between them. The apex of the tetrahedron may be directed up (T) or down (T') in the crystal. There are N tetrahedral holes of each type (giving $2N$ tetrahedral holes in all). In a model in which the atoms are modeled as hard spheres these holes can accommodate another atom of radius no greater than $0.225r$. Figure 4.6(b) shows the location of tetrahedral holes in an fcc lattice; it can be seen from the illustration that each hole has four nearest neighbor lattice points arranged tetrahedrally. Larger spheres may be accommodated in the holes if the close-packing of the parent structure is relaxed.

METALS

X-Ray diffraction studies (Box 3.1) reveal that many metals have close-packed structures. This finding can be rationalized if we suppose that metals have only a weak tendency toward directed covalency. As a result of this weak directional character, metal atoms tend to pack together efficiently and achieve high coordination numbers. One consequence of their close packing is that metals often have high densities. Indeed, the elements deep in the *d* block, near iridium and osmium, include the densest solids known under normal conditions of temperature and pressure. The low directionality of the bonds that metal atoms may form also accounts for the wide occurrence of polymorphism under different conditions of pressure and temperature. Iron, for example, shows several solid–solid phase transitions as it is heated and the atoms adopt new packing arrangements, generally (but not always) with the most closely packed phases stable at low temperatures and less closely packed structures stable at high temperatures. We shall deal with this aspect of metals shortly.

4.3 Metallic elements

The structures of metals are relatively simple to describe because all the atoms of a given element can be represented by spheres of the same size. Nevertheless, elemental metals show their own idiosyncrasies of structure, for a variety of polytypes can be formed in close-packed structures, and not all elemental metals have close-packed structures (Table 4.1).

Table 4.1 The crystal structures adopted by some metallic elements at 25 °C and 1 bar

Crystal structure	Element
Hexagonal close-packed (hcp)	Be, Cd, Co, Mg, Ti, Zn
Cubic close-packed (fcc)	Ag, Al, Au, Ca, Cu, Ni, Pb, Pt
Body-centered cubic (bcc)	Ba, Cr, Fe, W, alkali metals
Primitive cubic (cubic-P)	Po

Close-packed metals

Which common close-packed polytype—hcp or fcc—a metal adopts (if either) depends on the details of the elements, the interaction of atoms with second-nearest neighbors, and the residual effects of some directional character in their atomic orbitals. A close-packed structure need not be either of the regular ABAB... or ABCABC... polytypes, for these two common polytypes are only two of many possibilities. An infinite range of polytypes can in fact occur, for the planes may stack in a more complex manner.

Cobalt is an example of more complex polytypism. Above 500 °C, cobalt is fcc, but it undergoes a transition when cooled. The resulting metastable structure is a randomly stacked set (ABACBABABC...) of close-packed layers. In some samples of cobalt (and of SiC too), the polytypism is not random, for the sequence of planes repeats after several hundred layers. It is difficult to account for this behavior in terms of valence forces. The long range repeat may be a consequence of a spiral growth of the crystal that requires several hundred turns before a stacking pattern is repeated.

Structures that are not close-packed

Not all metals are close-packed, and some other packing patterns use space nearly as efficiently. Even metals that are close packed often undergo a phase transition to a less closely packed structure when they are heated and their atoms undergo large-amplitude vibrations.

One common structure is the **body-centered cubic** (cubic-I or bcc) structure in which there is a lattice point at the center of a cube with lattice points at each corner (Fig. 4.7). Metals with this structure have a coordination number of 8. Although a bcc structure is less closely packed than the ccp and hcp structures (for which the coordination number is 12), the difference is not very great because the central atom has six second-nearest neighbors only 15 percent further away. This arrangement leaves 32 percent of the space unfilled compared with 26 percent in the close-packed structures.

The least common metallic structure is the **primitive cubic** (cubic-P) structure (Fig. 4.8), in which atoms occupy lattice points located at the corners of a cube. The coordination number of a cubic-P structure is only 6. One form of polonium (α-Po) is the only example of this structure among the elements under normal conditions. Solid mercury, however, has a closely related structure which is obtained from the simple cubic arrangement by stretching the cube along one of its body diagonals.

Metals that have structures more complex than those described so far can sometimes be regarded as slightly distorted versions of simple structures. Zinc and cadmium, for instance, have almost hcp structures, but the planes of close-packed atoms are separated by a slightly greater distance than in pure hcp. This difference suggests stronger bonding between the atoms in the plane: the bonding draws these atoms together and, in doing so, squeezes out the atoms of the neighboring layers.

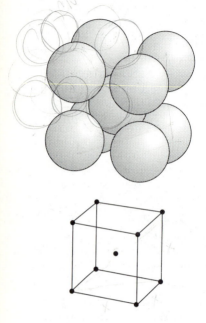

4.7 The body-centered cubic unit cell and its lattice-point representation.

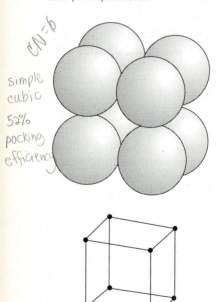

4.8 The primitive cubic unit cell and its lattice-point representation.

Polymorphism of metals

The polymorphs of metals are generally (but not always systematically) labeled α, β, γ,... with increasing temperature. Some metals revert to a low-temperature form at higher temperatures. Iron, for example, is polymorphic: α-Fe, which is bcc, is stable up to 906 °C, γ-Fe, which is fcc, is stable up to 1401 °C, and then α-Fe is stable again up to the melting point at 1530 °C. β-Fe, which is hcp, is formed at high pressures.

The room-temperature polymorph of tin is white tin (β-Sn): it undergoes a transition to gray tin (α-Sn) below 14.2 °C but the conversion occurs at an appreciable rate only after prolonged exposure to a much lower temperature. Gray tin has a diamond-like structure (Fig. 4.9). The structure of white tin is unusual in that each atom has four nearest neighbors that are more distant than in gray tin, as would be expected for the high-temperature form. However, white tin is the appreciably denser polymorph (7.31 g cm^{-3} compared with 5.75 g cm^{-3} for gray tin). The explanation is that in white tin the second-nearest neighbors are closer than in gray tin, so overall the solid is more compact. A further point of interest, which shows that crystal structure can influence chemical properties, is that when tin dissolves in concentrated hydrochloric acid, white tin forms tin(II) chloride whereas gray tin forms tin(IV) chloride.

The bcc structure is common at high temperatures for metals that are close-packed at low temperatures because the increased amplitude of atomic vibrations demand a less close-packed structure. For many metals (among them calcium, titanium, and manganese), the transition temperature is above room temperature. For others (among them lithium and sodium), the transition temperature is below room temperature. The observation that a bcc structure is favored by a small number of valence electrons per orbital suggests that a dense electron sea is needed to draw cations together into the close-packed arrangement and that the alkali metals do not have enough valence electrons for this to be achieved at room temperature.

Atomic radii of metals

An informal definition of the atomic radius of a metallic element is that it is half the distance between the nuclei of neighboring atoms in the solid at room temperature and pressure. However, it is found that this distance generally increases with the coordination number of the lattice. For example, the same atom in lattices with different coordination numbers may appear to have different radii, and an atom of an element in a twelve-coordinate lattice appears bigger than one in an eight-coordinate lattice. In an extensive study of internuclear separations in a wide variety of polymorphic elements and alloys, V. Goldschmidt found that the average relative radii are as shown in the margin.

It is desirable to put all elements on the same footing when comparing trends in their characteristics–that is, when comparing the *intrinsic*

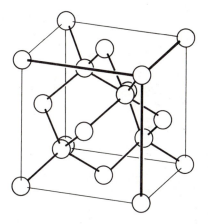

4.9 The structure of α-tin. Note its relation to the fcc unit cell in Fig. 4.6(b), with an additional Sn atom in half the tetrahedral holes. This structure is also that of diamond, silicon, and germanium.

Coordination number	Relative radius
12	1
8	0.97
6	0.96
4	0.88

properties of their atoms rather than the properties that stem from their environment. Therefore, it is common to adjust the empirical internuclear separation to the value that would be expected if the element were in fact close packed (with C.N. = 12). Thus, the empirical atomic radius of Na is 1.85 Å, but that is for a structure in which C.N. = 8; we therefore multiply this radius by $1/0.97 = 1.03$ and obtain 1.91 Å as the radius that an Na atom would have if it were close-packed.

The 'corrected' Goldschmidt radii of the elements were in fact the ones listed in Table 1.5. The Goldschmidt radii were used in the discussion of the periodicity of atomic radius (Section 1.9). The essential features of that discussion to bear in mind now are as follows:

Atomic radii generally increase down a group.
Atomic radii generally decrease from left to right across a period.

As remarked in Section 1.9, atomic radii reveal the presence of the lanthanide contraction in Period 6, with atomic radii of the elements that follow the lanthanides being found to be smaller than simple extrapolation from earlier periods would suggest. As remarked there, this contraction can be traced to the poor shielding effect of f electrons. A similar contraction occurs across each row of the d block.

4.4 Alloys

An **alloy** is a blend of metals prepared by mixing the molten components and then cooling the mixture. Alloys may be homogeneous solid solutions, in which the atoms of one metal are distributed randomly among the atoms of the other, or they may be compounds with a definite composition and internal structure. Solid solutions are sometimes classified as either substitutional or interstitial. A **substitutional solid solution** is a solid solution in which atoms of the solute metal occupy some of the locations of the solvent metal atoms (Fig. 4.10(a)). An **interstitial solid solution** is a solid solution in which the solute atoms occupy the gaps (the interstices) between the solvent atoms (Fig. 4.10(b)). However, this distinction is not particularly fundamental, because interstitial atoms often lie in a definite lattice (Fig. 4.10(c)), and hence can be regarded as a substitutional version of another lattice. A better viewpoint is that a solid solution is a new structure, and that its relation to the original lattice may be largely coincidental. Some of the classic examples of alloys are brass (up to 40 percent zinc in copper), bronze (a metal other than zinc or nickel in copper; casting bronze, for instance, is 10 percent tin and 5 percent lead), and stainless steel (over 12 percent chromium in iron).

Substitutional alloys

It is found that substitutional solid solutions are generally formed if three criteria are fulfilled:

1. The atomic radii of the elements are within about 15 percent of each other.

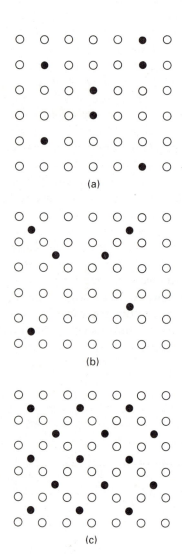

4.10 (a) Substitutional and (b) interstitial alloys. (c) In some cases an interstitial alloy may be regarded as a substitutional alloy derived from another lattice.

2. The crystal structures of the two pure metals are the same, for this indicates that the directional forces between the two types of atom are compatible with each other.
3. The electropositive character of the two components are similar, for otherwise compound formation would be more likely.

Thus, although sodium and potassium are chemically similar and have bcc structures, the atomic radius of sodium (1.91 Å) is 19 percent smaller than that of potassium (2.35 Å), and the two metals do not form a solid solution. On the other hand, copper and nickel, two neighbors late in the d block, have similar electropositive character, similar crystal structures (both fcc), and similar atomic radii (Ni 1.25 Å, Cu 1.28 Å, only 2.3 percent different), and form a continuous series of solid solutions, ranging from pure nickel to pure copper. Zinc, copper's other neighbor in Period 4, has a similar atomic radius (1.37 Å, 7 percent larger), but it is hcp, not fcc. In this instance, zinc and copper are partially miscible and form solid solutions over only a limited concentration range.

Interstitial solid solutions of nonmetals

Interstitial solid solutions are formed by solute nonmetals (such as boron and carbon) that are small enough to inhabit the interstices in the solvent structure. The small atoms enter the host solid with preservation of the crystal structure of the original metal, either in a simple whole number ratio of metal and interstitial atoms (as in Fe_3C), or the small atoms are distributed randomly in the available holes in the metal lattice. The former substances are true compounds and the latter are solid solutions.

Considerations of size can help to decide whether the formation of a solid solution is likely. Thus, the largest solute atom that can enter a close-packed solid without distorting the structure appreciably is one that just fits an octahedral hole, which as we have seen has radius $0.414r$ (if the structure is modeled in terms of hard spheres). Hence, on geometrical grounds and with no reconstruction of the crystal structure, to accommodate hard-sphere H, B, C, or N atoms, the atomic radii of the host metal atoms must be no less than 0.90 Å, 1.95 Å, 1.88 Å, or 1.80 Å, respectively. The observation that the Period 4 metals close to nickel (which have atomic radii close to 1.3 Å) form an extensive series of interstitial solid solutions with boron, carbon, and nitrogen must indicate that specific bonding occurs between the host and the interstitial atom. Hence, as electronegativity considerations should lead us to expect, these substances are best regarded as examples of compounds of the nonmetals, and they are treated as such in Chapters 11 and 12.

Intermetallic compounds

In contrast to the interstitial solid solutions of metals and nonmetals, where geometrical considerations are consistent with intuition, there is a class of solid solutions formed between two metals that are also better regarded as actual compounds despite the similarity of the electronegativities of the metals. For instance, when some liquid mixtures of metals are cooled, they form phases with definite structures that are often unrelated to the parent structure. These phases are called **intermetallic compounds**. They include β-brass (CuZn) and compounds of composition $MgZn_2$, Cu_3Au, and Na_5Zn_{21}. Chemical formulas such as these show the *limiting* compositions at the phase boundaries and are not necessarily applicable to a specific sample of the compound.

The **Zintl phases** (named for E. Zintl, who first characterized them) are intermetallic compounds formed between a strongly electropositive element (the alkali metals and alkaline earth metals) and a somewhat less electropositive metal (typically from the late *d* block or the early *p* block). Examples of Zintl phases include NaTl, Mg_2Sn, $CaZn_2$, and LiZn. Their electronic structures can be regarded (to a first approximation) as arising from the transfer of the valence electrons from the more electropositive metal to the less electropositive metal. As a result, in some cases the composition reflects the conventional valencies of the metals (as in Mg_2Sn, but not in $CaZn_2$).

A useful perspective on the structures of these compounds can be obtained by recognizing that some of them are isoelectronic with more familiar substances. As an illustration, consider sodium thalide, NaTl. The transfer of an electron from an atom of one metal, sodium, to an atom of another metal, thallium, results in the formation of Tl^-, which is isoelectronic (in its valence shell) with a C atom. Indeed, the Tl atoms are found to lie in a diamond lattice. Moreover, just as diamond has full bands, so too does NaTl, and it is a colorless nonmetallic solid.

As usual, isoelectronic analogies help to rationalize observations but must be used cautiously when making predictions. Thus, electron transfer to zinc in lithium zincide, LiZn, does not result in a C-atom configuration, but nevertheless, the Zn atoms do lie in a diamond-like lattice. Unlike in NaTl, the diamond-like zinc bands are not full and the compound is colored and a metallic conductor. Extensive transfer of electrons from Ca to the Zn bands also occurs in calcium zincide, $CaZn_2$, and in this case the Zn atoms form a graphite-like layer structure of hexagons of Zn atoms with Ca^{2+} ions between them.

IONIC SOLIDS

We now consider **ionic solids**, or solids that to a good approximation may be regarded as composed of cations and anions. Ionic solids, such as sodium chloride and ammonium nitrate, are often recognized by their brittleness, their moderately high melting points, and their solubility in polar solvents. However, there are exceptions: magnesium oxide, MgO, for example, is a high-melting ionic solid, but it is highly insoluble in water. Ammonium nitrate, NH_4NO_3, is ionic, but melts at only 170 °C. The existence of anomalies like MgO and NH_4NO_3 points to the need for a more fundamental definition of an 'ionic solid'.

An indication that a solid is ionic is provided by X-ray diffraction, for the coordination number of an ionic lattice is generally low: this low value is consistent with the low densities of ionic solids, as we remarked before. However, although coordination number helps to distinguish metallic bonding from ionic, it does not distinguish between ionic and covalent bonding: covalent solids also have low coordination numbers (diamond, for example, has C.N. = 4), and so we need a more fundamental criterion. The classification of a solid as ionic is based on comparison of its properties with those of a model. The **ionic model** of bonding treats a solid as an assembly of oppositely charged spheres that interact primarily by coulombic forces (together with repulsions between the complete shells of the ions in contact). If the thermodynamic properties of the solid calculated on this model agree with experiment, then the solid may be ionic. However, it should be noted that many examples of coincidental agreement with the ionic model are known, so numerical agreement may be misleading.

As we did for metals, we start by describing some common ionic structures in terms of the packing together of spheres but in this case the spheres have different sizes and opposite charges. After that, we shall see how to rationalize the structures in terms of the energetics of crystal formation. The structures we shall describe were obtained using X-ray diffraction and were among the first inorganic solids to be examined in this way.

4.5 Characteristic structures of ionic solids

The ionic structures described in this section are prototypes of a wide range of solids. For instance, although the rock-salt structure takes its name from a mineral form of NaCl, it is characteristic of numerous other solids (Table 4.2). Many of the structures we describe can be regarded as derived from arrays in which the anions (sometimes the cations) stack together in fcc or hcp patterns and the counterions (the ions of opposite charge) occupy the octahedral and tetrahedral holes in the lattice. Throughout the following discussion, it will be helpful

Table 4.2 Compounds with particular crystal structures

Crystal structure	Example*
Antifluorite	K_2O, K_2S, Li_2O, Na_2O, Na_2Se, Na_2S
Cesium chloride	**CsCl**, CaS, TlSb, CsCN, CuZn
Fluorite	**CaF$_2$**, UO_2, $BaCl_2$, HgF_2, PbO_2,
Nickel arsenide	**NiAs**, NiS, FeS, PtSn, CoS
Perovskite	**CaTiO$_3$**, $BaTiO_3$, $SrTiO_3$
Rock salt	**NaCl**, LiCl, KBr, RbI, AgCl, AgBr, MgO, CaO, TiO, FeO, NiO, SnAs, UC, ScN
Rutile	**TiO$_2$**, MnO_2, SnO_2, WO_2, MgF_2, NiF_2
Sphalerite (zinc blende)	**ZnS**, CuCl, CdS, HgS, GaP, InAs
Wurtzite	**ZnS**, ZnO, BeO, MnS, AgI,† AlN, SiC, NH_4F

*The substance in bold type is the one that gives its name to the structure.
†Silver iodide is also found with a sphalerite structure, which is metastable.

(margin annotations)
8–8 (Cesium chloride)
6–6 (Nickel arsenide)
6–6 (Rock salt)
4–4 (Sphalerite)
4–4 (Wurtzite)

to refer back to Fig. 4.6 to see how the structure being described is related to the hole patterns shown there. The close-packed layer usually needs to expand in order to accommodate the counterions, but this expansion is often a minor perturbation of the anion arrangement. Hence, the close-packed structure is often a good starting point for the discussion of ionic structures.

The rock-salt structure

The **rock-salt structure** (Fig. 4.11) is based on an fcc array of bulky Cl^- anions in which the cations occupy all the octahedral holes. Alternatively, it can be viewed as a structure in which the anions occupy all the octahedral holes in an fcc array of Na^+ ions. It can be seen from the diagram that each ion is surrounded by an octahedron of six counterions. The coordination number of each type of ion is therefore 6, and the structure is said to have '(6,6)-coordination'. In this notation, the first number in parentheses is the coordination number of the cation, and the second number is the coordination number of the anion.

To visualize the local environment of an ion in the rock-salt structure, we should note that the six nearest neighbors of the central ion of the cell shown in Fig. 4.11 lie at the centers of the faces of the cell and form an octahedron around the central ion. All six neighbors have a charge that is opposite to that of the central ion. The 12 second-nearest neighbors of the central ion, those next furthest away, are at the centers of the edges of the cell, and all have the same charge as the central ion. The eight third-nearest neighbors are at the corners of the unit cell, and have a charge opposite to that of the central ion.

When assessing the number of ions of each type in a unit cell we must take into account that any ions that are not fully inside the cell are shared by neighboring cells:

1. *Body*. An ion in the body of a cell belongs entirely to that cell and counts as 1.
2. *Face*. An ion on a face is shared by two cells and contributes $\frac{1}{2}$ to the cell in question.
3. *Edge*. An ion on an edge is shared by four cells and hence contributes $\frac{1}{4}$.
4. *Corner*. An ion on a corner is shared by eight cells that share the corner, and so contributes $\frac{1}{8}$.

In the unit cell shown in Fig. 4.11, there are four Na^+ ions and four Cl^- ions. Hence, each unit cell contains four NaCl formula units.

The cesium-chloride structure

Much less common than the rock-salt structure is the **cesium-chloride structure**, which is shown by CsCl, CsBr, and CsI as well as some other compounds formed of ions of similar radii to these, including NH_4Cl (see Table 4.2). The cesium-chloride structure (Fig. 4.12) has a cubic unit cell with each lattice point occupied by a

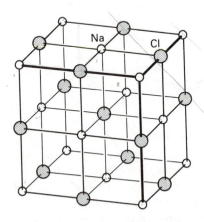

4.11 The rock-salt structure. Note its relation to the fcc structure in Fig. 4.6(a), with an anion in each octahedral hole. Alternatively, treat the lattice as the location of anions, in which case the cations occupy the octahedral holes.

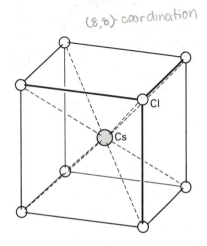

4.12 The cesium-chloride structure. Note that the corner ions, which are shared by eight cells, are surrounded by eight nearest-neighbor center atoms.

halide anion and a metal cation at the cell center (or vice versa). The coordination number of both types of ion is 8, for their radii are so similar that this energetically highly favorable (8,8) form of packing, with numerous counterions adjacent to a given ion, is feasible. Note that the cesium-chloride lattice is *not* body-centered, because a lattice is defined in terms of the locations of the fundamental asymmetric unit whereas in the cesium-chloride unit cell there are different atoms at the cell center and its vertices. The fundamental asymmetric unit is the Cs^+ ion together with its eight Cl^- neighbors.

The sphalerite structure

The **sphalerite structure** (ZnS), which is also known as the **zinc-blende structure** (Fig. 4.13), is based on an expanded fcc anion lattice, but now the cations occupy one type of tetrahedral hole. Each ion is surrounded by four neighbors, so the structure has (4,4)-coordination.

Example 4.1: *Counting the number of ions in a unit cell*

How many ions are there in the unit cell shown in the sphalerite structure shown in Fig. 4.13?

Answer. An ion in the body of the cell belongs entirely to that cell and counts as 1. An ion on the face is shared by two adjacent unit cells and counts $\frac{1}{2}$. An ion on an edge is shared by four cells and so counts $\frac{1}{4}$. An ion at a corner position is shared by eight cells and so counts $\frac{1}{8}$. In the sphalerite structure, the count is

Location (share)	Number of ions	Contribution
Body (1)	4	4
Face ($\frac{1}{2}$)	6	3
Edge ($\frac{1}{4}$)	0	0
Corner ($\frac{1}{8}$)	8	1
	Total 8	

Note that there are four cations and four anions in the unit cell. This ratio is consistent with the chemical formula ZnS.

Exercise E4.1. Count the ions in the cesium-chloride unit cell shown in Fig. 4.12.

The fluorite and antifluorite structures

The last of the three simple structures based on the expanded fcc lattice of anions is the **fluorite structure** (Fig. 4.14), which is shown by calcium fluoride, CaF_2. In it, the anions (which are twice as numerous as the cations) occupy both types of tetrahedral hole. (Recall that if there are N atoms, then there are $2N$ tetrahedral holes.) Fluorite,

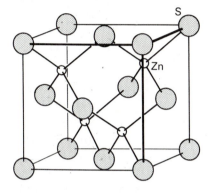

fcc with half tetrahedral holes filled with cations (4,4)-coordination

4.13 The sphalerite (zinc blende) structure. Note its relation to the fcc structure in Fig. 4.6(b), with half the tetrahedral holes occupied by Zn^{2+} ions.

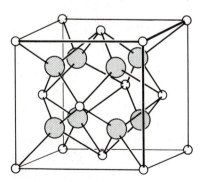

(8,4)-coordination fcc with all tetrahedral holes filled with anions

4.14 The fluorite structure has an fcc array of Ca^{2+} ions (○), and all the tetrahedral holes contain F^- ions (◐).

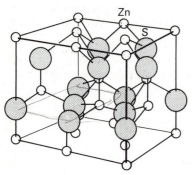

hcp
(4,4)-coordination

Zn
S

4.15 The wurtzite structure, which is derived from the hcp structure (Fig. 4.4(a)).

hcp with all octahedral holes filled with cations

As
Ni

Prismatic

anti-prismatic

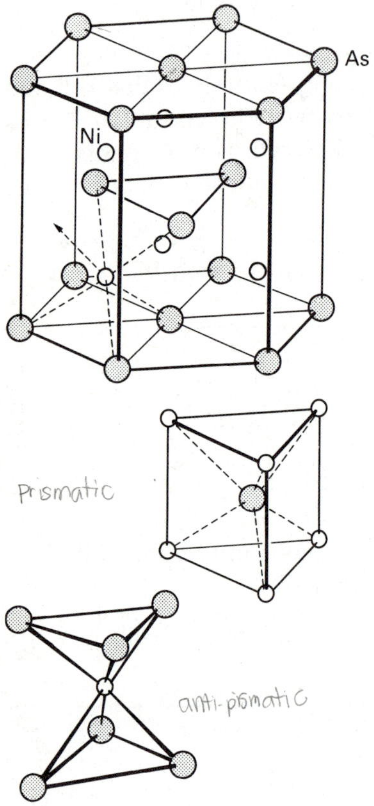

4.16 The nickel-arsenide structure, another structure derived from the hcp structure (Fig. 4.4(a)). Note the prismatic and trigonal antiprismatic local symmetries of the As and Ni atoms, respectively.

m.p. 1423 °C, is so called because it melts and flows on being heated with a mineralogist's blowpipe, and so can be distinguished from gemstones. The **antifluorite structure**, which is adopted by K_2O, is the inverse of the fluorite structure in the sense that the locations of cations and anions are reversed. In it, O^{2-} ions lie in an expanded fcc array and the K^+ cations occupy the two types of tetrahedral hole.

In the fluorite structure, the anions in their tetrahedral holes have four nearest neighbors. The cation site is surrounded by a cubic array of eight anions; hence, the lattice has (8,4) coordination. This coordination is consistent with there being twice as many anions as cations. In the antifluorite structure, the coordination is the opposite of this, namely (4,8).

The wurtzite structure

The **wurtzite structure**, which is shown in Fig. 4.15, takes its name from another polymorph of zinc sulfide. It differs from the sphalerite structure in being derived from an expanded hexagonally close-packed anion array rather than a face-centered cubic array, but as in sphalerite the cations occupy one type of tetrahedral hole. The wurtzite structure, which has (4,4)-coordination, is possessed by ZnO, AgI, and one polymorph of SiC as well as several other compounds (Table 4.2). When comparing wurtzite and sphalerite, note that the local symmetries of the cations and anions are identical toward their nearest neighbors but differ at the second-nearest neighbors.

The nickel-arsenide structure

The **nickel-arsenide structure** (NiAs, Fig. 4.16) is also based on an expanded, distorted hcp anion array, but the Ni atoms now occupy the octahedral holes and each As atom lies at the center of a trigonal prism of Ni atoms. It is also the structure of NiS, FeS, and a number of other sulfides. The structure is typical of MX compounds that contain soft cations in combination with soft anions, which suggests that the nickel-arsenide structure is favored by covalence. It should be noted, however, that in no compounds with this structure do the anions actually form a regular hcp array because the layers are drawn together as a result of, or to allow, metal–metal bonding.

The rutile structure

The **rutile structure**, which is shown in Fig. 4.17, takes its name from rutile, a mineral form of titanium(IV) oxide, TiO_2. It is also an example of an hcp anion lattice, but now the cations occupy only half the octahedral holes. This arrangement results in a tetragonal structure that reflects the strong tendency of a Ti atom to acquire octahedral coordination. The structure consists of TiO_6 octahedra, with the O atoms shared by neighboring Ti atoms. Each Ti atom is surrounded by six O atoms and each O atom is surrounded by three Ti atoms;

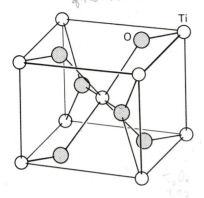

hcp with half of octahedral holes filled with cations

4.17 The rutile structure. Rutile itself is one polymorph of TiO_2.

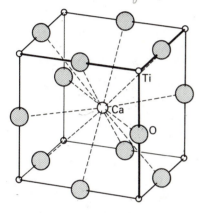

CN of Ti = 8
CN of Ca = 12
CN of O = 6

4.18 The perovskite structure. Perovskite itself is $CaTiO_3$. In general, Ca is A and Ti is B in an ABX_3 solid.

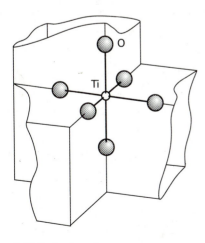

4.19 The local coordination environment of a Ti atom in perovskite.

hence the rutile structure has (6,3)-coordination. The principal ore of tin, cassiterite (SnO_2), has the rutile structure, as do a number of fluorides (Table 4.2).

The perovskite structure

The mineral perovskite $CaTiO_3$ has a structure that is the prototype of many ABX_3 solids (Table 4.2), particularly oxides. In its ideal form (Fig. 4.18), the **perovskite structure** is cubic with the A atoms surrounded by 12 O atoms and the B atoms surrounded by 6 O atoms. The sum of the charges on the A and B ions must be 6, but that can be achieved in several ways ($A^{2+}B^{4+}$ and $A^{3+}B^{3+}$ among them), including the possibility of mixed oxides of formula $A(B_{0.5}B'_{0.5})O_3$, as in $La(Ni_{0.5}Ir_{0.5})O_3$. The perovskite structure is closely related to the materials that show interesting electrical properties, such as piezo-electricity, ferroelectricity, and high-temperature superconductivity (see Chapter 18).

Example 4.2: *Interpreting a prototype structure*

What is the coordination number of a Ti atom in perovskite?

Answer. We need to imagine eight of the unit cells in Fig. 4.18 stacked together with a Ti atom shared by them all. A local fragment of the structure is shown in Fig. 4.19; we can see at once that there are six O atoms around the central Ti atom; so the coordination number of Ti in perovskite is 6.

Exercise E4.2. What is the coordination number of a Ca atom in perovskite?

4.6 The rationalization of structures

It was noted in Chapter 2 that the electronic structures of all solids can be expressed in terms of delocalized orbitals that form nearly continuous bands. However, as was remarked in Section 2.9, when those bands are full, the localized bonding description is equivalent and may be more convenient. In the case of ionic solids, the filled band (that composed largely of anion orbitals; recall the discussion of the band structure of NaCl in Section 2.9) is often very well approximated by electrons in the orbitals of the anions. The empty band (which is largely composed of cation orbitals), represents the vacancies in the cation orbitals. That is, in the conventional, localized picture, the valence electrons are regarded as having been transferred from the more electropositive atoms to the more electronegative atoms to form cations and anions, respectively.

The thermodynamic properties of such ionic solids can be treated very simply in terms of the ionic model. However, a model of a solid in terms of charged spheres interacting coulombically is likely to be quite crude, and we should expect significant departures from its predictions because many solids are more covalent than ionic. Even conventional 'good' ionic solids, such as the alkali metal halides, have significant covalent character. Nevertheless, the ionic model provides an attractively simple and effective scheme for correlating many properties.

Ionic radii

A difficulty that confronts us at the outset is the meaning of the term **ionic radius**. As remarked in Section 1.9, it is necessary to apportion the single internuclear separation of nearest-neighbor ions between the two different species (for example, an Na^+ ion and a Cl^- ion in contact).[1] The most direct way to solve the problem is to make an assumption about the radius of one ion, and then to use that value to compile a set of self-consistent values for all other ions. The O^{2-} ion has the advantage of being found in combination with a wide range of elements. It is also reasonably hard (that is, unpolarizable), so its size does not vary much as the identity of the accompanying cation is changed. In a number of compilations, therefore, the values are based on $r(O^{2-}) = 1.40$ Å. This value not only produces a consistent set of radii but also satisfies a number of theoretical criteria proposed by Linus Pauling. However, this value is by no means sacrosanct: a set of values compiled by Goldschmidt was based on $r(O^{2-}) = 1.32$ Å, and J. C. Slater even published radii that were self-consistent but used the value $r(O^{2-}) = 0.60$ Å.

The lesson at this stage is that for certain purposes (such as for predicting the sizes of unit cells and for predicting ion replacements in mineralogy) ionic radii can be helpful, but they are reliable only if they are all based on the same fundamental choice (such as the value 1.40 Å for O^{2-}). *It is very dangerous to mix values from different sources without verifying that they are based on the same convention.*

An additional complication that was first noted by Goldschmidt is that ionic radii increase with coordination number (Fig. 4.20, as happens with the atomic radius too). Hence, when comparing ionic radii, we should compare like with like, and use values for a single coordination number (typically 6).

The problems of the early workers have only partly been resolved by developments in X-ray diffraction. Thus, it is now possible to measure the electron density between two neighboring ions, locate the minimum, and identify that with the boundary between the two ions. However, as we see from Fig. 4.21, the electron density passes through a very broad minimum, and its exact location may be very sensitive to experimental uncertainties and to the identities of the two neighbors. That being so, and in the general spirit of inorganic chemistry, it is still probably more useful to express the sizes of ions in a self-consistent manner than to seek calculated values of individual radii in certain combinations. Very extensive lists of self-consistent values, which have

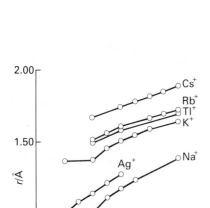

4.20 The variation of ionic radius with coordination number.

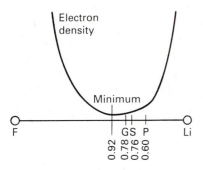

4.21 The variation in electron density along the Li—F axis in LiF. The point P indicates the Pauling radii of the ions, G the original (1927) Goldschmidt radii, and S the Shannon radii.

---

[1] A helpful survey of the history of the problem of defining and measuring ionic radii has been given by R. D. Shannon and C. T. Prewitt, *Acta Crystallogr.*, **B25**, 925 (1969).

Table 4.3 Ionic radii (in Å)*

Li⁺	Be²⁺	B³⁺			O²⁻	F⁻
0.59(4)	0.27(4)	0.12(4)			1.35(2)	1.28(2)
0.76(6)					1.38(4)	1.31(4)
					1.40(6)	1.33(6)
Na⁺	**Mg²⁺**	**Al³⁺**			1.42(8)	**Cl⁻**
0.99(4)	0.49(4)	0.39(4)				1.67(6)
1.02(6)	0.72(6)	0.53(6)				
1.16(8)	0.89(8)					
K⁺	**Ca²⁺**	**Ga³⁺**				
1.38(6)	1.00(6)	0.62(6)				
1.51(8)	1.12(8)					
1.59(10)	1.28(10)					
1.60(12)	1.35(12)					
Rb⁺	**Sr²⁺**	**In³⁺**	**Sn²⁺**	**Sn⁴⁺**		
1.49(6)	1.16(6)	0.79(6)		0.69(6)		
1.60(8)	1.25(8)	0.92(8)	1.22(8)			
1.73(12)	1.44(12)					
Cs⁺	**Ba²⁺**					
1.67(6)	1.49(6)					
1.74(8)	1.56(8)					
1.88(12)	1.75(12)					

*Numbers in parentheses are the coordination number of the ion.
Source: R. D. Shannon, *Acta Crystallogr.,* **A32**, 751 (1976). The radius of the six-coordinate NH_4^+ ion is approximately 1.46 Å.

been compiled by analyzing X-ray data on thousands of compounds, particularly oxides and fluorides, are now available, and some are given in Table 4.3.

The general trends for ionic radii are the same as for atomic radii. Thus

1. Ionic radii increase on going down a group:

    $$Li^+ < Na^+ < K^+ < Rb^+ < Cs^+$$

    (The lanthanide contraction, Section 1.9, restricts the increase among the heaviest ions of the 5*d* metal ions.)

2. The radii of ions of the same charge decrease across a period:

    $$Ca^{2+} > Mn^{2+} > Zn^{2+}$$

3. When an ion can occur in environments with different coordination numbers, its radius increases as the coordination number increases:

    $$4 < 6 < 8 < 10 < 12$$

4. If an element can exist in several different oxidation states, then for a given coordination number its ionic radius decreases with increasing oxidation number:

    $$Fe^{2+} > Fe^{3+}$$

5. Because a positive charge indicates a reduced number of electrons, and hence a more dominant nuclear attraction, cations are usually smaller than anions.

The radius ratio

A parameter that figures quite widely in the literature of inorganic chemistry, particularly in introductory texts, is the **radius ratio**, ρ, of the ions. The radius ratio is the ratio of the radius of the smaller ion ($r_<$) to that of the larger ($r_>$):

$$\rho = \frac{r_<}{r_>}$$

In most cases, $r_<$ is the cation radius and $r_>$ is the anion radius. The minimum radius ratio that can tolerate a given coordination number is then calculated by considering the geometrical problem of packing together spheres of different sizes. The results are listed below.

Coordination number	Radius ratio	Diagram
8	> 0.7	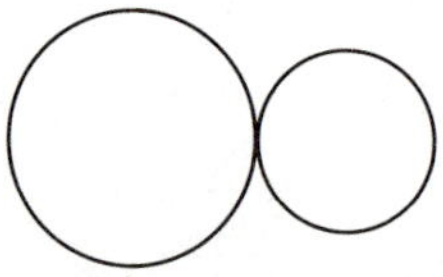 **2**
6	0.4–0.7	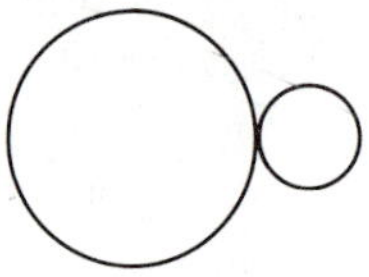 **3**
4	0.2–0.4	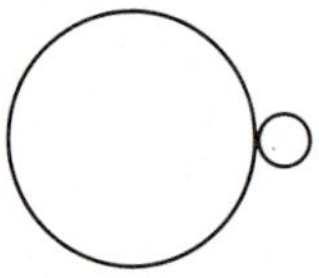 **4**
3	0.1–0.2	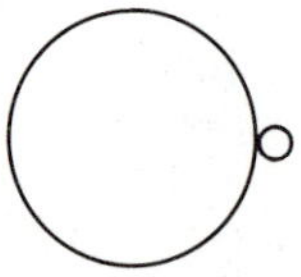 **5**

It is argued that if the radius ratio falls below the minimum given, then ions of opposite charge will not be in contact and ions of like charge will touch. According to a simple electrostatic argument, the lower coordination number, in which the contact of oppositely charged ions is restored, then becomes favorable. As the ionic radius of the M^+ ion increases, more anions can pack around it, as we have seen for CsCl with (8,8) coordination compared with NaCl and its (6,6) coordination.

The radius ratio has been used to predict the structure that some compounds are likely to adopt, but ionic radii are so arbitrary (and vary with coordination number), that what success the approach has had is probably coincidental. The relations are least reliable for the simplest structures, such as the alkali metal halides and the alkaline earth metal oxides. They are most reliable for complex fluorides and oxides and for the (very ionic) oxoanion salts. The radius ratio rules

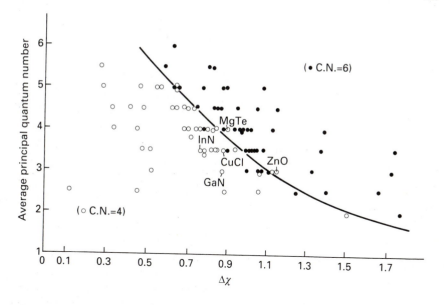

4.22 A structure map for compounds of formula MX. A point is defined by the electronegativity difference between anion and cation $\Delta\chi$ and the average principal quantum number n. Its location in the map indicates the coordination number expected for that pair of properties. From E. Mooser and W. B. Pearson, *Acta Crystallogr.*, **12**, 1015 (1959).

are probably a quantitative over-elaboration of the observation that large cations generally have large coordination numbers.

Structure maps

Granted that the use of radius ratios is hazardous, it is still possible to make progress with the rationalization of structures by collecting enough information empirically and looking for a pattern in it. This attitude has motivated the compilation of **structure maps**. A structure map is an empirically compiled map that depicts the dependence of crystal structure on the electronegativity difference $\Delta\chi$ between the elements involved and the average principal quantum number of the valence shells of the two atoms.[2] Because the ionic character of a bond increases with $\Delta\chi$, moving from left to right along the horizontal axis corresponds to an increase in ionic character in the bonding. Moreover, because the principal quantum number is an indication of the radius of an ion, traveling up the vertical axis corresponds to an increase in the average radius of the ions. Because atomic energy levels also become closer as the atom expands, the softness of the atom increases too (see the discussion in Section 1.13). Consequently the vertical axis of a structure map corresponds to increasing size and softness of the bonded atoms. In summary:

> A structure map is a representation of the variation in crystal structure with the character of the bonding, the electronegativity difference horizontally and the average softness vertically.

Figure 4.22 is an example of a structure map for MX compounds and Fig. 4.23 is an example of one for MX_2 compounds. We see from Fig. 4.22 that the structures we have been discussing fall in quite

---

[2] E. Mooser and W. B. Pearson, *Acta Crystallogr.*, **12**, 1015 (1959).

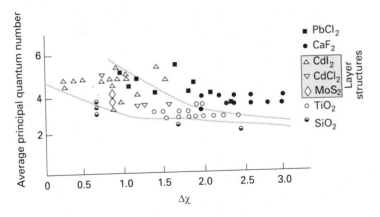

4.23 A structure map for compounds of formula MX$_2$. From E. Mooser and W. B. Pearson, *Acta Crystallogr.*, **12**, 1015 (1959).

distinct regions of the map. Elements with large $\Delta\chi$ have (6,6)-coordination, such as is found in the rock-salt and rutile structures; elements with small $\Delta\chi$ (and hence where there is the expectation of covalence) have a lower coordination number. In terms of a structure map representation, GaN is in a more covalent region of Fig. 4.22 than ZnO because $\Delta\chi$ is appreciably smaller.

Example 4.3: *Using a structure map*
What type of crystal structure should be expected for magnesium sulfide, MgS?

Answer. The electronegativities of magnesium and sulfur are 1.3 and 2.6 respectively, so $\Delta\chi = 1.3$. The average principal quantum number is 3 (both elements are in Period 3). The point $\Delta\chi = 1.3$, $n = 3$ lies just in the C.N. = 6 region of the structure map. This location is consistent with the observed rock-salt structure of MgS.

Exercise E4.3. Predict the coordination environment of rubidium chloride, RbCl.

4.7 Lattice enthalpies

The criterion of stability of a crystal lattice under conditions of constant temperature and pressure, and hence the thermodynamic criterion of why one structure is adopted rather than another, is the Gibbs free energy of lattice formation:

$$M^+(g) + X^-(g) \rightarrow MX(s) \quad \Delta G^\ominus = \Delta H^\ominus - T\Delta S^\ominus$$

If $\Delta G^\ominus$ is more negative for the formation of a structure A rather than B, then the transition from B to A is spontaneous under the prevailing conditions, and we can expect the solid to be found with structure A.

The process of lattice formation from the gas of ions is so exothermic that at and near room temperature the contribution of the entropy may be neglected (this is rigorously true at $T = 0$). Hence, discussions of the thermodynamic properties of crystal lattices normally focus, initially at least, on the lattice enthalpy. That being so, we look for the structure that is formed most exothermically and identify it as the thermodynamically most stable form.

Lattice enthalpy

The **lattice enthalpy**, $\Delta H_L^{\ominus}$, is the standard enthalpy change accompanying the formation of a gas of ions from the solid:

$$MX(s) \rightarrow M^+(g) + X^-(g) \quad \Delta H_L^{\ominus}$$

Because lattice disruption is always endothermic, lattice enthalpies are always positive. If entropy considerations are neglected, the most stable crystal structure of the compound is the structure with the greatest lattice enthalpy under the prevailing conditions.

Lattice enthalpies are determined from enthalpy data by using a **Born–Haber cycle**, such as that shown in Fig. 4.24. In this special case of a thermodynamic cycle, a cycle (a closed path) of steps is formed that includes lattice formation as one stage. The standard enthalpy of decomposition of a compound into its elements is the negative of its standard enthalpy of formation, $\Delta H_f^{\ominus}$, the standard enthalpy of lattice formation is the negative of the lattice enthalpy. For a solid element, the standard enthalpy of atomization, $\Delta H^{\ominus}$(atomization), is the standard enthalpy of sublimation, as in the process

$$K(s) \rightarrow K(g) \quad \Delta H^{\ominus} = +89 \text{ kJ}$$

For a gaseous element, the standard enthalpy of atomization is the standard enthalpy of dissociation, as in

$$Cl_2(g) \rightarrow 2Cl(g) \quad \Delta H^{\ominus} = +244 \text{ kJ}$$

The standard enthalpy of ionization is the ionization enthalpy (for the formation of cations) and the electron-gain enthalpy (for anions). Two examples are

$$K(g) \rightarrow K^+(g) + e^-(g) \quad \Delta H^{\ominus} = +425 \text{ kJ}$$

$$Cl(g) + e^-(g) \rightarrow Cl^-(g) \quad \Delta H^{\ominus} = -355 \text{ kJ}$$

The value of the lattice enthalpy—the only unknown in a well-chosen cycle—is found from the requirement that the sum of the enthalpy changes round a complete cycle is zero (because enthalpy is a state property):

$$\Delta H^{\ominus}(\text{atomization}) + \Delta H^{\ominus}(\text{ionization}) + (-\Delta H_L^{\ominus}) + (-\Delta H_f^{\ominus}) = 0$$

Therefore,

$$\Delta H_L^{\ominus} = \Delta H^{\ominus}(\text{atomization}) + \Delta H^{\ominus}(\text{ionization}) - \Delta H_f^{\ominus}$$

Example 4.4: *Using a Born–Haber cycle to determine a lattice enthalpy*
Calculate the lattice enthalpy of KCl(s) using a Born–Haber cycle and the following information:

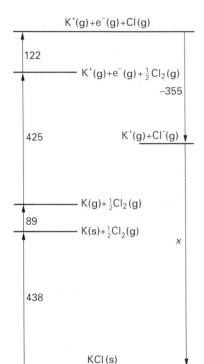

4.24 A Born–Haber cycle for KCl. The lattice enthalpy is equal to $-x$.

	$\Delta H^{\ominus}/(\text{kJ mol}^{-1})$
Sublimation of K(s)	+89
Ionization of K(g)	+425
Dissociation of Cl_2(g)	+244
Electron attachment to Cl(g)	−355
Formation of KCl(s)	−438

Answer. The required cycle is shown in Fig. 4.24. The first step is the sublimation of solid potassium: $\Delta H^{\ominus}/(\text{kJ mol}^{-1})$

$$K(s) \rightarrow K(g) \hspace{6cm} +89$$

followed by its ionization:

$$K(g) \rightarrow K^+(g) + e^-(g) \hspace{4.5cm} +425$$

Chloride ions are formed by dissociation of Cl_2:

$$\tfrac{1}{2}Cl_2(g) \rightarrow Cl(g) \hspace{5.5cm} +122$$

followed by electron gain:

$$Cl(g) + e^-(g) \rightarrow Cl^-(g) \hspace{4.5cm} -355$$

The solid is now formed:

$$K^+(g) + Cl^-(g) \rightarrow KCl(s) \hspace{3.5cm} -\Delta H_L^{\ominus}$$

and the cycle is completed by decomposing KCl(s) into its elements:

$$KCl(s) \rightarrow K(s) + \tfrac{1}{2}Cl_2(g) \hspace{4cm} +438$$

(the reverse of its formation). The sum of the enthalpies is $-\Delta H_L^{\ominus} + 719\,\text{kJ mol}^{-1}$; however, the sum must be equal to zero, so $\Delta H_L^{\ominus} = 719\,\text{kJ mol}^{-1}$.

Exercise E4.4. Calculate the lattice enthalpy of magnesium bromide from the following data:

	$\Delta H^{\ominus}/(\text{kJ mol}^{-1})$
Sublimation of Mg(s)	+148
Ionization of Mg(g) to Mg^{2+}(g)	+2187
Vaporization of Br_2(l)	+31
Dissociation of Br_2(g)	+193
Electron attachment to Br(g)	−331
Formation of $MgBr_2$(s)	−524

Once the lattice enthalpy is known, it can be used to judge the character of the bonding in the solid. If the value calculated on the assumption that the lattice consists of ions interacting coulombically is in good agreement with the measured value, then it may be appropriate to adopt a largely ionic model of the compound. A discrepancy indicates a degree of covalence. As we warned earlier, numerical coincidences can be misleading in this assessment, so the character of bonding in a solid is not proved by lattice enthalpy considerations.

Coulombic contributions to lattice enthalpies

To calculate the lattice enthalpy of a supposedly ionic solid we need to take into account several contributions to its energy, including the attraction between the ions and their repulsions. The kinetic energy of the vibrating atoms (apart from its contribution to the zero-point energy of the crystal) can be neglected if it is supposed that the solid is at the absolute zero of temperature.

The total coulombic potential energy of a crystal is the sum of the individual coulomb potential energy terms of the form

$$V_{AB} = \frac{(z_A e) \times (z_B e)}{4\pi\varepsilon_0 r_{AB}}$$

for ions of charge numbers z_A and z_B (with cations having positive charge numbers and anions negative charge numbers) separated by a distance r_{AB} (ε_0 is the vacuum permittivity, 8.85×10^{-12} C^2 J^{-1} m^{-1}; in a precise calculation, the actual permittivity of the medium should be used). The sum over all the pairs of ions in the solid may be carried out for any crystal structure, although in practice it converges very slowly because nearest neighbors contribute a large negative term, second-nearest neighbors an only slightly weaker positive term, and so on. The overall result is that the attraction between the cations and anions predominates and yields a favorable (negative) contribution to the energy of the solid.

For example, in a uniformly spaced one-dimensional line of alternating cations and anions with $z_A = +z$ and $z_B = -z$, the interaction of one ion with all the others is proportional to

$$-\frac{2z^2}{d} + \frac{2z^2}{2d} - \frac{2z^2}{3d} + \frac{2z^2}{4d} - \ldots = -\frac{2z^2}{d}(1 - \tfrac{1}{2} + \tfrac{1}{3} - \tfrac{1}{4} + \ldots)$$
$$= -\frac{2z^2}{d}\ln 2$$

(The factor of 2 comes from the fact that the same ions occur on both sides of the central ion.) As in this case, it is found that, apart from the explicit appearance of the ion charge numbers, the value of the sum depends only on the type of the lattice and on a single scale parameter which may be taken as the separation of the centers of nearest neighbors, d. We may write

$$V = -\frac{e^2}{4\pi\varepsilon_0} \times \frac{z^2}{d} \times 2\ln 2$$

The first factor is a collection of fundamental constants. The second is specific to the identities of the ions and to the scale of the lattice. The third term, $2\ln 2 = 1.386$, characterizes the symmetry of the lattice (in this case a straight line of ions) and is the simplest example of a **Madelung constant**, A. In simple solids, the Madelung constant is specific to the crystal type and independent of the interionic distances.

In general, the total potential energy per mole of formula units in an arbitrary crystal structure is

$$V = N_A \frac{e^2}{4\pi\varepsilon_0}\left(\frac{z_A z_B}{d}\right) \times A$$

where N_A is Avogadro's constant (in this case, the number of formula units per mole) and z_A and z_B are the charge numbers of the ions: because the charge number of cations is positive and that of anions is negative, V is negative overall, corresponding to a lowering in potential energy relative to the gas of widely separated ions. Some computed values of the Madelung constant for a variety of lattices are given in Table 4.4. The general trend is for the values to increase with coordination number. This trend reflects the fact that a large contribution comes from nearest neighbors, and such neighbors are more numerous when the coordination number is large. The values for the rock-salt structure (C.N. = 6) and for the cesium-chloride structure (C.N. = 8) illustrate the trend. However, a high coordination number does not necessarily mean that the interactions are stronger in the cesium-chloride structure, for the potential energy also depends on the scale of the lattice. Thus, d may be so large in lattices with ions big enough to adopt eightfold coordination that it overcomes the small increase in Madelung constant and results in a less favorable potential energy.

Another contribution to the lattice enthalpy arises from the **van der Waals attractions** between the ions and molecules, the weak intermolecular interactions that are responsible for the formation of condensed phases of neutral species. The dominant contribution of this kind is often the **dispersion interaction** (the 'London interaction'). The dispersion interaction arises from the transient fluctuations in electron density (and, consequently, instantaneous electric dipole moment) on one molecule driving a fluctuation in electron density (and dipole moment) on a neighboring molecule, and the attractive interaction between these two instantaneous electric dipoles. The potential energy of this interaction is inversely proportional to the sixth power of the separation, so in the crystal as a whole we can expect it to vary as the sixth power of the scale of the lattice:

$$V = -\frac{N_A C}{d^6}$$

The constant C depends on the substance. For ions of low polarizability, this contribution is only about 1 percent of the coulombic contribution and is ignored in elementary lattice enthalpy calculations of ionic solids.

Repulsions arising from overlap

When two closed-shell ions are in contact, another contribution to the total potential energy is the repulsion arising from the overlap of their electron distributions. Because orbitals decay exponentially toward zero at large distances from the nucleus, and repulsive interactions depend on the overlap of orbitals, it is plausible that their contribution

Table 4.4 Madelung constants*

Structural type	A
Cesium chloride	1.763
Fluorite	2.519
Rock salt	1.748
Rutile	2.408
Sphalerite	1.638
Wurtzite	1.641

*The values given are for the geometric factor A described in the text (Section 4.7). Some sources cite values that include the charge numbers of the ions (so, for instance, the value for CaF_2 is quoted as 5.039), and it is necessary to verify the definition before using them.

to the potential energy has the form

$$V = +N_A C' e^{-d/d*}$$

where C' and $d*$ are constants. We shall see in a moment that C' cancels and need not be known. The constant $d*$ can be estimated from measurements of compressibilities, which reflect the increase in potential energy that occurs when ions are pressed together by an applied force. Although the values of $d*$ measured in this way actually span a range, it is often found that setting $d* = 0.345$ Å gives reasonable agreement with experiment.

The Born-Mayer equation

The contribution of the attractive and repulsive components to the total potential energy of a solid is depicted by the curve in Fig. 4.25. The total potential energy passes through a minimum at the equilibrium lattice parameter of the solid, which can be found by looking for a minimum in the expression

$$V = N_A \frac{e^2}{4\pi\varepsilon_0} \left(\frac{z_A z_B}{d} \right) \times A + N_A C' e^{-d/d*}$$

The minimum potential energy is obtained where $dV/dd = 0$, which occurs when

$$N_A C' e^{-d/d*} = -N_A \frac{z_A z_B e^2}{4\pi\varepsilon_0} \times \frac{Ad*}{d^2}$$

Substitution of this relation into the preceding one results in the **Born–Mayer equation**:

$$V = \frac{N_A z_A z_B e^2}{4\pi\varepsilon_0 d} \left(1 - \frac{d*}{d} \right) A$$

The experimental scale of the lattice may now be used for d. At $T = 0$, where there is no contribution from the kinetic energy, we can identify this minimum potential energy as the molar internal energy of the crystal relative to the energy of the widely separated ions. It follows that the negative of V may be identified as the lattice enthalpy (strictly, its value at $T = 0$ with the zero-point vibrational energy of the lattice ignored).

As remarked previously, the agreement between the experimental lattice enthalpy and the value calculated using the ionic model of the solid (in practice, from the Born–Mayer equation) is used as a measure of the extent to which the solid is ionic. Some calculated and measured lattice enthalpies are given in Table 4.5.

As a general guide, the ionic model is reasonably valid if $\Delta\chi$ for the neutral atoms is more than about 2, and the bonding is thought of as predominantly covalent if $\Delta\chi$ is less than about 1. However, it should be remembered that the electronegativity criterion ignores the role of hardness of the ions that result from electron transfer between the atoms. Thus, the alkali metal halides give quite good agreement with

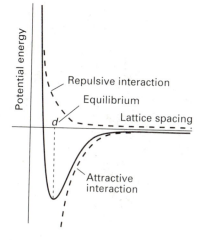

4.25 Contributions to the potential energy of ions in a crystal. The attractive interaction scales as $1/d$ where d represents the dimensions of the lattice. The repulsive interaction has a shorter range, but becomes strong very rapidly once the ions are in contact. The total potential energy, the sum of these two contributions, passes through a minimum that corresponds to the equilibrium dimension of the lattice.

Table 4.5 Measured and calculated lattice enthalpies

Compound*	$\Delta H_L/(kJ\ mol^{-1})$		Calc/expt (percent)
	(calc)	(expt)	
LiF[a]	1033	1037	99.6
LiCl[a]	845	852	99.2
LiBr[a]	798	815	97.9
LiI[a]	740	761	97.2
CsF[a]	748	750	99.7
CsCl[b]	652	676	96.4
CsBr[b]	632	654	96.6
CsI[b]	601	620	96.9
AgF[a]	920	969	94.9
AgCl[a]	833	912	91.3
AgBr[a]	816	900	90.7

[a] Rock-salt structure; [b] cesium-chloride structure.
Source: D. Cubicciotti, *J. Chem. Phys.*, **31**, 1646 (1951).
The calculated values use a more complete ionic model that includes terms beyond those in the Born–Mayer equation.

the ionic model, the best with the hard halide ions (F^-) formed from the highly electronegative F atom, and the worst with the soft halide ions (I^-) formed from the less electronegative I atom. It is not clear whether it is the electronegativity of the atoms or the hardness of the resultant ions that should be used as a criterion. The worst agreement with the ionic model is for soft cation–soft anion combinations, which are likely to be substantially covalent. Here again, though, the difference between the electronegativities of the parent elements is small and it is not clear whether electronegativity or hardness provides the better criterion.

The Kapustinskii equation

The Russian chemist A. F. Kapustinskii noticed that if the Madelung constants for a number of structures are divided by the number of ions per formula unit n, then approximately the same value is obtained for them all. Moreover, he also noticed that the order of values so obtained increases with the coordination number. Therefore, because ionic radius also increases with coordination number, the variation in A/nd from one structure to another can be expected to be quite small. This observation led Kapustinskii to propose that there exists a hypothetical rock-salt structure that is energetically equivalent to the true structure of any ionic solid. That being so, the lattice enthalpy can be calculated by using the Madelung constant and the appropriate ionic radii for (6,6)-coordination. The resulting expression is called the **Kapustinskii equation**:

$$\Delta H_L = -\frac{n z_A z_B}{d}\left(1 - \frac{d^*}{d}\right)K$$

In this equation, $d = r_A + r_B$, $K = 1.21$ MJ Å mol^{-1}, and n is the number of ions in each formula unit.

Table 4.6 The thermochemical radii of ions, $r/\text{Å}$

Main-group elements

BeF_4^{2-}	BF_4^-	CO_3^{2-}	NO_3^-	OH^-			
2.45	2.28	1.85	1.89	1.40			
		CN^-	NO_2^-	O_2^-			
		1.82	1.55	1.80			
			PO_4^{3-}	SO_4^{2-}	ClO_4^-	CrO_4^{2-}	MnO_4^-
			2.38	2.30	2.36	2.30	2.40
			AsO_4^{3-}	SeO_4^{2-}		MoO_4^{2-}	
			2.48	2.43		2.54	
			SbO_4^{3-}	TeO_4^{2-}	IO_4^-		
			2.60	2.54	2.49		
					IO_3^-		
					1.82		

Complex ions

$[TiCl_6]^{2-}$	$[IrCl_6]^{2-}$	$[SiF_6]^{2-}$	$[GeCl_6]^{2-}$
2.48	2.54	1.94	2.43
$[TiBr_6]^{2-}$	$[PtCl_6]^{2-}$	$[GeF_6]^{2-}$	$[SnCl_6]^{2-}$
2.61	2.59	2.01	2.47
$[ZrCl_6]^{2-}$			$[PbCl_6]^{2-}$
2.47			2.48

Source: A. F. Kapustinskii, *Q. Rev. Chem. Soc.*, **20**, 203 (1956).

The Kapustinskii equation can be used to ascribe a meaning to the 'radii' of nonspherical molecular ions, for their values can be adjusted until the calculated value of the lattice enthalpy matches that obtained experimentally from the Born–Haber cycle. The self-consistent set of parameters obtained in this way are called **thermochemical radii** (Table 4.6). They may be used to estimate lattice enthalpies, and hence enthalpies of formation, of a wide range of compounds.

Example 4.5: *Using the Kapustinskii equation*
Estimate the lattice enthalpy of potassium nitrate, KNO_3.

Answer. To use the Kapustinskii equation we need the number of ions per formula unit ($n=2$), their charge numbers $z(K^+) = +1$ and $z(NO_3^-) = -1$, and the sum of their thermochemical radii ($1.38\text{ Å} + 1.89\text{ Å} = 3.27\text{ Å}$). For monatomic species we use the ionic radii in Table 4.3. Then, with $d^* = 0.345\text{ Å}$

$$\Delta H_L = -\frac{2(+1)(-1)}{3.27\text{ Å}} \times \left(1 - \frac{0.345\text{ Å}}{3.27\text{ Å}}\right) \times 1.21\text{ Å MJ mol}^{-1}$$

$$= 662\text{ kJ mol}^{-1}$$

Exercise E4.5. Evaluate the lattice enthalpy of calcium sulfate, $CaSO_4$.

4.8 Consequences of lattice enthalpies

The Born–Mayer equation shows that for a given lattice type (a given value of A), the lattice enthalpy increases with increasing ion charge numbers (as $|z_A z_B|$). The lattice enthalpy also increases as the scale d

of the lattice decreases. Energies that vary as the **electrostatic parameter**, ξ, where

$$\xi = \frac{z^2}{d}$$

z being an ion charge number and d a scale length, are widely adopted in inorganic chemistry as indicative that an ionic model is appropriate.[3] In this chapter we consider three consequences of lattice enthalpy and its correlation with the electrostatic parameter.

Thermal stabilities of ionic solids

The particular aspect we consider here is the difference in the temperatures needed to bring about thermal decomposition of carbonates. Magnesium carbonate, for instance, decomposes when heated to about 300 °C whereas calcium carbonate decomposes only if the temperature is raised to over 800 °C. In general

> Large cations stabilize large anions (and vice versa); in particular, the decomposition temperatures of thermally unstable compounds (containing large anions, such as carbonates) increases with cation radius.

The aim of the following discussion is to demonstrate that the stabilizing influence of a cation on an unstable anion can be explained in terms of trends in lattice enthalpies. First, we note that the decomposition temperatures of solid inorganic compounds can be discussed in terms of their Gibbs free energies of decomposition into specified products. In many cases it is sufficient to consider only the reaction enthalpy, for the reaction entropy is almost constant for the substances being compared. Moreover, as we shall see, the calculations center on differences of lattice enthalpy between the solid reactants and the solid products, for these differences dominate the reaction enthalpy. The differences in lattice enthalpy in turn may be discussed, at least in a general way, in terms of the Kapustinskii equation. Hence the variation in stability may be correlated with the variation of the electrostatic parameter ξ.

As an illustration of what is involved, consider the thermal decomposition of carbonates:

$$MCO_3(s) \rightarrow MO(s) + CO_2(g)$$

The experimental observation that the decomposition temperature increases with increasing cation radius can be expressed in terms of the reaction free energy: the temperature at which $\Delta G^{\ominus}$ becomes negative and the reaction favorable, increases with cation radius.

---

[3] The correlation of properties with ξ is a very useful guide for many properties, but we should not take it as proof of the dominance of specifically charge–charge interactions. Often there is a correlation of a property with a wide range of parameters such as z^2/r^2 or z/r: indeed, a correlation can often be found with almost any expression with charge number in the numerator and radius in the denominator.

Table 4.7 Decomposition data on carbonates*

	Mg	Ca	Sr	Ba
$\Delta G^{\ominus}/(\text{kJ mol}^{-1})$	+ 48.3	+ 130.4	+ 183.8	+ 218.1
$\Delta H^{\ominus}/(\text{kJ mol}^{-1})$	+ 100.6	+ 178.3	+ 234.6	+ 269.3
$\Delta S^{\ominus}/(\text{J K}^{-1}\text{ mol}^{-1})$	+ 175.0	+ 160.6	+ 171.0	+ 172.1
$\theta/°\text{C}/(\text{J K}^{-1}\text{ mol}^{-1})$	300	840	1100	1300

*Data are for the reaction

$$MCO_3(s) \rightarrow MO(s) + CO_2(g) \quad \Delta X^{\ominus}(298\text{ K})$$

θ is the temperature required to reach 1 bar of CO_2, and has been estimated from the 298 K reaction enthalpy and entropy. The experimental values reported in the literature for calcium (and rubidium) are unreliable because the samples were probably damp.
Source: Gibbs free energy data calculated from NBS Thermochemical tables.

It follows from the thermodynamic relation

$$\Delta G^{\ominus} = \Delta H^{\ominus} - T\Delta S^{\ominus}$$

that the decomposition temperature is reached when

$$T = \frac{\Delta H^{\ominus}}{\Delta S^{\ominus}}$$

The decomposition entropy is almost constant for all carbonates because in each case it is dominated by the formation of gaseous carbon dioxide. Hence

> The higher the reaction enthalpy, the higher the decomposition temperature.

This trend can be verified by reference to the experimental data in Table 4.7.

The enthalpy of decomposition depends in part on the difference between the lattice enthalpy of the decomposition product MO, which is formed, and that of the parent carbonate MCO_3. The overall reaction enthalpy is positive (decomposition is endothermic), but it is less strongly positive if the lattice enthalpy of the oxide is markedly greater than that of the carbonate. Hence, the decomposition temperature will be low for oxides that have a relatively high lattice enthalpies compared to their parent carbonates. The compounds for which this is true are composed of small, highly charged cations, such as Mg^{2+}.

The pictorial explanation of why a small cation increases the lattice enthalpy of an oxide more than that of a carbonate is illustrated in Fig. 4.26. The illustration shows that the relative change in scale of the lattice is large when a compound composed of a small cation and a large anion becomes an oxide. The change in scale is relatively small when the parent compound has a large cation initially. As the illustration shows in an exaggerated way, when the cation is very big, the change in size of the anion barely affects the scale of the lattice. Therefore, the lattice enthalpy difference is more favorable to decomposition when the cation is small than when it is large.

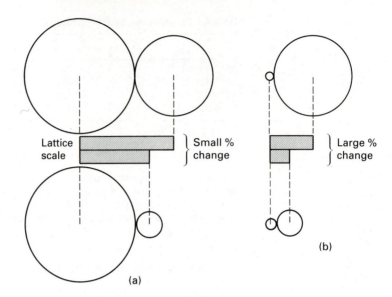

4.26 A greatly exaggerated representation of the change in lattice parameter for cations of different sizes. (a) When the anion changes size (when CO_3^{2-} decomposes to O^{2-} and CO_2, for instance), and the cation is large, the lattice scale changes by a relatively small amount. (b) If the cation is small, the relative change in lattice scale is large, and decomposition is thermodynamically more favorable.

The difference in lattice enthalpy between MO and MCO_3 is magnified by a large charge number on the cation. As a result, thermal decomposition of a carbonate will occur at lower temperatures if it contains a highly charged cation. One consequence of this dependence on charge number is that alkaline earth carbonates (M^{2+}) decompose at lower temperatures than the corresponding alkali metal carbonates (M^{+}).

Example 4.6: *Assessing the dependence of stability on ionic radius*

Present an argument to account for the fact that, when they burn in oxygen, lithium forms the oxide Li_2O but sodium forms the peroxide Na_2O_2.

Answer. Since the small Li^+ ion results in Li_2O having a more favorable lattice enthalpy than Na_2O, the decomposition reaction $M_2O_2 \rightarrow M_2O + \frac{1}{2}O_2$ is thermodynamically more favorable for Li_2O_2 than Na_2O_2.

Exercise E4.6. Predict the order of decomposition temperatures of alkaline earth metal sulfates in the reaction $MSO_4(s) \rightarrow MO(s) + SO_3(g)$.

High oxidation numbers and small anions

A similar argument can be used to account for the following general observation:

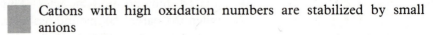

Cations with high oxidation numbers are stabilized by small anions

In particular, fluorine has a greater ability compared with the other halogens to stabilize the high oxidation states of metals. Thus, the only known halides of Ag(II), Co(III), and Mn(IV) are the fluorides.

Table 4.7 Decomposition data on carbonates*

	Mg	Ca	Sr	Ba
$\Delta G^{\ominus}/(\text{kJ mol}^{-1})$	+ 48.3	+ 130.4	+ 183.8	+ 218.1
$\Delta H^{\ominus}/(\text{kJ mol}^{-1})$	+ 100.6	+ 178.3	+ 234.6	+ 269.3
$\Delta S^{\ominus}/(\text{J K}^{-1}\text{ mol}^{-1})$	+ 175.0	+ 160.6	+ 171.0	+ 172.1
$\theta/{}^{\circ}\text{C}/(\text{J K}^{-1}\text{ mol}^{-1})$	300	840	1100	1300

*Data are for the reaction

$$MCO_3(s) \rightarrow MO(s) + CO_2(g) \quad \Delta X^{\ominus}(298\text{ K})$$

θ is the temperature required to reach 1 bar of CO_2, and has been estimated from the 298 K reaction enthalpy and entropy. The experimental values reported in the literature for calcium (and rubidium) are unreliable because the samples were probably damp.
Source: Gibbs free energy data calculated from NBS Thermochemical tables.

It follows from the thermodynamic relation

$$\Delta G^{\ominus} = \Delta H^{\ominus} - T\Delta S^{\ominus}$$

that the decomposition temperature is reached when

$$T = \frac{\Delta H^{\ominus}}{\Delta S^{\ominus}}$$

The decomposition entropy is almost constant for all carbonates because in each case it is dominated by the formation of gaseous carbon dioxide. Hence

> The higher the reaction enthalpy, the higher the decomposition temperature.

This trend can be verified by reference to the experimental data in Table 4.7.

The enthalpy of decomposition depends in part on the difference between the lattice enthalpy of the decomposition product MO, which is formed, and that of the parent carbonate MCO_3. The overall reaction enthalpy is positive (decomposition is endothermic), but it is less strongly positive if the lattice enthalpy of the oxide is markedly greater than that of the carbonate. Hence, the decomposition temperature will be low for oxides that have a relatively high lattice enthalpies compared to their parent carbonates. The compounds for which this is true are composed of small, highly charged cations, such as Mg^{2+}.

The pictorial explanation of why a small cation increases the lattice enthalpy of an oxide more than that of a carbonate is illustrated in Fig. 4.26. The illustration shows that the relative change in scale of the lattice is large when a compound composed of a small cation and a large anion becomes an oxide. The change in scale is relatively small when the parent compound has a large cation initially. As the illustration shows in an exaggerated way, when the cation is very big, the change in size of the anion barely affects the scale of the lattice. Therefore, the lattice enthalpy difference is more favorable to decomposition when the cation is small than when it is large.

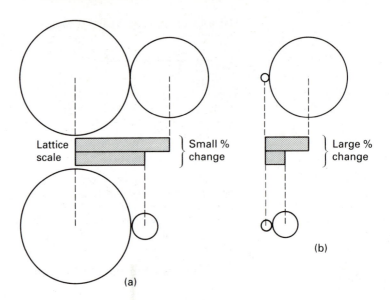

4.26 A greatly exaggerated representation of the change in lattice parameter for cations of different sizes. (a) When the anion changes size (when CO_3^{2-} decomposes to O^{2-} and CO_2, for instance), and the cation is large, the lattice scale changes by a relatively small amount. (b) If the cation is small, the relative change in lattice scale is large, and decomposition is thermodynamically more favorable.

The difference in lattice enthalpy between MO and MCO_3 is magnified by a large charge number on the cation. As a result, thermal decomposition of a carbonate will occur at lower temperatures if it contains a highly charged cation. One consequence of this dependence on charge number is that alkaline earth carbonates (M^{2+}) decompose at lower temperatures than the corresponding alkali metal carbonates (M^+).

Example 4.6: *Assessing the dependence of stability on ionic radius*
Present an argument to account for the fact that, when they burn in oxygen, lithium forms the oxide Li_2O but sodium forms the peroxide Na_2O_2.

Answer. Since the small Li^+ ion results in Li_2O having a more favorable lattice enthalpy than Na_2O, the decomposition reaction $M_2O_2 \rightarrow M_2O + \frac{1}{2}O_2$ is thermodynamically more favorable for Li_2O_2 than Na_2O_2.

Exercise E4.6. Predict the order of decomposition temperatures of alkaline earth metal sulfates in the reaction $MSO_4(s) \rightarrow MO(s) + SO_3(g)$.

High oxidation numbers and small anions

A similar argument can be used to account for the following general observation:

> Cations with high oxidation numbers are stabilized by small anions

In particular, fluorine has a greater ability compared with the other halogens to stabilize the high oxidation states of metals. Thus, the only known halides of Ag(II), Co(III), and Mn(IV) are the fluorides.

Another sign of the decrease in stability of the heavier halides of metals with high oxidation numbers is that the iodides of Cu(II) and Fe(III) decompose at room temperature. The oxygen atom is a very effective species for drawing out the high oxidation states of atoms because not only is it small, but it can accept up to two electrons.

To explain these observations, we consider the redox reaction

$$MX + \tfrac{1}{2}X_2 \rightarrow MX_2$$

where X is a halogen. Our aim is to show why this reaction is most strongly spontaneous for $X = F$. If we ignore entropy contributions, we must show that the reaction is most exothermic for fluorine.

One contribution to the reaction enthalpy is the conversion of $\tfrac{1}{2}X_2$ to X^-. Despite fluorine having a lower electron affinity than chlorine, this step is more exothermic for $X = F$ than for $X = Cl$ on account of the lower bond enthalpy of F_2 compared with that of Cl_2. The lattice enthalpies, however, play the major role. In the conversion of MX to MX_2, the charge number of the cation increases from $+1$ to $+2$, so the lattice enthalpy increases. As the radius of the anion increases, though, this difference in the two lattice enthalpies diminishes, and the exothermic contribution to the overall reaction decreases too. Hence, both the lattice energy and the X^- formation enthalpy lead to a less exothermic reaction as the halogen changes from F to I. So long as entropy factors are similar, we expect an increase in thermodynamic stability of MX relative to MX_2 on going from $X = F$ to $X = I$ down Group 17/VII.

Solubility

Lattice enthalpies play a role in solubilities, but one that is much more difficult to analyze than for reactions. One general rule that is quite widely obeyed is that

> Compounds that contain ions with widely different radii are generally soluble in water. Conversely, the least water-soluble salts are those of ions with similar radii.

That is, in general, *difference in size favors solubility in water*. Two familiar series of compounds illustrate these trends. In gravimetric analysis, Ba^{2+} is used to precipitate SO_4^{2-}, and the solubilities of the alkaline earth metal sulfates decrease from $MgSO_4$ to $BaSO_4$. In contrast, the solubility of the alkaline earth metal hydroxides increases down the group: $Mg(OH)_2$ is the sparingly soluble 'milk of magnesia' but $Ba(OH)_2$ is used as a soluble hydroxide for preparation of solutions of OH^-. The first case shows that a large anion requires a large cation for precipitation. The second case shows that a small anion requires a small cation for precipitation. It is found empirically that an ionic compound MX tends to be most soluble when the radius of M^+ is smaller than that of X^- by about 0.8 Å.

Before attempting to rationalize the observations, we should note that the solubility of an ionic compound depends on the Gibbs free

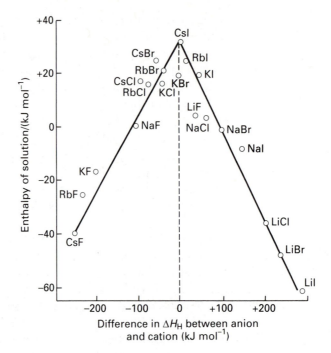

4.27 The correlation between enthalpies of solution of halides and the differences between the hydration enthalpies of the ions. Dissolution is most exothermic when the difference is large.

energy of the process

$$MX(s) \rightarrow M^+(aq) + X^-(aq)$$

In the process, the interactions responsible for the lattice enthalpy of MX are replaced by hydration (and by solvation in general) of the ions. However, the exact balance of enthalpy and entropy effects is delicate and difficult to assess, particularly because the entropy change also depends on the degree of order of the *solvent* molecules that is brought about by the presence of the dissolved solute. The data in Fig. 4.27 suggest that enthalpy considerations are important in some cases at least, for the graph shows that there is a correlation between the enthalpy of solution of a salt and the difference in hydration enthalpies of the two ions. If the cation has a larger hydration enthalpy than its anion partner (reflecting the difference in their sizes), or vice versa, then the dissolution of the salt is exothermic (reflecting the favorable solubility equilibrium).

The variation in enthalpy can be explained using the ionic model. The lattice enthalpy is inversely proportional to the distance between the ions:

$$\Delta H_L^{\ominus} \propto \frac{1}{r_+ + r_-}$$

whereas the hydration enthalpy is the sum of individual ion contributions:

$$\Delta H_H^{\ominus} \propto \frac{1}{r_+} + \frac{1}{r_-}$$

If the radius of one ion is small, then the term in the hydration enthalpy for that ion will be large. However, in the expression for the lattice enthalpy one small ion cannot make the denominator of the expression small by itself. Thus, one small ion can result in a large hydration enthalpy but not necessarily lead to a high lattice enthalpy, so ion size asymmetry can result in exothermic dissolution. However, if both ions are small, then both the lattice enthalpy and the hydration enthalpy may be large, and dissolution might not be very exothermic.

Example 4.7: *Accounting for trends in the solubility of s-block compounds*
What is the trend in the solubilities of the alkaline earth metal carbonates?

Answer. The CO_3^{2-} anion has a large radius and is of the same charge type (2) as the cations M^{2+} of the Group 2 elements. The least soluble carbonate of the group is predicted to be that of the largest cation, Ra^{2+}. The most soluble is expected to be the carbonate of the smallest cation, Mg^{2+}. Although magnesium carbonate is more soluble than radium carbonate, it is still only sparingly soluble: its solubility constant (its solubility-product constant) is only 3×10^{-8}.

Exercise E4.7. Which can be expected to be more soluble in water, $NaClO_4$ or $KClO_4$?

FURTHER READING
Some introductory texts on solid-state inorganic chemistry at about the level of this text are:

M. F. C. Ladd, *Structure and bonding in solid state chemistry.* Wiley, New York (1979).

D. M. Adams, *Inorganic solids.* Wiley, New York (1974).

M. H. B. Stiddard, *The elementary language of solid state physics.* Academic Press, New York (1975).

A. R. West, *Solid state chemistry and its applications.* Wiley, New York (1984).

P. A. Cox, *The electronic structure and chemistry of solids.* Oxford University Press (1987).

The standard reference book, which surveys the structures of a huge number of elements and compounds, is

A. F. Wells, *Structural inorganic chemistry.* Clarendon Press, Oxford (1985).

Two very useful, thoughtful introductory texts on the application of thermodynamic arguments to inorganic chemistry are:

W. E. Dasent, *Inorganic energetics.* Cambridge University Press (1982).

D. A. Johnson, *Some thermodynamic aspects of inorganic chemistry.* Cambridge University Press (1982).

KEY POINTS

1. Unit cells and crystal lattices
The structures of crystalline solids are discussed in terms of a unit cell, the fundamental unit from which the crystal may be regarded as constructed, and the pattern of atoms in a crystal is depicted in terms of the crystal lattice.

2. The hard-sphere model
The structures of simple solids can sometimes be expressed in terms of a model in which hard spheres representing ions are stacked together.

3. Close-packed structures
Many metals have close-packed structures in which the spheres pack together with least waste of space. Many other substances have structures that can be expressed in terms of less closely packed structures or in terms of the occupation of the tetrahedral and octahedral holes in a close-packed structure.

4. Alloys
Alloys may be either substitutional or interstitial. Nonmetals may also form interstitial solid solutions in metals.

5. Intermetallic phases
Some pairs of metals form intermetallic compounds which have definite structures unrelated to the parent compounds. The Zintl phases are a particular case in which a strongly electropositive metal combines with a less electropositive metal.

6. Typical crystal structures
A number of ionic solids have characteristic structures that include the rock-salt structure, the cesium-chloride structure, the sphalerite structure, the fluorite and antifluorite structures, the wurtzite structure, the nickel-arsenide structure, the rutile structure, and the perovskite structure.

7. Ionic radii
There are several conventions for the definition of ionic radii and values from different sources should be mixed with care. Large ionic radii tend to favor high coordination numbers.

8. Structure maps
An empirical rationalization of structure is in terms of structure maps, in which the axes denote difference in electronegativity and the sizes of the ions.

9. Lattice enthalpy
A measure of the strength of bonding in a solid is the lattice enthalpy, which is determined by using a Born–Haber cycle and thermodynamic data. If the observed lattice enthalpy agrees with that calculated on the basis of coulombic interactions between ions, then ionic bonding is suggested (but not guaranteed).

10. Trends in lattice enthalpies
The Born–Mayer equation can be used to rationalize trends in lattice enthalpies in terms of charge number and ionic radius: large charge number and small radii result in high lattice enthalpies.

11. Thermal stability–size correlations
Large cations stabilize large polyatomic anions (and vice versa); in particular, the decomposition temperatures of thermally unstable compounds (such as carbonates) increase with cation radius.

12. Oxidation number–size correlations
Species with high oxidation numbers are stabilized by small anions. In particular, fluorine has a greater ability compared with the other halogens to stabilize the high oxidation states of metals. Oxygen helps to stabilize species in high oxidation states.

13. Solubility–size correlations
Compounds that contain ions with widely different radii are generally soluble in water; conversely, the least water-soluble salts are those of ions with similar radii.

EXERCISES

4.1 Which of the following schemes for the repeating pattern of close-packed planes are not ways of generating close-packed lattices? (a) ABCABC... (b) ABAC... (c) ABBA... (d) ABCBC... (e) ABABC... (f) ABCCB.. ..

4.2 Draw one layer of close-packed spheres. On this layer mark the positions of the centers of the B layer atoms using the symbol ⊗ and, with the symbol ◯ mark the positions of the centers of the C layer atoms of an fcc lattice.

4.3 Show that the largest sphere that can fit into an octahedral hole in a close packed lattice has a radius equal to $\sqrt{2} - 1 = 0.414$ times that of the spheres composing the lattice. (Hint: Consider the four circles obtained by cutting a plane through the hole and the surrounding square array of spheres.)

4.4 (a) Distinguish between the terms polymorph and polytype. (b) Give an example of each.

4.5 The interstitial alloy tungsten carbide, WC, has the rock-salt structure. Describe it in terms of the holes in a close-packed structure.

4.6 Depending on temperature, RbCl can exist in either the rock-salt or cesium-chloride structure. (a) What is the coordination number of the anion and cation in each of these structures? (b) In which of these structures will Rb have the larger apparent radius?

4.7 The ReO_3 structure is cubic with Re at each corner of the unit cell and one O atom on each unit cell edge midway between the Re atoms. Sketch this unit cell and determine (a) the coordination number of the cation and anion and (b) the identity of the structure type that would be generated if a cation were inserted in the center of the ReO_3 structure.

4.8 Sketch the s and p blocks of the periodic table and mark the boxes for the elements that contribute monatomic cations and anions to solids that are described well by the ionic model. Identify the elements.

4.9 Consider the structure of rock salt. (a) What are the coordination numbers of the anion and cation? (b) How many Na^+ ions occupy second-nearest neighbor locations of an Na^+ ion? (c) Pick out the closest-packed plane of Cl^- ions. (Hint: this hexagonal plane will be perpendicular to a three-fold axis.)

4.10 Consider the structure of cesium chloride. (a) What is the coordination number of the anion and cation? (b) How many Cs^+ ions occupy second-nearest neighbor locations of a Cs^+ ion?

4.11 (a) How many Cs^+ ions and how many Cl^- ions are in the CsCl unit cell? (b) How many Zn^{2+} ions and how many S^{2-} ions are in the sphalerite unit cell?

4.12 Confirm that in rutile (TiO_2, Fig. 4.17) the stoichiometry is consistent with the structure.

4.13 Figure 4.18 shows the perovskite structure of $CaTiO_3$. Confirm that the stoichiometry is consistent with the structure.

4.14 Imagine the construction of an MX_2 structure from the bcc CsCl structure by removal of half the Cs^+ ions to leave tetrahedral coordination around each Cl^-. What is this MX_2 structure?

4.15 Given the following data for the length of a side of the unit cell for compounds that crystallize in the rock-salt structure, determine the cation radii: MgSe (5.45 Å), CaSe (5.91 Å), SrSe (6.23 Å), BaSe (6.62 Å). (To determine the Se^{2-} radius, assume that the Se^{2-} ions are in contact in MgSe.)

4.16 Use the structure map in Fig. 4.22 to predict the coordination numbers of the cations and anions in (a) LiF, (b) RbBr, (c) SrS, and (d) BeO. The observed coordination numbers are (6, 6) for LiF, RbBr, and SrS, and (4, 4) for BeO. Propose a possible reason for the discrepancies.

4.17 (a) Calculate the enthalpy of formation of the hypothetical compound KF_2 assuming a CaF_2 structure. Use the Born–Mayer equation to obtain the lattice enthalpy and estimate the radius of K^{2+} by extrapolation of trends in Table 4.3. Ionization enthalpies and electron gain enthalpies are given in Tables 1.7 and 1.8. (b) What factor prevents the formation of this compound despite the favorable lattice enthalpy?

4.18 The coulombic attraction of nearest-neighbor cations and anions accounts for the bulk of the lattice enthalpy of an ionic compound. With this fact in mind, estimate the order of increasing lattice enthalpy for the following solids, all of which crystallize in the rock-salt structure: (a) MgO, (b) NaCl, (c) LiF. Give your reasoning.

4.19 Which one of each of the following pairs of isostructural compounds is likely to undergo thermal decomposition at lower temperature? Give your reasoning. (a) $MgCO_3$ and $CaCO_3$ (decomposition products $MO + CO_2$). (b) CsI_3 and $N(CH_3)_4I_3$ (both compounds contain I_3^-; decomposition products $MI + I_2$; the radius of $N(CH_3)_4^+$ is much bigger than that of Cs^+).

4.20 Which member of each pair is likely to be the more soluble in water: (a) $SrSO_4$ or $MgSO_4$, (b) NaF or $NaBF_4$?

PROBLEMS

4.1 In the structure of MoS_2, the S atoms are arranged in close-packed layers that repeat themselves in the sequence AAA... The Mo atoms occupy holes with C.N. = 6. Show that each Mo atom is surrounded by a trigonal prism of S atoms.

4.2 Show that the maximum fraction of the available volume occupied by hard spheres on various lattices is (a) simple cubic: 0.52, (b) bcc: 0.68, (c) fcc: 0.74.

4.3 X-Ray diffraction experiments indicate that NH_4NO_3

has the CsCl structure at 150 °C. An NO_3^- ion can replace a Cl^- ion and preserve the symmetry of the lattice only if it has at least the symmetry of the Cl^- ion site or behaves as if it did. (a) What is the local point-group of the site it occupies? (b) What is the point-group symmetry of NO_3^-? (c) How might the high-temperature results be explained?

4.4 The common oxidation number for an alkaline earth metal is $+2$. Aided by the Born–Mayer equation for lattice enthalpy and a Born–Haber cycle, show that CaCl is an exothermic compound. Use a suitable analogy to estimate an ionic radius for Ca^+. The sublimation enthalpy of Ca(s) is 176 kJ mol^{-1}. Show that an explanation for the nonexist-

ence of CaCl can be found in the enthalpy change for the reaction

$$2CaCl(s) \rightarrow Ca(s) + CaCl_2(s)$$

4.5 At various times the following two proposals have been advanced for the elements to be included in Group 3: (a) Sc, Y, La, Ac; (b) Sc, Y, Lu, Lr. Because ionic radii strongly influence the chemical properties of the metallic elements, it might be thought that ionic radii could be employed as one criterion for the periodic arrangement of the elements. Using this criterion describe which of these sequences is preferred.

5

Acids and bases

The focus of this chapter is on species that are classified as acids and bases. The first part of the chapter deals with one particular class of acids and bases in which the criterion is their participation in proton transfer reactions. Proton transfer can account for a wide range of chemical reactions and can be discussed quantitatively in terms of the acidity constant of a species, which is a measure of the strength with which a species donates a proton. In the second part of the chapter, we broaden the definition of acids and bases to include reactions that involve electron pair sharing between a donor and an acceptor. Because of the greater diversity of these species, it is more difficult to devise a quantitative scale of strength. However, we introduce two approaches that have been adopted. One involves classifying such acids and bases as 'hard' or 'soft'. The other approach is based on fitting thermochemical data to a set of parameters characteristic of each species.

The chemical manufactured in greatest mass annually is sulfuric acid; the bases ammonia and lime (CaO) were in fifth and sixth places in annual production (in 1992). These figures emphasize the central role of acids and bases in industrial chemistry. Acids and bases are equally important in laboratory chemistry, where many of the synthetic and analytical reactions that are used routinely and in the development of new materials employ acid–base neutralizations of one kind or another. Acids and bases also have powerful ecological impact; different flora, for example, thrive on soils that differ in their acidic or basic character.

The first recognition of the existence of the classes of compound we now identify as acids and bases was based, hazardously, on criteria of taste and feel: acids were sour and bases felt soapy. A deeper chemical understanding of the distinction of the two types of substance emerged from Liebig's idea of reactive hydrogen and Arrhenius's conception of an acid as a compound that produced hydrogen ions in water. Modern definitions, the only ones we consider in this chapter, exploit general viewpoints that group together a wide range of experimental observations.

BRØNSTED ACIDITY

Johannes Brønsted in Denmark and Thomas Lowry in England proposed (in 1923) that the essential feature of an acid–base reaction is the transfer of a proton from one species to another. In the context of the Brønsted–Lowry definitions, a proton is a hydrogen ion, H^+. They suggested that any substance that acts as a proton donor should be classified as an **acid**, and any substance that acts as a proton acceptor should be classified as a **base**. Substances that act in this way are now called **Brønsted acids** and **Brønsted bases**, respectively. The definitions make no reference to the environment in which proton transfer occurs, and so apply to the gas phase and in any solvent.

An example of a Brønsted acid is HF, which can donate a proton to another molecule, such as H_2O, when it dissolves in water:

$$HF(g) + H_2O(l) \rightarrow H_3O^+(aq) + F^-(aq)$$

An example of a Brønsted base is NH_3, which can accept a proton from an acid:

$$HF(aq) + NH_3(aq) \rightarrow NH_4^+(aq) + F^-(aq)$$

Water is an example of substance that is **amphiprotic** and can act as both a Brønsted acid and a Brønsted base. Thus, it acts as an acid toward NH_3:

$$H_2O(l) + NH_3(aq) \rightarrow OH^-(aq) + NH_4^+(aq)$$

and as a base toward H_2S:

$$H_2S(aq) + H_2O(l) \rightarrow H_3O^+(aq) + HS^-(aq)$$

5.1 The proton in water

Because the most important environment for acid–base reactions is an aqueous solution, we set the stage by characterizing the behavior of the proton in water.

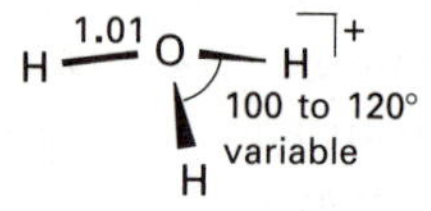

The hydronium ion

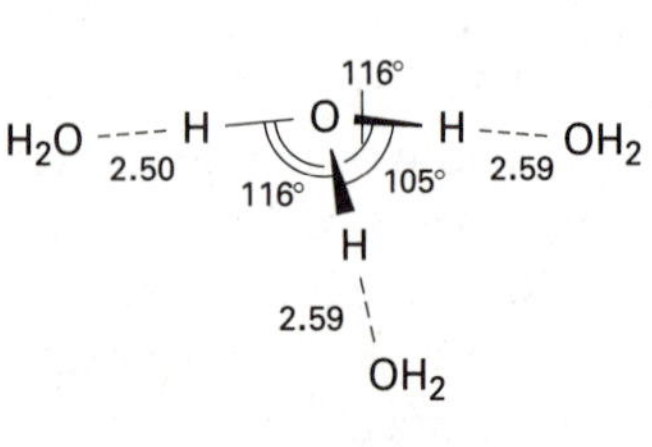

When an acid donates a proton to a water molecule, the product is the **hydronium ion**, H_3O^+. The dimensions of the hydronium ion (**1**) are taken from the crystal structure of $H_3O^+ClO_4^-$ in which it is the cation. However, the structure H_3O^+ is almost certainly an oversimplified description of the ion in water, for it participates in extensive hydrogen bonding. Manfred Eigen and his colleagues at Göttingen suggested that the hydronium ion in water is best represented as $H_9O_4^+$ (**2**), if a simple formula is required. Recent results from gas

1 H_3O^+

2 $H_9O_4^+$

Table 5.1 Rate constants* for aqueous proton transfer reactions at 25 °C

Reaction	$k_\rightarrow/(\text{L mol}^{-1}\text{ s}^{-1})$	$k_\leftarrow/(\text{L mol}^{-1}\text{ s}^{-1})$
$H_3O^+ + OH^- \rightleftharpoons H_2O + H_2O$	1.4×10^{11}	2.5×10^{-5}
$H_3O^+ + SO_4^{2-} \rightleftharpoons HSO_4^- + H_2O$	1×10^{11}	7×10^7
$H_3O^+ + NH_3 \rightleftharpoons NH_4^+ + H_2O$	4.3×10^{10}	8.4×10^5
$OH^- + NH_4^+ \rightleftharpoons H_2O + NH_3$	3.4×10^{10}	6×10^5

*$k_\rightarrow$ is the rate constant for the forward reaction and $k_\leftarrow$ is the rate constant for the reverse reaction.

phase studies of water clusters using mass spectrometry suggest that a cage of 20 H_2O molecules can condense around one H_3O^+ ion in a regular pentagonal dodecahedral arrangement, resulting in the formation of the species $H^+(H_2O)_{21}$.[1] As these structures indicate, the most appropriate description of a proton in water varies according to the environment and the experiment under consideration.

The mobility of hydrogen ions in water

No simple formula conveys the whole sense of the structure of the hydronium ion in water because a proton can transfer rapidly from one H_2O molecule to another. Early evidence for this ability to transfer rapidly came from the observation that the mobility of the proton in water (as measured from its electric conductivity) is about three times that of typical ions. The explanation, which is called the **Grotthus mechanism,** is that the migration of a proton is in fact not an actual movement of the ion through the solvent but a cooperative rearrangement of the atoms: a proton jumps from one O atom to the next along a hydrogen bond (Fig. 5.1), the receiving molecule becomes the cation, and one of its protons can now migrate to another neighbor in the same way. The migration is therefore a cooperative process that takes place through a network of several hydrogen-bonded H_2O molecules.

Eigen's work shows that a proton remains at one end of an O—H···O hydrogen bond for a half-life of 1 to 4 ps (1 ps = 10^{-12} s) before it jumps and forms O···H—O. The O—H bond is not the only example of a very labile (short-lived) structure: fast proton transfers from one electronegative atom (such as O, N, and Cl) to another are common. Table 5.1 shows some rate constants for such proton transfers.

5.1 The Grotthus mechanism for the effective migration of H^+ in water by rearrangement of covalent and hydrogen bonds.

5.2 Acid equilibria in water

Because proton-transfer reactions are fast in both directions, the dynamic equilibria

$$HF(aq) + H_2O(l) \rightleftharpoons H_3O^+(aq) + F^-(aq)$$

$$H_2O(l) + NH_3(aq) \rightleftharpoons OH^-(aq) + NH_4^+(aq)$$

give a more complete description of the behavior of the acid HF and the base NH_3 in water than the forward reaction alone. Brønsted

[1] A. W. Castleman, S. Wei, and Z. Shi, *J. Chem. Phys.*, **94**, 3268 (1991).

acid–base chemistry is mainly a matter of rapidly attained proton-transfer equilibria, and we concentrate on this aspect.

Conjugate acids and bases

In the first of the two equilibria above, H_3O^+ is a Brønsted acid in the reverse reaction because it donates a proton to F^-. The latter anion, being a proton acceptor, is a Brønsted base. Likewise, in the reverse reaction in the second equilibrium, NH_4^+ is a Brønsted acid because it donates a proton to OH^-. The fact that OH^- accepts a proton identifies it as a Brønsted base.

The symmetry of the forward and reverse reactions, both of which depend on the transfer of a proton from an acid to a base, is expressed by writing the general **Brønsted equilibrium** as

$$Acid_1 + Base_2 \rightleftharpoons Base_1 + Acid_2$$

The species $Base_1$ is called the **conjugate base** of $Acid_1$ and $Acid_2$ is the **conjugate acid** of $Base_2$. Thus, H_3O^+ is the conjugate acid of H_2O and F^- is the conjugate base of HF. Likewise, OH^- is the conjugate base of H_2O, ClO^- is the conjugate base of HClO, and NH_4^+ is the conjugate acid of NH_3. By 'acid' and 'base' we normally mean the reactants in the chemical equation as written; the products are their conjugates. However, there is no fundamental distinction between an acid and a conjugate acid or a base and a conjugate base: a conjugate acid is just another acid and a conjugate base is just another base. Thus, HF is the conjugate acid of the base F^- and NH_3 is the conjugate base of the acid NH_4^+.

The classification of a species as an acid or a base may depend on the reaction in which it is a participant, and a species classified as a base in one reaction may be able to act as an acid in another (and vice versa). An example of such an **amphiprotic species** is HSO_4^-, the conjugate base of the acid H_2SO_4. It is an acid in its own right because it can donate its remaining proton:

$$HSO_4^-(aq) + H_2O(l) \rightleftharpoons SO_4^{2-}(aq) + H_3O^+(aq)$$

Thus, the conjugate base of HSO_4^- is the sulfate ion SO_4^{2-}. Similarly, although OH^- is the conjugate base of H_2O, in nonaqueous solution it can donate the remaining proton and hence act as an acid. The conjugate base of the acid OH^- is the oxide ion O^{2-}.

Example 5.1: *Identifying acids and bases*

Identify the Brønsted acid and its conjugate base in

(a) $HSO_4^- + OH^- \rightleftharpoons SO_4^{2-} + H_2O$

(b) $PO_4^{3-} + H_2O \rightleftharpoons HPO_4^{2-} + OH^-$

(c) $H_2Fe(CO)_4 + CH_3OH \rightleftharpoons [HFe(CO)_4]^- + CH_3OH_2^+$

Answer. In (a), the hydrogensulfate ion (HSO_4^-) transfers a proton to hydroxide; it is therefore the acid and the SO_4^{2-} ion produced is its conjugate base. In (b), H_2O transfers a proton to the phosphate ion acting as a base; thus H_2O is the acid and the OH^- ion is its conjugate base. In (c), $H_2Fe(CO)_4$ is the acid that transfers a proton to methanol; its conjugate base is $[HFe(CO)_4]^-$.

Exercise E5.1. Identify the acid, base, conjugate acid, and conjugate base in (a) $HNO_3 + H_2O \rightleftharpoons H_3O^+ + NO_3^-$, (b) $CO_3^{2-} + H_2O \rightleftharpoons HCO_3^- + OH^-$, (c) $NH_3 + H_2S \rightleftharpoons NH_4^+ + HS^-$.

The strengths of Brønsted acids

The strength of a Brønsted acid HA in aqueous solution is expressed by its **acidity constant** (or **acid ionization constant**, K_a), the equilibrium constant for proton transfer between acid and water:

$$HA(aq) + H_2O(aq) \rightleftharpoons A^-(aq) + H_3O^+(aq) \quad K_a = \frac{a(H_3O^+)a(A^-)}{a(HA)a(H_2O)}$$

In this expression $a(X)$ is the activity of X, its effective thermodynamic concentration, in the solution at equilibrium. Because the activity of pure water (as for any pure liquid at 1 bar) is 1, and is close to 1 for dilute solutions, the approximation is normally made that $a(H_2O) = 1$. When thermodynamic precision is not required (as in approximate descriptive work) or when total ion concentrations are very low (less than about 10^{-3} mol L^{-1}), activities of solutes are often replaced by molar concentrations [X]. We shall make both approximations throughout this text, and use

$$K_a = \frac{[H_3O^+][A^-]}{[HA]} \tag{1}$$

The proton transfer from water acting as a Brønsted acid is described by the water **autoprotolysis constant** K_w:

$$H_2O(l) + H_2O(l) \rightleftharpoons H_3O^+(aq) + OH^-(aq) \quad K_w = [H_3O^+][OH^-] \tag{2}$$

Because molar concentrations and acidity constants span many orders of magnitude, it proves convenient to report them as their logarithms (to the base 10) by using

$$pH = -\log [H_3O^+] \quad pK_a = -\log K_a \quad pK_w = -\log K_w$$

At 25 °C, $pK_w = 14.00$.

The acidity constants of some common acids are given in Table 5.2. Substances with $pK_a < 0$ (corresponding to $K_a > 1$ and usually to $K_a \gg 1$) are classified as **strong acids**, in the sense that the proton transfer equilibrium lies strongly in favor of donation to water. Substances with $pK_a > 0$ (corresponding to $K_a < 1$) are classified as **weak acids**; for them, the proton transfer equilibrium lies in favor of nonionized HA. The conjugate bases of strong acids are **weak bases** because it is thermodynamically unfavorable for them to accept a proton from H_3O^+. A general feature to keep in mind, therefore, is that *the weaker the acid, the stronger its conjugate base.*

Table 5.2 Acidity constants for species in aqueous solution at 25 °C

Acid	HA	A$^-$	K_a	pK_a
Hydriodic	HI	I$^-$	10^{11}	-11
Perchloric	HClO$_4$	ClO$_4^-$	10^{10}	-10
Hydrobromic	HBr	Br$^-$	10^9	-9
Hydrochloric	HCl	Cl$^-$	10^7	-7
Sulfuric	H$_2$SO$_4$	HSO$_4^-$	10^2	-2
Hydronium ion	H$_3$O$^+$	H$_2$O	1	0.0
Sulfurous	H$_2$SO$_3$	HSO$_3^-$	1.5×10^{-2}	1.81
Hydrogensulfate ion	HSO$_4^-$	SO$_4^{2-}$	1.2×10^{-2}	1.92
Phosphoric	H$_3$PO$_4$	H$_2$PO$_4^-$	7.5×10^{-3}	2.12
Hydrofluoric	HF	F$^-$	3.5×10^{-4}	3.45
Pyridinium ion	HC$_5$H$_5$N$^+$	C$_5$H$_5$N	5.6×10^{-6}	5.25
Carbonic	H$_2$CO$_3$	HCO$_3^-$	4.3×10^{-7}	6.37
Hydrogen sulfide	H$_2$S	HS$^-$	9.1×10^{-8}	7.04
Boric acid*	B(OH)$_3$	B(OH)$_4^-$	7.2×10^{-10}	9.14
Ammonium ion	NH$_4^+$	NH$_3$	5.6×10^{-10}	9.25
Hydrocyanic	HCN	CN$^-$	4.9×10^{-10}	9.31
Hydrogencarbonate ion	HCO$_3^-$	CO$_3^{2-}$	4.8×10^{-11}	10.32
Hydrogenarsenate ion	HAsO$_4^{2-}$	AsO$_4^{3-}$	3.0×10^{-12}	11.53
Hydrogensulfide ion	HS$^-$	S^{2-}	1.1×10^{-12}	11.96
Hydrogenphosphate ion	HPO$_4^{2-}$	PO$_4^{3-}$	2.2×10^{-13}	12.67

*The proton transfer equilibrium is $B(OH)_3(aq) + 2H_2O(l) \rightleftharpoons H_3O^+(aq) + B(OH)_4^-(aq)$.

Polyprotic acids

The successive acidity constants of **polyprotic acids**, which are species that can donate more than one proton, are defined analogously. For a diprotic acid H$_2$A, such as H$_2$S, for example, there are two successive proton donations and two acidity constants:

$$H_2A(aq) + H_2O(aq) \rightleftharpoons HA^-(aq) + H_3O^+(aq) \quad K_{a1} = \frac{[H_3O^+][HA^-]}{[H_2A]}$$

$$HA^-(aq) + H_2O(aq) \rightleftharpoons A^{2-}(aq) + H_3O^+(aq) \quad K_{a2} = \frac{[H_3O^+][A^{2-}]}{[HA^-]}$$

The second acidity constant, K_{a2}, is almost always smaller than K_{a1} (and hence pK_{a2} is generally larger than pK_{a1}). For oxoacids in which both OH groups are attached to the same atom, pK_{a2} is typically about 5 units larger than pK_{a1}, corresponding to a decrease in equilibrium constant by a factor of about 10^5 (the decrease is largely independent of the identity of the central atom). The large nonspecific decrease is consistent with an electrostatic model of the acid in which, in the second ionization, a proton must separate from a center with one more unit of negative charge than in the first ionization. Because additional electrostatic work must be done to remove the positively charged proton, the ionization is less favorable.

The clearest representation of the concentrations of the species that are formed in the successive acid-base equilibria of polyprotic acids is a **distribution diagram**. Consider the specific case of the triprotic acid H$_3$PO$_4$, which releases three protons in succession to give

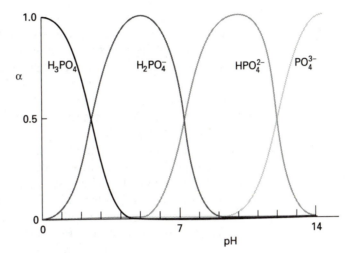

5.2 The distribution diagram for the various forms of the triprotic acid phosphoric acid as a function of pH.

$H_2PO_4^-$, HPO_4^{2-}, and PO_4^{3-}. The diagram in Fig. 5.2 shows the fraction, $\alpha(X)$, of a species X as a function of pH; for the fraction of H_3PO_4 molecules, for instance,

$$\alpha(H_3PO_4) = \frac{[H_3PO_4]}{[H_3PO_4] + [H_2PO_4^-] + [HPO_4^{2-}] + [PO_4^{3-}]}$$

The variation of α with pH reveals the relative importance of each acid and its conjugate base at each pH. Conversely, the diagram indicates the pH of the solution that contains a particular equilibrium concentration of the species. It can be seen that if $pH < pK_{a1}$, then the dominant species is the fully protonated H_3PO_4 molecule, and if $pH > pK_{a3}$, then the dominant species is the fully deprotonated PO_4^{3-} ion. The intermediate species are dominant at pH values that lie between the relevant pK_a values.

5.3 Solvent leveling

Because solvents may be acids or bases, there is a limit to the range of acidity that can be studied in a given solvent. For instance, acids that are weak in water may appear strong in a basic solvent, and it may not be possible to arrange them according to strength because all of them will be fully deprotonated. Similarly, bases that are weak in water may appear strong in an acidic solvent, and once again we cannot arrange them according to strength, for all of them will be effectively fully protonated. We shall now see that the autoprotolysis constant of a solvent plays a crucial role in determining the range of acid or base strengths that can be distinguished for species that are dissolved in it.

Discrimination in water

Any acid stronger than H_3O^+ in water donates a proton to H_2O and forms H_3O^+. Consequently, no acid significantly stronger than H_3O^+ can survive in water. No experiment conducted in water can tell us which is the stronger acid between HBr and HI because both transfer their protons essentially completely to give H_3O^+. Water is said to have a **leveling effect** that brings all stronger acids down to the acidity of H_3O^+. In water, HBr and HI have indistinguishable acid strengths. However, in a solvent that is less basic than water (for example, acetic acid), HBr and HI behave as weak acids and their strengths can be distinguished. It is found in this way that HI is stronger than HBr.

An analogous limit can be found for bases in water. Any base B that is strong enough to undergo complete protonation by donation from water, leaving OH^- ions in solution in place of every B molecule added initially, will be leveled. The OH^- ion is the strongest base that can exist in water because anything stronger immediately forms OH^- ions by reaction with water. For this reason, we cannot study NH_2^- or CH_3^- in water by dissolving alkali metal amides or methides because both anions generate OH^- ions quantitatively and are fully protonated to NH_3 and CH_4:

$$KNH_2(s) + H_2O(l) \rightarrow K^+(aq) + OH^-(aq) + NH_3(aq)$$

$$LiCH_3(s) + H_2O(l) \rightarrow Li^+(aq) + OH^-(aq) + CH_4(g)$$

Any negative value of pK_a ($K_a > 1$, signifying a proton transfer equilibrium that lies in favor of H_3O^+ and the conjugate base of the acid) indicates an acid that has been leveled in water. Similarly, the conjugate base of any acid with $pK_a \geq 14$ is a strong base and is leveled in water. The window of acid strengths that are not leveled in water is from $pK_a = 0$ to $pK_a = 14$. Note that the width of this range corresponds to pK_w.

Discrimination in nonaqueous solvents

The range over which acid and base strengths can be discriminated in a given solvent is related to the solvent's autoprotolysis constant. For water, as we have seen, $pK_w = 14$ and the range of discrimination spans 14 units. For liquid ammonia,

$$NH_3(l) + NH_3(l) \rightleftharpoons NH_4^+(am) + NH_2^-(am) \quad pK_{am} = 33$$

and the range of discrimination is considerably wider. (The *am* signifies solution in liquid ammonia.) The autoprotolysis constant for a solvent gives the width of the window. It characterizes the reaction of the strongest possible acid, the conjugate acid of the solvent, with the strongest possible base, the conjugate base of the solvent.

To determine the acidity constants of strong acids, such as HCl, we need data from acidic solvents, such as acetic acid. To study strong bases, such as NH_2^-, we need data from basic solvents, such as liquid ammonia. The discrimination windows of a number of solvents are shown in Fig. 5.3. The window for dimethylsulfoxide (DMSO, $(CH_3)_2SO$) is wide because the pK for the reaction

$$(CH_3)_2SO + (CH_3)_2SO \rightleftharpoons CH_3SOCH_2^- + (CH_3)_2SOH^+$$

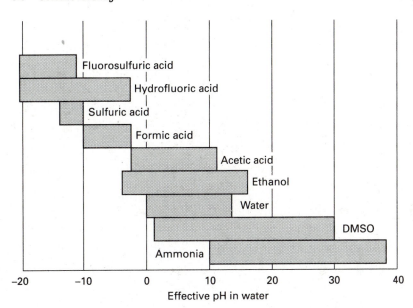

5.3 The acid–base discrimination windows of a variety of solvents. In each case, the width of the window is equal to the autoprotolysis constant pK of the solvent.

is very large (p$K = 37$). Consequently, DMSO can be used to study a wide range of acids (from H_2SO_4 to PH_3). Notice that water has a narrow window compared to some of the other solvents shown in Fig. 5.3. One reason is the high relative permittivity (dielectric constant) of water: the autoprotolysis creates two ions and high permittivity favors such a reaction.

PERIODIC TRENDS IN BRØNSTED ACIDITY

Now that we have seen the principles of Brønsted acidity, we can explore how it manifests itself in compounds formed by elements in different regions of the periodic table. This discussion will be limited to aqueous solutions. The largest class of acids in water are those that donate protons from an —OH group attached to a central atom. A donatable proton of this kind is called an **acidic proton** to distinguish it from other protons that may be present in the molecule.

There are three classes of hydroxyl group acids to consider:

1. **Aqua acids**, in which the acidic proton is on a water molecule coordinated to a central metal ion:

$$E(OH_2)(aq) + H_2O(l) \rightleftharpoons [E(OH)]^-(aq) + H_3O^+(aq)$$

An example is

$$[Fe(OH_2)_6]^{3+}(aq) + H_2O(l) \rightarrow [Fe(OH_2)_5(OH)]^{2+}(aq) + H_3O^+(aq)$$

The aqua acid, the hexaaquairon(III) ion, is shown as structure (**3**).

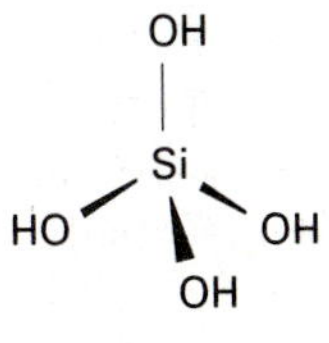

2. **Hydroxoacids**, in which the acidic proton is on a hydroxyl group without a neighboring oxo group (=O). An example is $Si(OH)_4$ (**4**), which is important in the formation of minerals.

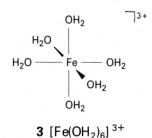

3 $[Fe(OH_2)_6]^{3+}$

4 $Si(OH)_4$

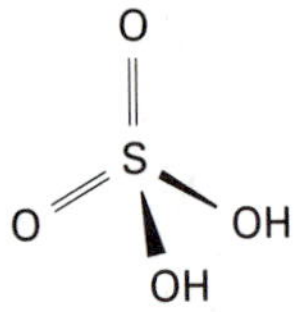

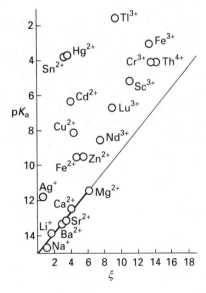

5 H_2SO_4

3. **Oxoacids,** in which the acidic proton is on a hydroxyl group with an oxo group attached to the same atom. Sulfuric acid, H_2SO_4 ($O_2S(OH)_2$; **5**), is an example of an oxoacid.

We can think of the three classes of acid as successive stages in the deprotonation of an aqua acid

$$H_2O-E-OH_2 \xrightleftharpoons{-2H^+} HO-E-OH^{2-} \xrightleftharpoons{-H^+} HO-E=O^{3-}$$

aqua acid   hydroxoacid   oxoacid

An example of these successive stages is provided by a *d*-block metal in an intermediate oxidation state, such as Ru(IV):

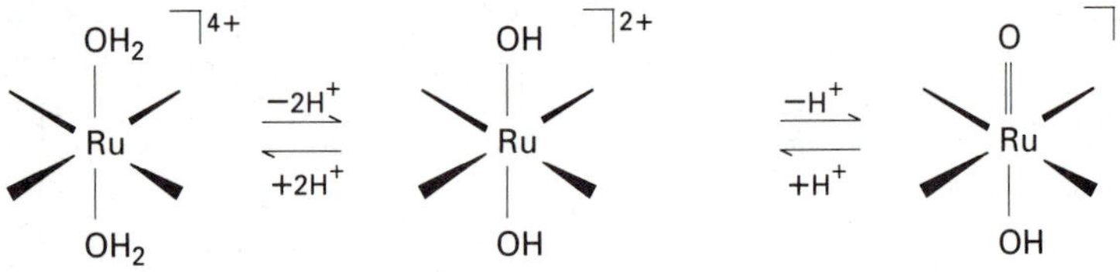

Aqua acids are characteristic of central atoms with low oxidation numbers, of *s* and *d* block metals, and of metals on the left of the *p* block. Oxoacids are found where the central element has a high oxidation number. In addition, an element from the right of the *p* block in one of its intermediate oxidation states may also produce an oxoacid ($HClO_2$ is an example).

5.4 Periodic trends in aqua acid strength

The strengths of aqua acids typically increase with increasing positive charge of the central metal ion and with decreasing ionic radius.

The variation of the strengths of aqua acids through the periodic table can be rationalized to some extent in terms of an ionic model, in which the metal cation is represented by a sphere carrying *z* units of positive charge. The gas-phase pK_a is proportional[2] to the work of removing a proton to infinity from a distance equal to the sum of the ionic radius, *r*, and the diameter of a water molecule, *d*. Because protons are more easily removed from the vicinity of cations of high charge and small radius, the model predicts that the acidity should increase with increasing *z* and with decreasing *r*, very roughly as the electrostatic parameter $\xi = z^2/(r+d)$. The trends this model predicts for the gas phase will also apply in solution if the effects of solvation are reasonably constant.

The validity of this ionic model can be judged from Fig. 5.4. Aqua ions of elements that form ionic solids (principally those from the *s* block) have pK_a values that are quite well described by the ionic model. Several *d*-block ions (such as Fe^{2+} and Cr^{3+}) lie reasonably near the same straight line, but many ions (particularly those with low pK_a,

5.4 The correlation between acidity constant and electrostatic parameter ξ of aqua ions. Note that only hard ions with low charge follow the correlation; all others are more acidic than the correlation suggests.

---

[2] This comes from $pK_a \propto \Delta G^{\ominus}$ coupled with the identification of ΔG and electrical work, so $pK_a \propto w_{electrical}$.

corresponding to high acid strength) deviate markedly from it. This deviation indicates that the metal ions repel the departing proton more strongly than is predicted by the ionic model. This enhanced repulsion can be rationalized by supposing that the cation's positive charge is not fully localized on the central ion but is delocalized over the ligands and hence is closer to the departing proton. The delocalization is equivalent to attributing covalence to the E—O bond. Indeed, the correlation is worst for ions that are disposed to covalent bond formation.

For the later d-block and p-block metals (such as Cu^{2+} and Sn^{2+}), the strengths of the aqua acids exceed the predictions of the ionic model by a large margin. For these species, covalent bonding is very important and the ionic model is unrealistic. The extent of overlap between metal orbitals and the orbitals of an oxygen ligand increases from left to right across a period. It also increases down a group, so aqua ions of heavier d-block metals tend to be stronger acids.

Example 5.2: *Accounting for trends in aqua acid strength*
Account for the trend in acidity

$$[Fe(OH_2)_6]^{2+} < [Fe(OH_2)_6]^{3+} < [Al(OH_2)_6]^{3+} \approx [Hg(OH_2)_n]^{2+}$$

Answer. The weakest acid is the Fe^{2+} complex on account of its relatively large radius and low charge. The increase of charge to $+3$ increases the acid strength. The greater acidity of Al^{3+} can be explained by its smaller radius. The anomalous ion in the series is the Hg^{2+} complex. This reflects the failure of an ionic model, for in this complex there is a large transfer of positive charge to oxygen as a result of covalent bonding.

Exercise E5.2. Arrange the following ions in order of increasing acidity:

$$[Na(OH_2)_n]^+, \ [Sc(OH_2)_6]^{3+}, \ [Mn(OH_2)_6]^{2+}, \ [Ni(OH_2)_6]^{2+}.$$

5.5 Simple oxoacids

The simplest oxoacids are the **mononuclear acids**, which contain one atom of the parent element. They include H_2CO_3, HNO_3, H_3PO_4, and H_2SO_4. These oxoacids are formed by the electronegative elements at the upper right of the periodic table and by other elements with high oxidation numbers. Table 5.3 summarizes some of their structures. One interesting feature in that table is the occurrence of planar $B(OH)_3$, H_2CO_3, and HNO_3 molecules but not their analogs in later periods. As we saw in Section 2.2, π bonding is more important among the Period 2 elements, so the atoms are more likely to be constrained to lie in a plane.

Substituted oxoacids

One or more hydroxyl groups of an oxoacid may be replaced by other groups to give a series of substituted oxoacids, which include fluoro-

Table 5.3 The structures and acidity constants of oxoacids*

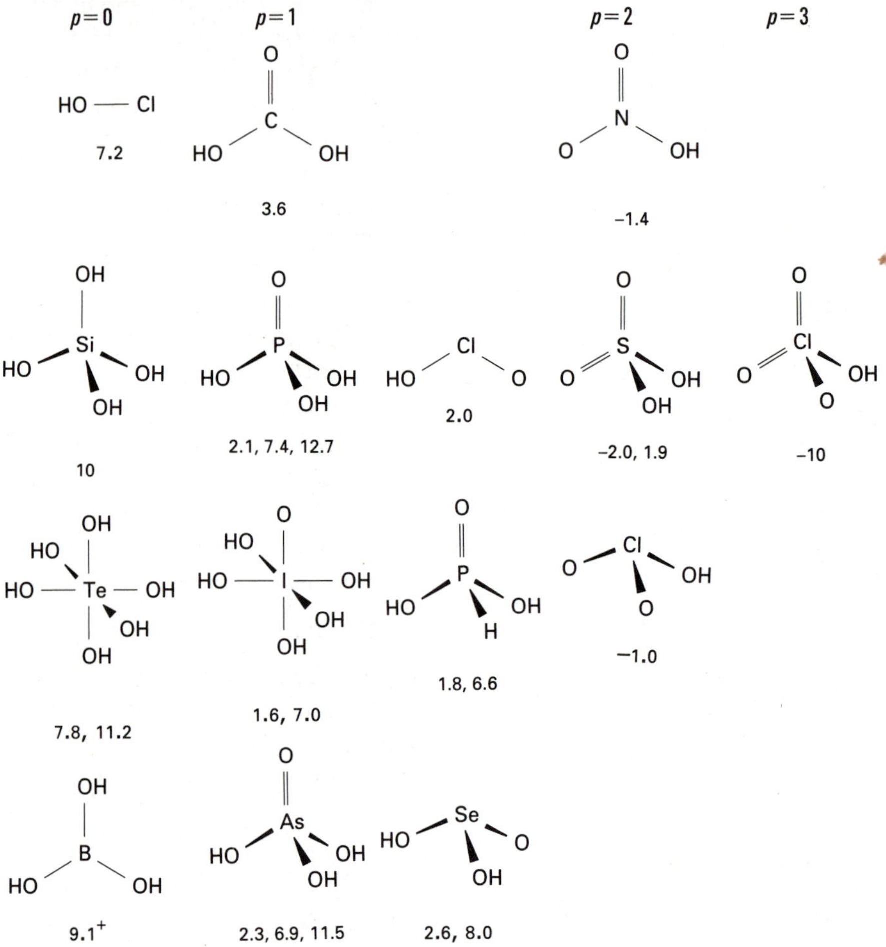

$p=0$	$p=1$	$p=2$	$p=3$
HO—Cl 　7.2	HO—C(=O)—OH (carbonic) 　3.6	O=N(O)(OH) 　−1.4	
Si(OH)$_4$: HO—Si(OH)(OH)—OH 　10	HO—P(=O)(OH)—OH 　2.1, 7.4, 12.7 ;　 HO—Cl(=O)—O 　2.0	O=S(=O)(OH)—OH 　−2.0, 1.9	O=Cl(=O)(OH)—O 　−10
Te(OH)$_6$: HO,HO—Te(OH)(OH)—OH 　7.8, 11.2	HO—I(=O)(OH)(OH)—OH 　1.6, 7.0	O=P(OH)(OH)—H 　1.8, 6.6 ;　 O—Cl(OH)—O 　−1.0	
B(OH)$_3$: HO—B(OH)—OH 　9.1$^+$	O=As(HO)(OH)—OH 　2.3, 6.9, 11.5	HO—Se(O)—O / with OH 　2.6, 8.0	

*p is the number of non-protonated O atoms.
†See Table 5.2.

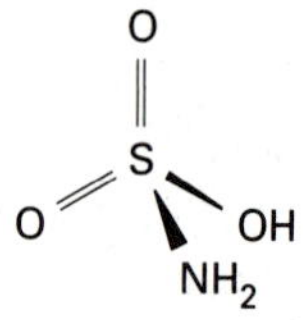

6 O$_2$S(NH$_2$)OH

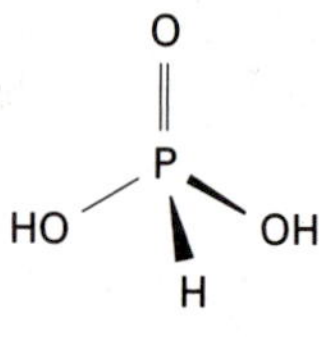

7 H$_3$PO$_3$

sulfuric acid, $O_2SF(OH)$, and aminosulfuric acid, $O_2S(NH_2)OH$ (**6**). Because fluorine is highly electronegative, it withdraws electrons from the central S atom and confers on it a higher effective positive charge, making it a stronger acid than H_2SO_4. In contrast, the NH$_2$ group, in which N is less electronegative than F, can donate electron density to S by π bonding. This transfer of charge reduces the positive charge of the central atom and weakens the acid. Another electron donor substituent is CH$_3$, as in methylsulfuric acid CH_3SO_3H, but its effect is smaller because CH$_3$ has no lone pair.

A trap for the unwary is that not all oxoacids follow the familiar structural pattern of a central atom surrounded by OH and O groups. Occasionally an H atom is attached directly to the central atom, as in phosphorous acid, H_3PO_3. Phosphorous acid is in fact only a *di*protic acid, for the substitution results in a P—H bond (**7**) and consequently a nonacidic proton. This structure is consistent with NMR and Raman

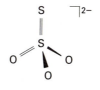

8 $S_2O_3^{2-}$

spectra, and the formula is best written as $OPH(OH)_2$. Substitution for an oxo group (as distinct from a hydroxyl group) is another example of a structural change that can occur. An important example is thiosulfate ion, $S_2O_3^{2-}$ (**8**), in which an S atom replaces an oxo-O atom of a sulfate ion.

Pauling's rules

The observed strengths of mononuclear oxoacids can be systematized by using two empirical rules devised by Linus Pauling:

1. For the oxoacid $O_pE(OH)_q$, $pK_a \approx 8 - 5p$.
2. The successive pK_a values of polyprotic acids (those with $q > 1$), increase by 5 units for each successive proton transfer.

Neutral hydroxoacids with $p = 0$ have $pK_a \approx 8$, acids with one oxo group have $pK_a \approx 3$, and acids with two oxo groups have $pK_a \approx -2$. For example, sulfuric acid, $O_2S(OH)_2$, has $p = 2$ and $q = 2$, and $pK_{a1} \approx -2$ (signifying a strong acid). Similarly, pK_{a2} is predicted to be $+3$, which is reasonably close to the experimental value of 1.9.

The success of these simple rules may be gauged by inspection of Table 5.3, in which acids are grouped according to p, the number of oxo groups. That the estimates are good to about ± 1 is pleasantly surprising. The variation in strengths down a group is not large, and the complex and perhaps canceling effects of changing structures allow the rules to work moderately well. The more important variation across the periodic table from left to right and the effect of change of oxidation number are taken into account by the number of oxo groups characteristic of the neutral acids. In Group 15/V, the oxidation number $+5$ requires one oxo group (as in $OP(OH)_3$) whereas in Group 16/VI the oxidation number $+6$ requires two (as in $O_2S(OH)_2$).

Structural anomalies

An interesting use of the Pauling rules is to detect structural anomalies. For example, carbonic acid, $OC(OH)_2$, is commonly reported as having $pK_{a1} = 6.4$, but the rules predict $pK_{a1} = 3$. The reduced acidity is the result of treating the concentration of dissolved CO_2 as if it were all H_2CO_3. However, in the equilibrium

$$CO_2(aq) + H_2O(l) \rightleftharpoons OC(OH)_2(aq)$$

only about 1 percent of the dissolved CO_2 is present as $OC(OH)_2$, so the actual concentration of acid is much less than the concentration of dissolved CO_2. When this difference is taken into account, the true pK_{a1} of H_2CO_3 is about 3.6, as the Pauling rules predict.

The value $pK_a = 1.8$ reported for sulfurous acid, H_2SO_3, suggests another anomaly, this time acting in the opposite direction. In fact, spectroscopic studies have failed to detect the molecule $OS(OH)_2$ in solution, and the equilibrium constant for

$$SO_2(aq) + H_2O(l) \rightleftharpoons H_2SO_3(aq)$$

is less than 10^{-9}. The equilibria of dissolved SO_2 are complex, and a simple analysis is inappropriate. The ions that have been detected include HSO_3^- and $S_2O_5^{2-}$, and there is evidence for an S—H bond in the solid salts of the hydrogensulfite ion.[3]

The pK_a values of aqueous CO_2 and SO_2 call attention to the important point that not all nonmetal oxides react fully with water to form acids. Carbon monoxide is another example: although it is *formally* the anhydride of formic acid, HCOOH, it does not in fact react with water at room temperature to give the acid. The same is true of some metal oxides: OsO_4, for example, can exist as dissolved neutral molecules.

Example 5.3: *Using Pauling's rules*
Identify the formulas that are consistent with the following pK_a values: H_3PO_4, 2.12; H_3PO_3, 1.80; H_3PO_2, 2.0.

Answer. All three values are in the range which Pauling's first rule associates with one oxo group. This observation suggests the formulas $(HO)_3P{=}O$, $(HO)_2HP{=}O$, and $(HO)H_2P{=}O$. The second and the third formulas are derived from the first by replacement of —OH by —H bound to P (as in structure 7).

Exercise E5.3. Predict the pK_a values of (a) H_3PO_4, (b) $H_2PO_4^-$, (c) HPO_4^{2-}. Experimental values are given in Table 5.3.

5.6 Anhydrous oxides

We have treated oxoacids as derived from the deprotonation of parent aqua acids. It is also useful to take the opposite viewpoint and to consider aqua acids and oxoacids as derived by hydration of the oxides of the central element. This approach emphasizes the acid and base properties of oxides and their correlation with the location of the element in the periodic table.

Acidic and basic oxides

An **acidic oxide** is one which, on dissolution in water, binds an H_2O molecule and releases a proton to the surrounding solvent:

$$CO_2(g) + H_2O(l) \rightarrow [OC(OH)_2](aq)$$

$$[OC(OH)_2](aq) + H_2O(l) \rightleftharpoons [O_2C(OH)]^-(aq) + H_3O^+(aq)$$

An equivalent interpretation is that an acidic oxide is an oxide that reacts with an aqueous base (an alkali):

$$CO_2(g) + OH^-(aq) \rightarrow [O_2C(OH)]^-(aq)$$

---

[3] Solution spectroscopic studies of the sulfite problem have been carried out by R. E. Connick and his collaborators. See *Inorg. Chem.*, **21**, 103 (1982) and **25**, 2414 (1986).

A **basic oxide** is an oxide to which a proton is transferred when it dissolves in water:

$$CaO(s) + H_2O(l) \rightarrow Ca^{2+}(aq) + 2OH^-(aq)$$

The equivalent interpretation in this case is that a basic oxide is an oxide that reacts with an acid:

$$CaO(s) + 2H^+(aq) \rightarrow Ca^{2+}(aq) + H_2O(l)$$

Because acidic and basic oxide character often correlates with other chemical properties, a wide range of properties can in fact be predicted from a knowledge of the character of oxides. In a number of cases the correlations follow from the basic oxides being largely ionic and of acidic oxides being largely covalent.

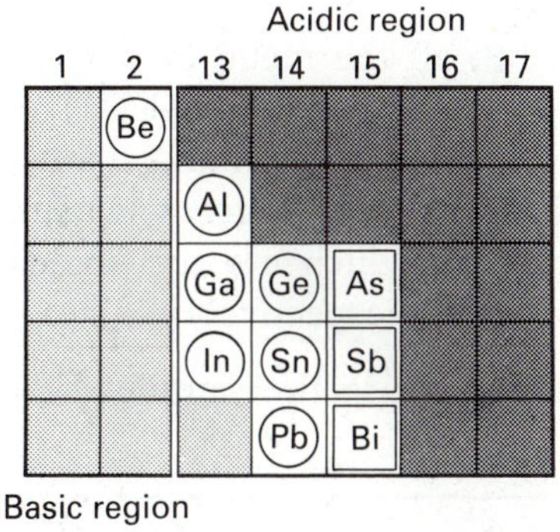

Amphoterism

An **amphoteric oxide** (from the Greek word meaning 'both') is an oxide that reacts with both acids and bases. Thus, aluminum oxide reacts with acids and alkalis:

$$Al_2O_3(s) + 6H_3O^+(aq) + 3H_2O(l) \rightarrow 2[Al(OH_2)_6]^{3+}(aq)$$

$$Al_2O_3(s) + 2OH^-(aq) + 3H_2O(l) \rightarrow 2[Al(OH)_4]^-(aq)$$

Amphoterism is observed for the lighter elements of Groups 2 and 13/III, as in BeO, Al_2O_3, and Ga_2O_3. It is also observed for some of the d-block elements, such as TiO_2 and V_2O_5, and some of the heavier elements of Groups 14/IV and 15/V, such as SnO_2 and Sb_2O_5. The location in the periodic table of elements that in their characteristic group oxidation states have amphoteric oxides is shown in Fig. 5.5. They lie on the frontier between acidic and basic oxides, and hence serve as an important guide to the character of an element.

An important issue in the d block is the oxidation number necessary for amphoterism. Figure 5.6 shows the oxidation number for which an element in the first row of the block has an amphoteric oxide. We see that on the left of the block, from titanium to manganese and perhaps iron, oxidation state +4 is amphoteric (with higher values on the border of acidic and lower values of the border of basic). On the

5.5 The elements in circles have amphoteric oxides even in their highest oxidation states. The elements in boxes have acidic oxides in their maximum oxidation states and amphoteric oxides at lower oxidation numbers.

5.6 The influence of oxidation number on the acid–base character of oxides of the elements in the first row of the d block. Mainly acidic oxidation states are represented in black. Mainly basic oxidation states are shown shaded. Oxidation numbers in the shaded region are amphoteric, with the center of the amphoteric zone indicated by unshaded circles.

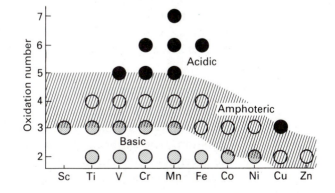

right of the block, amphoterism occurs at lower oxidation numbers: the oxidation states $+3$ for cobalt and nickel and $+2$ for copper and zinc are fully amphoteric. There is no simple way of predicting the onset of amphoterism, but it presumably reflects the ability of the metal cation to polarize the oxide ions that surround it, and this ability varies with the oxidation state of the metal.

Example 5.4: *Using oxide acidity in qualitative analysis*

In the traditional scheme of qualitative analysis, a solution of metal ions is oxidized and then aqueous ammonia is added to raise the pH. The ions Fe^{3+}, Ce^{3+}, Al^{3+}, Cr^{3+}, and V^{3+} precipitate as hydrous oxides. The addition of H_2O_2 and NaOH redissolves the aluminum, chromium, and vanadium oxides. Discuss these steps in terms of the acidities of oxides.

Answer. When the oxidation number is $+3$, all the metal oxides are sufficiently basic to be insoluble in a $pH \approx 10$ solution. Al(III) is amphoteric and redissolves in strong base to give aluminate ions, $[Al(OH)_4]^-$. V(III) and Cr(III) are oxidized by H_2O_2 to give vanadate ions, $[VO_4]^{3-}$, and chromate ions, $[CrO_4]^{2-}$, which are the anions derived from the acidic oxides V_2O_5 and CrO_3, respectively.

Exercise E5.4. If titanium(IV) ions were present in the sample, how would they behave?

Polyoxo compound formation

One of the most important aspects of the reactivity of acids containing the O—H group is the formation of condensation polymers. Polycation formation from simple aqua cations occurs with the loss of H_3O^+ ions:

$$2[Al(OH_2)_6]^{3+}(aq) \rightarrow [(H_2O)_5Al-OH-Al(OH_2)_5]^{5+}(aq) + H_3O^+(aq)$$

Polyanion formation from oxoanions occurs by protonation of an O atom and its departure as H_2O:

$$2[CrO_4]^{2-}(aq) + 2H_3O^+(aq) \rightarrow [O_3Cr-O-CrO_3]^{2-}(aq) + 3H_2O(l)$$

The importance of polyoxo anions can be judged by the fact that they account for most of the mass of oxygen in the Earth's crust, for they include almost all silicate minerals. They also include the phosphate polymers (such as ATP) used for energy storage in living cells. The silicates are so important that they are treated separately (Chapter 11).

5.7 Polymerization of aqua ions to polycations

As the pH of a solution is increased, the aqua ions of metals that have basic or amphoteric oxides generally undergo polymerization and precipitation. One application is to the separation of metal ions, because the precipitation occurs quantitatively at a different pH characteristic of each metal.

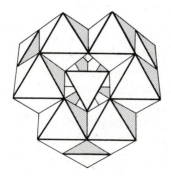

5.7 The structure of the $[AlO_4(Al(OH)_2)_{12}]^{7+}$ ion, with AlO_6 groups represented by octahedra around the central tetrahedron which represents the AlO_4 unit.

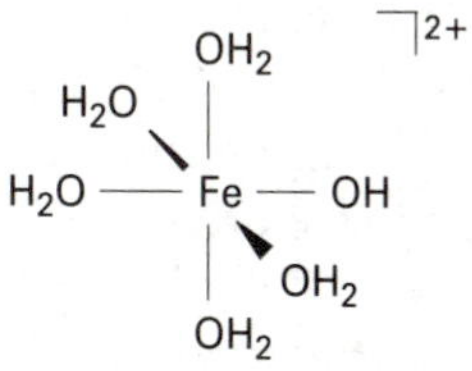

9 $[Fe(OH_2)_5OH]^{2+}$

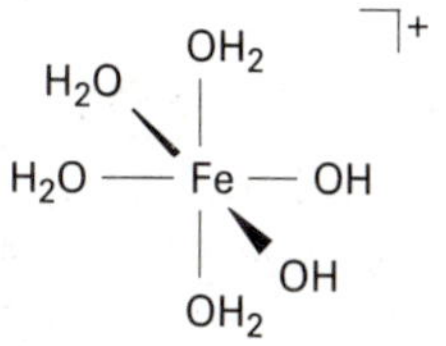

10 $[Fe(OH_2)_4(OH)_2]^+$

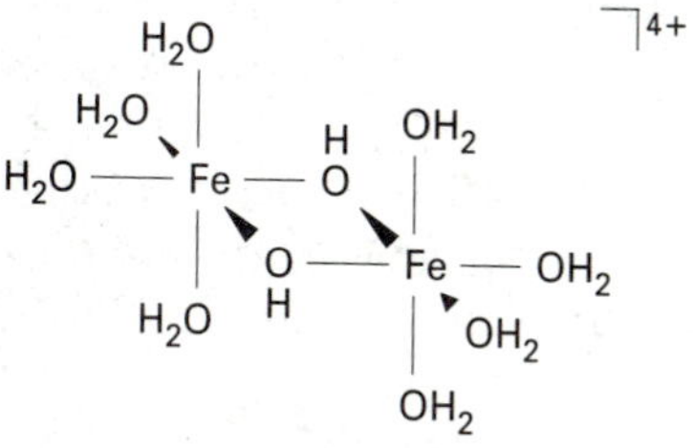

11 $[Fe_2(OH_2)_8(OH)_2]^{4+}$

12 $[AlO_4]^{5-}$

Only the most basic metals (of Groups 1 and 2) have no important solution species beyond the aqua ion. However, the solution chemistry becomes very rich as the amphoteric region of the periodic table is approached. The two most common examples are polymers formed by Fe(III) and Al(III), both of which are abundant in the Earth's crust. In acidic solutions, both form octahedral hexaaqua ions, $[Al(OH_2)_6]^{3+}$ and $[Fe(OH_2)_6]^{3+}$. In solutions of pH > 4, both precipitate as gelatinous hydrous oxides:

$$[Fe(OH_2)_6]^{3+}(aq) + nH_2O(l) \rightarrow Fe(OH)_3 \cdot nH_2O(s) + 3H_3O^+(aq)$$

$$[Al(OH_2)_6]^{3+}(aq) + nH_2O(l) \rightarrow Al(OH)_3 \cdot nH_2O(s) + 3H_3O^+(aq)$$

The precipitated polymers, which are often of colloidal dimensions, slowly crystallize to stable mineral forms.

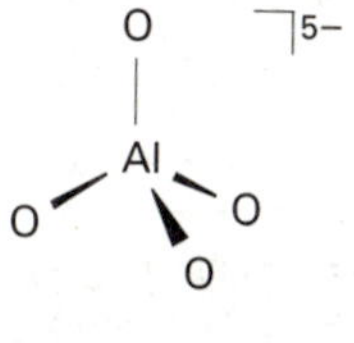

Aluminum and iron behave differently at intermediate pH, in the region between the existence of aqua ions and the occurrence of precipitation. Relatively few iron species have been characterized, but those that have been include two monomers (**9**, **10**), a dimer (**11**), and a polymer containing about 90 Fe atoms. In contrast, Al(III) forms a series of discrete polymeric cations in which the monomer unit consists of a central Al^{3+} ion surrounded tetrahedrally by four O atoms (**12**). A 'simple' polymeric cation of this kind is $[AlO_4(Al(OH)_2)_{12}]^{7+}$, with a charge of $+0.54$ per Al atom. An impression of the structure is given in Fig. 5.7, where the AlO_6 units are represented as octahedral blocks packed around the central tetrahedron; the larger Fe^{3+} ion does not fit so well in such a structure. This particular Al^{3+} polycation with 13 Al atoms has been shown to inhibit plant growth and to be extremely toxic to brook trout fry. It is now thought that polycations may be the most important toxic species leached by acid rain into lakes and soils. Studies using ^{27}Al-NMR show that the 13-Al polycation is a prominent species of aluminum in acidic organic soils.[4]

[4] D. Hunter and D. S. Ross, *Science*, **251**, 1056 (1991).

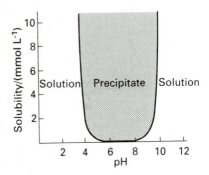

5.8 The variation of the solubility of Al_2O_3 with pH expressed as total concentration of Al. In the extreme acid region the aluminum is present as $[Al(OH_2)_6]^{3+}$. In the extreme basic region it is present as $[Al(OH)_4]^-$.

13 $[Ni_4(OH)_4]^{4+}$

14 $[VO_2(OH_2)_4]^+$

The extensive network structure of aluminum polymers, which are neatly packed in three dimensions, contrasts to the linear polymers of their iron analogs. Aluminum polycations and similar ions are used constructively in water treatment to precipitate anions (such as F^-) that are present as pollutants in effluents from aluminum refining plants.[5] As the pH is increased, H^+ ions are picked off these polycations and their charge is reduced. The pH at which the net charge is zero is known as the **point of zero charge**. Because both Fe(III) and Al(III) form amphoteric oxides, increasing the pH to a sufficiently high value can lead to the redissolution of their oxides as anions (Fig. 5.8).

The analogs of aqua ions in which H_2O is replaced by H_2S are not well known. However, just as sulfides like Fe_2S_3 can be thought of as analogs of oxides, there is an important family of sulfur-based polycations that can be considered to be the derivatives of the hydrogen sulfide analogs of aqua ions. For example, an extremely important structure involved in electron transport in living cells (Section 19.8) is based on the polysulfo ion $[Fe_4S_4]^{2+}$. This species is analogous to the $[Ni_4(OH)_4]^{4+}$ ion, which has alternating Ni^{2+} and OH^- groups on the eight corners of a distorted cube (**13**).

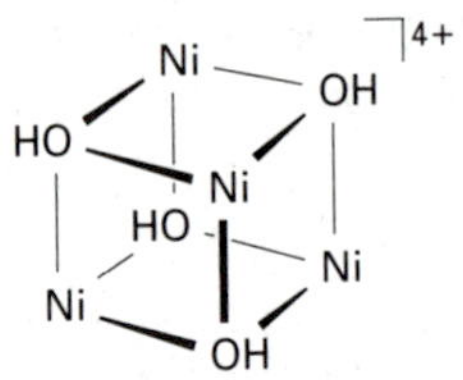

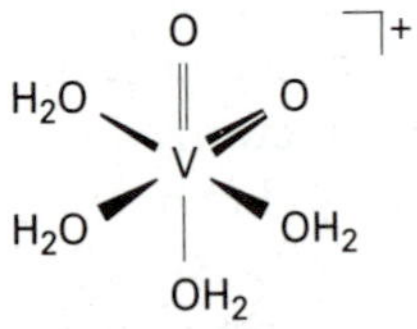

5.8 Polyoxoanions

Polyoxoanions form from the oxoacids of the early *d*-block elements in high oxidation states. Such polymerization is important for V(V), Mo(VI), W(VI), and (to a lesser extent) Nb(V), Ta(V), and Cr(VI); see Section 8.7.

A solution formed by dissolving the amphoteric oxide V_2O_5 in a strongly basic solution is colorless, and the dominant species is the tetrahedral $[VO_4]^{3-}$ ion (the analog of the colorless PO_4^{3-} ion). As the pH is decreased, the solution passes through a series of deeper colors from orange to red. This sequence indicates a complicated series of condensations and hydrolyses that yield ions including (in an order from basic to acidic solution) $[V_2O_7]^{4-}$, $[V_3O_9]^{3-}$, $[V_4O_{12}]^{4-}$, $[HV_{10}O_{28}]^{5-}$, and $[H_2V_{10}O_{28}]^{4-}$ (see Section 8.7). The successive decrease in anionic charge per V atom as the polyanions grow should be noted. The strongly acid solution is pale yellow and contains the hydrated $[VO_2]^+$ ion (**14**).

The polymerization of the nonmetal oxoanions is different from that of the *d*-metal analogs. The common species in solution are rings and chains. As we have remarked, the silicates are very important examples of polymeric oxoanions, and we discuss them in detail in Chapter 11. One example of a polysilicate mineral is $MgSiO_3$, which contains an infinite chain of SiO_3^{2-} units. In this section we illustrate some features of polyoxoanions using phosphates as examples.

[5] Some of the factors involved have been discussed by N. Parthasathey, J. Buffle, and W. Haerdi in *Canad. J. Chem.*, **64**, 24 (1986). They describe how coprecipitation by hydrous oxides depends on the different polycation structures that form under different conditions of addition of aluminum sulfate (alum) to the effluent and the changing initial effluent pH.

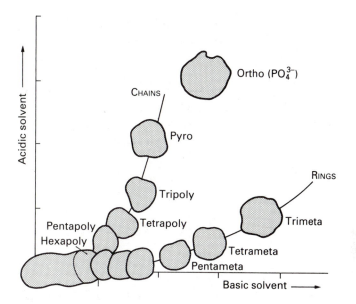

5.9 Two-dimensional paper chromatogram of a complex mixture of phosphates formed by condensation reactions. The sample spot was placed at the lower left, basic solvent separation was used first, followed by acidic solvent perpendicular to the first elution. The latter treatment separates open chain from cyclic phosphates.

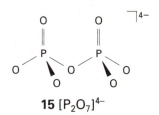

15 $[P_2O_7]^{4-}$

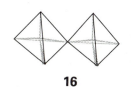

16

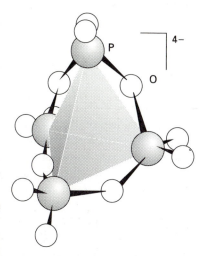

17 $[P_4O_{12}]^{4-}$

The simplest polymerization reaction, starting with the **orthophosphate ion**, PO_4^{3-}, is

$$2\,PO_4^{3-} \;+\; 2\,H^+ \;\longrightarrow\; \left[\, O-P-O-P-O \,\right]^{4-} \;+\; H_2O$$

Elimination of water consumes protons and reduces the average charge number of each P atom to -2. The diphosphate ion, $[P_2O_7]^{4-}$ (**15**), can be drawn as (**16**) if each phosphate group is represented as a tetrahedron with the O atoms located at the corners.

Phosphoric acid can be prepared by hydrolysis of the solid phosphorus(V) oxide, P_4O_{10}. An initial step using a limited amount of water produces a metaphosphate ion (**17**) with the formula $[P_4O_{12}]^{4-}$. This reaction is only the simplest among many, and the separation of products from the hydrolysis of phosphorus(V) oxide by chromatography reveals the presence of chain species with from 1 to 9 P atoms. Higher polymers are also present and can be removed from the column only by hydrolysis. A diagrammatic representation of a two-dimensional paper chromatogram is shown in Fig. 5.9. The upper spot sequence corresponds to linear polymers and the lower sequence corresponds to rings. Chain polymers P_n with $n = 10$ to 50 can be isolated as mixed amorphous glasses analogous to those formed by silicates and borates.

The biological importance of polyphosphates was mentioned at the outset. At physiological pH values (close to 7.4), the P—O—P bond is unstable with respect to hydrolysis. Consequently, its hydrolysis can serve as a mechanism for the delivery of free energy. Similarly, the formation of the P—O—P bond is a means of storing free energy.

The key to energy exchange in metabolism is the hydrolysis of ATP (**18**) to ADP (**19**):

$$ATP^{4-} + 2H_2O \rightarrow ADP^{3-} + HPO_4^{2-} + H_3O^+ \quad \Delta G^\ominus = -41 \text{ kJ at pH} = 7.4$$

Energy storage in metabolism depends on the subtle construction of pathways to make ATP from ADP. Energy is utilized metabolically by pathways that have evolved to exploit the delivery of driving force from the hydrolysis of ATP.

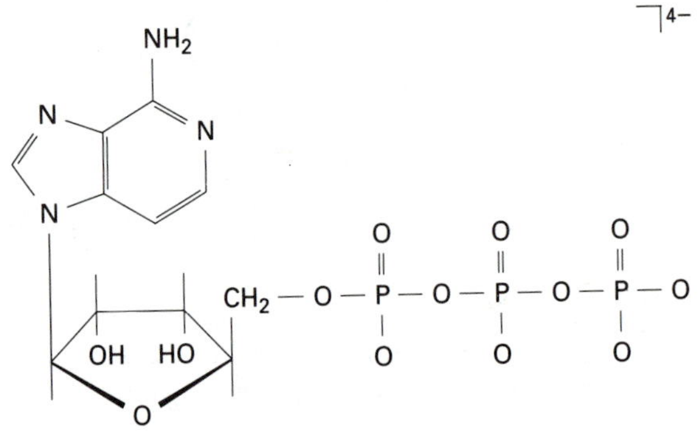

18 ATP^{4-}

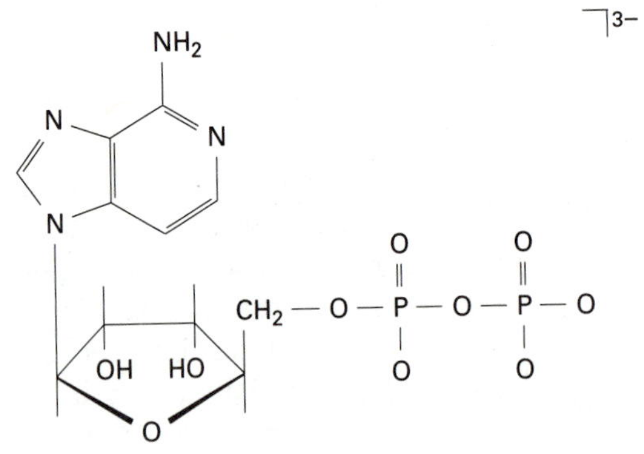

19 ADP^{3-}

THE LEWIS DEFINITIONS OF ACIDS AND BASES

The Brønsted–Lowry concept of acids and bases focuses on the transfer of a proton between species. While the approach is more general than any that preceded it, it still fails to take into account reactions between substances that show similar features but in which no protons are transferred. This deficiency was remedied by a more general concept of acidity introduced by G. N. Lewis in the same year as the Brønsted concepts (1923), but this approach became influential only in the 1930s.

A **Lewis acid** is a substance that acts as an electron pair acceptor. A **Lewis base** is a substance that acts as an electron pair donor. We denote a Lewis acid by A and a Lewis base by :B, often omitting any other lone pairs that may be present. The fundamental reaction of Lewis acids and bases is the formation of a **complex**, A—B, in which A and :B bond together by sharing the electron pair supplied by the base. It should be noted that the terms Lewis acidity and basicity are used in discussions of the equilibrium properties of reactions.[6]

5.9 Examples of Lewis acids and bases

The proton is a Lewis acid because it can attach to an electron pair. It follows that any Brønsted acid, since it provides protons, exhibits Lewis acidity too.[7] All Brønsted bases are Lewis bases, because a proton acceptor is also an electron pair donor. Therefore, the whole of the material presented in the preceding sections of this chapter can be regarded as a special case of Lewis's approach. However, because the proton is not a part of the definition, a wider range of substances can be classified as Lewis acids and bases than can be classified in the Brønsted scheme.

We meet many examples of Lewis acids later, but we should be alert to the following possibilities:

1. A metal cation can bond to an electron pair supplied by the base in a coordination compound.

This aspect of acids and bases will be treated at length in Chapter 6. An example is the hydration of Co^{2+}, where the O atom lone pairs of H_2O (acting as a Lewis base) donate to the central cation to give $[Co(OH_2)_6]^{2+}$. The cation is therefore the Lewis acid. We need to be alert for complex formation in which a cation, the acid, interacts with the π electrons of a base, as in the formation of the complex of Ag^+ and benzene (**20**).

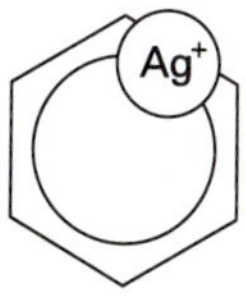

20 $[C_6H_6Ag]^+$

2. A molecule with an incomplete octet can complete its octet by accepting an electron pair.

A prime example is $B(CH_3)_3$, which can accept the lone pair of NH_3 and other donors:

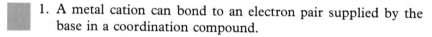

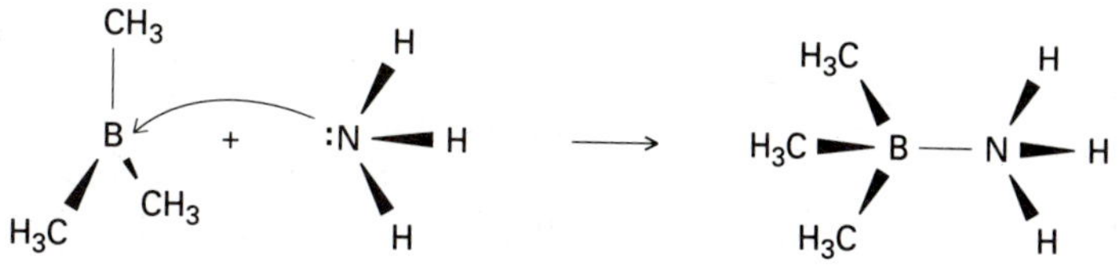

Hence, $B(CH_3)_3$ is a Lewis acid.

---

[6] In the context of reaction rates (Section 6.8), an electron-pair donor is called a nucleophile and an electron acceptor is called an electrophile.

[7] The Brønsted acid HA is the complex formed by the Lewis acid H^+ with the Lewis base A^-. This is why we said that a Brønsted acid 'exhibits' Lewis acidity rather than saying that a Brønsted acid *is* a Lewis acid.

3. A molecule or ion with a complete octet can rearrange its valence electrons and accept an additional electron pair.

For example, CO_2 acts as a Lewis acid when it forms HCO_3^- by accepting an electron pair from an O atom in an OH^- ion:

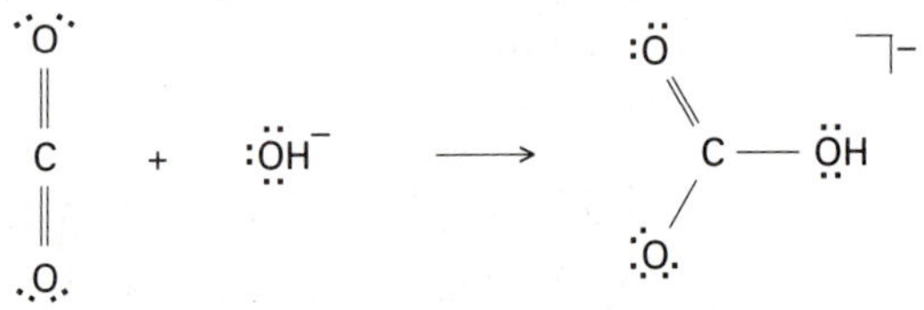

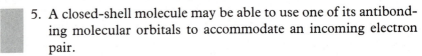

4. A molecule or ion may be able to expand its octet (or simply be large enough) to accept another electron pair.

An example is the formation of the complex $[SiF_6]^{2-}$ when two F^- ions (the Lewis bases) bond to SiF_4 (the acid).

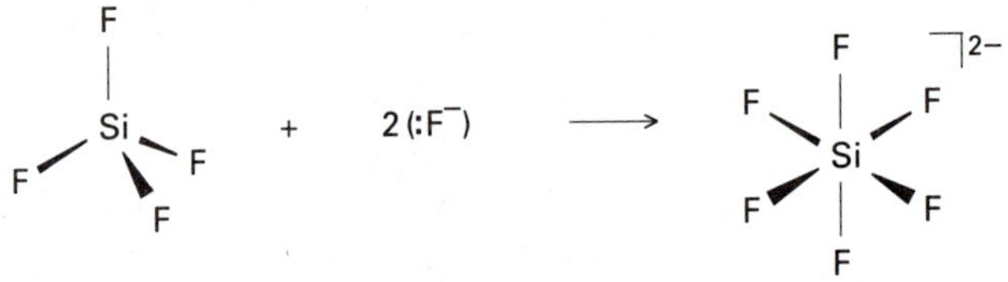

This type of Lewis acidity is common for the halides of the heavier *p*-block elements, such as SiX_4, AsX_3, and PX_5 (with X a halogen).

5. A closed-shell molecule may be able to use one of its antibonding molecular orbitals to accommodate an incoming electron pair.

An example of this behavior is the ability of tetracyanoethylene (TCNE, **21**) to accept a lone pair into its antibonding π^* orbital, and hence to act as an acid:

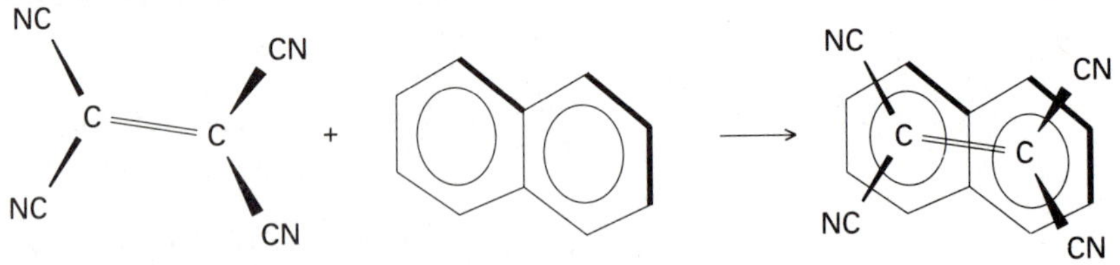

Two other examples of π-acids are tetracyanoxylylene (**22**) and picric acid (**23**):

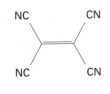

21 TCNE

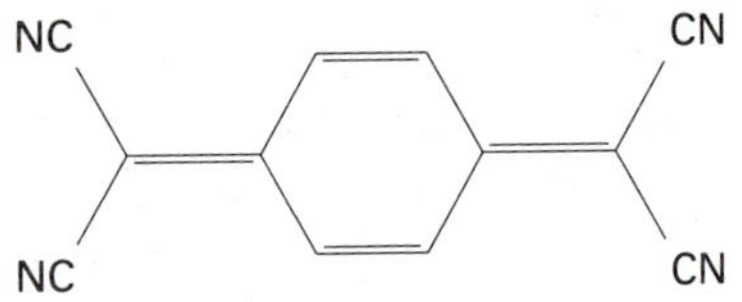

22 Tetracyanoxylylene

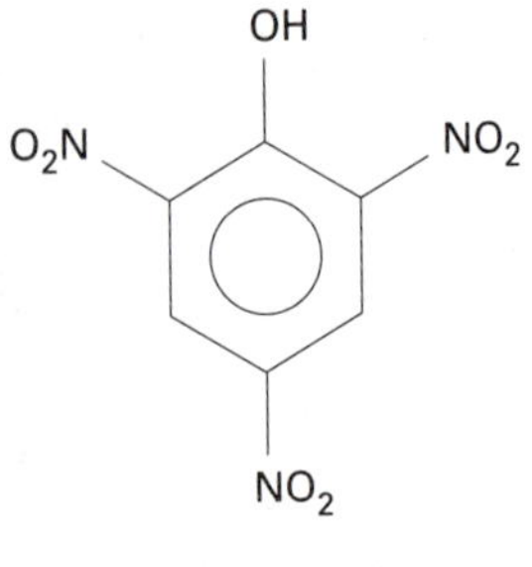

23 Picric acid

Example 5.5: *Identifying Lewis acids and bases*

Identify the Lewis acids and bases in the reactions (a) $BrF_3 + F^- \rightarrow [BrF_4]^-$, (b) $(CH_3)_2CO + I_2 \rightarrow (CH_3)_2CO—I_2$, (c) $KH + H_2O \rightarrow KOH + H_2$.

Answer. (a) The acid BrF_3 adds the base $:F^-$. (b) Acetone (propanone) acts as a base and donates a lone pair of electrons from O into an empty antibonding orbital of the I_2 molecule, which therefore acts as an acid. (c) The ionic hydride complex KH provides the base H^- to displace the acid H^+ from water to give H_2 along with KOH, in which the base OH^- is combined with the very weak acid K^+.

Exercise E5.5. Identify the acids and bases in the reactions (a) $FeCl_3 + Cl^- \rightarrow [FeCl_4]^-$, (b) $I^- + I_2 \rightarrow I_3^-$, (c) $[:SnCl_3]^- + (CO)_5MnCl \rightarrow (CO)_5Mn\!-\!SnCl_3 + Cl^-$.

All Brønsted bases are also Lewis bases; the metal ions in the coordination compounds that we meet in Chapter 6 are Lewis acids and the ligands are Lewis bases. That chapter will provide many examples and describe some periodic trends. For the remainder of this section we shall concentrate on the Lewis acids of *p*-block elements.

5.10 Boron and carbon group acids

The halides of boron and aluminum are among the most familiar Lewis acids. The planar BX_3 and AlX_3 molecules have incomplete octets, and the vacant *p* orbital perpendicular to the plane (**24**) can accept a lone pair from a Lewis base:

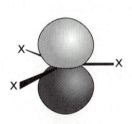

24 AlX_3 and BX_3

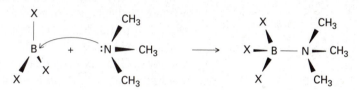

The acid molecule becomes pyramidal as the complex is formed.

Boron halides

The order of stability of complexes with BX_3 is

$$BF_3 < BCl_3 < BBr_3$$

This order is opposite to that expected on the basis of the relative electronegativities of the halogens. An electronegativity argument would suggest that fluorine, the most electronegative halogen, ought to leave the B atom in BF_3 most electron deficient and hence most acidic. The paradox is resolved by noting that the halogen atoms in the BX_3 molecule can form π bonds with the empty $B2p$ orbital (**25**), and that these π bonds must be broken to make the acceptor orbital available for complex formation. In a sense, complex formation is a subtle type of displacement reaction (Section 5.13) in which the incoming base, the amine, must displace the halogen ions from their role as π-donor bases. The small F atom is the best able to form π bonds with the $B2p$ orbital. (It is a general rule that *p–p* π bonding is strongest for Period 2 elements, largely on account of the small atomic radii of these elements and the significant overlap of their compact $2p$ orbitals.) Thus, the BF_3 molecule has the strongest π bond to be broken when the amine attacks and forms a complex.

Boron trifluoride is widely used as an industrial catalyst. Its role there is to extract bases bound to carbon and hence to generate carbocations:

25

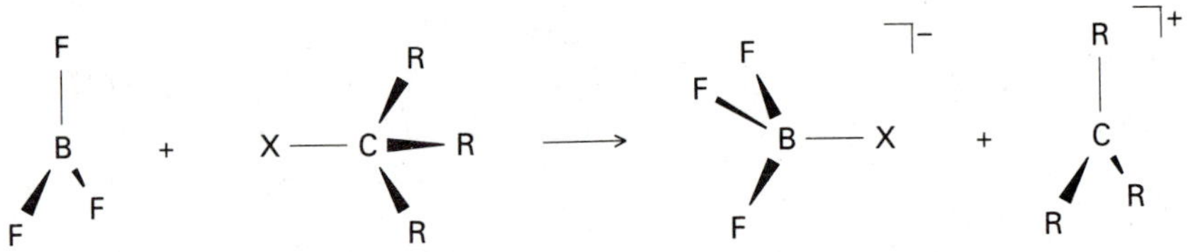

Boron trifluoride is a gas, but it dissolves in diethyl ether to give a solution that is convenient to use. Even this dissolution is an aspect of its Lewis acid character, for it forms a complex with the basic O atom of a solvent molecule.

Aluminum halides

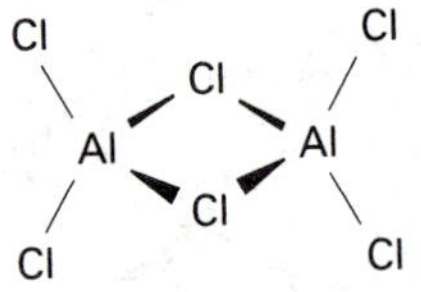

In contrast to the boron halides, the aluminum halides are dimers in the gas phase; aluminum chloride, for example, has molecular formula Al_2Cl_6 in the vapor (26). The existence of dimers illustrates the reduced tendency of p orbitals to participate in π bonding in Period 3 as compared with Period 2. The molecule is a 'self acid–base complex', because each Al atom acts as an acid toward a Cl atom initially belonging to the other Al atom.

Aluminum chloride is widely used as a Lewis acid catalyst of organic reactions. The classic examples are Friedel–Crafts alkylation (the attachment of R— to an aromatic ring) and acylation (the attachment of R—CO—):

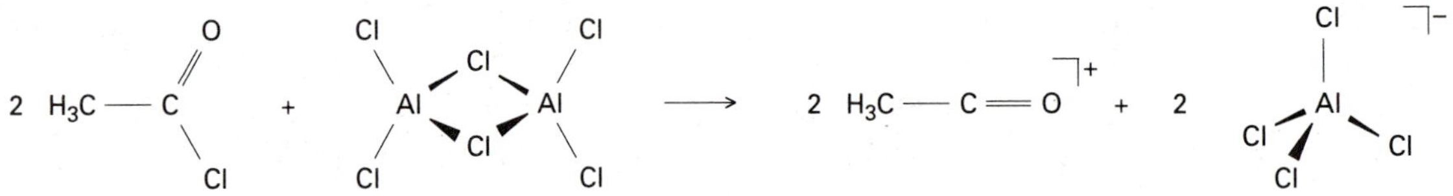

26 Al_2Cl_6

Silicon and tin complexes

Unlike carbon, an Si atom can expand its octet (or is simply large enough) to become hypervalent:

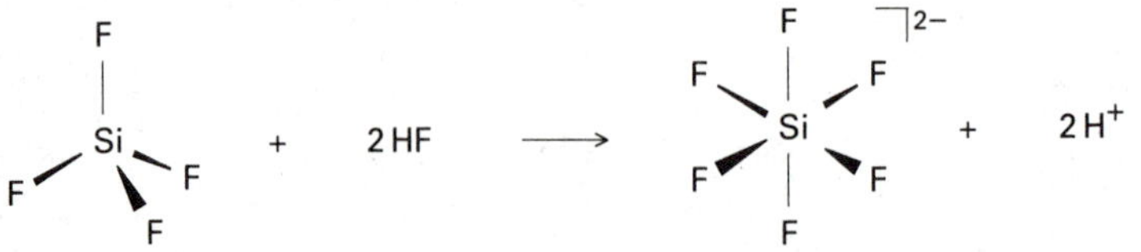

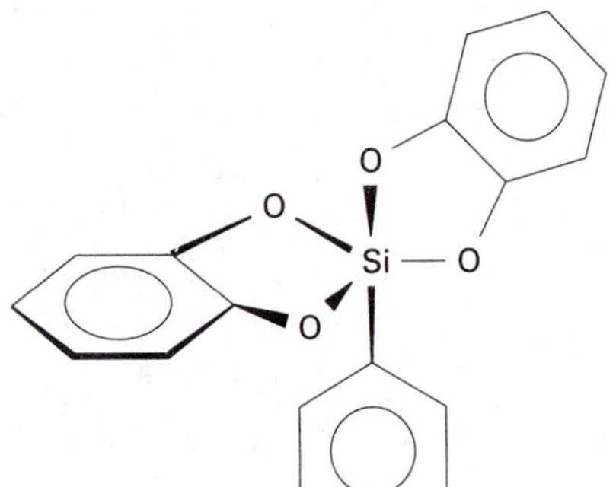

Because the Lewis base F^-, aided by a proton, can displace O^{2-} from Si, hydrofluoric acid is corrosive toward glass (SiO_2). The trend in acidity for SiX_4, which follows the order

$$SiI_4 < SiBr_4 < SiCl_4 < SiF_4$$

is the reverse of that for BX_3. This trend correlates with the increase in the electron-withdrawing power of the halogen from I to F. Coordination number 6, as in $[SiF_6]^-$, is not the only state of coordination for silicon above 4. For example, $C_6H_5—Si(—OC_6H_4O—)_2$ is a five-coordinate trigonal bipyramid (27).

27 $Si(C_6H_5)(C_6H_4O_2)_2$

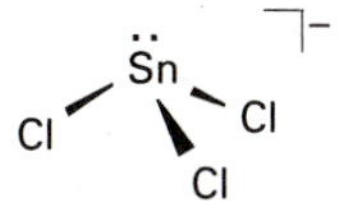

28 SnCl₃⁻

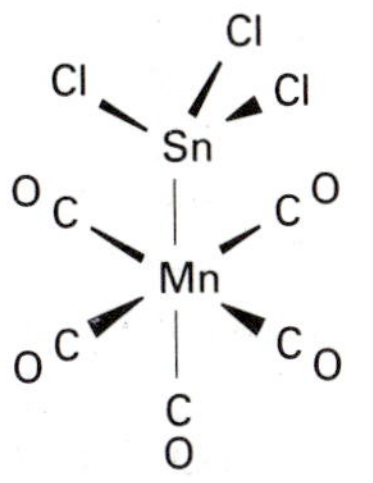

29 (CO)₅MnSnCl₃

$H_3Si \longrightarrow \overset{..}{\underset{..}{O}} \longrightarrow SiH_3$

$H_3Si = \overset{..}{O} \longrightarrow SiH_3$

$H_3Si \longrightarrow \overset{..}{O} = SiH_3$

30

Tin(II) chloride is both a Lewis acid and a Lewis base. As an acid, it forms $[SnCl_3]^-$ with a Cl^- ion (**28**). The complex retains a lone pair, and it is sometimes more revealing to write it as $:SnCl_3^-$. It acts as a base to give metal-metal bonds, such as in the complex $(CO)_5Mn—SnCl_3$ (**29**). Compounds like this are currently the focus of much attention in inorganic chemistry, as we see later in the text (Chapter 8).

Example 5.6: *Predicting the relative basicity of compounds*
Predict the relative Lewis basicity of (a) $(H_3Si)_2O$ and $(H_3C)_2O$; (b) $(H_3Si)_3N$ and $(H_3C)_3N$.

Answer. If delocalization of the O or N lone pairs occurs as shown in (**30**), the silyl ether and silyl amine should be the weaker Lewis base in each pair. This prediction is borne out by the observation that whereas complexes of $B(CH_3)_3$ with dimethyl ether and trimethylamine are readily observed, their silicon analogs do not form stable complexes.

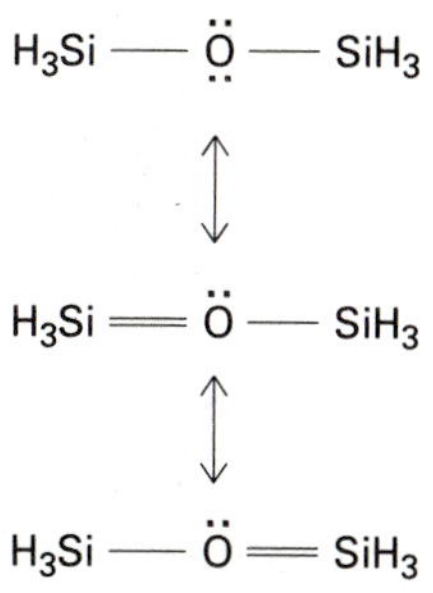

Exercise E5.6. If π bonding between Si and the lone pairs of N is important, what difference in structure between $(H_3Si)_3N$ and $(H_3C)_3N$ do you expect?

5.11 Nitrogen and oxygen group acids

The heavier elements of the nitrogen group (Group 15/V) form some of the most important Lewis acids, SbF_5 being among the most widely studied species. This Lewis acid can be used to produce some of the strongest Brønsted acids, as in the reaction

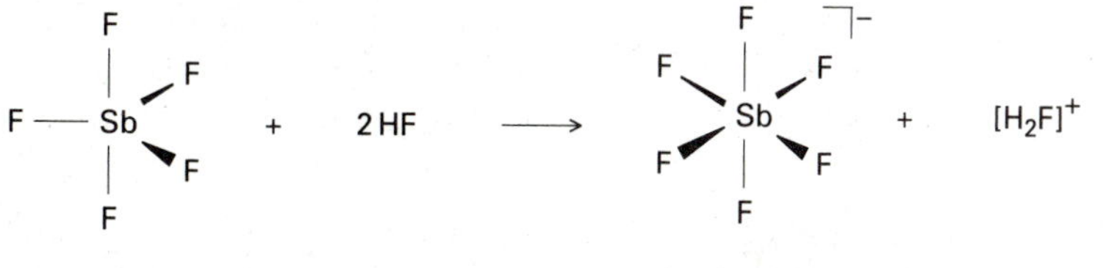

A **superacid** is a mixture that can protonate almost all organic compounds. One can be produced by dissolving SbF_5 in a mixture of HSO_3F and SO_3. The simplest of the many reactions occurring in this mixture is

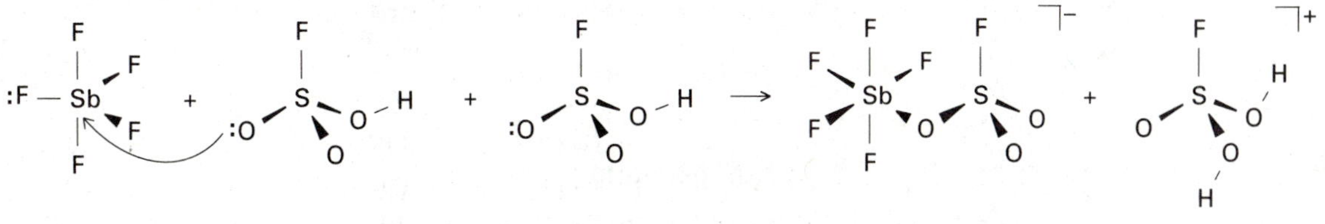

where the protonated fluorosulfuric acid acts as the powerful Brønsted acid.

Sulfur dioxide is both a Lewis acid and a Lewis base. Its acidity is weak and conventional:

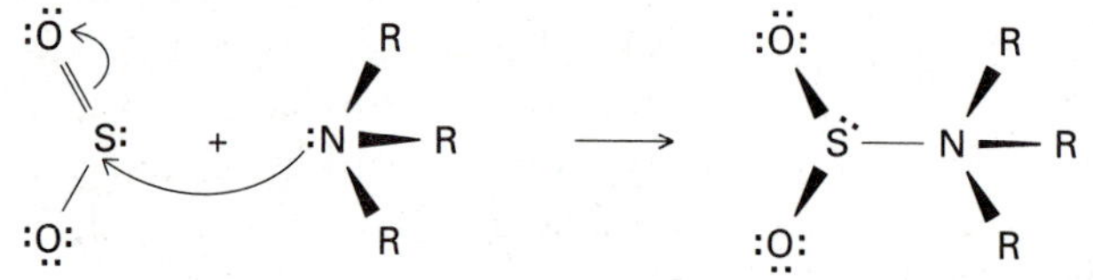

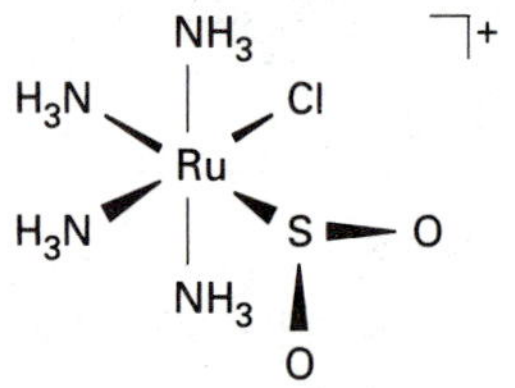

31 [RuCl(NH₃)₄(SO₂)]⁺

The SO_2 molecule acts as an ambidentate base because it can donate either its S or its O lone pairs to an appropriate acid. When SbF_5 is the acid, the O atom of SO_2 acts as the electron pair donor, but when Ru(II) is the acid, the S atom acts as donor (**31**).

Sulfur trioxide is a strong Lewis acid and a weak (O donor) Lewis base. Its acidity is illustrated by the reactions

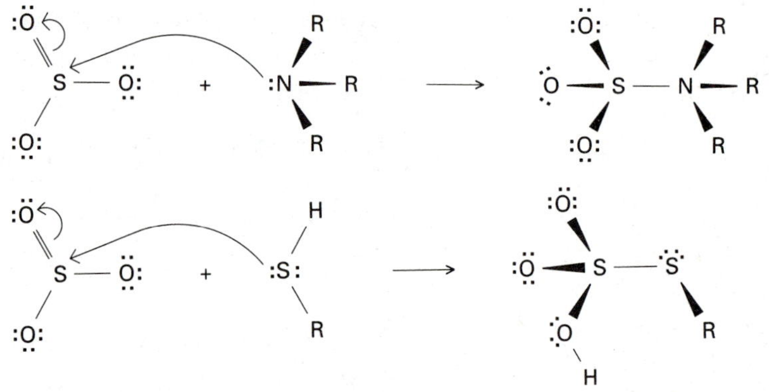

A classic aspect of the acidity of SO_3 is its highly exothermic reaction with water in the formation of sulfuric acid. The resulting problem of having to remove large quantities of heat from the reactor used for the commercial production of the acid is solved using another aspect of the trioxide's Lewis acidity. Prior to dilution, sulfur trioxide is dissolved in sulfuric acid to form the mixture known as oleum. This reaction is in fact an example of complex formation:

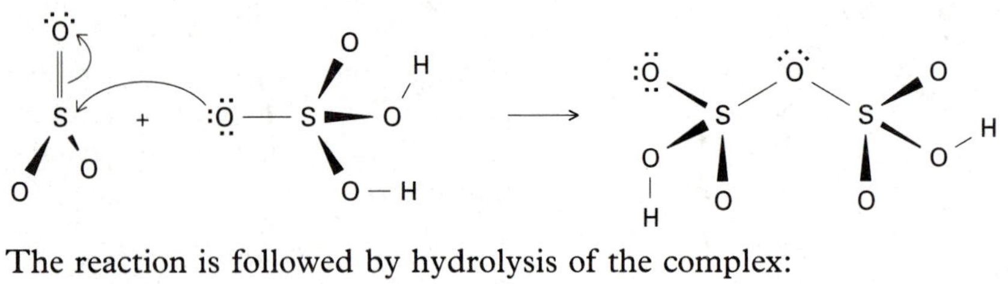

The reaction is followed by hydrolysis of the complex:

$$H_2S_2O_7 + H_2O \rightarrow 2H_2SO_4$$

5.12 Halogen acids

Lewis acidity is expressed in an interesting and subtle way by the dihalogen molecules, especially I_2 and Br_2. Iodine is violet in the solid and gas phases and in nondonor solvents such as trichloromethane; in

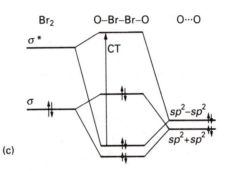

5.10 (a) The structure of $(CH_3)_2COBr_2$ as shown by X-ray diffraction. (b) The orbital overlap responsible for complex formation. (c) A partial molecular orbital energy level diagram for the interaction of the σ and σ^* orbitals of Br_2 with the appropriate symmetry-adapted combinations of sp^2 orbitals on the two O atoms. The donor–acceptor charge-transfer band in the near-UV is labeled CT.

water, acetone, or ethanol, all of which are Lewis bases, it is brown. Visible, UV, and IR spectroscopy of iodine dissolved in trichloromethane that also contains added donor molecules, such as $(CH_3)_2CO$, provide convincing evidence for 1:1 complex formation. The color changes because the solvent–solute complex (which is formed from the lone pair of donor molecule O atoms and a low-lying σ^* orbital of the dihalogen) has a strong optical absorption.

The strong visible absorption spectra of Br_2 and I_2 arise from transitions from filled to low-lying unfilled orbitals; they therefore provide a hint that the empty orbitals may be low enough in energy to serve as acceptor orbitals in Lewis acid–base complex formation.[8] The interaction of Br_2 with the carbonyl group of acetone is shown in Fig. 5.10. The illustration also shows the transition responsible for the new absorption band observed when a complex is formed. The orbital from which the electron originates in the transition is predominantly the lone pair orbital of the base (the ketone). The orbital to which the transition occurs is predominately the LUMO of the acid (the dihalogen). Thus, to a first approximation, the transition transfers an elec-

---

[8] The terms 'donor-acceptor complex' and 'charge-transfer complex' were at one time used to denote these complexes. However, the distinction between these and the more familiar Lewis acid–base complexes is arbitrary and in the current literature the terms are used interchangeably.

tron from the base to the acid and is therefore called a **charge–transfer transition**.

The tri-iodide ion (I_3^-) is an example of a complex between a halogen acid (I_2) and a halide base (I^-). One of the applications of its formation is to render molecular iodine water soluble so that it can be used as a titration reagent:

$$I_2(s) + I^-(aq) \rightarrow I_3^-(aq) \quad K = 725$$

The triiodide ion is one example of a large class of polyhalide ions (see Section 13.5).

THE REACTIONS OF LEWIS ACIDS AND BASES

In this section we establish the systematics of Lewis acid–base reactions, first by identifying the types of reaction that acids and bases can undergo, and then by assessing the driving force of the reactions in terms of the strength of the interaction between an acid and a base. Whereas proton-transfer reactions are somewhat limited in their variety (but of enormous importance), Lewis acid–base reactions, as befits a more general theory, show a much greater variety and need to be considered in terms of a number of classes of reaction.

5.13 The fundamental types of reaction

The simplest Lewis acid–base reaction in the gas phase is complex (or sometimes 'adduct') formation:

$$A + :B \rightarrow A{\rm —}B$$

Three examples are

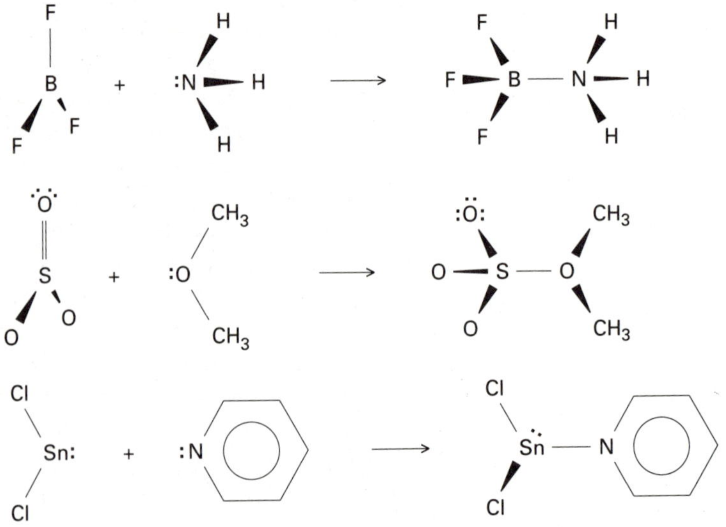

All three reactions involve Lewis acids and bases that are independently stable in the gas phase or in solvents that do not form complexes with them. Consequently, the individual species (as well as the complexes)

are easily studied experimentally. Figure 5.11 shows the interaction of orbitals responsible for the bonding in Lewis complexes. The exothermic character of the formation of the complex stems from the fact that the newly formed bonding orbital is populated by the two electrons supplied by the base and the newly formed antibonding orbital is left unoccupied.

Displacement reactions

A **displacement** of one Lewis base by another is a reaction of the form

$$B—A + :B' \rightarrow :B + A—B'$$

An example is

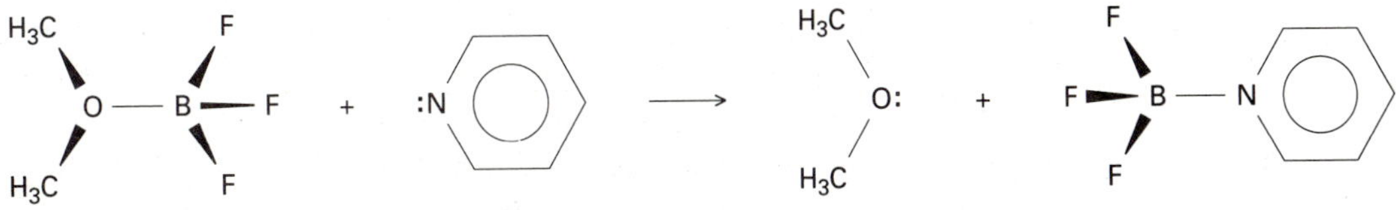

All Brønsted proton transfer reactions are of this type, as in

$$HS^-(aq) + H_2O(l) \rightarrow H_3O^+(aq) + S^{2-}(aq)$$

In this reaction, the Lewis base H_2O displaces the Lewis base S^{2-} from its complex with the acid H^+. Displacement of one acid by another is also possible, as in the reaction

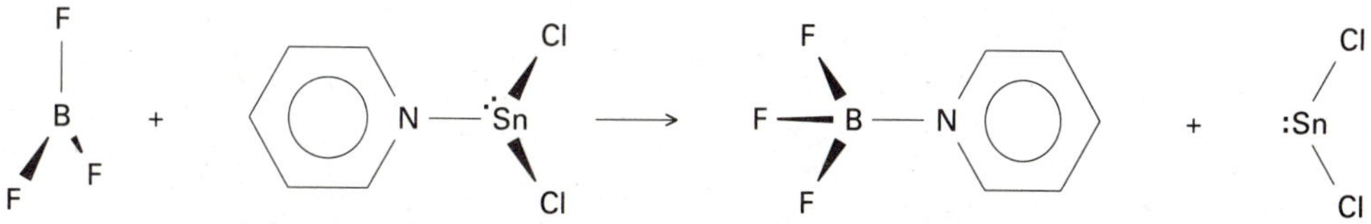

In the context of *d*-metal complexes, a displacement reaction (in which one ligand is driven out of the complex and is replaced by another), is generally called a **substitution reaction**.

Double displacement reactions

A **double displacement reaction** is the interchange of partners:

$$A—B + A'—B' \rightarrow A—B' + A'—B$$

The displacement of the base :B by :B' is assisted by the extraction of :B by the acid A'. An example is the reaction

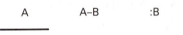

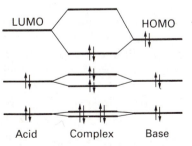

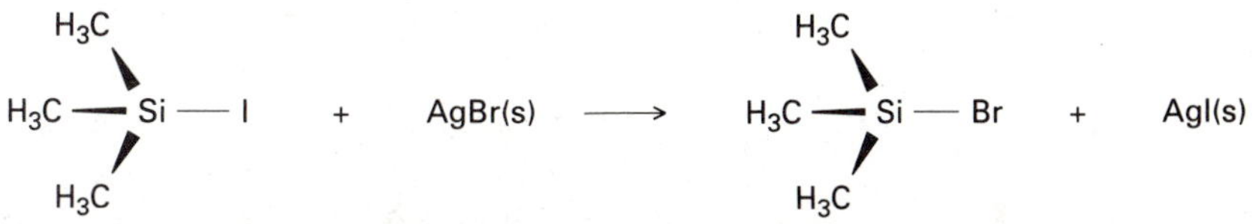

5.11 Localized molecular orbital representation of the interaction between frontier orbitals in the formation of a complex between a Lewis acid A and a Lewis base :B.

In this reaction, the base Br^- displaces the base I^-, the latter being extracted from its complex with Me_3Si^+ by the acid Ag^+.

5.14 Hard and soft acids and bases

The proton (H^+) was the key electron-pair acceptor in the discussion of Brønsted acid and base strengths. In the Lewis theory, we must allow for a greater variety of acceptors and hence a greater variety of factors that influence strength. When we consider more general acids, we find in many cases (among them Al^{3+}, Cr^{3+}, and BF_3) that there are excellent correlations between the order of affinity toward bases obtained with them and the order obtained when H^+ is used as the acid. However, a different order is observed for an acid such as Hg^{2+}.

The classification of acids and bases

To deal with the interactions of acids and bases containing elements drawn from throughout the periodic table, we need to consider at least two main classes of substance. These are the **hard and soft** acids and bases. This classification was introduced by R. G. Pearson and is a generalization—and a more evocative renaming—of the distinction between two types of behavior which originally were named simply 'class *a*' and 'class *b*', respectively, by Arland, Chatt, and Davies. The two classes are identified empirically by the opposite order of strengths (as measured by the equilibrium constant, K_f, for the formation of the complex) with which they form complexes with halide ion bases:

Class *a* bond in the order $I^- < Br^- < Cl^- < F^-$

Class *b* bond in the order $F^- < Cl^- < Br^- < I^-$

Figure 5.12 shows the trends in K_f for complex formation with a variety of halide ion bases. The equilibrium constants increase steeply from F^- to I^- when the acid is Hg^{2+}, indicating that Hg^{2+} belongs to class *b* and hence is a soft acid. The trend is less steep but in the same direction for Pb^{2+}, which indicates that the ion is a borderline soft acid. The trend is in the opposite direction for Zn^{2+}, so this ion belongs to class *a* and is classified as a borderline hard acid. The steep downward slope for Al^{3+} indicates that it is a hard acid.

For Al^{3+}, the binding strength increases as the electrostatic parameter, $\xi = z^2/r$, of the anion increases, which is consistent with an ionic model of the bonding. For Hg^{2+} the binding strength increases with increasing polarizability (the responsiveness of the electron distributions to perturbation). These two correlations suggest that hard acid cations form complexes in which simple coulombic interactions are dominant, and that soft acid cations form more complexes in which covalent bonding is important.

A similar classification can be applied to neutral molecular acids and bases. For example, the Lewis acid phenol forms a more stable complex by hydrogen bonding to $(C_2H_5)_2O$: than to $(C_2H_5)_2S$:. This behavior is analogous to the preference of Al^{3+} for F^- over Cl^-. In contrast, the Lewis acid I_2 forms a more stable complex with $(C_2H_5)_2S$:. We can conclude that phenol is hard (because it resembles Al^{3+}), whereas I_2 (which resembles Hg^{2+}), is soft.

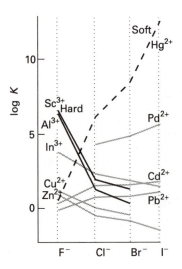

5.12 Trends in stability constants and the classification of cations as hard, borderline, and soft. Borderline ions are indicated by gray lines and may be borderline soft or borderline hard. This diagram, which is adapted from J. Burgess, *Metal ions in solution*, Ellis Horwood, Chichester (1988), emphasizes that there are degrees of hardness and softness.

Table 5.4 The classification of Lewis acids and bases*

Hard	Borderline	Soft
Acids H^+, Li^+, Na^+, K^+ Be^{2+}, Mg^{2+}, Ca^{2+} Cr^{2+}, Cr^{3+}, Al^{3+} SO_3, BF_3	Fe^{2+}, Co^{2+}, Ni^{2+} Cu^{2+}, Zn^{2+}, Pb^{2+} SO_2, BBr_3	Cu^+, Ag^+, Au^+, Tl^+, Hg^+ Pd^{2+}, Cd^{2+}, Pt^{2+}, Hg^{2+} BH_3
Bases F^-, OH^-, H_2O, NH_3 CO_3^{2-}, NO_3^-, O^{2-} SO_4^{2-}, PO_4^{3-}, ClO_4^-	$\underline{N}O_2^-$, SO_3^{2-}, Br^- N_3^-, N_2 C_6H_5N, $SC\underline{N}^-$	H^-, R^-, $\underline{C}N^-$, $\underline{C}O$, I^- $\underline{S}CN^-$, R_3P, C_6H_6 R_2S

*The underlined element is the site of attachment to which the classification refers.

In general, hard acids are identified empirically by their preferential binding of lighter basic atoms within a group:

$$\text{Hard acids } K_f: F^- \gg Cl^- > Br^-, I^- \quad R_2O \gg R_2S \quad R_3N \gg R_3P$$

The soft acids are identified empirically by exhibiting the opposite trend down each group:

$$\text{Soft acids } K_f: F^- \ll Cl^- < Br^- < I^- \quad R_2O \ll R_2S \quad R_3N \ll R_3P$$

It follows from the definition of hardness that

Hard acids tend to bind to hard bases.

Soft acids tend to bind to soft bases.

When species are analyzed with these rules in mind, it is possible to identify the classification summarized in Table 5.4.

The interpretation of hardness

The bonding between hard acids and bases can be described approximately in terms of ionic or dipole–dipole interactions. Soft acids and bases are more polarizable than hard acids and bases, and are more richly covalent. In Section 1.13 we saw that hard monatomic acids and bases are those with large energy separations between their frontier orbitals. In contrast, soft monatomic acids and bases have small frontier orbital energy separations. Molecular hardness and softness can be interpreted similarly. That is, when the frontier *molecular* orbital separation is small (Fig. 5.13), the electron distribution is easily distorted by an applied field (perhaps arising from a neighboring atom) and the molecule or ion is soft. When the frontier orbital separation is large, the electron distribution resists being distorted even when the perturbation is moderately strong. A hard acid does not have a low energy LUMO. A hard base has a low energy (strongly bound) HOMO. Because the electronic structures are barely perturbable, their interaction is primarily electrostatic. In contrast, the LUMO and HOMO of a soft acid–base pair rearrange substantially to give a covalent bond.

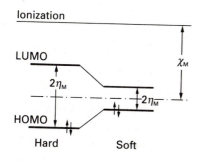

5.13 The relation between the frontier orbital separation in a molecule and the molecular hardness η_M. Both molecules have the same absolute molecular electronegativity χ_M.

Although this electronic interpretation of hardness in terms of the frontier orbitals is reasonably straightforward, it must never be forgotten that there are contributions to the free energy of complex formation other than the strength of the A—B bond. Among them are

1. The rearrangement of the substituents of the acid and base that may be necessary to permit formation of the complex.
2. Steric repulsion between substituents on the acid and the base.
3. In solution, the relative solvation energies of the acid, the base, and the complex.

Later in the chapter we shall see that these additional contributions can have a marked effect on the outcome of a reaction.

Chemical consequences of hardness

The concepts of hardness and softness help to rationalize a great deal of inorganic chemistry. For instance, they are useful for choosing preparative conditions and predicting the directions of reactions, and they help to rationalize the outcome of double-displacement reactions. However, the concepts must always be used with due regard for other factors that may affect the outcome of reactions. This deeper understanding of chemical reactions will grow in the course of the rest of the book. For the time being we shall limit the discussion to a few straightforward examples.

Hard and soft acids and bases are relevant to the terrestrial distribution of the elements. The tendency of soft acids to prefer soft bases and of hard acids to prefer hard bases explains certain aspects of the **Goldschmidt classification** of the elements into four types, a scheme widely used in geochemistry. Two of the classes are the **lithophile elements** and the **chalcophile elements**. The lithophile elements, which are found primarily in the Earth's crust (the lithosphere) in silicate minerals, include lithium, magnesium, titanium, aluminum, and chromium (as their cations). These cations are hard and are found in association with the hard base O^{2-}. On the other hand, the chalcophile elements are often found in combination with sulfide (and selenide and telluride) minerals, and include cadmium, lead, antimony, and bismuth. These elements (as their cations) are soft, and are found in association with the soft base S^{2-} (or Se^{2-} and Te^{2-}). Zinc cations are borderline hard, but softer than Al^{3+} and Cr^{3+}, and zinc is also often found as its sulfide.

Example 5.7: *Explaining the Goldschmidt classification*

The common ores of nickel and copper are sulfides. In contrast, aluminum is obtained from the oxide and calcium from the carbonate. Can these observations be explained in terms of hardness?

Answer. Both O^{2-} and CO_3^{2-} are hard bases; S^{2-} is a soft base. The cations Ni^{2+} and Cu^{2+} are considerably softer acids than Al^{3+} or Ca^{2+}. Hence the hard–hard and soft-soft rule accounts for the sorting observed.

Exercise E5.7. Of the metals cadmium, rubidium, chromium, lead, strontium, and palladium, which might be expected to be found in aluminosilicate minerals and which in sulfides?

An illustration on a somewhat smaller scale concerns the ability of the thiocyanate ion, SCN^-, to act as an ambidentate base, a molecule or ion that can donate an electron pair from more than one atom. The SCN^- ion is a base by virtue of both the harder N atom and the softer S atom; the ion binds to the hard Si atom through N. However, with a soft acid, such as a metal ion with a low oxidation number, the ion bonds through S. Platinum(II), for example, forms Pt—SCN in the complex $[Pt(SCN)_4]^{2-}$.

The Lewis viewpoint systematizes many reactions in solids and molten salt solutions. It recognizes that reactions often involve the transfer of a basic anion (typically O^{2-}, S^{2-}, or Cl^-) from one cationic acid center to another. For example, the reaction of CaO with SiO_2 to give the Ca^{2+} salt of the polyanion $[SiO_3^{2-}]_n$ can be regarded as the transfer of the base O^{2-} from the weak acid Ca^{2+} to the stronger acid 'Si^{4+}'. This reaction is a model for slag formation, which is used to remove silicates from the molten iron phase during reduction of iron ores in a blast furnace, and the floating of slag on top of iron is a microcosm of the core/mantle/crust division of the Earth. Similar molten salt and solid reactions are involved in the formation of glass and ceramics. In these reactions, alkali metal oxides or hydroxides transfer a basic O^{2-} ion to the acid silicate center.

Thermodynamic acidity parameters

An important alternative to the hard–soft classification of acids and bases makes use of an approach in which electronic, structural rearrangement, and steric effects are incorporated into a small set of parameters. A leading example of this approach parameterizes the reaction enthalpies of complex formation

$$A(g) + :B(g) \rightarrow A{-}B(g) \quad \Delta H^{\ominus}(A{-}B)$$

Many examples have been studied, among them

$$SO_3 + :NH_3 \rightarrow O_3S{-}NH_3$$

$$I_2 + :OEt_2 \rightarrow I_2{-}OEt_2$$

$$(CH_3)_3Al + :NC_5H_5 \rightarrow (CH_3)_3Al{-}NC_5H_5$$

$$Cl_2Sn + :N(CH_3)_3 \rightarrow Cl_2Sn{-}N(CH_3)_3$$

It has been found that the standard reaction enthalpies of reactions like these can be reproduced by the **Drago–Wayland equation**:

$$-\Delta H^{\ominus}(A{-}B)/kJ\ mol^{-1} = E_A E_B + C_A C_B$$

The parameters E and C were introduced with the idea that they

Table 5.5 Drago–Wayland parameters for some acids and bases*

	E	C
Acids		
Antimony pentachloride	15.1	10.5
Boron trifluoride	20.2	3.31
Iodine	2.05	2.05
Iodine monochloride	10.4	1.70
Phenol	8.86	0.904
Sulfur dioxide	1.88	1.65
Trichloromethane	6.18	0.325
Trimethylboron	12.6	3.48
Bases		
Acetone	2.02	4.67
Ammonia	2.78	7.08
Benzene	0.57	1.21
Dimethyl sulfide	0.702	15.26
Dimethylsulfoxide	2.76	5.83
Methylamine	2.66	12.0
p-Dioxane	2.23	4.87
Pyridine	2.39	13.10
Trimethylphosphane	1.72	13.40

*E and C parameters are often reported to give ΔH in kcal mol^{-1}; we have multiplied both by $\sqrt{(4.184)}$ to obtain ΔH in kJ mol^{-1}.

represented 'electrostatic' and 'covalent' factors, but in fact they must accommodate all factors except solvation. The compounds for which the parameters are listed in Table 5.5 satisfy the equation with an error of less than ± 3 kJ mol^{-1}, as do a much larger number of examples in the original papers.[9]

The Drago–Wayland equation is very successful and useful. In addition to giving the enthalpies of complex formation for over 1500 complexes, these enthalpies can be combined to calculate the enthalpies of displacement and double-displacement reactions. Moreover, the equation is useful for reactions of acids and bases in nonpolar, non-donor solvents as well as for reactions in the gas phase. The major limitation is that it is restricted to substances that can conveniently be studied in the gas phase or 'inert' solvents; hence, in the main it is limited to neutral molecules.

5.15 Solvents as acids and bases

Most solvents are either electron pair acceptors or donors and hence are either Lewis acids or bases. The chemical consequences of solvent acidity and basicity are considerable (see Box 5.1), for they help to account for the differences between reactions in aqueous and nonaqueous media. It follows that a displacement reaction often occurs when

[9] A good source of E and C parameters is R. S. Drago, N. Wong, C. Bilgrien, and C. Vogel, *Inorg. Chem.*, **26**, 9 (1987). For extensions of these concepts, see R. S. Drago, D. C. Ferris, and N. Wong, *J. Am. Chem. Soc.*, **112**, 8953 (1990) and R. S. Drago, N. Wong, and D. C. Ferris, *J. Am. Chem. Soc.*, **113**, 1970 (1990).

B1 Tetrahydrofuran

Box 5.1 Some useful nonaqueous solvents

A good example of a synthetically useful nonaqueous solvent is tetrahydrofuran, THF (**B1**), a cyclic nonpolar ether which boils at 66 °C. This weak hard base solvent can readily be dried and deoxygenated by distillation from sodium strips in a nitrogen atmosphere. After use in syntheses (including preparation of air-sensitive compounds) it can be easily removed from a reaction mixture by heating under reduced pressure. For these reasons, it is one of the most widely used solvents for the synthesis of organometallic compounds. The basicity of the O atom is sometimes important, for it can coordinate to cations. For example, the preparation of substituted metal carbonyls, $[M(CO)_5L]$ (L = phosphane, amine, etc.) can be accomplished by photolysis of $[M(CO)_6]$ in THF to give $[M(CO)_5(THF)]$ and carbon monoxide. This intermediate is then allowed to react with the entering ligand L. Other useful polar aprotic solvents for inorganic substances include methyl cyanide, CH_3CN, and dimethylsulfoxide, $(CH_3)_2SO$ (DMSO).

Ammonia is a strong hard base; this property is very important for its ability to coordinate to *d*-block acids and protons. It should not be forgotten that although ammonia is commonly regarded as a base, its own protons can act as centers of Lewis acidity by virtue of their ability to participate in hydrogen bonding. Liquid ammonia (b.p. −33 °C) is readily exploited as a solvent by using a Dewar flask. Despite a somewhat lower relative permittivity ($\varepsilon_r = 22$) than that of water, many salts of alkali metal cations with large anions are reasonably soluble in liquid ammonia. Organic compounds are often more soluble in liquid ammonia than they are in water.

One of the most remarkable reactions in liquid ammonia is the result of dissolution of alkali metals. Solutions of alkali metals are strongly reducing (Section 8.5). Electron paramagnetic resonance spectra (Section 14.9) show that the solutions contain unpaired electrons. The blue color typical of the solutions is the outcome of a very broad optical absorption band in the near IR with a maximum near 1500 nm. The metal is ionized in ammonia solution to give **solvated electrons**:

$$Na(s) + NH_3(l) \rightarrow Na^+(am) + e^-(am)$$

where am denotes ammonia solution. The blue solutions survive for long times at low temperature but decompose slowly to give hydrogen and sodium amide, $NaNH_2$. The exploitation of the blue solutions to produce compounds called **electrides** is discussed in Section 8.5.

Liquid hydrogen fluoride (b.p. 19.5 °C) is an acidic solvent with considerable Brønsted acid strength and a relative permittivity comparable to that of water. It is a good solvent for ionic substances. However, as it is both highly reactive and toxic, it presents handling problems, including its ability to etch glass. In practice, hydrogen fluoride is usually contained in polytetra-fluoroethylene and polychlorotrifluoroethylene vessels.

Although the conjugate base of HF is formally F^-, the ability of HF to form a strong hydrogen bond to F^- means that the conjugate base is better regarded as the bifluoride ion, FHF^-. Many fluorides are soluble in hydrogen fluoride as a result of the formation of this ion, for example

$$LiF(s) + HF(l) \rightarrow Li^+(hf) + [FHF]^-(hf)$$

where hf denotes solution in hydrogen fluoride. Because HF is quite acidic, protonation of the solute commonly accompanies bifluoride ion formation;

$$CH_3OH(l) + 2HF(l) \rightarrow [CH_3OH_2]^+(hf) + [FHF]^-(hf)$$

$$CH_3COOH(l) + 2HF(l) \rightarrow [CH_3COOH_2]^+(hf) + [FHF]^-(hf)$$

The second reaction is striking since acetic acid, an acid in water, is here acting as a base.

[References: W. L. Jolly, *The synthesis and characterization of inorganic compounds*, Waveland Press, Prospect Heights (1991); J. J. Lagowski (ed.), *The chemistry of nonaqueous solvents*, Vols. 1–5, Academic Press, New York (1966–78).]

a solute dissolves in a solvent, and that the subsequent reactions of the solution are also usually either displacements or double-displacements. For example, when antimony pentafluoride dissolves in bromine tri-fluoride, the following displacement reaction occurs:

$$SbF_5 + BrF_3 \rightarrow [BrF_2]^+ + [SbF_6]^-$$

In the reaction, the strong Lewis acid SbF_5 displaces the weaker Lewis acid BrF_2^+ from its complex with the Lewis base F^-. A more familiar example of the solvent as participant in a reaction is in Brønsted theory. In this theory, the acid (H^+) is always regarded as complexed with the solvent, as in H_3O^+ if the solvent is water, and reactions are treated as the transfer of the acid, the proton, from a basic solvent molecule to another base.

Example 5.8: *Accounting for properties in terms of the Lewis basicity of solvents*
Silver perchlorate, $AgClO_4$, is significantly more soluble in benzene than in alkane solvents. Account for this observation in terms of Lewis acid-base properties.

Answer. The π electrons of benzene, a soft base, are available for complex formation with the empty orbitals of the cation Ag^+, a soft acid (recall structure **20**). The species $[Ag\text{—}C_6H_6]^+$ is the complex of the acid Ag^+ with the base benzene using its π electrons.

Exercise E5.8. Boron trifluoride, BF_3, a hard acid, is used industrially as a solution in diethyl ether, $(C_2H_5)_2O:$, a hard base. Draw the structure of the complex that results from the dissolution of $BF_3(g)$ in $(C_2H_5)_2O(l)$.

Basic solvents

Solvents with Lewis base character are common. Most of the well known polar solvents, including water, alcohols, ethers, amines, dimethylsulfoxide (DMSO, $(CH_3)_2SO$), dimethylformamide (DMF, $(CH_3)_2NCHO$), and acetonitrile (CH_3CN), are hard Lewis bases. Dimethylsulfoxide is an interesting example of an ambidentate solvent that is hard on account of its O donor atom and soft on account of its S donor atom. Reactions of acids and bases in these solvents are generally displacements:

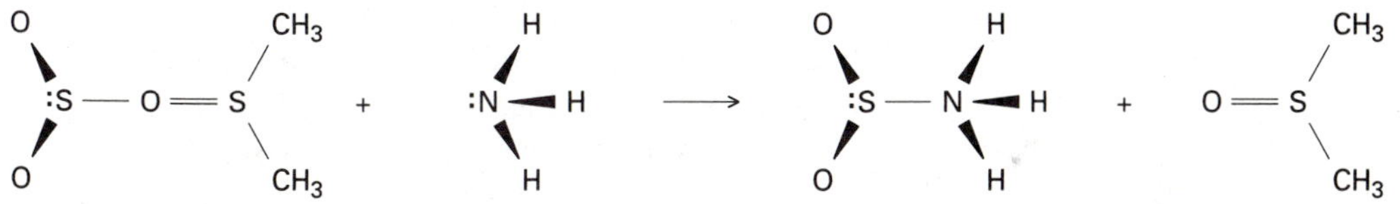

Acidic and neutral solvents

Hydrogen bonding is complex formation between A—H (the Lewis acid) and :B (the Lewis base) to give the complex conventionally denoted A—H···B. It is commonly the case that acidic solvents must be displaced if proton transfer to the base is to occur:

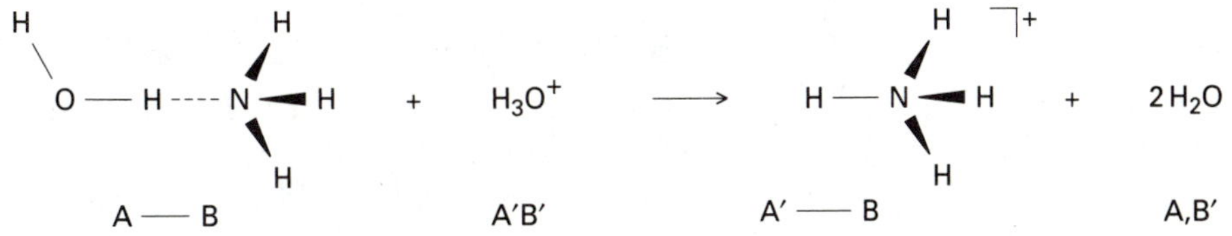

Liquid SO_2 is a good soft acidic solvent for dissolving the soft base benzene. Only the saturated hydrocarbons among common solvents lack significant Lewis acid or base character. Unsaturated hydrocarbons, however, may act as acids or bases by using their π or π^* orbitals as frontier orbitals. Electronegatively substituted alkanes such as haloalkanes (e.g. $CHCl_3$) are significantly acidic at the hydrogen atom.

Solvent parameters

The basicity of solvents can be expressed quantitatively in terms of the reaction enthalpy for the formation of a complex with a reference

Table 5.6 Donor and acceptor numbers and relative permittivities (dielectric constants) at 25 °C

Solvent	D.N.	A.N.	ε_r
Acetic acid		52.9	6.2
Acetone	17.0	12.5	20.7
Benzene	0.1	8.2	2.3
Carbon tetrachloride		8.6	2.2
Diethyl ether	19.2	3.9	4.3
Dimethylsulfoxide	29.8	19.3	45
Ethanol	19.0	37.1	24.3
Pyridine	33.1	14.2	12.3
Tetrahydrofuran	20.0	8.0	7.3
Water	18	54.8	81.7

Source: V. Gutmann, *Coordination chemistry in nonaqueous solutions.* Springer-Verlag, Berlin (1968).

acid. Victor Gutmann selected the strong Lewis acid $SbCl_5$ in 1,2-dichloroethane for this reference role, the relevant reaction being

$$SbCl_5 + :B \rightarrow Cl_5Sb\text{—}B \quad \Delta H^{\ominus}$$

The negative of $\Delta H^{\ominus}/\text{kcal mol}^{-1}$ (kilocalories are used for historical reasons) is called the **donor number** (D.N.) of the solvent. Some representative values are collected in Table 5.6: the higher the donor number, the stronger the Lewis base.

A corresponding parameter to measure solvent acidity, the **acceptor number** (A.N.), has been introduced. In this case triethylphosphane oxide, $(C_2H_5)_3PO:$, is used as a reference base and the NMR chemical shift of ^{31}P is measured for the base dissolved in the pure solvent. The zero of the scale is defined as the shift in hexane and the value 100 is ascribed to $SbCl_5$. On this arbitrary basis, numbers similar in magnitude to donor numbers are obtained (Table 5.6). As for donor numbers, the higher the acceptor number, the stronger the Lewis acid.

HETEROGENEOUS ACID–BASE REACTIONS

Some of the most important reactions involving the Lewis and Brønsted acidity of inorganic compounds occur at solid surfaces. For example, **surface acids**, which are solids with a high surface area and Lewis acid sites, are used as catalysts in the petrochemical industry for hydration and dehydration reactions. The surfaces of many materials that are important in the chemistry of soil and natural waters also have Brønsted and Lewis acid sites.[10]

Silica surfaces do not readily produce Lewis acid sites because —OH groups remain tenaciously attached at the surface of SiO_2 derivatives; as a result, Brønsted acidity is dominant. The Brønsted acidity of silica surfaces themselves is only moderate (and comparable to that

---

[10] An interesting account is found in G. Sposito, *The surface chemistry of soils*, Oxford University Press (1984).

of acetic acid). However, as we have already remarked, aluminosilicates display *strong* Brønsted acidity.

Surface reactions carried out using the Brønsted acid sites of silica gels are used to prepare thin coatings of a wide variety of organic groups using surface modification reactions such as

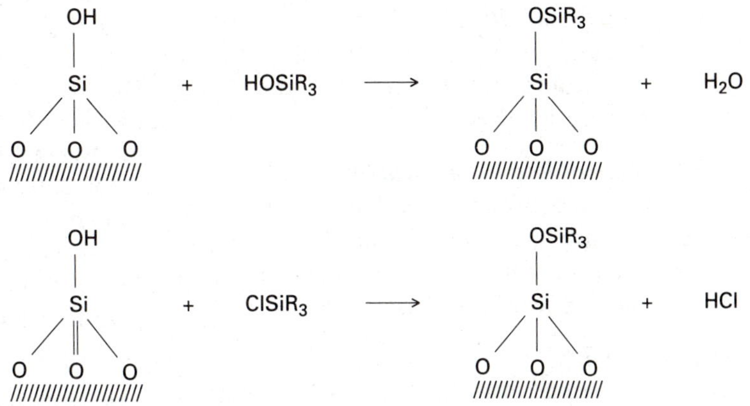

Thus silica gel surfaces can be modified to have affinities for specific classes of molecules. This greatly expands the range of stationary phases that can be used for chromatography. The surface OH groups on glass can be modified similarly, and glassware treated in this manner is sometimes used in the laboratory when proton-sensitive compounds are being studied.

FURTHER READING

R. P. Bell, *The proton in chemistry*. Cornell University Press, Ithaca (1969). A classic discussion of Brønsted acidity with many examples drawn from organic chemistry.

C. F. Baes, Jr and R. E. Messmer, *The hydrolysis of cations*. Wiley-Interscience, New York (1976). A survey of the acidities and the polymerization of aqua ions.

J. Burgess, *Ions in solution*. Ellis Horwood, Chichester, UK (1988). A readable account of ion solvation with an introduction to acidity and polymerization.

W. Stumm and J. J. Morgan, *Aquatic chemistry*. Wiley-Interscience, New York (1981). The standard text on the chemistry of natural waters.

Two general treatments of Lewis acids and bases are:

R. S. Drago and N. A. Matwiyoff, *Acids and bases*. Heath, Boston (1968).

W. B. Jensen, *The Lewis acid–base concepts*. Wiley, New York (1980).

More specialized books are:

R. G. Pearson, Chapter 1 in *Survey of progress in chemistry*, 1, ed. A. Scott. Academic Press, New York (1969). This is an account of the hard and soft classification from the originator of the terminology.

V. Gutmann, *Coordination chemistry in nonaqueous solutions*, Springer-Verlag, Berlin (1968). This monograph analyzes the role of solvents in detail.

KEY POINTS

1. Brønsted acids and bases
In the Brønsted definition, acids are proton donors and bases are proton acceptors.

2. Proton transfer equilibria
A Brønsted equilibrium exists in solution between conjugate acids and bases, and has the form $Acid_1 + Base_2 \rightleftharpoons Base_1 + Acid_2$.

3. Acidity and autoprotolysis constants
The strength of a Brønsted acid is expressed in terms of its acidity constant, K_a, and the extent of self-protonation of water is expressed in terms of the autoprotolysis constant of water, K_w.

4. Solvent leveling
Water has a leveling effect that brings the strengths of all stronger acids down to the acid strength of H_3O^+ and the strengths of all strong bases down to the strength of OH^-. Similar leveling effects are found in other solvents, such as liquid ammonia and methanol

5. Classes of oxoacid
Brønsted acids in which the acidic hydrogen atom is attached to an O atom are classified as aqua acids, hydroxoacids, and oxoacids. The observed strengths of mononuclear oxoacids can be systematized in terms of Pauling's rules.

6. Classes of oxides
Oxides are classified as acidic, amphoteric, and basic. The character of an oxide varies systematically through the periodic table and (particularly in the d block) with the oxidation state of the element.

7. Effect of pH on amphoteric oxides
As the pH of a solution is increased, the aqua ions of metals that have amphoteric oxides generally precipitate and then redissolve.

8. Lewis acids and bases
In the Lewis definition, acids are electron-pair acceptors and bases are electron-pair donors.

9. Varieties of Lewis acids
Lewis acids include metal cations in complexes, molecules with an incomplete octet, molecules with a complete octet that can rearrange their electrons to accept another pair, molecules or ions that can expand their octet, and closed shell molecules that can use their antibonding orbitals to accommodate electrons.

10. Reactions of Lewis acids and bases
The three important types of acid–base reaction are complex formation, displacement (substitution), and double displacement.

11. Strengths of Lewis acids and bases
Electronic and steric effects determine the strengths of Lewis acids and bases. Electronic effects are summarized by the distinction between hard and soft acids and bases.

12. E and C parameters
Quantitative, empirical correlation of the thermochemical aspects of complex formation is expressed by the E and C parameters.

13. Solvent properties
Solvents are usually Lewis acids or Lewis bases, and their abilities to act in this manner are summarized quantitatively by the donor and acceptor numbers of the solvent.

EXERCISES

5.1 Sketch an outline of the s and p blocks of the periodic table and indicate on it the elements that form (a) strongly acidic oxides and (b) strongly basic oxides, and (c) show the regions for which amphoterism is common.

5.2 Identify the conjugate bases corresponding to the following acids: $[Co(NH_3)_5(OH_2)]^{3+}$, HSO_4^-, CH_3OH, $H_2PO_4^-$, $Si(OH)_4$, HS^-.

5.3 Identify the conjugate acids of the bases C_5H_5N (pyridine), HPO_4^{2-}, O^{2-}, CH_3COOH, $[Co(CO)_4]^-$, CN^-.

5.4 List the bases HS^-, F^-, I^-, and NH_2^- in order of increasing proton affinity.

5.5 Aided by Fig. 5.3 (taking solvent leveling into account), identify which bases from the following lists are (a) too strong to be studied experimentally; (b) too weak to be studied experimentally; or (c) of directly measurable base strength.

   (i) CO_3^{2-}, O^{2-}, ClO_4^-, and NO_3^- in water

   (ii) HSO_4^-, NO_3^-, ClO_4^- in H_2SO_4

5.6 The aqueous solution pK_a values for HOCN, H_2NCN, and CH_3CN are approximately 4, 10.5, and 20 (estimated), respectively. Explain the trend in these $-CN$ derivatives of binary acids and compare them with H_2O, NH_3, and CH_4. Is $-CN$ electron donating or withdrawing?

5.7 The pK_a value of $HAsO_4^{2-}$ is 11.6. Is this value consistent with the two Pauling rules?

5.8 Draw the structures and indicate the charges of the tetraoxoanions of $X = Si$, P, S, and Cl. Summarize and account for the trends in the pK_a values of their conjugate acids.

5.9 Which of the following pairs is the stronger acid? Give reasons for your choice. (a) $[Fe(OH_2)_6]^{3+}$ or $[Fe(OH_2)_6]^{2+}$, (b) $[Al(OH_2)_6]^{3+}$ or $[Ga(OH_2)_6]^{3+}$, (c) $Si(OH)_4$ or $Ge(OH)_4$, (d) $HClO_3$ or $HClO_4$, (e) H_2CrO_4 or $HMnO_4$, (f) H_3PO_4 or H_2SO_4.

5.10 Arrange the oxides Al_2O_3, B_2O_3, BaO, CO_2, Cl_2O_7, SO_3 in order from the most acidic through amphoteric to the most basic.

5.11 Arrange the acids HSO_4^-, H_3O^+, H_4SiO_4, CH_3GeH_3, NH_3, HSO_3F in order of increasing acid strength.

5.12 The ions Na^+ and Ag^+ have similar radii. Which aqua ion is the stronger acid? Why?

5.13 Which of the elements Al, As, Cu, Mo, Si, B, Ti form oxide polyanions and which form oxide polycations?

5.14 When a pair of aqua cations forms an $M-O-M$ bridge with the elimination of water, what is the general rule for the change in charge per M atom on the ion?

5.15 Write a balanced equation for the formation of $P_2O_7^{4-}$ from PO_4^{3-}. Write a balanced equation for the dimerization of the complex $[Fe(OH_2)_6]^{3+}$ to give $[(H_2O)_4Fe(OH)_2Fe(OH_2)_4]^{4+}$.

5.16 Write balanced equations for the main reaction occurring when (a) H_3PO_4 and Na_2HPO_4 and (b) CO_2 and $CaCO_3$ are mixed in aqueous media.

5.17 Sketch the *p* block of the periodic table. Identify as many elements as you can that form Lewis acids in one of their oxidation states and give the formula of a representative Lewis acid for each element.

5.18 For each of the following processes, identify the acids and bases involved and characterize the process as complex formation or acid–base displacement. Identify the species that exhibit Brønsted acidity as well as Lewis acidity.
  (a) $SO_3 + H_2O \rightarrow HSO_4^- + H^+$
  (b) $CH_3[B_{12}] + Hg^{2+} \rightarrow [B_{12}]^+ + CH_3Hg^+$; $[B_{12}]$ designates the Co-macrocycle, vitamin B_{12}.
  (c) $KCl + SnCl_2 \rightarrow K^+ + [SnCl_3]^-$
  (d) $AsF_3(g) + SbF_5(l) \rightarrow [AsF_2]^+[SbF_6]^-(s)$
  (e) Ethanol dissolves in pyridine to produce a nonconducting solution.

5.19 Select the compound on each line with the named characteristic and state the reason for your choice.
  (a) Strongest Lewis acid:

    BF_3 BCl_3 BBr_3

    $BeCl_2$ BCl_3

    $B(n\text{-}Bu)_3$ $B(t\text{-}Bu)_3$

  (b) More basic toward $B(CH_3)_3$:

    Me_3N Et_3N

    $2\text{-}CH_3C_5H_4N$ $4\text{-}CH_3C_5H_4N$

5.20 Using hard–soft concepts, which of the following reactions are predicted to have an equilibrium constant greater than 1? Unless otherwise stated, assume gas phase or hydrocarbon solution and 25 °C.
  (a) $R_3PBBr_3 + R_3NBF_3 \rightleftharpoons R_3PBF_3 + R_3NBBr_3$
  (b) $SO_2 + (C_6H_5)_3PHOC(CH_3)_3 \rightleftharpoons (C_6H_5)_3PSO_2 + HOC(CH_3)_3$
  (c) $CH_3HgI + HCl \rightleftharpoons CH_3HgCl + HI$
  (d) $[AgCl_2]^-(aq) + 2CN^-(aq) \rightleftharpoons [Ag(CN)_2]^-(aq) + 2Cl^-(aq)$

5.21 The molecule $(CH_3)_2N-PF_2$ has two basic atoms, P and N. One is bound to B in a complex with BH_3, the other to B in an complex with BF_3. Decide which is which and state your reason.

5.22 The enthalpies of reaction of trimethylboron with NH_3, CH_3NH_2, $(CH_3)_2NH$, and $(CH_3)_3N$ are -58, -74, -81, and -74 kJ mol^{-1}, respectively. Why is trimethylamine out of line?

5.23 With the aid of the table of E and C values, discuss the relative basicity in (a) acetone and dimethylsulfoxide, (b) dimethylsulfide and dimethylsulfoxide. Comment on a possible ambiguity for DMSO.

5.24 Give the equation for the dissolution of SiO_2 glass by HF and interpret the reaction in terms of Lewis and Brønsted acid–base concepts.

5.25 Aluminum sulfide, Al_2S_3, gives off a foul odor characteristic of hydrogen sulfide when it becomes damp. Write a balanced chemical equation for the reaction and discuss it in terms of acid–base concepts.

5.26 Describe the solvent properties which would (a) favor displacement of Cl^- by I^- from an acid center, (b) favor basicity of R_3As over R_3N, (c) favor acidity of Ag^+ over Al^{3+}, (d) promote the reaction $2FeCl_3 + ZnCl_2 \rightarrow Zn^{2+} + 2[FeCl_4]^-$. In each case, suggest a specific solvent which might be suitable.

5.27 Why are strongly acidic solvents (e.g. SbF_5/HSO_3F) used in the preparation of cations like I_2^+ and Se_8^+, whereas strongly basic solvents are needed to stabilize anionic species such as S_4^{2-} and Pb_9^{4-}?

5.28 The Lewis acid $AlCl_3$ catalysis of the acylation of benzene was described in Section 5.10. Propose a mechanism for a similar reaction catalyzed by an alumina surface.

5.29 Use acid–base concepts to comment on the fact that the only important ore of mercury is cinnabar, HgS, whereas zinc occurs in nature as sulfides, silicates, carbonates, and oxides.

5.30 Write balanced Brønsted acid–base equations for the dissolution of the following compounds in liquid hydrogen fluoride: (a) CH_3CH_2OH, (b) NH_3, (c) C_6H_5COOH.

5.31 Is the dissolution of silicates in HF a Lewis acid–base reaction, a Brønsted acid–base reaction, or both?

5.32 The f-block elements are found as M(III) lithophiles in silicate minerals. What does this indicate about their hardness?

5.33 Consider the reaction forming metasilicates from carbonates:

$$nCaCO_3(s) + nSiO_2(s) \rightarrow [CaSiO_3]_n + nCO_2(g)$$

Identify the stronger acid between SiO_2 and CO_2.

5.34 The ores of titanium, tantalum, and niobium may be brought into solution near $800\,°C$ using sodium disulfate. A simplified version of the reaction is

$$TiO_2 + Na_2S_2O_7 \rightarrow Na_2SO_4 + TiO(SO_4)$$

Identify the acids and bases.

5.35 Sketch the shapes of AsF_5 and its complex with F^- (use VSEPR if necessary) and identify their point groups. What are the point groups of X_3BNH_3 and Al_2Cl_6?

PROBLEMS

5.1 In analytical chemistry a standard trick for improving the detection of the stoichiometric point in titrations of weak bases with strong acids is to use acetic acid as a solvent. Explain the basis of this approach.

5.2 In the gas phase, the base strength of amines increases regularly along the series $NH_3 < CH_3NH_2 < (CH_3)_2NH < (CH_3)_3N$. Consider the role of steric effects and the electron-donating ability of CH_3 in determining this order. In aqueous solution, the order is reversed. What solvation effect is likely to be responsible?

5.3 The hydroxoacid $Si(OH)_4$ is weaker than H_2CO_3. Write balanced equations to show how dissolving a solid M_2SiO_4 can lead to a reduction in the pressure of CO_2 over an aqueous solution. Explain why silicates in ocean sediments might limit the increase of CO_2 in the atmosphere.

5.4 The precipitation of $Fe(OH)_3$ discussed in this chapter is used to clarify waste waters, because the gelatinous hydrous oxide is very efficient at the coprecipitation of some contaminants and the entrapment of others. The solubility constant of $Fe(OH)_3$ is $K_s = [Fe^{3+}][OH^-]^3 \approx 10^{-38}$. Since the autoprotolysis constant of water links $[H_3O^+]$ to $[OH^-]$ by $K_w = [H_3O^+][OH^-] = 10^{-14}$, we can rewrite the solubility constant by substitution as $[Fe^{3+}]/[H^+]^3 = 10^4$. (a) Balance the chemical equation for the precipitation of $Fe(OH)_3$ when iron(III) nitrate is added to water. (b) If 6.6 kg of $Fe(NO_3)_3 \cdot 9H_2O$ is added to 100 L of water, what is the final pH of the solution and the molar concentration

of $Fe^{3+}(aq)$, neglecting other forms of dissolved Fe(III)? Give formulas for two most important Fe(III) species that have been neglected in this calculation.

5.5 The frequency of the symmetrical M—O stretching vibration of the octahedral aqua ions $[M(OH_2)_6]^{2+}$ increases along the series, $Ca^{2+} < Mn^{2+} < Ni^{2+}$. How does this trend relate to acidity?

5.6 An electrically conducting solution is produced when $AlCl_3$ is dissolved in the basic polar solvent CH_3CN. Give formulas for the most probable conducting species and describe their formation using Lewis acid–base concepts.

5.7 The complex anion $[FeCl_4]^-$ is yellow whereas $[Fe_2Cl_6]$ is reddish. Dissolution of 0.1 mol $FeCl_3(s)$ in 1 L of either $POCl_3$ or $PO(OR)_3$ produces a reddish solution which turns yellow on dilution. Titration of red solutions in $POCl_3$ with Et_4NCl solutions leads to a sharp color change (from red to yellow) at 1:1 mole ratio of $FeCl_3/Et_4NCl$. Vibrational spectra suggest that oxochloride solvents form adducts with typical Lewis acids via coordination of oxygen. Compare the following two sets of reactions as possible explanations of the observations.
  (a) $Fe_2Cl_6 + 2POCl_3 \rightleftharpoons 2[FeCl_4]^- + 2[POCl_2]^+$
        $POCl_2^+ + Et_4NCl \rightleftharpoons Et_4N^+ + POCl_3$
  (b) $Fe_2Cl_6 + 4POCl_3 \rightleftharpoons [FeCl_2(OPCl_3)_4]^+ + [FeCl_4]^-$
Both equilibria are shifted to products by dilution.

5.8 In the traditional scheme for the separation of metal ions from solution that is the basis of qualitative analysis, ions of Au, As, Sb, and Sn precipitate as sulfides but redissolve on addition of excess ammonium polysulfide. In contrast, ions of Cu, Pb, Hg, Bi, and Cd precipitate as sulfides but do not redissolve. In the language of this chapter, the first group is amphoteric for reactions involving SH^- in place of OH^-. The second group is less acidic. Locate the amphoteric boundary in the periodic table for sulfides implied by this information. Compare this boundary with the amphoteric boundary for hydrous oxides in Fig. 5.5. Does this analysis agree with describing S^{2-} as a softer base than O^{2-}?

5.9 The compounds SO_2 and $SOCl_2$ can undergo an exchange of radioactively labeled sulfur. The exchange is catalyzed by Cl^- and $SbCl_5$. Suggest mechanisms for these two exchange reactions with the first step being the formation of an appropriate complex.

5.10 In the reaction of t-butyl bromide with $Ba(NCS)_2$, the product is 91 percent S-bound t-butyl-SCN. However, if $Ba(NCS)_2$ is impregnated into solid CaF_2, the yield is higher and the product is 99 percent t-butyl-NCS. Discuss the effect of alkaline earth metal salt support on the hardness of the ambident nucleophile SCN^-. See T. Kimura, M. Fujita, and T. Ando, *J. Chem. Soc., Chem. Commun.*, 1213 (1990).

6

d-Metal complexes

The concepts of Lewis acids and bases and their combination as complexes are useful for describing the structures of molecules and polyatomic ions. This viewpoint is useful for molecules from all blocks of the periodic table, but it is especially useful for discussing d-block compounds because many of them can be regarded as being formed from molecular species that have an independent existence.

First, we introduce the systematics of the geometrical structures of d-metal complexes, and then analyze their electronic structures in terms of crystal field theory and a more sophisticated version—ligand field theory—based on the molecular orbital theory of polyatomic molecules introduced in Chapter 2. Both theories provide an excellent example of how a single parameter—we shall come to know it as the 'ligand field splitting parameter'—can be used to correlate a wide range of properties, including structure, spectra, magnetic properties, and some aspects of thermochemistry.

We shall use the term **complex** to mean a central metal atom or ion surrounded by a set of ligands. By **ligands** we mean ions or molecules that can have an independent existence. An example of a complex ion is $[Co(NH_3)_6]^{3+}$, in which the Co^{3+} ion is surrounded by six NH_3 ligands. We shall use the term **coordination compound** to mean a neutral complex or an ionic compound in which at least one of the ions is a complex. Thus, $[Ni(CO)_4]$ and $[Co(NH_3)_6]Cl_3$ are both coordination compounds.[1] A complex is a combination of a Lewis acid (the central metal atom) with a number of Lewis bases (the ligands). The atom in the Lewis base ligand that forms the bond to the central atom

---

[1] Although the term complex is widely used in the sense given here, IUPAC recommends its replacement by the term 'coordination entity' on the grounds that the word 'complex' has a variety of other meanings.

is called the **donor atom**, because (at least formally) it donates electrons used in bond formation. The metal atom or ion, the Lewis acid in the complex, is the **acceptor atom**. This chapter focuses on complexes that contain *d*-metal ions, but *s*- and *p*-metal ions also form complexes (see Chapter 8).

In this chapter we address two principal questions relating to complexes. One concerns their geometrical structures; the other concerns their electronic structures and the implications of these structures for the properties of the species. The credit for elucidating the principal features of the geometrical structures of *d*-metal complexes belongs to the Swiss chemist Alfred Werner (1866–1919) whose training was in organic stereochemistry. Werner combined the interpretation of optical and geometrical isomerism, patterns of reactions, and conductance data in work that remains a model of how to use chemical evidence effectively and imaginatively. The electronic structures of complexes, however, were a mystery to Werner, for he and his contemporaries were puzzled by the ability of a metal ion to bind large numbers of ligands. This feature was clarified only when the description of electronic structure in terms of orbitals was applied to the problem in the period from 1930 to 1960.

The geometrical structures of *d*-metal complexes, which we consider first, can now be studied in many more ways than Werner had at his disposal. When single crystals of a compound can be grown, X-ray diffraction (Box 3.1) gives precise shapes, bond distances, and angles. Nuclear magnetic resonance (see Further information 2) can be used to study complexes with lifetimes longer than microseconds. Very short-lived complexes, those with lifetimes comparable to diffusional encounters in solution (a few nanoseconds), can be studied by vibrational and electronic spectroscopy. It is possible to infer the geometries of complexes with long lifetimes in solution (such as the classic complexes of Co(III), Cr(III), and Pt(II) and many 4*d*, 5*d*, and organometallic compounds) by analyzing patterns of reactions and isomerism. This method was originally exploited by Werner, and it still teaches us much about the synthetic chemistry of the compounds as well as helping to establish their structures.

STRUCTURES AND SYMMETRIES

The **primary coordination sphere** of a complex consists of a set of ligands directly attached to a central metal ion. The number of ligands in the coordination sphere is called the **coordination number (C.N.)** of the complex. As in solids, a wide range of coordination numbers can occur, and the origin of the structural richness and chemical diversity of complexes is the ability of the coordination number to range up to 12.

Although we shall concentrate on the primary coordination sphere of the metal ion and its associated ligands, we should keep in mind that complex cations can associate electrostatically with anionic ligands (and by other weak interactions with solvent molecules) without dis-

H_2O

H_2O

H_2O —— Mn —— OH_2

SO_4^{2-} OH_2

OH_2

$\rceil$2+

1 $[Mn(OH_2)_6]SO_4$

placement of the ligands already present (**1**). The product of this association is an **outer-sphere complex**. For $[Mn(OH_2)_6]^{2+}$ and SO_4^{2-} ions, for instance, the equilibrium concentration of outer-sphere complex $[Mn(OH_2)_6]^{2+}SO_4^{2-}$ can exceed that of the **inner-sphere complex** $[Mn(OH_2)_5SO_4]$ in which the SO_4^{2-} ligand is directly attached to the metal ion. It is worth remembering that most methods of measuring complex formation equilibria do not distinguish outer-sphere from inner-sphere complex formation but simply detect the sum of all bound ligands. Outer-sphere complexation should be suspected whenever the metal and ligand have opposite charges.

The coordination number of a *d*-block metal ion is not always evident from the composition of the solid, for solvent molecules and species that are potentially ligands may simply fill spaces within the structure and not have any direct bonds to the metal ion. For example, X-ray diffraction shows that $CoCl_2 \cdot 6H_2O$ contains the neutral complex $[CoCl_2(OH_2)_4]$ and two uncoordinated H_2O molecules occupying well defined positions in the crystal. Such additional solvent molecules are called **lattice solvent** or **solvent of crystallization**.

6.1 Constitution

In this section we describe some examples of each class, indicating how some are prepared and their structures identified. Three factors determine the coordination number of a complex:

1. The size of the central atom.
2. The steric interactions between the ligands.
3. Electronic interactions.

In general, the large radii of atoms and ions in Periods 5 and 6 favor higher coordination numbers in the complexes these elements form. On the other hand, bulky ligands often favor low coordination numbers. Across a row of the *d* block, it is found that high coordination numbers are most common on the left of a period, especially if the metal ions have only a few *d* electrons. Lower coordination numbers are found on the right of a period, particularly if the ions are rich in *d* electrons. It is also found that low coordination numbers occur (even for small numbers of *d* electrons) if the ligands can form multiple bonds with the central metal.

Low coordination numbers

Complexes with C.N. = 1 (ML) and 2 (ML$_2$) are found in the gas phase at high temperatures. The best known complexes with low coordination numbers that are formed under ordinary laboratory conditions are linear compounds of the Group 11 and 12 ions, such as Cu(I), Ag(I), Au(I), and Hg(II); see Summary chart 1. The complex $[AgCl_2]^-$, which is responsible for the dissolution of solid silver chloride in aqueous solutions containing excess Cl^- ions, is one example. The toxic complex $[Hg(CH_3)_2]$, which is formed in bacteria by the

11	12
Cu(I) X —— Cu —— X⁻ X = Cl, Br	
Ag(I) H₃N —— Ag —— NH₃⁺	
Au(I) R₃P —— Au —— PR₃⁺	H₃C —— Hg —— CH₃

Summary chart 1:
Linear complexes.

action of methylating enzymes on $Hg^{2+}(aq)$, is another example. A series of linear Au(I) complexes of formula LAuX, where X is a halogen and L is a neutral Lewis base such as a substituted phosphane,[2] R_3P, or thioether, R_2S, is also known. The thioether ligand in $[(R_2S)AuCl]$ is easily displaced by the stronger donor SR^-.[3]

Many two-coordinate complexes readily gain additional ligands to form three- or four-coordinate complexes. In some cases, the formula that suggests two coordination in a solid compound conceals a polymer with a higher coordination number. This ambiguity emphasizes that there is not always an obvious correlation between a formula and a structure. The salt $K[Cu(CN)_2]$ in the solid state, for example, contains a chain-like anion (**2**) with three-coordinate Cu atoms. Three coordination is rare among d-metal complexes, and MX_3 compounds with X a halogen are usually chains or networks with a higher coordination number and shared ligands.

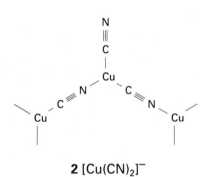

2 $[Cu(CN)_2]^-$

Four coordination

Four coordination is found in an enormous number of compounds. Tetrahedral complexes of approximately T_d symmetry (**3**) are favored over higher coordination numbers if the central atom is small or the ligands large (such as Cl^-, Br^-, and I^-), for then steric effects override the energy advantage of forming more metal–ligand bonds. Four-coordinate s- and p-block complexes with no lone pairs on the central atom are almost always tetrahedral, as in the examples $[BeCl_4]^{2-}$, $[BF_4]^-$, $[ZnCl_4]^{2-}$, and $[SnCl_4]$. Tetrahedral complexes are common for oxoanions of metal atoms on the left of the $3d$ series in high oxidation states (see Summary chart 2). The halide complexes of M^{2+} ions on the right of the $3d$ series are generally tetrahedral (see Summary chart 3).

[2] The current IUPAC recommendation names PH_3 as phosphane rather than phosphine The IUPAC recommendation is used throughout this text and is common in the current literature.

[3] Gold(I) complexes are exploited in the treatment of rheumatoid arthritis, and the interaction of Au(I) with thiol groups of proteins is believed to be involved. See S. J. Lippard (ed.), *Platinum, gold, and other chemotherapeutic agents. Chemistry and biochemistry*. ACS Symposium Series, No. 209. American Chemical Society, Washington, DC (1983).

3 Tetrahedral complex, T_d

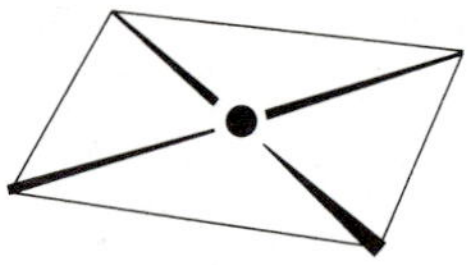

4 Square–planar complex, D_{4h}

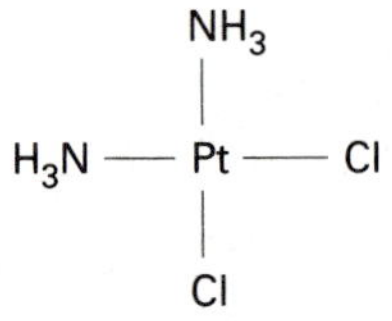

```
        NH3
         |
H3N —— Pt —— Cl
         |
         Cl
```

5a *cis*–[PtCl$_2$(NH$_3$)$_2$]

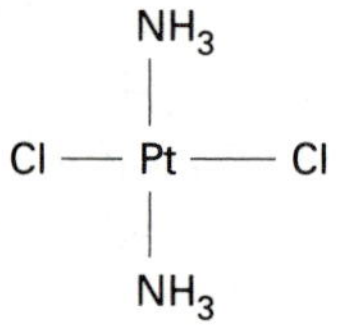

```
        NH3
         |
Cl —— Pt —— Cl
         |
        NH3
```

5b *trans*–[PtCl$_2$(NH$_3$)$_2$]

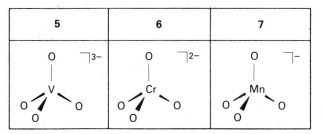

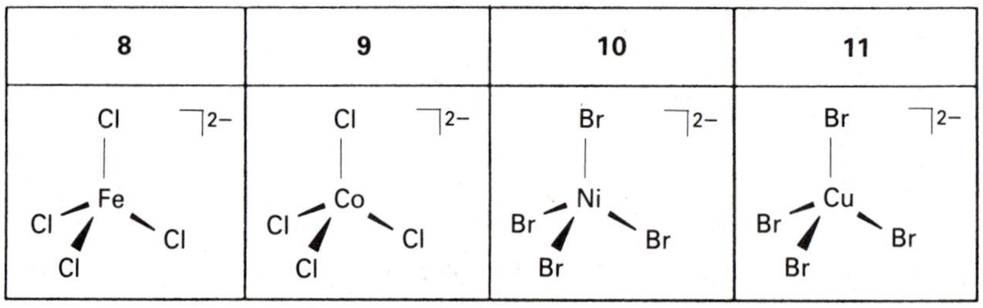

Summary chart 2:
Tetrahedral complexes.

Werner studied[4] a series of four-coordinate Pt(II) complexes formed by the reactions of PtCl$_2$ with NH$_3$ and HCl. For a complex of formula MX$_2$L$_2$, only one isomer is expected if the species is tetrahedral, but two isomers are expected if the species is square-planar (**4**). Because he was able to isolate *two* nonelectrolytes of formula [PtCl$_2$(NH$_3$)$_2$], he concluded that they cannot be tetrahedral and were, in fact, square-planar. The complex with like ligands on adjacent corners of the square is called a *cis* isomer (**5a**) and the complex with like ligands opposite is the *trans* isomer (**5b**). The existence of different spatial arrangements of the same ligands is called **geometrical isomerism**. Geometrical isomerism is far from being of only academic interest: platinum complexes are used in cancer chemotherapy[3], and it is found that only *cis*-Pt(II) complexes can bind to the bases of DNA and be effective. A useful generalization is that:

> Square-planar structures are common for complexes of the d^8 metal ions Rh$^+$, Ir$^+$, Pd^{2+}, Pt^{2+}, Au^{3+}, and sometimes Ni^{2+}.

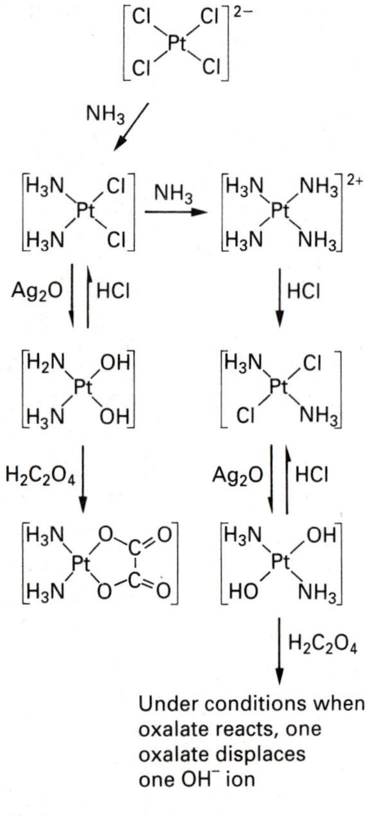

Example 6.1: *Identifying isomers from chemical evidence*
Use the reactions in Fig. 6.1 to show how the *cis* and *trans* geometries may be assigned.

Answer. The *cis* isomer reacts with Ag$_2$O to lose Cl$^-$, and the product adds one oxalic acid molecule (H$_2$C$_2$O$_4$) at neighboring positions. The *trans* isomer loses Cl$^-$, but the product cannot displace the two OH$^-$ ligands with only one H$_2$C$_2$O$_4$ molecule. A reasonable explanation is that the H$_2$C$_2$O$_4$ molecule cannot reach across the square plane to bridge two *trans* positions. This conclusion is supported by X-ray crystallography.

Under conditions when oxalate reacts, one oxalate displaces one OH$^-$ ion

6.1 The preparation of *cis*- and *trans*-diamminedichloroplatinum(II) and a chemical method for distinguishing the isomers.

[4] G. B. Kauffman gives a fascinating account of the history of structural coordination chemistry in *Inorganic coordination compounds*. Wiley, New York (1981).

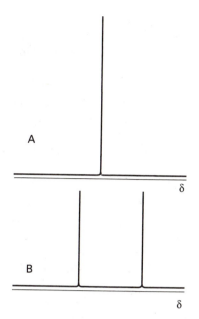

6.2 NMR spectra of the isomeric phosphane complexes of Pt(II).

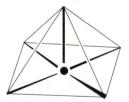

6 [Co(C$_2$S$_2$(CN)$_2$)$_2$]

7 Trigonal–bipyramidal complex, D_{3h}

8 Square–pyramidal complex, C_{4v}

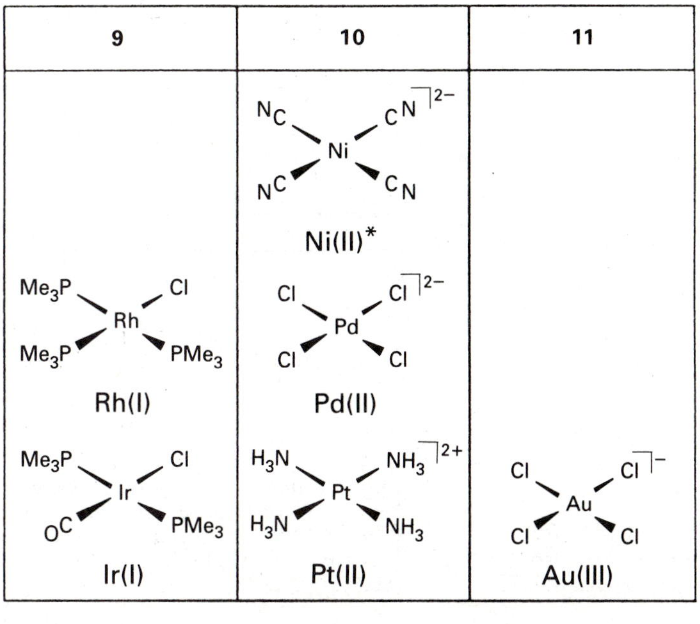

Exercise E6.1. The two square-planar isomers of [PtBrCl(PR$_3$)$_2$] (where —PR$_3$ is a trialkylphosphane group) have different ^{31}P-NMR spectra, which are illustrated in Fig. 6.2. One (A) shows a single ^{31}P group of lines; the other (B) shows two distinct ^{31}P resonances, each one being similar to the single resonance region of A. Which isomer is *cis* and which is *trans*? The Pt couplings have been eliminated.

For simple ligands, only d^8 configurations favor square-planar over tetrahedral geometry (see Summary chart 4). The examples of square-planar complexes in the first row of the *d* block are generally those in which the ligands can form π bonds with the metal by accepting electrons from it; two examples are [Ni(CN)$_4$]$^{2-}$ and complexes such as (**6**). The four-coordinate d^8 complexes of the elements belonging to the second and third rows of the *d* block are almost invariably square planar. Square-planar geometry can also be forced on a central atom by complexation with a ligand that contains a rigid ring of four donor atoms, much as in the formation of five-coordinate porphyrin complexes described below.

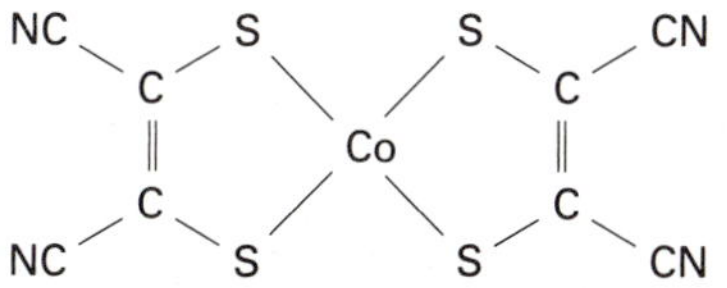

Five coordination

The geometry of five coordination is a delicate balance between trigonal bipyramidal (TBP, **7**) and square pyramidal (SPy, **8**). One shape is converted into the other by a simple distortion (Fig. 6.3) which brings a pair of ligands from the equatorial to the axial positions and vice versa. This mechanism for the transposition is called a **Berry pseudo-rotation**. The neutral complex [Fe(CO)$_5$] provides a good example: it is trigonal bipyramidal in the crystal and in solution as judged by IR spectroscopy, but ^{13}C-NMR of a sample in solution exhibits only

9	10	11
	NC—Ni—CN (with N$_2$C / C$_N$) $]^{2-}$ Ni(II)*	
Me$_3$P—Rh—Cl / PMe$_3$ Rh(I)	Cl—Pd—Cl $]^{2-}$ / Cl Pd(II)	
Me$_3$P—Ir—Cl / OC PMe$_3$ Ir(I)	H$_3$N—Pt—NH$_3$ $]^{2+}$ / H$_3$N NH$_3$ Pt(II)	Cl—Au—Cl $]^-$ / Cl Cl Au(III)

Summary chart 3: Planar complexes.

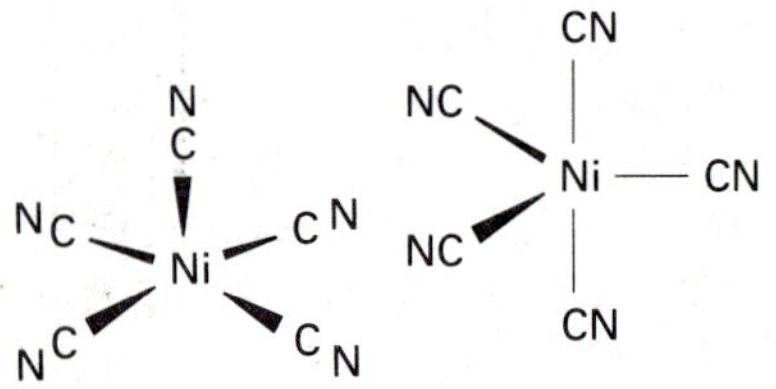

9a [Ni(CN)$_5$]$^{3-}$ **9b** [Ni(CN)$_5$]$^{3-}$

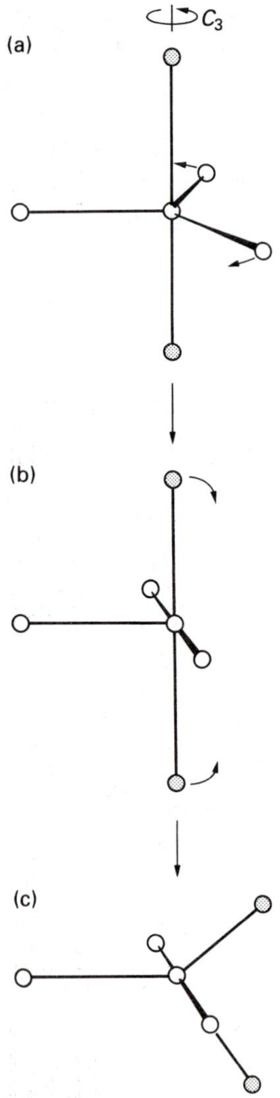

(a)

(b)

(c)

6.3 A Berry pseudorotation in which a trigonal bipyramidal (top complex) distorts into a square pyramidal isomer and then becomes trigonal bipyramidal again, but with two initially equatorial groups now axial. An example of a complex of this kind is [Fe(CO)$_5$].

one signal rather than the two expected for the axial and equatorial CO ligands of a trigonal-bipyramidal species. The explanation of the appearance of the NMR spectrum is the interchange of the axial and equatorial positions at a rate that is fast on an NMR timescale but slow on an IR timescale (see Section 3.2). It has been proposed that the complex undergoes a facile pseudorotation through a square-pyramidal conformation. The delicacy of the energy balances between square-pyramidal and trigonal-bipyramidal conformations is underlined by the fact that [Ni(CN)$_5$]$^{3-}$ (**9a** and **9b**) can exist with both geometries in the same crystal.

Five-coordinate complexes can be divided into five main groups according to their shapes and degrees of distortion from a particular ideal geometry.[5] These groupings are summarized in Table 6.1. Square-pyramidal five-coordination is found among the biologically important porphyrins, where the ligand ring enforces a square planar structure and a fifth ligand attaches above the plane. Structure (**10**) shows the active center of deoxymyoglobin, the oxygen transport protein; in this complex the Fe atom accepts O$_2$ as a sixth ligand, as we shall see in Section 19.2. Another effective method for inducing five-coordination is to use a ligand containing an atom that can bind to an axial location of a trigonal bipyramid, with the remainder of the molecule extending an arm that reaches down to each of the three equatorial positions (**11**).

Six coordination

In the *d*-block, six-coordination is the most common arrangement for electronic configurations ranging from d^0 to d^9. For example, complexes formed by the +3 ions of the 3*d* metals are usually octahedral

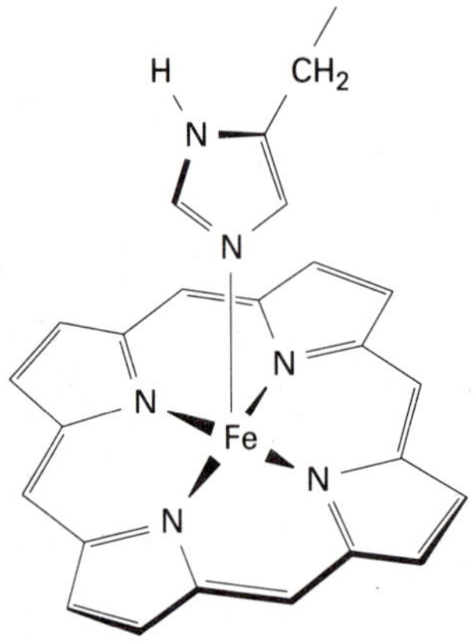

10 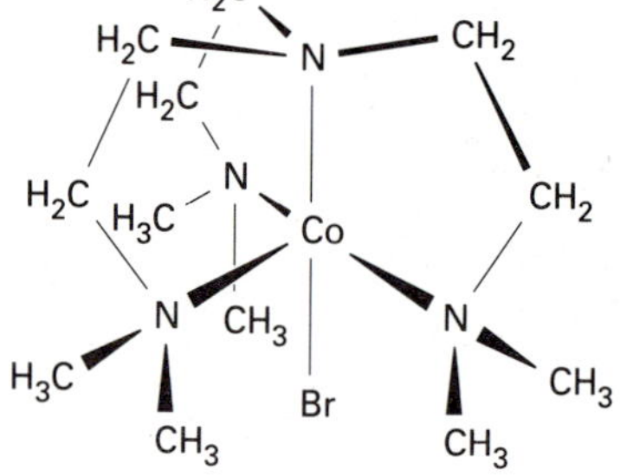   **11**

[5] E. L. Muetterties and L. J. Guggenberger, *J. Am. Chem. Soc.* **96**, 1748 (1974), identified a series of five-coordinate structures that show a smooth transition from ideal trigonal bipyramidal to ideal square pyramidal. The compounds in question ranged from (TBP) [CdCl$_5$]$^{3-}$, [P(C$_6$H$_5$)$_5$], [Co(C$_6$H$_7$NO)$_5$]$^{2+}$, [Ni(CN)$_5$]$^{3-}$, [Nb(NC$_5$H$_{10}$)$_5$], to [Sb(C$_6$H$_5$)$_5$] (SPy).

Table 6.1 The classification of five-coordinate complexes

Geometry	Description	Example
Regular trigonal bipyramid	All five ligands the same; no distortion	$[CuCl_5]^{3-}$
Slightly distorted trigonal bipyramid	Three equatorial ligands identical, but one axial ligand is part of a ring whereas the other is not	
Highly distorted structures	Intermediate between trigonal bipyramid and square pyramid	$[Ni(CN)_2(PPh(OEt)_2)_3]$
Regular square pyramid	Metal ion in square basal plane with fifth ligand above	
Distorted square pyramid	The ligand distribution is square-pyramidal, but the metal atom lies above the basal plane	

12 Octahedral complex, O_h

(**12**). Octahedral geometry is much less common for complexes formed by large metal ions, such as the +3 ions of *f*-block metals. A few examples representative of the wide range of six-coordinate complexes that can occur are $[Sc(OH_2)_6]^{3+}$ (d^0), $[Cr(NH_3)_6]^{3+}$ (d^3), $[Mo(CO)_6]$ (d^6), $[Fe(CN)_6]^{3-}$ (d^5), and $[RhCl_6]^{3-}$ (d^6). Even some halides of the *f*-block elements can display six coordination (although, higher coordination numbers, especially 8 and 9, are more common in the *f* block.)

The regular octahedron is especially important because it is also the starting point for discussions of complexes of lower symmetry, such as those shown in Fig. 6.4. The simplest distortion from O_h symmetry is tetragonal (D_{4h}), and occurs when two *trans* ligands are significantly

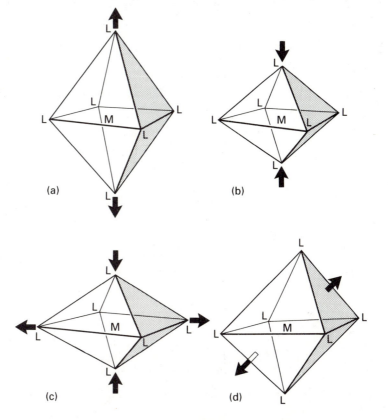

6.4 (a) and (b) Tetragonal (D_{4h}) distortions of a regular octahedron, (c) rhombic (D_{2h}), and (d) trigonal (D_{3d}) distortions. The last can lead to a trigonal prism (D_{3h}) by a further 60° rotation of the faces containing the arrows.

different from the other four. For the d^9 configuration (particularly Cu^{2+} complexes) significant distortions from O_h symmetry occur even when all ligands are identical. Rhombic (D_{2h}) and trigonal (D_{3d}) distortions are also common. Trigonal distortion gives rise to a large family of structures that are intermediate between regular octahedral and trigonal prismatic (D_{3h}).

The trigonal prism (**13**) itself is rare, but was first found in solid MoS_2 and WS_2; it is also the shape of several complexes of formula $[M(S_2C_2R_2)_3]$ (**14**, see Summary chart 5). Trigonal-prismatic d^0 complexes such as $[Zr(CH_3)_6]^{2-}$, have been isolated recently. Such structures require either very small σ-donor ligands or favorable ligand–ligand interactions that can constrain the complex into a trigonal prismatic geometry (and which are often provided by ligands that contain sulfur ligands, for covalent S—S interactions are attractive). However, the view that ligand–ligand attraction is not necessary

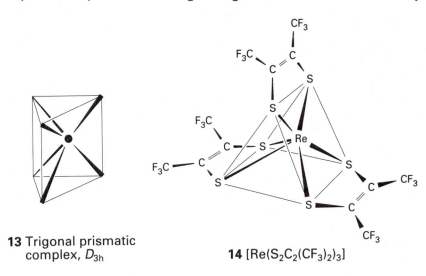

13 Trigonal prismatic
complex, D_{3h}

14 $[Re(S_2C_2(CF_3)_2)_3]$

	4	5	6
	H₃C, CH₃, CH₃ ⌐²⁻ Zr H₃C CH₃ CH₃ Zr(IV)		S S S Mo S S S Mo(VI)
	H₃C, CH₃, CH₃ ⌐²⁻ Hf H₃C CH₃ CH₃ Hf(IV)		S S S W S S S W(VI)

Summary chart 4:
Trigonal-prismatic
complexes.

15 Pentagonal–bipyramidal complex, D_{5h}

16 Capped octahedron

17 Capped trigonal prism

has received some theoretical support recently, for calculations indicate that d^0 systems with six compact ligands tend to be trigonal prismatic; this observation is consistent with the structure of $[Zr(CH_3)_6]^{2-}$.[6]

High coordination numbers

Seven coordination is encountered for a few $3d$ metal complexes and many more $4d$ and $5d$ complexes, where the larger central atom permits the close approach of more than six ligands. It resembles five coordination in the similarity in energy of its various geometries. These limiting 'ideal' geometries include the pentagonal bipyramid (**15**), a capped octahedron (**16**), and a capped trigonal prism (**17**); in each of the latter two, the seventh capping ligand occupies one face. There are a number of intermediate structures, and interconversions are facile. Examples include $[Mo(CNR)_7]^{2+}$, $[ZrF_7]^{3-}$, $[TaCl_4(PR_3)_3]$, and $[ReOCl_6]^{2-}$ from the d block and $[UO_2(OH_2)_5]^{2+}$ from the f block. A method to force seven coordination rather than six on the lighter elements is to synthesize a ring of five donor atoms (**18**) which then occupy the equatorial positions, leaving the axial positions free to accommodate two more ligands.

Stereochemical non-rigidity is also shown in eight coordination, for such complexes may be square antiprismatic (**19**) in one crystal but dodecahedral (**20**) in another. Two examples of complexes with these geometries are shown as (**21**) and (**22**), respectively.

Nine coordination is important in the structures of f-block complexes, for their relatively large ions can act as host to a large number

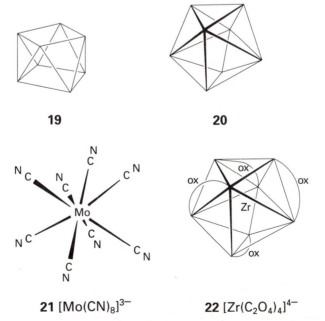

19

20

21 $[Mo(CN)_8]^{3-}$

22 $[Zr(C_2O_4)_4]^{4-}$

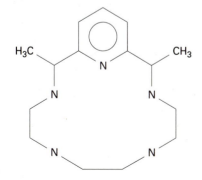

18

---

[6] The structures of six-coordinate trigonal prismatic zirconium and hafnium complexes are described and interpreted in P. M. Morse and G. S. Girolami, *J. Am. Chem. Soc.*, **111**, 4114 (1989).

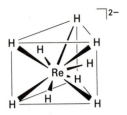

23 $[ReH_9]^{2-}$ D_{3h}

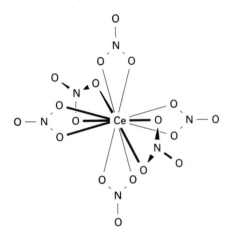

24 $[Ce(NO_3)_6]^{2-}$

of ligands. Examples include $[Nd(OH_2)_9]^{3+}$ and the MCl_3 solids, with M ranging from La to Gd. An example of nine coordination in the *d* block is $[ReH_9]^{2-}$ (**23**), which has small enough ligands for this coordination number to be feasible. Coordination numbers 10 and 12 are encountered in complexes of the *f*-block M^{3+} ions but are rare in the *d* block. Examples include $[Ce(NO_3)_6]^{2-}$ (**24**), which is formed in the reaction of Ce(IV) salts with nitric acid. Each NO_3^- ligand is bonded to the metal atom by two O atoms. An example of a ten-coordinate complex is $[Th(ox)_4(OH_2)_2]^{4-}$, in which each oxalate ion ligand (ox, or $C_2O_4^{2-}$) provides two O atoms as donors.

Polymetallic complexes

A considerable amount of attention has been given recently to the synthesis of **polymetallic complexes**, which are complexes that contain more than one metal atom (Fig. 6.5). In some cases, the metal atoms are held together by bridging ligands, in others there are direct metal–metal bonds, and in yet others there are both types of link. The term **metal cluster** is usually reserved for polymetallic complexes in which there are direct metal–metal bonds. When no metal–metal bonds are present polymetallic complexes are referred to as **cage complexes** (or cage compounds).[7]

Cage complexes may arise in the first *d* series through the formation of HO^-, O^{2-}, and $CH_3CO_2^-$ bridges. For example, two Cu^{2+} ions

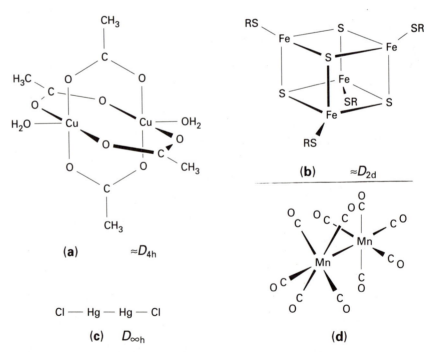

6.5 Representative types of polymetallic complexes. (a) The copper(II) acetate dimer in which there is almost negligible metal–metal bonding. (b) A synthetic Fe—S complex which models biochemically important electron-transfer agents. (c) Mercury(I) chloride, with a definite Hg—Hg bond. (d) The complex $[Mn_2(CO)_{10}]$ which is held together by an Mn—Mn bond.

---

[7] The term 'cage compound' has a variety of meanings in inorganic chemistry, and it is important to keep them distinct. For example, another use of the term is as a synonym of a clathrate compound (an inclusion compound, in which a species is trapped in a cage formed by molecules of another species).

can be held together with acetate-ion bridges (Fig. 6.5(a)). Figure 6.5(b) shows an example of a cubic structure formed from four Fe atoms bridged by S atoms. This structure is of great biological importance, for it is involved in a number of biochemical redox reactions (see Section 19.5).

With the advent of modern structural techniques, such as automated X-ray diffraction and multinuclear NMR, polymetallic clusters containing metal–metal bonds have been discovered and have given rise to an active area of research. A simple example is the mercury(I) cation Hg_2^{2+}, and complexes derived from it, such as $[Hg_2Cl_2]$ (Fig. 6.5(c)). A metal cluster containing CO ligands is illustrated in Fig. 6.5(d). More details about *d*-block cage and cluster compounds will be found in Sections 8.9, 15.16, and 16.10–16.12.

6.2 Representative ligands and nomenclature

We shall outline here a few key ideas of nomenclature and introduce a number of common ligands. More detailed guidance is given in Further information 1. We begin by considering complexes that contain only **monodentate ligands**, or ligands that have only one point of attachment to the metal atom, as distinct from **polydentate ligands**, which are ligands that have more than one point of attachment.

Nomenclature

Table 6.2 gives the names and formulas of a number of common simple ligands. Note that anionic ligands have names ending in -o. Complexes are named with their ligands in alphabetical order. The ligand names are followed by the name of the metal with either its oxidation number in parentheses, as in hexaamminecobalt(III), or with the overall charge on the complex specified in parentheses, as in hexaamminecobalt(3 +). The suffix -ate is added to the name of the metal (sometimes in its Latin form) if the complex is an anion, as in the name hexacyanoferrate(II) for $[Fe(CN)_6]^{4-}$.

The number of occurrences of a ligand in a complex is indicated by the prefixes mono-, di-, tri-, and tetra-. Where confusion with the names of ligands is likely (as with ethylenediamine) the alternative prefixes bis-, tris-, and tetrakis- are used, with the ligand name in parentheses. For example, dichloro- is unambiguous but bis(ethylenediamine) shows more clearly that there are two ligands of formula $H_2NCH_2CH_2NH_2$. Ligands that bridge two metal centers are denoted by a prefix μ- added to the name of the relevant ligand, as in μ-oxo-bis(pentaamminechromium(III)) (**25**). The number of metal atoms in a bridged complex is indicated by the prefixes di-, tri-, etc, as in octachlorodirhenate(III), $[Re_2Cl_8]^{2-}$ (**26**).

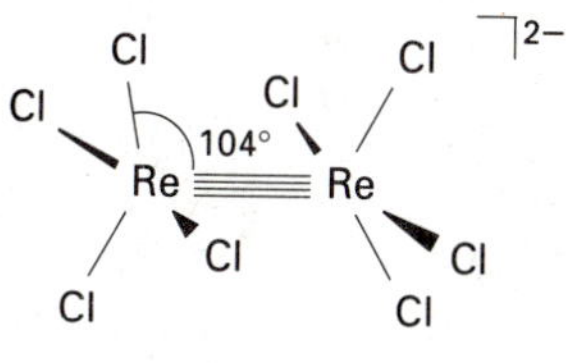

The formula of a complex should be enclosed within square brackets whether it is charged or not; however, in casual usage, neutral complexes and oxoanions are often written without brackets, as in $Ni(CO)_4$ and MnO_4^-. The metal symbol is given first, then the anionic ligands,

25 μ–Oxo–bis(penta–amminechromium(III))

26 $[Re_2Cl_8]^{2-}$ D_{4h}

Table 6.2 Typical ligands and their names

Name	Formula	Abbreviation	Classification*
Acetylacetonato	$(CH_3COCHCOCH_3)^-$	acac	B(O)
Ammine	NH_3		M(N)
Aqua	OH_2		M(O)
2,2-Bipyridine	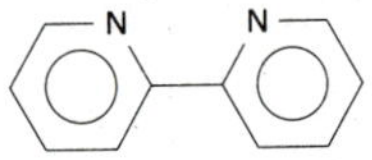	bipy	B(N)
Bromo	Br^-		M(Br)
Carbonato	CO_3^{2-}		M(O) or B(O)
Carbonyl	CO		M(C)
Chloro	Cl^-		M(Cl)
Cyano	CN^-		M(C)
Diethylenetriamine	$NH(C_2H_4NH_2)_2$	dien	T(N)
Ethylenediamine	$H_2NCH_2CH_2NH_2$	en	B(N)
Ethylenediaminetetraacetato	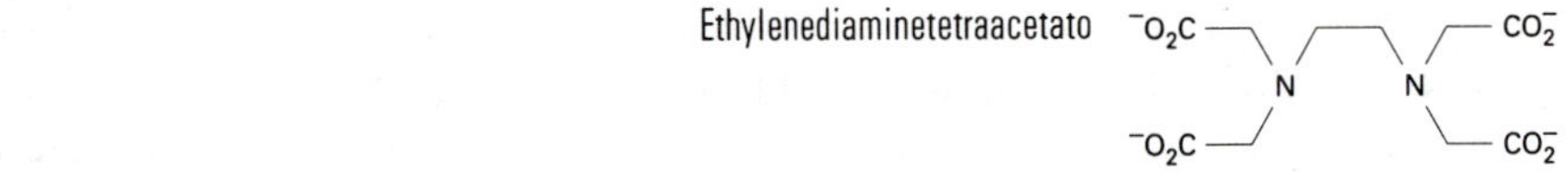	EDTA	S(N,O)
Glycinato	$NH_2CH_2CO_2^-$	gly	B(N,O)
Hydrido	H^-		M
Hydroxo	OH^-		M(O)
Maleonitriledithiolato	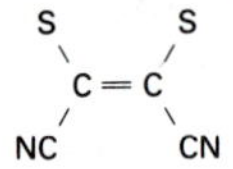	mnt	B(S)
Nitrilotriacetato	$N(CH_2CO_2^-)_3$	nta	Te(N,O)
Oxo	O^{2-}		M
Oxalato	$C_2O_4^{2-}$	ox	B(O)
Nitrito	NO_2^-		M(O)
Tetraazacyclotetradecane	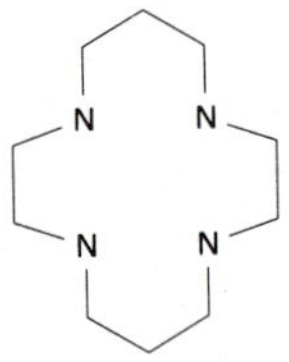	cyclam	Te(N)
Thiocyanato	SCN^-		M(S)
Isothiocyanato	SCN^-		M(N)
2,2′,2″-Triaminotriethylamine	$N(C_2H_4NH_2)_3$	trien	Te(N)

*M: monodentate, B: bidentate, T: tridentate, Te: tetradentate, S: sexidentate. The letters in parentheses identify the coordinating atom.

and finally the neutral ligands, as in $[CoCl_2(NH_3)_4]^+$. (This order is sometimes varied to clarify which ligand is involved in a reaction.) Polyatomic ligand formulas are sometimes written in an unfamiliar sequence (as for OH_2 in $[Fe(OH_2)_6]^{2+}$) to place the donor atom adjacent to the metal atom.

Ambidentate ligands

Ligands with more than one donor atom are called **ambidentate**. An example is the thiocyanate ion (NCS^-) which can attach to a metal atom either by the N atom, to give isothiocyanato complexes, or by the S atom, to give thiocyanato complexes. Compounds with S—Cr bonds can form in fast reactions in which Cr(II) attacks the S atom of a Co(III)—NCS complex, as in

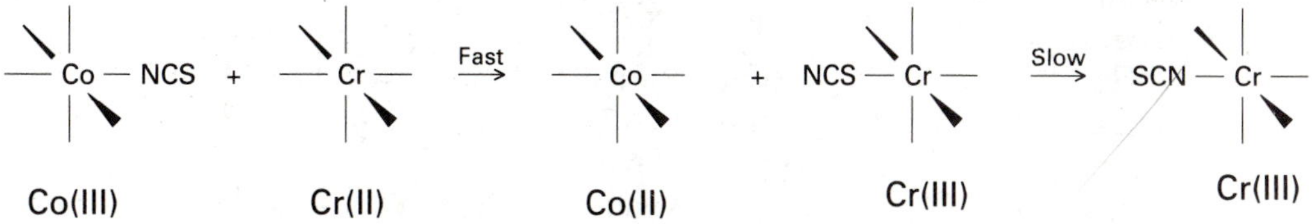

These Cr(III)—SCN thiocyanato complexes are short-lived and rearrange to give the Cr(III)—NCS isothiocyanato complex. Two more examples of ambidentate ligands are $-NO_2^-$, nitro, and $-ONO^-$, nitrito; and $-SO_3^{2-}$ and $-OSO_2^{2-}$.[8]

The existence of ambidentate character gives rise to the possibility of **linkage isomerism**, in which the same ligand may link through alternative atoms. This type of isomerism accounts for the red and yellow isomers of the formula $[Co(NO_2)(NH_3)_5]^{2+}$. The red compound has a nitrito Co—O link (**27**). The yellow isomer, which forms from the unstable red form on standing, has a nitro Co—N link (**28**).

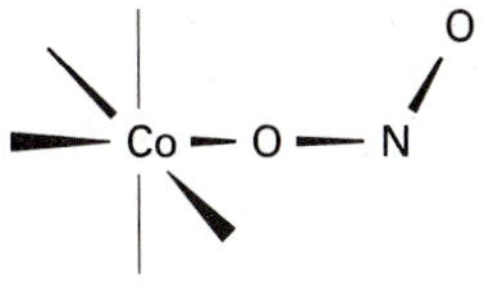

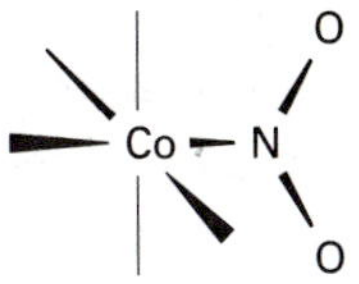

Chelating ligands

Polydentate ligands are often **chelating** (from the Greek for claw) in that they can form a ring that includes the metal atom. The resulting complex is called a **chelate**. An example is the bidentate ligand ethylenediamine ($NH_2CH_2CH_2NH_2$, en), which forms a five-membered ring (**29**). A more complex example is the hexadentate ligand ethylenediaminetetra-acetic acid as its anion ($EDTA^{4-}$) which is used to trap metal ions (**30**). Table 6.2 includes some of the most common chelating ligands.

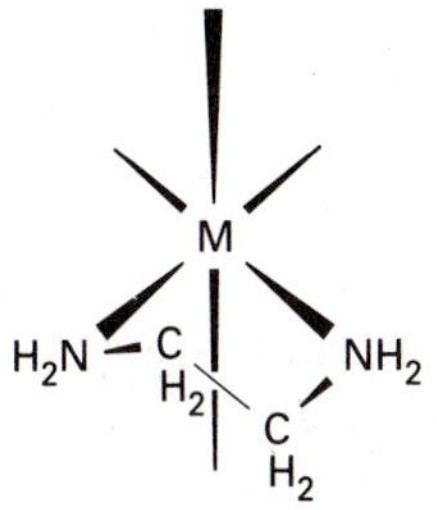

Example 6.2: *Naming complexes*

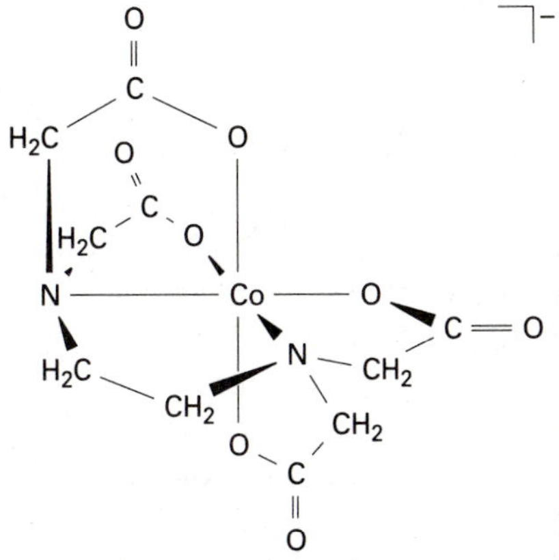

Name the complexes (a) *trans*-$[PtCl_2(NH_3)_4]^{2+}$, (b) $[Ni(CO)_3(py)]$, (c) $[Cr(EDTA)]^-$.

Answer. (a) The complex has two anionic ligands (Cl^-) and four neutral ligands (NH_3); hence the oxidation number of platinum must be $+4$. According to the alphabetical order rules, the name of the complex is *trans*-tetraamminedichloroplatinum(IV). (b) The ligands CO and py (pyridine) are neutral, so the oxidation number of nickel must be 0. It follows that the name of the complex is tricarbonylpyridylnickel(0). (c) This complex uses hexadentate $EDTA^{4-}$ as the sole ligand. The four negative charges of the ligand make an anion with a single negative charge if the central metal ion is Cr^{3+}. The complex is therefore ethylenediaminetetraacetatochromate(III).

Exercise E6.2. Write the formulas of the following complexes: (a) *cis*-diaquadichloroplatinum(II); (b) diamminetetra(isothiocyanato)chromate(III); (c) tris(ethylenediamine)rhodium(III).

---

[8] Many other interesting examples are given in the article 'Ambidentate ligands, the schizophrenics of coordination chemistry', by J. L. Burmeister, *Coord. Chem. Rev.*, **105**, 77 (1990).

27

28

29

30 [Co(edta)]⁻

31

32 Acetylacetonato ion

33

34

⌐ ⌐2−

35 Porphyrin

The rings formed by chelation may have more than one accessible conformation and may put constraints on the L—M—L angle, which in an octahedral complex is 90°. In a chelate formed from saturated C and N centers, such as (**29**), the five-membered ring can fold into a conformation that preserves the tetrahedral angles within the ligand and yet still achieve the L—M—L angle of 90°. Six-membered rings may be favored sterically or by electron delocalization (by conjugation of single and double bonds). β-Diketones, for example, coordinate as the anions of their enols in six-membered ring structures (**31**). An important example is the acetylacetonato anion, $CH_3COCHCOCH_3^-$ (**32**; acac). Since biochemically important amino acids can form five- or six-membered rings, they also chelate readily.

The degree of strain in a chelating ligand is often expressed in terms of the **bite distance** (**33**), the separation of the two donor atoms in a chelate ring. A small ratio of the bite distance to the metal–ligand distance is one of the main causes of distortion from octahedral toward trigonal prismatic geometry in six-coordinate complexes (**34**).

A special case of chelation occurs with a **macrocyclic ligand**, a polydentate ligand in which several donor atoms form a large ring even before the complex is formed. An example is the porphyrin ring (**35**), a tetradentate ligand, which is found (in modified forms) complexed with Fe at the O_2-binding site of hemoglobin and complexed with Mg in chlorophyll. A number of electron-transfer enzymes are iron porphyrins. A saturated tetradentate ligand is tetraazacyclotetradecane (called familiarly 'cyclam', Table 6.2).

The template effect

A metal ion such as Ni(II) can be used to assemble a group of ligands which then undergo a condensation reaction to form a macrocyclic ligand. (A condensation reaction is a reaction in which a bond is formed between two molecules, and a small molecule—often H_2O—is eliminated.) This trick, which is called the **template effect**, can be applied to produce a surprising variety of macrocyclic ligands. The most generally useful condensation reaction is the Schiff's base condensation of an amine and a ketone, of which an example is

$$(CH_3)_2C{=}O + H_2NCH_3 \rightarrow (CH_3)_2C{=}NCH_3 + H_2O$$

(A Schiff's base, an imine, is the product $R_2C{=}N{-}R'$.) The trouble with such condensation reactions is that they often give unwanted side products, and a metal ion can play a useful role in directing the reaction toward the desired ligand product or aiding in its isolation.

A good example of a synthesis that uses the template effect is the reaction that follows coordination of a 2,3-butanedione molecule and two 2-aminoethanethiol molecules to Ni^{2+}. The mixture undergoes condensation to give the square-planar Ni(II) complex shown as the first product in the equation below. The square-planar complex can then be condensed with 1,2-di(bromomethyl)benzene to form a macrocycle:

$$CH_3COCOCH_3 + 2H_2NCH_2CH_2SH + Ni^{2+} \xrightarrow{-2H^{\cdot}}$$

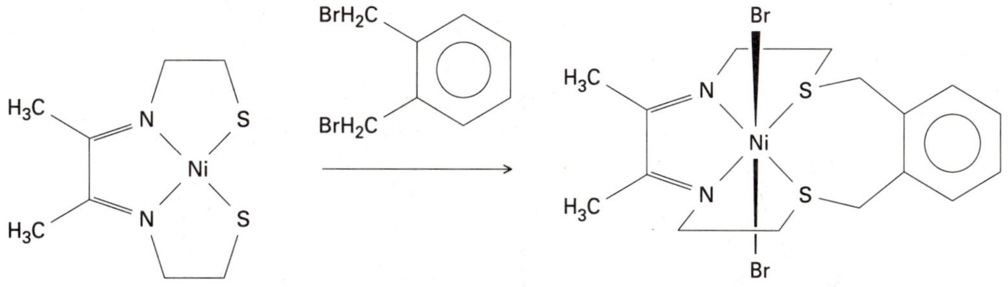

Any reaction of this type may be called a **template synthesis** because the ligand is assembled attached to the Ni(II).

The template effect can stem from either thermodynamics or kinetics. For example, the condensation may either involve enhancement of the rate of the reaction between coordinated ligands or it may arise from the added stability of the macrocycle end product.[9]

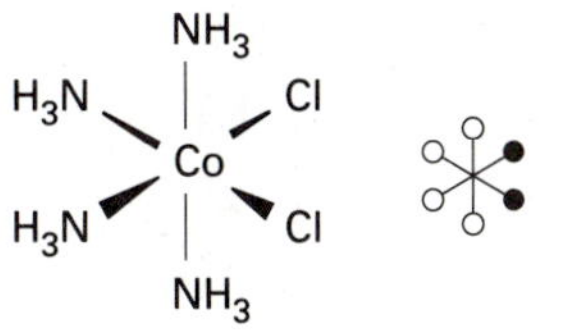

6.3 Isomerism and chirality

We have remarked that the existence and identification of isomerism has played a central role in the development of an understanding of structures of complexes. The essentially two-dimensional square-planar complexes exhibit *cis* and *trans* geometrical isomerism; however, in octahedral complexes the possibilities for isomerism are even greater because these complexes are essentially three-dimensional.

Geometrical isomerism in six coordination

One type of geometrical isomerism in six-coordinate complexes resembles that in square-planar complexes. For example, the two X ligands of an $[ML_2X_4]$ complex may be placed on adjacent octahedral positions to give a *cis* isomer (**36**) or on diametrically opposite positions to give a *trans* isomer (**37**). The *trans* isomer has D_{4h} symmetry and the *cis* isomer has C_{2v} symmetry.

A further type of geometrical isomerism is obtained when there are two groups of three ligands in a complex, as in $[MX_3Y_3]$. For example, one of the products of the oxidation of Co(II) in the presence of nitrite ions and ammonia is the yellow nonelectrolyte $[Co(NO_2)_3(NH_3)_3]$. There are two ways of arranging the ligands in $[MX_3Y_3]$ complexes. In one isomer (**38**), three X ligands lie in one plane and three Y ligands lie in a perpendicular plane. This complex is designated the **mer isomer** (for meridional) because each set of ligands can be regarded as lying on a meridian of a sphere. The second complex (**39**), where

36 *cis-* $[CoCl_2(NH_3)_4]^+$

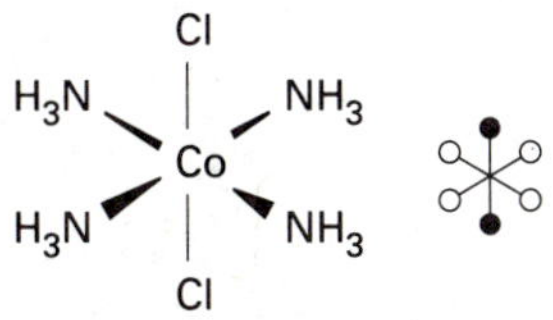

37 *trans-* $[CoCl_2(NH_3)_4]^+$

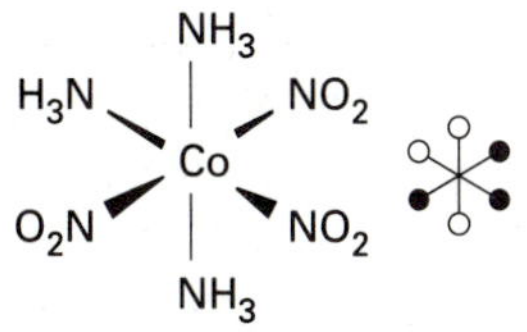

38 *mer-* $[Co(NO_2)_3(NH_3)_3]$

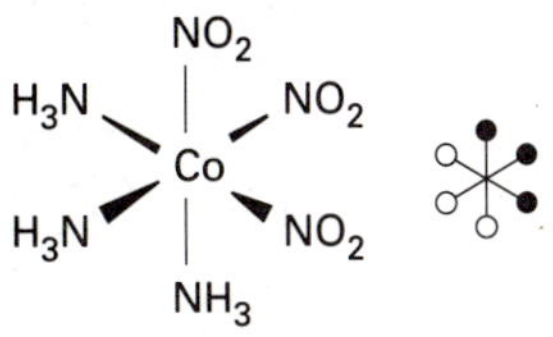

39 *fac-* $[Co(NO_2)_3(NH_3)_3]$

[9] A recent account of the origins of template effect in the case of Pt(II), in which relatively slow reactions of complex formation assist in the elucidation of the details, is D. J. Sheeran and K. B. Mertes, *J. Am. Chem. Soc.*, 1990, **112**, 1055.

all three X (or Y) ligands are adjacent and occupy the corners of one triangular face of the octahedron, is the **fac isomer** (for facial). An approach to the synthesis of specific isomers is summarized in Box 6.1.

Box 6.1 The synthesis of specific isomers

The synthesis of specific isomers often requires subtle changes in synthetic conditions. For example, the most stable Co(II) complex in ammoniacal solutions of Co(II) salts, $[Co(NH_3)_6]^{2+}$, is only slowly oxidized to $[Co(NH_3)_6]^{3+}$ by air. As a result, a variety of complexes containing other ligands as well as NH_3 can be prepared by bubbling air through a solution containing ammonia and a Co(II) salt. Starting with ammonium carbonate yields $[CoCO_3(NH_3)_4]^+$, in which CO_3^{2-} is a bidentate ligand that occupies two adjacent coordination positions. The complex *cis*-$[Co(NH_3)_4L_2]$ can be prepared by decomposition of the CO_3^{2-} ligand in acidic solution. When concentrated hydrochloric acid is used, the violet *cis*-$[CoCl_2(NH_3)_4]Cl$ compound (**B1**) can be isolated:

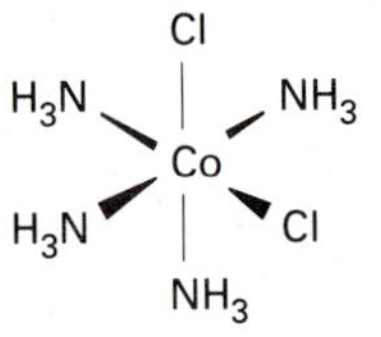

$$[Co(CO_3)(NH_3)_4]^+(aq) + 2H^+(aq) + 3Cl^-(aq) \rightarrow$$
$$cis\text{-}[CoCl_2(NH_3)_4]Cl(s) + H_2CO_3(aq)$$

In contrast, reaction with a mixture of HCl and H_2SO_4 gives the bright green *trans*-$[CoCl_2(NH_3)_4]Cl$ isomer (**B2**).

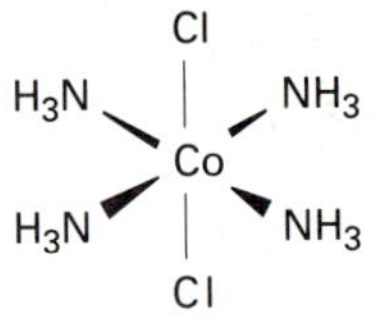

B1

B2

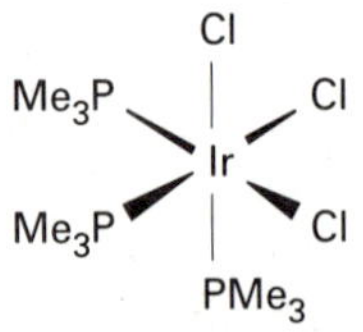

40 *fac*–$[IrCl_3(PMe_3)_3]$

41 *mer*-$[IrCl_3(PMe_3)_3]$

Example 6.3: *Identifying types of isomerism*

When the four-coordinate square-planar complex $[IrCl(PMe_3)_3]$ (where PMe_3 is trimethylphosphane) reacts with Cl_2, two six-coordinate products of formula $[IrCl_3(PMe_3)_3]$ are formed by a reaction known as 'oxidative addition'. ^{31}P-NMR spectra indicate one P environment in one of these isomers and two in the other. What isomers are possible?

Answer. Because the complexes have the formula $[ML_3X_3]$, we expect meridional and facial isomers. Structures (**40**) and (**41**) show the arrangement of the three Cl^- ions in the *fac* and *mer* isomers, respectively. Notice that, in agreement with the NMR data, all P atoms are equivalent in the *fac* isomer and two environments exist in the *mer* isomer.

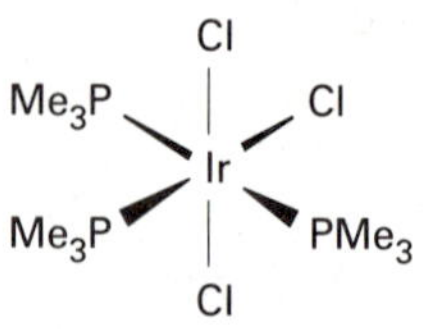

Exercise E6.3. When the anion of glycine $H_2NCH_2CO_2^-$ (gly$^-$) reacts with Co(III) oxide, both the N and an O atom of gly$^-$ coordinate and two Co(III) nonelectrolyte *mer* and *fac* isomers of $[Co(gly)_3]$ are formed. Sketch the two isomers.

Chirality and optical isomerism

A **chiral complex** is a complex that is not superimposable on its own mirror image. The existence of a pair of chiral complexes that are each

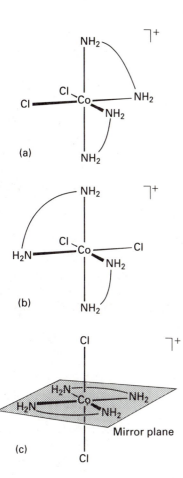

6.6 (a) and (b) Enantiomers of *cis*-[CoCl$_2$(en)$_2$] and (c) the achiral *trans* isomer. The curves represent the —CH$_2$CH$_2$— bridges in the en ligands. One of the mirror planes for testing whether there is an S_1 axis is also shown.

other's mirror image (like a right and left hand), and which have lifetimes that are long enough for them to be separable, is called **optical isomerism**. The two mirror-image isomers jointly make up an **enantiomeric pair**. Optical isomers are so called because they are optically active, in the sense that one enantiomer rotates the plane of polarized light in one direction and the other rotates it through an equal angle in the opposite direction.

As an example of optical isomerism, we can consider the products of the reaction of Co(III) and ethylenediamine. The products include a pair of dichloro complexes, one of which is violet and the other green; they are, respectively, the *cis* and *trans* isomers of dichlorobis-(ethylenediamine)cobalt(III), [CoCl$_2$(en)$_2$]$^+$. (The reaction also results in the slow formation of a yellow complex, tris(ethylenediamine)-cobalt(III) ion, [Co(en)$_3$]$^{3+}$.) As can be seen from Fig. 6.6, the *cis* isomer cannot be superimposed on its mirror image. It is therefore chiral and hence (because the complexes are long-lived) optically active. The *trans* isomer has a mirror plane and can be superimposed on its mirror image; it is achiral and optically inactive.

The formal criterion of chirality (Section 3.5) is the absence of an axis of improper rotation (S_n, an *n*-fold axis in combination with a horizontal mirror plane). The existence of such a symmetry element is implied by the presence of either a mirror plane through the central atom (which is equivalent to an S_1 axis) or a center of inversion (which is equivalent to an S_2 axis), and if either of these elements is present the complex is achiral. As remarked in Section 3.5, we must also be alert for higher-order axes of improper rotation (particularly S_4), because the presence of any S_n axis implies achirality (the absence of chirality). An example of a species with an S_4 axis but not an S_1 or S_2 axis was given as structure **18** in Chapter 3.

Example 6.4: *Recognizing chirality*

Which of the complexes (a) [Cr(edta)]$^-$, (b) [Ru(bipy)$_3$]$^{2+}$, (c) [PtCl(dien)]$^+$ are chiral?

Answer. The complexes are shown schematically in (**42**) to (**44**). Neither (**42**) nor (**43**) has a mirror plane or a center of inversion; so both are chiral (they also have no higher S_n axis); (**44**) has a plane of symmetry and hence is achiral. (Although the CH$_2$ groups in a dien ligand are not in the mirror plane, they oscillate rapidly above and below it.)

Exercise E6.4. Which of the complexes (a) *cis*-[CrCl$_2$(ox)$_2$]$^{3-}$, (b) *trans*-[CrCl$_2$(ox)$_2$]$^{3-}$, (c) *cis*-[RhH(CO)(PR$_3$)$_2$] are chiral?

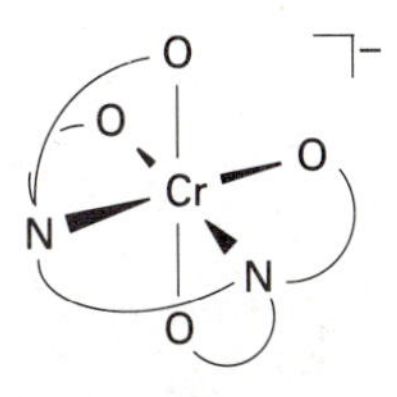

42a 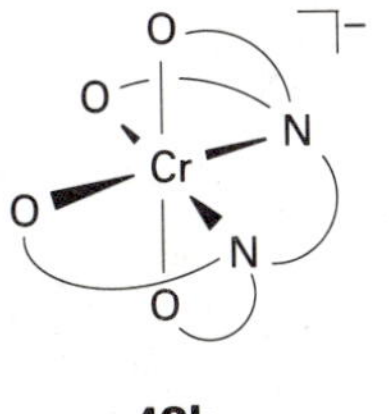   **42b** 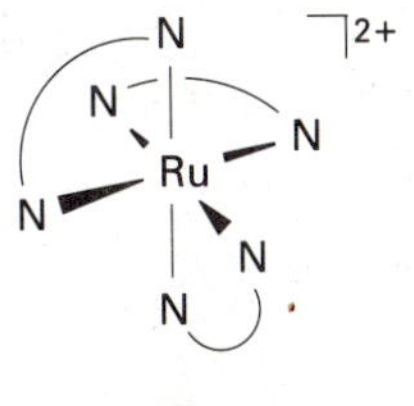   **43a** 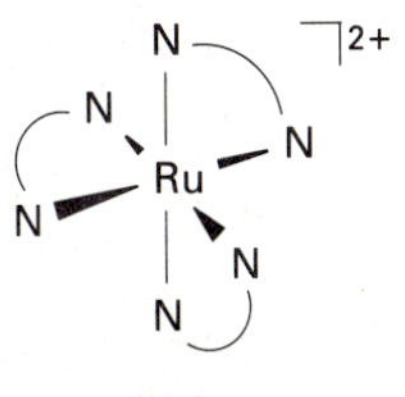   **43b** 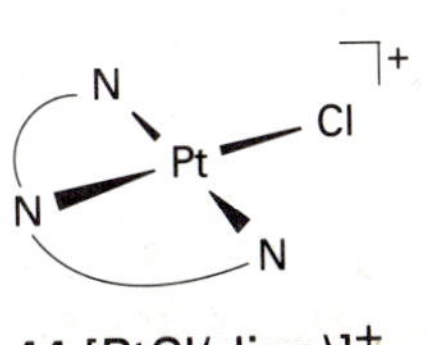   **44 [PtCl(dien)]$^+$**

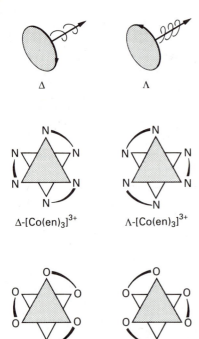

Δ-[Co(en)$_3$]$^{3+}$ Λ-[Co(en)$_3$]$^{3+}$

Δ-[Co(ox)$_3$]$^{3-}$ Λ-[Co(ox)$_3$]$^{3-}$

6.7 Absolute configurations of [M(L—L)$_3$] complexes; Δ is a right-hand screw and Λ is a left-hand screw, as is indicated in the diagrams at the top of the figure by the direction that a screw would turn when being driven in the direction shown.

The absolute configuration of a chiral complex is described by imagining a view along a threefold rotation axis of a regular octahedron (Fig. 6.7) and noting the handedness of the helix formed by the ligands. Right rotation of the helix is then designated Δ and left rotation Λ. The designation of the absolute configuration must be distinguished from the experimentally determined direction in which an isomer rotates polarized light: some Λ compounds rotate in one direction, others rotate in the opposite direction (and the direction may change with wavelength). The isomer that rotates to the right (when viewed into the oncoming beam) at a specified wavelength is designated the **d-isomer**, or the (+)-**isomer**; the one rotating the plane to the left is designated the **l-isomer**, or the (−)-**isomer**.

The resolution of enantiomers

Apart from tiny effects related to the weak interaction (an interaction between fundamental particles), optical activity is the only physical manifestation of chirality for a compound with a single chiral center. However, as soon as more than one chiral center is present, other physical properties—such as solubility and melting points—are affected because they depend on the strengths of intermolecular forces, which are different between different isomers (just as there are different forces between a given nut and bolts with left- and right-handed threads). One method of separating a pair of enantiomers into the individual isomers is therefore to prepare **diastereomers**. As far as we need be concerned, these are isomeric compounds that contain two chiral centers, one being of the same absolute configuration in both components and the other being enantiomeric between the two components. An example of diastereomers is provided by the two salts of an enantiomeric pair of cations with an optically pure anion, and hence of composition [Δ-A][Δ-B] and [Λ-A][Δ-B]. Because diastereomers differ in physical properties (such as solubility), they are separable by conventional techniques. The procedure is summarized in Box 6.2.

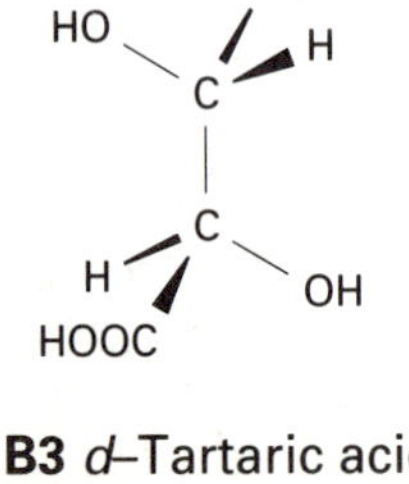

COOH

HO   H
    C

H   C   OH

HOOC

B3 *d*–Tartaric acid

Box 6.2 A typical resolution procedure

A classical resolution procedure begins with the isolation of a naturally optically active species from a biochemical source (many naturally occurring compounds are chiral). A convenient compound is *d*-tartaric acid (**B3**), a carboxylic acid obtained from grapes. This molecule is a chelating ligand for complexation of antimony, so a convenient resolving agent is the K$^+$ salt of the 'antimony *d*-tartrate' anion (**B4**). This anion is used for the resolution of [Co(NO$_2$)$_2$(en)$_2$]$^+$ as follows.

The enantiomeric mixture of the cobalt(III) complex is dissolved in warm water and a solution of potassium antimony *d*-tartrate is added. The mixture is cooled immediately to induce crystallization. The less soluble diastereomer {*l*-[Co(NO$_2$)$_2$(en)$_2$]}$_2${*d*-[Sb$_2$(OC$_4$H$_4$O$_6$)$_2$]} separates as fine yellow

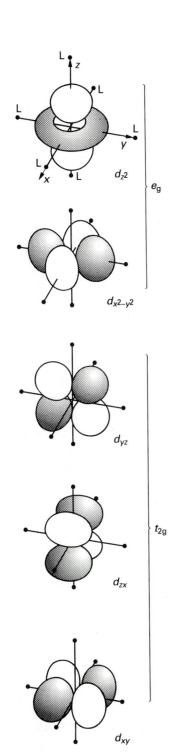

6.8 The orientation of the five *d* orbitals with respect to the ligands of an octahedral complex: (a) the degenerate e_g pair; (b) the degenerate t_{2g} triplet.

B4 $[Sb_2(d\text{-}C_4H_4O_6)_2]^{2-}$

crystals. The filtrate is reserved for isolation of the *d* enantiomer. The solid diastereomer is ground with water and sodium iodide. The sparingly soluble compound *l*-$[Co(NO_2)_2(en)_2]I$ separates, leaving sodium antimony tartrate in the solution. The *d* isomer is obtained from the filtrate by precipitation of the bromide salt.

[References: G. Pass and H. Sutcliffe, *Practical inorganic chemistry.* Chapman and Hall, London (1974); W. L. Jolly, *The synthesis and characterization of inorganic compounds.* Waveland Press, Prospect Heights (1991).]

BONDING AND ELECTRONIC STRUCTURE

The first theory of the electronic structure of complexes was developed to account for the properties of *d*-metal ions in ionic crystals. In this **crystal field theory**, a ligand lone pair is modeled as a point negative charge (or as the partial negative charge of an electric dipole) that repels electrons in the *d* orbitals of the central metal ion. This approach concentrates on the resulting splitting of the *d* orbitals into groups and then uses that splitting to account for the number of unpaired electrons on the ion and thence for the spectra, stability, and magnetic properties of complexes. The approach is simple, readily visualizable, and correctly identifies the importance of the symmetry of key orbitals in a complex. However, crystal field theory ignores bonding interactions between the ligand and the central metal ion, and the approach has been superseded by **ligand field theory**. This theory is an extension of molecular orbital theory that focusses on the role of the *d* orbitals on the central ion and their overlap with orbitals on the ligands.

6.4 Crystal field theory

In the model of an octahedral complex used in crystal field theory, the six ligands are placed on the cartesian axes (Fig. 6.8) centered on the metal ion. Electrons in the two *d* orbitals pointing along these axes, namely d_{z^2} and $d_{x^2-y^2}$ (which are jointly of symmetry type e_g), are repelled more strongly by the negative charge of ligands than electrons

Table 6.3. Ligand field stabilization energies*

d^n	Example	Octahedral					Tetrahedral	
			N	LFSE			N	LFSE
d^0	Ca^{2+}, Sc^{3+}		0	0			0	0
d^1	Ti^{3+}		1	0.4			1	0.6
d^2	V^{3+}		2	0.8			2	1.2
d^3	Cr^{3+}, V^{2+}		3	1.2			3	0.8
		Strong-field			Weak-field			
d^4	Cr^{2+}, Mn^{3+}	2	1.6		4	0.6	4	0.4
d^5	Mn^{2+}, Fe^{3+}	1	2.0		5	0	5	0
d^6	Fe^{2+}, Co^{3+}	0	2.4		4	0.4	4	0.6
d^7	Co^{2+}	1	1.8		3	0.8	3	1.2
d^8	Ni^{2+}		2	1.2			2	0.8
d^9	Cu^{2+}		1	0.6			1	0.4
d^{10}	Cu^+, Zn^{2+}		0	0			0	0

*N is the number of unpaired electrons; LFSE is in units of Δ_O for octahedra or Δ_T for tetrahedra; the calculated relation is $\Delta_T \approx 0.45\Delta_O$.

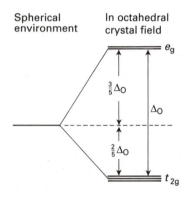

Spherical environment   **In octahedral crystal field**

e_g

$\frac{3}{5}\Delta_O$

Δ_O

$\frac{2}{5}\Delta_O$

t_{2g}

6.9 The relative energies of the *d* orbitals in an octahedral crystal field.

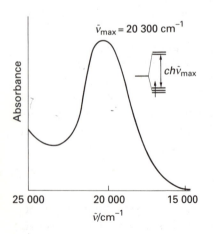

$\tilde{v}_{max} = 20\ 300\ cm^{-1}$

$ch\tilde{v}_{max}$

Absorbance

25 000   20 000   15 000
$\tilde{v}/cm^{-1}$

6.10 The optical absorption spectrum of $[Ti(OH_2)_6]^{3+}$.

in the three *d* orbitals that point between the ligands, namely d_{xy}, d_{yz}, and d_{zx} (symmetry type, t_{2g}). Group theory shows that the e_g orbitals are doubly degenerate (although this is not readily apparent from drawings), and the t_{2g} orbitals are triply degenerate. This simple model leads to an energy level diagram in which the triply degenerate group of t_{2g} orbitals lie lower in energy than the e_g orbitals (Fig. 6.9). The separation of the orbitals is called the **ligand-field splitting parameter**, Δ_O (where the subscript O signifies an octahedral crystal field).

The simplest property that can be interpreted by crystal field theory is the electronic spectrum of a complex. Figure 6.10 shows the optical absorption spectrum of the d^1 ion $[Ti(OH_2)_6]^{3+}$. Crystal field theory assigns the first absorption maximum at 20 300 cm^{-1} to the transition $e_g \leftarrow t_{2g}$. (In keeping with spectroscopic notation, the higher-energy orbital is shown first.) We can identify 20 300 cm^{-1} with Δ_O for the complex. It is more complicated to obtain values of Δ_O for complexes with more than one *d* electron because the energy of a transition then depends not only on orbital energies (which we wish to know) but also on the repulsion energies between the several electrons present. This aspect is treated more fully in Chapter 14, and the results from the analyses described there have been used to obtain the values of Δ_O in Table 6.3.

Ligand field splitting parameters

The ligand field splitting parameter varies systematically with the identity of the ligand. The empirical evidence was the observation, by the Japanese chemist R. Tsuchida, that there are certain regularities in the absorption spectra as the ligands of a complex are varied. For instance, in the series of complexes $[CoX(NH_3)_5]^{n+}$ with $X = I^-$, Br^-, Cl^-, H_2O, and NH_3, the colors range from deep purple (for $X = I^-$) through pink (for Cl^-) to yellow (with NH_3). This observation indi-

cates that there is an increase in the energy of the lowest energy electronic transition (and therefore in Δ_O) as the ligands are varied along the series. Moreover, this variation is quite general, for the same order of ligands is followed at all metal centers.

On the basis of these observations, Tsuchida proposed that ligands could be arranged in a **spectrochemical series** of ligands, in which the members are arranged in order of increasing energy of transitions that occur when they are present in a complex:

$$I^- < Br^- < S^{2-} < \underline{S}CN^- < Cl^- < NO_3^- < F^- < OH^- < C_2O_4^{2-} < H_2O$$

$$< \underline{N}CS^- < CH_3CN < NH_3 < en < bipy < phen < \underline{N}O_2^- < PPh_3$$

$$< \underline{C}N^- < \underline{C}O$$

(The donor atom in an ambidentate ligand is underlined.) For example, the series indicates that the optical absorption of a hexacyano complex will occur at much higher energy than that of a hexachloro complex of the same metal ion.

The values of Δ_O also depend in a systematic way on the metal ion, and it is not in general possible to say that a particular ligand exerts a strong or a weak ligand field without considering the metal ion too. In this connection, the most important trends to keep in mind are that

1. Δ_O increases with increasing oxidation number.
2. Δ_O increases down a group.

This series is thought to reflect the improved metal–ligand bonding of the more expanded $4d$ and $5d$ orbitals compared with the compact $3d$ orbitals. The spectrochemical series for metal ions is approximately:

$$Mn^{2+} < Ni^{2+} < Co^{2+} < Fe^{2+} < V^{2+} < Fe^{3+} < Co^{3+}$$

$$< Mn^{4+} < Mo^{3+} < Rh^{3+} < Ru^{3+} < Pd^{4+} < Ir^{3+} < Pt^{4+}$$

Ligand field stabilization energies

The kind of experimental information we need to explain can be exemplified by the variation in the enthalpies of hydration of M^{2+} ions,

$$M^{2+}(g) + 6H_2O(l) \rightarrow [M(OH_2)_6]^{2+}(aq) \qquad \Delta H^{\ominus}$$

across Period 4. The data are shown in Fig. 6.11. The nearly linear increase across a period shown by the filled circles represents the increasing strength of the bonding between H_2O ligands and the central metal ion: ionic radii decrease from left to right across a period. We shall concentrate here on the wave-like variation in the strength of bonding shown by the open circles and see that it arises from the occupation of the d orbitals of the complex.

It is conventional to take the *average* energy of the d orbitals as zero

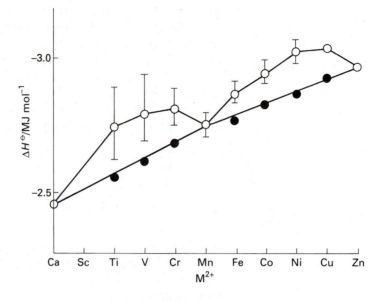

6.11 The hydration enthalpy of M^{2+} ions of the first row of the *d* block. The straight lines show the trend when the ligand field stabilization energy has been subtracted from the observed values. Note the general trend to greater hydration enthalpy (more exothermic hydration) on crossing the period from left to right.

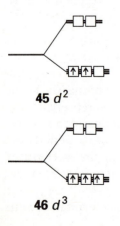

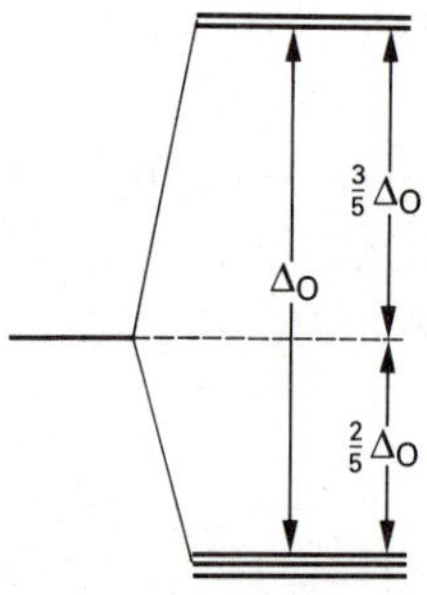

6.12 The orbital energy level diagram used in the application of the building-up principle in a crystal field analysis.

45 d^2

46 d^3

(not forgetting that, as shown by the data in Fig. 6.11, the average moves lower in energy across a period). Because there are three t_{2g} and two e_g orbitals, the t_{2g} orbitals lie $\frac{2}{5}\Delta_O$ below the average energy and the e_g orbitals lie $\frac{3}{5}\Delta_O$ above that average. Therefore, relative to the average energy, the energy of the t_{2g} orbitals is $-0.4\Delta_O$ and that of the e_g orbitals is $+0.6\Delta_O$. It follows that the net energy of a $t_{2g}^x e_g^y$ configuration relative to the average energy of the orbitals, which is called the **ligand field stabilization energy** (LFSE), is

$$\text{LFSE} = (-0.4x + 0.6y)\Delta_O$$

The values of the LFSE are given in Table 6.3.

   In the following sections we shall see how the ligand field splitting parameter Δ_O and the ligand field stabilization energy help to account for many aspects of *d*-block complexes. To establish the ground-state electron configurations of *d*-metal complexes, we use the *d*-orbital energy level diagram shown in Fig. 6.12 as a framework for the building-up principle. As usual, we look for the lowest energy configuration subject to the Pauli exclusion principle and (if more than one degenerate orbital is available) to the requirement that electrons first occupy separate orbitals and do so with parallel spins.

Weak-field and strong-field limits

The first three *d* electrons of a d^n complex occupy separate t_{2g} nonbonding orbitals, and do so with parallel spins. For example, the ions Ti^{2+} and V^{2+} have electron configurations d^2 and d^3, respectively. The *d* electrons occupy the lower t_{2g} orbitals (**45** and **46**, respectively) and the complexes are stabilized by $2 \times 0.4\Delta_O = 0.8\Delta_O$ (Ti^{2+}) and $3 \times 0.4\Delta_O = 1.2\Delta_O$ (V^{2+}). The next electron needed for the d^4 ion Cr^{2+} may enter one of the t_{2g} orbitals and pair with the electron already

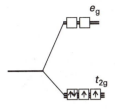

47 Strong-field d^4

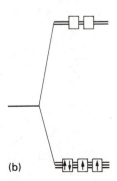

48 Weak-field d^4

(a)

(b)

6.13 (a) The weak-field, high-spin $t_{2g}^3 e_g^1$ and (b) strong-field, low-spin t_{2g}^4 configurations of a d^4 complex.

there (47); but if it does so, it experiences a strong coulombic repulsion, which is called the **pairing energy** P. Alternatively, the electron may occupy one of the e_g orbitals (48) and, although avoiding the pairing penalty, will have an energy higher by Δ_O. In the first case (t_{2g}^4), the LFSE is $1.6\Delta_O$, the pairing energy is P, and the net stabilization is $1.6\Delta_O - P$. In the second case ($t_{2g}^3 e_g^1$), the LFSE is $3 \times 0.4\Delta_O - 0.6\Delta_O = 0.6\Delta_O$, and there is no pairing energy to consider. Which configuration is adopted depends on which of $1.60\Delta_O - P$ and $0.60\Delta_O$ is the larger.

If $\Delta_O < P$, which is called the **weak-field case** (Fig. 6.13(a)), then occupation of the upper orbital is more favorable because the electron repulsion is minimized; the configuration adopted is then $t_{2g}^3 e_g^1$. If $\Delta_O > P$, which is called the **strong-field case** (Fig. 6.13(b)), pairing is more favorable, despite the repulsion, because it is energetically expensive to occupy the upper orbitals; the configuration adopted is then t_{2g}^4. For example, $[Cr(OH_2)_6]^{2+}$ has the ground-state configuration $t_{2g}^3 e_g^1$ (Fig. 6.13(a)) whereas $[Cr(CN)_6]^{4-}$ has the configuration t_{2g}^4 (Fig. 6.13(b)). The t_{2g}^4 configuration that is observed for the d^4 complex $[RuCl_6]^{2-}$ (Fig. 6.13(b)) illustrates the more general observation that $4d$ and $5d$ metals have larger values of Δ_O than $3d$ metals. Hence, complexes of these metals generally have electron configurations that are characteristic of strong ligand fields. For example, $[Ru(ox)_3]^{3-}$ has the configuration t_{2g}^5, in which all five electrons occupy the lower energy orbitals, whereas $[Fe(ox)_3]^{3-}$ has the configuration $t_{2g}^3 e_g^2$, in which two electrons decrease the pairing energy by occupying orbitals of slightly higher energy.

The ground state electron configurations of d^1, d^2, and d^3 complexes are unambiguous because there is no competition between LFSE and pairing energy: the configurations are t_{2g}^1, t_{2g}^2, and t_{2g}^3, respectively. A competition is possible for d^n complexes in which $n = 4$ and 5, because a large ligand field splitting parameter will favor the occupation of the lower orbitals (and give rise to t_{2g}^n configurations) whereas a small ligand field parameter will allow electrons to escape the pairing energy by occupying the upper orbitals. Then the configurations will be $t_{2g}^3 e_g^1$ and $t_{2g}^3 e_g^2$. Because in the latter case all the electrons occupy different orbitals, they will have parallel spins.

When alternative configurations are possible, the configuration with the smaller number of parallel electron spins is called a **low-spin configuration**, and the complex with the greater number of parallel electron spins is called a **high-spin configuration**. A d^4 complex has a low-spin configuration if Δ_O is large but it may have a high-spin configuration if Δ_O is small. The same applies to d^5 complexes (see Table 6.3). High and low spin configurations are also found for d^6 and d^7 complexes. In these cases, a large Δ_O results in the low-spin configurations t_{2g}^6 (no unpaired electrons) and $t_{2g}^6 e_g^1$ (one unpaired electron), respectively, and a small Δ_O results in the high-field configurations $t_{2g}^4 e_g^2$ (four unpaired electrons) and $t_{2g}^5 e_g^2$ (three unpaired electrons), respectively. The value of Δ_O depends on the identity of both the metal and the ligand, and the spin-pairing energy P varies with the metal, so it is not possible to specify the point in the spectrochemical series at which the complex changes from high spin to low spin. For the present, it is worth noting that complexes are low spin for ligands that are very high in the spectrochemical series (such as CN^-) in combination with $3d$ metal ions. They are generally high spin for ligands that are very low in the series (such as F^-) in combination with $3d$ metal ions. Low-spin complexes are common for $4d$ and $5d$ metal ions because Δ_O is larger and P is smaller than for $3d$ ions.

For octahedral d^n complexes with $n = 8$, 9, or 10 there is no ambiguity about the configuration (Table 6.3), and the terms high spin and low spin are not employed.

Magnetic measurements

The experimental distinction between high-spin and low-spin complexes is based on the determination of their magnetic properties. Complexes are classified as **diamagnetic** if they tend to move out of a magnetic field and **paramagnetic** if they tend to move into a magnetic field. The two can be distinguished using a Gouy balance (see Box 6.3).

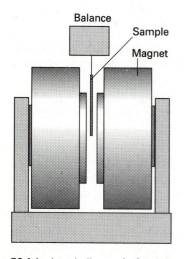

B6.1 A schematic diagram of a Gouy balance.

Box 6.3 The measurement of magnetic susceptibility

The attraction to or repulsion from a magnetic field can be measured by the change in the apparent weight of a sample when the magnetic field is turned on in an apparatus like that shown in the illustration, which is called a 'Gouy balance'. (Fig. B6.1) The observed change in apparent weight of the sample is the net outcome of a paramagnetic term arising from unpaired electrons and a diamagnetic term common to all matter. If unpaired electrons are present, the paramagnetic term is commonly much larger than the diamagnetic contribution. The force on the sample is proportional to the magnetic field gradient, so the sample usually protrudes from the field so that it is subject to a field gradient. The measurement yields a value for the magnetic susceptibility that must be corrected for diamagnetism using measurements on samples with similar total electron population but without unpaired electrons.

A more modern procedure for measuring magnetic susceptibility makes use of the solid-state device known as a superconducting quantum interference device (SQUID). A SQUID takes advantage of the quantization of magnetic flux and the property of current loops in superconductors that, as a part of the circuit, include a weakly conducting link through which electrons must tunnel. The current that flows in the loop in a magnetic field is determined by the value of the magnetic flux, and SQUIDs can be exploited as very sensitive magnetometers.

[References: W. E. Hatfield, Magnetic measurements. In *Solid state chemistry*, ed. A. K. Cheetham and P. Day. Oxford Univer-

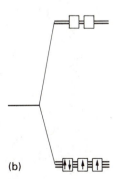

there (**47**); but if it does so, it experiences a strong coulombic repulsion, which is called the **pairing energy** P. Alternatively, the electron may occupy one of the e_g orbitals (**48**) and, although avoiding the pairing penalty, will have an energy higher by Δ_O. In the first case (t_{2g}^4), the LFSE is $1.6\Delta_O$, the pairing energy is P, and the net stabilization is $1.6\Delta_O - P$. In the second case ($t_{2g}^3 e_g^1$), the LFSE is $3 \times 0.4\Delta_O - 0.6\Delta_O = 0.6\Delta_O$, and there is no pairing energy to consider. Which configuration is adopted depends on which of $1.60\Delta_O - P$ and $0.60\Delta_O$ is the larger.

If $\Delta_O < P$, which is called the **weak-field case** (Fig. 6.13(a)), then occupation of the upper orbital is more favorable because the electron repulsion is minimized; the configuration adopted is then $t_{2g}^3 e_g^1$. If $\Delta_O > P$, which is called the **strong-field case** (Fig. 6.13(b)), pairing is more favorable, despite the repulsion, because it is energetically expensive to occupy the upper orbitals; the configuration adopted is then t_{2g}^4. For example, $[\text{Cr}(\text{OH}_2)_6]^{2+}$ has the ground-state configuration $t_{2g}^3 e_g^1$ (Fig. 6.13(a)) whereas $[\text{Cr}(\text{CN})_6]^{4-}$ has the configuration t_{2g}^4 (Fig. 6.13(b)). The t_{2g}^4 configuration that is observed for the d^4 complex $[\text{RuCl}_6]^{2-}$ (Fig. 6.13(b)) illustrates the more general observation that $4d$ and $5d$ metals have larger values of Δ_O than $3d$ metals. Hence, complexes of these metals generally have electron configurations that are characteristic of strong ligand fields. For example, $[\text{Ru}(\text{ox})_3]^{3-}$ has the configuration t_{2g}^5, in which all five electrons occupy the lower energy orbitals, whereas $[\text{Fe}(\text{ox})_3]^{3-}$ has the configuration $t_{2g}^3 e_g^2$, in which two electrons decrease the pairing energy by occupying orbitals of slightly higher energy.

The ground state electron configurations of d^1, d^2, and d^3 complexes are unambiguous because there is no competition between LFSE and pairing energy: the configurations are t_{2g}^1, t_{2g}^2, and t_{2g}^3, respectively. A competition is possible for d^n complexes in which $n = 4$ and 5, because a large ligand field splitting parameter will favor the occupation of the lower orbitals (and give rise to t_{2g}^n configurations) whereas a small ligand field parameter will allow electrons to escape the pairing energy by occupying the upper orbitals. Then the configurations will be $t_{2g}^3 e_g^1$ and $t_{2g}^3 e_g^2$. Because in the latter case all the electrons occupy different orbitals, they will have parallel spins.

When alternative configurations are possible, the configuration with the smaller number of parallel electron spins is called a **low-spin configuration**, and the complex with the greater number of parallel electron spins is called a **high-spin configuration**. A d^4 complex has a low-spin configuration if Δ_O is large but it may have a high-spin configuration if Δ_O is small. The same applies to d^5 complexes (see Table 6.3). High and low spin configurations are also found for d^6 and d^7 complexes. In these cases, a large Δ_O results in the low-spin configurations t_{2g}^6 (no unpaired electrons) and $t_{2g}^6 e_g^1$ (one unpaired electron), respectively, and a small Δ_O results in the high-field configurations $t_{2g}^4 e_g^2$ (four unpaired electrons) and $t_{2g}^5 e_g^2$ (three unpaired electrons), respectively. The value of Δ_O depends on the identity of both the metal and the ligand, and the spin-pairing energy P varies with the metal, so it is not possible to specify the point in the spectrochemical series

47 Strong-field d^4

48 Weak-field d^4

(a)

(b)

6.13 (a) The weak-field, high-spin $t_{2g}^3 e_g^1$ and (b) strong-field, low-spin t_{2g}^4 configurations of a d^4 complex.

at which the complex changes from high spin to low spin. For the present, it is worth noting that complexes are low spin for ligands that are very high in the spectrochemical series (such as CN^-) in combination with $3d$ metal ions. They are generally high spin for ligands that are very low in the series (such as F^-) in combination with $3d$ metal ions. Low-spin complexes are common for $4d$ and $5d$ metal ions because Δ_O is larger and P is smaller than for $3d$ ions.

For octahedral d^n complexes with $n = 8$, 9, or 10 there is no ambiguity about the configuration (Table 6.3), and the terms high spin and low spin are not employed.

Magnetic measurements

The experimental distinction between high-spin and low-spin complexes is based on the determination of their magnetic properties. Complexes are classified as **diamagnetic** if they tend to move out of a magnetic field and **paramagnetic** if they tend to move into a magnetic field. The two can be distinguished using a Gouy balance (see Box 6.3).

Box 6.3 The measurement of magnetic susceptibility

The attraction to or repulsion from a magnetic field can be measured by the change in the apparent weight of a sample when the magnetic field is turned on in an apparatus like that shown in the illustration, which is called a 'Gouy balance'. (Fig. B6.1) The observed change in apparent weight of the sample is the net outcome of a paramagnetic term arising from unpaired electrons and a diamagnetic term common to all matter. If unpaired electrons are present, the paramagnetic term is commonly much larger than the diamagnetic contribution. The force on the sample is proportional to the magnetic field gradient, so the sample usually protrudes from the field so that it is subject to a field gradient. The measurement yields a value for the magnetic susceptibility that must be corrected for diamagnetism using measurements on samples with similar total electron population but without unpaired electrons.

A more modern procedure for measuring magnetic susceptibility makes use of the solid-state device known as a superconducting quantum interference device (SQUID). A SQUID takes advantage of the quantization of magnetic flux and the property of current loops in superconductors that, as a part of the circuit, include a weakly conducting link through which electrons must tunnel. The current that flows in the loop in a magnetic field is determined by the value of the magnetic flux, and SQUIDs can be exploited as very sensitive magnetometers.

[References: W. E. Hatfield, Magnetic measurements. In *Solid state chemistry*, ed. A. K. Cheetham and P. Day. Oxford Univer-

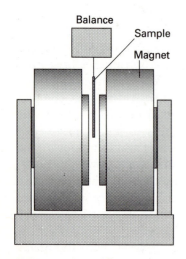

B6.1 A schematic diagram of a Gouy balance.

Balance

Sample

Magnet

sity Press (1987); L. N. Mulay and I. L. Mulay. Static magnetic techniques and applications. *Techniques of physical chemistry*, **IIIB**, 133 (1989).]

In a free atom or ion, both the orbital and the spin angular momenta give rise to a magnetic moment and contribute to the paramagnetism. When the atom or ion is part of a complex, its orbital angular momentum may be reduced—the technical term is **quenched**—as a result of the interactions of the electrons with their non-spherical environment. The electron spin angular momentum, however, survives, and gives rise to **spin-only paramagnetism**, which is characteristic of d-metal complexes. The spin-only magnetic moment μ of a complex with total spin quantum number S is

$$\mu = 2\{S(S+1)\}^{1/2}\mu_B$$

where μ_B is the Bohr magneton:

$$\mu_B = \frac{e\hbar}{2m_e} = 9.274 \times 10^{-24} \text{ J T}^{-1}$$

Because each unpaired electron has a spin quantum number of $\frac{1}{2}$, it follows that $S = \frac{1}{2}N$, where N is the number of unpaired electrons; therefore

$$\mu = \{N(N+2)\}^{1/2}\mu_B$$

A measurement of the magnetic moment of a d block complex can usually be interpreted in terms of the number of unpaired electrons it contains, and hence the measurement can be used to distinguish between high-spin and low-spin complexes. For example, magnetic measurements on a d^6 complex permit us to distinguish between a high-spin $t_{2g}^4 e_g^2$ ($S=2$) electronic configuration and a low-spin t_{2g}^6 ($S=0$) configuration.

The spin-only magnetic moments for the configurations $t_{2g}^x e_g^y$ are listed in Table 6.4 and compared there with experimental values for a number of complexes from the first row of the d block. For most $3d$ and some $4d$ complexes, experimental values lie reasonably close to spin-only predictions, so it becomes possible to identify correctly the

Table 6.4 Calculated spin-only magnetic moments

Ion	n	S	μ/μ_B	
			Calculated	Experiment
Ti^{3+}	1	$\frac{1}{2}$	1.73	1.7–1.8
V^{3+}	2	1	2.83	2.7–2.9
Cr^{3+}	3	$\frac{3}{2}$	3.87	3.8
Mn^{3+}	4	2	4.90	4.8–4.9
Fe^{3+}	5	$\frac{5}{2}$	5.92	5.9

number of unpaired electrons and thereby assign the ground state configuration of a complex. For instance, the complex $[Fe(OH_2)_6]^{3+}$ is paramagnetic with a magnetic moment of $5.3\mu_B$. As shown in Table 6.4, this is reasonably close to the value for five unpaired electrons and thus implies a high-spin $t_{2g}^3 e_g^2$ configuration.

Example 6.5: *Inferring an electron configuration from a magnetic moment*

The magnetic moment of an octahedral Co(II) complex is $4.0\mu_B$. What is its electron configuration?

Answer. A Co(II) complex is d^7. The two possible configurations are $t_{2g}^5 e_g^2$ (high spin) with three unpaired electrons or $t_{2g}^6 e_g^1$ (low spin) with one unpaired electron. The spin-only magnetic moments (Table 6.4) are $3.87\mu_B$ and $1.73\mu_B$ respectively. Therefore, the only consistent assignment is the high spin $t_{2g}^5 e_g^2$.

Exercise E6.5. The magnetic moment of the complex $[Mn(NCS)_6]^{4-}$ is $6.06\mu_B$. What is its electron configuration?

The interpretation of magnetic susceptibility measurements is sometimes less straightforward than Example 6.5 might suggest. For example, the potassium salt of $[Fe(CN)_6]^{3-}$ has $\mu = 2.3\mu_B$, which is between the spin-only values for one and two unpaired electrons ($1.7\mu_B$ and $2.8\mu_B$ respectively). In this case, the spin-only assumption has failed and the orbital magnetic contribution is substantial.

For orbital angular momentum to contribute, and hence for the paramagnetism to differ significantly from the spin-only value, there must be an unfilled or half-filled orbital similar in energy to that of the orbitals occupied by the unpaired spins. If that is so, the electrons can make use of the available orbital to circulate around the center of the complex and hence generate angular momentum and a magnetic moment (Fig. 6.14). More specifically, an orbital must be available that is related by rotational symmetry to the occupied orbital (as a d_{xy} orbital is related to $d_{x^2-y^2}$ by a 45° rotation about the z axis, Fig. 6.14) and it should not contain an electron with the same spin as the first electron. These conditions are fulfilled whenever any one or two of the three t_{2g} (d_{xy}, d_{yz}, d_{zx}) orbitals contain an odd number of electrons. Departure from spin-only values is generally large for low-spin d^5 and for high-spin $3d^6$ and $3d^7$ complexes. In other $3d$ complexes, the orbital angular momentum contributions of the electrons cancel each other and values near spin-only are observed.

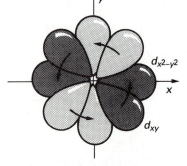

6.14 If there is a low-lying orbital of the correct symmetry, the applied field may induce the circulation of the electrons in a complex and hence generate orbital angular momentum. This diagram shows the way that circulation may arise when the field is applied perpendicular to the *xy* plane (perpendicular to this page).

Thermochemical correlations

We now return to Fig. 6.11 and see how it can be explained. We have noted that the increase in hydration enthalpy across the $3d$ series corresponds to a decrease in the ionic radius of the central atom and consequently a stronger interaction of that ion with the partial charges

Table 6.5 Ligand field splitting parameters Δ_0 of ML_6 complexes*

	Ions	Ligands				
		Cl^-	H_2O	NH_3	en	CN^-
d^3	Cr^{3+}	13.7	17.4	21.5	21.9	26.6
d^5	Mn^{2+}	7.5	8.5		10.1	30
	Fe^{3+}	11.0	14.3			(35)
d^6	Fe^{2+}		10.4			(32.8)
	Co^{3+}		(20.7)	(22.9)	(23.2)	(34.8)
	Rh^{3+}	(20.4)	(27.0)	(34.0)	(34.6)	(45.5)
d^8	Ni^{2+}	7.5	8.5	10.8	11.5	

*Values are in multiples of 1000 cm^{-1}; entries in parentheses are for low-spin complexes.
Source: H.B. Gray, *Electrons and chemical bonding*, Benjamin, Menlo Park (1965).

that represent the donor atoms of the ligands. The wave-like deviation of hydration enthalpies from a straight line reflects a similar variation in the ligand-field stabilization energies: as Table 6.3 shows:

> The LFSE increases from d^1 to d^3, decreases again to d^5, then rises to d^8.

(As we see in the next section, d^9 is a special case.) The filled circles in Fig. 6.11 were calculated by subtracting the high-spin LFSE from ΔH_H by using the spectroscopic values of Δ_O in Table 6.5. We see that the LFSE calculated from spectroscopic data nicely accounts for the additional ligand binding energy for the complexes shown in the illustration.

Example 6.6: *Using the LFSE to account for thermochemical properties*

The oxides of formula MO, which all have octahedral coordination of the metal ions, have the following lattice enthalpies:

CaO	TiO	VO	MnO
3460	3878	3913	3810 kJ mol^{-1}

Account for the trends in terms of the LFSE given that O^{2-} is low in the spectrochemical series.

Answer. The general trend across the d block is the increase from CaO (d^0) to MnO (d^5), both of which have an LFSE of zero. Since O^{2-} is a weak-field ligand, TiO (d^2) has an LFSE of $0.8\Delta_O$ and VO (d^3) has an LFSE of $1.2\Delta_O$ (Table 6.3). If we suppose that Δ_O is fairly constant across the series, the contribution of LFSE to lattice enthalpies is in the same order as the experimental values of the lattice enthalpy.

Exercise E6.6. Account for the variation in lattice enthalpy of the solid fluorides in which each metal ion is surrounded by an octahedral array of F^- ions: MnF_2 (2780 kJ mol^{-1}), FeF_2 (2926 kJ mol^{-1}), CoF_2 (2976 kJ mol^{-1}), NiF_2 (3060 kJ mol^{-1}), and ZnF_2 (2985 kJ mol^{-1}).

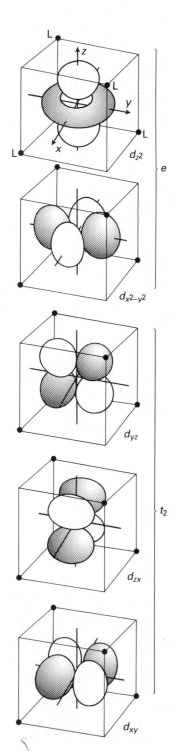

6.15 The effect of a tetrahedral crystal field on a set of *d* orbitals is to split them into two sets; the *e* pair (which point less directly at the ligands) lie lower in energy than the t_2 triplet.

Table 6.6 Values of Δ_T for representative tetrahedral complexes

Complex	Δ_T/cm^{-1}
VCl_4	9010
$[CoCl_4]^{2-}$	3300
$[CoBr_4]^{2-}$	2900
$[CoI_4]^{2-}$	2700
$[Co(NCS)_4]^{2-}$	4700

6.5 Four-coordinate complexes

Second only in abundance to octahedral complexes are the tetrahedral and square-planar four-coordinate complexes. As remarked earlier, the former are commonly observed for the tetrachloro, tetrabromo, and tetraiodo complexes of $3d$ metal ions with oxidation number $+2$, such as $[NiCl_4]^{2-}$. Square-planar complexes are common for $4d^8$ and $5d^8$ complexes, such as $[PtCl_4]^{2-}$.

Tetrahedral coordination compounds

A tetrahedral crystal field also splits *d* orbitals into two sets, one of which is doubly degenerate and the other triply degenerate. The principal difference from octahedral complexes is the reversal of the order of the two sets: in a tetrahedral crystal field, the doubly degenerate *e* set lies *below* the triply degenerate t_2 set. This difference can be understood from a detailed analysis of the spatial arrangement of the orbitals (Fig. 6.15), for the two *e* orbitals point between the positions of the ligands and their partial negative charges whereas the three t_2 orbitals point more directly toward the ligands.[10] A secondary difference is that the ligand field splitting parameter, Δ_T, in a tetrahedral complex is less than Δ_O, as might be expected for complexes with fewer ligands (but in fact, $\Delta_T < \frac{2}{3}\Delta_O$, which is harder to explain but is related to the fact that d_{z^2} and $d_{x^2-y^2}$ point directly at the ligands in an octahedral complex). Hence, only weak-field tetrahedral complexes are common, and they are the only ones we need consider.

Ligand field stabilization energies can be calculated in exactly the same way as for octahedral complexes, the only differences being the order of occupation (*e* before t_2) and the contribution of each orbital to the total energy ($-\frac{3}{5}\Delta_T$ for an *e* electron and $+\frac{2}{5}\Delta_T$ for a t_2 electron). Some calculated values are given in Table 6.3 and the values of Δ_T for a number of complexes are collected in Table 6.6.

Tetragonal and square-planar complexes

Copper(II) complexes usually depart considerably from octahedral symmetry and have lower energies than pure-octahedral ligand field stabilization predicts. Low spin d^7 complexes (of configuration $t_{2g}^6 e_g^1$) may show a similar distortion, but this case is rarer.

It is convenient to take the regular octahedron as a starting point

---

[10] Because there is no center of inversion in a tetrahedral complex, the orbital designation does not include the parity label g or u.

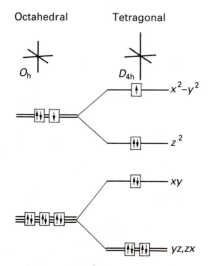

6.16 The effect of tetragonal distortions (compression along x and y and extension along z) on the energies of d orbitals. The electron occupation is for a d^9 complex.

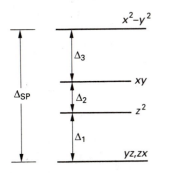

6.17 The orbital splitting parameters for a square-planar complex.

for the explanation of the existence of six-coordinate complexes with geometries that are distorted toward square planar. A tetragonal distortion, which corresponds to extension along the z axis and compression on the x and y axes, reduces the energy of the $e_g(d_{z^2})$ orbital (Fig. 6.16), but it increases the energy of the $e_g(d_{x^2-y^2})$ orbital. Therefore, if one, two, or three electrons occupy the e_g orbitals (as in d^7, d^8, and d^9 complexes) a tetragonal distortion may be energetically advantageous. For example, in a d^9 complex (with configuration $t_{2g}^6 e_g^3$), such a distortion leaves two electrons stabilized and one destabilized.

The distortion of d^8 ($t_{2g}^6 e_g^2$) complexes may be large enough to encourage the two e_g electrons to pair in the d_{z^2} orbital. This distortion actually often goes as far as the total loss of the ligands on the z axis and the formation of the d^8 square planar complexes, such as those found for Rh(I), Ir(I), Pt(II), Pd(II), and Au(III). The preponderance of square-planar complexes for the $4d^8$ and $5d^8$ metals correlates with the high values of the ligand field splitting parameter in these two series, which gives rise to a high ligand-field stabilization of the low-spin square-planar complexes. In contrast, $3d$ metal complexes such as $[NiX_4]^{2-}$, with X a halogen, are generally tetrahedral because the ligand field splitting parameter is generally quite small for members of this series. Only when the ligand is high in the spectrochemical series is the LFSE large enough to favor the formation of a square-planar complex, as for example with $[Ni(CN)_4]^{2-}$.

The d-orbital splitting for square-planar complexes is shown in Fig. 6.17. The sum of the three distinct orbital splittings is denoted Δ_{SP}. This sum is greater than Δ_O; simple theory predicts that $\Delta_{SP} = 1.3\Delta_O$ for complexes of the same metal and ligands.

The Jahn–Teller effect

The tetragonal distortions just described represent specific examples of the **Jahn–Teller effect**:

> If the ground electronic configuration of a nonlinear molecule is degenerate, then the molecule will distort so as to remove the degeneracy and achieve a lower energy.

An octahedral d^9 complex is degenerate because the odd electron can occupy either the $d_{x^2-y^2}$ orbital or the d_{z^2} orbital, and a tetragonal distortion lowers the energy of the latter. Distortion to square-planar, with the electrons paired in the d_{z^2} orbital, removes the degeneracy and lowers the energy of the complex, and can be regarded as an extreme case of the Jahn–Teller effect. Note that the effect does not refer to the degeneracies of orbitals but to the degeneracies of the electron configurations that arise when the orbitals are partially occupied.

The Jahn–Teller effect identifies an unstable geometry; it does not predict the preferred distortion. The examples we have cited involve the elongation of two axial bonds and the compression of the four bonds that lie in a plane. The alternative distortion, compression along an axis and elongation in a plane, would also remove the degeneracy. Which distortion occurs in practice is a matter of energetics, not symmetry. However, because the axial elongation weakens only two bonds whereas elongation in the plane would weaken four, axial elongation is more common than axial compression.

Apparent exceptions to the Jahn–Teller effect are sometimes encountered. Some Cu(II) complexes, for instance, seem to be undistorted. This symmetry may, however, only be apparent, for it could be a result of the time-scale of the measurement and the detection of a time-average of fluxional structures (Section 3.2). For example, the EPR signal from $[Cu(OH_2)_6]^{2+}$ in a $[Zn(OH_2)_6][SiF_6]$ host crystal at room temperature displays a single isotropic absorption line, which suggests a high symmetry octahedral environment. However, if a distortion alternates among the three equivalent axes faster than the difference in their resonance frequencies, then the observed spectral lines will lie at the average of the locations of the three lines of the distorted species. Because resonance frequency differences are of the order of 1 MHz in EPR spectroscopy, a fluxionally averaged spectrum is observed if a distortion changes orientation in less than about 1 μs.

The hopping of a distortion from one orientation to another is called the **dynamic Jahn–Teller effect**, and its rate depends on the temperature. In the EPR spectrum of $[Cu(OH_2)_6]^{2+}$, a static distortion (more precisely, one effectively stationary on the time-scale of the resonance experiment) occurs when the temperature is below 20 K; distinct signals are then observed from the different environments.

6.6 Ligand field theory

Crystal field theory provides a simple conceptual model and can be used to interpret spectra as long as we employ only empirical values of Δ_O. On closer inspection, however, the theory is defective; for example, a CO ligand gives rise to a large splitting even though it has no charge and only a very small dipole moment.

Ligand field theory, which is an application of molecular orbital theory, overcomes this objection and provides a satisfactory theory of Δ_O. As we shall see, ligand field theory is essentially an improvement of crystal field theory, and the concepts introduced by the latter need not be abandoned. The strategy for describing the molecular orbitals of a metal complex will follow procedures similar to those utilized in Chapter 2 for bonding in polyatomic molecules: identify the valence orbitals on the metal and ligand, determine the symmetry-adapted orbitals, and then—using empirical energy and overlap considerations—estimate the relative energies of the molecular orbitals. The estimated order can be verified and positioned more precisely by comparison with experimental data (particularly optical absorption and photoelectron spectroscopy[11]).

---

[11] High resolution PES data are available only for complexes in the gas phase, and are therefore restricted to uncharged complexes. For details, see J. H. D. Eland, *Photoelectron spectroscopy*. Open University Press, Milton Keynes (1983).

σ *Bonding*

We begin the systematic discussion of ligand field theory by imagining an octahedral complex in which each ligand has a single valence orbital directed toward the central metal atom with local σ symmetry around the M—L axis. Examples of such ligands include the isolobal NH_3 molecule and F^- ion.

In an octahedral (O_h) environment, the metal atom orbitals divide by symmetry into four sets (Fig. 6.18 and Appendix 4):

Metal atom orbital	Symmetry label
s	a_{1g} (non-degenerate)
p_x, p_y, p_z	t_{1u} (triply degenerate)
d_{xy}, d_{yz}, d_{zx}	t_{2g} (triply degenerate)
$d_{x^2-y^2}, d_{z^2}$	e_g (doubly degenerate)

Six symmetry-adapted linear combinations of the six ligand orbitals can also be formed. These combinations can be taken from Appendix 4 and are also shown in Fig. 6.18. One (unnormalized) ligand combination is a nondegenerate a_{1g} linear combination:

$$a_{1g}: \quad \sigma_1 + \sigma_2 + \sigma_3 + \sigma_4 + \sigma_5 + \sigma_6$$

where the σ_i denote ligand σ orbitals. Three form a t_{1u} set:

$$t_{1u}: \quad \sigma_1 - \sigma_3, \quad \sigma_2 - \sigma_4, \quad \sigma_5 - \sigma_6$$

The remaining two form an e_g pair:

$$e_g: \quad \sigma_1 - \sigma_2 + \sigma_3 - \sigma_4, \quad 2\sigma_6 + 2\sigma_5 - \sigma_1 - \sigma_2 - \sigma_3 - \sigma_4$$

These six combinations account for all the ligand orbitals of σ symmetry: there is no combination of ligand σ orbitals that has the symmetry of the metal t_{2g} orbitals, so the latter do not participate in σ bonding.[12]

Molecular orbitals are now formed by allowing symmetry adapted combinations of the same symmetry type on the metal ion and the surrounding ligands to overlap. Calculations of the resulting energies (adjusted to agree with experimental data) result in the molecular orbital energy level diagram shown in Fig. 6.19.

It should be recalled from Chapter 2 that the lowest energy linear combination of orbitals has principally the character of the lowest energy atomic orbitals that contribute to it. For NH_3, F^-, and most other ligands, the ligand σ orbitals are derived from atomic valence orbitals that lie in energy well below the metal d orbitals. As a result, the six bonding molecular orbitals of the complex are mainly ligand-

---

[12] The normalized forms of the symmetry-adapted linear combinations (with overlap neglected) are:

$$a_{1g}: \quad (\tfrac{1}{6})^{1/2}(\sigma_1 + \sigma_2 + \sigma_3 + \sigma_4 + \sigma_5 + \sigma_6)$$

$$t_{1u}: \quad (\tfrac{1}{2})^{1/2}(\sigma_1 - \sigma_3), (\tfrac{1}{2})^{1/2}(\sigma_2 - \sigma_4), (\tfrac{1}{2})^{1/2}(\sigma_5 - \sigma_6)$$

$$e_g: \quad (\tfrac{1}{4})^{1/2}(\sigma_1 - \sigma_2 + \sigma_3 - \sigma_4)$$

$$(\tfrac{1}{12})^{1/2}(2\sigma_6 + 2\sigma_5 - \sigma_1 - \sigma_2 - \sigma_3 - \sigma_4)$$

Metal   Ligands (σ)   Metal   Ligands (σ)

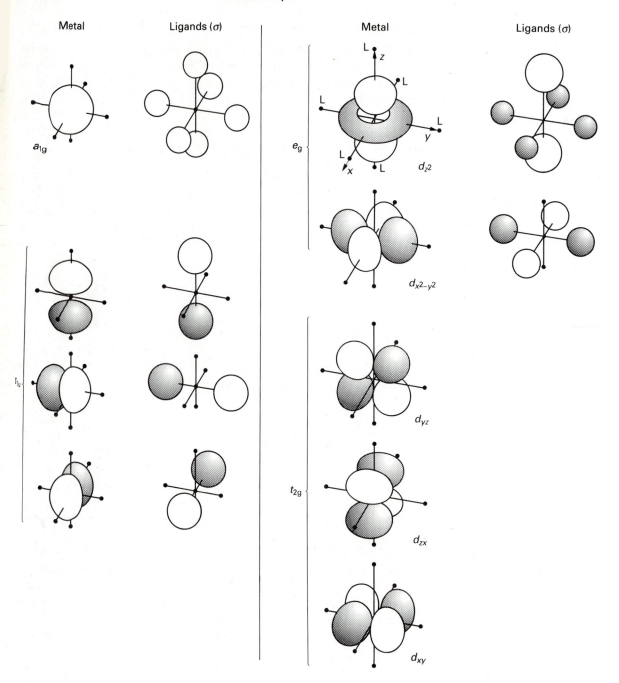

6.18 Symmetry-adapted combinations of ligand σ orbitals (represented here by spheres) in an octahedral complex. For symmetry adapted orbitals in other point groups, see Appendix 4.

orbital in character. These six bonding orbitals can accommodate the 12 electrons provided by the six ligand lone pairs.

The number of electrons to accommodate in addition to those supplied by the ligands depends on the number of d electrons supplied by the central metal ion. These additional electrons occupy the nonbonding d orbitals (the t_{2g} orbitals) and the antibonding combination (the upper e_g orbitals) of the d orbitals and ligand orbitals. Both these sets of orbitals are largely metal-ion in character, so the electrons

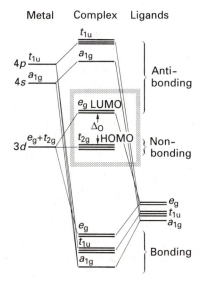

Metal   Complex   Ligands

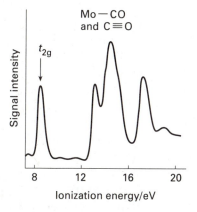

6.19 Molecular orbital energy levels of a typical octahedral complex. The frontier orbitals are inside the gray box.

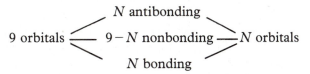

6.20 The He(II) (30.4 nm) photoelectron spectrum of $[Mo(CO)_6]$. With six electrons from Mo and twelve from :CO, the configuration indicated by Fig. 6.14 is $a_{1g}^2 t_{1u}^6 e_g^4 t_{2g}^6$. (From B. R. Higgenson, D. R. Lloyd, P. Burroughs, D. M. Gibson, and A. F. Orchard, *J. Chem. Soc., Faraday Trans., II*, **69**, 1659 (1973).)

supplied by the central ion remain largely on the metal ion. In summary, the frontier orbitals of the complex are the nonbonding t_{2g} orbitals (essentially purely metal in character when only σ bonding is considered) and the antibonding e_g orbitals (mainly metal in character). The octahedral ligand field splitting parameter, Δ_O, in this approach is the HOMO–LUMO separation. It is approximately the splitting of the metal d orbitals caused by the ligands, whereas in crystal field theory Δ_O is purely a metal d-orbital separation.

Some additional insight into bonding in complexes is obtained from photoelectron spectroscopy (recall Section 2.4). The photoelectron spectrum of gas-phase $[Mo(CO)_6]$ is shown in Fig. 6.20. With six electrons from Mo and twelve electrons from the six CO ligands (treated as :CO), the ground state electron configuration of the complex is expected to be $a_{1g}^2 t_{1u}^6 e_g^4 t_{2g}^6$. The HOMOs are the three t_{2g} orbitals that are largely confined to the Mo atom, and their energy can be identified by ascribing the peak of lowest ionization energy (close to 8 eV) to them. The group of ionization energies around 14 eV are assigned to the Mo—CO σ bonding orbitals. However, 14 eV is close to the ionization energy of CO itself, and so the peaks at that energy also arise from orbitals in CO.

Molecular orbitals for other coordination numbers

The preceding discussion leads to the conclusion that, in an octahedral complex, most of the electrons supplied by the ligands and the central metal ion can be accommodated in bonding molecular orbitals. It should then be easy to see that ligand field theory solves the puzzle that confronted Werner and his contemporaries concerning the ability of a metal ion to bind numerous ligands: with several metal d orbitals available in the valence shell, enough delocalized molecular orbitals can be built to accommodate all the electrons necessary to bind the ligands.[13]

More specifically, a d-metal atom can utilize up to nine valence orbitals (one s, three p, and five d orbitals) to form molecular orbitals. For coordination number N (and considering N ligand orbitals to be pointing at the metal atom, with N not greater than 9), N bonding and N antibonding orbitals can be formed, leaving $9 - N$ nonbonding d orbitals unused on the metal atom. Thus a complex with N metal–ligand bonds has the molecular orbital scheme

Metal orbitals   molecular orbitals   ligand orbitals

9 orbitals $\Big\langle$ N antibonding / $9-N$ nonbonding / N bonding $\Big\rangle$ N orbitals

For six coordination, this pattern corresponds to six bonding orbitals, three nonbonding orbitals, and six antibonding orbitals.

---

[13]   In fact we have already seen that hypervalence is possible without d orbitals. A contemporary account of the continuity between d and main group complexes and the role of d orbitals is given by M. Gerloch, *Coord. Chem. Rev.* **99**, 117 (1990).

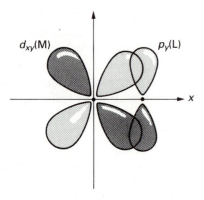

$d_{xy}(M)$ $p_y(L)$

x

6.21 The π overlap that may occur between a ligand p orbital perpendicular to the M—L axis and a metal d_{xy} orbital.

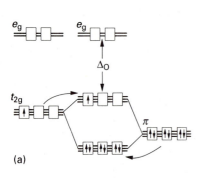

π^*

e_g e_g

Δ_O

t_{2g}

π

(a)

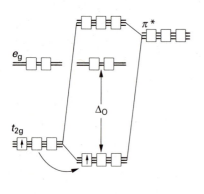

π^*

e_g

Δ_O

t_{2g}

(b) π

6.22 The effect of π bonding on the ligand field splitting parameter. (a) Ligands that act as π donors decrease Δ_O. (b) Ligands that act as π acceptors increase Δ_O.

π *Bonding*

If the ligands in a complex have orbitals with local π symmetry with respect to the M—L axis (as two of the p orbitals of a halide ligand have), then they may form π molecular orbitals with the t_{2g} metal orbitals (Fig. 6.21). The effect of this π bonding on the value of Δ_O depends on whether the π ligand orbitals act as electron donors or acceptors. In this section, we shall illustrate the essence of π bonding effects by a simplified argument based on π bonding in a single M—X fragment (in a fuller treatment, SALCs would be constructed from all the available π orbitals on all the ligand atoms). We shall employ two general principles described in Section 2.5. First, we use the idea that when atomic orbitals overlap strongly, they mix strongly, and the resulting bonding molecular orbitals are significantly lower in energy and the antibonding molecular orbitals are significantly higher in energy than the atomic orbitals. Second, we note that atomic orbitals with similar energies interact strongly, whereas those of very different energies mix only slightly even if their overlap is large.

By a π-**donor ligand** we mean a ligand that has filled orbitals that have π symmetry around the M—L axis. Furthermore, the energies of these orbitals are similar to those of the metal d orbitals, and the ligand has no low energy vacant π orbitals. Such ligands include Cl^-, I^-, and H_2O; it is significant that these ligands are low in the spectrochemical series, and shortly we shall see that considerations of π-bonding ability help to explain the order of ligands in that series. The full π orbitals of π-donor ligands lie lower in energy than the partially filled d orbitals of the metal. Therefore, when they form molecular orbitals with the t_{2g} orbitals of the central ion, the bonding combination lies lower than the ligand orbitals and the antibonding combination lies above the energy of the d orbitals of the free metal ion (Fig. 6.22(a)). The electrons supplied by the ligand lone pairs occupy and fill the bonding combinations, leaving the electrons originally in the d orbitals of the central metal atom to occupy the antibonding t_{2g} orbitals. The net effect is that the hitherto nonbonding metal-ion t_{2g} orbitals become antibonding and hence are raised closer in energy to the mainly metal antibonding e_g orbital. It follows that *strong π-donor ligand interactions decrease* Δ_O.

Now we turn to ligands that can accept electrons into their π orbitals, and we see that such ligands lie high in the spectrochemical series and give rise to a large ligand field splitting parameter. By a π-**acceptor ligand** we shall mean a ligand that (usually) has filled π orbitals at lower energies than metal t_{2g} orbitals and low energy empty π orbitals that are available for occupation. Typically, the π acceptor orbitals are vacant antibonding orbitals on the ligand, as in CO and N_2, and these orbitals lie above the metal d orbitals in energy. When these vacant π^* ligand orbitals are close in energy to the metal t_{2g} orbitals and the metal–ligand π overlap is strong, a little electron density will be delocalized from the metal to the ligand. For example, the π^* orbital of CO (Fig. 2.16) has its largest amplitude on the C atom and has the correct symmetry for overlap with the t_{2g} orbitals of the metal. In contrast, the full bonding π orbital of CO is low in energy and is largely localized on the O atom (because that is the more electronegative atom), so the π-donor character of CO is very low and in most (if not all) d-metal carbonyl complexes CO is a net π acceptor.

Because the π-acceptor orbitals on most ligands are higher in energy than the metal d orbitals, they form molecular orbitals with the metal in which t_{2g} orbitals of mainly metal d orbital character are lowered in energy (Fig. 6.22(b)). The net result is that Δ_O *is increased by the* π*-acceptor interaction.*

We can now put the role of π bonding in perspective. The order of ligands in the spectrochemical series is partly that of the strengths with which they can participate in M—L σ bonding. For example, both CH_3^- and H^- are very high in the spectrochemical series because they are very strong σ donors. However, when π bonding is significant it has a strong influence on Δ_O, and is responsible for CO (a strong π acceptor) being high on the spectrochemical series and OH^- (a strong π donor) being low in the series. The overall order of the spectrochemical series may be interpreted in broad terms as dominated by π effects (with a few important exceptions), and in general the series can be interpreted as follows:

$$\xrightarrow{\text{increasing } \Delta_O}$$

π donor $<$ weak π donor $<$ no π effects $<$ π acceptor

Representative ligands that match these classes are

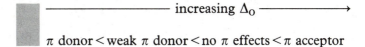

$$I^- < Br^- < Cl^- < F^- < H_2O < NH_3 < PR_3 < CO$$

π donor $<$ weak π donor $<$ no π effects $<$ π acceptor

Notable exceptions where the σ bonding effect is clearly important include CH_3^- and H^-, which are neither π donor nor π acceptor ligands.

REACTIONS OF COMPLEXES

The reactions of d-metal complexes are usually studied in solution. The solvent molecules compete for the central metal ion, so the formation of a complex with another ligand is a **substitution reaction**, a reaction in which an incoming group displaces a ligand (a solvent molecule) already present. The incoming group is called the **entering group** and the displaced ligand is the **leaving group**. We normally denote the leaving group as X and the entering group as Y. Then a substitution reaction is the Lewis displacement reaction

$$M—X + Y \rightarrow M—Y + X$$

Both the thermodynamics and the kinetics of complex formation lead to an understanding of the reaction of complexes, and we shall introduce both aspects here. A more thorough account of the kinetics of reactions is given in Chapter 15.

6.7 Coordination equilibria

A specific example of a coordination equilibrium is the reaction of Fe(III) with SCN^- to give the complex $[Fe(OH_2)_5(NCS)]^{2+}$, which is closely related to the analytically useful red complex used to detect either iron or the thiocyanate ion:

$$[Fe(OH_2)_6]^{3+} + NCS^- \rightleftharpoons [Fe(OH_2)_5(NCS)]^{2+} + H_2O$$

$$K_f = \frac{[Fe(OH_2)_5(NCS)^{2+}]}{[Fe(OH_2)_6^{3+}][NCS^-]}$$

The equilibrium constant K_f is the **formation constant** of the complex. The concentration of H_2O does not appear because it is taken to be constant in dilute solution and is absorbed into K_f. A ligand for which K_f is large is one that binds more tightly than H_2O. A ligand for which K_f is small may not be a weak ligand in an absolute sense, but merely weaker than H_2O.

The discussion of stabilities is more involved when more than one ligand may be replaced. In the series from $[Ni(OH_2)_6]^{2+}$ to $[Ni(NH_3)_6]^{2+}$, for instance, there are as many as six steps even if *cis-trans* isomerism is ignored. For the general case of the complex ML_n, the **stepwise formation constants** are

$$M + L \rightleftharpoons ML \qquad K_1 = \frac{[ML]}{[M][L]}$$

$$ML + L \rightleftharpoons ML_2 \qquad K_2 = \frac{[ML_2]}{[ML][L]}$$

$$ML_2 + L \rightleftharpoons ML_3 \qquad K_3 = \frac{[ML_3]}{[ML_2][L]}$$

$$\ldots \qquad\qquad \ldots$$

$$ML_{n-1} + L \rightleftharpoons ML_n \qquad K_n = \frac{[ML_n]}{[ML_{n-1}][L]}$$

These are the constants to consider when seeking to understand the relations between structure and reactivity. When we want to calculate the concentration of the final product (the complex ML_n) we use the **overall formation constant** β_n:

$$\beta_n = \frac{[ML_n]}{[M][L]^n}$$

The overall formation constant is the product of the stepwise constants:

$$\beta_n = K_1 K_2 K_3 \ldots K_n$$

The inverse of K_f, the **dissociation constant** K_d, is also sometimes useful:

$$ML \rightleftharpoons M + L \qquad K_d = \frac{[M][L]}{[ML]}$$

K_d has the same form as K_a for acids, which facilitates comparisons between metal complexes and Brønsted acids. The values of K_d and K_a can be tabulated together if the proton is considered to be simply another cation.

Trends in successive formation constants

It is commonly observed that stepwise formation constants lie in the order

$$K_1 > K_2 > K_3 \ldots > K_n$$

This general trend can be explained quite simply in terms of the numbers of ligands present and the number of opportunities for reaction. We can find a simple explanation by considering the decrease in the number of the ligand H_2O molecules available for replacement in the formation step, as in

$$M(OH_2)_5L + L \rightarrow M(OH_2)_4L_2 + H_2O$$

compared with

$$M(OH_2)_4L_2 + L \rightarrow M(OH_2)_3L_3 + H_2O$$

This sequence reduces the number of H_2O ligands available for replacement as n increases. Conversely, the increase in the number of bound L groups increases the importance of the reverse of these reactions as n increases. Therefore, so long as the reaction enthalpy is largely unaffected, the equilibrium constants lie progressively in favor of the reactants as n increases. That such a simple explanation in terms of entropy is more or less correct is illustrated by data for the successive complexes in the series from $[Ni(OH_2)_6]^{2+}$ to $[Ni(NH_3)_6]^{2+}$ (Table 6.7). The enthalpy changes for the six successive steps vary by less than 2 kJ mol^{-1} from 16.7 to 18.0 kJ mol^{-1}.

A reversal of the relation $K_n < K_{n+1}$ is usually an indication of a major change in the structure and bonding at the metal center as more ligands are added. An example is that the tris(bipyridyl) complex of Fe(II) is strikingly stable compared with the bis complex. This obser-

Table 6.7 Formation constants of Ni(II) ammines

n	pK_f	K_n/K_n-1	
		Experimental	Statistical*
1	− 2.72		
2	− 2.17	0.28	0.42
3	− 1.66	0.31	0.53
4	− 1.12	0.29	0.56
5	− 0.67	0.35	0.53
6	− 0.03	0.2	0.42

*Based on ratios of numbers of ligands available for replacement, with the reaction enthalpy assumed constant.

vation can be correlated with the change from a weak field $t_{2g}^4 e_g^2$ to a strong field t_{2g}^6 configuration. A contrasting example is the anomalously low value of K_3/K_2 (approximately $\frac{1}{7}$) for the halogeno complexes of Hg(II). This decrease is too large to be explained statistically and suggests the onset of four-coordination:

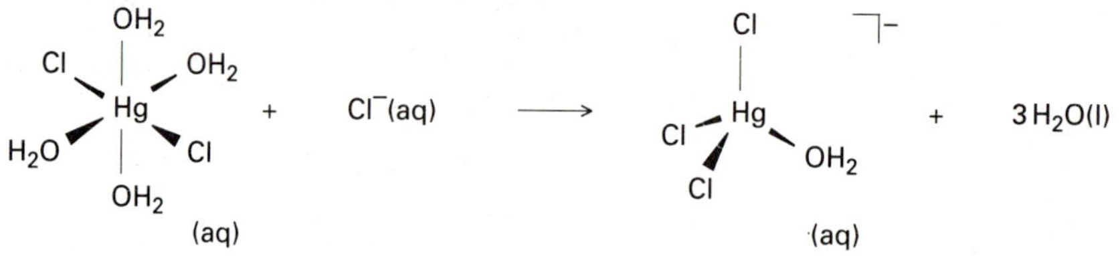

Example 6.7: *Interpreting irregular successive formation constants*
The formation of cadmium complexes with Br^- exhibits the successive equilibrium constants $K_1 = 1.56$, $K_2 = 0.54$, $K_3 = 0.06$, $K_4 = 0.37$. Suggest an explanation of why K_4 is larger than K_3.

Answer. The anomaly suggests a structural change. Aqua complexes are usually six coordinate whereas halo complexes are commonly tetrahedral. The reaction of the complex with three Br^- groups to add the fourth is

$$[CdBr_3(OH_2)_3](aq) + Br^-(aq) \rightarrow [CdBr_4]^{2-}(aq) + 3H_2O(l)$$

This step is entropically favored because of the release of three molecules of water from the relatively restricted coordination sphere environment. The result is an increase in K.

Exercise E6.7. A square-planar four-coordinate Fe(II) porphyrin complex may add two further ligands axially. The maximally coordinated complex $[FePL_2]$ is low spin whereas $[FePL]$ is high spin. Account for an increase of the second formation constant with respect to the first.

The chelate effect

When K_1 for a bidentate chelate ligand, such as ethylenediamine, is compared with the value of β_2 for the corresponding bisligand complex (a bisammine), it is found that the former is generally larger:

$$[Cu(OH_2)_6]^{2+} + en \rightleftharpoons [Cu(OH_2)_4(en)]^{2+} + 2H_2O$$

$$\log K_1 = 10.6 \quad \Delta H^\ominus = -54 \text{ kJ mol}^{-1} \quad \Delta S^\ominus = +23 \text{ J K}^{-1} \text{ mol}^{-1}$$

$$[Cu(OH_2)_6]^{2+} + 2NH_3 \rightleftharpoons [Cu(OH_2)_4(NH_3)_2]^{2+} + 2H_2O$$

$$\log \beta_2 = 7.7 \quad \Delta H^\ominus = -46 \text{ kJ mol}^{-1} \quad \Delta S^\ominus = -8.4 \text{ J K}^{-1} \text{ mol}^{-1}$$

Two similar Cu—N bonds are formed in each case, yet the formation of the chelate is distinctly more favorable. The **chelate effect** is this greater stability of chelated complexes compared with their non-chelated analogs.

We can trace the chelate effect primarily to differences in reaction entropies between chelated and nonchelated complexes in dilute solutions. The chelation reaction results in an increase in the number of independent molecules in solution, but the nonchelating reaction produces no net change (compare the two chemical equations above). The former therefore has the more positive entropy change and hence is the more favorable process. The entropy changes measured in dilute solution support this interpretation.

The chelate effect is of great practical importance. The majority of reagents used in complexometric titrations in analytical chemistry are multidentate chelates like EDTA. Most biochemical metal binding sites are chelating ligands. When a formation constant is measured as 10^{12} to 10^{25} it is generally a sign that the chelate effect is in operation.

Steric effects

Steric effects also have an important influence on formation constants. They are particularly important in chelate formation since ring completion may be difficult geometrically. Chelate rings with five members are generally the most stable, as we explained in Section 6.2. Six-membered rings are reasonably stable and may be favored if electron delocalization can occur.

An interesting trend is observed when the basic site on the ligand is crowded but the cavity remains large enough to accept a proton. The formation constants for the metal complexes correlate well with the Brønsted basicity if steric factors are considered. Figure 6.23, for instance, shows correlations for substituted pyridine bases. All lines have the same slope, but there is a reduction of stability for the metal complexes for each blocking R next to the donor N. The extreme example of such steric effects is a ligand called engagingly a 'Proton sponge' (**49**). This compound can strongly bind a proton but is sterically blocked from binding any other cations.

Diimine ligands (**50**), such as bipyridine (**51**) and phenanthroline (**52**), are constrained to form five-membered rings with the metal. The great stability of their complexes is probably a result of their ability to act as π acceptors as well as σ donors, and to form π bonds with full metal d orbitals and their vacant ring π^* orbitals. This bond formation is favored by electron population in the metal t_{2g} orbitals, which allows the metal to act as a π donor and donate to the ligand

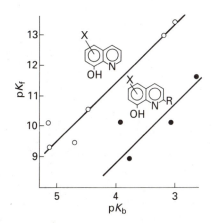

6.23 The correlation of the formation constant pK_f (for complexation with Cu^{2+}) and Brønsted basicity pK_b for two series of substituted 8-hydroxyquinoline ligands. The filled circles are for the more sterically hindered ligands.

49 **50** **51** bipy **52** phen

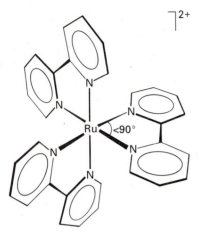

53 [Ru(bipy)$_3$]$^{2+}$

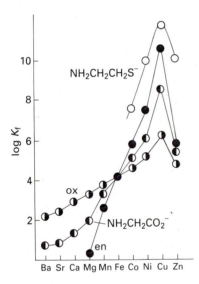

6.24 The variation of formation constants for the M^{2+} ions of the Irving–Williams series.

rings. An example is the complex [Ru(bipy)$_3$]$^{2+}$ (**53**). The small bite distance of these ligands distorts the complex from octahedral symmetry.

The Irving–Williams series

Figure 6.24 is obtained when log K_f is plotted along the first d series of divalent metal ions (M^{2+}). This variation is summarized by the **Irving–Williams series** for the order of formation constants. For divalent cations:

$$Ba^{2+} < Sr^{2+} < Ca^{2+} < Mg^{2+} < Mn^{2+}$$
$$< Fe^{2+} < Co^{2+} < Ni^{2+} < Cu^{2+} > Zn^{2+}$$

The order is relatively insensitive to the choice of ligands.

In the main, the Irving–Williams series reflects electrostatic effects. However, beyond Mn^{2+} there is a sharp increase in the value of K_f for Fe(II), d^6; Co(II), d^7; Ni(II), d^8; and Cu(II), d^9. These ions have additional stabilizations that are proportional to the ligand field stabilization energies shown for d^6 to d^9 in Table 6.3. However, there is one important exception: the stability of Cu(II) complexes is greater than that of Ni(II) despite the fact that Cu(II) has an additional antibonding e_g electron. This anomaly is a consequence of the stabilizing influence of the Jahn–Teller distortion, which enhances the value of K_f. In the tetragonally distorted complex, there is strong binding of the four ligands in the plane. Note that the remaining distorted axial positions are more weakly bound.

6.8 Rates and mechanisms of ligand substitution

Rates of reaction are as important as equilibria in coordination chemistry. The numerous isomers of the ammines of Co(III) and Pt(II), which were so important to the development of the subject, could not have been isolated if ligand substitutions and interconversion of the isomers had been fast.

Lability and inertness

Complexes that are thermodynamically unstable but survive for long periods (at least a minute) are called **inert**. Complexes that undergo more rapid equilibration are called **labile**. Octahedral complexes of the first d series show an interesting correlation with electron configuration: and strong-field d^3 and d^6 complexes (such as Cr(III) and Co(III) complexes respectively) are generally inert; all others are generally labile.

Figure 6.25 shows the characteristic lifetimes of octahedral complexes of the important aqua metal ions. We see a range of lifetimes starting at about 1 ns, which is approximately the time it takes for a molecule to diffuse one molecular diameter in solution. At the other end of the scale are lifetimes in years. Even so, the illustration does

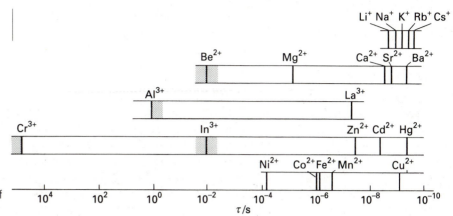

6.25 Characteristic lifetimes for exchange of water molecules in hexaaqua complexes.

not show the longest times that could be considered, which are comparable to geological eras.

There are a number of generalizations that help us to anticipate the lability of the complexes we are likely to meet. Thus:

> All complexes of *s*-block ions except the smallest (Be^{2+} and Mg^{2+}) are very labile. Across the first *d* series, complexes of *d*-block M(II) ions are moderately labile, with distorted Cu(II) complexes among the most labile. Complexes of lower oxidation number d^{10} ions (Zn^{2+}, Cd^{2+}, and Hg^{2+}) are also extremely labile. Complexes of M(III) ions are distinctly less labile. The M(III) ions of the *f* block are all extremely labile. In the first *d* series, the least labile M(II) and M(III) ions are those with greatest LFSE. Among the second and third *d* series complexes, inertness is quite common, which reflects the high LFSE and strength of the metal–ligand bonding.

Nucleophilicity

Because the formation of any complex in solution is a substitution reaction, we need a preliminary idea of the mechanisms of these reactions in order to understand lability and inertness.

The activation barrier in a substitution reaction is determined by two main factors. One is the energy required to break the bond between the metal ion and the leaving group. The other is the energy released as the metal ion forms a bond to the entering group. The latter assistance is the kinetic equivalent of basicity and is called **nucleophilicity** (an affinity for positive centers). Just as relative basicity is measured by comparing *equilibrium* constants for acid–base reactions, nucleophilicity is measured by comparing *rate* constants for substitution reactions, and the faster the reaction with an entering group, the greater its nucleophilicity.

The kinetic analog of acidity is **electrophilicity** (an affinity for negative centers): the faster the reaction of a Lewis acid with an entering group, the greater the electrophilicity of the acid. It turns out that the most telling information about mechanisms of ligand substitution comes from studies of the variation of the ligands and observations of their effect on the rate of substitution reactions, and we concentrate on this aspect.

Associative reactions

Provided the reaction does not proceed by slow solvent substitution, reactions of square-planar complexes of Pt(II) are first-order in the complex, first-order in the entering group, and second-order overall. The rate constants vary widely with the choice of the nucleophile. For example, *trans*-[PtCl$_2$(py)$_2$] reacts with various entering groups (Y) to give *trans*-[PtClY(py)$_2$] and the rate constant (in units of L mol^{-1} s^{-1}) varies from 4.7×10^{-4} for Y $=$ NH$_3$, through 3.7×10^{-3} for Br$^-$ and 0.107 for I$^-$, to 6.00 for thiourea, a factor of 10^4. The corresponding change in the rate constants of reactions in which the leaving group is changed from Cl$^-$ to I$^-$ is a factor of only 3.5. In general, entering group effects greatly exceed those of the leaving groups. The sensitivity of the reaction to the entering nucleophile in contrast to its insensitivity to the leaving group indicates that the nucleophile is the major factor determining the overall activation energy. This behavior is expected in an **associative substitution reaction**, a reaction in which the reactant passes through an activated complex with an increased coordination number. The observation that *cis* or *trans* isomers substitute with retention of the original stereochemistry suggests that the activated complex of square planar complexes is approximately a trigonal bipyramid:

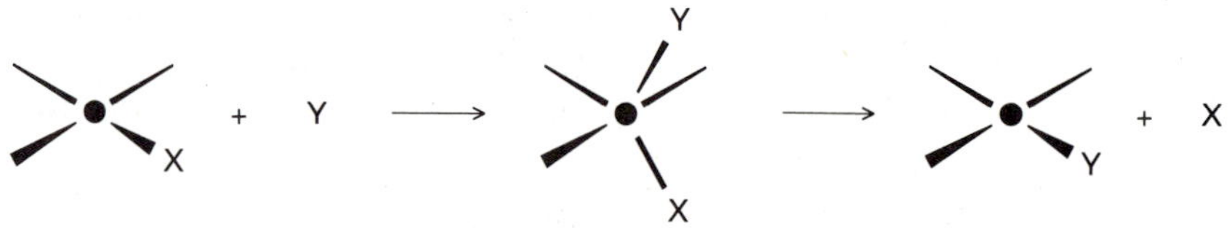

It is difficult to make generalizations about rates of associative substitutions because both the initial complex and the nucleophile must be specified.[14] However, it can be said that among the d^8 systems with similar entering and leaving groups, most are comparatively inert. Typically, lability decreases in the series

$$\text{Ni(II)} > \text{Pd(II)} \gg \text{Pt(II)} \approx \text{Rh(I)}$$

Dissociative substitutions

Octahedral Ni(II) complex formation reactions are representative of reactions of octahedral complexes. They are first order in Ni^{2+}(aq), first order in the entering group, and second order overall. However, in contrast to the square-planar case, there is only a comparatively small change in the rate constant as the entering group is varied: for

---

[14] One of the most interesting generalizations that can be made is that the ligand *trans* to the leaving group plays an important role in determining reactivity, a role comparable to the entering nucleophile. This '*trans* effect' is discussed in Section 15.2.

example, there is only a factor of 10 difference between the rate constants for the formation of $[Ni(SO_4)(OH_2)_4]$ and $[Ni(OH_2)_4(phen)]^{2+}$ (phen is 1,10-phenanthroline). The higher rate constant of SO_4^{2-} substitution arises from the electrostatic attraction between Ni^{2+} and SO_4^{2-} ions, which increases the probability of their encountering each other in solution (this topic is discussed in detail in Chapter 15).

The principal factor determining the activation energy for the replacement of a ligand in such an octahedral complex is the breaking of the bond between the metal and the leaving group. The entering group plays only a minor role in the substitution reaction. Reactions of this kind are called **dissociative substitution reactions;** they are reactions in which the activated complex has a lower coordination number than the reactant. The pathway can be written as follows:

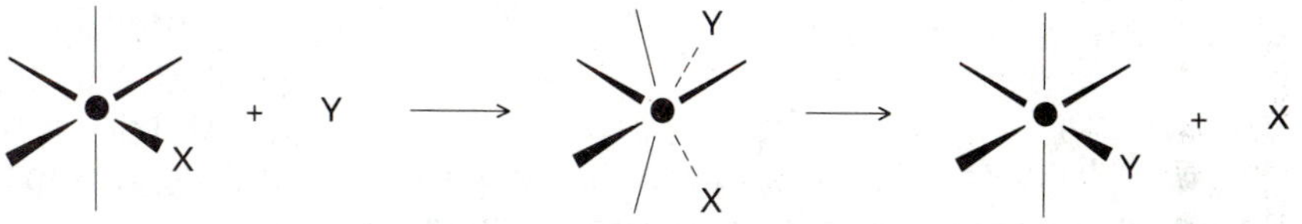

In this example, the dotted lines represent weak bonds. In practice, the dissociative character is identified from the insensitivity of the rate to variation of the entering group. Judged in this way, most octahedral substitutions seem more dissociative than associative. This is helpful because the lability of an octahedral coordination compound can be described without specifying the entering group: lability is a property of the complex itself. Among the aqua metal ions shown in Fig. 6.25, the ones that form 'weak' bonds on account of their relatively low charge and large ionic radius are more labile than those with higher charge and smaller radius. This is consistent with a dissociative mechanism of substitution, which involves the breaking of a bond.

That there is a correlation of inertness with LFSE has a complicated explanation, but the fact itself is worth remembering. The strong-field d^6 Co(III) and d^3 Cr(III) complexes are inert. The d^2 and d^8 V(II) and Ni(II) complexes are distinctly less labile than other high spin complexes. All four have large LFSE. We explore this feature further in Chapter 15.

FURTHER READING

Three introductory texts that cover much of the material in this chapter are:

R. DeKock and H. B. Gray, *Chemical structure and bonding.* Benjamin/Cummings, Menlo Park (1980).

F. Basolo and R. Johnson, *Coordination chemistry.* Science Reviews, Northwood (1987).

S. F. A. Kettle. *Coordination compounds.* Nelson, London (1987).

A useful general reference is:

G. Wilkinson, R. D. Gillard, and J. A. McCleverty (eds), *Comprehensive coordination chemistry.* Pergamon Press, Oxford (1987). Volume 1 of this seven-volume set provides chapters on history, coordination numbers and geometry, ligand field theory, and reactions.

KEY POINTS

1. Complexes

Assemblies of units (ligands) about a central atom are called complexes; they are characterized by their coordination numbers and characteristic geometries. The ligands act as Lewis bases and the central metal atom acts as a Lewis acid.

2. Coordination geometries

Important coordination numbers range from 2 to 12 with 4, 5, and 6 being most common for *d* block ions. The most common geometries include tetrahedral, square planar, trigonal bipyramidal, square pyramidal, and octahedral.

3. Isomerism

The possibility of different spatial arrangements of ligands about a metal atom gives rise to geometrical and optical isomerism.

4. Ligand field splitting

The bonding in complexes can be modeled by crystal field theory in which the ligands are represented by partial negative charges. The theory leads to the ligand field splitting parameter (Δ_O for octahedral complexes and Δ_T for tetrahedral complexes).

5. The spectrochemical series

The size of the ligand field splitting parameter leads to the ordering of ligands in the spectrochemical series.

6. Electron configurations of octahedral complexes

The electron configuration of a complex is predicted by applying the building-up principle to the *d* orbitals. If the ligand field splitting is large, then the lowest energy is obtained by filling the lower set of *d* orbitals (t_{2g} in octahedral complexes). If the ligand field splitting is small, then the lowest energy is obtained by occupying the upper set (e_g in octahedral symmetry) before pairing electrons in the lower set.

7. High- and low-spin complexes

When the ligand field splitting parameter is larger than the electron pairing energy, a low-spin complex results; when the opposite is true, a high-spin complex results. The number of unpaired spins can sometimes be determined by measuring the magnetic susceptibility of the complex.

8. Ligand-field splitting parameter

The energy of an electron configuration relative to the average energy of the *d* electrons in a complex is called the ligand field stabilization energy. It accounts for trends in the observed enthalpies of hydration of complexes and provides a means of correlating thermodynamic and spectroscopic data.

9. Electron configurations of tetrahedral complexes

The electron configurations of tetrahedral complexes can be explained in terms of a smaller ligand field splitting parameter and a splitting of the *d* orbitals into two sets with the *e* orbitals below the t_2 orbitals. Almost all tetrahedral complexes are high-spin complexes.

10. Ligand field theory

In ligand field theory, the bonding is modeled in terms of molecular orbitals that are formed by overlap between the *d* orbitals of the central metal ion and SALCs of the ligand orbitals. The ligand field splitting parameter is identified with the energy separation of the frontier orbitals of the complex.

11. π Bonding

Ligand field theory accommodates the effects of π bonding between the ligand and the metal ion. When the ligand acts as a π donor, the ligand field splitting parameter is decreased, and when the ligand acts as a π acceptor, the parameter is increased.

12. The spectrochemical series

The effects of π bonding correlate with the position of a ligand in the spectrochemical series, with π-donor ligands lying low in the series and π-acceptor ligands lying high in the series. Very strong σ donors, such as CH_3^- and H^-, are also high in the series.

13. Spectroscopic, magnetic, and thermochemical correlations

The ligand field splitting parameter can be used to correlate the trends in electronic spectra, magnetic properties, and some thermochemical properties of complexes of *d*-block metals.

14. Formation constants

The thermodynamic stabilities of complexes are expressed in terms of stepwise and overall formation constants. Stepwise formation constants K_n commonly decrease as n increases.

15. Chelate effects

Chelate complexes are generally more stable than their nonchelated analogs.

16. Irving–Williams series

The Irving–Williams series summarizes the variation in thermodynamic stability of complexes with change in the central metal ion.

17. Electrophiles and nucleophiles

Incoming groups are classified as electrophiles (electron seeking) or nucleophiles (positive charge seeking); their relative strengths are assessed by comparing rate constants for substitution reactions.

18. Associative substitution

In an associative substitution reaction the initial complex passes through an activated complex with an increased coordination number.

19. Dissociative substitution

In a dissociative substitution reaction the activated complex has a lower coordination number than the initial complex. Most octahedral substitutions are dissociative rather than associative.

20. Relative labilities

Among aqua metal ions, the ones that form relatively weak bonds (on account of their low charge and large ionic radius) are more labile than those with high charge and small radius.

EXERCISES

6.1 Preferably without using reference material, write out the $3d$ elements in their arrangement in the periodic table. Indicate the metal ions that commonly form tetrahedral complexes of formula $[MX_4]^{2-}$ where X^- is a halide ion.

6.2 (a) On a chart of the d-block elements in their periodic table arrangement, identify the elements and associated oxidation numbers that form square-planar complexes. (b) Give formulas for three examples of square-planar complexes.

6.3 (a) Sketch the two structures that describe most six-coordinate complexes. (b) Which one of these is rare? (c) Give formulas for three different d-metal complexes that have the more common six-coordinate structure.

6.4 Name and draw structures of the following complexes: (a) $[Ni(CO)_4]$; (b) $[Ni(CN)_4]^{2-}$; (c) $[CoCl_4]^{2-}$; (d) $[Ni(NH_3)_6]^{2+}$.

6.5 Draw the structures of representative complexes that contain the ligands (a) en, (b) ox^{2-}, (c) phen, and (d) $EDTA^{4-}$.

6.6 Draw the structure of (a) a typical square-planar four-coordinate complex; (b) a typical trigonal prismatic six-coordinate complex; (c) a typical complex of coordination number 2. Name each complex.

6.7 Give formulas for (a) pentaamminechlorocobalt(III) chloride; (b) hexaaquairon(3+) nitrate; (c) *cis*-dichlorobis-(ethylenediamine)ruthenium(II); (d) μ-hydroxobis-[pentaamminechromium(III)] chloride.

6.8 Name the octahedral complex ions (a) *cis*-$[CrCl_2(NH_3)_4]^+$, (b) *trans*-$[Cr(NCS)_4(NH_3)_2]^-$, and (c) $[Co(C_2O_4)(en)_2]^+$. Is the oxalato complex *cis* or *trans*?

6.9 Draw all possible isomers of (a) octahedral $[RuCl_2(NH_3)_4]$, (b) square-planar $[IrH(CO)(PR_3)_2]$, (c) tetrahedral $[CoCl_3(OH_2)]^-$, (d) octahedral $[IrCl_3(PEt_3)_3]$, and (e) octahedral $[CoCl_2(en)(NH_3)_2]^+$.

6.10 The compound Na_2IrCl_6 reacts with triphenylphosphane in diethylene glycol under an atmosphere of CO to give *trans*-$[IrCl(CO)(PPh_3)_2]$, known as Vaska's compound. Excess CO produces a five-coordinate species and treatment with $NaBH_4$ in ethanol gives $[IrH(CO)_2(PPh_3)_2]$. Draw and name the three complexes.

6.11 Which of the following complexes are chiral? (a) $[Cr(ox)_3]^{3-}$, (b) *cis*-$[PtCl_2(en)]$, (c) *cis*-$[RhCl_2(NH_3)_4]^+$, (d) $[Ru(bipy)_3]^{2+}$, (e) $[Co(edta)]^-$, (f) *fac*-$[Co(NO_2)_3(trien)]$, (g) *mer*-$[Co(NO_2)_3(trien)]$. Draw the enantiomers of the complexes identified as chiral and identify the plane of symmetry in the structures of the achiral complexes.

6.12 One pink solid has the formula $CoCl_3 \cdot 5NH_3 \cdot H_2O$. A solution of this salt is also pink and rapidly gives 3 mol AgCl on titration with silver nitrate solution. When the pink solid is heated, it loses 1 mol H_2O to give a purple solid with the same ratio of $NH_3 : Cl : Co$. The purple solid releases two of its chlorides rapidly; then, on dissolution and after titration with $AgNO_3$, releases one of its chlorides slowly. Deduce the structures of the two octahedral complexes and draw and name them.

6.13 The hydrated chromium chloride that is available commercially has the overall composition $CrCl_3 \cdot 6H_2O$. On boiling a solution, it becomes violet and has a molar electrical conductivity similar to that of $[Co(NH_3)_6]Cl_3$. In contrast, $CrCl_3 \cdot 5H_2O$ is green and has a lower molar conductivity in solution. If a dilute acidified solution of the green complex is allowed to stand for several hours, it turns violet. Interpret these observations with structural diagrams.

6.14 The complex first denoted β-[$PtCl_2(NH_3)_2$] was identified as the *trans* isomer. (The *cis* isomer was denoted α.) It reacts slowly with solid Ag_2O to produce $[Pt(NH_3)_2(OH_2)_2]^{2+}$. This complex does not react with ethylenediamine to give a chelated complex. Name and draw the structure of the diaqua complex.

6.15 The 'third isomer' (neither α nor β, see Exercise 6.14) of composition $PtCl_2 \cdot 2NH_3$ is an insoluble solid which, when ground with $AgNO_3$, gives a solution containing $[Pt(NH_3)_4](NO_3)_2$ and a new solid phase of composition $Ag_2[PtCl_4]$. Give the structures and names of each of the three Pt(II) compounds.

6.16 Phosphane and arsane analogs of $[PtCl_2(NH_3)_2]$ were prepared in 1934 by Jensen. He reported zero dipole moments for the β isomers, where the β designation represents the product of a synthetic route analogous to that of the ammines. Give the structures of the complexes.

6.17 Describe the differences in the ^{31}P-NMR spectra of the *cis* and *trans* isomers referred to in Exercise 6.16.

6.18 (a) Indicate what parameter of NMR spectra might be used to distinguish the *cis* and *trans* isomers of $[W(CO)_4(P(CH_3)_3)_2]$. (b) What features of NMR spectra would distinguish a complex $[M(CO)_3(PR_3)_2]$ of trigonal bipyramid geometry with the phosphane ligands in the axial positions from one with the phosphane ligands in the trigonal plane?

6.19 In *trans*-[$W(CO)_4(PR_3)_2$], the alkylphosphane ligands define the z-axis of the coordinate system. Sketch the symmetry-adapted linear combination of σ orbitals from the two P atoms that can combine with the metal d_{z^2} orbital. Show the bonding and antibonding orbitals that may be formed.

6.20 Determine the configuration (in the form $t_{2g}^m e_g^n$ or $e^m t_2^n$, as appropriate), the number of unpaired electrons, and the ligand field stabilization energy as a multiple of Δ_O

or Δ_T for each of the following complexes using the spectrochemical series to decide, where relevant, which are likely to be strong-field and which weak-field. (a) $[Co(NH_3)_6]^{3+}$, (b) $[Fe(OH_2)_6]^{2+}$, (c) $[Fe(CN)_6]^{3-}$, (d) $[Cr(NH_3)_6]^{3+}$, (e) $[W(CO)_6]$, (f) tetrahedral $[FeCl_4]^{2-}$, and (g) tetrahedral $[Ni(CO)_4]$.

6.21 Both H^- and $P(C_6H_5)_3$ are ligands of similar field strength high in the spectrochemical series. Recalling that phosphanes act as π acceptors, is π-acceptor character required for strong-field behavior? What orbital factors account for the strength of each ligand?

6.22 Estimate the spin-only contribution to the magnetic moment for each complex in Exercise 6.20.

6.23 Solutions of the complexes $[Co(NH_3)_6]^{2+}$, $[Co(H_2O)_6]^{2+}$ (both O_h), and $[CoCl_4]^{2-}$ are colored. One is pink, another is yellow, and the third is blue. Considering the spectrochemical series and the relative magnitudes Δ_T and Δ_O try to assign each color to one of the complexes.

6.24 Interpret the variation, including the overall trend across the $3d$ series, of the following values of oxide lattice enthalpies (in $kJ\,mol^{-1}$), All the compounds have the rock-salt structure. CaO (3460); TiO (3878); VO (3913); MnO (3810); FeO (3921); CoO (3988); NiO (4071).

6.25 A neutral macrocyclic ligand with four donor atoms produces a red diamagnetic low-spin d^8 complex of Ni(II) if the anion is the weakly coordinating perchlorate ion. When perchlorate is replaced by two thiocyanate ions, SCN^-, the complex turns violet and is high-spin with two unpaired electrons. Interpret the change in terms of structure.

6.26 Bearing in mind the Jahn–Teller theorem, predict the structure of $[Cr(OH_2)_6]^{2+}$.

6.27 The spectrum of d^1 Ti^{3+}(aq) is attributed to a single electronic transition $e_g \leftarrow t_{2g}$. The band shown in Fig. 6.10 is not symmetrical and suggests that more than one state is involved. Suggest how to explain this observation using the Jahn–Teller theorem.

6.28 The rate constants for the formation of $[CoX(NH_3)_5]^{2+}$ from $[Co(NH_3)_5OH_2]^{3+}$ for $X = Cl^-$, Br^-, N_3^-, and SCN^- differ by no more than a factor of two. What is the mechanism of the substitution?

6.29 If a substitution process is associative, why may it be difficult to characterize an aqua ion as labile or inert?

PROBLEMS

6.1 Air oxidation of Co(II) carbonate and aqueous ammonium chloride gives a pink chloride salt with a ratio of $4NH_3$:Co. On addition of HCl to a solution of this salt, a gas is rapidly evolved and the solution slowly turns violet

on heating. Complete evaporation of the violet solution yields $CoCl_3 \cdot 4NH_3$. When this is heated in concentrated HCl, a green salt can be isolated which analyzes as $CoCl_3 \cdot 4NH_3 \cdot HCl$. Write balanced equations for all the transformations occuring after the air oxidation. Give as much information as possible concerning the isomerism occuring and give the basis of your reasoning. If you know that the form of $[Co(en)_2Cl_2]^+$ that is resolvable into enantiomers is violet, is that helpful?

6.2 By considering the splitting of the octahedral orbitals as the symmetry is lowered, draw the symmetry-adapted linear combinations and the molecular orbital energy level diagram for σ bonding in a *trans*-$[ML_4X_2]$ complex. Assume that the ligand X is lower in the spectrochemical series than L.

6.3 By referring to Appendix 4, draw the appropriate symmetry adapted linear combinations and the molecular orbital diagram for σ bonding in a square-planar complex. The group is D_{4h}. Take note of the small overlap of the ligand with the d_{z^2} orbital. What is the effect of π bonding?

6.4 When cobalt(II) salts are oxidized by air in a solution containing ammonia and sodium nitrite, a yellow solid, $[Co(NO_2)_3(NH_3)_3]$, can be isolated. In solution it displays a conductivity small enough to be attributed to impurities only. Werner treated it with HCl to give a complex which, after a series of further reactions, he identified as *trans*-$[CoCl_2(NH_3)_3(OH_2)]^+$. It required an entirely different route to prepare *cis*-$[CoCl_2(NH_3)_3(OH_2)]^+$. Is the yellow substance *fac* or *mer*? What assumption must you make in order to arrive at a conclusion?

6.5 The reaction of $[ZrCl_4(dppe)]$ (dppe is a bidentate phosphane ligand) with $Mg(CH_3)_2$ gives $[Zr(CH_3)_4(dppe)]$. NMR spectra indicate that all methyl groups are equivalent. Draw octahedral and trigonal prism structures for the complex and show how the conclusion from NMR supports the trigonal prism assignment. (P. M. Morse and G. S. Girolami, *J. Am. Chem. Soc.*, **111**, 4114 (1989).)

6.6 Write balanced equations for the formation of the triaza (three N atoms) and tetraaza (four N atoms) macrocycles that can be formed by a template synthesis in which *o*-aminobenzaldehyde self-condenses in the presence of

Ni(II). (See: G. A. Melson and D. H. Busch, *J. Am. Chem. Soc.*, **86**, 4830 (1964).)

6.7 The resolving agent *d-cis*-$[Co(NO_2)_2(en)_2]Br$ can be converted to the soluble nitrate by grinding in water with $AgNO_3$. Outline the use of this species for resolving a racemic mixture of the *d* and *l* enantiomers of $K[Co(EDTA)]$. (The *l*-$[Co(EDTA)]^-$ enantiomer forms the less soluble diastereomer. See F. P. Dwyer and F. L. Garvan, *Inorg. Synth.*, **6**, 192 (1965).)

6.8 Figures 6.16 and 6.17 show the relation between the frontier orbitals of octahedral and square planar complexes. Construct similar diagrams for two-coordinate linear complexes.

6.9 In a fused magma liquid from which silicate minerals crystallize, the metal ions can be four coordinate. In olivine crystals, the M(II) coordination sites are octahedral. Partition coefficients, which are defined as $K_p = [M(II)]_{olivine}/[M(II)]_{melt}$, follow the order Ni(II) > Co(II) > Fe(II) > Mn(II). Account for this in terms of ligand field theory. (See I. M. Dale and P. Henderson, *24th Int. Geol. Congress, Sect.*, **10**, 105 (1972).)

6.10 The equilibrium constants for the successive reactions of ethylenediamine with Co^{2+}, Ni^{2+}, and Cu^{2+} are as follows.

$$[M(OH_2)_6]^{2+} + en \rightleftharpoons [M(en)(OH_2)_4]^{2+} + 2H_2O \qquad K_1$$

$$[M(en)(OH_2)_4]^{2+} + en \rightleftharpoons [M(en)_2(OH_2)_2]^{2+} + H_2O \qquad K_2$$

$$[M(en)_2(OH_2)_2]^{2+} + en \rightleftharpoons [M(en)_3]^{2+} + 2H_2O \qquad K_3$$

Ion	$\log K_1$	$\log K_2$	$\log K_3$
Co^{2+}	5.89	4.83	3.10
Ni^{2+}	7.52	6.28	4.26
Cu^{2+}	10.55	9.05	−1.0

Discuss whether these data support the generalizations in the text about successsive formation constants and the Irving–Williams series. How do you account for the very low value of K_3 for Cu^{2+}?

7

Oxidation and reduction

We close Part 1 with a discussion of the third major class of chemical reactions, one in which electrons are transferred from one species to another and oxidation and reduction take place. An understanding of these two processes, which are jointly called redox reactions, grew out of the early work on the extraction of elements from their natural sources, and this application continues to be of considerable technological importance. The thermodynamic analysis of industrially important redox reactions is well established, and we start with that. We shall see that to analyze redox reactions in solution we need to consider both kinetic and thermodynamic effects. In these cases we must decide first whether the reaction is thermo-dynamically feasible, and then whether it is kinetically practicable.

Redox reactions in solution can be studied by using electrochemical techniques. As a result, much of the data relating to these reactions is available in the form of electrode potentials, and we shall describe how such information is used. We shall introduce diagrammatic techniques that summarize thermodynamic relationships in a compact form. One applica-tion that we illustrate is the use of these diagrams to treat equilibria in natural water systems.

A very large class of reactions can be regarded as occurring by the formal loss of electrons associated with one atom and their gain by another. Since electron gain is called **reduction** and electron loss is **oxidation**, the joint process is called a **redox reaction**. The species that supplies electrons is the **reducing agent** (or reductant) and the species that removes electrons is the **oxidizing agent** (or oxidant). These remarks show that redox reactions are similar to proton transfer reactions, but instead of a proton being transferred from the Brønsted acid to the Brønsted base, one or more electrons are transferred (per-haps with accompanying atoms) from the reducing agent to the oxidizing agent. However, the electron migrations that accompany many redox reactions of covalent species are very small and quite unlike the actual, complete proton transfer that characterizes acid–base reactions. Hence it is safest to analyze redox reactions according to a set of *formal* rules expressed in terms of oxidation numbers (Section 2.1) and not to think in terms of actual electron transfers.

EXTRACTION OF THE ELEMENTS

The original definition of oxidation was a reaction in which an element reacts with oxygen and is converted to an oxide. Similarly, reduction originally meant the reverse reaction, in which an oxide of a metal was converted to the metal. Both terms have been generalized and expressed in terms of electron transfer, but these special cases are still the basis of a major part of chemical industry and laboratory chemistry.

7.1 Elements extracted by reduction

Oxygen has been a component of the atmosphere since photosynthesis became a dominant process over 10^9 years ago, and many metals are found as their oxides. After about 4000 BC, copper could be extracted from its ores at temperatures attainable in primitive hearths, and the process of **smelting** was discovered, in which ores are reduced by heating with a reducing agent such as carbon. Because many important ores of easily reduced metals are sulfides, smelting is often preceded by conversion of some of the sulfide to an oxide by 'roasting' in air, as in the reaction

$$2Cu_2S(s) + 3O_2(g) \rightarrow 2Cu_2O(s) + 2SO_2(g)$$

It was not until nearly 1000 BC, at the beginning of the Iron Age, that the temperatures needed for the reduction of less readily reduced elements, such as iron, could be attained. Carbon remained the dominant reducing agent until the end of the nineteenth century, and metals that need higher temperatures for their production could not be produced, even though their ores were reasonably abundant.

The technological breakthrough that resulted in the conversion of aluminum from a rarity to a major construction metal was the introduction of electrolysis. The availability of electric power also expanded the scope of carbon reduction, for electric furnaces can reach much higher temperatures than carbon-combustion furnaces such as the blast furnace. Thus, magnesium is a twentieth-century metal even though one of its modes of recovery, the **Pidgeon process**, the electrothermal reduction of the oxide, uses carbon as the reducing agent:

$$MgO(s) + C(s) \xrightarrow{\Delta} Mg(l) + CO(g)$$

The temperature dependence of carbon reductions can be rationalized by some simple thermodynamic considerations, which we discuss next.

Thermodynamics

Thermodynamic arguments can be used to identify which reactions are spontaneous (that is, have a natural tendency to occur) under the prevailing conditions and to choose the most economical reducing agents and reaction conditions. The criterion of spontaneity is that, at constant temperature and pressure, the Gibbs free energy of the reaction, ΔG, is negative. It is usually sufficient to consider the *standard* Gibbs free energy, $\Delta G^{\ominus}$, which is related to the equilibrium constant through

$$\Delta G^{\ominus} = -RT \ln K$$

Hence, a negative value of $\Delta G^{\ominus}$ corresponds to $K > 1$ and therefore to a 'favorable' reaction. Reaction rates are also relevant, but, at high temperature, reactions are often fast, and thermodynamically favorable reactions are likely to occur. A fluid phase (typically a gas) is usually required to facilitate what would be a sluggish reaction between coarse particles. It should be noted that equilibrium is rarely attained in commercial processes; even a process for which $K < 1$ can be viable if the products are swept out of the reaction chamber.

· To achieve a negative $\Delta G^{\ominus}$ for the reduction of a metal oxide with carbon, one of the reactions

(a) $C(s) + \frac{1}{2}O_2(g) \rightarrow CO(g)$ $\Delta G^{\ominus}(C,CO)$

(b) $\frac{1}{2}C(s) + \frac{1}{2}O_2(g) \rightarrow \frac{1}{2}CO_2(g)$ $\Delta G^{\ominus}(C,CO_2)$

(c) $CO(g) + \frac{1}{2}O_2(g) \rightarrow CO_2(g)$ $\Delta G^{\ominus}(CO,CO_2)$

must have a more negative standard free energy than does a reaction of the form

(d) $xM(s \text{ or } l) + \frac{1}{2}O_2(g) \rightarrow M_xO(s)$ $\Delta G^{\ominus}(M,M_xO)$

under the same reaction conditions. If that is so, then one of the reactions

(a–d) $M_xO(s) + C(s) \rightarrow xM(s \text{ or } l) + CO(g)$

$$\Delta G^{\ominus} = \Delta G^{\ominus}(C,CO) - \Delta G^{\ominus}(M,M_xO)$$

(b–d) $M_xO(s) + \frac{1}{2}C(s) \rightarrow xM(s \text{ or } l) + \frac{1}{2}CO_2(g)$

$$\Delta G^{\ominus} = \Delta G^{\ominus}(C,CO_2) - \Delta G^{\ominus}(M,M_xO)$$

(c–d) $M_xO(s) + CO(g) \rightarrow xM(s \text{ or } l) + CO_2(g)$

$$\Delta G^{\ominus} = \Delta G^{\ominus}(CO,CO_2) - \Delta G^{\ominus}(M,M_xO)$$

will have a negative reaction free energy, and be spontaneous. The relevant information is usefully summarized in an **Ellingham diagram**, which is a graph of $\Delta G^{\ominus}$ for each of these reactions against temperature (Fig. 7.1).

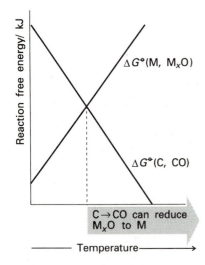

7.1 The variation of the standard free energies of formation of a metal oxide and carbon monoxide with temperature. The formation of carbon monoxide from carbon can reduce the metal oxide to the metal at temperatures higher than the point of intersection of the two lines. (More specifically, at the intersection K changes from less than 1 to more than 1.)

Some understanding of the appearance of an Ellingham diagram can be obtained by noting that

$$\Delta G^{\ominus} = \Delta H^{\ominus} - T\Delta S^{\ominus}$$

and using the fact that the enthalpy and entropy of reaction are, to a reasonable approximation, independent of temperature: the slope of a line in an Ellingham diagram should therefore be equal to $-\Delta S^{\ominus}$ for the relevant reaction. Because the standard entropies of gases are much larger than those of solids, the standard entropy of reaction (d), in which there is a net consumption of gas, is negative, and hence the Ellingham plot should have a positive slope, as shown in Fig. 7.2. The kink in the graph, where the slope of the metal oxidation line changes, is where the metal undergoes a phase change, particularly melting, and the reaction entropy changes accordingly.

The reaction entropy of (a), in which there is a net formation of gas (because 1 mol CO replaces $\frac{1}{2}$ mol O_2), is positive and its line in the Ellingham diagram will therefore have a negative slope. The reaction entropy of (b) is close to zero as there is no net change in the amount of gas, so its slope will be close to zero: its line is horizontal in the Ellingham diagram. Finally, (c) has a negative reaction entropy because $\frac{3}{2}$ mol of gas molecules is replaced by 1 mol of CO_2; hence the line in the diagram has a positive slope.

At temperatures for which the C,CO line lies above the metal oxide lines in Fig. 7.1, the difference $\Delta G^{\ominus}(C,CO) - \Delta G^{\ominus}(M,M_xO)$ is positive and so the reduction reaction (a)−(d) is not spontaneous. However, for temperatures for which the C,CO line lies below the metal oxide line, the reduction of the metal oxide by carbon is spontaneous. Similar remarks apply to the temperatures at which the other two carbon oxidation lines lie above or below the metal oxide lines in Fig. 7.2. In summary:

> For temperatures at which the C,CO line lies below the metal oxide line, carbon can be used to reduce the metal oxide and itself is oxidized to carbon monoxide.
>
> For temperatures at which the C,CO_2 line lies below the metal oxide line, carbon can be used to achieve the reduction, but is oxidized to carbon dioxide.
>
> For temperatures at which the CO,CO_2 line lies below the metal oxide line, carbon monoxide can reduce the metal oxide to the metal and is oxidized to carbon dioxide.

Figure 7.3 shows a combined Ellingham diagram for a selection of common metals. In principle, production of all of the metals shown in the diagram, even magnesium and calcium, could be accomplished by **pyrometallurgy**, heating with a reducing agent. However, there are severe practical limitations. Efforts to produce aluminum by pyrometallurgy (most notably in Japan, where electricity is expensive) were frustrated by the volatility of Al_2O_3 at the very high temperatures required. A difficulty of a different kind is encountered in the pyrometallurgical extraction of titanium, where titanium carbide, TiC, is

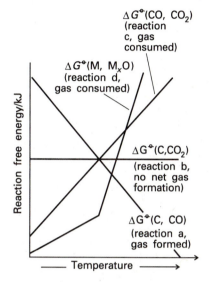

$\Delta G^{\oplus}(CO, CO_2)$ (reaction c, gas consumed)

$\Delta G^{\oplus}(M, M_xO)$ (reaction d, gas consumed)

$\Delta G^{\oplus}(C,CO_2)$ (reaction b, no net gas formation)

$\Delta G^{\oplus}(C, CO)$ (reaction a, gas formed)

Reaction free energy/kJ

⟶ Temperature ⟶

7.2 A fragment of an Ellingham diagram showing the free energy of formation of a metal oxide and the three carbon oxidation free energies. The slopes of the curves are determined largely by whether or not there is net gas formation or consumption in the course of the reaction. A phase change can result in a kink in the graph (because the entropy of the substance changes).

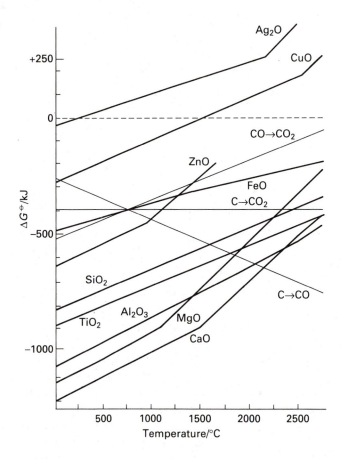

7.3 An Ellingham diagram for the reduction of metal oxides. The standard free energies are for the formation of the oxides from the metal and for the three carbon oxidations quoted in the text.

formed instead of the metal. In practice, pyrometallurgical extraction of metals is confined principally to magnesium, iron, cobalt, nickel, zinc, and to a variety of ferroalloys.

Example 7.1: *Using an Ellingham diagram*

What is the lowest temperature at which ZnO can be reduced to zinc metal by carbon? What is the overall reaction at this temperature?

Answer. The C,CO line in Fig. 7.3 lies below the ZnO line at approximately 950 °C; above this temperature reduction of the metal oxide is spontaneous. The contributing reactions are reaction (a) and

$$Zn(g) + \tfrac{1}{2}O_2(g) \rightarrow ZnO(s)$$

so the overall reaction is the difference, or

$$C(s) + ZnO(s) \rightarrow CO(g) + Zn(g)$$

The physical state of zinc is given as a gas because the element boils at 907 °C (the corresponding inflection in the ZnO line in the Ellingham diagram can be seen in Fig. 7.3).

Exercise E7.1. What is the minimum temperature for reduction of MgO by carbon?

Similar principles apply to reductions using other reducing agents. For instance, an Ellingham diagram can be used to explore whether a metal M′ can be used to reduce the oxide of another metal M. In this case, we note from the diagram whether at a temperature of interest the M′ oxide line lies below the M oxide line, for M′ is now taking the place of C. When

$$\Delta G^{\ominus} = \Delta G^{\ominus}(M', \text{oxide}) - \Delta G^{\ominus}(M, M_xO)$$

is negative, the reaction

$$MO(s) + M'(s \text{ or } l) \rightarrow M(s \text{ or } l) + M'O(s)$$

(and its analogs for MO_2 and so on) is feasible. For example, because in Fig. 7.3 the line for magnesium lies below the line for silicon at temperatures below 2200 °C, magnesium may be used to reduce SiO_2 below that temperature. This reaction has in fact been employed to produce low grade silicon.

A survey of chemical reductions

Industrial processes for achieving the reductive extraction of metals show a greater variety than the thermodynamic analysis might suggest. An important factor is that the ore and carbon are both solids, and a reaction between two coarse solids is rarely fast. Most processes exploit gas–solid or liquid–solid reactions. Current industrial processes are quite varied in the strategies they adopt to ensure good kinetics, exploit materials economically, and avoid environmental problems. We can explore these strategies by considering three important examples which reflect low, moderate, and extreme difficulty of reduction.

The least difficult reductions include that of copper ores. Roasting and smelting are still widely used in the pyrometallurgical extraction of copper. However, some recent techniques seek to avoid the major environmental problems caused by the production of the large quantity of SO_2 that accompanies roasting. One promising development is the **hydrometallurgical extraction** of copper, the extraction of a metal by reduction of aqueous solutions of its ions (with H_2 or scrap iron). In this process, Cu^{2+} ions leached from low grade ores by acid or bacterial action are reduced by hydrogen in the reaction

$$Cu^{2+}(aq) + H_2(g) \rightarrow Cu(s) + 2H^+(aq)$$

or by an equivalent reduction with iron. As well as being environmentally relatively benign, this process also allows economic exploitation of lower grade ores.

The extraction of iron is of intermediate difficulty (as shown by the fact that the Iron Age followed the Bronze Age). In economic terms, iron ore reduction is the most important application of carbon pyrometallurgy. In a blast furnace (Fig. 7.4), which is still the major source of the element, the charge of iron ores (Fe_2O_3, Fe_3O_4), coke (C), and limestone ($CaCO_3$) is heated with a blast of hot air. Combustion of coke in this air blast raises the temperature to 2000 °C, and the carbon

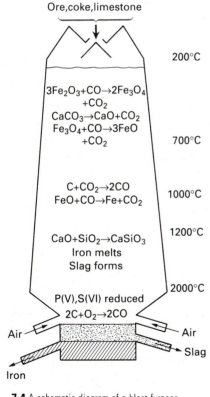

Ore, coke, limestone

200°C

$3Fe_2O_3 + CO \rightarrow 2Fe_3O_4$
$+CO_2$
$CaCO_3 \rightarrow CaO + CO_2$
$Fe_3O_4 + CO \rightarrow 3FeO$
$+CO_2$

700°C

$C + CO_2 \rightarrow 2CO$
$FeO + CO \rightarrow Fe + CO_2$

1000°C

$CaO + SiO_2 \rightarrow CaSiO_3$
Iron melts
Slag forms

1200°C

P(V), S(VI) reduced
$2C + O_2 \rightarrow 2CO$

2000°C

Air

Air

Slag

Iron

7.4 A schematic diagram of a blast furnace showing the typical composition and temperature profile.

burns to carbon monoxide in the lower part of the furnace. The supply of Fe_2O_3 from the top of the furnace meets the hot CO rising from below. The iron(III) oxide is reduced, first to magnetite (Fe_3O_4) and then to wüstite (FeO) at 500–700 °C, and the CO is oxidized to CO_2 (as in reaction (c)). At the same time, limestone ($CaCO_3$) is converted to lime (CaO), so increasing the CO_2 content of the exhaust gases. The final reduction to iron occurs between 1000 and 1200 °C in the central region of the furnace:

$$FeO(s) + CO(g) \rightarrow Fe(s) + CO_2(g)$$

A sufficient supply of CO is assured by the reaction

$$CO_2(g) + C(s) \rightarrow 2CO(g)$$

which augments the carbon monoxide formed by the incomplete combustion of carbon. This series of gas–solid reactions accomplishes the overall solid–solid reaction between the ore and the coke.

The function of the lime, CaO, formed by the thermal decomposition of calcium carbonate is to combine with the silicates present in the ore to form, in the hottest (lowest) part of the furnace, a molten layer of slag. The iron formed melts at about 400 °C below the melting point of the pure metal on account of the dissolved carbon it contains. The iron, the densest phase, settles to the bottom and is drawn off to solidify into 'pig iron', in which the carbon content is high (about 4 percent by mass). The manufacture of steel is then a series of reactions in which the carbon content is reduced and other metals are used to form alloys with the iron.

More difficult than the extraction of either copper or iron is the extraction of silicon from its oxide: indeed, silicon is very much an element of the twentieth century. Silicon of 96 to 99 percent purity is prepared by reduction of quartzite or sand (SiO_2) with high purity coke. The Ellingham diagram shows that the reduction is feasible only at temperatures in excess of about 1500 °C. This high temperature is achieved in an electric arc furnace in the presence of excess silica (to prevent the accumulation of SiC):

$$SiO_2(l) + 2C(s) \xrightarrow{1500\,°C} Si(l) + 2CO(g)$$

$$2SiC(s) + SiO_2(l) \rightarrow 3Si(l) + 2CO(g)$$

Very pure silicon (for semiconductors) is made by converting crude silicon to volatile compounds, such as $SiCl_4$. These compounds are purified by exhaustive fractional distillation and then reduction to silicon with pure hydrogen. The resulting semiconductor-grade silicon is melted and then large single crystals are pulled slowly from the melt (this procedure is called the Czochralski process).

The Ellingham diagram shows that the direct reduction of Al_2O_3 with carbon becomes feasible only above 2000 °C, which is uneconomically expensive and complicated by features common to high temperature chemistry, such as the volatility of the oxide. However, the reduction can be brought about electrolytically, and all modern production uses the **Hall–Héroult process**, which was invented in 1886 independently by Charles Hall in the United States and Paul Héroult in France.

Electrolytic reduction can be regarded as a technique for driving a reduction by coupling it (through electrodes and external circuitry) to a reaction or physical process with a more negative ΔG. The free energy available from the external source can be assessed from the potential difference E it produces across the electrodes using the thermodynamic relation

$$\Delta G = -nFE$$

where n is the amount (in moles) of electrons transferred and F is Faraday's constant ($F = 96.5 \text{ kC mol}^{-1}$). Hence, because the total Gibbs free energy change of the coupled internal and external process is

$$\Delta G + \Delta G(\text{external process}) = \Delta G - nFE$$

then if the applied potential difference is no less than

$$E = \frac{\Delta G}{nF}$$

the reduction is spontaneous.

In practice, the bauxite ore is a mixture of the acidic oxide SiO_2 and the amphoteric oxides Al_2O_3 and Fe_2O_3 (plus some TiO_2). The Al_2O_3 is extracted with aqueous sodium hydroxide, which separates the aluminum and silicon oxides from the iron(III) oxide (which needs a more concentrated alkali for significant reaction). The neutralization of the solution with CO_2 results in the precipitation of $Al(OH)_3$, leaving silicates in solution. The aluminum hydroxide is then dissolved in fused cryolite (Na_3AlF_6) and the melt is reduced electrolytically at a steel cathode with graphite anodes. The latter participate in the overall reaction, which is

$$2Al_2O_3 + 3C \rightarrow 4Al + 3CO_2$$

and hence need to be continuously replaced. The commercial process uses an electrode potential of about 4.5 V with a current density of about 1 A cm^{-2}. This current density might not seem much, but when multiplied by the total area of the electrodes, the current required is enormous; moreover, because the power (in watts) is the product of current and potential ($P = IV$) the power consumption of a typical plant is huge. Consequently, aluminum is often produced where electricity is cheap (in Quebec, for instance) and not where bauxite is mined (in Jamaica, for example).

7.2 Elements extracted by oxidation

The halogens are the most important elements extracted by oxidation. The standard reaction free energy for the oxidation of Cl$^-$ ions in water

$$2Cl^-(aq) + 2H_2O(l) \rightarrow 2OH^-(aq) + H_2(g) + Cl_2(g) \quad \Delta G^{\ominus} = +422 \text{ kJ}$$

is strongly positive, which suggests that electrolysis is required. The minimum potential difference that can achieve the oxidation of Cl^- is about 2.2 V (because $n = 2$ mol for the reaction as written, and $E(\text{applied}) = +\Delta G^{\ominus}/2F$).

It may appear that there is a problem with the competing reaction

$$2H_2O(l) \rightarrow 2H_2(g) + O_2(g) \quad \Delta G^{\ominus} = +414\,kJ \quad E^{\ominus} = -1.2\ V$$

which can be driven forward by a potential difference of only 1.2 V. However, the rate of oxidation of water is extremely slow at potentials at which it first becomes favorable thermodynamically. This slowness is expressed by saying that the reduction requires a high **overpotential**, the potential that must be applied in addition to the equilibrium value before a significant rate of reaction is achieved. (Overpotential is treated more fully in Section 7.4.) Consequently, the electrolysis of brine produces Cl_2, H_2, and aqueous NaOH, but not much O_2. One problem faced by the industry is therefore to try to develop uses for both Cl_2 and NaOH in a market balance as closely adjusted to the stoichiometry of the reaction as possible. The desire to reduce the consumption of chlorine (for environmental reasons) has resulted in the exploitation of alternative sources of NaOH, such as

$$Na_2CO_3(aq) + Ca(OH)_2(aq) \rightarrow CaCO_3(s) + 2NaOH(aq)$$

The sodium carbonate used in this process is mined and the calcium hydroxide is formed by roasting calcium carbonate and hydrating the resulting oxide.

Oxygen, not fluorine, is produced if aqueous solutions of fluorides are electrolyzed. Therefore, F_2 is prepared by the electrolysis of an anhydrous mixture of potassium fluoride and hydrogen fluoride, an ionic conductor which is molten above 72 °C. The more readily oxidizable halogens, Br_2 and I_2, are obtained by chemical oxidation of the aqueous halides with chlorine.

Because O_2 is available from fractional distillation of liquid air, chemical methods of oxygen production are not necessary (but may become necessary as we colonize other planets). Sulfur is an interesting mixed case. Elemental sulfur is either mined or produced by oxidation of the H_2S that is removed from 'sour' natural gas by trapping in ethanolamine ($HOCH_2CH_2NH_2$). The oxidation is accomplished by the **Claus process**, which consists of two stages. In the first, some hydrogen sulfide is oxidized to sulfur dioxide:

$$2H_2S + 3O_2 \rightarrow 2SO_2 + 2H_2O$$

In the second, this sulfur dioxide is allowed to react in the presence of a catalyst with more hydrogen sulfide:

$$2H_2S + SO_2 \xrightleftharpoons[]{\text{Oxide catalyst, 300 °C}} 3S + 2H_2O$$

The catalyst is typically Fe_2O_3 or Al_2O_3. The Claus process is environmentally benign, for otherwise it would be necessary to burn the toxic hydrogen sulfide to polluting sulfur dioxide.

The only important metals obtained by oxidation are the ones that occur native (that is, as the element). Gold is an example, because it is difficult to separate the granules of metal in low-grade ores by simple 'panning'. The dissolution of gold depends on the promotion of its oxidation, which is favored by complexation with CN^- ions, when it forms the $[Au(CN)_2]^-$ ion. This complex can be reduced to the metal by reaction with another reactive metal, such as zinc:

$$2[Au(CN)_2]^-(aq) + Zn(s) \rightarrow 2Au(s) + [Zn(CN)_4]^{2-}(aq)$$

REDUCTION POTENTIALS

One of the points we have made is that one reaction may be used to drive another if the *overall* reaction free energy is negative. In a blast furnace the driving reaction is the oxidation of carbon or carbon monoxide, and the oxidizing agent is in direct contact with the oxide that is being reduced. In electrochemical metallurgy, the oxidizing agent and the species undergoing reduction are physically separated. Moreover, the driving reaction (or physical process) for the nonspontaneous redox reaction is at a distant location, such as a power station.

7.3 Redox half-reactions

It is convenient to think of a redox reaction as the sum of two conceptual **half-reactions**. In a reduction half-reaction, a substance gains electrons, as in

$$2H^+(aq) + 2e^- \rightarrow H_2(g)$$

In an oxidation half-reaction, a substance loses electrons, as in

$$Zn(s) \rightarrow Zn^{2+}(aq) + 2e^-$$

A state is not ascribed to the electrons in the equation of a half-reaction because the division into half-reactions is only conceptual and may not correspond to an actual physical separation of the two processes.

The oxidized and reduced species in a half-reaction form a redox **couple**. A couple is written with the oxidized species before the reduced, as in H^+/H_2 and Zn^{2+}/Zn, and typically the phases are not shown. It is usually desirable to adopt the convention of writing *all* half-reactions as reductions. (This convention is reminiscent of discussing Brønsted bases in terms of the properties of their conjugate acids, so that all species are discussed as acids.) Because an oxidation is the reverse of a reduction, we can write the second of the two half-reactions above as

$$Zn^{2+}(aq) + 2e^- \rightarrow Zn(s)$$

The overall reaction is then the *difference* of the two reduction half-reactions with the number of electrons in the two half-reactions matching. This procedure is similar to the discussion of the Ellingham diagram, in which all the reactions are written as oxidations, and the overall reaction is the difference of reactions with matching numbers of O atoms.

Standard potentials

A reduction half-reaction can be thought of as contributing a certain value of $\Delta G^{\ominus}$ to the overall standard reaction free energy. It follows that the standard free energy of the overall reaction is the difference of the standard free energies of the two reduction half-reactions. The overall reaction is favorable in the direction that corresponds to a negative value of the resulting overall $\Delta G^{\ominus}$.

Because reduction half-reactions must always occur in pairs in any actual reaction, only the *difference* in their standard free energies has any significance. Therefore, we can chose one half-reaction as having $\Delta G^{\ominus} = 0$, and report all other values relative to it. By convention, the specially chosen half-reaction is the reduction of hydrogen ions:

$$2H^+(aq) + 2e^- \rightarrow H_2(g) \quad \Delta G^{\ominus} = 0 \text{ by definition}$$

With this choice, the standard free energy for the reduction of Zn^{2+} ions is found by determining experimentally that

$$Zn^{2+}(aq) + H_2(g) \rightarrow Zn(s) + 2H^+(aq) \quad \Delta G^{\ominus} = +147 \text{ kJ}$$

Then, because the H^+ reduction half-reaction makes zero contribution to the reaction free energy, it follows that

$$Zn^{2+}(aq) + 2e^- \rightarrow Zn(s) \quad \Delta G^{\ominus} = +147 \text{ kJ}$$

The standard free energies of overall reactions may be measured by setting up a galvanic cell (Fig. 7.5) in which the reaction driving the electric current through the external circuit is the reaction of interest. The potential difference between its electrodes is then measured[1] and, if desired, converted to a free energy using $\Delta G = -nFE$. Tabulated values—normally for standard conditions[2]—are usually kept in the units in which they were measured, namely volts (V).

The potential that corresponds to the $\Delta G^{\ominus}$ of a half-reaction is written $E^{\ominus}$ and is called the **standard potential** (or standard reduction potential, to emphasize that the half-reaction is a reduction). Because $\Delta G^{\ominus}$ for the reduction of H^+ is arbitrarily set at zero, the standard potential of the H^+/H_2 couple is also zero (at all temperatures):

$$2H^+(aq) + 2e^- \rightarrow H_2(g) \quad E^{\ominus}(H^+, H_2) = 0 \text{ by definition}$$

Similarly, for the Zn^{2+}/Zn couple, for which $n = 2$ mol, it follows from the measured value of $\Delta G^{\ominus}$ (and $\Delta G^{\ominus} = -nFE^{\ominus}$) that

$$Zn^{2+}(aq) + 2e^- \rightarrow Zn(s) \quad E^{\ominus}(Zn^{2+}, Zn) = -0.76 \text{ V at } 25\,°C$$

The value of $E^{\ominus}$ for an overall reaction is the difference of the two standard potentials of the reduction half-reactions into which the over-

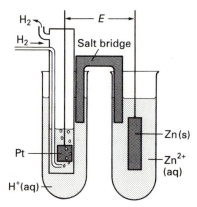

7.5 A schematic diagram of a galvanic cell. The standard cell potential $E^{\ominus}$ is the potential difference when the cell is not generating current and all the substances are in their standard states.

---

[1] In practice, we must ensure that the cell is acting reversibly in a thermodynamic sense, which means that the potential difference must be measured with negligible current flowing.

[2] Standard conditions are all substances at 1 bar pressure and unit activity. For our purposes, there is negligible difference using 1 atm in place of 1 bar. It is helpful to bear in mind that for reactions involving H^+ ions, standard conditions correspond to pH = 0, approximately 1 M acid.

Table 7.1 Selected standard potentials at 25 °C

Couple	$E^{\ominus}$/V
$F_2(g) + 2e^- \rightarrow 2F^-(aq)$	$+2.87$
$Ce^{4+}(aq) + e^- \rightarrow Ce^{3+}(aq)$	$+1.76$
$MnO_4^-(aq) + 8H^+(aq) + 5e^- \rightarrow Mn^{2+}(aq) + 4H_2O(l)$	$+1.51$
$Cl_2(g) + 2e^- \rightarrow 2Cl^-(aq)$	$+1.36$
$O_2(g) + 4H^+(aq) + 4e^- \rightarrow 2H_2O(aq)$	$+1.23$
$[IrCl_6]^{2-}(aq) + e^- \rightarrow [IrCl_6]^{3-}(aq)$	$+0.87$
$Fe^{3+}(aq) + e^- \rightarrow Fe^{2+}(aq)$	$+0.77$
$[PtCl_4]^{2-}(aq) + 2e^- \rightarrow Pt(s) + 4Cl^-(aq)$	$+0.60$
$I_3^-(aq) + 2e^- \rightarrow 3I^-(aq)$	$+0.54$
$[Fe(CN)_6]^{3-}(aq) + e^- \rightarrow [Fe(CN)_6]^{4-}(aq)$	$+0.36$
$AgCl(s) + e^- \rightarrow Ag(s) + Cl^-(aq)$	$+0.22$
$2H^+(aq) + 2e^- \rightarrow H_2(g)$	0
$AgI(s) + e^- \rightarrow Ag(s) + I^-(aq)$	-0.15
$Fe^{2+}(aq) + 2e^- \rightarrow Fe(s)$	-0.44
$Zn^{2+}(aq) + 2e^- \rightarrow Zn(s)$	-0.76
$Al^{3+}(aq) + 3e^- \rightarrow Al(s)$	-1.68
$Ca^{2+}(aq) + 2e^- \rightarrow Ca(s)$	-2.84
$Li^+(aq) + e^- \rightarrow Li(s)$	-3.04

all reaction can be divided:

(a) $2H^+(aq) + 2e^- \rightarrow H_2(g)$ $E^{\ominus} = 0$

(b) $Zn^{2+}(aq) + 2e^- \rightarrow Zn(s)$ $E^{\ominus} = -0.76$ V

The difference (a) − (b) is

$2H^+(aq) + Zn(s) \rightarrow Zn^{2+}(aq) + H_2(g)$ $E^{\ominus} = +0.76$ V

The consequence of the negative sign in $\Delta G^{\ominus} = -nFE^{\ominus}$ is that the rule about a reaction being favorable if $\Delta G^{\ominus} < 0$ is replaced by the rule that the reaction is favorable if $E^{\ominus} > 0$. Because $E^{\ominus} > 0$ for the reaction above ($E^{\ominus} = +0.76$ V), we know that zinc has a thermodynamic tendency to reduce H^+ ions under standard conditions (acidic solution, pH = 0, and Zn^{2+} at unit activity). The same is true for any couple with a negative standard potential.

The electrochemical series

A negative standard potential signifies a couple in which the reduced species (the Zn in Zn^{2+}/Zn) is a reducing agent for H^+ ions under standard conditions in aqueous solution. That is, if $E^{\ominus}(Ox,Red) < 0$, then the substance 'Red' is a strong enough reducing agent to reduce H^+ ions.

A short list of $E^{\ominus}$ values at 25 °C is given in Table 7.1. The list is arranged in the order of the **electrochemical series**:

Ox/Red couple with strongly positive $E^{\ominus}$ [Ox strongly oxidizing]

⋮

Ox/Red couple with strongly negative $E^{\ominus}$ [Red strongly reducing]

An important feature of the electrochemical series is that the reduced member of a couple can (thermodynamically, not always kinetically) reduce the oxidized member of any couple that lies above it in the series. In other words:

1. Oxidized species with high positive standard potentials are strong oxidizing agents.
2. Reduced species with large negative standard potentials are strong reducing agents.

In each case, the classification refers to the *thermodynamic* aspect of the reaction, its spontaneity, not its rate (and hence its practical usefulness).

Example 7.2: *Using the electrochemical series*

Among the couples in Table 7.1 is the permanganate ion, MnO_4^-, the common analytical reagent used in redox titrations of iron. Which of the ions Fe^{2+}, Cl^-, and Ce^{3+} can permanganate oxidize in acidic solution?

Answer. The standard potential of MnO_4^- to Mn^{2+} in acidic solution is $+1.51$ V. The potentials for the listed ions are $+0.77$, $+1.36$, and $+1.76$ V, respectively. It follows that permanganate ions are sufficiently strong oxidizing agents in acidic solution to oxidize the first two ions, which have less positive reduction potentials. Precautions should be taken to avoid the rapid oxidation of Cl^- ions in titrations of iron solutions with permanganate. Permanganate ions cannot oxidize Ce^{3+}, which has a more positive reduction potential. It should be noted that in practice the presence of other ions in the solution can modify reduction potentials and the conclusions. This is particularly the case with hydrogen ions, and the influence of pH is discussed in Section 7.10.

Exercise E7.2. Another common analytical oxidizing agent is an acidic solution of dichromate ions ($Cr_2O_7^{2-}$), for which $E^\ominus = +1.38$ V. Is the solution useful for titration of Fe^{2+}? Could there be a side reaction when Cl^- is present?

The Nernst equation

Standard potentials are signposts of spontaneous change under *standard* conditions. To judge the tendency of a reaction to run in a particular direction under *nonstandard* conditions, we need to know the sign and value of ΔG under the relevant conditions. For this information, we use the thermodynamic result that

$$\Delta G = \Delta G^\ominus + RT \ln Q$$

where Q is the **reaction quotient**[3]

$$aOx_A + bRed_B \rightarrow a'Red_A + b'Ox_B \qquad Q = \frac{[Red_A]^{a'}[Ox_B]^{b'}}{[Ox_A]^{a}[Red_B]^{b}}$$

---

[3] The molar concentrations [X] in the reaction quotient should be interpreted as molar concentrations *relative* to a standard molar concentration of 1 mol L^{-1}: that is, they are the numerical value of the molar concentration. For gas phase reactions, the relative molar concentration of a species is replaced by its partial pressure relative to the standard pressure $p^\ominus = 1$ bar.

The reaction is spontaneous under the stated conditions if $\Delta G < 0$. This criterion can be expressed in terms of potentials by substituting $E = -\Delta G/nF$ and $E^{\ominus} = -\Delta G^{\ominus}/nF$, which gives

$$E = E^{\ominus} - \frac{RT}{nF} \ln Q$$

This expression is called the **Nernst equation**.

A reaction is spontaneous if $E > 0$ (because that corresponds to $\Delta G < 0$). Because at equilibrium $E = 0$ and $Q = K$, we can also write the following very important relation between the standard potential of a reaction and its equilibrium constant at a temperature T:

$$E^{\ominus} = \frac{RT}{nF} \ln K$$

$E^{\ominus}/V$	K
+2	10^{34}
+1	10^{17}
0	1
−1	10^{-17}
−2	10^{-34}

The table in the margin lists the value of K that corresponds to potentials in the range -2 to $+2$ V, with $n = 1$ mol and at 25 °C.

Although electrochemical data are often compressed into the range -2 to $+2$ V, that brief range corresponds to 68 orders of magnitude in the value of the equilibrium constant.

If we regard E as the difference of two potentials, just as $E^{\ominus}$ is the difference of two standard potentials, then the potential of each couple can be written as

$$E = E^{\ominus} - \frac{RT}{nF} \ln Q \quad Q = \frac{[\text{Red}]^{a'}}{[\text{Ox}]^{a}}$$

A useful numerical form of this expression is that at 25 °C,

$$E = E^{\ominus} - \frac{0.059 \text{ V}}{n} \log Q$$

where $\log Q$ is the common logarithm (to the base 10) of Q. It follows that each increase of an order of magnitude in concentration ratio ($Q \to 10Q$) corresponds to a $(59/n)$ mV decrease in potential.

Example 7.3: *Using the Nernst equation*

What is the dependence of the reduction potential of the H^+/H_2 couple on pH when the hydrogen pressure is 1.00 bar and the temperature is 25 °C?

Answer. The reduction half-reaction is

$$2H^+(aq) + 2e^- \to H_2(g)$$

The Nernst equation for the couple is

$$E = E^{\ominus} - \frac{0.059 \text{ V}}{2} \log \left(\frac{p(H_2)}{[H^+]^2} \right)$$

Therefore, with $p(H_2) = 1$ bar,

$$E(H^+, H_2) = -59 \text{ mV} \times pH$$

That is, the reduction potential becomes more negative by 59 mV for each unit increase in pH.

Exercise E7.4. What is the potential for reduction of MnO_4^- to Mn^{2+} in neutral (pH = 7) solution?

7.4 Kinetic factors

We have already remarked that thermodynamic considerations can be used only to analyze the spontaneity of a reaction under the prevailing conditions: thermodynamics is silent on the rates with which a spontaneous reaction occurs. In this section we summarize some of the information that is available about the rates of redox processes in solution.

Overpotential

A negative standard potential for a metal ion–metal couple indicates that the metal can reduce H^+ ions or any more positive couple under standard conditions in aqueous solution; it does not assure us that a mechanistic pathway exists for this reduction to be realized. There are no fully general rules that predict when reactions are likely to be fast, for the factors that control them are diverse (as we see in Chapter 15). However, a useful rule-of-thumb (but one with notable exceptions) is that couples with potentials that are more negative than about -0.6 V bring about hydrogen ion reduction to H_2 at a significant rate. Similarly, potentials more positive than 1.23 V + 0.60 V = 1.83 V bring about O_2 evolution at a significant rate (the standard potential of the couple $O_2, H^+/H_2O$ is $+1.23$ V, Table 7.1.) The additional 0.6 V is an example of an **overpotential,** the potential beyond the zero-current reduction potential that must exist before the reaction proceeds at a significant rate.

The existence of an overpotential explains why some metals reduce acids and not water itself. Such metals (which include iron and zinc) have negative reduction potentials, but these potentials are not low enough to achieve the necessary overpotential, about -0.6 V, for H^+ reduction in neutral solution (at pH = 7, $E(H^+, H_2) = -0.41$V). However, the difference $E(H^+, H_2) - E(Fe^{2+}, Fe)$ can be increased if $E(H^+, H_2)$ is made more positive, which can be done by lowering the pH from 7 toward a more acidic value. When the difference in potential exceeds about 0.6 V, the metal brings about reduction at an appreciable rate. The commercial (electrolytic) concentration of D_2O depends on the greater overpotential for the evolution of D_2 than for the evolution of H_2.

Example 7.4: *Taking overpotential into account*
Is iron likely to be rapidly oxidized to Fe^{2+}(aq) by water at 25 °C?

Answer. We need to judge the difference in potentials of the couples at pH = 7. The Nernst equation for the hydrogen couple (supposing that the partial pressure of hydrogen is about 1 bar) is

$$H^+(aq) + e^- \rightarrow \tfrac{1}{2}H_2(g)$$

$$E = 0 + (0.059 \text{ V}) \log [H^+] = -(0.059 \text{ V}) \times pH$$

Therefore, because pH = 7.0, $E \approx -0.41$ V. Similarly, the Nernst equation for the iron couple is

$$Fe^{2+}(aq) + 2e^- \rightarrow Fe(s) \quad E = -0.44 \text{ V} + \tfrac{1}{2}(0.059 \text{ V}) \log [Fe^{2+}]$$

If the concentration of Fe^{2+} approaches 1 mol L^{-1}, $E \approx -0.44$ V. Although the iron couple is more positive than the hydrogen couple, the difference (0.03 V) is less than the 0.6 V typically required for a significant rate of hydrogen evolution. This small value suggests that the reaction will be slow.

Exercise E7.4. If an exposed surface is maintained, magnesium can undergo rapid oxidation in water at pH = 7 and 25 °C. Is the value of E for this reaction in line with the generalization about overpotential stated in the text?

One of the techniques that is used to study inorganic reaction mechanisms is outlined in Box 7.1.

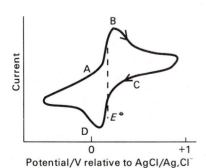

B7.1 The arrangement for a cyclic voltammetry experiment. Each period of the time-varying potential (the first is marked by the vertical lines) corresponds to one complete cycle of a cyclic voltammogram (as in the following two illustrations).

B7.2 A typical cyclic voltammogram. The points are described in the text.

Box 7.1 Cyclic voltammetry

In the course of investigations of inorganic reactions, it is often very important to identify the species (aqua ions, complexes, and so on) that are involved in redox reactions and to identify any reaction intermediates. The technique that has become popular with inorganic chemists is **cyclic voltammetry**.

Cyclic voltammetry makes use of a small **working electrode** of about 1 mm or 2 mm in diameter. The current toward the electrode is limited by the rate at which an electrochemically active species can diffuse toward it. The experiment (Fig. B7.1) is conducted in a three-electrode cell in which the potential of the platinum working electrode is held constant with respect to a reference electrode by controlling the current through a counter electrode. The potential is applied to the working electrode in a sawtooth manner. A **cyclic voltammogram** is a plot of the current through the electrode as the potential is increased linearly, then reduced linearly, and so on. Figure B7.2 shows a cyclic voltammogram of a solution containing the redox couple $[Fe(CN)_6]^{3-}/[Fe(CN)_6]^{4-}$ at a platinum electrode.

The experiment begins by setting the potential at a highly cathodic (negative) value (on the left in Fig. B7.2). After a reasonable period of time, all the $[Fe(CN)_6]^{3-}$ near the working electrode will be reduced to $[Fe(CN)_6]^{4-}$. The current will now be very small because the concentration of the oxidized species will be very low near the electrode. A potential scan is then initiated. As the potential approaches the value of $E^{\ominus}$ for the couple, the oxidation of $[Fe(CN)_6]^{4-}$ at the electrode surface begins. As a result, at point A in the illustration, a current begins to flow as some iron(II) complex is oxidized to iron(III). As the value of $E^{\ominus}$ is passed, a condition is approached at which all the complex near the electrode has been oxidized to iron(III). When the oxidizable iron(II) species becomes scarce near the electrode, the current falls again as it becomes limited by diffusion of the oxidizable iron(II) complex from a distance far from the electrode surface. The result is that the current returns to a small value at highly positive potentials (on the right of the illustration). There is a peak in the current–voltage curve near (but not at) $E^{\ominus}$ for the couple.

When the current has reached a low value on the right, the direction of the potential sweep is reversed. As the potential becomes more negative, a current begins to flow in the reverse direction (point C in the illustration). The remainder of the curve is the reverse of the scan in the anodic direction. Notice that the peak current (point B) in the first sweep is to the right (positive) of the peak in the second sweep (point D). These positions reflect the fact that the concentration of the product species becomes small only *after* $E^{\ominus}$ has been passed. In every case, the peak current will appear at a potential beyond $E^{\ominus}$ in the direction of the sweep of the potential. The two peak current values in the two directions bracket $E^{\ominus}$, so it is easy to recognize the couple responsible for the peaks. In the illustration $E^{\ominus} \approx 0.17$ V relative to AgCl/Ag,Cl$^-$. Because the latter has $E^{\ominus} = 0.22$ V, it follows that $E^{\ominus}$ for the iron couple is $+0.39$ V relative to the SHE, which is recognizable as that of the $[Fe(CN)_6]^{4-}/[Fe(CN)_6]^{3-}$ couple.

The difference in potential between the peak current in the two potential sweeps is directly derivable from theory if the electron transfer reaction at the electrode is rapid and the couple is reversible (in the sense that the concentrations at the surface of the electrode rapidly adopt the values dictated by the Nernst equation). If electron transfer at the electrode is slow, the peak currents are separated by more than the reversible value, so a cyclic voltammogram also provides information about electron transfer kinetics.

In some cases, the reaction in one direction is simple but the reaction in the reverse direction is complex. For example, oxidation of RCO_2^- produces the radical $RCO_2\cdot$ which decomposes rapidly to $R\cdot$ and CO_2. The result of the decomposition is that the oxidation is irreversible. Neither $R\cdot$ nor CO_2 is reduced at a potential near the standard potential for oxidation of RCO_2^-. For a couple like this, the 'wave' for oxidation will appear in the cyclic voltammogram, but the reduction wave will be absent in the potential region near $E^{\ominus}$ (Fig. B7.3).

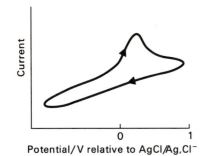

B7.3 A cyclic voltammogram in which the oxidation is simple but the oxidation product decomposes, so its reduction wave is absent.

The electron transfer couple is characterized by peaks near $E^{\ominus}$ that can be used to identify the electron transfer reaction and to measure the reduction potential. Whether or not the electron transfer reaction is rapid at the electrode is identified by the separation of the anodic and cathodic peaks. The absence of a peak in one direction shows that the product of the electrode reaction undergoes a reasonably rapid chemical transformation before the potential is swept back in the reverse direction. Often it is possible to recover the peak in the back sweep direction by increasing the rate at which the potential is scanned. This recovery implies that the potential cycle is being completed faster than the irreversible reaction, and hence its onset can be interpreted in terms of the rate of that reaction.

Cyclic voltammetry can be used to identify the specific electron transfer reaction occurring at the electrode and to decide whether or not it is electrochemically reversible. In fact, considerations beyond those discussed here allow additional inferences to be extracted from cyclic voltammograms. The technique is very powerful for the identification of redox reaction pathways.

[References: P. H. Rieger, *Electrochemistry*. Prentice-Hall, Englewood Cliffs (1987); A. J. Bard and L. R. Faulkner, *Electrochemical methods*. Wiley, New York (1980).]

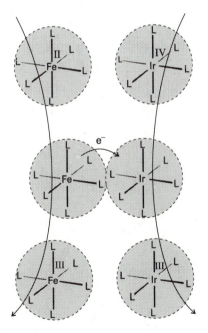

7.6 Schematic diagram of an outer-sphere redox reaction. (a) The reacting ions diffuse together through the solvent and, when they are in contact, an electron is transferred from one to the other. (b) The products then diffuse apart with their original coordination spheres (represented by the shaded spheres) intact. A small readjustment of the metal–ligand bond lengths may occur to accommodate the changes in radius of the central metal ions.

Electron transfer

In some cases, the need for an overpotential can be interpreted mechanistically. The transfer of an electron in bulk solution often takes place by **outer-sphere electron transfer**, a process in which a minimal change occurs in the coordination spheres of the redox centers (the atoms undergoing a change of oxidation number). This process is depicted in Fig. 7.6. The redox process is accompanied by only small changes in the metal–ligand distance. The electron transfer can be fast, and it is often found that the rate is proportional to the exponential of the difference in the standard potentials of the two couples. That is, the more favorable the equilibrium, the faster the reaction. The quantitative form of the dependence is discussed in Section 15.12.

The next simplest case is an **inner-sphere electron transfer**, in which a change does occur in the coordination sphere of the redox center. A classic example is the reduction of $[CoCl(NH_3)_5]^{2+}$ by $[Cr(OH_2)_6]^{2+}$, for which the reaction is

$$[CoCl(NH_3)_5]^{2+}(aq) + [Cr(OH_2)_6]^{2+}(aq) + 5H_2O(l) + 5H^+(aq)$$

$$\rightarrow [Co(OH_2)_6]^{2+}(aq) + [CrCl(OH_2)_5]^{2+}(aq) + 5NH_4^+(aq)$$

Because it is known that both replacement of NH_3 or Cl^- at Co(III) is a slow process, the change of ligands on Co must take place after the formation of the more labile Co(II) complex. The observation of

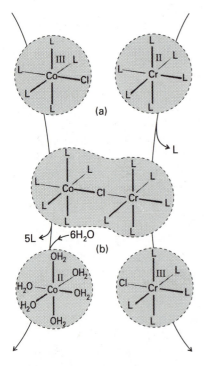

7.7 Schematic diagram of an inner-sphere redox reaction. (a) The reacting ions diffuse together through the solvent and, when they are in contact, a ligand substitution reaction occurs to create a ligand-bridged species. (b) The bridging ligand transfers (with all its electrons, as :$\ddot{C}l$:⁻, in this example) from one complex to the other, so resulting in a change in oxidation state of the metal atoms in the complexes. In the present example, a substitution-labile Co(II) complex is formed, and it quickly undergoes ligand substitution with the surrounding solvent.

Cl⁻ coordinated to Cr(III) indicates that a bridging Cl atom is involved in the mechanism because the incorporation of Cl⁻ into the coordination sphere of an aqua Cr(III) ion, $[Cr(OH_2)]^{3+}$, is very slow (Fig. 7.7). It is frequently found in this kind of reaction that the greater the difference in standard potentials of the couples, then the faster the reaction.

Noncomplementary redox reactions are reactions in which the change in oxidation numbers of the oxidizing and reducing agents are unequal. Such reactions are often slow because the reaction cannot be accomplished in a single electron-transfer step. The reaction may be slow if any one of the steps is slow or the concentration of an essential intermediate is low. Noncomplementary reactions are characteristic of the oxidation of p-block oxoanions by d-block ions because the oxidation numbers of the element in such oxoanions often differ by steps of 2 (as in NO_2^-, with oxidation number +3, and NO_3^-, with oxidation number +5) whereas redox reactions of d-block ions frequently occur in one-electron steps. Even if the change in oxidation number is 1, more than one step may be involved. For example, the oxidation of sulfite ions, SO_3^{2-} (oxidation number +4) to dithionate ions, $S_2O_6^{2-}$ (oxidation number +5) by Fe^{3+} ions requires two steps. In the first step, Fe(III) oxidizes a sulfite ion to a radical, and then the reaction is completed by dimerization:

$$2(SO_3 \cdot)^-(aq) \rightarrow S_2O_6^{2-}(aq)$$

Empirical generalizations

Inner-sphere mechanisms provide an important alternative to radical pathways for multiple electron transfer reactions and are common for oxoanions. It is found that

> The rate of oxoanion reduction varies strongly with the oxidation number of the central atom in an oxoanion: the lower the oxidation number, the faster the reaction.

For example, rates of oxoanion reduction lie in the order

$$ClO_4^- < ClO_3^- < ClO_2^- < ClO^-$$

$$ClO_4^- < SO_4^{2-} < HPO_4^{2-}$$

The radius of the central atom is also important, and the larger the central atom the greater the rate:

$$ClO_3^- < BrO_3^- < IO_3^-$$

Iodate reactions are fast and equilibrate rapidly enough to be useful in titrations.

Another very useful empirical rule is that

> The formation and decomposition of the common diatomic molecules O_2, N_2, and H_2 have complex mechanisms and are usually slow.

REDOX STABILITY IN WATER

An ion or molecule in solution may be destroyed by oxidation or reduction by any of the other species present. Therefore, when assessing the stability of a species in solution, we must bear in mind all possible reactants: the solvent, other solutes, the solute itself, and dissolved oxygen. In the following discussion, we shall stress the types of reaction that result from the thermodynamic instability of a solute. We shall also comment on rates, but the trends they show are generally less systematic than those shown by stabilities.

7.5 Reactions with water

Water may act as an oxidizing agent; when it acts in this way it is reduced to H_2. It may also act as a reducing agent, when it is oxidized to O_2. Species that can survive in water must have reduction potentials that lie between the limits defined by these two processes.

Oxidation by water

The reaction of metals with water or aqueous acid is actually the oxidation of the metal by water or hydrogen ions, because the overall reaction is one of the following processes (and their analogs for more highly charged metal ions):

$$M(s) + H_2O(l) \rightarrow M^+(aq) + \tfrac{1}{2}H_2(g) + OH^-(aq)$$

$$M(s) + H^+(aq) \rightarrow M^+(aq) + \tfrac{1}{2}H_2(g)$$

These reactions are thermodynamically favorable when M is an *s*-block metal other than beryllium or a first *d*-series metal from Group 4 through at least Group 7 (Ti, V, Cr, Mn). A number of other metals undergo similar reactions but with different numbers of electrons being transferred. An example from Group 3 is

$$2Sc(s) + 6H^+(aq) \rightarrow 2Sc^{3+}(aq) + 3H_2(g)$$

When the standard potential for the reduction of a metal ion to the metal is negative, the metal should undergo oxidation in 1 M acid with the evolution of H_2. However, the reaction may be slow, in which case it is appropriate to consider the role of the overpotential, as we have already explained.

Although the reactions of magnesium and aluminum with moist air are spontaneous, both metals can be used for years in the presence of water and oxygen. They survive because they are **passivated**, or protected against reaction, by an impervious film of oxide. Magnesium oxide and aluminum oxide both form a protective skin on the parent metal beneath. A similar passivation occurs with iron, copper, and zinc. The process of 'anodizing' a metal, in which the metal is made an anode (the site of oxidation) in an electrolytic cell, is one in which partial oxidation produces a smooth, hard passivating film on its surface. Anodization is especially effective for the protection of aluminum.

Reduction by water

Water can act as a reducing agent through the half-reaction

$$2H_2O(l) \rightarrow 4H^+(aq) + O_2(g) + 4e^-$$

This oxidation half-reaction is the reverse of the reduction half-reaction

$$O_2(g) + 4H^+(aq) + 4e^- \rightarrow 2H_2O(l) \quad E^\ominus = +1.23 \text{ V}$$

The strongly positive reduction potential shows that acidified water is a poor reducing agent except toward strong oxidizing agents. An example of the latter is $Co^{3+}(aq)$, for which $E^\ominus(Co^{3+},Co^{2+}) = +1.82$ V. It is reduced by water with the evolution of O_2:

$$4Co^{3+}(aq) + 2H_2O(l) \rightarrow 4Co^{2+}(aq) + O_2(g) + 4H^+(aq)$$

$$E^\ominus = +0.59 \text{ V}$$

This $E^\ominus$ is very close to the notional overpotential needed for a significant reaction rate. Because H^+ ions are produced in the reaction, changing from acid to neutral or basic solution favors the oxidation, since lowering the concentration of H^+ ions favors the products.

Only a few oxidizing agents (Ag^{2+} is another example) can oxidize water rapidly enough to give appreciable rates of O_2 evolution: the necessary overpotentials are not attained by many reagents that are thermodynamically capable of acting. Indeed, standard potentials greater than $+1.23$ V occur for several redox couples that are regularly used in aqueous solution, including Ce^{4+}/Ce^{3+} ($+1.72$ V), the acidified dichromate ion couple $Cr_2O_7^{2-}/Cr^{3+}$ ($+1.38$ V), and the acidified permanganate couple MnO_4^-/Mn^{2+} ($+1.51$ V). The origin of the barrier to reaction is the need to transfer four electrons and to form an oxygen–oxygen double bond.

Given that the rates of redox reactions are often controlled by the slowness with which an oxygen–oxygen bond can be formed, it remains a challenge for inorganic chemists to find good catalysts for O_2 evolution. Some progress has been made using ruthenium complexes. Existing catalysts include the relatively poorly understood coatings that are used in the anodes of cells for the commercial electrolysis of water. They also include the enzyme system that is found in the O_2 evolution apparatus of the plant photosynthetic center. This system is based on manganese with four identifiable levels of oxidation (see Section 19.9). Although Nature is elegant and efficient, it is also complex, and the photosynthetic process is only slowly being elucidated by biochemists and bioinorganic chemists.

The stability field of water

A reducing agent that can reduce water to H_2 rapidly, or an oxidizing agent that can oxidize water to O_2 rapidly, cannot survive in aqueous solution. The **stability field** of water (Fig. 7.8) is the range of values of reduction potential and pH for which water is thermodynamically stable toward both oxidation and reduction.

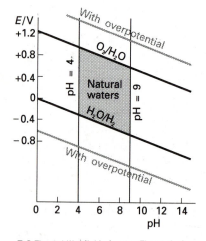

7.8 The stability field of water. The vertical axis is the reduction potential of redox couples in water: those above the upper line can oxidize water; those below the lower line can reduce water. The gray lines are the boundaries when overpotential is taken into account, and the vertical lines represent the normal range for natural waters. Hence, the gray area is the stability field for natural waters.

The upper and lower boundaries of the stability field are identified by finding the dependence of E on pH for the relevant half-reactions. The half-reaction for the $O_2,H^+/H_2O$ couple is

$$O_2(g) + 4H^+(aq) + 4e^- \rightarrow 2H_2O(aq) \quad n=4 \quad Q = \frac{1}{p_{O_2}[H^+]^4}$$

where p_{O_2} is the partial pressure of oxygen. It follows that

$$E = E^{\ominus} + \frac{RT}{4F}\ln\left(p_{O_2}[H^+]^4\right)$$

For a partial pressure of oxygen of 1 bar and at 25 °C,

$$E = 1.23 \text{ V} - 0.059 \text{ V} \times \text{pH}$$

Any species with a reduction potential *higher* than this value can be reduced by water, with the production of O_2. Hence, this expression defines the upper boundary of the stability field. The reduction of $H^+(aq)$ to H_2 occurs by the half-reaction

$$2H^+(aq) + 2e^- \rightarrow H_2(g) \quad n=2 \quad Q = \frac{p_{H_2}}{[H^+]^2}$$

so

$$E = E^{\ominus} - \frac{RT}{2F}\ln\frac{p_{H_2}}{[H^+]^2}$$

If the partial pressure of hydrogen is about 1 bar and the temperature 25 °C, then

$$E = -0.059 \text{ V} \times \text{pH}$$

Any species with a reduction potential that is *lower* than this value can reduce $H^+(aq)$ to H_2, so the line gives the lower boundary of the stability field. To simplify discussions, chemists generally refer to the reduction of hydrogen ions in water (at any pH) as 'the reduction of water'.

Couples that are thermodynamically stable in water lie between the limits defined by the sloping lines in Fig. 7.8. A couple that lies outside the field is unstable. The couple consists either of too strong a reducing agent (below the H_2 production line) or too strong an oxidizing agent (above the O_2 production line). The stability field in 'natural' water is represented by the addition of two vertical lines at pH = 4 and pH = 9 which mark the limits on pH that are commonly found in lakes and streams. Diagrams like that shown in the illustration are widely used in geochemistry, as we shall see in Section 7.10.

7.6 Disproportionation

Because $E^{\ominus}(Cu^+,Cu) = +0.52$ V and $E^{\ominus}(Cu^{2+},Cu^+) = +0.16$ V, and both potentials lie within the stability field of water, Cu^+ ions neither oxidize nor reduce water. Despite this, Cu(I) is not stable in aqueous solution because it can undergo **disproportionation**, a reaction in which the oxidation number of an element is simultaneously raised and lowered. In other words, the element undergoing disproportionation serves as its own oxidizing and reducing agent:

$$2Cu^+(aq) \rightarrow Cu^{2+}(aq) + Cu(s)$$

This reaction is spontaneous because $E^\ominus = 0.52$ V $- 0.16$ V $= +0.36$ V. We can obtain a more quantitative picture of the equilibrium position by employing the relation $E^\ominus = (RT/nF) \ln K$ in the form

$$0.36 \text{ V} = \frac{0.059 \text{ V}}{n} \log K$$

Because one electron is transferred in the reaction, $K = 1.3 \times 10^6$.

Hypochlorous acid is also subject to disproportionation:

$$5HClO(aq) \rightarrow 2Cl_2(g) + ClO_3^-(aq) + 2H_2O(l) + H^+(aq)$$

This reaction is the difference of the following two half reactions:

$$4HClO(aq) + 4H^+(aq) + 4e^- \rightarrow 2Cl_2(g) + 4H_2O(l) \quad E^\ominus = +1.63 \text{ V}$$

$$ClO_3^-(aq) + 5H^+(aq) + 4e^- \rightarrow HClO(aq) + 2H_2O(l) \quad E^\ominus = +1.43 \text{ V}$$

So overall $E^\ominus = 1.63$ V $- 1.43$ V $= +0.20$ V, implying that $K = 3 \times 10^{13}$.

Example 7.5: *Assessing the importance of disproportionation*

Show that manganate(VI) ions are unstable with respect to disproportionation into Mn(VII) and Mn(II) in acidic aqueous solution.

Answer. The overall reaction must involve reduction of one Mn(VI) to Mn(II) balanced by four Mn(VI) oxidizing to Mn(VII). The equation is therefore

$$5MnO_4^{2-}(aq) + 8H^+(aq) \rightarrow 4MnO_4^-(aq) + Mn^{2+}(aq) + 4H_2O(l)$$

This reaction can be expressed as the difference of the reduction half-reactions

$$MnO_4^{2-}(aq) + 8H^+(aq) + 4e^- \rightarrow Mn^{2+}(aq) + 4H_2O(l)$$

$$E^\ominus = +1.75 \text{ V}$$

$$4MnO_4^-(aq) + 4e^- \rightarrow 4MnO_4^{2-}(aq) \quad E^\ominus = +0.56 \text{ V}$$

The difference of the standard potentials is $+1.19$ V, so the disproportionation is essentially complete ($K = 5 \times 10^{80}$, because $n = 4$ mol). A practical consequence of the disproportionation is that high concentrations of MnO_4^{2-} ions cannot be obtained in acidic solution but must be prepared in a basic solution.

Exercise E7.6. The standard potentials for Fe^{2+},Fe and Fe^{3+},Fe^{2+} are -0.41 V and $+0.77$ V, respectively. Can Fe^{2+} disproportionate under standard conditions?

The reverse of disproportionation is **comproportionation**. In comproportionation, two species with the same element in different oxidation states form a product in which the element is in an intermediate oxidation state. An example is

$$Ag^{2+}(aq) + Ag(s) \rightarrow 2Ag^+(aq) \quad E^\ominus = +1.18 \text{ V}$$

The large positive potential indicates that Ag(II) and Ag(0) are completely converted to Ag(I) in aqueous solution ($K = 9 \times 10^{19}$).

7.7 Oxidation by atmospheric oxygen

The possibility of reaction between the solutes and dissolved O_2 must be considered when a solution is contained in an open beaker or is otherwise exposed to air. As an example, consider a solution containing Fe^{2+}. The standard potentials of the Fe^{2+}/Fe and Fe^{3+}/Fe^{2+} couples are -0.44 V and $+0.77$ V, which lie within the stability field of water and hence suggests that Fe^{2+} should survive in water. Moreover, we can also infer that the oxidation of metallic iron by $H^+(aq)$ should not proceed beyond Fe(II), because further oxidation to Fe(III) is unfavorable (by 0.77 V) under standard conditions. However, the picture changes considerably in the presence of O_2. In fact, Fe(III) is the most common form of iron in the Earth's crust, and most iron in sediments that have been deposited from aqueous environments is present as Fe(III).

The potential for the reaction

$$4Fe^{2+}(aq) + O_2(g) + 4H^+(aq) \rightarrow 4Fe^{3+}(aq) + 2H_2O(l)$$

$$E^\ominus = +0.46 \text{ V}$$

shows that the oxidation of Fe^{2+} by O_2 is spontaneous. However, $+0.46$ V is not a large enough overpotential for rapid reaction, and atmospheric oxidation of Fe(II) in aqueous solution is slow in the absence of catalysts. As a result, it is possible to use Fe^{2+} aqueous solutions in laboratory procedures without elaborate precautions.

Example 7.6: *Judging the importance of atmospheric oxidation*

The oxidation of copper roofs to a substance with a characteristic green color is another example of atmospheric oxidation in a damp environment. Estimate the potential for oxidation of copper by oxygen.

Answer. The reduction half-reactions are

$$O_2(g) + 4H^+(aq) + 4e^- \rightarrow 2H_2O(l) \quad E^\ominus = +1.23 \text{ V}$$

$$Cu^{2+}(aq) + 2e^- \rightarrow Cu(s) \quad E^\ominus = +0.34 \text{ V}$$

The difference is $E^\ominus = +0.89$ V, so atmospheric oxidation by the reaction

$$2Cu(s) + O_2(g) + 4H^+(aq) \rightarrow 2Cu^{2+}(aq) + 2H_2O(l)$$

is spontaneous. Nevertheless, copper roofs do last for more than a few minutes: their familiar green surface is a passive layer of an almost impenetrable hydrated copper(II) carbonate and sulfate formed from oxidation in the presence of atmospheric CO_2 and SO_2. Near the sea, the copper patina contains substantial amounts of chloride.

Exercise E7.6. The standard potential for the conversion of sulfate ions, $SO_4^{2-}(aq)$, to $SO_2(aq)$ by the reaction

$$SO_4^{2-}(aq) + 4H^+(aq) + 2e^- \rightarrow SO_2(aq) + 2H_2O(l)$$

is $+0.16$ V. What is the thermodynamically expected fate of SO_2 emitted into fog or clouds?

THE DIAGRAMMATIC PRESENTATION OF POTENTIAL DATA

There are several useful diagrammatic summaries of the relative thermodynamic stabilities of a series of species in which one element exists with a series of different oxidation numbers. We shall describe two of them: Latimer diagrams, which are useful for summarizing quantitative data element by element, and Frost diagrams, which are useful for qualitative portrayals of the relative stabilities of oxidation states.

7.8 Latimer diagrams

The simplest type of diagram was introduced by Wendell Latimer, one of the pioneers in the application of thermodynamics to inorganic solution chemistry. In a **Latimer diagram** for an element, the value of the standard potential (in volts) is written over a horizontal line connecting species with the element in different oxidation states. The most highly oxidized form of the element is on the left, and in species to the right the element is in successively lower oxidation states. The Latimer diagram for chlorine in acidic solution, for instance, is

$$ClO_4^- \xrightarrow{+1.20} ClO_3^- \xrightarrow{+1.18} ClO_2^- \xrightarrow{+1.65} HClO \xrightarrow{+1.63} Cl_2 \xrightarrow{+1.36} Cl^-$$
$$+7 +5 +3 +1 0 -1$$

As in this example, oxidation numbers are sometimes written under (or over) the species. The notation

$$ClO_4^- \xrightarrow{+1.20} ClO_3^-$$

denotes

$$ClO_4^-(aq) + 2H^+(aq) + 2e^- \rightarrow ClO_3^-(aq) + H_2O(l) \quad E^\ominus = +1.20 \text{ V}$$

Similarly,

$$HClO \xrightarrow{+1.63} Cl_2$$

denotes

$$2HClO(aq) + 2H^+(aq) + 2e^- \rightarrow Cl_2(g) + 2H_2O(l) \quad E^{\ominus} = +1.63 \text{ V}$$

In this example, the conversion of a Latimer diagram to a half-reaction involves balancing elements by including the predominant species present in acidic aqueous solution (H^+ and H_2O). Electric charge is then balanced by the inclusion of the appropriate number of e^-.

In basic aqueous solution, the Latimer diagram for chlorine is

$$ClO_4^- \xrightarrow{+0.37} ClO_3^- \xrightarrow{+0.20} ClO_2^- \xrightarrow{+0.68} ClO^- \xrightarrow{+0.42} Cl_2 \xrightarrow{+1.35} Cl^-$$

with ClO^- to Cl^- at $+0.89$.

Note that the value for the Cl_2/Cl^- couple is the same as in acidic solution, since its half-reaction does not involve the transfer of protons. Under basic conditions, the predominant species present in solution are OH^- and H_2O, and so they are used to balance the half-reactions. As an example, the half-reaction for the ClO^-/Cl_2 couple is

$$2ClO^-(aq) + 2H_2O(l) + 2e^- \rightarrow Cl_2(g) + 4OH^-(aq) \quad E^{\ominus} = +0.42 \text{ V}$$

The Latimer diagram given above includes the reduction potential for two nonadjacent species (the couple ClO^-/Cl^-). This information is redundant (see below), but it is often included for commonly used couples as a convenience.

As we can see from these examples, a Latimer diagram summarizes a great deal of information in a compact form and (as we shall explain) shows the relationships between the various species in a particularly clear manner. The potential data in Appendix 2 are presented in the form of Latimer diagrams. Such diagrams are compact and their arrangement by element makes it easy to locate the information required and place it in the context of periodic trends.

A Latimer diagram contains sufficient information to derive the standard potentials of nonadjacent couples. The connection is obtained from the relation $\Delta G^{\ominus} = -nFE^{\ominus}$, and the fact that the overall $\Delta G^{\ominus}$ for two successive steps is the sum of the individual values:

$$\Delta G^{\ominus} = \Delta G^{\ominus\prime} + \Delta G^{\ominus\prime\prime}$$

To find the standard potential of the composite process, we convert the individual $E^{\ominus}$ values to $\Delta G^{\ominus}$ values by multiplication by the relevant $-nF$ factor, add them together, and then convert the sum back to $E^{\ominus}$ for the nonadjacent couple by division by $-nF$ for the overall electron transfer:

$$-nFE^{\ominus} = -n'FE^{\ominus\prime} - n''FE^{\ominus\prime\prime}$$

Because the factors $-F$ cancel and $n = n' + n''$, the net result is

$$E^{\ominus} = \frac{n'E^{\ominus\prime} + n''E^{\ominus\prime\prime}}{n' + n''}$$

Example 7.7: *Extracting $E^{\ominus}$ for nonadjacent oxidation numbers*
Use the Latimer diagram to calculate the value of $E^{\ominus}$ for reduction of HClO to Cl$^-$ in aqueous acidic solution.

Answer. The oxidation number of chlorine changes from $+1$ to 0 in one step ($n' = 1$) and from 0 to -1 ($n'' = 1$) in the subsequent step. From the Latimer diagram given in the text above, we can write

$$\text{HClO(aq)} + \text{H}^+\text{(aq)} + \text{e}^- \rightarrow \tfrac{1}{2}\text{Cl}_2\text{(g)} + \text{H}_2\text{O(l)} \quad E^{\ominus\prime} = +1.63 \text{ V}$$

$$\tfrac{1}{2}\text{Cl}_2\text{(g)} + \text{e}^- \rightarrow \text{Cl}^-\text{(aq)} \quad E^{\ominus\prime\prime} = +1.36 \text{ V}$$

It follows that the standard potential of the HClO/Cl$^-$ couple is

$$E^{\ominus} = \frac{E^{\ominus\prime} + E^{\ominus\prime\prime}}{2} = \frac{1.63 \text{ V} + 1.36 \text{ V}}{2} = +1.50 \text{ V}$$

Exercise E7.7. Calculate $E^{\ominus}$ for the reduction of ClO$_3^-$ to HClO in aqueous acidic solution.

A Latimer diagram shows the species for which disproportionation is spontaneous: *a species has a thermodynamic tendency to disproportionate into its neighbors if the potential on the right of the species is higher than the potential on the left.* Thus, H_2O_2 has a tendency to disproportionate into O_2 and H_2O under acid conditions:

$$\text{O}_2 \xrightarrow{+0.70} \text{H}_2\text{O}_2 \xrightarrow{+1.76} \text{H}_2\text{O}$$

The basis of this general rule can be appreciated by considering the two half-reactions:

$$\text{H}_2\text{O}_2\text{(aq)} + 2\text{H}^+\text{(aq)} + 2\text{e}^- \rightarrow 2\text{H}_2\text{O(l)} \quad E^{\ominus} = +1.76 \text{ V}$$

$$\text{O}_2\text{(g)} + 2\text{H}^+\text{(aq)} + 2\text{e}^- \rightarrow \text{H}_2\text{O}_2\text{(aq)} \quad E^{\ominus} = +0.70 \text{ V}$$

and forming the difference:

$$2\text{H}_2\text{O}_2\text{(aq)} \rightarrow 2\text{H}_2\text{O(l)} + \text{O}_2\text{(g)} \quad E^{\ominus} = +1.06 \text{ V}$$

Because $E^{\ominus} > 0$, the disproportionation is spontaneous.

7.9 Frost diagrams

A **Frost diagram** of an element X is a plot of $nE^{\ominus}$ for the couple X(N)/X(0) against the oxidation number N of the element. An example is shown in Fig. 7.9. Because $nE^{\ominus}$ is proportional to the standard reaction free energy for the conversion of the species X(N) to the element, a Frost diagram can also be regarded as a plot of standard formation free energy against oxidation number. Therefore, the most stable oxidation state of an element corresponds to the species that lies lowest in its Frost diagram (**1**).

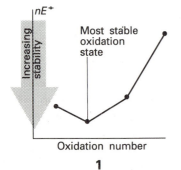

1

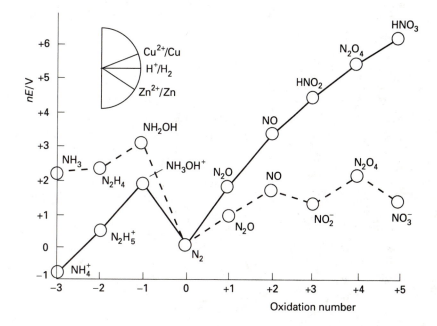

7.9 The Frost diagram for nitrogen: the steeper the slope of the line, the higher the standard reduction potential for the couple. The solid line refers to standard conditions (pH = 0), the broken line to pH = 14. The insert on the upper left emphasizes the zero slope for the H^+/H_2 couple and indicates the slope for two common metal couples.

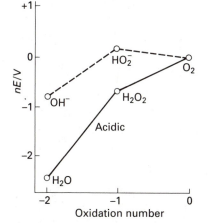

7.10 The Frost diagram for oxygen in acidic solution (full line) and basic solution (broken line).

Example 7.8: *Constructing a Frost diagram*

Construct a Frost diagram for oxygen from the Latimer diagram:

$$O_2 \xrightarrow{+0.70} H_2O_2 \xrightarrow{+1.76} H_2O$$
$$\underleftarrow{\qquad +1.23 \qquad}$$

Answer. The oxidation numbers of O are 0, -1, and -2 in the three species. For the change of oxidation number from 0 to -1 (O_2 to H_2O_2), $E^{\ominus} = +0.70$ V and $n = -1$, so $nE^{\ominus} = -0.70$ V. Because the oxidation number of O in H_2O is -2, $n = -2$, and $E^{\ominus}$ for the formation of H_2O is $+1.23$ V, $nE^{\ominus} = -2.46$ V. These results are plotted in Fig. 7.10.

Exercise E7.8. Construct a Frost diagram from the Latimer diagram for Tl:

$$Tl^{3+} \xrightarrow{+1.25} Tl^+ \xrightarrow{-0.34} Tl$$
$$\underleftarrow{\qquad +0.72 \qquad}$$

The slope of the line joining any two points in a Frost diagram is equal to the standard potential of the couple formed by the two species the points represent. To see that this is so, refer to the oxygen diagram. At the point corresponding to oxidation number -1 (for H_2O_2), $nE^{\ominus} = -0.70$ V, and at oxidation number -2 (for H_2O) it is -2.46 V, a difference of -1.76 V. The change in oxidation number of oxygen on going from H_2O_2 to H_2O is -1. Therefore, the slope of the line is

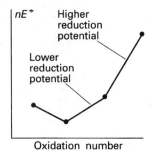

$nE^{\oplus}$　Higher reduction potential

Lower reduction potential

Oxidation number

2

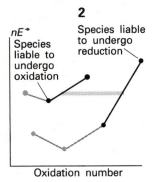

$nE^{\oplus}$

Species liable to undergo oxidation

Species liable to undergo reduction

Oxidation number

3

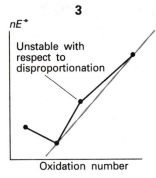

$nE^{\oplus}$

Unstable with respect to disproportionation

Oxidation number

4

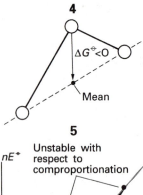

$\Delta G^{\oplus} < 0$

Mean

5

$nE^{\oplus}$　Unstable with respect to comproportionation

Oxidation number

6

$(-1.76\ \text{V})/(-1) = +1.76\ \text{V}$, in accord with the value for the H_2O_2/H_2O couple in the Latimer diagram.

The most useful aspect of Frost diagrams is the quick *qualitative* impression they provide for trends in the chemical properties of elements. We shall use them frequently in this context in the following chapters to convey the sense of trends in the redox properties of the members of a group. For numerical calculations, however, Latimer diagrams and other explicit tabulations of standard potentials are more convenient. To interpret the qualitative information contained in a Frost diagram it will be useful to keep the following features in mind.

First, *the steeper the line joining two points in a diagram, the higher the potential of the corresponding couple* (**2**). It follows that we can infer the spontaneity of the reaction between any two couples by comparing the slopes of the corresponding lines. Then two points to bear in mind (**3**) when considering a redox reaction involving two couples are:

1. The oxidizing agent in the couple with the more positive slope (the more positive $E^{\ominus}$) is liable to undergo reduction.
2. The reducing agent of the couple with the less positive slope (the less positive $E^{\ominus}$) is liable to undergo oxidation.

For example, the steep slope connecting HNO_3 to lower oxidation numbers in Fig. 7.9 shows that nitric acid is a good oxidizing agent under standard conditions. By comparing the slope of the Cu^{2+}/Cu couple in the upper left corner of the illustration with the nitrogen couples, we see that HNO_3/NO has a more positive slope than Cu^{2+}/Cu; hence, HNO_3 can oxidize copper to Cu^{2+}.

A third rule of interpretation (**4**) is as follows:

3. An ion or molecule in a Frost diagram is unstable with respect to disproportion if it lies above the line connecting two adjacent species.

The origin of this rule is illustrated in (**5**), where we show geometrically that the reaction free energy of an intermediate species lies above the average value of the two terminal species, implying that the former is unstable with respect to disproportionation into the two terminal species. A specific example is NH_2OH in Fig. 7.9, which is unstable with respect to disproportionation into NH_3 and N_2. However, it is one of the intricacies of nitrogen chemistry that slow formation of N_2 often prevents its production.

A fourth rule (**6**) relates to the tendency for comproportionation to occur:

4. Two species will tend to comproportionate into an intermediate species that lies below the straight line joining the terminal species.

A substance that lies below the line connecting its neighbors in a Frost diagram is more stable than they are because their average free energy

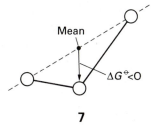

Mean

$\Delta G^{\ominus} < 0$

7

is higher (**7**). Accordingly, comproportionation is thermodynamically favorable. The nitrogen in NH_4NO_3, for instance, has two ions with oxidation numbers -3 (NH_4^+) and $+5$ (NO_3^-). Because N_2O, in which the oxidation number of N has the average of these two values, $+1$, lies below the line joining NH_4^+ to NO_3^-, their comproportionation is spontaneous:

$$NH_4^+(aq) + NO_3^-(aq) \rightarrow N_2O(g) + 2H_2O(l)$$

The reaction is kinetically inhibited in solution and does not ordinarily occur. However, it does occur in the solid state, when it can be explosively fast: indeed, ammonium nitrate is often used in place of dynamite for blasting rocks.

When the points for three substances lie approximately on a straight line, no one species will be the exclusive product. The three reactions

$$NO(g) + NO_2(g) + H_2O(l) \rightarrow 2HNO_2(aq) \text{ (rapid)}$$

$$3HNO_2(aq) \rightarrow HNO_3(aq) + 2NO(g) + H_2O(l) \text{ (rapid)}$$

$$2NO_2(g) + H_2O(l) \rightarrow HNO_3(aq) + HNO_2(aq)$$
$$\text{(slow and normally rate determining)}$$

are examples of this behavior. They are all important in the industrial synthesis of nitric acid by the oxidation of ammonia.

Example 7.9: *Using a Frost diagram to judge the stability of ions*
Figure 7.11 shows a Frost diagram for manganese. Comment on the stability of Mn^{3+} in aqueous solution.

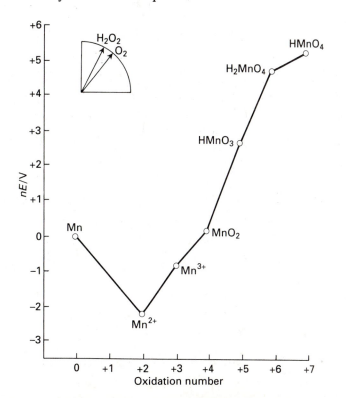

7.11 The Frost diagram for manganese species in acidic solution.

Answer. Because Mn^{3+} lies above the line joining Mn^{2+} to MnO_2, it should disproportionate. It is a one-electron transfer, and is therefore likely to be reasonably rapid. Indeed, Mn^{3+} is unstable in aqueous solution.

Exercise E7.9. What is the oxidation number of Mn in the product when MnO_4^- is used as an oxidizing agent in aqueous acid?

7.10 pH dependence

So far, we have discussed Latimer and Frost diagrams only for aqueous solution at pH = 0, but they can equally well be constructed for other conditions. As is done in Appendix 2, Latimer diagrams are commonly presented for both pH = 0 and pH = 14.[4]

Conditional Latimer diagrams

'Basic Latimer diagrams' are expressed in terms of reduction potentials at pH = 14 (pOH = 0). The potentials are denoted $E_B^\ominus$ and the dotted line in Fig. 7.9 is a representation of their values. The important difference from the behavior in acidic solution is the stabilization of NO_2^- against disproportionation: its point no longer lies above the line connecting its neighbors. The practical outcome is that metal nitrites can be isolated whereas HNO_2 cannot.

In some cases, there are marked differences between strongly acidic and basic solutions, as for the phosphorus oxoanions:

$$E^\ominus$$
$$H_3PO_4 \xrightarrow{-0.28} H_3PO_3 \xrightarrow{-0.50} H_3PO_2 \xrightarrow{-0.51} P \xrightarrow{-0.06} PH_3$$

$$E_B^\ominus$$
$$PO_4^{3-} \xrightarrow{-1.12} HPO_3^{2-} \xrightarrow{-1.56} HPO_2^- \xrightarrow{-2.05} P \xrightarrow{-0.89} PH_3$$

This example illustrates an important general point about oxoanions: when their reduction requires removal of oxygen, the reaction consumes H^+ ions, and *all oxoanions are stronger oxidizing agents in acidic than in basic solution.*

Reduction potentials in neutral solution (pH = 7) are denoted $E_W^\ominus$. These potentials are particularly useful in biochemical discussions because cell fluids are buffered near pH = 7.

Example 7.10: *Using conditional diagrams*

Potassium nitrite is stable in basic solution, but when the solution is acidified, a gas is evolved that turns brown on exposure to air. What is the reaction?

Answer. Figure 7.9 shows that the NO_2^- ion in basic solution lies below the line joining NO to NO_3^-; the ion therefore is not liable to dispropor-

---

[4] Limited potential data are available for nonaqueous solvents. For example, data are available for liquid ammonia solutions. See W. L. Jolly, *J. Chem. Educ.*, **33**, 512 (1956).

tionation. On acidification, the HNO_2 point rises and the straightness of the line through NO, HNO_2, and N_2O_4 implies that all three species are present at equilibrium. The brown gas is NO_2 formed from the reaction of NO evolved from the solution with air. In solution, the species of oxidation number $+2$ (NO) tends to disproportionate. However, the escape of NO from the solution prevents its disproportionation to N_2O and HNO_2.

Exercise E7.10. By reference to Fig. 7.9, compare the strength of NO_3^- as an oxidizing agent in acidic and basic solution.

Pourbaix diagrams

Diagrams can also be used for discussing the general relations between redox activity and Brønsted acidity. In a **Pourbaix diagram**, the regions indicate the conditions of pH and potential under which each species is thermodynamically stable. Figure 7.12 is a simplified Pourbaix diagram for iron, omitting such minor species as oxygen-bridged dimers. This diagram is useful for the discussion of iron species in natural water because the total iron concentration is low and dimers tend to dissociate.

We can see how the diagrams are constructed by considering some of the reactions involved. The reduction half-reaction

$$Fe^{3+}(aq) + e^- \rightarrow Fe^{2+}(aq) \quad E^\ominus = +0.77 \text{ V}$$

does not involve H^+ ions, and so (if we ignore the effect of pH on the activity of the iron ions) its potential is independent of pH, giving a horizontal line on the diagram. If the environment contains a couple with a reduction potential above this line (a more positive, more oxidizing potential), then the oxidized species, Fe^{3+}, will be the major species. Hence, the horizontal line toward the top left of the diagram is a boundary that separates the regions where Fe^{3+} and Fe^{2+} dominate.

Another reaction to consider is

$$Fe^{3+}(aq) + 3H_2O(l) \rightarrow Fe(OH)_3(s) + 3H^+(aq)$$

This reaction is not a redox reaction (there is no change in oxidation number of any element), and the boundary between the region where Fe^{3+} is soluble and the region where $Fe(OH)_3$ precipitates and limits $Fe^{3+}(aq)$ to very low concentration is independent of any redox couples. However, this boundary does depend on pH, with $Fe^{3+}(aq)$ favored by low pH and the $Fe(OH)_3$ favored by high pH. We shall accept that Fe^{3+} is the dominant species in the solution if its concentration exceeds 10^{-5} mol L^{-1} (a typical freshwater value). The equilibrium concentration of Fe^{3+} varies with pH, and the vertical boundary represents the pH at which the Fe^{3+} becomes dominant according to this definition.

A third reaction we need to consider is

$$Fe(OH)_3(s) + 3H^+(aq) + e^- \rightarrow Fe^{2+}(aq) + 3H_2O(l)$$

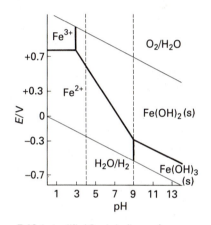

7.12 A simplified Pourbaix diagram for some important naturally occurring compounds of iron. The broken vertical lines represent the normal pH range in natural waters.

This reaction depends on the pH, as we can see by noting that H^+ ions appear in the half-reaction, and as is shown explicitly by writing the Nernst equation:

$$E = E^\ominus - 0.059 \text{ V} \log \frac{[Fe^{2+}]}{[H^+]^3}$$
$$= E^\ominus - (0.059 \text{ V}) \log [Fe^{2+}] - (0.177 \text{ V}) \times pH$$

This expression shows that the potential falls linearly as the pH increases, as is shown in Fig. 7.12. (The line has been drawn for an Fe^{2+} concentration of 10^{-5} mol L^{-1}.) The region of potential and pH that lie above this line corresponds to conditions in which the oxidized species $(Fe(OH)_3)$ is stable; those below it correspond to conditions in which the reduced species (Fe^{2+}) is stable. We see that a stronger oxidizing agent is required to oxidize and precipitate Fe^{2+} ions in acidic media than in basic media.

It is a general rule for coupled redox-acid reactions that presence of H^+ on the left of the reduction couple implies a slope down from left to right of a boundary in a Pourbaix diagram. Conversely, H^+ on the right implies an upward slope of a boundary.

The vertical line at $pH = 9$ divides the regions in which either the reactants or the products are stable in the reaction

$$Fe^{2+}(aq) + 2H_2O(l) \rightarrow Fe(OH)_2(s) + 2H^+(aq)$$

This reaction is not a redox reaction, and the vertical line is drawn at a pH that corresponds to an Fe^{2+} equilibrium concentration of 10^{-5} mol L^{-1}.

Another line separates the regions in which $Fe(OH)_2$ and $Fe(OH)_3$ are stable:

$$Fe(OH)_3(s) + H^+(aq) + e^- \rightarrow Fe(OH)_2(s) + H_2O(l)$$

The potential of this reduction depends on the pH, but when we write the Nernst equation

$$E = E^\ominus - \frac{RT}{F} \ln \frac{1}{[H^+]} = E^\ominus - (0.136 \text{ V}) \times pH$$

we see that it has a less steep slope than for the $Fe(OH)_3/Fe^{2+}$ couple (because the number of H^+ ions involved is smaller than in the latter).

Finally, we have drawn on the diagram the two sloping lines that act as boundaries for the stability field of water (recall Fig. 7.8). As we saw earlier, any couples present with potentials more positive than the upper line will oxidize water to O_2, and any with potentials more negative than the lower line will reduce it to H_2. All the couples that we need consider when discussing the iron redox reactions in water therefore lie within the stability field.

Natural waters

The chemistry of natural waters can be rationalized by using Pourbaix diagrams of the kind we have just constructed. Thus, where fresh water is in contact with the atmosphere, it is saturated with O_2, and many species may be oxidized by this powerful oxidizing agent ($E^{\ominus} = +1.23$ V). More fully reduced forms are found in the absence of oxygen, especially where there is organic matter to act as a reducing agent. The major acid system that controls the pH of the medium is the $CO_2/H_2CO_3/HCO_3^-/CO_3^{2-}$ diprotic system, where atmospheric CO_2 provides the acid and dissolved carbonate minerals provide the base. Biological activity is also important, because respiration consumes oxygen and releases CO_2. This acidic oxide reduces the pH and hence makes the reduction potential more negative. The reverse process, photosynthesis, consumes CO_2 and releases O_2. This consumption of acid raises the pH and makes the reduction potential less negative. The condition of typical natural waters—their pH and the potentials of the redox couples they contain—is summarized in Fig. 7.13.

From Fig. 7.12 we see that Fe^{3+} can exist in water if the environment is oxidizing; hence, where O_2 is plentiful and the pH is low (below 4), iron will be present as Fe^{3+}. Because few natural waters are so acidic, Fe^{3+} is very unlikely to be found in them. The iron in insoluble Fe_2O_3 can enter solution as Fe^{2+} if it is reduced, which occurs when the condition of the water lies below the sloping boundary in the diagram. We should observe that as the pH rises, Fe^{2+} can form only if there are strong reducing couples present, and its formation is very unlikely in oxygen-rich water. If we compare Figs 7.12 and 7.13, we see that iron will be reduced and dissolved in the form of Fe^{2+} in bog waters and organic-rich waterlogged soils (pH near 4.5 and E near $+0.03$ V and -0.1 V, respectively).

It is instructive to analyze a Pourbaix diagram in conjunction with an understanding of the physical processes that occur in water. As an example, consider a lake where the temperature gradient tends to prevent vertical mixing. At the surface, the water is fully oxygenated and the iron must be present in particles of the insoluble $Fe(OH)_3$; these particles will tend to settle. At greater depth, the O_2 content is low. If the organic content or other sources of reducing agents are sufficient, the oxide will be reduced and iron will dissolve as Fe^{2+}. The Fe(II) ions will then diffuse toward the surface where they will encounter O_2 and be oxidized to insoluble $Fe(OH)_3$ again.

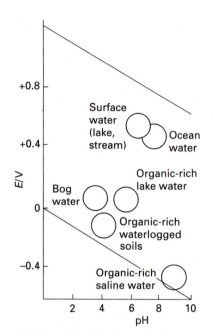

7.13 The stability field of water showing regions typical of various natural waters.

Example 7.11: *Using a Pourbaix diagram*
Figure 7.14 is part of a Pourbaix diagram for manganese. Identify the environment in which the solid MnO_2 or its corresponding hydrous oxides are important.

Answer. Manganese dioxide is thermodynamically favored only if $E > +0.6$ V. Under mildly reducing conditions, the stable species is $Mn^{2+}(aq)$. Thus, MnO_2 is important in well aerated waters (near the air–water boundary) where E approaches the value for the O_2/H_2O couple.

Exercise E7.11. Use Figs 7.12 and 7.13 to evaluate the possibility of finding $Fe(OH)_3(s)$ in a waterlogged soil.

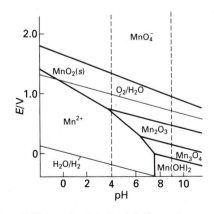

7.14 A section of the Pourbaix diagram for manganese. The broken vertical lines represent the normal pH range in natural waters.

7.11 The effect of complex formation on potentials

The formation of metal complexes affects reduction potentials because the ability of the complex ML_6 to accept or release an electron is different from that of the corresponding aqua ion with $L=H_2O$. An example is

$$[Fe(OH_2)_6]^{3+}(aq)+e^- \rightarrow [Fe(OH_2)_6]^{2+}(aq) \quad E^\ominus = +0.77 \text{ V}$$

compared with

$$[Fe(CN)_6]^{3-}(aq)+e^- \rightarrow [Fe(CN)_6]^{4-}(aq) \quad E^\ominus = +0.36 \text{ V}$$

We see that the hexacyanoferrate(III) complex is more resistant to reduction than the aqua complex. To analyze this difference, we treat the reduction of the cyano complex as the sum of the following three reactions:

1. $[Fe(CN)_6]^{3-}(aq)+6H_2O(l) \rightarrow [Fe(OH_2)_6]^{3+}(aq)+6CN^-(aq)$

2. $[Fe(OH_2)_6]^{3+}(aq)+e^- \rightarrow [Fe(OH_2)_6]^{2+}(aq)$

3. $[Fe(OH_2)_6]^{2+}(aq)+6CN^-(aq) \rightarrow [Fe(CN)_6]^{4-}(aq)+6H_2O(l)$

The replacement of aqua ligands by cyano ligands stabilizes the hexacyanoferrate(III) complex more than it stabilizes the hexacyanoferrate(II) complex, so the net contribution of reactions 1 and 2 is to favor the reactants in the overall reaction

$$[Fe(CN)_6]^{3-}(aq)+e^- \rightarrow [Fe(CN)_6]^{4-}(aq)$$

As a result, the hexacyanoferrate(III) complex is less susceptible to reduction than the hexaaquairon(III) complex.

The central lesson of this example is that the formation of a more stable complex when the metal has the higher oxidation number favors oxidation and makes the reduction potential more negative. The formation of a more stable complex when the metal has the lower oxidation number favors reduction and the reduction potential becomes more positive.

Example 7.12: *Assessing the effect of complexation on potential*
In general, CN^- forms complexes that are thermodynamically more stable than those formed by Br^-. Which complex, $[Ni(CN)_4]^{2-}$ or $[NiBr_4]^{2-}$, is expected to have the more positive potential for reduction to $Ni(s)$?

Answer. The reduction half-reaction is

$$[NiX_4]^{2-}(aq)+2e^- \rightarrow Ni(s)+4X^-(aq)$$

The more stable the complex, the higher its resistance to reductive decomposition. Because $[Ni(CN)_4]^{2-}$ is more stable, it is more difficult to reduce than $[NiBr_4]^{2-}$, and therefore it will have the more negative reduction potential.

Exercise E7.12. Which couple has the higher reduction potential, $[Ni(en)_3]^{2+}/Ni$ or $[Ni(NH_3)_6]^{2+}/Ni$?

FURTHER READING

T. W. Swaddle, *Applied inorganic chemistry*. University of Calgary Press (1990). This book provides a good introductory account of practical aspects of extractive metallurgy. It also treats redox reactions in solution, including kinetics, electrolysis, and corrosion.

A. J. Bard, R. Parsons, and R. Jordan, *Standard potentials in aqueous solution*. M. Dekker, New York (1985). The most recent critical collection of potential data. Comments on rates are also included.

S. G. Bratsch. Standard electrode potentials and temperature coefficients in water at 298.15 K, *J. Phys. Chem. Ref. Data*, **18**, 1 (1989).

D. M. Stanbury. Potentials for radical species, *Adv. Inorg. Chem.*, **33**, 69 (1989).

I. Baran, *Thermochemical data of pure substances*, Vols 1 and 2. VCH, Weinheim (1989). A comprehensive and reliable source of thermodynamic data as a function of temperature for inorganic substances.

P. H. Rieger, *Electrochemistry*. Prentice-Hall, Englewood Cliffs (1987). A good introduction to electrochemical methods and their interpretation.

M. Pourbaix, *Atlas of electrochemical equilibria in aqueous solution*. Pergamon Press, Oxford (1966). The source of Pourbaix diagrams for many elements.

R. M. Garrels and C. L. Christ, *Solutions, minerals, and equilibria*. Harper and Row, New York (1965). A standard text on aqueous geochemistry, which makes use of Pourbaix diagrams throughout and analyzes the implications of chemical equilibria in geology.

W. Stumm and J. J. Morgan, *Aquatic chemistry*. Wiley, New York (1981). The standard reference on natural water chemistry.

D. C. Harris, *Introduction to analytical chemistry*. W. H. Freeman, New York (1987). Chapters 14 and 16 are good introductions to electrochemical cells and electrochemical measurements.

The following are two recent textbooks on inorganic reaction mechanisms:

R. B. Jordan, *Reaction mechanisms of inorganic and organometallic systems*. Oxford University Press (1991).

R. G. Wilkins, *Kinetics and mechanism of reactions of transition metal complexes*. VCH, New York (1991).

KEY POINTS

1. Oxidation and reduction

Oxidation is the loss of electrons; reduction is the gain of electrons. These definition are generalizations of the original definitions based on reaction with oxygen to produce an oxide.

2. Thermodynamic aspects of metal production

Many elements exist in nature as oxides or as sulfides that are easily converted to oxides. An Ellingham diagram summarizes the temperature variation of the thermodynamic aspects of the reduction of metal oxides to the element.

3. Varieties of industrial element production
Industrial extraction of elements from ores uses low temperature carbon reduction for readily reduced species (zinc), higher temperature carbon reduction for less readily reduced species (iron), and high temperature (electric furnace) carbon reduction for some (magnesium) or electrolysis (aluminum) for least readily reduced species. Several elements (e.g. halogens) exist in a reduced state and are extracted by oxidation, which is commonly electrolytic.

4. Reduction half-reactions and standard potentials
The thermodynamics of redox reactions in solution are conveniently tabulated in terms of reduction half-reactions and their standard potentials $E^{\ominus}$. The overall standard potential of an electrochemical cell is proportional to $\Delta G^{\ominus}$ for the cell reaction. The Nernst equation summarizes the dependence of potentials on concentrations.

5. Redox stability in solution
The stability of a species in solution refers to its thermodynamic tendency toward oxidation or reduction by the medium or to disproportionation.

6. Overpotential
The kinetics of redox reactions often determine the observed processes. An overpotential is often required to achieve a significant reaction rate.

7. Reaction mechanisms
The mechanisms of redox reactions include outer-sphere electron transfer, inner-sphere electron-transfer, and atom or group transfer.

8. Latimer, Frost, and Pourbaix diagrams
Latimer diagrams provide compact portrayal of $E^{\ominus}$ values for an element. Frost diagrams provide a striking representation of trends in the stabilities of oxidation states of elements. Pourbaix diagrams provide a summary of the influence of pH and potentials on the identity of the predominant species.

9. The effect of pH and complexation
Where proton transfer or complexation occurs together with redox processes, potentials depend on pH and complex formation.

10. Natural waters
The behavior of natural water systems is conveniently organized in terms of dependence on potential and pH (and expressed in terms of Pourbaix diagrams).

EXERCISES
7.1 Consult the Ellingham diagram in Fig. 7.3 and determine if there are any conditions under which aluminum might be expected to reduce MgO. Comment on these conditions.

7.2 Standard potential data suggest that gaseous oxygen rather than gaseous chlorine should be the anode product in electrolysis of aqueous NaCl. In fact, only traces of oxygen are produced. Suggest an explanation.

7.3 Using data from Appendix 2, suggest chemical reagents which would be suitable for carrying out the following transformations and write balanced equations: (a) oxidation of HCl to chlorine gas, (b) reducing Cr(III)(aq) to Cr(II)(aq), (c) reducing Ag^+(aq) to Ag(s), (d) reducing I_2 to I^-.

7.4 Use standard potential data from Appendix 2 as a guide to write balanced equations for the reactions that might be expected for each of the following species in aerated aqueous acid. If the species is stable, write 'no reaction'. (a) Cr^{2+}, (b) Fe^{2+}, (c) Cl^-, (d) HClO, (e) Zn(s).

7.5 Use the information in Appendix 2 to write balanced equations for the reactions, including disproportionations, that can be expected for each of the following species in aerated acidic aqueous solution: (a) Fe^{2+}, (b) Ru^{2+}, (c) $HClO_2$, (d) Br_2(l).

7.6 Write the Nernst equation for (a) the reduction of O_2:

$$O_2(g) + 4H^+(aq) + 4e^- \rightarrow 2H_2O(l)$$

(b) the reduction of Fe_2O_3(s):

$$Fe_2O_3(s) + 6H^+(aq) + 6e^- \rightarrow 2Fe(s) + 3H_2O(l)$$

and in each case express the formula in terms of pH. What is the potential for the O_2 reduction at pH = 7 and $p(O_2)$ = 0.20 bar (the partial pressure of oxygen in air)?

7.7 Given the following potentials in basic solution

$$CrO_4^{2-}(aq) + 4H_2O(l) + 3e^- \rightarrow$$
$$Cr(OH)_3(s) + 5OH^-(aq) \quad E^\ominus = -0.11 \text{ V}$$

$$[Cu(NH_3)_2]^+(aq) + e^- \rightarrow Cu(s) + 2NH_3(aq)$$
$$E^\ominus = -0.10 \text{ V}$$

and assuming that a reversible reaction can be established on a suitable catalyst, calculate $E^\ominus$, $\Delta G^\ominus$, and K for the reduction of CrO_4^{2-} and $[Cu(NH_3)_2]^+$ in basic solution. Comment on why $\Delta G^\ominus$ and K are so different between the two cases despite the values of $E^\ominus$ being so similar.

7.8 Answer the following questions using the Frost diagram in Fig. 7.15. (a) What are the consequences of dissolving Cl_2 in aqueous basic solution? (b) What are the consequences of dissolving Cl_2 in aqueous acid? (c) Is the failure of $HClO_3$ to disproportionate in aqueous solution a thermodynamic or a kinetic phenomenon?

7.9 Use standard potentials as a guide to write equations for the main net reaction which you would predict in the

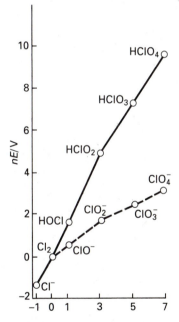

7.15 A Frost diagram for chlorine. The solid line refers to standard conditions (pH $= 0$) and the broken line to pH $= 14$.

following experiments: (a) N_2O is bubbled into aqueous NaOH solution, (b) zinc metal is added to aqueous sodium triiodide, (c) I_2 is added to excess aqueous $HClO_3$.

7.10 Characterize the condition of acidity or basicity which would most favor the following transformations in aqueous solution: (a) $Mn^{2+} \rightarrow MnO_4^-$, (b) $ClO_4^- \rightarrow ClO_3^-$, (c) $H_2O_2 \rightarrow O_2$, (d) $I_2 \rightarrow 2I^-$.

7.11 Comment on the likelihood that the following reactions occur by a simple outer-sphere electron transfer, simple atom transfer, or a multistep mechanism: (a) HIO $(aq) + I^-(aq) \rightarrow I_2(aq) + OH^-(aq)$, (b) $[Co(phen)_3]^{3+}(aq) + [Cr(bipy)_3]^{2+}(aq) \rightarrow [Co(phen)_3]^{2+}(aq) + [Cr(bipy)_3]^{3+}(aq)$, (c) $IO_3^-(aq) + 8I^-(aq) + 6H^+(aq) \rightarrow 3I_3^-(aq) + 3H_2O(l)$.

7.12 Use the Latimer diagram for chlorine to determine the potential for reduction of ClO_4^- to Cl_2. Write a balanced equation for this half-reaction.

7.13 Calculate the equilibrium constant of the reaction $Au^+(aq) + 2CN^-(aq) \rightarrow [Au(CN)_2]^-(aq)$ from the standard potentials

$$Au^+(aq) + e^- \rightarrow Au(s) \quad E^\ominus = +1.68 \text{ V}$$

$$[Au(CN)_2]^-(aq) + e^- \rightarrow Au(s) + 2CN^-(aq)$$
$$E^\ominus = -0.6 \text{ V}$$

7.14 Use Fig. 7.13 to find the approximate potential of an aerated lake at pH $= 6$. With this information and Latimer diagrams from Appendix 2, predict the species at equilibrium for the elements (a) iron, (b) manganese, (c) sulfur.

7.15 The species Fe^{2+} and H_2S are important at the bottom of a lake where O_2 is scarce. If the pH $= 6$, what is the maximum value of E characterizing the environment?

7.16 The ligand EDTA forms stable complexes with hard acid centers. How will complexation with EDTA affect the reduction of M^{2+} to the metal in the first d series?

7.17 In Fig. 7.12, which of the boundaries depend on the choice of Fe^{2+} concentration as 10^{-5} mol L^{-1}?

7.18 How could a cyclic voltammogram be used to determine whether a metal ion in aqueous solution is present in a complex with ligands other than water?

7.19 Explain why water with high concentrations of dissolved carbon dioxide and open to atmospheric oxygen is very corrosive toward iron.

PROBLEMS

7.1 Using standard potential data, suggest why permanganate is not a suitable oxidizing agent for the quantitative estimation of Fe^{2+} in the presence of HCl but becomes so

if sufficient Mn^{2+} and phosphate ion are added to the solution. (Hint: Phosphate complexes Fe^{3+}, thereby stabilizing it.)

7.2 R. A. Binstead and T. J. Meyer have described (*J. Am. Chem. Soc.* **109**, 3287 (1987)) the reduction of $[Ru^{IV}O(bipy)_2(py)]^{2+}$ to $[Ru^{III}(OH)(bipy)_2(py)]^{2+}$. The study revealed that the rate of the reaction is strongly influenced by the change of the solvent from H_2O to D_2O. What does this suggest about the mechanism of the reaction? Comment on the relation of this result to the cases of simple electron transfer and simple atom transfer described in the chapter.

7.3 It is often found that O_2 is a slow oxidizing agent. Suggest a mechanistic explanation that takes into account the two standard potentials.

$$O_2(g) + 4H^+(aq) + 4e^- \rightarrow 2H_2O(l) \quad E^\ominus = +1.23 \text{ V}$$

$$O_2(g) + 2H^+(aq) + 2e^- \rightarrow H_2O_2(aq) \quad E^\ominus = +0.70 \text{ V}$$

7.4 A number of anaerobic bacteria utilize oxidizing agents other than O_2 as an energy source. Some important examples include SO_4^{2-}, NO_3^-, and Fe^{3+}. One half-reaction is

$$FeO(OH)(s) + HCO_3^-(aq) + 2H^+(aq) + e^- \rightarrow$$

$$FeCO_3(s) + 2H_2O(l)$$

for which $E^\ominus = +1.67$ V. What mass of iron gives the same free energy yield as 1.00 g of oxygen?

7.5 Many of the tabulated data for standard potentials have been determined from thermochemical data rather than direct electrochemical measurements of cell potentials. Carry out a calculation to illustrate this approach for the half reaction

$$Sc_2O_3(s) + 3H_2O(l) + 6e^- \rightarrow 2Sc(s) + 6OH^-(aq)$$

	Sc^{3+}(aq)	OH^-(aq)	H_2O(l)
$\Delta H^\ominus$/kJ mol^{-1}	-614.2	-230.0	-285.8
$S^\ominus$/J K^{-1} mol^{-1}	-255.2	-10.75	$+69.91$

	Sc_2O_3(s)	Sc(s)
$\Delta H^\ominus$/kJ mol^{-1}	-1908.7	0
$S^\ominus$/J K^{-1} mol^{-1}	$+77.0$	$+34.76$

7.6 The reduction potential of an ion such as OH^- can be strongly influenced by the solvent. (a) From the review article by D. T. Sawyer and J. L Roberts, *Acc. Chem. Res.*, **21**, 469 (1988), describe the magnitude of the change in the $OH\cdot/OH^-$ potential on changing the solvent from water to acetonitrile, CH_3CN. (b) Suggest a qualitative interpretation of the difference in solvation of the OH^- ion in these two solvents.

7.7 Reactions that involve atom transfer may be expressed either in terms of the redox language of changes in oxidation number or from the standpoint of nucleophilic substitution. Describe, from these two viewpoints, the mechanism of the reaction of NO_2^- with ClO^- in water to produce NO_3^- and Cl^-. See D. W. Johnson and D. W. Margerum, *Inorg. Chem.*, **30**, 4845 (1991).

PART TWO

Systematic chemistry of the elements

We focus in this part on periodic trends in the properties of the elements. Some physical and chemical properties vary in a simple manner down a group or across a period, but often the trend is less straightforward. Despite this complexity, the periodic table remains a useful framework for remembering the individuality of an element or a group of elements. In addition, we shall see that many trends can be rationalized in terms of models of structure, bonding, and reactivity that have been developed in previous chapters. It is important to appreciate, however, that explanations widely presented in inorganic chemistry range from sound deductions to speculations. Some of these explanations will be overthrown by new experimental results or more accurate theories. Facts relating to reactions and structures will, for the most part, survive.

The opening chapter of Part 2 gives a survey of the systematics of the metallic elements. These elements are the most numerous, and their chemical properties are encountered in many contexts, including the chapters on complexes, organometallic compounds, catalysis, and bioinorganic chemistry. The remaining chapters in Part 2 deal with the p-block elements, particularly the nonmetals. Although fewer in number, the nonmetallic elements span a wide range of chemical properties and play a major role in all aspects of chemistry, the geosphere, and the biosphere.

8

The metals

The metallic elements are the most numerous of the elements, and their chemical properties are central to both industry and contemporary research. We shall see many systematic trends in properties of the metals within each of the blocks of the periodic table, and some striking differences between these blocks. In this chapter we provide a survey of the periodic trends in their chemical properties. One of the most important is the trend in stabilities of the oxidation states within each block of the periodic table. These stabilities are intimately related to the ease of recovery of metals from their ores and to the way the various metals and their compounds are handled in the laboratory. Other important properties are ligand preference and the arrangement of ligands around metal ions in complexes. It will be seen that simple binary compounds, such as metal halides and oxides, often follow systematic trends. However, when the metal is in a low oxidation state, interesting variations such as metal–metal bonding, may be encountered.

All the elements in the *s*, *d*, and *f* blocks of the periodic table are metals, and seven of the thirty *p*-block elements (aluminum, gallium, indium, thallium, tin, lead, and bismuth) are usually considered to be metals (Fig. 8.1). We say usually because the diagonal borderline between metals and nonmetals in the *p* block is ill-defined, and germanium (Ge) and polonium (Po) are also sometimes described as metals.

GENERAL PROPERTIES

Most metals have high electric and thermal conductivities and are malleable and ductile. However, there is a wide range of properties within this general uniformity. One aspect of this diversity is the

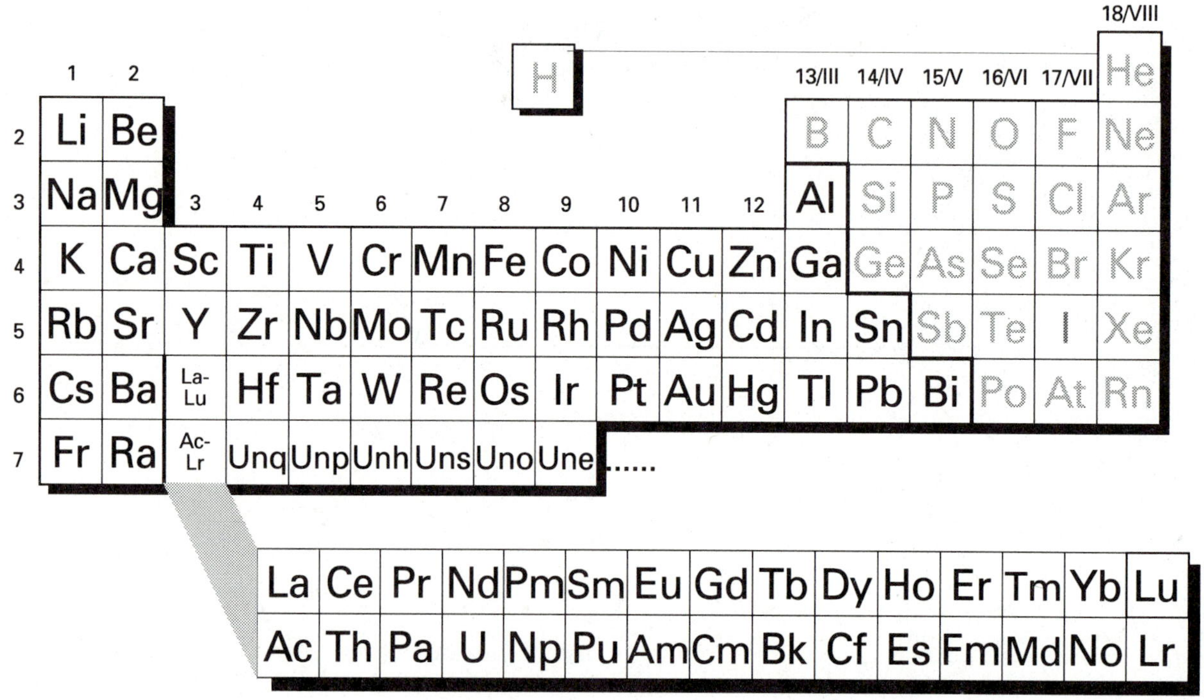

8.1 The distribution of metallic elements (shown bold) in the periodic table.

strength of the cohesion between atoms as indicated by the range of enthalpies of sublimation (Fig. 8.2). A practical application of the low enthalpies of sublimation in Groups 1 and 12 (the alkali metals, and zinc, cadmium, and mercury) is the use of sodium and mercury vapor in electric discharge lamps, such as fluorescent lamps and street lamps. The metals with the highest enthalpies of sublimation are found in the middle of the 4*d* and 5*d* series. Tungsten, with the highest enthalpy of sublimation of any element, is used as the filament material in incandescent lights because it volatilizes only very slowly at high temperatures.

Two characteristic chemical properties of many metals are the formation of basic oxides and hydroxides when the metal is in the +1 or +2 oxidation state and the formation of simple (hydrated) cations in acidic aqueous solution. Most metals react with oxygen. However, the rate and thermodynamic spontaneity of this reaction vary greatly from cesium, which ignites on contact with air, to metals such as aluminum and iron that survive in air. The latter metals are commercially useful because they react only slowly with the atmosphere at ordinary temperatures on account of a thin protective layer of oxide that forms on their surface. The only metals that have no tendency to form oxides under standard conditions are the small group of noble metals in the lower right of the *d* block (Fig. 8.3, with the exception of osmium which smells of OsO_4 even at room temperature); the most familiar examples are gold and platinum. The elements in the middle of the *d* block have highly varied chemical properties on account of their wide range of accessible oxidation states and ability to form a multitude of complexes.

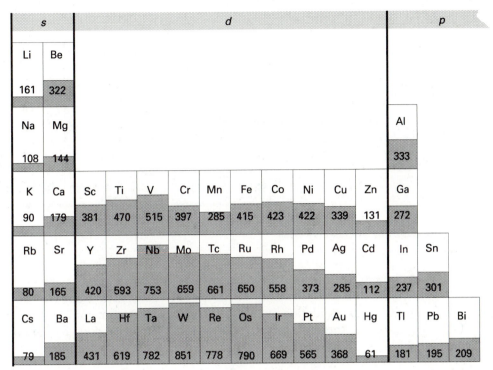

8.2 The enthalpies of vaporization (in kJ mol^{-1}) for the metallic elements in the *s*, *d*, and *p* blocks.

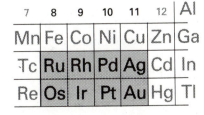

8.3 The location of the noble metals in the periodic table.

The traditional view of the chemical properties of the metallic elements is from the perspective of ionic solids and complexes containing a single central metal ion. However, with the development of improved techniques for the determination of structure it has been recognized that there are also many *d*-metal compounds that have metal–metal (M—M) bond distances comparable to or shorter than those in the pure metals. These findings have stimulated additional X-ray structural investigations of compounds that might contain M—M bonds, and have inspired research into syntheses designed to broaden the range of such cluster compounds. Examples of metal (and nonmetal) cluster compounds are now known in every block of the periodic table (Fig. 8.4), but they are most numerous in the *d* block.

The cluster compounds in Fig. 8.4 are classified according to ligand type. Although this classification is not accurate in detail for all cluster compounds, it provides information on the predominant ligand types and some insight into the bonding. For example, alkyllithium compounds (see Chapter 10) often have alkyl groups attached to a metal cluster by two-electron multicenter bonds. Clusters of the early *d*-block elements and the lanthanides generally contain π-donor ligands such as Br$^-$. These ligands can fill some of the low-lying metal orbitals of these electron-poor metal atoms by π-electron donation. In contrast, electron-rich metal clusters on the right of the *d* block generally contain π-acceptor ligands such as CO which remove some of the electron density from the metal. Many *p*-block elements do not require ligands to complete the valence shells of the individual atoms in clusters and can exist as **naked clusters**. The ionic clusters Pb_5^{2-} and Sn_9^{4-} are two examples of naked clusters formed by metals. (Section 4.4.)

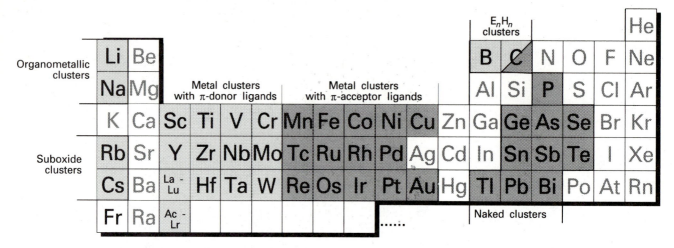

8.4 The major classes of cluster forming elements. Note that carbon is in two classes (for example, it occurs as C_4H_4 and C_{60}). Adapted from D. M. P. Mingos and D. J. Wales, *Introduction to cluster chemistry*. Prentice Hall, Englewood Cliffs (1990).

Although the chemical properties of the metallic elements are very rich, each block of elements displays fairly consistent trends in the stability of the oxidation states, the tendency to form complexes, and the nature of the halides and chalcogenides of the various oxidation states.

THE *s*-BLOCK METALS

Alkali metal (Group 1) and alkaline earth metal (Group 2) cations are commonly found in minerals and natural waters and some are important constituents of biological fluids such as blood. The cheaper metals (lithium, sodium, potassium, and calcium) are often used as powerful reducing agents for chemical reactions in nonaqueous solvents. The characteristic oxidation numbers of the *s*-block elements are the same as their group numbers: $+1$ for the alkali metals and $+2$ for the alkaline earth metals. When water and air are excluded, some unusual compounds with the metals in lower oxidation states can be prepared (including sodides, containing Na^-). The weakness with which *s*-block metal ions bind ligands meant that until recently very few *s*-metal complexes had been characterized. However, the development of polydentate ligands that can entrap cations has led to a considerable resurgence of research into *s*-block complexes.

8.1 Occurrence and isolation

The abundance of Group 1 and 2 metals in the Earth's crust spans a broad range from calcium (the fifth most abundant metal), sodium (sixth), magnesium (seventh), and potassium (eighth), to the relatively rare metals cesium and beryllium (Fig. 8.5). We saw in Section 1.1 that the low abundances of lithium and beryllium arise from the details of nucleosynthesis. The low abundances of the heavy alkali and alkaline earth metals are associated with the decline in nuclear binding energies of elements beyond iron.

Group 1

Li	Na	K	Rb	Cs
1.30	4.36	4.23	1.85	0.20

Group 2

Be	Mg	Ca	Sr	Ba
0.30	4.45	4.71	2.65	2.58

8.5 The crustal abundances of Group 1 and 2 elements. The abundances are given as their logarithms (to the base 10) of grams of metal per 100 kg of sample. Because the vertical scale is logarithmic, the differences are much greater than they appear.

Table 8.1 Mineral sources and methods of recovery of some commercially important *s*-block metals

Metal	Natural source	Method of recovery
Lithium	Spodumene, $LiAl(SiO_3)_2$	Electrolysis of molten $LiCl + KCl$
Sodium	Rock salt, $NaCl$, sea water and brines	Electrolysis of molten $NaCl$
Potassium	Silvite, KCl brines	Action of sodium on KCl at 850°C
Beryllium	Beryl, $Be_3AlSi_6O_{18}$	Electrolysis of molten $BeCl_2$
Magnesium	Dolomite, $CaMg(CO_3)_2$	$2MgCaO_2(l) + FeSi(l) \xrightarrow{1150°C} Mg(g) + Fe(l) + Ca_2SiO_4(l)$
Calcium	Limestone, $CaCO_3$	Electrolysis of molten $CaCl_2$

Table 8.1 lists the principal natural sources and methods of isolation of the commercially important Group 1 and 2 metals. Because these elements are all highly reducing, their recovery requires relatively expensive technology, such as the electrolysis of molten salts, the use of other alkali metals as reducing agents, or high temperature metal reduction.

8.2 Redox reactions

The standard potentials of the alkali and alkaline earth metals (Table 8.2) suggest that they are all capable of being oxidized by water:

Group 1: $\quad M(s) + H_2O(l) \rightarrow M^+(aq) + OH^-(aq) + \frac{1}{2}H_2(g)$

Group 2: $\quad M(s) + 2H_2O(l) \rightarrow M^{2+}(aq) + 2OH^-(aq) + H_2(g)$

Table 8.2 Standard potentials for the *s*-block elements ($E^{\ominus}/V$)

Group 1		Group 2	
Li	− 3.04	Be	− 1.97
Na	− 2.71	Mg	− 2.36
K	− 2.93	Ca	− 2.84
Rb	− 2.92	Sr	− 2.89
Cs	− 2.92	Ba	− 2.92

This reaction is so rapid and exothermic for sodium and its heavier congeners that the hydrogen is ignited. The vigor of these reactions is associated with the low melting points of the metals, because once molten a clean metal surface is more readily exposed and rapid reaction ensues. In Group 2, both beryllium and magnesium are protected from further oxidation by a thin oxide coating and therefore survive in the presence of water and air.

The standard potentials of the alkali metals are surprisingly uniform (and close to − 3 V). This uniformity stems from compensating terms in the thermochemical steps (Fig. 8.6) into which the reduction half-reaction may be analyzed. The enthalpies of sublimation and ionization both decrease down the group (making oxidation more favorable); however, this trend is counteracted by a smaller enthalpy of hydration as the radii of the ions increase (making oxidation less favorable).

8.3 Binary compounds

The aqueous potentials often give an indication of the tendency of the *s*-block elements to form compounds. However the interactions of these cations with anions in the solid state may differ significantly from the interaction between metal ions and H_2O molecules, so the tendency toward compound formation does not correlate with the standard potentials in all cases. For example, despite the relative uniformity of

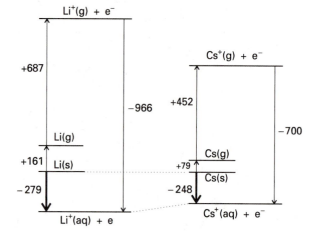

8.6 Thermochemical cycles (enthalpy changes) for the oxidation half-reaction $M(s) \rightarrow M^+(aq) + e^-$ for lithium and cesium. Note that the large difference in the formation of $M^+(g)$ is compensated by a similar difference in M^+ hydration enthalpies.

standard potentials down a group, the only stable nitride of the Group 1 metals is lithium nitride, Li_3N. We will also see chemical individuality in the reactions of the *s*-block elements with O_2.

Trends in chemical properties are much simpler for the halides, and examples are known of all binary combinations of an alkali metal and a halogen. The rock-salt and cesium-chloride structures of the alkali halides are typical of ionic compounds and will be familiar from the discussion in Section 4.5. Most of the alkali halides have a (6,6)-coordinated rock-salt structure, but CsCl, CsBr, and CsI have the more closely packed, (8,8)-coordinated cesium-chloride structure. At high pressures the halides of sodium, potassium, and rubidium undergo a transition from the rock-salt to the more closely packed cesium-chloride structure.

The great majority of the oxides and sulfides of the *s*-block metals also display the group oxidation number, with the added possibility of catenation (chain formation) of the anion. For example, Na_2O_2 contains the peroxide ion, O_2^{2-}, and KO_2 contains the superoxide ion, O_2^-. In addition, the larger Group 2 metal ions form peroxides. Barium peroxide, BaO_2, which forms at moderate temperatures and decomposes at higher temperatures, was at one time used as a renewable source of oxygen:

$$BaO(s) + \tfrac{1}{2}O_2(g) \xrightarrow{\text{below } 500\,°C} BaO_2(s)$$

$$BaO_2(s) \xrightarrow{\text{above } 700\,°C} BaO(s) + \tfrac{1}{2}O_2(g)$$

The stabilization of the large peroxide and superoxide anions by the larger alkali metal and alkaline earth metal cations was used in Section 4.8 as an example of how large cations can help to stabilize large anions.[1]

---

[1] The instability of ionic solids containing large anions in combination with small and highly charged cations, does not carry over to many covalent systems having a metal atom in a high oxidation state. For example Cr(VI) peroxo complexes are well known and many of them have been isolated and characterized. One example, which contains Cr(VI), is $[Cr(O_2)_2O(py)]$. Other examples include the *d*-block polysulfide complexes, which are discussed later in the chapter.

A distinctive aspect of the alkali metal cations is the high solubility in water of most simple salts. The major exceptions to this rule are the larger cations, K^+ through Cs^+, in combination with large anions: this property was also mentioned in Section 4.8 as an example of the variation of lattice enthalpies with the relative radii of the cations and anions in a compound. For instance, the solubility of heavy alkali metal perchlorates is much lower than that of the light alkali metal perchlorates: the molar concentration of saturated $CsClO_4$ is 0.09 mol L^{-1} whereas that of $LiClO_4$ is 4.5 mol L^{-1}. Tetraphenylborate salts of potassium and the heavier alkali metals are even less soluble in water. Similarly, the heavier alkaline earth metal cations form insoluble salts with large dinegative ions: a familiar example is hydrated calcium sulfate, the major constituent of plaster. The trend toward lower solubility in water down the group is very striking: $MgSO_4$ is highly soluble but the solubility of $CaSO_4 \cdot 2H_2O$ is 5×10^{-2} mol L^{-1} and that of $BaSO_4$ is only 10^{-5} mol L^{-1}.

X-Ray structure determinations show that the Li^+ ion is often found in sites ranging from four- to six-coordination. Examples include Li_2O, in which it is four-coordinate (in the antifluorite structure), and LiF, in which it is six-coordinate (in the rock-salt structure). The larger alkali metal ions display even more variation in their common coordination numbers.

The structures of beryllium compounds are distinctive on account of the element's tendency toward covalent bond formation. A recurring motif in beryllium compounds is a tetrahedral unit with Be at its center. For example, the low temperature form of BeO has the wurtzite structure, with tetrahedral coordination of the Be atom. Were it not for the high toxicity of beryllium compounds, ceramic BeO might find many applications because it displays the uncommon combination of very low electrical conductivity with high thermal conductivity. In contrast with BeO, the simple oxides of all the larger alkali metal ions are six-coordinate in the rock-salt structure.

8.4 Complex formation

Most of the complexes of *s*-block ions arise from coulombic interactions with small, hard donors, such as those possessing O or N atoms. Therefore, the Group 1 and 2 ions are classified as chemically hard. In general, the smaller the ion and the greater the charge, the more stable the complex. For example, Be^{2+} and Mg^{2+} complexes with hard ligands are often stable with respect to decomposition, but complexes of the other *s*-block metal ions are not. For these complexes lability parallels stability. Indeed, only for $Be^{2+}(aq)$ and $Mg^{2+}(aq)$ is the formation of simple complexes slower than the rate of encounter in solution.

The most notable complexes of the cations of Group 1 and the heavier Group 2 metals (calcium through barium) are formed by polydentate ligands. Monodentate ligands are only weakly bound on account of the weak coulombic interactions and lack of significant

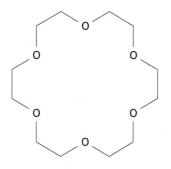

1 18–crown–6

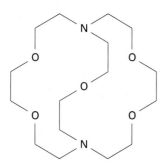

2 2.2.1 crypt

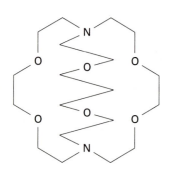

3 2.2.2 crypt

covalent bonding by these ions. Crown ethers such as 18-crown-6 (**1**) form complexes with alkali metal ions that survive indefinitely in nonaqueous solution. Bicyclic cryptate (or cryptand) ligands, such as 2.2.1 (**2**) and 2.2.2 (**3**), form even more stable complexes with alkali metals, and they can survive even in aqueous solution. Cryptand ligands are sterically selective for a particular metal ion. The dominant factor in the selectivity is the geometrical fit between the cation and the cavity in the ligand that accommodates it (Fig. 8.7).

The Group 2 cations form complexes with crown and crypt ligands. More stable complexes are formed with charged polydentate ligands, such as the analytically important ethylenediaminetetraacetate ion $((^-O_2CH_2)_2NCH_2CH_2N(CH_2CO_2^-)_2$, EDTA). The formation constants of EDTA complexes of the alkali metal complexes lie in the order $Ca^{2+} > Mg^{2+} > Sr^{2+} > Ba^{2+}$. The analysis of this odd trend is quite involved and beyond the scope of the present discussion. In the solid state the structure of the Mg^{2+} EDTA complex is seven-coordinate (**4**), with H_2O filling a coordination site. The calcium complex is either seven- or eight-coordinate, depending on the counterion, with one or two H_2O molecules serving as ligands. A large number of complexes of Ca^{2+} and Mg^{2+} occur naturally, the most familiar being chlorophyll, which contains Mg^{2+} (see Section 19.9).

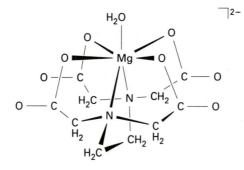

4 $[Mg(EDTA)(OH_2)]^{2-}$

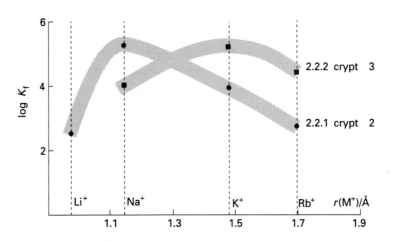

8.7 The stability of complexes of metals with cryptand ligands shown as a plot of the logarithm of the formation constant in water versus cation radius. Note that the smaller 2.2.1 crypt favors complex formation with Na^+ and the larger 2.2.2 crypt favors K^+.

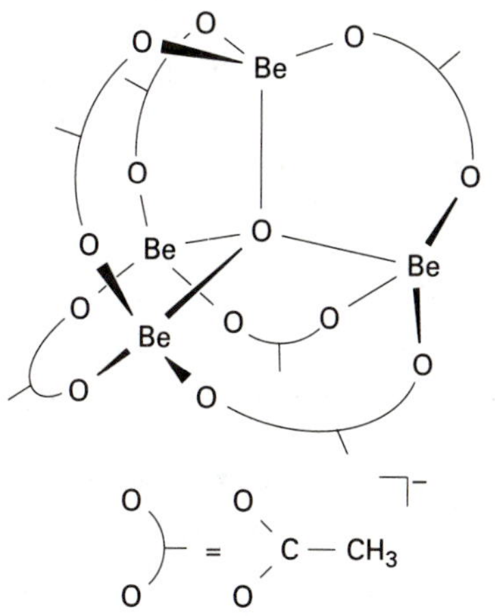

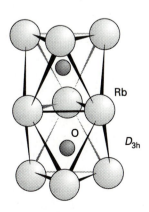

5 $Be_4O(O_2CCH_3)_6$

Compounds of beryllium show properties consistent with a greater covalent character than those of its congeners, and some if its complexes with ordinary ligands are quite stable. For example, basic beryllium acetate (beryllium oxoacetate, $Be_4O(O_2CCH_3)_6$) consists of a central O atom surrounded by a tetrahedron of four Be atoms, which in turn are bridged by acetate groups (**5**). It may be prepared by the reaction of acetic acid with beryllium carbonate:

$$4BeCO_3(s) + 6CH_3COOH(l) \rightarrow$$

$$4CO_2(g) + 3H_2O(l) + Be_4O(O_2CCH_3)_6(s)$$

Basic beryllium acetate is a colorless, sublimable, molecular compound; it is soluble in chloroform, from which it can be recrystallized.

Example 8.1: *Explaining trends in s-block chemistry*

Use simple bonding models to explain why barium forms a peroxide but beryllium does not.

Answer. Large anions are generally stabilized by large cations (Section 4.8). Therefore, barium peroxide should be more stable than beryllium peroxide. In fact, the peroxide compound forms spontaneously on exposure of barium to air. Beryllium forms only BeO.

Exercise E8.1. Choose the most appropriate ligand for each of the two groups of metal ions and give an approximate order of stability within each group. *Ligand*: 2.2.2 crypt, EDTA, OH^-; *Metal cation*: (a) Li^+, Na^+, Rb^+, Cs^+; (b) Cu^{2+}, Fe^{3+}.

8.5 Metal-rich oxides, electrides, and alkalides

Special conditions are needed to form compounds of the *s*-block metals in which the elements occur with oxidation numbers lower than their group numbers. These remarkable compounds are formed only when air, water, and other oxidizing agents are rigorously excluded. For example, a series of metal-rich oxides is formed by the reaction of rubidium or cesium with a limited supply of oxygen. These compounds, which are classified in Fig. 8.4 as cluster compounds, are dark, highly reactive metallic conductors with seemingly strange formulas such as Rb_6O, Rb_9O_2, Cs_4O, and Cs_7O.[2] A clue to the explanation of these constitutions is that Rb_9O_2 consists of O atoms surrounded by octahedra of six Rb atoms, with two neighboring octahedra sharing faces (Fig. 8.8). These clusters are thought to owe their existence to coulombic M^+O^{2-} interactions together with weaker M—M bonds delocalized over the metal system. The metallic conduction of the compounds suggests that the valence electrons are delocalized beyond the individual Rb_9O_2 clusters.

Another interesting set of compounds has been characterized in the

8.8 The structure of Rb_9O_2. Each O atom in this cluster is surrounded by an octahedron of Rb atoms.

[2] A. Simon, Clusters of valence-poor metals: structure, bonding, and properties. *Angew. Chem., Int. Edn. Encl.*, **27**, 159 (1988).

course of studying sodium in liquid ammonia. Sodium dissolves in pure, anhydrous liquid ammonia (without hydrogen evolution) to give solutions that are deep blue when dilute. The color of these **metal–ammonia solutions** originates from the tail of a strong absorption band that peaks in the near infrared. Other electropositive metals with low enthalpies of sublimation, including calcium and europium, dissolve in liquid ammonia to give solutions with a color that is independent of the metal. A variety of experiments indicate that the absorption responsible for the color arises from transitions of an electron in a cavity formed by a group of NH_3 molecules and with energy levels that resemble those of a particle in a spherical well (Section 1.4).

The dissolution of sodium in liquid ammonia to give a very dilute solution is represented by the equation

$$Na(s) \rightarrow Na^+(am) + e^-(am)$$

where 'am' denotes solution in ammonia. These solutions survive for long periods at the temperature of boiling ammonia ($-33\,°C$) and in the absence of air. However, they are only metastable, and their decomposition is catalyzed by some *d*-block compounds:

$$Na^+(am) + e^-(am) + NH_3(l) \xrightarrow{\text{iron oxides}} NaNH_2(am) + \tfrac{1}{2}H_2(g)$$

More concentrated solutions have a bronze appearance. Optical spectra and electrical conductivity measurements indicate that the electrons are delocalized throughout the solution, much as they are in a metal.

The more dilute blue metal–ammonia solutions are excellent reducing agents. For example, the unusual Ni(I) complex $[Ni_2(CN)_6]^{4-}$ may be prepared by the reduction of Ni(II) with potassium in liquid ammonia:

$$2K_2[Ni(CN)_4] + 2K^+(am) + 2e^-(am) \rightarrow$$

$$K_4[Ni_2(CN)_6](am) + 2KCN$$

The reaction is performed in the absence of air in a vessel cooled to the boiling point of ammonia.

Alkali metals also dissolve in ethers and alkylamines to give solutions with absorption spectra that depend on the alkali metal.[3] The metal dependence suggests that the spectrum is associated with charge transfer from an **alkalide ion**, M^- (such as a sodide ion, Na^-), to the solvent. Further evidence for the occurrence of alkalide ions is the diamagnetism associated with the species assigned as M^-, which would have the spin-paired s^2 valence electron configuration. Another observation in agreement with this interpretation is that when sodium–potassium alloy is dissolved, the metal-dependent band is the same as for solutions of sodium itself. When ethylenediamine is used as a solvent (denoted 'en'), the metal-independent band is not observed, so

[3] J. L. Dye, *J. Chem. Educ.*, **54**, 323 (1977); S. B. Dawes, D. L. Ward, O. Fussa-Rydal, R.-H. Huang, and J. L. Dye, *Inorg. Chem.*, **28**, 2132 (1989).

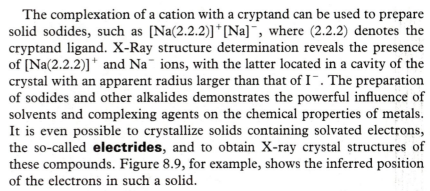

the dissolution equation is written as

$$2Na(s) \rightarrow Na^+(en) + Na^-(en)$$

$$NaK(l) \rightarrow K^+(en) + Na^-(en)$$

The complexation of a cation with a cryptand can be used to prepare solid sodides, such as $[Na(2.2.2)]^+[Na]^-$, where (2.2.2) denotes the cryptand ligand. X-Ray structure determination reveals the presence of $[Na(2.2.2)]^+$ and Na^- ions, with the latter located in a cavity of the crystal with an apparent radius larger than that of I^-. The preparation of sodides and other alkalides demonstrates the powerful influence of solvents and complexing agents on the chemical properties of metals. It is even possible to crystallize solids containing solvated electrons, the so-called **electrides**, and to obtain X-ray crystal structures of these compounds. Figure 8.9, for example, shows the inferred position of the electrons in such a solid.

The *s*-block elements also form a range of organometallic compounds that are useful in organic and inorganic synthesis. Two familiar examples are the Grignard reagents, such as CH_3MgBr, and methyllithium, $Li_4(CH_3)_4$. These compounds are treated at length in Chapter 10.

8.9 An ORTEP diagram of the crystal structure of $[Cs(18\text{-crown-}6)_2]^+e^-$. The symbols $\odot$ mark the sites where the e^- appears to reside. (From S. B. Dawes, D. L. Ward, R. H. Huang and J. L. Dye, *J. Am. Chem. Soc.*, **108**, 3534 (1986).)

THE *d*-BLOCK METALS

The *d*-block elements have rich and interesting chemical properties because of their variety of oxidation states, their extensive ability to form complexes (Chapter 6), organometallic compounds (Chapter 16), and useful solid compounds (Chapter 18), and their ability to participate in catalysis (Chapter 17) and biochemical processes (Chapter 19). In this section we focus on trends in the stabilities of oxidation states within the *d* block and on the properties of some representative compounds.

8.6 Occurrence and recovery

The elements on the left of the 3*d* block occur in nature primarily as metal oxides or as metal cations in combination with oxoanions. This characteristic is illustrated in Table 8.3 for the mineral sources of the commercially most important 3*d* metals. Of these elements, titanium is the most difficult to reduce, and until new electrochemical techniques were developed it was widely produced by using the expensive but strong reducing agents sodium or magnesium. The oxides of chromium, manganese, and iron are reduced with carbon (see Section 7.1), a much cheaper reagent. To the right of iron in the 3*d* series, cobalt, nickel, and copper occur as sulfides and arsenides, which is consistent with the increasingly soft acid character of the dipositive ions toward the right of the series. Copper is used in large quantities for electrical conductors, and electrolysis is used to refine crude copper so as to achieve the high purity needed for high electrical conductivity (Fig. 8.10).

Table 8.3 Mineral sources and methods of recovery of some commercially important d-block metals

Metal	Principal minerals	Method of recovery	Note
Titanium	Ilmenite, $FeTiO_3$ Rutile, TiO_2	$TiO_2 + 2C + 2Cl_2 \rightarrow TiCl_4 + 2CO$ followed by reduction of $TiCl_4$ with Na or Mg	
Chromium	Chromite, $FeCr_2O_4$	$FeCr_2O_4 + 4C \rightarrow Fe + 2Cr + 4CO$	(a)
Molybdenum	Molybdenite, MoS_2	$2MoS_2 + 7O_2 \rightarrow 2MoO_3 + 4SO_2$ followed by either: $MoO_3 + 2Fe \rightarrow Mo + Fe_2O_3$ or: $MoO_3 + 3H_2 \rightarrow Mo + 3H_2O$	(b)
Tungsten	Scheelite, $CaWO_4$ Wolframite, $FeMn(WO_4)_2$	$CaWO_4 + 2HCl \rightarrow WO_3 + CaCl_2 + H_2O$ followed by $2WO_3 + 6H_2 \rightarrow 2W + 6H_2O$	
Manganese	Pyrolusite, MnO_2	$MnO_2 + C \rightarrow Mn + CO_2$	(c)
Iron	Hematite, Fe_2O_3 Magnetite, Fe_3O_4 Limonite, $FeO(OH)$	$Fe_2O_3 + 3CO \rightarrow 2Fe + 3CO_2$	
Cobalt	Cobaltite, $CoAsS$ Smaltite, $CoAs_2$ Linnaeite, Co_3S_4	By-product of copper and nickel production	
Nickel	Pentlandite, $(Fe,Ni)_6S_8$	$2NiS + 3O_2 \rightarrow 2NiO + 2SO_2$	(d)
Copper	Chalcopyrite, $CuFeS_2$ Chalcocite, Cu_2S	$2CuFeS_2 + 2SiO_2 + 4O_2 \rightarrow Cu_2S + 2FeSiO_3$ $+ 3SO_2$	

(a) The iron–chromium alloy is used directly for stainless steel.
(b) The iron–molybdenum alloy is used in cutting steel.
(c) The reaction is carried out in a blast furnace with Fe_2O_3 to produce alloys.
(d) NiS is formed by melting the mineral and separated by physical processes. NiO is used in a blast furnace with iron oxides to produce steels. Nickel is purified by electrolysis or the Mond process via $Ni(CO)_4$.

The difficulty of reducing the early $4d$ and $5d$ metals molybdenum and tungsten is apparent from Table 8.3. It reflects the tendency of these elements to have stable high oxidation states, as discussed later in this section. The platinum metals (Ru and Os, Rh and Ir, and Pd and Pt) which are found on the lower right of the d block, occur as

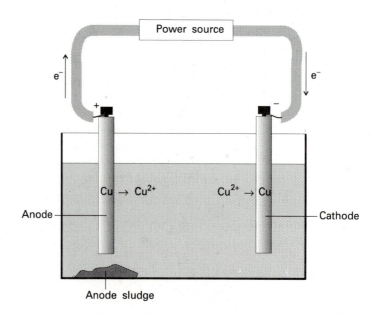

8.10 A schematic diagram of the arrangement for the electrolytic refinement of copper. Crude copper from the anode is deposited as nearly pure copper on the cathode. Noble metals are collected in the anode sludge. More electropositive elements, such as nickel and cobalt, collect in the electrolyte.

sulfide and arsenide ores, usually in association with larger quantities of copper, nickel, and cobalt. They are collected from the sludge that forms during the electrolytic refinement of copper and nickel.

8.7 High oxidation states

For the most part, we shall confine this discussion of the compounds of d metals in high oxidation states to those containing electronegative ligands such as the halogens, oxygen, and sulfur. Because of the convention of assigning a negative oxidation number to any nonmetal in combination with a metal, high formal oxidation numbers may be encountered, such as Re(VII) in $[ReH_9]^{2-}$ and W(VI) in $W(CH_3)_6$. However, these compounds are not oxidizing agents in the usual sense, and they are best discussed at the same time as organometallic compounds (Chapter 16).

Oxidation states across the 3d series

If all the outermost s and d electrons of a d-block element are chemically accessible, then the maximum oxidation number is equal to the group number (in the 1 to 18 system) and the element is said to have its **group oxidation number**. A general observation is that the group oxidation number may be achieved in elements that lie toward the left of the block but not by elements on the right. For example, scandium, yttrium, and lanthanum in Group 3 are found in aqueous solution only with oxidation number $+3$, and the great majority of their complexes contain the metals in these oxidation states. The group oxidation number is never achieved from Group 9 (Co, Rh, and Ir) onward (that is, it is not achieved in Group 10 by Ni, Pd, and Pt nor in Group 11 by Cu, Ag, and Au).

The trend in thermodynamic stability of the group oxidation states of the 3d metals is clearly illustrated by the Frost diagram in Fig. 8.11. We see that the group oxidation states of scandium, titanium, and vanadium fall in the lower part of the diagram. This location indicates that the metal and any intermediate oxidation states are readily oxidized to the group oxidation state. In contrast, the group oxidation states for chromium and manganese ($+6$ and $+7$, respectively), lie in the upper part of the diagram. This location indicates that they are susceptible to reduction. The illustration also shows that the group oxidation state is not achieved in Groups 8 through 11 of Period 4 (iron, cobalt, nickel, copper, and zinc).

The binary compounds of the 3d elements with halogens and oxygen also illustrate the trend in stability of the group oxidation state. Chlorides can be prepared for the earliest metals in their group oxidation states ($ScCl_3$ and $TiCl_4$) but the more strongly oxidizing halogen fluorine is necessary to achieve the group oxidation states of vanadium (Group 5) and chromium (Group 6), which form VF_5 and CrF_6 respectively. Beyond Group 6 in Period 4, even fluorine cannot bring out the group oxidation state, and MnF_7 and FeF_8 have never been prepared. Oxygen brings out the group oxidation state for many metals

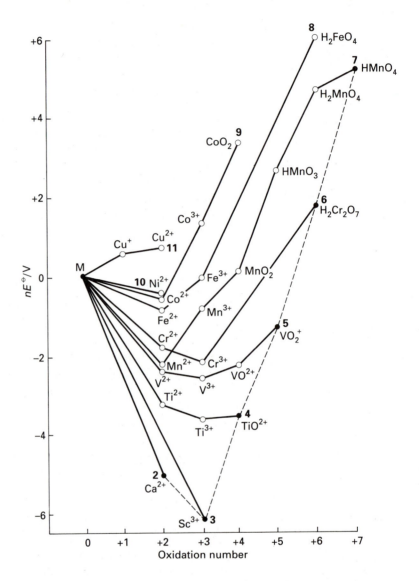

8.11 A Frost diagram for the first series of *d*-block elements in acidic solution (pH = 0). The bold numbers designate the group numbers and the broken line connects species in their group oxidation states.

more readily than does fluorine: fewer O atoms than F atoms need to surround the central metal atom to induce the same change in oxidation number. For example, the group oxidation state of $+7$ for manganese is achieved in permanganate salts such as potassium permanganate, $KMnO_4$, and there are claims for the existence of FeO_4.

As can be inferred from the Frost diagram, these oxoanions are strong oxidizing agents. The synthesis and properties of the oxoanions chromate, CrO_4^{2-}, permanganate, MnO_4^-, and ferrate, FeO_4^{2-} (see Box 8.1), illustrates the decreasing stability of the maximum attainable oxidation state for Groups 6, 7, and 8.

The air oxidation of MnO_2 in molten potassium hydroxide does not take manganese to its group oxidation state but instead yields the deep green compound potassium manganate, K_2MnO_4, which contains Mn(VI). This is another example of the greater difficulty of oxidizing an element to the right of chromium to its maximum oxidation state.

Box 8.1 The preparation of dichromate

To prepare the Cr(VI) oxoanion, chromite ore, $FeCr_2O_4$, which contains Fe(II) and Cr(III), is dissolved in molten potassium hydroxide and oxidized with atmospheric oxygen:

$$FeCr_2O_4(s) + 6KOH(l) + \tfrac{5}{2}O_2(g) \rightarrow K_2FeO_4(melt) + 2K_2CrO_4(melt) + 3H_2O(g)$$

In this process, both the iron and chromium are converted to the +6 state, FeO_4^{2-} and CrO_4^{2-}, so chromium reaches its group oxidation state but iron does not. Dissolution in water followed by filtration leads to a solution of the two oxoanions. These ions can be separated by taking advantage of the greater oxidizing power of Fe(VI) relative to that of Cr(VI) in acidic solution. Acidification leads to the reduction of FeO_4^{2-} and the conversion of CrO_4^{2-} to dichromate $Cr_2O_7^{2-}$; the latter conversion is a simple acid–base reaction, and not a redox reaction:

$$2FeO_4^{2-}(aq) + 4H^+(aq) + 2e^- \rightarrow 2Fe^{3+}(aq) + 3O_2(g) + 2H_2O(l) \qquad \text{reduction}$$

$$2CrO_4^{2-}(aq) + 2H^+(aq) \rightarrow O_3CrOCrO_3^{2-}(aq) + H_2O(l) \qquad \text{acid–base}$$

The disproportionation of MnO_4^{2-} in acidic aqueous solution yields manganese(IV) oxide, MnO_2, and the deep purple permanganate ion, MnO_4^-, containing Mn(VII):

$$3MnO_4^{2-}(aq) + 4H^+(aq) \rightarrow 2MnO_4^-(aq) + MnO_2(s) + 2H_2O(l)$$

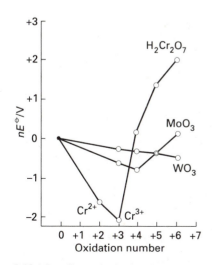

8.12 A Frost diagram for the chromium group in the *d* block (Group 6) in acidic solution (pH = 0).

Oxidation states down a group

The general trend in oxidation states down a group can be summarized as follows:

> In Groups 4 through 10, the highest oxidation state of an element becomes more stable on descending a group, with the largest change in stability occurring between the first two rows of the *d* block.

This trend is illustrated for the chromium group in Fig. 8.12. Note that the Mo(VI) and W(VI) species lie below Cr(VI) in $H_2Cr_2O_7$, indicating that the maximum oxidation state is more stable for molybdenum and tungsten. The relative positions of Cr(VI), Mo(VI) and W(VI) in the diagram also illustrates the point about the change between the first and second rows of the *d* block.

The increasing stability of high oxidation states for the heavier *d*-block metals is also seen in the formulas of their halides (Table 8.4).

Table 8.4 Highest oxidation state *d*-block binary halides*

Group							
4	5	6	7	8	9	10	11
TiI_4	VF_5	CrF_5†	MnF_4	$FeBr_3$	CoF_3	NiF_4	$CuBr_2$
ZrI_4	NbI_5	MoF_6	$TcCl_6$	RuF_6	RhF_6	PdF_4	AgF_3
HfI_4	TaI_5	WBr_6	ReF_7	OsF_6	IrF_6	PtF_6	$AuCl_5$

*The formulas show the least electronegative halide that brings out the highest oxidation state.
†CrF_6 exists for several days at room temperature in a passivated Monel container.

Fluorine is more effective in bringing out the high oxidation states, but the formulas MnF_4, TcF_6, and ReF_7 show the greater ease of oxidizing the 4*d* and 5*d* metals. The hexafluorides of the heavier *d*-block elements have been prepared from Group 6 through Group 10 (as in PtF_6). As expected from the periodic trends in oxidation numbers, WF_6 is not a significant oxidizing agent, but the oxidizing character of the hexafluorides increases to the right, and PtF_6 is so potent that it can oxidize O_2 to O_2^+:

$$O_2(g) + PtF_6(s) \rightarrow [O_2][PtF_6](s)$$

The ability to achieve the highest oxidation state does not correlate with the ease of oxidation of the bulk metal to an intermediate oxidation state. For example, elemental iron is thermodynamically susceptible to oxidation by $H^+(aq)$ under standard conditions,

$$Fe(s) + 2H^+(aq) \rightarrow Fe^{2+}(aq) + H_2(g) \qquad E^\ominus = +0.44 \text{ V}$$

but, as already noted, it is not certain that any oxidizing agent has been found that will take it to its group oxidation state. The two heavier metals in Group 8 (ruthenium and osmium) are not oxidized in acidic aqueous solution:

$$Os(s) + 2H_2O(l) \rightarrow OsO_2(s) + 2H_2(g) \qquad E^\ominus = -0.81 \text{ V}$$

but they can be oxidized by oxygen to the +8 state, as RuO_4 and OsO_4:

$$Os(s) + 2O_2(g) \rightarrow OsO_4(s)$$

Ruthenium tetroxide and osmium tetroxide are low melting, highly volatile toxic molecular compounds that are used as selective oxidizing agents. For example, osmium tetroxide (as well as the permanganate ion) is used to oxidize alkenes to diols:

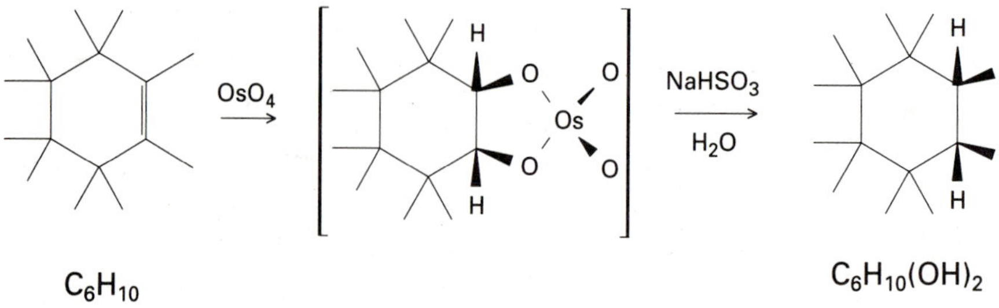

C_6H_{10} $\qquad\qquad\qquad\qquad\qquad\qquad\qquad$ $C_6H_{10}(OH)_2$

Table 8.5 Coordination numbers of some early d-block fluoro and cyano complexes

Group	Complex (coordination number)		
	$3d$	$4d$	$5d$
3	$[NH_4]_3[ScF_6]$ (6)	$NaYF_4$ (9)	$NaLaF_4$ (9)
4	$Na_2[TiF_6]$ (6)	$Na_3[ZrF_7]$ (7)	$Na_3[HfF_7]$ (7)
5	$K[VF_6]$ (6)	$K_2[NbF_7]$ (7)	$K_3[TaF_8]$ (8)
	$K_2[V(CN)_7] \cdot 2H_2O$ (7)	$K_5[Nb(CN)_8]$ (8)	

The Frost diagrams for the positive oxidation states of molybdenum and (especially) tungsten are quite flat (Fig. 8.12). This flatness indicates that neither element exhibits the marked tendency to form the $+3$ oxidation state that is so characteristic of chromium. Mononuclear molybdenum and tungsten complexes are common in the $+2$, $+3$, $+4$, $+5$, or $+6$ oxidation states; as we shall see in more detail later, M—M bonded dinuclear or polynuclear complexes are numerous for the $+2$ and $+3$ oxidation states.

Structural trends down a group

As may be anticipated from considerations of atomic and ionic radii, the $4d$ and $5d$ elements often have higher coordination numbers than their $3d$ congeners. This trend is illustrated in Table 8.5 for the fluoro and cyano complexes of the early d-block metals. Note that with the compact F^- ligand, the early $3d$ metals tend to form six-coordinate complexes but that the larger $4d$ and $5d$ metals in the same oxidation state tend to form seven- and nine-coordinate complexes. The octacyanomolybdate complex illustrates the tendency toward high coordination numbers with compact ligands. These complexes are readily interconverted electrochemically or chemically:

$$[Mo(CN)_8]^{3-}(aq) + e^- \rightarrow [Mo(CN)_8]^{4-}(aq) \qquad E^{\ominus} = +0.73 \text{ V}$$

Complexes of oxo and related ligands

Because oxygen is readily available in an aqueous environment, in the atmosphere, and as the donor atom in many organic molecules, it is not surprising that oxygen-containing ligands play a major role in the chemistry of the metallic elements, particularly the hard, high oxidation state metals on the left of the d block. Our focus in this section is on the oxo ligand and its ability to bring out the high oxidation states of metals. Another important issue will be the relation between oxo and aqua ligands through acid–base equilibria, and the relation of the oxo ligand to ligands that are isoelectronic with it. First we consider simple oxo complexes having single metal centers, and then the more complex polyoxometallates, which consist of two or more metal centers bridged by oxo ligands.

Mononuclear oxo complexes

Metals in high oxidation states typically occur as oxoanions in aqueous solution, such as permanganate, MnO_4^-, which contains mangan-

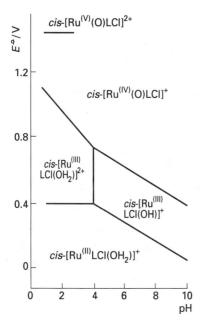

8.13 A Pourbaix diagram for *cis*-[RuLCl(OH₂)]²⁺ and related species. The ligand L is tetradentate, and its structure is shown in (**6**). Adapted from C.-K. Li, W.-T. Tang, C.-M. Chi, K.-Y. Wong, R.-J. Wang, and T. C. W. Mak, *J. Chem. Soc., Dalton Trans.*, 1909 (1991).

ese(VII), and chromate, CrO_4^{2-}, which contains chromium(VI). These oxoanions contrast with the simple aqua ions for these metals in lower oxidation states, such as $[Mn(OH_2)_6]^{2+}$ for manganese(II) and $[Cr(OH_2)_6]^{2+}$ for chromium(II), These and related observations can be summarized by the following generalization:

> The conversion of an aqua ligand to an oxo ligand is favored by a high pH and by a high oxidation state of the central metal atom.

The occurrence of aqua complexes for low oxidation state metal cations can be rationalized by noting the relatively small electron-withdrawing effect that these cations exert on the O atoms of the water molecule. In contrast, high oxidation state metal ions form strong bonds which deplete the electron density on the attached O atoms and thereby decrease their Brønsted basicity. The combined influence of pH and oxidation state is summarized by Pourbaix diagrams (Section 7.10). The example shown in Fig. 8.13 involves a complex containing a stabilizing tetradentate ligand, L, which results in a similar coordination geometry for a wide range of conditions. The aqua complex *cis*-[RuLCl(OH₂)]⁺ (**6**) in solution at pH = 2 is stable up to +0.40 V. Just above this potential, simple oxidation (that is, electron removal) occurs to give the Ru(III) aqua complex *cis*-[RuLCl(OH₂)]²⁺. Under even more oxidizing conditions (at +0.95 V) and pH = 2, both oxidation and deprotonation occur and result in the formation of a Ru(IV) oxo species, *cis*-[RuLCl(O)]⁺. At an even higher potential (about +1.4 V), further oxidation yields the Ru(V) oxo complex *cis*-[RuLCl(O)]²⁺. The influence of more basic conditions is to deprotonate an H₂O ligand and thus to favor hydroxo or oxo complexes. For example, at pH = 8, the Ru(III) species is deprotonated to the hydroxo complex *cis*-[RuLCl(OH)]⁺, which in turn is converted to the oxo-Ru(IV) complex at a lower potential than for the Ru(III)-Ru(IV) transformation at pH = 2.

A list of simple complexes containing oxo ligands is given in Table 8.6. Many complexes are known that contain the vanadyl moiety,[4] VO^{2+}, with vanadium in its penultimate oxidation state (+4). Vanadyl complexes generally contain four additional ligands and are square-pyramidal (**7**). Many of these d^1 complexes are blue (as a result

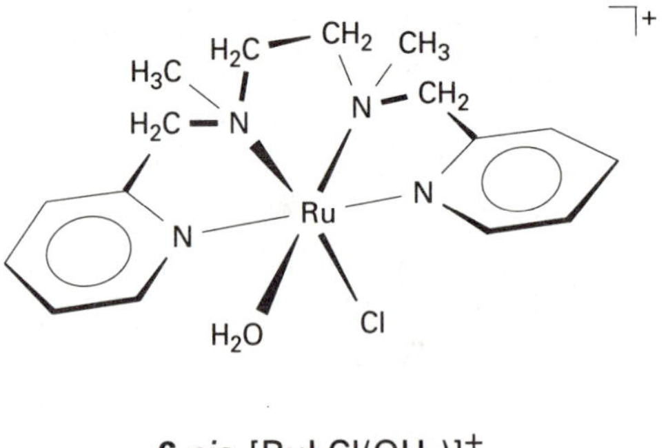

6 *cis*–[RuLCl(OH₂)]⁺   **7** [VOCl₄]²⁻

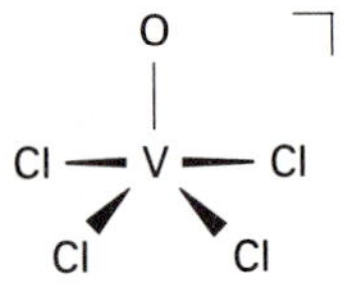

---

[4] Although the term moiety strictly means one half of an entity, it is widely used in the looser sense of one of two parts into which an entity may be divided.

Table 8.6 Some common monoxo and dioxo complexes

Group	Element	Structure	Formula
5	V(IV), d^1	square pyramidal	$[V(O)(acac)_2]$, $[V(O)Cl_4]^{2-}$
	V(V), d^0	*cis*-octahedral	$[V(O)_2(OH_2)_4]^+$
6	Mo(VI), d^0; W(VI), d^0	tetrahedral	$[M(O)_2(Cl)_2]$
7, 8	Re(V), d^2; Os(VI), d^2	*trans*-octahedral	$[Re(O)_2(py)_4]^+$, $[Re(O)_2(CN)_4]^{3-}$; $[Os(O)_2Cl_4]^{2-}$

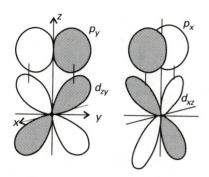

8.14 The d–p π-bonding between oxygen and vanadium in the vanadyl moiety, VO^{2+}. It is conventional to think of the nonmetal oxygen as having oxidation number -2 and vanadium as having oxidation number $+4$. From this point of view, both electron pairs for the two π bonds are donated from the oxide ligand to the metal d_{zx} and d_{yz} orbitals.

of a ligand field d–d transition) and may take on a weakly bound sixth ligand *trans* to the oxo ligand. The vanadyl V—O bond length (1.58 Å) in $[VO(acac)_2]$ is short compared with the four V—O bond lengths to the acac ligand (1.97 Å). The short bond distances in vanadyl complexes, together with high VO stretching wavenumbers (940–980 cm^{-1}) provide strong evidence for VO multiple bonding. The d–p orbital overlap involved in a multiple bond is illustrated in Fig. 8.14. This strong multiple bonding with the oxygen appears to be responsible for the *trans* influence of the oxo ligand, which disfavors attachment of a ligand *trans* to O.[5]

Vanadium in its highest oxidation state forms an extensive series of oxo compounds, many of which are the polyoxo species discussed later. The simplest oxo complex, $[V(O)_2(OH_2)_4]^+$ exists in the acidic solution formed when the sparingly soluble vanadium pentoxide, V_2O_5, dissolves in water. This pale yellow complex has a *cis*-geometry (**8**). Here again we see the *trans* influence of an oxo ligand. Apparently the *cis* geometry minimizes competition between the oxo ligands for π bonding. As shown in Fig. 8.15, the two O^{2-} ligands in the *trans* configuration would π bond with the same two d orbitals, but in the *cis* configuration they compete for only one d orbital and each O interacts independently with a second d orbital.

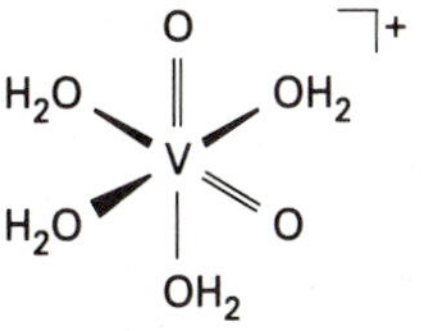

In contrast with the *cis* structure of the d^0 complex $[V(O)_2(OH_2)_4]^+$, many *trans*-dioxo complexes (**9**) are known for the d^2 metal centers Re(V) and Os(VI); some examples are given in Table 8.6. It is thought that this configuration is favored because the *trans* configuration leaves a vacant low energy orbital to be occupied by the two d electrons. According to this explanation, the avoidance of the destabilizing effect of the d^2 electrons more than offsets the loss in energy involved in the added competition for d orbitals for the *trans*-oxo configuration.

8 $[V(O)_2(OH_2)_4]^+$

9 $[Re(O)_2(py)_4]^+$

Nitrido and alkylidyne complexes

The N^{3-} and CR^{3-} groups are isolobal with O^{2-}. Both nitrido and alkylidyne complexes are known to have short M—N and M—C bond lengths, which suggests the existence of multiple bonding as with their oxo counterparts. As with the oxo ligand, nitrido and alkylidyne ligands often weaken the *trans* metal–ligand bond. The order of this *trans* influence is RC≡ > N≡ > O≡. (Because O≡M bonding is weaker

[5] The *trans* influence of the oxo ligand is reviewed by E. M. Shustrovitch, M. A. Pori-Koshits, and Yu. A. Buslaev, *Coord. Chem. Rev.*, **17**, 1 (1975).

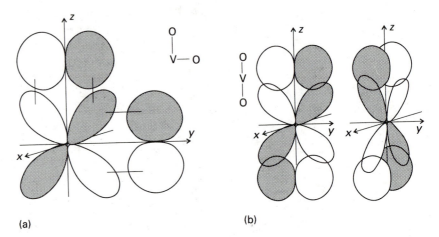

8.15 Comparison of the competition between *cis* and *trans* oxygen ligands attached to vanadium. (a) In the *cis* configuration the metal has only one *d* orbital in common with the π orbitals on the O atoms. (b) In the *trans* configuration the metal has two *d* orbitals in common with the π orbitals on the two O atoms.

(a)   (b)

than in other members of this series it is usually depicted O=M.) The highly reactive compound $(Me_3CO)_3W \equiv W(OCMe_3)_3$ cleaves the $C \equiv N$ bond in phenylcyanide to give both a nitrido and an alkylidyne ligand:

$$(Me_3CO)_3W \equiv W(OCMe_3)_3 + PhC \equiv N \rightarrow$$

$$(Me_3CO)_3W \equiv CPh + (Me_3CO)_3W \equiv N$$

The remarkable feature of this reaction is that the $C \equiv N$ bond is broken despite its considerable strength (890 kJ mol^{-1}), so the $W \equiv C$ and $W \equiv N$ bond enthalpies must be substantial.[6]

Many nitrido complexes are known for the elements in Groups 5 through 8. They are most numerous for molybdenum and tungsten in Group 6, rhenium in Group 7, and ruthenium and osmium in Group 8. Square-pyramidal nitrido complexes of formula $[MNX_4]^-$ are known for M = Mo(VI), Re(VI), Ru(VI), and Os(VI), and X = F, Cl, Br, and (in some cases) I. These complexes have square-pyramidal structures (**10**) with short $M \equiv N$ bond lengths (1.57 Å to 1.66 Å). In a few cases a ligand is attached in the sixth site, but the resulting bond is long and weak, as with oxo complexes discussed above. Examples of M=N=M species are known, such as $[Ta_2NBr_8]^{3-}$ (**11**), but nitrogen analogs of the polyoxometallates described below are as yet unknown.

10 $[Mn(N)X_4]^-$, C_{4v}

Polyoxometallates

The H_2O ligand created by protonation of an oxo ligand at low pH can be eliminated from the central metal atom and thus lead to the condensation of mononuclear oxometallates. A familiar example is the reaction of a basic chromate solution, which is yellow, with excess acid to bring about dehydration and formation of the oxo bridged di-

11 $[Ta_2(N)Br_8]^{3-}$

[6] Metathesis of tungsten–tungsten triple bonds with acetylenes and nitriles to give alkylidyne and nitrido complexes, R. R. Schrock, M. L. Listmann, and L. G. Sturgeoff, *J. Am. Chem. Soc.*, **104**, 4291 (1982).

chromate ion, which is orange:

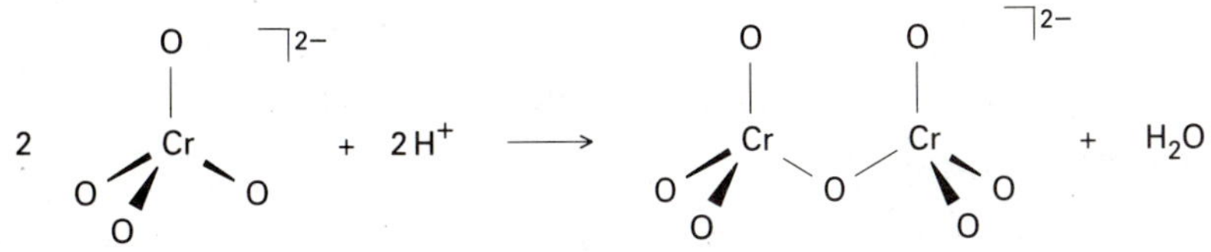

$$2CrO_4^{2-}(aq) + 2H^+(aq) \rightarrow Cr_2O_7^{2-}(aq) + H_2O(l)$$

In highly acidic solution, longer chain oxo-bridged Cr(VI) species are formed. The tendency for Cr(VI) to form polyoxo species is limited by the fact that the O tetrahedra link through vertices: edge and face bridging would result in too close an approach of the metal centers. It is found that five- and six-coordinate metal oxo complexes can share an oxo ligand between either vertices or edges, and these structural possibilities lead to a rich variety of polyoxometallates.

Chromium's neighbors in Groups 5 and 6 (Fig. 8.16) form six-coordinate polyoxo complexes. In Group 5 the polyoxometallates are most numerous for vanadium, which forms many V(V) complexes and a few V(IV) or mixed oxidation state V(IV)–V(V) polyoxo complexes. In short, polyoxometallate formation is most pronounced in Groups 5 and 6 for vanadium(V), molybdenum(VI), and tungsten(VI).

It is often convenient to represent the structures of the polyoxometallate ions by polyhedra, with the metal atom understood to be in the center and O atoms at the vertices. The sharing of O atom vertices in the dichromate ion, $Cr_2O_7^{2-}$, may be depicted in either the traditional way (**12**) or in the polyhedral representation (**13**). Similarly, the important M_6O_{19} structure of $[Nb_6O_{19}]^{8-}$, $[Ta_6O_{19}]^{8-}$, $[Mo_6O_{19}]^{2-}$, and $[W_6O_{19}]^{2-}$, can be depicted by the conventional or polyhedral structures shown in Fig. 8.17. The structures for this series of polyoxometallates contain terminal O atoms (those projecting outward from a single metal) and two types of bridging O atoms: two-metal bridges, M—O—M, and one O atom in the center of the structure and common to all six metal atoms. Viewed from the perspective of polyhedra, the structure consists of six MO_6 octahedra, each sharing an edge with four neighbors. The overall symmetry of the M_6O_{19} array is O_h.

Typically, the polyoxometallate anions are prepared by carefully adjusting pH and concentrations.[7] Polyoxomolybdates and polyoxotungstates are formed by acidification of solutions of the simple molybdate or tungstate:

$$6[MoO_4]^{2-}(aq) + 10H^+(aq) \rightleftharpoons [Mo_6O_{19}]^{2-}(aq) + 5H_2O(l)$$

$$8[MoO_4]^{2-}(aq) + 12H^+(aq) \rightleftharpoons [Mo_8O_{26}]^{4-}(aq) + 6H_2O(l)$$

Another example of a polyoxometallate is $[W_{12}O_{40}(OH)_2]^{10-}$ (**14**). As shown in this formula, the polyoxocation is protonated. Proton transfer

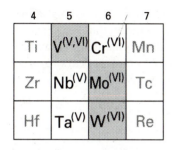

8.16 Elements in the *d* block that form polyoxometallates. The shaded elements form a greater variety of polyoxometallates.

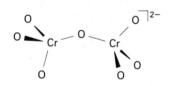

12 $[Cr_2O_7]^{2-}$

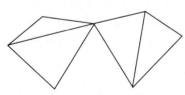

13 $[Cr_2O_7]^{2-}$

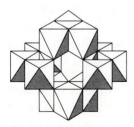

14 $[W_{12}O_{40}(OH)_2]^{10-}$

---

[7] For excellent summaries of this area, see M. T. Pope, *Heteropoly and isopoly oxometallates*. Springer Verlag, Berlin (1983), and *Comprehensive coordination chemistry*, Vol. 3. Pergamon Press, Oxford (1987).

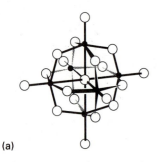

(a)

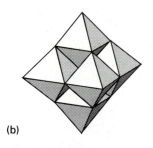

(b)

8.17 (a) Conventional representation and (b) polyhedral representation of the six edge-shared octahedra as found in $[M_6O_{19}]^{2-}$.

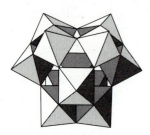

15 $[PMo_{12}O_{40}]^{3-}$

equilibria are common for the polyoxometallates, and may occur together with condensation and fragmentation reactions.[8]

In addition to the large number of polyoxomolybdates and polyoxo-tungstates, there is a large class of heteropolymolybdates and tungstates that incorporate phosphorus, arsenic, and other heteroatoms. For example, $[PMo_{12}O_{40}]^{3-}$ contains a PO_4^{3-} tetrahedron that shares O atoms with surrounding octahedral MoO_6 groups (**15**). Many different heteroatoms can be incorporated into this structure, and the general formulation is $[X^{(n+)}Mo_{12}O_{40}]^{(8-n)-}$ where $(n+)$ represents the oxidation state of the heteroatom, which may be As(V), Si(IV), Ge(IV), or Ti(IV). An even broader range of heteroatoms is observed with the analogous tungsten heteropolyanion. Heteropolymolybdates and tungstates can undergo one-electron reduction with no change in structure but with the formation of a deep blue color. The color appears to arise from the transition of the added electron from a Mo(V) or W(V) site to an adjacent Mo(VI) or W(VI) site.

8.8 Intermediate oxidation states

Simple $+1$ cations (M^+) in general disproportionate (to M and M^{2+}) because the metal binding energies are so large; with the exception of copper, silver, and gold such cations are not observed. Therefore, the $+2$ oxidation state is generally the lowest one to consider for the d-metal ions in aqueous solution and in combination with hard ligands. Exceptions to this behavior are mainly confined to metal–metal bonded and organometallic compounds, such as the 0 oxidation state metal carbonyls, $Ni(CO)_4$ and $Mo(CO)_6$, discussed in Chapter 16.

Oxidation state $+2$ of the 3d metals

The dipositive aqua ions M^{2+}(aq) (specifically, the octahedral complexes $[M(OH_2)_6]^{2+}$) play an important role in the chemistry of the $3d$ metals. Many of these ions are colored as a result of d–d transitions in the visible region of the spectrum. For example, Mn^{2+}(aq) is pale pink, Fe^{2+}(aq) is pale green, Co^{2+}(aq) is wine red, Ni^{2+}(aq) is green, and Cu^{2+}(aq) is blue.

A general feature is that:

> The $+2$ oxidation state becomes increasingly common from left to right across the period.

For example, among the early members of this series, Sc^{2+}(aq) (Group 3) is unknown and Ti^{2+}(aq) (Group 4) is not readily accessible. For Groups 5 and 6, V^{2+}(aq) and Cr^{2+}(aq) are thermodynamically unstable with respect to oxidation by H^+:

$$2V^{2+}(aq) + 2H^+(aq) \rightarrow 2V^{3+}(aq) + H_2(g) \qquad E^{\ominus} = +0.26 \text{ V}$$

However, the slowness of H_2 evolution makes it possible to work with solutions of these dipositive ions in the absence of air, and as a result

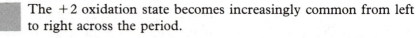

[8] V. W. Day and W. G. Klemperer, Metal oxide chemistry in solution: the early transition metal polyoxyanions. *Science*, **228**, 533 (1985).

they are useful reducing agents. Beyond chromium (for Mn^{2+}, Fe^{2+}, Co^{2+}, Ni^{2+}, and Cu^{2+}), the $+2$ state is stable with respect to reaction with water, and only Fe^{2+} is oxidized by air. The only stable aqua ions of copper and nickel are M(II). The same is nearly true for cobalt, because $Co^{3+}(aq)$ is reduced to $Co^{2+}(aq)$ by water:

$$4Co^{3+}(aq) + 2H_2O(l) \rightarrow 4Co^{2+}(aq) + O_2(g) + 4H^+(aq)$$

$$E^{\ominus} = +0.69 \text{ V}$$

The trend toward increased stability of the lower oxidation states can be understood from the general increase in ionization energies from left to right across a period in the d block (Section 1.10).

The dipositive ions of the $3d$ metals form four-coordinate complexes with halide ions, such as the brown–yellow complex $[NiBr_4]^{2-}$. The absorption intensities of these tetrahedral complexes are greater than those of the simple octahedral hexaaqua complexes because the lack of a center of symmetry in the tetrahedral array means that the d–d transition is allowed. The structures of $[CuX_4]^{2-}$ complexes are distorted because the tetrahedral d^9 configuration is orbitally degenerate and therefore susceptible to Jahn–Teller distortion (Section 6.5). The magnetic moments of the $3d$ tetrahalo complexes demonstrate that they are all high spin, which is in line with the low values of Δ_T.

A wider range of M^{2+} ions is known in solids than in aqueous solution because water can serve as an oxidizing agent. For example, $TiCl_2$ can be prepared by the reduction of $TiCl_4$ with hexamethyldisilane

$$(CH_3)_3SiSi(CH_3)_3(l) + TiCl_4(l) \rightarrow TiCl_2(s) + 2(CH_3)_3SiCl(l)$$

but it is oxidized by both water and air. With the exception of $ScCl_2$, these dihalides are well described by structures containing isolated M^{2+} ions (Table 8.7). As we see from the table, the fluorides have a rutile structure and most heavier halides are layered; in both cases the metal is in an octahedral site. This shift in structure is understandable because the rutile structure is associated with ionic bonding and the layered structures are associated with more covalent bonding. The metal–metal bonded $ScCl_2$ and related M—M bonded dihalides of the $4d$ and $5d$ elements will be discussed in Section 8.9.

Monoxides are known for many of the $3d$ metals (Table 8.8). They

Table 8.7 Structures of dihalides*

	Ti	V	Cr	Mn	Fe	Co	Ni	Cu
F	R	R†	R†	R	R	R	R†	
Cl	L	L	R†	L	L	L	L	L
Br	L	L	L	L	L	L	L	L
I	L	L	L	L	L	L	L	L

*R = rutile, L = layered (CdI_2, $CdCl_2$ or related structures).
†The structure is distorted from the ideal type.
Source: Adapted from A. F. Wells, *Structural inorganic chemistry.* Oxford University Press (1984).

Table 8.8 *d*-block MO compounds

	Group						
	4	5	6	7	8	9	10
Rock-salt structure (shaded)	Ti Zr	V		Mn	Fe	Co	Ni Pd*

*PtS structure (square-planar, four-coordinate metal)
Source: A. F. Wells, *Structural inorganic chemistry.* p. 537. Oxford University Press (1984).

have the rock-salt structure characteristic of ionic solids, but their properties, which will be discussed in more detail in Chapter 18, indicate significant deviations from the simple ionic $M^{2+}O^{2-}$ model. For example, TiO has metallic conductivity and FeO is always iron deficient. The early *d*-block monoxides are strong reducing agents. Thus, TiO is easily oxidized by water or oxygen, and MnO is a convenient oxygen scavenger that is used to lower the concentration of O_2 in inert gases down to the parts-per-billion range.

Example 8.2: *Judging trends in redox stability in the d block*
On the basis of trends in the 3*d* series, suggest suitable M^{2+} aqua ions for use as reducing agents, and write a balanced chemical equation for the reaction of one of these ions with O_2 in acidic solution.

Answer. Because the $+2$ oxidation state is most stable for the late 3*d* elements, strong reducing agents include ions of the metals on the left of the 3*d* series: such ions include $Ti^{2+}(aq)$, $V^{2+}(aq)$, and $Cr^{2+}(aq)$. The ion $Fe^{2+}(aq)$ is only weakly reducing. The ions $Co^{2+}(aq)$, $Ni^{2+}(aq)$, and $Cu^{2+}(aq)$ are not oxidizable in water. The Latimer diagram for iron indicates that Fe^{3+} is the only accessible higher oxidation state of iron in acid solution:

$$Fe^{3+} \xrightarrow{0.770} Fe^{2+} \xrightarrow{-0.44} Fe$$

The chemical equation is then

$$4Fe^{2+}(aq) + O_2(g) + 4H^+(aq) \rightarrow 4Fe^{3+}(aq) + 2H_2O(l)$$

Exercise E8.2. Refer to the appropriate Latimer diagram in Appendix 2 and identify the oxidation state and formula of the species that is thermodynamically favored when an acidic aqueous solution of V^{2+} is exposed to oxygen.

Oxidation state $+2$ of the 4d and 5d elements

In contrast to the 3*d* series, the 4*d* and 5*d* metals only rarely form simple $M^{2+}(aq)$ ions. A few examples have been characterized recently; including $[Ru(OH_2)_6]^{2+}$, $[Pd(OH_2)_4]^{2+}$, and $[Pt(OH_2)_4]^{2+}$. The 4*d* and 5*d* metals form many M(II) complexes with ligands other than

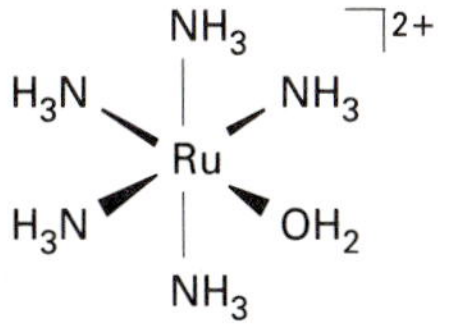

NH₃ structure label:

16 $[Ru(OH_2)(NH_3)_5]^{2+}$

17 $[RuCl_2(PPh_3)_3]$

18 $[Re_2Cl_8]^{2-}$

H_2O; they include the very stable d^6 octahedral complexes, such as (**16**), and the much rarer square-pyramidal d^6 complexes, such as (**17**), that form with bulky ligands.[9] Palladium(II) and platinum(II) form many square-planar d^8 complexes, such as $[PtCl_4]^{2-}$, and they will be discussed in Section 8.10. The Ru(II) complex in (**16**) is obtained by reducing $RuCl_3 \cdot 3H_2O$ with zinc in the presence of ammonia; it is a useful starting material for the synthesis of a range of ruthenium(II) pentaammine complexes that have π-acceptor ligands, such as CO, as the sixth ligand:

$$[Ru(NH_3)_5(OH_2)]^{2+}(aq) + L(g) \rightarrow [Ru(NH_3)_5L]^{2+}(aq) + H_2O(l)$$

$$L = CO,\ N_2,\ N_2O$$

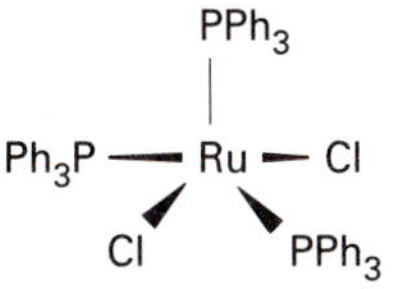

These ruthenium and the related osmium pentaammine species are strong π-donors, and consequently the complexes they form with the π-acceptors CO and N_2 are stable. Ammine ligands in conjunction with a $4d$ or $5d$ metal ion result in a strong ligand field, so the configuration is approximately t_{2g}^6; the electrons in two of the t_{2g} orbitals have π symmetry with respect to the M—CO bond and so can back-donate into that π-acceptor ligand.[10]

8.9 Metal–metal bonded *d*-metal compounds[11]

A general feature of the low oxidation state metals from the early part of the d block is their formation of metal–metal bonded species (Fig 8.4). These clusters are stabilized by π-donor ligands, such as halides and alkoxides. They may involve M—M bonds within a molecular cluster or M—M bonds in extended solid state compounds. When no bridging ligands are present, as in $[Re_2Cl_8]^{2-}$ (**18**), the presence of a metal–metal bond is unambiguous. Although this compound may be prepared by the reduction of ReO_4^- by a conventional reducing agent (such as zinc in dilute acid, or H_3PO_2), it is best prepared by the reduction of ReO_4^- with benzyl chloride.[12] When bridging ligands are present, careful observations and measurement (typically of bond lengths and magnetic properties) are needed to identify direct metal–metal bonding. We shall see in Chapter 16 that there is an extensive range of metal cluster compounds of the middle to late d-block elements that are stabilized by π-acceptor ligands, particularly CO.

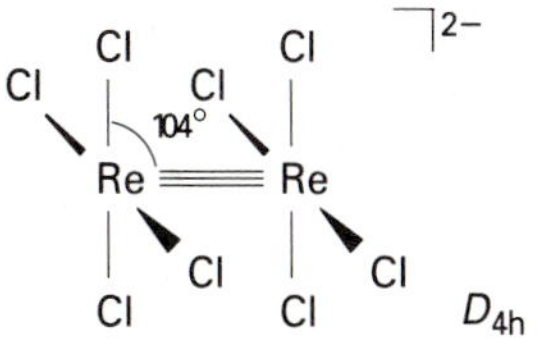

As a consequence of the intricate bonding patterns in most metal cluster compounds, metal–metal bond strengths cannot be determined with great precision. Some evidence, however, such as the stability of

[9] P. R. Hoffman and K. G. Caulton, *J. Am. Chem. Soc.*, **97**, 4221 (1975).

[10] The t_{2g} designation applies strictly to an O_h complex, but the terminology is used here to facilitate the discussion. The precise orbital designation for the C_{4v} symmetry group of the complex are b_2 and e, and the configuration is $b_2^2e^4$. The e electrons form the π bond.

[11] F. A. Cotton and R. A. Walton, *Multiple bonds between metal atoms.* Wiley, New York (1993); M. H. Chisholm, The $\sigma^2\pi^4$ triple bond between molybdenum and tungsten atoms: developing the chemistry of an inorganic functional group. *Angew. Chem., Int. Edn. Engl.*, **25**, 21 (1986).

[12] T. J. Border and R. A. Walton, *Inorg. Synth.*, **23**, 116 (1985).

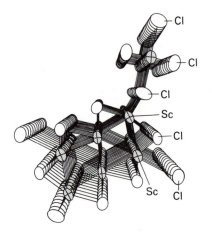

8.18 The structure of Sc_7Cl_{10} showing the single chain of Sc atoms at the top and multiple chains at the bottom. (From J. D. Corbett, *Acc. Chem. Res.*, **14**, 239 (1981).)

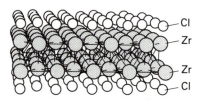

8.19 The structure of ZrCl consists of layers of metal atoms in graphite-like hexagonal nets.

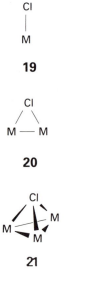

compounds and M—M force constants, indicates that there is an increase in M—M bond strength down a group, perhaps on account of the greater spatial extension of *d* orbitals down a group. This trend may be the reason why there are so many more metal–metal bonded compounds for the 4*d* and 5*d* metals than for their 3*d* counterparts. Figure 8.2 shows that bulk metal–metal bonds in the *d* block are strongest in the 4*d* and 5*d* series, and this feature carries over into their compounds. In contrast, element-element bonds weaken down a group in the *p* block.

Not all the early *d*-block metal–metal bonded compounds are discrete clusters. Many extended metal–metal bonded compounds exist, such as the multiple chain scandium subhalides, Sc_7Cl_{10} and Sc_5Cl_6 (Fig. 8.18) and layered compounds such as ZrCl (Fig. 8.19). In the latter compound, the Zr atoms are within bonding distance in two adjacent layers of metal atoms sandwiched between Cl^- layers. We have already seen that zirconium and scandium adopt their group oxidation states (+3 and +4, respectively) when exposed to air and moisture, so these metal–metal bonded compounds are prepared out of contact with air and moisture. For example, ZrCl is prepared by reducing zirconium tetrachloride with zirconium metal in a sealed tantalum tube at high temperatures:

$$3Zr(s) + ZrCl_4(g) \xrightarrow{600-800\,°C} 4ZrCl(s)$$

True clusters (as distinct from the extended M—M bonded solid state compounds just described) are often soluble and can be manipulated in solution. The π-donor ligands in these clusters typically occur in one of several locations, namely in a terminal position (**19**), bridging two metals (**20**), or bridging three metals, (**21**). Clusters may be linked by bridging halides or chalcogenides in the solid state. For example, solid $MoCl_2$ consists of octahedral Mo clusters linked by Cl bridges. It may be prepared in a sealed glass tube by the reaction of $MoCl_5$ in a mixture of molten $NaAlCl_4$ and $AlCl_3$ with aluminum as the reducing agent:

$$MoCl_5(s) + Al(s) \xrightarrow{NaAlCl_4/AlCl_3(l),\ 200\,°C} MoCl_2(s) + AlCl_3(l)$$

The compound is fairly robust with respect to oxidation under mild conditions, and when treated with hydrochloric acid it forms the anionic cluster $[Mo_6Cl_{14}]^{2-}$. This cluster contains an octahedral array of Mo atoms with a Cl atom bridging each triangular face and a terminal Cl atom on each Mo vertex (**22**). These terminal Cl atoms can be replaced by other halides, alkoxides, and phosphanes. An analogous series of tungsten cluster compounds is known. Once formed, these molybdenum and tungsten compounds can be handled in air and water at room temperature on account of their kinetic barriers to decomposition. The oxidation number of the metal is +2, so the metal valence electron count is $(6-2) \times 6 = 24$. One-electron oxidation and reduction of the clusters is possible.

Similar octahedral clusters are known for niobium and tantalum in

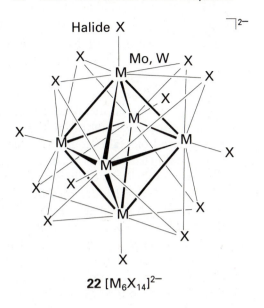

22 $[M_6X_{14}]^{2-}$

Group 5, and for zirconium in Group 4. The cluster $[Nb_6Cl_{12}L_6]^{2+}$ and its tantalum analog have an octahedral framework with edge-bridging Cl atoms and six terminal ligands (**23**). The metal valence electron count in this case is 16 but these clusters can be oxidized in several steps. One example from Group 4 is the cluster $[Zr_6Cl_{18}C]^{4-}$, which has the same metal and Cl atom array together with a C atom in the center of the octahedron. Here the metal electron count is 14. Clearly, there is not a high preference for a specific electron count in these clusters.

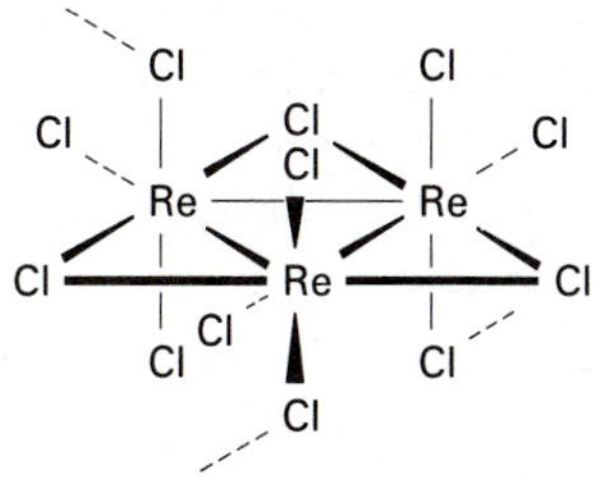

Rhenium trichloride, which consists of Re_3Cl_9 clusters linked by weak halide bridges (Fig. 8.20), provides the starting material for the preparation of a set of three-metal clusters that have been studied thoroughly. As with molybdenum dichloride, the intercluster bridges

8.20 The structure of $ReCl_3$ in the solid state. The dotted bonds correspond to interaction with a Cl atom belonging to an adjacent cluster.

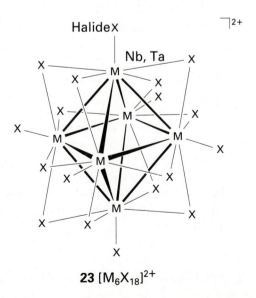

23 $[M_6X_{18}]^{2+}$

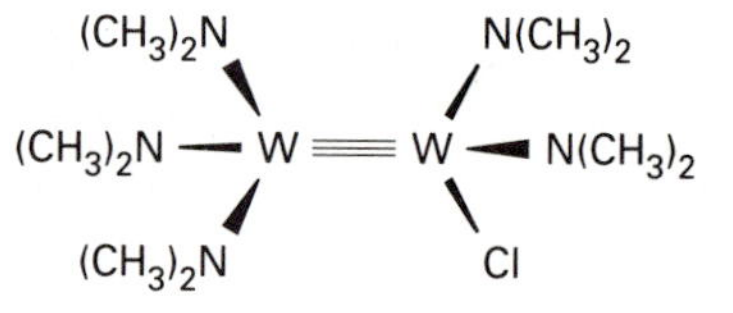

24 [((CH₃)₂N)₃WWCl(N(CH₃)₂)₂]

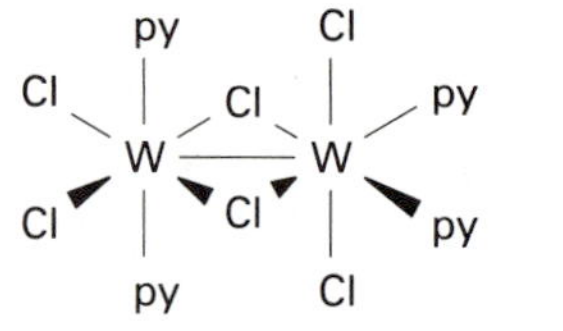

25 [W₂(py)₄Cl₆]

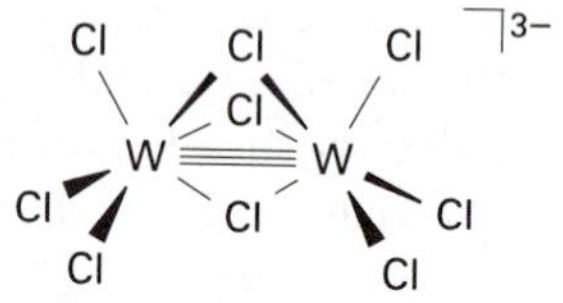

26 [W₂Cl₉]³⁻

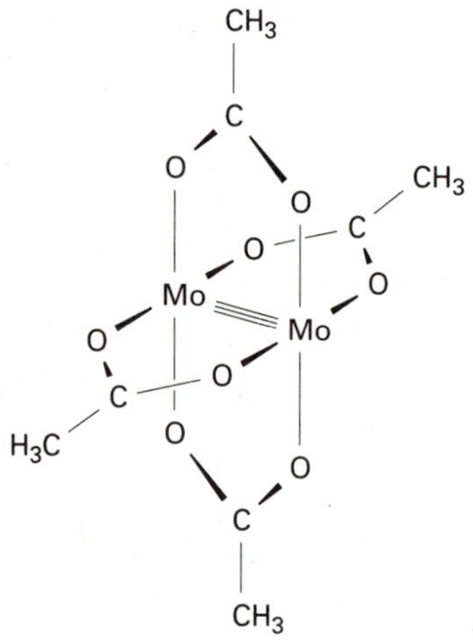

27 [Mo₂(CH₃CO₂)₄]

in the solid state can be broken by reaction with potential ligands. For example, treatment of Re_3Cl_9 with Cl^- ions produces the discrete complex $[Re_3Cl_{12}]^{3-}$. Neutral ligands, such as alkylphosphanes, can also occupy these coordination sites and result in clusters of the general formula $Re_3Cl_9L_3$.

Many two-metal metal–metal bonded compounds are known; some of the common structural motifs are an ethane-like structure (**24**), an edge-shared bioctahedron (**25**), a face-shared bioctahedron (**26**), and a tetragonal prism, which we have already encountered with $[Re_2Cl_8]^{2-}$ (**18**). We shall focus on bonding in the last of these structural types, in which the bond order may be considered as lying between 1 and 4. Figure 8.21 shows that a σ bond between two metals can arise from the overlap of a d_{z^2} orbital from each metal, π bonds can arise from the overlap of d_{zx} or d_{yz} orbitals (two such π bonds are possible), and a δ bond can be formed from the overlap of two face-to-face d_{xy} orbitals on two different metal atoms (the remaining $d_{x^2-y^2}$ orbital is used in the M—L σ bonds). A quadruple M≡M bond results when all the bonding orbitals are occupied (Fig. 8.22) and the configuration is $\sigma^2\pi^4\delta^2$. Evidence for quadruple bonding comes from the observation that $[Re_2Cl_8]^{2-}$ has an eclipsed array of Cl ligands, which is sterically unfavorable. It is argued that the δ bond, which is formed only when the d_{xy} orbitals are confacial, locks the complex in the eclipsed configuration. Another well known quadruply bonded compound is molybdenum(II) acetate (**27**), which is prepared by heating the molybdenum(0) compound $Mo(CO)_6$ with acetic acid:

$$2Mo(CO)_6 + 4CH_3COOH \rightarrow Mo_2(O_2CCH_3)_4 + 2H_2 + 12CO$$

The quadruply bonded molybdenum acetato complex is an excellent starting material for other Mo≡Mo compounds. For example, the quadruply bonded chloro complex is obtained when the acetato complex is treated with concentrated hydrochloric acid at below room temperature:

$$Mo_2(OCCH_3)_4 + 4H^+(aq) + 8Cl^-(aq) \rightarrow$$

$$[Mo_2Cl_8]^{4-}(aq) + 4CH_3COOH(aq)$$

As shown in Table 8.9, triply-bonded M≡M systems arise for tetragonal prismatic complexes when δ and δ^* orbitals are both occupied. These complexes are more numerous than the quadruply bonded complexes and, because δ bonds are weak, M≡M bond lengths are often similar to those of quadruply bonded systems. Many of the triply bonded systems also include bridging ligands (Table 8.9). The table shows that singly occupied δ or δ^* orbitals lead to a formal bond order of $3\frac{1}{2}$. Once the δ and δ^* orbitals are fully occupied, successive filling of the two higher lying π^* orbitals leads to further reduction in the bond order from $2\frac{1}{2}$ to 1.

Like carbon–carbon multiple bonds, metal–metal multiple bonds are centers of reaction, but the variety of structures resulting from the reactions of metal–metal multiply bonded compounds is more diverse

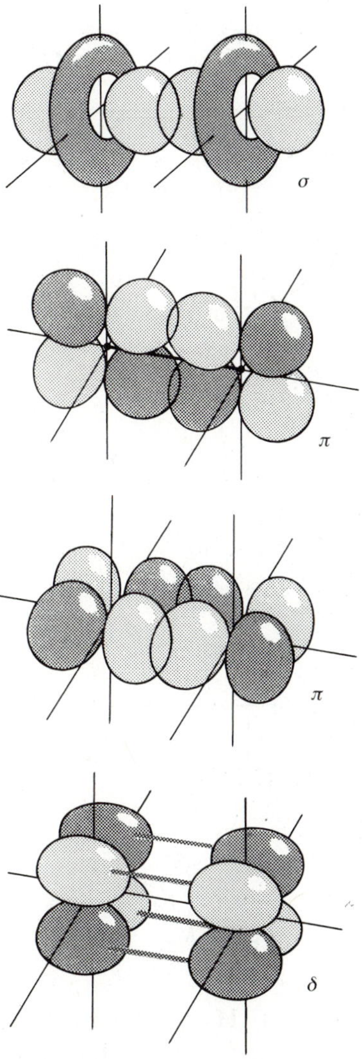

8.21 The origin of σ, π, and δ interactions between two d-block metal atoms situated along the z-axis. Only the bonding combinations are shown.

Table 8.9 Examples of metal–metal bonded tetragonal prismatic complexes

Complex*	Configuration	Bond order	M–M bond length/Å
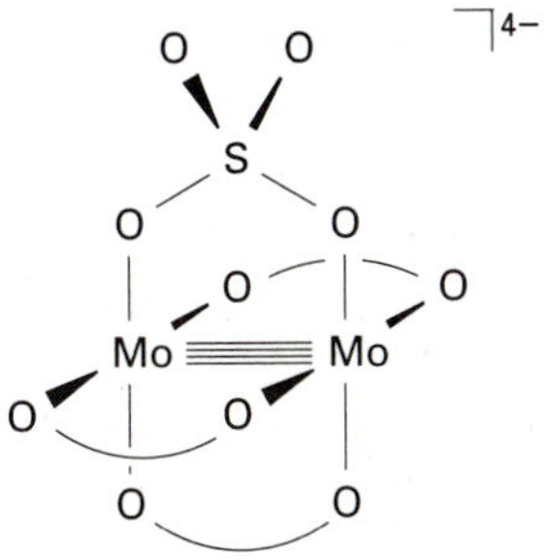	$\sigma^2\pi^4\delta^2$	4	2.11
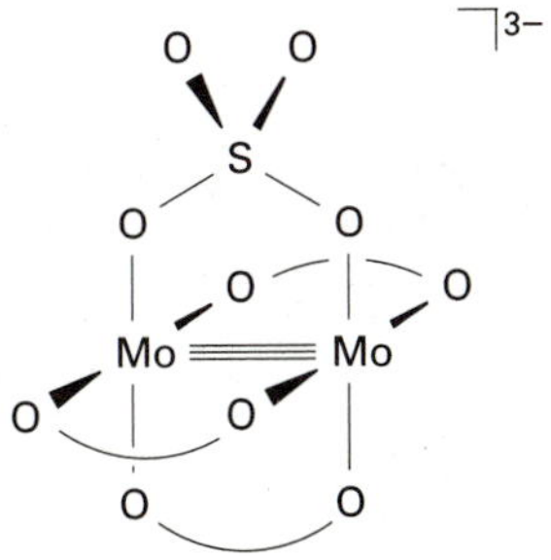	$\sigma^2\pi^4\delta^1$	3.5	2.17
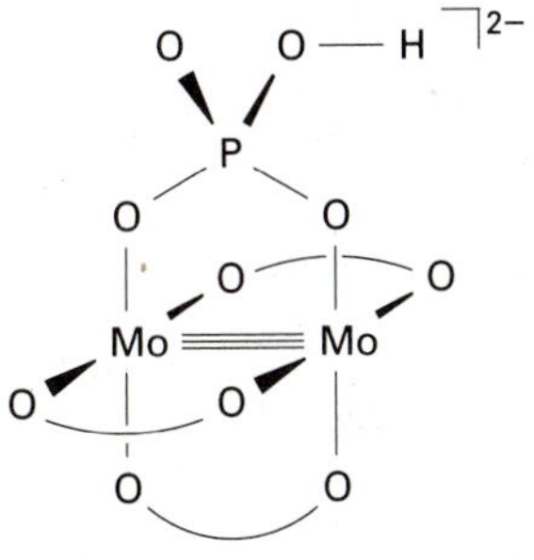	$\sigma^2\pi^4\delta^2\delta^{*2}$	3	2.22
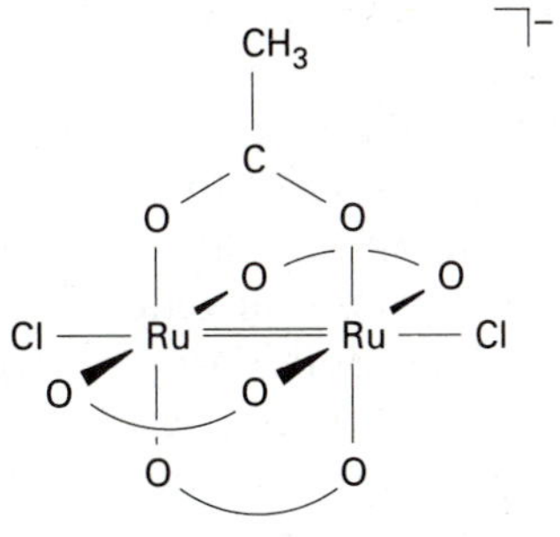	$\sigma^2\pi^4\delta^2\delta^*\pi^{*2}$	2.5	2.27
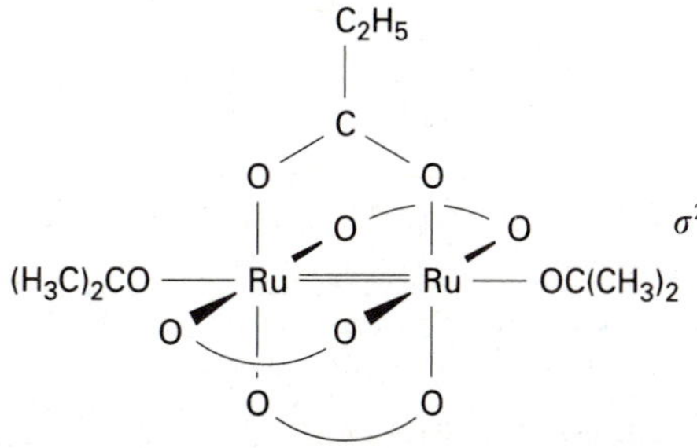	$\sigma^2\pi^4\delta^2\delta^{*2}\pi^{*2}$	2	2.26

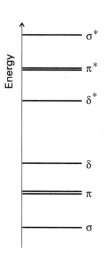

8.22 Approximate molecular orbital energy level scheme for the M–M interactions in tetragonal prismatic two-metal clusters.

Table 8.9 continued

Complex*	Configuration	Bond order	M–M bond length/Å
	$\sigma^2\pi^4\delta^2\delta^{*2}\pi^{*3}$	1.5	2.32
	$\sigma^2\pi^4\delta^2\delta^{*2}\pi^{*4}$	1	2.39

Source: F. A. Cotton and G. Wilkinson, *Advanced inorganic chemistry*. Wiley, New York (1988) and F. A. Cotton. *Chem. Soc. Rev.*, **12**, 35 (1983).
*When multiple bridging ligands are present only one is shown in detail.

than for organic compounds.[13] For example,

$$Cp(OC)_2Mo \equiv Mo(CO)_2Cp \quad + \quad HI \quad \longrightarrow \quad Cp(OC)_2Mo \underset{I}{\overset{H}{\diagdown\diagup}} Mo(CO)_2Cp$$

where Cp denotes the cyclopentadienyl group, C_5H_5. In this reaction, HI adds across a triple bond but both the H and I bridge the metal atoms; the outcome is quite unlike the addition of HX to an alkyne, which results in a substituted alkene. The reaction product can be regarded as containing a 3c,2e M—H—M bridge (Section 2.7) and an I atom bonded by two conventional 2c,2e bonds, one to each Mo atom.

It is possible to build larger metal clusters by addition to a metal–metal multiple bond. For example, [Pt(PPh$_3$)$_4$] loses two triphenylphosphane ligands when it adds to the Mo–Mo triple bond, resulting in a three-metal cluster:

$$Cp(OC)_2Mo \equiv Mo(CO)_2Cp + Pt(PPh_3)_4 \longrightarrow Cp(CO)_2Mo = Mo(CO)_2Cp + 2\,PPh_3$$

[13] For several reviews see *Reactivity of metal–metal bonds* (ed. M. H. Chisholm), ACS Symposium Series 155, American Chemical Society, Washington, DC (1981); R. E. McCarley, T. R. Ryan, and C. M. C. Torardi, p. 41; R. A. Walton, p. 207; M. D. Curtis, L. Messerle, N. A. Fotinus, and R. G. Gerlach, p. 221; and A. F. Dyke, S. R. Finnimore, S. A. R. Knox, P. J. Naish, A. G. Orpen, G. H. Riding, and G. E. Taylor, p. 259.

Example 8.3: *Deciding probable structures of metal halides*

List the major structural classes of *d*-metal halides and decide the most likely class for (a) MnF_2, (b) WCl_2, (c) RuF_6, and (d) FeI_2.

Answer. The difluorides of the 3*d* metals (MnF_2) have the simple rutile structure characteristic of ionic AB_2 compounds; heavier halides (FeI_2) are commonly layered. The chlorides, bromides, and iodides of the low oxidation state early 4*d* and 5*d* metals have metal–metal bonds (WCl_2 is in this class; in fact it contains a W_6 cluster). The hexahalides are molecular (RuF_6).

Exercise E8.3. Describe the probable structure of the compound formed when Re_3Cl_9 is dissolved in a solvent containing PPh_3.

8.10 Noble character

Another general feature of the *d* block is that metals on the right of the block are resistant to oxidation. This resistance is most evident for silver, gold, and the 4*d* and 5*d* metals in Groups 8 through 10 (Fig. 8.23). The latter are referred to as the **platinum metals** because they occur together in platinum-bearing ores. In recognition of their traditional use, copper, silver, and gold are referred to as the **coinage metals**. Gold occurs as the metal; silver, gold, and the platinum metals are also recovered in the electrolytic refining of copper. The prices of the individual platinum metals vary widely because they are recovered together but their consumption is not proportional to their abundance. Rhodium is by far the most expensive metal in this group because it is widely used in industrial catalytic processes and in automotive catalytic converters (Chapter 17). For example, rhodium is about 40 times more costly than the less catalytically useful metal palladium even though they occur in similar abundance.

Copper, silver, and gold are not susceptible to oxidation by hydrogen ions under standard conditions, and this noble character accounts for their use, together with platinum, in jewelry and ornaments. *Aqua regia*, a 3:1 mixture of concentrated hydrochloric and nitric acids, is an old but effective reagent for the oxidation of gold and platinum. Its function is twofold: the NO_3^- ions provide the oxidizing power and the Cl^- ions act as complexing agents. The overall reaction is

$$Au(s) + 4H^+(aq) + NO_3^-(aq) + 4Cl^-(aq) \rightarrow$$
$$[AuCl_4]^-(aq) + NO(g) + 2H_2O(l)$$

The active species in solution are thought to be Cl_2 and $NOCl$, which are generated in the reaction

$$3HCl(aq) + HNO_3(aq) \rightarrow Cl_2(aq) + NOCl(aq) + 2H_2O(l)$$

Oxidation state preferences are erratic in Group 11. For copper, the +1 and +2 states are most common, but for silver +1 is typical and for gold +1 and +3 are common. The simple aqua ions $Cu^+(aq)$ and

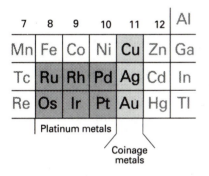

8.23 The locations of the platinum and coinage metals in the periodic table.

Au$^+$(aq) undergo disproportionation in aqueous solution:

$$2Cu^+(aq) \rightarrow Cu(s) + Cu^{2+}(aq)$$

$$3Au^+(aq) \rightarrow 2Au(s) + Au^{3+}(aq)$$

Complexes of Cu(I), Ag(I) and Au(I) often have a linear structure. For example, [H$_3$NAgNH$_3$]$^+$ forms in aqueous solution and linear [XAgX]$^-$ complexes have been identified by X-ray crystallography. The currently preferred explanation for the tendency toward linear coordination is the similarity in energy of the outer ns, np, and $(n-1)d$ orbitals, which permits the formation of collinear spd hybrids (Fig. 8.24).

The soft character of Cu$^+$, Ag$^+$, and Au$^+$, which also results from the relatively small energy difference between highest occupied and lowest unoccupied orbitals of these ions, is illustrated by their affinity order, which is I$^-$ > Br$^-$ > Cl$^-$. Complex formation, as in the formation of [Cu(NH$_3$)$_2$]$^+$ and [AuI$_2$]$^-$, provides a means of stabilizing the +1 oxidation state of these metals in aqueous solution. Many tetrahedral complexes are also known for Cu(I), Ag(I), and Au(I).

Square-planar complexes are common for the platinum metals and gold in oxidation states that yield the d^8 electronic configuration, which include Rh(I), Ir(I), Pd(II), Pt(II), and Au(III). An example is [Pt(NH$_3$)$_4$]$^{2+}$. Characteristic reactions for these complexes are ligand substitution (Section 6.8) and, except for gold(III) complexes, **oxidative addition**. In the latter reaction, a ligand XY is cleaved and the components X and Y fill two coordination sites to give a six-coordinate complex:

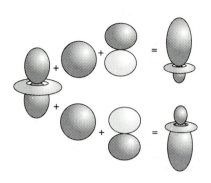

8.24 The hybridization of s, p_z, and d_{z^2} with the choice of phases shown here produces a pair of collinear orbitals that can be used to form strong σ bonds.

The classification of the reaction as 'oxidative addition' stems from the formal increase in oxidation state of the metal center (from Pt(II) to Pt(IV), and from Ir(I) to Ir(III), in the two examples above) upon addition of a nonmetal species of formula XY. Because negative oxidation numbers are assigned to nonmetal atoms attached to a metal, the addition of H$_2$, CH$_3$I, or HCl leads to an increase of 2 in the oxidation number of the metal. This type of reaction is frequently implicated in the mechanisms for catalytic reactions of platinum metal complexes (see Section 17.3). The properties of platinum complexes with organic ligands will be discussed in Chapter 16. An interesting series of Pt(I) and Pd(I) complexes has been found in which two M—M bonded metal atoms are bridged by phosphane ligands (**28**). These compounds undergo reactions in which a two-electron ligand inserts into the M–M bond:

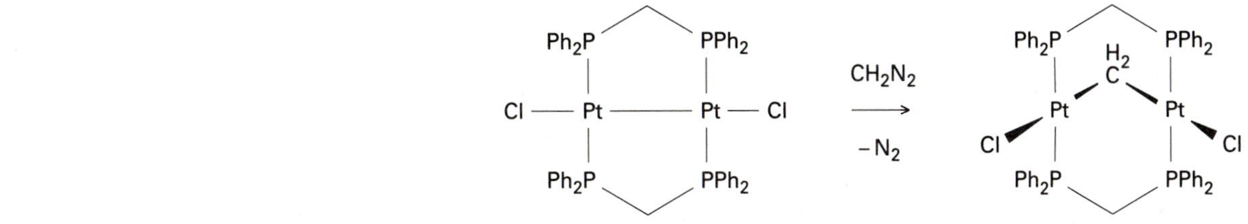

28   **29**

In this example, the CH_2 group inserts into the M—M bond and the resulting compound (**29**), which is informally called an 'A-frame complex', now contains two square-planar Pt(II) groups bridged by CH_2 as well as the phosphane ligands.

8.11 Metal sulfides and sulfido complexes

Sulfur is softer and less electronegative than oxygen; it therefore has a broader span of oxidation states and a stronger affinity for the softer metals on the right of the *d* block. For example, zinc(II) sulfide is readily precipitated when $Zn^{2+}(aq)$ is added to an aqueous solution containing H_2S and NH_3 (to adjust the pH) but $Sc^{3+}(aq)$ gives $Sc(OH)_3(s)$ instead. The covalent contribution to the lattice enthalpy of the soft metals (e.g. Ag^+) sulfides cannot be overcome in water, and in general these sulfides are only very sparingly soluble. Another striking feature of sulfur is its tendency to form catenated sulfide ions. Thus, in combination with alkali metals a broad set of compounds containing S_n^{2-} ions can be prepared, and the smaller polysulfides can serve as chelating ligands toward *d*-block metal ions.

Monosulfides

As with the *d*-metal monoxides, the monosulfides are most common in the first series of the *d* block (Table 8.10). In contrast to the monox-

Table 8.10 Structures of *d*-block MS compounds*

	Group						
	4	5	6	7	8	9	10
Nickel-arsenide structure (shaded)	Ti	V		Mn†	Fe	Co	Ni
Rock-salt structure (unshaded)	Zr	Nb					

*Metal monosulfides not shown here for Group 6; some of the heavier metals have more complex structures.
†MnS has two polymorphs; one has a rock-salt structure, the other has a wurtzite structure.
Source: A. F. Wells, *Structural inorganic chemistry.* p. 752. Oxford University Press (1984).

Table 8.11 Structures of *d*-block MS$_2$ compounds*

	Group							
	4	5	6	7	8	9	10	11
Layered (shaded)	Ti			Mn	Fe	Co	Ni	Cu
Pyrite or marcasite (unshaded)	Zr	Nb	Mo		Ru	Rh		
	Hf	Ta	W	Re	Os	Ir	Pt	

*Metals not shown either do not form disulfides or have disulfides with complex structures.
Source: A. F. Wells, *Structural inorganic chemistry.* p. 757. Oxford University Press (1984).

ides, most of the monosulfides have the nickel-arsenide structure (Fig. 4.16). The different structures are consistent with the rock-salt structure of the monoxides being favored by the ionic (harder) cation–anion combinations. The nickel-arsenide structure is favored by more covalent (softer) combinations and is found only when there is appreciable metal–metal bonding and correspondingly shorter metal–metal distances.

Disulfides

The disulfides of the *d* metals fall into two broad classes (Table 8.11). One class consists of layered compounds with the CdI$_2$ or MoS$_2$ structure and the other of compounds containing discrete S$_2^{2-}$ groups.

The layered disulfides are built from a sulfide layer, a metal layer, and then another sulfide layer (Fig. 8.25). These sandwiches stack together in the crystal with sulfide layers in one slab adjacent to a sulfide layer in the next. Clearly, this crystal structure is not in harmony with a simple ionic model and its formation is a sign of covalence in the bonds between the soft sulfide ion and *d*-metal cations. The metal ion in these layered structures is surrounded by six S atoms. Its coordination environment is octahedral in some cases (PtS$_2$, for instance) and trigonal prismatic in others (MoS$_2$). The layered MoS$_2$ structure is favored by S—S bonding as indicated by short S—S distances within each of the MoS$_2$ slabs. The common occurrence of the trigonal prismatic structure in many of these compounds is in striking contrast to isolated metal complexes, where the octahedral arrangement of ligands is by far the most common.

Some of the layered metal sulfides readily undergo **intercalation reactions** in which ions or molecules penetrate between adjacent sulfide layers, often with accompanying redox reactions:

$$0.6Na(am) + TaS_2(s) \rightarrow Na_{0.6}TaS_2(s)$$

In this reaction, sodium dissolved in liquid ammonia (Section 8.5) gives up an electron to a vacant band in TaS$_2$, and the Na$^+$ ion worms its way into positions between the sulfide layers. Similar intercalation reactions are common for graphite (Section 11.5) and various *d*-block

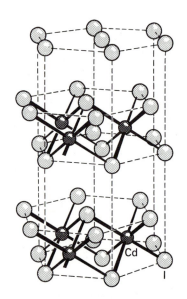

8.25 The CdI$_2$ structure adopted by many disulfides. The latter have adjacent sulfide layers without an intervening layer of metal ions.

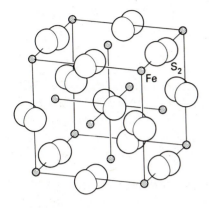

8.26 The structure of pyrite, FeS$_2$.

oxides and sulfides (Section 18.7). The intercalation reaction between lithium and TiS$_2$ has been investigated for use in lightweight automobile batteries that can be recharged rapidly because the solid-state reaction involves no essential structural change.

Compounds containing discrete S$_2^{2-}$ ions adopt the pyrite (Fig. 8.26) or marcasite structure. The stability of the formal S$_2^{2-}$ ion in metal sulfides is much greater than that of the O$_2^{2-}$ ion in peroxides, and there are many more metal sulfides in which the anion is S$_2^{2-}$ than there are peroxides.

Example 8.4: *Contrasting the structures of two different d-block disulfides*
Contrast the structures of MoS$_2$ and FeS$_2$ and explain their existence in terms of the oxidation states of the metal ions.

Answer. The point to be decided is whether the metals are likely to be in the $+4$ oxidation state, in which case the two S atoms would be present as S^{2-} ions. If the $+4$ oxidation state is relatively inaccessible, the metal might be in the $+2$ oxidation state, and the S atoms present as the S—S bonded species S$_2^{2-}$. Because it is a fair reducing agent, S^{2-} will only be found with a metal ion in an oxidation state that is not easily reduced. As with many of the d metals in Periods 5 and 6, molybdenum is easily oxidized to Mo(IV); therefore Mo(IV) can coexist with S^{2-}. Molybdenum(IV) sulfide (MoS$_2$) has the layered structure typical of metal disulfides. Iron is readily oxidized to Fe(II) but not to Fe(IV). Therefore, Fe(IV) cannot coexist with S^{2-}. The compound is therefore likely to contain Fe(II) and S$_2^{2-}$. The mineral name for FeS$_2$ is pyrite; its common name 'fools gold' indicates its misleading color.

Exercise E8.4. Molybdenum(IV) sulfide is a very effective lubricant. Present a plausible reason for this property.

Sulfido complexes

The coordination chemistry of sulfur is quite different from that of oxygen. Much of this difference is connected with sulfur catenation and preference for metal centers that are not highly oxidizing. Research on Fe—S cluster complexes has flourished recently as a result of the discovery that they are present in electron-transfer and nitrogen fixing enzymes (Section 19.8).[14] The structure of one such model compound, [Fe$_4$S$_4$(SR)$_4$]$^{2-}$, was shown in Fig. 6.5. It is readily prepared from simple starting materials in the absence of air:

$$4FeCl_3 + 4HS^- + 6RS^- + 4CH_3O^- \xrightarrow{\text{Methanol}}$$

$$[Fe_4S_4(SR)_4]^{2-} + RS-SR + 12Cl^- + 4CH_3OH$$

The observation that this reaction is successful with many different R

---

[14] R. H. Holm, S. Ciurli, and J. A. Weigel, *Prog. Inorg. Chem.*, **38**, 1 (1990).

groups, and its good yield, indicates that the 4Fe–4S cage is thermo-dynamically more stable than other possibilities. The HS^- ion provides the sulfide ligands for the cage, the RS^- ion serves both as a ligand and a reducing agent, and the CH_3O^- ion acts as a base. The cubic cluster contains Fe and S atoms on alternate corners, so each S atom bridges three Fe atoms. Each of the thiolate groups, RS^-, occupies a terminal position on an Fe atom. The cluster remains intact upon one-electron reduction to $[Fe_4S_4(SR)_4]^{3-}$, and analogous clusters are implicated in the redox reactions of the electron transfer protein fer-redoxin.

Simple thiometallate complexes such as $[MoS_4]^{2-}$ can be synthesized easily by passing H_2S gas through a strongly basic aqueous solution of the molybdate or tungstate ion:

$$[MoO_4]^{2-}(aq) + 4H_2S(g) \rightarrow [MoS_4]^{2-}(aq) + 4H_2O(l)$$

These tetrathiometallate anions are building blocks for the synthesis of complexes containing more metal atoms. For example, they will coordinate to many dipositive metal ions, such as Co^{2+} and Zn^{2+}:

$$Co^{2+}(aq) + 2[MoS_4]^{2-}(aq) \rightarrow [S_2MoS_2CoS_2MoS_2]^{2-}(aq)$$

The polysulfides such as S_2^{2-} and S_3^{2-}, which are formed by addition of elemental sulfur to a solution of ammonium sulfide, can also act as ligands. An example is $[Mo_2(S_2)_6]^{2-}$ (**30**), which is formed from ammonium polysulfide and MoO_4^{2-}; it contains side-bonded S_2^{2-} ligands. The larger polysulfides bond to metal atoms forming chelate rings, as in $[MoS(S_4)_2]^{2-}$ (**31**), which contains chelating S_4 ligands.

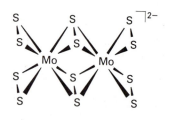

30 $[Mo_2(S_2)_6]^{2-}$

31 $[MoS(S_4)_2]^{2-}$

THE ELEMENTS OF GROUP 12

We shall see in this section that Group 12 displays trends in oxidation states that contrast strongly with those just described for the other *d*-block metals. Thus, the noble character—the resistance to oxidation—that developed across the *d* block is suddenly lost at Group 12. We shall see below that there is a connection between the greater ease of oxidation of these metals and an abrupt lowering of *d*-orbital energies at the end of the *d*-block.

8.12 Occurrence and recovery

Zinc is by far the most abundant element in Group 12. It ranks twenty-third in crustal abundance, just ahead of copper. Cadmium and mer-cury are much less abundant; they are less abundant even than most of the lanthanides. Sulfides are the principal ores for the elements in the group, with zinc and cadmium often occurring together (Table 8.12). Zinc sulfide is roasted in air to produce the oxide;

$$ZnS(s) + \tfrac{3}{2}O_2(g) \rightarrow ZnO(s) + SO_2(g)$$

The oxide is then reduced in a blast furnace charged with carbon. The reduction occurs primarily by CO in a hot portion of the blast furnace,

Table 8.12 Occurrence and methods of recovery of the Group 12 metals

Metal	Principal minerals	Method of recovery
Zinc	Sphalerite, ZnS	$ZnS + \frac{3}{2}O_2 \rightarrow ZnO + SO_2$ followed by $2ZnO + C \rightarrow 2Zn + CO_2$
Cadmium	Traces in zinc ores	
Mercury	Cinnabar, HgS	$HgS + O_2 \rightarrow Hg + SO_2$

as was shown in Fig. 7.4 for the reduction of iron. When cadmium and zinc occur together, the sulfide ore is roasted in air to produce a mixture of the metal sulfates and oxides. This mixture is dissolved in sulfuric acid and reduced. The separation is based on the greater ease of reduction of Cd^{2+} compared with Zn^{2+}.

Mercury occurs in the bright red mineral cinnabar, HgS, which was at one time used as a pigment (vermilion) by artists, a practice which has been discontinued because of the toxicity of mercury. Elemental mercury is recovered from the sulfide by roasting in air.

8.13 Redox reactions

Zinc and cadmium are much more readily oxidized than their neighbors copper and silver in Group 11. This difference is apparent in the standard potentials, which are much lower for Zn^{2+} (-0.76 V) and Cd^{2+} (-0.40 V) than for Cu^{2+} ($+0.34$ V) and Ag^+ ($+0.80$ V). The origin of the thermodynamic difference between the two groups is the lower sublimation enthalpy, and to a lesser extent the lower ionization enthalpy, of Group 12 elements compared with those of Group 11: for $\Delta H_{sub}^{\ominus}$ is 338 kJ mol^{-1} for Cu but only 131 kJ mol^{-1} for Zn (Fig. 8.2). This decrease in sublimation enthalpy in turn reflects a decrease in metal–metal bond strength from Group 11 to 12. A variety of evidence suggests that the weaker metal–metal bonds arise from the lack of d-orbital contribution to the bonding in Group 12. The decrease in energy of the d orbitals beyond Group 12 was illustrated in Fig. 1.23.

Another difference from other d-block elements is the substantial change in chemical properties between the lightest d-block element (Zn) and its congeners (Cd and Hg). For example, mercury is much less electropositive than zinc and cadmium and is alone in having a $+1$ oxidation state (the Hg_2^{2+} ion) that survives in aqueous solution. Partly as a result of this stability, Hg_2^{2+} is a far more important species than Zn_2^{2+} and Cd_2^{2+}.

The chemical, spectroscopic, and X-ray structural evidence that Hg(I) occurs as the dinuclear cation Hg_2^{2+} provided the first example of a metal–metal bonded species. Much later Cd_2^{2+} was identified in $Cd_2(AlCl_4)_2$, and there is spectroscopic evidence for the formation of Zn_2^{2+} in the reaction of zinc metal with molten $ZnCl_2$. The Cd_2^{2+} compound $Cd_2(AlCl_4)_2$ is well characterized. It appears to owe its existence in part to the presence of the large anion, which helps to stabilize large cations (Section 4.8). The synthesis of the Cd_2^{2+} compound is performed in a strictly nonaqueous molten salt medium:

$$CdCl_2(l) + 2AlCl_3(l) + Cd(s) \rightarrow Cd_2[AlCl_4]_2(s)$$

Cadmium(I) rapidly disproportionates in the presence of water:

$$Cd_2^{2+}(aq) \rightarrow Cd(s) + Cd^{2+}(aq)$$

The substantial hydration enthalpy of Cd^{2+} provides a driving force for this reaction. The situation with mercury(I) ions is not radically different because it has a measurable equilibrium constant for disproportionation:

$$Hg_2^{2+}(aq) \rightleftharpoons Hg(l) + Hg^{2+}(aq) \quad K = 6.0 \times 10^{-3}$$

This reaction is driven to the right by ligands that strongly complex or precipitate Hg^{2+}: the addition of CN^- yields a stable Hg(II) complex:

$$Hg_2^{2+}(aq) + 2CN^-(aq) \rightarrow Hg(l) + Hg(CN)_2(aq)$$

and OH^- gives mercury(II) oxide:

$$Hg_2^{2+}(aq) + 2OH^-(aq) \rightarrow Hg(l) + HgO(aq) + H_2O(l)$$

8.14 Coordination chemistry

Mercury(II) forms some linear two-coordinate complexes, such as $Hg(CN)_2$, $Hg(CH_3)_2$, and O—Hg—O in solid mercury(II) oxide, HgO. Zinc and cadmium usually exhibit higher coordination numbers, 4 to 6. (One major exception to this rule is that zinc, cadmium, and mercury all form linear alkyl compounds, $M(CH_3)_2$, Section 10.8.) The coordination of zinc in ZnO is tetrahedral, and cadmium in CdO is octahedral, in the rock-salt structure. Both the reduced tendency toward linear coordination and the harder character of Zn^{2+} compared with Cu^+, and of Cd^{2+} compared with Ag^+, can be traced to the much larger energy difference between filled d orbitals and empty s and p orbitals that occurs at Group 12 (see Section 1.7).

The increase in amphoterism of lower oxidation states on going from left to right across Period 4 was illustrated in Fig. 5.6. In Group 12 we find that Zn^{2+} is amphoteric:

$$Zn(OH)_2(s) + 2H^+(aq) \rightarrow Zn^{2+}(aq) + 2H_2O(l)$$

$$Zn(OH)_2(s) + OH^-(aq) \rightarrow Zn(OH)_3^-(aq)$$

Cadmium hydroxide also reacts with hydroxide ions, but it requires more concentrated base to dissolve. Mercury(II) hydroxide is not amphoteric.

In keeping with their closed subshell (d^{10}) electronic configurations, the Group 12 cations Zn^{2+}, Cd^{2+}, and Hg^{2+} are colorless in aqueous solution. However, some of their heavy halides and chalcogenides are colored. For example, $ZnCl_2$ and ZnI_2 are colorless, but only the chlorides of cadmium and mercury are colorless; CdI_2 is yellow and

HgI_2 occurs as two polymorphs, one of which is yellow and the other red. The presence of heavy halides shifts the energy of these charge-transfer transitions from the ultraviolet (for the colorless materials) to the visible region of the spectrum. Compounds that owe their color to charge-transfer transitions are very useful in practice (for pigments) because their absorptions are more intense than those of d–d transitions.

THE p-BLOCK METALS

To provide comparisons and contrasts with the chemical properties of the other metallic elements, we shall describe the redox stability and some coordination chemistry of the p-block elements aluminum, gallium, indium, and thallium in Group 13/III, of tin and lead in Group 14/IV, and of bismuth in Group 15/V. Further information on these elements in relation to the p-block nonmetals is presented in Chapters 11 and 12.

A general summary is that, in contrast to the elements in the d block:

> The heavier p-block metals favor low oxidation states.

This trend is illustrated in Fig. 8.27(a) for the Group 13/III metals, where the maximum oxidation state is easily achieved for gallium but not for thallium. The latter favors the +1 oxidation state. This tendency to favor an oxidation state 2 less than the group oxidation number also occurs in Groups 14/IV (Fig. 8.27(b)) and 15/V and is an example of the inert pair effect (Section 1.10). When the inert-pair effect was first introduced, we warned that it might not have a simple explanation. It is probably best attributed to the low M—X bond enthalpies for the heavy p-block elements and the fact that it requires less energy to oxidize an element to a low oxidation state than to a higher oxidation state. This energy has to be supplied by ionic or covalent bonds, so if bonding to a particular element is weak, the high oxidation state may be inaccessible.

The most commonly encountered oxidation states for the three heaviest elements in these groups are Tl(I), Pb(II), and Bi(III), and compounds containing these elements in their group oxidation states, Tl(III), Pb(IV), and Bi(V), are easily reduced. The highly oxidizing character of the heavy p-block metals in their group oxidation states is also reflected in the limited range of binary compounds of that state of the metal. For example, TlX_3 exists for X = F, Cl, and Br, but TlI_3 contains the Tl^+ cation in combination with the I_3^- anion. Similarly, PbF_4 is known, $PbCl_4$ does not survive slightly above room temperature, and neither $PbBr_4$ nor PbI_4 are known.

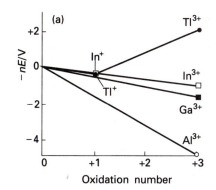

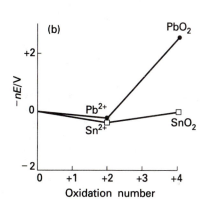

8.27 Frost diagrams for the early p-block metals in acid solution. (a) Group 13/III metals, (b) Group 14/IV metals.

8.15 Occurrence and recovery

The abundances of the p-block metals in the Earth's crust vary widely from aluminum, which is the third most abundant element (behind oxygen and silicon), to bismuth, which is the heaviest element with a stable isotope. Gallium, which is more abundant than lithium, boron,

Table 8.13 Occurrence and methods of recovery of commercially important metals in Groups 13/III to 15/V

Metal	Principal minerals	Method of recovery
Group 13		
Aluminum	Bauxite, $Al_2O_3 \cdot xH_2O$	Electrolytic (Hall process)
Gallium	Traces in aluminum and zinc ores	
Group 14		
Tin	Cassiterite, SnO_2	$SnO_2 + C \rightarrow Sn + CO_2$
Lead	Galena, PbS	$PbS + \frac{3}{2}O_2 \rightarrow PbO + SO_2$
		followed by $2PbO + C \rightarrow 2Pb + CO_2$
Group 15		
Bismuth	Traces in zinc, copper, and lead sulfide ores	

lead, and many other familiar elements, is an expensive element because it is widely dispersed in aluminum- and iron-containing minerals. Gallium is difficult to recover because the chemical properties of Ga^{3+} and Al^{3+} and Fe^{3+} are similar owing to their similar radii and acid–base properties. The recovery of these elements is summarized in Table 8.13. The lighter, more electropositive metals aluminum and gallium are recovered from oxides. The energy demands are high for the high-temperature electrolytic recovery of the most electropositive element, aluminum (Section 7.1), but the price is moderated by the abundance of bauxite and the economics of large-scale production.

The least abundant elements considered here are thallium and bismuth. The occurrence in this group of three elements from Period 6 is in harmony with the general observation of lower nuclear binding energies with increasing atomic number. Lead does not follow this trend, and is more abundant than the lanthanide elements and germanium. As with mercury and cadmium, lead is environmentally hazardous because it is highly toxic. Unfortunately, lead, mercury, and cadmium are uniquely suited for and used in a variety of consumer products, including batteries and electrical switches.

Although aluminum is the third most abundant element in the Earth's crust, most of it is distributed in clays and aluminosilicate minerals that are not economically attractive sources of the metal. The primary ore for aluminum is bauxite, a hydrated oxide. Gallium, which occurs as a trace component of bauxite, is produced as a byproduct of aluminum refining. The heavier and chemically softer *p*-block elements in Group 13/III (indium and thallium) together with germanium in Group 14/IV, are recovered as byproducts of the refining of sulfide ores of more abundant elements (Table 8.13). Bismuth is sometimes recovered from the minerals bismuthinite, Bi_2S_3, and bismite, Bi_2O_3, but like the other heavy *p*-block metals most is recovered during copper, zinc, or lead smelting.

8.16 Group 13/III

The Group 13/III metals have a silver luster and erratic variations in melting points down the group: Al ($660\,°C$), Ga ($30\,°C$), In ($157\,°C$), and Tl (303 °C). The low melting point of gallium is reflected in the unusual structure of the metal, which contains Ga_2 units that persist into the melt. Gallium, indium, and thallium are all mechanically soft metals.

The group oxidation state (+3)

Although direct reaction of aluminum or gallium with a halogen yields a halide, both these electropositive metals also react with HCl or HBr gas, and this is usually a more convenient route:

$$2Al(s) + 6HCl(g) \xrightarrow{100\,^{\circ}C} 2AlCl_3(s) + 3H_2(g)$$

In the laboratory, a 'hot tube' reactor (one is illustrated in Fig. B12.2) is often used. The halides of both elements are available commercially, but it is common to synthesize them in the laboratory when a material free of hydrolysis products is required. Halides of Ga(I) and Ga(II), such as GaCl and $GaCl_2$, can be prepared by a comproportionation reaction in which the Ga(III) halide is heated with gallium metal:

$$2GaX_3 + Ga \xrightarrow{\Delta} 3GaX_2 \qquad X = Cl,\ Br,\ or\ I\ but\ not\ F$$

Because the F^- ion is so small, the fluorides AlF_3 and GaF_3 are hard solids that have much higher melting points and sublimation enthalpies than the other halides. Their high lattice enthalpies also result in their having very limited solubility in most solvents, and they do not act as Lewis acids to simple donor molecules. In contrast, the heavier halides are soluble in a wide variety of polar solvents and are excellent Lewis acids. However, despite their low reactivity toward most donors, AlF_3 and GaF_3 form salts of the type Na_3AlF_6 and Na_3GaF_6 which contain octahedral $[MF_6]^{3-}$ complex ions.

The relative Lewis acidities of the halides reflect the relative chemical hardness of the two elements. Thus, toward a hard Lewis base (such as ethyl acetate, which is hard on account of its O atoms), the Lewis acidities of the halides weaken as the softness of the element increases:

$$BCl_3 > AlCl_3 > GaCl_3$$

In contrast, toward a soft Lewis base (such as dimethylsulfide, which is soft on account of its S atom), the Lewis acidities strengthen as the softness of the element increases:

$$GaX_3 > AlX_3 > BX_3 \qquad (X = Cl\ or\ Br)$$

In keeping with the general tendency toward higher coordination numbers for the heavier p-block elements, halides of aluminum and its heavier congeners may take on more than one Lewis base and become hypervalent:

$$AlCl_3 + N(CH_3)_3 \rightarrow Cl_3AlN(CH_3)_3$$

$$Cl_3AlN(CH_3)_3 + N(CH_3)_3 \rightarrow Cl_3Al(N(CH_3)_3)_2$$

The stable form of Al_2O_3, α-alumina, is a very hard and refractory material. In its mineral form it is known as corundum and as a gemstone it is sapphire. The blue of the latter arises from a charge-transfer transition from Fe^{2+} to Ti^{4+} ion impurities. The structure of α-alumina and gallia, Ga_2O_3, consists of an hcp array of O^{2-} ions with the metal ions occupying two-thirds of the octahedral holes in an ordered array. Ruby is α-alumina in which a few percent of the Al^{3+} is replaced by Cr^{3+}. The Cr(III) is red rather than the characteristic violet of $[Cr(OH_2)_6]^{3+}$ or the green of Cr_2O_3 because when Cr^{3+} substitutes for the smaller Al^{3+} ion, the O ligands are compressed about Cr^{3+}. This compression increases the ligand field parameter Δ_O and shifts the spin-allowed *d–d* bands.

Dehydration of aluminum hydroxide at temperatures below 900 °C leads to the formation of γ-alumina, a metastable polycrystalline form with a very high surface area. Partly on account of its surface acid and base sites, this material is used as a solid phase in chromatography and as a catalyst and catalyst support (Section 17.5).

Low oxidation state gallium, indium, and thallium

Oxidation states of aluminum lower than +3 are rarely encountered. Thus, AlCl(g) is encountered only at very high temperatures and Al^+ is not stable in solid compounds. The oxidation state +1 is encountered more often for the heavier element gallium in compounds such as the solids GaI and $Ga(AlCl_4)$. Gallium(I) has much in common with indium(I); for example, both disproportionate when dissolved in water:

$$3MX(s) \rightarrow 2M(s) + M^{3+}(aq) + 3X^-(aq) \qquad M = Ga, In$$

On the other hand, Tl^+ is stable with respect to disproportionation in water because Tl^{3+} is difficult to achieve, as discussed above. Mixed oxidation state compounds are known for the heavier metals, such as $GaCl_2$, $InCl_2$, and $TlBr_2$; these compounds all appear to contain M^+ and M^{3+}. For instance, structural data indicate that the compound $GaCl_2$ is in fact $Ga^+[GaCl_4]^-$. The presence of M^{3+} ions is indicated by the existence of $[MX_4]^-$ complexes in these salts with short M—X distances, and the presence of M^+ ions is indicated by longer and less regular separation from the halide ions.

We have seen that the *d*-block elements become soft Lewis acids in their low oxidation states; however, the trend for the heavy *p*-block elements is just the opposite. The relative affinities for hard and soft donors suggest that Tl^+ is harder than Tl^{3+}. However, the Tl^+ ion is in fact still borderline between hard and soft. For example, it is transported into cells along with the hard K^+ ion, and like the alkali metal hydroxides TlOH is soluble in water; unlike K^+, however, thallium(I) chloride, thallium(I) bromide, and sulfide are insoluble. The transport through cell walls and reactions with soft donors appear to account for the observation that thallium is highly toxic to mammals.

The monohalides GaX, InX, and TlX are known for X = Cl, Br, and I. Under ordinary conditions the Tl(I) halides are insulators, as is typical of ionic compounds. However, at high pressures a new phase is formed with a significant electrical conductivity that decreases with increasing temperature. This behavior signifies metallic conduction (Section 2.9). An interesting point is that TlI_3 is not a halide of $Tl(III)$ but of $Tl(I)$, for structural data show that it contains the triiodide ion I_3^-. This compound forms when solutions of Tl^{3+} and I^- are mixed because Tl^{3+} can oxidize the I^- ion.

There is only a fine line between the formation of a mixed oxidation state ionic compound containing both M^+ and M^{3+} ions and the formation of a compound that contains M—M bonds. For example, mixing $Ga(I)[Ga(III)Cl_4]$ with a solution of $[N(CH_3)_4]Cl$ in a non-aqueous solvent yields the compound $[N(CH_3)_4]_2[Cl_3Ga—GaCl_3]$, in which the anion has an ethane-like structure with a Ga—Ga bond.

Example 8.5: *Proposing reactions of Group 13/III halides*

Propose balanced chemical equations (or indicate no reaction) for reactions between (a) $AlCl_3$ and $(C_2H_5)_3NGaCl_3$ in toluene, (b) $(C_2H_5)_3NGaCl_3$ and GaF_3 in toluene, and (c) TlCl and NaI in water.

Answer. (a) Al(III) is a stronger and harder Lewis acid than Ga(III); therefore, the following reaction can be expected:

$$AlCl_3 + (C_2H_5)_3NGaCl_3 \rightarrow (C_2H_5)_3NAlCl_3 + GaCl_3$$

(b) No reaction, because GaF_3 has a very high lattice enthalpy and thus is not a good Lewis acid. (c) Tl(I) is chemically soft, so it combines with the softer I^- ion rather than Cl^-:

$$TlCl(s) + NaI(aq) \rightarrow TlI(s) + NaCl(aq)$$

Like silver halides, Tl(I) halides have low solubility in water, so the reaction will probably proceed very slowly.

Exercise E8.5. Propose, with reasons, the chemical equation (or indicate no reaction) for reactions between (a) $(CH_3)_2SAlCl_3$ and $GaBr_3$ and (b) $TlCl_3$ and formaldehyde (CH_2O) in acidic aqueous solution. (Hint: formaldehyde is easily oxidized to CO_2 and H^+.)

8.17 Tin and lead

Aqueous and nonaqueous solutions of tin(II) salts are useful mild reducing agents, but they must be stored under an inert atmosphere because air oxidation is spontaneous and rapid:

$$Sn^{2+}(aq) + \tfrac{1}{2}O_2(g) + 2H^+(aq) \rightarrow Sn^{4+}(aq) + H_2O(l)$$

$$E^{\ominus} = +1.08 \text{ V}$$

Tin dihalides and tetrahalides are both well known. The tetrachloride, bromide, and iodide are molecular compounds, but the tetrafluoride has a structure consistent with it being an ionic solid because the small F^- ion permits a six-coordinate structure. Lead tetrafluoride also has a structure consistent with being an ionic solid but, because of the

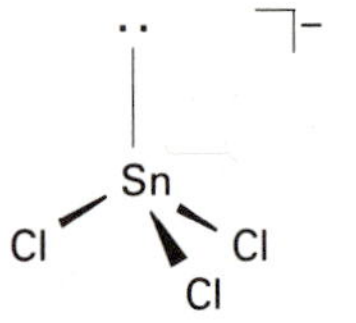

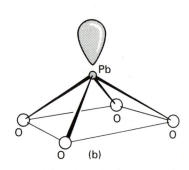

32 [SnCl$_3$]$^-$

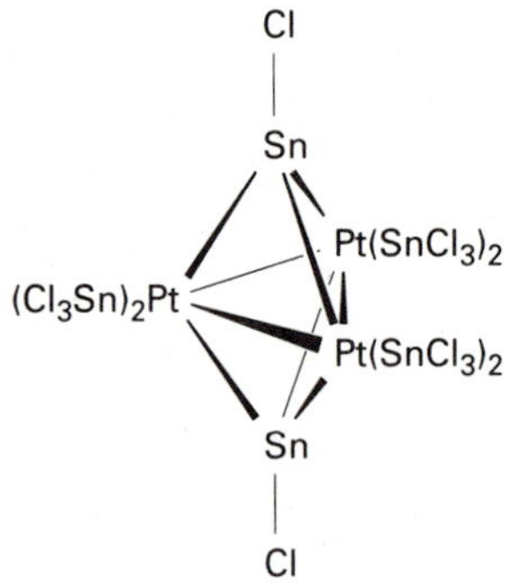

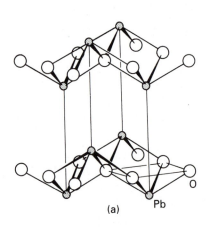

33 [(SnCl)$_2$(Pt(SnCl$_3$)$_2$)$_3$]

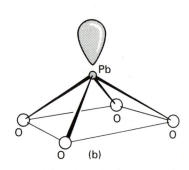

(a)

(b)

8.28 (a) The structure of PbO. (b) The square pyramidal array of PbO showing the possible orientation of a stereochemically active lone pair of electrons.

inert pair effect, PbCl$_4$ is an unstable compound that decomposes into PbCl$_2$ and Cl$_2$ at room temperature. Lead tetrabromide and tetraiodide are unknown and the dihalides dominate the halogen chemistry of lead. The arrangement of halogen atoms around the central metal atom in the dihalides of tin and lead often deviates from simple tetrahedral or octahedral coordination, and is attributed to the presence of a stereochemically active lone pair. The tendency to achieve the distorted structure is more pronounced with the small F$^-$ ion, and less distorted structures are observed with larger halides.

Both Sn(IV) and Sn(II) form a variety of complexes. Thus, SnCl$_4$ forms complex ions such as [SnCl$_5$]$^-$ and [SnCl$_6$]$^{2-}$ in acidic solution. In nonaqueous solution, a variety of donors interact with the moderately acidic SnCl$_4$ to form complexes such as *cis*-SnCl$_4$(OPMe$_3$)$_2$. In aqueous and nonaqueous solutions Sn(II) forms trihalo complexes, such as [SnCl$_3$]$^-$, where the pyramidal structure indicates the presence of a stereochemically active lone pair (**32**). The [SnCl$_3$]$^-$ ion can serve as a soft donor to *d*-metal ions. One unusual example of this ability is the red cluster compound Pt$_3$Sn$_8$Cl$_{20}$, which has a trigonal bipyramidal structure (**33**).

The oxides of lead are very interesting from both fundamental and technological standpoints. In the red form of PbO, the Pb(II) ions are four-coordinate (Fig. 8.28), but the O^{2-} ions around the Pb(II) lie in a square. As for the halides, this structure can be rationalized by the presence of a stereochemically active lone pair on the metal atom. Lead also forms mixed oxidation state oxides. The best known is red lead, Pb$_3$O$_4$, which contains Pb(IV) in an octahedral environment and Pb(II) in an irregular six-coordinate environment. The assignment of different oxidation numbers to the lead in these two sites is based on the shorter Pb—O distances for the atom identified as Pb(IV). The maroon form of lead(IV) oxide, PbO$_2$, crystallizes in the rutile structure. This oxide is a component of the cathode of a lead-acid battery (Box 8.2).

8.18 Bismuth

The chemical properties of bismuth strikingly illustrate the inert pair effect. For instance, Bi gives up its full complement of five valence electrons with great difficulty, and most of its chemical properties relate to the +3 state. Bismuth(III) can be regarded as borderline between hard and soft. An indication of this classification is the insolubility of both Bi(OH)$_3$ and Bi$_2$S$_3$. The approximate standard potentials for bismuth in acidic aqueous solution illustrate the strongly oxidizing character of the +5 oxidation state and the mildly electropositive character of the element:

$$Bi^{3+}(aq) + 3e^- \rightarrow Bi(s) \qquad E^{\ominus} = +0.32 \text{ V}$$

$$Bi^{5+}(aq) + 2e^- \rightarrow Bi^{3+}(aq) \qquad E^{\ominus} \approx +2 \text{ V}$$

Bismuth(V) is prepared by heating Bi$_2$O$_3$ with sodium peroxide:

$$Bi_2O_3(s) + 2Na_2O_2(s) \rightarrow 2NaBiO_3(s) + Na_2O(s)$$

Box 8.2 The lead-acid battery

The chemistry of the lead-acid battery is noteworthy because, as well as being the most successful rechargeable battery, it illustrates the role of both kinetics and thermodynamics in the operation of cells.

In its fully charged state, the active material on the cathode is PbO_2 and at the anode it is lead metal; the electrolyte is dilute sulfuric acid. One feature of this arrangement is that the lead-containing reactants and products at both electrodes are insoluble. When the cell is producing current, the reaction at the cathode is the reduction of Pb(IV) as PbO_2 to Pb(II), which in the presence of sulfuric acid is deposited on the electrode as insoluble $PbSO_4$:

$$PbO_2(s) + HSO_4^-(aq) + 3H^+(aq) + 2e^- \rightarrow$$
$$PbSO_4(s) + 2H_2O(l)$$

At the anode, lead is oxidized to Pb(II), which is also deposited as the sulfate:

$$Pb(s) + SO_4^{2-}(aq) \rightarrow PbSO_4(s) + 2e^-$$

The overall reaction is

$$PbO_2(s) + 2HSO_4^-(aq) + 2H^+(aq) + Pb(s) \rightarrow$$
$$2PbSO_4(s) + 2H_2O(l)$$

The potential difference of about 2 V is remarkably high for a cell in which an aqueous electrolyte is used, and exceeds by far the potential for the oxidation of water to O_2, which is 1.23 V. The success of the battery hinges on the high overpotentials (and hence low rates) of oxidation of H_2O on PbO_2 and of reduction of H_2O on lead.

When the sodium bismuthate product is dissolved in an aqueous solution of a noncoordinating acid, such as $HClO_4$, a poorly characterized metastable Bi(V) species is produced. This species is shown as Bi^{5+} in the half-reaction above: it may be $[Bi(OH)_6]^-$.

The coordination chemistry of Bi(III) reflects the borderline affinity for hard and soft ligands and the tendency toward distorted coordination environments. Distorted structures are often attributed to a stereochemical lone pair. The tendency for a low oxidation state p-block element to exhibit a distorted structure follows the trends already discussed for Pb(II) and Sn(II). Thus:

1. Low coordination numbers favor a distorted structure; for example, BiF_3 is pyramidal in the gas phase.

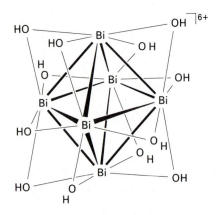

34 $Bi_2(OR)_6$ $R = C_2H_4OCH_3$

35 $[Bi_6(OH)_{12}]^{6+}$

2. The lighter *p*-block central atom has the greater tendency toward a distorted structure; for instance, Sb(III) compounds are more often distorted than Bi(III) compounds.
3. Small ligands promote a distorted structure; thus, fluoride or alkoxide ligands are more likely to lead to a distorted structure.

For example, a recently characterized bismuth alkoxide with the empirical formula $Bi(OC_2H_4OCH_3)_3$ turns out to be a dimer in the solid state with square-bipyramidal coordination around bismuth (**34**). A similar structure is observed for the anionic complex $[Bi_2Cl_8]^{2-}$.[15] It is interesting to note that Bi(III) forms a hydroxide complex in aqueous solution, with the formula $[Bi_6(OH)_{12}]^{6+}$, which consists of an octahedral array of Bi^{3+} ions with the OH^- ligands bridging the edges of the octahedron (**35**).

THE *f*-BLOCK METALS

The span of 15 elements across the lanthanide series and again across the actinide series corresponds to the filling of the seven 4*f* and 5*f* orbitals, respectively (from f^0 to f^{14}). There is a striking uniformity in the properties of the 4*f* elements, the **lanthanides**, and greater diversity in the chemistry of the 5*f* elements, the **actinides**. A general lanthanide is represented by the symbol Ln and an actinide by An.

8.19 Occurrence, recovery, and applications

Other than promethium, which has no stable isotopes, the least abundant lanthanide, thulium, is similar to iodine in crustal abundance. The principal mineral source for the early lanthanides is monazite, which contains mixtures of lanthanides and thorium $(Ln,Th)PO_4$. Another phosphate mineral, xenotime, is the principal source of yttrium and the heavier lanthanides. (Although yttrium is not strictly a lanthanide, its radius and chemical properties are similar to those of the heavier lanthanides.) The common oxidation state for the lanthanides is $+3$ (Table 8.14). Cerium, which can be oxidized to the $+4$ oxidation state, and europium, which can be reduced to Eu^{2+}, are chemically separable from the other lanthanides. Separation of the remaining Ln^{3+} ions is accomplished on a large scale by multistep liquid–liquid extraction in which the ions distribute themselves between an aqueous phase and an organic phase containing complexing agents. Ion-exchange chromatography, which is described in greater detail later in this section, is used to separate the individual lanthanide ions when high purity is required. Pure and mixed lanthanide metals are prepared by the electrolysis of molten lanthanide halides.

A mixture of the early lanthanide metals, including cerium, is referred to in commerce as *mischmetal*. It is used in steel making to remove impurities such as oxygen, hydrogen, sulfur, and arsenic, which reduce the mechanical strength and ductility of steel. Compounds of

---

[15] Soluble and volatile alkoxides of bismuth, M. A. Matchett, M. Y. Chang, and W. E. Buhro, *Inorg. Chem.*, **29**, 359 (1990).

Table 8.14 Names, symbols, and properties of the lanthanides

Atomic number	Name	Symbol	Configuration of M^{3+}	$E^{\ominus}/V$	$r(M^{3+})/Å$*	O.N.†
57	Lanthanum	La	[Xe]	−2.38	1.16	**3**
58	Cerium	Ce	[Xe]$4f^1$	−2.34	1.14	**3** 4
59	Praseodymium	Pr	[Xe]$4f^2$	−2.35	0.96	**3**, 4
60	Neodymium	Nd	[Xe]$4f^3$	−2.32	1.11	2(n), **3**
61	Promethium	Pm	[Xe]$4f^4$	−2.29	1.09	**3**
62	Samarium	Sm	[Xe]$4f^5$	−2.30	1.08	2(n), **3**
63	Europium	Eu	[Xe]$4f^6$	−1.99	1.07	2(a), **3**
64	Gadolinium	Gd	[Xe]$4f^7$	−2.28	1.05	**3**
65	Terbium	Tb	[Xe]$4f^8$	−2.31	1.04	**3**, 4
66	Dysprosium	Dy	[Xe]$4f^9$	−2.29	1.03	2(n), **3**
67	Holmium	Ho	[Xe]$4f^{10}$	−2.33	1.02	**3**
68	Erbium	Er	[Xe]$4f^{11}$	−2.32	1.00	**3**
69	Thulium	Tm	[Xe]$4f^{12}$	−2.32	0.99	2(n), **3**
70	Ytterbium	Yb	[Xe]$4f^{13}$	−2.22	0.99	2(a), **3**
71	Lutetium	Lu	[Xe]$4f^{14}$	−2.30	0.98	**3**

*Ionic radii for C.N. = 8 from R. D. Shannon, *Acta Crystallogr.*, **A32**, 751 (1976).
†Oxidation numbers in bold type indicate the most stable states; other states that can be achieved in aqueous (a) and nonaqueous (n) solution are also included.

the lanthanides find a wide range of applications, many of which are associated with their *f–f* electronic transitions: europium oxide or orthovanadate is used as a red phosphor in television and computer-terminal displays, and neodymium (Nd^{3+}), samarium (Sm^{3+}), and holmium (Ho^{3+}) are employed in solid-state lasers.

Beyond bismuth ($Z = 83$) none of the elements has stable isotopes, but two of the actinides, thorium (Th, $Z = 90$) and uranium (U, $Z = 92$), have very long-lived isotopes and occur in significant quantities in nature (Table 8.15). The primary source of the rest is synthesis by nuclear reactions, and all of these are more radioactive than thorium and uranium (Table 8.15). One major ore for uranium, uranite (also called pitchblende), has the approximate formula UO_2. The current primary use for uranium is in nuclear reactors for electric power generation at hundreds of reactors throughout the world.

8.20 Lanthanides[16]

The lanthanides are a family of highly electropositive metals in Period 6 that intervene between the *s* and *d* blocks. They are sometimes referred to as the **rare earths**; however, that name is inappropriate because they are not particularly rare, except for promethium, which has no stable isotopes. The lanthanides mark the first appearance of *f* orbitals in the ground state configurations of the elements. In contrast to the wide variation in properties across each series of *d* elements, the chemical properties of the lanthanides are highly uniform. These properties may be summarized as follows:

[16] There is an ongoing controversy over whether the lanthanides should include 14 elements from La through Yb or be displaced one position to the right, Ce through Lu. We shall include 15 elements in this discussion, La to Lu. Similar controversy extends to the actinides. See W. B. Jensen, *J. Chem. Educ.*, **59**, 634 (1982).

Table 8.15 Names, symbols, and properties of the actinides

Z	Name	Symbol	Mass number	$t^*_{1/2}$	$r(M^{3+})/Å†$	O.N.‡
89	Actinium	Ac	227	21.8 y	1.26	**3**
90	Thorium	Th	232	1.41×10^{10} y	–	**4**
91	Protactinium	Pa	231	3.28×10^4 y	1.18	4, **5**
92	Uranium	U	238	4.47×10^9 y	1.17	3, 4, 5, **6**
93	Neptunium	Np	237	2.14×10^6 y	1.15	3, 4, **5**, 6, 7
94	Plutonium	Pu	244	8.1×10^7 y	1.14	3, **4**, 5, 6
95	Americium	Am	243	7.38×10^3 y	1.12	**3**, 4, 5, 6
96	Curium	Cm	247	1.6×10^7 y	1.11	**3**, 4
97	Berkelium	Bk	247	1.38×10^3 y	1.10	**3**, 4
98	Californium	Cf	249	350 y	1.09	**3**, 4
99	Einsteinium	Es	254	277 d	(1.07)	**3**, 4
100	Fermium	Fm	257	100 d		2, **3**
101	Mendelevium	Md	258	55 d	(1.04)	2, **3**
102	Nobelium	No	259	1.0 h		**2**, 3
103	Lawrencium	Lr	260	3 min	(1.02)	**3**

*Half life of the most long-lived isotope.
†Effective ionic radii for C.N. = 6, from R. D. Shannon, *Acta Crystallogr.*, **A32**, 751 (1976). Estimates in parenthesis are from W. Brüchle, M. Schädel, U.W. Sherer, J.V. Kratz, K.E. Gregorich, D. Lee, R.M. Chasteler, H.L. Hall, R.A. Henderson, and D.L. Hoffman, *Inorg. Chim. Acta*, **146**, 267 (1988).
‡Oxidation states in aqueous solution, the predominant oxidation state is in bold face. From G. T. Seaborg and W. D. Loveland, *The elements beyond uranium*, p. 84. Wiley Interscience, New York (1990).

The elements La through Yb favor the +3 oxidation state with a uniformity that is unprecedented in the periodic table.

Their common adoption of the +3 oxidation state can probably be traced to the high sensititivy of the 4f electrons to the charge outside the inner core of the atom, so any increase in charge beyond +3 results in the 4f electrons becoming too firmly held to be generally available in chemical reactions. It should be noted that various relevant properties of the elements vary significantly. For example, the eight coordinate radii of the M^{3+} ions (Table 8.14) contract greatly, from 1.16 Å for La^{3+} to 0.98 Å for Lu^{3+}, and this 28 percent decrease in radius leads to a great increase in the hydration enthalpy across the series. A detailed analysis shows in fact that there is fortuitous cancellation of the various terms for sublimation, solvation, and ionization in the Born–Haber cycle for aqua ion formation; as a result of this coincidence, the potential for the reduction of La^{3+} to the metal, -2.38 V, is close to that for Lu^{3+}, -2.30 V, at the other end of the block.

Superimposed on this uniformity there are some atypical oxidation states that are most prevalent when the ion can attain an empty (f^0), half-filled (f^7), or filled (f^{14}) subshell (Table 8.14). Thus, Ce^{3+}, which is an f^1 ion, can be oxidized to the f^0 ion Ce^{4+}, a strong and useful oxidizing agent. The next most common of the atypical oxidation states is Eu^{2+}, which is an f^7 ion that readily reduces water.

A Ln^{3+} ion has hard acid properties, as indicated by its preference for F^- and oxygen-containing ligands and its occurrence with PO_4^{3-} in the mineral monazite. The decrease in ionic radius from La^{3+} (1.16 Å) to Lu^{3+} (0.98 Å) is attributed to in part to the increase in Z_{eff} as

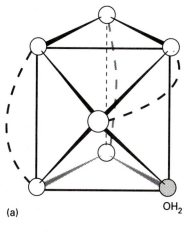

(a)

OH₂

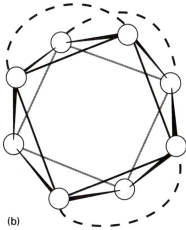

(b)

8.29 (a) Capped trigonal prism of donor atoms around ytterbium in [Yb(acac)₃(OH₂)]. Positions of acac chelate rings are indicated by dashed lines. (b) The square antiprism of donor atoms around lanthanum in [La(acac)₃(OH₂)₂]. Acac chelate rings are indicated by dashed lines.

$$(CH_3)_3C \overset{O \quad\quad O^-}{\diagdown\diagup} CF_2CF_2CF_3$$

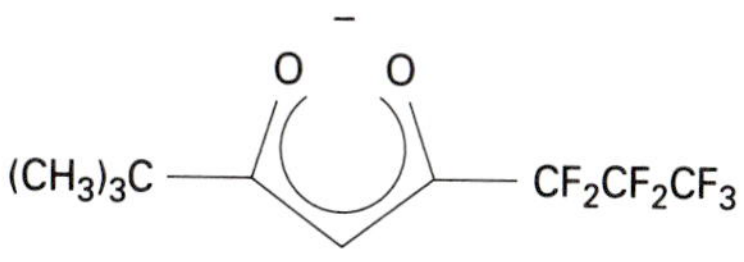

36 fod

electrons are added to the 4f subshell (Section 1.6). Detailed calculations indicate that relativistic effects also make a substantial contribution to the decrease in radius across the series. Most lanthanide ions are colored, and their spectra in complexes of solids generally show much narrower and more distinct absorption bands than those for d-metal complexes. These spectra are associated with weak f–f electronic transitions. Both the narrowness of the spectroscopic features and their insensitivity to the nature of coordinated ligands indicate that the f orbitals have a smaller radial extension than the filled 5s and 5p orbitals. Similarly, the magnetic properties of the ions (which are dealt with more fully in Chapter 14) can be explained on the assumption that the f electrons in Ln^{3+} ions are only slightly perturbed by ligands because they are buried so deeply. Ligand field stabilization plays no part in the chemical properties of the lanthanide complexes.

Two other general features of lanthanide complexes are that they often have high coordination numbers and that their electronic configurations vary widely. The variation in structure is in harmony with the view that the spatially buried f electrons have no significant stereochemical influence, and consequently that the ligands adopt positions that minimize ligand–ligand repulsion. In addition, polydentate ligands must satisfy their own stereochemical constraints, much as for the s-block and Al^{3+} complexes. For example, many lanthanide complexes have been formed with crown ether and β-diketonate ligands. The coordination numbers for $[Ln(OH_2)_n]^{3+}$ in aqueous solution are thought to be 9 for the early lanthanides and 8 for the later, smaller members of the series, but these ions are highly labile and the measurements are subject to considerable uncertainty. Similarly, a striking variation is observed for the coordination numbers and structures of lanthanide salts and complexes. For example, the small ytterbium cation, Yb^{3+}, forms the seven-coordinate complex $[Yb(acac)_3(OH_2)]$, and the larger La^{3+} is eight-coordinate in $[La(acac)_3(OH_2)_2]$. The structures of these two complexes are approximately a capped trigonal prism and a square antiprism respectively (Fig. 8.29).

The partially fluorinated β-diketonate ligand nicknamed fod (**36**) produces complexes with Ln^{3+} that are volatile and soluble in organic solvents. Because of their volatility these complexes are used as precursors for the synthesis of lanthanum-containing superconductors by vapor deposition (Section 18.5). Moreover, their solubility in organic solvents and the presence of available coordination sites mean that they are useful as NMR shift reagents, where the resonance signals for the magnetic nuclei in the attached ligand are spread out by the local magnetic field of the lanthanide ion (Fig. 8.30). This technique can be applied to a variety of molecules having a donor group that can serve as a ligand toward the lanthanide ion. The signal is most shifted for the H nuclei that are closest to the lanthanide.

Charged ligands generally have the highest affinity for the smallest Ln^{3+} ion, and the resulting increase in formation constants from large, lighter Ln^{3+} (on the left of the series) to small, heavier Ln^{3+} (on the right of the series) provides a convenient method for the chromato-

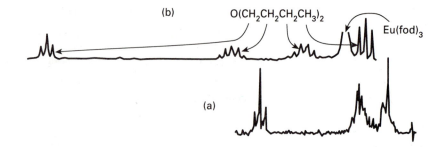

8.30 The influence of coordination to a paramagnetic Eu(III) center on the ^{1}H NMR spectrum of an ether. (a) NMR spectrum of $O(C_4H_9)_2$. (b) Spectrum of this ether coordinated to $[Eu(fod)_3]$. Note the greatest shift for the CH_2 groups attached to oxygen. These groups are closest to the paramagnetic Eu(III) in the complex. (From R. E. Sievers (ed.), *Nuclear magnetic resonance shift reagents.* Academic Press, New York (1973).)

graphic separation of these ions (Fig. 8.31 and Box 8.3). In the early days of lanthanide chemistry, before ion-exchange chromatography was developed, tedious repetitive crystallizations of lanthanide salts were used to separate these elements.

8.21 Actinides

The 14 elements from actinium (Ac, $Z=89$) through nobelium (No, $Z=102$) involve the filling of the $5f$ subshell, and in this sense are analogs of the lanthanides. However, the actinides do not exhibit the chemical uniformity of the lanthanides. A common oxidation state of the actinides (which for the present discussion will include Ac through Lr) is $+3$. Unlike the lanthanides, however, the early members of the series occur in a rich variety of oxidation states. The Frost diagrams (Fig. 8.32) and the data in Table 8.15 show that:

> Oxidation states higher than $+3$ are preferred for the early elements of the block (Th, Pa, U, and Np).
> The linear or nearly linear MO_2^+ and MO_2^{2+} ions are the dominant aqua species for oxidation numbers $+5$ and $+6$.

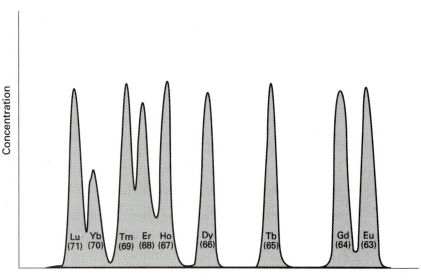

8.31 Elution of the heavy lanthanide ions from an ion exchange column using ammonium 2-hydroxyisobutyrate as the eluant. Note that the higher atomic number lanthanides elute first because they have smaller radii and are more strongly complexed by the 2-hydroxyisobutyrate ligand.

Volume of eluant

Box 8.3 Ion-exchange

To perform an ion-exchange separation, a solution of lanthanide ions is introduced at the top of a cation-exchange column in the sodium form (typically sodium polystyrene sulfonate). The Ln^{3+} ions readily undergo ion exchange displacing Na^+ ions, thus forming a band of the lanthanide ions bound to the top of the cation exchange column. To move these ions down the column and effect a separation, a solution consisting of an anionic ligand (citrate, lactate, or 2-hydroxyisobutyrate) is slowly passed through the column. These anionic chelating ligands form complexes with lanthanides. Moreover, because the complexes possess a lower positive charge than the initial Ln^{3+}, they are less tightly held by the resin than Ln^{3+}, and are displaced from the ion exchange material into the surrounding solution. The equilibria that are established between cations on the ion exchange resin (res) and neutral or anionic complexes in solution can be summarized as follows.

Initial ion exchange displacement of Na^+ by La^{3+} in a band at the top of column:

$$Ln^{3+}(aq) + 3Na^+(res) \rightleftharpoons Ln^{3+}(res) + 3Na^+(aq)$$

Subsequent elution with a solution of a complexing agent leads to formation of neutral or negatively charged lanthanide complexes. To maintain electroneutrality within the ion exchange resin sodium cations take the place of the neutral or negative lanthanide complex.

$$3Na^+(aq) + Ln^{3+}(res) + 3RCO_2^-(aq) \rightleftharpoons$$
$$3Na^+(res) + Ln(RCO_2)_3(aq)$$

The Ln^{3+} cations with smallest radius are most strongly bound to the anionic ligand so these ions have the greatest tendency to be eluted first (see Fig. 8.31).

A further general point is that

> The actinides have large atomic radii, and as a result often have high coordination numbers.

For example, uranium in solid UCl_4 is eight-coordinate and in solid UBr_4 it is seven-coordinate in a pentagonal bipyramidal array.

Thorium and uranium

Because of their ready availability and low level of radioactivity, the chemical manipulation of thorium and uranium can be carried out with ordinary laboratory techniques. As indicated by Fig. 8.32 and Table 8.15, the only stable oxidation state of thorium in aqueous solu-

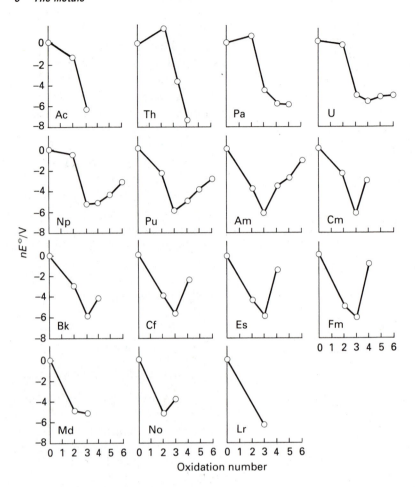

8.32 Frost diagrams for the actinides in acidic solution. (From J. J. Katz, G. T. Seaborg, and L. Morss, *Chemistry of the actinide elements*, 2nd edn. Chapman and Hall, London, (1986).)

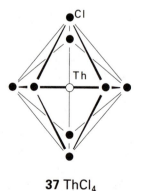

37 ThCl$_4$

tion is $+4$. This oxidation state also dominates the solid state chemistry of the element. Eight-coordination is common in simple thorium(IV) compounds. For example, ThO$_2$ has the fluorite structure (in which a Th atom is surrounded by a cube of O^{2-} ions) and in ThCl$_4$ the coordination number is again 8 with dodecahedral symmetry (**37**). The coordination number of Th in Th(NO$_3$)$_4$(OPPh$_3$)$_2$ is 10 with the bidentate NO$_3^-$ ions and triphenylphosphane oxide groups arranged in an octahedral array around the metal (**38**).

The chemical properties of uranium are more varied than those of thorium because it has access to oxidation states from $+3$ through $+6$ (Fig. 8.32), with the $+4$ and $+6$ states the most common. Uranium halides are known for the full range of oxidation states $+3$ through $+6$, with a trend toward decreasing coordination number with increase in oxidation number. The uranium atom is nine-coordinate in solid UCl$_3$, eight-coordinate in UCl$_4$, and six-coordinate for the U(V) and U(VI) chlorides U$_2$Cl$_{10}$ and UCl$_6$, both of which are molecular. The high volatility of UF$_6$ (it sublimes at 57 °C) together with the occurrence of fluorine in a single isotopic form accounts for the use of this compound in the separation of the uranium isotopes by gaseous diffusion or gaseous centrifugation.

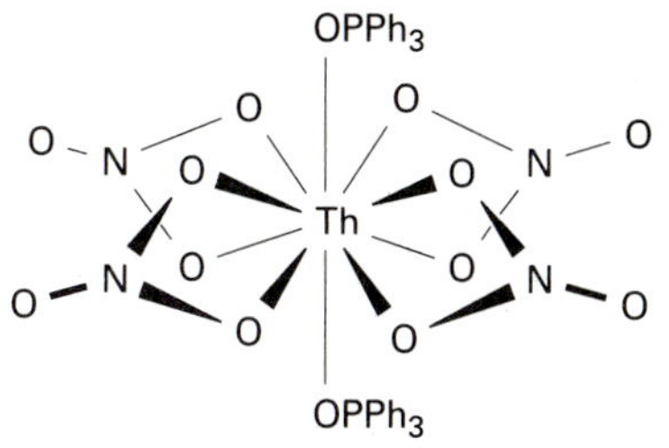

OPPh₃

38 [Th(NO₃)₄(OPPh₃)₂]

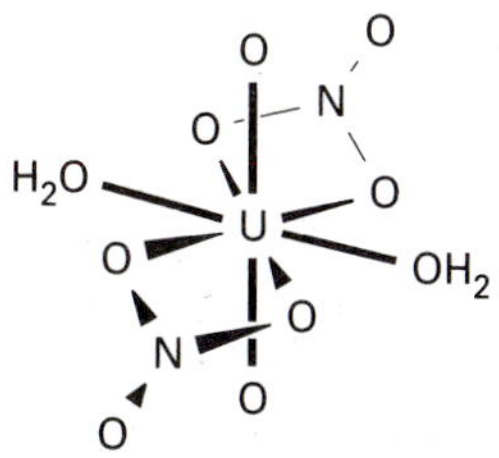

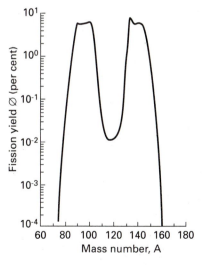

8.33 The structure of the uranyl nitrate diaqua complex. Note the six ligand O atoms in a plane perpendicular to the OUO axis. The $[UO_2Cl_4]^{2-}$ complex has four atoms in this plane, and many uranyl complexes have five ligating atoms in this plane.

Fission yield ∅ (per cent)

60   80   100   120   140   160   180
Mass number, A

8.34 Smoothed mass distribution for the products from the thermal neutron induced fission of ^{235}U. (From G. T. Seaborg and W. D. Loveland, *The elements beyond uranium*. Wiley-Interscience, New York, (1990).)

Uranium metal does not form a passive oxide coating, and so it is consumed on prolonged exposure to air to give a complex mixture of oxides. The most important oxide is UO_3, which dissolves in acid to give the uranyl ion, UO_2^{2+}. In water, this bright iridescent yellow ion forms complexes with many anions, such as NO_3^- and SO_4^{2-} (Fig. 8.33). In contrast to the angular shape of the VO_2^+ ion and similar d^0 complexes, the AnO_2^{2+} moiety, with An = U, Np, Pu, and Am maintains its linearity in all complexes. Both f-orbital bonding and relativistic effects have been invoked to explain this linearity. Unlike the lanthanides, f orbitals extend into the bonding region for the early actinides, so the complex actinide spectra are strongly affected by ligands.

The separation of uranium from most other metals is accomplished by the extraction of the neutral uranyl nitrato complex $[UO_2(NO_3)_2(OH_2)_4]$ from the aqueous phase into a polar organic phase, such as a solution of tributylphosphate dissolved in a hydrocarbon solvent. These kinds of solvent extraction processes are used to separate actinides from other fission products in spent nuclear fuel.

The fission of heavy elements, such as ^{235}U, occurs upon bombardment by neutrons. Thermal neutrons (neutrons with low velocities) bring about the fission of ^{235}U to produce two medium mass nuclides, and a large amount of energy is released because the binding energy per nucleon decreases for atomic numbers beyond 26 (see Fig. 1.2). That unsymmetrical fission of the uranium nucleus has a high probability of occurring is shown by the double-humped distribution of fission products (Fig. 8.34), which has maxima close to mass numbers 95 (Mo) and 135 (Ba). Almost all the fission products are unstable nuclides, and the most troublesome are those with half-lives in the range of years to centuries: these nuclides decay fast enough to be highly radioactive but not sufficiently fast to disappear in a convenient time.

The difficult tasks of separating, immobilizing, and storing unwanted fission products has not been satisfactorily solved. One proposal is to extract uranium and plutonium and other fissionable materials, immobilize the unwanted nuclides in a glass, and deposit that glass in a geological stratum that is stable and out of communication with ground water.

Example 8.6: *Assessing the redox stability of actinide ions*
Use the Frost diagram for thorium (Fig. 8.32) to describe the relative stability of its +2 and +3 oxidation states.

Answer. The initial slope in the Frost diagrams indicates that the Th^{2+} ion might be readily attained with a mild oxidant. However, Th^{2+} is above lines connecting Th(0) with the higher oxidation states, so it is susceptible to disproportionation. The +3 state is readily oxidized to Th(IV) and the steep negative slope indicates that it would be oxidized by water:

$$Th^{3+}(aq) + H^+(aq) \rightarrow Th^{4+}(aq) + \tfrac{1}{2}H_2(g)$$

We can confirm, from data in Appendix 2, that this reaction is highly favored, $E^{\ominus} = +3.8$ V. Thus, Th^{4+} will dominate in aqueous solution.

Exercise E8.6. Use the Frost diagrams and data in Appendix 2 to determine the most stable uranium ion in acid aqueous solution in the presence of air, and give its formula.

The transamericium elements

For americium (Am, $Z = 95$) and beyond, the properties of the actinides begins to converge with those of the lanthanides:

> With increasing atomic number, M(III) becomes progressively more stable relative to higher oxidation states.

Oxidation state $+3$ is dominant for curium (Cm), berkelium (Bk), californium (Cf), and einsteinium (Es); these elements therefore resemble the lanthanides. The striking difference between the chemical properties of the lanthanides and early actinides led to controversy about the most useful placement of the actinides in the periodic table. For example, before 1945, periodic tables usually showed uranium below tungsten because both elements have a maximum oxidation state of $+6$. The emergence of the $+3$ oxidation state for the later actinides was a key point in the way that chemists thought about the location of the actinides in the periodic table. The similarity of the heavy actinides and the lanthanides is illustrated by their similar elution behavior in ion exchange separation (compare Figs 8.35 and 8.31).

Because of the small quantities of material available in most cases and the intense radioactivity, most of the chemical properties of the transamericium elements have been established by experiments carried out on a microgram scale or even a few hundreds of atoms. For example, the actinide ion complexes have been absorbed on and eluted

8.35 Elution of the heavy actinide ions from an ion exchange column by ammonium 2-hydroxyisobutyrate as the eluent. Note the similarity in elution sequence with Fig. 8.31. Like the Ln^{3+} ions, heavy (smaller) An^{3+} ions elute first. (From J. J. Katz, G. T. Seaborg, and L. Morss, *Chemistry of the actinide elements*, 2nd edn. Chapman and Hall, London (1986).)

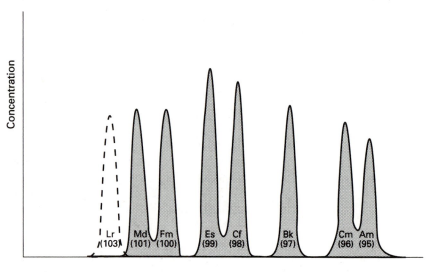

from a single bead of ion exchange material of about 0.2 mm in diameter. For the heaviest and most unstable actinides, such as hassium ($Z = 108$), the lifetimes are too short for chemical separation, and the identity of the element is based exclusively on the properties of the radiation it emits.

FURTHER READING

G. Wilkinson, R. D. Gillard, and J. McCleverty (ed.), *Comprehensive coordination chemistry*. Pergamon Press, Oxford (1987 et seq.). Volume 1 of this set provides a general introduction and succeeding volumes cover the particulars for each metallic element.

A. F. Wells, *Structural inorganic chemistry*, 5th edn. Oxford University Press (1984). This book is a good single-volume reference for information on the structures of compounds of the metals.

Ullmann's encyclopedia of industrial chemistry, 5th edn. VCH, Weinheim (1985 et seq.); *Kirk–Othmer Encyclopedia of chemical technology*, 3rd edn. Wiley-Interscience, New York (1978). These volumes provide general information on the metallic elements, with emphasis on mineral sources, methods of extraction, and applications.

J. C. Bailar, Jr., H. J. Emeleus, R. Nyholm, and A. F. Trotman-Dickenson (ed.), *Comprehensive inorganic chemistry*, Vols 1–5. Pergamon Press, Oxford (1973). Although out of date, these volumes provide much useful information on the chemical properties of the metallic elements.

F. A. Cotton and G. Wilkinson, *Advanced inorganic chemistry*, 6th edn. Wiley-Interscience, New York (1992). Parts 3 and 4 of this advanced text cover *d*- and *f*-block chemistry.

G. T. Seaborg and W. D. Loveland, *The elements beyond uranium*. Wiley-Interscience, New York (1990). This primer for the transuranium elements describes their discovery and provides a good overview of their separation, detection, and chemical properties.

J. Katz, G. Seaborg, and L. Morss, *The chemistry of the actinide elements*, 2nd edn. Longman, London (1986). A comprehensive two-volume account of the actinides.

KEY POINTS

1. Group 1 and 2 metal properties
The *s*-block elements are low-melting and highly electropositive metals with the group oxidation state dominant. The principal exceptions are beryllium and magnesium, which are physically harder and kinetically less reactive.

2. Nonaqueous solutions of Group 1 and 2 metals
Alkali metals dissolve in low acidity solvents (such as liquid ammonia) to yield highly reducing solutions that contain the metal cation and e^-(solv) or sometimes the M^-(solv) anion. Metal solutions also can be prepared for the heavier Group 2 metals.

3. Oxidation state stability in the d block
In aqueous redox chemistry, the M^{2+} ion is favored toward the right of the $3d$ metals, and the highest oxidation state becomes more stable on descending each of the Groups 4 through 8.

4. Coordination number in the d block
In contrast to the common occurrence of 4 and 6-coordinate $3d$ cations, those in the $4d$ and $5d$

series are larger and often display higher coordination numbers.

5. Polyoxometallates

The polyoxometallates observed for Group 5 and 6 metals, especially vanadium, molybdenum, and tungsten, usually consist of edge- and vertex-shared $[MO_6]^{n-}$ octahedra.

6. Metal–metal bonding in the d block

Metal–metal bonded compounds are encountered for low-oxidation state metals on the left of the d block in conjunction with π-donor ligands such as halide and for metals on the right of the d block combined with π-acceptor ligands.

7. Noble character

The noble metals and the coinage metals are not oxidized by hydrogen ions in acids. Square-planar

structures are the rule for their d^8 complexes and octahedral structures are common for their d^6 complexes.

8. Oxidation states of p block metals

The heavy p-block metals thallium, lead, and bismuth are most stable in an oxidation state that is two less than the group oxidation state.

9. Oxidation states in the f block

The $+3$ oxidation state predominates for the lanthanides and the heavy actinides. For uranium, oxidation states $+4$ and $+6$ are most important.

10. f-Block metal complexes

Because of their large size, high coordination numbers are common for $4f$ and $5f$ cations. Linear dioxo complexes are characteristic of the actinides in $+5$ and $+6$ oxidation states.

EXERCISES

8.1 Without using reference material, sketch the s block of the periodic table, including the chemical symbols for the elements, and indicate the trends in (a) melting point, (b) radii for the common cations, (c) the tendency of peroxides to decompose thermally to simple oxides.

8.2 Which of the following pairs are most likely to form the desired compound or undergo the process mentioned? Describe the periodic trend and the physical basis for your answer in each case. (a) Cs^+ or Mg^{2+}, form an acetate complex; (b) Be or Sr, dissolve in liquid ammonia in the absence of air; (c) Li^+ or K^+, form a complex with crypt 2.2.2.

8.3 Contrast the coordination environment of a metal ion in each of the following pairs of compounds and give a plausible explanation for the difference: (a) CaF_2 versus MoS_2; (b) CdI_2 versus $MoCl_2$; (c) BeO versus CaO; (d) molybdenum(II) acetate and the beryllium acetate compound crystallized from slightly basic solution.

8.4 Without reference to a periodic table, sketch the first series of the d block, including the symbols of the elements. Indicate those elements for which the group oxidation number is common by C; those for which the group oxidation number can be reached but is a powerful oxidizing agent by O; and those for which the group oxidation number is not achieved by N.

8.5 State the trend in the stability of the group oxidation number on descending a group of metallic elements in the d and p blocks. Illustrate the trend using standard potentials in acidic solution for Groups 5, 6, and 13/III.

8.6 For each part, give balanced chemical equations or NR (for no reaction) and rationalize your answer in terms of trends in oxidation states.

    (a) $Cr^{2+}(aq) + Fe^{3+}(aq) \rightarrow$

    (b) $CrO_4^{2-}(aq) + MoO_2(s) \rightarrow$

    (c) $MnO_4^-(aq) + Cr^{3+}(aq) \rightarrow$

8.7 (a) Which ion, $Ni^{2+}(aq)$ or $Mn^{2+}(aq)$, is more likely to form a sulfide in the presence of H_2S? (b) Rationalize your answer with the trends in hard and soft character across Period 4. (c) Give a balanced equation for the reaction.

8.8 Preferably without reference to the text (a) write out the d block of the periodic table, (b) indicate the metals that form difluorides with the rutile or fluorite structures, and (c) indicate the region of the periodic table in which metal–metal bonded halide compounds are formed, and give one example.

8.9 Write a balanced chemical equation for the reaction that occurs when cis-$[RuLCl(OH_2)]^+$ (see Fig. 8.13) in acidic solution at 0.2 V is made strongly basic at the same potential. Write a balanced equation for each of the successive reactions when this same complex at pH = 6 and 0.2 V is exposed to progressively more oxidizing environments up to 1.0 V. Give other examples and a reason for the redox state of the metal center affecting the extent of protonation of coordinated oxygen.

8.10 Give plausible balanced chemical reactions (or NR for no reaction) for the following combinations, and state the basis for your answer: (a) MoO_4^{2-}(aq) plus Fe^{2+}(aq) in acidic solution; (b) the preparation of $[Mo_6O_{19}]^{2-}$(aq) from K_2MoO_4(s); (c) $ReCl_5$(s) plus aqueous $KMnO_4$; (d) $MoCl_2$ plus warm HBr(aq).

8.11 Speculate on the structures of the following species and present bonding models to justify your answers. (a) $[Re(O)_2(py)_4]^+$, (b) $[V(O)_2(ox)_2]^{3-}$, (c) $[Mo(O)_2(CN)_4]^{4-}$, (d) $[VOCl_4]^{2-}$.

8.12 Which of the following are likely to have structures that are typical of (a) predominantly ionic, (b) significantly covalent, (c) metal–metal bonded compounds: NiI_2, $NbCl_4$, FeF_2, PtS, and WCl_2? Rationalize the difference and speculate on the structure.

8.13 From trends in the chemical properties of the metals, write plausible balanced chemical equations (or NR for no reaction) for the following and state your reasoning: (a) TiO with aqueous HCl under an inert atmosphere; (b) Ce^{4+}(aq) with Fe^{2+}(aq); (c) Rb_9O_2 with water; (d) Na(am) with CH_3OH.

8.14 Indicate the probable occupancy of σ, π, and δ bonding and antibonding orbitals and the bond order for the following tetragonal prismatic complexes: (a) $[Mo_2(O_2CCH_3)_4]$, (b) $[Cr_2(O_2CC_2H_5)_4]$, (c) $[Cu_2(O_2CCH_3)_4]$.

8.15 The acid–base chemistry of liquid ammonia often parallels that of aqueous solutions. Assuming this to be the case, write balanced chemical reactions for the reaction of solid $Zn(NH_2)_2$ with (a) NH_4^+ in liquid ammonia and (b) with KNH_2 in liquid ammonia.

8.16 Give balanced chemical equations (or NR for no reaction) for the following cases and give a rationale for the reaction or lack of reaction: (a) Hg^{2+}(aq) added to Cd(s); (b) Tl^{3+}(aq) plus Ga(s); (c) $[AlF_6]^{3-}$(aq) with Tl^{3+}(aq).

8.17 (a) Summarize the trends in relative stabilities of the oxidation states of the elements of Groups 13/III and 14/IV, and indicate the elements that display the inert pair effect. (b) With this information in mind, write balanced chemical reactions or NR (for no reaction) for the following combinations, and explain how the answer fits the trends.

(i) Sn^{2+}(aq) + PbO_2(s) (excess) → (air excluded)

(ii) Tl^{3+}(aq) + Al(s) (excess) → (air excluded)

(iii) In^+(aq) → (air excluded)

(iv) Sn^{2+}(aq) + O_2(air) →

(v) Tl^+(aq) + O_2(air) →

8.18 Use data from Appendix 2 to determine the standard potential for each of the reactions in Exercise 8.17(b). In each case, comment on the agreement or disagreement with the qualitative assessment you gave for reactions (i) through (v).

8.19 (a) Give a balanced equation for the reaction of any of the lanthanide metals with aqueous acid. (b) Justify your answer with redox potentials and with a generalization about the most stable positive oxidation states for the lanthanides. (c) Name two lanthanides that have the greatest tendency to deviate from the usual positive oxidation state and correlate this deviation with electronic structure.

8.20 From a knowledge of their chemical properties, speculate on why cerium and europium were the easiest lanthanides to isolate before the development of ion-exchange chromatography.

8.21 Describe the general nature of the distribution of the elements formed in the thermal neutron fission of ^{235}U, and decide which of the following highly radioactive nuclides are likely to present the greatest radiation hazard in the spent fuel from nuclear power reactors: (a) ^{39}Ar, (b) ^{228}Th, (c) ^{90}Sr, (d) ^{144}Ce.

8.22 Account for the similar electronic spectra of Eu^{3+} complexes with various ligands and the variation of the electronic spectra of Am^{3+} complexes as the ligand is varied.

PROBLEMS

8.1 It is wrong to think of the periodic table as a fixed and unambiguous arrangement of the elements that was put forward by Mendeleev. In fact, there were many forerunners of Mendeleev's periodic table. One of these, put forward by William Olding in 1864, placed H, Ag, and Au in the same group. Mendeleev considered many variations on the periodic table. In one of these variations he placed Na over Cu, Ag, and Au. Discuss the arguments that might be made for and against each of these arrangements from the standpoints of, (i) chemical properties of the elements (their accessible oxidation states, and formulas and physical properties of their simple halides and oxides), and (ii) modern

information on the electronic configurations of the elements and their ions.

8.2 An amateur chemist claimed the existence of a new metallic element, grubium (Gr), which has the following characteristics. Metallic Gr reacts with 1 M H^+(aq) in the absence of air to produce Gr^{3+}(aq) and H_2(g). In the absence of air $GrCl_2$(s) dissolves in 1 M H^+(aq) and very slowly yields H_2(g) plus Gr^{3+}(aq). When Gr^{3+}(aq) is exposed to air GrO^{2+}(aq) is produced. From this information, estimate the range of potentials for (a) the reduction of Gr^{3+}(aq) to Gr(s), (b) the reduction of Gr^{3+}(aq) to

$GrCl_2(s)$, and (c) the reduction of $GrO^{2+}(aq)$ to $Gr^{3+}(aq)$. Suggest a known element that fits the description of grubium.

8.3 The following arrangements have been used for Group 3 in the periodic table at various times:

B
Al
Sc   Sc   Sc
Y   Y   Lu
La   La   Lr

Discuss the relative merits of each classification assuming that the following chemical trends are the criteria for the classification: (a) aqueous oxidation state stability, (b) radii of the atoms and $+3$ ions, (c) electronic configuration of the atoms.

8.4 Many metal salts can be vaporized to a small extent at high temperatures and their structures studied in the vapor phase by electron diffraction. Speculate on the structures that you might expect to find for the following gas-phase species and present your reasoning: (a) TaF_5; (b) MoF_6.

8.5 The bonding in the linear uranyl ion, OUO^{2+}, is often explained in terms of significant π bonding utilizing $5f$ orbitals on the metal. Using the f orbitals illustrated in Fig. 1.20, construct a reasonable molecular orbital diagram for π-bonding with the appropriate oxygen p orbitals.

8.6 In the presence of catalytic amounts of $[Pt(P_2O_5H_2)_4]^{4-}$ and light, 2-propanol produces H_2 and acetone (E. L. Harley, E. Stiegman, A. Vlcek Jr., and H.B. Gray, *J. Am. Chem. Soc.*, **109**, 5233 (1987); D. C. Smith and H. B. Gray, *Coord. Chem. Rev.*, **100**, 169 (1990)). (a) Give the equation for the overall reaction. (b) Give a plausible molecular orbital scheme for the metal–metal bonding in this tetragonal prismatic complex and indicate the nature of the excited state that is thought to be responsible for the photochemistry. (c) Indicate the metal complex intermediates and the evidence for their existence.

8.7 The initial extraction of a uranium ore with 2 M sulfuric acid contains 0.2 mol L^{-1} UO_2^{2+} and 0.2 mol L^{-1} Fe^{3+} in equilibrium with their sulfate complexes as the primary metal constituents. The equilibrium constants for coordina-

tion of UO_2^{2+} with sulfate are log $K_1 = 3.3$, log $\beta_2 = 4.3$ and for Fe^{3+} the formation constants with SO_4^{2-} as a ligand are log $K_1 = 2.2$, log $\beta_2 = 2.5$. The equilibrium mixture is adsorbed on the top of an *anion* exchange column and the column is then eluted with 2 M $HClO_4(aq)$. (The perchlorate ion is an extremely weak ligand.) (a) What are the major metal containing species in the initial solution? (b) Write chemical equations for the interaction of the initial solution with the ion exchange resin which is initially in the perchlorate form. (c) What is the effect of the 2 M perchloric acid solution on metal complexes in solution? (d) Sketch a qualitative concentration versus time profile for the two metal constituents in the eluent, and explain your reasoning.

8.8 Identify the incorrect statements in the following description and provide corrections and an explanation of the trend. (a) Sodium dissolves in ammonia and amines to produce the sodium cation and solvated electrons or the sodide ion. (b) Sodium dissolved in liquid ammonia will not react with NH_4^+ because of strong hydrogen bonding with the solvent. (c) The ion $Fe^{2+}(aq)$ is a poorer reducing agent than $V^{2+}(aq)$, in keeping with the trends in oxidation state stability across the $3d$ series. (d) The compound WBr_2 has a layered CdI_2 structure, in keeping with the tendency for bromides to be significantly covalent.

8.9 The existence of a maximum oxidation number of $+6$ for both uranium ($Z = 92$) and tungsten ($Z = 74$) prompted the placement of uranium under tungsten in early periodic tables. When the element after uranium, neptunium ($Z = 93$), was discovered in 1940 its properties did not correspond to those of rhenium ($Z = 75$), and this cast doubt on the original placement of uranium. (G. T. Seaborg and W. D. Loveland, *The elements beyond uranium*, p. 9 et seq Wiley-Interscience, New York (1990).) Using standard potential data from Appendix 2, discuss the differences in oxidation state stability between Np and Re.

8.10 The extraction of uranium involves both chemical and physical separation techniques. Consult a general source (for instance, Kirk–Othmer) and prepare a summary of the steps involved in the separation of nuclear fuel grade uranium from its ore.

9

Hydrogen and its compounds

Despite its simplicity, hydrogen has a very extensive chemistry; it was in fact at the forefront of much of the discussion in Part 1, most importantly in the discussion of acids and bases. Here we summarize some of the properties of the element and its binary compounds. The latter range from solid ionic compounds and solids with metallic properties to molecular compounds. We shall see that many of the reactions of these compounds are understandable in terms of their ability to supply an H^- or H^+ ion, and that it is often possible to anticipate when one tendency is likely to dominate. The chapter concludes with a brief survey of one of the striking characteristics of certain compounds of hydrogen, their ability to participate in the formation of hydrogen bonds.

Hydrogen is the most abundant element in the universe and the fifteenth most abundant in the Earth. The partial depletion of hydrogen from the Earth reflects its volatility during the formation of the planet (Section 1.1). It is one of the most important elements in the Earth's crust, where it is found in minerals, the oceans, and in all living things.

Because an H atom has only one electron, it might be thought that the element's chemical properties will be mundane, but this is far from the case. Hydrogen has richly varied chemical properties, for despite its single electron it can bond to more than one atom simultaneously. Moreover, it ranges in character from a strong Lewis base (as the hydride ion, H^-) to a strong Lewis acid (as the hydrogen cation, H^+, the proton).

THE ELEMENT

Hydrogen does not fit neatly into the periodic table. It is sometimes placed at the head of the alkali metals in Group 1, with the rationale that hydrogen and the alkali metals have only one valence electron, but that position is not entirely satisfactory in terms of either the chemical or physical properties of the elements; in particular, hydrogen is not a metal.[1] Less frequently, hydrogen is placed above the halogens in Group 17/VII, with the rationale that, like the halogens, hydrogen requires one electron to complete its valence shell, and, as we shall see, there are indeed some parallels between its properties and those of the halogens.

9.1 Nuclear properties

There are three isotopes of hydrogen, with mass numbers 1, 2 (deuterium, ^{2}H or D), and 3 (tritium, ^{3}H or T). The justification for giving these isotopes their own names is the large relative differences in their masses, which make their chemical and physical differences readily observable.

The lightest isotope, ^{1}H—which is very occasionally called protium—is by far the most abundant. Deuterium has variable natural abundance with an average value of about 0.016 percent. Neither ^{1}H nor ^{2}H is radioactive, but tritium decays by the loss of an electron (β^-) from the nucleus to yield a rare but stable isotope of helium:

$$\begin{smallmatrix}3\\1\end{smallmatrix}\text{H} \rightarrow \begin{smallmatrix}3\\2\end{smallmatrix}\text{He} + \beta^- \qquad t_{\frac{1}{2}} = 12.4 \text{ y}$$

Tritium's abundance of 1 in 10^{21} hydrogen atoms in surface water reflects a steady state between its production by bombardment of cosmic rays on the upper atmosphere and its loss by radioactive decay. However, the low natural abundance of the isotope is augmented artificially because it is manufactured on an undisclosed scale for use in thermonuclear weapons. This synthesis uses the neutrons from a fission reactor with Li as a target:

$$\begin{smallmatrix}1\\0\end{smallmatrix}\text{n} + \begin{smallmatrix}6\\3\end{smallmatrix}\text{Li} \rightarrow \begin{smallmatrix}3\\1\end{smallmatrix}\text{H} + \begin{smallmatrix}4\\2\end{smallmatrix}\text{He}$$

Deuterium and tritium are used in studies of reaction mechanisms as **tracers**, isotopes that can be followed through a series of reactions. Thus, it is possible to trace the transfer of hydrogen between different molecules by following the isotopic composition of the products and any intermediates using NMR, IR, or mass spectroscopy. Nuclear magnetic resonance spectroscopy is isotope specific, and the vibrational frequencies of molecules are influenced strongly by changing a CH_3 group to CD_3. Tritium is sometimes preferred as a tracer because it can be detected by its radioactivity, which is a far more sensitive probe than spectroscopy.

Isotope effects

The physical and chemical properties of isotopically substituted molecules are usually very similar. For example, the bond enthalpies, vapor

---

[1] At very high pressures hydrogen is expected to be metallic. The necessary pressures have not yet been achieved in the laboratory, but currently attainable pressures would appear to be approaching the values required. Hydrogen may be metallic at the core of Jupiter.

Table 9.1 The effect of deuteration on physical properties

	H_2	D_2	H_2O	D_2O
Normal boiling point/°C	−252.8	−249.7	100.00	101.42
Mean bond enthalpy/(kJ mol^{-1})	436.0	443.3	463.5	470.9

pressures, and Lewis acidities of $^{10}BF_3$ and $^{11}BF_3$ are nearly identical. However, the same is not true when D is substituted for H, for the mass of the substituted atom is doubled. For instance, Table 9.1 shows that the differences in boiling points and bond enthalpies are easily measurable for H_2 and D_2. The difference in boiling point between H_2O and D_2O reflects the greater strength of the O⋯D–O hydrogen bond compared with that of the O⋯H–O bond.

Much of the difference between E—D and E—H bonds (and between the analogous hydrogen bonds) arises from differences in their zero-point vibrational energies. The solution of the Schrödinger equation for a harmonic oscillator shows that the minimum permitted vibrational energy is

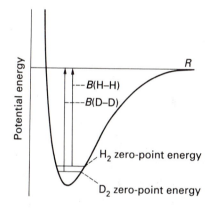

9.1 Molecular potential energy curve for H_2 and D_2 and the differences in zero-point energy and, because of that, the differences in dissociation energy.

$$E_0 = \tfrac{1}{2}\hbar\omega \qquad \text{with } \omega = \sqrt{\frac{k}{\mu}}$$

In this expression k is the **force constant** of the bond (a measure of its stiffness to stretching) and μ is the **effective mass**:

$$\frac{1}{\mu} = \frac{1}{m_A} + \frac{1}{m_B}$$

m_A and m_B are the masses of the two atoms. For two identical atoms of mass m, $\mu = \tfrac{1}{2}m$. The D_2 molecule has a larger effective mass than H_2, and therefore a lower zero-point energy and a higher bond enthalpy (Fig. 9.1). The difference in zero-point energy also accounts for the differences between the strengths of E—H and E—D bonds.

Reaction rates are also measurably different for processes in which E—H and E—D bonds are broken, made, or rearranged. The detection of this **kinetic isotope effect** can often help to support a proposed reaction mechanism. Kinetic isotope effects are frequently observed when an H atom is transferred from one atom to another in an activated complex. The effect on rate when this occurs is referred to as a **primary isotope effect**, and D transfer can be as much as ten times slower than H transfer. For example, the electrochemical reduction of $H^+(aq)$ to $H_2(g)$ occurs with a substantial isotope effect, with H_2 being liberated more rapidly; a practical consequence of the difference in rates of formation of H_2 and D_2 is to concentrate D_2O electrolytically. Substantial **secondary isotope effects** may occur when H is not transferred. For example, the rate of hydrolysis of the Cr—NCS link in *trans*-$[Cr(NCS)_4(NH_3)_2]^-$ is twice as fast when the ligands are ND_3 as when they are NH_3 even though no N—H bonds are broken in the reaction. This secondary isotope effect is ascribed to the change in strength of the N–H⋯O hydrogen bond between the complex and the solvent and the effect of this bonding on the ease with which the NCS^- ligand is able to depart from the Cr(III) site. The rate is greater for D than for H because N–D···O bonds are stronger than N–H···O bonds.

Isotopic substitution and IR spectra

Because the frequencies of infrared transitions between vibrational levels depend on the masses of atoms, they are strongly influenced by substitution of D for H. The heavier isotope results in the lower frequency. Inorganic chemists often take advantage of this isotope effect by comparing the IR spectra of a compound with normal hydrogen and with deuterium to determine whether a particular infrared transition involves significant motion of hydrogen. To a first approximation (because hydrogen is so much lighter than other atoms), the ratio of the frequencies is given by the expression

$$\frac{v_D}{v_H} = \sqrt{\frac{m_H}{m_D}}$$

Example 9.1: *Identifying an IR band by isotopic substitution*

For a dialkylarsane R_2AsH, an IR band, thought to be due to an As—H stretch, is observed at 2080 cm^{-1}. For the deuterated compound R_2AsD the band is shifted to 1475 cm^{-1}. Do the data support the assignment of the band as an As—H stretch?

Answer. The expected wavenumber for As—D is 2080 cm$^{-1} \times (1/2)^{\frac{1}{2}} =$ 1471 cm^{-1}, in reasonable agreement with the observed 1475 cm^{-1}. Thus, the assignment is supported.

Exercise E9.1. An IR band thought to be the Co—H stretch is observed at 1840 cm^{-1} for $[Co(CN)_5(H)]^{3-}$. What is the expected wavenumber of the corresponding band in $[Co(CN)_5(D)]^{3-}$?

Nuclear spin and NMR spectra

Another important property of the hydrogen nucleus is its spin, which makes possible the technique of ^{1}H-NMR spectroscopy (see *Further information 2*). Although the nuclei of many different elements can be observed readily with modern NMR spectrometers, ^{1}H-NMR is still the most widely used version of the technique.

Proton-NMR is useful for detecting the presence of H atoms in a compound and for identifying the atoms to which the hydrogen is bound by observation of the chemical shift or the spin–spin coupling to another nucleus (Fig. 9.2). A hydrogen ligand on a late *d*-block element has an NMR signal that is much more shielded than the usual standard (tetramethylsilane). Theoretical considerations suggest that the strong shielding is not simply the consequence of high electron density around the proton. Only in *s*-block hydrides is the electron

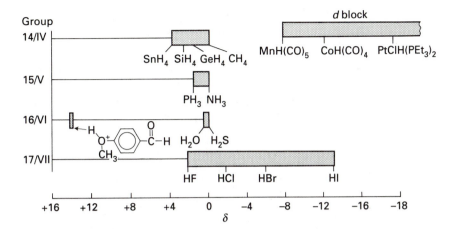

9.2 Typical ^{1}H-NMR chemical shifts. The tinted boxes group families of elements together but have no other significance.

density great enough for H^- to be a reasonably accurate portrayal of the entity present in the compound. As can be seen by inspection of Fig. 9.2, strong acids have strongly shielded ^{1}H-NMR signals, and in this case the simple picture of depletion of electron density around the proton does appear to be appropriate. The example of spin–spin coupling between ^{1}H and ^{73}Ge given in *Further information 2* provides an unambiguous demonstration that a compound such as $Ge(CH_3)_3H$ contains a GeH bond.

9.2 Hydrogen atoms and ions

The H atom has a high ionization enthalpy ($+1310\,kJ\,mol^{-1}$) and a low but positive electron affinity ($77\,kJ\,mol^{-1}$). The Pauling electronegativity is 2.2. This value is similar to the electronegativities of B, C, and Si, so their E—H bonds are not expected to be very polar. With *s*-block elements, hydrogen forms salt-like compounds in which the H atom acquires electron density and can be regarded as the hydride ion H^-. Conversely, when combined with the highly electronegative elements from the right of the periodic table, the E—H bond is best regarded as covalent and polar, with the H atom carrying the very small partial positive charge. In keeping with these trends:

> H is normally assigned oxidation number -1 when in combination with metals and $+1$ when in combination with nonmetals.

Some examples are:

Oxidation number:	+1 −1	+3 −1	−3 +1	+1 −1
Compound:	NaH	AlH$_3$	NH$_3$	HCl

Although the oxidation number is not the charge on an atom, it does provide an approximate guide to the variation of the polarities of hydrogen compounds through the periodic table.

When a high-voltage discharge is passed through hydrogen gas at low pressure, the molecules dissociate, ionize, and recombine to form a plasma that contains spectroscopically observable amounts of H, H^+, H_2^+, and H_3^+. The free hydrogen cation (H^+, the proton) has a very high charge/radius ratio and so it is not surprising to find that it is a very strong Lewis acid. In the gas phase it readily attaches to other molecules and atoms; for example, H^+ attaches to He to form HeH^+. In condensed phases, H^+ is always found in combination with a Lewis base, and its ability to transfer between Lewis bases gives it the special role in chemistry that we explored in detail in Chapter 5.

The molecular cations H_2^+ and H_3^+ have only a transitory existence in the gas phase and are unknown in solution. Their electronic structures were described in Section 2.7, where we saw that spectroscopic data indicate that $H_3^+(g)$ is an equilateral triangle. The H_3^+ ion is the simplest example of a three-center, two-electron bond (a 3c,2e-bond) in which three nuclei are bonded by only two electrons.

9.3 Properties and reactions of dihydrogen

The stable form of elemental hydrogen under normal conditions is **dihydrogen**, H_2, more informally and henceforth plain 'hydrogen'. The H_2 molecule has a high bond enthalpy ($436 \, kJ \, mol^{-1}$) and a short bond length (0.74 Å). Because it has so few electrons, the forces between H_2 molecules are weak, and at 1 atm the gas condenses to a liquid only when cooled to 20 K.

The production of hydrogen

Molecular hydrogen is not present in significant quantities in the Earth's crust but is produced in huge quantities to satisfy the needs of industry. The most economical sources of hydrogen are the reactions of carbon, methane, or carbon monoxide with water at high temperatures. The main commercial process is currently **steam reforming**, the catalyzed reaction of water and hydrocarbons (typically methane from natural gas):

$$CH_4(g) + H_2O(g) \xrightarrow{1000\,°C} CO(g) + 3H_2(g)$$

A similar reaction, but with coke as the reducing agent, is sometimes called the **water-gas reaction**:

$$C(s) + H_2O(g) \xrightarrow{1000\,°C} CO(g) + H_2(g)$$

This reaction was once a primary source of H_2 and it may become important again when natural hydrocarbons are depleted. As with the reduction of metal oxides by CO (Section 7.1), the negative free energy of both reactions results in part from the strength of the bond in CO and from the favorable entropy change. The yield of H_2 can be increased by means of a subsequent **shift reaction**:

$$CO(g) + H_2O(g) \rightarrow CO_2(g) + H_2(g)$$

Hydrogen is also produced as a by-product of the electrolysis of brine during the manufacture of chlorine and sodium hydroxide (Sec-

Chart 9.1

1

2

tion 13.1), but electrolytic production will become the primary source only if cheap electricity becomes available. Another source is as a by-product of the dehydrogenation of alkanes to produce alkenes for the petrochemical industry and arenes for high-octane gasolines:

$$CH_3CH_3(g) \xrightarrow{\Delta} CH_2{=}CH_2(g) + H_2(g)$$

The hydrogen produced in one of these ways is normally used directly in an integrated chemical plant, such as an ammonia synthesis plant or a petroleum refinery (Chart 9.1).

Because of its low mass and high enthalpy of combustion, hydrogen is an excellent fuel for large rockets. Its more general use as a fuel has been analyzed seriously since the early 1970s when petroleum prices rose sharply. Strategies have been devised for a **hydrogen economy** in which hydrogen is the primary fuel. One scenario is to collect solar energy with photovoltaic cells and employ the resulting electricity to electrolyze water. The hydrogen from this electrolysis would be a form of stored solar energy that could be burned as a fuel when needed or used as a feed-stock in chemical processes. The current prices of petroleum, natural gas, and coal are too low to make such a hydrogen economy viable, but the idea might have its day.[2] Quite aside from the eventual scarcity of fossil fuels, when hydrogen burns it produces water, which is not a greenhouse gas[3] like CO_2, so a hydrogen economy would not necessarily lead to the global warming that now seems to confront us. The availability of photochemically produced H_2 would also eliminate the CO_2 resulting from steam reforming and the shift reaction. Thus a first step toward a hydrogen economy might be the use of electrolytic or photochemical hydrogen in the chemical and petrochemical industries.

Mechanistic aspects

The reaction of molecular hydrogen with most other elements is slow partly because of its high bond enthalpy and hence high activation energy for reaction. Under special conditions, however, the reactions are rapid. These conditions include:

1. The activation of the molecule by homolytic dissociation on a metal surface or a metal complex.
2. Heterolytic dissociation by a surface or metal ion.
3. Initiation of a radical-chain reaction.

Two examples of **homolytic dissociation** are chemisorption on platinum (**1**) and coordination to the Ir atom in a complex (**2**). The former accounts for the use of finely divided platinum metal to catalyze

---

[2] Research on the generation of hydrogen by advanced photochemical methods is described in M. Grätzel (ed.), *Energy resources through photochemistry and catalysis*, Academic Press, New York (1983).

[3] A greenhouse gas is a species in the atmosphere that is relatively transparent to incoming ultraviolet and visible radiation from the Sun, but absorbs outgoing infrared radiation. Thus it acts rather like the glass in a greenhouse, and the atmosphere becomes warmed.

the hydrogenation of alkenes and of platinum as the electrode material used to interconvert H_2 and H^+ electrochemically. Thus the over-potential for H_2 formation is much smaller at a platinum electrode, on which a surface hydride can form, than on a mercury electrode, where the formation of a surface hydride is unfavorable. Similarly, H_2 readily bonds to the Ir atom in complexes such as $[IrClCO(PPh_3)_2]$ to give a complex containing two hydrido (H) ligands:

This type of reaction is an example of oxidative addition because the metal center undergoes a formal increase in oxidation number when the reactant adds to it (see Section 8.10, where this type of reaction was introduced). Oxidative addition of H_2 is common for Rh(I), Ir(I), and Pt(0) complexes. When considering hydrido complexes, we should recognize that H^- is a high-field ligand. There are many d-block hydrido complexes, some of which are formed by oxidative addition, and they are discussed in more detail in Chapters 16 and 17.

An H_2 molecule can coordinate to a metal atom without cleavage of the H—H bond.[4] The first such compound was $[W(CO)_3(H_2)(P^iPr_3)_2]$ (**3**; ^iPr denotes isopropyl, $-CH(CH_3)_2$) and over a hundred compounds of this type are now known. The significance of these compounds is that they provide examples of species intermediate between molecular H_2 and a dihydrido complex. Molecular orbital calculations indicate that the bonding may be considered in terms of donation of electron density from the H—H bond into a vacant metal d orbital and a simultaneous back π bonding from a different d orbital into the empty σ^* orbital of H_2 (**4**). This bonding pattern is similar to that of CO and ethylene with metal atoms (Sections 6.6 and 16.3). The back-donation of electron density from the metal to a σ^* orbital on H_2 is consistent with the observation that the H—H bond is cleaved when the metal center is made electron-rich by highly basic ligands. No dihydrogen complexes are known for the early d-block (Groups 3, 4, and 5), the f-block, or the p-block metals.

An example of **heterolytic dissociation** is the reaction of H_2 with a ZnO surface, which appears to produce a Zn(II)-bound hydride and an O-bound proton:

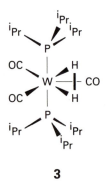

3

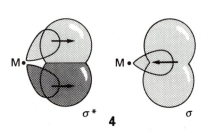

σ^* **4** σ

[4] An informative account of this discovery and its implications is given by G. Kubas, *Acc. Chem. Res.*, **21**, 120 (1988) and the detection of dihydrogen complexes by NMR is described by R. H. Crabtree, *Acc. Chem. Res.*, **23**, 95 (1990).

This reaction is thought to be involved in the catalytic hydrogenation of carbon monoxide to methanol

$$CO(g) + 2H_2(g) \xrightarrow{Cu/Zn} CH_3OH(g)$$

which is carried out on a very large scale world-wide. Another example is the sequence

$$H_2(g) + Cu^{2+}(aq) \rightarrow [CuH]^+(aq) + H^+(aq) \rightarrow Cu(s) + 2H^+(aq)$$

which is important in the hydrometallurgical reduction of Cu^{2+} (Section 7.1). The $[CuH]^+$ intermediate has only a transitory existence.

Radical chain mechanisms account for the thermally or photochemically initiated reactions between H_2 and the halogens. Initiation is by thermal or photochemical dissociation of the dihalogen molecules to give atoms that act as radical chain carriers:

Initiation: $Br_2 \xrightarrow{\Delta \text{ or } h\nu} 2Br\cdot$

Propagation: $Br\cdot + H_2 \rightarrow HBr + H\cdot$
   $H\cdot + Br_2 \rightarrow HBr + Br\cdot$

The activation energy of radical attack is low because a new bond is formed as one bond is lost, and so once initiated the formation and consumption of radicals is self-sustaining and the production of HBr is very rapid. Chain termination occurs when the radicals recombine:

Termination: $2H\cdot \rightarrow H_2$ $2Br\cdot \rightarrow Br_2$

Termination becomes more important toward the end of the reaction when the concentrations of H_2 and Br_2 are low.

Example 9.2: *Recognizing initiation steps in a radical reaction*
Suggest a mechanistic explanation for the observation that the addition of NO gas to a mixture of H_2 and Cl_2 leads to an explosion.

Answer. The NO radical initiates a very fast reaction between H_2 and Cl_2 by generation of $Cl\cdot$ atoms:

$$NO\cdot + Cl_2 \rightarrow ClNO + Cl\cdot$$

The $Cl\cdot$ produced then reacts with H_2 in a radical chain mechanism in which the propagation steps are similar to those already given for bromine:

$$H_2 + Cl\cdot \rightarrow HCl + H\cdot$$

$$H\cdot + Cl_2 \rightarrow HCl + Cl\cdot$$

Exercise E9.2. Suggest why, under some conditions, a radical chain reaction will slow down when a foreign radical is introduced.

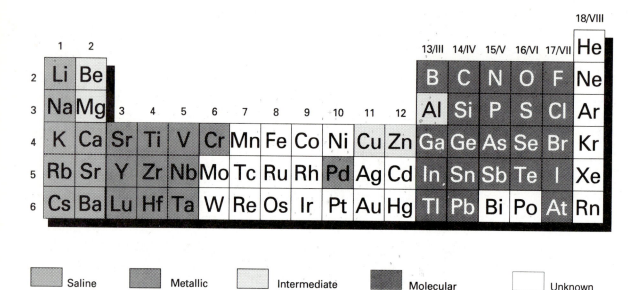

9.3 Classification of the binary hydrogen compounds of the *s*-, *p*-, and *d*-block elements. Although some *d*-block elements, such as iron and ruthenium, do not form binary hydrides, they do form metal complexes containing the hydride ligand.

CLASSIFICATION AND STRUCTURE OF COMPOUNDS

Most binary compounds of hydrogen fall into one of three categories. The hydrogen compounds of the highly electropositive *s*-block metals are nonvolatile, electrically nonconducting, crystalline solids. These properties as well as their structures lead to their classification as **saline hydrides** (or 'salt-like hydrides'). On the other hand, the *d*-block and *f*-block metals form hydrides that are often nonstoichiometric, electrically conducting solids, called **metallic hydrides**. Most of the *p*-block binary hydrogen compounds are volatile molecular compounds, and accordingly are classified as **molecular hydrides**. The distribution of these hydrides in the periodic table is shown in Fig. 9.3. As so often in chemistry, the main point of this classification scheme is to emphasize the major trends in properties, and in fact there is a continuum of structural types. We shall see, for instance, that some elements (among them beryllium and aluminum) form hydrides that cannot be classified as either strictly saline or strictly molecular.

9.4 The saline hydrides

The Group 1 hydrides have rock-salt structures, and the Group 2 hydrides have crystal structures like those of some heavy metal halides (Table 9.2). These structures (and their chemical properties) are the basis for classifying hydrogen compounds of the *s*-block elements other than beryllium as saline. The ionic radius of H^- determined from the dimensions of these compounds varies quite widely, from 1.26 Å in LiH to 1.54 Å in CsH. This wide variability reflects the loose control that the single charge of the proton has on its two surrounding electrons and the resulting high compressibility of H^-.

The saline hydrides are insoluble in common nonaqueous solvents

Table 9.2 Structures of *s*-block hydrides

Compound	Crystal structure
LiH, NaH, KH, RbH, CsH	Rock salt
MgH_2	Rutile
CaH_2, SrH_2, BaH_2	Distorted $PbCl_2$

Source: A. F. Wells, *Structural inorganic chemistry*. Oxford University Press (1984).

but they do dissolve in molten alkali halides. Electrolysis of this molten-salt solution produces hydrogen gas at the anode (the site of oxidation)

$$2H^-(melt) \rightarrow H_2(g) + 2e^-$$

thus providing chemical evidence for the existence of H^-. The reaction of saline hydrides with water is dangerously vigorous:

$$NaH(s) + H_2O(l) \rightarrow H_2(g) + NaOH(aq)$$

Indeed, finely divided sodium hydride can ignite if it is left exposed to humid air. Such fires are difficult to extinguish because even carbon dioxide is reduced when it comes into contact with hot metal hydrides (water, of course, forms more of the flammable hydrogen); they may be blanketed with an inert solid, such as silica sand. The reaction between metal hydrides and water is employed in the laboratory to remove traces of water from solvents and inert gases such as nitrogen and argon:

$$CaH_2(s) + 2H_2O(g) \rightarrow Ca(OH)_2(s) + 2H_2(g)$$

Calcium hydride is preferred in this application because it is the cheapest of the saline hydrides, and is available in granular form, which is easy to handle. Large amounts of water should not be removed from solvent in this manner because the strongly exothermic reaction evolves flammable hydrogen.

The absence of convenient solvents limits the use of saline hydrides as reagents, but this problem is partially overcome by the availability of commercial dispersions of finely divided NaH in oil.[5] This material is reactive by virtue of the dispersion's high surface area. Alkali metal hydrides are convenient reagents for making other hydride compounds. For example, the reaction with a trialkylboron compound yields a hydride complex which is soluble in polar organic solvents and therefore a useful reducing agent and H^- source:

$$NaH(s) + B(C_2H_5)_3(et) \rightarrow Na[HB(C_2H_5)_3](et)$$

where et denotes solution in diethyl ether.

9.5 Metallic hydrides

Metallic hydrides are known for all d-block metals of Groups 3 through 5 and for the f-block metals (Fig. 9.4). However, the only hydride in Group 6 is CrH and no hydride phases are known for the unalloyed metals in Groups 7 through 9. (Claims have been made, however, that hydrogen dissolves in iron at very high pressure, and that iron hydride is abundant at the center of the Earth.) The region of the periodic table covered by Groups 7 through 9 is sometimes referred to as the **hydride gap**. The f-block elements form hydrides with the limiting

[5] Even more finely divided and reactive alkali metal hydrides can be prepared from the metal alkyl and hydrogen: P. A. A. Klusener, L. Brandsma, H. D. Verkruijsse, P. von Rague Schleyer, T. Friedl, and R. Pi, *Angew. Chem., Int. Ed. Engl.*, **25**, 465 (1986).

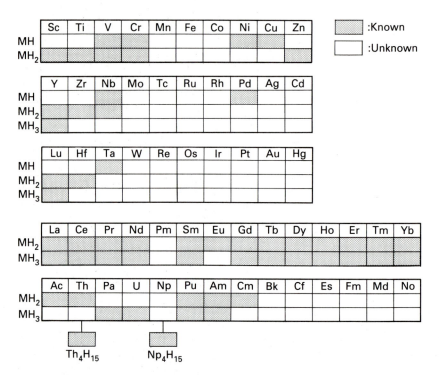

9.4 Hydrides formed by *d*- and *f*-block elements. The formulas are limiting stoichiometries based on the structural type. In many cases they are not attained. For example, PdH_x never attains $x = 1$. (From G. G. Libowitz, *Solid state chemistry of the binary metal hydrides.* Benjamin (1965).)

formulas MH_2 and MH_3, but large deviations from the ideal stoichiometric composition are generally found.

The Group 10 metals, especially nickel and platinum, are often used as hydrogenation catalysts in which surface hydride formation is implicated (Section 17.6). However, somewhat surprisingly, at moderate pressures only palladium forms a stable bulk phase; its composition is PdH_x, with $x < 1$. Nickel forms hydride phases at very high pressures but platinum does not form any at all. Apparently the Pt—H bond enthalpy is sufficiently great to disrupt the H—H bond but not to offset the loss of Pt—Pt bonding, which would occur upon formation of a bulk platinum hydride. In agreement with this interpretation, the enthalpies of sublimation, which reflect MM bond enthalpies, lie in the order Pd (378 kJ mol^{-1}) < Ni (430 kJ mol^{-1}) < Pt (565 kJ mol^{-1}).

Most metallic hydrides are metallic conductors (hence their name) and have variable composition. For example, at $550\,°C$ a compound ZrH_x exists over a composition range from $ZrH_{1.30}$ to $ZrH_{1.75}$; it has the fluorite structure (Fig. 4.14) with a variable number of anion sites unoccupied. We can understand the variable stoichiometry and metallic conductivity of these hydrides in terms of a model in which the band of delocalized orbitals responsible for the conductivity accommodates electrons supplied by arriving H atoms. In this model, the H^+ ions as well as the metal cations take up equilibrium positions in this electron sea. The conductivities of metallic hydrides typically vary with hydrogen content, and this variation can be correlated with the extent to which the conduction band is filled or emptied as hydrogen is added or removed. Thus, CeH_{2-x} is a metallic conductor, whereas CeH_3 (which has a full conduction band) is an insulator.

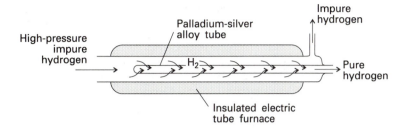

9.5 Schematic diagram of a hydrogen purifier. Because of a pressure differential and the mobility of H atoms in palladium, hydrogen diffuses through the palladium-silver tube as H atoms but impurities do not.

A striking property of many metallic hydrides is the high rate of hydrogen diffusion through the solid at slightly elevated temperatures. This mobility is utilized in the ultrapurification of H_2 by diffusion through a palladium–silver alloy tube (Fig. 9.5). The high mobility of the hydrogen they contain, and their variable composition, also make the metallic hydrides potential hydrogen storage media. Thus the intermetallic compound $LaNi_5$ forms a hydride phase with a limiting composition $LaNi_5H_6$, and at this composition it contains more hydrogen per unit volume than liquid H_2. A less expensive system with the composition $FeTiH_x$ ($x < 1.95$) is now commercially available for low pressure hydrogen storage, and it has been used as a source of energy in vehicle trials.

9.6 Simple molecular compounds

Hydrogen forms molecular compounds with the *p*-block elements. They include the familiar Period 2 compounds methane, ammonia, water, and hydrogen fluoride, as well as the corresponding compounds of the heavier elements in each group.

Nomenclature and classification

The systematic names of the molecular hydrogen compounds are usually formed from the name of the element and the suffix -ane, as in phosphane for PH_3. Older usage, however, employs the suffix -ine (as in phosphine, Table 9.3). In an earlier terminology, compounds of hydrogen were often named as though they were ionic, as in hydrogen sulfide (H_2S, sulfane). Many of these traditional names are still widely used. The nonsystematic names ammonia and water are universally used rather than their systematic names, azane and oxidane.

A useful classification of the molecular hydrogen compounds takes note of the relative numbers of electrons and bonds in their Lewis structures. An **electron-deficient compound** is one with too few electrons for a Lewis structure to be written. An **electron-precise compound** is one with the correct number of electron pairs for bond formation with none left over as nonbonding electron pairs on the central atom. An **electron-rich compound** is one with more electron pairs than are needed for bond formation, the extra electron pairs being present as nonbonding electron pairs on the central atom.

Diborane (B_2H_6) is a classic example of an electron-deficient com-

Table 9.3 Some molecular hydrogen compounds

Group	Formula	Traditional name	IUPAC name
13/III	B_2H_6	Diborane	Diborane(6)
14/IV	CH_4	Methane	Methane
	SiH_4	Silane	Silane
	GeH_4	Germane	Germane
	SnH_4	Stannane	Stannane
15/V	NH_3	Ammonia	Azane
	PH_3	Phosphine	Phosphane
	AsH_3	Arsine	Arsane
	SbH_3	Stibine	Stibane
16/VI	H_2O	Water	Oxidane
	H_2S	Hydrogen sulfide	Sulfane
	H_2Se	Hydrogen selenide	Sellane
	H_2Te	Hydrogen telluride	Tellane
17/VII	HF	Hydrogen fluoride	Hydrogen fluoride
	HCl	Hydrogen chloride	Hydrogen chloride
	HBr	Hydrogen bromide	Hydrogen bromide
	HI	Hydrogen iodide	Hydrogen iodide

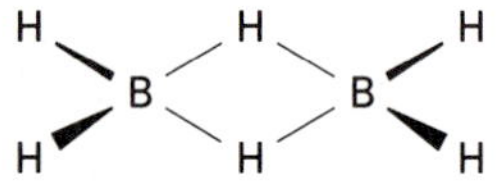

5 B_2H_6

6 CH_4

7 NH_3

8 H_2O

pound (Section 2.8). Its Lewis structure would require at least 14 valence electrons to bind the eight atoms together, but the molecule in fact has only 12 electrons. A simple description of its structure is one in which a bridging hydrogen and two boron atoms are joined by a 3c,2e (three-center, two-electron) bond (**5**): each B atom has an octet of electrons and each H atom has a pair of electrons, so the valence shells of all atoms are satisfied. That the 3c,2e hydrogen bridge bond is weaker than the terminal 2c,2e bonds is reflected in diborane's chemical properties.

Methane is an example of an electron-precise compound. Electron-precise binary compounds are formed by the elements in Group 14/IV (each of which has four valence electrons). The molecules are tetrahedral (**6**) with the expected increase in E—H bond length down the group.

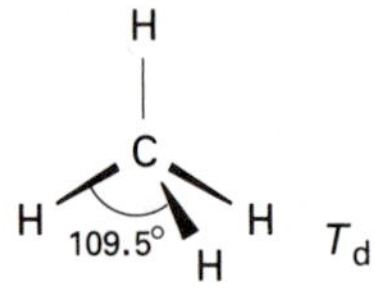

Ammonia is an example of an electron-rich compound. Its four atoms could be bonded by only six valence electrons, whereas in fact it has eight, the extra two forming a nonbonding lone pair. A part of the reason for the existence of electron-rich compounds is that H has too low an electronegativity to draw out the highest oxidation number of Group 15/V to 17/VII elements. For example, in combination with hydrogen, phosphorus forms the electron-rich compound phosphane (PH_3) and not the electron precise compound PH_5. With the more electronegative chlorine it forms both PCl_3 and PCl_5.

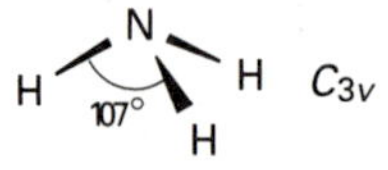

General aspects of properties

The shapes of the electron-precise and electron-rich compounds can all be predicted by the VSEPR rules (Section 3.1). Thus, CH_4 is tetrahedral (**6**), NH_3 is pyramidal (**7**), H_2O angular (**8**), and HF (necessarily) linear. However, the simple VSEPR rules do not indicate the considerable change in bond angle between NH_3 and its heavier analogs

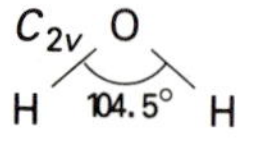

Table 9.4 Bond angles (in degrees) for Group 15/V and 16/VI hydrogen compounds

NH_3	106.6	H_2O	104.5
PH_3	93.8	H_2S	92.1
AsH_3	91.8	H_2Se	91
SbH_3	91.3	H_2Te	89

Source: A. F. Wells, *Structural inorganic chemistry.* Oxford University Press (1984).

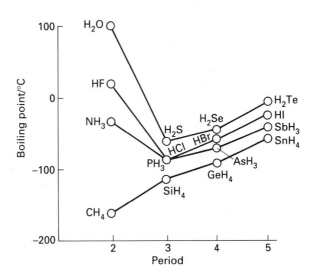

9.6 Normal boiling points of *p*-block binary hydrogen compounds.

9 $(HF)_n$

or between H_2O and its analogs in Group 16/VI. As noted in Table 9.4, the bond angles for the Period 2 molecules NH_3 and H_2O are slightly less than the tetrahedral angle, but for their heavier analogs the bond angle is as small as 90°.

An important consequence of the simultaneous presence of highly electronegative atoms (N, O, and F) and lone pairs in the electron-rich compounds is the possibility of forming hydrogen bonds. The idea of hydrogen bonding originally came from the observation of unusually high relative permittivities of water and similar fluids.[6] Striking evidence for hydrogen bonding is provided by the trends in boiling points, which are unusually high for the strongly hydrogen bonded molecules water, ammonia, and hydrogen fluoride (Fig. 9.6). Some of the most persuasive evidence for hydrogen bonding comes from structural data, such as the open network structure of ice (Fig. 9.7 and Section 9.16), the existence of chains in solid HF (**9**) that survive partially even in the vapor, and the location of hydrogen atoms in solids by diffraction and NMR techniques.[7] As indicated by boiling point trends, H_2S, PH_3, HCl, and the heavier *p*-block hydrogen compounds do not form strong hydrogen bonds.

[6] W. M. Latimer and W. H. Rhodebush, *J. Am. Chem. Soc.*, **42**, 1635 (1912). It took many years after this publication for the full significance of hydrogen bonding to be appreciated.
[7] W. C. Hamilton and J. A. Ibers, *Hydrogen bonding in solids.* W. A. Benjamin, New York (1968); J. Emsley, *Chem. Soc. Rev.*, **9**, 91 (1990).

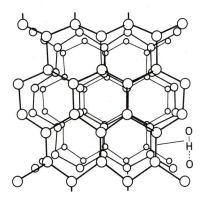

9.7 The structure of ice. The large spheres represent the O atoms; the H atoms lie on the lines joining O atoms.

Example 9.3: *Correlating the classification and properties of hydrogen compounds*

Classify $HfH_{1.5}$, PH_3, CsH, and B_2H_6 and discuss their physical properties. For the molecular compounds specify their subclassification (electron-deficient, electron-precise, or electron-rich).

Answer. The two hydrides $HfH_{1.5}$ and CsH will be solids. The former is an example of a metallic hydride, which are common for *d*- and *f*-block metals, and it exhibits significant electrical conductivity. CsH is a saline hydride typical of the *s*-block metals. It is an electrical insulator with the rock-salt crystal structure. The *p*-block molecular hydrides PH_3 and B_2H_6 have low molar mass and are therefore expected to be highly volatile. (They are in fact gases under normal conditions.) The Lewis structure indicates that PH_3 has a lone pair on P and that it is therefore an electron-rich molecular hydride. As explained in Section 2.8, diborane is electron deficient.

Exercise E9.3. Give examples of hydrogen compounds of elements in Groups 2, 11, and 16/VI; classify them, and discuss whether any of them are likely to participate in hydrogen bonding.

SYNTHESIS AND REACTIONS OF HYDROGEN COMPOUNDS

Many of the chemical properties of hydrogen compounds can be systematized by noting the periodic trends in their thermodynamic stabilities and their polarities (specifically, whether the partial charge on hydrogen is $H^{\delta+}$ or $H^{\delta-}$). Furthermore, the less predictable but important radical chain reactions play a major role in the gas-phase reactions of hydrogen with oxygen and the halogens.

9.7 Stability and synthesis

A negative Gibbs free energy of formation (signifying an exoergic compound) is a clue that the direct combination of hydrogen and an element may be the preferred synthetic route for a hydrogen compound. When a compound is thermodynamically unstable with respect to its elements (that is, when it is endoergic), an indirect synthetic route can often be found, but each step in the indirect route must be thermodynamically favorable. Our principal objective in this section is to convey the intuition that suggests which reactions are likely to be thermodynamically favorable.

Thermodynamic aspects

The standard free energies of formation of the hydrogen compounds of *s*- and *p*-block elements (Table 9.5) reveal a regular variation in stability. With the possible exception of BeH_2, for which good data are not available, all the *s*-block hydrides are exoergic ($\Delta G_f^{\ominus} < 0$) and therefore thermodynamically stable with respect to their elements at room temperature. The trend is erratic in Group 13/III, in that only AlH_3 is exoergic. In all the other groups of the *p* block, the hydrogen

Table 9.5 Standard Gibbs energy of formation, $\Delta G_f^{\ominus}/(\text{kJ mol}^{-1})$, of binary *s*- and *p*-block hydrogen compounds at 25 °C

Period	Group 1 I	2 II	13 III	14 IV	15 V	16 VI	17 VII
2	LiH(s) −68.4	BeH$_2$(s) (+20)	B$_2$H$_6$(g) +86.7	CH$_4$(g) −50.7	NH$_3$(g) −16.5	H$_2$O(l) −237.1	HF(g) −273.2
3	NaH(s) −33.5	MgH$_2$(s) −35.9	AlH$_3$(s) (−1)	SiH$_4$(g) +56.9	PH$_3$(g) +13.4	H$_2$S(g) −33.6	HCl(g) −95.3
4	KH(s) (−36)	CaH$_2$(s) −147.2	GaH$_3$ >0	GeH$_4$(g) +113.4	AsH$_3$(g) +68.9	H$_2$Se(g) +15.9	HBr(g) −53.5
5	RbH(s) (−30)	SrH$_2$(s) (−141)		SnH$_4$(g) +188.3	SbH$_3$(g) +147.8	H$_2$Te(g) >0	HI(g) +1.7
6	CsH(s) (−32)	BaH$_2$(s) (−140)					

Data from *J. Phys. Chem. Ref. Data*, **11**, Supplement 2 (1982). Values in parenthesis are based on $\Delta H_f^{\ominus}$ data from this source and entropy contributions estimated by the method of W. M. Latimer, p. 359 of *Oxidation potentials*. Prentice Hall, Englewood Cliffs, NJ (1952).

compounds of the first members of the groups (CH$_4$, NH$_3$, H$_2$O, and HF) are exoergic and become progressively less stable down the group. This general trend in stability is largely a reflection of the decreasing E—H bond strength down a group in the *p* block (Fig. 9.8). The weak bonds formed by the heavier elements are often attributed to poor overlap between the relatively compact H1*s* orbital and the more diffuse *s* and *p* orbitals of the heavier atoms in these groups. The heavier members become more stable on going from Group 14/IV across to the halogens. For example, SnH$_4$ is highly endoergic whereas HI is barely so.

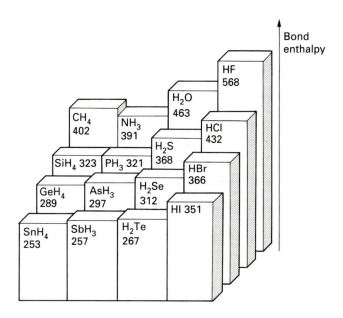

9.8 Average bond enthalpies (in kJ mol^{-1} at 298 K).

Synthesis

There are three common methods for synthesizing binary hydrogen compounds:

1. *Direct combination of the elements:*

$$2E + H_2(g) \rightarrow 2EH$$

2. *Protonation of a Brønsted base:*

$$E^- + H_2O(aq) \rightarrow EH + OH^-$$

3. *Double replacement (metathesis) of a halide or pseudohalide with a hydride:*

$$E'H + EX \rightarrow E'X + EH$$

(In such general equations, the symbol E could also denote an element with higher valence, with corresponding changes of detail in the formulas and stoichiometric numbers.)

An example of a direct synthesis is

$$2Li(l) + H_2(g) \xrightarrow{\Delta} 2LiH(s)$$

This method is used commercially for the synthesis of exoergic compounds, including NH_3 and the hydrides of lithium, sodium, and calcium. However, in some cases forcing conditions (high pressure, high temperature, and a catalyst) are necessary to overcome the unfavorable kinetics. The high temperature used for the lithium reaction is an example: it melts the metal and hence helps to break up the surface layer of hydride that would otherwise passivate it. This inconvenience is avoided in many laboratory preparations by adopting one of the other methods (2 or 3 above), which may also be used to prepare endoergic compounds.

An example of protonation of a Brønsted base is

$$Li_3N(s) + 3H_2O(l) \rightarrow 3LiOH(aq) + NH_3(g)$$

The Brønsted base in this reaction is the nitride ion, N^{3-}. The starting materials are too expensive for the reaction to be suitable for the commercial production of NH_3, but it is very useful in the laboratory for the preparation of ND_3. The success of the reaction depends on the proton donor being stronger than the conjugate acid of the N^{3-} anion (NH_3 in this case). Water is a sufficiently strong acid to protonate the very strong base N^{3-}, but a stronger acid, such as H_2SO_4, is required to protonate the weak base Cl^-:

$$NaCl(s) + H_2SO_4(l) \rightarrow NaHSO_4(s) + HCl(g)$$

A synthesis by a double replacement reaction is

$$LiAlH_4 + SiCl_4 \rightarrow LiAlCl_4 + SiH_4$$

This reaction involves (at least formally) the exchange of Cl^- ions for H^- ions in the coordination sphere of the Si atom. Hydrides of the more electropositive elements (LiH, NaH, and $[AlH_4]^-$) are the most active H^- sources. The favorite sources are often the $[AlH_4]^-$ and $[BH_4]^-$ ions in salts such as $LiAlH_4$ and $NaBH_4$, these compounds are soluble in ether solvents (such as $CH_3OC_2H_4OCH_3$) which solvate the alkali metal ion. Of these anion complexes, $[AlH_4]^-$ is much the strongest hydride donor (Section 9.11).

9.8 Reaction patterns for hydrogen compounds

Three types of reaction can lead to the scission of an E—H bond. They are similar to the processes already considered for H—H bond cleavage:

1. *Heterolytic cleavage by hydride transfer:*

$$E—H \rightarrow E^+ + {:}H^-$$

2. *Homolytic cleavage:*

$$E—H \rightarrow E{\cdot} + H{\cdot}$$

3. *Heterolytic cleavage by proton transfer:*

$$E—H \rightarrow {:}E^- + H^+$$

A free H^- or H^+ ion is unlikely to be formed in reactions that take place in solution. Instead, the ion is transferred via a complex that involves a hydrogen bridge.

It is often not clear which of these three primary processes occurs in practice. For example, any of them can lead to the addition of EH across a double bond:

$$E—H + H_2C{=}CH_2 \rightarrow \overset{\displaystyle E}{\underset{\textstyle |}{H_2C}}{-}\overset{\displaystyle H}{\underset{\textstyle |}{CH_2}}$$

However, there are some patterns of behavior that at least suggest which one is likely to occur.

Heterolytic cleavage and hydridic character

Compounds with hydridic character (that is, compounds that can transfer an H^- ion and hence participate in Process 1) react vigorously with Brønsted acids, with the evolution of H_2:

$$NaH(s) + H_2O(l) \rightarrow NaOH(aq) + H_2(g)$$

A compound that takes part in this **protolysis reaction** with a weak proton donor (such as water, as in this example) is said to be **strongly hydridic**. If a compound requires a strong proton donor, then it is classified as **weakly hydridic** (an example is germane, GeH_4). Hydridic character is most pronounced toward the left of a period where the element is most electropositive (in the s-block) and decreases rapidly after Group 13/III. For example, the saline hydrides of Groups

1 and 2 are strongly hydridic. In addition to protolysis, the saline hydrides display hydridic behavior, by transferring an H^-, in the following synthetically useful reactions:

1. *Double replacement with a halide*, such as the reaction of finely divided lithium hydride with silicon tetrachloride in dry ether (et):

$$4LiH(s) + SiCl_4(et) \rightarrow 4LiCl(s) + SiH_4(g)$$

2. *Addition to a Lewis acid*:

$$LiH(s) + B(CH_3)_3(g) \rightarrow Li[BH(CH_3)_3](et)$$

3. *Reaction with a proton source*, to produce H_2 as one of the products:

$$NaH(s) + CH_3OH(l) \rightarrow NaOCH_3(s) + H_2(g)$$

Note that the hydrogen bonded to the O atom, but not the hydrogen bonded to the C atom, is sufficiently acidic to undergo this reaction.

Combinations of these reactions are also common. For example the following reaction may be thought of as metathesis (to form B_2H_6) to which $2H^-$ may add (to form BH_4^-):

$$4LiH(s) + BF_3(et) \rightarrow Li[BH_4](et) + 3LiF(s)$$

Homolytic cleavage and radical character

Homolytic cleavage (Process 2) appears to occur readily for the hydrogen compounds of some *p*-block elements, especially the heavier elements. For example, the use of a radical initiator ($Q\cdot$ in the following equations, typically a peroxide) greatly facilitates the reaction of trialkylstannanes (R_3SnH) with haloalkanes, R—X. The overall reaction is

$$R_3SnH + R'X \rightarrow R'H + R_3SnX$$

and the proposed mechanism is

Initiation: $R_3Sn\text{-}H + Q\cdot \rightarrow R_3Sn\cdot + QH$

Propagation: $R_3Sn\cdot + XR' \rightarrow R_3SnX + R'\cdot$
(rate-determining halogen abstraction)
$R_3SnH + R'\cdot \rightarrow R'H + R_3Sn\cdot$

Termination: $R'\cdot + R'\cdot \rightarrow R'_2$
$R'\cdot + R_3Sn\cdot \rightarrow R_3SnR'$
$R_3Sn\cdot + R_3Sn\cdot \rightarrow R_3Sn\text{—}SnR_3$

The order of reactivity for various haloalkanes with trialkylstannanes is

$$RF < RCl < RBr < RI$$

Thus fluoroalkanes do not react with R_3SnH, chloroalkanes require heat, photolysis, or chemical radical initiators, and bromoalkanes and iodoalkanes react spontaneously at room temperature. This trend indicates that the rate-determining step is halogen abstraction. Similarly, the tendency toward radical reactions increases toward the heavier elements in each group, and Sn—H compounds are in general more prone to radical reactions than Si—H compounds. The ease of homolytic E—H bond cleavage mirrors the decrease in E—H bond strength down a group and the decrease in stretching frequencies (or wavenumbers):[8]

H_2O	H_2S	H_2Se
3652 and 3756 cm^{-1}	2611 and 2684 cm^{-1}	2260 and 2350 cm^{-1}

Heterolytic cleavage and protonic character

Compounds reacting by deprotonation (Process 3) are said to show **protic behavior**: in other words, they are Brønsted acids. We saw in Section 5.2 that Brønsted acid strength increases from left to right across a period in the *p* block and down a group. One striking example of the former trend is the increase in acidity from CH_4 to HF.

Example 9.4: *Using hydrogen compounds in synthesis*

Suggest a procedure for synthesizing lithium tetraethoxyaluminate from $Li[AlH_4]$ and reagents and solvents of your choice.

Answer. The reaction of the slightly acidic compound ethanol with the strongly hydridic lithium tetrahydroaluminate should yield the desired alkoxide and hydrogen. The reaction might be carried out by dissolving $LiAlH_4$ in tetrahydrofuran and slowly dropping ethanol into this solution:

$$LiAlH_4(thf) + 4C_2H_5OH(l) \rightarrow Li[Al(OEt)_4](thf) + 4H_2(g)$$

This type of reaction should be carried out under a blanket of inert gas (N_2 or Ar) and with a bubbler to permit the evolved hydrogen to escape into a hood.

Exercise E9.4. Suggest a way of making Et_3MeSn from Et_3SnH and a reagent of your choice.

THE ELECTRON-DEFICIENT HYDRIDES OF THE BORON GROUP

Boron hydrides were first prepared in a pure form by the German chemist Alfred Stock, who investigated them from 1912 to the early 1930s. Stock was motivated by the thought that boron, as carbon's neighbor in the periodic table, was likely to form analogs of the hydrocarbons. He was served well by this analogy, since it led him to prepare and characterize six boron hydrides. His achievement was all the more noteworthy because he had to find techniques to deal with these highly reactive compounds, and in particular he needed to develop new

---

[8] There are two EH_2 stretches for each molecule, a symmetric stretch in which both bonds stretch and contract in phase, and an antisymmetric stretch in which one bond contracts as the other lengthens.

vacuum-line procedures (Box 9.1). Nevertheless, although the analogy was fruitful, it is not perfect, and we shall see that boron chemistry is unlike carbon chemistry in many interesting ways.

Periodic trends and analogies with known compounds form the basis for many **exploratory synthetic** projects. Exploratory synthesis is a strong theme in inorganic chemistry and it continues to yield novel inorganic compounds. To be successful, it requires good chemical intuition to motivate the project and keen observation to capitalize on unexpected results.

Box 9.1. Chemical vacuum lines

Chemical vacuum line techniques provide a method for handling gases or volatile liquids without exposure to the air. They enable a wide range of operations necessary for the preparation and characterization of such air-sensitive compounds to be performed.

A vacuum line, such as the one in Fig. B9.1, is evacuated by a high vacuum pumping system. Once the apparatus is evacuated, condensable gases can be moved from one site into another by low-temperature condensation. For example, a sample of gas from a storage bulb on the right might be condensed onto a solid reactant in reaction vessel on the left. Products of different volatility can be separated by passing the vapors through U-traps. The first trap in the series is held at the highest temperature (perhaps $-78\,°C$, by means of a carbon dioxide/acetone slush bath surrounding the trap) to capture the least volatile component; the next trap might be held at $-196\,°C$ by means of a Dewar flask of liquid nitrogen; noncondensable gases, such as H_2, might be pumped away.

As well as excluding air and enabling the manipulation of vapors with ease, the vacuum line provides a closed system in

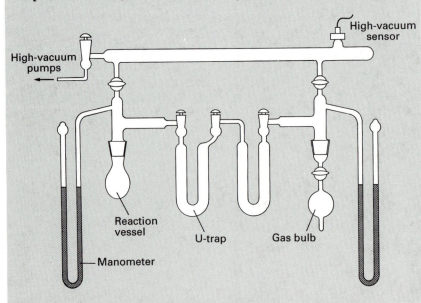

9B.1 A simple chemical vacuum line suitable for the preparation, characterization, and quantitative measurement of condensable gases, particularly gases that are sensitive to air and moisture.

which gaseous reactants or products can be measured quantitatively. For instance, a sample of gas in the bulb could be allowed to expand into the adjacent manometer where its pressure is measured. If the volume of the system has been previously calibrated, the amount of gas present can be calculated using the gas laws. The identity and purity of a compound can be determined on the vacuum line by measuring its vapor pressure at one or more fixed temperatures, and comparison with tabulated vapor pressures. NMR, IR, and other spectroscopic cells can also be attached to the vacuum line and loaded by condensation of a vapor.

Glass vacuum systems such as the one illustrated here are widely used for volatile hydrogen compounds, organometallic compounds, and halides. Since elemental fluorine and highly reactive fluorides corrode glass, they are generally handled in a vacuum line constructed from nickel tubing, with metal valves in place of the stopcocks, and electronic pressure transducer gauges in place of the mercury filled manometers shown here.

[References: R. J. Angelici, *Synthesis and techniques in inorganic chemistry*. Saunders, Philadelphia (1977); W. L. Jolly, *The synthesis and characterization of inorganic compounds*, Waveland Press, Prospect Heights, IL (1971); D. F. Shriver and M. A. Drezdzon, *The manipulation of air-sensitive compounds*. Wiley, New York (1986).]

9.9 Diborane

In this section, we explore some of the chemical properties of diborane for comparison with other molecular hydrogen compounds. The structures, bonding, and reactions of the higher boron hydrides will be covered in Chapter 11.

Synthesis

As shown in Table 9.5, diborane is an endoergic compound at 25 °C, so direct combination of the elements is not feasible as a synthetic route. Stock's entry into the world of boron hydrides was the protolysis of magnesium boride, but this route has been superseded by syntheses that give better yields of specific compounds. Thus, the simplest stable boron hydride, diborane (B_2H_6), may be prepared in the laboratory by metathesis of a boron halide with either $LiAlH_4$ or $LiBH_4$ in ether:

$$3LiEH_4 + 4BF_3 \rightarrow 2B_2H_6 + 3LiEF_4 \qquad (E = B, Al)$$

Both $LiBH_4$ and $LiAlH_4$, like LiH, are good reagents for the transfer of H^-, but they are generally preferred over LiH and NaH because of their solubility in ethers. The synthesis is carried out with the strict exclusion of air (typically, in a vacuum line, because diborane ignites on contact with air). Diborane decomposes very slowly at room temperature, forming higher boron hydrides and a nonvolatile and insoluble yellow solid, the boron chemist's counterpart of the organic chemist's 'black tar'.

Stock prepared six different boron hydrides,[9] which form two classes with the formulas B_nH_{n+4} and B_nH_{n+6}, the former being the more stable. Examples include pentaborane(11) B_5H_{11}, tetraborane(10) B_4H_{10}, and pentaborane(9) B_5H_9. The nomenclature, in which the number of B atoms is specified by a prefix and the number of H atoms is given in parentheses, should be noted. Thus, the systematic name for diborane is diborane(6); however, as there is no diborane(8), the simpler term diborane is almost always used.

All the boranes are colorless, diamagnetic substances. They range from gases (B_2 and B_4 hydrides), through volatile liquids (B_5 and B_6 hydrides), to the sublimable solid $B_{10}H_{14}$.

Oxidation

All the boron hydrides are flammable, and several of the lighter ones, including diborane, react spontaneously with air, often with explosive violence and a green flash (an emission from an excited state of the reaction intermediate BO). The final product of the reaction is the hydrated oxide:

$$B_2H_6(g) + 3O_2(g) \rightarrow 2B(OH)_3(s)$$

Air oxidation is fairly general for the p-block hydrides. Only HF and H_2O do not burn in air; some (including B_2H_6) ignite spontaneously as soon as they come into contact with air.

The lighter boranes are readily hydrolyzed by water:

$$B_2H_6(g) + 6H_2O(l) \rightarrow 2B(OH)_3(aq) + 6H_2(g)$$

As described below, BH_3 is a Lewis acid and the mechanism of this hydrolysis reaction involves coordination of H_2O acting as a Lewis base. Molecular hydrogen forms through the combination of the partially positively charged H atom on O with the partially negatively charged H atom on B.

Lewis acidity

As implied by the mechanism of hydrolysis, diborane and many other light boron hydrides act as Lewis acids and are cleaved by reaction with Lewis bases. Two different cleavage patterns have been observed, namely symmetric cleavage and unsymmetric cleavage.

---

[9] Stock had good but inconclusive evidence for a seventh boron hydride, B_6H_{12}, which (together with many other boron hydrides) was characterized only after his death. In 1933 he wrote an interesting account of his research, which involved the development of vacuum line apparatus for handling these highly air-sensitive compounds. See *Hydrides of boron and silicon*, Cornell University Press, Ithaca (1957).

In **symmetric cleavage**, B_2H_6 is broken symmetrically into two BH_3 fragments, each of which forms a complex with a Lewis base:[10]

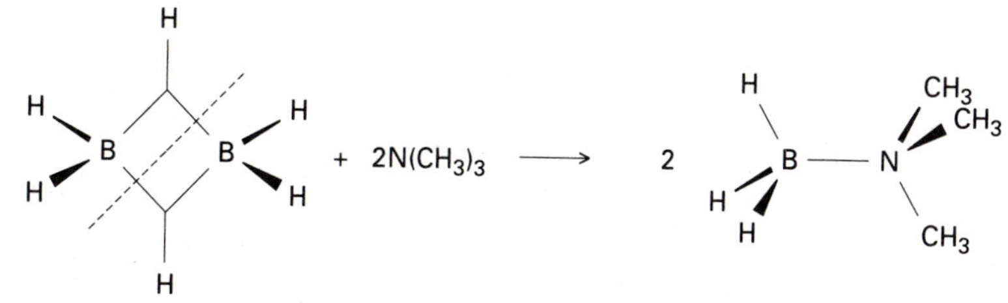

Many complexes of this kind exist. They are interesting partly because they are isoelectronic with hydrocarbons. For instance, the product of the reaction above is isoelectronic with 2,2-dimethylpropane (neopentane, $C(CH_3)_4$). Stability trends indicate that BH_3 is a soft Lewis acid, as illustrated by the following favorable interchange reaction, where BH_3 transfers to the soft S donor atom and the harder Lewis acid, BF_3, combines with the hard N donor atom:

$$H_3BN(CH_3)_3 + F_3BS(CH_3)_2 \rightarrow H_3BS(CH_3)_2 + F_3BN(CH_3)_3$$

The direct reaction of diborane and ammonia results in **unsymmetrical cleavage**, which is cleavage leading to an ionic product:

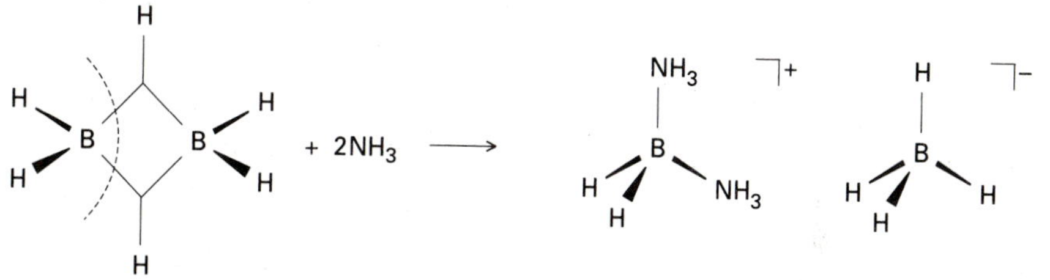

Unsymmetrical cleavage of this kind is generally observed when diborane and a few other boron hydrides react with strong, sterically uncrowded bases at low temperatures. The steric repulsion is such that only two small ligands can attack one B atom in the course of the reaction.

Hydroboration

An important component of a synthetic chemist's repertoire of reactions is **hydroboration**, the addition of H—B across a multiple bond:

---

[10] Wherever possible, we use the symbol

to depict a 3c,2e bond. Unlike in the representation of the structures of organic molecules by lines, there is no atom at the intersection of the three spokes. In large structures, where this symbol would clutter the diagram, we shall simply connect the central H atom to its neighbors by lines, as in

$$H_3B-OR_2 + H_2C=CH_2 \xrightarrow{\text{Ether}} CH_3CH_2BH_2 + R_2O$$

From the viewpoint of an organic chemist, the C—B bond in the primary product of hydroboration is an intermediate stage in the formation of a C—H or C—OH bond, to which it can be converted. From the viewpoint of the inorganic chemist, the reaction is a convenient method for the preparation of a wide variety of organoboranes. The hydroboration reaction is one of a class of reactions in which E—H adds across the multiple bond; hydrosilylation (Section 9.12) is another important example. Much of the driving force for this type of reaction stems from the greater strength of the C—H bond compared with that of B—H and Si—H.

9.10 The tetrahydroborate ion

The simplest borohydride anion is the tetrahydroborate ion BH_4^-, which may be prepared by the reaction of B_2H_6 with LiH in ether. We can view this reaction as another example of the Lewis acidity of BH_3 toward the strong Lewis base H^-. The BH_4^- ion is isoelectronic with CH_4 and NH_4^+, and these show a correlation between the electronegativity of the central atom and the reactivity of the attached H atom:

	BH_4^-	CH_4	NH_4^+
Character:	hydridic	—	protic

CH_4 is neither acidic nor basic under the conditions prevailing in aqueous solution.

Sodium tetrahydroborate, more informally 'sodium borohydride', and the corresponding lithium and potassium borohydrides are very useful laboratory and commercial reagents. They are often used as a mild source of H^- ions, as general reducing agents, and as precursors for most boron–hydrogen compounds. Most of these reactions are carried out in polar nonaqueous solvents. The preparation of diborane mentioned previously,

$$3LiEH_4 + 4BF_3 \rightarrow 2B_2H_6 + 3LiEF_4 \quad (E = B, Al)$$

is one example. Although BH_4^- is thermodynamically unstable with respect to hydrolysis, the reaction is very slow at high pH and some synthetic applications have been devised in water. For example, germane can be prepared by dissolving GeO_2 and KBH_4 in aqueous potassium hydroxide and then acidifying the solution:[11]

$$HGeO_3^-(aq) + BH_4^-(aq) + 2H^+(aq) \rightarrow GeH_4(g) + B(OH)_3(aq)$$

Aqueous BH_4^- also can serve as a simple reducing agent, as in the reduction of aqua ions such as Ni^{2+} or Cu^{2+} to the metal or metal boride. With halo complexes of $4d$ and $5d$ elements that also have stabilizing ligands such as phosphanes, tetrahydroborate ions can be

---

[11] The detailed procedure is given by W. L. Jolly, *The synthesis and characterization of inorganic substances*, p. 496. Waveland Press, Prospect Heights, IL (1991).

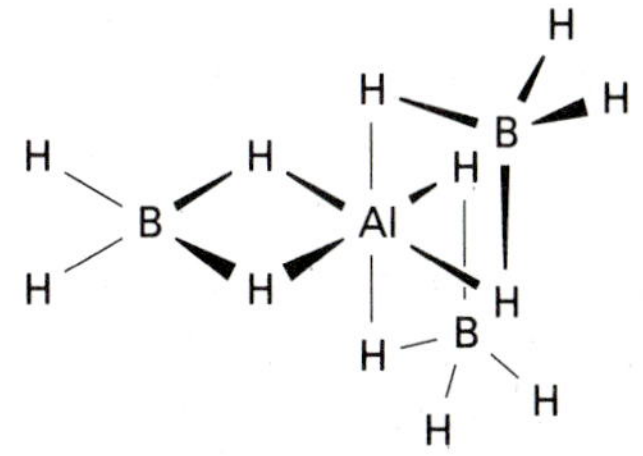

10 [Al(BH$_4$)$_3$]

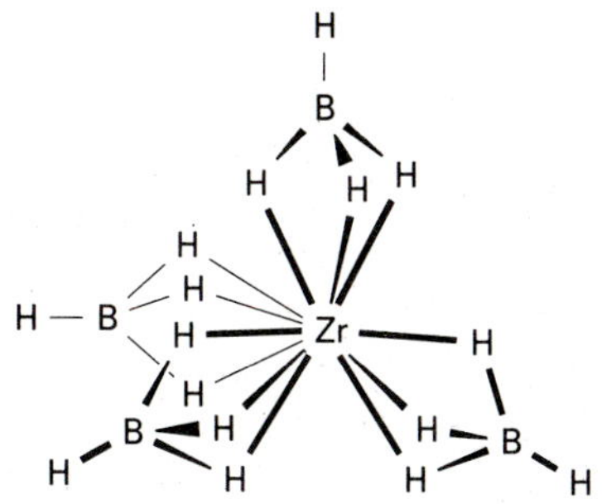

11 [Zr(BH$_4$)$_4$]

used to introduce a hydride ligand by a metathesis reaction in a non-aqueous solvent:

$$RuCl_2(PPh_3)_3 + NaBH_4 + PPh_3 \xrightarrow{\text{Benzene/ethanol}} RuH_2(PPh_3)_4 + \text{other products}$$

It is probable that many of these metathesis reactions proceed via a transient BH$_4^-$ complex. Indeed, many borohydride complexes are known, especially with highly electropositive metals: they include Al(BH$_4$)$_3$ (**10**), which contains a diborane-like double hydride bridge, and Zr(BH$_4$)$_4$ (**11**), in which triple hydride bridges are present. We see from these examples that many compounds can be described in terms of 3c,2e bonds.

Example 9.5: *Predicting the reactions of boron-hydrogen compounds*

By means of a chemical equation, indicate the products resulting from the interaction of equal amounts of [HN(CH$_3$)$_3$]Cl with LiBH$_4$ in tetrahydrofuran (THF).

Answer. The interaction of the hydridic BH$_4^-$ ion with the protonic [HN(CH$_3$)$_3$]$^+$ ion will evolve hydrogen to produce trimethylamine and BH$_3$. In the absence of other Lewis bases, the BH$_3$ would coordinate to THF; however the stronger Lewis base trimethylamine is produced in the initial reactions, so the overall reaction will be:

$$[HN(CH_3)_3]Cl + LiBH_4 \rightarrow H_2 + H_3BN(CH_3)_3 + LiCl$$

The boron in the H$_3$BN(CH$_3$)$_3$ product has four tetrahedrally attached groups.

Exercise E9.5. Write plausible equations for the reaction of B$_2$H$_6$ with propene in THF in the stoichiometric ratio 1:2.

9.11 The hydrides of aluminum and gallium

The hydrides of indium and thallium are very labile. Pure GaH$_3$ has been prepared only recently[12] but derivatives have been known for some time. The range of binary aluminum hydrides is much more limited than that of boron. The metathesis of their halides with LiH leads to lithium tetrahydroaluminate, LiAlH$_4$, or the analogous tetra-hydrogallate, LiGaH$_4$:

$$4LiH + ECl_3 \xrightarrow{\text{Ether}} LiEH_4 + 3LiCl \qquad (E = Al, Ga)$$

The direct reaction of lithium, aluminum, and hydrogen leads to the formation of either LiAlH$_4$ or Li$_3$AlH$_6$, depending on the conditions of the reaction. Their formal analogy with halo complexes such as [AlCl$_4$]$^-$ and [AlF$_6$]$^{3-}$ should be noted.

[12] For a description of this experimental *tour de force*, see C. R. Pulham, A. J. Downs, M. J. Goode, D. W. H. Rankin, and H. E. Robertson, *J. Am. Chem. Soc.*, **113**, 5149 (1991).

The $[AlH_4]^-$ and $[GaH_4]^-$ ions are tetrahedral, and are much more hydridic than $[BH_4]^-$. The latter property is consistent with the higher electronegativity of boron compared with aluminum and gallium. Thus they are also much stronger reducing agents; $LiAlH_4$ is commercially available and widely used as a strong hydride source and as a reducing agent. With the halides of many nonmetallic elements AlH_4^- serves as a hydride source in double-replacement reactions, such as the reaction of lithium tetrahydroaluminate with silicon tetrachloride in tetrahydrofuran solution (denoted thf) to produce silane:

$$LiAlH_4(thf) + SiCl_4(thf) \rightarrow LiAlCl_4(thf) + SiH_4(g)$$

The general rule in this important type of reaction is that H^- *transfers from the element of lower electronegativity* (Al in the example) *to the one of greater electronegativity* (Si in the example). That tetrahydridoaluminate is a more powerful hydride source than tetrahydroborate is in line with the lower electronegativity of aluminum than boron. For example, $NaAlH_4$ reacts violently with water, but basic aqueous solutions of $NaBH_4$ are useful in synthetic chemistry, as described above.

Under conditions of controlled protolysis, both AlH_4^- and GaH_4^- lead to complexes of aluminum or gallium hydride:

$$LiEH_4 + [(CH_3)_3NH]Cl \rightarrow (CH_3)_3N-EH_3 + LiCl + H_2$$

$$(E = Al, Ga)$$

In striking contrast to BH_3 complexes, these complexes will add a second molecule of base to form five-coordinate complexes of aluminum or gallium hydride (**12**):

$$(CH_3)_3N-EH_3 + N(CH_3)_3 \rightarrow ([(CH_3)_3N]_2EH_3 \qquad (E = Al, Ga)$$

This behavior is consistent with the trend for Period 3 and higher *p*-block elements to form 5- and 6-coordinate hypervalent compounds (Section 2.1).

Aluminum hydride, AlH_3, is a solid that is best regarded as saline, like the hydrides of the *s*-block metals. Unlike CaH_2 and NaH, which are more readily available commercially, AlH_3 does not find many applications in the laboratory. The alkylaluminum hydrides, such as $Al_2(C_2H_5)_4H_2$ (**13**), are well known molecular compounds and contain $Al-H-Al$ 3c,2e bonds. Hydrides of this kind are used to couple alkenes, the initial step being the addition of $Al-H$ across the $C=C$ double bond, as in hydroboration.

12 $[(CH_3)_3N]_2AlH_3$, R = CH_3

13 $Al_2(C_2H_5)_4H_2$

ELECTRON-PRECISE HYDRIDES OF THE CARBON GROUP

The electron-precise hydrides of the carbon group (Group 14/IV) have no lone pairs of electrons and so are not Lewis bases. Their lack of lone pairs also means that they are not associated by hydrogen bonding. Moreover, the C atom in hydrocarbons has no vacant, energetically accessible orbitals, and so is not a Lewis acid. The numerous hydrocarbons are best regarded from the viewpoint of organic chemistry, and this section concentrates mainly on the **silanes**, the silicon hydrides. The extensive range of metals and metalloids bound to hydrocarbon ligands is discussed in Chapters 11, 16, and 17.

9.12 Silanes

The silanes, with their greater number of electrons and stronger inter-molecular forces, are less volatile than their hydrocarbon analogs: whereas propane (C_3H_8) boils at $-44\,°C$ and is a gas under normal conditions, its analog trisilane (Si_3H_8) is a liquid and boils at $53\,°C$. The chemical properties of silanes are less fully documented than those of the alkanes and other hydrocarbons. This is partly because there have been few practical or theoretical reasons to prepare many of them, and their greater reactivity makes their investigation more demanding. Stable unsaturated analogs of alkenes, alkynes, and arenes are unknown for the simple silanes.

Synthesis

Partly because he happened to use magnesium boride contaminated with magnesium silicide, Stock also found himself studying the silicon hydrides. He identified four silicon analogs of the alkanes, specifically SiH_4, Si_2H_6, Si_3H_8, and Si_4H_{10}. Separation of these products by modern gas chromatography suggests that Si_4H_{10} in fact consists of a mixture of a linear and a branched isomer analogous to butane and methylpropane. Indeed, gas chromatography suggests that there are silicon analogs of all the straight and branched-chain alkanes, at least up to Si_9H_{20}.

The silanes are thermally less stable than the hydrocarbons. Thus, silanes are cracked at moderate temperatures, when they form a mixture of silane itself and higher silanes:

$$Si_2H_6 \xrightarrow{400\,°C} H_2 + SiH_4 + \text{higher silanes}$$

Complete decomposition occurs above about $500\,°C$, when the silane decomposes to silicon and hydrogen. These **pyrolysis reactions**, reactions in which a compound is degraded by heat, have considerable technological utility because silane is used as a source of pure crystalline silicon by the semiconductor industry:

$$SiH_4 \xrightarrow{500\,°C} Si(s, \text{crystalline}) + 2H_2$$

In this process a thin film of silicon can be laid down on a heated substrate.[13] A related reaction, the decomposition of silane in an electric discharge, yields amorphous silicon:

$$SiH_4 \xrightarrow{\text{Electric discharge}} Si(s, \text{amorphous}) + 2H_2$$

[13] J. M. Jasinski and S. M. Gates, *Acc. Chem. Res.*, **24**, 9 (1991).

Amorphous silicon is used in photovoltaic devices, such as power sources for pocket calculators. 'Amorphous silicon' is in fact a misnomer, for this material contains a significant proportion of hydrogen, which IR spectroscopy shows to be bound by Si—H bonds. The amorphous structure arises because Si—H bonds prevent the formation of the ordered Si—Si bonds that are responsible for the diamond-like structure of pure crystalline silicon.

Stock's protonation of a silicide is no longer of much importance in the preparation of Si—H bonds. Instead, the laboratory preparation is generally based on the double replacement reactions of Si—Cl or Si—Br compounds with $LiAlH_4$ and on reactions such as

$$4R_3SiCl + LiAlH_4 \xrightarrow{\text{Ether}} 4R_3SiH + LiAlCl_4$$

Although useful in the laboratory, this method is too expensive for the commercial production of silane. Instead, the commercial production of silane employs the less expensive reagents HCl and either silicon or iron silicide, which form trichlorosilane, $HSiCl_3$. The trichlorosilane is then heated in the presence of a catalyst, such as aluminum chloride, to yield silane:

$$4SiHCl_3 \xrightarrow{\Delta} SiH_4 + 3SiCl_4$$

This reaction is endoergic and becomes more so at higher temperatures ($\Delta S^\ominus = -58.5 \text{ J K}^{-1} \text{ mol}^{-1}$), but it is possible to drive the reaction forward by removing the highly volatile SiH_4. Silane itself is also prepared commercially by the reduction of SiO_2 with aluminum under a high pressure of hydrogen in a molten salt mixture of NaCl and $AlCl_3$. An idealized equation for this reaction is

$$6H_2(g) + 3SiO_2(s) + 4Al(s) \rightarrow 3SiH_4(g) + 2Al_2O_3(s)$$

Reactions

The silanes are generally more reactive than their hydrocarbon analogs, and ignite spontaneously on contact with air. They also react explosively with fluorine, chlorine, and bromine. Silane itself is also a reducing agent in aqueous solution. For example, when it is bubbled through an oxygen-free aqueous solution containing Fe^{3+}, reduction to Fe^{2+} results.

Bonds between silicon and hydrogen are not readily hydrolyzed in neutral water, but the reaction is rapid in strong acid or in the presence of traces of base. Likewise, alcoholysis is accelerated by catalytic amounts of alkoxide:

$$SiH_4 + 4ROH \xrightarrow{\text{OR}^-} Si(OR)_4 + 4H_2$$

Kinetic studies indicate that the reaction proceeds through a structure in which OR^- attacks the Si atom while H_2 is being formed via a kind of H···H hydrogen bond between hydridic and protic hydrogen atoms (**14**). Although the silicon tetrahalides form many stable complexes with amines and similar donors, the proposed intermediate cannot be isolated because silane is a far weaker Lewis acid than the tetrahalides. This difference in Lewis acidity is reasonable in view of the low electronegativity of hydrogen relative to the halogens, which results in a more positive Si atom and therefore a stronger Lewis acid site in $SiCl_4$ than in SiH_4.

The silicon analog of hydroboration is **hydrosilylation**, the addition of Si—H across the multiple bonds of alkenes and alkynes. This reaction, which is used in both industrial and laboratory syntheses, can be carried out under conditions (300 °C or UV irradiation) that produce a radical intermediate. In practice it is usually performed under far milder conditions using a platinum complex as catalyst:

$$CH_2 {=} CH_2 + SiH_4 \xrightarrow[\text{2-propanol}]{H_2PtCl_6} CH_3CH_2SiH_3$$

The current view is that this reaction proceeds through an intermediate in which both the alkene and silane are attached to the metal atom (**15**).

9.13 Germane, stannane, and plumbane

The decreasing stability of the hydrides on going down the group severely limits the accessible chemical properties of stannanes and plumbane. Germane (GeH_4) and stannane (SnH_4) can be synthesized by the reaction of the appropriate tetrachloride with $LiAlH_4$ in THF solution. It has been claimed that plumbane has been synthesized in trace amounts by the protolysis of a magnesium/lead alloy. The presence of alkyl or aryl groups stabilizes the hydrides of all three elements. For example, trimethylplumbane, $(CH_3)_3PbH$, begins to decompose at $-30\,°C$, but it manages to survive for several hours at room temperature.

ELECTRON-RICH COMPOUNDS OF GROUPS 15/V TO 17/VII

The neutral binary hydrogen compounds in Groups 15/V to 17/VII all have lone electron pairs on the central atom. As we have already remarked, this is consistent with the angular structure of H_2O and its analogs and with the pyramidal structure of NH_3. The other consequences of the these lone pairs is Lewis basicity and ability to take part in hydrogen bonding. The Brønsted acidity of the electron-rich hydrogen compounds, such as the hydrogen halides, has already been discussed in Section 5.2 and will not be covered here.

We shall consider here the production of the commercially important electron-rich hydrogen compounds. We shall also discuss the mechanism for the reaction of hydrogen and oxygen to form water, because it is interesting mechanistically and presents a serious safety problem whenever hydrogen gas escapes into the atmosphere.

14

15

9.14 Ammonia

Ammonia (NH_3) is produced in huge quantities worldwide for use as a fertilizer and as a primary source of nitrogen in the production of many chemicals. Virtually only one method, the **Haber process**, is used for the entire global production. This process is the direct combination of N_2 and H_2 over promoted iron as catalyst (the promoters required include SiO_2, MgO, and other oxides). The reaction is carried out at high temperature (around 450 °C) to overcome the kinetic inertness of N_2, and at high pressure (around 270 atm) to overcome the thermodynamic effect of an unfavorable equilibrium constant at 450 °C. So novel and great were the chemical and engineering problems arising from the then (early twentieth century) uncharted area of large-scale high-pressure technology, that two Nobel Prizes were awarded. One went to Fritz Haber (in 1918), who developed the chemical process. The other went to Carl Bosch (in 1931), the chemical engineer who designed the first plants to realize Haber's process.

Ammonia is the primary source of nitrogen for fertilizers and commercially important compounds that contain nitrogen. An indication of the range of its uses is given in Chart 9.2; more details on the synthesis and individual reactions of ammonia may be found in Section 12.2.

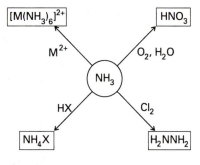

Chart 9.2

9.15 Phosphane, arsane, and stibane

In contrast to the commanding role that ammonia plays in nitrogen chemistry, the highly poisonous hydrides of the heavier elements of Group 15/V (phosphane, PH_3, arsane, AsH_3, stibane, SbH_3, and bismuthane, BiH_3) are of minor importance in the chemistry of their respective *p*-block elements. Both phosphane and arsane find use in the semiconductor industry for the doping of silicon or the preparation of semiconductor compounds such as GaAs by chemical vapor deposition. The thermal deposition of the element is made possible by the thermal fragility of these compounds.

Perhaps the best route to PH_3 is the one used for its commercial synthesis, the disproportionation of white phosphorus in basic solution:

$$P_4(s) + 3OH^-(aq) + 3H_2O(l) \rightarrow PH_3(g) + 3H_2PO_2^-(aq)$$

Arsane and stibane may be prepared by the protolysis of compounds that contain an electropositive metal in combination with arsenic or antimony:

$$Zn_3E_2(s) + 6H^+(aq) \rightarrow 2EH_3(g) + 3Zn^{2+}(aq) \qquad E = As, Sb$$

It should be evident from Fig. 9.6 that PH_3, AsH_3, and SbH_3 are subject to little if any hydrogen bonding with themselves, but PH_3, and AsH_3 can be protonated by strong acids, such as HI. All the Group 15/V hydrides are pyramidal, but an interesting change in bond angle occurs down the series:

$$NH_3, 107.8° \qquad PH_3, 93.6° \qquad AsH_3, 91.8° \qquad SbH_3, 91.3°$$

The large change in bond angle and basicity from NH_3 to PH_3 is in harmony with the idea that the N—H bonds and lone pair of NH_3 can be regarded as approximately sp^3-hybrid orbitals, whereas the the lone pair in PH_3 appears to have much more s-orbital character and the P—H bonds correspondingly more p-orbital character.

9.16 Water

The bonding and properties of water have been discussed extensively in a variety of contexts already. The standard Gibbs free energy of formation of $H_2O(l)$ is strongly negative ($-237.1\ kJ\ mol^{-1}$), so the compound is thermodynamically stable with respect to its elements.

The mechanism of reaction of hydrogen gas with oxygen gas to produce water has been extensively studied. A hint of its complexity is evident from the pressure dependence of the rate of the reaction (Fig. 9.9). This diagram shows that at 550 °C an increase in total pressure can turn an explosive reaction into a smooth reaction and that a further pressure increase leads to an explosion again. It is now recognized that the complexity of the reaction stems from the coexistence of a branched-chain mechanism as well as a simple chain propagation step. In the simple chain, a radical carrier ($\cdot OH$) is consumed with the production of another carrier ($H\cdot$),

$$\cdot OH + H_2 \rightarrow H_2O + H\cdot$$

In a branching chain, more than one radical carrier is produced when one radical reacts:

$$H\cdot + O_2 \rightarrow \cdot OH + \cdot O\cdot$$

$$\cdot O\cdot + H_2 \rightarrow \cdot OH + H\cdot$$

Under normal reaction conditions, the chain carriers are scavenged by collisions with the walls of the reactor or other chain-terminating gas-phase encounters. However, the branching steps allow the rate of carrier production to outstrip quenching, producing a cascade of radical carriers, an increase in rate, and an explosion. Hydrogen in air poses a serious explosion hazard because of the very wide range of hydrogen partial pressures under which its combustion is explosive.

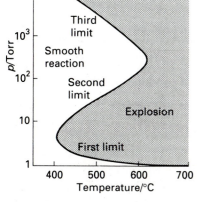

9.9 The variation of the character of the $H_2 + O_2$ reaction with pressure and temperature.

9.17 Hydrogen sulfide, selenide, and telluride

The decrease in E—H bond enthalpy down Group 16/VI from H_2S to H_2Te (Fig. 9.8) is reflected in a sharp change in Gibbs free energy of formation to more positive values; indeed, only H_2S is exoergic. All three compounds can be prepared by protonation of their metal salts:

$$Na_2E(s) + 2H^+(aq) \rightarrow 2Na^+(aq) + H_2E(g)$$

The bond angles are smaller than in water:

$$H_2O,\ 104.5° \qquad H_2S,\ 92.1° \qquad H_2Se,\ 91° \qquad H_2Te,\ 89°$$

This trend indicates that, as in Group 15/V, there is less p-orbital character in the lone pair of the heavier members of this series. The Brønsted basicity of H_2S and heavier members of this series is negligible in aqueous solution (but H_2S can be protonated to H_3S^+ by superacids).

In addition to the simple hydrogen compounds of formula EH_2, a series of hydrogen polysulfides may be prepared by the protonation of a metal polysulfide:

$$Na_2S_n + 2H^+(aq) \rightarrow 2Na^+(aq) + H_2S_n \qquad (n = 4 \text{ to } 6)$$

Hydrogen polysulfide molecules consist of zigzag chains of S atoms that are capped at the ends by H atoms.

9.18 Hydrogen halides

The hydrogen halides may be formed by direct reaction of the elements in a radical-chain reaction, such as that for the reaction between Br_2 and H_2. With the lighter halogens (F_2 and Cl_2), the reaction is explosive under a wide range of conditions. However, all commercial HF and most HCl is synthesized by the protonation of halide ions:

$$CaF_2(s) + H_2SO_4(l) \rightarrow CaSO_4(s) + 2HF(g)$$

This reaction cannot be used for the preparation of HI because concentrated sulfuric acid oxidizes the product to iodine; it can be used to prepare HBr if the concentration of H_2SO_4 is carefully controlled.

9.19 Hydrogen bonding

A **hydrogen bond** consists of an H atom between atoms of more electronegative nonmetallic elements. This definition includes the widely recognized N–H···N and O–H···O hydrogen bonds but excludes the B—H—B bridges in boron hydrides. It also excludes the W—H—W links present in $[(OC)_6WHW(CO)_6]^-$, for tungsten is a metal.

Structural and energetic aspects

Although hydrogen bonds are much weaker than conventional bonds (Table 9.6) they have important consequences for the properties of the electron-rich hydrogen compounds of Period 2, including their densities, viscosities, vapor pressures, and acid–base characters (Section 5.3). Hydrogen bonding is readily detected by the shift to lower wavenumber and broadening of E—H stretching bands in infrared spectra (Fig. 9.10). The origin of the broadening appears to be the thermal disorder of the hydrogen-bonded system at ordinary temperatures, which gives rise to slightly different local environments for the covalent E—H bonds. Thermodynamic data on hydrogen-bonded systems reveal a wide range of hydrogen bond strengths (Table 9.6). The majority of hydrogen bonds are weak. In these cases, the H atom is

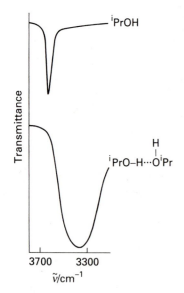

9.10 Infrared spectra of 2-propanol. In the upper curve, 2-propanol is present as unassociated molecules in dilute solution. In the lower curve the pure alcohol is associated through hydrogen bond. The association lowers the frequency and broadens the O–H stretching absorption band. (From N. B. Colthup, L. H. Daly, and S. E. Wiberley, *Introduction to infrared and Raman spectroscopy*, Academic Press, New York (1975).)

Table 9.6 Comparison of hydrogen bond enthalpies with the corresponding E—H covalent bond enthalpies (kJ mol^{-1})

	Hydrogen Bond ($\cdots$)		Covalent Bond (—)	
HS—H$\cdots$SH$_2$	7	S—H	363
H$_2$N—H$\cdots$NH$_3$	17	N—H	386
HO—H$\cdots$OH$_2$	22	O—H	464
F—H$\cdots$F—H	29	F—H	565
HO—H$\cdots$Cl$^-$	55	Cl—H	428
F$\cdots$H$\cdots$F$^-$	165	F—H	565

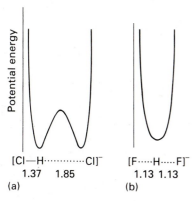

[Cl—H$\cdots\cdots\cdots$Cl]$^-$ [F$\cdots\cdots$H$\cdots\cdots$F]$^-$
1.37   1.85   1.13 1.13
(a)   (b)

9.11 The variation of the potential energy with the position of the proton between two atoms in a hydrogen bond. (a) The double-minimum potential characteristic of a weak hydrogen bond. (b) The single-minimum potential characteristic of the strong hydrogen bond in [FHF]$^-$.

not midway between the two nuclei, even when the heavier linked atoms are identical. For example, the [ClHCl]$^-$ ion is linear but the H atom is not midway between the Cl atoms. In contrast, the bifluoride ion, [FHF]$^-$, has a strong hydrogen bond and the H atom appears to be midway between the F atoms, and the F—F separation (2.26 Å) is significantly less than twice the van der Waals radius of the F atom (2 × 1.35 Å). Very strong hydrogen bonds, such as that in [FHF]$^-$, are generally thought to have a single minimum in their potential energy curve, whereas weak hydrogen bonds are thought to have double minima (Fig. 9.11). The latter may be homonuclear (as in the [ClHCl]$^-$ ion) or heteronuclear (as in hydrogen bonds between water and an amine).

The structures of hydrogen-bonded complexes have been observed in the gas phase through the use of microwave spectroscopy.[14] To a first approximation,[15] the lone-pair orientation of electron rich compounds determined on the basis of the VSEPR model shows good agreement with the HF orientation. For example, HF is oriented along the threefold axis of NH$_3$ and PH$_3$, out of the H$_2$O plane in its complex with H$_2$O, and off the HF axis in the HF dimer (Fig. 9.12). X-ray single crystal structure determinations often show the same patterns, as for example in the ice structure (Fig. 9.7) and in solid HF (**9**), but packing forces in solids may have a strong influence on the orientation of the relatively weak hydrogen bond.

Ice and clathrate hydrates

One of the most interesting manifestations of hydrogen bonding is the structure of ice. There are seven different phases of ice above 2 kbar but only one at lower pressures. The low pressure phase crystallizes in a hexagonal unit cell with each O atom surrounded tetrahedrally by four others. These O atoms are held together by hydrogen bonds with O–H$\cdots$O and O$\cdots$H–O bonds largely randomly distributed through the solid. The resulting structure is quite open, which accounts for

---

[14] A. C. Legon and D. J. Millen, *Acc. Chem. Res.*, **20**, 39 (1987), describe the structures of gas-phase hydrogen bonded dimers and the simple interpretation of their electronic origin.

[15] A more detailed treatment is based on the analysis of the gradient and curvature of the charge density of the donor molecule, see: M. T. Carroll, C. Chang, and R. F. W. Bader, *Mol. Phys*, **63**, 387 (1988).

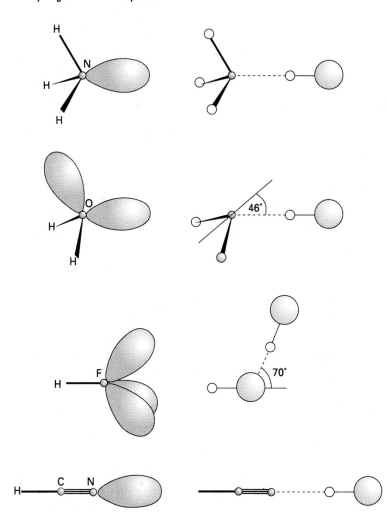

9.12 The orientation of lone pairs as indicated by VSEPR theory (left) compared with the structure of the gas-phase hydrogen-bonded complex formed with HF (right). (After A. C. Legon and D. J. Millen, *Acc. Chem. Res.*, **20**, 39 (1987).)

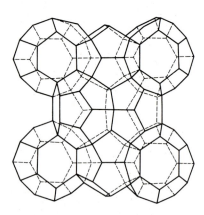

9.13 The cages of water molecules in clathrate hydrates, such as $Cl_2 \cdot (H_2O)_{7.25}$. Each intersection is the location of an O atom, and H atoms lie on the lines joining these O atoms. Two of the central 14-hedra are about 80 percent occupied by Cl_2.

the density of ice being lower than that of water. When ice melts, the network of hydrogen bonds partially collapses.

Water can also form **clathrate hydrates**, consisting of hydrogen-bonded cages of water molecules surrounding foreign molecules or ions. One example is the clathrate hydrate with the composition $Cl_2 \cdot (H_2O)_{7.25}$ (Fig. 9.13). In this structure, the cages with O atoms defining their corners consist of 14-faced and 12-faced polyhedra in the ratio 3:2. These O atoms are held together by hydrogen bonds, and Cl_2 molecules occupy the interiors of the 14-faced polyhedra. Similar clathrate hydrates are formed at elevated pressures and low temperatures with Ar, Xe, and CH_4, but in these cases all the polyhedra appear to be occupied. Aside from their interesting structures—which illustrate the organization that can be enforced by hydrogen bonding—clathrate hydrates are often used as models for the way in which water appears to become organized around nonpolar groups, such as those in proteins. Methane clathrate hydrates occur in the Earth at high pressures, and it is estimated that huge quantities of natural gas are trapped in these formations.

Some ionic compounds form clathrate hydrates in which the anion is incorporated into the framework by hydrogen bonding. This type of clathrate is particularly common with the very strong hydrogen bond acceptors F^- and OH^-. One such example is $N(CH_3)_4F \cdot 4H_2O$.

FURTHER READING

A. G. Massey, *Main group chemistry*. Ellis Horwood, Chichester (1990). Chapter 2 provides a good introductory summary of hydrogen chemistry.

N. N. Greenwood and A. Earnshaw, *Chemistry of the elements*. Pergamon, Oxford (1984). Chapter 3 of this comprehensive text gives a good summary of hydrogen chemistry.

A. F. Wells, *Structural inorganic chemistry*. Oxford University Press (1984). Chapter 5 gives structural data and correlations relating to many hydrogen compounds, and Chapter 15 describes water and hydrates.

Kirk–Othmer encyclopedia of chemical technology, Wiley–Interscience, New York (1980). Vol. 12, p. 930 is a description of aspects of the industrial production of hydrogen. Metal hydrides are also covered on p. 772.

Ullmann's encyclopedia of industrial chemistry, VCH, Weinheim (1985), Vol. 20, p. 143, provides a description of the synthesis of ammonia.

E. A. Evans, *Tritium*. Wiley, New York (1974).

K. M. Mackay, *Hydrogen compounds of the metallic elements*. Spon, London (1966). Although this text is dated, no more recent text is so comprehensive.

E. L. Muetterties (ed.), *Boron hydride chemistry*. Academic Press, New York (1975).

E. L. Muetterties (ed.), *Transition metal hydrides*. Marcel Dekker, New York (1971). This book deals mainly with *d*-block hydrido complexes but some broader aspects are covered too.

Two very good general accounts of hydrogen bonding are:
M. D. Joesten and L. J. Schaad, *Hydrogen bonding*. Marcel Dekker, New York (1974).
J. Emsley, Very strong hydrogen bonds. In *Chem. Soc. Rev.*, **9**, 91 (1980).

KEY POINTS

1. Isotope effects
The effects of isotopic substitution on spectra and reactions are more pronounced for hydrogen compounds than for others on account of the large percentage mass change when H is replaced by D or T.

2. Production of H_2
The commercial production of hydrogen is currently based mainly on steam reforming and dehydrogenation of hydrocarbons.

3. Classification
The hydrogen compounds are classified as saline, metallic, and molecular. Saline hydrides are formed by *s*-block elements, metallic hydrides by many *d*- and *f*-block elements, and molecular hydrides by many *p*-block elements.

4. Stabilities
The stabilities of the *p*-block hydrides relative to their elements decreases markedly down a group. The endoergic character of many *p*-block hydrides necessitates indirect synthetic routes for their preparation.

5. Reaction patterns
The reactions of hydrogen include homolytic

cleavage on metal surfaces and some electron-rich metal complexes, heterolytic cleavage on surfaces such as ZnO, and radical chain reactions with many nonmetals in the gas phase

6. Synthesis of hydrogen compounds.

The synthesis of compounds of hydrogen is generally based on one of three strategies: direct combination of the elements, protonation of a basic-ide salt, or double replacement between a metal hydride and halide of a more electronegative element.

7. Hydrogen bonding

Hydrogen bonding is responsible for the low volatility and high electric permittivities of NH_3, H_2O, and HF. It also results in the organization of water molecules in ice and clathrate hydrates.

EXERCISES

9.1 Assign oxidation numbers to the elements in (a) H_2S, (b) KH, (c) $[ReH_9]^{2-}$, (d) H_2SO_4, (e) $H_2PO(OH)$.

9.2 Write balanced chemical equations for three major industrial preparations of hydrogen gas. Propose two different reactions that would be convenient for the preparation of hydrogen in the laboratory.

9.3 Preferably without consulting reference material, construct the periodic table, identify the elements, and (a) indicate positions of salt-like, metallic, and molecular hydrides, (b) add arrows to indicate trends in $\Delta G_f^{\ominus}$ for the hydrogen compounds of the p-block elements, (c) delineate the areas where the molecular hydrides are electron-deficient, electron-precise, and electron-rich.

9.4 Name and classify the following hydrogen compounds: (a) BaH_2, (b) SiH_4, (c) NH_3, (d) AsH_3, (e) $PdH_{0.9}$, (f) HI.

9.5 Identify the compounds from Exercise 9.4 that provide the most pronounced example of the following chemical characteristics and give a balanced equation that illustrates each of the characteristics: (a) hydridic character, (b) Brønsted acidity, (c) variable composition, (d) Lewis basicity.

9.6 Divide the compounds in Exercise 9.4 into those that are solids, liquids, or gases at room temperature and pressure. Which of the solids are likely to be good electrical conductors?

9.7 Use Lewis structures and the VSEPR model to predict the shapes of H_2Se, P_2H_4, and H_3O^+ and to assign point groups. Assume a skew structure for P_2H_4.

9.8 Identify the reaction that is most likely to give the highest proportion of HD and give your reasoning: (a) $H_2 + D_2$ equilibrated over a platinum surface, (b) $D_2O + NaH$, (c) electrolysis of HDO.

9.9 Identify the compound in the following list that is most likely to undergo radical reactions with alkyl halides, and describe the reason for your choice: H_2O, NH_3, $(CH_3)_3SiH$, $(CH_3)_3SnH$.

9.10 Arrange H_2O, H_2S, and H_2Se in order of (a) increasing acidity, (b) increasing basicity toward a hard acid such as the proton.

9.11 Describe the three different common methods for the synthesis of binary hydrogen compounds and illustrate each one with a balanced chemical equation.

9.12 Give balanced chemical equations for practical laboratory methods of synthesizing (a) H_2Se, (b) SiD_4, and (c) $Ge(CH_3)_2H_2$ from $Ge(CH_3)_2Cl_2$, and for the commercial synthesis of (d) SiH_4 from elemental silicon and HCl.

9.13 Does B_2H_6 survive in air? If not, write the equation for the reaction. Describe a step-by-step procedure for transferring B_2H_6 quantitatively from a gas bulb at 200 Torr into a reaction vessel containing diethyl ether.

9.14 What is the trend in hydridic character of $[BH_4]^-$, $[AlH_4]^-$, and $[GaH_4]^-$? Which is the strongest reducing agent? Give the equations for the reaction of $[GaH_4]^-$ with excess 1 M HCl(aq).

9.15 Given $NaBH_4$, a hydrocarbon of your choice, and appropriate ancillary reagents and solvents, give formulas and conditions for the synthesis of (a) $B(C_2H_5)_3$, (b) Et_3NBH_3.

9.16 Write balanced chemical equations for the formation of pure silicon from crude silicon via silane.

9.17 Describe the important physical differences and a chemical difference between each of the hydrogen compounds of the p-block elements in Period 2 and their counterparts in Period 3.

9.18 What type of compound is formed between water and krypton at low temperatures and elevated krypton pressure? Describe the structure in general terms.

9.19 Sketch the approximate potential energy surfaces for the hydrogen bond between H_2O and the Cl^- ion and contrast this with the potential energy surface for the hydrogen bond in $[FHF]^-$.

PROBLEMS

9.1 What is the expected infrared stretching wavenumber of gaseous ^3HCl given that the corresponding value for ^1HCl is 2991 cm^{-1}?

9.2 Consult *Further information 2* and then sketch the qualitative splitting pattern and relative intensities within each set for the ^1H- and ^{31}P-NMR spectra of PH_3.

9.3 (a) Sketch a qualitative molecular orbital energy level diagram for the HeH^+ molecule-ion and indicate the correlation of the molecular orbital levels with the atomic energy levels. The ionization energy of H is 13.6 eV and the first ionization energy of He is 24.6 eV. (b) Estimate the relative contribution of H1s and He1s orbitals to the bonding orbital and predict the location of the partial positive charge of the polar molecule. (c) Why do you suppose that HeH^+ is unstable on contact with common solvents and surfaces?

9.4 Borane exists as the molecule B_2H_6 and trimethylborane exists as a monomer $B(CH_3)_3$. In addition, the molecular formulas of the compounds of intermediate compositions are observed to be $B_2H_5(CH_3)$, $B_2H_4(CH_3)_2$, $B_2H_3(CH_3)_3$, and $B_2H_2(CH_3)_4$. Based on these facts, describe the probable structures and bonding in this series of mixed alkyl hydrides.

9.5 Observations on a complex in which HD replaces H_2 were used to show that H_2 is bound in $[W(CO)_3(P^iPr_3)_2(H_2)]$ without H—H bond rupture (G. Kubas, R. R. Ryan, B. I. Swanson, P. J. Vergamini, and H. J. Wasserman, *J. Am. Chem. Soc.*, **106**, 451 (1984)). If the dihydrogen complex has an H–H stretch at 2695 cm^{-1}, what would be the expected wavenumber of the HD complex? What should be the pattern of the ^1H-NMR signal arising from coupling with D in the HD complex?

9.6 Hydrogen bonding can influence many reactions, including the rates of O_2 binding and dissociation from metalloproteins (G. D. Armstrong and A. G. Sykes, *Inorg.*

Chem., **25**, 3135 (1986)). Describe the evidence for (or against) hydrogen bonding with O_2 in the metalloproteins hemerythrin, myoglobin, and hemocyanin.

9.7 (a) Compare the structures of borane and aluminum hydride as described in this text with gallane (see reference in footnote 12 of this chapter). (b) Compare the compounds formed from diborane and digallane with trimethylamine and rationalize any differences.

9.8 Spectroscopic evidence has been obtained for the existence of $[Ir(C_5H_5)(H_3)(PR_3)]^+$, a complex in which one ligand is formally H_3^+. Devise a plausible molecular orbital scheme for the bonding in the complex, assuming that an angular H_3 unit occupies one coordination site and interacts with the e_g and t_{2g} orbitals of the metal. An alternative formulation of the structure of the complex, however, is as a trihydro species with very large coupling constants (see *J. Am. Chem. Soc.*, **113**, 6074 (1991) and the references therein, and especially *J. Am. Chem. Soc.*, **112**, 909 and 920 (1990)). Review the evidence for this alternative formulation.

9.9 Correct the faulty statements in the following description of hydrogen compounds. 'Hydrogen, the lightest element, forms thermodynamically stable compounds with all the nonmetals and most metals. The isotopes of hydrogen have mass numbers of 1, 2, and 3, and the isotope of mass number 2 is radioactive. The structures of the hydrides of the Group 1 and 2 elements are typical of ionic compounds because the H^- ion is compact and has a well-defined radius. The structures of the hydrogen compounds of the nonmetals are adequately described by VSEPR theory. The compound $NaBH_4$ is a versatile reagent because it has greater hydridic character than the simple Group 1 hydrides such as NaH. Heavy element hydrides such as the tin hydrides frequently undergo radical reactions, in part because of the low E—H bond energy. The boron hydrides are called electron-deficient compounds because they are easily reduced by hydrogen.'

10

Main-group organometallic compounds

There are many similarities between the chemical properties of hydrogen compounds and the alkyl derivatives of the main-group elements. This is due in part to the similar electronegativities of hydrogen and carbon and the resulting similar strengths and polarities of element–carbon and element–hydrogen bonds. Here we consider the organometallic compounds of the main-group elements and include the zinc group, which is very similar. The organometallic compounds of the d and f blocks are treated in Chapter 16 and the role of organometallic compounds in catalysis is treated in Chapter 17.

We shall see that most organometallic compounds can be synthesized by using one of four $M—C$ bond-forming reactions. Moreover, many of their properties can be rationalized in terms of a small number of classes of reaction. One feature that we shall constantly emphasize is the important role of steric congestion around the central atom. This congestion is believed to be responsible for the ability of some organometallic compounds to withstand hydrolysis, and has allowed the synthesis of compounds containing multiple bonds between heavy elements.

One of the pioneers of main-group organometallic chemistry was the English chemist E. C. Frankland, who learned about organoarsenic compounds during his doctoral studies in Germany. In 1848 he was the first person to synthesize dimethylzinc, and over the next fourteen years he discovered $Zn(C_2H_5)_2$, $Hg(CH_3)_2$, $Sn(C_2H_5)_4$, and $B(CH_3)_3$. He also developed the use of organozinc reagents in organic synthesis and introduced the term 'organometallic' into the vocabulary of chemistry.

The organic derivatives of lithium, magnesium, boron, aluminum, and silicon received considerable attention in the early part of this century. These and other developments led to several major industrial

Table 10.1 Names of common organic substituents in organometallic chemistry*

Formula	Systematic name	Alternative name (and abbreviation)
CH_3—	Methyl	(Me)
CH_3CH_2—	Ethyl	(Et)
$(CH_3)_2CH$—	1-Methylethyl	Isopropyl (^iPr)
$CH_2{=}CHCH_2$—	2-Propenyl	Allyl
$(CH_3)_2CHCH_2$—	2-Methylpropyl	Isobutyl (^iBu)
$(CH_3)_3C$—	1,1-Dimethylethyl	*tert*-Butyl (^tBu)
C_5H_5	Cyclopentadienyl	(Cp)
C_6H_5—	Phenyl	(Ph)
$H_2C{=}$	Methylene	
$HC{\equiv}$	Methylidyne	

*For a longer list, see Further information 1.

applications for organometallic compounds such as alkene polymerization catalysts and silicone polymers. Since the 1960s, exploratory synthetic research in organometallic compounds has been dominated by studies of *d*-block compounds, but a recent revival of interest in exploratory synthesis and physical studies of main-group organometallic compounds has produced new classes of compounds and has broadened our knowledge of the range of types of bonding and reactions of the *s*- and *p*-block elements.

CLASSIFICATION, NOMENCLATURE, AND STRUCTURE

A compound is regarded as organometallic if it contains at least one carbon–metal (C—M) bond. In this context the suffix 'metallic' is interpreted broadly to include metalloids such as boron, silicon, and arsenic as well as true metals. To emphasize the distinctions between the members of this broad class of compounds, we shall employ the notation E—C bonds when E is usually a *p*-block metalloid, and M—C bonds when the element M is specifically an electropositive metal.

The *s*-block organometallic compounds are commonly named according to the substituent names used in organic chemistry (Table 10.1), as in methyllithium for CH_3Li. Compounds that have considerable ionic character are named as salts, as for sodium naphthalide, $Na[C_{10}H_8]$. The *p*-block organometallic compounds are also often named as simple organic species, such as either boron trimethyl or trimethylboron, $B(CH_3)_3$. They may also be named as derivatives of their hydrogen counterparts, as in trimethylborane for $B(CH_3)_3$, tetramethylsilane for $Si(CH_3)_4$, and trimethylarsane for $As(CH_3)_3$.

The nomenclature we shall use for main group organometallic compounds does not include an explicit statement of oxidation number, but it is sometimes useful to take note of the oxidation number of the metallic element when considering the properties of a compound. As elsewhere in inorganic chemistry, the oxidation number of an element is useful for keeping track of electrons and inferring the type of reaction it is likely to undergo, but it must not be interpreted as the actual charge on the atom. The convention employed in assigning an oxidation number to a metal atom in an organometallic compound is that *the organic moiety is considered to be present as an anion.* For example, in dimethylmercury, $Hg(CH_3)_2$, each CH_3 group is assigned a charge of -1, which (because the sum of the oxidation numbers of all the atoms in a neutral compound is zero) implies that the oxidation number of Hg is $+2$. In bis(cyclopentadienyl)magnesium, $Mg(\eta^5\text{-}C_5H_5)_2$, the C_5H_5 group is assigned a charge of -1, and so the oxidation number of Mg is $+2$. (The notation η^5 indicates that the ligand has five points of attachment to the metal atom; this notation is explained in more detail in the introduction to Chapter 16.)

10.1 Comparison with hydrogen compounds

The polarities and strengths of many E—C bonds are similar to those of E—H bonds. As a result, there are a number of structural and chemical similarities between alkyl compounds and their simple hydrogen analogs. This may be seen by comparing Fig. 10.1 and Fig. 9.4. These trends can be summarized as follows:

> Alkyl compounds of the *s*-block elements and aluminum have highly polar $M^{\delta+}$—$C^{\delta-}$ bonds. The electron-precise compounds of Group 14/IV and the electron-rich organometallic compounds of Groups 15/V and 16/VI have M—C bonds of relatively low polarity.

10.2 Structure and bonding

There are some subtle differences between organometallic compounds and binary hydrogen compounds, which stem in part from the tendency of alkyl groups to avoid ionic bonding. For example, the molecular structures of trimethylaluminum (1) and methyllithium (2) contrast with the saline character of their solid hydrides, AlH_3 and LiH. Even the more ionic methylpotassium crystallizes in the nickel-arsenide

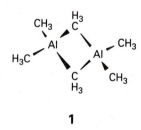

1

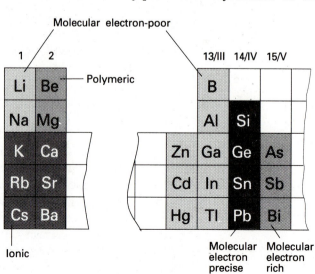

10.1 The classification of methyl compounds of the metals and metalloids.

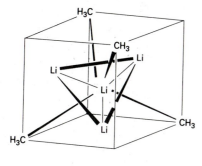

2 $Li_4(CH_3)_4$

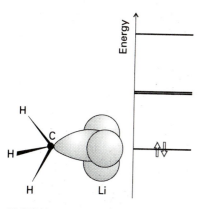

10.2 The interaction between an sp^3 orbital from a face-bridging methyl group and three $2s$ orbitals of the Li atoms on a triangular face of $Li_4(CH_3)_4$ to form a totally symmetric 4c,2e bonding orbital. The next higher orbital is doubly degenerate and nonbonding. The uppermost orbital is antibonding.

3 M = Be or Mg

structure rather than the rock-salt structure adopted by potassium chloride. It will be recalled that the nickel-arsenide structure is typical of soft-cation, soft-anion combinations (Section 4.5). As well as the saline compounds, such as methylpotassium, KCH_3, and electron-precise alkyl compounds, such as tetramethylsilane, $Si(CH_3)_4$, there is a series of electron-deficient compounds, such as hexamethyldialuminum (**1**), that may contain 3c,2e bonds analogous to the B—H—B bridges in diborane (Section 9.9).

s-Block organometallic compounds

Methyllithium in nonpolar solvents consists of a tetrahedron of Li atoms with each face bridged by a methyl group (**2**). As with $Al_2(CH_3)_6$, it is convenient to describe the bonding in terms of a set of semi-localized molecular orbitals. Thus, a totally symmetric combination of three Li$2s$ orbitals on each face of the Li_4 tetrahedron and one sp^3 hybrid orbital from CH_3 gives an orbital that can accommodate one electron pair (Fig. 10.2), leading to a **4-center, 2-electron bond** (a 4c,2e bond). The lower energy of the C orbital compared with the Li orbitals indicates that the bonding pair of electrons will be associated primarily with the CH_3 group, in agreement with the observed carbanionic character of the molecule. Some analyses of the bonding in these compounds indicate about 90 percent ionic character for the Li—CH_3 interaction. Dimethylberyllium and dimethylmagnesium exist in a polymeric structure with two 3c,2e-bonding CH_3 bridges between each metal atom (**3**). In summary:

> The methyl compounds of lithium, sodium, beryllium, magnesium, and aluminum are associated through alkyl bridges, and multi-center two-electron bonding may be involved.

Compounds with greater ionic character are known in which the alkali metal is combined with cyclic or polycyclic arenes (which can accept an electron into their π^* orbitals). One such compound, sodium naphthalide, has already been mentioned: it contains the radical anion $[C_{10}H_8]^-$ in which a single unpaired electron occupies an antibonding π orbital of naphthalene. Because of the delocalization of the charge, the effective radius of the anion is large and the coulombic interaction between cation and anion is weak. As a result, compounds that contain such aromatic anion salts are slightly dissociated in polar aprotic solvents, such as tetrahydrofuran or tris(dimethylamine)phosphane oxide, $((CH_3)_2N)_3PO$.

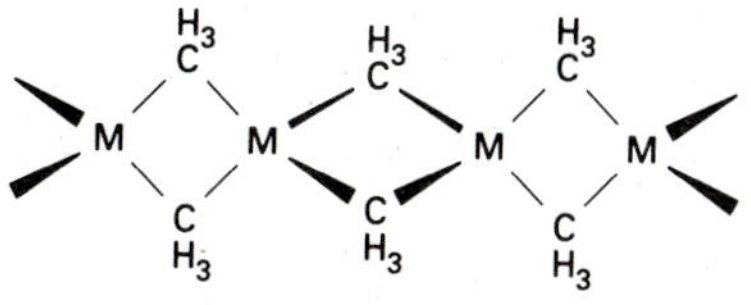

The zinc group

A general feature to keep in mind is that the metals of Group 12 form linear molecular compounds, such as $Zn(CH_3)_2$, $Cd(CH_3)_2$, and $Hg(CH_3)_2$, that are not associated in the solid, liquid, or gaseous states or in hydrocarbon solution.

The elements in Group 12 have two valence electrons and their linear monomeric structures indicate that they employ these electrons

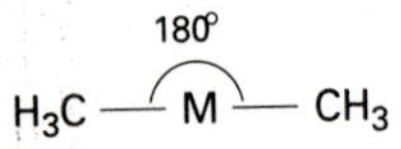

$$H_3C \overset{180°}{\underset{}{—\,M\,—}} CH_3$$

4 M = Zn,Cd,Hg

5

6

to form molecules with localized 2c,2e bonds (**4**). Unlike their beryllium and magnesium analogs, the elements do not complete their valence shells by association through alkyl bridges. To put these monomeric linear compounds into perspective we should note the general tendency toward linear coordination of metals with the d^{10} configuration, such as Cu(I), Ag(I), and Au(I) in Group 11 (for example, $[N{\equiv}C{-}M{-}C{\equiv}N]^-$, M = Ag, Au), which is sometimes rationalized by invoking pd hybridization in the M^+ ion, leading to orbitals that favor linear attachment of the two electron donor ligands (much like the analogous spd hybridization illustrated in Fig. 8.24).

The boron group

Unlike BH_3, which exists as the dimer, trimethylboron is a monomeric molecular compound, probably on account of the steric repulsion that would result between methyl groups in the bridged structure. In support of this explanation, the bigger Al atoms in alkylaluminum compounds do allow dimerization: they form 3c,2e bonds composed of orbitals supplied by the two Al atoms and bridging CH_3 groups (**1**). However, $Ga(CH_3)_3$ is monomeric even though Ga is larger than Al, so size is not the only factor.

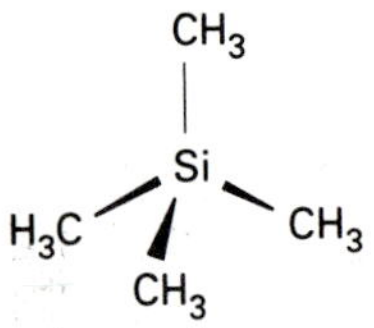

The carbon and nitrogen groups

The electron-precise compounds of the carbon group, such as tetramethylsilane (**5**), and the electron-rich compounds of the nitrogen group, such as trimethylarsane (**6**), are adequately described by conventional, localized 2c,2e bonds, and their shapes can be rationalized by VSEPR theory. However, that the simplest form of VSEPR theory cannot account for everything is shown by the fact that C—E—C bond angles for trimethyl compounds of the heavy elements (arsenic and antimony) are close to 90°. This is reminiscent of the bond angles of the corresponding hydrogen compounds, such as AsH_3 and SbH_3 (Table 9.4).

10.3 Stability

As reliable Gibbs free energies of formation of organometallic compounds are not widely available, it is necessary to resort to the enthalpies of formation in Table 10.2 for an approximate indication of the stability of the compounds with respect to their elements. It will be noted that the trends are similar to those encountered for hydrogen compounds of the p block:

> The methyl compounds of the light elements of Groups 13/III to 15/V are exothermic with respect to the elements, but the exothermic character decreases down the groups, and the compounds of the heaviest elements are endothermic.

The relative instability of organometallic compounds of the heavy members of the group is a consequence in part of the decrease in

Table 10.2 Standard molar enthalpies of formation of gaseous methyl derivatives of the *p*-block elements ($\Delta H_f^\ominus/\text{kJ mol}^{-1}$)

Period	12	13/III	14/IV	15/V
2		$B(CH_3)_3$ -124	$C(CH_3)_4$ -167	$N(CH_3)_3$ -24
3		$Al(CH_3)_3$ -74	$Si(CH_3)_4$ -239	$P(CH_3)_3$ -101
4	$Zn(CH_3)_2$ $+53$	$Ga(CH_3)_3$ -45	$Ge(CH_3)_4$ -71	$As(CH_3)_3$ $+13$
5	$Cd(CH_3)_2$ $+101$	$In(CH_3)_3$?	$Sn(CH_3)_4$ $+21$	$Sb(CH_3)_3$ $+32$
6	$Hg(CH_3)_2$ $+94$	$Tl(CH_3)_3$ >0	$Pb(CH_3)_4$ $+136$	$Bi(CH_3)_3$ $+194$

Sources: D. D. Wagman, W. H. Evans, V. B. Parker, R. H. Shumm, I. Halow, S. M. Baiky, K. L. Churney, and R. L. Nuttall, *J. Phys. Chem. Ref. Data*, **11**, Supplement 2 (1982); M. E. O'Neill and K. Wade in *Comprehensive organometallic chemistry* (ed. G. Wilkinson, F. G. A. Stone, and E. W. Abel), Pergamon, Oxford (1982).

M—C bond strength (Fig. 10.3). The trend toward decreasing E—C bond strengths down a main group is also encountered for other types of bonds, such as E—H bonds (recall Fig. 9.8). As with the analogous hydrogen compounds, the endothermic heavy metal organometallic compounds are susceptible to homolytic cleavage of the M—C bond. For example, when $Pb(CH_3)_4$ is heated in the gas phase, it undergoes homolytic Pb—C bond cleavage to produce CH_3 radicals. Similarly, dimethylcadmium can decompose explosively.

10.4 Synthesis

Most organometallic compounds can be synthesized using one of four M—C bond-forming reactions: reaction of a metal with an organic halide, metal displacement, double replacement, and hydrometallation.

The net reaction of an electropositive metal M and a halogen-substituted hydrocarbon is

$$2M + RX \rightarrow MR + MX$$

where RX is an alkyl or aryl halide. For example, methyllithium is

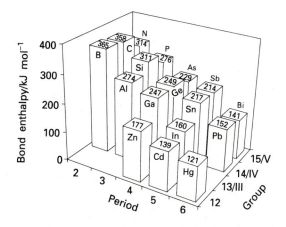

10.3 Average M—CH$_3$ bond enthalpies (in kJ mol^{-1}) at 298 K. From the data in M. E. O'Neill and K. Wade, *Comprehensive organometallic chemistry* (ed. G. Wilkinson, F. G. A. Stone, and E. W. Abel), Vol. 1, 1. Pergamon Press, Oxford (1982).

produced commercially by the reaction

$$8Li + 4CH_3Cl \rightarrow Li_4(CH_3)_4 + 4LiCl$$

With other active metals, such as magnesium, aluminum, and zinc, the reaction generally yields the organometal halide. A familiar example is the synthesis of a **Grignard reagent**, an alkylmagnesium halide:

$$Mg + CH_3Br \xrightarrow{\text{diethyl ether}} CH_3MgBr$$

In a **transmetallation reaction**, one metal atom takes the place of another, as in

$$M + M'R \rightarrow M' + MR$$

This reaction is favorable when the metal M is higher in the electrochemical series than M′ (which is typically Hg), so the reaction is analogous to a simple inorganic reaction in which one metal is displaced from a compound by a metal with a more negative standard potential. The probable outcome of a transmetallation reaction correlates with the electrochemical series even though the conditions for the reaction are very far from those used to determine standard potentials. With somewhat lower reliability, the outcome of the reaction can also be predicted from relative electronegativities: a metal with low electronegativity tends to displace a metal of higher electronegativity from its organometallic compounds.

Because all the metals of Groups 1, 2, and 13/III have more negative reduction potentials than mercury, transmetallation can be carried out between them and dimethylmercury, as in

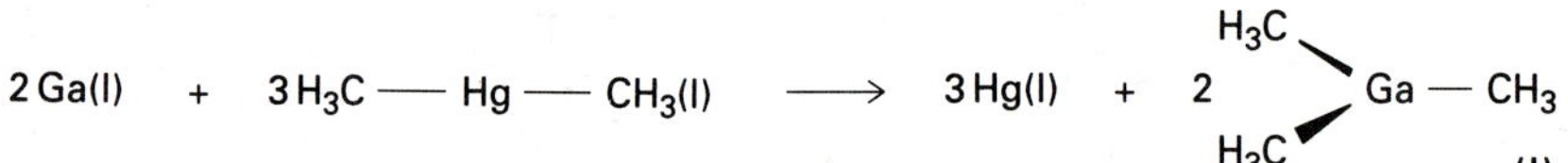

$$2\,Ga(l) \quad + \quad 3\,H_3C - Hg - CH_3(l) \quad \longrightarrow \quad 3\,Hg(l) \quad + \quad 2 \quad \begin{array}{c} H_3C \\ \\ H_3C \end{array}\!\!\!\!\!> Ga - CH_3$$

(I)

In this reaction gallium ($E^{\ominus}(Ga^{3+}/Ga) = -0.53$ V) drives mercury ($E^{\ominus}(Hg^{2+}/Hg) = +0.85$ V) out of dimethylmercury.

Double replacement of an organometallic compound MR and a binary halide EX provides a widely used synthetic procedure in organometallic chemistry:

$$MR + EX \rightarrow MX + ER$$

An approximate but instructive view of the reaction is that it takes place by the exchange of a formal carbanion (R^-) and a halide (X^-). Double replacement is an effective way of preparing a large number of organoelement compounds: the most common reagents are alkyllithium, alkylmagnesium, and alkylaluminum compounds and the halides of the Group 13/III to 15/V elements:

$$Li_4(CH_3)_4 + SiCl_4 \rightarrow 4LiCl + Si(CH_3)_4$$

$$Al_2(CH_3)_6 + 2BF_3 \rightarrow 2AlF_3 + 2B(CH_3)_3$$

The tendency for reaction can frequently be predicted by either electronegativity concepts or hard and soft acid–base concepts. From the standpoint of electronegativity, hydrocarbon groups form stronger covalent than ionic bonds, so the hydrocarbon group tends to prefer the more electronegative element; the halogen favors the formation of ionic compounds with the more electropositive metal. In brief, the alkyl or aryl group tends to migrate from the less to the more electronegative element:

$$M\text{—}R \quad + \quad E\text{—}X \rightarrow MX + R\text{—}E$$

Li	Mg	Al	Zn		Si	B	As	P
χ: 0.98	1.31	1.61	1.65		1.90	2.04	2.18	2.19

When the electronegativities are similar, the correct outcome can usually be predicted as the outcome of the combination of the softer element with the organic group and the harder element with fluoride or chloride. However, predictions from electronegativities and hard–soft characters must be used with some care. For example, an insoluble product or reactant may change the outcome, as in the reaction

$$SnPh_4(thf) + HgBr_2(thf) \rightarrow HgPhBr(s) + Ph_3SnBr(thf)$$

where HgPhBr turns out to be insoluble in tetrahydrofuran.

Double replacements involving the *same* central element are often referred to as **redistribution reactions**. For example, silicon tetrachloride and tetramethylsilicon (tetramethylsilane) undergo redistribution when heated and produce a variety of chloromethylsilanes:

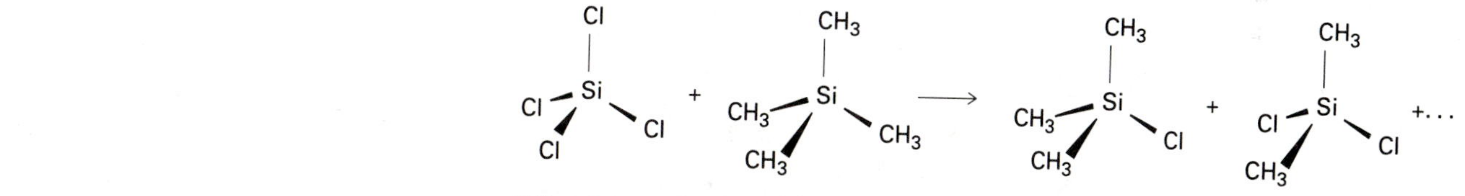

The formation of compounds with mixed substituents is favored by entropy considerations, for a random distribution of ligands is a more disorderly arrangement than the ligand-specific starting compounds.

Example 10.1: *Classifying and predicting a potential M—C bond forming reaction*
Decide on the type of reaction that might occur between $Al_2(CH_3)_6$ and $GeCl_4$.

Answer. We are presented with an organometallic compound and a metal halide, so we should explore the possibility that these compounds can undergo a double replacement reaction. The organometallic compound has a metal (aluminum) that is more electropositive than the central atom in the halide (germanium); therefore the reaction

$$3GeCl_4 + 2Al_2(CH_3)_6 \rightarrow 3Ge(CH_3)_4 + 4AlCl_3$$

should be thermodynamically favorable. This expectation is borne out in practice and it provides a convenient synthetic procedure.

Exercise E10.1. What is the likely outcome of the reaction between magnesium with dimethylmercury?

The net outcome of the addition of a metal hydride to an alkene is an alkylmetal compound:

$$E-H + H_2C=CH_2 \rightarrow ECH_2-CH_3$$

This reaction is driven mainly by the high strength of the C—H bond relative to that of most E—H bonds, and occurs with a wide variety of compounds that contain E—H bonds. We have already encountered several examples in Chapter 9, the most important being

(1) *hydroboration* (Section 9.9):

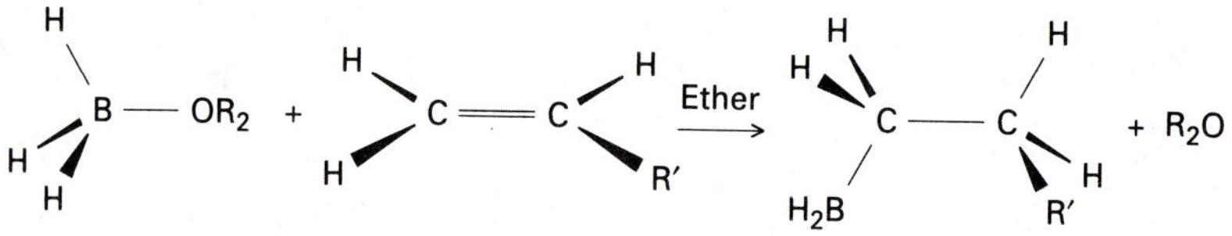

(2) *hydrosilylation* (Section 9.12):

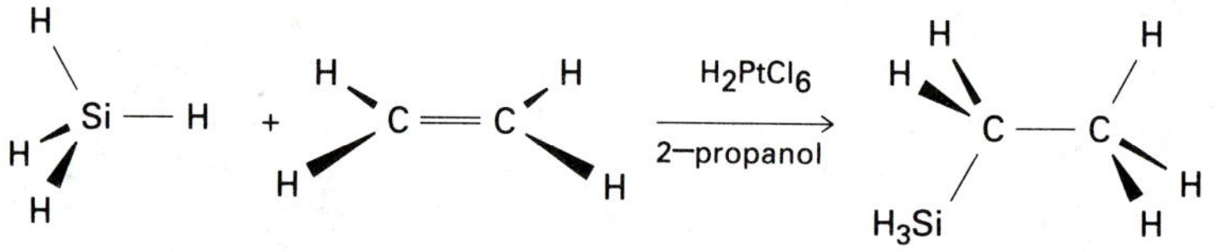

In the hydroboration and hydrosilylation of unsymmetrical alkenes, the bulkier B or Si group adds to the sterically less hindered C atom, and the smaller H atom adds to the more hindered C atom.

10.5 Reaction patterns

The reactions of organometallic compounds of electropositive elements are dominated by factors such as the carbanion character of the organic moiety and presence or absence of a potential coordination site on the central metal atom.

Oxidation

All organometallic compounds are potentially reducing agents; those of the electropositive elements are in fact very strong reducing agents. It is important to be mindful of the potential fire hazard this character implies, for many organometallic compounds are **pyrophoric** and ignite spontaneously on contact with air. The strong reducing character also presents a potential explosion hazard if the compounds are mixed with large amounts of oxidizing agents.

All organometallic compounds of the electropositive metals that have unfilled valence orbitals, or that readily dissociate to fragments with unfilled orbitals, are pyrophoric. These compounds include $Li_4(CH_3)_4$,

$Zn(CH_3)_2$, $B(CH_3)_3$, and $Al_2(CH_3)_6$. Volatile pyrophoric compounds, such as $B(CH_3)_3$, may be handled in a vacuum line (Box 9.1) and inert atmosphere techniques are used for less volatile but air-sensitive organometallic compounds (Box 10.1). Compounds such as $Si(CH_3)_4$ and $Sn(CH_3)_4$, which do not have low-lying empty orbitals, require elevated temperatures to initiate combustion, and can be handled in air. The combustion of many organometallic compounds takes place by a radical chain mechanism (Section 9.8).

Nucleophilic (carbanion) character

The partial negative charge of an organic group attached to an electropositive metal results in the group being a strong nucleophile and Lewis base. This is frequently referred to as its **carbanion character** even though the compound itself is not ionic.

Alkyllithium and alkylaluminum compounds and Grignard reagents are the most common carbanion reagents in laboratory-scale synthetic chemistry; carbanion character is greatly diminished for the less metallic elements boron and silicon. We described this nucleophilic character of organometallic compounds in Section 10.3 in connection with the formation of M—C bonds. As may be inferred from the summary in Fig. 10.4, carbanion character has many other synthetic applications. For instance, an $R^{\delta-}$ group from the organometallic reagent attacks the carbonyl C atom of a ketone and upon hydrolysis a tertiary alcohol results. Similarly, aldehydes can be converted to secondary alcohols by reaction with an organometallic compound followed by hydrolysis. There are many examples of the use of double replacement reactions for the preparation of p-block organometallic compounds and sulfones and sulfoxides by treatment of SO_2Cl_2 or $SOCl_2$ with an alkyllithium or Grignard reagent. We shall amplify on many of these reactions in this chapter.

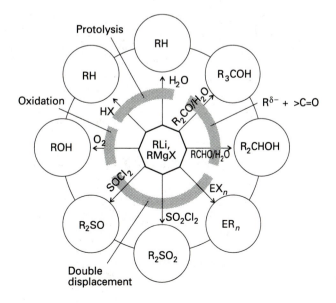

10.4 Typical protolysis, attack of carbonyl, double replacement, and oxidation reactions of alkyllithium compounds and Grignard reagents; X = halide, E = B, Si, Ge, Sn, Pb, As, and Sb.

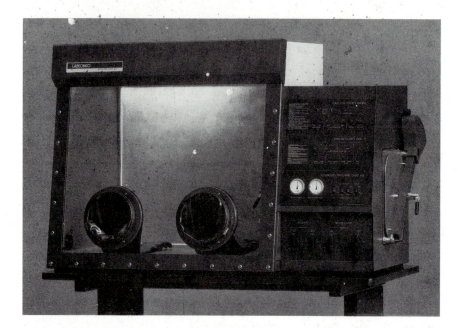

B10.1 Inert-atmosphere glove box. The door to the airlock is on the right. A vacuum pump is used to evacuate the transfer lock, which is refilled with inert gas before items are brought into the main chamber. The system also contains a circulation pump and columns, which remove water and oxygen from the glove box atmosphere. (Reproduced by permission of LabConco, Kansas City, MO.)

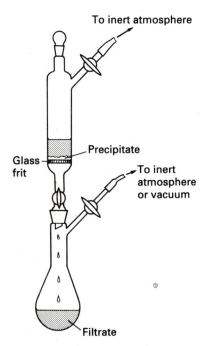

B10.2 Typical inert-atmosphere Schlenk ware. In this example, a reaction mixture has been filtered in the upper part of the apparatus and the filtrate collected in the Schlenk tube shown in the lower part of the illustration.

Box 10.1 Inert atmosphere techniques

Because they react so readily with oxygen, moisture, and carbon dioxide, many organometallic compounds are handled in an inert atmosphere. The simplest of these techniques (conceptually, at least) is the **inert-atmosphere glove box** (Fig. B10.1). This apparatus allows manipulations to be carried out in a large metal enclosure filled with nitrogen, argon, or helium. The enclosure has a window, and a tightly attached pair of rubber gloves are used to manipulate chemicals inside the box. Two other important components of the box are a source of pure inert gas and an antechamber—an airlock—which permits items to be brought into or out of the box while avoiding the influx of air. Because air slowly diffuses into the box through the rubber gloves, it is common to maintain the purity of atmosphere inside the chamber by a recirculating gas purifier or a constant flush of inert gas.

Another common method of handling air-sensitive compounds uses standard glassware constructed so that an inert atmosphere can be maintained inside it. Typically this apparatus is equipped with sidearms for pumping out the air and introducing inert gas (Fig. B10.2). When apparatus containing air-sensitive compounds must be reconfigured, it is opened under a flush of inert gas. This type of apparatus can be employed for all the standard synthetic operations, such as reactions in solution, filtration, and crystallization. The apparatus is often referred to as **Schlenk ware** in recognition of the German chemist Wilhelm Schlenk, who introduced this general design during his pioneering research in organometallic chemistry during the first quarter of the twentieth century.

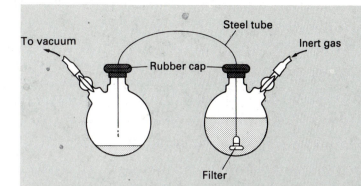

B10.3 Filtration and liquid transfer by means of a pressure differential and stainless steel tube (a cannulum).

A variation on the above method for transferring air-sensitive solutions utilizes syringes or metal cannula (flexible, small diameter metal tubing). For these operations, a flask is fitted with a rubber serum bottle cap (a septum) which can be pierced with a syringe needle or sharpened cannula. Figure B10.3 illustrates the transfer of a solution from one flask to the other by a cannula and a pressure differential between the two flasks.

[References: J. J. Eisch, *Organometallic synthesis*, Vol. 2. Academic Press, New York (1981); A. L. Wayda and M. Y. Darensbourg (ed.), *Experimental organometallic chemistry: A practicum in synthesis and characterization*. ACS Symposium Series 357, American Chemical Society, Washington, DC (1988); D. F. Shriver and M. A. Drezdzon, *The manipulation of air-sensitive compounds*, Wiley, New York (1986).]

One noteworthy consequence of high carbanion character is the **protolysis** reaction that takes place with very weak Brønsted acids, including water and alcohols. The criteria for judging when this reaction can be rapid are similar to those for judging whether reaction with O_2 (combustion) is likely.[1] Thus, organometallic compounds with low-lying vacant orbitals have low kinetic barriers for the formation of an initial complex, which can then undergo proton transfer. According to this mechanism, the lone pairs on the O atom of an alcohol coordinate to the Ga atom in triethylgallium. This complex formation is followed by transfer of the proton to the ethyl group and hence leads to the evolution of ethane, as illustrated below for $Al(CH_3)_3$:

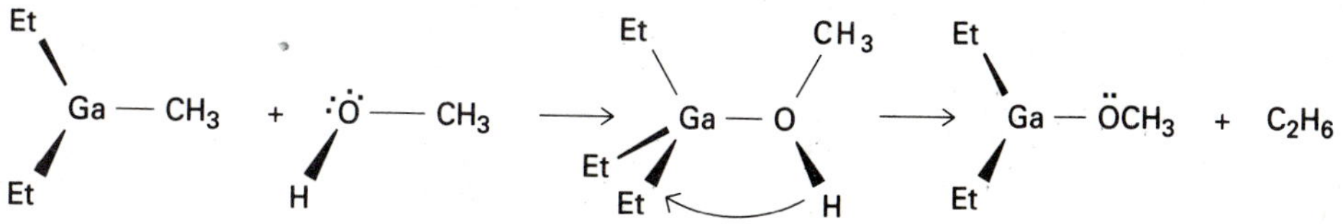

$$Et_2Ga{-}CH_3 + \ddot{O}(H){-}CH_3 \longrightarrow Et_2Ga{-}O \longrightarrow Et_2Ga{-}\ddot{O}CH_3 + C_2H_6$$

---

[1] As with most chemical generalizations, this statement has limitations. For example, $B(CH_3)_3$ is oxidized rapidly by air but it is resistant to hydrolysis.

Similarly, alkylaluminum compounds react vigorously with excess alcohol to produce alkoxyaluminum compounds:

$$Al_2(CH_3)_6 + 6C_2H_5OH \rightarrow 2Al(OC_2H_5)_3 + 6CH_4$$

Again, it appears that protolysis occurs by prior coordination:

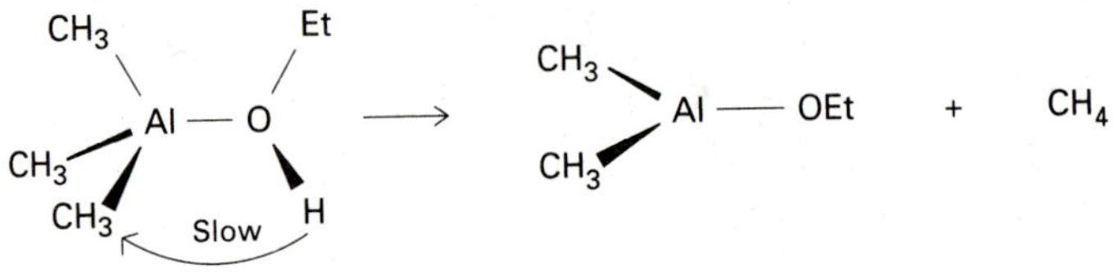

In keeping with the lengthening of the O—H bond implied by this mechanism, the reaction shows a significant kinetic isotope effect, and on substitution $k(H-OCH_3)/k(R-OCH_3) = 2.9$ when R is an octyl group.

Because the protolysis of organometallic compounds liberates hydrocarbon, water should not be used to extinguish a fire involving trialkylaluminum compounds or any other organometallic compounds of electropositive metals. Protolysis of electropositive organometallic compounds with sterically hindered alcohols, such as 2-propanol or *tert*-butanol (2-methyl-2-propanol) proceeds at a moderate rate and therefore provides a convenient way of destroying reactive organometallic wastes.

Lewis acidity

As a result of the presence of unoccupied orbitals on the metal atom, electron-deficient organometallic compounds are observed to be Lewis acids. An illustration of this Lewis acidity is the synthesis of organometallic anions, such as the tetraphenylborate ion:

$$B(C_6H_5)_3 + Li(C_6H_5) \rightarrow Li[B(C_6H_5)_4]$$

This reaction may be viewed as the transfer of the strong base $C_6H_5^-$ from the weak Lewis acid Li^+ to the stronger acid B(III).

Organometallic species that are bridged by organic groups can also serve as Lewis acids and, in the process, undergo bridge cleavage. For example, $Al_2(CH_3)_6$ is cleaved by tertiary amines to form a simple Lewis acid–base complex:

$$Al_2(CH_3)_6 + 2N(C_2H_5)_3 \rightarrow 2(CH_3)_3AlN(C_2H_5)_3$$

This reaction illustrates once again the weakness of the 3c,2e Al—CH_3—Al bond. Solvents such as THF coordinate to the Li atoms in $Li_4(CH_3)_4$, but such a mild base does not disrupt the metal cluster.

β-hydrogen elimination

In the process known as **β-hydrogen elimination**, a metal atom extracts an H atom from the next-nearest neighbor (β) carbon atom. An example is the reaction

$$M-CH_2-CH_3 \rightarrow M-H + H_2C{=}CH_2$$

The reaction is the reverse of the addition of an $M-H$ bond to an alkene (hydrometallation) and under some conditions significant equilibrium concentrations of both reactants and products are observed. The reaction mechanism is believed to involve the formation of β-hydrogen bridges to an open coordination site on the metal atom:

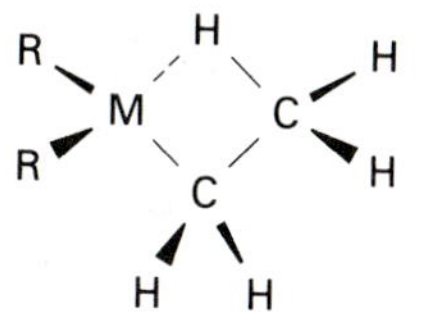

As might be expected from the proposed mechanism, compounds in which the central atom has a low coordination number tend to undergo this reaction. For example, it occurs with trialkylaluminum compounds but not with tetraalkylgermanium compounds.

Example 10.2: *Predicting the products of thermal decomposition*

Summarize the stabilities of (a) $Bi(CH_3)_3$ and (b) $Al_2({}^iBu)_6$, with respect to their thermal decomposition and give chemical equations for their decomposition (^iBu denotes the $(CH_3)_2CHCH_2-$ group).

Answer. (a) As with the other heavy *p*-block elements, bismuth–carbon bonds are weak and readily undergo homolytic cleavage. The resulting methyl radicals will react with surrounding molecules or form ethane:

$$2Bi(CH_3)_3 \xrightarrow{\ \Delta\ } 2Bi(s) + 3H_3C-CH_3$$

(b) The $Al_2({}^iBu)_6$ dimer readily dissociates. At elevated temperatures dissociation is followed by β-hydrogen elimination, which is common for organometallic compounds that (1) have alkyl groups with β-hydrogens, (2) can form stable $M-H$ bonds, and (3) can provide a coordination site on the central metal. The decomposition reaction is

$$Al_2({}^iBu)_6 \xrightarrow{\ \Delta\ } 2Al({}^iBu)_3 \xrightarrow{\ \Delta\Delta\ } Al({}^iBu)_2H + (CH_3)_2C{=}CH_2$$

In this equation, Δ implies moderate heating and $\Delta\Delta$ implies stronger heating.

Exercise E10.2. Describe the probable mode of thermal decomposition of $Pb(CH_3)_4$.

IONIC AND ELECTRON-DEFICIENT COMPOUNDS OF GROUPS 1, 2, AND 12

In this and the following sections we discuss the chemical properties of certain organometallic compounds in more detail. We shall concentrate on the compounds formed by lithium, magnesium, zinc, and mercury because of their interesting properties and utility in syntheses.

10.6 Alkali metals

Organometallic derivatives have been made for all the Group 1 metals. Of the simple alkyl compounds the alkyllithium compounds are by far the most thoroughly studied and useful synthetic reagents.

Organolithium compounds

Organolithium compounds are available commercially as solutions. Methyllithium is generally handled in ether solution, but alkyllithium compounds with longer chains are soluble in hydrocarbons. Because the commercial preparation of alkyllithium compounds is by the reaction of the metal with the organic halide, they are often contaminated by halide. This contamination can be avoided by a preparation of the form

$$HgR_2 + 2Li \rightarrow 2LiR + Hg$$

Although methyllithium exists as a tetrahedral cluster in the solid state and in solution, many of its higher homologs exist in solution as hexamers or equilibrium mixtures of aggregates ranging up to hexamers. These larger aggregates can be broken down by strong Lewis bases, such as chelating amines. For example, the interaction of TMEDA[2] with phenyllithium produces a complex containing two Li atoms bridged by phenyl groups with each Li atom coordinated by the chelating diamine (7).

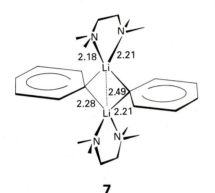

7

In addition to the common organolithium compounds with one Li atom per organic group, a variety of **polylithiated** organic molecules, or organometallic compounds containing several Li atoms per molecule, are known.[3] The simplest example is dilithiomethane, Li_2CH_2, which can be prepared by the pyrolysis of methyllithium. This compound crystallizes in a distorted antifluorite structure (Section 4.5), but the finer details of the orientation of the CH_2 groups are as yet unknown.

Radical anion salts

Sodium naphthalide is an example of an organometallic salt that contains a delocalized radical anion, the $C_{10}H_8^-$ ion. Such compounds are readily prepared by the direct reaction of the aromatic compound with an alkali metal in a polar aprotic solvent. Thus, naphthalene dissolved in THF reacts with sodium metal to produce a dark green solution of sodium naphthalide:

$$Na(s) + C_{10}H_8(thf) \rightarrow Na[C_{10}H_8](thf)$$

EPR spectra show that the odd electron is delocalized in an antibonding orbital of $C_{10}H_8$.

---

[2] TMEDA is *N,N,N',N'*-tetramethylethylenediamine. It is a favorite chelating ligand in *p*-block organometallic chemistry because it lacks the mildly acidic N—H bonds of ethylenediamine, and therefore does not undergo protolysis with carbanionic organometallic compounds such as the alkyllithium compounds.

[3] A description of X-ray structures of lithium compounds is given by W. N. Setzer and P. von R. Schleyer, *Adv. Organometal. Chem.*, **24**, 1385 (1985).

Table 10.3 Standard potentials of some conjugated hydrocarbons*

Compound		$E^{\ominus}/V$
Biphenyl	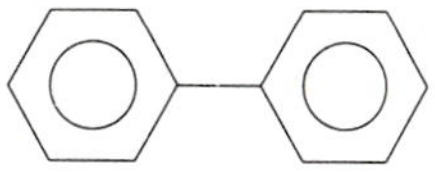	+ 0.00
Naphthalene	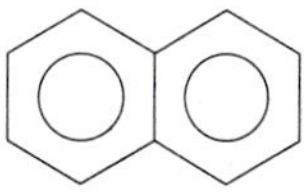	+ 0.09
Phenanthrene	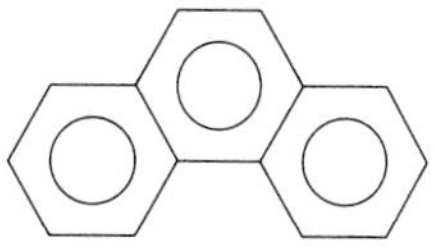	+ 0.17
Anthracene	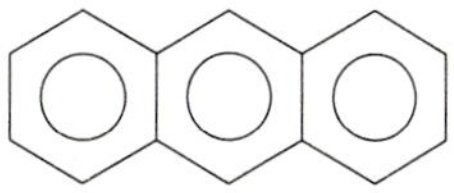	+ 0.78

*Relative to the value for biphenyl in 1,2-dimethoxyethane. *Source*: E. de Boer, *Adv. Orgmet. Chem.*, **2**, 115 (1964).

The formation of the radical anion is more favorable when the π LUMO of the arene is low in energy. Simple molecular orbital theory predicts that the LUMO occurs at progressively lower energies on going from benzene to more extensively conjugated hydrocarbons. This is analogous to the lowering of energy levels of an electron in a box as the length of the box is increased (Section 1.4). The prediction is borne out by the reduction potentials of aromatic hydrocarbons (Table 10.3). For this reason, the radical anion of benzene cannot be formed in most solvents, but naphthalene and more extensively conjugated arenes readily yield alkali metal salts.

Sodium naphthalide and similar compounds are highly reactive reducing agents and are often preferred to sodium because—unlike sodium itself—they are readily soluble in ethers. The resulting homogeneous reaction is generally faster and easier to control than a heterogeneous reaction between one component in solution and pieces of sodium metal, which are often coated with unreactive oxide or with insoluble reaction products. As indicated in Table 10.3, an additional advantage of the radical anion reagents is that by proper choice of the aromatic group the reduction potential of the reagent can be chosen to match the requirements of a particular synthetic task.

Another route to delocalized anions is the reductive cleavage of acidic C—H bonds by an alkali metal or alkylmetallic compound. Thus, in the presence of a good coordinating ligand such as TMEDA, butyllithium reduces dihydronaphthalene to produce a diamagnetic dinegative anion:

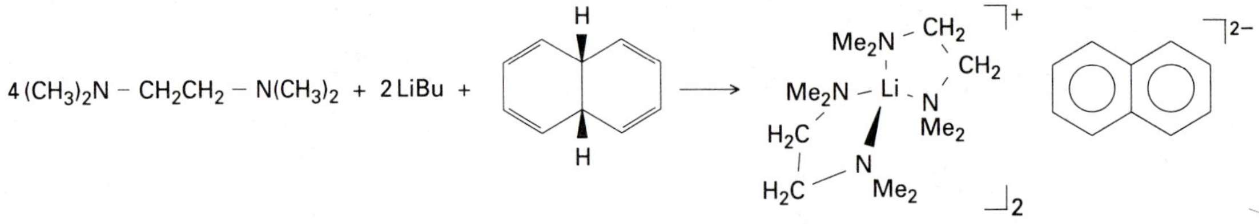

$$4\,(CH_3)_2N - CH_2CH_2 - N(CH_3)_2 + 2\,LiBu +$$

TMEDA contributes to the favorable reaction free energy through its affinity for the Li^+ produced in the reaction.

The planar aromatic cyclopentadienide ion $C_5H_5^-$ can be prepared by the reductive cleavage of a C—H bond in cyclopentadiene, using either sodium metal or NaH:

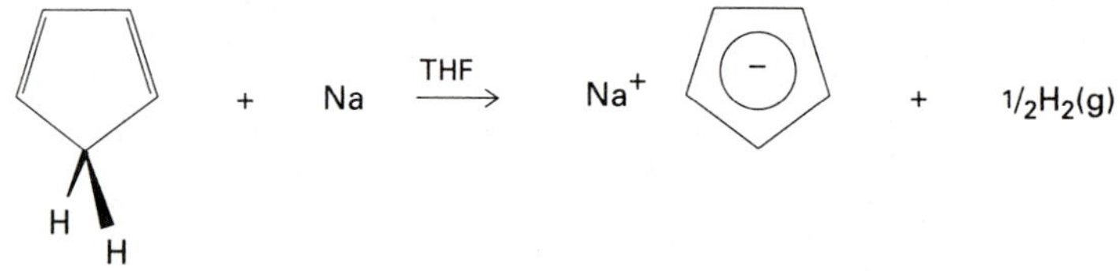

The product, sodium cyclopentadienide, is an important reagent in organometallic synthesis. It readily participates in double replacement with a variety of halides of *p*-block elements to produce either σ- or π-bonded cyclopentadienyl compounds (which we shall discuss later in this chapter). It is also used to synthesize a wide variety of *d*-block organometallic compounds, as we shall see in Chapter 16.

10.7 Alkaline earth metals

Organoberyllium and organomagnesium compounds have significant covalent character, whereas the analogous compounds of the heavier congeners are more ionic. The latter have not yet been thoroughly investigated.

A characteristic of organoberyllium and organomagnesium compounds is the tendency of the metal atoms to adopt four-coordination; three-coordination is also observed for beryllium. The dimethyl compounds, for instance, appear to be bridged species (**3**). The fact that more bulky groups lead to decreased association is shown by the dimeric structure of diethylberyllium in benzene solution compared with the monomeric character of $Be(^tBu)_2$.

Synthesis and structure

Because beryllium is more electronegative than magnesium, ether-solvated dialkylberyllium compounds can be prepared by trans-metallation with a Grignard reagent:

$$BeCl_2 + 2RMgX + (C_2H_5)_2O \xrightarrow{\text{Diethyl ether}}$$

$$R_2Be - O(C_2H_5)_2 + 2MgXCl$$

When an ether-free product is needed, transmetallation with dialkyl-mercury can be used because beryllium is higher than mercury in the electrochemical series:

$$Be + H_3C - Hg - CH_3 \longrightarrow \quad Be \begin{smallmatrix} CH_3 \\ \\ CH_3 \end{smallmatrix} + Hg$$

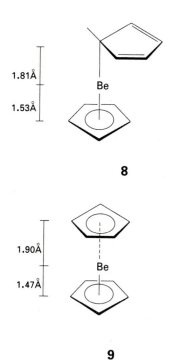

8

9

Bis(cyclopentadienyl)beryllium is readily prepared by a double replacement reaction between beryllium halide and $Na[C_5H_5]$. In agreement with the usual trend, the $[C_5H_5]^-$ ion transfers to the more electronegative Be(II) atom:

$$BeCl_2 + 2NaC_5H_5 \rightarrow Be(C_5H_5)_2 + 2NaCl$$

Data from X-ray crystal structure determinations demonstrate that in the solid state the compound has a mixed η^5-C_5H_5,η^1-C_5H_5 coordination (**8**). The symbol η^1 indicates the attachment of a C_5H_5 group through only one C atom and denotes **monohapto bonding**. The notation η^5 for the other C_5H_5 ring indicates that all five C atoms bond to beryllium; this arrangement is called **pentahapto bonding**. (The term hapto, is derived from the Greek word 'to fasten'. More examples of this type are given in the introduction to Chapter 16.) The structure in the gas phase is less clear. It may be the same unsymmetrical sandwich (**8**) with the molecule undergoing rearrangements on a time scale too fast to resolve by electron diffraction, or it may be an antisymmetric sandwich (**9**), where both C_5H_5 groups are pentahapto, but one ring is further than the other from the Be atom. This ambiguity illustrates the difficulty of obtaining precise structural data on large, possibly fluxional molecules in the gas phase.

Reactions

Simple organoberyllium compounds react with oxygen, water, and other weak Brønsted acids:

$$Be(CH_3)_2 + CH_3OH \rightarrow (CH_3)Be(OCH_3) + CH_4$$

Dimethylberyllium also readily forms complexes with Lewis bases:

$$Be(CH_3)_2 + N(CH_3)_3 \rightarrow (CH_3)_2BeN(CH_3)_3$$

$$Be(CH_3)_2 + (CH_3)_2NCH_2CH_2N(CH_3)_2 \longrightarrow$$

As beryllium compounds are highly toxic, and there appear to be no compelling reasons to use them as reagents, organoberyllium compounds do not have significant commercial applications. Their use as synthetic intermediates in the laboratory is confined to the synthesis of other beryllium compounds.

Grignard reagents

Organomagnesium compounds are familiar to any student of organic chemistry as useful carbanion reagents. The most common of these compounds are the alkylmagnesium halides, or **Grignard reagents**, prepared by the reaction of a haloalkane with magnesium metal.[4] This reaction is carried out in ether, and because a coating of oxide on the magnesium acts as a kinetic barrier, a trace of iodine is often added to initiate the reaction.

From the early work carried out by Wilhelm Schlenk, it is known that redistribution equilibria occur in ether solution. The simplest of these, often called the **Schlenk equilibrium**, is

$$2RMgX \rightleftharpoons MgR_2 + MgX_2$$

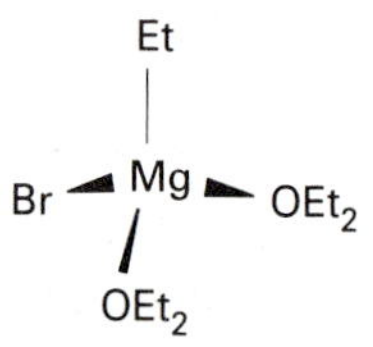

The addition of dioxane to the equilibrium mixture in diethyl ether leads to the precipitation of a dioxane complex of the magnesium halide, $MgX_2(C_4H_8O_2)$, and the filtrate can be evaporated to yield the dialkylmagnesium. More recent spectroscopic studies indicate a rather complex set of equilibria between alkylmagnesium halides in ether solution. In line with these observations, the species crystallized from ether solution include a wide variety of structures, such as monomeric four-coordinate magnesium complexes (**10**) and larger clusters (**11**). An important feature of the latter structure is that halide bridges, where conventional 2c,2e bonds can be formed by donation of halide lone electron pairs, are preferred over alkyl bridges, which would require two-electron multicenter bonds. We shall encounter this preference for 2c,2e bridge bonds in aluminum chemistry.

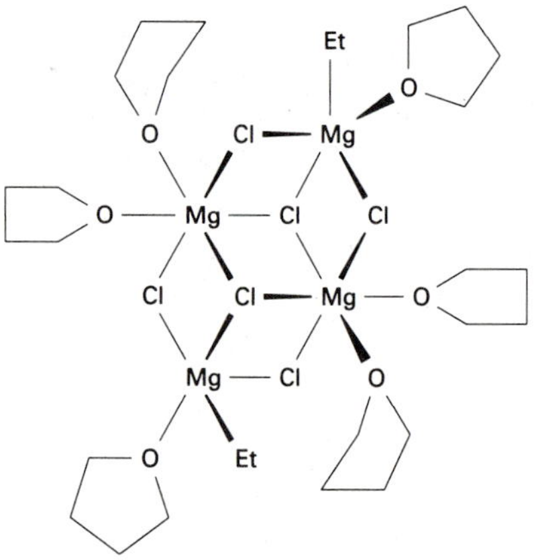

Because Grignard reagents are produced in ether solution, their use is limited to reactions in which that mildly Lewis basic solvent is not objectionable. The ether may form an unwanted complex with Lewis acid products, such as $(CH_3)_3BO(C_2H_5)_2$ in the reaction of BF_3 with CH_3MgBr in diethyl ether. In these cases, alkylaluminum or alkyllithium compounds are preferred because they can be used in hydrocarbon solution.

10

11

10.8 The zinc group

As mentioned earlier in the chapter, the dialkyl compounds of zinc, cadmium, and mercury are remarkable for their lack of association through alkyl bridges. Another feature to bear in mind is that dialkylzinc compounds are only weak Lewis acids, organocadmium compounds are even weaker, and organomercury compounds do not act as Lewis acids except under special circumstances.

Organozinc and organocadmium compounds

A convenient synthesis of organometallic compounds of zinc is double replacement with alkylaluminum or alkyllithium compounds. With

---

[4] A translation of Grignard's first full-length paper on this subject is available in *J. Chem. Educ.*, **47**, 290 (1970).

alkyllithium compounds as reactants this reaction conforms to the correlation of electronegativity that is often characteristic of double replacement reactions, but for alkylaluminum reactions the correlation is not decisive because the electronegativities of aluminum and zinc are so similar (1.61 and 1.65, respectively); however, hardness considerations correctly predict the formation of the softer $ZnCH_3$ and harder AlCl pairs:[5]

$$ZnCl_2(s) + \text{(structure)} \longrightarrow H_3C - Zn - CH_3 + \text{(structure)}$$

Alkylzinc compounds are pyrophoric and readily hydrolyzed, whereas alkylcadmium compounds react more slowly with air. Because of their mild Lewis acidity, dialkylzinc and dialkylcadmium compounds form stable complexes with amines, especially chelating amines (**12**). The C—Zn bond has greater carbanionic character than the C—Cd bond. One example is the addition of alkylzinc compounds across the carbonyl group of a ketone, $R_2C{=}O$,

$$Zn(CH_3)_2 + (CH_3)_2C{=}O \rightarrow (CH_3)_3C - O - ZnCH_3$$

This reaction does not take place when the less polar alkylcadmium and alkylmercury compounds are used instead. The reaction also occurs for organolithium, organomagnesium, and organoaluminum compounds, all of which contain metals of even lower electronegativity than zinc.

   Yet again, the cyclopentadienyl compounds are structurally unusual. (Cyclopentadienyl)methylzinc is monomeric in the gas phase with a pentahapto C_5H_5 group (**13**). In the solid it is associated in a zigzag chain (**14**), each C_5H_5 group being pentahapto with respect to *two* Zn atoms.

Organomercury compounds

Organomercury compounds are readily prepared by double displacement reactions between mercury(II) halides and strong carbanion reagents, such as Grignard reagents or trialkylaluminum compounds:

$$2RMgX + HgX_2 \rightarrow HgR_2 + 2MgX_2$$

This reaction is in line with both electronegativity and hardness considerations. As we have already remarked, dialkylmercury compounds are versatile starting materials for the synthesis of many organometallic compounds of more electropositive metals by transmetallation. How-

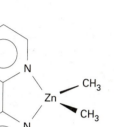

12

13

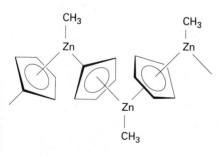

14

[5] For examples where hard and soft acid–base principles are more reliable for the prediction of the outcome of a reaction than are electronegativities; see R. G. Pearson, *J. Chem. Educ.*, **45**, 643 (1968). The great convenience of using electronegativity as a first approximation to reactivity is the availability of numerical values that follow fairly simple trends in the periodic table.

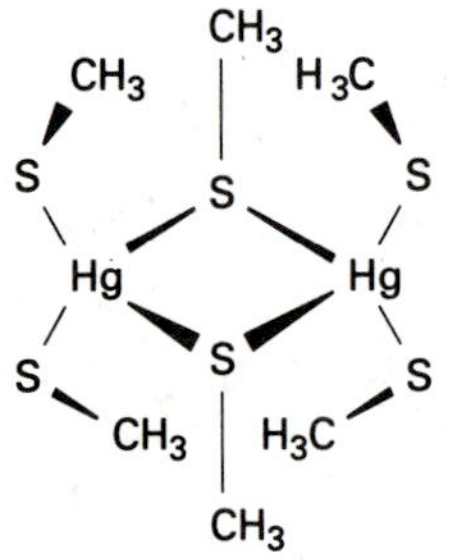

B1

Box 10.2 The toxicity of organomercury compounds

The toxicity of mercury arises from the very high affinity of the soft Hg atom for sulfhydryl (—SH) groups in enzymes. Simple mercury-sulfur compounds have been studied as potential analogs of natural systems. The Hg atoms are most commonly four-coordinated, as in $[Hg_2(SMe_2)_6]^{2-}$ (B1).

Mercury poisoning was a serious problem for early scientists, including Isaac Newton in the eighteenth century and Alfred Stock in the early twentieth century, both of whom worked with mercury in poorly ventilated laboratories. More recently, it became a major public concern following the incidence of brain damage and death it caused among the inhabitants in Minamata, Japan. This incident arose because mercury from a plastics factory was allowed to escape into a bay where it found its way into fish that were later eaten. Research since that time has shown that bacteria found in sediments are capable of methylating mercury, and that species such as $Hg(CH_3)_2$ and $[HgCH_3]^+$ enter the food chain because they readily penetrate cell walls. The bacteria appear to produce $Hg(CH_3)_2$ as a means of eliminating toxic mercury through their cell walls and into the environment.

[References: P. J. Craig (ed.), *Organometallic compounds in the environment*. Wiley, New York (1986); J. G. Wright, M. J. Natan, F. M. MacDonnell, D. Ralston, and T. V. O'Halloran, Mercury(II) thiolate chemistry and the mechanism of the heavy metal biosensor, *Progr. Inorg. Chem.*, **38**, 323 (1990).]

ever, owing to the high toxicity of alkylmercury compounds (Box 10.2), other syntheses are often preferred. In striking contrast to the high sensitivity of dimethylzinc to oxygen, dimethylmercury survives exposure to air.

ELECTRON-DEFICIENT COMPOUNDS OF THE BORON GROUP

We shall illustrate two general points in this section. One is that electron-deficient organometallic compounds of Group 13/III are molecular. The second is that for compounds in which the metallic element has oxidation number +3, only the organoaluminum compounds are significantly associated through 3c,2e bonded organic bridges.

10.9 Organoboron compounds

Trimethylboron is colorless, gaseous (b.p. −22 °C), and monomeric. It is pyrophoric but is not rapidly hydrolyzed by water. The alkylboranes can be synthesized by double replacement between boron halides and organometallic compounds of metals with low electronegativity, such as Grignard reagents or organoaluminum compounds:

$$BF_3 + 3CH_3MgBr \xrightarrow{\text{dibutyl ether}} B(CH_3)_3 + 3MgBrF$$

Dibutyl ether is used here rather than diethyl ether because it has a much lower vapor pressure than trimethylboron, which facilitates separation by trap-to-trap distillation on a vacuum line. Another key to the success of this separation is the very weak association of the dibutyl ether–trimethylboron complex.

Although trialkyl- and triaryl-boron compounds are mild Lewis acids, strong carbanion reagents lead to anions of the type $[BR_4]^-$. The best known of these anions is the tetraphenylborate ion, in which $R = C_6H_5$. This bulky anion hydrolyzes very slowly in neutral or basic water and is useful for the precipitation of large monopositive cations. For example, the addition of an aqueous solution of $Na[BPh_4]$ to a solution containing K^+ ions leads to the formation of insoluble $K[BPh_4]$. This precipitation reaction, which can be used for the gravimetric determination of potassium, is an example of the low solubility of large cation, large anion salts in water (Section 4.8).

The incorporation of boron into heterocycles is common, as in the protolysis of a triarylboron with 1,2-dihydroxybenzene at elevated temperatures:

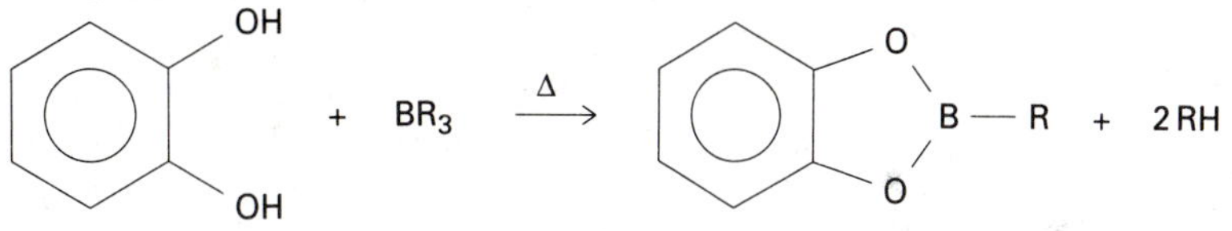

Organohaloboron compounds are more reactive than simple trialkylboron compounds. One preparative route is by the reaction of boron trichloride with a stoichiometric amount of an alkylaluminum in a hydrocarbon solvent:

$$3BCl_3 + 6AlR_3 \rightarrow 3R_2BCl + 6AlR_2Cl$$

Another procedure is the redistribution of boron trihalide and trialkylborane in the presence of diborane as catalyst:

$$2BCl_3 + BMe_3 \xrightarrow{\text{Diborane}} 3BMeCl_2$$

The products may be subjected to the full range of protolysis reactions with ROH, R_2NH, and other reagents, and to displacement of the halide by a carbanion reagent:

$$B(CH_3)_2Cl + 2HNR_2 \rightarrow B(CH_3)_2NR_2 + [NR_2H_2]Cl$$

$$B(CH_3)_2Cl + Li(C_4H_9) \rightarrow B(CH_3)_2(C_4H_9) + LiCl$$

Among the many other organoboranes, an especially interesting series contains B—N linkages. A BN fragment is isoelectronic with a CC fragment and B—N and C—C analogs generally have the same structures. However, despite this formal similarity, their chemical and physical properties are often quite different (see Section 11.2).

10.10 Organoaluminum compounds

One of the distinguishing features of the methyl bridge bond in alkylaluminum compounds (**1**) is the small Al—C—Al angle, which is approximately 75°. The weakness of these 3c,2e bridge bonds is indicated by the dissociation of trialkylaluminum compounds in the pure liquid to an extent that increases with the bulkiness of the alkyl group:

$$Al_2(CH_3)_6 \rightleftharpoons 2Al(CH_3)_3 \qquad K = 1.52 \times 10^{-8}$$

$$Al_2(C_4H_9)_6 \rightleftharpoons 2Al(C_4H_9)_3 \qquad K = 2.3 \times 10^{-4}$$

(The equilibrium constants refer to 25 °C and are expressed in mole fractions.) With very bulky groups, dissociation is virtually complete: trimesitylaluminum (**15**), for instance, is a monomer.[6] These examples provide clear evidence for the powerful role of steric effects on the structures of the alkylaluminum compounds.

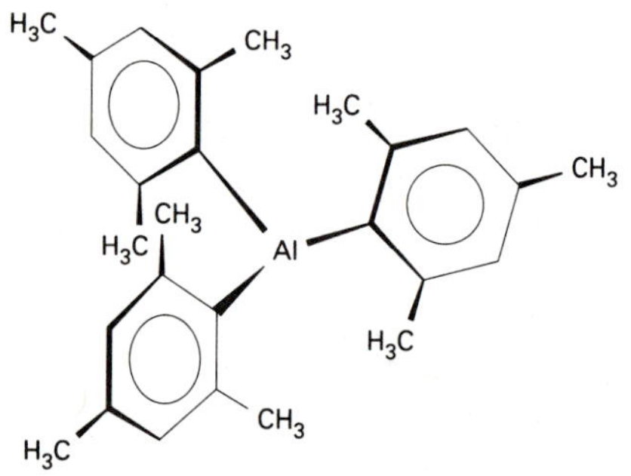

With bridging halides, alkoxides, and amides the Al—X—Al angle is close to 90° (Fig. 10.5). In contrast to a bridging alkyl group, a halogen atom has more orbitals and electrons available and can form a bridge utilizing two 2c,2e bonds. Similarly, triphenylaluminum exists as a dimer with bridging η^1-phenyl groups lying in a plane perpendicular to the line joining the two Al atoms (Fig. 10.6). This structure is favored partly on steric grounds and partly by supplementation of the Al—C—Al bond by electron donation from the phenyl π orbitals to the Al atoms. These bonding arguments are consistent with the preference for the bridging site, which falls in the order (OR, halide) > phenyl > alkyl.

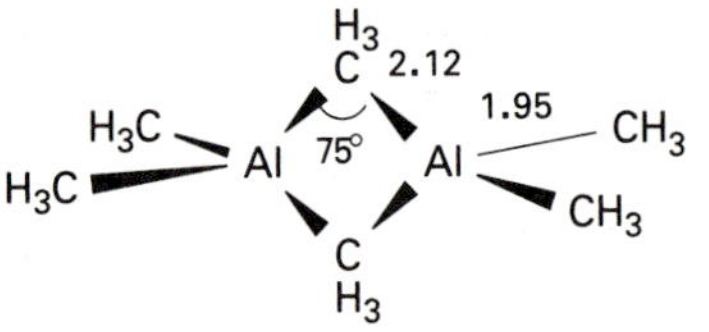

Alkylaluminum compounds have been extensively investigated on account of their use as alkene polymerization catalysts and chemical intermediates. They are relatively inexpensive carbanion reagents for the replacement of halogens with organic groups by double replacement. A general laboratory-scale preparation of trimethylaluminum is transmetallation with dimethylmercury:

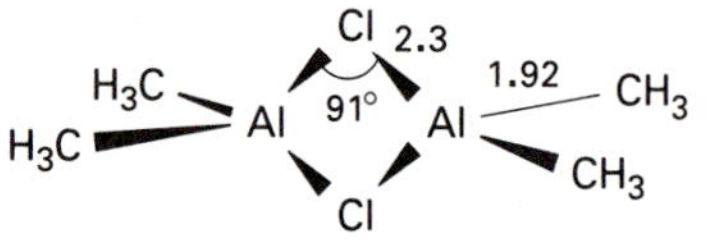

$$2Al + 3Hg(CH_3)_2 \rightarrow Al_2(CH_3)_6 + 3Hg$$

The commercial synthesis uses the reaction of aluminum metal with chloromethane to produce $Al_2Cl_2(CH_3)_4$, which is then reduced with sodium:

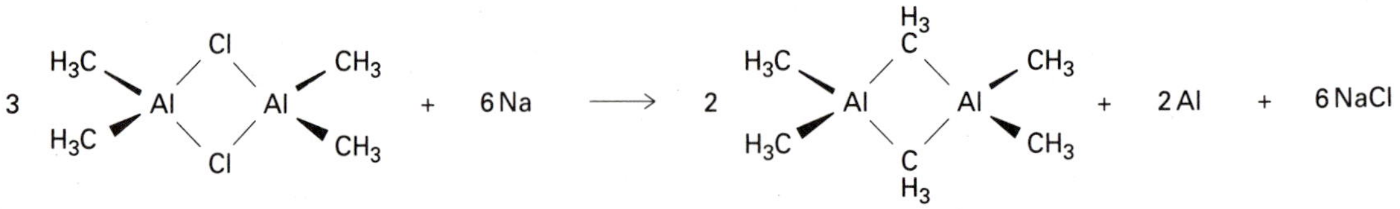

15 Al(mes)₃

(a)

(b)

10.5 Comparison of the bond angles between the 3c,2e bond for the Al₂CH₃ bridge and the two conventional 2c,2e bonds for the Al₂Cl bridge.

[6] The mesityl (2,4,6-trimethylphenyl) group is commonly used on account of its bulk. It plays a prominent role in the discussion of organosilicon chemistry in Section 10.12.

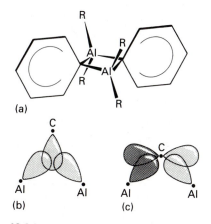

10.6 Structure and bonding of the phenyl bridge in Ph$_6$Al$_2$. (a) Structure illustrating the perpendicular orientation of the bridging phenyl relative to the AlCAlC plane. (b) The 3c,2e bond formed by a symmetric combination of C and Al orbitals. (c) An additional interaction between the $p\pi$ orbital on C and an antisymmetric combination of Al orbitals.

The commercial synthesis of triethylaluminum and higher homologs is by the addition of H$_2$ and the appropriate alkene to aluminum metal at elevated pressures and temperatures:

$$2Al + 3H_2 + 6RHC{=}CH_2 \xrightarrow[100\text{–}200\ \text{atm}]{60\text{–}110\,°C} Al_2(CH_2CH_2R)_6$$

It is probable that this reaction proceeds by the formation of a surface Al—H species that adds across the C=C bond of the alkene in a hydrometallation reaction. The commercial use of alkylaluminum compounds would be very limited without this relatively economical synthesis.

As might be expected from the more electropositive nature of aluminum, alkylaluminum compounds have much greater carbanionic character than alkylboron compounds. As a result, the former are very sensitive to water and oxygen and most are pyrophoric; both the pure liquid and solutions must be handled using inert atmosphere or vacuum line techniques. However, this reactivity can be turned to advantage, for the susceptibility of the Al—R bond to protolysis provides a simple method for the preparation of aluminum alkoxides and amides:

$$2\,AlR_3 \ + \ 2\,HOR' \ \longrightarrow \qquad\qquad\qquad +\quad 2\,RH$$

$$2\,AlR_3 \ + \ 2\,HNR'_2 \ \longrightarrow \qquad\qquad\qquad +\quad 2\,RH$$

Alkylaluminum compounds are mild Lewis acids and form complexes with ethers, amines, and anions.

β-Hydrogen elimination (Section 10.5) yields dialkylaluminum hydride when triethylaluminum and higher alkylaluminum compounds are heated. Tri(isobutyl)aluminum has a strong tendency to undergo this reaction:

$$2\,Al(^iC_4H_9)_3 \ \longrightarrow \qquad\qquad\qquad +\quad 2\,H_2C{=}C(CH_3)_2$$

The general preference of hydride for the bridging position indicates that the 3c,2e bond is stronger for H than an alkyl group, probably because the small H atom can lie more readily between the atoms. (See the footnote on p. 397 concerning the notation for 3c,2e bonds.)

Example 10.3: *Proposing structures for some boron and aluminum organometallic compounds*

Based on your knowledge of the bonding in organoboron and organo-aluminum compounds, propose structures for compounds having the empirical formulas (a) $B(^iPr)_3$, (b) $Al(C_2H_5)_2Ph$, (c) $Al(C_2H_5)_2P(CH_3)_2$.

Answer. (a) Triisopropylboron, as with all the simple alkylboranes, will be monomeric and the B atom and the three C atoms bonded to it should lie in a plane. (b) If the attached groups have moderate bulk, the alkyl and arylaluminum compounds are associated into dimers through multicenter bonds. Recalling that the tendency toward bridge structures is

$$PR_2^- > X^- > H^- > Ph^- > R^-$$

(where R is alkyl), the structures will be

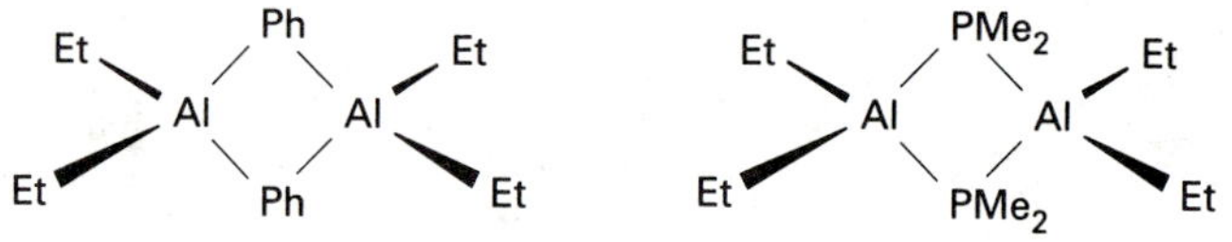

Exercise E10.3. Propose a structure for $Al_2(^iBu)_4H_2$

10.11 Organometallic compounds of gallium, indium, and thallium

There is a striking alternation of structures for the alkyl compounds of the first three elements in Group 13/III. As we have seen, the first member of the series, trimethylboron, is a monomer with three-coordinate B; the next, trimethylaluminum, is a dimer with four-coordinate Al; all the succeeding members (trimethylgallium, tri-methylindium and trimethylthallium) are monomeric in solution and in the gas phase. Carbanion character, as indicated by the tendency to hydrolyze and undergo double replacement reactions, follows the order indicated by aqueous potentials: $Al > Ga > In > Tl$. This behavior can be summarized as follows:

> Judging from the tendency to hydrolyze, carbanion character is greatest for aluminum and decreases in the order $Al_2R_6 > GaR_3 > InR_3 > TlR_3 > BR_3$.

Thus alkylaluminum compounds are rapidly and completely hydrolyzed by water:

$$Al_2(CH_3)_6 + 6H_2O \rightarrow 2Al(OH)_3 + 6CH_4$$

Hydrolysis of the analogous gallium, indium, and thallium compounds under mild conditions yields $M(CH_3)_2^+$ which itself is susceptible to hydrolysis in acid solution in the order Ga > In > Tl. As we have seen (Section 10.9), trialkylboranes require much more forcing conditions to undergo hydrolysis.

Gallium

Trialkylgallium compounds can be synthesized by the reaction of alkyllithium compounds with gallium chloride in a hydrocarbon solvent:

$$3Li_4(C_2H_5)_4 + 4GaCl_3 \rightarrow 12LiCl + 4Ga(C_2H_5)_3$$

Trialkylgallium compounds are mild Lewis acids, and so the corresponding double-replacement reaction in ether produces the complex $(C_2H_5)_2OGa(C_2H_5)_3$. Similarly, the use of excess methyllithium leads to the uptake of a fourth alkyl group by the Ga atom to form a salt:

$$Li_4(C_2H_5)_4 + GaCl_3 \rightarrow 3LiCl + Li[Ga(C_2H_5)_4]$$

Trialkylgallium compounds are pyrophoric and react with weak Brønsted acids such as water, alcohols, and thiols. In keeping with the decrease in carbanion character with the more electronegative metal atom, their reactivity toward Brønsted acids is somewhat less than that of alkylaluminum compounds. The general (unbalanced) form of the reactions they undergo is

$$Ga(CH_3)_3 + H_2O \rightarrow [Ga(CH_3)_2OH]_n + CH_4 \quad n = 2, 3, \text{ or } 4$$

The GaR$_2$ group is somewhat resistant to protolysis, and complexes such as $[Ga(CH_3)_2(en)]^+$ (**16**) and even $[Ga(CH_3)_2(OH_2)_2]^+$ can be handled in aqueous solution.

Indium and thallium and group trends

Alkylindium and alkylthallium compounds may be prepared by reactions analogous to those used to make the alkylgallium compounds. Trimethylindium is monomeric in the gas phase and in the solid the bond lengths indicate that association is weak (if present at all). Partial hydrolysis of Tl(CH$_3$)$_3$ yields the linear $[CH_3TlCH_3]^+$ ion, which is isoelectronic and isostructural with CH$_3$HgCH$_3$.

A new aspect of organometallic chemistry becomes evident this deep in the boron group, for now the inert pair effect can play a role and give rise to stable indium(I) and thallium(I) compounds. Two examples of these compounds are $(\eta^5\text{-}C_5H_5)In$ and $(\eta^5\text{-}C_5H_5)Tl$, which exist as monomers in the gas phase but are associated as solids. Cyclopentadienylthallium is useful as a synthetic reagent in organometallic chemistry

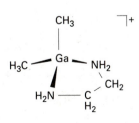

16

because it is not as highly reducing as $Na[C_5H_5]$ and the insolubility of TlCl provides an added driving force for double replacement reactions. A drawback is that thallium is even more poisonous than mercury, so the disposal of the reaction byproducts must be done with care.

ELECTRON-PRECISE COMPOUNDS OF THE CARBON GROUP

In Group 14/IV we meet organometallic compounds that are formed by carbon with its own congeners. Because the electronegativity of carbon is similar to that of its congeners, the bonds these elements form to one another are not very polar. This low polarity in conjunction with the greater steric protection of the four-coordinate central atom and the lack of a low energy LUMO appear to be responsible for their greater resistance to hydrolysis compared with the organometallic compounds of the boron group.

10.12 Organosilicon compounds

The chemistry of organosilicon compounds is very extensive. This is partly because the compounds have been studied for a long time but also because they have wide commercial use as water repellents, lubricants, and sealants.

Structures and properties

The general feature that will be illustrated in this section is that

> In contrast to the M—C bonds of more electropositive elements, such as aluminum, Si—C bonds are resistant both to hydrolysis and to air oxidation.

Many oxo-bridged organosilicon compounds can be synthesized; one example is hexamethyldisiloxane $(CH_3)_3Si-O-Si(CH_3)_3$. As with all simple organosilicon compounds, this compound is resistant to moisture and air. Two properties of these materials, the very weak Lewis basicity of the O atom and ready deformation of the Si—O—Si bond angle, have been rationalized by a model in which lone pairs on O are partially delocalized into vacant σ^* or d orbitals of Si (**17**). The delocalization reduces the directionality of the Si—O single bond and hence makes the structure more flexible. This flexibility permits silicone elastomers to remain rubber-like down to very low temperatures. Delocalization also accounts for the low basicity of an O atom attached to silicon, as in $(CH_3)_3Si-O-Si(CH_3)_3$, for the electrons that are needed for the O atom to act as a base are partially removed. The planarity of trisilylamine, $(SiH_3)_3N$, is also explained by the delocalization of the lone pair on N; moreover, this compound is only very weakly basic.[7]

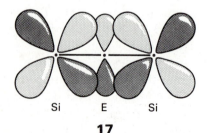

Si   E   Si

17

---

[7] The interesting shapes of silyl molecules containing Si—N and Si—O bonds have been reviewed by E. A. V. Ebsworth, *Acc. Chem. Res.*, **20**, 295 (1987).

Another closely related observation is the relative ease of deprotonation of tetramethylsilane by very strong bases, such as strong carbanion reagents:

$$LiBu + Si(CH_3)_4 \rightarrow Li[CH_2Si(CH_3)_3] + BuH$$

Once again, the $-CH_3$ group is thought to exhibit this mild Brønsted acidity because the resulting conjugate base, the $-CH_2^-$ group, can delocalize electron density on to the neighboring Si atom.

Example 10.4: *Predicting the properties of silyl ethers and amines*
Predict the extent of hydrogen bonding of ethanol with (a) $(H_3Si)_2O$ compared with $(H_3CCH_2)_2O$ and (b) $(H_3Si)_3N$ compared with $(H_3C)_3N$.

Answer. Strong Lewis bases form the stronger hydrogen bonds with a given reference hydrogen donor, such as ethanol. We have seen that O or N attached to silicon has reduced Lewis basicity as compared with the carbon analogs. So we expect that for (a) diethyl ether and for (b) trimethyl amine will form the stronger hydrogen bonds.

Exercise E10.4. For the compounds in (a), which do you expect to have the lower force constant, $Si-O-Si$ bending or $C-O-C$ bending?

The formation of Si—C bonds

A convenient way of linking alkyl groups to Group 14/IV elements is by double replacement reactions between $E-Cl$ and Grignard or alkyllithium reagents:

$$Li_4(CH_3)_4 + SiCl_4 \rightarrow 4LiCl + Si(CH_3)_4$$

Large quantities of chloromethylsilanes are needed in industry to synthesize silicone rubber and oils, but Grignard and lithium reagents are expensive when used on such a scale. It was to satisfy this need more cheaply that led Eugene Rochow (in the early 1940s at General Electric in the USA) to develop a process for their direct synthesis from elemental silicon and an alkyl or aryl halide in the presence of copper as a catalyst. The reaction has the (unbalanced) form

$$Si + RX \xrightarrow{250-550\,°C,\ Cu} R_nSiX_{4-n}$$

The conditions are usually adjusted to favor the formation of dimethyl-dichlorosilane, but other useful halosilanes are also produced. This relatively inexpensive direct process transformed silicone polymers from being expensive laboratory curiosities to widely used materials.[8]

---

[8] An interesting personal account of his discovery of the direct process is given by E. G. Rochow in his book: *Silicon and silicones.* Springer-Verlag, Berlin (1987).

Redistribution reactions

Redistribution reactions are useful for the synthesis of a variety of Group 14/IV compounds. The reaction is readily performed in the laboratory and is employed commercially to prepare halosilanes:

$$2SiCl_4 + Si(CH_3)_4 \rightarrow xSiCl(CH_3)_3 + ySiCl_2(CH_3)_2 + zSiCl_3(CH_3)$$

Traces of Lewis acids such as $AlCl_3$ are effective catalysts for this reaction. It might be expected that because Si—Cl and Si—CH_3 bonds are being broken and then reformed, the product distribution from an initial 1:1 mixture of $SiCl_4$ and $Si(CH_3)_4$ should yield a statistical mixture of products: $SiCl_4$, $SiCl_3(CH_3)$, $SiCl_2(CH_3)_2$, $SiCl(CH_3)_3$, and $Si(CH_3)_4$ in the ratio $1:4:6:4:1$. However, subtle aspects of the bonding and steric interactions often lead to nonstatistical mixtures.

Double replacement reactions that utilize haloorganosilane, germane, and stannane compounds are very useful for laboratory-scale syntheses, especially for preparing compounds with mixed organic substituents:

$$4SiCl_3(C_6H_5) + 3Li_4(CH_3)_4 \rightarrow 4Si(CH_3)_3(C_6H_5) + 12LiCl$$

Studies of the kinetics of substitution reactions on silicon suggest that the mechanisms of these reactions are associative (Section 6.8). Thus, the reaction is generally second-order overall

$$Rate = k[SiABCX][Y]$$

and k depends strongly on the identity of the entering group Y. That redistribution reactions on silicon and heavier members of the group are more facile than those on carbon also suggests that a Si atom forms a five-coordinate transient intermediate much more readily than a C atom does. This difference is in line with the existence of many more five-coordinate silicon inorganic compounds than five-coordinate carbon compounds. (The latter are largely confined to carboranes, metal cluster compounds, and solid metal carbides.)

Stereochemical investigations reveal that substitution reactions may proceed with either inversion or retention of configuration. A mechanism that accounts for inversion is very similar to that proposed for associative substitutions on C (that is, a mechanism resembling an S_N2 substitution in organic chemistry), with the entering and leaving groups occupying the axial positions in a trigonal bipyramidal activated complex:

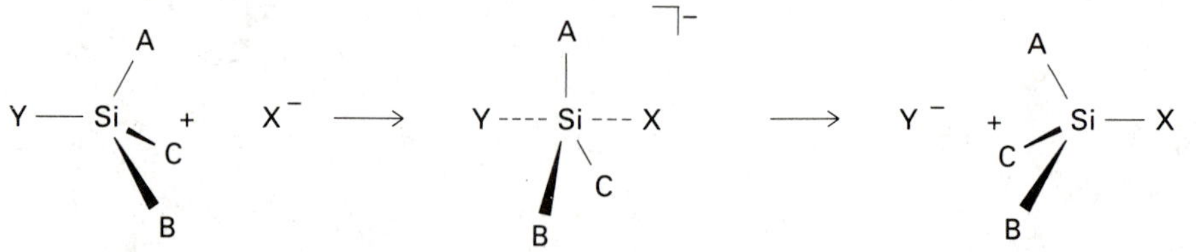

Retention is common when Y is a poor leaving group, such as H^- or OR^-. This kind of reaction is believed to involve a pseudorotation (Fig. 6.3) of a five-coordinate intermediate that places Y on the pseudo threefold axis and *trans* to one of the original substituents:

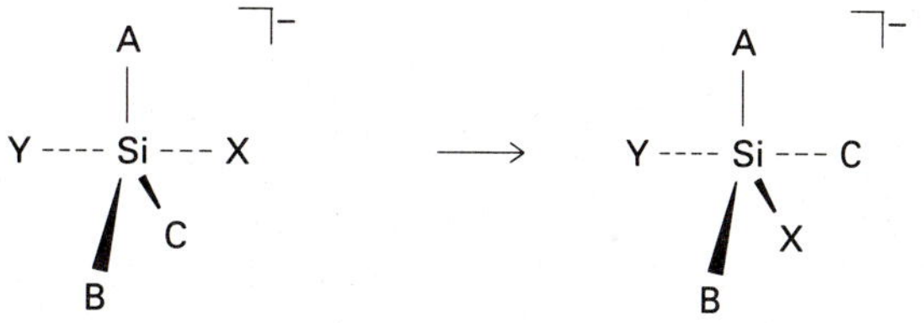

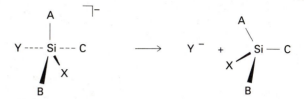

The pseudorotation is then followed by departure of the leaving group:

The ability of the Si atom to form a five-coordinate intermediate with sufficiently long lifetime to rearrange results in retention of configuration in some nucleophilic displacement reactions even though they proceed by an associative mechanism. On the other hand, the smaller C atom, which has a much less stable five-coordinate activated complex, immediately undergoes inversion.

Protolysis of halosilanes

The protolysis of Si—X bonds is a convenient route to a myriad of compounds, particularly **siloxanes**, which are compounds with Si—O bonds. Thus, polar Si—Cl bonds are susceptible to protolysis by species with H—O, H—N, and H—S bonds, but the much less polar Si—C bonds are more resistant. This difference accounts, for instance, for the formation of hexamethyldisiloxane in the reaction of water with chlorotrimethylsilane. The initial reaction is the hydrolysis of the Si—Cl bond:

$$(CH_3)_3SiCl + H_2O \rightarrow (CH_3)_3SiOH + HCl$$

This step is followed by a slower reaction, the elimination of water to form the Si—O—Si link:

$$2(CH_3)_3SiOH \rightarrow (CH_3)_3Si—O—Si(CH_3)_3 + H_2O$$

The condensation of the SiOH compound is analogous to the transformation of metal hydroxide complexes to polycations (Section 5.7) and to the polymerization of $Si(OH)_4$ in aqueous solution to produce silica gel. These reactions demonstrate the general tendency of Si—OH groups to eliminate water.

When dimethyldichlorosilane is hydrolyzed, both cyclic (**18**) and long-chain compounds are produced:

$$(CH_3)_2SiCl_2 + H_2O \rightarrow HO[Si(CH_3)_2O]_nH + [(CH_3)_2SiO]_4 + etc$$

When $RSiCl_3$ is used as a starting material and contains bulky organic

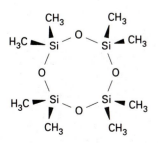

18

19

groups, more elaborate structures are possible, including cage compounds (**19**). Similar reactions occur with ammonia and with primary and secondary amines to produce a variety of **silazanes**, compounds with Si—N bonds. For example, excess $(CH_3)_3SiCl$ reacts with NH_3 to produce $((CH_3)_3Si)_2NH$. In this case, the protolysis is incomplete on account of the protection afforded by the bulky groups around the Si atom; the less hindered $HSi(CH_3)_2Cl$ yields $((CH_3)_2HSi)_3N$.

Siloxanes undergo redistribution reactions to form polymers, the **silicones**, in the presence of sulfuric acid as a catalyst:

$$n[cyclo\text{-}(CH_3)_2SiO]_4 + (CH_3)_3SiOSi(CH_3)_3 \xrightarrow{H_2SO_4}$$

$$(CH_3)_3SiO[Si(CH_3)_2O]_{4n}Si(CH_3)_3$$

In this reaction, the hexamethyldisiloxane provides the $Si(CH_3)_3$ end groups, so the higher the proportion of it present, the lower the molar mass of the resulting polymer. Silicones, including poly(dimethylsiloxane), are fluids, waxes, and, when cross-linked, elastomers. They are produced on a large scale. In general, the polysiloxanes are highly pliable materials that retain their flexibility at low temperatures on account of the low bending force constants of Si—O—Si bonds. Their useful properties include their flexibility at low temperatures, their ability to repel water, and their resistance to oxidation by air. Additionally, their low toxicity leads to their use in medical and cosmetic implants.

Example 10.5: *Contrasting the ease of hydrolytic cleavage of Al—CH₃ and Si—CH₃ bonds*

Give chemical equations for the possible reactions of trimethylaluminum and tetramethylsilane with water, and identify the difference in bonding in the two compounds that accounts for the difference in behavior.

Answer. Organoaluminum compounds have considerable carbanion character, which correlates with the highly electropositive character of aluminum. This bonding description is consistent with the experimental observation that trimethylaluminum hydrolyzes rapidly on contact with water:

$$Al_2(CH_3)_6 + 6H_2O \rightarrow 2Al(OH)_3 + 6CH_4$$

As mentioned in Section 10.5, this reaction probably occurs by initial coordination of the O atom in H_2O to the relatively low energy acceptor orbitals of the Al atom of the alkylaluminum compound. In contrast, silicon and carbon have similar electronegativity and they form covalent bonds of low polarity. Their low polarities, relative steric congestion, and low Lewis acidities prevent coordination prior to hydrolysis. This resistance to hydrolysis is illustrated by the use of silicone polymers as waterproofing agents on cloth or leather and the use of silicone cements to isolate electrical and electronic devices from water.

Exercise E10.5. Illustrate how the difference in reactivity between Al—C and Si—C bonds with O—H groups leads to the choice of different strategies for the synthesis of aluminum and silicon alkoxides.

Single bonds between Si atoms

Although not as extensive as the simple carbon–carbon bonded organic compounds, there are many catenated silicon organometallic compounds. (Catenation means the formation of chains, such as R_3Si—SiR_3 and R_3Si—SiR_2—SiR_3.) Considerable progress has been made in the synthesis of compounds containing the Si—Si bond, which is weaker than the C—C bond but not dramatically so. Acyclic, cyclic, bicyclic, and cage alkylsilicon compounds are known (**20** and **21**). These compounds are prepared by reductive halide elimination reaction:

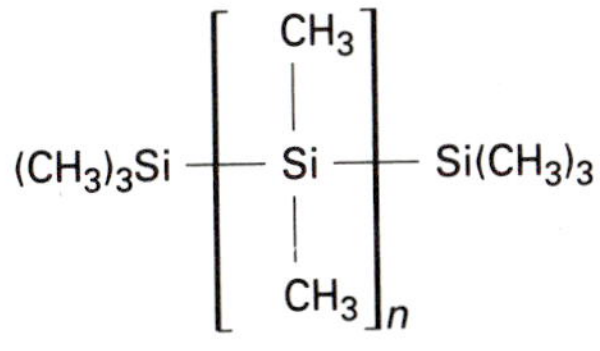

$$3(Xyl)_2SiCl_2 + 6Li[C_{10}H_8] \xrightarrow[-78°C]{CH_3OC_2H_4OCH_3}$$

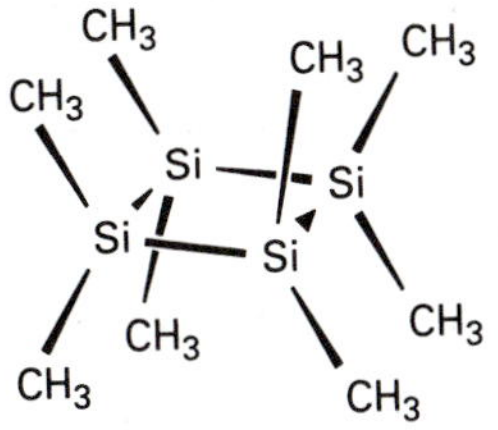

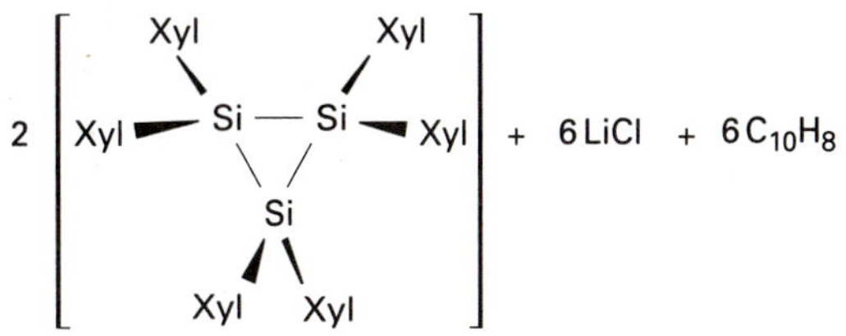

20

21

The abbreviation Xyl represents the bulky xylyl group (2,6-dimethylphenyl). The strong reducing agent lithium naphthalide reduces a Si—X halogen bond with the expulsion of the X^- halide ion and the formation of Si—Si bonds. The product of this reaction will be important in our discussion of Si=Si bonds in the following section.

Spectroscopic and chemical evidence suggests that the catenated compounds of silicon and germanium have fairly low lying vacant orbitals that are delocalized over the chain or ring. Thus, they have near-UV absorption bands that decrease in energy with increasing chain length. This absorption is attributed to the promotion of an electron from a σ orbital of the Si—Si or Ge—Ge backbone into an excited σ^* orbital that is also delocalized along the Si—Si or Ge—Ge chain. In keeping with this interpretation, silane chains undergo photolysis when exposed to UV radiation. For example, poly(dimethylsilane), $H_3C(-Si(CH_3)_2-)_nCH_3$ is cleaved when irradiated with UV radiation.[9]

Another aspect of the chemical properties of silicon that emphasizes the delocalized nature of the σ orbitals is the formation, by one-electron reduction of a saturated silicon compound, of a radical anion in which

---

[9] This photolysis forms the basis for the potential use of polydimethylsilane in the photolithographic production of integrated circuits. In this application the polydimethylsilane appears to be superior to the photosensitive polymers that are currently used. See: J. Michl, J. W. Downing, T. Karatsu, A. J. Mckinley, G. Poggi, G. W. Wallraff, R. Sooriyakumaran, and R. D. Miller, *Pure Appl. Chem.*, **60**, 959 (1988).

the odd electron occupies a delocalized Si—Si σ^* orbital (represented by the ring, as in benzene, but here signifying σ delocalization, not π delocalization):

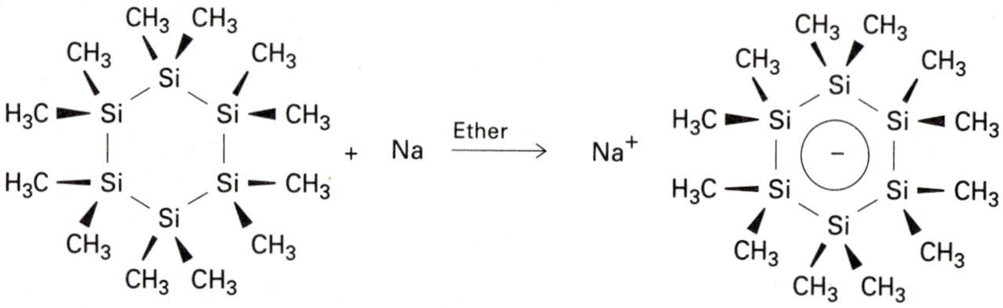

The EPR spectrum of the anion shows that the unpaired electron occurs with equal probability on all six Si atoms, proving that the electron is fully delocalized (at least on the timescale of this experiment, which is about 1 μs).

Multiple bonds between Si atoms

All attempts to synthesize compounds having Si=C or Si=Si bonds were unsuccessful until quite recently. As is the case for carbon, two single Si—Si bonds are stronger than one Si=Si double bond, so there is an energy advantage for Si=C or Si=Si groups to couple together. Carbon C=C bonds are protected from coupling by the high activation energy of the reaction, but there is much less kinetic protection for multiple bonds involving silicon or germanium.

A clue that multiply bonded silicon compounds might exist came from the detection of transient compounds such as $(CH_3)_2Si=CH_2$ in the gas phase and in low temperature inert gas matrices. However, few of the chemical properties of the molecules were established because they rapidly form dimers and polymers:

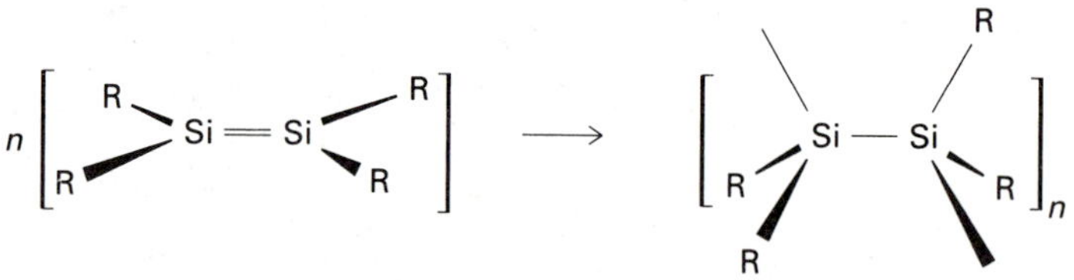

The key to success in the formation of Si=Si and double-bonded compounds of germanium, tin, phosphorus, and arsenic was the discovery that bulky substituents block dimerization and polymerization.

Compounds containing Si=C bonds, the **silaethenes**, and Si=Si bonds, the **disilaethenes**, are now known. These new silaethenes and disilaethenes not only extend our understanding of chemical bonding but also exhibit interesting chemical properties.[10] Photochemical cleavage of Si—Si bonds, following excitation into an antibonding Si—Si σ^* orbital, provides a convenient route to disilaethenes. An

[10] Silaethene chemistry is reviewed by A. G. Brook and K. M. Baines, *Adv. Organometal. Chem.*, **25**, 1 (1986). A helpful survey of disilaethenes is given by R. West in *Angew. Chem. Int. Ed. Engl.*, **26**, 1201 (1986).

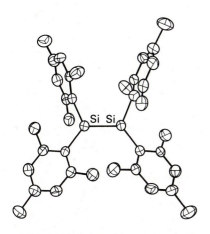

10.7 An ORTEP diagram of the structure of $(Mes)_2Si=Si(Mes)_2$ as determined by X-ray single crystal diffraction. (From M. J. Fink, M. J. Michalczyk, K. J. Haller, R. West, and J. Michl, *Organometallics*, **3**, 793 (1983).)

example is the photolysis of a cyclic trisilane with bulky substituents:

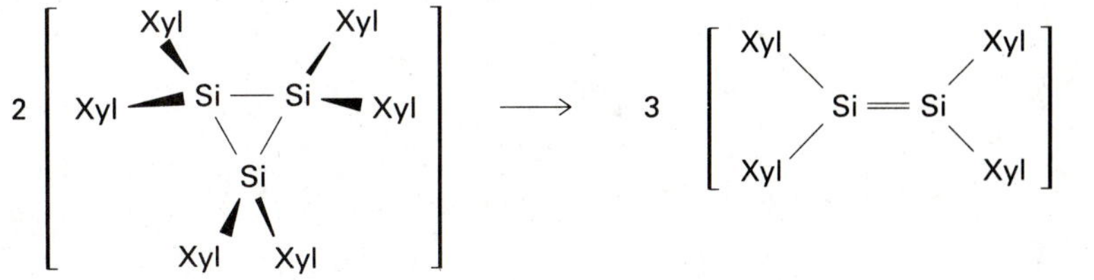

$$2\begin{bmatrix} \text{Xyl} & & \text{Xyl} \\ \text{Xyl}-\!\!\!\blacktriangleright\text{Si}-\text{Si}\blacktriangleleft\!\!\!-\text{Xyl} \\ & \text{Si} & \\ \text{Xyl} & & \text{Xyl} \end{bmatrix} \longrightarrow 3\begin{bmatrix} \text{Xyl} & & \text{Xyl} \\ & \text{Si}=\text{Si} & \\ \text{Xyl} & & \text{Xyl} \end{bmatrix}$$

The bulky groups on the Si atoms prevent cyclization and polymerization; they also obstruct reaction of the disilaethene product.

X-Ray structure determinations show the expected bond length decrease from Si—Si to Si=Si. For example, tetramesityldisilaethene (Fig. 10.7) has a silicon–silicon bond length of 2.16 Å, which is about 0.20 Å shorter than a typical Si—Si single bond. A significant difference between the disilaethenes and alkenes, however, is the greater ease of distorting the $R_2Si=SiR_2$ group from planarity. We shall see that this tendency is even more pronounced in heavier Group 14/IV analogs, and an explanation in terms of bonding will be given there.

The electronic absorption spectra of the disilaethenes contain a band in the visible or near-UV region, and the compounds are often brightly colored. This coloration indicates that the π and π^* orbitals are closer in energy than in the alkenes, for which the corresponding absorption is well into the UV (Fig. 10.8).

The energies of the bonding and antibonding π orbitals can be inferred from reduction potentials in solution. It is found that the oxidation of disilaethenes occurs at less positive potentials and reduction at less negative potentials than for corresponding alkenes. As depicted in Fig. 10.8, this observation indicates that the energies of π and π^* orbitals of the disilaethenes are between those of the corresponding alkenes. As a result, disilaethenes are better π donors (using electron donation from the filled π orbital) and better π acceptors (accepting electrons into the π^* orbital).

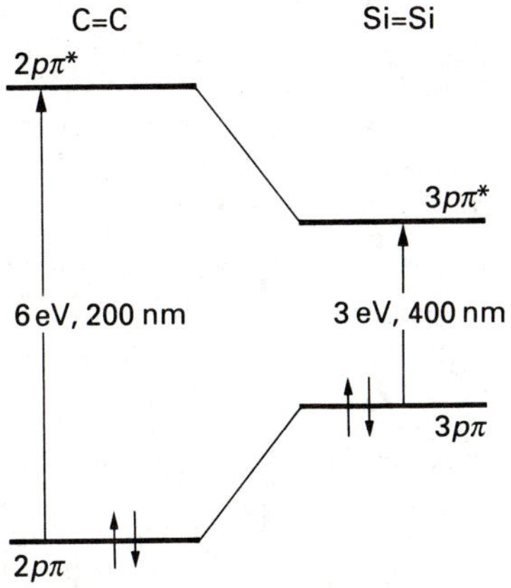

Molecules that are not sterically encumbered can penetrate a molecule's stereochemical defences and have access to the Si=Si bond in disilaethenes. Some of the resulting reactions are very similar to those found for alkenes. For example, hydrogen halides and halogens add across the Si=Si double bond:

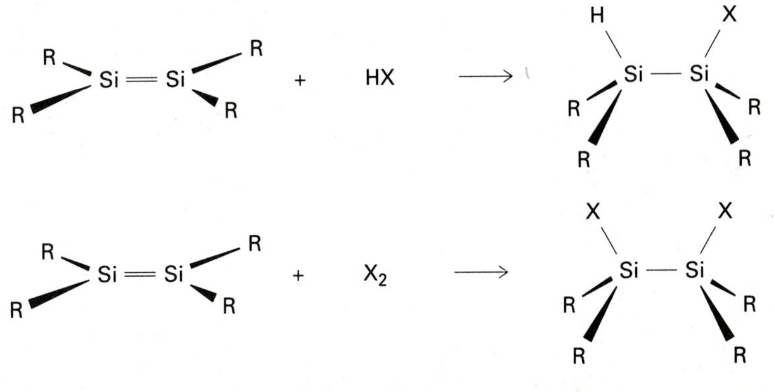

10.8 Approximate energy levels for alkenes and disilenes. The energy separations are obtained from the analysis of UV absorption spectra. (From R. West, *Angew. Chem. Int. Ed. Engl.*, **26**, 1201 (1983).)

C=C

$2p\pi^*$

Si=Si

$3p\pi^*$

6 eV, 200 nm   3 eV, 400 nm

$3p\pi$

$2p\pi$

$$\begin{array}{c} R \\ R \end{array}\!\!\!>\!\!\text{Si}=\text{Si}\!<\!\!\!\begin{array}{c} R \\ R \end{array} + \text{HX} \longrightarrow \begin{array}{cc} H & X \\ \text{Si}-\text{Si} \end{array}$$

$$\begin{array}{c} R \\ R \end{array}\!\!\!>\!\!\text{Si}=\text{Si}\!<\!\!\!\begin{array}{c} R \\ R \end{array} + X_2 \longrightarrow \begin{array}{cc} X & X \\ \text{Si}-\text{Si} \end{array}$$

where $X = Cl$ or Br. More interesting, perhaps, are the contrasts with the properties of alkenes. Unlike alkenes, disilenes undergo addition of ROH across the $Si = Si$ bond:

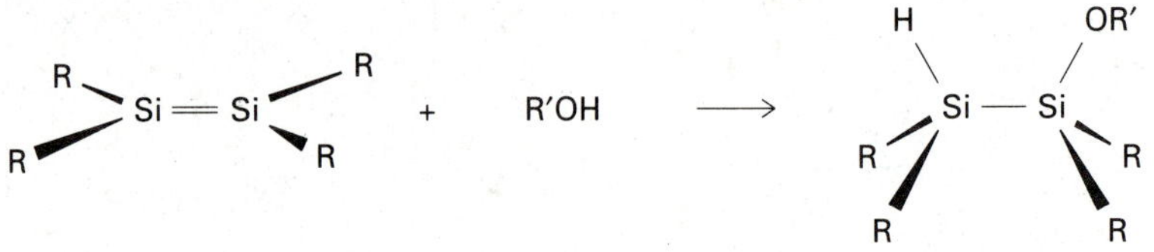

$$R_2Si = SiR_2 + R'OH \longrightarrow R_2Si(H) - Si(OR')R_2$$

They also undergo $2 + 2$ additions with some alkynes:

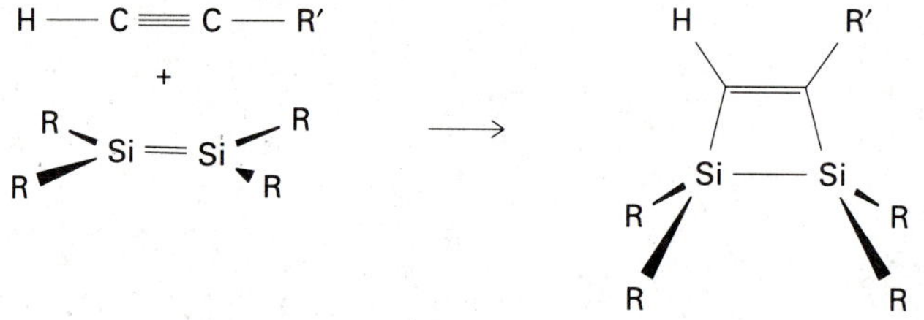

$$HC \equiv C - R' + R_2Si = SiR_2 \longrightarrow$$

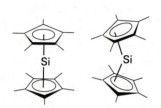

10.9 Two conformations of bis(pentamethylcyclopentadienyl)silicon in the solid state. In the bent structure the rings are canted and the Si atom is offset relative to the center of each C_5 ring.

It is possible to attach a pentamethylcyclopentadiene ligand to a Si atom. This possibility has led to the synthesis of the first formal Si(II) organometallic compound, bis(pentamethylcyclopentadienyl)silicon.[11] An X-ray crystal structure determination of the compound reveals the presence of two different configurations (Fig. 10.9). Apparently the energy difference between the two configurations is small enough for packing forces in the solid to influence the structure.

Thus we have seen a full range of silicon compounds from the common R_4Si, to the recently discovered disilaethenes with moderately bulky ligands and the formally divalent R_2Si compounds with the highly bulky pentamethylcyclopentadiene ligand.

10.13 Organometallic compounds of germanium, tin, and lead

Many of the reactions of organotin compounds (organostannanes) and organolead compounds (organoplumbanes) are similar to those already mentioned for silicon and germanium. One major difference from organosilicon compounds, however, is the existence (because of the inert-pair effect) of some Ge(II), Sn(II), and Pb(II) compounds. Another contrast with organosilicon compounds results from the rapid decrease in E—C bond strength down the group (Fig. 10.3). Because of the latter, organolead compounds generally decompose when heated above 100°C.

The gas-phase decomposition of alkyllead compounds occurs with the formation of alkyl radicals:

$$Pb(CH_3)_4(l \text{ or } g) \xrightarrow{\Delta} Pb(s) + 4 \cdot CH_3(g)$$

This decomposition was the reason why, for many years, tetramethyl-

[11] P. Jutzi, Main-group metallocenes: Recent developments, *Pure Appl. Chem.*, **61**, 1731 (1989).

lead and tetraethyllead were added to gasoline to improve its octane rating (a measure of its smoothness of combustion). The alkyl radicals generated by the decomposition of the tetraalkyllead in the hot combustion chamber of the engine are radical chain terminators which reduce the likelihood of explosive combustion. The toxicity of lead, however, and its deactivation of catalytic converters for pollution control, has prompted the elimination of organolead additives from gasoline in many parts of the world.

Organotin compounds find many different applications, ranging from stabilizers for poly(vinyl chloride) plastics to fungicides and antifouling paints for the hulls of boats. Recently, though, some of these applications have come under close scrutiny because they may harm benign and desirable organisms: organotin compounds kill not only barnacles but oysters too.

Germaethenes ($R_2Ge{=}CR_2$), stannaethenes ($R_2Sn{=}CR_2$), digermaethenes ($R_2Ge{=}GeR_2$), and distannaethenes ($R_2Sn{=}SnR_2$) have recently all been prepared. As with the corresponding silicon compounds, it is necessary for the R groups to be bulky to prevent association. The digermaethenes and distannaethenes are distinctly nonplanar and several bonding explanations have been proposed. In one of them, the nonplanarity is attributed to an unconventional pattern of multiple bonding, which is between the $sp^2\sigma$ orbital on one atom with a $p\pi$ orbital on the other (**22**).

Germanium–germanium and tin–tin multiple bonds are long, and dissociation has been observed in solution to form the germylene and stannylene divalent compounds, GeR_2 and SnR_2. As expected from the increased stability of the divalent state going down the group, the SnR_2 species are more stable than their GeR_2 counterparts. The divalent state is also seen in $Sn(\eta^5\text{-}C_5H_5)_2$ and $Pb(\eta^5\text{-}C_5H_5)_2$ which have angular structures in the gas phase, indicating that a stereochemically active lone pair may be present on the metal atom (**23**). The very bulky pentaphenylcyclopentadienyl ligands in $Sn(C_5Ph_5)_2$ suppress the stereochemical activity of the electron pair on the Sn atom, and X-ray structure determination reveals that the two C_5 rings are coplanar.

In addition to the simple monomeric organogermanium and organotin compounds, catenated compounds (**24**) and cyclic compounds (**25**) are known. The synthetic routes to these compounds resemble those used for their silicon analogs, but the reactions are often more facile with organogermanium and organotin compounds. One possible reason for this greater reactivity is that these larger atoms may be more accessible to potential reactant species. We have seen already that the heavier members of this group more readily participate in radical reactions.

Polyhedral compounds of carbon, such as cubane and its derivatives, R_4C_4, are thermodynamically less stable than arenes, but the same is not true for analogous compounds of carbon's congeners, and these heavier members of Group 14/IV have never been found to form benzene-like structures. Instead, a series of closed polyhedral com-

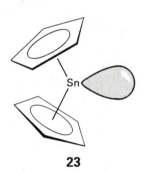

22

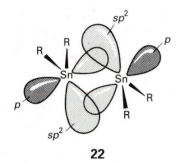

23

$$(CH_3)_3Ge \;-\!\!\!\left[\begin{array}{c} CH_3 \\ | \\ Ge \\ | \\ CH_3 \end{array}\right]_n\!\!\!-\; Ge(CH_3)_3$$

24

25 R = CH_3, CH_2CH_3

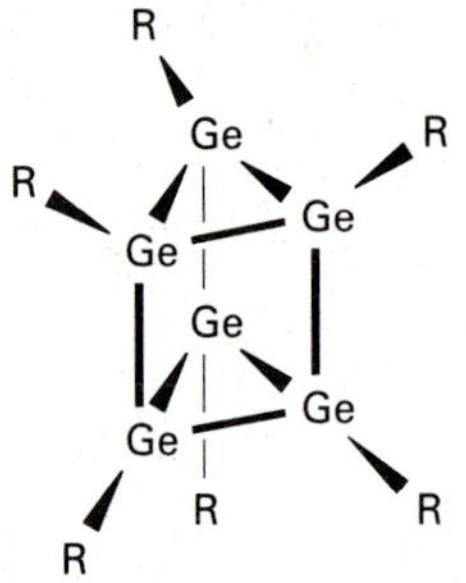

26 R = $CH(Si(CH_3)_3)_2$

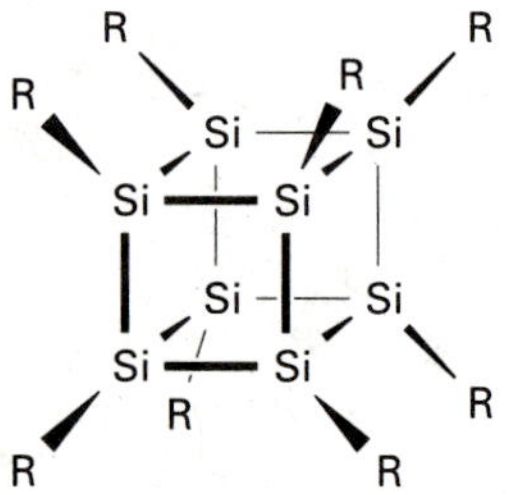

27

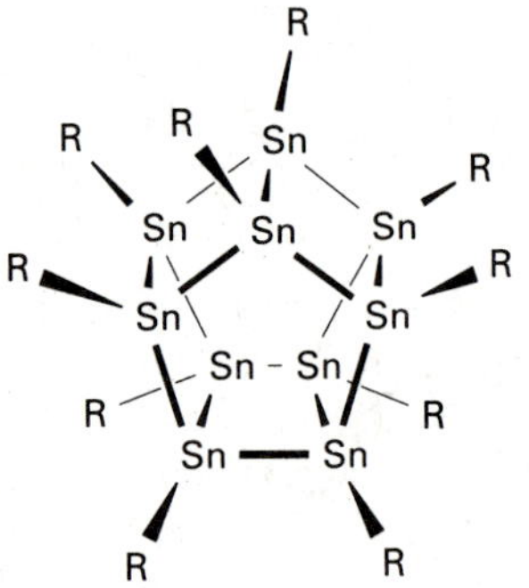

28 R = $(C_2H_5)_2C_6H_3$

pounds has been prepared recently, including trigonal-prismatic Ge_6R_6 (**26**), square-prismatic Si_8R_8 and Sn_8R_8, (**27**), and pentagonal-prismatic $Sn_{10}R_{10}$ (**28**).[12] It appears that the bulkiness of the R group plays a role in determining the particular structure, but the general systematics of its role have not been identified. The preparation of a germanium prismane compound is representative of the simple reactions that lead to these polyhedral compounds:

$$6RGeCl_3 + 18Li \rightarrow Ge_6R_6 + 18LiCl \qquad R = -CH_2Si(CH_3)_3$$

We can view this reaction as the reduction of the Ge(IV) compound $RGeCl_3$ to a formal Ge(I) cluster compound.

ELECTRON-RICH COMPOUNDS OF THE NITROGEN GROUP

Some new features of organometallic compounds are encountered in Group 15/V, including the Lewis basicity arising from the lone pair on the central atom (as in $:AsR_3$). Moreover, the central atom may exist in oxidation states $+3$ or $+5$, as in AsR_3 and $AsPh_4^+$, respectively.

10.14 Organometallic compounds of arsenic, antimony, and bismuth

Because the first two members of the nitrogen group (N and P) are unambiguously nonmetals, we do not consider them here. Recently, there has been significant work on the organic derivatives of the heavier elements in the group. It has resulted in interesting new compounds with unusual bonding patterns.

Oxidation states $+3$ and $+5$ are encountered in many of the organometallic compounds of arsenic, antimony, and bismuth. An example of a compound with an element in the $+3$ oxidation state is $As(CH_3)_3$ (**6**) and an example of the $+5$ state is $As(C_6H_5)_5$ (**29**). Compounds in which the elements have oxidation number $+3$ contain lone pairs and therefore can be considered electron rich; those with elements in the $+5$ oxidation state are electron precise.

Organoarsenic compounds were once widely used to treat bacterial infections and as herbicides and fungicides. However, because of their high toxicity arsenic, antimony, and bismuth organometallic compounds no longer have major commercial applications. Nevertheless, they can be handled in the laboratory with proper techniques, and many interesting compounds have been prepared.

Oxidation state $+3$

Here we consider a broad class of compounds of these elements with the general formulas ER_3, and related organo halides and cyclopentadiene compounds.

---

[12] A. Sekiguchi, C. Kabuto, and H. Sakuri, *Angew. Chem. Int. Ed. Engl.*, **28**, 55 (1989), and L. R. Sita and I. Kinoshita, *J. Am. Chem. Soc.*, **113**, 1856 (1991).

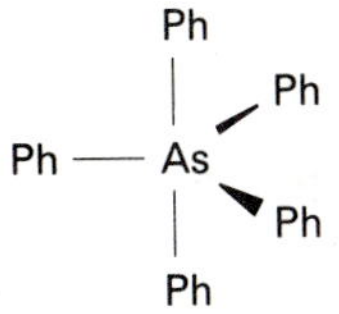

29

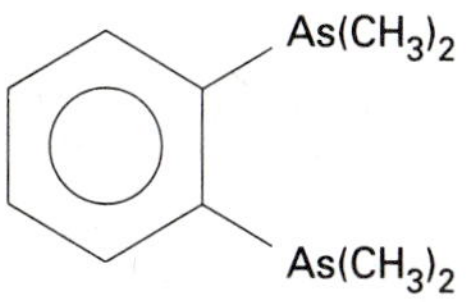

30 $C_6H_4(As(CH_3)_2)_2$, diars

The double replacement reaction of a Grignard or organolithium reagent with EX_3 ($E = As$, Sb, Bi; $X = Cl$, Br) provides a convenient route to the trialkyl and triaryl compounds:

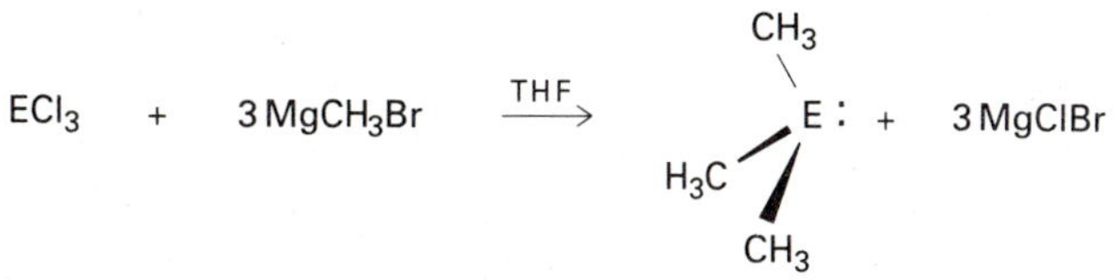

The $E(CH_3)_3$ molecules and their analogs are trigonal pyramidal, which is as expected in view of the lone pair on the central atom. Alkylarsanes, such as trimethylarsane are volatile, malodorous, and toxic. The arylarsanes are more air-stable and less volatile.

The aryl- and alkyl-arsanes, such as trimethylarsane (**6**), are encountered as ligands in *d*-metal complexes. The order of affinity of these soft Lewis bases for a *d*-metal ion generally follows the order $PR_3 > AsR_3 > SbR_3 \gg BiR_3$. Many complexes of alkyl- and aryl-arsanes have been prepared but fewer stibane complexes are known. A useful ligand, for example, is the bidentate compound known as diars (**30**). Because of their soft-donor character, many aryl- and alkyl-arsane complexes of the soft species Rh(I), Ir(I), Pd(II), and Pt(II) have been prepared. However, hardness criteria are only approximate, so we should not be surprised to see phosphane and arsane complexes of some metals in higher oxidation states. For example, the unusual palladium(IV) oxidation state is stabilized by the diars ligand (**31**).

The synthesis of diars provides a good illustration of some common reactions in the synthesis of organoarsenic compounds. The starting material is $(CH_3)_2AsI$. This compound is not conveniently prepared by double replacement reaction between AsI_3 and a Grignard or similar carbanion reagent because that reaction is not selective to partial substitution on the As atom when the organic group is compact. Instead, the compound can be prepared by the direct action of a haloalkane, CH_3I, on the metallic allotrope of arsenic:

$$4As + 6CH_3I \rightarrow 3(CH_3)_2AsI + AsI_3$$

In the next step, the action of sodium on $(CH_3)_2AsI$ is used to produce $[(CH_3)_2As]^-$:

$$(CH_3)_2AsI + 2Na \rightarrow Na[(CH_3)_2As] + NaI$$

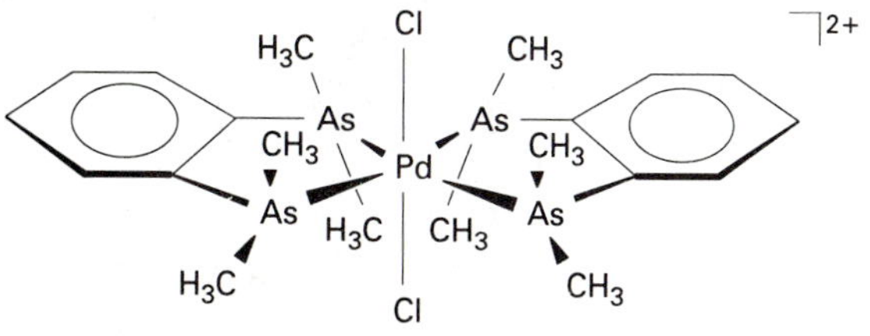

31

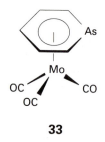

32 Arsabenzene

33

The resulting powerful nucleophile $[(CH_3)_2As]^-$ is then employed to displace chlorine from *ortho*-dichlorobenzene:

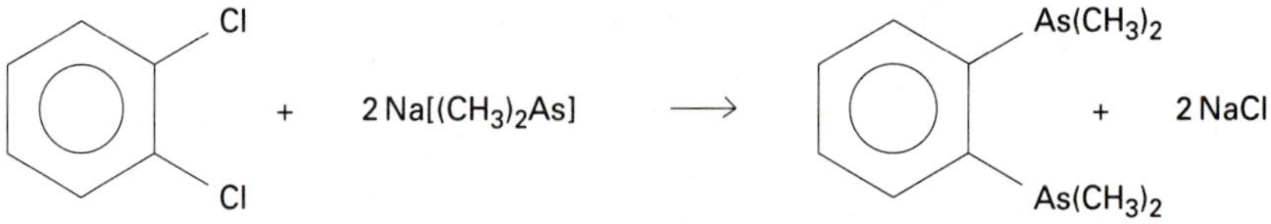

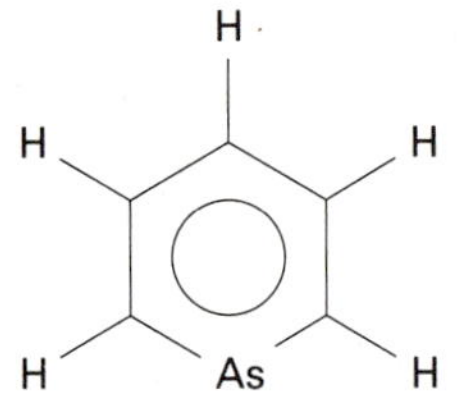

Another interesting compound of As(III), arsabenzene (**32**), is an analog of pyridine. It lacks the σ-donor ability of pyridine but forms π-complexes with *d*-block metals (**33**) analogous to those formed by benzene.[13]

Oxidation state +5

The trialkylarsanes act as nucleophiles toward haloalkanes to produce tetraalkylarsonium salts, which contain As(V):

$$As(CH_3)_3 + (CH_3)Br \rightarrow [As(CH_3)_4]Br$$

This type of reaction cannot be used for the preparation of the tetraphenylarsonium ion, $[AsPh_4]^+$, because triphenylarsane is a much weaker nucleophile than trimethylarsane. Instead, a suitable synthetic reaction is:

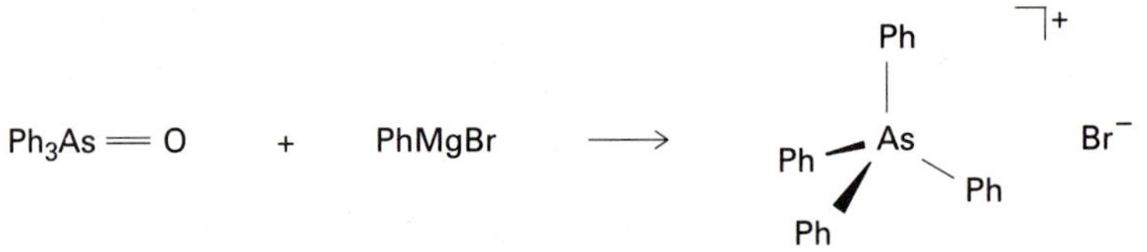

$$Ph_3As \!=\! O \;+\; PhMgBr \longrightarrow [AsPh_4]^+ \, Br^-$$

This reaction may look unfamiliar, but it is simply a double replacement in which the Ph^- anion replaces the formal O^{2-} ion attached to the As atom, resulting in a compound in which the arsenic retains its $+5$ oxidation state. The formation of the highly exoergic compound MgO also contributes to the Gibbs free energy of this reaction, and its formation drives the reaction forward.

The tetraphenylarsonium, tetraalkylammonium, and tetraphenylphosphonium cations are used in synthetic inorganic chemistry as bulky cations to stabilize bulky anions. The tetraphenylarsonium ion is also a starting material for the preparation of other As(V) organometallic compounds. For instance, the action of phenyllithium on a tetraphenylarsonium salt produces pentaphenylarsenic (**28**), a compound (formally) of As(V):

$$[AsPh_4]Br + LiPh \rightarrow AsPh_5 + LiBr$$

Pentaphenylarsenic is trigonal bipyramidal, as expected from VSEPR considerations. We have seen (Section 3.1) that a square-pyramidal

---

[13] A. J. Ashe, The Group 5 heterobenzenes, *Acc. Chem. Res.*, **11**, 153 (1978).

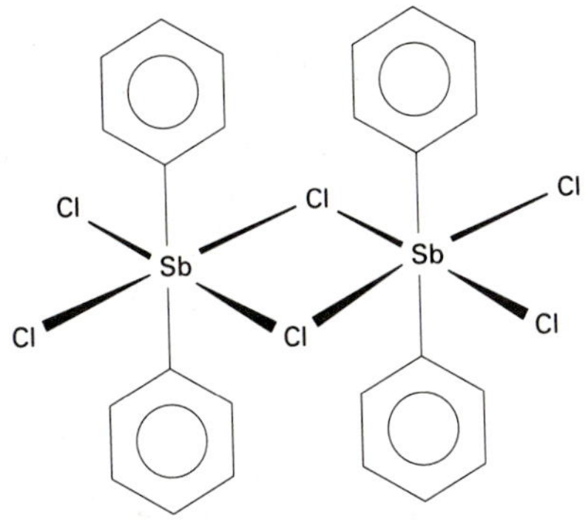

structure is often close in energy to the trigonal-bipyramidal structure, and the antimony analog, $SbPh_5$, is in fact square pyramidal. A similar reaction under carefully controlled conditions yields the unstable compound $As(CH_3)_5$.

Mixed-substituent compounds are common for the $+5$ oxidation state and we have already seen one example in $OAsR_3$; similarly, halides such as $Sb_2Ph_4Cl_6$ are also known. Molecules of the latter compound have an edge-shared structure with octahedral coordination around each Sb atom (**34**).

34

Example 10.6: *Correlating oxidation numbers and stabilities*
Describe the stability of the alkyl compounds of the adjacent elements germanium and arsenic with their group oxidation number and the group oxidation number minus 2.

Answer. The most common oxidation number for germanium in organogermanium compounds is $+4$, the group oxidation number. Only a few organogermanium(II) compounds are known. In contrast, As(V) compounds such as $As(CH_3)_5$ are unstable with respect to organoarsenic(III) compounds, such as $As(CH_3)_3$. One way of explaining this trend is that on going to the right along a period, the elements become more electronegative and it becomes progressively more difficult to bring out their high oxidation states. This difficulty is particularly strong with a substituent, such as CH_3, with a low electronegativity. These trends in the stability of oxidation states follow the trends for simple inorganic compounds of germanium and arsenic.

Exercise E10.6. Compare formulas of the most stable hydrogen compounds of germanium and arsenic (Chapter 9) with those of their methyl compounds. Can the similarities and differences be explained in terms of the relative electronegativities of C and H?

10.15 Catenated and multiply bonded compounds

Tetramethyldiarsane, $(CH_3)_2AsAs(CH_3)_2$, a catenated compound, was one of the first organometallic compounds to be made. One convenient synthesis is the reaction of $As(CH_3)_2Br$ with zinc:

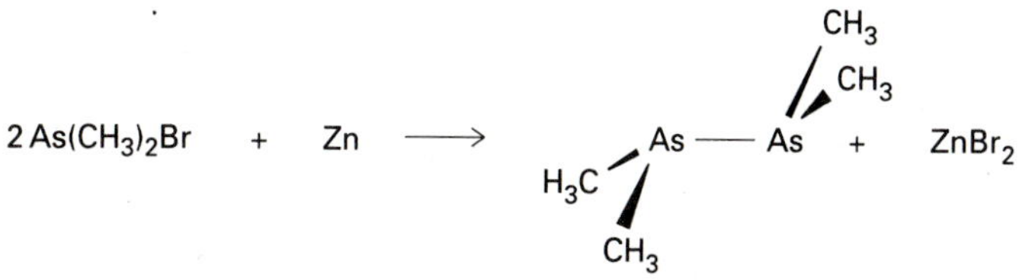

$$2\,As(CH_3)_2Br \;+\; Zn \;\longrightarrow\; (CH_3)_2As\!-\!As(CH_3)_2 \;+\; ZnBr_2$$

Cyclic arsanes are also known, including $(PhAs)_6$. The synthesis of $(PhAs)_6$ is accomplished by abstracting iodine from phenylarsenic diiodide:

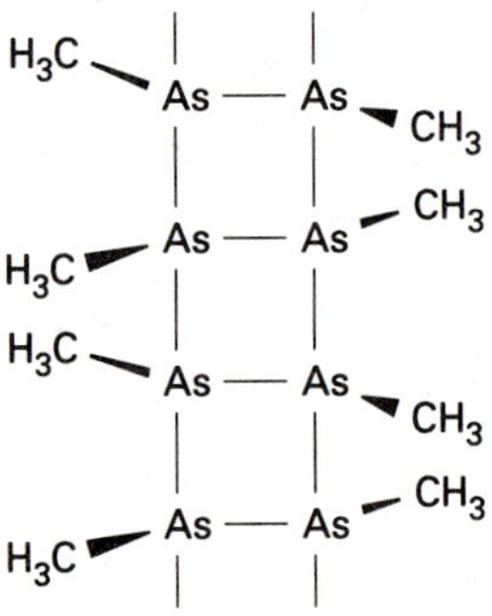

35

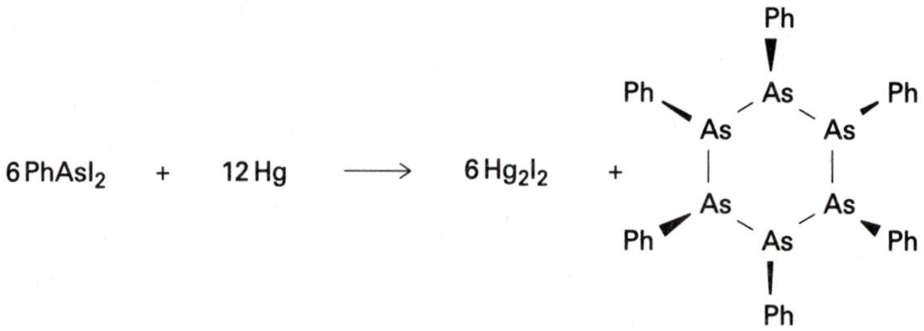

36 (E,E′) = (P,As),(P,Sb),(As,P)

$$6\,PhAsI_2 \quad + \quad 12\,Hg \quad \longrightarrow \quad 6\,Hg_2I_2 \quad +$$

There are also many other arsenic–arsenic bonded compounds. Among them is one containing a ladder structure (**35**), which is obtained by reducing CH_3AsI_2 with $Sb(C_4H_9)_3$:

$$nCH_3AsI_2 + nSb(C_4H_9)_3 \rightarrow [AsCH_3]_n + nSb(C_4H_9)_3I_2$$

As with the disilenes, it has been found recently that bulky substituents make it possible to prevent the polymerization and cyclization of E—E. This strategy leads to formal double bonded systems of formula RE=ER, where E may be P, As, or Sb, including the series of compounds shown (**36**) that contain P=As and P=Sb bonds.

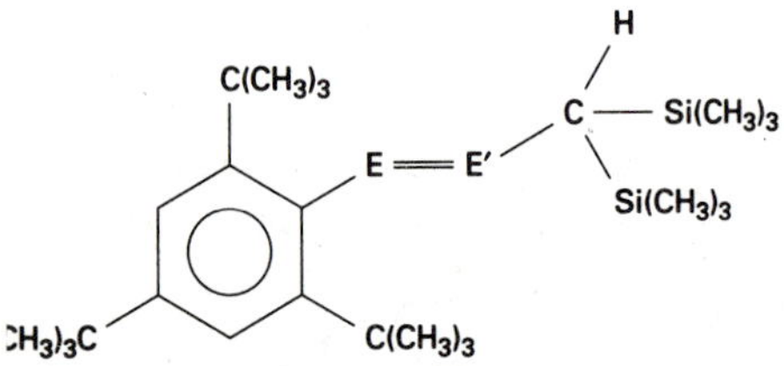

FURTHER READING

Ch. Elschenbroich and A. Salzer, *Organometallics*. VCH, Weinheim and New York (1992). This concise and well illustrated book provides an excellent introduction to the chemistry of both main-group and *d*-metal organometallic compounds.

G. Wilkinson, F. G. A. Stone, and E. W. Abel (ed.), *Comprehensive organometallic chemistry*. Pergamon, Oxford (1982). Volumes 1 and 2 are devoted to the *s* and *p* blocks. Chapter 1, Main group structure and bonding relationships (Vol. 1, p. 1) by M. E. O'Neill and K. Wade, provides a good introduction, and later chapters provide thorough coverage. A new edition is in preparation.

B. J. Aylett, *Organometallic compounds*. Chapman and Hall, London (1979). Group 14/IV and 15/V organometallic compounds.

J. J. Eisch, *Organometallic syntheses*, Vol. 2. Academic Press, New York (1981). The volume provides a thorough description of methods and detailed procedures for the synthesis of important *p*-block organoelement compounds.

KEY POINTS

1. Classification
The organometallic compounds of the *s*-block elements range from molecular or polymeric lithium, beryllium, and magnesium compounds, which have highly polar bonds $M^{\delta+}C^{\delta-}$, to the even more ionic compounds of the heavier metals. The more covalent compounds of the *p* block are classified as electron poor through Group 13/III, electron precise for Group 14/IV, and electron rich for many compounds in Group 15/V.

2. Structures
The highly useful lithium organometallic compounds form cluster compounds in non-coordinating or weakly coordinating solvents.

3. Stability
The thermodynamic stability of organometallic compounds (as revealed by Gibbs free energies of formation) decreases down a group.

4. Synthesis

The primary routes to organometallic compounds are:

(1) Active metal plus an organohalide: $M + RX$ (X = halogen);

(2) Transmetallation: $M + M'R$ (where M is more electropositive than M');

(3) Double replacement (metathesis): $MR + EX$ (where M is more electropositive than E);

(4) Hydrometallation: $MH + RHC = CH_2$ (where MH is a molecular hydride, often borane or silane).

5. s-Block reactivity

The *s*-block organometallic compounds are highly reactive with proton sources and are good carbanion sources in metathesis reactions with the halides of more electronegative elements. Organolithium and Grignard reagents are the most frequently used carbanion reagents in synthesis.

6. Group 13/III structures

BR_3 and GaR_3 compounds are unassociated trigonal and planar. Triorganoaluminum compounds are usually associated though alkyl or aryl bridges.

7. Group 13/III reactivity

Carbanion character follows the order $AlR_3 > GaR_3 \gg BR_3$; accordingly, trialkylaluminum compounds are frequently used in double replacement reactions with the halides of more electronegative elements.

8. Group 14/IV structures

Silicon, germanium, and lead form tetrahedral MR_4 compounds. With bulky groups, $R_2E = ER_2$ (E = Si, Ge, or Sn) is formed.

9. Group 14/IV reactivity

Associative substitution reactions are characteristic of organometallic compounds such as SiR_3Br and the greater access to the larger Si atom accounts for the greater reactivity of silicon compounds than their carbon analogs.

10. Group 15/V structures

The compounds AsR_3, SbR_3, and BiR_3 of Group 15/V are trigonal pyramidal, with a stereochemically active lone pair. The less common ER_5 organometallic compounds of the group are either trigonal bipyramidal or square pyramidal.

11. Group 15/V reactivity

The compounds AsR_3, and to a lesser extent SbR_3, form complexes with many soft *d*-block metals. These formal $+3$ oxidation state compounds can be converted to formal $+5$ species such as $(C_6H_5)_3AsO$ and $[(C_6H_5)_4As]^+$.

EXERCISES

10.1 Explain why certain of the following compounds qualify as organometallic compounds whereas others do not: (a) $B(CH_3)_3$, (b) $B(OCH_3)_3$, (c) $(NaCH_3)_4$, (d) $SiCl_3(CH_3)$, (e) $N(CH_3)_3$, (f) sodium acetate, (g) $Na[B(C_6H_5)_4]$.

10.2 Preferably without consulting reference material, construct the periodic table for the *s*- and *p*-block elements and the Group 12 elements. Head each group with the formula of a representative methyl compound and indicate (a) the positions of the salt-like methyl compounds, (b) the positions of the electron-deficient, electron-precise, and electron-rich methyl compounds, (c) trends in $\Delta H_f^{\ominus}$ in the *p* block.

10.3 Write formulas for each of the following compounds. For nonionic compounds give alternative names as derivatives of their hydrogen compounds if they are named as organometallic compounds of the element, and vice versa:

(a) trimethylbismuth, (b) tetraphenylsilane, (c) tetraphenylarsonium bromide, (d) potassium tetraphenylborate.

10.4 Name each of the following compounds and classify them: (a) $SiH(C_2H_5)_3$, (b) $BCl(C_6H_5)_2$, (c) $Al_2Cl_2(C_6H_5)_4$, (d) $Li_4(C_2H_5)_4$, (e) $RbCH_3$.

10.5 Sketch the structures of: (a) methyllithium, (b) trimethylboron, (c) hexamethyldialuminum, (d) tetramethylsilane, (e) trimethylarsane, and (f) tetraphenylarsonium.

10.6 Describe the tendency toward association through methyl bridges for the trimethyl compounds of boron, aluminum, gallium, and indium. Give a plausible explanation for the difference between boron and aluminum.

10.7 For each of the following compounds, indicate those that may serve as (1) a good carbanion nucleophile reagent,

(2) a mild Lewis acid, (3) a mild Lewis base at the central atom, (4) a strong reducing agent. (A compound may have more than one of these properties.) (a) $Li_4(CH_3)_4$, (b) $Zn(CH_3)_2$, (c) $(CH_3)MgBr$, (d) $B(CH_3)_3$, (e) $Al_2(CH_3)_6$, (f) $Si(CH_3)_4$, (g) $As(CH_3)_3$.

10.8 For an appropriate compound from Exercise 10.7, give balanced chemical equations for: (a) the reactions of one of the carbanion reagents with $AsCl_3$ and $SiPh_2Cl_2$, (b) the reaction of a Lewis acid with NH_3, (c) the reaction of a Lewis base with the Lewis acid $[HgCH_3][BF_4]$.

10.9 Give examples of the synthesis of organometallic compounds by each of the following reaction types and describe the factors that favor reaction in each case: (a) reaction of a metal with an organic halide, (b) transmetallation, (c) double replacement.

10.10 Determine which compound in each pair is likely to be the stronger reducing agent and explain the physical basis for your answer: (a) $Na[C_{10}H_8]$ and $Na[C_{14}H_{10}]$, (b) $Na[C_{10}H_8]$ and $Na_2[C_{10}H_8]$ (where $C_{10}H_8$ is naphthalene and $C_{14}H_{10}$ is anthracene).

10.11 Determine the likely reaction type, write a reasonable balanced chemical equation (or NR for no reaction), and explain the systematics of the reaction type that led you to decide whether or not a reaction is likely.
   (a) Calcium with dimethylmercury.
   (b) Mercury with diethylzinc.
   (c) Methyllithium with triphenylchlorosilane in ether.
   (d) Tetramethylsilane with zinc chloride in ether.
   (e) Trimethylsilane with ethene in a solution of chloroplatinic acid in 2-propanol.

10.12 Give balanced chemical equations that illustrate: (a) direct synthesis of $SiCl_2(CH_3)_2$; (b) redistribution of $SiCl_2(CH_3)_2$.

10.13 Using silicon and a chloromethane as the primary starting materials, give equations and conditions for the synthesis of a poly(dimethylsiloxane).

10.14 Disregarding compounds with metal–metal bonds, describe the trends in the oxidation states for the organometallic compounds of Groups 13/III through 15/V.

10.15 For the simple organometallic compounds, briefly describe the periodic trends in (a) metal–carbon bond enthalpies, (b) Lewis acidity, (c) Lewis basicity for Groups 1, 2, 12, 13/III, and 14/IV.

10.16 Summarize the trend in each process listed below, giving your reasoning:
   (a) The relative ease of pyrolysis of $Si(CH_3)_4$ and $Pb(CH_3)_4$ at 300°C.
   (b) The relative Lewis acidity of $Li_4(CH_3)_4$, $B(CH_3)_3$, $Si(CH_3)_4$, and $Si(CH_3)Cl_3$.
   (c) The relative Lewis basicity of $Si(CH_3)_4$ and $As(CH_3)_3$.
   (d) The tendency of $Li_4(CH_3)_4$ and $Hg(CH_3)_2$ to displace halide from $GeCl_4$.

10.17 Taking into account sensitivity to oxygen and moisture and the volatility of the compound, indicate the general techniques (vacuum line, inert atmosphere, Schlenk ware, or open flasks) that are most appropriate for handling the following organometallic compounds: (a) $Li_4(CH_3)_4$ in ether, (b) trimethylboron, (c) triisobutylaluminum, (d) $AsPh_3$ (solid), (e) $(CH_3)_3SiOSi(CH_3)_3$ (liquid).

10.18 The 'saturated' cyclic compound $Si_6(CH_3)_{12}$, which has a near-UV absorption band, can be reduced by sodium to produce a cyclic radical anion $[Si_6(CH_3)_{12}]^-$. What types of molecular orbitals are thought to be involved in these processes?

10.19 Give equations and conditions for the synthesis of $R_2Si=SiR_2$ from R_2SiCl_2 and indicate the type of R group that is necessary to yield a stable product.

PROBLEMS

10.1 The bulk price per mole of trimethylaluminum is an order of magnitude greater than that of triethylaluminum. Describe the methods of synthesis available for these two compounds and rationalize the price difference.

10.2 Correct any errors and explain the nature of the errors in the following statements. (a) The methyl compounds of lithium, zinc, and germanium are associated through multicenter M—C—M bonds because they each contain fewer methyl groups than valence orbitals on the metal. (b) Substitution reactions of organometallic compounds of silicon such as chloroethylmethylphenylsilane generally occur by an associative mechanism with complete loss of stereochemistry. (c) The tendency toward radical reaction by ER_4 compounds increases on going down the group

from $E = Si$ to Pb, which correlates with decreasing E–C bond enthalpies.

10.3 The compound $(C_6H_5)_3Pb—Pb(C_6H_5)_3$ survives up to about 300°C. Write plausible reactions for this compound with (a) sodium in liquid ammonia, (b) HBr, and (c) $[Pt(PPh_3)_4]$ (this compound readily loses up to two triphenylphosphane ligands). Give your reasoning in each case.

10.4 Based on general trends in atomic energy levels, explain the probable order of the π–π^* separation in $R_2Si=SiR_2$ and $R_2Ge=GeR_2$. Give specific examples of how this energy order might be reflected in the reactions of digermene in comparison with disilenes.

10.5 Although tetraalkyltin compounds are not Lewis acids, the haloalkylstannanes are Lewis acids. C. Yoder and coworkers have determined thermodynamic parameters of complex formation of triphenylphosphane oxide with a series of halotrialkystannanes (*Organometallics*, **5**, 118 (1986)). Summarize the trends in enthalpies of complex formation and discuss reasonable explanations of these trends.

10.6 The synthesis of the first stable germaethene was reported by C. Couret, J. Escudie, J. Satge, and M. Lazraq, *J. Am. Chem. Soc.*, **109**, 441 (1987). Summarize the reactions involved in this synthesis and speculate on why each reaction is favorable. Describe the interaction of Lewis bases with this germaethene and propose reasons for the interaction of the germaethene with a Lewis base and lack of similar interaction for a simple alkene.

10.7 Given $Ge(CH_3)_3Cl$, $H_2C=CH(C_6H_5)$, and other reagents of your choice, and drawing on analogies with the chemical properties of silicon, give balanced chemical equations and conditions for the synthesis of $(CH_3)_3GeCH_2CH_2(C_6H_5)$.

11

The boron and carbon groups

This chapter and the following two introduce the general chemical properties of the p-block elements with emphasis on the nonmetals. The p block is a very rich region of the periodic table, for its members show a much greater variation in properties than those of the s and d blocks, and range from metals such as aluminum to the highly electronegative nonmetals such as fluorine. A single viewpoint cannot cope adequately with this great diversity, and we shall see that we need to adjust our perspective as we travel across the block. For instance, as we travel toward the right, the number of oxidation states available to the elements increases, so redox properties become more important. This character is in contrast to that shown by the elements on the left of the block (boron, carbon, and silicon) for which redox reactions are less important. However, the latter elements make up for lack of richness in their redox properties by their ability to bind to themselves to form chains, rings, and clusters. Compounds of the elements with oxygen are important throughout the block, and will frequently recur in the discussion.

The elements of Groups 13/III and 14/IV have interesting and diverse physical and chemical properties as well as considerable importance in industry and nature. Carbon, of course, plays a central role in organic chemistry, but it also forms many binary and organometallic compounds (as we saw in Chapter 10 and will see again in Chapter 16). In combination with oxygen, carbon's congener silicon is a dominant component of the Earth's crust, just as carbon in combination with hydrogen and oxygen is dominant in the biosphere. The other elements of these two groups are vital to modern high technology, particularly as semiconductors.

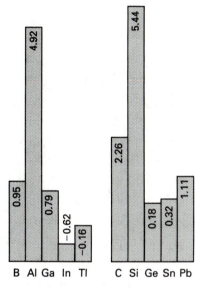

11.1 Abundances of the Group 13/III and Group 14/IV elements. The numbers are logarithms of the abundance in parts per million.

THE ELEMENTS

The elements of the two groups show a wide variation in abundance in crustal rocks, the oceans, and the atmosphere. Carbon, aluminum, and silicon are all abundant (Fig. 11.1), but the low cosmic and terrestrial abundance of boron, like that of lithium and beryllium, reflects how these light elements are sidestepped in nucleosynthesis (Section 1.1). The low abundance of heavier members of both groups is in keeping with the progressive decrease in nuclear stability of the elements that follow iron. As Fig. 11.1 indicates, the carbon-group elements are generally more abundant than the members of the boron group; they are also more abundant than their neighbours in Group 15/V. This difference stems from the greater stability of nuclei that have even numbers of protons (and therefore even atomic number) compared with those having odd numbers.

A large variation in chemical and physical properties is observed on descending Group 13/III and Group 14/IV. The lightest members of each group are nonmetals and the heaviest are metals. Physical similarities are particularly pronounced between boron and its diagonal neighbor silicon. In compounds, boron and silicon are chemically hard, and in their elemental form they are mechanically hard, semiconducting solids. The occurrence of two or more significantly different polymorphs is a common characteristic of *p*-block elements, and is well illustrated by elemental boron and carbon (as we describe below). Thallium and lead are the only members of the two groups that crystallize in the close-packed structures typical of many metals.

The chemical properties of boron, carbon, silicon, and germanium are distinctly those of nonmetals. Their electronegativities are similar to hydrogen's and, as we saw in Chapters 9 and 10, they form many covalent hydrogen and alkyl compounds. The light elements, boron, aluminum, carbon, and silicon are strong **oxophiles** and **fluorophiles**, in the sense that they have high affinities for oxygen and fluorine, respectively. Their oxophilic character is evident in the existence of an extensive series of oxoanions, the borates, aluminates, carbonates, and silicates. The similar oxophilicity and fluorophilicity of boron and silicon is an example of chemical similarity that is observed for diagonal neighbors in the periodic table. In contrast, the heavy elements thallium and lead have higher affinities for soft anions, such as I^- and S^{2-} ions, than for hard anions. Thallium and lead are therefore classified as chemically soft.

Table 11.1 shows that, for most of the members of the two groups, the dominant oxidation number is the group oxidation number, which is $+3$ for Group 13/III and $+4$ for Group 14/IV. The major exceptions are thallium and lead, for which the most common oxidation number is 2 less than the group maximum, being $+1$ for thallium and $+2$ for lead. This relative stability of the low oxidation state is an example of the inert pair effect (Sections 8.16 and 8.17).

Table 11.1 Properties of the boron and carbon group elements

Element	(I/kJ mol^{-1})	χ^*	r_{cov}/Å†	r_{ion}/ţ	Appearance and properties	Common oxidation numbers§
Group 13/III						
B	899	2.04	0.85		Dark Semiconductor	**3**
Al	578	1.61	1.43	0.54	Metal	**3**
Ga	579	1.81	1.53	0.62	Metal m.p. 30°C	1,**3**
In	558	1.78	1.67	0.80	Soft metal	1,**3**
Tl	589	2.04	1.71	0.89	Soft metal	**1**,3
Group 14/IV						
C	1086	2.55	0.77		Hard insulator (diamond) Semimetal (graphite)	**4**
Si	786	1.90	1.17	0.40	Hard semiconductor	**4**
Ge	760	2.01	1.22	0.53	Metal	2,**4**
Sn	708	1.96	1.40	0.69	Metal	**2,4**
Pb	715	2.33	1.75	0.92	Soft metal	**2,4**

*Pauling values recalculated by A. L. Allred, *J. Inorg. Nucl. Chem.*, **17**, 215 (1961).
†Covalent radii from M. C. Ball and A. H. Norbury, *Physical data for inorganic chemists.* Longman, London (1974).
‡Ionic radii from R. D. Shannon, *Acta. Crystallogr.*, **A32**, 751 (1975) for elements with C.N. = 6 and in their maximum group oxidation states.
§Most common oxidation number in bold type.

THE BORON GROUP (GROUP 13/III)

The boron group shows considerable structural diversity. Boron itself, for instance, exists in several hard and refractory polymorphs. The three solid phases for which crystal structures are available contain the icosahedral (20-faced) B_{12} unit as a building block (Fig. 11.2). This icosahedral unit is a recurring motif in boron chemistry and we shall meet it again in the structures of metal borides and boron hydrides. All boron's congeners are metals, and their chemical properties are described in Chapter 8. Only gallium, which has one nearest neighbor in the solid (and hence resembles the structure of solid iodine), is structurally unusual.

11.1 Occurrence and recovery

The light members of Group 13/III are found in nature in combination with oxygen. The primary sources for boron are hydrated sodium borates, such as the mineral borax ($Na_2B_4O_5(OH)_4 \cdot 8H_2O$). Bauxite, the primary ore of aluminum, consists of various hydrates of aluminum oxide, such as $Al_2O_3 \cdot H_2O$. The difficulty of reducing aluminum by carbon is evident from an Ellingham diagram (Fig. 7.3). In particular, the relative positions of the lines for the oxides of these elements indicate that they have a more negative free energy of formation than their heavier congeners. Because of its abundance and wide use as a

(a)

(b)

11.2 A view of the B_{12} icosahedron in α-rhombohedral boron (a) along and (b) perpendicular to the threefold axis of the crystal. The individual icosahedra are linked by 3c,2e bonds.

B₁₂ icosahedron (20)

Table 11.2 Mineral sources and methods of recovery of the nonmetals in Groups 13/III and 14/IV*

Element	Natural source	Recovery
Boron	Borax	Reduction by magnesium
Carbon	Coal, hydrocarbons, graphite	Pyrolysis
Silicon	Silica	Reduction by carbon $SiO_2 + 2C \xrightarrow{\Delta} Si + 2CO$
Germanium	By-product of zinc refining	Reduction of GeO_2 by hydrogen

$$GeO_2 + 4H_2 \xrightarrow{\Delta} Ge + 2H_2O$$

*Information on the recovery of *p*-block metals is available in Table 8.13.

structural metal, the recovery of elemental aluminum by the Hall process (Section 7.1) is carried out on by far the largest scale of any element in its group. An outline of the occurrence and recovery of the elements is given in Table 11.2.

11.2 Compounds of boron with the electronegative elements

In this section we introduce the boron halides, which are very useful reagents and Lewis acid catalysts, and the numerous boron oxides and oxoanions. We also introduce the structures and properties of compounds of boron with nitrogen. The latter are intriguing largely because the BN unit is isoelectronic with the CC unit, and hence compounds that contain it are structural analogs of the hydrocarbons. Chemically, though, the two classes are quite different.

Halides

All the boron trihalides except BI_3 may be prepared by direct reaction of the halogens with elemental boron. However, the preferred method for BF_3 is the reaction of B_2O_3 with CaF_2 in H_2SO_4, which is driven in part by the favorable reaction of the strong acid H_2SO_4 with oxides and the affinity of the hard boron atom for fluorine:

$$B_2O_3(s) + 3CaF_2(s) + 6H_2SO_4(l) \rightarrow$$
$$2BF_3(g) + 3[H_3O][HSO_4](soln) + 3CaSO_4(s)$$

Boron trihalides consist of trigonal planar BX_3 molecules. Unlike the halides of the other elements in the group, they are monomeric in the gas, liquid, and solid states. However, halogen exchange does occur and it may well go through the formation and dissociation of a bridged dimer (**1**) like those characteristic of the gaseous aluminum halides. Boron trifluoride and boron trichloride are gases, the tribromide is a volatile liquid, and the triiodide is a solid (Table 11.3). This trend in volatility is consistent with the increase in strength of dispersion forces with the number of electrons in the molecules.

The trihalides are Lewis acids. We have already drawn attention to the order of their strengths in this role, which is $BF_3 < BCl_3 \leq BBr_3$ and contrary to the order of electronegativity of the attached halogens (Section 5.10). This trend is thought to stem from greater X—B π bonding for the lighter, smaller halogens giving rise to the occupation

1 Br_3BBCl_3

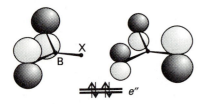

11.3 The bonding π orbitals of boron trihalide are largely localized on the electronegative halogen atoms, but overlap with a p orbital of boron is significant in the a_1'' orbital.

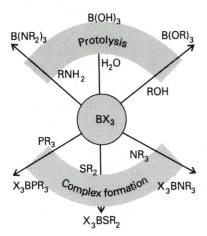

11.4 The reactions of boron–halogen compounds (X = halogen).

Table 11.3 Properties representative of the boron trihalides

	Halide	m.p./°C	b.p./°C	$\Delta G_f^{\ominus}$/(kJ mol^{-1})*
BX$_3$, X =	F	-127	-100	-1112
	Cl	-107	12	-339
	Br	-46	91	-232
	I	49	210	$+21$

*For the formation of the gaseous trihalide at 25 °C.

of the p orbital on the B atom by electrons supplied by the halogen atoms (Fig. 11.3). All the boron trihalides form simple Lewis complexes with suitable bases, as in the reaction

$$BF_3(g) + :NH_3(g) \rightarrow F_3B\!-\!NH_3(s)$$

However, the chlorides, bromides, and iodides are susceptible to protolysis by mild proton sources such as water, alcohols, and even amines. As shown in the chart in Fig. 11.4, this reaction, along with double displacement reactions, is very useful in preparative chemistry. An example is the hydrolysis of BCl$_3$:

$$BCl_3(g) + 3H_2O(l) \rightarrow B(OH)_3(aq) + 3HCl(aq)$$

It is probable that a first step in this reaction, which occurs rapidly, is the formation of the complex Cl$_3$B$-$OH$_2$, which then eliminates HCl and reacts further with water.

Example 11.1: *Predicting the products of reactions of the boron trihalides*
Predict the probable products of the following reactions, and write the balanced chemical equations.

(a) BF$_3$ and excess NaF in acidic aqueous solution.

(b) BCl$_3$ and excess NaCl in acidic aqueous solution.

(c) BBr$_3$ and excess NH(CH$_3$)$_2$ in a hydrocarbon solvent.

Answer. (a) The F$^-$ ion is a hard and fairly strong base; BF$_3$ is a hard and strong Lewis acid with a high affinity for the F$^-$ ion. Hence, the reaction should result in the formation of a complex:

$$BF_3(g) + F^-(aq) \rightarrow [BF_4]^-(aq)$$

Excess F$^-$ and acid prevent the formation of hydrolysis products such as [BF$_3$OH]$^-$, which are formed at high pH.

(b) Unlike B$-$F bonds, which are only mildly susceptible to hydrolysis, the other boron–halogen bonds are vigorously hydrolyzed by water. We can anticipate that BCl$_3$ will undergo hydrolysis rather than coordinate to aqueous Cl$^-$:

$$BCl_3(g) + 3H_2O(l) \rightarrow B(OH)_3(aq) + 3HCl(aq)$$

(c) Boron tribromide will undergo protolysis with formation of a B$-$N bond:

$$BBr_3(g) + 3NH(CH_3)_2 \rightarrow B(N(CH_3)_2)_3 + 3HBr(g)$$

The HBr will protonate any excess dimethylamine.

Exercise E11.1. Write and justify balanced equations for plausible reactions between (a) BCl_3 and ethanol, (b) BCl_3 and pyridine in hydrocarbon solution, (c) BBr_3 and $F_3BN(CH_3)_3$.

The tetrafluoroborate anion, BF_4^-, which is mentioned in Example 11.1, is a very weak Lewis base. It is used in preparative chemistry when a relatively large noncoordinating anion is needed. The other tetrahaloborate anions, BCl_4^- and BBr_4^-, can be prepared in nonaqueous solvents. However, because of the ease with which B—Cl and B—Br bonds undergo solvolysis, they are stable in neither water nor alcohols.

Boron halides containing B—B bonds have been prepared. The best known of these compounds are those with the formula B_2X_4, with X = F, Cl, and Br, and the tetrahedral cluster compound B_4Cl_4. One route to B_2Cl_4 is to pass an electric discharge through BCl_3 gas in the presence of a Cl atom scavenger, such as mercury vapor. Spectroscopic data indicate that BCl is produced by electron impact on BCl_3:

$$BCl_3(g) \xrightarrow{\text{electron impact}} BCl(g) + 2Cl(g)$$

The Cl atoms are scavenged by mercury vapor and removed as $Hg_2Cl_2(s)$, and the BCl fragment is thought to combine with BCl_3 to yield B_2Cl_4.

Double-displacement reactions can be used to make B_2X_4 derivatives from B_2Cl_4. The thermal stability of these derivatives increases with increasing tendency of the X group to form a π bond with B:

$$B_2Cl_4 < B_2F_4 < B_2(OR)_4 \ll B_2(NR_2)_4$$

For a long time it was thought that X groups with lone pairs were essential for the existence of B_2X_4 compounds, but recently diboron compounds with alkyl groups have been prepared. When the alkyl groups are bulky, compounds that survive at room temperature are obtained, such as $B_2(^tBu)_4$.

Diboron tetrachloride is a highly reactive volatile molecular liquid. The B_2Cl_4 molecules are planar (2) in the solid state but staggered (3) in the gas. This difference suggests that rotation about the B—B bond is quite easy, as is expected for a B—B single bond. One of the interesting reactions of B_2Cl_4 is its addition across a C=C double bond:

$$B_2Cl_4 + C_2H_4 \xrightarrow{\text{low temperature}} Cl_2BCH_2CH_2BCl_2$$

A secondary product in the synthesis of B_2Cl_4 is B_4Cl_4, a pale yellow solid composed of molecules with the four B atoms forming a tetrahedron. This compound, like B_2Cl_4, does not have a formula

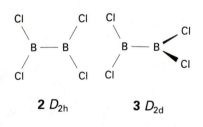

2 D_{2h} **3** D_{2d}

that conforms to those of the boranes discussed below. Part of this difference may lie in the tendency of halogens to donate electrons into B—X π bonds.

Oxides and oxo compounds

Boric acid $B(OH)_3$ is a very weak Brønsted acid in aqueous solution. However, the equilibria are more complicated than the simple Brønsted proton transfer reactions characteristic of the later *p*-block oxoacids. Boric acid is in fact primarily a *Lewis* acid, and the complex it forms with H_2O is the actual source of protons:

$$B(OH)_3(aq) + 2H_2O(l) \rightleftharpoons H_3O^+(aq) + [B(OH)_4]^-(aq) \quad pK_a = 9.2$$

As is typical for many of the lighter elements in the two groups, there is a tendency for the anion to polymerize by condensation with the loss of H_2O. Thus, in concentrated neutral or basic solution, equilibria such as

$$3B(OH)_3(aq) \rightleftharpoons [B_3O_3(OH)_4]^-(aq) + H^+(aq) + 2H_2O(l)$$
$$K = 1.4 \times 10^{-7}$$

occur to yield polynuclear anions (4).

The reaction of boric acid with an alcohol in the presence of sulfuric acid leads to the formation of simple **borate esters**, which are compounds of the form $B(OR)_3$:

$$B(OH)_3 + 3CH_3OH \xrightarrow{H_2SO_4} B(OCH_3)_3 + 3H_2O$$

Borate esters are much weaker Lewis acids than the boron trihalides, presumably because the O atom acts as an intramolecular π donor, like the F atom in BF_3, and donates electron density to the *p* orbital of the B atom. Hence, judging from Lewis acidity, an O atom is more effective than an F atom as a π donor toward boron. 1,2-Diols have a particularly strong tendency to form borate esters on account of the chelate effect, for they produce a cyclic borate ester (as in **5**).

As with silicates and aluminates, there are many polynuclear borates, and both cyclic and chain species are known. An example is the cyclic polyborate anion $[B_3O_6]^{3-}$ (**6**), the conjugate base of (**4**). A notable feature of borate formation is the possibility of both three-coordinate B atoms, as in (**6**), and four-coordinate B atoms, as in $[B(OH)_4]^-$ and for one of the B atoms in (**4**). Polyborates form by sharing one O atom with a neighboring B atom, as in (**4**) and (**5**); structures in which two adjacent B atoms share two or three O atoms are unknown.

The rapid cooling of molten B_2O_3 or metal borates often leads to the formation of borate glasses. Although these glasses themselves have little technological significance, the fusion of sodium borate with silica leads to the formation of borosilicate glasses (such as Pyrex), which have low thermal expansivities and hence little tendency to crack when heated or cooled rapidly (Section 18.6). Borosilicate glass is used extensively for cooking- and laboratory-ware.

4 $[B_3O_3(OH)_4]^-$

5

6 $[B_3O_6]^{3-}$

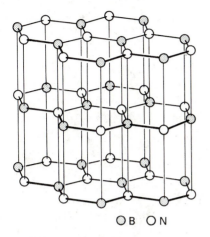

11.5 The structure of layered hexagonal boron nitride. Note that the atoms are in register between layers.

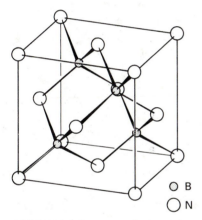

11.6 The sphalerite structure of cubic boron nitride.

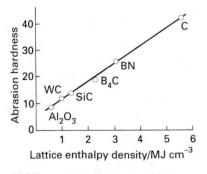

11.7 The correlation of hardness with lattice enthalpy density (the lattice enthalpy divided by the molar volume of the substance). The point for carbon represents diamond and for boron nitride the diamond-like sphalerite structure.

Compounds with nitrogen

The simplest binary compound of boron and nitrogen is easily synthesized by heating boron oxide with a nitrogen compound:

$$B_2O_3(l) + 2NH_3(g) \xrightarrow{1200\,°C} 2BN(s) + 3H_2O(g)$$

The form of boron nitride this reaction produces, and the thermodynamically stable phase under normal laboratory conditions, consists of planar sheets of atoms like those in graphite (Section 11.5). The planar sheets of alternating B and N atoms consist of edge-shared hexagons, and the B—N distance within the sheet (1.45 Å) is much shorter than the distance between the sheets (3.33 Å, Fig. 11.5). The difference between the structures of graphite and boron nitride, however, lies in the register of the atoms of neighboring sheets. In BN, the hexagonal rings are stacked directly over each other, with B and N atoms alternating in successive layers; in graphite, the hexagons are staggered. Molecular orbital calculations suggest that the stacking in BN stems from a partial positive charge on B and partial negative charge on N. This charge distribution is consistent with the greater electronegativity of nitrogen compared with boron.

As with impure graphite, layered boron nitride is a slippery material that is used as a lubricant. Unlike graphite, it is a colorless electrical insulator, for there is a large energy gap between the filled and vacant π bands. The size of the band gap is consistent with the very small number of BN intercalation compounds analogous to those formed by graphite (Section 11.5).

Layered boron nitride changes into a cubic phase (Fig. 11.6) at high pressures and temperatures (60 kbar and 2000 °C). This phase is a hard crystalline analog of diamond, but as it has a lower lattice enthalpy, it has a slightly lower mechanical hardness (Fig. 11.7). Cubic boron nitride is manufactured and used as an abrasive for certain high temperature applications in which diamond cannot be used because it forms carbides with the material being ground.

There are many molecular compounds that contain BN bonds. Here too the isoelectronic character of the units BN and CC invites comparison with hydrocarbons. Many **amine-boranes**, the boron–nitrogen analogs of saturated hydrocarbons, can be synthesized by reaction between a nitrogen Lewis base and a boron Lewis acid:

$$\tfrac{1}{2}B_2H_6 + N(CH_3)_3 \rightarrow H_3B - N(CH_3)_3$$

Although amine-boranes are isoelectronic with hydrocarbons, their properties are significantly different. The most striking difference in physical properties is found between the two simplest analogs, ammoniaborane ($H_3N - BH_3$) and ethane ($H_3C - CH_3$). The former is a solid at room temperature with a vapor pressure of a few torr, whereas ethane is a gas that condenses at −89 °C. This difference can be traced to the difference in polarity of the two molecules: ethane is nonpolar, whereas ammoniaborane has a large dipole moment (5.2 D). The H

7 NH₃BH₃

atoms attached to the electronegative N atom have a partial positive charge and those on the less electronegative B atom have a partial negative charge (**7**).

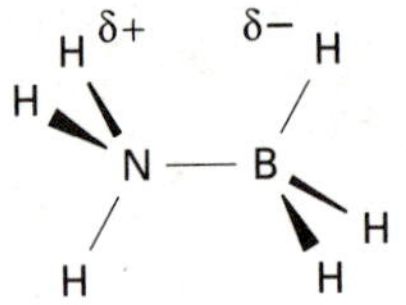

Several BN analogs of the amino acids have been prepared, including ammoniacarboxyborane, H_3NBH_2COOH, the analog of glycine, H_2NCH_2COOH.[1] These compounds display significant physiological activity, including tumor inhibition and reduction of serum cholesterol.

The simplest unsaturated boron-nitrogen compound is aminoborane, $H_2N{=}BH_2$, which is isoelectronic with ethylene. It has only a transient existence in the gas phase because it readily forms cyclic ring compounds such as the cyclohexane analog (**8**). However, the aminoboranes do survive as monomers when the double bond is shielded from reaction by bulky alkyl groups on the N atom and by Cl atoms on the B atom (**9**). For instance, monomeric aminoboranes can be synthesized readily by the reaction of a dialkylamine, which is protic, and a boron halide:

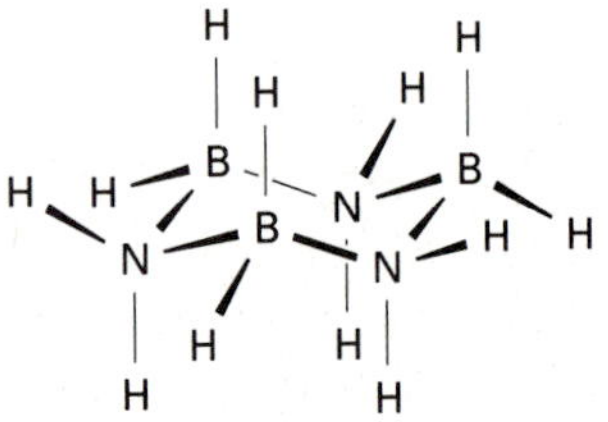

8 N₃B₃H₁₂

$$((CH_3)_2CH)_2NH + BCl_3 \rightarrow \underset{\textbf{9 } Cl_2B{=}N(^iPr)_2}{\begin{array}{c} Cl \\ \diagdown \\ B{=}N \\ \diagup \quad \diagdown \\ Cl \quad\quad CH(CH_3)_2 \end{array}} \begin{array}{c} CH(CH_3)_2 \\ \diagup \\ \\ \diagdown \\ CH(CH_3)_2 \end{array} + HCl$$

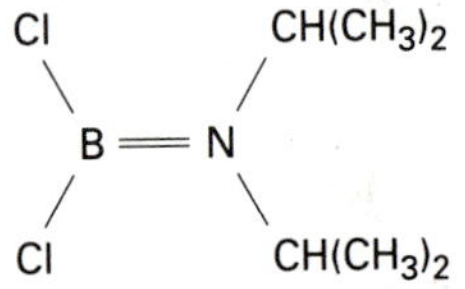

(The reaction also occurs with xylyl groups in place of isopropyl.)

Apart from layered boron nitride, the best known unsaturated compound of boron and nitrogen is borazine, $B_3N_3H_6$ (**10**), which is isoelectronic and isostructural with benzene. Borazine was first prepared in Alfred Stock's laboratory (see Section 9.9) in 1926 by the reaction between diborane and ammonia. Many symmetrically trisubstituted derivatives can be made by more recent procedures that depend on the protolysis of BCl bonds of BCl_3 by an ammonium salt:

borazine

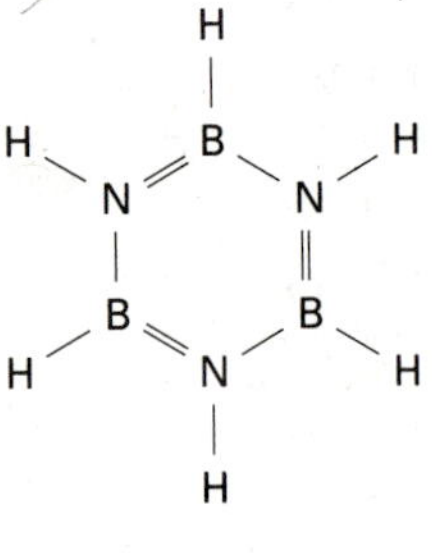

10 H₃B₃N₃H₃

$$3NH_4Cl + 3BCl_3 \xrightarrow{\Delta,\ C_6H_5Cl} \text{(cyclic structure)} + 9HCl$$

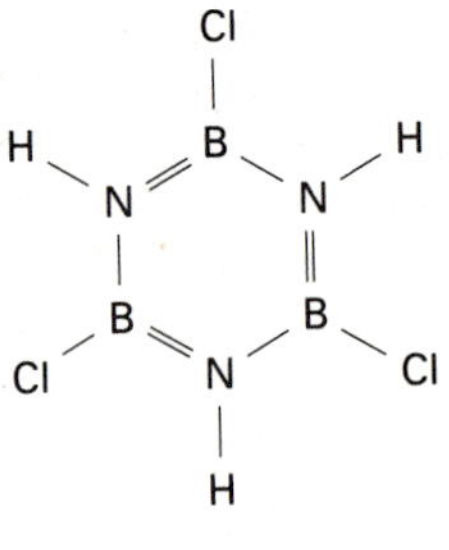

11 Cl₃B₃N₃H₃

The use of alkylammonium chloride yields *N*-alkyl substituted *B*-trichloroborazines.

Although borazine is isoelectronic with benzene, there is little resemblance between them chemically. Once again, the difference in

[1] B. F. Spielvogel, F. U. Ahmed, and A. T. McPhail, *Inorg. Chem.*, **25**, 4395 (1986).

electronegativities of boron and nitrogen is influential, and BCl bonds in trichloroborazine are much more labile than the CCl bonds in chlorobenzene because the π electrons are concentrated on the N atoms. This charge distribution leaves partial positive charges on the B atoms, which are therefore open to attack by nucleophiles. A sign of the difference is that the reaction of a chloroborazine with a Grignard reagent or hydride source results in the substitution of Cl by alkyl, aryl, or hydride groups. Another example of the difference is the ready addition of HCl to borazine to produce a trichlorocyclohexane analog (**12**):

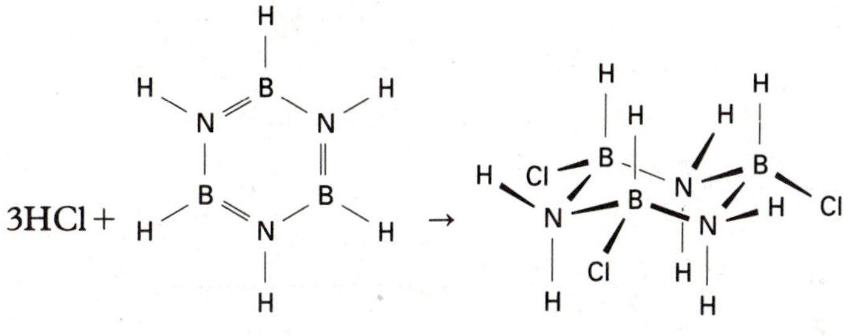

10 $H_3B_3N_3H_3$ **12** $Cl_3B_3N_3H_9$

The electrophile H^+ in this reaction attaches to the partially negative N atom and the nucleophile Cl^- attaches to the partially positive B atom.

Ultraviolet spectra indicate that the separation of the π and π* orbitals in borazine is greater than in benzene. This difference recalls the much greater separation of the π and π* bands in layered BN compared with their separation in graphite. The greater size of the splitting in borazine is not due to a stronger p–p interaction but to differences in the separations of the atomic orbitals on the nitrogen, boron, and carbon atoms.

Example 11.2: *Preparing borazine derivatives*
Give balanced chemical equations for the synthesis of borazine starting with NH_4Cl and other reagents of your choice,

Answer. The reaction of NH_4Cl with BCl_3 in refluxing chlorobenzene will yield the *B*-trichloroborazine:

$$3NH_4Cl + 3BCl_3 \rightarrow H_3N_3B_3Cl_3 + 9HCl$$

The Cl atoms in *B*-trichloroborazine can be displaced by hydride ions from reagents such as $LiBH_4$, to yield borazine:

$$3LiBH_4 + H_3N_3B_3Cl_3 \xrightarrow{THF} H_3N_3B_3H_3 + 3LiCl + 3THF \cdot BH_3$$

Exercise E11.2. Suggest a reaction or series of reactions for the preparation of *N,N′,N″*-trimethyl-*B,B′,B″*-trimethylborazine starting with methylamine and boron trichloride.

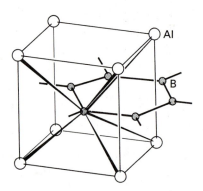

11.8 The AlB_2 structure. To give a clear picture of the hexagonal layer B atoms outside the Al cube are displayed.

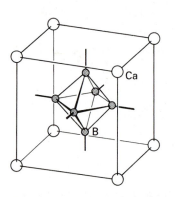

11.9 The CaB_6 structure. Note that the B_6 octahedra are connected by bonds between vertices of adjacent B_6 octahedra. The crystal is a simple analog of CsCl. Thus eight Ca atoms surround the central B_6 octahedron.

13

11.3 Boron clusters

The boron subhalide B_4Cl_4 discussed in the previous section provides a hint of boron's ability to form cluster compounds. The first recognition of these clusters was a direct outgrowth of improvements in X-ray crystallography which gave the first accurate indication of the structures of metal borides and of boranes more complex than B_2H_6. The neutral boranes and the anionic borohydrides were the first molecular cluster compounds to be studied in depth. More recently, cluster chemistry has been developed for elements in every block of the periodic table (see Fig. 8.4).

Metal borides

The direct reaction of elemental boron and a metal at high temperatures provides a useful route to many metal borides. An example is the reaction of calcium and some other highly electropositive metals with boron to produce a phase of composition MB_6:

$$Ca(l) + 6B(s) \rightarrow CaB_6(s)$$

The wide range of compositions of metal borides stem from the existence of a wide variety of structures ranging from isolated B atoms to chains, planar and puckered nets, and clusters. The simplest metal borides are metal-rich compounds that contain species resembling isolated B^{3-} ions. The most common examples of these compounds have the formula M_2B, where M may be one of the middle to late $3d$ metals (manganese through nickel). Another important class of metal borides contain planar or puckered hexagonal nets and have the composition MB_2 (Fig. 11.8). These compounds are formed primarily by electropositive metals, including aluminum, the early d-block metals (such as scandium through manganese), and uranium.

The boron-rich borides, typically MB_6 and MB_{12} where M is an electropositive metal, are of even greater structural interest. In them, the B atoms link to form an intricate network of interconnecting cages. In MB_6 compounds (which are formed by metals such as sodium, potassium, calcium, barium, strontium, europium, and ytterbium) the B_6 octahedra are linked by their vertices to form a cubic framework (Fig. 11.9). The similarity between these anionic linked octahedra and the structure of *closo*-$[B_6H_6]^{2-}$ should be noted. The linked B_6 clusters bear a charge of -1, -2, or -3 depending on the cation with which it is associated. In the MB_{12} compounds the B-atom networks are based on linked cuboctahedra (**13**) rather than the more familiar icosahedron. This type of compound is formed by some of the heavier electropositive metals, particularly those of the f block.

Bonding and structure of higher boranes and borohydrides

We have already discussed (in Section 9.9) the synthesis and chemical properties of B_2H_6 and the existence of some higher boranes. In this section we describe the structures and properties of the cage-like

Table 11.4 Structure of some boranes and borohydrides

closo-$B_nH_n^{2-}$ nido-B_nH_{n+4} arachno-B_nH_{n+6}

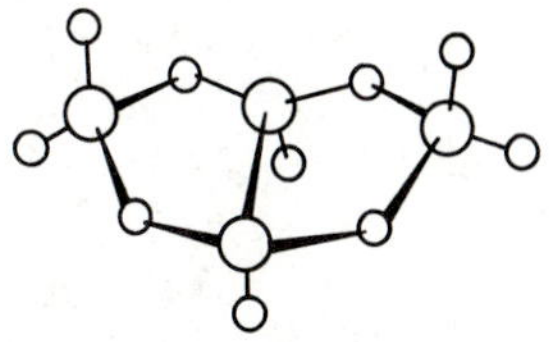

B_4H_{10} Tetraborane(10) C_{2v}

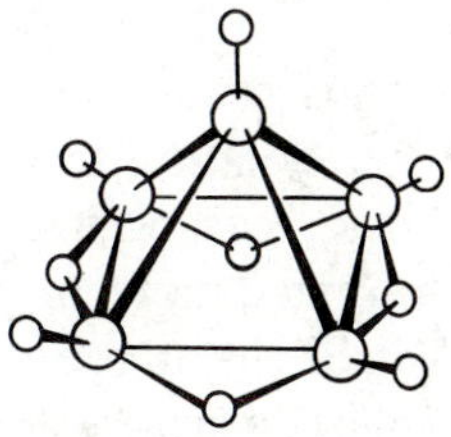

B_5H_9 Pentaborane(9) C_{4v}

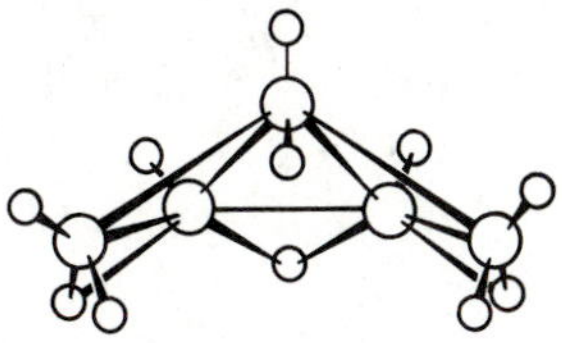

B_5H_{11} Pentaborane(11) C_s

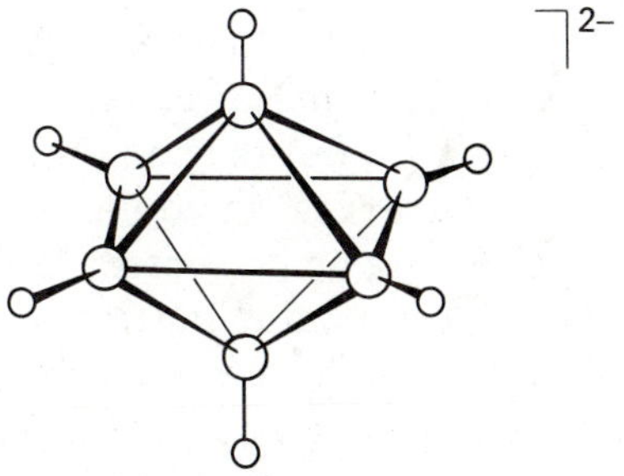

closo-$[B_6H_6]^{2-}$ O_h

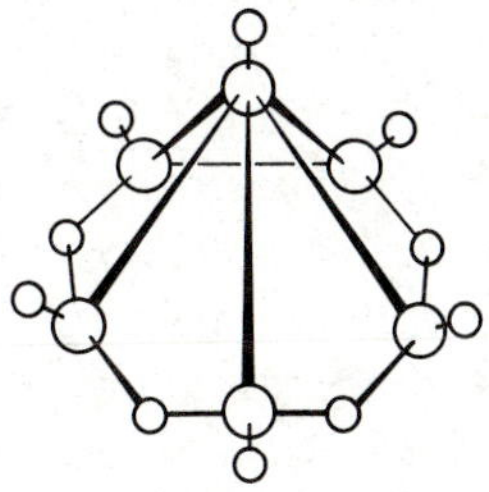

B_6H_{10} Hexaborane(10) C_s

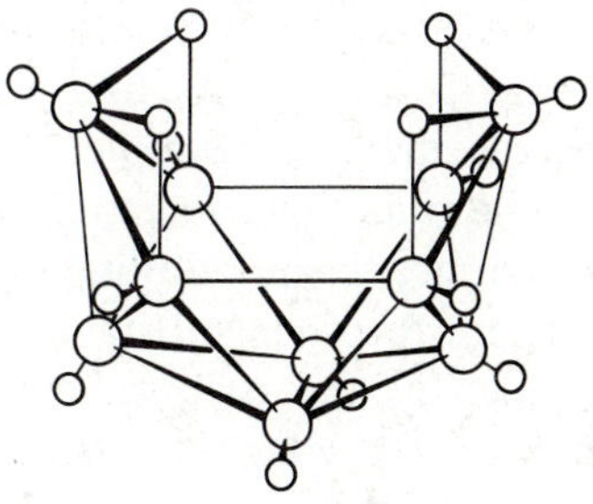

$B_{10}H_{14}$ Decaborane(14) C_{2v}

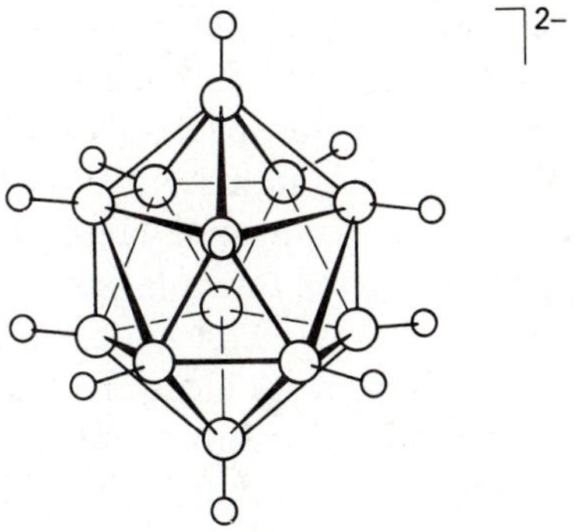

closo-$[B_{12}H_{12}]^{2-}$ I_h

boranes and borohydrides, which include Alfred Stock's series B_nH_{n+4} and B_nH_{n+6} as well as the more recently discovered $[B_nH_n]^{2-}$ closed polyhedra. Boranes and borohydrides occur with a variety of shapes, some resembling nests and others (to the imaginative eye) butterflies. Some of their structures are shown in Table 11.4. They are all electron deficient in the sense that they cannot be described by simple Lewis electronic structures.[2]

The modern interpretation of the bonding in boranes and borohydrides derives from the work of Christopher Longuet-Higgins who, as an undergraduate at Oxford, published a seminal paper in which he introduced the concept of 3c,2e bonds. He later developed a fully delocalized molecular orbital treatment for boron polyhedra and predicted the stability of the icosahedral ion $[B_{12}H_{12}]^{2-}$, which was subsequently verified. William Lipscomb and his students in the USA used single crystal X-ray diffraction to determine the structures of a large number of boranes and borohydrides, and extended the concept of multicenter bonding to these more complex species.

Boron cluster compounds are best considered from the standpoint of fully delocalized molecular orbitals containing electrons that contribute to the stability of the entire molecule. However, it is sometimes fruitful to identify groups of three atoms and to regard them as bonded together by versions of the 3c,2e bonds of the kind that occur in diborane itself (14). In the more complex boranes, the three centers of the 3c,2e bonds may be B—H—B bridge bonds, but they may also be bonds in which three B atoms lie at the corners of an equilateral triangle with their sp^3 hybrid orbitals overlapping at its center (15). To reduce the complexity of the structural diagrams, the illustrations that follow will not in general indicate the 3c,2e bonds in the structures.

A correlation between the number of electrons (counted in a specific way), the formula, and the shape of the molecule was established by the British chemist Kenneth Wade in the 1970s.[3] Wade's rules apply to a class of polyhedra called **deltahedra** (because they are made up of triangular faces each resembling a Greek delta, Δ) and can be used in two ways. For boranes and borohydrides, they enable us to predict the general shape of the molecule or anion from its formula. However, because the rules are also expressed in terms of the number of electrons that contribute to the bonding of the framework of the species, we can extend them to other substances in which there are atoms other than boron, such as the carboranes and other p-block clusters. Here we concentrate on the boron clusters, where knowing the formula is sufficient for predicting the shape. However, so that we can cope with other clusters we shall show how to count the framework electrons too.

For boranes and borohydrides, the building block from which the deltahedron is constructed is assumed to be one BH group (16). The electrons in the B—H bond are ignored in the counting procedure, but all others are included whether or not it is obvious that they help to hold the skeleton together. By the 'skeleton' is meant the framework of the cluster with each BH group counted as a unit. If a B atom happens to carry two H atoms, only one of the B—H bonds is treated as a unit. (This rather odd feature is just a part of Wade's rules: it

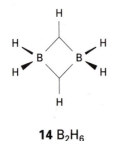

14 B_2H_6

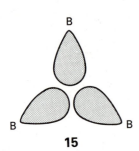

15

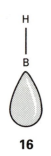

16

[2] An introduction to this subject is provided by C. E. Housecroft, *Boranes and metalloboranes.* Ellis Horwood, Chichester (1990).

[3] Wade has given a very clear account of his rules in K. Wade, The key to cluster shapes. *Chem. Brit.*, **11**, 177 (1975).

works. That is, counting electrons in this way provides a parameter that helps to correlate properties.) For instance, in B_5H_{11}, one of the B atoms has two 'terminal' H atoms, but only one BH entity is treated as a unit, the other pair of electrons being treated as part of the skeleton and hence classified as **skeletal electrons**. A BH group makes two electrons available to the skeleton (the B atom provides three electrons and the H atom provides one, but of these four, two are used for the B—H bond).

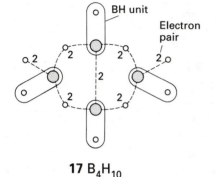

17 B_4H_{10}

Example 11.3: *Counting skeletal electrons*
Count the number of skeletal electrons in B_4H_{10} (Table 11.4).

Answer. Four BH units contribute $4 \times 2 = 8$ electrons, and the six additional H atoms contribute a further 6 electrons, giving 14 in all. The resulting seven pairs are distributed as shown in (**17**): two are used for the additional terminal B—H bonds, four are used for the four B—H—B bridges, and one is used for the central B—B bond.

Exercise E11.3. How many skeletal electrons are present in B_5H_{11}?

According to Wade's rules (Table 11.5), species of formula $[B_nH_n]^{2-}$ will be found to have the *closo* structure (from the Greek for 'cage') with a B atom at each corner of a closed deltahedron and no B—H—B bonds. Such structures have $n + 1$ pairs of skeletal electrons. This series of anions is known for $n = 5$ to 12, and examples include the trigonal bipyramidal $[B_5H_5]^{2-}$ ion, the octahedral $[B_6H_6]^{2-}$ ion, and the icosahedral $[B_{12}H_{12}]^{2-}$ ion. The *closo*-borohydrides and their carborane analogs (see below) are typically thermally stable and moderately unreactive.

Boron clusters of formula B_nH_{n+4} have the *nido* structure (the name is derived from the Latin for 'nest'). They can be regarded as derived from a *closo*-borane that has lost one vertex but may have B—H—B bonds as well as B—B bonds. The compounds in this series contain $n + 2$ pairs of skeletal electrons. An example is B_5H_9 (Table 11.4). In general, the thermal stability of the *nido*-boranes is intermediate between that of *closo*- and *arachno*-boranes.

Clusters of formula B_nH_{n+6} have the *arachno* structure (from the Greek for 'spider') and are *closo*-borane polyhedra less two vertices

Table 11.5 Classification and electron count of boron hydrides

Type	Formula*	Skeletal electron pairs	Examples
Closo	$[B_nH_n]^{2-}$	$n+1$	$[B_5H_5]^{2-}$ to $[B_{12}H_{12}]^{2-}$
Nido	B_nH_{n+4}	$n+2$	B_2H_6, B_5H_9, B_6H_{10}
Arachno	B_nH_{n+6}	$n+3$	B_4H_{10}, B_5H_{11}
Hypho†	B_nH_{n+8}	$n+4$	None‡

*In some cases, protons can be removed; thus $[B_5H_8]^-$ is a *nido* analog of B_5H_9.
†The name comes from the Greek word for 'net'.
‡Some derivatives are known.

(and may have B—H—B bonds); they are so called because they resemble an untidy spider's web. The *arachno*-boranes have $n+3$ skeletal electron pairs. One example of an *arachno*-borane is pentaborane(11) (B_5H_{11}; Table 11.4). As with most *arachno*-boranes, pentaborane(11) is thermally unstable at room temperature and is highly reactive.

Example 11.4: *Using Wade's rules*

Infer the structure of $[B_5H_5]^{2-}$ from its formula and from its electron count.

Answer. We note that the formula $[B_5H_5]^{2-}$ belongs to a class of borohydrides having the formula $[B_nH_n]^{2-}$, which is characteristic of a *closo* species. Alternatively, we can count the number of skeletal electron pairs and from that deduce the structural type. Assuming one B—H bond per B atom, there are five BH units to take into account and therefore ten skeletal electrons plus two from the overall -2 charge: $5 \times 2 + 2 = 12$, or $2(n+1)$ with $n = 5$. This number is characteristic of *closo* clusters. The closed polyhedron must contain triangular faces and five vertices; therefore a trigonal bipyramidal structure is indicated.

Exercise E11.4. How many framework electron pairs are present in B_5H_9 and to what structural category does it belong? Sketch its structure.

A very useful structural correlation between *closo*-, *nido*-, and *arachno*- species is based on the observation that clusters with the same numbers of skeletal electrons are related by removal of successive BH groups and the addition of the appropriate numbers of electrons and H atoms. This conceptual process provides a good way to think about the structures of the various boron clusters. The idea is amplified in Fig. 11.10 where the removal of a BH unit and two electrons and the addition of four H atoms converts the octahedral *closo*-$[B_6H_6]^{2-}$ anion to the square-pyramidal *nido*-B_5H_9 borane. A similar process (removal of a BH unit and addition of two H atoms) converts *nido*-B_5H_9 into a butterfly-like *arachno*-B_4H_{10} borane. Each of these three boranes has 14 skeletal electrons, but as the number of skeletal electrons per B atom increases, the structure becomes more open. A more schematic correlation of this type is indicated for many different boranes in Fig. 11.11. Although this approach gives a good way of deducing the shapes of the molecules, it is purely conceptual and does not represent how they are interconverted chemically.

Wade's rules have been justified by molecular orbital calculations. We shall illustrate the reasoning involved by considering the first of them (the $n+1$ rule). In particular we shall show that $[B_6H_6]^{2-}$ has a low energy if it has an octahedral *closo* structure, as predicted by the rules.

A B—H bond utilizes one electron and one orbital of the B atom, leaving three orbitals and two electrons for the skeletal bonding. One

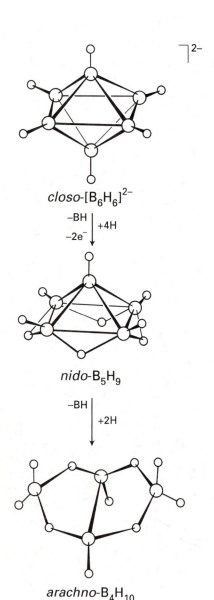

closo-$[B_6H_6]^{2-}$

$\begin{array}{c} -BH \\ -2e^- \end{array}\Bigg| +4H$

nido-B_5H_9

$-BH \Bigg| +2H$

arachno-B_4H_{10}

11.10 Structural correlations between a B_6 *closo* octahedral structure, a B_5 *nido* square pyramid, and a B_4 *arachno* butterfly.

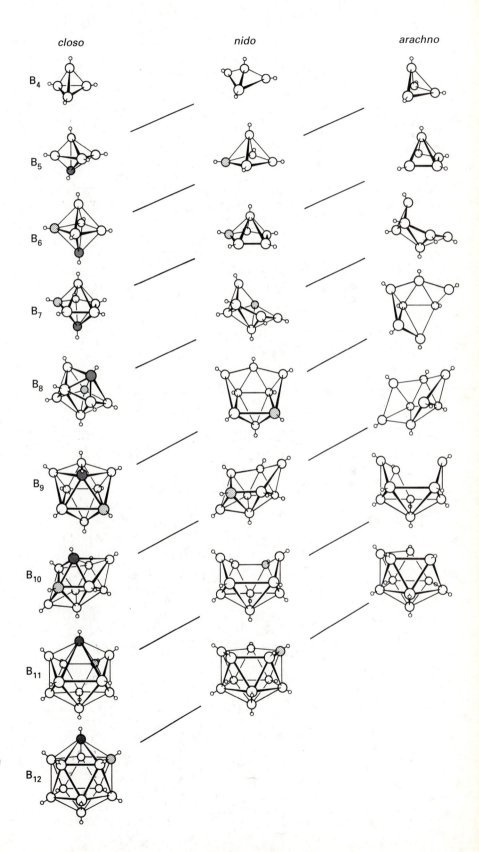

11.11 Structural relations between *closo*, *nido*, and *arachno* boranes and heteroatomic boranes. Diagonal lines connect species that have the same number of skeletal electrons. Hydrogen atoms beyond those in the B—H framework and charges have been omitted. The dark tinted atom is removed first, and the pale tinted atom is removed second. (Based on R. W. Rudolph, *Acc. Chem. Res.*, **9**, 446 (1976).)

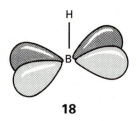

18

of these orbitals, which is called a **radial orbital**, can be considered to be a boron sp hybrid pointing toward the interior of the fragment (as in **16**). The remaining two boron p orbitals, the **tangential orbitals**, are perpendicular to the radial orbital (**18**). The shapes of the 18 symmetry adapted linear combinations of these 18 orbitals in an octahedral B_6H_6 cluster can be inferred from the drawings in Appendix 4, and we show the ones with net bonding character in Fig. 11.12.

The lowest energy orbital is totally symmetric (a_{1g}) and arises from in-phase contributions from all the radial orbitals. Calculations show that the next higher orbitals are the t_{1u} orbitals, each one of which is a combination of four tangential and two radial orbitals. Above these three degenerate orbitals lie another three t_{2g} orbitals, giving seven bonding orbitals in all. Hence, there are seven net bonding orbitals delocalized over the skeleton, and they are separated by a considerable gap from the remaining eleven largely antibonding orbitals (Fig. 11.13).

11.12 Radial and tangential bonding molecular orbitals for $[B_6H_6]^{2-}$. The relative energies are $a_{1g} < t_{1u} < t_{2g}$.

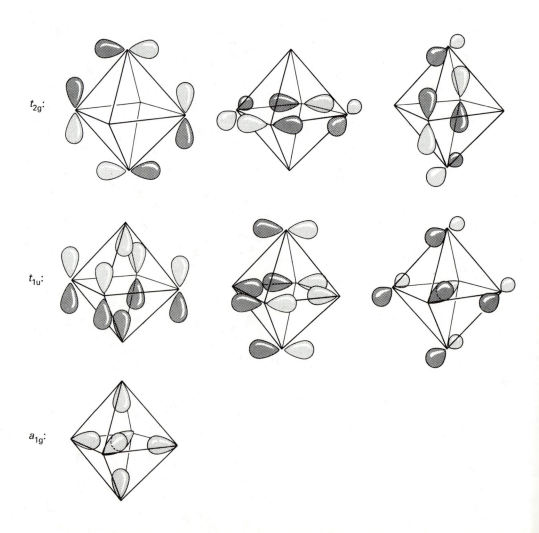

t_{2g}:

t_{1u}:

a_{1g}:

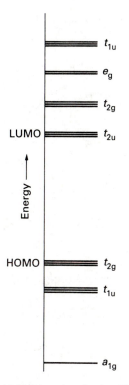

11.13 Schematic molecular orbital energy levels of the B-atom skeleton of $[B_6H_6]^{2-}$. The form of the bonding orbitals is shown in Fig. 11.12.

There are seven electron pairs to accommodate, one pair from each of the six B atoms and one pair from the overall -2 charge. These seven pairs can all enter and fill the seven bonding skeleton orbitals, and hence give rise to a stable structure, in accord with the $n+1$ rule. Note that the unknown neutral octahedral B_6H_6 molecule would have too few electrons to fill the t_{2g} bonding orbitals and would be expected to be unstable.

Synthesis of higher boranes and borohydrides

As discovered by Alfred Stock and perfected by many subsequent workers, the controlled pyrolysis of B_2H_6 in the gas phase provides a route to most of the higher boranes and borohydrides, including B_4H_{10}, B_5H_9, and $B_{10}H_{14}$. A key first step in the proposed mechanism is the dissociation of B_2H_6 and the condensation of the resulting BH_3 with borane fragments. For example, the mechanism of tetraborane(10) formation by the pyrolysis of diborane appears to be

$$B_2H_6(g) \rightarrow 2BH_3(g)$$

$$B_2H_6(g) + BH_3(g) \rightarrow B_3H_7(g) + H_2(g)$$

$$BH_3(g) + B_3H_7(g) \rightarrow B_4H_{10}(g)$$

The synthesis of tetraborane(10), B_4H_{10}, is particularly difficult because it is highly thermally unstable, in keeping with the instability of the B_nH_{n+6} (*arachno*) series. To improve the yield, the product that emerges from the hot reactor is immediately quenched on a cold surface. Pyrolytic syntheses to form species belonging to the more stable B_nH_{n+4} (*nido*) series proceed in higher yield, without the need for a rapid quench. Thus B_5H_9 and $B_{10}H_{14}$ are readily prepared by the pyrolysis reaction. More recently, these brute force methods of pyrolysis have given way to more specific methods that are described below (see reaction 3 below).

The characteristic reactions of boranes and borohydrides

The characteristic reactions of boron clusters are Lewis base cleavage, deprotonation, cluster enlargement, and displacement of protons by electrophiles. We shall consider each one in turn.

1. **Lewis base cleavage** reactions have already been introduced in Section 9.9 in connection with diborane. An example is

with the robust higher borane B_4H_{10}, cleavage may break some $B-B$ bonds leading to partial fragmentation of the cluster:

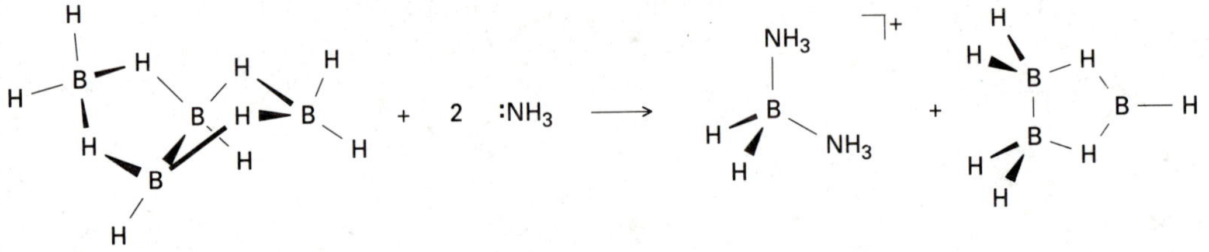

2. **Deprotonation**, rather than cleavage, occurs readily with the large borane $B_{10}H_{14}$:

$$B_{10}H_{14} + N(CH_3)_3 \rightarrow [HN(CH_3)_3]^+ [B_{10}H_{13}]^-$$

The structure of the product anion indicates that deprotonation occurs from a 3c,2e BHB bridge, leaving the electron count on the boron cluster unchanged. This deprotonation of a $B-H-B$ 3c,2e bond to yield a 2c,2e $B-B$ bond occurs without major disruption of the bonding:

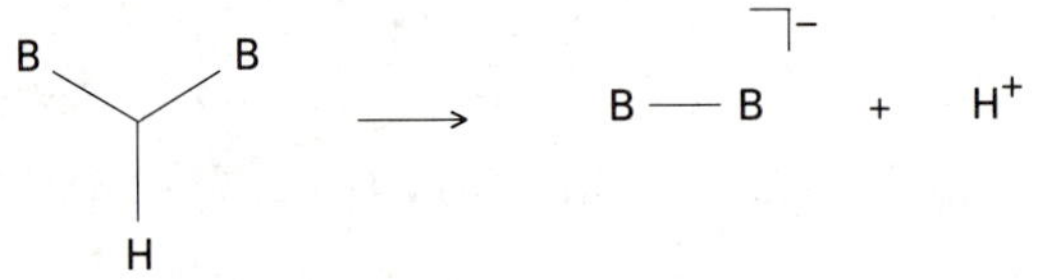

The Brønsted acidity of boron hydrides increases approximately with size:

$$B_4H_{10} < B_5H_9 < B_{10}H_{14}$$

This trend correlates with the greater delocalization of charge in the larger clusters, in much the same way that delocalization accounts for the greater acidity of phenol than of methanol. The variation in acidity is illustrated by the observation that, as shown above, the weak base trimethylamine deprotonates decaborane(10), but the much stronger base methyllithium is required to deprotonate B_5H_9:

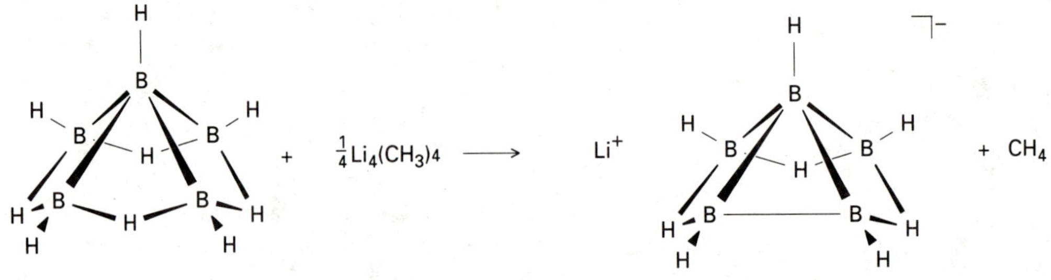

Hydridic character is most characteristic of small anionic borohydrides. As an illustration, whereas BH_4^- readily surrenders a hydride ion in the reaction

$$BH_4^- + H^+ \rightarrow \tfrac{1}{2}B_2H_6 + H_2$$

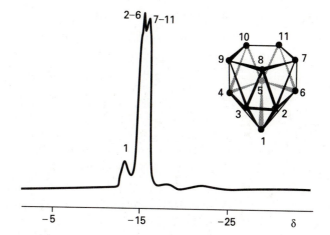

11.14 The proton-decoupled ^{11}B-NMR spectrum of $[B_{11}H_{14}]^-$. The *nido* structure (truncated icosahedron) is indicated by the 1:5:5 pattern. (From: N. S. Hosmane, J. R. Wermer, Z. Hong, T. D. Getman, and S. G. Shore, *Inorg. Chem.*, **26**, 3638 (1987).)

the $[B_{10}H_{10}]^{2-}$ ion survives even in strongly acidic solution. Indeed, the hydronium salt $(H_3O)_2B_{10}H_{10}$ can even be crystallized.

3. The **cluster building reaction** between a borane and a borohydride provides a convenient route to higher borohydride ions:[4]

$$5K[B_9H_{14}] + 2B_5H_9 \xrightarrow{\text{polyether, 85 °C}} 5K[B_{11}H_{14}] + 9H_2$$

Similar reactions are used to construct other borohydrides, such as $[B_{10}H_{10}]^{2-}$. This type of reaction has been employed to synthesize a wide range of polynuclear borohydrides. Boron-11 NMR spectroscopy (Fig. 11.14) reveals that the boron skeleton in $[B_{11}H_{14}]^-$ consists of an icosahedron with a missing vertex.

4. The **electrophilic displacement** of H^+ provides a route to alkylated and halogenated species. As with Friedel-Crafts reactions, the electrophilic displacement of H is catalyzed by a Lewis acid, such as aluminum chloride, and the substitution generally occurs on the closed portion of the boron clusters.

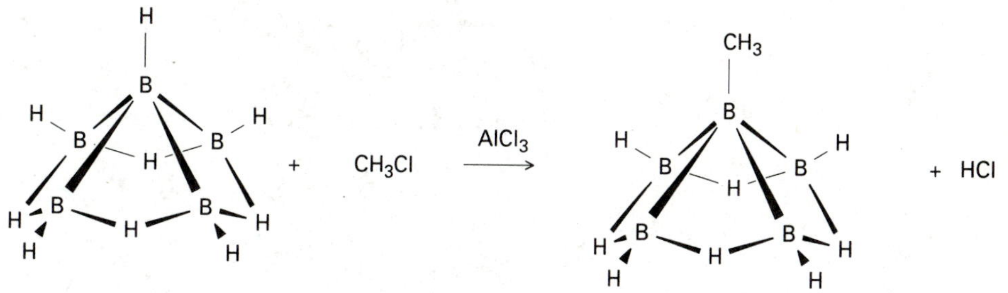

[4] KB$_9$H$_{14}$ is produced from KH and B$_5$H$_9$, so by adjusting the ratio of B$_5$H$_9$ to KH a 'single pot reaction' of KH and B$_5$H$_9$ gives the desired product. The 'pot' is a reaction flask attached to a chemical high vacuum line (Box 8.1). Further details are given by N. S. Hosmane, J. R. Wermer, Z. Hong, T. D. Getman, and S. G. Shore, *Inorg. Chem.*, **26**, 3638 (1987).

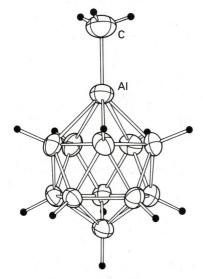

19 [Fe(CO)$_3$B$_4$H$_8$]

11.15 The structure of *closo-*[B$_{11}$H$_{11}$AlCH$_3$]$^{2-}$. (From: T. D. Getman and S. G. Shore, *Inorg. Chem.*, **27**, 3440 (1988).)

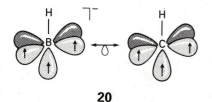

20

Metalloboranes

Many **metalloboranes**, or metal-containing boron clusters, have been characterized. In some cases the metal is appended to a borohydride ion through hydrogen bridges. A more common and generally more robust group of metalloboranes have direct metal–boron bonds. An example of a main-group metal borane with an icosahedral framework is shown in Fig. 11.15. It is prepared by interaction of the protic hydrogens in Na$_2$[B$_{11}$H$_{13}$] with trimethylaluminum:

$$2[B_{11}H_{13}]^{2-} + Al_2(CH_3)_6 \xrightarrow{\Delta} 2[B_{11}H_{11}AlCH_3]^{2-} + 4CH_4$$

When B$_5$H$_9$ is heated with Fe(CO)$_5$, a metallated analog of pentaborane is formed (**19**).

Example 11.5: *Proposing a structure for a boron-cluster reaction product*
Propose a structure for the product of the reaction of B$_{10}$H$_{14}$ with LiBH$_4$ in a refluxing polyether, CH$_3$OC$_2$H$_4$OCH$_3$.

Answer. The prediction of the probable outcome for the reactions of a boron cluster is difficult because several products are often plausible and the actual outcome is often sensitive to conditions of the reaction. In the present case we note that an acidic borane, B$_{10}$H$_{14}$, is brought into contact with the hydridic anion BH$_4^-$ under rather vigorous conditions. Therefore, we might expect the evolution of hydrogen:

$$B_{10}H_{14} + Li[BH_4] \xrightarrow{\text{Ether}} Li[B_{10}H_{13}] + R_2OBH_3 + H_2$$

This set of products suggests the further possibility of condensation of the neutral BH$_3$ complex with B$_{10}$H$_{13}^-$ to yield a larger borohydride. That is in fact the observed outcome under these conditions:

$$Li[B_{10}H_{13}] + R_2OBH_3 \rightarrow Li[B_{11}H_{14}] + H_2 + R_2O$$

It turns out that in the presence of excess LiBH$_4$ the cluster building continues to give the very stable icosahedral B$_{12}$H$_{12}^{2-}$ anion:

$$Li[B_{11}H_{14}] + Li[BH_4] \rightarrow Li_2[B_{12}H_{12}] + 3H_2$$
    (*nido*)   (*closo*)

Exercise E11.5. Propose a plausible product for the interaction of Li[B$_{10}$H$_{13}$] with Al$_2$(CH$_3$)$_6$.

Carboranes

Closely related to the polyhedral boranes and borohydrides are the **carboranes**, a large family of clusters that contain both B and C atoms. Now we begin to see the full generality of Wade's electron counting rules, for BH$^-$ is isoelectronic and isolobal with CH (**20**), and we can expect the polyhedral borohydrides and carboranes to be related. Thus, an analog of $[B_6H_6]^{2-}$ (**21**) is the neutral carborane $B_4C_2H_6$ (**22**).

One entry into the interesting and diverse world of carboranes is the conversion of decaborane(14) to *closo*-1,2-$B_{10}C_2H_{12}$ (**23**). The first reaction in this preparation is the displacement of an H_2 molecule from decaborane by a thioether:

$$nido\text{-}B_{10}H_{14} + 2SEt_2 \rightarrow arachno\text{-}B_{10}H_{12}(SEt_2)_2 + H_2$$

In this reaction the *nido*-$B_{10}H_{14}$ molecule loses the two-electron molecule H_2 and gains two two-electron donors, SEt_2; so this net two-electron gain results in an *arachno* product. That product in turn may be converted to a *closo* species by the reaction

$$arachno\text{-}B_{10}H_{12}(SEt_2)_2 + C_2H_2 \rightarrow closo\text{-}B_{10}C_2H_{12} + 2SEt_2 + 2H_2$$

The electron bookkeeping for this second reaction is as follows: the *arachno* starting material gains the four-electron C_2H_2 but loses eight electrons in the two Et_2S and two H_2 molecules, for a net four-electron loss by the cluster. The *closo* product contains C atoms in adjacent (1,2) positions (**23**) reflecting their origin from acetylene. This *closo* carborane survives in air and can be heated without decomposition. At 500 °C in an inert atmosphere it undergoes isomerization into 1,7-$B_{10}C_2H_{12}$ (**24**) which in turn isomerizes at 700 °C to the 1,12-isomer (**25**).

The H atoms attached to carbon in *closo*-$B_{10}C_2H_{12}$ are very mildly acidic in the clusters, so it is possible to lithiate these compounds with butyllithium:

$$B_{10}C_2H_{12} + 2LiC_4H_9 \rightarrow B_{10}C_2H_{10}Li_2 + 2C_4H_{10}$$

These dilithiocarboranes are good nucleophiles and undergo many of the reactions characteristic of organolithium reagents (Section 10.6). Thus, a wide range of carborane derivatives can be synthesized. For example, reaction with CO_2 gives a dicarboxylic acid carborane:

$$B_{10}C_2H_{10}Li_2 + 2CO_2 \xrightarrow{\text{(1) react, (2) } 2H_2O} B_{10}C_2H_{10}(COOH)_2$$

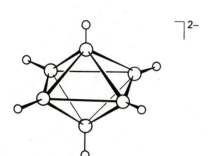

21 *closo*-$[B_6H_6]^{2-}$

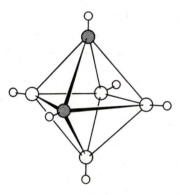

22 *closo*-1,2-$B_4C_2H_6$

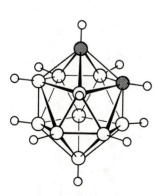

23 *closo*-1,2-$B_{10}C_2H_{12}$

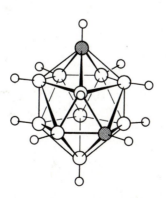

24 *closo*-1,7-$B_{10}C_2H_{12}$

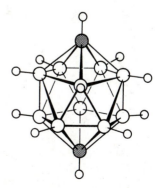

25 *closo*-1,12-$B_{10}C_2H_{12}$

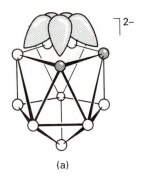

(a)

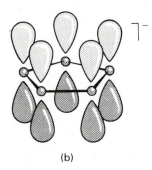

(b)

11.16 The isolobal relation between (a) $[B_9C_2H_{11}]^{2-}$ and (b) $[C_5H_5]^-$. The H atoms have been omitted for clarity.

Similarly, I_2 leads to the diiodocarborane and NOCl yields $B_{10}C_2H_{10}(NO)_2$.

Although $1,2\text{-}B_{10}C_2H_{12}$ is very stable, the cluster can be partially fragmented in strong base, and then deprotonated with NaH to yield $nido\text{-}[B_9C_2H_{11}]^{2-}$:

$$B_{10}C_2H_{12} + OEt^- + 2EtOH \rightarrow [B_9C_2H_{12}]^- + B(OEt)_3 + H_2$$

$$Na[B_9C_2H_{12}] + NaH \rightarrow Na_2[B_9C_2H_{11}] + H_2$$

The importance of these reactions is that $nido\text{-}[B_9C_2H_{11}]^{2-}$ (Fig. 11.16(a)) is an excellent ligand. In this role it mimics the isolobal cyclopentadienide ligand ($C_5H_5^-$; Fig. 11.16(b)) which is widely employed in organometallic chemistry:

$$2Na_2[B_9C_2H_{11}] + FeCl_2 \xrightarrow{\text{THF}} 2NaCl + Na_2[Fe(B_9C_2H_{11})_2]$$

$$2Na[C_5H_5] + FeCl_2 \xrightarrow{\text{THF}} 2NaCl + Fe(C_5H_5)_2$$

Although we shall not go into the details of their synthesis, a wide range of d-metal-coordinated carboranes can be synthesized in this manner. A notable feature is the ease of formation of multi-decker sandwich compounds containing carborane ligands (**26** and **27**).[5] The highly negative $[B_3C_2H_5]^{4-}$ ligand has a much greater tendency to form stacked sandwich compounds than the less negative and therefore poorer donor $C_5H_5^-$.

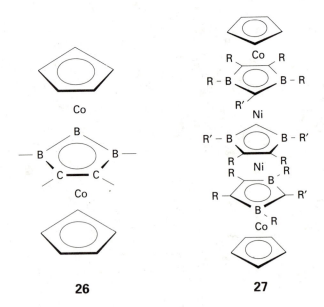

26   **27**

[5] R. N. Grimes, Boron–carbon ring ligands in organometallic synthesis. *Chem. Rev.*, **92**, 251 (1992).

Example 11.6: *Planning the synthesis of a carborane derivative*

Give balanced chemical equations for the synthesis of 1,2-$B_{10}C_2H_{10}(Si(CH_3)_3)_2$ starting with decaborane(10) and other reagents of your choice.

Answer. The attachment of substituents to C atoms in 1,2-*closo*-$B_{10}C_2H_{12}$ is most readily carried out using the dilithium derivative $B_{10}C_2H_{10}Li_2$. We first prepare 1,2-$B_{10}C_2H_{12}$ from decaborane:

$$B_{10}H_{14} + 2SR_2 \rightarrow B_{10}H_{12}(SR_2)_2 + H_2$$

$$B_{10}H_{12}(SR_2)_2 + C_2H_2 \rightarrow B_{10}C_2H_{12} + 2SR_2 + H_2$$

The product is then lithiated:

$$B_{10}C_2H_{12} + 2LiC_4H_9 \rightarrow B_{10}C_2H_{10}Li_2 + 2C_4H_{10}$$

The resulting nucleophilic carborane is then employed in a nucleophilic displacement on $Si(CH_3)_3Cl$ to yield the desired product:

$$B_{10}C_2H_{10}Li_2 + 2Si(CH_3)_3Cl \rightarrow B_{10}C_2H_{10}(Si(CH_3)_3)_2 + 2LiCl$$

Exercise E11.6. Propose a synthesis for the polymer precursor 1,7-$B_{10}C_2H_{10}(Si(CH_3)_2Cl)_2$ from 1,2-$B_{10}C_2H_{12}$ and other reagents of your choice.

Table 11.6 Band gaps at 25°C of Group 14/IV elements and compounds and some III/V compounds

Material	E_g/eV
C (diamond)	5.47
SiC	3.00
Si	1.12
Ge	0.66
Sn	0
BN	7.5 (approx.)
BP	2.0
GaN	3.36
GaP	2.26
GaAs	1.42
InAs	0.36

Source: S. A. Schwartz, *Kirk-Othmer encyclopedia of chemical technology*, Vol. 20, p. 601. Wiley-Interscience, New York (1982).

THE CARBON GROUP (GROUP 14/IV)

We discuss carbon in many contexts throughout this text, including organometallic compounds in Chapters 10 and 16 and catalysis in Chapter 17. In this section we shall focus attention on the more classically 'inorganic' chemistry of carbon and its congeners.

All the elements of the group except lead have at least one solid phase with the diamond structure. That of tin, which is called **gray tin**, is not stable at room temperature; the more stable phase, **white tin**, has six nearest neighbors in a highly distorted octahedral array. This distortion is reminiscent of the unsymmetrical environment within tin's neighbors gallium and indium. The band gap energy between the top of the valence band and bottom of the conduction band (Section 2.9) decreases steadily from diamond, which is commonly regarded as an insulator, to gray tin, which behaves like a metal just below its transition temperature (Table 11.6).

11.4 Occurrence and recovery

Two reasonably pure forms of carbon, diamond and graphite, are mined. There are many less pure forms, such as coke, which is made by the pyrolysis of coal, and lamp black, which is the product of incomplete combustion of hydrocarbons. As these examples imply, carbon is polymorphous, and some of its forms will be discussed in more detail in Section 11.5.

The recovery of elemental silicon from silica, SiO_2, by very high temperature reduction with carbon in an electric arc furnace (Section 7.1) is the first step in the production of pure silicon for the manufacture of modern semiconducting devices. Germanium is low in abundance and generally not concentrated in nature. Most germanium is recovered in the processing of zinc ores. Tin is produced by the reduction of the mineral cassiterite, SnO_2, with coke in an electric furnace. Lead is obtained from its sulfide ores, which are converted to oxide and reduced by carbon in a blast furnace.

11.5 Carbon allotropes

Apart from its central role in organic chemistry, carbon forms numerous inorganic compounds and an extensive series of organometallic compounds (Chapters 10 and 16). We shall first consider the elemental forms of carbon, the most notable of which are graphite, diamond, and the fullerenes.

Diamond and graphite

Diamond and graphite, the two common crystalline forms of elemental carbon, are strikingly different. Diamond is effectively an electrical insulator; graphite is a good conductor. Diamond is the hardest known substance and hence the ultimate abrasive; impure (partially oxidized) graphite is slippery and is frequently used as a lubricant. Because of its durability, clarity, and high refractive index, diamond is one of the most highly prized gemstones; graphite is soft and black with a slightly metallic luster and is neither durable nor particularly attractive. The origin of these widely different physical properties can be traced to the very different structures and bonding in the two polymorphs.

In diamond (Fig. 11.17), each C atom forms single bonds (of length 1.54 Å) with four adjacent C atoms at the corners of a regular tetrahedron. The result is a tightly bonded, covalent, three-dimensional crystal. On the other hand, graphite (Fig. 11.18) consists of stacks of planar layers within which each C atom has three nearest-neighbors at 1.42 Å. The σ bonds between neighbors within the sheets are formed from the overlap of sp^2 hybrids, and the remaining perpendicular p orbitals overlap to form π bonds that are delocalized over the plane. The planes themselves are widely separated from each other (at 3.35 Å), which indicates that there are weaker forces between them. These forces are sometimes, but not very appropriately, called **van der Waals forces** (because in the common impure form of graphite, graphitic oxide, they are weak, like intermolecular forces), and the region between the planes is called the **van der Waals gap**. The ready cleavage of graphite parallel to the planes of atoms (which is enhanced by the presence of impurities) accounts for its slipperiness. Diamond can be cleaved, but this ancient craft requires considerable expertise since the forces in the crystal are more symmetrical.

The transformation of diamond to graphite at room temperature and pressure is spontaneous ($\Delta G^{\ominus} = -2.90$ kJ mol^{-1}) but does not occur

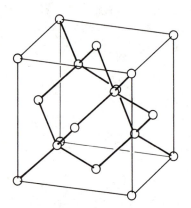

11.17 The cubic diamond structure.

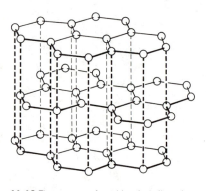

11.18 The structure of graphite. As indicated by the vertical dotted lines, the C atoms are in register in *alternate* planes, not adjacent planes.

at an observable rate under ordinary conditions. Diamond is the denser phase, so it is favored by high pressures, and large quantities of diamond abrasive are manufactured commercially by a *d*-metal catalyzed high-temperature, high-pressure process. The *d*-metal (typically nickel) dissolves the graphite at high temperature and pressure (1800 °C and 70 kbar), and the less soluble diamond phase crystallizes from it. The synthesis of gem-quality diamonds is possible but not yet economical.

Because the high-pressure synthesis of diamond is costly and cumbersome, a low-pressure process would be highly attractive. It has in fact been known for a long time that microscopic diamond crystals mixed with graphite can be formed by depositing C atoms on a hot surface in the absence of air. The C atoms are produced by the pyrolysis of methane, and the atomic hydrogen also produced in the pyrolysis plays an important role in favoring diamond over graphite.[6] One property of the atomic hydrogen is that it reacts more rapidly with the graphite than with diamond to produce volatile hydrocarbons, so the unwanted graphite is swept away. Although it is not fully perfected, synthetic diamond films are already finding applications ranging from the hardening of surfaces subjected to wear to the construction of electronic devices. Microscopic diamonds are also formed by subjecting graphite to pressure created by an explosive charge.

The electrical conductivity and many of the chemical properties of graphite are closely related to the structure of its conjugated π bonds. Its conductivity perpendicular to the planes is low (5 S cm^{-1} at 25 °C) and increases with increasing temperature, signifying that graphite is a semiconductor in that direction. The electrical conductivity is much higher parallel to the planes (3×10^4 S cm^{-1} at 25 °C) but decreases as the temperature is raised. This behavior indicates metallic conduction in that direction.[7] The anisotropy of the conductivity is consistent with a simple band model in which the mobile electrons are in a half-full π band extending over the sheets (Fig. 11.19).

A chemical consequence of the small separation of the valence and conduction bands in graphite is that it may serve as either an electron donor or an electron acceptor toward atoms and ions that intercalate between its sheets. Thus, K atoms reduce graphite by donating their valence electron to the empty orbitals of the π band and the resulting K$^+$ ions penetrate between the layers. The electrons added to the band are mobile, and therefore alkali metal graphite intercalates have high electrical conductivity. The stoichiometry of the compound depends on the quantity of potassium and on the reaction conditions. The different stoichiometries are associated with an interesting series of structures, where the alkali metal ion may insert between every other carbon layer, every third layer, and so on (Fig. 11.20).

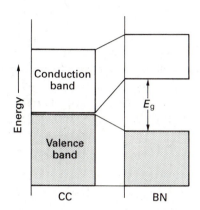

11.19 A schematic diagram of the π valence and conduction bands of graphite and layered boron nitride.

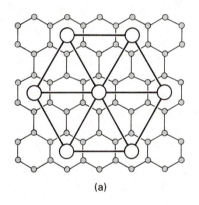

11.20 Potassium graphite compounds. (a) Viewed perpendicular to the plane of graphite. (b) Viewed parallel with graphite planes (schematic).

[6] J. C. Anderson and C. C. Hayman, Low-pressure metastable growth of diamond and 'diamond-like' phases. *Science*, **241**, 913 (1988) and M. N. Geis and J. C. Angus, Thin diamond films. *Scientific American*, **267** (4), 64 (1992).

[7] More precisely, graphite is a semimetal in that direction (Section 2.9).

An example of an oxidation of graphite by removal of electrons from the π band is the formation of the substances called **graphite bisulfates** by heating graphite with a mixture of sulfuric and nitric acids. In this reaction, electrons are removed from the π band, and HSO_4^- ions penetrate between the sheets to give substances of approximate formula $(C_{24})^+ SO_3(OH)^-$. In this oxidative intercalation reaction, the removal of electrons from the full π band leads to a higher conductivity than that of pure graphite. This process is analogous to the formation of p-type silicon by electron accepting dopants (Section 2.10). The layered graphite compounds are examples of a broader class of **intercalation compound** (Section 18.7) in which an atom or molecule is inserted between the layers of a solid.

Carbon clusters

Metal and nonmetal cluster compounds have been known for decades, but the recent discovery of the soccer-ball shaped C_{60} cluster created a great amount of excitement in the scientific community and in the popular press. Some of this interest undoubtedly stemmed from the fact that carbon is a common element and there had seemed little likelihood that new molecular carbon structures would be found.[8]

When an electric arc is struck between carbon electrodes in an inert atmosphere, a large quantity of soot is formed together with significant quantities of C_{60} and much smaller quantities of related **fullerenes**[9] such as C_{70}, C_{76}, and C_{84}. The fullerenes can be dissolved in a hydrocarbon solvent and separated by chromatography on an alumina column. The structure of C_{60} has been determined by X-ray crystallography on the solid at low temperature and electron diffraction in the gas phase.[10] It consists of five- and six-membered carbon rings, and the over-all symmetry is icosahedral in the gas phase (Fig. 11.21).

The fullerene molecule is readily reduced electrochemically. The process is attributed to the addition of electrons to antibonding π orbitals, in much the same way that naphthalene is reduced (Section 10.6). Similarly fullerene reacts with alkali metals to produce solids having compositions such as K_3C_{60}. The structure of K_3C_{60} consists of an fcc array of C_{60} clusters in which K^+ ions occupy the one octahedral and two tetrahedral sites available per C_{60} (Fig. 11.22). The compound is a superconductor below 18 K.

In the two years following the discovery of large-scale synthesis of C_{60} dozens of reactions were discovered. Many of them are similar to the organometallic chemistry of alkenes (Section 16.7). Electron-rich platinum complexes attach to two carbon atoms of C_{60} (**28**). In this and similar reactions, the two C atoms mimic the ethene ligand. Another indication of this reactivity pattern is the addition of osmium

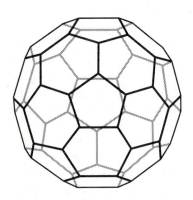

11.21 The structure of buckminsterfullerene, C_{60}. Note that each pentagon is surrounded by hexagons and that each hexagon is linked to three pentagons and three hexagons.

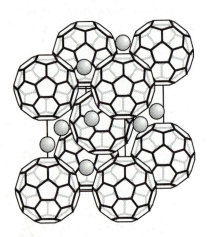

11.22 The fcc structure of K_3C_{60}. Only a fragment of the complete unit cell is shown (the full cell is fcc).

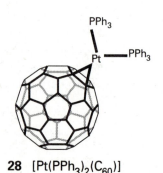

28 $[Pt(PPh_3)_2(C_{60})]$

---

[8] An informative series of articles on the discovery, structure, and chemistry of C_{60} and related carbon clusters appears in *Acc. Chem. Res.*, **25**, 98 et seq. (1992).

[9] The name fullerne (or sometimes buckminsterfullerene) is applied to these carbon clusters because they resemble the geodesic domes devised by the architect Buckminster Fuller.

[10] S. Lu, Y. Lu, M. M. Kappes, and J. A. Ibers, *Science*, **254**, 408 (1991); K. Hedberg, L. Berg, D. S. Gethune, C. S. Brown, and D. R. Dorn, *Science*, **254**, 410 (1991).

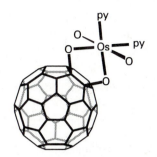

29 [Os(O)$_2$(py)$_2$(OC$_{60}$O)]

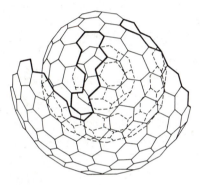

11.23 A proposed structure for soot resulting from imperfect closure of a curved C-atom network. Graphite-like structures have also been proposed.

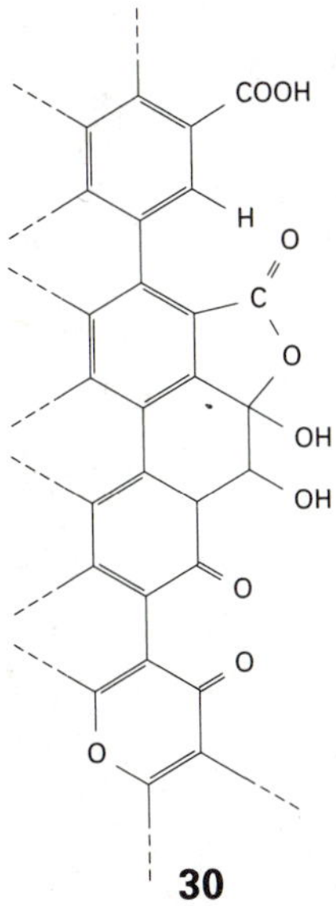

30

tetroxide to a C$_2$ unit in C$_{60}$ (**29**). New forms of carbon with tubular shapes have also been synthesized.[11] Indeed, it appears likely that a large number of metastable forms of carbon will be found.

Partially crystalline carbon

There are many forms of carbon that have a low degree of crystallinity. These partially crystalline materials have considerable commercial importance, for they include **carbon black**, **activated carbon**, and **carbon fibers**. Since single crystals suitable for complete X-ray analyses of these materials are not available, their structures are uncertain. However, what information there is suggests that their structures are similar to that of graphite, but the degree of crystallinity and shapes of the particles differ.

Carbon black is a very finely divided form of carbon. It is prepared (on a scale that exceeds 8×10^9 kg annually) by the combustion of hydrocarbons under oxygen-deficient conditions. Planar stacks, like those of graphite, and multilayer balls, reminiscent of the fullerenes, have both been proposed for the structure of carbon black (Fig. 11.23). The major uses of carbon black are as a pigment, in printer's ink (as on this page), and as a filler for rubber goods, including automobile tires, where it greatly improves the strength and wear resistance of the rubber and helps to protect it from degradation by sunlight.

Activated carbon is prepared from the controlled pyrolysis of organic material, including coconut shells. It has a high surface area (in some cases exceeding 1000 m^2 g^{-1}) which arises from the small particle size. It is therefore a very efficient adsorbent for molecules, including organic pollutants from drinking water, noxious gases from the air, and impurities from reaction mixtures. There is evidence that the parts of the surface defined by the edges of the hexagonal sheets are covered with oxidation products, including carboxyl and hydroxyl groups (**30**). This structure may account for some of its surface activity.

Carbon fibers are made by the controlled pyrolysis of asphalt fibers or synthetic fibers and are incorporated into a variety of high-strength plastic products, such as tennis rackets and aircraft components. Their structure bears a resemblance to that of graphite, but in place of the extended sheets the layers consist of ribbons parallel to the axis of the fiber. The strong in-plane bonds (which resemble those in graphite) give the fiber its very high tensile strength.

[11] The discovery of these so-called 'bucky tubes' is described by S. Iijima, *Nature*, **354**, 56 (1991) and S. Iijima, T. Ichihashi, and Y. Ando, *Nature*, **356**, 776 (1992).

Example 11.7: *Comparing bonding in diamond and boron*
Each B atom in elemental boron is often bonded to five other B atoms but each C atom in diamond is bonded to four nearest neighbors. Suggest an explanation of this difference.

Answer. The B and C atoms both have four orbitals available for bonding (one *s* and three *p*). However, a C atom has four valence electrons, one for each orbital, and it can therefore use all its electrons and orbitals in forming 2c,2e bonds with the four neighboring C atoms. In contrast, B has one less electron; therefore, to use all its orbitals, it forms 3c,2e bonds. The formation of these three-center bonds brings another B atom into binding distance.

Exercise E11.7. Describe how the electronic structure of graphite is altered when it reacts with (a) potassium, (b) bromine.

11.6 Compounds of carbon with electronegative elements

Carbon has a high affinity for highly electronegative elements such as O and F, with which it forms many important compounds. For example, CO_2 is essential for plant life, and carbon–fluorine compounds are widely utilized in industry. From the viewpoint of a synthetic chemist, many of the compounds with heavier halide and chalcogenide elements are important starting materials for the synthesis of other compounds.

Halides

The tetrahalomethanes, the simplest halocarbons, vary from the highly stable and volatile CF_4 to the thermally unstable solid CI_4 (Table 11.7). These tetrahalomethanes and analogous partially halogenated alkanes provide a route to a wide variety of derivatives, mainly by nucleophilic displacement of one or more halogens. Some useful and interesting reactions from an inorganic perspective are outlined in the chart in Fig. 11.24. Note in particular the metal–carbon bond-forming reactions, which take place either by complete displacement of halide or by oxidative addition. The rates of nucleophilic displacement increase

Table 11.7 Properties of tetrahalomethanes

	CF_4	CCl_4	CBr_4	CI_4
m.p./°C	−187	−23	90	171 *dec*
b.p./°C	−128	77	190	*c.* 130 *sub*
$\Delta G_f^{\ominus}$/kJ mol^{-1}	−879(g)	−65.2(l)	+47.7(s)	>0

dec = decomposes; *sub* = sublimes.

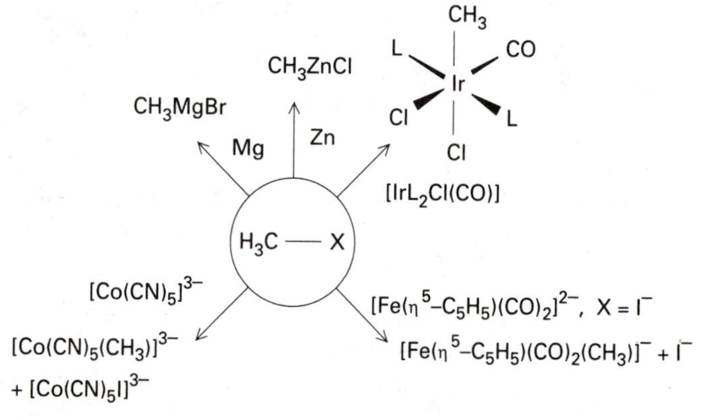

11.24 Some characteristic reactions of carbon–halogen bonds (X = halogen).

Table 11.8 Properties of carbonyl halides

	COF_2	$COCl_2$	$COBr_2$
m.p./°C	-114	-128	
b.p./°C	-83	8	65
$\Delta G_f^{\ominus}$/kJ mol^{-1}	-619(g)	-205(g)	-111(g)

greatly from fluorine to iodine, and lie in the order $F < Cl < Br < I$. All tetrahalomethanes are thermodynamically unstable with respect to hydrolysis,

$$CX_4(l \text{ or } g) + 2H_2O(l) \rightarrow CO_2(g) + 4HX(aq)$$

but the rate for C—F bonds is extremely slow, and therefore fluorocarbon polymers such as poly(tetrafluoroethylene) (e.g. Teflon) or fluorochlorocarbon polymers are highly resistant to attack by water. Halocarbons are oxidizing agents toward strong reducing agents such as alkali metals. For example, the reaction of carbon tetrachloride with sodium is highly exoergic:

$$CCl_4(l) + 4Na(s) \rightarrow 4NaCl(s) + C(s) \quad \Delta G^{\ominus} = -249 \text{ kJ mol}^{-1}$$

This reaction can occur with explosive violence with CCl_4 and other chlorocarbons, so alkali metals such as sodium should never be used to dry them. Analogous reactions occur on the surface of poly-(tetrafluoroethylene) when it is exposed to alkali metals or strongly reducing organometallic compounds. Fluorocarbons, together with other fluorine-containing molecules exhibit many interesting properties, such as high volatility and strong electron withdrawing character (Section 13.7).

The **carbonyl halides** (Table 11.8) are planar molecules and useful chemical intermediates. The simplest of these compounds ($OCCl_2$, phosgene) is a highly toxic gas. It is prepared on a large scale by the reaction of chlorine with carbon monoxide:

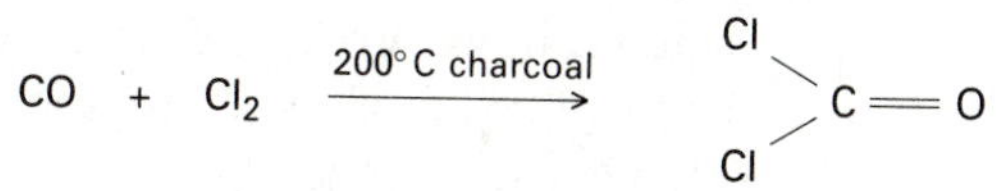

$$CO + Cl_2 \xrightarrow{200°C \text{ charcoal}} \begin{array}{c} Cl \\ \diagdown \\ \diagup \\ Cl \end{array} C = O$$

The utility of phosgene lies in the ease of nucleophilic displacement of chlorine to produce carbonyl compounds and isocyanates (Fig. 11.25). The fact that hydrolysis leads to carbon dioxide and hydrogen chloride, rather than carbonic acid, $(HO)_2CO$, can be traced to the latter's instability with respect to loss of water.

Oxygen and sulfur compounds

The two familiar oxides of carbon, CO and CO_2, have already been mentioned in several contexts; among its less familiar oxides is carbon suboxide, $O=C=C=C=O$. Physical data on all three compounds are summarized in Table 11.9. It should be noted that the bond length in carbon monoxide is short and the force constant high; both features

Table 11.9 Properties of some oxides of carbon

Oxide	m.p./°C	b.p./°C	$\bar{\nu}(CO)/cm^{-1}$	$k(CO)/(N\ m^{-1})$	Bond length/Å CC	Bond length/Å CO
CO	−199	−191.5	2145	1860		1.13
OCO	−78*		2349, 1318	1550		1.16
OCCCO	−111	7	2200, 2290		1.28	1.16

*Sublimes; at 5 atm, CO_2 melts at −57 °C.

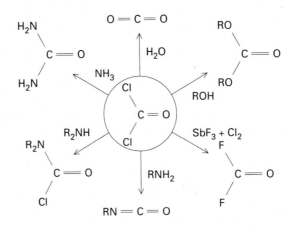

11.25 Characteristic reactions of Cl_2CO.

are in accord with its possession of a triple bond, as in the Lewis structure $:C\equiv O:$.

The uses of CO that we have described elsewhere include the reduction of metal oxides in blast furnaces (Sections 7.1 and 8.6) and the shift reaction (Section 9.3) for the production of H_2:

$$CO(g) + H_2O(g) \rightarrow CO_2(g) + H_2(g)$$

In Chapter 17, where we deal with catalysis, we shall describe the conversion of carbon monoxide to methanol, acetic acid, and aldehydes. The CO molecule has very low Brønsted basicity and negligible Lewis acidity toward neutral electron pair donors. Despite its weak Lewis acidity, however, CO is attacked by strong Lewis bases at high pressure and somewhat elevated temperatures. Thus, the reaction with hydroxide yields the formate ion, HCO_2^-:

$$CO(g) + OH^-(s) \rightarrow HCO_2^-(s)$$

Similarly, the reaction with methoxide ions (CH_3O^-) yields the acetate ion, $CH_3CO_2^-$.

Carbon monoxide is an excellent ligand toward metal atoms in low oxidation states, as we shall see in some detail in Chapter 16. Its well-known toxicity is an example of this behavior: it binds to the Fe atom in hemoglobin, so excluding the attachment of O_2, and the victim suffocates. An interesting point is that H_3BCO can be prepared from B_2H_6 and CO at high pressures in a rare example of the coordination of CO to a simple Lewis acid. A complex of similar stability is not formed by BF_3, and this observation prompts the classification of BH_3 as a soft acid and BF_3 as a hard acid.

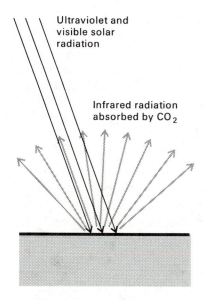

11.26 An illustration of the thermal parameters involved in the greenhouse effect. The IR absorbing molecules retain some of the heat that might otherwise radiate into space.

11.27 Characteristic reactions of CO₂.

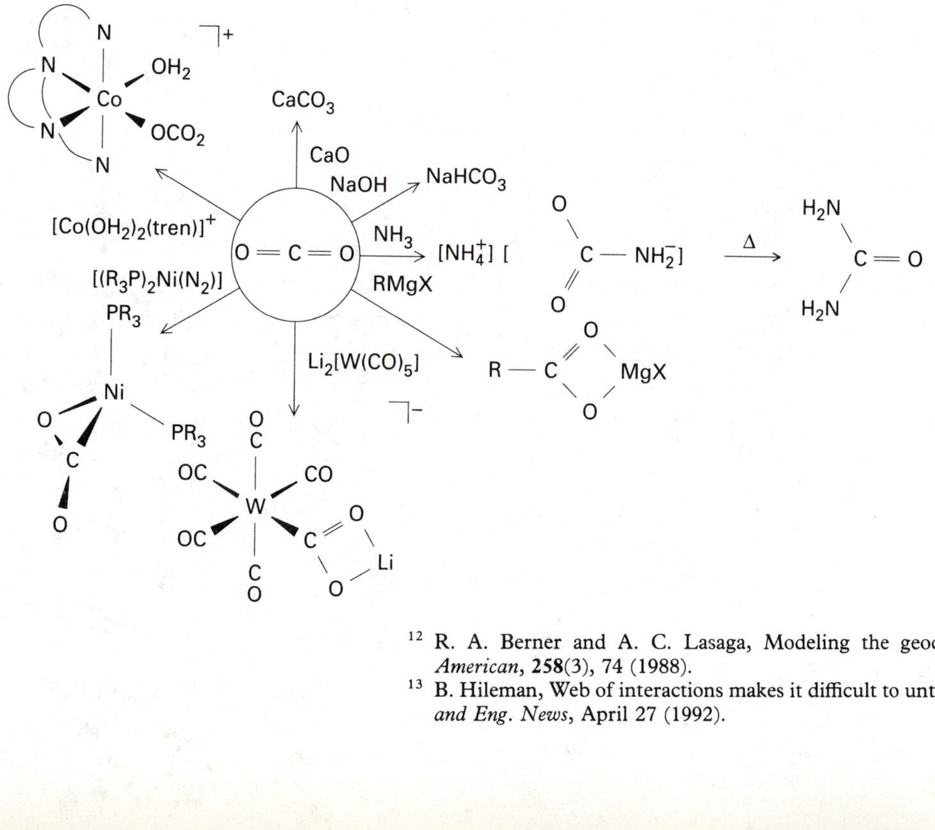

Carbon dioxide shows a number of subtle but significant differences from carbon monoxide. The CO bond is longer and the stretching force constants smaller in CO_2 than in CO, which is consistent with the CO bond in CO_2 being only a double bond. Carbon dioxide is only a very weak Lewis acid. For example, it is only weakly solvated in acidic aqueous solution (Section 5.10) but at higher pH the stronger base OH^- coordinates to the C atom, so forming the hydrogencarbonate (bicarbonate) ion, HCO_3^-.

Carbon dioxide is one several polyatomic molecules that are implicated in the **greenhouse effect**. In this effect, polyatomic molecules in the atmosphere permit the passage of visible light, but because of their vibrational infrared absorption they block the immediate radiation of heat from the Earth (Fig. 11.26). There is strong evidence for a significant increase in atmospheric CO_2 since the industrialization of society. In the past, nature has managed to stabilize the concentration of atmospheric CO_2, in part by precipitation of calcium carbonate in the deep oceans, but it seems that the rate of diffusion of CO_2 into the deep waters is now too slow to compensate for the increased influx of CO_2 into the atmosphere.[12] Although there is convincing evidence for increasing concentrations of the greenhouse gases CO_2, CH_4, N_2O, and chlorofluorocarbons, it is not clear whether they are having an impact on global temperatures. The difficulty with detecting global warming is associated with large natural short-term and long-term variations in temperature, which may mask the effects of greenhouse gases.[13]

The principal chemical properties of CO_2 are summarized in the chart in Fig. 11.27. These properties are based on its mild Lewis

[12] R. A. Berner and A. C. Lasaga, Modeling the geochemical carbon cycle. *Scientific American*, **258**(3), 74 (1988).

[13] B. Hileman, Web of interactions makes it difficult to untangle global warming data. *Chem. and Eng. News*, April 27 (1992).

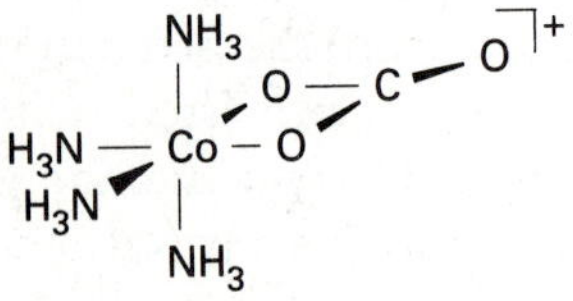

31 $[Co(NH_3)_4(CO_3)]^+$

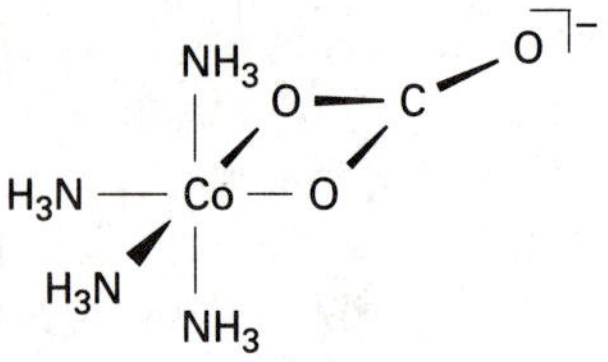

32 $[Co(NH_3)_4CO_3]^+$

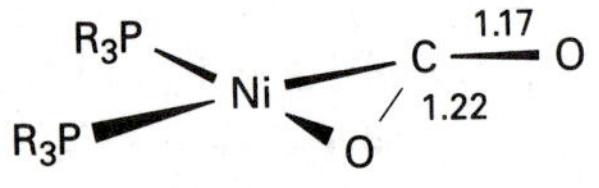

33 $[Ni(PR_3)_2CO_2]$

34

35

acidity toward hard donors, as in the formation of CO_3^{2-} ions in strongly basic solution. The ability of CO_2 to form limestone and thereby moderate atmospheric CO_2 has been mentioned above. Similarly, carbonato complexes of metals can be formed (**31**) in which CO_3^{2-} is a ligand. These complexes are often useful intermediates because CO_3^{2-} can be displaced in acidic solution to yield complexes that are otherwise difficult to prepare:

$$[Co(NH_3)_5(CO_3)]^+ + 2HF \rightarrow [Co(NH_3)_5F]^{2+} + CO_2 + H_2O + F^-$$

Carbonato complexes can be either monodentate, as shown above, or bidentate; in the latter case the CO_3^{2-} ion has a small bite angle (**32**).

From an economic perspective, an important reaction is CO_2 with ammonia to yield ammonium carbonate, $(NH_4)_2CO_3$, which at elevated temperatures is directly converted to urea, $CO(NH_2)_2$, a useful fertilizer, feed supplement for cattle, and chemical intermediate. In organic chemistry, a common synthetic reaction is that between CO_2 and carbanion reagents to produce carboxylic acids.

Metal complexes of CO_2 (**33**) are known, but they are rare and far less important than the metal carbonyls. In its interaction with a low-oxidation state, electron-rich metal center, the neutral CO_2 molecule acts as a Lewis acid and the bonding is dominated by electron donation from the metal atom into an antibonding π orbital of CO_2.

The sulfur analogs of carbon monoxide and carbon dioxide, CS and CS_2, are known. The former is an unstable transient molecule and the latter is endoergic ($\Delta G_f^{\ominus} = +65$ kJ mol^{-1}). Some complexes of CS (**34**) and CS_2 (**35**) exist, and their structures are similar to those formed by CO and CO_2. In basic aqueous solution, CS_2 undergoes hydrolysis and yields a mixture of carbonate ions, CO_3^{2-}, and trithiocarbonate ions, CS_3^{2-}.

Example 11.8: *Proposing a synthesis that uses the reactions of carbon monoxide*
Propose a synthesis of $CH_3CO_2^-$ that uses ^{13}CO, a primary starting material for many ^{13}C labeled compounds.

Answer. We first note that CO_2 is readily attacked by strong nucleophiles such as $Li_4(CH_3)_4$ to produce acetate ions. Therefore an appropriate procedure would be to oxidize ^{13}CO to $^{13}CO_2$ and then to react the latter with $Li_4(CH_3)_4$. A strong oxidizing agent such as solid MnO_2 can be used in the first step, to avoid the problem of excess O_2 in the direct oxidation.

$$^{13}CO(g) + 2MnO_2(s) \xrightarrow{\Delta} {}^{13}CO_2(g) + Mn_2O_3(s)$$

$$4{}^{13}CO_2(g) + Li_4(CH_3)_4(et) \rightarrow 4Li[CH_3{}^{13}CO_2](et)$$

where et denotes solution in ether.

Exercise E11.8. Propose a synthesis of $D^{13}CO_2^-$ starting from ^{13}CO.

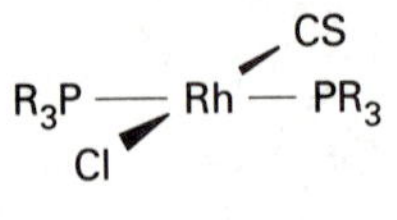

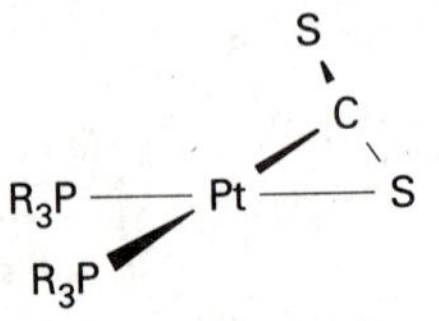

Compounds with nitrogen

Hydrogen cyanide, HCN, is produced in large amounts by the high-temperature catalytic combination of methane and ammonia, and is used as an intermediate in the synthesis of common polymers such as poly(methyl methacrylate) and polyacrylonitrile. It is highly volatile (b.p. 26°C) and, like the CN^- ion, very poisonous. In some respects the toxicity of the CN^- ion is similar to that of the isoelectronic CO molecule, because both form complexes with iron porphyrin molecules. However, whereas CO attaches to the Fe in hemoglobin and causes oxygen starvation, CN^- has a higher affinity for the Fe in cytochrome-c, and blocks energy transfer.

Unlike the neutral ligand CO, the negatively charged CN^- ion is a strong Brønsted base ($pK_a = 9.4$) and a much poorer Lewis π acid. Its coordination chemistry is therefore mainly associated with metal ions in positive oxidation states, as with Fe^{2+} in the hexacyanoferrate(II) complex, $[Fe(CN)_6]^{4-}$.

11.7 Carbides

Carbon forms binary compounds with most other elements. Many of them are important to daily life, for they include CO_2 and Fe_3C, the latter a major constituent of steel. It proves helpful to classify the binary compounds of carbon with metals and metalloids, the **carbides**, into three main categories:

1. **Saline carbides**, which may be considered to be largely ionic solids. These compounds are formed by the elements of Groups 1 and 2 and by aluminum.
2. **Metallic carbides**, which have a metallic conductivity and luster. These compounds are formed by *d*- and *f*-block elements.
3. **Metalloid carbides**, which are hard covalent solids formed by boron and silicon.

Figure 11.28 summarizes the distribution of the different types in the periodic table. This classification is very useful for the purpose of

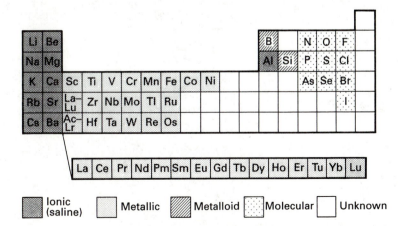

11.28 The distribution of carbides in the periodic table.

Ionic (saline)　　Metallic　　Metalloid　　Molecular　　Unknown

correlating chemical and physical properties, but (as so often in inorganic chemistry) the borderlines are sometimes indistinct.

Saline carbides

The saline carbides of Group 1 and 2 metals may be divided into three categories: **graphite intercalation compounds**, such as KC_8, **dicarbides** or **acetylides**, which contain the C_2^{2-} anion, and **methides**, which contain formally the C^{4-} anion.

The graphite intercalation compounds are formed by the Group 1 metals. They are formed by a redox process, and specifically by the reaction of graphite with alkali metal vapor or with metal–ammonia solution. For example, contact between graphite and potassium vapor in a sealed tube at $300\,°C$ leads to the formation of KC_8. The alkali metal ions lie in an ordered array (Fig. 11.20). A series of alkali metal–graphite intercalation compounds can be prepared with different metal:carbon ratios, including KC_8 and KC_{16}.

The dicarbides are formed by a broad range of electropositive metals, including those from Groups 1 and 2 and the lanthanides. The C_2^{2-} ion has a very short CC distance in some dicarbides (for example, 1.19 Å in CaC_2), which is consistent with it being a triply bonded $[C{\equiv}C]^{2-}$ ion isoelectronic with $[C{\equiv}N]^-$ and $N{\equiv}N$. Some dicarbides have a structure related to rock salt, but replacement of the spherical Cl^- ion by the dumbbell-shaped $[C{\equiv}C]^{2-}$ ion leads to an elongation of the crystal along one axis, and a resulting tetragonal symmetry (Fig. 11.29). The CC bond is significantly longer in the lanthanide dicarbides, which suggests that for them the simple triply bonded structure is not a good approximation.

Carbides such as Be_2C and Al_4C_3 are borderline between saline and metalloid, and the isolated C ion is only formally C^{4-}. The existence of directional bonding to the C atom in carbides (as distinct from the nondirectional character expected of purely ionic bonding) is indicated by the crystal structures of methides, which are not those expected for the simple packing of spherical ions.

The principal synthetic routes to the saline carbides and acetylides of Groups 1 and 2 are very straightforward:

1. *Direct reaction* of the elements at high temperatures:

$$Ca(l) + 2C(s) \xrightarrow{\ >2000\,°C\ } CaC_2(s)$$

The formation of graphite intercalation compounds is another example of a direct reaction, but is carried out at much lower temperatures. The intercalation reaction is more facile because no CC covalent bonds are broken when an ion slips between the graphite layers.

2. *Reaction of a metal oxide and carbon* at a high temperature:

$$CaO(l) + 3C(s) \xrightarrow{\ 2200\,°C\ } CaC_2(l) + CO(g)$$

Crude calcium carbide is prepared in electric arc furnaces by this

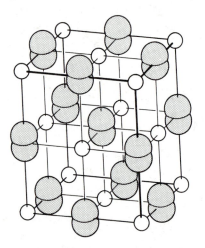

11.29 The calcium carbide structure. Note that this structure bears a similarity to the rock-salt structure. Because C_2^{2-} is not spherical, the cell is elongated along one axis. This crystal is therefore tetragonal rather than cubic.

method. The carbon serves both as a reducing agent to remove the oxygen and as a source of carbon to form the carbide.

3. Reaction of acetylene with a metal–ammonia solution:

$$2Na(am) + C_2H_2(g) \rightarrow Na_2C_2(s) + H_2(g)$$

This reaction occurs under mild conditions and leaves the carbon–carbon bonds of the starting material intact. As the acetylene molecule is a very weak Brønsted acid, the reaction can be regarded as a redox reaction between a highly active metal and a weak acid (with H^+ the oxidizing agent) to yield H_2 and the metal dicarbide.

The saline carbides have high electron density on carbon, so they are readily oxidized and protonated. For example, calcium carbide reacts with the weak acid water to produce acetylene in which the carbon–carbon bond is left intact:

$$CaC_2(s) + 2H_2O(l) \rightarrow Ca(OH)_2(s) + HC{\equiv}CH(g)$$

This reaction is readily understood as the transfer of a proton from a Brønsted acid (H_2O) to the conjugate base (C_2^{2-}) of a weaker acid ($HC{\equiv}CH$). Similarly, the controlled hydrolysis or oxidation of the graphite intercalation compound KC_8 restores the graphite and produces a hydroxide or oxide of the metal:

$$2KC_8(s) + 2H_2O(g) \rightarrow 16C(graphite) + 2KOH(s) + H_2(g)$$

Metallic carbides

Most metallic carbides are formed by the *d* metals. They are sometimes referred to as **interstitial carbides** because their structures are often related to those of metals by the insertion of C atoms in octahedral holes. However, this nomenclature gives the erroneous impression that the metallic carbides are not legitimate compounds. In fact the hardness and other properties of metallic carbides demonstrate that strong metal–carbon bonding is present in them. Some of these carbides are economically useful materials. Tungsten carbide (WC), for example, is used for cutting tools and high pressure apparatus such as that used to produce diamond. Cementite (Fe_3C) is a major constituent of steel and cast iron.

The metallic carbides of composition MC have an fcc or hcp arrangement of metal atoms with the C atoms in the octahedral holes. The fcc arrangement results in a rock-salt structure (Fig. 4.11). The C atoms in carbides of composition M_2C occupy only half the octahedral holes between the close-packed metal atoms. In these structures the C atom has only four valence orbitals to bond with six surrounding metal atoms, and so it is hypervalent. We must therefore adopt a delocalized bonding model using orbitals on C atoms and on the surrounding metal atoms. One such linear combination (**36**) consists of a p_x orbital and two adjacent *d* orbitals.

It has been found empirically that the formation of simple com-

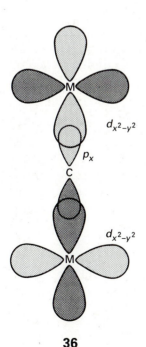

36

pounds in which the C atom resides in an octahedral hole of a close-packed structure occurs when

$$\frac{r_C}{r_M} < 0.59$$

In this expression, r_C is the covalent radius of C and r_M the metallic radius of M. This relation, which is known as **Hägg's rule**, also applies to metal compounds containing nitrogen or oxygen. When the ratio is exceeded the compound normally has a low-symmetry structure.

11.8 Silicon and germanium

The band gap and therefore conductivity of silicon is ideal for many semiconductor applications. However germanium was the first widely used material for the construction of transistors because it was easier to purify than silicon. When methods for purifying silicon were developed in the 1960s, germanium was largely eclipsed for the construction of semiconductors. The principal method of synthesizing semiconductor-grade silicon is to reduce high purity silicon tetrachloride with hydrogen:

$$SiCl_4(g) + 2H_2(g) \rightarrow Si(s) + 4HCl(g)$$

The semiconducting material used for solar cells to power calculators is called 'amorphous silicon,' but it is better regarded as a solid silicon hydride SiH_x ($x \leq 0.5$; Section 9.12).

11.9 Compounds of silicon with electronegative elements

Some of the most important compounds of silicon and germanium contain the electronegative halogens, oxygen, or nitrogen. The oxides are so extensive that they deserve their own section. In this section we consider the halides and the nitrides.

Compounds with halogens

The full range of tetrahalides is known for silicon and germanium; all of them are volatile molecular compounds. Lower down the group, germanium shows signs of an inert pair effect in that it also forms nonvolatile dihalides. Among the silicon tetrahalides, the most important is the tetrachloride, which is prepared by direct reaction of the elements:

$$Si(s) + 2Cl_2(g) \rightarrow SiCl_4(l)$$

Silicon and germanium halides are mild Lewis acids. They display this behavior when they add one or two ligands to yield complexes with a five-coordinate or six-coordinate central atom:

$$SiF_4(g) + 2F^-(aq) \rightarrow [SiF_6]^{2-}(aq)$$

$$GeCl_4(l) + N \equiv CCH_3(l) \rightarrow Cl_4GeN \equiv CCH_3(aq)$$

The ease of forming hypervalent silicon and germanium complexes accounts for the much faster hydrolysis of the silicon and germanium halides compared with CCl_4. The process is represented schematically as follows:

$$MX_4 + 2H_2O \rightarrow MX_4(OH_2)_2 \rightarrow MO_2 + 4HX$$

The substitution reactions of halosilanes have been studied extensively. The reactions are more facile than for their carbon analogs because a Si atom can readily expand its coordination sphere to accommodate the incoming nucleophile in an associative mechanism. The stereochemistry of these substitution reactions[14] indicates that a five-coordinate intermediate is formed with the most electronegative substituents adopting the axial position. Moreover, substituents leave from the axial position. The H^- ion is a poor leaving group, and alkyl groups are even poorer:

Note that in this example, the R^4 substituent replaces H with retention of configuration.

Compounds with nitrogen

The direct reaction of silicon and nitrogen gas at high temperatures produces Si_3N_4. This substance is very hard and inert, and may find use in high-temperature ceramic materials. Current industrial research projects focus on the use of suitable organosilicon–nitrogen compounds that might undergo pyrolysis to yield silicon nitride fibers and other shapes.

Trisilylamine, $(H_3Si)_3N$, the silicon analog of trimethylamine, has very low basicity and a planar structure. These features are rationalized by the delocalization of the lone pair from the $N2p_z$ orbital into a d or π^* orbital on Si (**37**). This effect possibly occurs in conjunction with a greater p-orbital character for the lone pair as a result of a flattening of the molecule by steric repulsions between the large silyl groups. On the other hand, $(H_3Ge)_3N$ is pyramidal.

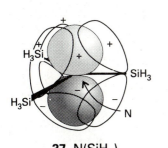

37 $N(SiH_3)_3$

11.10 Extended silicate structures

The high affinity of silicon for oxygen accounts for the existence of a vast array of silicate minerals and synthetic silicon–oxygen compounds, which are important in mineralogy, industrial processing, and the laboratory. Aside from rare high-temperature phases, the structures of silicates are confined to tetrahedral four-coordinate Si. The complicated silicate structures are often easier to comprehend if the SiO_4 unit

[14] R. R. Holmes, Stereochemistry of nucleophilic substitution at tetracoordinate silicon. *Chem. Rev.*, **90**, 17 (1990).

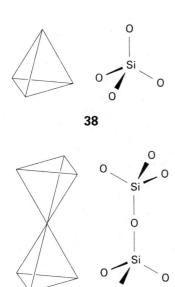

38

39

is drawn as a tetrahedron with the Si atom at the center and O atoms at the vertices. The representation is often cut to the bone by drawing the SiO_4 unit as a simple tetrahedron with the atoms omitted. In general, these tetrahedra share vertices and (much more rarely) edges or faces. Each terminal O atom contributes -1 to the charge of the SiO_4 unit, but each shared O atom contributes zero. Thus, orthosilicate (**38**) is $[SiO_4]^{4-}$, disilicate (**39**) is $[O_3SiOSiO_3]^{6-}$, and the SiO_2 unit of silica has no net charge because all O atoms are shared.

With the above principles of charge balance in mind, it should be clear that an endless single-stranded chain or a ring of SiO_4 units, which has two shared O atoms for each Si atom, will have the formula and charge $[(SiO_3)^{2-}]_n$. An example of a compound containing such a cyclic metasilicate ion is the mineral beryl, $Be_3Al_2Si_6O_{18}$, which contains the $[Si_6O_{18}]^{12-}$ ion (**40**). Beryl is a major source of beryllium. Emerald is beryl in which Cr^{3+} ions are substituted for some Al^{3+} ions. A chain metasilicate (**41**) is present in the mineral jadeite, $NaAl(SiO_3)_2$, one of two different minerals sold as jade. In addition to other configurations for the single chain, there are double-chain silicates which include the family of minerals known commercially as asbestos.[15]

Example 11.9: *Determining the charge on a cyclic silicate*
Draw the structure and determine the charge on the cyclic silicate anion $[Si_3O_9]^{n-}$.

Answer. The ion is a six-membered ring with alternating Si and O atoms and 6 terminal O atoms, two on each Si atom. Because each terminal O atom contributes -1 to the charge, the overall charge is -6. From another perspective, the conventional oxidation states of silicon and oxygen, $+4$ and -2, respectively, also indicate a charge of -6 for the anion.

Exercise E11.9. Repeat the question for the cyclic anion $[Si_4O_{12}]^{n-}$.

Silica and many silicates crystallize slowly. By cooling the melt at an appropriate rate amorphous solids known as **glasses** can be obtained instead of crystals. In some respects these glasses resemble liquids. As with liquids, their structures are ordered over distances of only a few interatomic spacings (such as within a single SiO_4 tetrahed-

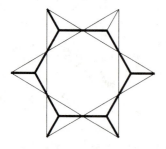

40 $[Si_6O_{18}]^{12-}$

41

---

[15] Asbestos has many commercially desirable qualities, but the fine asbestos particles are readily airborne and workers who handle the mineral are subject to a degeneration of the lung tissue which may manifest itself many years after the exposure. See L. Michaelis and S. S. Chissick (eds), *Asbestos: Properties, applications, and hazards.* Wiley, New York (1979).

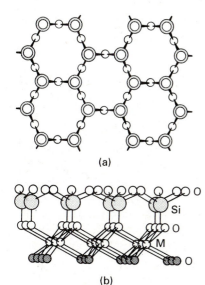

(a)

(b)

11.30 (a) A net of SiO_4 tetrahedra with one O atom on top of each Si atom projecting toward the viewer. (b) Edge view of the above net, with one O/Si incorporated into a net of MO_6 octahedra. This structure is for the mineral chrysotile, for which M is Mg. When M is Al^{3+} and each of the atoms on the bottom is replaced by an OH group this structure is close to that of the 1:1 clay mineral kaolinite.

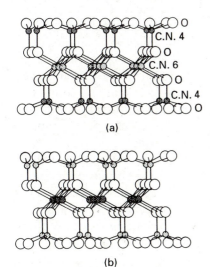

(a)

(b)

11.31 (a) The structure of 2:1 clay minerals such as muscovite mica $KAl_2(OH)_2Si_3AlO_{10}$ where K^+ resides between the charged layers (exchangeable cation sites), Si^{4+} resides in sites of C.N.=4 and Al^{3+} in sites of C.N.=6. (b) In talc, Mg^{2+} resides in the octahedral sites and O atoms on the top and bottom are replaced by OH groups.

ron). Unlike liquids, their viscosities are very high, and for many practical purposes they behave like solids.

The composition of silicate glasses has a strong influence on their physical properties. For example, fused quartz (amorphous SiO_2) softens at about 1600 °C, borosilicate glass (which contains boron oxide, as we have already seen) softens at about 800 °C, and sodalime glass softens at even lower temperatures. The variation in softening point can be understood by appreciating that the Si—O—Si links in silicate glasses form the framework that imparts rigidity. When basic oxides such as Na_2O and CaO are incorporated (as in sodalime glass), they react with the SiO_2 melt and convert Si—O—Si links into terminal SiO groups and hence lower the softening temperature.

11.11 Aluminosilicates

Even greater structural diversity is possible when Al atoms replace some of the Si atoms. The resulting **aluminosilicates** are largely responsible for the rich variety of the mineral world. We have already seen that in γ-alumina, Al^{3+} ions are present in both octahedral or tetrahedral holes. This versatility carries over into the aluminosilicates, where Al may substitute for Si in tetrahedral sites, enter an octahedral environment external to the silicate framework, or, more rarely, occur with other coordination numbers. Because aluminum occurs as Al(III), its presence in place of Si(IV) in an aluminosilicate renders the overall charge negative by one unit. An additional cation, such as H^+, Na^+, or $\frac{1}{2}Ca^{2+}$, is therefore required for each Al atom that replaces an Si atom. As we shall see, these additional cations have a profound effect on the properties of the materials.

Layered aluminosilicates

Many important minerals are varieties of layered aluminosilicates that also contain metals such as lithium, magnesium, and iron: they include clays, talc, and various micas. In one class of layered aluminosilicate, the repeating unit consists of a silicate layer with the structure shown in Fig. 11.30. An example of a simple aluminosilicate of this type (simple, that is, in the sense of there being no additional elements) is the mineral kaolinite, $Al_2(OH)_4Si_2O_5$, which is used commercially as china clay. The electrically neutral layers are held together by rather weak hydrogen bonds, so the mineral readily cleaves and incorporates water between the layers.

A larger class of aluminosilicates has Al^{3+} ions sandwiched between silicate layers (Fig. 11.31). One such mineral is pyrophyllite, $Al_2(OH)_2Si_4O_{10}$. The mineral talc $Mg_3(OH)_2Si_4O_{10}$ is obtained when three Mg^{2+} ions replace two Al^{3+} ions in the octahedral sites. In talc (and in pyrophyllite) the repeating layers are neutral, and as a result talc readily cleaves between them. This structure accounts for its familiar slippery feel.

Muscovite mica, $KAl_2(OH)_2Si_3AlO_{10}$, has charged layers because one Al(III) atom substitutes for one Si(IV) atom in the pyrophyllite structure. The resulting negative charge is compensated by a K^+ ion that lies between the repeating layers. Because of this electrostatic cohesion, muscovite is not soft like talc, but it is readily cleaved into sheets. More highly charged layers with dipositive ions between the layers lead to greater hardness.

Three-dimensional aluminosilicates

There are many minerals based on a three-dimensional aluminosilicate framework. The **feldspars**, for instance, which are the most important class of rock-forming minerals (and contribute to granite), belong to this class. The aluminosilicate frameworks of feldspars are built up by sharing all vertices of SiO_4 or AlO_4 tetrahedra. The cavities in this three-dimensional network accommodate ions such as K^+ and Ba^{2+}. Two examples are the feldspars orthoclase, $KAlSi_3O_8$, and albite, $NaAlSi_3O_8$.

Molecular sieves

The **molecular sieves** are crystalline aluminosilicates having open structures with apertures of molecular dimensions. These substances represent a major triumph of solid state chemistry, for their synthesis and our understanding of their properties combine challenging determinations of structures, imaginative synthetic chemistry, and important practical applications.

The name 'molecular sieve' is prompted by the observation that these materials adsorb only molecules that are smaller than the aperture dimensions and so can be used to separate molecules of different sizes. A subclass of molecular sieves, the **zeolites**,[16] have an aluminosilicate framework with cations (typically from Groups 1 or 2) trapped inside tunnels or cages. In addition to their function as molecular sieves, zeolites can exchange their ions for those in a surrounding solution.

The cages are defined by the crystal structure, so they are highly regular and of precise size. Consequently, molecular sieves capture molecules with greater selectivity than high surface area solids such as silica gel or activated carbon, where molecules may be caught in irregular voids between the small particles. Zeolites are also used for shape-selective heterogeneous catalysis. For example, the molecular sieve ZSM-5 is used to synthesize 1,2-dimethylbenzene (*o*-xylene) for use as an octane booster in gasoline. The other xylenes are not produced because the catalytic process is controlled by the size and shape of the zeolite cages and tunnels. This and other applications are summarized in Table 11.10 and discussed in Chapter 17.

Synthetic procedures have added to the many naturally occurring zeolite varieties that have specific cage sizes and specific chemical properties within the cages. These synthetic zeolites are sometimes made at atmospheric pressure, but more often they are produced in a high pressure autoclave. Their open structures appear to form around

[16] The name zeolite is derived from the Greek word for boiling stone. Geologists found that certain rocks appeared to boil when subjected to the flame of a blowpipe.

Table 11.10 Some uses of zeolites

Function	Application
Ion exchange	Water softeners in washing detergents
Absorption of molecules	Selective gas separation Industrial processes, gas chromatography, laboratory experiments
Solid acid	Cracking of high molar mass and selectivity hydrocarbons for fuel and petrochemical intermediates
	Alkylation and isomerization shape selectivity of aromatics for gasoline and polymer intermediates

Table 11.11 Composition and properties of some molecular sieves

Molecular sieve	Composition	Bottleneck diameter/Å	Chemical properties
A	$[Na_{12}[(AlO_2)_{12}(SiO_2)_{12}] \cdot xH_2O$	4	Absorbs small molecules; ion exchanger, hydrophilic
X	$Na_{86}[(AlO_2)_{86}(SiO_2)_{106}] \cdot xH_2O$	8	Absorbs medium sized molecules; ion exchanger, hydrophilic
Chabazite	$Ca_2[(AlO_2)_4(SiO_2)_8] \cdot xH_2O$	4-5	Absorbs small molecules; ion exchanger, hydrophilic
ZSM-5	$Na_3[(AlO_2)_3(SiO_2)] \cdot xH_2O$	5.5	Moderately hydrophilic
ALPO-5	$AlPO_4 \cdot xH_2O$	8	Moderately hydrophilic
Silicalite	SiO_2	6	Hydrophobic

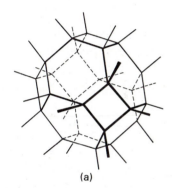

(a)

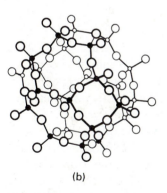

(b)

11.32 (a) Framework representation of a truncated octahedron (truncation perpendicular to the fourfold axes of the octahedron). (b) Relation of Si and O atoms to the framework. Note that a Si atom is at each vertex of the truncated octahedron and an O atom is approximately along each edge.

hydrated cations or other large cations such as NR_4^+ ions introduced into the reaction mixture. For example, a synthesis may be performed by heating colloidal silica to between 100 °C and 200 °C in an autoclave with an aqueous solution of tetrapropylammonium hydroxide. The microcrystalline product, which has the typical composition $\{[N(C_3H_7)_4][OH]\}(SiO_2)_{48}$, is converted into the zeolite by burning away the C, H, and N of the quaternary ammonium cation at 500 °C in air. Aluminosilicate zeolites are made by including high surface area alumina in the starting materials.

A wide range of zeolites has been prepared with varying cage and bottleneck sizes (Table 11.11). Their structures are based on approximately tetrahedral MO_4 units, which in the great majority of cases are SiO_4 and AlO_4. Because the structures involve many such tetrahedral units, it is common practice to abandon the polyhedral representation in favor of one that emphasizes the position of the Si and Al atoms. In this scheme, the Si or Al atom lies at the intersection of four line segments (Fig 11.32) and the O atom bridge lies on the line segment. This **framework representation** has the advantage of giving a clear impression of the shapes of the cages and channels in the zeolite.[17]

---

[17] The combination of X-ray crystallography, solid state ^{29}Si-NMR and ^{27}Al-NMR with high resolution electron microscopy has revolutionized the structural determination of zeolites and other aluminosilicates. An interesting account is given by J. M. Thomas and C. R. A. Catlow, *Prog. Inorg. Chem.*, **35**, 1 (1987).

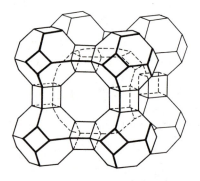

11.33 Framework representation of a type-A Zeolite. Note the sodalite cages (truncated octahedra), the small cubic cages, and the central supercage.

A large and important class of zeolites is based on the **sodalite cage**. This is a truncated octahedron (Fig. 11.33), a geometric shape that can be formed by slicing off each vertex of an octahedron. The truncation leaves a square face in the place of each vertex and the triangular faces of the octahedron are transformed into regular hexagons. The substance known as 'Zeolite type A' is based on sodalite cages that are joined by O bridges between the square faces. Eight such sodalite cages are linked in a cubic pattern with a large central cavity called an α cage. The α cages share octagonal faces, with an open diameter of 4.20 Å. Thus water or other small molecules can fill them and diffuse through octagonal faces. However, these faces are too small to permit the entrance of molecules with van der Waals diameters larger than 4.20 Å.

Example 11.10: *Analyzing the structure of the sodalite cage*

Identify the fourfold and sixfold axes in the truncated octahedral polyhedron used to describe the sodalite cage.

Answer. There is one fourfold axis running through each pair of opposite square faces. This gives a total of six fourfold axes. Similarly, a set of four sixfold axes runs through opposite sixfold faces.

Exercise E11.10. How many Si and Al atoms are there in one sodalite cage?

The charge on the aluminosilicate zeolite framework is neutralized by cations lying within the cages. In the type-A zeolite, Na^+ ions are present and the formula is $Na_{12}(AlO_2)_{12}(SiO_2)_{12} \cdot xH_2O$. Numerous other ions, including *d*-block cations and NH_4^+, can be introduced by ion exchange with aqueous solutions, and another of the major uses of zeolites is for ion exchange. In this connection, zeolites are widely used in laundry detergent to remove di- and tri-positive ions that decrease the effectiveness of the surfactant. Zeolites have largely replaced polyphosphates because the latter, which are plant nutrients, find their way into natural waters and stimulate the growth of algae.

In addition to the control of properties by selecting a zeolite with the appropriate cage and bottleneck size, the zeolite can be chosen for its affinity for polar or nonpolar molecules according to its polarity (Table 11.11). The aluminosilicate zeolites, which always contain charge-compensating ions, have high affinities for polar molecules such as H_2O and NH_3. In contrast, the nearly pure silica molecular sieves bear no net electric charge and are nonpolar to the point of being mildly hydrophobic. Another group of hydrophobic zeolites is based on the aluminum phosphate frameworks, for $AlPO_4$ is isoelectronic with Si_2O_4 and the framework is similarly uncharged.

One interesting aspect of zeolite chemistry is that large molecules can be synthesized from smaller molecules inside the zeolite cage. The result is like a ship in a bottle, because once assembled the large molecule is too large to escape. For example, Na^+ ions in a Y-type

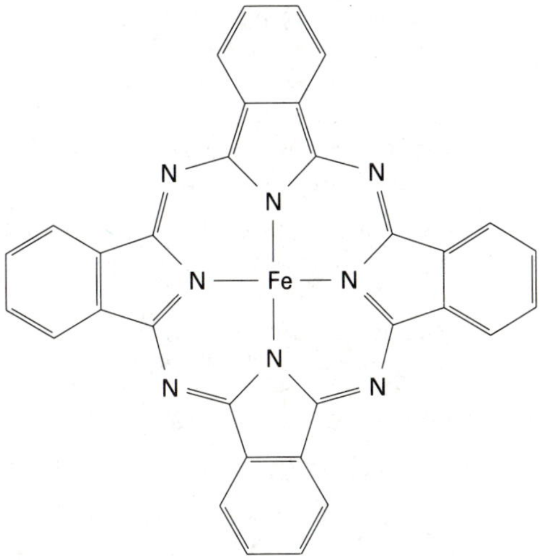

42

zeolite may be replaced by Fe^{2+} ions (by ion exchange). The resulting Fe^{2+}–Y zeolite is heated with phthalonitrile, which diffuses into the zeolite and condenses around the Fe^{2+} to form iron phthalocyanine (**42**) which is too large to escape from the cage.

11.12 Silicides

Silicon, like its neighbors boron and carbon, forms a wide variety of binary compounds with metals. Some of them contain isolated Si atoms. The structure of ferrosilicon Fe_3Si, for instance, which plays an important role in steel manufacture, can be viewed as an fcc array of Fe atoms with some replaced by Si. Compounds such as K_4Si_4 contain isolated tetrahedral cluster anions $[Si_4]^{4-}$ that are isoelectronic with P_4. Many of the *f*-block elements form compounds with the formula MSi_2 that have the layered AlB_2 structure (Fig. 11.34).

Silicon carbide (SiC) is the most common of the compounds of silicon with nonmetals because it is synthesized on a large scale for use as an abrasive. One polymorph of SiC mimics diamond itself and has the diamond-like sphalerite structure (Fig. 4.13).

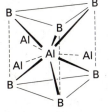

11.34 The AlB_2 structure. To give a clear picture of the hexagonal B layer, some B atoms outside the Al cube are displayed.

FURTHER READING
Boron

A good introduction to boron hydrides and carboranes is given by K. Wade, *Electron deficient compounds*. Nelson, London (1971).

A series of articles that cover major areas of borane and carborane chemistry is available in: G. A. Olah, K. Williams, and R. E. Williams (eds). *Electron deficient boron and carbon clusters*. Wiley-Interscience, New York (1991).

The incorporation of metals into boranes is described in R. N. Grimes (ed.), *Metal interactions with boron*. Plenum Press, New York (1982).

The metallic face of boron. T. P. Fehlner, *Adv. Inorg. Chem.*, **35**, 199 (1990).

An excellent series of reviews on boranes and carboranes can be found in *Chem. Rev.*, **92**, 177–362 (1992).

The carbon group

Kirk-Othmer encyclopedia of chemical technology, especially: carbon, semi-conductors, silicon.

A series of useful reviews on fullerene chemistry is available in *Acc. Chem. Res.*, **92**, 97 et seq. (1992).

Aqueous chemistry, silicate minerals, and aluminosilicates

C. F. Baes and R. E. Mesmer, *The hydrolysis of cations*. Wiley, New York (1976).

W. Stumm and J. J. Morgan, *Aquatic chemistry*. Wiley, New York (1981).

A. Dyer, *An introduction to zeolite molecular sieves*. Wiley, Chichester (1988).

R. M. Barrer, *Hydrothermal chemistry of zeolites*. Academic Press, New York (1982).

KEY POINTS

1. The elements

In Groups 13/III and 14/IV the light elements are distinctly nonmetals and the heavy elements are metals. The borderline occurs between boron and aluminum in Group 13/III and between germanium and tin in Group 14/IV.

2. Extraction

Elemental boron (Group 13/III) is not important commercially. In Group 14/IV, carbon from graphite mining or pyrolysis of hydrocarbons or coal, finds many uses. Reduction by carbon is used in the recovery of the other commercially important elements. Silica requires very high temperatures for its reduction, whereas the heavy metal oxides of tin and lead are reduced at moderate temperatures.

3. Boron halides

The simple trihalides are notable for their Lewis acidity and by the facile halide nucleophilic displacement reactions they undergo, to produce BR_3 and $[BR_4]^-$, where R = alkyl, aryl, alkoxy, amido, and the like. A series of boron subhalides is known, the most important of which have the formula B_2X_4.

4. Boron oxides

Boron has a high affinity for oxygen and is mined in the form of metal borates. Boron oxide, B_2O_3, in combination with SiO_2 and alkali metal oxides is widely used to make glass which is tolerant to thermal shock. The coordination number of B in oxides may be either 3 or 4.

5. Boron nitrogen compounds

Boron nitrogen compounds are often structural analogs of carbon compounds because BN is isoelectronic with CC. Thus, one form of BN has a structure similar to graphite and another resembles diamond. Other BN analogs of carbon compounds include Lewis acid–base complexes (such as H_3NBF_3) and analogs of alkenes and amino acids.

6. Boranes and carboranes

The boranes constitute a large group of compounds which are conveniently represented by delocalized bonding, such as the 3-center, 2-electron BHB bond. Most of them fall into one of three classes: *closo*, $[B_nH_n]^-$; *nido*, B_nH_{n+4}; and *arachno*, B_nH_{n+6}. An isoelectronic series of carboranes is known, which formally represents the replacement of BH^- by the isolobal CH group.

7. Allotropes of carbon

Elemental carbon exists in many forms, ranging from the extended crystalline solids, diamond and graphite, to semicrystalline analogs of graphite and clusters such as C_{60}.

8. Carbides

Carbon compounds include the familiar hydrocarbons and halogenated hydrocarbons, and simple oxygen and nitrogen compounds. In addition, carbon forms a series of compounds with metals and metalloids: saline carbides, which are formed by highly electropositive metals display ionic character in many of their reactions; metallic carbides, formed with many *d*-block metals are often hard materials with high electronic conductivity; metalloid carbides are often hard covalent insulating or semiconducting solids, such as SiC.

9. Silicon and germanium halides
The halides of silicon and germanium are notable for their mild Lewis acidity, which is achieved by hypervalence, and their utility as starting materials by the nucleophilic displacement of halide to form compounds such as $Si(CH_3)_4$ and SiH_4.

10. Silicates
The silicates are metal–silicon–oxygen compounds that contain the tetrahedral SiO_4 building block that may share one or two of its O atoms with an adjacent SiO_4. These arrangements correspond to shared tetrahedral vertices or edges respectively.

In SiO_2, the four O atoms around a Si atom have four surrounding SiO_4 units. The progressive introduction of alkali metal and alkaline earth metal oxides to form metal silicates leads to decreased sharing of O atoms between the SiO_4 tetrahedra.

11. Aluminosilicates
Aluminosilicates, together with silicates, constitute the bulk of the minerals in the Earth's crust. A vertex- or edge-shared MO_4 tetrahedron is the building block, where M is either Si or Al. Aluminosilicate minerals include the three-dimensional feldspars; many two-dimensional minerals, such as clays, talc, and mica; and a few with one-dimensional topology.

12. Molecular sieves
Most molecular sieves are aluminosilicate minerals or synthetic compounds that have open structures through which small molecules can diffuse. These compounds find use as desiccants through their water absorbing properties, ion exchange materials, and solid acid catalysts.

EXERCISES

11.1 Preferably without consulting reference material, list the elements in Groups 13/III and 14/IV and indicate (a) the metals and nonmetals, (b) those that can crystallize in the diamond structure, (c) the most oxophilic elements in these groups.

11.2 Draw the B_{12} unit that is a common motif of boron structures; take a viewpoint along a C_2 axis.

11.3 Give balanced chemical equations and conditions for the recovery of boron, silicon, and germanium from their ores. Which of these is likely to be the most energy efficient? Explain your answer.

11.4 Arrange the following in order of increasing Lewis acidity: BF_3, BCl_3, SiF_4, $AlCl_3$. (b) In the light of this order, write balanced chemical reactions (or NR) for

(i) $SiF_4N(CH_3)_3 + BF_3 \rightarrow$?
(ii) $BF_3N(CH_3)_3 + BCl_3 \rightarrow$?
(iii) $BH_3CO + BBr_3 \rightarrow$?

11.5 Using BCl_3 as a starting material and other reagents of your choice, devise a synthesis for the Lewis acid chelating agent, $F_2B-C_2H_4-BF_2$.

11.6 State the coordination numbers of boron, carbon, and silicon in their oxoanions and devise plausible explanations for the differences based on Lewis electron dot structures.

11.7 Which boron hydride would you expect to be more thermally stable, B_6H_{10} or B_6H_{12}? Give a generalization by which the thermal stability of a borane can be judged.

11.8 (a) Give a balanced chemical equation (including the state of each reactant and product) for the air oxidation of pentaborane(9). (b) Describe the probable disadvantages, other than cost, for the use of pentaborane as a fuel for an internal combustion engine.

11.9 (a) From its formula, classify $B_{10}H_{14}$ as *closo*, *nido*, or *arachno*. (b) Use Wade's rules to determine the number of framework electron pairs for decaborane(14). (c) Verify by detailed accounting of valence electrons that the number of cluster valence electrons of $B_{10}H_{14}$ is the same as that determined in (b).

11.10 Starting with $B_{10}H_{14}$ and other reagents of your choice, give the equations for the synthesis of [Fe(*nido*-$B_9C_2H_{11}$)$_2$]$^{2-}$, and sketch the structure of this species.

11.11 (a) What are the similarities and differences in the structures of layered BN and graphite? (b) Contrast their reactivity with Na and Br_2. (c) Suggest a rationalization for the differences in structure and reactivity.

11.12 Devise a synthesis for the borazines (a) $Ph_3N_3B_3Cl_3$ and (b) $Me_3N_3B_3H_3$, starting with BCl_3 and other reagents of your choice. Draw the structures of the products.

11.13 Give the structures and names of B_4H_{10}, B_5H_9, and $1,2-B_{10}C_2H_{12}$.

11.14 Arrange the following boron hydrides in order of increasing Brønsted acidity, and draw a structure for the probable structure of the deprotonated form of one of them: B_2H_6, $B_{10}H_{14}$, B_5H_9.

11.15 (a) Describe the trend in band gap energy, E_g, for the elements carbon (diamond) through tin (gray), and for cubic BN, AlP, and GaAs. (b) Does the electrical conductivity of silicon increase or decrease when its temperature is changed from $20\,°C$ to $40\,°C$? (c) State the functional dependence of conductivity on temperature for a semiconductor and decide whether the conductivity of AlP or GaAs will be more sensitive to temperature.

11.16 Preferably without consulting reference material, draw a periodic table and indicate the elements that form saline, metallic, and metalloid carbides.

11.17 Describe the preparation, structure, and classification of (a) KC_8, (b) CaC_2, (c) K_3C_{60}.

11.18 There are major commercial applications for semicrystalline and amorphous solids, many of which are formed by Group 14/IV elements or their compounds. List four different examples of amorphous or partially crystalline solids described in this chapter and briefly state their useful properties.

11.19 The lightest *p*-block elements often display different physical and chemical properties from the heavier members. Discuss the similarities and differences by comparison of:
(a) The structures and electrical properties of (i) boron and aluminum and of (ii) carbon and silicon.
(b) The physical properties and structures of the oxides of carbon and silicon.
(c) The Lewis acid–base properties of the tetrahalides of carbon and silicon.
(d) The structures of the halides of boron and aluminum.

11.20 Write balanced chemical equations for the reactions of K_2CO_3 with HCl(aq) and of Na_4SiO_4 with aqueous acid.

11.21 Describe in general terms the nature of the $[SiO_3]_n^{2n-}$ ion in jadeite and the silica-alumina framework in kaolinite.

11.22 (a) Determine the number of bridging O atoms are in the framework of a single sodalite cage? (b) Describe the (super-cage) polyhedron at the center of the Zeolite A structure in Fig. 11.33.

11.23 Describe the physical properties of pyrophyllite and muscovite mica and explain how these properties arise from the composition and structures of these closely related aluminosilicates.

PROBLEMS

11.1 Boron-11 NMR is an excellent spectroscopic tool for inferring the structures of boron compounds. Under conditions in which the $^{11}B-^{11}B$ coupling is absent, it is possible to determine the number of attached H atoms by the multiplicity of a resonance: BH gives a doublet, BH_2 a triplet, and BH_3 a quartet. Also, B atoms on the closed side of *nido* and *arachno* clusters are generally more shielded than those on the open face. Assuming no B–B or B–H–B coupling, predict the general pattern of the ^{11}B-NMR spectra of (a) BH_3CO, (b) $[B_{12}H_{12}]^{2-}$, and (c) B_4H_{10}.

11.2 Identify the incorrect statements in the following description of Group 13/III chemistry and provide corrections along with explanations of principles or chemical generalizations that apply. (a) All the elements in Group 13/III are nonmetals. (b) The increase in chemical hardness on going down the group is illustrated by greater oxophilicity and fluorophilicity for the heavier elements. (c) The Lewis acidity increases for BX_3 from X = F to Br and this may be explained by stronger Br—B π bonding. (d) *Arachno*-boron hydrides have a $2(n+3)$ skeletal electron count and are more stable than *nido*-boron hydrides. (e) In a series of *nido*-boron hydrides acidity increases with increasing size. (f) Layered boron nitride is similar in structure to graphite

and, because it has a small separation between HOMO and LUMO, is a good electrical conductor.

11.3 Correct any inaccuracies in the following descriptions of Group 14/IV chemistry. (a) None of the elements in this group is a metal. (b) At very high pressures, diamond is a thermodynamically stable phase of carbon. (c) Both CO_2 and CS_2 are weak Lewis acids and the hardness increases from CO_2 to CS_2. (d) Zeolites are layered materials exclusively composed of aluminosilicates. (e) The reaction of calcium carbide with water yields acetylene and this product reflects the presence of a highly basic C_2^{2-} ion in calcium carbide.

11.4 Acetylcholine, $[(CH_3)_3N(CH_2)_2OC(O)CH_3]^+$, is an important neurotransmitter, and the physiological properties of the neutral boron analog $(CH_3)_2(BH_3)N(CH_2)_2OC(O)CH_3$ are of interest. (B. F. Spielvogel, F. U. Ahmed, and A. T. McPhail, *Inorg. Chem.*, 25, 4395 (1986).) Devise a method of synthesis for this analog starting with $[(CH_3)_2HN(CH_2)_2OC(O)CH_3]^+$ and other reagents of your choice.

11.5 Two reactions used in commercial processes for the production of hydrogen are the reduction of $H_2O(g)$ with

CO to give CO_2 and H_2, often referred to as the 'water–gas shift reaction', and the analogous reduction of water with methane. Both reactions are carried out with reactants and products in the gas phase at elevated temperatures in the presence of a catalyst. (a) Using a reliable source of thermodynamic data,[18] determine $\Delta G^{\ominus}$, and $\Delta S^{\ominus}$ per mole of H_2 for each of these reactions assuming gas-phase species at 1 bar. (b) Are these reactions thermodynamically favorable under the above conditions? (c) From qualitative considerations and also from the thermodynamic data determine whether each of these reactions is more, or less favored, thermodynamically, by (i) higher temperature, (ii) higher total pressure.

11.6 Use a suitable extended Hückel molecular orbital program[19] to calculate the wavefunctions and energy levels for *closo*-$[B_6H_6]^{2-}$. From that output, draw a molecular orbital energy diagram for the orbitals primarily involved in BB bonding and sketch the form of the orbitals. How do these orbitals compare qualitatively with the description for this anion in Section 11.3? Is BH bonding neatly separated from the BB bonding in the extended Hückel wavefunctions?

11.7 The layered silicate compound $CaAl_2Si_2O_8$, contains a double aluminosilicate layer, with both Si and Al in four-coordinate sites. Sketch an edge-on view of a reasonable structure for the double layer, involving only vertex sharing between the SiO_4 and AlO_4 units. Discuss the likely sites occupied by Ca^{2+} in relation to the silica–alumina double layer.

[18] D. D. Wagman, *et al.*, The NBS Tables of Thermodynamic Properties, *J. Phys Chem. Ref. Data,* **11,** Supplement 2 (1982).

[19] Suitable programs include QCMP 001 from Quantum Chemistry Program Exchange, Chemistry Dept., Indiana University, Bloomington, IN; CACAO by C. Meali and D. M. Proserpio, *J. Chem. Educ.*, **67,** 3399 (1990); PLOT3D, J. A. Bertrand and M. R. Johnson, School of Chemistry, Georgia Institute of Technology, Atlanta, GA.

12

The nitrogen and oxygen groups

As in the rest of the p block, the elements at the head of these two groups, nitrogen and oxygen, differ significantly from their congeners. Their coordination numbers are generally lower in their compounds and they are the only members of the group to exist as diatomic molecules under normal conditions. The chemical properties of nitrogen, oxygen, phosphorus, and sulfur are intricate because these elements possess a wide range of oxidation states. Moreover, their reactions are strongly influenced by kinetic factors. A good approach to the mastery of their properties is to learn the structures of important species for each oxidation state and the general thermodynamic trends for their redox reactions. With these features in mind, the rates and mechanisms of their interconversion reactions complete the picture.

The nitrogen and oxygen groups, specifically Groups 15/V and 16/VI of the periodic table, contain some of the most important elements for geology, life, and industry. However, the heaviest elements of the groups are much less abundant than the lighter elements, for they lie at the frontier of nuclear stability. Indeed, the Group 15/V element bismuth ($Z=83$) is the heaviest element to have stable isotopes. The most common isotope of its neighbor, the Group 16/VI element polonium ($Z=84$), ^{210}Po, is an intense α emitter with a half-life of 138 d.

THE ELEMENTS

All the members of the two groups other than nitrogen and oxygen are solids under normal conditions (Table 12.1) and metallic character generally increases down the groups. However, the trend is not clear cut, because the electrical conductivities of the heavier elements decrease from arsenic to bismuth (Fig. 12.1). The usual trend (an

14	15	16	17
C	N	O	F
Si	P	S	Cl
Ge	As	Se	Br
Sn	Sb	Te	I
Pb	Bi	Po	At
IV	V	VI	VII

Table 12.1 Properties of the nitrogen and oxygen group elements

	I/kJ mol^{-1}	χ_P	Radius		Appearance and properties	Common oxidation numbers
			r_{cov}/Å	r_{ion}/Å		
Group 15/V						
N	1410	3.04	0.75		Gas b.p. −196 °C	**−3**, 0, +1, **+3**, **+5**
P	1020	2.06	1.10		Polymorphic solid	−3, +3, **+5**
As	953	2.18	1.22		Dark solid	**+3**, **+5**
Sb	840	2.05	1.43		Solid metallic luster brittle	**+3**, **+5**
Bi	710	2.02	1.52		Solid metallic luster brittle	**+3**, +5
Group 16/VI						
O	1320	3.44	0.73	1.24†	Paramagnetic gas b.p. −183 °C	**−2**, −1, 0
S	1005	2.44	1.02	1.70†	Yellow polymorphic solid	**−2**, 0, +4, **+6**
Se	947	2.55	1.17	1.84†	Polymorphic solid	−2, +4, **+6**
Te	875	2.10	1.35	2.07†	Solid silvery brittle	−2, +4, **+6**
Po					Solid all isotopes radioactive	−2, +2, +4, +6

† For oxidation number −2.

Sources: Ionization enthalpies from D. D. Wagman, W. H. Evans, V. B. Parker, R. H. Schumm, I. Halow, S. M. Bailey, K. L. Churney, and R. L. Nuttall, *J. Phys. Chem. Ref. Data* **11** (Suppl. 2) (1982). Electronegativities from A. L. Allred, *J. Inorg. Nucl. Chem.,* **17**, 215 (1961). Covalent radii from L. C. Allen and J. E. Huheey, *J. Inorg. Nucl. Chem.,* **42**, 1523 (1980). Ionic radii from R. D. Shannon, *Acta Crystallogr., A32,* 1751. (1976).

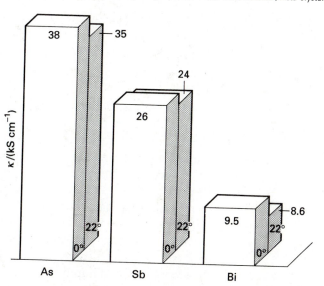

12.1 Electrical conductivities of the heavier elements of Group 15/V and their variation with (Celsius) temperature.

increase down a group) reflects the closer spacing of the atomic energy levels in heavier elements and hence a smaller separation of the valence and conduction bands. The opposite trends in these two groups may reflect the greater directional character of the bonding between their atoms.

In addition to their distinctive physical properties, nitrogen and oxygen are significantly different chemically from the other members of the groups. For one thing, they are among the most electronegative elements in the periodic table and significantly more electronegative than their congeners. Although oxygen never achieves the group maximum oxidation state of $+6$, the slightly less electronegative nitrogen does achieve the maximum for its group ($+5$), but only under much stronger oxidizing conditions than are necessary to achieve this state for its congeners. The small radii of the N and O atoms also contribute to their distinctive chemical character. Thus, nitrogen and oxygen seldom have coordination numbers greater than 4 in simple molecular compounds, but their heavier congeners frequently reach 5 and 6, as in PCl_5, AsF_6^-, and SeF_6. The heaviest member of Group 15/V, bismuth, requires a very strong oxidizing agent to achieve the group oxidation number, and in most of its compounds it has oxidation number $+3$. This behavior is a consequence of the inert pair effect, which is also important (as we saw in Chapter 11) for the preceding heavy elements thallium and lead.

THE NITROGEN GROUP (GROUP 15/V)

The members of the nitrogen group are sometimes referred to collectively as the **pnictides** but this name is neither widely used nor officially sanctioned.

12.1 Occurrence and recovery of the elements

As usual, we begin the discussion with an account of the elements themselves.

Nitrogen

Nitrogen is readily available as dinitrogen, N_2, for it is the principal constituent of the atmosphere and is obtained from it on a massive scale by the distillation of liquid air. Although the bulk price of N_2 is low, the scale on which it is used is an incentive for the development of less expensive processes than liquefaction and distillation. Membrane materials that are more permeable to O_2 than to N_2 are used in laboratory-scale separations of oxygen and nitrogen at room temperature (Fig. 12.2), and an active area of research is the development of membranes that can be used for large-scale commercial separations.

The major nonchemical use of nitrogen gas is as an inert atmosphere in metal processing, petroleum refining, and food processing. We saw in Box 10.1 how nitrogen is used to provide an inert atmosphere in the laboratory, and liquid nitrogen (b.p. $-196\,°C$, 77 K) is a convenient

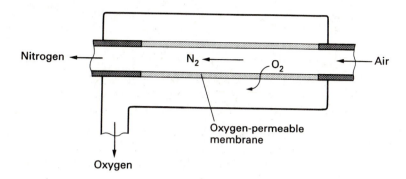

12.2 Schematic diagram of a membrane separator for nitrogen and oxygen.

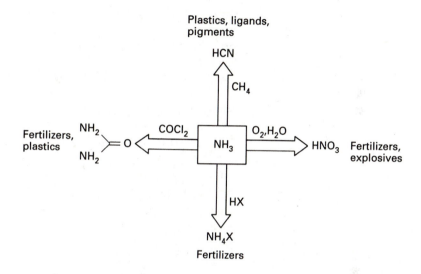

Chart 12.1 Uses of ammonia.

refrigerant in both industry and the laboratory. Nitrogen enters the chain of industrial and agricultural chemicals through its conversion into ammonia by the Haber process, which we describe later. Once 'fixed' in this way it can be converted to a wide range of compounds. The destinations of fixed nitrogen are summarized in Chart 12.1, and later in the chapter we shall see something of the chemistry involved in the reactions mentioned there.

Phosphorus

Phosphorus (along with nitrogen and potassium) is an essential plant nutrient. However, as a result of the low solubility of phosphates it is often depleted in soil, and hence hydrogenphosphates are important components of balanced fertilizers.[1] Approximately 85 percent of the phosphoric acid produced goes into fertilizer manufacture.

The principal raw material for the production of elemental phosphorus and phosphoric acid is phosphate rock, the insoluble, crushed, and compacted remains of ancient organisms, which consists primarily of the minerals fluorapatite, $Ca_5(PO_4)_3F$, and hydroxyapatite,

---

[1] The availability of soluble phosphate nutrients is slightly dependent on the pH of the soil because metal salts of PO_4^{3-} are much less soluble than those of HPO_4^{2-} and $H_2PO_4^{-}$.

Ca$_5$(PO$_4$)$_3$OH. Phosphoric acid can be produced by an acid-base reaction between the rock and concentrated sulfuric acid:

$$Ca_5(PO_4)_3F(s) + 5H_2SO_4(l) \rightarrow 3H_3PO_4(l) + 5CaSO_4(s) + HF(aq)$$

The hydrogen fluoride from the fluoride component of the rock is scavenged by reaction with SiO$_2$ to yield the less reactive [SiF$_6$]$^{2-}$ complex ion.

The product of the treatment of phosphate rock with acid contains *d*-metal contaminants that are difficult to remove, so its use is largely confined to fertilizers and metal treatment. However, ion exchange processes have been developed to remove the offending ions. Most pure phosphoric acid and phosphorus compounds are still produced from the element because it can be purified by sublimation. The production of elemental phosphorus starts with crude calcium phosphate (as calcined phosphate rock) which is reduced with carbon in an electric arc furnace. (This is another example of the very high temperature carbon reduction of a highly oxophilic element, Section 7.1.) Silica is added (as sand) to produce a slag of calcium silicate:

$$2Ca_3(PO_4)_2 + 6SiO_2 + 10C \xrightarrow{1500\,°C} 6CaSiO_3 + 10CO + P_4$$

The slag is molten at these high temperatures and so can be easily removed from the furnace. The phosphorus itself vaporizes and is condensed to the solid, which is stored under water to protect it from reaction with air. Most phosphorus produced in this way is burned to form P$_4$O$_{10}$, which is then hydrated to yield pure phosphoric acid.

The solid elements of Group 15/V (as well as those in Group 16/VI) exist as a number of allotropes.[2] White phosphorus, for example, is a solid consisting of tetrahedral P$_4$ molecules (**1**). Despite the small P—P—P angle, the molecules persist in the vapor up to about 800 °C, but above that temperature the equilibrium concentration of P$_2$ becomes appreciable. As with the N$_2$ molecule, P$_2$ has a formal triple bond and a short bond length. White phosphorus is thermodynamically less stable than the other solid phases under normal conditions. However, in contrast to the usual practice of choosing the most stable phase of an element as the reference phase for thermodynamic calculations, white phosphorus is adopted since it is better characterized than the other forms.

Red phosphorus is obtained by heating white phosphorus at 300 °C in an inert atmosphere for several days. It is normally obtained as an amorphous solid, but crystalline materials can be prepared that have very complex three-dimensional network structures. Unlike white phosphorus, red phosphorus does not spontaneously ignite in air. When phosphorus is heated under high pressure, a series of phases of

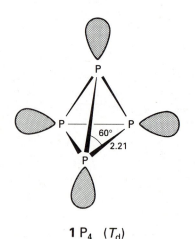

1 P$_4$ (*T*$_d$)

<hr>

[2] The term *allotrope* refers to different molecular structures of the same element; thus ozone and dioxygen are allotropes of oxygen. The word *polymorph* denotes the same substance (an element or a compound) in different crystal forms. For example, phosphorus occurs in several solid polymorphs (including white phosphorus and black phosphorus). Examples of allotropes of phosphorus are the molecules P$_4$ and P$_2$. For a further discussion of the use of the terms, see B. D. Sharma, *J. Chem. Educ.*, **64**, 404 (1987).

black phosphorus are formed. One of these phases consists of tubes composed of pyramidal three-coordinate P atoms (Fig. 12.3).

Arsenic, antimony, and bismuth

The chemically softer elements arsenic, antimony, and bismuth are often found in sulfide ores. Arsenic is usually present in copper and lead sulfide ores, and most of its production is from the flue dust of copper and lead smelters. The solubilities of arsenates, AsO_4^{3-}, are similar to those of phosphates, so traces of these ions are present in phosphate rock.

Arsenic, antimony, and bismuth exist as several allotropes. The most stable structures at room temperature for all three elements are built from puckered hexagonal nets in which each atom has three nearest neighbors. The nets stack in a way that gives three more distant neighbors in the adjacent net (Fig. 12.4). Arsenic vapor resembles phosphorus and consists of tetrahedral As_4 molecules. Bismuth, in common with the α-forms of arsenic and antimony, has a metallic luster and is one of a small class of substances that, like water, expand upon solidification. The electrical conductivity of bismuth is not as high as for most metals, and its structure is not typical of the isotropic bonding normally found. The band structure of bismuth suggests a low density of conduction electrons and holes and it is best classified as a semimetal (Section 2.9) rather than as a semiconductor or a true metal.

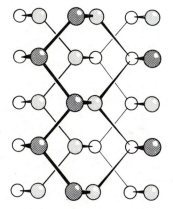

12.3 Two of the puckered layers of black phosphorus. Note the trigonal pyramidal coordination of the atoms. Dark atoms are closest to the viewer and open circles are furthest away.

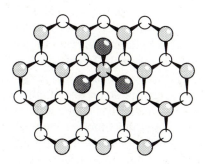

12.4 The puckered network structure of bismuth. Each Bi atom has three nearest neighbors; the dark shaded atoms indicate next-nearest neighbors from an adjacent puckered sheet. Unshaded atoms are furthest from the viewer.

Example 12.1: *Electronic structure and chemistry of P₄*
Draw the Lewis structure of P_4, and discuss its possible role as a ligand.

Answer. In the Lewis structure of P_4, there is a lone pair of electrons on each P atom:

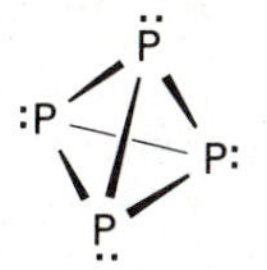

This structure, together with the fact that the electronegativity of phosphorus is moderate ($\chi = 2.06$), suggests that P_4 might be a moderately good donor ligand. Indeed, though rare, P_4 complexes are known.

Exercise E12.1. Consider the Lewis structure of a segment of bismuth (Fig. 12.4). Is this puckered structure consistent with VSEPR theory?

12.2 Nitrogen activation

Nitrogen occurs in many compounds, but N_2 itself is strikingly unreactive. A few strong reducing agents can transfer electrons to the N_2 molecule at room temperature, leading to scission of the $N \equiv N$ bond, but usually the reaction needs extreme conditions. The prime example of this reaction is the slow reaction of lithium metal at room temperature, which yields Li_3N. Similarly, when magnesium (lithium's diagonal neighbor) burns in air it forms the nitride as well as the oxide.

The slowness of the reactions of N_2 appears to be the result of several factors. One is the strength of the $N\equiv N$ bond and hence the high activation energy for breaking it. (The strength of this bond also accounts for the lack of nitrogen allotropes.) Another factor is the size of the HOMO–LUMO gap in N_2, which makes the molecule resistant to simple electron-transfer redox processes. A third factor is the low polarizability of N_2, which does not encourage the formation of the highly polar transition states that are often involved in electrophilic and nucleophilic displacement reactions.

Cheap methods of nitrogen fixation are highly desirable because they would have a profound effect on the economy. In the Haber process for the production of ammonia (which we discuss in detail in Section 17.7), H_2 and N_2 are combined at high temperatures and pressures over an iron catalyst. Although very successful and used throughout the world, the process requires a costly high temperature, high pressure plant, and research continues into more economical alternatives. As we shall see later in the chapter, ammonia is the starting material for a wide variety of commercial syntheses and it is also a useful laboratory solvent.

A clue to the direction nitrogen fixation is likely to take in the future is the observation that bacteria have achieved the room-temperature conversion of atmospheric N_2 to a wide variety of compounds. The dominant process for the fixation of nitrogen in soils is the catalytic conversion of dinitrogen to NH_4^+ by the enzyme nitrogenase, which occurs in the root nodules of legumes. The identity of this metal-containing enzyme is the topic of considerable research. In this connection, dinitrogen complexes of metals were discovered in 1965, and at about the same time the partial elucidation of the structure of nitrogenase indicated that the active site contains Fe and Mo atoms (Chapter 19). These developments led to optimism that efficient catalysts might be devised in which metal ions will coordinate to N_2 and promote its reduction, and in fact many N_2 complexes have been prepared. In some cases the preparation is as simple as bubbling N_2 through an aqueous solution of a complex:

$$[Ru(NH_3)_5(OH_2)]^{2+}(aq) + N_2(g) \rightarrow$$
$$[Ru(NH_3)_5(N_2)]^{2+}(aq) + H_2O(l)$$

As with the isoelectronic CO molecule, end-on bonding (2) is typical of N_2 when it acts as a ligand. The $N\equiv N$ bond is only slightly altered from that in the free molecule in the Ru(II) complex. However, when N_2 is coordinated to a more strongly reducing metal center, the nitrogen–nitrogen bond is considerably lengthened by back-donation of electron density into the π^* orbitals of N_2. Although new catalysts for N_2 reduction have not yet emerged from this research, there are high

2 $[Ru(NH_3)_5N_2]^{2+}$

hopes of it, because it is possible to convert bound N_2 in some of these complexes into NH_4^+:[3]

$$\textit{cis-}[W(N_2)_2\{P(CH_3)_2(C_6H_5)\}_4] \xrightarrow{H_2SO_4}$$

$$N_2 + NH_4^+ + W(VI) \text{ products}$$

12.3 Halides

Halogen compounds of phosphorus, arsenic, and antimony are numerous and important in synthetic chemistry. This richness stems from the fact that not only do these elements form compounds with all the halogens, but also that they do so with oxidation numbers of $+3$ and $+5$ (for example, AsF_3 and AsF_5). The halogen chemistry of nitrogen and bismuth is less rich because the $+5$ oxidation state is less accessible. For example, BiF_5 exists, but $BiCl_5$ and $BiBr_5$ do not. The difficulty of oxidizing Bi(III) to Bi(V) by chlorine or bromine is an example of the inert-pair effect (Section 1.10). Nitrogen does not reach its group oxidation state in neutral binary halogen compounds, but it does achieve it in NF_4^+. Presumably, an N atom is too small for NF_5 to be sterically feasible.

Nitrogen halides

The most extensive series of halogen compounds of nitrogen are the fluorides, and NF_3 is in fact its only exoergic binary halogen compound. This pyramidal molecule is highly unreactive. Thus, unlike NH_3, it is not a Lewis base because the strongly electronegative F atoms make the lone pair of electrons unavailable: whereas the polarity in NH_3 is $^{\delta-}NH_3^{\delta+}$, in NF_3 it is $^{\delta+}NF_3^{\delta-}$. Nitrogen trifluoride can be converted into the N(V) species NF_4^+ by the following reaction:[4]

$$NF_3 + 2F_2 + SbF_3 \rightarrow [NF_4^+][SbF_6^-]$$

Nitrogen trichloride, NCl_3, is a highly endoergic, explosive, and volatile liquid. It is prepared commercially by the electrolysis of a solution of ammonium chloride, and the resulting gas is used directly as an oxidizing bleach for flour. The extremely unstable nitrogen tribromide, NBr_3, has been prepared, but NI_3 is known only in the form of the highly explosive ammoniate $NI_3 \cdot NH_3$. In complete contrast to NF_3, X-ray diffraction indicates a polymeric structure for this compound, with iodine-bridged NI_4 tetrahedra.

Other halides

The trihalides and pentahalides of nitrogen's congeners are used extensively in synthetic chemistry, and their simple empirical formulas conceal an interesting and varied structural chemistry.

---

[3] A clever approach to the reduction of dinitrogen in metal complexes is described by J. Collman, J. E. Hutchinson, M. A. Lopez, R. Guilard, and R. A. Reed, *J. Am. Chem. Soc.*, **113**, 2794 (1991).

[4] The possible existence of NF_5 has been discussed by K. O. Christie, *J. Am. Chem. Soc.* **114**, 9934 (1992).

The trihalides range from gases and volatile liquids, such as PF_3 (b.p. $-102\,°C$) and AsF_3 (b.p. $63\,°C$), to solids, such as BiF_3 (m.p. $649\,°C$). A common method of preparation is direct reaction of the element and halogen. For phosphorus, the trifluoride is prepared by double replacement of the trichloride and a fluoride:

$$2PCl_3(l) + 3CaF_2(s) \rightarrow 2PF_3(g) + 3CaCl_2(s)$$

The trichlorides PCl_3, $AsCl_3$, and $SbCl_3$ are useful starting materials for the preparation of a variety of alkyl, aryl, alkoxy, and amino derivatives (Section 10.14) because they are susceptible to protolysis and double replacement:

$$ECl_3 + 3EtOH \rightarrow E(OEt)_3 + 3HCl \quad E = P, \text{ As, Sb}$$

$$ECl_3 + 6Me_2NH \rightarrow E(NMe_2)_3 + 3[Me_2NH_2]Cl \quad E = P, \text{ As, Sb}$$

Phosphorus trifluoride, PF_3, is an interesting ligand because in some respects it resembles CO. Like that molecule, it is a weak σ donor but a strong π acceptor, and complexes of PF_3 exist that are the analogs of carbonyls, such as $Ni(PF_3)_4$ the analog of $Ni(CO)_4$. The π-acceptor character is attributed to a P—F antibonding LUMO, which has mainly phosphorus p-orbital character when P is bonded to the electronegative F atom. The trihalides also act as mild Lewis acids toward bases such as trialkylamines and halides. Many halide complexes have been isolated, ranging from simple mononuclear species, such as $[AsCl_4]^-$ (3) and $[SbF_5]^{2-}$ (4), to more complex dinuclear and polynuclear anions linked by halide bridges, such as the polymeric chain $([BiBr_3]^{2-})_n$ in which Bi(I) is surrounded by a distorted octahedron of Br atoms.

The pentahalides vary from volatile substances, such as PF_5 (b.p. $-85\,°C$) and AsF_5 (b.p. $-53\,°C$), to solids, such as PCl_5 (sublimes at $162\,°C$) and BiF_5 (m.p. $154\,°C$). The five-coordinate gas phase molecules are trigonal bipyramidal, and we saw in Section 3.2 that PF_5 is fluxional. In contrast to the volatile PF_5 and AsF_5, SbF_5 is a highly viscous liquid in which the molecules are associated through F-atom bridges. In solid SbF_5 these bridges result in a cyclic tetramer (5), which reflects the tendency of Sb(V) to achieve six-coordination. A related phenomenon occurs with PCl_5, which in the solid state exists as $[PCl_4^+][PCl_6^-]$. In this case, the ionic contribution to the lattice enthalpy provides the driving force for the transfer of a Cl^- ion from one molecule to another. The pentafluorides of phosphorus, arsenic, and bismuth are stronger Lewis acids than the other known pentahalides of these elements. As we remarked in Section 5.15, SbF_5 is a very strong Lewis acid; it is much stronger, for example, than the aluminum halides.

The pentahalides of phosphorus and antimony are very useful in syntheses. Phosphorus pentachloride, PCl_5, is widely used in the laboratory and in industry as a starting material for syntheses, and some of its characteristic reactions are shown in Chart 12.2. Note, for example, that reactions with Lewis acids yield PCl_4^+ salts, simple Lewis bases

3 $AsCl_4^-$ **4** SbF_5^{2-} (C_{4v})

5 $(SbF_5)_4$

Chart 12.2 Uses of PCl_5.

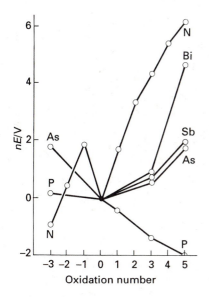

12.5 Frost diagram for the elements of the nitrogen group in acidic solution. The species with oxidation number − 3 are NH_3, PH_3, and AsH_3, and those with oxidation numbers − 2 and − 1 are N_2H_4 and NH_2OH respectively. The positive oxidation states refer to the most stable oxo or hydroxo species in acidic solution, and may be oxides, oxoacids, or oxoanions.

like F^- give six-coordinate complexes such as PF_6^-, compounds containing the NH_2 group lead to the formation of P=N bonds, and the interaction with either H_2O or P_4O_{10} yields O=PCl_3.

12.4 Oxides and aqueous redox chemistry

We can infer the redox properties of nitrogen-group compounds in acidic aqueous solution from the Frost diagram shown in Fig. 12.5. The steepness of the slopes of the lines on the far right of the diagram show the thermodynamic tendency for reduction of the +5 oxidation states of the elements. They show, for instance, that Bi_2O_5 is potentially a very strong oxidizing agent, which is consistent with the inert pair effect and the tendency of Bi(V) to form Bi(III). The next strongest oxidizing agent is NO_3^-. Both As(V) and Sb(V) are milder oxidizing agents, and P(V), in the form of phosphoric acid, is extremely weak.

The redox properties of nitrogen are important because of its widespread occurrence in the atmosphere, biosphere, industry, and the laboratory. Nitrogen chemistry also is quite intricate, partly because of the large number of oxidation states attainable, but also because reactions that are thermodynamically favorable are often slow or have rates that depend critically on the identity of the reactants. Carbon chemistry is similarly intricate, as we know from organic and organometallic chemistry, but the mechanisms of the reactions of N_2 and its oxides and oxoanions have turned out to be more difficult to rationalize.

The main points to keep in mind include the fact that, as the N_2 molecule is so inert, redox reactions that consume N_2 are slow. Moreover, for mechanistic reasons that seem to be different in each case, the formation of N_2 is also slow and often sidestepped in aqueous solution (Chart 12.3). As with several other p-block elements, the

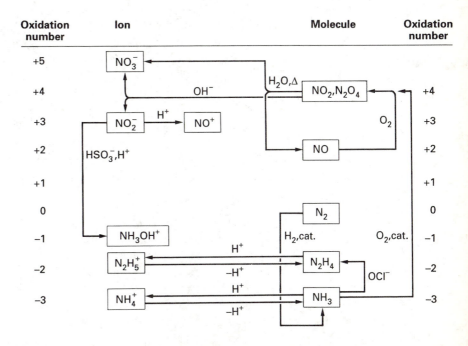

Chart 12.3 Interconversion of important nitrogen species.

Table 12.2 Oxides of nitrogen

Oxidation number	Formula	Name	Structure (gas phase)	Remarks
$+1$	N_2O	Nitrous oxide (Dinitrogen oxide)	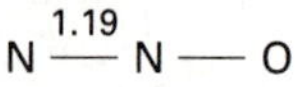 $N \overset{1.19}{—} N — O$ $C_{\infty v}$	Colorless gas, not very reactive
$+2$	NO	Nitric oxide (Nitrogen monoxide)	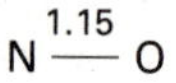 $N \overset{1.15}{—} O$ $C_{\infty v}$	Colorless, paramagnetic gas
$+3$	N_2O_3	Dinitrogen trioxide	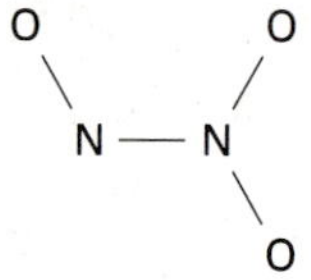 Planar, C_s	Forms blue solid (m.p. -101 °C) and dissociates into NO and NO_2 in the gas phase
$+4$	NO_2	Nitrogen dioxide	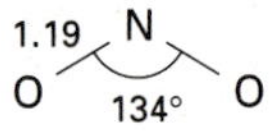 C_{2v}, 1.19, 134°	Brown, reactive, paramagnetic gas
	N_2O_4	Dinitrogen tetroxide	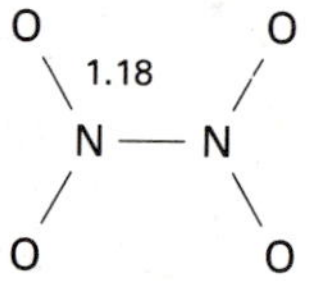 Planar, D_{2h}, 1.18	Forms colorless liquid (m.p. -11 °C); in equilibrium with NO_2 in the gas phase
$+5$	N_2O_5	Dinitrogen pentoxide	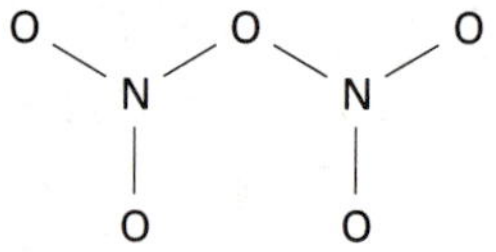 Planar, C_{2v}	Colorless ionic solid $[NO_2][NO_3]$ (m.p. 32 °C); unstable in gas phase

barriers to reaction of high oxidation state oxoanions, such as NO_3^-, are greater than for low oxidation state oxoanions, such as NO_2^-. We should also remember that low pH enhances the oxidizing power of oxoanions (Section 7.10). Low pH also often accelerates their oxidizing reactions by protonation, for this is thought to facilitate N—O bond breaking. Finally, it should be remembered that the reactions of the oxocompounds of nitrogen commonly take place by atom or ion transfer, and outer-sphere electron transfer is rare.

Table 12.2 summarizes some of the properties of the nitrogen oxides, and Table 12.3 does the same for the nitrogen oxoanions. Both tables will help us to pick our way through the details of their properties.

Nitrogen(V) oxoanions

The most common source of N(V) is nitric acid, HNO_3, which is a major industrial chemical used in the production of fertilizers, explosives, and a wide variety of nitrogen-containing chemicals. It is produced by modern versions of the **Ostwald process**, which make use of an indirect route from N_2 to the highly oxidized compound HNO_3 via the fully reduced compound NH_3. Thus, after nitrogen has been

Table 12.3 Nitrogen oxoions

Oxidation number	Formula	Common name	Structure	Remarks
+1	$N_2O_2^{2-}$	Hyponitrite	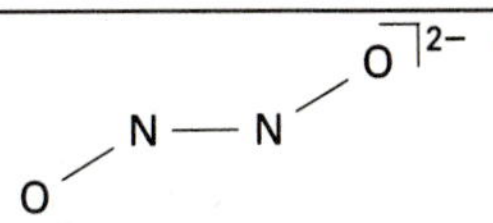 C_{2h}	Usually acts as a reducing agent
+3	NO_2^-	Nitrite	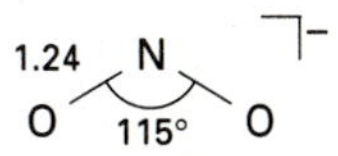 1.24, 115° C_{2v}	Weak base; acts as an oxidizing and a reducing agent
+3	NO^+	Nitrosonium	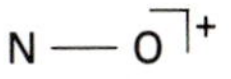 $N-O^+$ $C_{\infty v}$	Oxidizing agent and Lewis acid
+5	NO_3^-	Nitrate	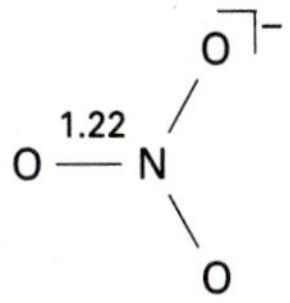 1.22 Planar, D_{3h}	Very weak base; an oxidizing agent
+5	NO_2^+	Nitronium	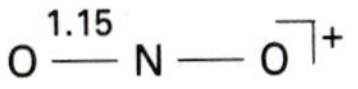 $O\overset{1.15}{-}N-O^+$ Linear, $D_{\infty h}$	Oxidizing agent, nitrating agent, and a Lewis acid

reduced to the -3 state as NH_3 by the Haber process, it is oxidized to the $+4$ state:

$$4NH_3(g) + 7O_2(g) \rightarrow 6H_2O(g) + 4NO_2(g)$$

$$\Delta G^{\ominus} = -308.0 \text{ kJ (mol } NO_2)^{-1}$$

The NO_2 then undergoes disproportionation into N(II) and N(V) in water at elevated temperatures:

$$3NO_2(aq) + H_2O(l) \rightarrow 2HNO_3^-(aq) + NO(g)$$

$$\Delta G^{\ominus} = -5.0 \text{ kJ (mol } HNO_3)^{-1}$$

Both steps are thermodynamically favorable. The byproduct NO is oxidized with O_2 to NO_2 and recirculated. Such an indirect route is employed because the direct oxidation of N_2 to NO_2 is thermodynamically unfavorable, with $\Delta G_f^{\ominus}(NO_2) = +51 \text{ kJ mol}^{-1}$. In part, this endoergic character is due to the great strength of the $N{\equiv}N$ bond.

Standard potential data imply that the NO_3^- ion is a fairly strong oxidizing agent; however, the kinetic aspects of its reactions are important, and they are generally slow in dilute acid solution. Because protonation of oxygen promotes N—O bond breaking, concentrated HNO_3 (in which NO_3^- is protonated) undergoes more rapid reactions than the dilute acid (in which HNO_3 is fully ionized). It is also a thermodynamically more potent oxidizing agent at low pH. A sign of this oxidizing character is the yellow color of the concentrated acid, which indicates its instability with respect to decomposition into NO_2:

$$4HNO_3(aq) \rightarrow 4NO_2(aq) + O_2(g) + 2H_2O(l)$$

This decomposition is accelerated by light and heat.[5]

The reduction of NO_3^- ions rarely yields a single product, as so many lower oxidation states of nitrogen are available at similar potentials, and there are substantial kinetic barriers to their interconversion. For example, a strong reducing agent such as zinc can reduce a substantial proportion of dilute HNO_3 as far as oxidation state -3:

$$HNO_3(aq) + 4Zn(s) + 9H^+(aq) \rightarrow NH_4^+(aq) + 3H_2O(l) + 4Zn^{2+}(aq)$$

A weaker reducing agent, such as copper, proceeds only as far as oxidation state $+4$ in the concentrated acid:

$$2HNO_3(aq) + Cu(s) + 2H^+(aq) \rightarrow 2NO_2(g) + Cu^{2+}(aq) + 2H_2O(l)$$

With the dilute acid, the $+2$ oxidation state is favored, and NO is formed:

$$2NO_3^-(aq) + 3Cu(s) + 8H^+(aq) \rightarrow 2NO(g) + 3Cu^{2+}(aq) + 4H_2O(l)$$

Example 12.2: *Correlating trends in the stabilities of N(V) and Bi(V)*

Compounds of $N(V)$ and $Bi(V)$ are stronger oxidizing agents than the $+5$ oxidation states of the three intervening elements. Correlate this observation with trends in the periodic table.

Answer. The light *p*-block elements are more electronegative than the elements immediately below them in the periodic table, accordingly these light elements are generally less easily oxidized. Nitrogen is generally a good oxidizing agent in its positive oxidation states. Bismuth is much less electronegative, but favors the $+3$ oxidation state in preference to the $+5$ state on account of the inert-pair effect.

Exercise E12.2. From trends in the periodic table, decide whether phosphorus or sulfur is likely to be the stronger oxidizing agent.

Nitrogen(IV) and nitrogen(III)

Nitrogen(IV) oxide, which is commonly called nitrogen dioxide, exists as an equilibrium mixture of the brown NO_2 radical and its colorless dimer N_2O_4 (dinitrogen tetroxide):

$$N_2O_4(g) \rightleftharpoons 2NO_2(g) \qquad K = 0.115 \text{ at } 25\,^\circ C$$

This readiness to dissociate is consistent with the N—N bond in N_2O_4 (6) being significantly longer and weaker than the C—C bond in the

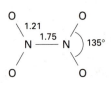

6 N_2O_4

[5] Useful comments on the explosion hazards of NO_3^- in combination with reducible cations, such as NH_4^+ in ammonium nitrate, will be found in L. Bretherick, *J. Chem. Educ.*, **66**, A220 (1989).

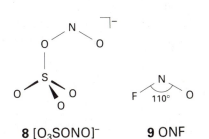

7 $C_2O_4^{2-}$

isoelectronic oxalate ion ($C_2O_4^{2-}$, **7**), which does not equilibrate with CO_2^- radicals in the same way. The unpaired electron occupies an antibonding orbital in NO_2 and the transient CO_2^- radical, and so is *less* localized on the more electronegative N atom in NO_2 than it is on C in CO_2^-.[6] Hence the bond in N_2O_4 is relatively weak and N_2O_4 is more likely to dissociate than the oxalate ion.

Nitrogen(IV) oxide is a poisonous oxidizing agent that is present in low concentrations in the atmosphere, especially in photochemical smog. In basic aqueous solution it disproportionates into N(III) and N(V), forming NO_2^- and NO_3^- ions:

$$2NO_2(aq) + 2OH^-(aq) \rightarrow NO_2^-(aq) + NO_3^-(aq) + H_2O(l)$$

In acidic solution (as in the Ostwald process) the reaction product is N(II) in place of N(III) because nitrous acid readily disproportionates:

$$3HNO_2(aq) \rightarrow NO_3^-(aq) + 2NO(g) + H_3O^+(aq)$$

$$E^{\ominus} = +0.05 \text{ V}, K = 50$$

Nitrous acid, HNO_2, is a strong oxidizing agent:

$$HNO_2(aq) + H^+(aq) + e^- \rightarrow NO(g) + H_2O(l) \qquad E^{\ominus} = +1.00 \text{ V}$$

and its reactions as an oxidizing agent are often more rapid than its disproportionation. The rate at which nitrous acid oxidizes is increased by acid as a result of its conversion to the nitrosonium ion, NO^+:

$$HNO_2(aq) + H^+(aq) \rightarrow NO^+(aq) + H_2O(l)$$

The nitrosonium ion is a strong Lewis acid and associates rapidly with anions and other nucleophiles. These species may not themselves be susceptible to oxidation (as in the case of SO_4^{2-} and F^- ions, which form SO_3ONO^- (**8**) and FNO (**9**), respectively), but the association is often the initial step in an overall redox reaction. Thus there is good experimental evidence that the reaction of HNO_2 with I^- ions leads to the rapid formation of INO:

$$I^-(aq) + NO^+(aq) \rightarrow INO(aq)$$

followed by the rate-determining second-order reaction between two INO molecules:

$$2INO(aq) \rightarrow I_2(aq) + 2NO(g)$$

Nitrosonium salts containing poorly coordinating anions, such as $[NO][BF_4]$, are useful reagents in the laboratory as facile oxidizing agents and as a source of NO^+.

8 $[O_3SONO]^-$ **9** ONF

Nitrogen(II) oxide

Nitrogen(II) oxide, more commonly nitric oxide, is an odd-electron molecule. However, unlike NO_2 it does not form a stable dimer in the

---

[6] It was remarked in Section 2.5 that bonding electrons are more likely to be found near the more electronegative atom in a bond, and antibonding electrons are more likely to be found near the less electronegative atom.

gas phase. This difference reflects the greater delocalization of the odd electron in the π^* orbital in NO than in NO_2. Nitric oxide reacts with O_2 to generate NO_2, but in the gas phase the rate law is second-order in NO, because a transient dimer, $(NO)_2$, is produced that subsequently collides with an O_2 molecule. Because the reaction is second-order, atmospheric NO (which is produced in low concentrations by coal-fired power plants and by gasoline and diesel engines) is slow to convert to NO_2. Until the late 1980s, no beneficial biological roles were known for NO. However, since then it has been found that NO is generated *in vivo*, and that it performs functions such as the reduction of blood pressure, neurotransmission, and the destruction of microbes.[7]

Because NO is endoergic, it should be possible to find a catalyst to convert the pollutant NO to the natural atmospheric gases N_2 and O_2 at its source in exhausts. A practical catalyst of this type is yet to be discovered, and the search for one is an interesting and socially important chemical challenge.[8]

Low oxidation states

The average oxidation number of nitrogen in dinitrogen oxide, N_2O (specifically, NNO), which is commonly called nitrous oxide, is $+1$; that of the azide ion, N_3^-, is $-\frac{1}{3}$. Both species are isoelectronic with CO_2, but the resemblance barely extends beyond their common linear shapes.

Dinitrogen oxide, a colorless unreactive gas, is produced by the comproportionation of molten ammonium nitrate. Care must be taken to avoid an explosion in this reaction, in which the cation is oxidized by the anion:

$$NH_4NO_3(l) \xrightarrow{250\,^{\circ}C} N_2O(g) + 2H_2O(g)$$

Its standard potentials suggest that N_2O should be a strong oxidizing agent in acidic and basic solutions:

$$N_2O(g) + 2H^+(aq) + 2e^- \rightarrow N_2(g) + H_2O(l)$$

$$E^{\ominus} = +1.77 \text{ V at pH} = 0$$

$$N_2O(g) + H_2O(l) + 2e^- \rightarrow N_2(g) + 2OH^-(aq)$$

$$E = +0.94 \text{ V at pH} = 14$$

However, kinetic considerations are paramount, and the gas is unreactive toward many reagents at room temperature. One sign of this inertness is that N_2O has been used as the propellant gas for instant whipping cream. Similarly, N_2O was used for many years as a mild

---

[7] See E. Culotta and D. E. Koshland, Jr, Molecule of the year: NO news is good news. *Science*, **258**, 1862 (1992), and J. S. Stamler, D. J. Singel, and J. Loscalzo, Biochemistry of nitric oxide and its redox active forms. *Science*, **258**, 1898 (1992).

[8] A successful catalyst for NO decomposition under laboratory conditions was found not long ago. It consists of Cu(II) in a zeolite. M. Iwamoto, H. Furukawa, Y. Mire, F. Umemura, S. Mikuriya, and S. Kagawa, *J. Chem. Soc., Chem Commun.*, 1272 (1986).

anesthetic; however, this practice has been discontinued because of some undesirable physiological side-effects.

Azide ions may be synthesized by the oxidation of sodium amide with either NO_3^- ions or N_2O at elevated temperatures:

$$3NH_2^- + NO_3^- \xrightarrow{175\,°C} N_3^- + 3OH^- + NH_3$$

$$2NH_2^- + N_2O \xrightarrow{190\,°C} N_3^- + OH^- + NH_3$$

The azide ion is a reasonably strong Brønsted base, the pK_a of its conjugate acid, hydrazoic acid (HN_3), being 4.77. It is also a good ligand toward *d*-block ions. However, heavy metal complexes or salts, such as $Pb(N_3)_2$ and $Hg(N_3)_2$, are shock-sensitive detonators. Ionic azides such as NaN_3 are thermodynamically unstable but kinetically inert; they can be handled at room temperature. When alkali metal azides are heated they liberate N_2 smoothly; this reaction is employed in the inflation of automobile 'air bags'.

Hydrazine and hydroxylamine

Hydrazine, N_2H_4 (**10**), in which the oxidation number of nitrogen is -2, and hydroxylamine, NH_2OH (**11**), in which it is -1, are isoelectronic molecules that are formally related to NH_3 by the replacement of an H atom by an NH_2 group or an OH group, respectively. Both compounds are liquids at room temperature. In each case, the electronegative substituent makes the nitrogen lone pair less readily available and results in weaker Brønsted bases (and hence stronger acidity for the conjugate acids) than NH_3:

	NH_4^+	$N_2H_5^+$	NH_3OH^+
pK_a:	9.26	7.93	5.82

The Lewis base strength is reduced in the same way.

Most commercial hydrazine is prepared by the oxidation of NH_3 by ClO^- ions:

$$2NH_3(aq) + ClO^-(aq) \rightarrow N_2H_4(aq) + Cl^-(aq) + H_2O(l)$$

This is formally a redox reaction. However, the mechanism is more intricate than simple electron transfer because it proceeds through the formation of the intermediate NH_2Cl. Once formed, the NH_2Cl is attacked by the nucleophile NH_3, which results in the displacement of Cl^- and the formation of a N—N bond:

$$NH_2Cl(aq) + 2NH_3(aq) \rightarrow H_2NNH_2(aq) + Cl^-(aq) + NH_4^+(aq)$$

Hydrazine (m.p. 2 °C, b.p. 113 °C) is strongly associated through hydrogen bonding. It is an endoergic compound ($\Delta G_f^\ominus = +149$ kJ mol^{-1}), kinetically fairly inert, and widely used as a reducing agent. For example, it is used to reduce dissolved oxygen in boiler water to suppress corrosion. Similarly, alkylhydrazines are used as the

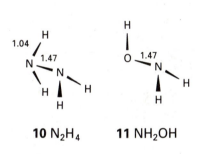

10 N_2H_4 **11** NH_2OH

reducing component in some rocket fuels. When hydrazine reacts with oxidizing agents, it yields a variety of nitrogen-containing products. One common product is N_2, as in the reactions

$$N_2H_4(aq) + O_2(g) \rightarrow N_2(g) + 2H_2O(l)$$

$$N_2H_4(aq) + 2Cl_2(g) \rightarrow N_2(g) + 4HCl(aq)$$

Hydrazine is a much stronger reducing agent in basic solution than in acidic solution:

$$N_2(g) + 5H^+(aq) + 4e^- \rightarrow N_2H_5^+(aq)$$

$$E^{\ominus} = -0.23 \text{ V at pH} = 0$$

$$N_2(g) + 4H_2O(l) + 4e^- \rightarrow N_2H_4(aq) + 4OH^-(aq)$$

$$E = -1.16 \text{ V at pH} = 14$$

Hydroxylamine is an unstable endoergic compound ($\Delta G_f^{\ominus} = +23 \text{ kJ mol}^{-1}$) which is prepared by the reduction of NO_2^- ions by HSO_3^- ions in neutral solution, followed by acidification and heating. In the first stage, HSO_3^- attacks NO_2^- and forms an N—S bond; in the second stage, this N—S bond is hydrolyzed

$$NO_2^-(aq) + 2HSO_3^-(aq) \xrightarrow{0\,^{\circ}C} N(OH)(SO_3)_2^{2-}(aq) + OH^-(aq)$$

$$N(OH)(SO_3)_2^{2-}(aq) + H_3O^+(aq) + H_2O(l) \xrightarrow{50\,^{\circ}C}$$
$$H_3NOH^+(aq) + 2HSO_4^-(aq)$$

Although the standard potentials indicate that hydroxylamine can serve as either an oxidizing agent or a reducing agent, the latter reactions generally occur more readily, as in

$$4Fe^{3+}(aq) + 2NH_3OH^+(aq) \rightarrow$$
$$4Fe^{2+}(aq) + N_2O(g) + 6H^+(aq) + H_2O(l)$$

This reaction sidesteps N_2, despite the latter's great stability, and the nitrogen is carried directly from oxidation state -1 in H_2NOH to the $+1$ state in N_2O by a mechanistically complex reaction.

Example 12.3: *Comparing the redox properties of nitrogen oxoanions and oxo compounds*

Compare (a) NO_3^- and NO_2^- as oxidizing agents; (b) NO_2, NO, and N_2O with respect to their ease of oxidation in air; (c) N_2H_4 and H_2NOH as reducing agents.

Answer. (a) NO_3^- and NO_2^- ions are both strong oxidizing agents. The reactions of the former are often sluggish but are generally faster in acidic solution. The reactions of NO_2^- ions are generally faster and become even faster in acidic solution, where the NO^+ is a common identifiable intermediate. (b) NO_2 is stable with respect to oxidation in air. N_2O and NO are thermodynamically susceptible to oxidation. However, the reaction of N_2O with oxygen is slow, and at low NO concentrations the reaction between NO and O_2 is slow because the rate law is second-order in NO. (c) Hydrazine and hydroxylamine are both good reducing agents. In basic solution hydrazine becomes a stronger reducing agent.

Exercise E12.3. Summarize the reactions that are employed for the synthesis of hydrazine and hydroxylamine. Are these reactions best described as electron transfer processes or nucleophilic displacements?

Chart 12.3 is a summary of the reactions of some important nitrogen species. All the reactions shown in solution occur by electrophile-nucleophile interaction rather than simple electron transfer. These reactions include the disproportionation of NO_2 in acidic or basic solution and the disproportionation of NO_2^- in acidic solution. Similarly, the conversion of NO_2^- ions to NH_2OH occurs by nucleophilic attack by HSO_3^- on NO_2^- followed by acid hydrolysis:

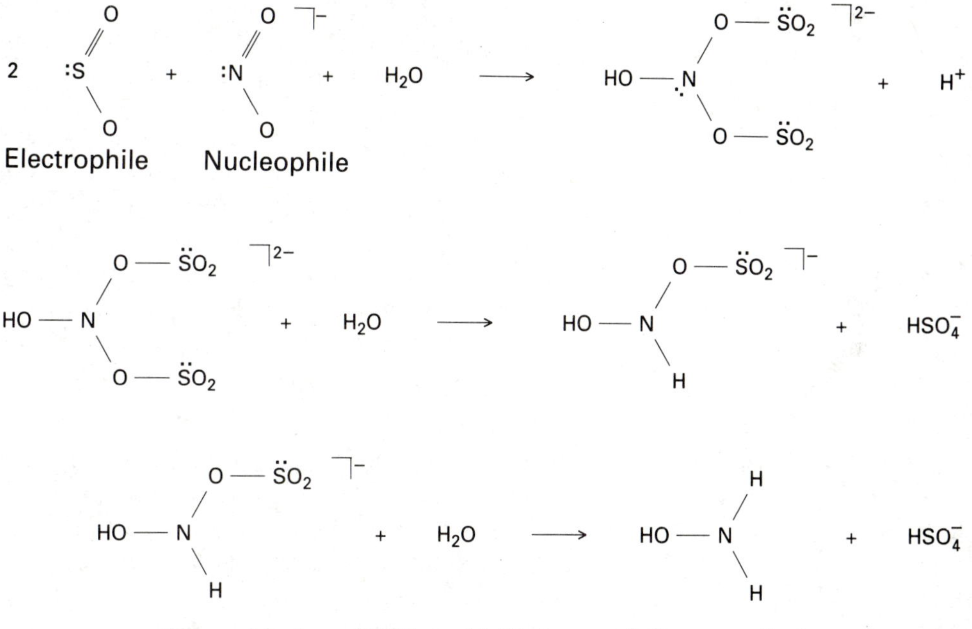

The oxidation of NH_3 to N_2H_4 by carefully controlled reaction with ClO^- involves the initial formation of NH_2Cl followed by nucleophilic attack of NH_3 on NH_2Cl:

$$ClO^- + :NH_3 \longrightarrow H_2NCl + OH^-$$
Electrophile Nucleophile

$$H_2NCl + NH_3 + OH^- \longrightarrow H_2N-NH_2 + H_2O + Cl^-$$

Other oxides

The complete combustion of phosphorus yields phosphorus(V) oxide, P_4O_{10}. Each P_4O_{10} molecule has a cage structure in which a tetrahed-

Table 12.4 Latimer diagrams for phosphorus

Acidic solution

$$\text{H}_3\text{PO}_4 \xrightarrow{-0.93} \text{H}_4\text{P}_2\text{O}_6 \xrightarrow{0.38} \text{H}_3\text{PO}_3 \xrightarrow{-0.50} \text{H}_3\text{PO}_2 \xrightarrow{-0.51} \text{P} \xrightarrow{-0.06} \text{PH}_3$$

with -0.28 bridging $\text{H}_4\text{P}_2\text{O}_6$ to H_3PO_3, and -0.50 bridging H_3PO_3 to P.

Basic solution

$$\text{PO}_4^{3-} \xrightarrow{-1.12} \text{HPO}_3^{2-} \xrightarrow{-1.57} \text{H}_2\text{PO}_2^{-} \xrightarrow{-2.05} \text{P} \xrightarrow{-0.89} \text{PH}_3$$

with -1.73 bridging HPO_3^{2-} to P.

12 P_4O_{10} (T_d)

13 P_4O_6 (T_d)

ron of P atoms is held together by bridging O atoms, and each P atom has a terminal O atom (**12**). Combustion in a limited supply of oxygen results in the formation of phosphorus(III) oxide, P_4O_6, instead; this molecule has the same O-bridged framework of P_4O_{10}, but lacks the terminal O atoms (**13**). (It is also possible to isolate the intermediate compositions having one, two, or three O atoms terminally attached to the P atoms.) Both oxides can be hydrated to yield the corresponding acids, the P(V) oxide giving phosphoric acid, H_3PO_4, and the P(III) oxide giving phosphorous acid, H_3PO_3. As remarked in Section 5.5, phosphorous acid has one H atom attached directly to the P atom; it is therefore a diprotic acid and better represented as $(\text{HO})_2\text{PHO}$.

In contrast to the high stability of phosphorus(V) oxide (P_4O_{10}), arsenic, antimony, and bismuth more readily form oxides with oxidation state $+3$, specifically As_2O_3, Sb_2O_3, and Bi_2O_3. In the gas phase, the arsenic(III) and antimony(III) oxides have the molecular formula E_4O_6, with the same tetrahedral structure as P_4O_6. Arsenic, antimony, and bismuth do form oxides with oxidation state $+5$, but Bi(V) oxide is unstable and has not been structurally characterized.

Oxoanions

It can be seen from the Latimer diagram in Table 12.4 that elemental phosphorus and most of its compounds other than P(V) are strong reducing agents. White phosphorus disproportionates into phosphane, PH_3, (oxidation number -3) and hypophosphite ions (oxidation number $+1$) in basic solution:

$$\text{P}_4(\text{s}) + 3\text{OH}^-(\text{aq}) + 3\text{H}_2\text{O}(\text{l}) \rightarrow \text{PH}_3(\text{g}) + 3\text{H}_2\text{PO}_2^-(\text{aq})$$

Some common phosphorus oxoanions are listed in Table 12.5. The approximately tetrahedral environment of the P atom in their structures should be noted, as should the existence of P—H bonds in the hypophosphite and phosphite anions. The synthesis of various P(III) oxoacids and oxoanions, including HPO_3^{2-} and alkoxophosphanes, is conveniently performed by solvolysis of phosphorus(III) chloride under mild conditions, such as in cold CCl_4 solution:

$$\text{PCl}_3(\text{l}) + 3\text{H}_2\text{O}(\text{l}) \rightarrow \text{H}_3\text{PO}_3(\text{aq}) + 3\text{HCl}(\text{aq})$$

$$\text{PCl}_3(\text{l}) + 3\text{ROH}(\text{sol}) + 3\text{N}(\text{CH}_3)_3(\text{sol}) \rightarrow$$

$$\text{P}(\text{OR})_3(\text{sol}) + 3[\text{HN}(\text{CH}_3)_3]\text{Cl}(\text{sol})$$

Table 12.5 Some phosphorus oxoanions

Oxidation number	Formula	Name	Structure	Remarks
+1	$H_2PO_2^-$	Hypophosphite (dihydrodioxophosphate)	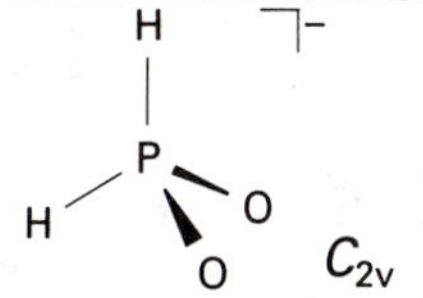 C_{2v}	Facile reducing agent
+3	HPO_3^{2-}	Phosphite (hydrotrioxophosphate)	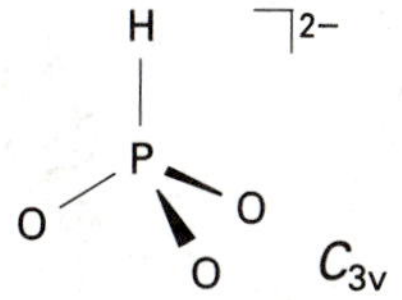 C_{3v}	Facile reducing agent
+4	$P_2O_6^{4-}$	Hypophosphate	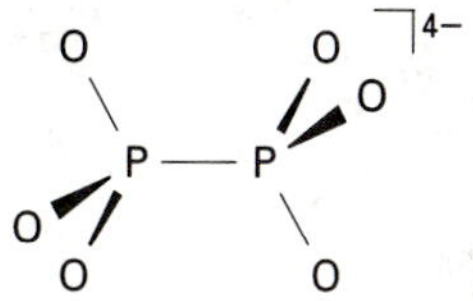	Basic
+5	PO_4^{3-}	Phosphate	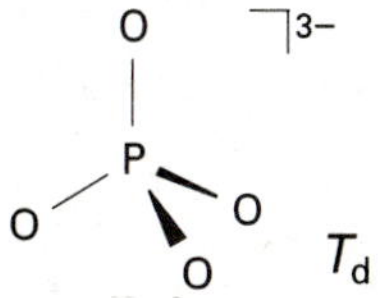 T_d	Strongly basic
+5	$P_2O_7^{4-}$	Diphosphate	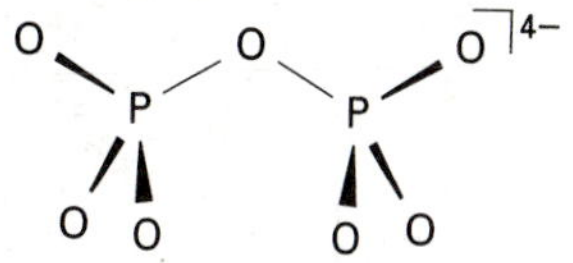	Basic; longer-chain analogs are known

Reductions with $H_2PO_2^-$ and HPO_3^{2-} are usually fast. One of the commercial applications is the use of $H_2PO_2^-$ to reduce Ni^{2+}(aq) ions and so coat surfaces with metallic nickel in the process called 'electrodeless plating'.

The Frost diagram for the elements (Fig. 12.5) reveals similar trends in aqueous solution, with oxidizing character following the order $PO_4^{3-} \ll AsO_4^{3-} < Sb(OH)_6^- \ll Bi(V)$. The thermodynamic tendency and kinetic ease of reducing AsO_4^{3-} is thought to be key to its toxicity toward animals. Thus, As(V) as AsO_4^{3-} readily mimics PO_4^{3-}, and so may be incorporated into cells. There, unlike phosphorus, it is reduced yielding an As(III) species, which is thought to be the real toxic agent. This toxicity may stem from the affinity of As(III) for sulfur-containing amino acids.

12.5 Compounds of nitrogen with phosphorus

Many analogs of phosphorus-oxygen compounds exist in which the O atom is replaced by the isolobal NR or NH group, such as $P_4(NR)_6$ (**14**), the analog of P_4O_6. Other compounds exist in which OH or OR groups are replaced by the isolobal NH_2 or NR_2 groups. An example is $P(NMe_2)_3$, the analog of $P(OMe)_3$. Another indication of the scope

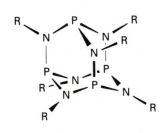

14 $P_4(NR)_6$

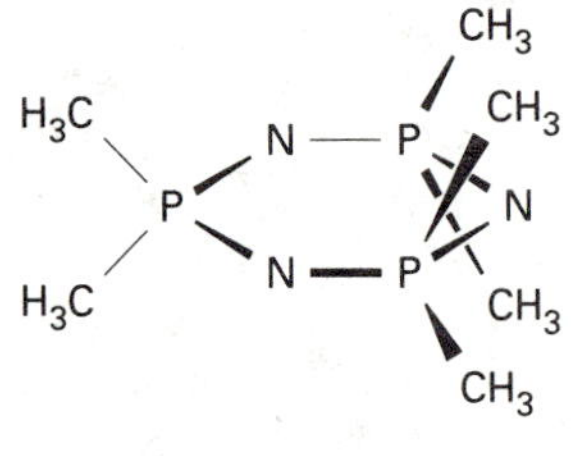

of PN chemistry, and a useful point to remember, is that PN is isoelectronic with SiO. For example, various phosphazenes, which are chains and rings containing R_2PN units (**15**), are analogous to the siloxanes (Section 10.12) and their R_2SiO units (**16**).

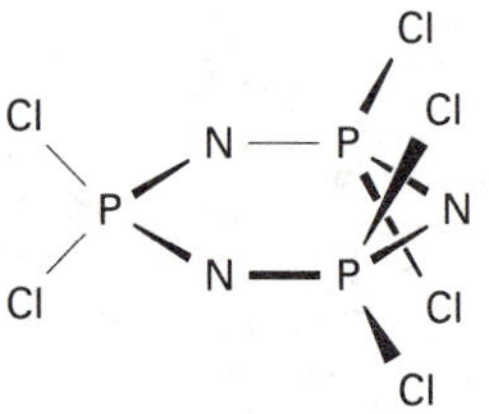

The cyclic phosphazene dichlorides, which are good starting materials for the preparation of the more elaborate phosphazenes, are easily synthesized:

$$nPCl_5 + nNH_4Cl \rightarrow (Cl_2PN)_n + 4nHCl \qquad n = 3 \text{ or } 4$$

A chlorocarbon solvent and temperatures near 130 °C produce the cyclic trimer (**17**) and tetramer (**18**), and when the trimer is heated to about 290 °C it changes to polyphosphazene. The P—Cl bonds in this rubbery material make it liable to hydrolysis, but very stable materials can be obtained by substitution:

$$(Cl_2PN)_n + 2nC_2F_5O^- \rightarrow [(F_5C_2O)_2PN]_n + 2nCl^-$$

Like silicone rubber, the polyphosphazenes remain rubbery at low temperatures because, as with the isoelectronic SiOSi group, PNP groups are highly flexible.

Example 12.4: *Devising a synthetic route to an alkoxy substituted cyclophosphazene*

Give balanced equations for the preparation of $[NP(OCH_3)_2]_4$ from PCl_5, NH_4Cl, and $NaOCH_3$.

Answer. The cyclic chlorophosphazene can be synthesized first:

$$4PCl_5 + 4NH_4Cl \xrightarrow{130\,°C} (Cl_2PN)_4 + 16HCl$$

The Cl atoms are readily replaced by strong nucleophiles such as alkoxides, in the present case to give the desired product:

$$(Cl_2PN)_4 + 8NaOCH_3 \rightarrow [(CH_3O)_2PN]_4 + 8NaCl$$

Exercise E12.4. Give the equations for the preparation of a high polymer phosphazene containing a PN backbone with two $N(CH_3)_2$ side groups attached to each P atom.

THE OXYGEN GROUP (GROUP 16/VI)

The elements of the oxygen group are often (and officially) called the **chalcogens**. The name derives from the Greek word for bronze, and refers to the association of sulfur and its congeners with copper.

12.6 Occurrence and recovery of the elements

The principal elements of the group are oxygen and sulfur, which are both found in native form.

15 [(CH₃)₂)PN]₃

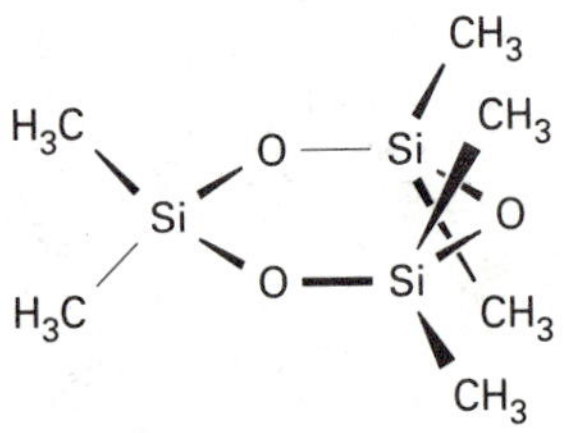

16 [(CH₃)₂SiO]₃

17 [Cl₂PN]₃   (D₃ₕ)

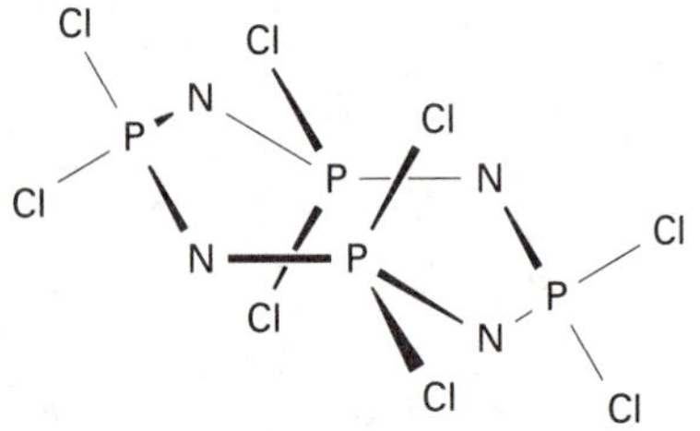

18 [Cl₂PN]₄

Oxygen

Oxygen is readily available as O_2 from the atmosphere and is obtained on a massive scale by the distillation of liquid air. The main commercial motivation is to recover O_2 for steel making, in which it reacts exothermically with coke (carbon) to produce carbon monoxide and heat. The heat is necessary to achieve a fast reduction of iron oxides by CO and carbon (Sections 7.1 and 8.6). Pure oxygen, rather than air, is advantageous in this process because heat is not wasted in warming nitrogen. About 1 ton of oxygen is needed to make 1 ton of steel.

The common allotrope of oxygen, formally dioxygen (O_2), boils at $-183\,°C$ and has a faint blue color in the liquid state (which arises from electronic transitions involving pairs of neighboring molecules). The molecular orbital description implies the existence of a double bond; however, as we saw in Section 2.4, the outermost two electrons occupy different antibonding π orbitals with parallel spins ($\uparrow,\uparrow$). The molecular term symbol for the ground state is $^3\Sigma$, and henceforth the molecule will be denoted $O_2(^3\Sigma)$ when it is appropriate to specify the spin state.[9] The singlet state $^1\Sigma$, with paired electrons in the same two π orbitals as in the ground state ($\uparrow,\downarrow$) is higher in energy by $155\ kJ\ mol^{-1}$, and another singlet state $^1\Delta$ ('singlet delta'), with electrons paired in one orbital ($\uparrow\downarrow$) lies between these two terms at $92\ kJ\ mol^{-1}$ above the ground state. Of the two singlet states, the latter has much the longer excited state lifetime. This $O_2(^1\Delta)$ survives long enough to participate in chemical reactions. When it is needed for reactions, $O_2(^1\Delta)$ can be generated in solution by energy transfer from a photoexcited molecule. Thus $[Ru(bipy)_3]^{2+}$ can be excited by absorption of blue light (452 nm) to give an electronically excited state, denoted $*[Ru(bipy)_3]^{2+}$, and this state transfers energy to $O_2(^3\Sigma)$:

$$*[Ru(bipy)_3]^{2+} + O_2(^3\Sigma) \rightarrow [Ru(bipy)_3]^{2+} + O_2(^1\Delta)$$

Another efficient way to generate $O_2(^1\Delta)$ is through the thermal decomposition of an ozonide:[10]

In contrast to the radical character of many $O_2(^3\Sigma)$ reactions, $O_2(^1\Delta)$ reacts as an electrophile. This mode of reaction is plausible because $O_2(^1\Delta)$ has a vacant π^* orbital. For example, $O_2(^1\Delta)$ adds across a

[9] Term symbols are described in Section 14.1. With linear molecules such as dioxygen, the symbols Σ, Π, and Δ are used in place of S, P, and D used for atoms. The Greek letters represent the magnitude of the total orbital angular momentum around the internuclear axis.

[10] P. D. Bartlett and C. M. Lonzetta, *J. Am. Chem. Soc.*, **105**, 1984 (1983).

diene, thus mimicking the Diels–Alder reaction of butadiene with an electrophilic alkene:

$$O_2(^1\Delta) \quad + \quad \text{(butadiene)} \quad \longrightarrow \quad \text{(endoperoxide)}$$

Singlet oxygen is implicated as one of the biologically hazardous products of photochemical smog.

The other allotrope of oxygen, ozone (O_3), boils at $-112\,°C$ and is an explosive and highly reactive endoergic blue gas ($\Delta G_f^\ominus = +163\ kJ\ mol^{-1}$). The O_3 molecule is angular, in accord with the VSEPR model (**19**), and has bond angle $117°$; it is diamagnetic.

19 O_3 (C_{2v})

Sulfur

Sulfur is obtained from deposits of the native element, metal sulfide ores, and liquid or gaseous hydrocarbons with a high sulfur content.

Unlike oxygen, all the heavier members of the group favor single bonds over double bonds. As a result, they aggregate into larger molecules or extended structures and hence are solids at room temperature. Sulfur vapor, which is formed at high temperatures, consists partially of paramagnetic disulfur S_2 molecules that resemble O_2 in having a triplet ground state and a formal double bond.

All the crystalline forms of sulfur that can be isolated at room temperature consist of S_n rings. The common orthorhombic polymorph, α-S_8, consists of crown-like eight-membered rings (**20**), but it is possible to synthesize and crystallize rings with from 6 to 20 S atoms. Orthorhombic sulfur melts at $113\,°C$; the yellow liquid darkens above $160\,°C$ and becomes more viscous as the sulfur rings break open and polymerize. The resulting helical S_n polymers (**21**) can be drawn from the melt and quenched to form metastable rubber-like materials that slowly revert to α-S_8 at room temperature. Sulfur is an important element for plant and animal life. For example, RSH and RSR groups are present in the amino acids cysteine and methionine.

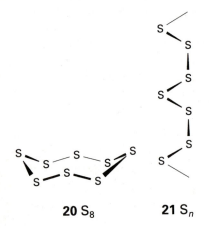

20 S_8 **21** S_n

Selenium, tellurium, and polonium

The chemically soft elements selenium and tellurium occur in metal sulfide ores, and their principal source is the electrolytic refining of copper.

Selenium exists as several different polymorphs. As with sulfur, a nonmetallic allotrope of selenium exists that contains Se_8 rings, but the most stable form at room temperature is gray selenium, a crystalline material composed of helical chains. The photoconductivity of gray selenium, which arises from the ability of incident light to excite electrons across its reasonably small band gap (2.6 eV in the crystalline material, 1.8 eV in the amorphous), accounts for its use in photocells. The common commercial form of the element is amorphous black selenium. Another amorphous form of selenium, obtained by deposition of the vapor, is used as the photoreceptor in the xerographic photocopying process. Selenium is an essential element for humans, but there is only a narrow range of concentration between the minimum daily requirement and toxicity.

Tellurium crystallizes in a chain structure like that of gray selenium. Polonium crystallizes in a primitive cubic structure and a closely related higher temperature form above 36 °C. We remarked in Section 4.3 that a simple cubic array represents inefficient packing of atoms, and polonium is the only element that adopts this structure under normal conditions. Tellurium and polonium are both highly toxic; the toxicity of polonium is enhanced by its intense radioactivity.

12.7 Halides

Oxygen forms many halogen oxides and oxoanions, and we shall discuss them in Chapter 13. Oxygen difluoride, OF_2, is the highest fluoride of oxygen and hence contains oxygen in its highest oxidation state (+2).

Sulfur, selenium, tellurium, and polonium have a very rich halogen chemistry, and some of the most common halides are summarized in Table 12.6. The more electronegative element sulfur forms very unstable iodides, but the more electropositive elements selenium, tellurium, and polonium form more robust compounds with iodine. Of the halogens, the small, electronegative F atom alone brings out the maximum group oxidation state of the elements, but it does not form stable binary compounds of selenium, tellurium, and polonium in low oxidation states (+1 and +2). A series of catenated subhalides exist for the heavy members of the group. For example, Te_2I and Te_2Br consist of a ribbon of edge-shared Te hexagons with halogen bridges (**22**). The inability of the less electronegative halogens to bring out the higher oxidation states is understandable on the basis that they are less electronegative than fluorine, and their single bond strengths to other elements are generally weaker too.[11] The lack of low oxidation state fluorides may be a consequence of their instability toward disproportionation into the element and a higher oxidation state fluoride.

The structures of the sulfur halides S_2F_2, SF_4, SF_6, and S_2F_{10} (Table 12.6) are all in line with the VSEPR model. Thus, SF_4 has 10 valence electrons around the S atom, two of which form a lone pair in an equatorial position of a trigonal bipyramid. We have already mentioned the theoretical evidence that the molecular orbitals bonding the F atoms to the central atom in SF_6 primarily utilize the sulfur $4s$ and $4p$ orbitals, with the $3d$ orbitals playing a relatively unimportant role (Section 3.1). The same seems to be true of SF_4 and S_2F_{10}.

Sulfur hexafluoride is a gas at room temperature. Its inertness stems from the suppression (presumably by steric protection of the central S atom) of thermodynamically favorable reactions such as the hydrolysis

$$SF_6(g) + 4H_2O(l) \rightarrow 6HF(aq) + H_2SO_4(aq)$$

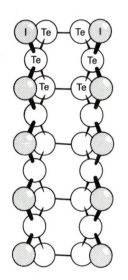

22 Te_2I

[11] J. Passmore, *Acc. Chem. Res.*, **22**, 234 (1989).

Table 12.6 Some halides of sulfur, selenium, and tellurium

Oxidation number	Formula	Structure	Remarks
$+\frac{1}{2}$	Te_2X (X = Br, I)	Halide bridges	Silver-gray
$+1$	S_2F_2	Two isomers:	Reactive
	S_2Cl_2	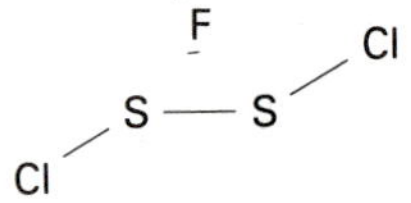	Reactive
$+2$	SCl_2	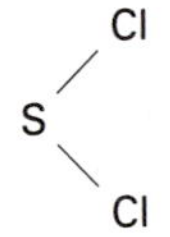	Reactive
$+4$	SF_4	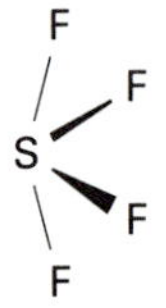	Gas
	SeX_4 (X = F, Cl, Br) TeF_4 (X = F, Cl, Br, I)		SeF_4 liquid TeF_4 solid
$+5$	S_2F_{10} Se_2F_{10}	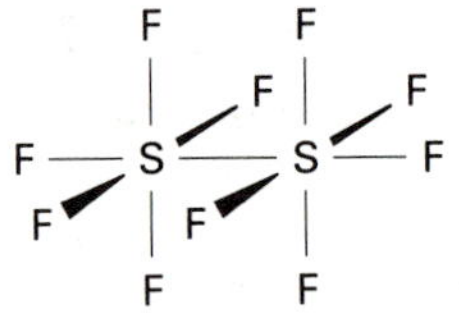	Reactive
$+6$	SF_6, SeF_6 TeF_6	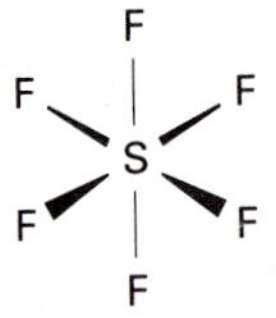	Colorless gases Liquid (b.p. 36 °C)

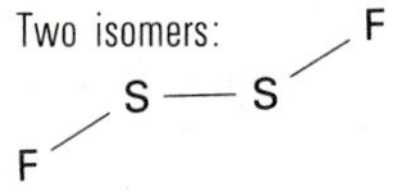

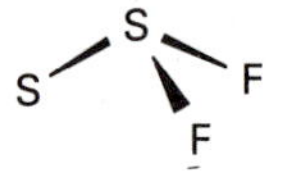

The less sterically crowded SeF_6 molecule is easily hydrolyzed and is generally more reactive than SF_6. Similarly, the sterically less hindered molecule SF_4 is reactive and undergoes rapid partial hydrolysis:

$$SF_4 + H_2O \rightarrow OSF_2 + 2HF$$

Both SF_4 and SeF_4 are very selective fluorinating agents for the conversion of C=O and P=O groups into CF_2 and PF_2 groups:

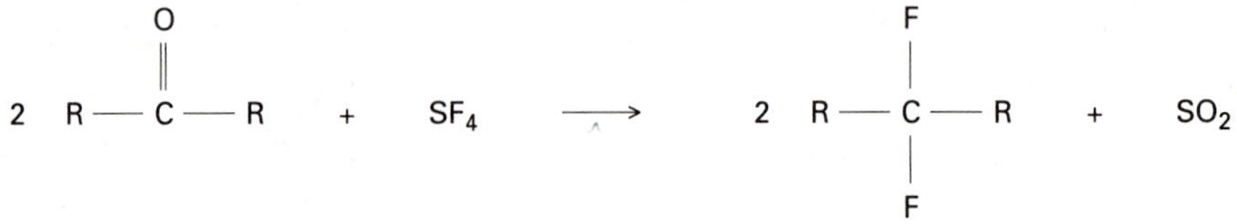

Other halides of the chalcogens exist, and are summarized in Table 12.6. Sulfur chlorides are commercially important. The reaction of molten sulfur with Cl_2 yields the foul-smelling and toxic substance disulfur dichloride, S_2Cl_2, which is a yellow liquid at room temperature (b.p. 138 °C). Disulfur dichloride and its further chlorination product SCl_2 (an unstable red liquid) are produced on a large scale for use in the vulcanization of rubber.

12.8 Oxygen and the *p*-block oxides

Oxygen is by no means an inert molecule, and yet many of its reactions are sluggish (a point first made in connection with overpotentials in Section 7.4). For example, a solution of Fe^{2+} is only slowly oxidized by air even though the potential is favorable.

There are several factors that contribute to the appreciable activation energy of many reactions of O_2. One is that, with weak reducing agents, single electron transfer to O_2 is mildly unfavorable thermodynamically:

$$O_2(g) + H^+(aq) + e^- \rightarrow HO_2(aq) \qquad E^{\ominus} = -0.13 \text{ V at pH} = 0$$

$$O_2(g) + e^- \rightarrow O_2^-(aq) \qquad E = -0.33 \text{ V at pH} = 14$$

Therefore a single-electron reducing reagent must exceed these potentials to achieve a significant reaction rate. Secondly, the ground state of O_2, with both π^* orbitals singly occupied, is neither an effective Lewis acid nor an effective base, and therefore has little tendency to undergo displacement reactions with *p*-block electrophiles or nucleophiles. Finally, the high bond energy of O_2 (463 kJ mol^{-1}) results in a high activation energy for reactions that depend on its dissociation. We saw in Section 9.8 that radical-chain mechanisms can provide reaction paths that circumvent some of these activation barriers in combustion processes at elevated temperatures, and radical oxidations also occur in solution.

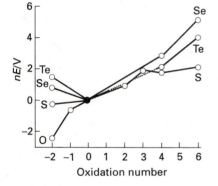

Hydrogen peroxide

The Frost diagram for oxygen (Fig. 12.6) shows that H_2O_2 is unstable with respect to disproportionation. In practice, however, it is not very labile and survives reasonably well at moderate temperatures unless traces of some ions are present, for these act as catalysts. Some insight into the mechanism is obtained by noting that effective catalysts have standard potentials in the range bounded by +1.76 V, the value for the reduction of H_2O_2 to H_2O, and +0.70 V, the value for the reduction of O_2 to H_2O_2. It is believed that the catalyzing ion shuttles back and forth between two oxidation states as it alternately oxidizes and reduces H_2O_2.

12.6 Frost diagram for the elements of the oxygen group in acidic solution. The species with oxidation number − 2 are H_2E and for oxidation number − 1 the compound is H_2O_2. The positive oxidation states refer to oxides or oxoacids. The species for S(II) is thiosulfate, $S_2O_3^{2-}$, and that for S(III) is hydrogendithionite, $HO_2SSO_2^-$.

Example 12.5: *Deciding whether an ion can catalyze H_2O_2 disproportionation*
Is Fe^{3+} thermodynamically capable of catalyzing the decomposition of H_2O_2?

Answer. The standard potential for the reduction of Fe^{3+} to Fe^{2+} is $+0.77$ V. This value falls between the potentials for the reduction of H_2O_2 to H_2O and for the reduction of O_2 to H_2O_2, so catalytic decomposition is expected. We can verify that the potentials are favorable. Subtraction of the equations and potentials for reduction of O_2 to H_2O_2 from that for reduction of $2Fe^{3+}$ gives:

$$2Fe^{3+}(aq) + H_2O_2(aq) \rightarrow 2Fe^{2+}(aq) + O_2(g) + 2H^+(aq)$$

$$E^{\ominus} = +0.07 \text{ V}$$

Because $E^{\ominus} > 0$, the reaction is thermodynamically favorable. Next we subtract the equation and potential for the reduction of $2Fe^{3+}$ from those for the reduction of H_2O_2 and obtain

$$2Fe^{2+}(aq) + H_2O_2(aq) + 2H^+(aq) \rightarrow 2Fe^{3+}(aq) + 2H_2O(l)$$

$$E^{\ominus} = +0.99 \text{ V}$$

This reaction is also spontaneous, so catalytic decomposition is thermodynamically favored. In fact, the rates also are high, so Fe^{3+} is an effective catalyst for the decomposition of H_2O_2, and in its manufacture great pains are taken to minimize iron contamination.

Exercise E12.5. Determine whether either Br^- or Cl^- are candidates for the catalytic decomposition of H_2O_2.

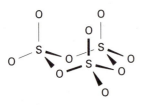

23 SO_2 (C_{2v}) **24** SO_3 (D_{3h})

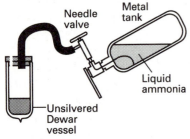

25 $(SO_3)_3$ (C_{3v})

Sulfur trioxide and sulfur dioxide

The molecules of the two common oxides of sulfur, SO_2 (b.p. $-10\,°C$) and SO_3 (b.p. $44.8\,°C$), are angular (**23**) and trigonal planar in the gas phase (**24**), respectively. They are both Lewis acids, with the S atom the acceptor site, but SO_3 is much the stronger and harder acid. The high Lewis acidity of SO_3 accounts for its occurrence as a polymeric O-bridged solid at room temperature and pressure (**25**).

Sulfur dioxide forms weak complexes with simple *p*-block Lewis bases. For example, it does not form a stable complex with H_2O. However, it does form stable complexes with stronger Lewis bases, such as trimethylamine and F^- ions. Sulfur dioxide is a useful solvent for acidic substances (see Box 12.1).

B12.1 The transfer of liquid ammonia from a pressurized tank. (Based on W. L. Jolly, *The synthesis and characterization of inorganic compounds*, Waveland Press, Prospect Heights (1991).)

Box 12.1: Synthesis in nonaqueous solvents

The nonaqueous solvents liquid ammonia, liquid sulfur dioxide, and sulfuric acid show an interesting set of contrasts as they range from a good Lewis base with a resistance to reduction (ammonia) to a strong Brønsted acid with resistance to oxidation (sulfuric acid).

Liquid ammonia (b.p. $-33\,°C$) and the solutions it forms may be handled in an open Dewar flask in an efficient hood (Fig. B12.1), in a vacuum line when the liquid is kept below its boiling point, or in sealed heavy-walled glass tubes at room temperature

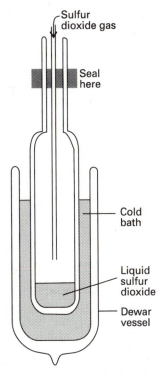

B12.2 The condensation of sulfur dioxide into a Carius tube.

(the vapor pressure of ammonia is about 10 atm at room temperature). Liquid ammonia solutions of the *s*-block metals are excellent reducing agents (Section 8.5). One example of their application is in the preparation of a complex of nickel in its unusual +1 oxidation state:

$$2K(am) + 2[Ni(CN)_4]^{2-}(am) \rightarrow$$
$$[Ni_2(CN)_6]^{4-}(am) + 2KCN(am)$$

Liquid sulfur dioxide (b.p. −10 °C) can be handled like liquid ammonia. An example of its use is the reaction

$$NH_2OH + SO_2(l) \rightarrow H_2NSO_2OH$$

This reaction is performed in a sealed thick-walled borosilicate glass tube, which is sometimes called a Carius tube (Fig. B12.2). The hydroxylamine is introduced into the tube and the sulfur dioxide is condensed into the tube after it has been cooled to about −45 °C. After the tube has been sealed, it is allowed to warm to room temperature behind a blast shield in a hood. The reaction is complete after several days at room temperature; the tube is then cooled (to reduce the pressure) and broken open.

Sulfuric acid and solutions of SO_3 in sulfuric acid (fuming sulfuric acid, $H_2S_2O_7$) are used as oxidizing acidic media for the preparation of polychalcogen cations:

$$8Se + 5H_2SO_4 \rightarrow Se_8^{2+} + 2H_3O^+ + 4HSO_4^- + SO_2$$

[References: W. M. Burgess and J. W. Estes, *Inorg. Synth.*, **5**, 197 (1957); G. Pass and H. Sutcliffe, *Practical inorganic chemistry*, p. 150. Chapman and Hall, London (1974); R. J. Gillespie and J. Passmore, *Acc. Chem. Res.*, **4**, 413 (1971).]

---

26 SO_2

27 $[SO_2F]^-$

28 NR_3SO_2

29 HSO_3^-

30 HSO_3^- (C_{3v})

Example 12.6: *Deducing the structures and chemistry of SO_2 complexes*
Suggest the probable structures of SO_2F^- and $(CH_3)_3NSO_2$, and predict their reactions with OH^-.

Answer. Although the Lewis structure of SO_2 (**26**) has an electron octet around the S atom, that atom can act as an acceptor (Section 5.9). Both resulting complexes have a lone pair on S, and the four electron pairs form a trigonal pyramid around S in both complexes (**27, 28**). The OH^- ion is a stronger Lewis base than either F^- or $N(CH_3)_3$, so exposure of either complex to OH^- will yield the hydrogensulfite ion, HSO_3^-, which has been found to exist in two isomers (**29**) and (**30**).

Exercise E12.6 Give the Lewis structures and point groups of (a) $SO_3(g)$ and (b) SO_3F^-.

Many oxohalides are known for the chalcogens. The most important of these are the thionyl dihalides, OSX_2, and the sulfuryl dihalides,

O_2SX_2. One laboratory application of thionyl dichloride that we shall describe in Box 13.2 is the dehydration of metal chlorides. The compound $F_5TeOTeF_5$ and its selenium analog are known, and the $[OTeF_5]^-$ ion, which is known informally as 'teflate', is one of the largest anions with the properties of very low basicity and resistance to oxidation. Accordingly, it has been employed to stabilize large unstable cations.[12]

Redox properties of sulfur oxoanions

A wide range of sulfur oxoanions exist, and many of them are important in the laboratory and in industry. The peroxodisulfate anion, $[O_3SOOSO_3]^{2-}$, for example, is a powerful and useful oxidizing agent:

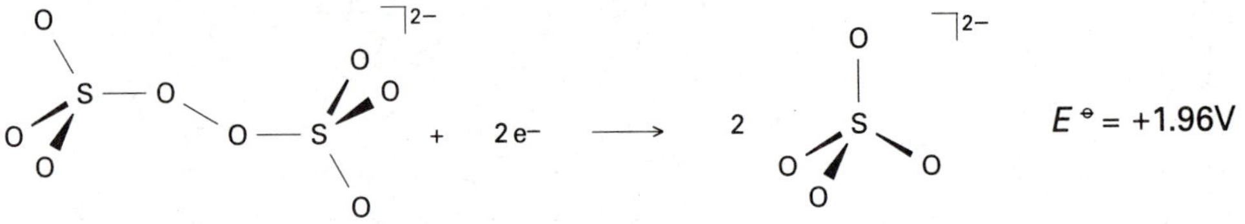

$$E^\ominus = +1.96\text{V}$$

Table 12.7 lists some of the most stable examples of the sulfur oxoanions.

Sulfur's common oxidation numbers are -2, 0, $+2$, $+4$, and $+6$, but there are also many S—S bonded species that are assigned odd and fractional average oxidation numbers. A simple example is the thiosulfate ion, $S_2O_3^{2-}$, in which the average oxidation number of S is $+2$, but in which the environments of the two S atoms are quite different (Table 12.7). The thermodynamic relations between the oxidation states are summarized by the Frost diagram (Fig. 12.6). As with many other p-block oxoanions, many of the thermodynamically favorable reactions are slow when the element is in its maximum oxidation state ($+6$), as in SO_4^{2-}. Another kinetic factor is suggested by the fact that oxidation numbers of compounds containing a single S atom generally change in steps of 2, which alerts us to look for an O atom transfer path for the mechanism. In some cases a radical mechanism operates, as in the oxidation of thiols and alcohols by peroxodisulfate, in which O—O bond cleavage produces the transient radical anion SO_4^-.

We saw in Section 7.10 that the pH of a solution has a marked effect on the redox properties of oxoanions. This is true for SO_2 and SO_3^{2-}, because the former is easily reduced in acidic solution and is therefore a good oxidizing agent, whereas the latter in basic solution is primarily a reducing agent:

$$SO_2(aq) + 4H^+(aq) + 4e^- \rightarrow S(s) + 2H_2O(l)$$

$$E^\ominus = +0.50\text{ V}$$

$$SO_4^{2-}(aq) + H_2O(l) + 4e^- \rightarrow SO_3^{2-}(aq) + 2OH^-(l)$$

$$E^\ominus = -0.94\text{ V}$$

---

[12] P. K. Hurlburt, J. J. Rack, S. F. Dec, O. P. Anderson, and S. H. Strauss, *Inorg. Chem.*, **32**, 373 (1993).

Table 12.7 Some sulfur oxoanions

Oxidation number	Formula	Name	Structure	Remarks
One S atom				
+4	SO_3^{2-}	sulfite	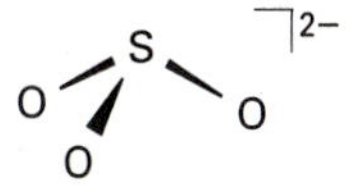	Basic, reducing agent
+6	SO_4^{2-}	sulfate	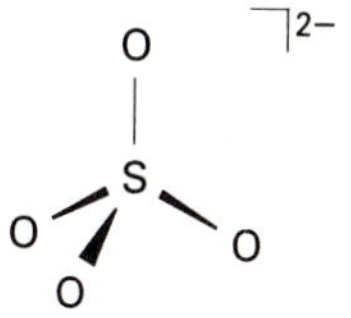	Weakly basic
Two S atoms				
+2	$S_2O_3^{-}$	thiosulfate	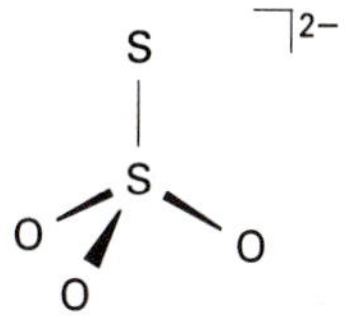	Mild reducing agent
+3	$S_2O_4^{2-}$	dithionite	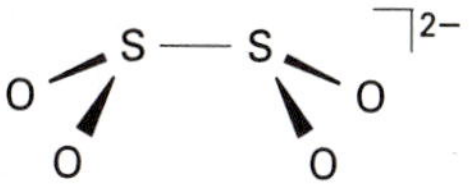	Strong and facile reducing agent
+5	$S_2O_6^{2-}$	dithionate	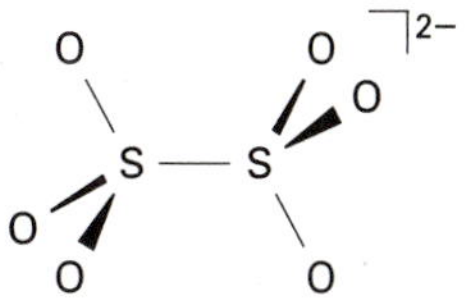	Resists oxidation and reduction
Polysulfur oxoanions				
Variable	$S_nO_6^{2-}$ $3 \leq n \leq 20$	$n = 3$, trithionate	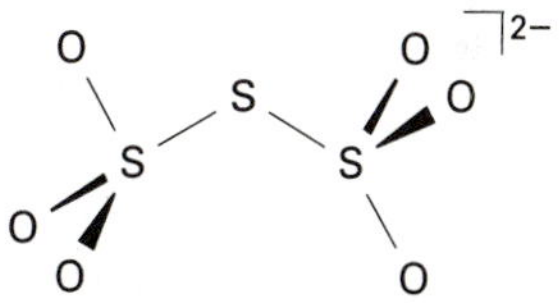	

The principal species present in acidic solution is SO_2, not H_2SO_3 (Section 5.5), but in more basic solution, HSO_3^- exists as an equilibrium mixture of $H—SO_3^-$ and $H—OSO_2^-$. The oxidizing character of SO_2 accounts for its use as a mild disinfectant and bleach for foodstuffs, such as dried fruit and wine.

The known oxoanions of selenium and tellurium are a much less diverse and extensive group. Selenic acid is thermodynamically a strong oxidizing acid:

$$SeO_4^{2-}(aq) + 4H^+(aq) + 2e^- \rightarrow H_2SeO_3(aq) + H_2O(l)$$
$$E^{\ominus} = +1.1 \text{ V}$$

However, like SO_4^{2-} and in common with the behavior of oxoanions of other elements in high oxidation states, the reduction of SeO_4^{2-} is generally slow. Telluric acid exists as $Te(OH)_6$ and also as $(HO)_2TeO_2$

in solution. Again, its reduction is thermodynamically favorable but kinetically sluggish.

12.9 Metal oxides

The O_2 molecule readily takes on electrons from metals to form a variety of metal oxides containing the anions O^{2-} (oxide), O_2^- (superoxide), or O_2^{2-} (peroxide). Even though these solids can be expressed in these terms, and the existence of O^{2-} can be rationalized in terms of a closed-shell noble gas electron configuration, the formation of $O^{2-}(g)$ from $O(g)$ is endothermic, and the ion is stabilized in the solid state by favorable lattice energies (Section 4.8).

Alkali metals and alkaline earth metals often form peroxides or superoxides (see Section 8.3). Among the metals, only some of the noble metals do not form thermodynamically stable oxides. However, even where no bulk oxide phase is formed, an atomically clean metal surface (which can be prepared only in an ultra high vacuum) is quickly covered with a surface layer of oxide when it is exposed to traces of oxygen.

An important trend in the chemical properties of metal oxides is the high basicity of oxides with metal ions having low charge and large radius (low ξ) and the progression through amphoteric to acidic oxides as the charge to radius ratio increases (high ξ). These trends are examined in Sections 5.6 and 8.7. Structural trends in the metal oxides are not so readily summarized, but for oxides in which the metal has oxidation state $+1$, $+2$, or $+3$, the oxide ion is generally in a site of high coordination number. Thus, oxides of M^{2+} ions usually have the rock-salt structure (6:6 coordination, Section 4.5), oxides of M^{3+} ions (of formula M_2O_3) often have 6:4 coordination, and oxides with the formula MO_2 often occur in the rutile or fluorite structures (6:3 and 8:4 coordination, respectively). At the other extreme, MO_4 compounds are molecular: the tetrahedral compound osmium tetraoxide, OsO_4, is an example. Deviations from these simple structures are common with *p*-block metals, where the less symmetric packing of oxide around the metal can often be rationalized by proposing the existence of a stereochemically active lone pair, as in PbO (Section 8.17).

The structures of the metal oxides contrast with those of the nonmetal oxides, in which an O atom may be attached with some multiple bond character (E=O). Another common structure with nonmetals and some metals in high oxidation state is two-coordinate oxygen, E—O—E.

12.10 Metal sulfides, selenides, and tellurides

Sulfide, selenide, and telluride ions are soft ligands and often occur together in nature in copper and zinc deposits. As with oxygen, Se and Te atoms can serve as bridging ligands between metal centers.[13]

---

[13] For a review see: W. A. Herrmann, *Angew. Chem., Int. Ed. Engl.*, **25**, 56 (1986).

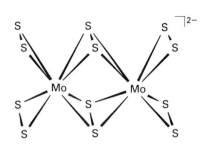

31 $[Mo_2(S_2)_6]^{2-}$

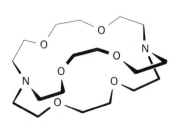

32 $[Ti(Cp)_2Se_5]$

33 $[WS(Se_4)_2]$

34 2,2,2–crypt

For example, S, Se, and Te atoms are found to bridge both two or three metal atoms (**31**). Judging from their structures, the trend toward multiple bonding with a metal atom in a low oxidation state is greater for sulfur, selenium, and tellurium than for oxygen, for which M=O multiple bonding is generally confined to metals in high oxidation states.

In Section 8.11 we saw that sulfur has an extensive coordination chemistry as the formal S^{2-} anion and higher polysulfide anions.[14] The extensive series of polysulfide ions, S_n^{2-}, with $n = 2$ to 22, is not matched by either selenium or tellurium, for which only the isolated ions Se_3^{2-} and Te_3^{2-} and *d*-metal complexes of larger polyselenides and polytellurides are known, such as Cp_2TiSe_5 (**32**).

It appears that in polysulfides, polyselenides, and polytellurides electron density is concentrated at the ends of an E_n^{2-} chain, which accounts for coordination through the terminal atoms, as shown in (**32**) and (**33**).

12.11 *p*-Block ring and cluster compounds

Cluster compounds are known for elements from across the periodic table (see Fig. 8.4).[15] We have already discussed boron clusters (the boranes and carboranes, Chapter 11) and clusters formed by *d* metals (Chapter 8); the latter will be met again in Chapter 16. In this section we concentrate on the clusters formed by the heavier *p*-block elements. These are frequently called **naked clusters** because they generally have no attached groups, such as the H atoms in boron clusters and the CO ligands in *d*-metal clusters. Many of the *p*-block clusters are ions.

Some *p*-block metals and metalloids react with alkali metals to form compounds such as K_2Pb_5. These compounds were investigated by Eduard Zintl in Germany in the 1930s, who found that some could be dissolved in liquid ammonia, a hospitable solvent to strong reducing agents such as these. More recently, John Corbett in the United States discovered that by complexing the alkali metal cation with 2,2,2-crypt (**34**), it was possible to prepare species in ethylenediamine solution that can be crystallized and investigated by X-ray diffraction. The crypt ligand encapsulates the alkali metal ion and thus creates a large cation complex:

$$K_2Pb_5 + 2crypt \xrightarrow{50\,°C} [Kcrypt]_2[Pb_5]$$

The stability of the product stems from the stabilizing effect that large cations have on large anions. Similar techniques have led to a variety of reduced species that have been characterized in the solid state by X-ray diffraction and in some cases in solution by NMR.

[14] Additional examples are given by: A. Muller, *Coord. Chem. Rev.*, **46**, 245 (1982).

[15] R. J. Gillespie, *Chem. Soc. Rev.*, 315 (1979); J. D. Corbett, *Chem. Rev.*, **85**, 383 (1985); H. G. von Schnering, *Angew. Chem., Int. Ed. Engl.*, **20**, 33 (1981); A. H. Cowley (ed.), *Rings, polymers, and clusters of main group elements*, ACS Symposium Series 232. American Chemical Society, Washington (1983).

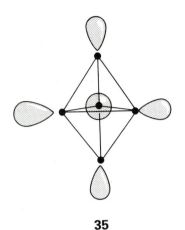

35

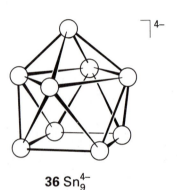

36 Sn_9^{4-}

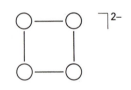

37 Bi_4^{2-} (D_{4h})

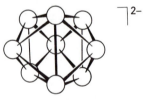

38 Ge_9^{2-}

Se —— Se $^{2+}$
| |
Se —— Se

39 Se_4^{2+}

The electron-counting correlations (Wade's rules) introduced for boranes and carboranes in Section 11.3 are successful with many of these naked *p*-block cluster anions, and we can use Table 11.5 to predict the shapes of deltahedra. As an example, the number of valence electrons available in the Pb_5^{2-} ion is $5 \times 4 = 20$ from the Pb atoms and 2 more from the charge, giving 22 electrons, or 11 pairs. Of these 11 pairs, one pair on each Pb atom, and so 5 pairs in all, are unavailable for bonding since, like the electron pair of a B—H unit in boranes, they are assumed to be directed away from the skeleton (**35**). Thus the total number of skeletal electron pairs is $11 - 5 = 6$. This number agrees with the count expected for a 5-atom ($n = 5$) *closo* cluster, for $5 + 1 = 6$.

Example 12.7: *Correlating the electron count and structure for Zintl clusters*

Determine the electron count and the structural classification for (a) the Sn_9^{4-} ion and (b) the diamond shaped Bi_4^{2-} ion.

Answer. (a) The Sn_9^{4-} ion contains $4 \times 9 + 4 = 40$ valence electrons, and hence 20 pairs. Subtracting one nonbonding pair on each atom (a total of 9) leaves 11 pairs. This fits the number expected for a *nido* cluster. In agreement with this conclusion, the structure is a truncated M_{10} deltahedron (**36**) characteristic of a *nido* cluster. (b) The Bi_4^{2-} anion contains $(4 \times 5) + 2 - (4 \times 2) = 14$ skeletal electrons, or 7 pairs. The count suggests an *arachno* cluster, but it is square planar (**37**) rather than the expected butterfly; see the isoelectronic Se_4^{2+} below.

Exercise E12.7. Determine the classification from electron count for Ge_9^{2-} and determine whether this is consistent with the structure (**38**).

12.12 Polycations

A large number of cationic chain, ring, and cluster compounds of the *p*–block elements have been prepared. The majority of them contain sulfur, selenium, or tellurium, but they may also contain elements from mercury in Group 12 to the halogens in Group 17/VII. Because these cations are oxidizing agents and Lewis acids, the preparative conditions are quite different from those used to synthesize the highly reducing polyanions. For example, S_8 is oxidized by AsF_5 in liquid SO_2 to yield the S_8^{2+} ion:

$$S_8 + 2AsF_5 \xrightarrow{\text{SO}_2} [S_8][AsF_6]_2 + AsF_3$$

A strong acid solvent, such as fluorosulfuric acid, is used and the strongly oxidizing peroxide compound FO_2SOOSO_2F (that is, $S_2O_6F_2$) oxidizes Se to Se_4^{2+}:

$$4Se + S_2O_6F_2 \xrightarrow{\text{HSO}_3\text{F}} [Se_4][SO_3F]_2$$

The Se_4^{2+} cation has a square-planar structure (**39**). In the molecular orbital model of the bonding, this square cation has a closed-shell configuration in which the delocalized π-bonding a_{1g} and nonbonding e_g orbitals are filled, and a higher energy nonbonding orbital is vacant. In contrast, most of the larger ring systems can be understood using localized 2c,2e bonds. For these larger rings, the removal of two electrons brings about the formation of an additional 2c,2e bond, thereby preserving the local electron count on each element. This is readily seen for the oxidation of S_8 to S_8^{2+} (**40**). An X-ray single crystal structure determination shows that the transannular bonds in S_8^{2+} are long compared with the other bonds. Long transannular bonds are common in these types of compounds.

12.13 Neutral heteroatomic ring and clusters

The P_4 molecule is a good example of a cluster that can be described by local 2c,2e bonds. It turns out that several chalcogen derivatives are structurally related to P_4 by the insertion of atoms into P—P bonds to give analogs of the P_4O_n compounds we have already described. Some sulfur compounds in this series are P_4S_3 (**41**), which apparently does not have an oxygen analog, and P_4S_{10}, the analog of P_4O_{10} (**12**).

Sulfur–nitrogen compounds have structures that can be related to the polycations discussed above. The oldest known, and perhaps easiest to prepare, is the pale yellow to orange tetrasulfurtetranitride (S_4N_4, **42**) which is made by passing ammonia through a solution of S_2Cl_2:

$$6S_2Cl_2 + 16NH_3 \rightarrow S_4N_4 + S_8 + 12NH_4Cl$$

S_4N_4 is endoergic ($\Delta G_f^{\ominus} = +536 \text{ kJ mol}^{-1}$) and may decompose explosively. The molecule is an eight-membered ring with the four N atoms in a plane and bridged by S atoms that project above and below the plane. The short S—S distance (2.58 Å) suggests that there is a weak interaction between pairs of S atoms. Lewis acids such as BF_3, SbF_5, and SO_3 form 1:1 complexes with one of the nitrogen atoms, and in the process the S_4N_4 ring rearranges (**43**).

Disulfurdinitride (S_2N_2, **44**) is formed (together with Ag_2S and N_2) when S_4N_4 vapor is passed over hot silver wool. It is even more touchy than its precursor, for it explodes above room temperature. When allowed to stand at 0 °C for several days, it transforms into a bronze-colored polymer (($SN)_x$, **45**). The chains have a zigzag shape and the compound exhibits metallic conductivity along the chain axis. Below 0.3 K $(SN)_x$ is superconducting. The discovery of this superconductivity was important because it was the first example of a superconductor that had no metal constituents.

40 S_8^{2+}

41 P_4S_3

42 S_4N_4

43 $S_4N_4SO_3$

44 S_2N_2

45 $(SN)_x$

FURTHER READING

J. Emsley and D. Hall, *The chemistry of phosphorus*. Harper and Row, New York (1976).

D. E. C. Corbridge, *Phosphorus*. Elsevier, New York (1980): Vol. 1, Phosphorus compounds; Vol. 2, Organophosphorus compounds.

H. G. Heal, *The inorganic heterocyclic chemistry of sulfur, nitrogen, and phosphorus*. Academic Press, New York (1980).

A. J. Bard, R. Parsons, and J. Jordan, *Standard potentials in aqueous solution*. Dekker, New York (1985). See Chapters 4, 6, and 9 particularly. In addition to a discussion of thermodynamic data, these chapters give qualitative information on rates.

A. F. Wells, *Structural inorganic chemistry*. Oxford University Press (1984); especially Chapters 11–12 and 16–20.

A. E. Martell and D. T. Sawyer (ed.), *Oxygen complexes and oxygen activation by transition metals*. Plenum, New York (1988). This volume presents a range of informative review papers on O_2 complexes, and O_2 activation by complexes.

D. T. Sawyer, *Oxygen chemistry*. Oxford University Press, New York (1991).

Kirk-Othmer encyclopedia of chemical technology. Wiley, New York (1979). Because of the great industrial importance of N, O, P, and S compounds, they receive extensive coverage in these volumes. Vol. 15: nitric acid, nitrides, nitrogen, and nitrogen fixation; Vol. 16: oxygen and ozone; Vol. 17: phosphoric acid, phosphates, phosphorus, phosphides, phosphorus compounds, and peroxides; Vol. 22: sulfur, sulfur recovery, sulfuric acid, sulfur trioxide, and sulfur compounds.

KEY POINTS

1. Physical state

In contrast to the gaseous diatomic species N_2 and O_2, the heavier elements in Groups 15/V and 16/VI are solids under normal conditions. Atmospheric nitrogen and oxygen are separated on a massive scale by low temperature distillation for use as a refrigerant and inert gas (N_2), and in steel making (O_2).

2. Elemental nitrogen

Nitrogen has low reactivity but bacteria manage to reduce it at room temperature. The commercial Haber process requires high temperatures and pressures to yield ammonia, which is a major ingredient in fertilizers and an important chemical intermediate.

3. Elemental phosphorus

Elemental phosphorus is recovered from the mineral apatite by carbon arc reduction. The resulting white phosphorus is a molecular solid (P_4). Treatment of apatite with sulfuric acid yields phosphoric acid, which is converted to fertilizers and other chemicals.

4. Halides

The halides of nitrogen and oxygen have limited stability, but their heavier congeners form an extensive series of compounds. Typical formulas are EX_3 and EX_5 for Group 15/V and EX_2, EX_4, and EX_6 for Group 16/VI. Nitrogen, oxygen, and bismuth have somewhat limited halide chemistry because of their resistance to oxidation.

5. Positive oxidation states of nitrogen

Such species are produced by the oxidation of ammonia rather than elemental nitrogen on account of the kinetic inertness of the latter.

6. Nitrogen oxoanions

Nitrate, NO_3^-, and nitrite, NO_2^-, are the most important oxoanions of nitrogen. Their reactions can be sluggish, especially those of nitrate. The mechanisms often involve atom transfer and rates are facilitated by low pH.

7. Phosphorus oxides and oxoanions

The oxides of phosphorus include P_4O_6 and P_4O_{10}, both of which are cage compounds with T_d symmetry. Important oxoanions are the P(I) species hypophosphite, $H_2PO_2^-$, the P(III) species phosphite, HPO_3^{2-}, and the P(V) species phosphate, PO_4^{3-}. The existence of P—H bonds and highly reducing character of the two lower oxidation states is notable. Phosphorus(V) also forms an extensive series of O-bridged polyphosphates. In contrast to N(V), P(V) species are not strongly oxidizing.

8. Arsenic

Arsenic(V) is more easily reduced than P(V), which may account for the toxicity of AsO_4^{3-}, which mimics PO_4^{3-} sufficiently to enter biological cells.

9. Phosphorus-nitrogen compounds

The range of PN compounds is extensive, and includes cyclic and polymeric phosphazenes, $(-PX_2{=}N-)_n$.

10. Oxygen allotropes

Oxygen has two allotropes, dioxygen (O_2) and ozone (O_3). Dioxygen has a triplet ground state which oxidizes hydrocarbons by a radical chain mechanism. Reaction with an excited state molecule can produce a fairly long-lived singlet state species, which is found in photochemical smog and often reacts as an electrophile. Ozone, is an unstable and highly aggressive oxidizing agent.

11. Allotropes and polymorphs of sulfur

Sulfur has many allotropic forms including rings and a metastable polymer; its solid form is polymorphic.

12. Sulfur oxoanions

The oxoanions of sulfur include the good reducing agent sulfite ion, SO_3^{2-}, the rather unreactive sulfate ion, SO_4^{2-}, and the strongly oxidizing peroxosulfate ion, $O_3S-OO-SO_3^{2-}$.

13. Metal oxides

The oxides formed by metals include the basic oxides with high oxygen coordination number that are formed with most M^+ and M^{2+} ions. Oxides of metals in intermediate oxidation states often have more complex structures and are amphoteric. Ruthenium and osmium tetroxides, MO_4, are molecular and not basic.

14. Cyclic and cluster compounds

Cyclic and cluster compounds are known for many of the heavier p-block elements. The valence electron count and structures of Group 14/IV anionic clusters can often be correlated by Wade's rules.

EXERCISES

12.1 List the elements in Groups 15/V and 16/VI and indicate the ones that are (a) diatomic gases, (b) nonmetals, (c) metalloids, (d) true metals. Also indicate those elements that do not achieve the maximum group oxidation number and indicate the elements that display the inert pair effect.

12.2 (a) Give complete and balanced chemical equations for each step in the synthesis of H_3PO_4 from hydroxyapatite to yield (a) high purity phosphoric acid and (b) fertilizer grade phosphoric acid. (c) Account for the large difference in costs between these two methods.

12.3 Ammonia can be prepared by (a) the hydrolysis of Li_3N or (b) the high-temperature high-pressure reduction of N_2 by H_2. Give balanced chemical equations for each method starting with N_2, Li, and H_2, as appropriate. (c) Account for the lower cost of the second method.

12.4 Compare and contrast the formulas and stabilities of the oxidation states of the common nitrogen chlorides with the phosphorus chlorides.

12.5 Use Lewis structures and the VSEPR model to predict the probable structure of (a) PCl_4^+, (b) PCl_4^-, (c) $AsCl_5$.

12.6 Give balanced chemical equations for each of the following reactions: (a) oxidation of P_4 with excess oxygen; (b) reaction of the product from part (a) with excess water; (c) reaction of the product from part (b) with a solution of $CaCl_2$ and name the product.

12.7 Select isoelectronic species from O_2^+, O_2, O_2^-, O_2^{2-}, N_2, NO, NO^+, CN^-, and $N_2H_5^+$. Within these isoelectronic groups, describe the probable relative oxidizing strengths and Lewis basicities.

12.8 Give the formula and name of a carbon-containing molecule or molecular ion that is isoelectronic and isostructural with (a) NO_3^-, (b) NO_2^-, (c) N_2O_4, (d) N_2, (e) NH_3.

12.9 Starting with $NH_3(g)$ and other reagents of your choice, give the chemical equations and conditions for the synthesis of (a) HNO_3, (b) NO_2^-, (c) NH_2OH, (d) N_3^-.

12.10 Write the balanced chemical equation corresponding to the standard enthalpy of formation of $P_4O_{10}(s)$. Specify the structure, physical state (s, l, or g), and allotrope of the reactants. Do either of these reactants differ from the usual practice of taking as reference state the most stable form of an element?

12.11 Without reference to the text, sketch the general form of the Frost diagrams for phosphorus (oxidation states 0 to +5) and bismuth (0 to +5) in acidic solution and discuss the relative stabilities of the +3 and +5 oxidation states of both elements.

12.12 Are reactions of NO_2^- as an oxidizing agent generally faster or slower when pH is lowered? Give a mechanistic explanation for the pH dependence of NO_2^- oxidations.

12.13 When equal volumes of nitric oxide (NO) and air are mixed at atmospheric pressure a rapid reaction occurs, to form NO_2 and N_2O_4. However, nitric oxide from an automobile exhaust, which is present in the parts per million concentration range, reacts slowly with air. Give an explanation for this observation in terms of the rate law and the probable mechanism.

12.14 Give balanced chemical equations for the reactions of the following reagents with PCl_5 and indicate the structures of the products: (a) water (1:1), (b) water in excess, (c) $AlCl_3$, (d) NH_4Cl.

12.15 Use standard potentials (Appendix 2) to calculate the standard potential of the reaction of H_3PO_2 with Cu^{2+}. Are HPO_3^{2-} and $H_2PO_2^{2-}$ useful as oxidizing or reducing agents?

12.16 (a) Use standard potentials (Appendix 2) to calculate the standard potential of the disproportionation of H_2O_2 and HO_2 in acid solution. (b) Is Cr^{2+} a likely catalyst for the disproportionation of H_2O_2? (c) Given the Latimer diagram

$$O_2 \xrightarrow{-0.13} HO_2 \xrightarrow{1.51} H_2O_2$$

in acidic solution, calculate $\Delta G^{\ominus}$ for the disproportionation of hydrogen superoxide (HO_2) into O_2 and H_2O_2, and compare the result with its value for the disproportionation of H_2O_2.

12.17 Which of the solvents ethylenediamine (which is basic and reducing) or SO_2 (which is acidic and oxidizing) might not react with (a) Na_2S_4, (b) K_2Te_3, (c) $Cd_2(Al_2Cl_7)_2$. Explain your reasons.

12.18 Rank the following species from the strongest reducing agent to the strongest oxidizing agent and give a plausible balanced reaction illustrating the strongest reducing agent and strongest oxidizing agent: SO_4^{2-}, SO_3^{2-}, $O_3SO_2SO_3^{2-}$

12.19 (a) Give the formula for Te(VI) in acidic aqueous solution and contrast it with the formula for S(VI). (b) Offer a plausible explanation for this difference. (c) If there is precedent for this trend in Group 15/V, identify and describe it.

12.20 Use isoelectronic analogies to infer the probable structures of (a) Sb_4^{2-} and (b) P_7^{3-}. Describe the bonding in these ions.

PROBLEMS

12.1 The tetrahedral P_4 molecule may be described in terms of localized 2c,2e bonds. Determine the number of skeletal valence electrons and from this decide whether P_4 is *closo, nido,* or *arachno*. If it is not *closo*, determine the parent *closo* polyhedron from which the structure of P_4 could be formally derived by the removal of one or more vertices.

12.2 On account of their slowness at electrodes, the potentials of most redox reactions of nitrogen compounds cannot be measured in an electrochemical cell. Instead, the values must be determined from other thermodynamic data. Illustrate such a calculation using $\Delta G_f^{\ominus}(NH_3,\ aq) = -26.5\ \mathrm{kJ\ mol^{-1}}$ to calculate the standard potential of the N_2/NH_3 couple in basic aqueous solution.

12.3 Correct any inaccuracies in the following statements and, after correction, provide examples to illustrate each statement. (a) Elements in the middle of Groups 15/V and 16/VI are easier to oxidize to the group oxidation state than are the lightest and heaviest members. (b) In its ground state, O_2 is a triplet and it undergoes Diels–Alder electrophilic attack on dienes. (c) The diffusion of ozone from the stratosphere into the troposphere poses a major environmental problem. (d) The starting material for the production of sodium nitrite is ammonia.

12.4 A mechanistic study of reaction between chloramine and sulfite has been reported (B. S. Yiin, D. M. Walker, and D. W. Margerum, *Inorg. Chem.*, **26**, 3435 (1987)). Summarize the observed rate law and the proposed mechanism. Accepting the proposed mechanism, why should $SO_2(OH)^-$ and HSO_3^- display different rates of reaction? Explain why it was not possible to distinguish the reactivity of $SO_2(OH)^-$ from that of HSO_3^-.

12.5 A compound containing pentacoordinate nitrogen has been characterized (A. Frohmann, J. Riede, and H. Schmidbaur, *Nature*, **345**, 140 (1990)). Describe (a) the synthesis and (b) the structure of the compound and (c) give a description of the bonding.

12.6 Tetramethyltellurium, $Te(CH_3)_4$ was prepared in 1989 (R. W. Gedrige, D. C. Harris, K. R. Higa, and R. A. Nissan, *Organometallics*, **8**, 2817 (1989)), and its synthesis was soon followed by the preparation of the hexamethyl compound (L. Ahmed, and J. A. Morrison, *J. Am. Chem. Soc.*, *112*, 7411 (1990)). Explain why these compounds are so unusual, give equations for their syntheses and speculate on why these synthetic procedures are successful. In the latter connection, speculate on why reaction of TeF_4 with a methyllithium does not yield the tetramethyltellurium.

12.7 The bonding in the square-planar Se_4^{2+} ion is briefly described in Section 12.12. Explore this description in more detail by carrying out an extended Hückel calculation on S_4^{2+} with S—S bond distances of 2.0 Å (sulfur is recommended because its extended Hückel parameters are more reliable than those of Se).[16] From the output (a) draw the molecular orbital energy level diagram, (b) assign the symmetry of each level, and (c) sketch the highest energy molecular orbital. Is a closed-shell molecule predicted?

13

The halogens and the noble gases

We now turn to the last two groups in the p block, and see that many of the systematic themes that were helpful for discussing the two preceding groups will again prove useful. For instance, we shall see that the VSEPR model can be used to predict the structures of a wide range of the compounds that the halogens form among themselves, with oxygen, and with the noble gases. As with the elements in Groups 15/V and 16/VI, we shall see that the oxoanions of the halogens and of xenon are oxidizing agents that often react by atom transfer. Another similarity between the groups is the useful correlation between the oxidation number of the central atom and the rates of redox reactions.

Although the range of compounds of the noble gases is much less extensive than that of the halogens, there are similarities between the two groups in the patterns of their reactions and in the structures of their compounds.

The **halogens**, the members of Group 17/VII, are among the most reactive nonmetallic elements; the **noble gases**, their neighbors in Group 18/VIII, are the least. Despite this contrast, there are some resemblances between the two groups, particularly in the structures of their compounds. The two groups are also related in the sense that the first compounds of xenon were fluorides, and these fluorides are the most common starting materials for other compounds of the noble gases.

The chemical properties of the halogens are extensive and have been mentioned many times already. Therefore, in this chapter we highlight their systematic features and pay particular attention to trends within the group.

Table 13.1 Atomic properties of the halogens and noble gases

Element	Ionization enthalpy/ kJ mol^{-1}	Electron affinity/ kJ mol^{-1}	χ_P	Ionic radius/ Å	Common oxidation numbers
Halogens					
F	1687	334	3.98	1.17	−1
Cl	1257	355	3.16	1.67	−1,1,3,5,7
Br	1146	325	2.96	1.82	−1,1,3,5
I	1015	295	2.66	2.06	−1,1,3.5,7
At		270			
Noble gases					
He	2378	−48			0
Ne	2087	−120			0
Ar	1527	−96			0
Kr	1357	−96			0,2
Xe	1177	−77	2.6		0,2,4,6,8
Rn	1043				

Sources: As for Table 12.1. Electronegativity of Xe from L. C. Allen and J. E. Huheey, *J. Inorg. Nucl. Chem.*, **42**, 1523 (1980).

THE ELEMENTS

The atomic properties of the halogens and the noble gases are listed in Table 13.1. The features to note include their high ionization energies and (for the halogens) their high electronegativities and electron affinities. The halogens have high electron affinities because the incoming electron can occupy an orbital of an incomplete valence shell and experience a strong nuclear attraction: recall that Z_{eff} increases progressively across the period (Section 1.9). The noble gases have negative electron affinities because their valence shells are full and an incoming electron must enter the orbitals of a new shell.

We have seen when discussing the earlier groups in the *p* block that the element at the head of each group has structures and properties that are quite different from those of its heavier congeners. The anomalies are much less striking for the halogens, and the most notable difference is that fluorine has a lower electron affinity than chlorine. Intuitively, this feature seems to be at odds with the high electronegativity of fluorine, but stems from the larger electron–electron repulsion in the compact F atom as compared with the larger Cl atom. Despite this difference in electron affinity, however, the enthalpies of formation of metal fluorides are generally much greater than those of metal chlorides. The explanation of this observation is that the low electron affinity of fluorine is more than offset by the high lattice enthalpies of ionic compounds containing the small F$^-$ ion (Fig. 13.1) and the strengths of bonds in covalent species (for examples, the fluorides of metals in high oxidation states).

Because fluorine is the most electronegative of all reactive elements, it is never found in a positive oxidation state (except in the transient gas-phase species F$_2^+$). With the possible exception of astatine, the other halogens occur with oxidation numbers ranging from −1 to +7.

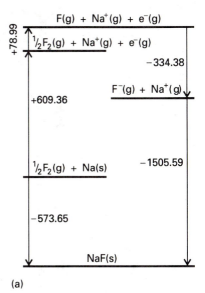

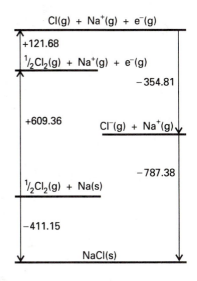

13.1 Thermochemical cycles for the formation of (a) sodium fluoride and (b) sodium chloride (values are in kilojoules).

The dearth of chemical information on astatine stems from its lack of any stable isotopes and the relatively short half-life (8.3 h) of its most stable isotope. Because of this short half-life, astatine solutions are intensely radioactive and may be studied only in high dilution. Astatine appears to exist as the anion At^- and as $At(I)$ and $At(V)$ oxoanions; no evidence for $At(VII)$ has been obtained.

The noble gas with the most extensive chemical properties is xenon. Compounds with Xe—F, Xe—O, Xe—N, and Xe—C bonds are known, and its most important oxidation numbers are $+2$, $+4$, and $+6$. The chemical properties of xenon's lighter neighbor, krypton, are much more limited. The study of radon chemistry, like that of astatine, is inhibited by the high radioactivity of the element.

THE HALOGENS (GROUP 17/VII)

Because of the high electronegativities and abundances of the lighter halogens, their compounds are important in practically every area of chemistry and are encountered throughout the text. Our attention here will be on the halogens themselves, halogen-rich compounds (the interhalogens), and oxides of the halogens.

13.1 Occurrence and recovery

The halogens are so reactive that they are found naturally only as compounds. Their abundances in the Earth's crust decrease steadily with atomic number from fluorine to iodine. All the dihalogens (except for the radioactive At_2) are produced commercially on a large scale, with chlorine production by far the greatest, followed by fluorine. Chlorine is widely used in industry to make chlorinated hydrocarbons and in applications in which a strong and effective oxidizing agent is needed (including bleaching and water purification). These applications are under close scrutiny, however, because some organic chlorine compounds are carcinogenic and others are implicated in the destruction of ozone in the stratosphere (Section 12.8).

The elements are found mainly as halides in nature, but the most easily oxidized element, iodine, is also found as iodates. Because many chlorides, bromides, and iodides are soluble, these anions occur in the oceans and in brines. The primary source of fluorine is calcium fluoride, which is highly insoluble and often found in sedimentary deposits (for example, as fluorite).

The principal method of production of the elements is oxidation of the halides (Section 7.2). The strongly positive standard potentials $E^{\ominus}(F_2, F^-) = +2.87$ V and $E^{\ominus}(Cl_2, Cl^-) = +1.36$ V indicate that the oxidation of F^- and Cl^- ions requires a strong oxidizing agent. Commercially only electrolytic oxidation is feasible. An aqueous electrolyte cannot be used for fluorine production both because water is oxidized at a much lower potential ($+1.23$ V) and because any fluorine produced reacts rapidly with water. The isolation of elemental fluorine eluded chemists for most of the nineteenth century until in 1886 the French chemist Henri Moisson prepared it by the electrolysis of a solution of potassium fluoride in liquid hydrogen fluoride using a cell very much like the one still used today (Fig. 13.2).

Most commercial chlorine production is by the electrolysis of aqueous sodium chloride solution in a **chloroalkali cell** (Fig. 13.3). The half-reactions are

Anode half-reaction: $2Cl^-(aq) \rightarrow Cl_2(g) + 2e^-$

Cathode half-reaction: $2H_2O(l) + 2e^- \rightarrow 2OH^-(aq) + H_2(g)$

The oxidation of water at the anode is suppressed by employing an electrode material that has a higher overpotential for O_2 evolution than for Cl_2 evolution. The best anode material appears to be RuO_2 (see Chapter 18).

In modern chloroalkali cells, the anode and cathode compartments are separated by a polymer ion-exchange membrane. The membrane can exchange cations, and hence permits Na^+ ions to migrate from the anode to the cathode compartment. The flow of cations maintains electroneutrality in the two compartments because, during electrolysis, negative charge is removed at the anode (by conversion of $2Cl^-$ to Cl_2) and supplied at the cathode (by formation of OH^-). The migration of OH^- in the opposite direction would also maintain electroneutrality, but then OH^- would react with Cl_2 and spoil the process. Hydroxide ion migration is suppressed because the membrane does not exchange anions.[1] A final detail of the membrane material is that the C atoms in the polymer backbone are fluorinated: fluorination protects the membrane because the C—F bonds are resistant both to the strong oxidizing agent Cl_2 and to the strong nucleophile OH^-.

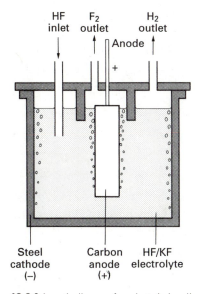

13.2 Schematic diagram of an electrolysis cell for the production of fluorine from potassium fluoride dissolved in liquid hydrogen fluoride.

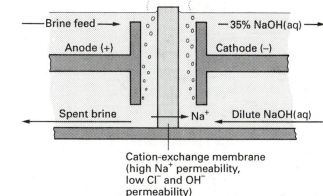

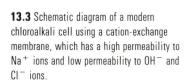

13.3 Schematic diagram of a modern chloroalkali cell using a cation-exchange membrane, which has a high permeability to Na^+ ions and low permeability to OH^- and Cl^- ions.

---

[1] From the standpoint of electrostatic interactions it might seem unlikely that the OH^- ion would migrate away from the positive electrode and toward the negative electrode. It must be remembered however that a solute will diffuse from regions of higher concentration to regions of lower concentration; see P. W. Atkins, *Physical chemistry*. Oxford University Press and W. H. Freeman, New York (1994). Even the cation exchange nature of the membrane material does not totally suppress the diffusion of Cl^- into the negative electrode compartment.

Bromine is obtained by the chemical oxidation of Br^- ions in sea water. A similar process is used to recover iodine from certain natural brines that are rich in I^-. The more strongly oxidizing halogen chlorine is used as the oxidizing agent in both processes, and the resulting Br_2 and I_2 are driven from the solution in a stream of air:

$$Cl_2(g) + 2X^-(aq) \xrightarrow{\text{air}} 2Cl^-(aq) + X_2(g) \quad X = Br \text{ or } I$$

Example 13.1: *Recovery of Br_2 from brine*

Give the chemical equation and potential for the commercial conversion of Br^- in brine to Br_2. Show that from a thermodynamic standpoint bromide could be oxidized to Br_2 by O_2, and offer a plausible reason for why O_2 is not used for this purpose.

Answer. Chlorine is used to oxidize Br^-:

$$Cl_2(g) + 2Br^-(aq) \rightarrow 2Cl^-(aq) + Br_2(g)$$

$$E^\ominus = 1.35\ V - 1.07\ V = +0.26\ V$$

and the resulting volatile Br_2 is removed in a steam–air mixture. Oxygen would be thermodynamically capable of carrying out this reaction in acidic solution:

$$O_2(g) + 4Br^-(aq) + 4H^+(aq) \rightarrow 2H_2O(l) + 2Br_2(l)$$

$$E^\ominus = 1.23\ V - 1.07\ V = +0.16\ V$$

but the reaction is not favorable at pH $= 7$, when $E^\ominus = -0.15\ V$. Even though the reaction is thermodynamically favorable in acidic solution, it is doubtful that the rate would be adequate, since an overpotential of about 0.6 V is associated with the reactions of O_2 (Section 7.4). Even if the oxidation by O_2 in acidic solution were kinetically favorable, the process would be unattractive because of the cost of acidifying large quantities of brine and then neutralizing the effluent.

Exercise E13.1. One source of iodine is sodium iodate, $NaIO_3$. Which of the reducing agents $SO_2(aq)$ or $Sn^{2+}(aq)$ would seem practical from the standpoints of thermodynamic feasibility and plausible judgements about cost? Standard potentials are given in Appendix 2.

13.2 Trends in properties

Unlike the structures of the elements of the preceding groups in the p block, the structures of the halogens are strikingly similar. They are all diatomic and many properties change in a continuous fashion down the group.

Molecular structure and properties

Among the most striking physical properties of the halogens are their colors. In the vapor they range from the almost colorless F_2, through yellowish-green Cl_2 and red–brown Br_2, to purple I_2. The progression

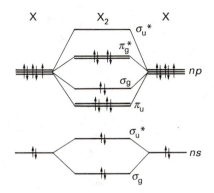

13.4 Schematic molecular orbital energy diagram for Cl_2, Br_2, and I_2. For F_2, the order of the π_u and the upper σ_g orbitals is reversed.

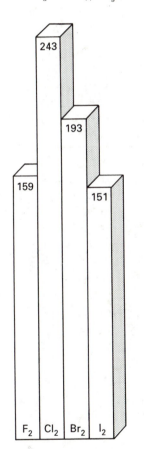

13.5 Bond dissociation enthalpies of the halogens (in $kJ\ mol^{-1}$).

13.6 Dissociation enthalpies of (a) carbon–halogen, (b) hydrogen–halogen, and (c) halogen–halogen bonds. Note the weakness of the X—F bond. (From P. Politzer, *J. Amer. Chem. Soc.*, **91**, 6235 (1969).)

Table 13.2 Bonding and shortest nonbonding distances for solid dihalogens

Element	Temperature/ °C	Bond length/ Å	Nonbonding distance/ Å	Ratio
Cl_2	−160	1.98	3.32	1.68
Br_2	−106	2.27	3.32	1.46
I_2	−163	2.72	3.50	1.29

Data from J. Donahue, *The structure of the elements*. Wiley, New York (1974).

of the maximum absorption to longer wavelengths reflects the decrease in the HOMO–LUMO gap on descending the group. In each case, the optical absorption spectrum arises primarily from transitions in which an electron is promoted from the highest filled σ and π^* orbitals into the vacant antibonding σ^* orbital (Fig. 13.4).

Except for F_2, the analysis of the UV absorption spectra gives precise values for the dihalogen bond dissociation enthalpies (Fig. 13.5). It is found that bond strengths decrease smoothly down the group from Cl_2. The UV spectrum of F_2, however, is a broad continuum that lacks structure because the absorption of radiation is accompanied by dissociation of the F_2 molecule. The lack of discrete absorptions make it difficult to estimate the dissociation energy spectroscopically, and thermochemical methods are complicated by the highly corrosive nature of this reactive halogen. When these corrosion problems were solved, the F—F bond enthalpy was found to be less than that of Br_2 and thus out of line with the trend in the group. However, fluorine's low bond enthalpy is consistent with the low single-bond enthalpies of N—N, O—O, and various combinations of N, F, and O (Fig. 13.6). The simplest explanation (like the explanation of the low electron affinity) is that the bond is weakened by the strong repulsions between nonbonding electrons in the small F_2 molecule.

Chlorine, bromine, and iodine all crystallize in lattices of the same symmetry (Fig. 13.7), so it is possible to make a detailed comparison of distances between bonded and nonbonded neighboring atoms (Table 13.2). The important conclusion is that the latter do not increase as rapidly as the bond lengths. This observation suggests the presence of weak intermolecular bonding interactions that strengthen on going

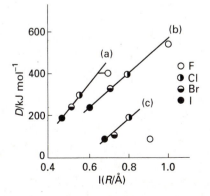

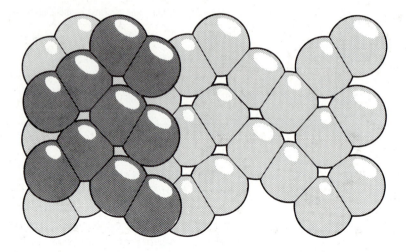

13.7 Solid chlorine, bromine, and iodine have similar structures; however the closest nonbonded interactions are relatively less compressed in Cl_2 and Br_2 than in I_2. (Based on J. Donohue, *The structures of the elements.* Wiley, New York (1974).)

from Cl_2 to I_2. The interaction also leads to a weakening of the I—I bond within the I_2 molecules, as is demonstrated by the lower I—I stretching frequency and greater I—I bond length in the solid as compared with that in the gas phase. Moreover, solid iodine is a semiconductor, and under high pressure exhibits metallic conductivity.

Reactivity trends

Fluorine, F_2, is the most reactive nonmetal and is the strongest oxidizing agent among the halogens. The rapidity of many reactions of fluorine with other elements may in part be due to a low kinetic barrier associated with the weak F—F bond. Despite the thermodynamic stability of most metal fluorides, fluorine can be handled in metal containers because a substantial number of common metals form a passive metal fluoride surface film upon contact with fluorine gas. Nickel and a nickel–molybdenum alloy are particularly resistant to fluorine and fluorides once the passive film is formed. Fluorocarbon polymers, such as polytetrafluoroethylene, are also useful materials for the construction of apparatus to contain fluorine and oxidizing fluorine compounds (Fig. 13.8).

13.8 A typical metal vacuum system for handling fluorine and reactive fluorides. Nickel tubing is used throughout; (A) monel valves, (B) nickel U-traps for the condensation of gases, (C) monel pressure gauge, (D) nickel container for gas storage, (E) polytetrafluoroethylene reaction tube, (F) nickel reaction vessel, (G) nickel canister filled with soda lime (a mixture of sodium and calcium hydroxides) to neutralize HF and react with F_2 and oxidizing fluorine compounds.

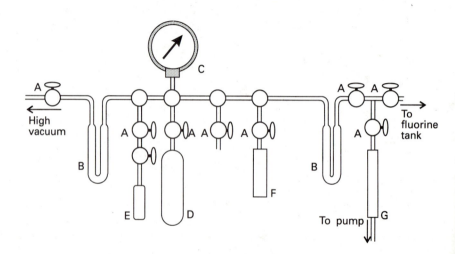

Table 13.3 Normal boiling points (in °C) of compounds of fluorine and their analogs*

F_2	-188.2	H_2	-252.8	Cl_2	-34.0
CF_4	-127.9	CH_4	-161.5	CCl_4	76.7
PF_3	-101.5	PH_3	-87.7	PCl_3	75.5

*Hydrogen-bonded molecules such as HF and NH_3 and their analogs have been omitted.

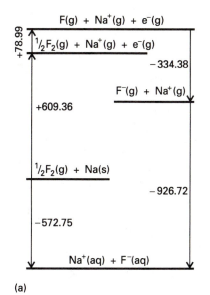

(a)

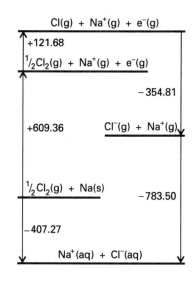

(b)

13.9 Thermochemical cycles for the enthalpy of formation of (a) aqueous sodium fluoride and (b) aqueous sodium chloride. Note that the enthalpy of hydration is much more favorable for F^-.

Reduction potentials for the halides indicate that F_2 ($E^\ominus = +2.87V$) is a much stronger oxidizing agent than Cl_2 ($E^\ominus = +1.36V$). The decrease in oxidizing strength continues in more modest steps from Cl_2 through Br_2 ($+1.07$ V) to I_2 ($+0.54$ V). Although the half-reaction

$$X_2(g) + 2e^- \rightarrow 2X^-(aq)$$

is favored by a high electron affinity (which suggests that fluorine should have a *lower* reduction potential than chlorine), the process is favored by the low bond enthalpy of F_2 and by the highly exothermic hydration of the small F^- ion (Fig. 13.9). The net outcome of these three competing effects of size is that fluorine is the most strongly oxidizing element of the group.

Special characteristics of compounds of fluorine

The boiling points in Table 13.3 demonstrate that molecular fluorine compounds tend to be highly volatile, in some cases even more volatile than the corresponding hydrogen compounds (compare, for example, PF_3 and PH_3) and in all cases much more volatile than their chlorine analogs. The volatilities of the compounds are a result of variations in the strength of the dispersion interaction (the interaction between instantaneous transient electric dipole moments) which are strongest for highly polarizable molecules. The electrons in the small F atoms are held tightly by the nuclei, and consequently fluorine compounds have low polarizabilities and hence weak dispersion interactions.

Another characteristic of fluorine is the ability of an F atom in a compound to withdraw electrons from the atoms to which it is attached. This withdrawal leads to enhanced Brønsted acidity, as for example the three orders of magnitude increase in acidity of trifluoromethane-sulfonic acid, HSO_3CF_3 ($pK_a = 3.0$ in nitromethane) over that of methanesulfonic acid, HSO_3CH_3, ($pK_a = 6.0$ in nitromethane). The presence of F atoms in a molecule also results in high Lewis acidity. For example, we saw in Section 5.15 that SbF_5 is one of the strongest Lewis acids of its type (it is much stronger than $SbCl_5$).

The ability of fluorine to stabilize high oxidation states is second only to that of oxygen: some examples of high oxidation state fluorine compounds are IF_7, PtF_6, BiF_5, and $KAgF_4$. All these compounds are examples of the highest oxidation state attainable for these elements, the rare oxidation state Ag(III) being perhaps the most notable. Another example is the existence of the stable Pb(IV) fluoride, PbF_4, and the instability of all other Pb(IV) halides. A related phenomenon is the tendency of fluorine to disfavor low oxidation states. Thus solid

Table 13.4 Pseudohalides, pseudohalogens, and corresponding acids

Pseudohalide	Pseudohalogen	$E^{\ominus}$/V	Acid	pK_a^*
CN⁻ cyanide	NCCN cyanogen	+ 0.27	HCN hydrogen cyanide	9.2
SCN⁻ thiocyanate	NCSSCN dithiocyanogen	+ 0.77	HNCS hydrogen isothiocyanate	− 1.9
OCN⁻ cyanate			HNCO isocyanic acid	3.5
CNO⁻ fulminate			HCNO fulminic acid	
NNN⁻ azide			HNNN hydrazoic acid	4.92

*A. Albert and E. R. Serjeant, *The determination of ionization constants*. Chapman and Hall, London (1984).

copper(I) fluoride, CuF, is unknown but CuCl, CuBr, and CuI are well known. These trends were discussed in Section 4.8 in terms of a simple ionic model in which the small size of the F^- ion results in a high lattice enthalpy with a small highly charged cation. As a result, there is a thermodynamic tendency for CuF to disproportionate and form CuF_2 (because Cu^{2+} is double charged and its ionic radius is smaller than that of Cu^+).

13.3 Pseudohalogens

A number of compounds have properties so similar to those of the halogens that they are called **pseudohalogens** (Table 13.4). For example, like the dihalogens, cyanogen, $(CN)_2$, undergoes thermal and photochemical dissociation in the gas phase; the resulting CN radicals are isolobal with halogen atoms and undergo similar reactions. Like the dihalogens, cyanogen takes part in a chain reaction with hydrogen:

$$NC—CN \xrightarrow{\text{Heat or light}} 2CN\cdot$$
$$H_2 + CN\cdot \rightarrow HCN + H\cdot$$
$$H\cdot + NC—CN \rightarrow HCN + CN\cdot$$

Overall: $$H_2 + C_2N_2 \rightarrow 2HCN$$

Another similarity is the reduction of a pseudohalogen:

$$NC—CN(aq) + 2e^- \rightarrow 2CN^-(aq)$$

The anion formally derived from a pseudohalogen is called a **pseudohalide ion**. An example is the cyanide anion, CN^-. We say 'formally derived' because pseudohalide ions are known in many cases even though the neutral dimer (the pseudohalogen) is not. Covalent pseudohalides similar to the covalent halides of the *p*-block elements are also common. They are often structurally similar to the corresponding covalent halides (compare **1** and **2**), and undergo similar double-displacement reactions.

As with all analogies, the concepts of pseudohalogen and pseudohalide have many limitations For example, pseudohalogen ions are not spherical, so structures of ionic compounds often differ (for example, NaCl is fcc but NaSCN is not). The pseudohalides are generally less electronegative than the lighter halides, and some of the pseudohalides have more versatile donor properties. The thiocyanate ion, SCN^-, for instance, acts as an ambidentate ligand with a soft base site, S, and a hard base site, N (see Section 6.2).

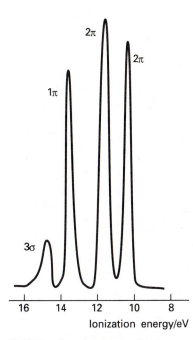

1 $(CH_3)_3SiCN$ **2** $(CH_3)_3SiCl$

13.4 Interhalogens

There are many compounds with halogen–halogen bonds. They are of special importance as highly reactive intermediates and for providing useful insights into bonding. The binary molecular **interhalogens** are compounds with the formulas XY, XY_3, XY_5, and XY_7, where the heavier, less electronegative halogen X is the central atom.

Physical properties and structure

The diatomic interhalogens XY have been made for all combinations of the elements, but many of them are thermally unstable. The most stable is ClF, but ICl and IBr can also be obtained in pure crystalline form. Their physical properties are intermediate between those of their component molecules. For example, the deep red ICl (m.p. 27 °C, b.p. 97 °C) is intermediate between yellowish-green Cl_2 (m.p. −101 °C, b.p. −35 °C) and black I_2 (m.p. 114 °C, b.p. 184 °C). Photoelectron spectra (Fig. 13.10) indicate that the molecular orbital energy levels in the mixed dihalogen molecules lie in the order $3\sigma < 1\pi < 2\pi$, which is the same as in the homonuclear dihalogen molecules. An interesting historical note is that ICl was discovered before Br_2 in the early nineteenth century, and when later the first samples of the dark red-brown Br_2 (m.p. −7 °C, b.p. 59 °C) were prepared, they were mistaken for ICl.

Most of the higher interhalogens are fluorides (Table 13.5). The only neutral interhalogen with the central atom in a +7 oxidation state is IF_7, but the cation ClF_6^+, a compound of Cl(VII), is known. The absence of a neutral ClF_7 reflects the destabilizing effect of F···F nonbonding repulsions between lone pairs on neighbouring atoms (indeed, coordination numbers greater than six are not observed for other *p*-block central atoms in Period 3). The lack of BrF_7 might be rationalized in a similar way, but in addition we shall see later that bromine is reluctant to achieve its maximum oxidation state. In this respect, it resembles some other Period 4 *p*-block elements, notably arsenic and selenium.

13.10 Photoelectron spectrum of ICl. The π levels give rise to two peaks because of spin-orbit interaction in the positive ion. (S. Evans and A. F. Orchard, *Inorg. Chem. Acta.*, **5**, 81 (1971).)

Table 13.5 Representative interhalogens

XY	XY$_3$	XY$_5$	XY$_7$
ClF colorless b.p. −100 °C	ClF$_3$ colorless b.p. 12 °C	ClF$_5$ colorless b.p. −13 °C	
BrF* light brown b.p. *c.* 20 °C	BrF$_3$ yellow b.p. 126 °C	BrF$_5$ colorless b.p. 41 °C	
IF*	(IF$_3$)$_n$ yellow dec. −28 °C	IF$_5$ colorless b.p. 105 °C	IF$_7$ colorless subl. 5 °C
BrCl*† red brown b.p. *c.* 5 °C			
ICl red solid	I$_2$Cl$_6$ bright yellow m.p. 101 °C(16 atm)		
IBr black solid			

*Very unstable.
†The pure solid is known at low temperatures.

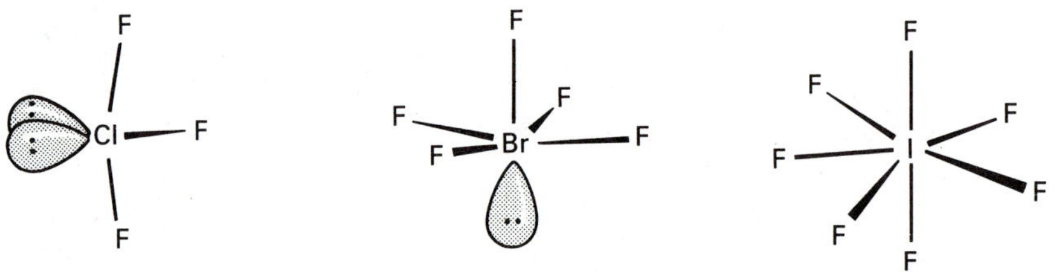

13.11 The structures of typical interhalogens. The shapes are in accord with the predictions of the VSEPR model.

ClF$_3$(C_{2v})   BrF$_5$(C_{4v})   IF$_7$($\approx D_{5h}$)

The shapes of interhalogen molecules are largely as the VSEPR model predicts (Fig. 13.11). For example, the XY$_3$ compounds (such as ClF$_3$) have five valence electron pairs around the X atom, and they adopt a trigonal bipyramidal arrangement. The Y atoms attach to the two axial pairs and one of the three equatorial pairs, and then the two axial bonding pairs move away from the two equatorial lone pairs. As a result, XY$_3$ molecules have a C_{2v} drooping-T shape. There are some discrepancies: for example, ICl$_3$ is a Cl-bridged dimer.

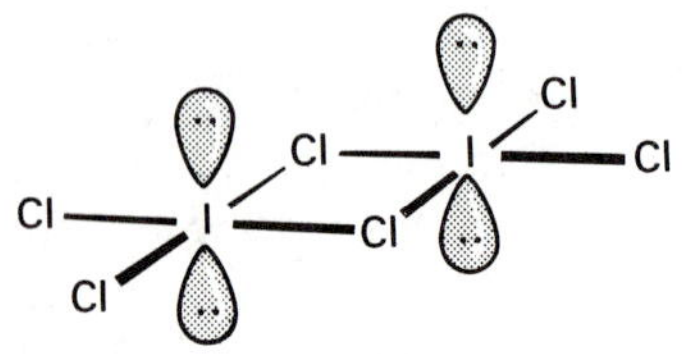

Example 13.2: *Predicting the shape of an interhalogen molecule*
Predict the shape of I$_2$Cl$_6$ using VSEPR model and assign the point group.

Answer. There are seven valence electrons on each atom in I$_2$Cl$_6$, giving 56 in all, and hence 28 pairs. Given the fact cited above that the molecule has two bridging Cl atoms, a reasonable Lewis structure is (**3**) with six electron pairs on each I atom. According to the VSEPR

3 I$_2$Cl$_6$

model, these pairs adopt an octahedral configuration, with the lone pairs in nonadjacent trans positions. Thus the bonds to surrounding Cl atoms take on a square planar configuration and the molecule should be planar. This shape is verified by X-ray diffraction on single crystals. To assign the point group, we may choose (in accord with convention) the principal axis as the twofold axis through the two I atoms. There are two additional twofold axes and a mirror plane perpendicular to this principal axis. These symmetry elements identify the point group as D_{2h}.

Exercise E13.2. Predict the structure and identify the point group of ClO_2F.

The Lewis structure of XF_5 puts five bonding pairs and one lone pair on the central halogen atom, and as expected from the VSEPR model, the XF_5 molecules are square-pyramidal (Fig. 13.11). As already mentioned, the only known XY_7 compound is IF_7, which is predicted to be pentagonal-bipyramidal. The most recent experimental evidence for its actual structure shows that it is a distorted pentagonal bipyramid with a puckered, fluxional IF_5 plane. As with other hypervalent molecules (Section 2.8), the bonding in IF_7 can be explained without invoking d-orbital participation by adopting a molecular orbital model in which bonding and nonbonding orbitals are occupied but antibonding orbitals are not.

Chemical properties

All the interhalogens are oxidizing agents. As with all the known halogen fluorides, the Gibbs free energy of formation of ClF_3 is negative, so it is thermodynamically a weaker fluorinating agent than F_2. However, the rate at which it fluorinates substances generally exceeds that of fluorine, so it is in fact an aggressive fluorinating agent toward many elements and compounds. In general, the rates of oxidation of interhalogens do not have a simple relation to their thermodynamic stabilities. Thus, ClF_3 and BrF_3 are much more aggressive fluorinating agents than BrF_5, IF_5, and IF_7; iodine pentafluoride, for instance, is a convenient mild fluorinating agent that can be handled in glass apparatus. One use of ClF_3 as a fluorinating agent is in the formation of a passive metal fluoride film on the inside of nickel apparatus used in fluorine chemistry.

Both ClF_3 and BrF_3 react vigorously (often explosively) with organic matter, burn asbestos, and expel oxygen from many metal oxides:

$$2Co_3O_4(s) + 6ClF_3(g) \rightarrow 6CoF_3(s) + 3Cl_2(g) + 4O_2(g)$$

They are produced on a significant scale for the production of UF_6 for ^{235}U enrichment:

$$UF_4(s) + ClF_3(g) \rightarrow UF_6(s) + ClF(g)$$

Bromine trifluoride autoionizes in the liquid state:

$$2 \quad F-\overset{\cdot\cdot}{\underset{|}{Br}}\overset{\cdot\cdot}{\underset{F}{}}\quad \longrightarrow \quad \left[\overset{\cdot\cdot}{:Br}-\underset{|}{F}\right]^{+} \quad + \quad \left[\overset{F}{\underset{F}{\diagdown}}\overset{\cdot\cdot}{\underset{F}{Br}}\overset{F}{\diagup}\right]^{-}$$

This Lewis acid–base behavior is shown in its ability to dissolve a number of halide salts:

$$CsF(s) + BrF_3(l) \rightarrow CsBrF_4(soln)$$

Bromine trifluoride is a useful solvent for ionic reactions that must be carried out under highly oxidizing conditions. The Lewis acid character of BrF_3 is shared by other interhalogens, which react with alkali metal fluorides to produce anionic fluoride complexes (Section 13.5).

Cationic polyhalogens

Under special strongly oxidizing conditions, such as in fuming sulfuric acid, I_2 is oxidized to the blue paramagnetic diiodonium cation, I_2^+; the dibrominium cation, Br_2^+, is also known. The bond lengths of these cations are shorter than those of the corresponding neutral dihalogens, which is the expected result for loss of an electron from a π^* orbital (Fig. 13.4) which increases the bond order from 1 to $1\frac{1}{2}$. Three higher polyiodonium cations, Br_5^+, I_3^+, and I_5^+, are known, and X-ray diffraction studies of the iodine species have established the structures shown in (4) and (5). The angular shape of I_3^+ is in line with the VSEPR model because the central I atom has two lone pairs of electrons.

Another class of polyhalogen cations of formula XF_n^+ is obtained when a strong Lewis acid, such as SbF_5, abstracts F^- from interhalogen fluorides.

$$ClF_3 + SbF_5 \rightarrow [ClF_2^+][SbF_6^-]$$

Table 13.6 lists a variety of interhalogen cations that are prepared in the same manner. X-Ray diffraction of solid compounds that contain these cations indicate that the F^- abstraction from the cations is incomplete and that the anions remain weakly associated with the cations by fluorine bridges (6).

13.5 Halogen complexes and polyhalides

The Lewis acidity of diiodine and the other heavy dihalogens toward electron pair donor molecules was mentioned in Section 5.12. This Lewis acidity is also evident in the interaction of halogen molecules with ion donors to give the range of ions known as **polyhalides**.

Polyiodides

A deep brown color develops when I_2 is added to a solution of I^- ions. This color is characteristic of the homoatomic polyiodides, which

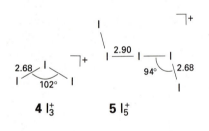

4 I_3^+ **5** I_5^+

Table 13.6 Representative interhalogen cations

$[XF_2]^+$	$[XF_4]^+$	$[XF_6]^+$
$[ClF_2]^+$	$[ClF_4]^+$	$[ClF_6]^+$
$[BrF_2]^+$	$[BrF_4]^+$	$[BrF_6]^+$
	$[IF_4]^+$	$[IF_6]^+$

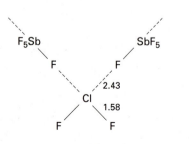

6 $(ClF_2)(SbF_6)$

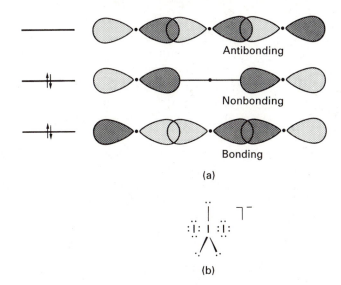

13.12 Some representations of the polyiodide ion. (a) The σ interaction in an I_3^- ion, (b) Lewis and VSEPR rationalizations of the linear structure, where the five electron pairs around the central atom are arranged in a trigonal bipyramidal array with the two bonding pairs along the threefold axis and the three nonbonding pairs in the equatorial plane.

include triiodide ions, I_3^-, and pentaiodide ions, I_5^-. These polyiodides are Lewis acid–base complexes in which I^- and I_3^- act as the bases and I_2 acts as the acid (Fig. 13.12). The Lewis structure of I_3^- has three equatorial lone pairs on the central I atom and two axial bonding pairs in a trigonal bipyramidal arrangement. This hypervalent Lewis structure is consistent with the observed linear structure of I_3^-, which is described in more detail below.

An I_3^- ion can interact with an additional I_2 molecule to yield larger mononegative polyiodides of composition, $[(I_2)_n(I^-)]$. The I_3^- ion is the most stable member of this series. In combination with a large cation, such as $[N(CH_3)_4]^+$, it is symmetrical and linear with a longer I—I bond than in I_2. However, the structure of the triiodide ion, like that of the polyiodides in general, is highly sensitive to the identity of the counterion. For example Cs^+, which is smaller than the tetramethylammonium ion, distorts the I_3^- ion and produces one long and one short I—I bond (7). The ease with which the ion responds to its environment is a reflection of the weakness of the delocalized bonds that just manage to hold the atoms together. A more extreme example of sensitivity to the cation is provided by NaI_3, which can be formed in aqueous solution but decomposes when the water is evaporated:

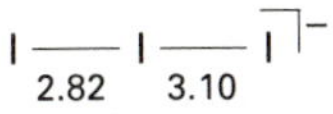

$$Na^+(aq) + I_3^-(aq) \xrightarrow{\text{remove water}} NaI(s) + I_2(s)$$

This behavior is another example of the instability of large anions in combination with small cations which, as we have seen (Section 4.8), can be rationalized by the ionic model.

The existence and structures of the higher polyiodides are sensitive to the counterion for similar reasons, and large cations are necessary to stabilize them in the solid state. In fact, entirely different shapes are observed for polyiodide ions in combination with various large

I —— I —— I⁻
  2.82   3.10

7 I_3^- in CsI_3

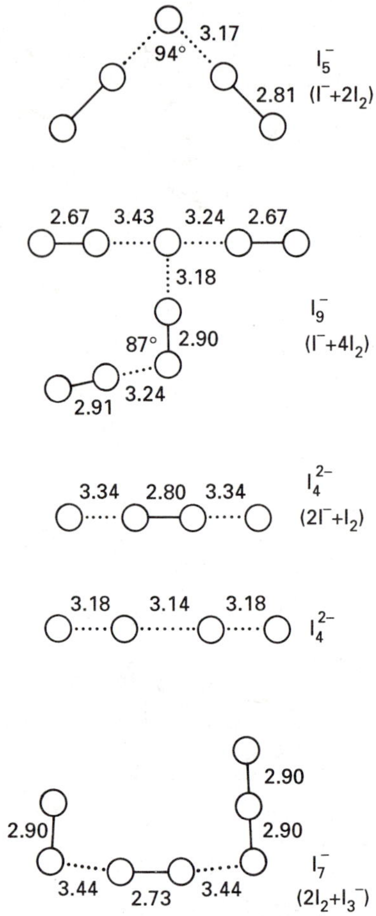

13.13 Some representative polyiodide structures and their approximate description in terms of I^-, I_3^-, and I_2 building blocks. The bond lengths and angles vary with the identity of the cation.

8 $[BrF_4]^-$

cations, for the structure of the anion is determined in large measure by the manner in which the ions pack together in the crystal. The bond lengths in a polyiodide ion often suggest that it can be regarded as a chain of associated I^-, I_2, I_3^- and sometimes I_4^{2-} units (Fig. 13.13). Solids containing polyiodides exhibit electrical conductivity which may arise from the hopping of electrons (or holes) along the polyiodide chain or by an ion relay along the polyiodide chain (Fig. 13.14). This proposed ion relay is similar to the Grotthus mechanism for the transport of H^+ through water and ice (Section 5.1).

Some dinegative polyiodides are known. These contain an even number of I atoms and their general formula is $[I^-(I_2)_n I^-]$. They have the same sensitivity to the cation as their mononegative counterparts.

Other polyhalides

Although polyhalide formation is most pronounced for iodine, other polyhalides are also known. They include Cl_3^-, Br_3^-, and BrI_2^-, which are known in solution and (in partnership with large cations) as solids too. Even F_3^- has been detected spectroscopically at low temperatures in an inert matrix. This technique, which is known as **matrix isolation**, employs the codeposition of the reactants with a large excess of noble gas at very low temperatures (in the region of 4 K to 14 K). The solid noble gas forms an inert matrix within which the F_3^- ion can sit in chemical isolation.

In addition to complex formation between dihalogens and halide ions, some interhalogens can act as Lewis acids toward halide ions. The reaction results in the formation of polyhalides which, in contrast to the chain-like polyiodides, are assembled around a central halogen acceptor atom in a high oxidation state. For example, BrF_3 reacts with CsF to form $CsBrF_4$, which contains the square-planar BrF_4^- anion (**8**). Recently an extensive series of these interhalogen anions has been synthesized (Table 13.7). Many of their shapes agree with the VSEPR model, but there are some interesting exceptions. Two such exceptions are ClF_6^- and BrF_6^-, in which the central halogen has a lone pair of electrons, but the apparent structure is octahedral.

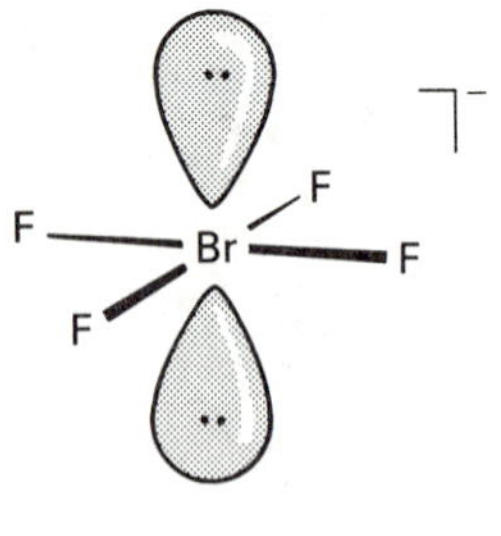

Example 13.3: *A bonding model for I^+ complexes*

In some cases the interaction of I_2 with strong donor ligands leads to the formation of cationic complexes such as bis(pyridyl)iodine($+1$), $[py-I-py]^+$. Propose a bonding model for this linear complex from (a) the standpoint of the VSEPR model and (b) simple molecular orbital considerations.

Answer. The (hypervalent) Lewis electron structure places 10 electrons around the central I^+ in $[py-I-py]^+$, six from the iodine cation and four from the lone pairs on the two pyridine ligands. According to the VSEPR model, these pairs should form a trigonal bipyramid. The lone pairs will occupy the equatorial positions and consequently the complex should be linear. (b) From a molecular orbital perspective, the orbitals of the N—I—N array can be pictured as formed from an iodine $5p$ orbital and an orbital of σ symmetry from each of the two ligand atoms. Three orbitals can be constructed: 1σ (bonding), 2σ (nearly nonbonding), and 3σ (antibonding). There are four electrons to accommodate (two from each ligand atom; the iodine $5p$ orbital is empty). The resulting configuration is $1\sigma^2 2\sigma^2$, which is net bonding.

Exercise E13.3. From the perspective of structure and bonding, indicate several polyhalides that are analogous to $[py—I—py]^+$, and describe their bonding.

13.6 Compounds of the halogens with oxygen

In contrast to the rather simple formulas and structures of halogen fluorides, the molecular compounds of the halogens with oxygen are a diverse group. The aqueous chemistry of the halogen oxoanions is the most uniform and important, so we shall concentrate on them after a brief discussion of the neutral oxides.

Halogen oxides

Many binary compounds of the halogens and oxygen are known, but most are unstable and not commonly encountered in the laboratory. We shall mention only a few of the most important.

Oxygen difluoride (FOF; m.p. $-224\,°C$, b.p. $-145\,°C$), the most stable binary compound of oxygen and fluorine, is prepared by passing

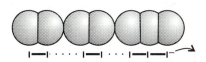

13.14 One possible mode of charge transport along a polyiodide chain is the shift of long and short bonds resulting in the effective migration of an I^- ion along the chain. Three successive stages of the migration are shown. As with the transport of the proton in ice or water, the iodine atom from the I_3^- on the left is not the same as the one emerging on the right.

Table 13.7 Representative interhalogen anions

Compound	Shape
IF_8^-	square antiprism[a]
ClF_6^-	octahedral[b]
BrF_6^-	octahedral[c]
IF_6^-	trigonally distorted octahedron[d]

[a] *Chem. Commun.*, 837 (1991); *Angew. Chem., Int. Ed. Engl.*, **30**, 876 (1991).
[b] *Inorg. Chem.*, **29**, 3506 (1990).
[c] *Angew. Chem., Int. Ed. Engl.*, **28**, 1526 (1989).
[d] *Angew. Chem., Int. Ed. Engl.*, **30**, 323 (1991).

fluorine through dilute aqueous hydroxide solution:

$$2F_2(g) + 2OH^-(aq) \rightarrow OF_2(g) + 2F^-(aq) + H_2O(l)$$

The pure difluoride survives in the gas phase above room temperature and does not react with glass. It is a strong fluorinating agent but less so than fluorine itself. As suggested by the VSEPR model (or by a Walsh diagram, by analogy with H_2O), the OF_2 molecule is angular.

Dioxygen difluoride (FOOF; m.p. $-154\,°C$, b.p. $-57\,°C$), in laboratory jargon 'foof', can be synthesized by photolysis of a liquid mixture of the two elements. It is unstable in the liquid state and decomposes rapidly above $-100\,°C$, but may be transferred (with some decomposition) as a low-pressure gas in a metal vacuum line. Dioxygen difluoride is an even more aggressive fluorinating agent than ClF_3. For example, it oxidizes plutonium metal and its compounds to PuF_6 in a reaction that ClF_3 cannot accomplish:

$$Pu(s) + 3O_2F_2(g) \rightarrow PuF_6(g) + 3O_2(g)$$

The interest in this reaction is in removing plutonium as its volatile hexafluoride from spent nuclear fuel.

Chlorine occurs with many different oxidation numbers in its oxides, for they include:

Oxidation number	+1	+4	+6	+7
Formula	Cl_2O	ClO_2	Cl_2O_6	Cl_2O_7
State	brown–yellow gas	yellow gas	dark red liquid	colorless liquid

Some of these oxides are odd-electron species, including ClO_2, in which chlorine has the unusual oxidation number $+4$, and Cl_2O_6, a compound of chlorine(VI). The latter exists in an ionic form, as $[ClO_2][ClO_4]$, in the solid in which the oxidation states are $+5$ and $+7$.

Chlorine dioxide is the only halogen oxide produced on a large scale. The reaction used is the reduction of ClO_3^- with HCl or SO_2 in strongly acidic solution:

$$2ClO_3^-(aq) + SO_2(g) \xrightarrow{\text{acid}} 2ClO_2(g) + SO_4^{2-}(aq)$$

It is a strongly endoergic compound ($\Delta_f G^\ominus = +121\ kJ\ mol^{-1}$) and must be kept dilute to avoid explosive decomposition; it is therefore consumed at the site of production. Its major uses are to bleach paper pulp and to disinfect sewage and drinking water. Some controversy surrounds these applications because the action of chlorine (or its product of hydrolysis, HClO) and chlorine dioxide on organic matter produces low concentrations of chlorocarbon compounds, which are potential carcinogens. On the other hand, the disinfection of water undoubtedly saves many more lives than the carcinogenic byproducts may take.

Table 13.8 Halogen oxoanions

Oxidation number	Formula	Name*	Point group	Shape	Remarks
+1	ClO^-	Hypochlorite [monoxochlorate(I)]	$C_{\infty v}$	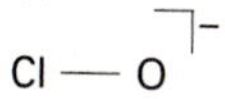 Cl — O	Good oxidizing agent
+3	ClO_2^- †	Chlorite [dioxochlorate(III)]	C_{2v}	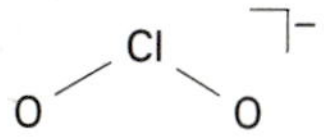	Strong oxidizing agent, disproportionates
+5	ClO_3^-	Chlorate [trioxochlorate(V)]	C_{3v}	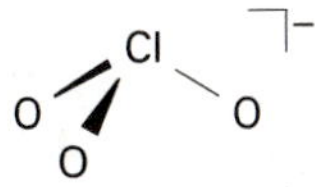	Oxidizing agent
+7	ClO_4^- ‡	Perchlorate [tetraoxochlorate(VII)]	T_d	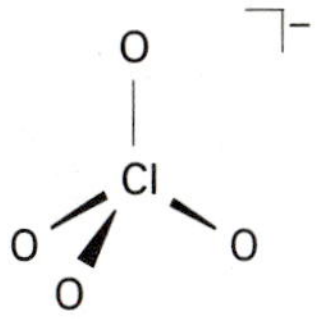	Oxidizing agent and weak ligand

*IUPAC names in square brackets.
†Iodite has not been isolated.
‡For iodine, IO_4^- and $H_3IO_6^{2-}$ are present in basic solution, and H_5IO_6 is present in acidic solution.

Oxoacids and oxoanions

The wide range of oxoanions and oxoacids of the halogens presents a challenge to those who devise systems of nomenclature. We shall use the most commonly employed names, such as chlorate for ClO_3^-, rather than the systematic names, such as trioxochlorate(V). Table 13.8 includes a brief dictionary for converting between the two systems of nomenclature.

The strengths of the oxoacids vary systematically with the number of O atoms on the central atom (Table 13.9; see Pauling's rules in Section 5.5, specifically $pK_a = 8 - 5p$ for the acid $O_pE(OH)_q$). At first sight, periodic acid, the I(VII) analog of perchloric acid, appears to be out of line, for it is weak ($pK_{a1} = 3.29$). However, as soon as we note that its formula is $(HO)_5IO$ with $p = 1$, we see that it is its structure, not its strength, that is anomalous. The O atoms in the conjugate base $H_4IO_6^-$ are very labile on account of the rapid equilibration

$$H_4IO_6^-(aq) \rightleftharpoons IO_4^-(aq) + 2H_2O(l) \quad K = 40$$

and in basic solution IO_4^- is the dominant ion. The tendency to have an expanded coordination shell is shared by the oxoacids of the neighboring Group 16/VI element tellurium, which in its maximum oxidation state forms the weak acid $Te(OH)_6$.

The halogen oxoanions, like many oxoanions, form metal complexes, including the metal perchlorates and periodates to be discussed here. In this connection we note that because $HClO_4$ is a very strong acid and H_5IO_6 is a weak acid, it follows that ClO_4^- ions are very weak bases and $H_4IO_6^-$ ions are stronger bases.

In view of the low Brønsted basicity and single negative charge of the perchlorate ion, ClO_4^-, it is not surprising that it is a weak Lewis base with little tendency to form complexes with cations in aqueous

Table 13.9 Acidities of chlorine oxoacids

Acid	q/p	pK_a
HOCl	0	7.53 (weak)
HOClO	1	2.00
$HOClO_2$	2	−1.2
$HOClO_3$	3	−10 (strong)

solution. Therefore, metal perchlorates are often used to study the properties of hexaaqua ions in solution. The ClO_4^- ion is used as a weakly coordinating ion that can readily be displaced from a complex by other ligands, or as a medium sized anion that might stabilize solid salts containing large cationic complexes with easily displaced ligands.

However, the ClO_4^- ion is a treacherous ally. Because it is a powerful oxidizing agent, solid compounds of perchlorate should be avoided whenever there are oxidizable ligands or metal ions present (which is commonly the case). In some cases the danger lies in wait, for the reactions of ClO_4^- are generally slow and it is possible to prepare many metastable perchlorate complexes or salts that may be handled with deceptive ease. However, once reaction has been initiated by mechanical action, heat, or static electricity, these compounds can detonate with disastrous consequences.[2] Such explosions have injured chemists who may have handled a compound many times before it unexpectedly exploded. Some readily available and more docile weakly basic anions may be used in place of ClO_4^-; they include trifluoromethanesulfonate $[SO_3CF_3]^-$, tetrafluoroborate $[BF_4]^-$, and hexafluorophosphate $[PF_6]^-$.

In contrast to perchlorate, periodate is a rapid oxidizing agent and a stronger Lewis base than perchlorate. These properties lead to the use of periodate as an oxidizing agent and stabilizing ligand for metal ions in high oxidation states. Some of the high oxidation states it can be used to form are very unusual: they include Cu(III) in a salt containing the $[Cu(HIO_6)_2]^{5-}$ complex and Ni(IV) in an extended complex containing the $[Ni(IO_6)]^-$ unit. In these complexes the periodate ligand forms a bridge between Ni(IV) ions in a layered lattice.

Thermodynamic aspects of redox reactions

The thermodynamic tendencies of the halogen oxoanions and oxoacids to participate in redox reactions are well understood. As we shall see, we can summarize their behavior with a Frost diagram that is quite easy to rationalize. It is a very different story with the rates of the reactions, for these vary widely. Their mechanisms are only partly understood despite many years of investigation. Recent progress in the understanding of some of these mechanisms stems from advances in techniques for fast reactions and interest in oscillating reactions (Box 13.1).

We saw in Section 7.9 that if a species lies above the line joining its two neighbors to each other in a Frost diagram, then it is unstable with respect to disproportionation into them. From the Frost diagram for the halogen oxoanions and oxoacids in Fig. 13.15 we can see that many of the oxoanions in intermediate oxidation states are susceptible to disproportionation. Chlorous acid, $HClO_2$, for instance, lies above

---

[2] Two examples are the explosion in a rocket-fuel plant of ammonium perchlorate with the intensity of an earthquake, *Chem. and Eng. News*, May 16, 1988 p. 5, and the injury of a chemist by the explosion of the perchlorate salt of a lanthanide complex, *Chem. and Eng. News*, December 5, 1983 p. 4.

Box 13.1 Oscillating reactions

Clock reactions and oscillating reactions provide fascinating lecture demonstrations and are an active topic of research. Most examples of oscillating reactions are based on reactions of halogen oxoanions, apparently because of the variety of oxidation states and their sensitivity to changes in pH. In 1895, H. Landot discovered that a mixture of sulfite, periodate, iodate, and starch in acidic solution remains nearly colorless for an initial period and then suddenly switches to the dark purple of the I_2–starch complex. When the concentrations are properly adjusted the reaction oscillates between nearly colorless and opaque blue. The reactions leading to this oscillation are the reduction of periodate to iodide by sulfite, where all reactants are colorless:

$$IO_4^-(aq) + 4SO_3^{2-}(aq) \rightarrow I^-(aq) + 4SO_4^{2-}(aq)$$

A comproportionation reaction between I^- and IO_3^- then produces I_2, which forms an intensely colored complex with starch.

$$IO_3^-(aq) + 6H^+(aq) + 5I^-(aq) \rightarrow 3H_2O(l) + 3I_2(starch)$$

Under some conditions this is the final state, but adjustment of concentrations may lead to bleaching of the I_2–starch complex by sulfite reduction of iodine to the colorless $I^-(aq)$ ion:

$$3I_2(starch) + 3SO_3^{2-}(aq) + 3H_2O(l) \rightarrow$$
$$6I^-(aq) + 6H^+(aq) + 3SO_4^{2-}(aq)$$

The reaction may then oscillate between colorless and blue as the I_2/I^- ratio changes.

The detailed analysis of the kinetic conditions for oscillating reactions is pursued by chemists and chemical engineers. In the former case the challenge is to employ kinetic data determined separately for the individual steps to model the observed oscillations with a view to testing the validity of the overall scheme. Since oscillating reactions have been observed in commercial catalytic processes, the concern of the chemical engineer is to avoid large fluctuations or even chaotic reactions that might degrade the process.

[References: I. Epstein, *Chem. Eng. News*, **65**, p. 25, 30 March 1987; R. J. Field and F. W. Schneider, *J. Chem. Educ.*, **66**, 195 (1989) and other papers in that issue.]

the line joining its two neighbors, and is liable to disproportionation:

$$2HClO_2(aq) \rightarrow ClO_3^-(aq) + HClO(aq) + H^+(aq) \quad E^\ominus = +0.52 \text{ V}$$

Although BrO_2^- is well characterized, the corresponding I(III) species is so unstable that it does not exist in solution, except perhaps as a transient intermediate.

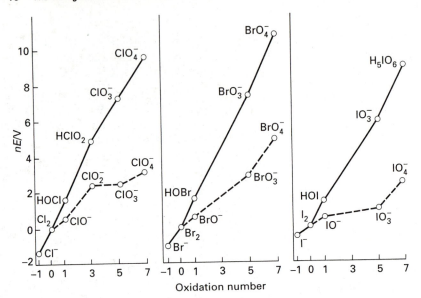

13.15 Frost diagrams for chlorine, bromine, and iodine in acidic solution (solid line) and basic solution (dashed line).

We also saw in Section 7.9 that the more positive the slope for the line from a lower to a higher oxidation state species in a Frost diagram, the stronger the oxidizing power of the couple. A glance at Fig. 13.15 shows that all three diagrams have steep positively sloping lines, which immediately shows that all the oxidation states except the lowest (Cl^-, Br^-, and I^-) are strongly oxidizing.

Finally, basic conditions decrease reduction potentials for oxoanions as compared with their conjugate acids (Section 7.10). This decrease is evident in the less steep slopes of the lines in the Frost diagrams for the oxoanions in basic solution (Fig. 13.15). The numerical comparison for ClO_4^- ions in 1 M acid compared with 1 M base makes this clear:

$$At \ pH = 0: \ ClO_4^-(aq) + 2H^+(aq) + 2e^- \rightarrow ClO_3^-(aq) + H_2O(l)$$

$$E^{\ominus} = +1.20 \ V$$

$$At \ pH = 14: \ ClO_4^-(aq) + H_2O(l) + 2e^- \rightarrow ClO_3^-(aq) + 2OH^-(aq)$$

$$E_B^{\ominus} = +0.37 \ V$$

The potentials show that perchlorate is thermodynamically a weaker oxidizing agent in basic solution than in acidic solution.

Trends in rates of redox reactions

Mechanistic studies show that the redox reactions of halogen oxoanions are intricate. Nevertheless, despite this complexity there are a few discernable patterns that help to correlate the trends in rates of reaction. These correlations have practical value and give some clues about the mechanisms that may be involved.

The oxidation of many molecules and ions by halogen oxoanions becomes progressively faster as the oxidation number of the halogen decreases. Thus the rates observed are often in the order

$$ClO_4^- < ClO_3^- < ClO_2^- \approx ClO^- \approx Cl_2$$

$$BrO_4^- < BrO_3^- \approx BrO^- \approx Br_2$$

$$IO_4^- < IO_3^- < I_2$$

For example, aqueous solutions containing Fe^{2+} and ClO_4^- are stable for many months in the absence of dissolved oxygen, but an equilibrium mixture of aqueous ClO^- and Cl_2 rapidly oxidizes Fe^{2+}.

Oxoanions of the heavier halogens tend to react most rapidly. This is particularly striking for the elements in their highest oxidation states:

$$ClO_4^- < BrO_4^- < IO_4^-$$

As we have remarked, perchlorates in dilute aqueous solution are usually unreactive, but periodate oxidations are fast enough to be employed for titrations. The mechanistic details are often complex, but the fast reaction of periodate appears to arise from access of the reductant to the electrophilic I atom. The existence of both four-coordinate and six-coordinate periodate ions shows that the I atom in periodate is accessible to nucleophiles. Thus H_2O may add to the I atom in IO_4^- to yield $IO_2(OH)_4^-$ and protons. This reaction is facile, as can be seen from the rapid exchange of labeled water ($H_2{}^{18}O$) with periodate:

$$IO_2(OH)_4^- + H_2{}^{18}O \rightleftharpoons IO_2(OH)_3({}^{18}OH) + H_2O$$

Oxidations by oxoanions are often accelerated by acid. The oxidation of halides by BrO_3^- ions, for instance, is second-order in H^+:

$$\text{Rate} = k[BrO_3^-][X^-][H^+]^2$$

The acid is thought to act by protonating the oxo group in the oxoanion, so aiding oxygen–halogen bond scission. Another role of protonation is to increase the electrophilicity of the halogen as in HClO where, as described below, the Cl atom may be viewed as an electrophile toward an incoming reducing agent (**9**). We have already seen that the thermodynamic tendency of oxoanions to act as oxidizing agents increases as the pH is lowered, and now we see that their rates are increased too. Thus, kinetics and equilibria unite to bring about otherwise difficult oxidations. An example is the use of a mixture of H_2SO_4 and $HClO_4$ in the *final* stages of the oxidation of organic matter in certain analytical procedures. It is important to be especially alert to the risk of an explosion when concentrated perchloric acid is mixed with organic compounds or other oxidizable substances.

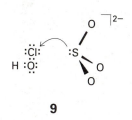

9

Reactions of individual oxidation states

With the general redox properties of the halogens now established, we can consider the characteristic properties and reactions of specific oxidation states, with emphasis on mechanisms of reaction. Although we are dealing here with halogen oxides, it is convenient to mention, for the sake of completeness, the redox properties of halogen(-1) and halogen(0) species. Figure 13.16 summarizes some of the reactions that

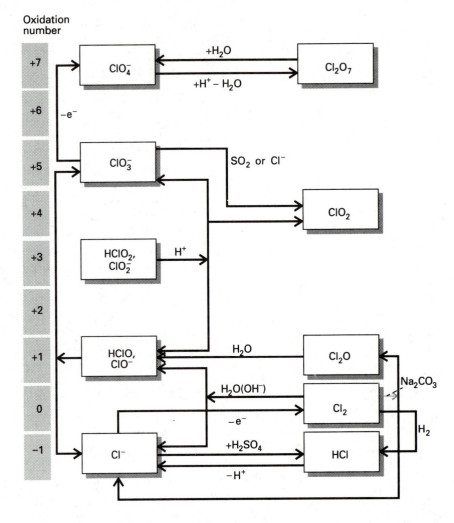

13.16 The interconversion of oxidation states of some important chlorine compounds.

interconvert the oxoions and oxoacids of chlorine in its various oxidation states. One point to note is the major role of disproportionation and electrochemical reactions in the scheme. For example, the chart includes the production of Cl_2 by the electrochemical oxidation of Cl^-, which was discussed in Section 13.1. Then, starting with Cl_2, the chart indicates that two successive disproportionations lead to the important Cl(V) species, ClO_3^-. The highest oxidation state Cl(VII) is achieved by the electrochemical oxidation of ClO_3^- to ClO_4^-.

Halogen(−1). The halides are familiar ions found in natural waters and mineral deposits such as rock salt (NaCl), sylvite (KCl), and fluorite (CaF_2). They also are useful starting materials and in this application are often needed in their anhydrous form. The removal of residual water is not straightforward for metal ions in the +3 or higher oxidation state, because simple heating leads to hydrolysis:

$$FeCl_3 \cdot H_2O(s) \xrightarrow{\Delta} FeCl_2(OH)(s) + HCl(g)$$

$$\xrightarrow{\Delta} FeOCl(s) + 2HCl(g)$$

Several methods are available to dehydrate metal salts without hydrolysis (see Box 13.2). The halides become thermodynamically easier to oxidize to their elemental form on descending the group from fluorine to iodine.

Box 13.2 Preparation of anhydrous metal halides

Because anhydrous halides are very common starting materials for inorganic syntheses, their preparation and the removal of water from impure commercial halides is of considerable practical importance. The principal synthetic methods are the direct reaction of a metal and a halogen and the reaction of halogen compounds with metals or metal oxides.

An example of direct reaction is shown in Fig. B13.1 where the halogen in the gas phase reacts with the metal contained in a heated tube. When the halide is volatile at the temperature of the reaction, it sublimes to the exit end of the tube.

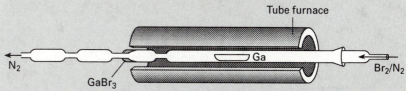

B13.1 A tube furnace used for the production of anhydrous gallium bromide.

It is sometimes advantageous to use a halogen compound for the synthesis of a metal halide. For example, the reaction of ZrO_2 with CCl_4 vapor in a heated tube is a good method for the preparation of $ZrCl_4$. A contribution to the Gibbs free energy of this reaction comes from the production of the $C{=}O$ bond in phosgene ($COCl_2$):

$$ZrO_2(s) + 2CCl_4(g) \rightarrow ZrCl_4(s) + 2COCl_2(g)$$

Anhydrous halides can often be prepared by the removal of water from hydrated metal chlorides.* Simply heating the sample in a dry gas stream is usually not satisfactory for metals with high charge to radius ratio because significant hydrolysis may lead to the oxide or oxohalide:

$$2CrCl_3 \cdot 6H_2O(s) \rightarrow Cr_2O_3(s) + 6HCl(g) + 9H_2O(g)$$

This hydrolysis can be suppressed by carrying out the dehydration in a heated tube in a stream of hydrogen halide, or by dehydration with thionyl chloride (b.p. 79°C), which produces the desired anhydrous chloride and volatile HCl and SO_2 by-products:

$$FeCl_3 \cdot 6H_2O(s) + 6SOCl_2(l) \rightarrow$$

$$FeCl_3(s) + 6SO_2(g) + 12HCl(g)$$

---

* The procedures for dehydration of metal chloride hydrates by thionyl chloride and references to other methods are given by A. R. Pray, *Inorg. Synth.*, **28**, 321 (1990).

Halogen(0). One of the most characteristic redox reactions of the halogens is disproportionation. In general, disproportionation is thermodynamically favorable for basic solutions of Cl_2, Br_2, and I_2. The equilibria in basic aqueous solution are:

$$X_2(aq) + 2OH^-(aq) \rightleftharpoons XO^-(aq) + X^-(aq) + H_2O(l)$$

$$K = 7.5 \times 10^{15} \text{ (Cl)}, \ 2 \times 10^8 \text{ (Br)}, \ 30 \text{ (I)}$$

In writing the associated equilibrium expressions in aqueous solution, we conform to the usual convention that the activity of water is 1, so $[H_2O]$ does not appear in the equilibrium constant:

$$K = \frac{[XO^-][X^-]}{[X_2][OH^-]^2}$$

Disproportionation is much less favorable in acidic solution, as would be expected from the fact that H^+ is a product of the reaction:

$$Cl_2(aq) + H_2O(l) \rightarrow HClO(aq) + H^+(aq) + Cl^-(aq)$$

$$K = 3.9 \times 10^{-4}$$

The mechanism of this reaction appears to involve attack of H_2O on the Cl_2 molecule:

$$H_2O + Cl_2 \rightarrow [H_2O{-}Cl{-}Cl] \rightarrow H^+ + [HO{-}Cl{-}Cl]^-$$

$$\rightarrow H^+ + HOCl + Cl^-$$

It is not yet known if the Lewis complex $H_2O{-}Cl_2$ is actually formed. According to one interpretation, proton transfer is so rapid that the reaction yields $HOCl_2^-$ directly. The $[HO{-}Cl{-}Cl]^-$ complex is at most transitory, for Cl^- is rapidly displaced by the strong nucleophile OH^-. Thus we see that the reaction which is ordinarily viewed as a redox (because the oxidation number of chlorine changes), is mechanistically a nucleophilic displacement.

Because the redox reactions of HClO and Cl_2 are often fast, Cl_2 in water is widely used as an inexpensive and powerful oxidizing agent. A part of the reason for its usefulness is that the rapid equilibration between ClO^- and Cl_2 gives rise to a variety of possible reaction pathways. For example, the nonlabile complex $[Fe(phen)_3]^{2+}$ reacts more rapidly with Cl_2 than with HClO because the former can more readily undergo an outer-sphere electron transfer, whereas ClO^- or HClO appear to require dissociation of phen prior to an inner-sphere redox process. In other situations HClO is a more facile oxidizing agent than Cl_2. In general, labile aqua complexes such as $[Cr(OH_2)_6]^{2+}$ react faster with ClO^- than with Cl_2.

The equilibrium constants for the hydrolysis of Br_2 and I_2 in acid solution are smaller than for Cl_2, and both elements are unchanged when dissolved in slightly acidified water. Because F_2 is a much stronger oxidizing agent than the other halogens, it produces mainly O_2 and H_2O_2 when in contact with water, so hypofluorous acid, HFO, was discovered only long after the other hypohalous acids. The first indication of its existence came from IR spectroscopic studies of H_2O and F_2 trapped in a matrix of frozen N_2 at about 20 K. With this encouragement, HFO (which is structurally H—O—F) was finally isolated and chemically characterized in the early 1970s by the controlled reaction of F_2 with ice at $-40\,°C$, which leads to the highly unstable HFO in approximately 50 percent yield:

$$F_2(g) + H_2O(s) \rightarrow HFO(g) + HF(g)$$

Halogen(I). The aqueous Cl(I) species hypochlorous acid, HClO (the angular molecular species H—O—Cl)[3] and hypochlorite ions, ClO^-, are facile oxidizing agents that are used as household bleach and disinfectant and as useful laboratory oxidizing agents. The mechanism invoked for oxidation reactions of HClO was thought for many years to involve O atom transfer, but recent mechanistic studies strongly implicate Cl^+ transfer in the reactions of ClO^- and HClO.[4] For example, kinetic data indicate that the first step in the mechanism of reaction of HClO with SO_3^{2-} is Cl^+ transfer to produce the chlorosulfate anion:

$$HClO(aq) + SO_3^{2-}(aq) \rightarrow OH^-(aq) + ClSO_3^-(aq)$$

The activated complex is pictured as a Cl bridged species $[HO—Cl—SO_3]^{2-}$ (**9**). To account for the outcome of this reaction, the Cl atom is thought to carry only six electrons with it as it transfers to SO_3^{2-}; so in effect a Cl^+ ion is transferred even though an isolated Cl^+ cation is not involved. This step is followed by the hydrolysis of chlorosulfate, to yield the sulfate and chloride ions:

$$ClSO_3^-(aq) + H_2O(l) \rightarrow SO_4^{2-}(aq) + Cl^-(aq) + 2H^+(aq)$$

A similar Cl^+ ion transfer is indicated for the reactions of HClO with other nucleophilic anions and the rate of reaction, which lies in the order

$$Cl^- < Br^- < I^- < SO_3^{2-} < CN^-$$

correlates with the nucleophilicity of the anion. The ready access to the electrophilic Cl atom in HClO appears to be one feature that leads to the very fast redox processes of this compound, which contrast with much slower redox reactions of perchlorate ions, in which access of nucleophiles to the Cl atom is blocked by the surrounding O atoms.

Hypohalite ions are subject to disproportionation. For instance, the ClO^- ion forms Cl^- and ClO_3^- ions:

$$3ClO^-(aq) \rightarrow 2Cl^-(aq) + ClO_3^-(aq) \quad K = 1.5 \times 10^{27}$$

oxidation number: $\quad +1 \qquad -1 \qquad +5$

This reaction (which is used for the commercial production of chlorates) is slow at or below room temperature for ClO^- but is much faster

---

[3] Hypochlorous acid is widely denoted HOCl to emphasize its structure. We have adopted the formula HClO to emphasize its family relationship to the other oxoacids, $HClO_n$.

[4] C. Gerritsen and D. W. Margerum, *Inorg. Chem.*, **29**, 2757 (1990) and references therein.

for BrO^-. It is so fast for IO^- that this ion has been detected only as a reaction intermediate.

Halogen(III). Chlorite ions, ClO_2^-, and bromite ions, BrO_2^-, are both susceptible to disproportionation. However, the rate is strongly dependent on pH, and ClO_2^- (and to a lesser extent BrO_2^-) can be handled in basic solution with only slow decomposition. In contrast, chlorous acid, $HClO_2$, and bromous acid, $HBrO_2$, both disproportionate rapidly. As the following potentials indicate, disproportionation is thermodynamically quite favorable in acidic solution.

$$2HClO_2(aq) \rightarrow H^+(aq) + HClO(aq) + ClO_3^-(aq) \quad E^\ominus = +0.51 \text{ V}$$

$$2HBrO_2(aq) \rightarrow H^+(aq) + HBrO(aq) + BrO_3^-(aq) \quad E^\ominus = +0.50 \text{ V}$$

Iodine(III) is even more elusive, and HIO_2 was only recently identified as a transient species in aqueous solution.

These primary reactions are usually accompanied by further reaction. For example, the disproportionation of $HClO_2$ under a broad set of conditions is quite well described by the equation

$$4HClO_2(aq) \rightarrow$$
$$2H^+(aq) + 2ClO_2(aq) + ClO_3^-(aq) + Cl^-(aq) + H_2O(l)$$

This complex disproportionation is shown in the center of Fig. 13.16. Because two oxidized products (ClO_2 and ClO_3^-) are produced, the equation can be balanced differently by changing their ratio. The ratio actually observed reflects the relative rates of the alternative reactions.

Halogen(V). The Frost diagrams for chlorine (Fig. 13.15) indicate that chlorate ions, ClO_3^-, are unstable with respect to disproportionation in both acidic and basic solution:

$$4ClO_3^-(aq) \rightarrow 3ClO_4^-(aq) + Cl^-(aq) \quad E^\ominus = +0.25 \text{ V}$$

This value of $E^\ominus$ corresponds to $K = 1.4 \times 10^{25}$; however, as $HClO_3$ is a strong acid, and this reaction is slow at both low and high pH, ClO_3^- ions can be handled readily in aqueous solution. Bromates and iodates are thermodynamically stable with respect to disproportionation.

Example 13.4: *The decomposition of HFO*
It was remarked that HFO is unstable in water. Write plausible equations for the decomposition of HFO(aq) and discuss why it is different from that for HClO.

Answer. Oxidation states 0, +1, and +3 are accessible for halogens except fluorine. In part this is simply an artifact of our way of assigning the most negative oxidation number to the most electronegative element, but the higher electronegativity of fluorine is also reflected in the chemical properties of HFO. We might for example expect this thermodynamically unstable compound to yield more stable compounds, such as HF, O_2, H_2O_2 and possibly OF_2. In fact some of these products are observed:

$$2HFO(aq) \rightarrow 2HF(aq) + O_2(g)$$

$$HFO(aq) + H_2O(l) \rightarrow H_2O_2(aq) + HF(aq)$$

In the presence of excess fluorine a further reaction occurs:

$$F_2(aq) + HFO(aq) \rightarrow OF_2(aq) + HF(aq)$$

Exercise E13.4. Use standard potentials to decide which dihalogens are thermodynamically capable of oxidizing H_2O to O_2. Which of these reactions are facile?

Halogen(VII). The perchlorate ion, ClO_4^-, and periodate ion, IO_4^-, were known in the nineteenth century, but the perbromate ion BrO_4^- was not discovered until the late 1960s.[5] This difference is another example of the instability of the maximum oxidation state of the Period 4 elements arsenic, selenium, and bromine. Although perbromate is not available commercially, it is fairly readily synthesized by the action of fluorine on bromate ions in basic aqueous solution:

$$BrO_3^-(aq) + F_2(g) + 2OH^-(aq) \rightarrow BrO_4^-(aq) + 2F^-(aq) + H_2O(l)$$

Of the three XO_4^- ions, BrO_4^- is the strongest oxidizing agent. That perbromate is out of line with its adjacent halogen congeners, fits a general pattern for anomalies in the chemistry of *p*-block elements of Period 4.

The reduction of BrO_4^- is faster than that of ClO_4^- because perbromate ions undergo electron transfer more rapidly (by an outer-sphere mechanism). However, reduction of periodate ions in dilute acid solution is by far the fastest. Periodates are therefore used in analytical chemistry as oxidizing titrants and also in syntheses, such as the oxidative cleavage of diols. As we remarked earlier, the speed of periodate reactions appear to stem from the ease with which the reducing agent can coordinate to the electrophilic I(VII) atom, and thereby undergo inner-sphere redox, as in the diol oxidation:

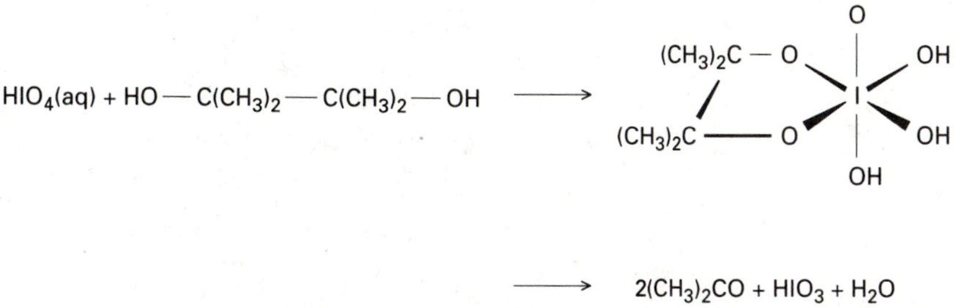

$$HIO_4(aq) + HO - C(CH_3)_2 - C(CH_3)_2 - OH \longrightarrow$$

$$\longrightarrow 2(CH_3)_2CO + HIO_3 + H_2O$$

[5] Appelman's discovery of BrO_4^- (and of HFO) is described in 'Nonexistent compounds: Two case histories', *Acc. Chem. Res.*, 4, 113 (1973). His strategy was first to synthesize, detect, and study the important chemical properties of traces of the compound. This information enabled him to devise practical syntheses for larger quantities.

We have already commented on the low basicity of perchlorate, which leads to very weak coordination with metal aquaions, and the fact that perchlorate in combination with many reducing cations forms metastable compounds that may not react under mild condition but may explode when provoked by friction or heat. With certain reducing agents, the ClO_4^- ion reacts smoothly in aqueous solution. The order of reactivity observed for a series of metal ion reducing agents is

$$Ru^{2+} > Ti^{3+} > Mo^{3+} > V^{2+} \approx V^{3+} > Cr^{2+} \gg Fe^{2+}$$

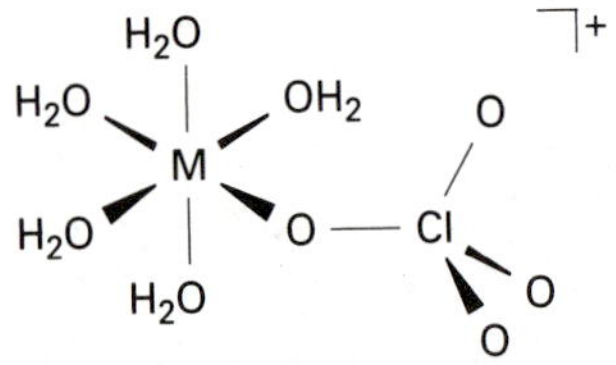

For the most part, this series correlates with the value of the ligand field splitting parameter Δ_O but not with the values of $E^{\ominus}$ for the oxidation of the individual ions. The parallel between rates and Δ_O indicates that stronger bonding between the metal ion and perchlorate O atom may facilitate O atom transfer from Cl(VII) to the metal ion, thereby reducing the chlorine to Cl(V) (**10**).

Anhydrous perchloric acid is a particularly aggressive oxidizing medium, which must not be allowed to come into accidental contact with organic matter or similar oxidizable material. The presence of Cl_2O_7 as a result of the equilibrium

$$3HClO_4 \rightleftharpoons Cl_2O_7 + H_3O^+ + ClO_4^-$$

and at elevated temperatures of the radical species $\cdot OH$, $\cdot ClO_3$, and $\cdot ClO_4$, appear to be responsible for its rapid oxidation reactions.

10 $[(H_2O)_5MOClO_3]^+$

13.7 Fluorocarbons

The synthesis of fluorocarbon compounds is of great technological importance because they are useful in applications ranging from coatings for non-sticking cooking ware and halogen-resistant laboratory vessels to the volatile chlorofluorocarbons, which are used as the refrigerant in most air conditioners and refrigerators. Fluorocarbon derivatives also have been the topic of considerable exploratory synthetic research because they often have unusual properties.

The direct reaction of an aliphatic hydrocarbon with an oxidizing metal fluoride leads to the formation of C—F bonds and produces HF as a byproduct:

$$RH + 2CoF_3 \rightarrow RF + 2CoF_2 + HF \qquad (R = \text{alkyl or aryl})$$

When R is an aryl, CoF_3 yields the cyclic saturated fluoride:

$$C_6H_6 + 18CoF_3 \rightarrow C_6F_{12} + 18CoF_2 + 6HF$$

The strongly oxidizing fluorinating agent used in these reactions, CoF_3, is regenerated by the reaction of CoF_2 with fluorine:

$$2CoF_2 + F_2 \rightarrow 2CoF_3$$

Another important route to C—F bond formation is halogen exchange by the reaction of a nonoxidizing fluoride, such as HF, with a chlorocarbon in the presence of a catalyst, such as SbF_3:

$$CHCl_3 + 2HF \rightarrow CHClF_2 + 2HCl$$

This process is performed on a large scale to produce the chlorofluoro-carbons that have been used as refrigerant fluids, the propellant in spray cans, and in the blowing agent in plastic foam products. When heated, difluorochloromethane is converted to the useful monomer, C_2F_4:

$$2CHClF_2 \xrightarrow{600-800\,°C} C_2F_4 + 2HCl$$

The polymerization of tetrafluoroethylene is carried out with a radical initiator:

$$nC_2F_4 \xrightarrow{ROO\cdot} (-CF_2-CF_2-)_n$$

Polytetrafluoroethylene is sold under many trade names, one of which is Teflon (DuPont). It can be depolymerized at high temperatures and this is the most convenient method of preparing tetrafluoroethylene in the laboratory:

$$(-CF_2-CF_2-)_n \xrightarrow{600°C} nC_2F_4$$

Although tetrafluoroethylene is not highly toxic, a by-product, per-fluoroisobutylene, is toxic and its presence dictates care in handling the crude tetrafluoroethylene.

THE NOBLE GASES (GROUP 18/VIII)

The elements in Group 18/VIII of the periodic table have been given and have lost various collective names over the years as different aspects of their properties have been identified and then disproved. Thus they have been called the rare gases and the inert gases, and are currently called the noble gases. The first is inappropriate because argon is by no means rare (it is more abundant than CO_2 in the atmosphere). The second has been inappropriate since the discovery of compounds of xenon. The appellation 'noble gases' is now accepted because it gives the sense of low but significant reactivity.

13.8 Occurrence and recovery

Because they are so unreactive and are rarely concentrated in nature, the noble gases eluded recognition until the end of the nineteenth century. Indeed, Mendeleev did not make a place for them in his periodic table because the chemical regularities of the other elements, upon which his table was based, did not suggest their existence. How-ever, in 1868 a new frequency was observed in the spectrum of the Sun that did not correspond to a known element. This frequency was eventually attributed to a new element, helium, and in due course helium and its congeners were found on Earth.

All the noble gases occur in the atmosphere. The crustal and atmo-spheric abundances of argon (0.94 percent by volume) and neon (1.5×10^{-3} percent) make these two elements more plentiful than many

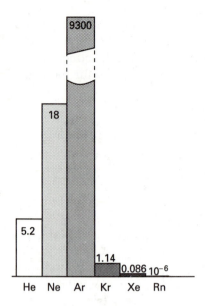

13.17 Abundance of the noble gases in the Earth's crust (atmospheric ppm by volume). Source: J. Emsley, *The elements.* Oxford University Press (1991).

familiar elements, such as arsenic and bismuth, in the Earth's crust (Fig. 13.17). Xenon and radon are the rarest elements of the group; the latter is a product of radioactive decay and, since it lies beyond bismuth, is itself unstable. The elements neon through xenon are extracted from liquid air by low-temperature distillation. Unlike the atoms of the other noble gases, the light He atom is not retained by the Earth's gravitational field, so most of the helium on Earth is the product of α-emission in the decay of radioactive elements and has been trapped by rock formations. High helium concentrations are found in certain natural gas deposits (mainly in the USA and eastern Europe) from which it can be recovered by low-temperature distillation. Some helium arrives from the Sun as a solar wind of α particles.

The low boiling points of the lighter noble gases (He, 4.2 K; Ne, 27.1 K; Ar 87.3 K) follow from the weakness of the dispersion forces between the atoms and absence of other forces. Helium has the lowest boiling point of any substance and is widely used in cryogenics as a very low temperature refrigerant; it is the coolant for superconducting magnets used for NMR spectroscopy and NMR imaging. When ^{4}He is cooled below 2.178 K it undergoes a transformation into a second liquid phase known as **helium-II**. This phase is classed as a superfluid, for it flows without viscosity.

On account of its low density and nonflammability, helium is used in balloons and lighter-than-air craft. The greatest use of argon is as inert atmospheres for the production of air-sensitive compounds and as an inert gas blanket to suppress the oxidation of metals when they are welded. The noble gases are also extensively used in various light sources, including conventional sources (neon signs, fluorescent lamps, xenon flash lamps) and lasers (helium–neon, argon-ion, and krypton-ion lasers). In each case, an electric discharge through the gas ionizes some of the atoms and promotes both ions and neutral atoms into excited states which then emit light upon return to a lower state.

Radon, which is a product of nuclear power plants and of the radioactive decay of naturally occurring uranium and thorium, is a health hazard because of the ionizing nuclear radiation it produces. It is usually a minor contributor to the background radiation arising from cosmic rays and terrestrial sources. However, in regions where the soil, underlying rocks, or building materials contain significant concentrations of uranium, excessive amounts of the gas have been found in buildings.

13.9 Compounds

The reactivity of the noble gases has been investigated sporadically ever since their discovery, but all early attempts to coerce them into compound formation were unsuccessful. However, in March 1962, Neil Bartlett, then at the University of British Columbia, observed the reaction of a noble gas.[6] Bartlett's report, and another from Rudolf

---

[6] For an enjoyable account of the early failure and final success in the quest for compounds of the noble gases, see P. Lazlo and G. J. Schrobilgen, *Angew. Chem., Int. Ed. Engl.*, **27**, 479 (1988).

Hoppe's group at the University of Munster a few weeks later, set off a flurry of activity throughout the world. Within a year, a series of xenon fluorides and oxo compounds had been synthesized and characterized.[7]

Synthesis and structure of xenon fluorides

Bartlett's motivation for studying xenon was based on the observations that PtF_6 can oxidize O_2 to form a solid, and that the ionization energy of xenon is similar to that of molecular oxygen. Indeed, reaction of xenon with PtF_6 did give a solid, but the reaction is complex and the formulation of the product (or products) is still in doubt. The direct reaction of xenon and fluorine leads to a series of compounds with oxidation numbers $+2$ (in XeF_2), $+4$ (in XeF_4), and $+6$ (in XeF_6).

The molecules XeF_2, XeF_4, and XeF_6 are linear (11), square-planar (12), and distorted octahedral (13), respectively. The first two structures are consistent with the VSEPR model for a Xe atom with five and six electron pairs respectively and they bear a molecular and electronic structural resemblance to the isoelectronic polyhalide anions I_3^- and ClF_4^- (Section 13.5). The structures of XeF_2 and XeF_4 are well established over a wide timescale by diffraction and spectroscopic methods. Similar measurements on XeF_6 in the gas phase, however, lead to the conclusion that this molecule is fluxional (Section 3.2). Infrared spectra and electron diffraction on XeF_6 show that a distortion occurs about a threefold axis such that one triangular face of F atoms is opened up to accommodate a lone pair of electrons (13). It is thought that the fluxional process is the migration of the distortion from one triangular face to another.

The xenon fluorides are synthesized by direct reaction of the elements, usually in a nickel reaction vessel that has been passivated by exposure to F_2 to form a thin protective NiF_2 coating. This treatment also removes surface oxide, which would react with the xenon fluorides. The synthetic conditions indicated in the following equations show that formation of the higher halides is favored by a higher proportion of fluorine and higher total pressure:

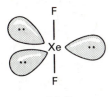

11 XeF_2

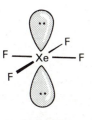

12 XeF_4

$$Xe(g) + F_2(g) \xrightarrow{400\,°C,\ 1\ atm} XeF_2(g) \quad \text{(Xe in excess)}$$

$$Xe(g) + 2F_2(g) \xrightarrow{600\,°C,\ 6\ atm} XeF_4(g) \quad (Xe{:}F_2 = 1{:}5)$$

$$Xe(g) + 3F_2(g) \xrightarrow{300\,°C,\ 60\ atm} XeF_6(g) \quad (Xe{:}F_2 = 1{:}20)$$

A simple 'windowsill' synthesis is also possible. Xenon and fluorine are sealed in a glass bulb (previously rigorously dried to prevent the formation of HF and the attendant etching of the glass) and the bulb is exposed to sunlight, whereupon beautiful crystals of XeF_2 slowly

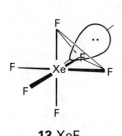

13 XeF_6

---

[7] John Holloway provides a nice summary of the development of this field in 'Twenty-five years of noble gas chemistry', *Chem. Brit.*, 658 (1987).

form in the bulb. It will be recalled that fluorine undergoes photodissociation (Section 13.1), and in this synthesis the photochemically generated F atoms react with Xe atoms.

Reactions of xenon fluorides

The reactions of the xenon fluorides are similar to those of the high oxidation number interhalogens, and redox and metathesis reactions dominate. One important reaction of XeF_6 is double-displacement with oxides:

$$XeF_6(s) + 3H_2O(l) \rightarrow XeO_3(aq) + 6HF(g)$$

$$2XeF_6(s) + 3SiO_2(s) \rightarrow 2XeO_3(s) + 3SiF_4(g)$$

Xenon trioxide presents a serious hazard since this endoergic compound is highly explosive. In basic aqueous solution the Xe(VI) oxoanion $HXeO_4^-$ slowly decomposes in a coupled disproportionation and water oxidation to yield a Xe(VIII) perxenate ion, XeO_6^{4-} and xenon. Another important chemical property of the xenon fluorides is their strong oxidizing power, as illustrated by the following examples:

$$2XeF_2(s) + 2H_2O(l) \rightarrow 2Xe(g) + 4HF(g) + O_2(g)$$

$$XeF_4(s) + Pt(s) \rightarrow Xe(g) + PtF_4(s)$$

As with the interhalogens, the xenon fluorides react with strong Lewis acids to form xenon fluoride cations:

$$XeF_2(s) + SbF_5(l) \rightarrow [XeF^+][SbF_6^-](s)$$

These cations are associated with the counterion by F^- bridges. Compounds with Xe—N (**14**) and Xe—C bonds, as in $[Xe(C_6F_5)][C_6F_5BF_3]$,[8] have also been synthesized.[9] Evidence also exists for the formation of the paramagnetic species Xe_2^+.

Another similarity with the interhalogens is the reaction of XeF_4 with the Lewis base fluoride in acetonitrile solution to produce the XeF_5^- ion:

$$XeF_4 + [N(CH_3)_4]F \rightarrow [N(CH_3)_4]XeF_5$$

The XeF_5^- ion has a planar pentagonal shape (**15**), and in the VSEPR model the two electron pairs on Xe occupy axial positions.[10] Similarly, it has been known for many years that reaction of XeF_6 with a F^- source produces the XeF_7^- or XeF_8^{2-} ions depending on the proportion of fluoride. Only the shape of XeF_8^{2-} is known; it is a square antiprism (**16**), which is difficult to reconcile with the simple VSEPR model because this shape does not provide a site for the lone pair on Xe.[11]

14 $FXeN(SO_2F)_2$

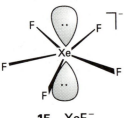

15 XeF_5^-

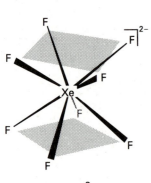

16 XeF_8^{2-}

[8] H. J. Frohn and S. Jakobs, *J. Chem. Soc., Chem. Commun.*, 625 (1989).

[9] In each of these cases, strongly electron withdrawing groups are attached to the C or N, but this is not the case for $HCNXeF^+$, A. A. A. Emara and G. J. Schrobilgen, *J. Chem. Soc., Chem. Commun.*, 1644 (1987).

[10] K. O. Christe, E. C. Curtis, D. A. Dixon, H. Mercier, J. C. P. Sanders, and G. J. Schrobilgen. The pentafluoroxenate(IV) anion, XeF_5^-: the first example of a pentagonal planar AX_5 species. *J. Am. Chem. Soc.*, **113**, 3351 (1991).

[11] A.-R. Mahjoub and K. Seppelt, The structure of $[IF_8]^-$. *Angew. Chem.*, **30**, 876 (1991).

Example 13.5: *Synthesis and structure of a noble gas compound*

(a) Describe a procedure for the synthesis of potassium perxenate, starting with xenon and other reagents of your choice. (b) Using the VSEPR model, determine the probable structure of the perxenate ion.

Answer. (a) Xe–O compounds are endoergic, so they cannot be prepared by direct reaction of xenon and oxygen. The hydrolysis of XeF_6 yields XeO_3 and as described in the text, the latter in basic solution undergoes disproportionation yielding perxenate, XeO_6^{4-}. Thus the reaction of xenon and excess F_2 would be carried out at 300 °C and 60 atm in a stout nickel container. The resulting XeF_6 might then be converted to the perxenate in one step by exposing it to aqueous KOH solution. The resulting potassium perxenate (which turns out to be a hydrate) could then be crystallized. (b) A Lewis structure of the perxenate ion is shown in (**17**). With six electron pairs around the Xe atom, the VSEPR model predicts an octahedral arrangement of bonding electron pairs and an octahedral overall structure.

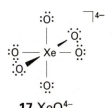

17 XeO_6^{4-}

Exercise E13.5. Write a balanced equation for the decomposition of xenate ions in basic solution for the production of perxenate ions, xenon, and oxygen.

Compounds of other noble gases

Radon has a lower ionization energy than xenon, so we can expect it to form compounds even more readily. Evidence exists for the formation of RnF_2 and of cationic compounds such as $[RnF^+][SbF_6^-]$, but detailed characterization is frustrated by their radioactivity.[12] Krypton has a much higher ionization energy than xenon (Table 13.1) and its ability to form compounds is much more limited. Krypton difluoride is prepared by passing an electric discharge or ionizing radiation through a fluorine–krypton mixture at low temperatures (-196°C). As with XeF_2, the krypton compound is a colorless volatile solid and the molecule is linear. It is an endoergic and highly reactive compound that must be stored at low temperatures.

FURTHER READING

A. J. Bard, R. Parsons, and J. Jordan, *Standard potentials in aqueous solution.* Dekker, New York (1985). See the chapters on the halogens and the 'inert' gases. *Halides of the transition elements*: D. Brown, Vol. 1 *Halides of the lanthanides and actinides*; J. H. Canterford and R. Colton, Vol. 2, *Halides of the first row transition metals*; Vol. 3, *Halides of the second and third row transition metals.* Wiley, New York (1968).

H. Selig and J. H. Holloway, Cationic and anionic complexes of the noble gases. In *Topics in Current Chemistry*, **124**, 33 (1984).

R. C. Thompson, Reaction mechanisms of the halogens and oxohalogen species in acidic, aqueous solution. In *Advances in inorganic and bioinorganic mechanisms* (ed. A. G. Sykes), Vol. 4. Academic Press, New York (1986).

[12] L. Stein, *Inorg. Chem.*, **23**, 2670 (1984).

K. Seppelt and D. Lentz, Novel developments in noble gas chemistry. In *Prog. Inorg. Chem.*, **29**, 167 (1982).

Kirk–Othmer encyclopedia of chemical technology. Wiley, New York (1980). This series of volumes contains many reviews of the individual halogens and their compounds, as well as of the noble gases, with emphasis on practical applications.

KEY POINTS

1. Structures of the elements

There is greater structural uniformity for the elements within the halogen group and those within the noble gas group than for any other *p*-block groups. The former are diatomic molecules and uniformly nonmetals, whereas the latter are monatomic gases with low reactivity.

2. Recovery and use of halogens

Chlorine, bromine, and a high proportion of iodine is produced by oxidation of the halides. Chlorine is by far the most commercially important halogen, although its use in the preparation of chlorocarbon compounds is coming under close scrutiny for environmental reasons.

3. Recovery and use of noble gases

Except for helium, which is extracted from gas wells, and radon, which is radioactive, the noble gases are extracted from the atmosphere by distillation. Helium and argon are used as inert gases and helium is used as a refrigerant.

4. Special properties of fluorine

The distinctive properties of fluorine include the stabilization of metal ions in their high oxidation states, the high volatility of many molecular fluorine compounds, and strong electron withdrawing effects in covalent compounds, which give rise to strong proton and Lewis acidity of fluorine-containing acids.

5. Polyiodides and interhalogens

Polyiodides most of which have the formula I_n^-, with n odd, may be viewed as loosely associated aggregates of I_3^-, I_2, and I^-; the most important species is I_3^-. Many interhalogen compounds exist, most of which are composed of a central Cl, Br, or I atom bound to peripheral F atoms. The structures are generally predictable from the VSEPR model. Some interhalogen compounds, such as ClF_3 and BrF_3, are aggressive fluorinating agents.

6. Halogen oxoanions

The halogen oxoanions are the most important oxygen-containing halogen compounds.

7. Redox reactions of halogen compounds

Halogen oxides and oxoanions are often thermodynamically potent oxidizing agents. Many of the intermediate oxidation states undergo redox disproportionation.

8. Reaction mechanisms of halogen oxoanions

Atom transfer mechanisms are common, such as Cl transfer from the facile oxidant OCl^- and O transfer from the sluggish but potent oxidant ClO_4^-.

9. Compounds of noble gases

Far more compounds are known for xenon than for any of the other noble gases. The fluorides XeF_n, with $n = 2$, 4, and 6, are the most important, but Xe—O, Xe—C, and Xe—N bonds are known.

EXERCISES

13.1 Preferably without consulting reference material, write out the halogens and noble gases as they appear in the periodic table, and indicate the trends in (a) physical state (s, l, or g) at room temperature and pressure, (b) electronegativity, (c) hardness of the halide ion, and (d) color.

13.2 Describe how the halogens are recovered from their naturally occurring halides and rationalize the approach in terms of standard potentials. Give balanced chemical equations and conditions where appropriate.

13.3 Sketch the chloroalkali cell. Show the half-cell reactions and indicate the direction of diffusion of the ions. Give the chemical equation for the unwanted reaction that would occur if OH^- migrated through the membrane and into the anode compartment.

13.4 Sketch the form of the vacant σ^* orbital of a dihalogen molecule and describe its role in the Lewis acidity of the dihalogens.

13.5 Nitrogen trifluoride, NF_3, boils at $-129\,°C$ and is devoid of Lewis basicity. By contrast the lower molecular weight compound NH_3 boils at $-33\,°C$ and is well known as a Lewis base. (a) Describe the origins of this very large difference in volatility. (b) Describe the probable origins of the difference in basicity.

13.6 Based on the analogy between halogens and pseudohalogens: (a) Write the balanced equation for the probable reaction of cyanogen $(CN)_2$ with aqueous sodium hydroxide. (b) Write the equation for the probable reaction of excess thiocyanate with the oxidizing agent $MnO_2(s)$ in acidic aqueous solution. (c) Write a plausible structure for trimethylsilyl cyanide.

13.7 (a) Use the VSEPR model to predict the probable shapes of IF_6^+, and IF_7. (b) Give a plausible chemical equation for the preparation of $[IF_6][SbF_6]$.

13.8 Describe whether each of the following solutes is likely to make liquid BrF_3 a stronger Lewis acid or a stronger Lewis base: (a) SbF_5, (b) SF_6, (c) CsF.

13.9 Describe whether each of the following compounds is likely to be dangerously explosive in contact with BrF_3, and explain your answer: (a) SbF_5, (b) CH_3OH, (c) F_2, (d) S_2Cl_2.

13.10 The formation of Br_3^- from a tetraalkylammonium bromide and Br_2 is only slightly exoergic. Write an equation (or NR for no reaction) for the interaction of $[NR_4][Br_3]$ with excess I_2 in CH_2Cl_2 solution, and give your reasoning.

13.11 Explain why $CsI_3(s)$ is stable with respect to the elements but $NaI_3(s)$ is not.

13.12 Write plausible Lewis structures for (a) ClO_2 and (b) I_2O_6 and predict their shapes and point groups.

13.13 (a) Give the formulas and the probable relative acidities of perbromic acid and periodic acid. (b) Which is the more stable?

13.14 (a) Describe the expected trend in the standard potential of an oxoanion in a solution with decreasing pH. (b) Demonstrate this phenomenon by calculating the reduction potential of ClO_4^- at $pH = 7$ and comparing it with the tabulated value at $pH = 0$.

13.15 With regard to the general influence of pH on the standard potentials of oxoanions, explain why the disproportionation of an oxoanion is often promoted by low pH.

13.16 Which oxidizing agent reacts more readily in dilute aqueous solution, perchloric acid or periodic acid? Give a mechanistic explanation for the difference.

13.17 (a) For which of the following anions is disproportionation thermodynamically favorable in acidic solution: ClO^-, ClO_2^-, ClO_3^-, and ClO_4^-? (If you do not know the properties of these ions, determine them from a table of standard potentials.) (b) For which of the favorable cases is the reaction very slow at room temperature?

13.18 Which of the following compounds present an explosion hazard? (a) NH_4ClO_4, (b) $Mg(ClO_4)_2$, (c) $NaClO_4$, (d) $[Fe(OH_2)_6][ClO_4]_2$. Explain your reasoning.

13.19 Explain why helium is present in low concentration in the atmosphere even though it is the second most abundant element in the universe.

13.20 Which of the noble gases would you choose as (a) the lowest temperature liquid refrigerant, (b) an electric discharge light source requiring a safe gas with the lowest ionization energy, (c) the least expensive inert atmosphere.

13.21 By means of balanced chemical equations and a statement of conditions, describe a suitable synthesis of (a) xenon difluoride, (b) xenon hexafluoride, (c) xenon trioxide.

13.22 Give the formula and describe the structure of a noble gas species that is isostructural with (a) ICl_4^-, (b) IBr_2^-, (c) BrO_3^-, (d) ClF.

13.23 (a) Give Lewis formulas and formal charges for ClO^-, Br_2, and XeF^+. (b) Are these species isolobal? (c) Describe the chemical similarities as judged by their reactions with nucleophiles. (d) Rationalize the trends in electrophilicity and basicity.

13.24 (a) Give a Lewis structure for XeF_7^-. (b) Speculate on its possible structures by using the VSEPR model and analogy with other xenon fluoride anions.

PROBLEMS

13.1 Many of the acids and salts corresponding to the positive oxidation numbers of the halogens are not listed in the catalog of a major international chemical supplier:

(a) $KClO_4$ and KIO_4 are available but $KBrO_4$ is not; (b) $KClO_3$, $KBrO_3$, and KIO_3 are all available; (c) $NaClO_2$ and $NaBrO_2 \cdot 3H_2O$ are available but salts of IO_2^- are not;

(d) ClO^- salts are available but the bromine and iodine analogs are not. Describe the probable reasons for the missing salts of oxoanions.

13.2 Determine the incorrect statements among the following descriptions and provide correct statements. (a) Oxidation of the halides is the only commercial method of preparing the halogens F_2 through I_2. (b) ClF_4^- and I_5^- are isolobal and isostructural. (c) Atom-transfer processes are common in the mechanisms of oxidations by the halogen oxoanions, and an example is the O atom transfer in the oxidation of SO_3^{2-} by ClO^-. (d) Periodate appears to be a more facile oxidizing agent than perchlorate because the former can coordinate to the reducing agent at the $I(VII)$ center, whereas the $Cl(VII)$ center in perchlorate is inaccessible to reducing agents.

13.3 The reaction of I^- ions is often used to titrate ClO^-, giving deeply colored I_3^- ions, along with Cl^- and H_2O. Although never proved, it was thought that the initial reaction proceeds by O atom transfer from Cl to I. However, it is now believed that the reaction proceeds by Cl atom transfer to give ICl as the intermediate (K. Kumar, R. A. Day, and D. W. Margerum, *Inorg. Chem.*, **25**, 4344 (1986)). Summarize the evidence for Cl atom transfer.

13.4 Given the bond lengths and angles in I_5^+ (5), describe the bonding in terms of two-center and three-center σ bonds and account for the structure in terms of the VSEPR model.

13.5 Until the work of K. O. Christe (*Inorg. Chem.*, **25**, 3721 (1986)), F_2 could be prepared only electrochemically. Give chemical equations for Christe's preparation and summarize the reasoning behind it.

13.6 The first compound containing an Xe—N bond was reported by R. D. LeBlond and K. K. DesMarteau (*J. Chem. Soc., Chem. Commun.*, 554 (1974)). Summarize the method of synthesis and characterization. (The proposed structure was later confirmed in an X-ray crystal structure determination.)

PART THREE

Advanced topics

Our final set of topics deals with more advanced aspects of inorganic chemistry. The first two chapters (on spectroscopy and reaction mechanisms) look in more detail at topics that have been introduced earlier and introduce some more advanced material to show how a detailed understanding of structure and mechanism can be obtained. The remaining four chapters (on d-block organometallic compounds, catalysis, solid state chemistry, and bioinorganic chemistry) span areas that are currently being pursued vigorously in research laboratories and which are the focus of considerable scientific and technological excitement. Research in these fields often combines the talents of inorganic chemists with those of researchers in other disciplines, such as condensed matter physics, biology, and electrical and chemical engineering. Chapter 17 follows on naturally from Chapter 16 because it shows how certain reactions of organometallic compounds can be used in catalysis. In many cases, such as in the development of an enantioselective catalyst for the synthesis of a drug to combat Parkinson's disease, the mechanism is subtle. In Chapter 18 we extend the discussion of the solid state to include new structural types, nonstoichiometry, and the role of defects. In this way we will be able to discuss high temperature superconductors. In Chapter 19, the final chapter of the book, we shall see how metal ions and metal complexes are involved in some vital biological processes, such as oxygen and ion transport.

14

Electronic spectra of complexes

The overall aim of this chapter is to see how to analyze the electronic spectra of complexes (particularly those of the d block) and hence to enrich our understanding of their bonding. Many of the topics we treat are elaborations of the material we introduced in Chapter 6 on the molecular orbital theory of complexes. The principal new work in the first part of the chapter is the problem of interelectronic repulsions, which we largely ignored in Chapter 6. To see how to take repulsions into account, we must step back from the concepts introduced in that earlier chapter and discuss free atoms, for atomic spectra provide the parameters that we need to take interelectronic repulsion into account quantitatively. We can then go on to see how repulsions compete with the ligand field when the atoms and ions are present as a part of a complex. This chapter also extends the material in Chapter 6 by describing how to apply molecular orbital techniques to systems of lower symmetry than we treated there. In particular, we shall see how to build up a description of the molecular orbitals of complexes from the orbitals of simpler fragments. This sets the stage for the discussion of cluster compounds in Chapters 16 and 17.

We shall set the scene for the material we discuss in this chapter by examining Fig. 14.1, which shows the electronic absorption spectrum of the d^3 complex $[Cr(NH_3)_6]^{3+}$ in aqueous solution in the range $50\,000\ \text{cm}^{-1}$ (200 nm, the ultraviolet) to $10\,000\ \text{cm}^{-1}$ (1000 nm, the infrared). The band at lowest energy is very weak, and later we shall see that it is an example of a 'spin forbidden' transition. Next are two bands with intermediate intensities, which will turn out to be 'spin allowed' HOMO–LUMO transitions between the t_{2g} and e_g orbitals of the complex ($t_{2g}^2 e_g^1 \leftarrow t_{2g}^3$) and which are mainly derived from the metal d orbitals. Closer analysis shows that there is also a third transition hidden under the very intense band marked 'CT' and which can

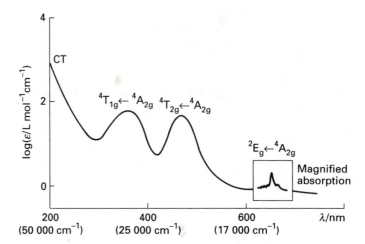

14.1 The spectrum of the d^3 complex $[Cr(NH_3)_6]^{3+}$, which illustrates the features studied in this chapter, and the assignments of the transitions as explained in the text.

be ascribed to the transition $t_{2g}^1 e_g^2 \leftarrow t_{2g}^3$; we shall not consider this high frequency transition further. However, we do need to explore why we obtain two absorptions at different energies from the apparently single transition $t_{2g}^2 e_g^1 \leftarrow t_{2g}^3$. The splitting into different bands is in fact an outcome of the electron–electron repulsions that will be one focus of this chapter. The intense band at high energy (labeled CT for 'charge-transfer'; we see only the tail of it in the illustration) has an entirely different origin from the other three. We shall see that it is an example of a transition in which charge is transferred from the ligands to the central metal atom.

THE ELECTRONIC SPECTRA OF ATOMS

We have been content so far to describe atoms and molecules by giving their electronic **configurations**, the statement of the number of electrons in each orbital (as in $1s^2 2s^1$ for Li). However, a configuration is an incomplete description of the arrangement of electrons in atoms. In the configuration $2p^2$, for instance, the two electrons might occupy orbitals with different orientations of their orbital angular momenta (that is, with different values of m_l from among the possibilities $+1$, 0, and -1 that are available when $l = 1$). Similarly, the designation $2p^2$ tells us nothing about the spin orientations of the two electrons ($m_s = +\frac{1}{2}$ or $-\frac{1}{2}$). The atom may in fact have several different states of total orbital angular momentum, and each one corresponds to an occupation of orbitals with different values of m_l and m_s. The different ways in which the electrons can occupy the orbitals specified by the configuration are called the **microstates** of the configuration. For example, one microstate of a $2p^2$ configuration is $(1^+, 1^-)$ in a notation that signifies that both electrons occupy a $2p$ orbital with $m_l = +1$ but do so with opposite spins (the superscript $+$ indicates $m_s = +\frac{1}{2}$ and $-$ indicates $m_s = -\frac{1}{2}$). Another microstate of the same configuration is $(-1^+, 0^-)$.

14.1 Spectroscopic terms

The microstates of a given configuration have the same energy only if we ignore interelectronic repulsions. However, interelectron repulsions are in fact strong and cannot always be ignored. As a result, microstates that correspond to different relative spatial distributions of electrons have different energies.

Permissible microstates and terms

The 15 possible microstates of a $2p^2$ configuration are

$$(1^+,1^-)\,(1^+,0^+)\,(1^+,0^-)\,(1^-,0^+)\,(1^-,0^-)\,(0^+,0^-)$$

$$(1^+,-1^+)\,(1^+,-1^-)\,(1^-,-1^+)\,(1^-,-1^-)$$

$$(0^+,-1^+)\,(0^+,-1^-)\,(0^-,-1^+)\,(0^-,-1^-)\,(-1^+,-1^-)$$

Microstates such as $(1^+,1^-)$ and $(1^-,1^+)$ are identical and not listed separately because it is not possible to distinguish individual electrons. Microstates such as $(1^+,1^+)$ are excluded by the Pauli principle for they would correspond to two electrons with the same spin in the same orbital (the orbital with $m_l = +1$).

The microstates $(0^+,0^-)$ and $(1^+,1^-)$ both correspond to two electrons in the same orbital. The only difference is that the doubly occupied orbital has $m_l = 0$ in one case and $m_l = +1$ in the other; consequently, the two microstates have the same energy. However, they both differ from $(0^+,1^-)$ in which the two electrons occupy different orbitals. If we group together the microstates that have the same energy when electron repulsions are taken into account, we obtain the spectroscopically distinguishable energy levels called **terms**.

Our first task is to classify all the microstates of a given configuration into terms. The next is to identify transitions between these terms in the spectrum of the atom. The differences in energies of the transitions will then tell us the interelectron repulsion energies within the atom, which is the information we need if we are to make sense of spectra like that in Fig. 14.1.

Russell–Saunders coupling

It turns out that the most important property of a microstate for helping us to decide the energy of a term is the relative orientation of the spins of the electrons. Next in importance is the relative orientation of the orbital angular momenta of the electrons. Thus we can identify the terms of lighter atoms and put them in order of energy by noting first their total spin (which is determined by the relative orientation of the individual spins) and then their total orbital angular momentum (which is determined by the relative orientation of the individual orbital angular momenta of the electrons). The process of combining electron angular momenta by summing first the spins, then the orbital momenta, and finally the two resultants, is called **Russell–Saunders coupling**. An alternative procedure, called ***jj*-coupling**, is more appropriate when the spin–orbit coupling of electrons is strong (in heavy atoms).

The total spin S of an atom is the vector sum of the spin angular momenta of the individual electrons:

$$S = s_1 + s_2 + \ldots$$

The magnitude of this total spin is given by the total spin quantum number S:

Magnitude of $S = \{S(S+1)\}^{1/2}\hbar$

$S = 1$ if the two spins are parallel ($\uparrow\uparrow$), and $S = 0$ if they are antiparallel ($\uparrow\downarrow$). How S is ascribed in atoms with more than two electrons is explained in Further information 4. Similarly, the total orbital angular momentum L is the vector sum of the individual orbital momenta of the electrons:

$$L = l_1 + l_2 + \ldots$$

Its magnitude is described by the quantum number L:

Magnitude of $L = \{L(L+1)\}^{1/2}\hbar$

For two p electrons with $l = 1$, L may be 2 (if the two orbital angular momenta are in the same directions), 0 (if the orbital momenta are in opposite directions), and 1 (if the relative orientation is intermediate).[1] Further information 4 also explains how these values of L are obtained and how to deal with more complicated cases.

By analogy with the notation s, p, d, ... for orbitals with $l = 0$, 1, 2,..., the total orbital angular momentum of an atomic term is denoted by the upper case equivalent:

$$
\begin{array}{ccccc}
L = & 0 & 1 & 2 & 3 & 4 \ldots \\
 & \text{S} & \text{P} & \text{D} & \text{F} & \text{G then alphabetical (omitting J)}
\end{array}
$$

The total spin quantum number (S) of an atom is normally reported as the **multiplicity** of the term, the value of $2S + 1$:

$$
\begin{array}{cccccc}
S = & 0 & \frac{1}{2} & 1 & \frac{3}{2} & 2 \ldots \\
2S+1 = & 1 & 2 & 3 & 4 & 5 \ldots
\end{array}
$$

The multiplicity is written as a left superscript on the letter representing the value of L, and the entire label of a term is called a **term symbol**. Thus, the term symbol ^3P denotes a term (a collection of degenerate states) with $L = 1$ and $S = 1$, and is called a 'triplet' term.

[1] In classical terms we could think of the intermediate case as one electron orbiting round the equator and the other as orbiting over the poles.

Example 14.1: *Deriving term symbols*

Give the term symbols for an atom with the configurations (a) s^1, (b) p^1, and (c) s^1p^1.

Answer. (a) The single s electron has $l = 0$ and $s = \frac{1}{2}$. Because there is only one s electron, $L = 0$ (an S term), $S = \frac{1}{2}$, and $2S + 1 = 2$ (a doublet term). The term symbol is therefore ^{2}S. (b) For a single p electron, $l=1$ so $L=1$ and the term is ^{2}P. (These terms arise in the spectrum of an alkali metal atom, such as Na.) With one s and one p electron, $L=0+1=1$, a P term. The electrons may be paired ($S=0$) or parallel ($S=1$). Hence both ^{1}P and ^{3}P terms are possible.

Exercise E14.1. What terms arise from an $s^1 d^1$ configuration?

14.2 Terms of a d^2 configuration

The Pauli principle restricts the microstates that can occur in a configuration, and consequently it affects the terms that can occur. For example, two electrons cannot both have the same spin and be in an orbital with $m_l = +1$; as a result, a p^2 configuration cannot give rise to a term with $L=2$ and $S=1$, a ^{3}D term. On the other hand, a p^2 configuration can give rise to a ^{1}D term, for in a singlet term the electrons have opposite spins and so may occupy the same orbital. Our first problem is therefore to decide which terms are allowed by the Pauli principle. We shall illustrate the ideas involved by considering a d^2 configuration, for then the outcome will be useful in the discussion of the complexes encountered later in the chapter. An example of a species with a d^2 configuration is a Ti^{2+} ion.

We start the analysis by setting up a table of microstates of the d^2 configuration (Table 14.1). The strategy we then adopt is to identify the values of L and S to which the microstates belong.

The classification of microstates

We need two facts in order to be able to ascribe each microstate to a term. The first is based on the fact that each d electron may have $m_l = +2, +1, 0, -1,$ or -2 and $m_s = +\frac{1}{2}$ or $-\frac{1}{2}$. It follows that if one electron has the quantum number m_{l1} and the other has m_{l2}, the total orbital angular momentum will have a z component given by the quantum number

$$M_L = m_{l1} + m_{l2}$$

Table 14.1 Microstates of the d^2 configuration

M_L	-1	M_S 0	$+1$
$+4$		$(2^+,2^-)$	
$+3$	$(2^-,1^-)$	$(2^+,1^-)(2^-,1^+)$	$(2^+,1^+)$
$+2$	$(2^-,0^-)$	$(2^+,0^-)(2^-,0^+)$ $(1^+,1^-)$	$(2^+,0^+)$
$+1$	$(2^-,-1^-)(1^-,0^-)$	$(2^+,-1^-)(2^-,-1^+)$ $(1^+,0^-)(1^-,0^+)$	$(2^+,-1^+)(1^+,0^+)$
0	$(1^-,-1^-)(2^-,-2^-)$	$(1^+,-1^-)(1^-,-1^+)$ $(2^+,-2^-)(2^-,-2^+)$ $(0^+,0^-)$	$(1^+,-1^+)(2^+,-2^+)$
-1 to -4*			

*The lower half of the diagram is a reflection of the upper half.

The same applies to the total spin:

$$M_S = m_{s1} + m_{s2}$$

Thus $(0^+, -1^-)$ is a microstate with $M_L = -1$ and $M_S = 0$ and may contribute to any term for which these two quantum numbers apply.

The second fact we need is that a term with quantum number L has $2L + 1$ states distinguished by the value of M_L. Thus a G term (with $L = 4$) has the nine states with $M_L = +4, +3, \ldots -4$. Similarly, a term with quantum number S (and multiplicity $2S + 1$) has $2S + 1$ states distinguished by different values of M_S. Thus a triplet term, with $S = 1$, has three states with $M_S = +1$, 0, and -1 for each value of M_L. A ^{3}G term, for instance, has three spin states for each of its nine orbital states, and therefore consists of 27 states. All these states have the same energy, but they differ in energy from the states of any other term.

It follows from these remarks that by picking out the values of M_L that can arise, we can infer the values of L to which the states belong. Similarly, by picking out the values of M_S that can go with a given value of L, we can infer the value of S that is associated with that value of L. The strategy is to identify the state with the largest value of M_L and M_S, and hence to identify the largest values of L and S that may stem from the configuration. We then know that the term accounts for $2L + 1$ values of M_L and $2S + 1$ values of M_S for each value of M_L, and so can assign $(2L + 1) \times (2S + 1)$ of the microstates to this term. When that is done, we identify the next largest values of L and S, and continue until all the microstates have been assigned. Keeping track of this process is made easy by crossing out one microstate in Table 14.1 as the terms are identified, as we illustrate below.

First, we note the largest value of M_L, which for a d^2 configuration is $+4$. This state must belong to a term with $L = 4$ (a G term). Table 14.1 shows that the only value of M_S that occurs for this term is $M_S = 0$, so the G term must be a singlet. Moreover, since there are nine values of M_L when $L = 4$, one of the microstates in each of the boxes in the column below $(2^+, 2^-)$ must belong to this term.[2] We can therefore strike out one microstate from each row in the central column of Table 14.1, which leaves 36 microstates to classify.

The next largest value is $M_L = +3$, which must stem from $L = 3$ and hence belong to an F term. That row contains one microstate in each column (that is, each box contains one unassigned combination for $M_S = +1$, 0, and -1), which signifies a triplet term. Hence the microstates belong to a ^{3}F term. The same is true for one microstate in each of the rows down to $M_L = -3$, which accounts for a further $3 \times 7 = 21$ microstates. If we strike out one state in each of the 21 boxes, we are left with 15 to be assigned.

There is one unassigned microstate in the row with $M_L = +2$ ($L = 2$) and the column under $M_S = 0$ ($S = 0$), which must therefore

[2] In fact, it is unlikely that one of the microstates itself will correspond to one of these states: in general, a state will be a linear combination of microstates. However, as N linear combinations can be formed from N microstates, each time we cross off one microstate, we are taking one linear combination into account, so the book-keeping is correct even though the detail may be wrong.

belong to a ^1D term. This term has five values of M_L, which removes one microstate from each row (in the column headed $M_S = 0$) down to $M_L = -2$, leaving 10 microstates unassigned. Because these unassigned microstates include one with $M_L = +1$ and $M_S = +1$, nine of them must belong to a ^3P term. There now remains only one microstate in the central box of the table, with $M_L = 0$ and $M_S = 0$. This microstate (or, in fact, a linear combination of all the microstates in that box; see footnote 2) must be the one and only state of a ^1S term.

At this point we can conclude that the terms of a $3d^2$ configuration are ^1G, ^3F, ^1D, ^3P, and ^1S, which account for all 45 permitted states (see table in the margin).

Term	Number of states
^1G	$9 \times 1 = 9$
^3F	$7 \times 3 = 21$
^1D	$5 \times 1 = 5$
^3P	$3 \times 3 = 9$
^1S	$1 \times 1 = 1$
	Total: 45

The energies of the terms

Once the values of S and L that can arise from a given configuration are known, it is possible to predict the ground term by using **Hund's rules**. The first of these empirical rules was introduced in Section 1.7. There it was expressed as 'the lowest energy configuration is achieved if the electrons are parallel'. Because a high value of S stems from parallel electron spins, an alternative statement is

> 1. For a given configuration, the term with the greatest multiplicity lies lowest in energy.

The rule implies that a triplet term of a configuration (if one is permitted) has a lower energy than a singlet term of the same configuration.

Hund also proposed a second rule for predicting the relative energies of the terms of a given multiplicity:

> 2. For a term of given multiplicity, the greater the value of L, the lower the energy.

When L is high, each electron can stay clear of its partner and hence have a lower coulombic interaction with it because (in a classical picture of the atom) the electrons are then orbiting in the same direction. If L is low, the electrons are orbiting in opposite directions and meet more often (and hence repel each other more strongly). The second rule implies that if a configuration can give rise to both a ^3F and a ^3P term, then the ^3F term is lower in energy. It follows that the ground term of Ti^{2+} is ^3F.

Hund's rules are reasonably reliable for predicting which term has lowest energy (the ground state), but they are not very reliable for predicting the order in which terms of higher energy lie. Thus, for Ti^{2+} the rules predict the order

$$^3F < {}^3P < {}^1G < {}^1D < {}^1S$$

but the observed order is

$$^3F < {}^1D < {}^3P < {}^1G < {}^1S$$

The spin multiplicity rule is fairly reliable for predicting the ordering of terms, but the 'greatest L' rule is reliable only for predicting the ground terms; there is generally little correlation of L with the order of the higher terms.

All the previous work can be shortened considerably if all we want to know is the identity of the ground term of an atom or ion. The procedure may be summarized as follows:

■■■ 1. Identify the microstate that has the highest value of M_S.

This step tells us the highest multiplicity of the configuration.

■■■ 2. Identify the highest permitted value of M_L for that multiplicity.

This step tells us the highest value of L consistent with the highest multiplicity.

Example 14.2: *Identifying the ground term of a configuration*
What is the ground term for the configurations (a) $3d^5$ of Mn^{2+} and (b) $3d^3$ of Cr^{3+}

Answer. (a) The d^5 configuration permits occupation of each d orbital singly, so the maximum multiplicity is $2 \times \frac{5}{2} + 1 = 6$, a sextet term. If each of the electrons is to have the same spin quantum number, all must have different m_l values in order to obey the Pauli principle. Thus the m_l values will be $+2, +1, 0, -1$, and -2. The sum of these is 0, so $L = 0$ and the term is ^6S. (b) For the configuration d^3, the maximum multiplicity corresponds to all three electrons having the same spin quantum number to give $S = 2 \times \frac{3}{2} + 1 = 4$, a quartet. The three m_l values must be different if all spins are the same, which allows a maximum value of $M_L = 2 + 1 + 0 = 3$, indicating $L = 3$, an F term. Hence, the ground term of d^3 is ^4F.

Exercise E14.2. Identify the ground terms of (a) $2p^2$ and (b) $3d^9$. (Hint: because d^9 is one electron short of a closed shell with $L = 0$ and $S = 0$, treat it on the same footing as a d^1 configuration.)

Racah parameters

We have remarked several times that the different terms of a configuration have different energies on account of the coulombic repulsion between electrons. These interelectron repulsion energies are given by complicated integrals over the orbitals occupied by the electrons. Mercifully, though, all the integrals for a given configuration can be collected together in three specific combinations, and the repulsion energy of any term of a configuration can be expressed as a sum of these three quantities. The three combinations of integrals are called the **Racah parameters** A, B, and C. We do not even need to know the theoretical values of the parameters or the theoretical expressions for them, because it is more reliable to use A, B, and C as empirical quantities obtained from spectroscopy.

Each term stemming from a given configuration has an energy that

Table 14.2 Racah parameters for some *d*-block ions*

	1 +	2 +	3 +	4 +
Ti		720(3.7)		
V		765(3.9)	860(4.8)	
Cr		830(4.1)	1030(3.7)	1040(4.1)
Mn		960(3.5)	1130(3.2)	
Fe		1060(4.1)		
Co		1120(3.9)		
Ni		1080(4.5)		
Cu	1220(4.0)	1240(3.8)		

*The table gives the B parameter with the value of C/B in parentheses.

may be expressed as a linear combination of the three Racah parameters. For a d^2 configuration a detailed analysis shows that

$$E(^1S) = A + 14B + 7C$$

$$E(^1G) = A + 4B + 2C$$

$$E(^1D) = A - 3B + 2C$$

$$E(^3P) = A + 7B \qquad E(^3F) = A - 8B$$

The values of A, B, and C can be found by fitting these expressions to the observed energies of the terms. Note that A is common to all the terms; hence, if we are interested only in their relative energies, we do not need to know its value. Likewise, if we are interested only in the relative energies of the two triplet terms, we also do not need to know the value of C.

All three Racah parameters are positive (they represent repulsions). Therefore, provided $C > 5B$, the energies of the terms of the d^2 configuration lie in the order

$$^3F < {}^3P < {}^1D < {}^1G < {}^1S$$

This order is the same as we obtained by using Hund's rules. However, if $C < 5B$, the advantage of having an occupation of orbitals that corresponds to a high orbital angular momentum is greater than the advantage of having a high multiplicity, and the ^3P term lies above ^1D (as is in fact the case for the Ti^{2+} ion). Some experimental values of B and C are given in Table 14.2. It should be noted that $C \approx 4B$, so the ions listed there are in the regime where Hund's rules are not valid.

The importance of Racah parameters is that they summarize the energies of all the terms that may arise from a single configuration. The parameters are the quantitative expression of the ideas in Hund's rules, and are more powerful than the rules alone since they also account for deviations from them.

THE ELECTRONIC SPECTRA OF COMPLEXES

We now return to the spectrum of $[Cr(NH_3)_6]^{3+}$ shown in Fig. 14.1. The two central bands with intermediate intensities are HOMO–

LUMO transitions with energies that differ on account of the interelectronic repulsions (as we shall shortly explain). Because both the HOMO and LUMO of an octahedral complex are predominantly metal d orbital in character, with a separation characterized by the strength of the ligand field splitting parameter, these two transitions are called **d–d transitions** or **ligand-field transitions**.

In contrast, the ultraviolet absorption is very sensitive to ligand substitutions and the polarity of the solvent, which implies that the transition is accompanied by a substantial redistribution of charge. This observation suggests that it is a **charge-transfer transition** in which an electron is moved between orbitals that are predominantly ligand in character to orbitals that are predominantly metal in character. The transition is therefore classified as a **ligand-to-metal charge transfer** (LMCT) transition. In some complexes the charge migration occurs in the opposite direction, in which case it is classified as a **metal-to-ligand charge transfer** (MLCT) transition. An example of an MLCT transition is the one responsible for the red color of tris(bipyridyl)iron(II), the complex used for the colorimetric analysis of Fe(II). In this case, an electron makes a transition from a d orbital of the central metal into a π^* orbital of the ligand. Charge-transfer transitions are usually much more intense than ligand field transitions.

14.3 Ligand-field transitions

According to the discussion in Chapter 6, we would expect the octahedral d^3 complex $[Cr(NH_3)_6]^{3+}$ to have the ground configuration t_{2g}^3. We can identify the absorption spectrum in the region near $25\,000\ \mathrm{cm}^{-1}$ as arising from the excitation $t_{2g}^2 e_g^1 \leftarrow t_{2g}^3$ because that wavenumber is typical of ligand field splittings in complexes. Because there are three t_{2g} orbitals and two e_g orbitals, there are in fact six possible transitions because any of the three t_{2g} electrons can migrate to either of the two e_g orbitals. In the absence of interelectron repulsions, all six transitions occur at the same energy. However, because there are interelectron repulsions, the transition energies depend on which orbitals are specifically involved in the transitions.

Before we embark on a Racah-like analysis of the splitting, it may be helpful to see qualitatively from the viewpoint of simple molecular orbital theory why there are two bands. The $d_{z^2} \leftarrow d_{xy}$ transition promotes an electron from the xy plane into the already electron-rich z direction (it is electron rich because both d_{yz} and d_{zx} are occupied). However, the $d_{z^2} \leftarrow d_{zx}$ transition merely relocates an electron that is already largely concentrated along the z axis (Fig. 14.2). In the former case, but not in the latter, there is a distinct increase in electron repulsion. As a result, the two $e_g \leftarrow t_{2g}$ transitions lie at different energies. All the other transitions resemble one or other of these two cases, and three transitions fall into one group and the other three fall into the second group.

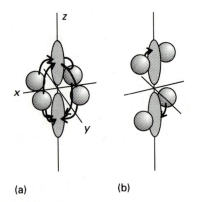

(a)   (b)

14.2 The shifts in electron density that accompany the two transitions discussed in the text. There is a considerable relocation of electron density toward the ligands on the z axis in (a), but a much less substantial relocation in (b).

The spectroscopic terms

The two bands we are discussing in Fig. 14.1 are labeled $^4T_{2g} \leftarrow {}^4A_{2g}$ (at $21\,550$ cm^{-1}) and $^4T_{1g} \leftarrow {}^4A_{2g}$ (at $28\,500$ cm^{-1}), where in each case the upper state precedes the lower. The labels are **molecular term symbols** and serve a similar purpose to the atomic term symbols we have already encountered. The left superscript denotes the multiplicity, so the superscript 4 denotes a quartet state with $S = \frac{3}{2}$, as expected when there are three unpaired electrons. However, as L is not a good quantum number in an octahedral environment (because orbital angular momentum is not well-defined unless a system has spherical symmetry), the letter denotes the symmetry species of the overall orbital state and not its total orbital angular momentum.

The letters in the term symbols are the upper-case versions of the symmetry species of individual orbitals (just as in atoms upper case letters are used to denote total angular momenta). Thus, the nearly totally symmetric ground state (with an electron in each of the three t_{2g} orbitals) is denoted A_{2g}. We say only *nearly* totally symmetric, because close inspection of the behavior of the three occupied t_{2g} orbitals shows that all the operations of the O_h point group transform the product $t_{2g} \times t_{2g} \times t_{2g}$ into ± 1 times itself, which identifies it as an A symmetry species (see the character table in Appendix 3). Likewise, since each d orbital has even parity (g), the overall parity is also g. However, each C_4 rotation transforms one t_{2g} orbital into the negative of itself (**1**) and the other two t_{2g} orbitals into each other (**2**), so overall there is a change of sign under this operation and its character is -1. Thus the term is A_{2g} rather than the totally symmetrical A_{1g} of the closed shell.

It is more difficult to establish that the term symbols which can arise from the quartet $t_{2g}^2 e_g^1$ excited configuration are $^4T_{2g}$ and $^4T_{1g}$, and we shall not consider this aspect here. The superscript 4 implies that the upper configuration continues to have the same number of unpaired spins as in the ground state, and the subscript g stems from the even parity of all the contributing orbitals. We have seen that the six excitations fall into two groups of three, which is consistent with the symmetry species T in each case.

The energies of the terms

In a free atom we needed to consider only the electron–electron repulsions in order to arrive at the relative ordering of the terms arising from a given d^n configuration. In an octahedral complex the problem has another aspect because as well as the repulsions it is necessary to take into account the effect of Δ_O, the difference in energy between the t_{2g} and e_g orbitals.

We can illustrate the considerations involved by treating first the simplest case of an atom or ion with a single valence electron. Thus, an s orbital in a free atom becomes an a_{1g} orbital in an octahedral field: a totally symmetrical orbital in one environment becomes a totally symmetrical orbital in another environment. As in Section 3.3 in

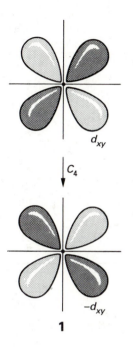

2

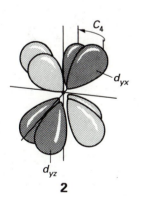

1

Table 14.3 The correlation of spectroscopic terms for *d* electrons in O_h complexes

Atomic term	Number of states	Terms in O_h symmetry
S	1	A_{1g}
P	3	T_{1g}
D	5	$T_{2g} + E_g$
F	7	$T_{1g} + T_{2g} + A_{2g}$
G	9	$A_{1g} + E_g + T_{1g} + T_{2g}$

connection with Walsh diagrams, we express the change by saying that the *s* orbital of the atom correlates with the a_{1g} orbital of the complex. Similarly, the totally symmetrical overall S term of a many-electron atom correlates with the totally symmetrical A_{1g} term of an octahedral environment. We have seen that the five *d* orbitals of a free atom split into the triply degenerate t_{2g} and doubly degenerate e_g sets in an octahedral complex. In the same way, an atomic D term splits into a T_{2g} term and an E_g term in O_h symmetry. Table 14.3 summarizes the correlations between free ion terms and their terms in an octahedral complex.

Example 14.3: *Identifying correlations between terms*

What terms in a complex with O_h symmetry correlate with the ^3P term of a bare ion with a d^2 configuration?

Answer. The three *p* orbitals of a free atom become the triply degenerate t_{1u} orbitals of an octahedral complex. Therefore, if we disregard parity for the moment, a P term of a many-electron atom becomes a T_1 term in the point group O_h. Because *d* orbitals have even parity, the term overall must be g. The multiplicity is unchanged in the correlation, so the ^3P term becomes the $^3T_{1g}$ term.

Exercise E14.3. What terms in a d^2 complex of O_h symmetry correlate with the ^3F and ^1D terms of a free ion?

Weak and strong field limits

Electron–electron repulsions are difficult to take into account, but the discussion is simplified by considering two extreme cases initially. In the **weak field limit** the ligand field is so weak that only electron repulsions are important and the relative energies of the terms are determined by the Racah parameters. The other extreme case is the **strong field limit** in which the ligand field is so strong that the repulsions can be ignored. Then, with the two extremes established, we can consider intermediate cases by drawing a correlation diagram between the two. We shall illustrate what is involved by considering two simple cases, namely d^1 and d^2. Then we show how the same ideas are used to treat more complicated cases.

The only term of a d^1 configuration in the free atom is ^2D. In an octahedral complex the configuration is either t_{2g}^1, which gives rise to a $^2T_{2g}$ term, or e_g^1, which gives rise to a ^2E term. Because there are no electron-electron repulsions to worry about, the separation of the $^2T_{2g}$ and ^2E_g terms is the same as the separation of the t_{2g} and e_g orbitals, which is Δ_O. The correlation diagram for the d^1 configuration will therefore resemble that shown in Fig. 14.3(a).

For a d^2 configuration, the triplet terms are ^3F and ^3P in the free atom, as we have seen, and their energies relative to the lower term (^3F) are

$$E(^3F) = 0 \qquad E(^3P) = 15B$$

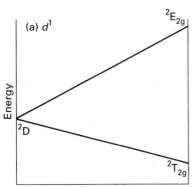

(a) d^1

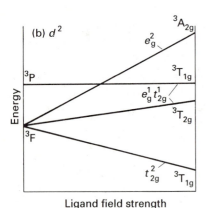

(b) d^2

14.3 Correlation diagram between a free ion (left) and the strong-field terms (right) of (a) d^1 and (b) d^2 configurations.

These two terms are shown on the left of Fig. 14.3(b). In the very strong field limit, a d^2 ion has the configurations

$$t_{2g}^2 < t_{2g}^1 e_g^1 < e_g^2$$

Each configuration corresponds to one term, and they are separated in energy by an amount that depends on Δ_O. From the information in Table 6.3 we can write their relative energies as

$$E(t_{2g}^2, \mathrm{T}_{1g}) = -0.8\Delta_O \quad E(t_{2g}^1 e_g^1, \mathrm{T}_{2g}) = +0.2\Delta_O$$

$$E(e_g^2, \mathrm{A}_{2g}) = +1.2\Delta_O$$

Therefore, relative to the energy of the lowest term (t_{2g}^2) their energies are

$$E(t_{2g}^2, \mathrm{T}_{1g}) = 0 \quad E(t_{2g}^1 e_g^1, \mathrm{T}_{2g}) = 1.0\Delta_O$$

$$E(e_g^2, \mathrm{A}_{2g}) = 2.0\Delta_O$$

Because we are ignoring electron repulsions in this limit, the Racah parameters play no role. These configurations give rise to the terms shown on the right in Fig. 14.3(b). The $^3\mathrm{T}_{1g}$ term of $e_g^1 t_{2g}^1$ is unaffected by the ligand field.

Our problem now is to account for the energies of the intermediate cases, where neither the ligand field nor the electron repulsion terms are dominant. We do this by identifying the correlation of the terms between the two extreme cases. To do so, we identify the term that each free atom term becomes in the very high field limit. This correlation must be done using group theory, because we need to know how one term becomes another as the spherical symmetry of the free atom is replaced by the lower octahedral symmetry of the complex. However, tracing the effects of this descent in symmetry is a standard problem in group theory, and has been worked out and tabulated once and for all; therefore, all we need to know are the results.

Tanabe–Sugano diagrams

Once the symmetry analysis has been done, diagrams like those shown in Fig. 14.4 can be constructed for any complex and quantitative calculations can be carried out for any strength of ligand field. The most widely used versions of these diagrams are called **Tanabe–Sugano diagrams** after the Japanese workers who devised them. Tanabe–Sugano diagrams for O_h complexes with the configurations d^2 through d^8 are given in Appendix 5. Figure 14.3 is in fact a simplified version of a Tanabe–Sugano diagram for a d^2 configuration, the full diagram for which is shown in Fig. 14.4. The term energies E are expressed as E/B and plotted against Δ_O/B. Because $C \approx 4B$, terms with energies that depend on both B and C can be plotted on the same diagrams. The zero of energy in a Tanabe–Sugano diagram is always taken as that of the lowest term. Hence the lines in some diagrams are discontinuous (see that for d^4) when there is a change in the ground

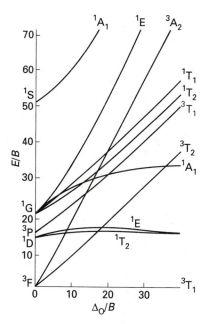

14.4 The Tanabe–Sugano diagram for the d^2 configuration. A complete collection of diagrams for d^n configurations is given in Appendix 5 at the back of the book.

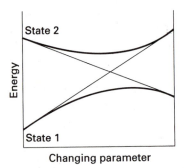

14.5 The noncrossing rule states that if two states of the same symmetry are likely to cross as a parameter is changed (as shown by the thin lines), they will in fact mix together and avoid the crossing (as shown by the heavy lines).

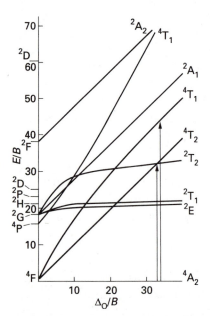

14.6 The Tanabe–Sugano diagram for the d^3 configuration.

term, when the ligand field becomes strong enough to favor electron pairing.

Some lines in a Tanabe–Sugano diagram are curved on account of the mixing of terms of the same symmetry type. States of the same symmetry obey the **noncrossing rule**, which says that if the increasing ligand field causes two weak field terms of the same symmetry to approach, they do not cross but bend apart from each other (Fig. 14.5).

We shall use the d^3 case that opened this chapter to illustrate how the relevant Tanabe–Sugano diagram (Fig. 14.6) is used to interpret spectroscopic data even in the absence of a detailed theoretical analysis. We have seen that two low-energy ligand-field transitions occur at 21 550 cm^{-1} ($^4T_{2g} \leftarrow {}^4A_{2g}$) and 28 500 cm^{-1} ($^4T_{1g} \leftarrow {}^4A_{2g}$), a ratio of 1.32. The only point in Fig. 14.6 where this ratio is satisfied is on the far right. Hence, we can read off the values of Δ_O and B from the location of this point.

Example 14.4:   *Determining Δ_O and B using a Tanabe–Sugano diagram*
Deduce the values of Δ_O and B for $[Cr(NH_3)_6]^{3+}$ from the spectrum in Fig. 14.1 and the Tanabe–Sugano diagram in Fig. 14.6 (or Appendix 5).

Answer. The transition energy ratio of 1.32 falls at $\Delta_O/B = 32.8$ at the point where the two arrows fit. The tip of the arrow representing the lower energy transition lies at $32.8B$ vertically, so equating $32.8B$ and 21 550 cm^{-1} gives $B = 657$ cm^{-1} and hence $\Delta_O = 21\,550$ cm^{-1}.

Exercise E14.4. Use the same Tanabe-Sugano diagram to predict the wavenumbers of the first two spin-allowed quartet bands in the spectrum of $[Cr(OH_2)_6]^{3+}$ for which $\Delta_O = 17\,600$ cm^{-1} and $B = 918$ cm^{-1}.

The nephelauxetic series

In Example 14.4 we found that $B = 657$ cm^{-1} for $[Cr(NH_3)_6]^{3+}$. This value is only 64 percent of the value for a free Cr^{3+} ion in the gas phase, and we conclude that electron repulsions are weaker in the complex than in the free ion. This weakening is to be expected, because the occupied molecular orbitals are delocalized over the ligands and away from the metal. The delocalization increases the average separation of the electrons and hence reduces their mutual repulsion.

The reduction of B from its free ion value is normally reported in terms of the **nephelauxetic parameter**, β:

$$\beta = \frac{B(\text{complex})}{B(\text{free ion})}$$

(The name is from the Greek words for 'cloud expanding'.) The values of β depend on the ligand and vary along the **nephelauxetic series**:

$$F^- > H_2O > NH_3 > CN^-, Cl^- > Br^-$$

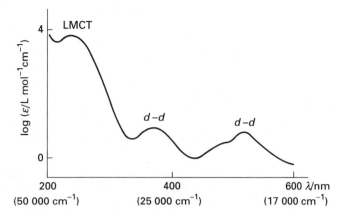

14.7 The absorption spectrum of $[CrCl(NH_3)_5]^{2+}$ in water in the visible and ultraviolet regions. The peak corresponding to the transition $^2E \leftarrow ^4A$ is not visible on this magnification.

A small β indicates a large measure of d-electron delocalization on to the ligands and hence a significant covalent character in the complex. Thus the series shows that a Br^- ligand results in a greater reduction in electron repulsions in the complex than an F^- ion, which is consistent with a greater covalent character in bromo complexes than in analogous fluoro complexes. Another way of expressing the trend represented by the series is that the softer the ligand, the smaller the nephelauxetic parameter.

The nephelauxetic character of a ligand is different for electrons in t_{2g} and e_g orbitals. Because the σ overlap of e_g is usually larger than the π overlap of t_{2g}, the cloud expansion is larger in the former case. The measured nephelauxetic parameter of an $e \leftarrow t$ transition is an average of the effects on both types of orbital.

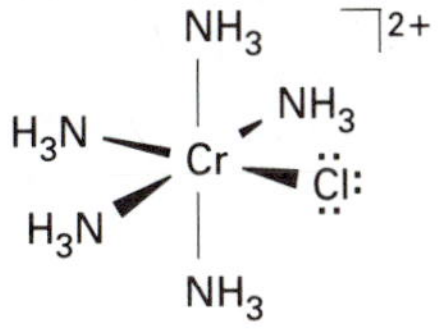

3 $[CrCl(NH_3)_5]^{2+}$

14.4 Charge-transfer bands

Figure 14.7 shows the visible and UV spectrum of $[CrCl(NH_3)_5]^{2+}$. It resembles the spectrum shown in Fig. 14.1, and we can recognize that the two ligand field bands are the pair in the visible region. The replacement of one NH_3 ligand by one weaker field Cl^- ligand moves the lowest energy band to lower energy than that for $[Cr(NH_3)_6]^{3+}$ in Fig. 14.1, but although the reduction in symmetry from O_h to C_{4v} does produce an additional splitting, the effect is small. The new feature in the spectrum is the strong band in the ultraviolet near $42\,000$ cm^{-1} (240 nm). This band is at lower energy than the corresponding band in the spectrum of $[Cr(NH_3)_6]^{3+}$ and arises because the Cl^- ligands have π lone pair electrons that are not directly involved in bonding (**3**). The band is an example of an LMCT transition in which a lone-pair electron of Cl^- is promoted into a predominantly metal orbital. Figure 14.8 summarizes the transitions we classify as charge transfer. The LMCT character of similar bands in $[CoX(NH_3)_6]^{2+}$ is confirmed by the decrease in energy in steps equivalent to about 8000 cm^{-1} as X is varied from Cl to Br to I.

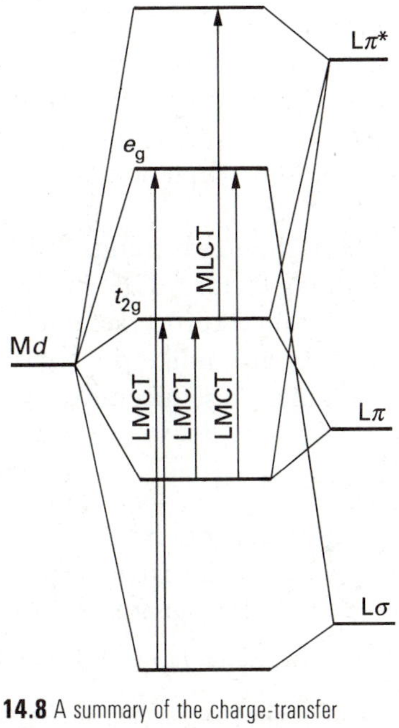

The charge transfer character of a transition is most often demonstrated (and distinguished from $\pi^* \leftarrow \pi$ transitions on ligands)

14.8 A summary of the charge-transfer transitions in an octahedral complex.

by demonstrating **solvatochromism,** the variation of the transition energy with changes in solvent permittivity. The observation of solvatochromism indicates a large change in the molecular dipole accompanying the transition.[3]

LMCT transitions

As in Fig. 14.7, charge-transfer transitions are generally intense compared with ligand-field transitions. Charge-transfer bands in the visible region of the spectrum (and hence contributing to the intense colors of complexes) may occur if the ligands have lone pairs of relatively high energy (as in sulfur and selenium) or if the metal has low lying empty orbitals. The color of the artists' pigment 'cadmium yellow', CdS, for instance, is due to the transition $Cd^{2+}(5s) \leftarrow S^{2-}(\pi)$. Similarly, HgS is red as a result of the transition $Hg^{2+}(6s) \leftarrow S^{2-}(\pi)$ and ochers are iron oxides that are colored red and yellow by the transition $Fe(3de_g^*) \leftarrow O^{2-}(\pi)$.

The tetraoxoanions of metals with high oxidation numbers (such as MnO_4^-) provide what are probably the most familiar examples of LMCT bands. In them, an O lone pair electron is promoted into a low-lying empty e metal orbital. High metal oxidation numbers correspond to a low d-orbital population (many are d^0), so the acceptor level is available and low in energy. The trend in LMCT energies is:

Oxidation number	
+7	$MnO_4^- < TcO_4^- < ReO_4^-$
+6	$CrO_4^{2-} < MoO_4^{2-} < WO_4^{2-}$
+5	$VO_4^{3-} < NbO_4^{3-} < TaO_4^{3-}$

The energies of the transitions correlate with the order of the electrochemical series, with the lowest energy transitions taking place to the most easily reduced ions. This correlation is consistent with the transition being the transfer of an electron from the ligands to the metal, corresponding, in effect, to the reduction of the metal ion by the ligands.

Polymeric and monomeric oxoanions follow the same trends, with the oxidation number of the metal the determining factor.[4] The similarity suggests that these LMCT transitions are localized processes which take place on molecular fragments and that they are unlike the transitions between delocalized orbitals that are common in semiconductors.

Optical electronegativity

The variation in the position of LMCT bands can be expressed in terms of molecular orbital energy level diagrams and in particular the **optical electronegativities** of the metal and the ligands. In the

[3] A useful discussion of solvatochromism is provided in A. B. P. Lever, *Inorganic electronic spectroscopy.* Elsevier, Amsterdam (1984).

[4] E. S. Dodsworth and A. B. P. Lever, Solvatochromism of dinuclear complexes. *Inorg. Chem.,* **29**, 499 (1990).

Table 14.4 Optical electronegativities

Metal	O_h	T_d	Ligand	π	σ
Cr(III)	1.8–1.9		F$^-$	3.9	4.4
Co(III)*	2.3		Cl$^-$	3.0	3.4
Ni(II)		2.0–2.1	Br$^-$	2.8	3.3
Co(II)		1.8–1.9	I$^-$	2.5	3.0
Rh(III)*	2.3		H$_2$O	3.5	
Mo(VI)	2.1		NH$_3$	3.3	

*Low-spin complexes.

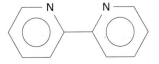

4 2,2′–Bipyridine (bipy)

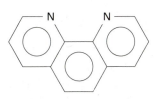

5 Phenthroline (phen)

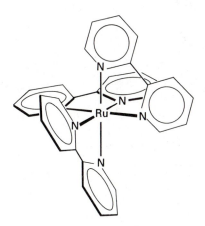

6 [Ru(bipy)$_3$]$^{2+}$

7 Dithiolene

simplest case of a complex with an anionic ligand, we focus on the electronegativities χ_{Ligand} and χ_{Metal} of the full ligand π orbitals and of the t_{2g} metal orbitals that are destined to become the π^* orbitals in the complex. The wavenumber of the transition is then written as the difference between the two electronegativities:

$$\tilde{\nu} = C(\chi_{\text{Ligand}} - \chi_{\text{Metal}})$$

Optical electronegativities have values comparable to Pauling electronegativities if we take $C = 30\,000$ cm^{-1}. If the LMCT transition terminates in an e_g orbital, Δ_O must be added to the energy predicted by this equation.

Table 14.4 lists some optical electronegativities. The values for metals are different in complexes of different symmetry and the ligand values are different if the transition originates from a π orbital rather than a σ orbital.

MLCT transitions

A transfer of charge from metal to ligand is most commonly observed in complexes with ligands that have low lying π^* orbitals, especially aromatic ligands. If the metal ion has a low oxidation number, in which case its d orbitals will be relatively high in energy, the transition will occur at low energy. The family of ligands most commonly involved in MLCT transitions are the diimines, which have two N donor atoms: two important examples are bipyridine (**4**) and phenanthroline (**5**). Complexes of diimines with strong MLCT bands may be *tris*(diimine) species such as *tris*(bipyridyl)ruthenium(II) (**6**), which is orange on account of its MLCT band. A diimine ligand may also be easily substituted into a complex with other ligands that favor a low oxidation state. Two examples are [W(CO)$_4$(phen)] and [Fe(CO)$_3$(bipy)]. The occurrence of MLCT transitions is not limited to diimine ligands. Another important ligand type that shows typical MLCT transitions is dithiolene, S$_2$C$_2$R$_2$ (**7**). Recently, resonance Raman spectroscopy (Box 14.1) has emerged as a powerful technique for the study of MLCT transitions, particularly of diimine complexes.

The MLCT excitation of *tris*(bipyridyl)ruthenium(II) has been the

Box 14.1: Resonance Raman spectroscopy

Resonance Raman spectroscopy is a powerful technique for the assignment and characterization of charge transfer bands. The conventional Raman technique described in Box 3.1 involves the inelastic scattering of radiation by a molecule. The frequency of the scattered radiation differs from that of the incident radiation by an amount corresponding to a vibrational excitation energy of the molecule, so the observation of the frequency shift can be interpreted as a vibrational spectrum. Conventional Raman spectroscopy depends upon the variation of the polarizability of the molecule as it vibrates, and the incident radiation is typically not at an absorption frequency of the molecule. In contrast, resonance Raman spectroscopy uses incident radiation that has a frequency within an electronic absorption band of the molecule. Under such circumstances, there is an enhancement of the Raman effect for the vibration modes that are affected by the electronic excitation.

Resonance Raman bands are observed when the electronic transition results in a change in the electron distribution that causes a shift in nuclear positions (bond lengths, bond angles). Thus, the vibrations that give intense resonance Raman bands at a particular incident frequency are vibrations that occur in the part of the molecule that is strongly affected by the electronic transition. As such, resonance Raman enhancements aid in the identification of the transition and its assignment.

Figure B14.1 shows the resonance Raman spectrum of

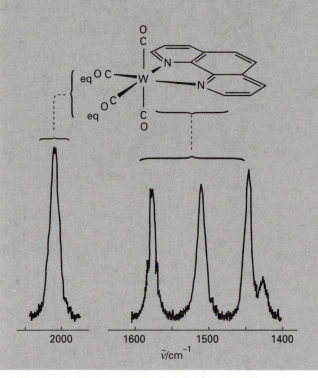

B14.1 Resonance Raman spectrum of $W(CO)_4(phen)$. The sample was irradiated by a 488.0 nm (blue) laser line, which falls in the lowest energy electronic absorption band of the complex. Only the symmetric stretch of the equatorial CO ligands, near 2000 cm^{-1} and the phen modes, between 1600 and 1400 cm^{-1}, are enhanced by the resonance Raman effect, indicating that the electronic transition is confined to the $(OC)_2W(phen)$ plane. (R. Balk, T. Snoeck, D. Stufkens, and A. Oskam, *Inorg. Chem.*, **19**, 3015 (1980).)

[W(CO)$_4$(phen)].[†] The incident radiation excites near the maximum of the electronic absorption band that has been assigned on the basis of solvatochromism as the π^*(phen) $\leftarrow$ d(W) MLCT transition. Notice that the two internal stretching vibrations of the phen ring labeled ν(phen II) and ν(phen III) give strong lines. This observation provides clear confirmation of the assignment of the electronic transition.

A surprise in this spectrum is the strong line for the CO stretch (marked eq), which is also resonance enhanced. The increasing intensity of such signals suggests that the π^* levels of both phen and the coplanar CO(eq) participate in the electronic transition.

The resonance enhancement of the phen ring vibrations decreases as the metal d and phen ligand π^* orbitals approach each other in energy along a series of analogous complexes. This decrease indicates a reduction in the extent of charge transfer in the transition as both the initial and final orbital become more mixed in character. There is a very good correlation between the extent of resonance Raman enhancement and solvatochromism, the simplest indicator of charge transfer.

[†] R. W. Balk, Th. L. Snoeck, D. J. Stufkens, and A. Oskam, *Inorg. Chem.*, **19**, 3015 (1980).

subject of intense research efforts because the excited state that results from the charge transfer has a lifetime of microseconds, and the complex is a versatile photochemical redox reagent (Section 15.14). The photochemical behavior of a number of related complexes has also been studied on account of their relatively long excited state lifetimes.

14.5 Selection rules and intensities

The contrast in intensities between typical charge transfer bands and typical ligand field bands raises the question of the factors that control the intensity of absorption bands. In an octahedral, nearly octahedral, or square planar complex, the maximum molar absorption coefficient ε_{max} (which measures the strength of the absorption)[5] for ligand-field transitions is typically less than or close to 100 L mol^{-1} cm^{-1}. In tetrahedral complexes, which have no center of symmetry, ε_{max} might exceed 250 L mol^{-1} cm^{-1}. In many cases, charge-transfer bands are allowed by symmetry rules and have intensities governed by the extent of overlap of ground-state and excited-state wavefunctions. Such transitions have ε_{max} of between 1000 and 50 000 L mol^{-1} cm^{-1}.

[5] The molar absorption coefficient is the constant in the Beer–Lambert law for the transmittance $\tau = I_f/I_i$ when light passes through a length l of solution of molar concentration [X] and is attenuated from an intensity I_i to an intensity I_f: $\log_{10} \tau = -\varepsilon l[X]$. Its earlier name is the 'extinction coefficient'.

Spin selection rules

The electromagnetic field of the incident radiation cannot change the *relative* orientations of the spins of the electrons in a complex. For example, an initially antiparallel pair ↑↓ cannot be converted to ↑↑, so a singlet ($S = 0$) cannot undergo a transition to a triplet ($S = 1$). This restriction is summarized by the rule $\Delta S = 0$ for **spin-allowed transitions**.

The coupling of spin and orbital angular momenta can relax the spin selection rule, but such **spin forbidden** $\Delta S \neq 0$ transitions are generally much weaker than spin-allowed transitions. The intensity of spin-forbidden bands increases as the atomic number increases because the strength of the spin–orbit coupling is greater for heavy atoms than for light atoms. The breakdown of the spin selection rule by spin–orbit coupling is often called the **heavy-atom effect**. In the first d series, in which spin–orbit coupling is weak, spin-forbidden bands have ε_{max} less than about 1 L mol^{-1} cm^{-1}; however, spin-forbidden bands are a significant feature in the spectra of heavy d-metal complexes.

The Laporte selection rule

The **Laporte selection rule** is a statement about the change in parity that accompanies a transition:

> In a centrosymmetric molecule or ion, the only allowed transitions are those accompanied by a change in parity.

That is, transitions between g and u terms are permitted, but a g term cannot combine with another g term and a u term cannot combine with another u term:

$$g \leftrightarrow u \qquad g \nleftrightarrow g \qquad u \nleftrightarrow u$$

The Laporte selection rule is based on the concept of an **electric dipole transition**, in which the transition gives rise to a transient electric dipole moment. The intensity of such a transition from a state with wavefunction ψ_i to one with wavefunction ψ_f is proportional to the square of the **transition dipole moment**, which is defined as the integral

$$\mu_{fi} = \int \psi_f^* \mu \psi_i \, d\tau$$

where μ is the electric dipole moment operator $-e\mathbf{r}$. The transition dipole moment can be regarded as a measure of the impulse that a transition imparts to the electromagnetic field: a large impulse corresponds to an intense transition; zero impulse corresponds to a forbidden transition. Because $\mathbf{r}$ changes sign under inversion (and is therefore u), the entire integral will also change sign under inversion if ψ_i and ψ_f have the same parity, because

$$g \times u \times g = u \quad \text{and} \quad u \times u \times u = u$$

Therefore, because the value of an integral cannot depend on the

Table 14.5 Intensities of spectroscopic bands in $3d$ complexes

Band type	$\varepsilon_{max}/(\text{L mol}^{-1}\text{cm}^{-1})$
Spin-forbidden	<1
Laporte forbidden d–d	20–100
Laporte allowed d–d	$c.$ 250
Symmetry allowed (e.g. CT)	1000–50 000

choice of coordinates,[6] it vanishes if ψ_i and ψ_f have the same parity. However, if they have opposite parity, the integral does not change sign under inversion of the coordinates because

$$g \times u \times u = g$$

and therefore need not vanish.

In a centrosymmetric complex, ligand-field d–d transitions are g–g and are therefore forbidden. Their forbidden character accounts for the relative weakness of these transitions in octahedral complexes compared with those in tetrahedral complexes, on which the Laporte rule is silent (as they have no center of symmetry).

The question remains why d–d transitions occur at all, albeit weakly. The Laporte selection rule may be relaxed in two ways. First, a complex may depart slightly from perfect centrosymmetry, perhaps on account of a distortion imposed by the environment of a complex packed into a crystal or on account of an intrinsic asymmetry in the structure of polyatomic ligands. Alternatively, the complex might undergo an asymmetrical vibration, which also destroys its center of inversion. In either case, a Laporte-forbidden d–d band tends to be much more intense than a spin-forbidden transition. As already mentioned, charge-transfer transitions are generally more intense than either kind of d–d transition.

Table 14.5 summarizes typical intensities of electronic transitions of complexes of the first series of d elements.

Example 14.5: *Assigning a spectrum using selection rules*
Assign the bands in the spectrum in Fig. 14.7 by considering their intensities.

Answer. Like the hexaammine spectrum in Fig. 14.1, which has d–d bands at similar wavenumbers, the complex is moderately strong field ($\Delta_O/B > 20$). If we assume that the complex is approximately octahedral, examination of the Tanabe–Sugano diagram again reveals that the ground state is $^4A_{2g}$. The first excited state is ^2E_g. The next two are $^2T_{1g}$ and $^2T_{2g}$. Transitions to these three will not have ε_{max} greater than 1 L mol^{-1} cm^{-1} since they are spin forbidden. Very weak bands at low energy are predicted. They may be too high in energy to be resolved from the more intense bands. The spectrum shows one such

---

[6] An integral is an area, and areas are independent of the coordinates used for their evaluation.

weak band which is probably the transition to the lowest energy excited term, ^2E_g, the next two higher terms are $^4T_{2g}$ and $^4T_{1g}$. These are reached by spin allowed but Laporte-forbidden d–d transitions, and have ε_{max} near 100 L mol^{-1} cm^{-1}. In the near UV, the band with ε_{max} near 10 000 L mol^{-1} cm^{-1} corresponds to the LMCT transitions in which the chlorine π lone pair is promoted into a molecular orbital that is principally metal d orbital.

Exercise E14.5. The spectrum of $[Cr(NCS)_6]^{3-}$ has a very weak band near 16 000 cm^{-1}, a band at 17 700 cm^{-1} with $\varepsilon_{max} = 160$ L mol^{-1} cm^{-1}, a band at 23 800 cm^{-1} with $\varepsilon_{max} = 130$ L mol^{-1} cm^{-1}, and a very strong band at 32 400 cm^{-1}. Assign these transitions using the d^3 Tanabe–Sugano diagram and selection rule considerations. (Hint: NCS$^-$ has low lying π^* orbitals.)

14.6 Luminescence

A complex is **luminescent** if it emits light after it has been electronically excited by the absorption of radiation. For luminescence to occur, the rate of radiative decay by photon emission must compete with thermal degradation of energy through heat loss to the surroundings. Relatively fast radiative decay is not especially common at room temperature for d-metal complexes, so strongly luminescent systems are comparatively rare. Nevertheless, they do occur, and we can distinguish two types of process. Traditionally, rapidly decaying luminescence was called **fluorescence** and luminescence that persists after the exciting illumination is extinguished was called **phosphorescence**. However, because the lifetime criterion is not reliable, the modern definitions of the two kinds of luminescence are based on the distinctive mechanisms of the processes. Fluorescence occurs when an excited state of the same multiplicity as the ground state decays radiatively into the ground state. The transition is spin-allowed and is commonly fast (a matter of nanoseconds). Phosphorescence is a spin-forbidden process, as we explain below, and hence is often slow.

Because light absorption usually populates a state by a spin allowed transition, the mechanism of phosphorescence involves a nonradiative **intersystem crossing**, or conversion from the initial excited state to another excited state of different multiplicity without the emission of radiation. This second state acts as an energy reservoir because radiative decay to the ground state is spin forbidden. However, just as spin–orbit coupling allows the intersystem crossing to occur, it also breaks down the spin-selection rule, so the radiative decay can occur, but usually only slowly. A phosphorescent state of a d-metal complex may survive for microseconds or longer.

Phosphorescent complexes

An important example of phosphorescence is provided by ruby, which consists of a low concentration of Cr^{3+} ions substituted for Al^{3+} in

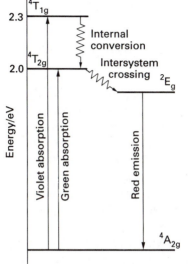

14.9 The transitions responsible for the absorption and luminescence of Cr^{3+} ions in ruby.

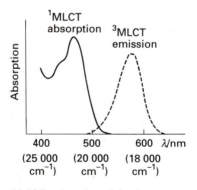

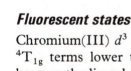

14.10 The absorption and phosphorescence spectra of $[Ru(bipy)_3]^{2+}$.

alumina. Each Cr^{3+} ion is surrounded octahedrally by six O^{2-} ions and the initial excitations are the spin-allowed processes

$$t_{2g}^2 e_g^1 \leftarrow t_{2g}^3: \quad {}^4T_{2g} \leftarrow {}^4A_{1g} \quad \text{and} \quad {}^4T_{1g} \leftarrow {}^4A_{1g}$$

These absorptions occur in the green and violet regions of the spectrum and are responsible for the red color of the gem (Fig. 14.9). Intersystem crossing to a ${}^2E_g$ term of the t_{2g}^3 configuration occurs in a few picoseconds or less, and red 627 nm phosphorescence occurs as this doublet decays back into the quartet ground state. This red emission adds to the red perceived by the subtraction of green and violet light from white light, and adds luster to the gem's appearance. When the optical arrangement is such that the emission is stimulated by 627 nm photons reflecting back and forth between two mirrors, it grows strongly in intensity. This stimulated emission in a resonant cavity was the process used by Theodore Maiman in the first successful laser (in 1960).

A similar ${}^2E \rightarrow {}^4A$ phosphorescence can be observed from a number of essentially octahedral Cr(III) complexes in solution. The emission is always in the red and close to the wavelength of ruby emission. The ${}^2E$ term belongs to the t_{2g}^3 configuration, which is the same as the ground state, and the strength of the ligand field is not important. If the ligands are rigid, as in $[Cr(bipy)_3]^{3+}$, the ${}^2E$ term may live for several microseconds in solution.

Another interesting example of a phosphorescent state is found in $[Ru(bipy)_3]^{2+}$. The excited singlet term produced by a spin allowed MLCT transition of this d^6 complex undergoes intersystem crossing to the lower energy triplet term of the same configuration $t_{2g}^5 \pi^{*1}$. Bright orange emission then occurs with a lifetime of about 1 μs (Fig. 14.10). The effects of other molecules (quenchers) on the lifetime of the emission may be used to monitor the rate of electron transfer from the excited state.

Fluorescent states

Chromium(III) d^3 complexes with weak-field ligands may have their ${}^4T_{1g}$ terms lower than the corresponding phosphorescent ${}^2E$ terms because the ligand field splitting is smaller than the electron repulsion energies. Such complexes may be fluorescent. Because the transition populates an antibonding e_g orbital, bond distances change in the excited state. As a result, the emission is considerably shifted to the red from the position of the absorption band. This red shift is called a **Stokes shift**.

14.7 Spectra of f-block complexes

The partially filled f shell can also participate in transitions in the visible region of the spectrum. In one respect, their description is more complex than for the analogous transitions in the d block because there are seven f orbitals. However, it is simplified by the fact that f orbitals are relatively deep inside the atom and overlap only weakly with ligand

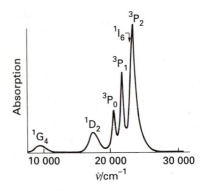

14.11 The spectrum of the f^2 Pr^{3+} (aq) ion in the visible region.

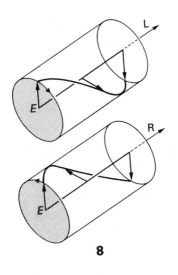

8

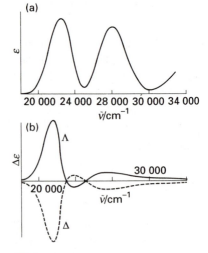

14.12 (a) The absorption spectrum of $[Co(en)_3]^{3+}$ and (b) the CD spectra of the two optical isomers.

orbitals. Hence, as a first approximation, their spectra can be discussed in the free-ion limit.

The colors of the aqua ions from $La^{3+}(f^0)$ to $Gd^{3+}(f^7)$ tend to repeat themselves in the reverse sequence $Lu^{3+}(f^{14})$ to $Gd^{3+}(f^7)$:

f electrons:	0	1	2	3	4	5	6	7
Color:	Colorless	green	red	pink	yellow	pink	colorless	
f electrons:	14	13	12	11	10	9	8	

This sequence suggests that the absorption maxima are related in a simple manner to the number of unpaired f electrons. Unfortunately, though, the apparent simplicity is deceptive and is not fully borne out by detailed analysis.

Figure 14.11 shows the spectrum of f^2 Pr^{3+}(aq) from the near infrared to the near ultraviolet regions. The four bands are labeled with the free-ion term symbols, which is appropriate when the interaction with ligands is very weak. The Russell–Saunders coupling scheme remains a good approximation despite the elements having high atomic numbers, because the f electrons penetrate only slightly through the inner shells. As a result, they do not sample the electric field produced by the nucleus very strongly and hence their spin–orbital coupling is weak.

The other noteworthy feature of the spectrum is the narrowness of the bands, which indicates that the electronic transition does not excite much molecular vibration when it occurs. The narrowness implies that there is little difference in the molecular potential energy surface when an electron is excited, which is consistent with an f electron interacting only weakly with the ligands. Because the excited electron interacts only weakly with its environment, the nonradiative lifetime of the excited state is quite long and luminescence can be expected. These complexes do tend to be strongly luminescent, and for that reason are used as phosphors on TV screens.

14.8 Circular dichroism

The photon is a particle with spin 1 (that is, it carries one unit of spin angular momentum), and in a state of circular polarization it has a definite **helicity**, or component of angular momentum along its direction of propagation. Left-circularly polarized light consists of photons of one helicity and right-circularly polarized light of photons of the opposite helicity (**8**).

A chiral molecule (one lacking an improper rotation axis, Section 3.5) displays **circular dichroism**. That is, the molecule has different absorption coefficients of right and left circularly polarized light at any given wavelength. A **circular dichroism spectrum** (a CD spectrum) is a plot of the difference of the molar absorption coefficients for right and left circularly polarized light against wavelength. As we see in Fig. 14.12, enantiomers have CD spectra that are mirror-images of each other. The usefulness of CD spectra can be appreciated by comparing the CD spectra of two enantiomers of $[Co(en)_3]^{3+}$ (Fig. 14.12(b)) with

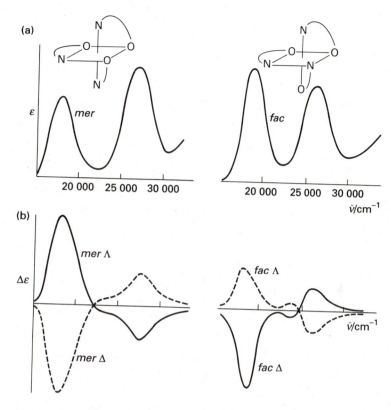

14.13 (a) The absorption spectra of the *fac* and *mer* isomers of [Co(ala)₃], where ala is alanine (specifically, its conjugate base) and (b) the CD spectra of the isomers and the assignment of their absolute configuration by comparison with the CD spectrum of [Co(en)₃]³⁺.

their conventional absorption spectra (Fig. 14.12(a)). The latter show two ligand field bands, just as though the complex were an octahedral complex, and give no sign that we are dealing with a complex of lower (D_3) symmetry. That difference, however, shows up in the CD spectra where an additional band is seen near $24\,000$ cm^{-1}.

Circular dichroism spectra are also important because they help to identify the absolute configurations of chiral complexes. The same absolute configuration of two complexes with similar electronic configurations give CD spectra of the same sign. Hence CD spectra can be used to relate large families of chiral complexes to the small number of primary cases, including $[Co(en)_3]^{3+}$, for which absolute configurations have been established by X-ray diffraction.

Circular dichroism spectra are useful in the study of isomerism. For example, the amino acid alanine $CH_3CH(NH_2)COOH$ forms *fac* and *mer* tris complexes with Co(III), and each form can have Λ or Δ isomers. The problem of identifying the arrangement of ligands is first to decide between the *fac* and *mer* isomers, and then between their Λ and Δ optical isomers. The two geometrical isomers have different absorption spectra (Fig. 14.13(a)) because the immediate symmetry of ligands about the Co(III) center differs. In the *mer* isomer, there is no axis of rotational symmetry; the *fac* isomer has a C_3 axis passing through the faces surrounded by N and O atoms, respectively. The lower symmetry of the *mer* isomer is likely to lead to a somewhat broader absorption band. Hence a plausible assignment is that the spectrum indicated in Fig. 14.13(a) is that of the *mer* isomer. The corresponding CD spectra are shown in Fig. 14.13(b). The two pairs of curves with peaks of opposite sign clearly indicate which isomers are enantiomers of the same geometrical isomer. Furthermore, we can assign the Δ and Λ absolute configuration labels by comparing these spectra with the CD spectra of $[Co(en)_3]^{3+}$, for its absolute configuration is known from X-ray diffraction.

Example 14.6: *Using CD spectra to assign a configuration*

The enantiomer commonly labeled $(-)546$ $[Co(edta)]^-$ because it rotates (to the left) the plane of linearly polarized 546 nm light has the CD spectrum shown in Fig. 14.14. Assign its absolute configuration by comparison with $[Co(en)_3]^{3+}$.

Answer. The complex is similar to the alanine complex with one more O atom bound to Co. Therefore it seems reasonable to continue to correlate the spectra with the two d–d transitions of $[Co(en)_3]^{3+}$. The lower energy band has a positive sign followed by a weak negative peak as energy increases. The higher energy absorption band corresponds to weak CD absorption with a larger positive than negative area. These features match the Λ absolute configuration of $[Co(en)_3]^{3+}$.

Exercise E14.6. What evidence do you find in Fig. 14.14 that $[Co(EDTA)]^-$ has lower symmetry at Co than $[Co(en)_3]^{3+}$?

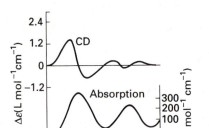

14.14 The absorption and CD spectra of $(-)546$ $[Co(edta)]^-$ referred to in Example 14.6.

14.9 Electron paramagnetic resonance

The electronic analog of nuclear magnetic resonance is called **electron paramagnetic resonance** (EPR) and is a powerful technique for studying complexes with unpaired electrons (those in other than singlet states). EPR can be used to map the distribution of an unpaired electron in a molecule and help to decide the extent to which electrons are delocalized over the ligands. The technique can also provide some information about the energy levels of complexes.

The g-value

Electron paramagnetic resonance relies on the fact that an electron spin can adopt two orientations along the direction defined by an applied magnetic field B. The energy difference between the states $m_s = +\frac{1}{2}$ and $-\frac{1}{2}$ is

$$\Delta E = g\mu_B B$$

where μ_B is the Bohr magneton and g is a constant characteristic of the complex. For a free electron, g has the 'free-spin' value $g_e = 2.0023$. If the sample is exposed to electromagnetic radiation of frequency v, a strong absorption—a resonance—occurs when the magnetic field satisfies the condition

$$hv = g\mu_B B$$

In a typical spectrometer operating near 0.3 T (where T denotes the unit tesla), resonance occurs in the 3 cm band of the microwave region of the electromagnetic spectrum (at about 9 GHz).

The g-factor in a complex differs from g_e by an amount that depends on the ability of the applied field to induce local magnetic fields. We can think of $(g/g_e)B$ as a modification of the applied field that takes into account any locally induced fields. If $g > g_e$, the local field is larger than that applied; if $g < g_e$, it is smaller. The sign and magnitude of the induced local fields depends on the separation of the energy levels of the complex. The closer they are together, the easier it is for the applied field to induce orbital circulation of the electrons and hence produce a local magnetic field.

The value of g can be determined by noting the applied field needed to achieve resonant absorption for a given microwave frequency. If the complex is part of a single crystal, then g may be measured along different directions and hence used to infer the symmetry of the complex. For example, the tetragonally distorted D_{4h} $[Cu(OH_2)_6]^{2+}$ complex has $g = 2.08$ in the direction parallel to the C_4 axis and $g = 2.40$ perpendicular to it. The fact that two g values are measured confirms the existence of the distortion, because an octahedral complex has an isotropic g value (one that is the same in all directions). The two numerical values of g give information about the energy levels of the complex.

Hyperfine structure

Magnetic nuclei—those with nonzero spin—give rise to an additional magnetic field experienced by an electron in a complex. Because a given magnetic nucleus of spin quantum number I can adopt $2I + 1$ different orientations, it can give rise to $2I + 1$ different contributions to the local field and hence result in the resonance condition being fulfilled at $2I + 1$ different values of the applied field. Hence, EPR spectra are often split into multiplets called their **hyperfine structure**.

The SiH_3 radical with three equivalent protons ($I = \frac{1}{2}$) is expected to exhibit four lines with an intensity pattern $1:3:3:1$ as a result of the coupling of the electron spin to the three protons. A further splitting occurs in the 4.7 percent of radicals that have a ^{29}Si nucleus ($I = \frac{1}{2}$): the four lines are split into doublets by this interaction.

A classic study of the delocalization of electrons on to ligands is the work on *bis*(salicylaldehyde-imine)copper(II) with ^{63}Cu (**9**).[7] There

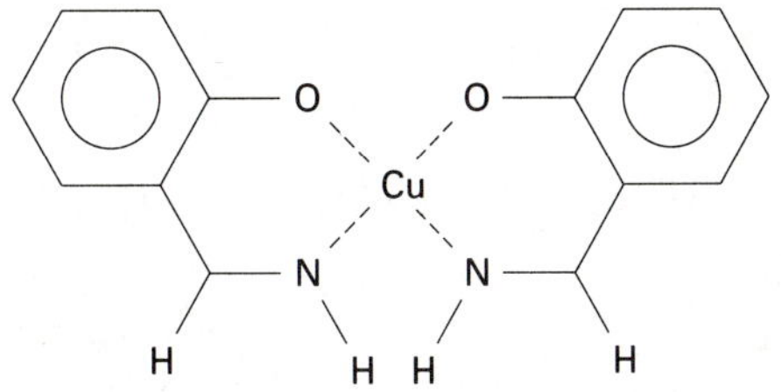

9 Bis(salicylaldehyde–imine)copper(II)

[7] See A. H. Maki and B. R. McGarvey, Electron spin resonance in transition metal chelates. II. Copper(II) bis-salicylaldehyde-imine. *J. Chem. Phys.*, **29**, 35 (1958).

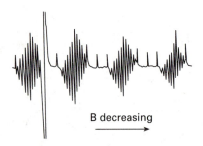

B decreasing

14.15 The EPR spectrum of copper(II) bis(salicylaldehyde-imine) diluted 1:200 with the isomorphous nickel(II) chelate. The copper is isotopically pure copper–63. The strong absorption at applied field (on the left) is a radical marker (DPPH) used to calibrate the spectrum. As is common in EPR, the spectrum is displayed as the first derivative of the absorption with respect to applied field, so each line has a shape like that of the strong DPPH absorption. (A. H. Maki and B. R. McGarvey, *J. Chem. Phys.*, **29**, 35 (1958).)

are four main signals in the EPR spectrum, and each one is split into eleven lines (Fig. 14.15). Substitution of H atoms of the NH group by deuterium (for which $I=1$) does not alter the spectrum whereas replacement of H of the neighboring CH group by CH_3 groups (the spin of ^{12}C is $I=0$) results in the collapse of each of the four groups into lines with only five components. The original $4 \times 11 = 44$ hyperfine lines correspond to coupling to ^{63}Cu ($I=\frac{3}{2}$, to give four lines of equal intensity) and then to two equivalent ^{14}N nuclei ($I=1$, jointly giving five lines) and two equivalent protons (jointly giving three lines). The fact that only eleven lines are seen whereas this analysis requires $5 \times 3 = 15$ is a consequence of partial overlap of the lines. When the proton coupling is removed by substitution with CH_3 groups, only the five lines of the ^{14}N splitting survive. The results indicate that electron delocalization is important to the H of C—H but not as far as the H of N—H.

The hyperfine splitting constant

We can write the $2I+1$ resonance conditions responsible for the hyperfine structure as

$$h\nu = g\mu_B(B + Am_I)$$

where A is the **hyperfine splitting constant**. This constant is a characteristic of the nucleus itself (its magnetic moment) and the probability that an unpaired electron will be found near it. By measuring the value of A, we can deduce the probability that an unpaired electron occupies an orbital on that atom and hence build up a description of the molecular orbital it occupies.

The value of A for either the magnetic nuclei of the central nucleus or of the ligands is a very good indication of the degree to which electrons are delocalized from the central metal atom to the surrounding ligands. For instance, for a Cu(II) complex, the hyperfine splitting of the Cu nucleus varies as follows:

Ligand	F^-	H_2O	O^{2-}	S^{2-}
A/mT	9.8	9.8	8.5	6.9

These values show that as the energy of the ligand orbitals approach that of the metal and their interaction increases, the orbital occupied by the unpaired electron becomes more ligand-like and interacts more weakly with the metal nucleus.

BONDING AND SPECTRA OF SIMPLE CLUSTERS

In this section we discuss the molecular orbitals of some complexes that we meet again in later chapters, particularly when we discuss the *d*-block organometallic compounds (Chapter 16). In particular, we deal here with complexes of lower symmetry than those considered in Chapter 6. We shall show that one very useful technique for constructing molecular orbitals of complicated molecules is to build them up from the valence orbitals of subunits. The technique is especially useful for organometallic molecules because they frequently consist of ML$_n$ subunits in which L is a σ-donor, π-acceptor ligand bound to an organic molecule.

14.10 The ML$_5$ fragment

We shall illustrate the technique by deriving the molecular orbitals for the fragment ML$_5$ from the octahedral molecular orbital diagram. Once we have established the orbitals of this C_{4v} unit, we can analyze some aspects of the spectra of C_{4v} and D_{4h} complexes. We shall also be able to use the fragment orbitals to construct an orbital diagram of an M$_2$L$_{10}$ binuclear metal–metal bonded complex.

ML$_5$ molecular orbitals

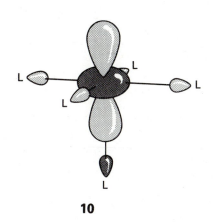

10

The HOMO and LUMO of an ML$_6$ octahedral complex are shown on the left of Fig. 14.16. Suppose we remove one z-axis ligand to give ML$_5$. There is little effect on the octahedral t_{2g} levels, for they are not σ with respect to the z axis. The $d_{x^2-y^2}$ orbital, which becomes b_1, is largely unaffected by removal of the ligand. The principal change occurs to the d_{z^2} orbital, which becomes of a_1 symmetry, and falls sharply in energy as it loses its antibonding character toward the departing ligand (**10**). As d_{z^2} becomes a_1, it can mix with s and p_z, which are also a_1 in C_{4v} symmetry. We can therefore expect the resulting hybrid orbital to have a large amplitude in the vacant coordination position, which makes it an ideal orbital for acting as a σ acceptor to a new donor ligand.

In C_{4v} symmetry the t_{2g} orbitals are split into a doubly degenerate pair (e) consisting of d_{zx} and d_{yz}, and one other orbital of b_2 symmetry, which is the d_{xy} orbital. The splitting will be smaller than that between the d_{z^2} and $d_{x^2-y^2}$ orbitals because the three orbitals are less directly affected by the ligand that is being removed. However, if the d_{zx} and d_{yz} orbitals are involved in π bonding with the departing ligand there may be an additional effect. Thus, if the ligand is a π donor (like Cl$^-$), then the antibonding effect will be reduced and the new e pair will lie below the b_2 orbital (as we have shown in Fig. 14.16). On the other hand, if the departing ligand is a π acceptor (like CO), then its removal will lead to an increase in the energy of the e orbitals above the b_2 orbital.

Square-pyramidal complexes

We need one more detail before we can use the orbital diagram of the ML$_5$ fragment to interpret the spectra of five-coordinate complexes. A square pyramidal five-coordinate complex can have its central M atom above the plane of the four ligands of the pyramid's base. Figure 14.17 shows an example of the energy variations of the HOMO and LUMO as the basal ligands bend away from the metal and the angle ϕ increases from 90° to 120°. This increase in angle reduces the overlap of the ligands with the d_{z^2} and $d_{x^2-y^2}$ orbitals and the net

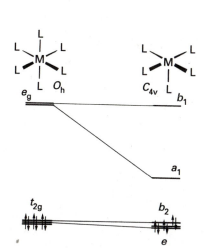

14.16 An orbital correlation diagram for the conversion of an ML$_6$ octahedral complex an ML$_5$ fragment. Only the frontier orbitals are shown.

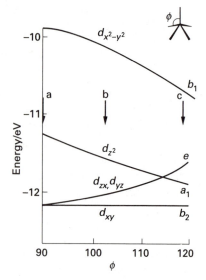

14.17 The correlation diagram for the metal *d* orbitals in an ML$_5$ fragment as the basal ligands bend away from the metal and the angle ϕ increases from 90° to 120°. The arrows show the location of (a) low-spin d^6 oxyhemoglobin, (b) low-spin d^8 [Ni(CN)$_5$]$^{3-}$, and (c) a d^{10} tetradentate macrocyclic complex of Cu(I) with an axial CO ligand.

antibonding character is reduced. At the same time, σ overlap with the d_{zx} and d_{yz} orbitals increases and they become more antibonding.

Figure 14.17 can be used to correlate the structures of square-pyramidal complexes with their electron counts. A low-spin d^6 complex should be expected to be a relatively flat pyramid because d_{zx} and d_{yz}, which are filled, rise in energy as ϕ increases from 90°. In contrast, a low-spin d^8 pyramid should have ϕ well above 90°, for then d_{z^2} (which is now occupied) is stabilized. In support of this conclusion, the d^8 complex [Ni(CN)$_5$]$^{3-}$ has $\phi = 101°$ whereas the d^6 oxyhemoglobin system has $\phi \approx 90°$.

14.11 Metal–metal bonding and mixed valence complexes

The molecular orbitals of polynuclear complexes of the form M$_2$L$_{10}$, such as Mn$_2$(CO)$_{10}$, can be constructed from the orbitals of the two ML$_5$ fragments. Because a reasonable first approximation is to suppose that the only important interactions are between the frontier orbitals of each complex, we need consider only them.

Metal–metal bonding

Figure 14.18 shows the formation of an M—M bond when the two ML$_5$ frontier orbitals interact to give σ, π, and δ orbitals. There are two quite different cases. If the metal in ML$_5$ has a low oxidation

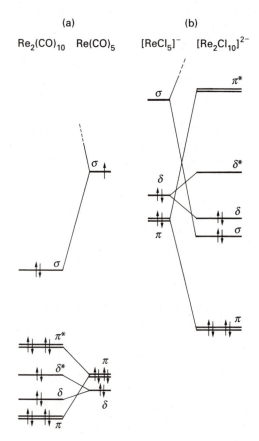

14.18 The correlation of Re$_2$L$_{10}$ with two ReL$_5$ units (a) when the ligands have π-acceptor character (L = CO) and (b) when they have π-donor character (L = Cl$^-$).

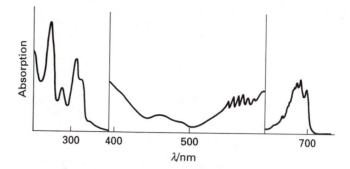

14.19 The absorption spectrum of [Bu$_4$N]$_2$[Re$_2$Cl$_8$] at 5 K (the outer two sections) and at 15 K (the inner section). Note the strong $\delta^* \leftarrow \delta$ band in the red near 15 000 cm^{-1} (670 nm).

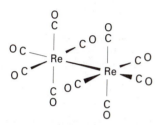

11 [Re$_2$(CO)$_{10}$]

12 [Re$_2$Cl$_{10}$]$^{2-}$

number (and hence is electron rich), as in Fig. 14.18(a), and the ligands are π acceptors, then the π and δ orbitals lie well below the σ orbitals and the complex is stabilized by back-donation from the metal to the ligands. If the ligands are π donors, as in Fig. 14.18(b), then the σ orbital is not so far above the π and δ orbitals. Molecular orbital diagrams for the M—M bond in Re$_2$(CO)$_{10}$ (**11**) and the hypothetical anion [Re$_2$Cl$_{10}$]$^{2-}$ (**12**) are shown[8] in Fig. 14.18, which also shows the electron populations for Re(0) and Re(III). Decacarbonyldirhenium has a net single σ bond; in contrast, the analogous chloro complex is regarded as having a quadruple bond consisting of a σ bond, two π bonds, and a δ bond; the HOMO is a δ bond.

Figure 14.19 shows the spectrum of [Bu$_4$N]$_2$[Re$_2$Cl$_8$], where Bu is the butyl group, and its remarkably strong band in the red near 15 000 cm^{-1} (near 650 nm). This band is too intense for a Laporte forbidden d–d transition and has been ascribed to a $\delta^* \leftarrow \delta$ transition of the metal–metal bond.[9] The δ^* excited state has a long lifetime and initiates several very interesting photochemical processes (Section 15.16).

Mixed-valence complexes

Another type of binuclear complex with interesting spectroscopic properties is a **mixed-valence complex** with identical metal atoms in different oxidation states. One of the earlier and more important examples is the 'Creutz–Taube ion', [(NH$_3$)$_5$Ru–pyz–Ru(NH$_3$)$_5$]$^{5+}$ (**13**, pyz is a bridging pyrazine ligand). The question this formula

13 [Ru$_2$(NH$_3$)$_{10}$(pyz)]$^{5+}$

[8] This hypothetical species is related to the well known quadruply bonded complex [Re$_2$Cl$_8$]$^{2-}$ which was illustrated in structure **26** of Chapter 6.
[9] C. D. Cowman and H. B. Gray, *J. Am. Chem. Soc.*, **95**, 8177 (1973). This is probably the only spectroscopic transition to have been featured in a detective novel, Joseph Wambaugh's *The delta star*. Bantum Books, New York (1974).

Table 14.6 The Robin and Day classification of IT bands

Class I	Class II	Class III
M and M' very different, different ligand fields, etc.	M and M' in similar environments, but not equivalent	M and M' not distinguishable
Orbital 'trapped'	Orbital distinguishable but not fully localized	Orbital delocalized
IT bands at high energy	IT in visible or near IR	Bands in visible or near IR
Electronic spectra of constituent ions seen	Electronic spectra of constituent ions seen but modified	Electronic spectra of constituent ions not distinguishable
Insulator	Semiconductor	IIIA: Usually insulator IIIB: Usually metallic conductor
Magnetic properties of complex	Magnetic properties of isolated complex except at low temperature	IIIA: Magnetic isolated properties of isolated complex IIIB: Ferromagnetic or paramagnetic

poses is whether the two Ru atoms are a localized Ru(II) and a localized Ru(III) or whether the electrons are sufficiently delocalized to give two equivalent Ru atoms in a $+2.5$ oxidation state. The problem can be expressed (but not solved) by writing a wavefunction ψ for Ru(II)–Ru(III) and another wavefunction ψ' for Ru(III)–Ru(II) and expressing the overall delocalized ground state by the wavefunction

$$\Psi = c\psi + c'\psi'$$

The values $c=0$ or $c'=0$ correspond to complete localization and $c^2 = c'^2$ corresponds to complete delocalization.

A classification of mixed valence compounds that reflects the differences in the strength of the interaction has been proposed by M. B. Robin and Peter Day[10]:

Class I: Fully localized electrons
Class II: Intermediate
Class III: Fully delocalized electrons

Class III is subdivided into IIIA for pairs or clusters of metals and IIIB for infinitely delocalized solids. The classification is summarized in Table 14.6.

Class I includes a large number of metal oxide and sulfide solids that seem curious from the standpoint of oxidation numbers. For example, the apparent oxidation state of lead in Pb_3O_4 may be regarded as a mixture of the two common oxidation states Pb(IV) and Pb(II). As outlined in Table 14.6, compounds in this class contain metal ions in ligand fields of different symmetry or strength, and there is definite structural evidence for localized oxidation states (corresponding to $c^2 \gg c'^2$ or vice versa). Spectroscopic and magnetic properties near room

[10]   M. B. Robin and P. Day, Mixed valence chemistry: A survey and classification. *Adv. Inorg. Chem. Radiochem.*, **10**, 248 (1967).

temperature are characteristic of the coordination environment of the individual ions.

Class III includes cases where formation of covalent or metallic metal–metal bonds lead to the delocalization of charge ($c^2 \approx c'^2$). The bronze colored metallic conductor Ag_2F, for example, contains sheets of metal–metal bonded Ag atoms bearing a positive charge, and is a member of class IIIB. In contrast, Ta_6Cl_{15} contains cluster cations of the formula $[Ta_6Cl_{12}]^{3+}$ containing six equivalent Ta atoms. Discrete clusters of this kind belong to class IIIA. The solid is an electrical insulator and does not have a metallic appearance.

Class II includes the compounds for which **intervalence transitions** (IT) are observed in which an electron is excited from the valence shell of one atom into that of the other. Thus, they correspond to metal-to-metal charge transfer transitions. As is the case in the Creutz–Taube ion, intervalence transitions are strong if there is an appropriate bridging ligand between the two centers. If the metal ions do not share a bridging ligand, the IT band may be too weak to be observed. The classic example of this type of compound is the pigment Prussian blue. Its structure is a cubic lattice with Fe^{2+} octahedrally coordinated by the C atom of the CN^- ligands and Fe^{3+} octahedrally coordinated by the N atom of the ligand. The bridging cyanide provides the conduit for the strong IT transition and the resulting blue color. The magnetic properties of Class II compounds at room temperature give little indication of differences from localized complexes but magnetic interactions between sites become evident at low temperatures.

Example 14.7: *Assigning a compound to a Robin–Day class*
At room temperature magnetite (Fe_3O_4) has half the Fe(III) in tetrahedral sites and the rest in half the octahedral sites; the Fe(II) also resides in the octahedral sites. At this temperature the compound is a metallic conductor. What is its probable Robin–Day class?

Answer. The occurrence of both Fe(II) and Fe(III) in the octahedral sites and the metallic conductivity suggest a delocalized system, Robin–Day Type IIIB.

Exercise E14.7. The compound Mn_3O_4 contains Mn(III) in octahedral sites and Mn(II) in tetrahedral sites. What are the possible Robin–Day classifications? What additional information would help to decide between these possibilities?

FURTHER READING

A. B. P. Lever, *Inorganic electronic spectroscopy.* Elsevier, Amsterdam (1984). A comprehensive, reasonably up to date treatment with a good collection of data.

T. A. Albright, J. K. Burdett, and M. H. Whangbo, *Orbital interactions in chemistry.* Wiley, New York (1985). A good source for the treatment of construction of molecular orbitals by combination of fragments.

B. N. Figgis, in *Comprehensive coordination chemistry* (ed. G. Wilkinson, R. D. Gillard, and J. A. McCleverty), Vol. 1, p. 213. Pergamon, Oxford (1987).

T. J. Meyer, Excited state electron transfer. *Prog. Inorg. Chem.*, **30**, 389 (1983).

KEY POINTS

1. Atomic energy levels
Atomic orbital configurations do not completely specify the states and energy levels of an atom: interelectronic repulsions must be taken into account. Angular momentum quantum numbers aid in grouping microstates that have similar energies.

2. Racah parameters
Interelectronic repulsion effects can be summarized with a small number of parameters called Racah parameters, which are known from atomic spectra.

3. Ligand field transitions
Electronic transitions between the HOMO and LUMO of complexes can often be assigned as ligand field transitions between the d orbitals of the metal atom. A consideration of ligand field stabilization energies and Racah parameters is used to identify the transitions. In complexes of low symmetry, the HOMO and LUMO are not the only orbitals involved in transitions.

4. Charge transfer transitions
The other major source of low energy electronic transitions is charge transfer between the metal atom and the ligands (CT bands).

5. Selection rules
Spectral intensities depend upon selection rules. These rules are based on the conservation of spin of the complex and on considerations of symmetry by using group theory.

6. Spectroscopic techniques
The spectroscopic techniques for investigation of d-block complexes include electronic absorption spectra, luminescence, circular dichroism, Raman scattering, and EPR.

7. Metal–metal bonding
The electronic structure of metal–metal bonded systems may be expressed conceptually by the consideration of the frontier orbitals of fragments that contain a single metal atom.

8. Mixed-valence complexes
Intense intervalence charge transfer absorption bands are found in multinuclear systems with metal atoms in more than one oxidation state.

EXERCISES

14.1 Write the Russell–Saunders term symbols for states with the angular momentum quantum numbers (L,S): (a) $(0,\frac{5}{2})$, (b) $(3,\frac{3}{2})$, (c) $(2,\frac{1}{2})$, (d) $(1,1)$.

14.2 Identify the ground term from each set of terms: (a) ^1G, ^3F, ^3P, ^1P; (b) ^3H, ^3P, ^5D, ^1I, ^1G; (c) ^6S, ^4G, ^4P, ^2I.

14.3 Give the Russell–Saunders terms of the configurations: (a) $4s^1$, (b) $3p^2$. Identify the ground term.

14.4 Identify an atom or an ion corresponding to the configuration $3p^5$ and $4d^9$. What are the ground terms for each?

14.5 Identify the ground terms of the gas phase species B^+, Na, Ti^{2+}, and Ag^+.

14.6 The free gas phase ion V^{3+} has a ^3F ground term. The ^1D and ^3P terms lie respectively $10\,642\ cm^{-1}$ and $12\,920\ cm^{-1}$ above it. The energies of the terms are given in terms of Racah parameters as $E(^3F) = A - 8B$, $E(^3P) = A + 7B$, $E(^1D) = A - 3B + 2C$. Calculate the values of B and C for V^{3+}.

14.7 Write the d orbital configurations and use the Tanabe–Sugano diagrams (Appendix 5) to identify the ground term of (a) low-spin $[Rh(NH_3)_6]^{3+}$, (b) $[Ti(OH_2)_6]^{3+}$, (c) high-spin $[Fe(OH_2)_6]^{3+}$.

14.8 Using the Tanabe–Sugano diagrams in Appendix 5, estimate Δ_O and B for (a) $[Ni(OH_2)_6]^{2+}$ (absorptions at

8500, 15 400, and 26 000 cm^{-1}) and (b) $[Ni(NH_3)_6]^{2+}$ (absorptions at 10 750, 17 500, and 28 200 cm^{-1}).

14.9 If an octahedral Fe(II) complex has a large paramagnetic susceptibility, what is the ground state label according to the Tanabe–Sugano diagram? What is the description of the states involved in the spin-allowed electronic transition?

14.10 The spectrum of $[Co(NH_3)_6]^{3+}$ has a very weak band in the red and two moderate intensity bands in the visible to near-UV. How should these transitions be assigned?

14.11 Explain why $[FeF_6]^{3-}$ is colorless whereas $[CoF_6]^{3-}$ is colored but exhibits only a single band in the visible.

14.12 The Racah parameter B is 460 cm^{-1} in $[Co(CN)_6]^{3-}$ and 615 cm^{-1} in $[Co(NH_3)_6]^{3+}$. Consider the nature of bonding with the two ligands and explain the difference in nephelauxetic effect.

14.13 An approximately 'octahedral' complex of Co(III) with ammine and chloro ligands gives two bands with ε_{max} between 60 and 80 L mol^{-1} cm^{-1}, one weak peak with $\varepsilon_{max} = 2$ L mol^{-1} cm^{-1} and a strong band at higher energy with $\varepsilon_{max} = 2 \times 10^4$ L mol^{-1} cm^{-1}. What do you suggest for the origins of these transitions?

14.14 Ordinary bottle glass appears nearly colorless when viewed through the wall of the bottle but green when viewed from the end so that the light has a long path through the glass. The color is associated with the presence of Fe^{3+} in the silicate matrix. Describe the transitions.

14.15 $[Cr(OH_2)_6]^{3+}$ ions are pale blue-green but the chromate ion CrO_4^{2-} is an intense yellow. Characterize the origins of the transitions and explain the relative intensities.

14.16 Classify the symmetry type of the d_{z^2} orbital in a tetragonal C_{4v} symmetry complex, such as $[CoCl(NH_3)_5]^+$ where the Cl is on the z axis. (a) Which orbitals will be displaced from their position in the octahedral molecular orbital diagram by π interactions with the lone pairs of the Cl$^-$ ligand? (b) Which orbital will move because the Cl$^-$ ligand is not as strong a σ-donor as NH$_3$? (c) Sketch the qualitative molecular orbital diagram for the C_{4v} complex.

14.17 Consider the molecular orbital diagram for a tetrahedral complex (based on Fig. 6.15) and the relevant d orbital configuration. Show that the purple color of MnO_4^- ions cannot arise from a ligand field transition.

14.18 Given that the wavenumbers of the two transitions in MnO_4^- are 18 500 cm^{-1} and 32 200 cm^{-1}, explain how to estimate Δ_T from an assignment of the two charge transfer transitions, even though Δ_T cannot be observed directly.

14.19 Prussian blue was one of the first synthetic pigments. It has the overall composition $K[Fe_2(CN)_6]$ in which each Fe(II) is surrounded by six carbon coordinated cyanide ligands which bridge to Fe(III) bound to the nitrogen. Assign the intense absorption bands at 15 000 and 25 000 cm^{-1}.

14.20 The compound $Ph_3Sn\text{-}Re(CO)_3(t\text{-butDAB})$ (t-butDAB = R—N=CH—CH=N—R where R is a t-butyl group) has a metal-to-diimine ligand MLCT band as the lowest energy transition in its spectrum. Irradiation of this band gives an $\cdot Re(CO)_3(t\text{-butDAB})$ radical as a photochemical product. The EPR spectrum shows extensive hyperfine splitting by ^{14}N and ^1H. The radical is assigned as a complex of the DAB radical anion with Re(I). Explain the argument (see D. J. Stufkens, *Coord. Chem. Rev.*, **104**, 39 (1990)).

PROBLEMS

14.1 Consider the trigonal prismatic six-coordinate ML$_6$ complex with D_{3h} symmetry. Use the D_{3h} character table to divide the d orbitals of the metal atom into sets of defined symmetry type. Assume that the ligands are at the same angle relative to the xy plane as in a tetrahedral complex.

14.2 The absorption spectra of the two isomers of $[CoCl_2(en)_2]^+$ are shown in Fig. 14.20. Considering the higher symmetry of the trans isomer, suggest which is *cis* and which is *trans*. Supposing that this assignment is correct, predict the CD spectrum of the Λ isomer of *cis*-$[CoCl_2(en)_2]^+$ by comparison with $[Co(en)_3]^{3+}$ shown in Fig. 14.20.

14.3 Vanadium(IV) species that have the V=O group have quite distinct spectra. What is the d-electron configuration of V(IV)? The most symmetrical of such complexes are

VOL$_5$ with C_{4v} symmetry with the O atom on the z axis. What are the symmetry species of the five d orbitals in VOL$_5$ complexes? How many d–d bands are expected in the spectra of these complexes? A band near 24 000 cm^{-1} in these complexes shows vibrational progressions of the V=O vibration, implicating an orbital involving V=O bonding. Which d–d transition is a candidate? (See C. J. Ballhausen and H. B. Gray, *Inorg. Chem.*, **1**, 111 (1962).)

14.4 Fig. 14.21 shows the EPR spectra of the anion produced by reduction of the square planar iron tetraphenylporphyrin [Fe(TPP)]$^-$, of labeled [^{57}Fe(TPP)]$^-$, and of [Fe(TPP)]$^-$ in the presence of excess pyridine. Account for the form of the observed spectra. (G. S. Srivatsa, D. T. Sawyer, N. J. Boldt, and D. F. Bocian, *Inorg. Chem.*, **24**, 2123 (1985).)

(a)

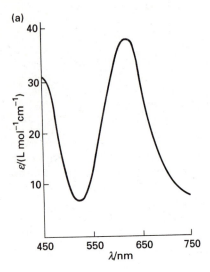

(b)

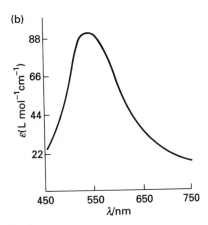

14.20 (a) The absorption and (b) the CD spectra of two isomers of $[CoCl_2(en)_2]^+$.

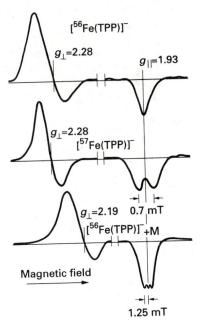

14.21 The EPR spectra of the anion produced by reduction of the square-planar iron tetraphenylporphyrin $[Fe(TPP)]^-$, of labeled $[^{57}Fe(TPP)]^-$, and of $[Fe(TPP)]^-$ in the presence of excess pyridine.

(a)

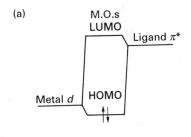

(b)

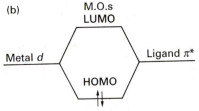

14.22 Representation of the orbitals involved in MLCT transitions for cases in which the energy of the ligand π^* orbital varies with respect to the energy of the metal d orbital. See Problem 14.7.

14.5 The complex $[Cr(CN)_6]^-$ has absorption bands at 264 nm, 310 nm, and 378 nm. Evaluate Δ_O and B using a Tanabe–Sugano diagram. First decide how to assign the observed transitions, noting that the ligand has empty π^* orbitals. The complex also luminesces at 725 nm. Is this luminescence fluorescence or phosphorescence?

14.6 The M(II) oxidation state is not common among the lanthanides but a 'normal' M(II) chemistry does exist for Sm^{2+}, Eu^{2+}, and Yb^{2+}. Write the f-electron configurations for these species. By comparison with Table 14.3, identify the ground terms.

14.7 It was remarked that MLCT bands can be recognized by the fact that the energy is a sensitive function of the polarity of the solvent (because the excited state is more polar than the ground state). Two simplified molecular orbital diagrams are shown in Fig. 14.22. In (a) is a case

with a ligand π level higher than the metal d orbital. In (b) is a case in which the metal d orbital and the ligand level are at the same energy. Which of the two MLCT bands should be more solvent sensitive? These two cases are realized by $[W(CO)_4(phen)]$ and $[W(CO)_4(iPr-DAB)]$, where $DAB = 1,4$-diaza-1,3-butadiene, respectively. (See P. C.

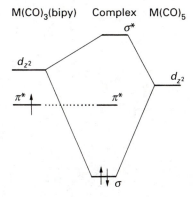

M(CO)$_3$(bipy)   Complex   M(CO)$_5$

14.23 Simplified MO diagram for a complex with an M—M bond formed from the combination of a fragment M(CO)$_5$ with a fragment with a π acceptor ligand, M'(CO)$_3$(bipy). See Problem 14.8.

Servas, H. K. van Dijk, T. L. Snoeck, D. J. Stufkens, and A. Oskam, *Inorg. Chem.*, **24**, 4494 (1985).) Comment on the CT character of the transition as a function of the extent of back donation by the metal.

14.8 Figure 14.23 shows a simplified scheme for the interaction of a fragment M(CO)$_5$ with M'(CO)$_3$(bipy). The M—M bond is simplified to the interaction of the d_{z^2} orbital of M and M'. The spectrum of [(CO)$_5$ReRe(CO)$_3$(phen)] has a strong band in the visible (at 18 940 cm^{-1}) and a second of similar intensity in the near ultraviolet (at 28 570 cm^{-1}). Propose assignments. (See D. J. Stufkens, *Coord. Chem. Revs.*, **104**, 39 (1990).)

14.9 Predict the number of lines in the EPR spectrum of bis(benzene)vanadium(0) given that the abundance of the isotope ^{51}V ($I = \frac{7}{2}$) is 99.8 percent. Comment on the interpretation of the spectrum given in S. M. Mattar and G. A. Ozin, *J. Chem. Phys.*, **90**, 1037 (1986).

15

Reaction mechanisms of *d*-block complexes

We encountered the distinction between labile and inert complexes in Chapter 6, and in Chapter 7 we saw some of the factors that influence the rates of electron transfer. Since then, many aspects of reactivity have been interpreted in terms of reaction mechanisms. Now we look in more detail at the evidence and experiments that are used in the analysis of reaction pathways and so develop a deeper understanding of the mechanisms of the reactions of d-block metal complexes. Because the mechanism is rarely known finally and completely, the nature of the evidence for a mechanism should always be kept in mind in order to recognize what other possibilities might also be consistent with it. In the first part of this chapter we describe how reaction mechanisms are classified, and distinguish between the steps by which the reaction takes place and the details of the formation of the activated complex. Then these concepts are used to describe the currently accepted mechanisms for the substitution reactions of complexes, their redox reactions, and the reactions of photoexcited states of the kind mentioned in Section 14.6.

We have seen the central importance of the interplay of equilibrium and kinetics for determining the outcome of inorganic reactions. It is often helpful to understand the mechanisms of the reactions in cases where kinetic considerations are dominant, and in this chapter we see how to achieve at least partial understanding for *d*-block complexes. The first two classes of reactions to be closely analyzed were ligand substitution and redox reactions, and we concentrate mainly on them. As we shall see, the quantitative understanding of the rates of inorganic reactions is far from secure, and in most cases all that it is possible to do is to distinguish reasons for differences in the order of magnitude of rate constants.

LIGAND SUBSTITUTION REACTIONS

A **ligand substitution reaction** is a reaction in which one Lewis base displaces another from a Lewis acid:

$$Y + M—X \rightarrow M—Y + X$$

This class of reactions includes complex formation reactions, in which the **leaving group**, the displaced base X, is a solvent molecule and the **entering group**, the displacing base Y, is some other ligand. An example is

$$[Co(OH_2)_6]^{2+} + Cl^- \rightarrow [CoCl(OH_2)_5]^+ + H_2O$$

The majority of the reactions we consider in this chapter take place in water, and we shall not write the phase explicitly in equations.

We saw in Section 6.7 that the equilibrium constants of displacement reactions can be used to rank ligands in order of their strength as Lewis bases. However, a different order may be found if bases are ranked according to the *rates* at which they react. Therefore, for kinetic considerations we replace the equilibrium concept of basicity by the kinetic concept of **nucleophilicity** (from the Greek for 'nucleus loving'). By nucleophilicity we mean the ratio of the rate of attack on a complex by a given base Y to the rate of attack by a standard base $Y^{\ominus}$. The shift from equilibrium to kinetic considerations is emphasized by referring to the displacements as **nucleophilic substitutions**.

We have already seen (in Section 6.8) that the rates of substitution reactions span a very wide range, and that to some extent there is a pattern in their behavior. For instance, aqua complexes of Group 1, 2, and 12 metal ions, lanthanide ions, and some $3d$-metal ions, particularly those in low oxidation states, have half-lives as short as nanoseconds. In contrast, half-lives are years long for complexes of heavier d-metals in high oxidation states, such as those of Ir(III) or Pt(IV). Intermediate half-lives are also observed. Two examples are $[Ni(OH_2)_6]^{2+}$, which has a half-life of the order of milliseconds, and $[Co(NH_3)_5(OH_2)]^{3+}$, in which H_2O survives for several minutes as a ligand before it is replaced by a stronger base. Table 15.1 puts the time scales of reactions into perspective.

15.1 The classification of mechanisms

One kinetic aspect of a reaction is its **stoichiometric mechanism**, the sequence of elementary steps by which the reaction takes place. A second kinetic aspect is its **intimate mechanism**, the details of the activation process and the energetics of formation of an activated complex in the rate determining step. The distinction reflects the two types of experiments that are available for studying mechanisms.

Stoichiometric mechanisms

The first stage in the kinetic analysis of a reaction is to study how its rate changes as the concentrations of reactants are varied. This type

Table 15.1 Representative time-scales of chemical and physical processes

Time-scale*	Process	Example
10^2 s	Ligand exchange (inert complex)	$[Cr(OH_2)_6]^{3+} - H_2O$ (*ca.* 10^8 s)
60 s	Ligand exchange (inert complex)	$[V(OH_2)_6]^{3+} - H_2O$ (50 s)
1 ms	Ligand exchange (labile complex)	$[Pt(OH_2)_4]^{2+} - H_2O$ (0.4 ms)
1 μs	Intervalence charge transfer	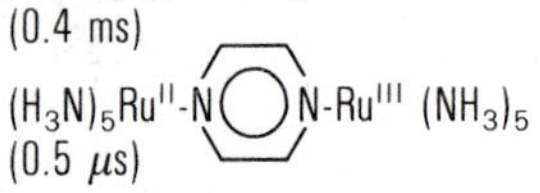 $(H_3N)_5Ru^{II}-N\bigcirc N-Ru^{III}$ $(NH_3)_5$ (0.5 μs)
1 ns	Hydrogen bond rearrangement	C_4H_9OH (l) (0.4 ns)
1 ps	Rotation time in liquid	CH_3CN (10 ps)
1 fs	Molecular vibration	Sn–Cl stretch (300 fs)

*Approximate time at room temperature.

of investigation leads to the identification of **rate laws**, the differential equations governing the rate of change of reactant or product concentration. For example, the observation that the rate of formation of $[Ni(OH_2)_5NH_3]^{2+}$ from $[Ni(OH_2)_6]^{2+}$ in the presence of excess ammonia is proportional to the concentration of NH_3 and to the concentration of $[Ni(OH_2)_6]^{2+}$ implies that the reaction is first-order in these two reactants, and that the overall rate law is

$$\text{rate} = k[\{Ni(OH_2)_6\}^{2+}][NH_3]$$

It is a fundamental assumption in kinetics that the species appearing in a rate law contribute to the formation or reaction of the activated complex in the rate-determining stage, and therefore that an acceptable mechanism must accommodate this contribution. Therefore, together with stereochemical and isotope tracer studies, the establishment and interpretation of the rate law is the route to the elucidation of the stoichiometric mechanism of the reaction.

Three types of stoichiometric mechanism for ligand substitutions have been recognized:

Dissociative *D*   Interchange *I*   Associative *A*

The first and third are two-step mechanisms in which an intermediate is formed and then transformed into the product.

A **dissociative mechanism** involves a step in which an intermediate of reduced coordination number is formed after the departure of the leaving group. Thus, it is proposed that the substitution of hexacarbonyltungsten (a d^6, 18-electron complex) by a phosphane takes place by the following dissociative mechanism:

$$W(CO)_6 \rightarrow W(CO)_5 + CO$$

$$W(CO)_5 + PPh_3 \rightarrow W(CO)_5PPh_3$$

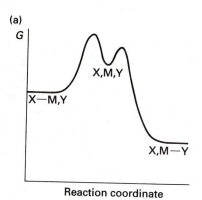

(a)

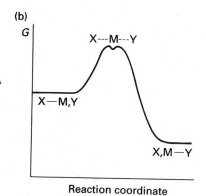

(b)

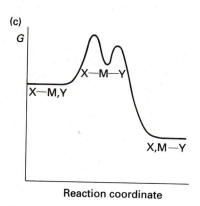

(c)

15.1 Reaction profiles for (a) dissociative *D*, (b) interchange *I*, and (c) associative mechanisms *A*. A true intermediate exists in (a) and (c) but not in (b).

Under the conditions in which this reaction is usually performed in the laboratory, the intermediate $W(CO)_5$ is rapidly captured by an ether solvent, such as THF, to form $[W(CO)_5(THF)]$, which in turn is converted to the phosphane product, presumably by a second dissociative process. The generalized reaction profile is shown in Fig. 15.1(a).

An **associative mechanism** (Fig. 15.1(c)) involves a step in which an intermediate is formed with a higher coordination number than the original complex. This mechanism is suspected for many reactions of square-planar Pt(II), Pd(II), and Ir(I) complexes. The exchange of $^{14}CN^-$ with the ligands in the square-planar complex $[Ni(CN)_4]^{2-}$ appears to be associative, and in the reaction the coordination number of Ni^{2+} is increased to form a transient species:

$$[Ni(CN)_4]^{2-} + {}^{14}CN^- \rightarrow [Ni(CN)_4({}^{14}CN)]^{3-}$$

$$[Ni(CN)_4({}^{14}CN)]^{3-} \rightarrow [Ni(CN)_3({}^{14}CN)]^{2-} + CN^-$$

The radioactivity of carbon-14 provides a means of monitoring this reaction. In the presence of excess CN^- ions, the proposed intermediate $[Ni(CN)_5]^{3-}$ can be isolated from solution and it can also be detected spectroscopically in solution.

An **interchange mechanism**, for which the reaction profile is given in Fig. 15.1(b), takes place in one step. In an interchange mechanism, the leaving and entering groups exchange in a single step by forming an activated complex but not a true intermediate. The interchange mechanism is illustrated by reactions of the six-coordinate d^8 complex $[Ni(OH_2)_6]^{2+}$.

Intimate mechanisms

We saw in Section 6.8 that it is useful to distinguish two types of intimate mechanism (the details of the formation of the activated complex), especially when dealing with an interchange mechanism. If the rate constant for the formation of the activated complex depends strongly on the identity of the entering group Y, then we can infer that the activated complex must involve significant bonding to Y. If the rate constant is little influenced by the identity of Y, then the value of the rate constant must be controlled largely by the rate at which the bond between the metal atom and the leaving group X can break. The distinction between reactions with rate constants that do not depend strongly on the nature of the entering group and those that do depend strongly on Y leads to the classification of reactions according to their intimate mechanisms. The activation process falls somewhere between the extremes called **associative** (*a*) and **dissociative** (*d*).

An **associative reaction** is identified by observing a rate constant that is sensitive to the entering group. Examples are found among reactions of the d^8 square-planar complexes of Pt(II), Pd(II), and Au(III), including

$$[PtCl(dien)]^+ + X^- \rightarrow [PtX(dien)]^+ + Cl^-$$

Table 15.2 Relation between intimate and stoichiometric mechanisms of ligand substitution

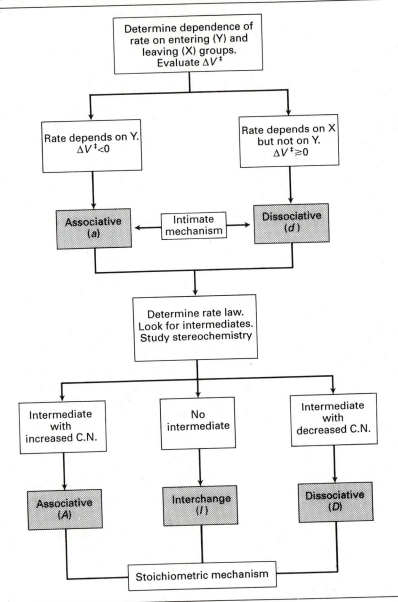

(where dien is diethylene triamine, $NH_2CH_2CH_2NHCH_2CH_2NH_2$), in which changing X^- from I^- to Br^- decreases the rate constant by an order of magnitude. A **dissociative reaction** is identified by observing a rate constant that is insensitive to changes in the entering group. This category includes some of the classic examples of ligand substitution in octahedral d-metal complexes, including

$$[Ni(OH_2)_6]^{2+} + L \rightarrow [Ni(OH_2)_5L]^{2+} + H_2O$$

Changing L from NH_3 to pyridine in this reaction changes the rate by at most a few percent.

Table 15.2 summarizes the relation between intimate and stoichiometric mechanisms. Dissociative and associative reactions (*d* and *a*) correspond approximately to the S_N1 and S_N2 mechanisms of nucleophilic substitution in organic chemistry.[1] However, *d* and *a* refer specifically to the *intimate* mechanism of the activation process, and for many substitution reactions only the intimate mechanism is known. The composite labels A, I_a, I_d, and D are used if both the stoichiometric and intimate mechanisms are known.

SUBSTITUTION IN SQUARE-PLANAR COMPLEXES

We saw in Section 6.8 that the substitution reactions of nonlabile d^8 complexes of Pt(II), Pd(II) and Au(III) are associative reactions (*a*). We emphasized the effect of the entering group and saw that the rate constants for substitution reactions lie in the order

$$Y: H_2O < Cl^- < I^- < H^- < PR_3 < CO, CN^-$$

It is quite easy to summarize the dependence of the rate on the leaving group too, because all we need do is to reverse the order:

$$X: CN^-, CO < PR_3 < H^- < I^- < Cl^- < H_2O$$

Thus, good entering groups (good nucleophiles) are usually poor (nonlabile) leaving groups.

In this section we shall also consider the role of a **spectator ligand**, a ligand that is present in the complex but not lost in the reaction. A general feature is that

> Most good nucleophiles are good reaction accelerators if they occupy a *trans* position (T) as a spectator ligand.

Thus X, Y, and T play related roles, and knowing how a ligand affects the rate when it is Y usually allows us to predict how it affects the rate when it is X or T.

15.2 Intimate mechanisms

The bulk of the evidence on the energetics of the activated complex involved in the substitution reactions of square-planar complexes support the view that their intimate mechanism is associative (*a*).

Entering group effects

If a reaction of the form

$$[PtCl(dien)]^+ + I^- \rightarrow [PtI(dien)]^+ + Cl^-$$

is first-order in the complex and independent of I^-, then the rate of reaction will be equal to $k_1[PtCl(dien)^+]$. On the other hand, if there were a pathway in which the rate law is overall second-order, first-order in the complex and first-order in the incoming group, then the

---

[1] A recent report of the IUPAC commission on physical organic chemistry has endorsed a system of nomenclature for organic reaction mechanisms that is readily mapped on to the *A, I, D* scheme. See R. D. Guthrie and W. P. Jencks, *Acc. Chem. Res.*, **22**, 343 (1989).

rate would be given by $k_2[\text{PtCl(dien)}^+][\text{I}^-]$. If both reaction pathways occurred at comparable rates, the rate law would have the form

$$\text{rate} = (k_1 + k_2[\text{I}^-])[\{\text{PtCl(dien)}\}^+]$$

where k_1 and k_2 are respectively first- and second-order rate constants. Experimental evidence for this rate law comes from a study of substitution in the presence of excess I^- ions, when the rate law is pseudofirst-order with effective rate constant

$$k_{\text{obs}} = k_1 + k_2[\text{I}^-]$$

When the observed pseudofirst-order rate constant is plotted against $[\text{I}^-]$, the slope of the straight line gives k_2 and the intercept gives k_1.

We shall deal first with the path that leads to second-order kinetics. To do so, we note that we can express the reactivity of the entering group Y (for instance, I^- in the reaction above) in terms of a **nucleophilicity parameter** n_{Pt}:

Table 15.3 A selection of n_{Pt} values for a range of nucleophiles

$$n_{\text{Pt}}(\text{Y}) = \log \frac{k_2(\text{Y})}{k_2^\circ}$$

or, equivalently,

Nucleophile	Donor atom	n_{Pt}
Cl^-	Cl	3.04
$\text{C}_6\text{H}_5\text{SH}$	S	4.15
CN^-	C	7.00
$(\text{C}_6\text{H}_5)_3\text{P}$	P	8.79
CH_3OH	O	0
I^-	I	5.42
NH_3	N	3.06

$$\log k_2(\text{Y}) = n_{\text{Pt}}(\text{Y}) + \log k_2^\circ$$

where $k_2(\text{Y})$ is the second-order rate constant of the reaction

$$trans\text{-}[\text{PtCl}_2(\text{py})_2] + \text{Y} \rightarrow trans\text{-}[\text{PtClY}(\text{py})_2]^+ + \text{Cl}^-$$

and k_2° is the rate constant for the same reaction with a standard base, methanol. We give some values of n_{Pt} in Table 15.3.

One striking feature of the data in Table 15.3 is that, although the entering groups in the table are all quite simple, the rate constants span nearly nine orders of magnitude. Another feature is that the nucleophilicity of the entering group toward Pt appears to correlate with soft Lewis basicity, with $\text{Cl}^- < \text{I}^-$, $\text{O} < \text{S}$, and $\text{NH}_3 < \text{PR}_3$.

The rates of reaction of platinum complexes show a range of different sensitivities toward changes in the entering group. To take these differences in sensitivity into account in a reaction of the form

$$[\text{L}_3\text{PtX}] + \text{Y} \rightarrow [\text{L}_3\text{PtY}] + \text{X}$$

we write

$$\log k_2(\text{Y}) = S n_{\text{Pt}}(\text{Y}) + C$$

in place of the equation above. The parameter S, which characterizes the sensitivity of the rate constant to the nucleophilicity parameter, is called the **nucleophilic discrimination factor**. We see from Fig. 15.2 that the straight line obtained by plotting $\log k_2(\text{Y})$ against n_{Pt} for reactions of Y with $trans\text{-}[\text{PtCl}_2(\text{PEt}_3)_2]$ is steeper than that for reactions with $cis\text{-}[\text{PtCl}_2(\text{en})]$. Hence, S is larger for the former reaction, which indicates that rate of the reaction is more sensitive to changes in nucleophilicity.

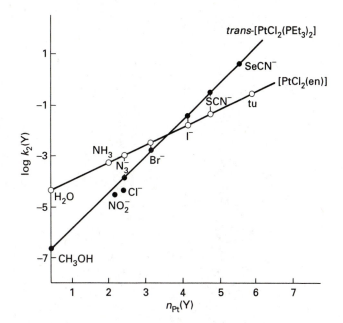

15.2 The slope of the straight line obtained by plotting log k_2 (Y) against the nucleophilicity parameter n_{Pt} (Y) for a series of ligands is a measure of the responsiveness of the complex to the nucleophilicity of the entering group.

Table 15.4 Nucleophilic discrimination factors

	S
trans-[PtCl$_2$(PEt$_3$)$_2$]	1.43
trans-[PtCl$_2$(py)$_2$]	1.00
[PtCl$_2$(en)]	0.64
[PtCl(dien)]$^+$	0.65

Some values of S are given in Table 15.4. Notice that S is close to 1 in all cases, so all the complexes are quite sensitive to n_{Pt}. This sensitivity is what we expect for associative reactions. Another feature to note is that larger values of S are found for complexes of platinum with softer base ligands.

Example 15.1: *Using the nucleophilicity parameter*

The second-order rate constant for the reaction of I$^-$ with *trans*-[PtCl(CH$_3$)(PEt$_3$)$_2$] in methanol at 30 °C is 40 L mol^{-1} s^{-1}. The corresponding reaction with N$_3^-$ has $k_2 = 7.0$ L mol^{-1} s^{-1}. Estimate S and the first-order rate constant for reaction of this complex with methanol (k_2°) given the n_{Pt} values 5.42 and 3.58 for the two nucleophiles.

Answer. Substituting the two values of n_{Pt} into the preceding equation gives

$$1.60 = 5.42S + C \quad \text{(for I}^-\text{)}$$

$$0.85 = 3.58S + C \quad \text{(for N}_3^-\text{)}$$

Solving these two simultaneous equations gives $S = 0.41$ and $C = -0.61$. The value of S is rather small, showing that the discrimination among different nucleophiles is not great. This lack of sensitivity is related to the rather large value of C, which corresponds to the rate constant being large and hence to the complex being reactive. It is commonly found that high reactivity correlates with low selectivity.

Exercise E15.1. Calculate the second-order rate constant for the reaction of the same complex with NO$_2^-$, for which $n_{Pt} = 3.22$.

Table 15.5 The effect of the *trans* ligand in reactions of *trans*-[PtCl(PEt$_3$)$_2$L]

L	k_1/s^{-1}	$k_2/(L\ mol^{-1}\ s^{-1})$
CH$_3^-$	1.7×10^{-4}	6.7×10^{-2}
C$_6$H$_5^-$	3.3×10^{-5}	1.6×10^{-2}
Cl$^-$	1.0×10^{-6}	4.0×10^{-4}
H$^-$	1.8×10^{-2}	4.2
PEt$_3$	1.7×10^{-2}	3.8

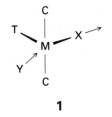

1

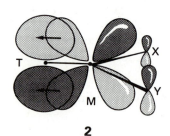

2

The *trans* effect

The ligands T that are *trans* to the leaving group in square planar complexes may influence the rate of substitution. This phenomenon is called the **trans effect**. If T is a strong σ-donor ligand or π-acceptor ligand, then it greatly accelerates substitution of a ligand that lies *trans* to itself (Table 15.5):

For T a σ-donor:

$$OH^- < NH_3 < Cl^- < Br^- < CN^-, CO,$$
$$CH_3 < I^- < SCN^- < PR_3 < H^-$$

For T a π-acceptor:

$$Br^- < I^- < NCS^- < NO_2^- < CN^- < CO, C_2H_4$$

These orders are roughly the order of increasing overlap of the ligand orbitals with either a σ or a π Pt6p orbital. That is, the greater the overlap, the stronger the *trans* effect. It should be noticed that the same factors contribute to a large ligand field splitting.

The significant degree of correlation of the strength of the *trans* effect with the nucleophilicity of the entering group and the nonlability of the leaving group provides a strong clue to the form of the activated complex involved in this associative reaction. If the two *cis* spectator ligands have little influence on the rate of substitution, and if the *trans*, entering, and leaving ligands have complementary influences on the reaction rate, then we can suspect that a trigonal bipyramidal structure is likely for the activated complex. In this model there are two locations of one kind (the axial locations, occupied by the two *cis* ligands) and three of another kind (the equatorial positions, occupied by X, Y, and T).

The process that might occur when Y attacks from in front of the initial complex and X is destined to leave from behind it, is shown in (**1**). (It is interesting to note that the stereochemistry is quite different from that of *p*-block central atoms, such as Si(IV) and P(V), where the leaving group departs from an axial position; see Section 11.9.) It is speculated that the π-acceptor ligands facilitate nucleophilic attack on the metal atom by removal of some d electron density from that atom (**2**).

Example 15.2: *Using the trans effect synthetically*
Use the *trans* effect series to suggest synthetic routes to *cis*- and *trans*-[PtCl$_2$(NH$_3$)$_2$] from [Pt(NH$_3$)$_4$]$^{2+}$ and [PtCl$_4$]$^{2-}$.

Answer. Reaction of [Pt(NH$_3$)$_4$]$^{2+}$ with HCl leads to [PtCl(NH$_3$)$_3$]$^+$. Because the *trans* effect of Cl$^-$ is greater than that of NH$_3$, substitution reactions will occur preferentially *trans* to Cl$^-$, so that the further action of HCl gives *trans*-[PtCl$_2$(NH$_3$)$_2$]. When the starting complex is [PtCl$_4$]$^{2-}$, reaction with NH$_3$ leads first to [PtCl$_3$(NH$_3$)]$^-$. A second step should substitute one of the two mutually *trans* Cl$^-$ ligands with NH$_3$ to give *cis*-[PtCl$_2$(NH$_3$)$_2$].

Exercise E15.2. Given the reactants triphenylphosphane (PPh_3), NH_3, and $[PtCl_4]^-$, propose efficient routes to *cis*- and *trans*-$[PtCl_2(NH_3)(PPh_3)]$.

Steric effects

Steric crowding at the reaction center is usually assumed to inhibit associative reactions (*a*) and to facilitate dissociative reactions (*d*). Bulky groups can block the approach of attacking nucleophiles, but the reduction of coordination number in a dissociative reaction can relieve the crowding in the activated complex. The rate constants for the hydrolysis of *cis*-$[PtClL(PEt_3)_2]$ complexes (the replacement of Cl^- by H_2O) at $25\,°C$ illustrate the point:

L=	pyridine	2-methylpyridine	2,6-dimethylpyridine
k/s^{-1}	8×10^{-2}	2.0×10^{-4}	1.0×10^{-6}

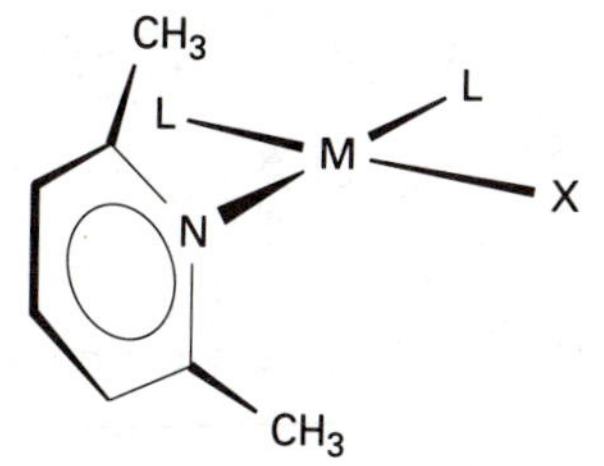

The observation that CH_3 groups adjacent to the N donor atom greatly decrease the rate is explained by their blocking of positions either above or below the plane of the 2-methylpyridine complex and both above and below the plane of the 2,6-dimethylpyridine complex (**3**), and so increasingly hindering attack by H_2O. The effect is smaller if L is *trans* to Cl^-. This difference is explained by the CH_3 groups then being further from the entering and leaving groups in the trigonal bipyramidal activated complex if the pyridyl ligand is in the trigonal plane (**4**).

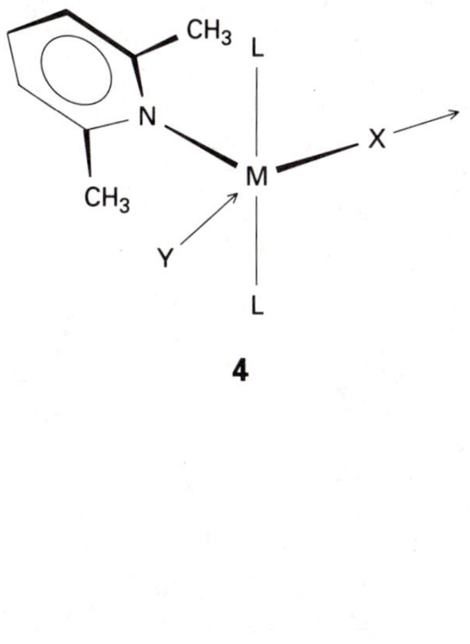

3

4

Stereochemistry

Further insight into the nature of the activated complex is obtained from the observation that substitution of a square-planar complex conserves the original geometry. That is, a *cis* complex gives a *cis* product and a *trans* complex gives a *trans* product. This behavior is readily explained on the assumption that the activated complex is approximately trigonal bipyramidal (**1**) with the entering (Y), leaving (X), and *trans* (T) groups in the trigonal plane. The steric course of the reaction is shown in Fig. 15.3. We can expect a *cis* ligand to

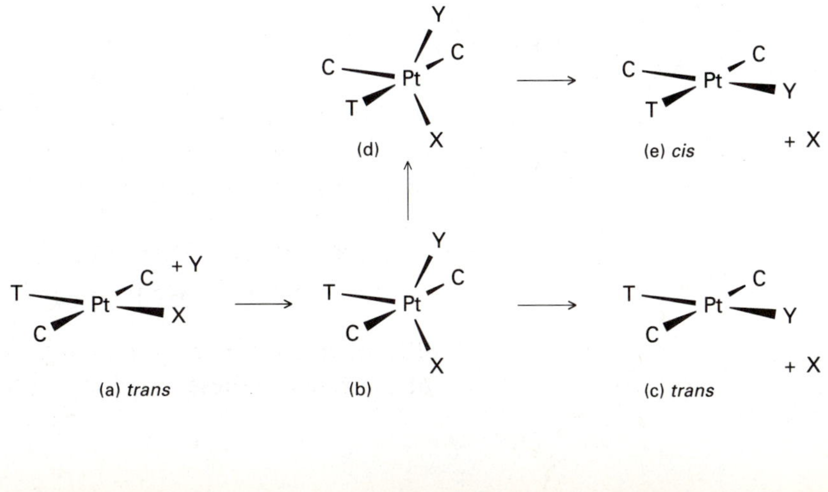

15.3 The stereochemistry of substitution in a square planar complex. The normal path (resulting in retention) is from (a) to (c). However, if the intermediate (b) is sufficiently long lived, it can undergo pseudorotation to (d), which leads to the isomer (e).

Table 15.6 Activation parameters for substitution in square planar complexes (in methanol as solvent)

Reaction	k_1			k_2		
	$\Delta H^{\ddagger}$	$\Delta S^{\ddagger}$	$\Delta V^{\ddagger}$	$\Delta H^{\ddagger}$	$\Delta S^{\ddagger}$	$\Delta V^{\ddagger}$
trans-[PtCl(NO$_2$)(py)$_2$] + py				50	−100	−38
trans-[PtBrP$_2$(mes)]* + SC(NH$_2$)$_2$	71	−84	−46	46	−138	−54
cis-[PtBrP$_2$(mes)]* + I$^-$	84	−59	−67	63	−121	−63
cis-[PtBrP$_2$(mes)]* + SC(NH$_2$)$_2$	79	−71	−71	59	−121	−54
[AuCl(dien)]$^{2+}$ + Br$^-$				54	−17	

*[PtBrP$_2$(mes)] is [PtBr(PEt$_3$)$_2$ (2,4,6-(CH$_3$)$_3$C$_6$H$_2$)].
Enthalpy in kJ mol^{-1}, entropy in J K^{-1} mol^{-1}, and volume in cm^3 mol^{-1}.

exchange places with the T ligand in the trigonal plane only if the intermediate lives long enough to be stereomobile. That is, it must be a long-lived associative (A) intermediate.

Temperature and pressure dependence

The values of the enthalpies, entropies, and volumes of activation for some reactions of Pt(II) and Au(III) complexes are given in Table 15.6. The two striking aspects of the data are the consistently strongly negative values of the entropies of activation and the volumes of activation (the changes in entropy and volume that accompany the formation of the activated complex). The simplest explanation is that the entering ligand is being incorporated into the activated complex without release of the leaving group. That is, we can conclude that the intimate mechanism is associative.

15.3 Stoichiometric mechanisms

The k_1 pathway

The first issue we must address is the first-order pathway in the rate equation and decide whether k_1 in the rate law

$$\text{rate} = (k_1 + k_2[\text{Y}])[\text{PtL}_4]$$

does indeed represent the operation of an entirely different reaction mechanism. It turns out that it does not, for we can account for k_1 by an associative reaction that involves the solvent. Thus, the substitution of Cl$^-$ by pyridine in methanol as solvent proceeds in two steps:

$$[\text{PtCl(dien)}]^+ + \text{CH}_3\text{OH} \rightarrow [\text{Pt(dien)(CH}_3\text{OH)}]^{2+} + \text{Cl}^- \quad \text{(slow)}$$

$$[\text{Pt(dien)(CH}_3\text{OH)}]^{2+} + \text{py} \rightarrow [\text{Pt(dien)(py)}]^{2+} + \text{CH}_3\text{OH} \quad \text{(fast)}$$

The evidence for this two-step mechanism comes from a correlation of the rates of these reactions with the nucleophilicity parameters of the solvent molecules. It has also been demonstrated explicitly that reactions of entering groups with solvent complexes are rapid compared to the step in which the solvent displaces a ligand. Supporting evidence for the similarity of the k_1 and k_2 pathways is found in Table 15.6, which shows that the activation parameters have the same signs and display similar trends.

Associative intermediates

The A and I_a mechanisms are very difficult to distinguish from each other because the composition of the activated complex is the same in both cases. The distinction hinges on whether or not the intermediate persists long enough to be detectable. The dip in the Gibbs free energy surface that distinguishes Fig. 15.1(b) from Fig. 15.1(c) must be deep if the intermediate is to be detectable chemically and hence justify an A rather than an I_a classification. Three types of evidence have been considered.

One type of evidence is the isolation of an intermediate in another related reaction or under different conditions. If an argument by extrapolation to the actual reaction conditions suggests that a moderately long-lived intermediate might exist during the reaction in question, then the A path is indicated. For example, the successful synthesis of the first trigonal bipyramidal Pt(II) complex, $[Pt(SnCl_3)_5]^{3-}$, indicated that a five-coordinate platinum complex may be plausible in substitution reactions of square planar Pt(II) ammine complexes. The fact that $[Ni(CN)_5]^{3-}$ is observed spectroscopically in solution, and that it has been isolated in the crystalline state, thus provides support for the view that it is involved when CN^- exchanges with the square-planar tetracyanonickelate(II) ion.

A second possible indication of the persistence of an intermediate is the observation of a stereochemical change, which would imply that the intermediate has lived long enough to undergo rearrangement (Fig. 15.3). A *cis* → *trans* isomerization is observed in the substitution reactions of certain square-planar phosphane Pt(II) complexes in contrast to the complete retention of original configuration usually observed. This difference implies that the trigonal bipyramid lives long enough for an exchange between the axial and equatorial ligand positions to occur.

Direct spectroscopic detection of the intermediate may be possible if a sufficient amount accumulates. Such direct evidence, however, requires an unusually stable intermediate with favorable spectroscopic characteristics and has not yet been obtained for Pt(II) complexes.

SUBSTITUTION IN OCTAHEDRAL COMPLEXES

The interchange mechanism plays a central role in the discussion of octahedral substitution. The analysis of rate laws for reactions that take place by such a mechanism helps to formulate the precise conditions for distinguishing associative and dissociative activation.

15.4 Rate laws and their interpretation

The rate law of an octahedral substitution by an interchange mechanism is usually consistent with an **Eigen–Wilkins mechanism**. In this mechanism, an encounter complex is formed in a pre-equilibrium step. This complex then rearranges to form the products.

The Eigen–Wilkins mechanism

The first step in the Eigen–Wilkins mechanism is an encounter in which the complex C and the entering group Y diffuse together and come into contact. They may also separate at diffusion-limited rates (that is, at a rate governed by their ability to migrate by diffusion through the solvent), and the equilibrium

$$C + Y \rightleftharpoons \{CY\} \quad K_E = \frac{[\{CY\}]}{[C][Y]}$$

is established. Because in ordinary solvents the lifetime of a diffusional encounter is approximately 1 ns, the formation of an encounter complex can be treated as a pre-equilibrium in all reactions that take longer than a few nanoseconds. A special case arises if Y is the solvent. Then the encounter equilibrium is saturated in the sense that a solvent molecule is always available to take the place of one that leaves the complex because the complex is always surrounded by solvent.

The second step is the rate-determining reaction of the encounter complex to give products:

$$\{CY\} \rightarrow \text{Product} \quad \text{rate} = k[\{CY\}]$$

When the equilibrium equation is solved for the concentration of $\{CY\}$ and the result substituted into the rate law for the rate determining step, the overall rate law is found to be

$$\text{rate} = \frac{kK_E[C][Y]}{1 + K_E[Y]}$$

An example is the reaction

$$[Ni(OH_2)_6]^{2+} + NH_3 \rightleftharpoons \{[Ni(OH_2)_6]^{2+}, NH_3\} \qquad K_E$$

$$\{[Ni(OH_2)_6]^{2+}, NH_3\} \rightarrow [Ni(OH_2)_5 NH_3]^{2+} + H_2O \qquad k$$

for which

$$\text{rate} = \frac{kK_E[\{Ni(OH)_2)_6\}^{2+}][NH_3]}{1 + K_E[NH_3]}$$

It is difficult to conduct experiments over a range of concentrations wide enough to test the rate law exhaustively, but this has been done in a few cases (see Problem 15.1). At such low concentrations of the entering group that $K_E[Y] \ll 1$, the rate law reduces to the second-order form

$$\text{rate} = k_{obs}[C][Y] \qquad k_{obs} = kK_E$$

Table 15.7 Complex formation by the $[Ni(OH_2)_6]^{2+}$ ion

Ligand	$k_{obs}/(L\ mol^{-1}\ s^{-1})$	$K_E/L\ mol^{-1}$	$(k_{obs}/K_E)/s^{-1}$
$CH_3CO_2^-$	1×10^5	3	3×10^4
F^-	8×10^3	1	8×10^3
HF	3×10^3	0.15	2×10^4
H_2O^*			3×10^3
NH_3	5×10^3	0.15	3×10^4
$[NH_2(CH_2)_2NH_3]^+$	4×10^2	0.02	2×10^3
SCN^-	6×10^3	1	6×10^3

*The solvent is always in encounter with the ion so that K_E is undefined and all rates are inherently first-order.

Because k_{obs} can be measured and K_E estimated as we describe below, the rate constant k can be found from k_{obs}/K_E. The results for reactions of Ni(II) hexaaqua complexes with various nucleophiles are shown in Table 15.7. The very small variation in k is in sharp contrast to the wide variation typical of associative reactions. This example is a model I_d reaction with very slight response to the nucleophilicity of the entering group.

In the special case that Y is the solvent and encounter equilibration is saturated, $K_E[Y] \gg 1$ and $k_{obs} = k$. Thus, reactions with the solvent can be directly compared to reactions with other entering ligands without needing to estimate the value of K_E.

The Fuoss–Eigen equation

The equilibrium constant for the encounter complex (K_E) can be estimated theoretically by using a simple equation proposed independently by R. M. Fuoss of Yale University and M. Eigen of Göttingen. Both sought to take the particle size and charge into account, expecting that larger, oppositely charged particles would meet more frequently than small ions of the same charge. Fuoss used an approach based on statistical thermodynamics and Eigen used one based on kinetics. Their result, which is called the **Fuoss–Eigen equation**, is

$$K_E = \frac{4\pi a^3}{3} N_A e^{-V/RT}$$

In this expression a is the distance of closest approach, V the Coulombic potential energy of the ions at that distance, and N_A is Avogadro's constant. The equilibrium constant favors the encounter if the reactants are large (so a is large) and oppositely charged (V negative).

15.5 Intimate mechanisms

Many studies of substitution in octahedral complexes support the view that they are dissociative reactions (d), and we summarize these studies first. However, octahedral substitutions can acquire a distinct associative (a) character in the case of large central ions (as in the second and third series of the d block) or where the d-electron population at the metal is low (the early members of the d block). More room for attack

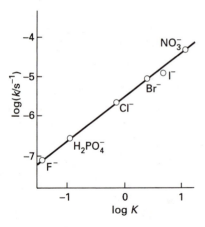

15.4 The straight line obtained when the logarithm of a rate constant is plotted against the logarithm of an equilibrium constant shows the existence of a linear free energy relation. This graph is for the reaction $[CoX(NH_3)_5]^{2+} + H_2O \rightarrow [Co(NH_3)_5(OH_2)]^{3+} + X^-$ with different leaving groups X.

or lower π^* electron density appears to facilitate nucleophilic attack and hence associative reaction.

Leaving-group effects

We can expect a large effect of the leaving group X in dissociatively activated reactions because their activation energies are determined by the unassisted scission of the M—X bond. When the identity of X can be isolated as the only variable, as in the reaction

$$[CoX(NH_3)_5]^{2+} + H_2O \rightarrow [Co(NH_3)_5(OH_2)]^{3+} + X^-$$

it is found for these complexes that there is a linear relation between the logarithms of the rate constants and equilibrium constants of the reaction (Fig. 15.4), and that specifically

$$\ln k = \ln K + c$$

Because both logarithms are proportional to Gibbs free energy changes ($\ln k$ is proportional to the activation Gibbs energy, $\Delta G^{\ddagger}$, and $\ln K$ is proportional to the standard reaction Gibbs energy, $\Delta G^{\ominus}$), we can write

$$\Delta G^{\ddagger} = p\Delta G^{\ominus} + b$$

with p and b constants (and $p \approx 1$). The existence of this **linear free energy relation** of unit slope shows that changing X has the same effect on $\Delta G^{\ddagger}$ for the conversion of Co–X to the activated complex as it has on $\Delta G^{\ominus}$ for the complete elimination of X^- (Fig. 15.5). This observation in turn suggests that in a dissociative reaction the leaving group (an anionic ligand) has become a solvated ion in the activated complex.

A similar linear free energy relation (with a slope smaller than 1.0, indicating some *a* character) is observed for the corresponding complexes of Rh(III). However, as we should expect for the softer Rh(III) center, the order of leaving groups is reversed, and the reaction rates are in the order

$$I^- < Br^- < Cl^-$$

because the softer acid, Rh(III), forms stronger bonds to I^-.

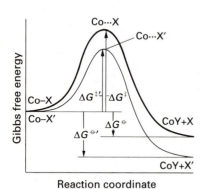

15.5 The existence of a linear free energy relation with unit slope shows that changing X has the same effect on $\Delta G^{\ddagger}$ for the conversion of M–X to the activated complex as it has on $\Delta G^{\ominus}$ for the complete elimination of X^-. This reaction profile shows the effect of changing the leaving group from X to X'.

The effects of spectator ligands

Ligands *cis* to a leaving group usually have nonspecific effects on rates. In Co(III), Cr(III), and related octahedral complexes, both *cis* and *trans* ligands affect rates of substitution in proportion to the strength of the bonds they form with the metal. There is no important *trans* effect. For instance, hydrolysis reactions such as

$$[NiXL_5]^+ + H_2O \rightarrow [NiL_5(OH_2)]^{2+} + X^-$$

are much faster when L is NH_3 than when it is H_2O. This difference can be explained on the grounds that as NH_3 is a stronger σ-donor

Table 15.8. Tolman cone angle for various ligands

Ligand	$\theta/°$	Ligand	$\theta/°$
CH_3	90	$P(OCH_3)_3$	128
CO	95	$P(OC_2H_5)_3$	134
Cl, Et	102	$\eta^5\text{-}C_5H_5$ (Cp)	136
PF_3	104	PEt_3	137
Br, Ph	105	PPh_3	145
I, $P(OCH_3)_3$	107	$C_5(CH_3)_5$ (Cp*)	165
$P(CH_3)_3$	118	$2,4\text{-}(CH_3)_2C_5H_3$	180
t-butyl	126	$P(t\text{-Bu})_3$	182

Source: C. A. Tolman, *Chem. Rev.*, **77**, 313 (1975); L. Stahl and R. D. Ernst, *J. Am. Chem. Soc.*, **109**, 5673 (1987).

than H_2O, it increases the electron density at the metal and hence facilitates the scission of the M—X bond and the formation of X^-. In the activated complex, the good donor stabilizes the reduced coordination number.

Steric effects

Steric effects on dissociative reactions can be illustrated by considering the rate of hydrolysis of the first Cl^- ligand in two complexes of the type $[CoCl_2(bn)_2]^+$:

$$[CoCl_2(bn)_2]^+ + H_2O \rightarrow [CoCl(bn)_2(OH_2)]^{2+} + Cl^-$$

The ligand bn is 2,3-butanediamine, and may be either chiral (5) or achiral (6). The important observation is that the complex formed with the chiral form of the ligand hydrolyzes 30 times more slowly than the complex of the achiral form. The two ligands have very similar electronic effects, but the CH_3 groups are on the opposite sides of the chelate ring in (5) and adjacent and crowded in (6). The latter is more reactive because the strain is relieved in the dissociatively activated complex with its reduced coordination number. In general:

> Steric crowding favors dissociative reaction because the five-coordinate activated complex can relieve strain.

In recent years, quantitative treatments of the steric effects of ligands have been developed using molecular modeling computer programs that take into account van der Waals interactions.[2] A more pictorial semiquantitative approach was introduced by C. A. Tolman (of Dupont). It assesses the extent to which various ligands (especially phosphanes) crowd each other by approximating the volume occupied by a ligand by a cone with an angle determined from a space-filling model and, for phosphane ligands, an M—P bond length of 2.28 Å (Fig. 15.6 and Table 15.8). The ligand CO is small in the sense of having a small cone angle; $P(t\text{-Bu})_3$ is regarded as bulky because it has

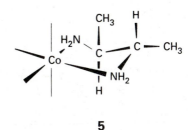

5

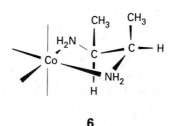

6

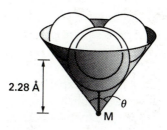

15.6 The determination of ligand cone angles from space-filling molecular models of the ligand and an assumed M—P bond length of 2.28 Å.

[2] J. F. Endicott, K. Kumar, C. L. Schwarz, W. Perkovic, and W. Lin, *J. Am. Chem. Soc.*, **111**, 7411 (1989).

Table 15.9 Activation parameters for the H_2O exchange reactions $[M(OH_2)_6]^{2+}$ + $H_2^{17}O \rightarrow [M(OH_2)_5(^{17}OH_2)]^{2+}$ + H_2O

	$\Delta H^{\ddagger}$/ kJ mol^{-1}	LFSE*/ Δ_0	LFSE†/ Δ_0	LFAE/ Δ_0	$\Delta V^{\ddagger}$/ cm^3 mol^{-1}
Ti^{2+} (d^2)		8	9.1	-1.1	
V^{2+} (d^3)	68.6	12	10	2	-4.1
Cr^{2+} $(d^4,$ hs)		6	9.1	-3.1	
Mn^{2+} $(d^5,$ hs)	33.9	0	0	0	-5.4
Fe^{2+} $(d^6,$ hs)	31.2	4	4.6	-0.6	$+3.8$
Co^{2+} $(d^7,$ hs)	43.5	8	9.1	-1.1	$+6.1$
Ni^{2+} (d^8)	58.1	12	10	2	$+7.2$

*Octahedral.
†Square pyramidal.
hs: High spin.

a large cone angle. Bulky ligands have considerable steric repulsion with each other when packed around a metal center. They favor dissociative reaction and inhibit associative reaction.

The rate of the reaction of $[Ru(SiCl_3)_2(CO)_3(PR_3)]$ with L to give $[Ru(SiCl_3)_2(CO)_2L(PR_3)]$ is independent of the identity of L, which suggests that the reaction is dissociative. Furthermore, in the case of ligands with cone angle 145° but which have significantly different values of pK_a, it has been found that there is only a small change in rate. This observation supports the assignment of the rate changes to steric effects because changes in pK_a should correlate with changes in electron distributions in the ligands.

Activation energetics

Among the most fruitful attempts to isolate the factors affecting activation parameters is the identification of a correlation between the enthalpy of activation, $\Delta H^{\ddagger}$, and the **ligand field activation energy** (LFAE). The latter is the difference in ligand field stabilization energies of the reactant and a model of the activated complex:

$$LFAE = (LFSE)^{\ddagger} - (LFSE)$$

Table 15.9 shows the results of a calculation in which the dissociatively activated complex is assumed to be a five-coordinate square pyramid. The data show some correlation between large $\Delta H^{\ddagger}$ and large LFAE. Through this approach we can begin to see why Ni^{2+} and V^{2+} complexes are not very labile and why they have higher experimental activation enthalpies. In these two cases there is a significant loss of LFSE on going from the six-coordinate complex to the five-coordinate activated complex.

Associative activation

Activation volumes reflect the changes in compactness (including that of the surrounding solvent) when the activated complex forms from

Table 15.10 Kinetic parameters for anion attack on Cr(III)*

X	L = H₂O			L = NH₃
	$k/(10^8\ \text{L mol}^{-1}\ \text{s}^{-1})$	$\Delta H^{\ddagger}$	$\Delta S^{\ddagger}$	$k/(10^4\ \text{L mol}^{-1}\ \text{s}^{-1})$
Br^-	1.0	122	8	3.7
Cl^-	2.9	126	38	0.7
NCS^-	180	105	4	4.2

*Enthalpy in kJ mol^{-1}, entropy in J K^{-1} mol^{-1}

the reactants. The last column in Table 15.9 gives $\Delta V^{\ddagger}$ for some H_2O ligand exchange reactions. We see that $\Delta V^{\ddagger}$ increases from $-4\ \text{cm}^3$ mol^{-1} for V^{2+} to $+7.2\ \text{cm}^3$ mol^{-1} for Ni^{2+}. We can interpret the negative volume of activation as the result of shrinkage on the incorporation of the entering H_2O molecule into the activated complex, which suggests that the reactions have significant associative character in the activation process. The increase in $\Delta V^{\ddagger}$ follows the increase in the number of nonbonding d electrons from d^3 to d^8 across the first row of the d block. In the earlier part of the d block, associative reaction appears to be favored by a low population of d electrons.

Negative volumes of activation are also observed in the second and third rows of the block, such as for Rh(III), and also indicate association of the entering group into the activated complex in the reactions of these complexes. Associative reaction begins to dominate when the metal center is more accessible to nucleophilic attack, either because it is large or has a low (nonbonding or π^*) d-electron population, and the mechanism shifts from I_d toward I_a. Table 15.10 shows some data for the formation of Br^-, Cl^-, and NCS^- complexes from $[Cr(NH_3)_5(OH_2)]^{3+}$ and $[Cr(OH_2)_6]^{3+}$. The pentaammine complex shows only a weak nucleophile dependence in contrast to the strong dependence of the hexaaqua complex. The two complexes probably mark the transition from I_d to I_a.

In the replacement of H_2O in $[Cr(OH_2)_6]^{3+}$ or $[Cr(NH_3)_5(OH_2)]^{3+}$ by Cl^-, Br^-, or NCS^-, the rate constants for the aqua complexes are all smaller than those for ammine complexes by a factor of about 10^4 (see Table 15.10). This difference suggests that the NH_3 ligands, which are stronger σ-donors than H_2O, promote dissociation of the sixth ligand more effectively. As we saw above, this is to be expected in dissociative reactions. Apparently in the ammines the reduced coordination number of the activated complex is stabilized more effectively; the aqua complex needs nucleophilic assistance.

Example 15.3: *Interpreting kinetic data in terms of a mechanism*
The second-order rate constants for formation of $[VX(OH_2)_5]^+$ from $[V(OH_2)_6]^{2+}$ and X^- for $X^- = Cl^-$, NCS^-, and N_3^- are in the ratio 1:2:10. What do the data suggest about the intimate mechanism of substitution?

Answer. Because all three complexes are monoanions of similar size, the encounter equilibrium constants are similar, and second-order rate constants (which are equal to $K_E k_2$) are proportional to first-order rate constants for substitution in the encounter complex, k_2. The greater rate constants for NCS^- than Cl^-, and especially the fivefold difference of SCN^- from its close structural analog N_3^-, suggest some contribution from nucleophilic attack and an associative reaction. (It turns out that there is no such systematic pattern for the same anions reacting with Ni(II), for which the reaction is believed to be dissociative.)

Exercise E15.3. Using the data in Table 15.9, estimate an appropriate value for K_E and calculate k for the reactions of V(II) with Cl^- if the observed second-order rate constant is 1.2×10^2 L mol^{-1} s^{-1}.

15.6 Stereochemistry

The stereochemical consequences of substitution of octahedral complexes are more involved than those of square-planar complexes. Classic examples of octahedral substitution stereochemistry are provided by Co(III) complexes, and in Table 15.11 we give some data for the hydrolysis of *cis* and *trans* $[CoAX(en)_2]^{2+}$ where X is the leaving group (either Cl^- or Br^-) and A is OH^-, NCS^-, or Cl^-. The *cis* complexes do not undergo isomerization when substitution occurs. The *trans* complexes show a tendency to isomerize to *cis* that increases in the order

$$A = NO_2^- < Cl^- < NCS^- < OH^-$$

We can understand the data within the framework of an I_d mechanism by recognizing that the five-coordinate metal center in the activated complex may resemble either of the two stable geometries for five-coordination, namely square-pyramidal or trigonal-bipyramidal. As we can see from Fig. 15.7, reaction through the square-pyramidal complex results in retention of the original geometry, but reaction through the trigonal-bipyramidal complex can lead to isomerization. The results suggest that a good π-donor ligand *trans* to the leaving group Cl^-

Table 15.11 Stereochemical course of hydrolysis reactions of $[CoAX(en)_2]^+$

	A	X	Percentage *cis* in product
cis	OH^-	Cl^-	100
	Cl^-	Cl^-	100
	NCS^-	Cl^-	100
	Cl^-	Br^-	100
trans	NO_2^-	Cl^-	0
	NCS^-	Cl^-	50–70
	Cl^-	Cl^-	35
	OH^-	Cl^-	75

X is the leaving group.

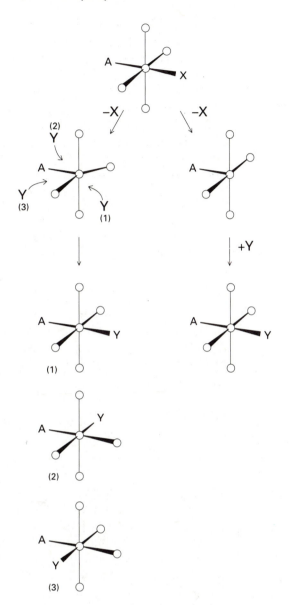

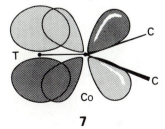

15.7 Reaction through a square-pyramidal complex (right hand path) results in retention of the original geometry but reaction through a trigonal-bipyramidal complex (left-hand path) can lead to isomerization.

7

favors isomerization, because it can stabilize a trigonal bipyramid (**7**) by participating in π bonding.

15.7 Base hydrolysis

Octahedral substitution is commonly greatly accelerated by OH^- ions when ligands with acidic protons are present. A typical rate law for such reactions is

$$\text{rate} = k[\{CoCl(NH_3)_5\}^{2+}][OH^-]$$

However, an extended series of studies has shown that the mechanism is not a simple bimolecular attack by OH^- on the complex.

The stereochemistry of the reaction indicates that a strong π-donor ligand is present and that the steric effects observed are characteristic of a dissociative reaction (d). There is a considerable body of indirect evidence relating to the problem, but one elegant experiment makes the essential point. This conclusive evidence comes from a study of ^{18}O isotope distribution in the product $[Co(OH)(NH_3)_5]^{2+}$, where it is found that the $^{18}O/^{16}O$ ratio differs between water and hydroxide ion at equilibrium. The isotope ratio in the product matches that for water, not that for the OH^- ions. Therefore, an H_2O molecule and not an OH^- ion appears to be the entering group.

The mechanism that takes these observations into account supposes that the role of OH^- is to act as a Brønsted base, not an entering group:

$$[CoCl(NH_3)_5]^{2+} + OH^- \rightleftharpoons [CoCl(NH_2)(NH_3)_4]^+ + H_2O$$

$$[CoCl(NH_2)(NH_3)_4]^+ \rightarrow [Co(NH_2)(NH_3)_4]^{2+} + Cl^- \quad \text{(slow)}$$

$$[Co(NH_2)(NH_3)_4]^{2+} + H_2O \rightarrow [Co(OH)(NH_3)_5]^{2+} \quad \text{(fast)}$$

In the first step, an NH_3 ligand acts as a Brønsted acid, resulting in the formation of its conjugate base, the NH_2^- ion, as a ligand. Because NH_2^- is a strong π-donor, it greatly accelerates Cl^- ion loss (see Example 15.4).

15.8 Isomerization reactions

Isomerization reactions are closely related to substitution reactions; indeed, a major pathway for isomerization is often via substitution. The square-planar Pt(II) and octahedral Co(III) complexes we have discussed can both form five-coordinate trigonal-bipyramidal activated complexes. The interchange of the axial and equatorial ligands in a trigonal-bipyramidal complex can be pictured as occurring by a pseudorotation through a square pyramid as shown in (8). As we saw above, when a trigonal bipyramid adds a ligand to produce a six-coordinate complex, a new direction of attack of the entering group can lead to an isomerization.

If a chelate ligand is present, isomerization can occur as a consequence of metal–ligand bond breaking without substitution. The exchange of the 'outer' CD_3 group with the 'inner' CH_3 group during

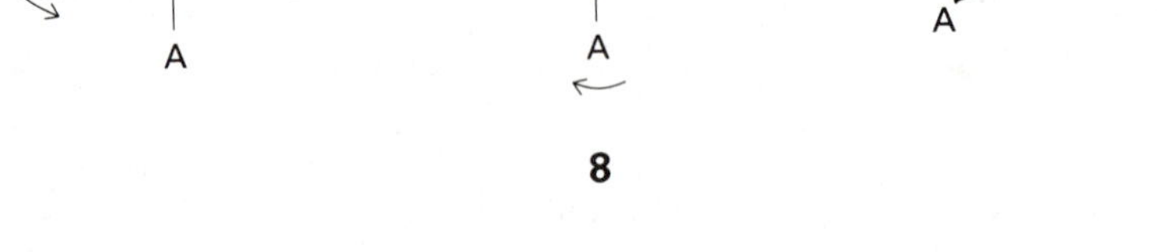

8

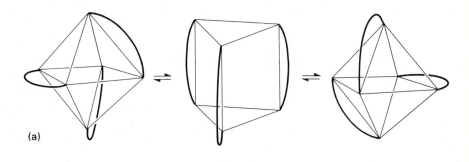

15.8 (a) The Bailar twist and (b) the Ray–Dutt twist by which an octahedral complex can also undergo isomerization without losing a ligand or breaking a bond.

the isomerization of tris(acetylacetylacetonato)cobalt(III) $(9 \rightarrow 10)$ shows that ring opening accompanies the isomerization.

An octahedral complex can also undergo isomerization by an intramolecular twist without loss of a ligand or breaking of a bond. There is evidence, for example, that racemization of $[Ni(en)_3]^{2+}$ occurs by such an internal twist. Two possible paths are the **Bailar twist** and the **Ray–Dutt twist**, which are illustrated in Fig. 15.8.

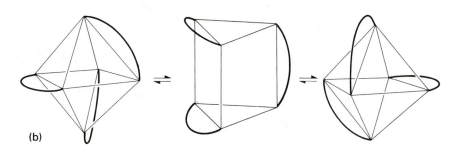

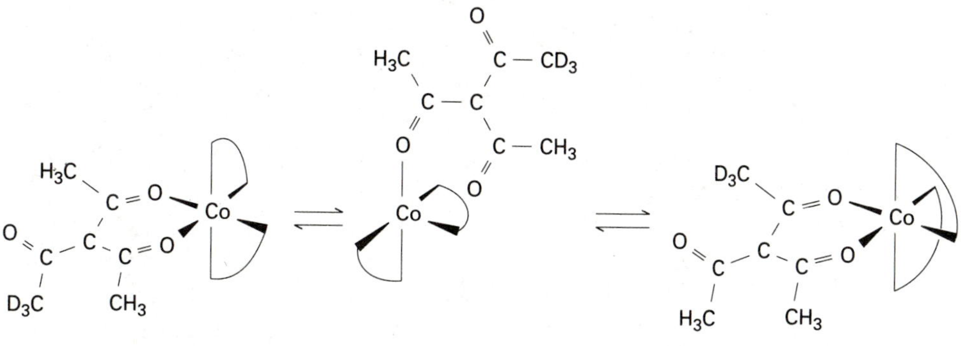

9   **10**

Example 15.4: *Interpreting the stereochemical course of a reaction*
Substitution of Co(III) complexes of the type $[CoAX(en)_2]^+$ results in *trans* to *cis* isomerization, but only when the reaction is catalyzed by a base. Show that this observation is consistent with the model in Fig. 15.8.

Answer. For base hydrolysis, the conjugate base is the strong π-donor ligand :NHR^-. A trigonal bipyramid of the type shown in the lower path in Fig. 15.8 is possible, and it may be attacked in the way shown there.

Exercise E15.4. If the directions of attack on the above trigonal bipyramid are random, what *cis–trans* isomer distribution does the analogy with Fig. 15.8 suggest for products of the type $[CoA(NH_3)_4(OH_2)]^{2+}$?

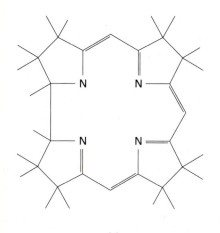

11

15.9 Stoichiometric mechanisms: the *D* intermediate

We may be able to distinguish a dissociative *D* mechanism from an interchange I_d mechanism if the intermediate lives long enough for competitive reactions to occur. An example is the substitution reaction of the corrin (**11**) complexes of Co(III), which are related to vitamin B_{12}.

The process begins with reversible intermediate formation:

$$[Co(corrin)(OH_2)_2] \underset{k_a'}{\overset{k_a}{\rightleftharpoons}} [Co(corrin)(OH_2)] + H_2O$$

The forward reaction is first-order and the reverse reaction is pseudofirst-order because one reactant is the solvent (water). This step is followed by an attack on the intermediate by the entering group:

$$[Co(corrin)(OH_2)] + Y \xrightarrow{k_b} [Co(corrin)(OH_2)Y]$$

The overall rate law implied by this mechanism (on the basis of the steady-state assumption for the intermediate) is

$$\text{rate} = \frac{k_a k_b [Co(corrin)(OH_2)_2][Y]}{k_a' + k_b[Y]}$$

If the reaction is studied at low enough entering group concentrations for $k_b[Y] \ll k_a'$, the rate law reduces to second order:

$$\text{rate} = k_{obs}[Co(corrin)(OH_2)_2][Y], \qquad k_{obs} = \frac{k_a k_b}{k_a'}$$

The result that the rate for any given nucleophile Y depends on the ratio k_b/k_a' shows that it is the outcome of the competition between Y and H_2O for capture of the five-coordinate intermediate.

Table 15.12 gives some examples of the variation of k_{obs} with the identity of Y for $[Co(corrin)(OH_2)_2]^{2-}$. In contrast to the I_d case, there is a pronounced dependence on the entering group. However, there is

Table 15.12 Second-order rate constants for reactions of Co(corrin)

Y	SCN^-	I^-	Br^-	HSO_3^-
$k_{obs}/(\text{L mol}^{-1}\text{ s}^{-1})$:	0.23	0.14	0.10	0.017

k_a for H_2O exchange is 14 s^{-1}.

a difference from the I_a case in that at high entering group concentrations ($k_b[Y] \ll k_a'$) the rate law approaches

$$\text{rate} = k_a[\text{Co(corrin)(OH}_2)_2]$$

so the observed rate constant approaches the common limit k_a independent of the identity of Y. This behavior is uncharacteristic of an I_a process. The 20-fold increase in the rate of product formation at low concentration associated with the change of Y from HSO_3^- to SCN^- reflects the increase in the rate of the attack on the five-coordinate intermediate. When that attack is fast compared to the rate of re-entry of the leaving group, formation of the intermediate becomes rate-limiting and all entering groups react at the same rate.

Probably the first D intermediate to be suggested was $[\text{Co(NH}_3)_5]^{3+}$; however, there is now conclusive evidence against its importance. The effect of the leaving group in $[\text{CoX(NH}_3)_5]^{2+}$ on the proportions of the linkage isomers $[\text{Co(SCN)(NH}_3)_5]^{2+}$ and $[\text{Co(NCS)(NH}_3)_5]^{2+}$ in solutions containing relatively high concentrations of NCS^- have provided the critical evidence. They show unambiguously that the leaving group influences the product distribution, and therefore that it is a part of the activated complex. Hence $[\text{Co(NH}_3)_5]^{3+}$ does not survive long enough to qualify as a true intermediate and for the reaction to be D.[3]

15.10 Involvement of the ligand; alkyl migration and CO insertion

Substitution reaction pathways are more complex when an atom belonging to a ligand other than an entering or leaving group becomes directly involved. In substitution reactions of alkyl carbonyl complexes such as

$$[\text{CH}_3\text{Mn(CO)}_5] + \text{PPh}_3 \rightarrow [\text{CH}_3\text{Mn(CO)}_4\text{PPh}_3] + \text{CO}$$

there is evidence that an **acyl intermediate** (**12**) is involved, where Sol is a solvent molecule. This intermediate arises from the cleavage of the M—CH$_3$ bond and its replacement by an H$_3$CC(O)—bond (henceforth, simply CH$_3$CO—). The substitution reaction is commonly completed by replacement of Sol by a more strongly binding ligand and restoration of the CH$_3$—M bond with loss of CO. When the entering ligand is CO itself, the net outcome of the reaction is **CO insertion** of the form

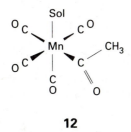

12

[3] W. G. Jackson, B. C. McGregor, and S. S. Jurisson, *Inorg. Chem.*, **26**, 1286 (1987). These authors argue that $[\text{Co(NH}_3)_5]^{3+}$ cannot live longer than a few picoseconds.

The mechanism of the reaction can be similar to the PPh$_3$ substitution above, up to the formation of the acyl intermediate. The most attractive mechanistic hypothesis for the formation of the acyl intermediate is **methyl migration** (13), in which the CH$_3$ group engages in nucleophilic attack on the C atom of a neighboring CO ligand. This interpretation of the reaction is consistent with the activation parameters for the reaction, especially $\Delta S^{\ddagger} = -88.2$ J K^{-1} mol^{-1}, the negative value indicating incorporation of an additional ligand into the activated complex. Moreover, electron-withdrawing groups substituted into CH$_3$ dramatically slow the reaction. This effect would be expected if CH$_3$ functioned as a nucleophile. For example, [Mn(CH$_2$NO$_2$)(CO)$_5$] is much less reactive than [MnCH$_3$(CO)$_5$].

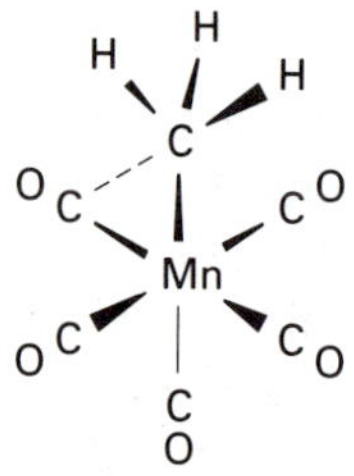

A definitive study exploited the reverse reaction with one CO position labeled with ^{13}CO:

$$[\text{Mn(CH}_3\text{CO)(CO)}_4(^{13}\text{CO)}] \rightarrow [\text{Mn(CH}_3\text{)(CO)}_4(^{13}\text{CO)}] + \text{CO}$$

Because the activated complex for a reverse reaction must be the same as that of the forward reaction, the distribution of products with CH$_3$ *cis* or *trans* to ^{13}CO confirms that the methyl group migrates to a *cis* CO in the forward reaction.[4]

The two possible paths for the reaction are CH$_3$ migration and CO insertion into the M—CH$_3$ bond. In the reverse reaction these paths correspond to migration of CH$_3$ to a position from which CO is lost, and migration of CO out of the M(COCH$_3$) group to the position from which a CO group is lost. The product distribution is as follows:

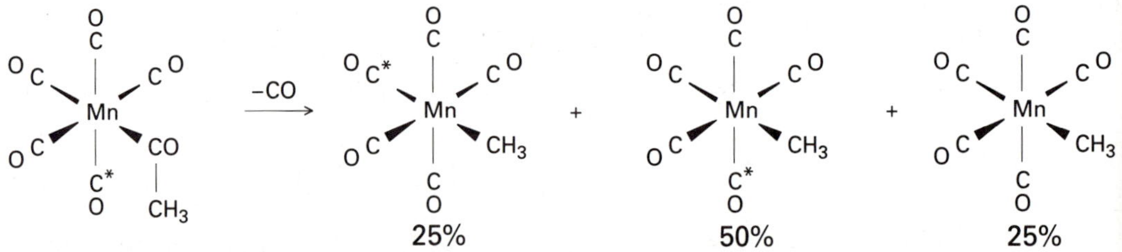

13

The four CO ligands adjacent to CH$_3$CO are equally likely to be lost. If the CO ligand *trans* to the labeled CO ligand is lost, then CH$_3$ migration results in the formation of a complex in which ^{13}CO and CH$_3$ are *trans* to each other. Loss of a CO ligand that is *cis* to both ^{13}CO and CH$_3$CO followed by CH$_3$ migration results in a complex in which ^{13}CO and CH$_3$ are *cis*. Finally, loss of the labeled ligand results in an unlabeled product. These considerations suggest that in the case of CH$_3$ migration the products should be in the ratio 1:2:1, as observed.

REDOX REACTIONS

As remarked in Chapter 7, redox reactions can occur by the direct transfer of electrons (as in some electrochemical cells). Alternatively, they may occur by the transfer of atoms and ions, as in the transfer of O atoms in reactions of oxoanions. Because redox reactions involve

[4] T. C. Flood, J. E. Jensen, and J. A. Slater, *J. Am. Chem. Soc.*, **103**, 4410 (1981).

both an oxidizing and a reducing agent, they are usually bimolecular in character. (The rare exceptions are reactions in which one molecule has both oxidizing and reducing centers.) The stoichiometric mechanisms depend on whether the centers react without modification of their coordination spheres or whether a ligand substitution must accompany the reaction.

15.11 The classification of redox reactions

In the 1950s, the American chemist Henry Taube[5] identified two stoichiometric mechanisms of redox reactions. One is the **inner-sphere mechanism**, which includes atom transfer processes. In an inner-sphere mechanism, the coordination spheres of the reactants share a ligand transitorily and form a bridged intermediate. The other is an **outer-sphere mechanism**, which includes many simple electron transfers. In an outer-sphere mechanism, the complexes come into contact without sharing a bridging ligand. We encountered both mechanisms briefly in Section 7.4.

Inner-sphere mechanisms

The inner-shell mechanism was first confirmed for the reduction of the nonlabile complex $[CoCl(NH_3)_5]^{2+}$ by Cr^{2+}(aq), because the products of the reaction included both Co^{2+} and $[CrCl(OH_2)_5]^{2+}$. Moreover, addition of $^{36}Cl^-$ to the solution did not lead to the incorporation of any of the isotope in the Cr(III) product. Furthermore, the reaction is faster than reactions that remove Cl^- from Co(III) or introduce Cl^- into the nonlabile $[Cr(OH_2)_6]^{3+}$ complex. These observations must mean that Cl^- has moved directly from the coordination sphere of one complex to that of the other during the reaction.

The Cl^- attached to Co(III) can easily enter into the labile coordination sphere of $[Cr(OH_2)_6]^{2+}$ to produce a bridged intermediate (**14**). Good bridging ligands are those with more than one pair of electrons available to donate to the two metal centers; they include

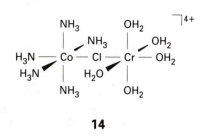

14

$$:\ddot{Cl}:^- \quad :S-C\equiv N:^- \quad :N\equiv N: \quad :\ddot{N}=N=\ddot{N}:^- \quad :C\equiv N:^-$$

Outer-sphere mechanism

An example of an outer-shell mechanism is the reduction of $[Fe(phen)_3]^{3+}$ by $[Fe(CN)_6]^{4-}$. Both reactants have nonlabile coordination spheres, and no ligand substitution can occur on the very short time scale of the redox reaction. The only possible mechanism is an electron transfer between the two complex ions in outer-sphere contact.

---

[5] Taube has given an excellent retrospective account in his 1983 Nobel Prize lecture reprinted in *Science*, **226**, 1028 (1984). A collection of informative chapters on the mechanisms of the redox reactions of coordination compounds is also available in *Prog. Inorg. Chem.*, **30** (1983).

An understanding of ligand substitution is a prerequisite to the full analysis of a redox reaction mechanism. If a redox reaction is faster than ligand substitution, then the reaction has an outer-sphere mechanism. Similarly, there is no difficulty in assigning an inner-sphere mechanism when the reaction involves ligand transfer from an initially nonlabile reactant to a nonlabile product. Unfortunately, it is difficult to make unambiguous assignments when the reactants and products are labile. Much of the study of well defined examples is directed toward the identification of the parameters that differentiate the two paths in the hope of being able to making correct assignments in more difficult cases.

15.12 The theory of redox reactions

The theory of the intimate mechanisms of redox reactions is well established. This is especially so for outer-sphere reactions where it is even possible to calculate rate constants in favorable cases.

The concepts needed to discuss outer-sphere processes

The analysis of the outer-sphere mechanism depends on two concepts. One is the **Franck–Condon principle**, which states that electron rearrangements occur so rapidly that nuclei can be considered as stationary until the rearrangement is complete. This separation of the motion of electrons and nuclei is based on the great difference of mass between electrons and nuclei, and hence the former's ability to respond almost instantaneously to rearrangements of nuclei and, conversely, for the nuclei to respond only sluggishly to movement of the electrons. The second concept we need is that electron transfer is most facile when the nuclei in the two complexes have positions that ensure that the electron has the same energy on each site. Taken together with the Franck–Condon principle, it follows that the rate of electron transfer, and the activation energy for the process, will be governed by the ability of nuclei to adopt arrangements that achieve this matching of energies.

To understand how an outer-sphere reaction occurs, we shall consider the exchange reaction

$$[Fe(OH_2)_6]^{2+} + [Fe*(OH_2)_6]^{3+} \rightarrow [Fe(OH_2)_6]^{3+} + [Fe*(OH_2)_6]^{2+}$$

where the * indicates a radioactive isotope of iron that acts as a tracer. The second-order rate constant is 3.0 L mol^{-1} s^{-1} at 25 °C and the activation energy is 32 kJ mol^{-1}. With the two theoretical points in mind we must consider the following three factors. First, bond lengths in Fe(II) are longer than those in Fe(III). Hence a part of the activation energy will arise from their adjustment to a common value in both complexes. This adjustment requires a change in Gibbs free energy that is called the **inner-sphere rearrangement energy**, $\Delta G^{\ddagger}_{IS}$. Secondly, the solvation shell must be reorganized, which results in a change in free energy called the **outer-sphere reorganization**

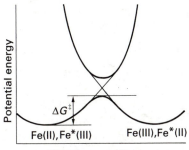

15.9 A simplified reaction profile for electron exchange in a symmetrical reaction. On the left of the graph, the nuclear coordinates correspond to Fe(II) and Fe*(III); on the right, the ligands and solvent molecules have adjusted locations and the nuclear coordinates correspond to Fe(III) and Fe*(II), where * denotes the isotope label.

energy, $\Delta G^{\ddagger}_{OS}$. Thirdly, there is the electrostatic interaction energy between the two reactants ($\Delta G^{\ddagger}_{ES}$). The total activation Gibbs energy is therefore

$$\Delta G^{\ddagger} = \Delta G^{\ddagger}_{IS} + \Delta G^{\ddagger}_{OS} + \Delta G^{\ddagger}_{ES}$$

The potential energy curves for reaction

The reactants initially have their normal bond lengths for Fe(II) and Fe*(III) respectively, and the reaction corresponds to motion in which the Fe(II) bonds shorten and the Fe*(III) bonds simultaneously lengthen (Fig. 15.9). The potential energy curve for the products of this symmetrical reaction is the same as for the reactants, the only difference being the interchange of the roles of the two Fe atoms. We have assumed that the metal–ligand stretching motions resemble a harmonic vibration and so have drawn them as parabolas.

The activated complex is located at the intersection of the two curves. However, the noncrossing rule (Section 14.3) states that molecular potential energy curves of states of the same symmetry do not cross but instead split into an upper and a lower curve (as seen in Fig. 15.9). The noncrossing rule implies that if the reactants in their ground states slowly distort, then they follow the path of minimum energy and transform into products in their ground states.

More general redox reactions correspond to a nonzero reaction free energy, so the parabolas representing the reactants and the products lie at different heights. If the product surface is higher (Fig. 15.10(a)), then the crossing point moves up and the reaction has a higher activation energy. Conversely, moving the product curve down (Fig. 15.10(c)) leads to lower crossing points and lower activation

(a)

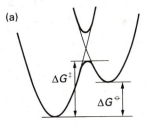

(b)

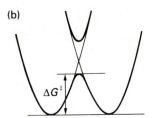

(c)

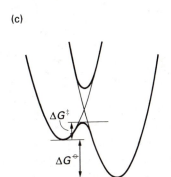

(d)

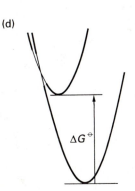

15.10 The effect of a change of reaction Gibbs free energy for electron transfer on the activation energy when the shape of the potential surfaces remain constant (corresponding to equal self-exchange rates).

energy, at least until the crossing begins to occur at the left. At the extreme of exergonic reaction (Fig. 15.10(d)), the crossing point rises and rates may become slower again.[6]

The Marcus equation

The diagrams in Fig. 15.10 suggest that there are two factors that determine the rate of electron transfer. The first is the shape of the potential curves. If the parabolas rise steeply, indicating a rapid increase in energy with increasing bond extension, then their crossing points will be high and activation energies will be high too. Shallow potential curves, in contrast, imply low activation energies. Similarly, large changes in the equilibrium internuclear distance mean that the equilibrium points are far apart, and the crossing point will not be reached without large distortions. The second factor is the standard reaction free energy, and the more negative it is, the lower the activation energy of the reaction.

These considerations were expressed quantitatively by R. A. Marcus (and the work was recognized by the award of the Nobel Prize in 1992). He derived an equation for predicting the rate constant for an outer-sphere reaction from the exchange rate constants for each of the redox couples involved and the equilibrium constant for the overall reaction. The Marcus equation for the rate constant k is

$$k^2 = f k_1 k_2 K$$

where k_1 and k_2 are the rate constants for the two exchange reactions and K is the equilibrium constant for the overall reaction. The factor f is a complex parameter composed of the rate constants and the encounter rate; it may be taken as near unity for approximate calculations. The idea of a weighted average of the two rates of self exchange is emphasized by the name **Marcus cross relation**, which is sometimes given to the expression above.

[6] We shall not consider this last point further here. The slowing of rate at high exothermicity is referred to as the 'Marcus inverted region'. In an experimental *tour de force*, appropriate highly energetic oxidizing agents have recently been prepared photochemically and direct evidence for this inverted region has been obtained. See, for example, L. S. Fox, M. Kozik, J. R. Winkler, and H. B. Gray, *Science*, **297**, 1069 (1990).

Example 15.5: *Using the Marcus equation*

Calculate the rate constant at $0\,°C$ for the outer-sphere reduction of $[\mathrm{Co(bipy)_3}]^{3+}$ by $[\mathrm{Co(terpy)_2}]^{2+}$.

Answer. The required exchange reactions are

$$[\mathrm{Co(bipy)_3}]^{2+} + [\mathrm{Co^*(bipy)_3}]^{3+} \xrightarrow{k_1}$$

$$[\mathrm{Co(bipy)_3}]^{3+} + [\mathrm{Co^*(bipy)_3}]^{2+}$$

$$[\mathrm{Co(terpy)_2}]^{2+} + [\mathrm{Co^*(terpy)_2}]^{3+} \xrightarrow{k_2}$$

$$[\mathrm{Co(terpy)_2}]^{3+} + [\mathrm{Co^*(terpy)_2}]^{2+}$$

where $k_1 = 9.0$ L mol^{-1} s^{-1}, $k_2 = 48$ L mol^{-1} s^{-1}, and $K = 3.57$, all at 0 °C. Setting $f = 1$ gives

$$k = (9.0 \times 48 \times 3.57)^{1/2} \text{ L mol}^{-1} \text{ s}^{-1} = 39 \text{ L mol}^{-1} \text{ s}^{-1}$$

This result compares reasonably well with the experimental value, which is 64 L mol^{-1} s^{-1}.

Exercise E15.5. For reactions with k_1 and k_2 as given in the example, what is the rate of electron transfer if the overall reaction has $E^{\ominus} = +1.00$ V?

The Marcus equation can be expressed as a linear free energy relation because the logarithm of the rate constants are proportional to the free energies of activation. Thus

$$2 \ln k = \ln k_1 + \ln k_2 + \ln K$$

implies that

$$2\Delta G^{\ddagger} = \Delta G_1^{\ddagger} + \Delta G_2^{\ddagger} + \Delta G^{\ominus}$$

Figure 15.11 shows the correlation between rates and overall free energy change for the oxidation of a series of substituted phenanthroline complexes of Fe(II) by Ce(IV) and Fe^{2+}(aq). The fit to the straight line is a sign of the success of the approximate Marcus equation.[7] Departures from the Marcus equation are taken as a sign of special features in outer-sphere reactions. These may include the barrier created by the change from high to low spin during electron transfer or a change in symmetry, such as from octahedral to distorted tetragonal. An important example is that of cobalt ammine complexes, in which electron exchange between Co(II) and Co(III) is quite slow. The spin factor may play a role because Co(II) complexes are high-spin d^7 and Co(III) are low-spin d^6. However, transfer of an electron into the σ antibonding LUMO of a low spin d^6 complex also requires a substantial tetragonal distortion.

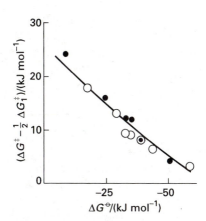

15.11 The correlation between rates and overall free energy change for the oxidation by Ce(IV) of a series of phenanthroline complexes of Fe(II) in sulfuric acid (open circles) and Fe^{2+}(aq) in perchloric acid (closed circles).

Inner-sphere reactions

Inner-sphere reactions are much more difficult to analyze because whereas an outer-sphere reaction must be an electron transfer, an inner-sphere reaction may be either electron or group transfer. In many cases, it is very difficult to distinguish the two.

The mechanism of any inner-sphere reaction must consist of three steps:

(1) Formation of the bridged (μ) complex:

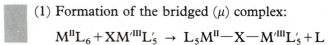

$$M^{II}L_6 + XM'^{III}L'_5 \rightarrow L_5M^{II}-X-M'^{III}L'_5 + L$$

---

[7] Tests of the Marcus equation are summarized in D. A. Pennington, in *Coordination chemistry*, Vol. 2, ACS Monograph 174, (ed. A. E. Martell). American Chemical Society, Washington (1978).

(2) The redox process:

$$L_5M^{II}-X-M'^{III}L'_5 \rightarrow L_5M^{III}-X-M'^{II}L'_5$$

(3) Decomposition of the successor complex to final products:

$$L_5M^{III}-X-M'^{II}L'_5 \rightarrow products$$

The rate determining step of the overall redox reaction may be any one of these processes, but the most common one is the redox step.

The reactions in which break-up of the successor complex is the rate determining step are common only if the configuration of both metals after electron transfer results in substitutional inertness, and then an inert binuclear complex is a characteristic product. A good example is the reduction of $[RuCl(NH_3)_5]^{2+}$ by $[Cr(OH_2)_6]^{2+}$ in which the rate determining step is the dissociation of $[Ru^{II}(NH_3)_5(\mu\text{-}Cl)Cr^{III}(OH_2)_5]^{4+}$.

Reactions in which the formation of the bridged complex is rate determining tend to be quite similar for a series of partners of a given species. For example, the oxidation of $V^{2+}(aq)$ proceeds at more or less the same rate for a long series of Co(III) oxidants with different bridging ligands. The explanation is that the rate determining step is the substitution of an H_2O molecule from the coordination sphere of V(II). As is typical for octahedral aqua ions, this reaction has a rate that depends only weakly on the identity of the entering group.

The large class of reactions in which the redox step is rate determining does not display such simple regularities. Rates vary over a wide range as metal ions and bridging ligands are varied. Because the change of oxidation states requires reorganization of ligands and solvent just as is the case for outer-sphere reactions, the trends indicated by the Marcus relation will also find some application to inner-sphere reactions. A very interesting way to study the role of the reorganization barriers is the synthesis of bridged complexes of the type

$$[(bipy)_2ClRu^{II} - N \bigcirc N - Ru^{III}(bipy)_2Cl]^{3+}$$

as models for inner sphere precursor complexes. The data in Table 15.13 show some typical variations as bridging ligand, oxidizing metal, and reducing metal are changed.

All the reactions in Table 15.13 result in the change of oxidation number by ± 1. Such reactions are still often called **one-equivalent processes**, the name reflecting the largely outmoded term 'chemical equivalent'. Their mechanisms require transfer of electrons or radicals. Similarly, reactions that result in the change of oxidation number by ± 2 are often called **two-equivalent processes** and may resemble nucleophilic substitutions. This resemblance can be seen by considering the reaction

$$[Pt^{II}Cl_4]^{2-} + [{}^*Pt^{IV}Cl_6]^{2-} \rightarrow [Pt^{IV}Cl_6]^{2-} + [{}^*Pt^{II}Cl_4]^{2-}$$

which occurs through a Cl^- bridge (15). The reaction depends on the

Table 15.13 Second order rate constants for selected inner-sphere reactions with variable bridging ligands

Oxidant	Reductant	Bridging ligand	$k/(\text{L mol}^{-1}\,\text{s}^{-1})$
$[\text{Co(NH}_3)_6]^{3+}$	$[\text{Cr(OH}_2)_6]^{2+}$		8×10^{-5}
$[\text{CoF(NH}_3)_5]^{2+}$	$[\text{Cr(OH}_2)_6]^{2+}$	F^-	2.5×10^5
$[\text{CoCl(NH}_3)_5]^{2+}$	$[\text{Cr(OH}_2)_6]^{2+}$	Cl^-	6.0×10^5
$[\text{CoI(NH}_3)_5]^{2+}$	$[\text{Cr(OH}_2)_6]^{2+}$	I^-	3.0×10^6
$[\text{Co(NCS)(NH}_3)_5]^{2+}$	$[\text{Cr(OH}_2)_6]^{2+}$	NCS^-	1.9×10
$[\text{Co(SCN)(NH}_3)_5]^{2+}$	$[\text{Cr(OH}_2)_6]^{2+}$	SCN^-	1.9×10^5
$[\text{Co(NH}_3)_5(\text{OH}_2)]^{3+}$	$[\text{Cr(OH}_2)_6]^{2+}$	H_2O	0.1
$[\text{CrF(OH}_2)_5]^{2+}$	$[\text{Cr(OH}_2)_6]^{2+}$	F^-	7.4×10^{-3}
$[\text{Co(NH}_3)_5\text{O}_2\text{C}-\!\!\bigcirc\!\!-\text{N}-\text{Ru(NH}_3)_4(\text{OH}_2)]$		$\text{O}_2\text{C}-\!\!\bigcirc\!\!-\text{N}$	1×10^2
$[\text{Co(NH}_3)_5\text{O}_2\text{C}$ ... $\bigcirc\!\!-\text{N}-\text{Ru(NH}_3)_4(\text{OH}_2)]$		O_2C ... $\bigcirc\!\!-\text{N}$	1.6×10^{-3}

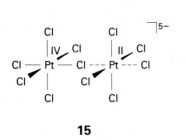

15

transfer of a Cl^- ion in the break-up of the successor complex. Note that an extra Cl^- is essential to make the activated complex symmetrical. It enters by adding to the four-coordinate Pt(II) center.

15.13 Oxidative addition

A reaction of the type

$$\text{IrCl(CO)L}_2 + \text{RI} \rightarrow \text{Ir(R)(I)(Cl)(CO)L}_2$$

where $L = \text{PPh}_3$ and R is an alkyl group is called an **oxidative addition**. (H_2 or HX, where X is a halogen, may replace the iodoalkane.) In the reaction, the d^8 square-planar complex uses the nonbonding axial electron pair to bond to the incoming Lewis acid (R^+ or H^+ in this case) and the lone pair of the base (X^- or H^-) coordinates to the metal acting as an acid. The net effect is a metal oxidized to the extent of two electrons, from d^8 to d^6. Because the alkyl group and the anion are both more electronegative than the metal, they formally acquire all four electrons involved in M—R and M—I bonding, so the oxidation number of the metal changes by $+2$ (corresponding to oxidation). Similar reactions are exhibited by the d^{10} species $[\text{Pd(PPh}_3)_4]$ and $[\text{Ni(P(OEt)}_3)_4]$.

There are three important mechanisms of oxidative addition. The first is a concerted electrophilic attack by the acid and nucleophilic attack by the base (Fig. 15.12(a)). Another mechanism involves nucleophilic attack on a C atom by the lone pair electrons of the metal atom (Fig. 15.12(b)). A third mechanism involves radicals; it is accelerated by O_2 and light and occurs one electron at a time (Fig. 15.12(c)). Loss of stereochemistry, inhibition by radical scavengers, and sensitivity to O_2 are common indicators of radical pathways.

Oxidative addition of H_2 (as depicted in Fig. 15.12(a)) always produces a *cis* dihydride.[8] The activation enthalpy for the reaction, which typically lies in the range 20 to 40 kJ mol^{-1}, is much smaller than the

[8] A. J. Kunin, C. E. Johnson, J. A. Maguire, W. D. Jones, and R. Eisenberg, *J. Am. Chem. Soc.*, **109**, 2963 (1987).

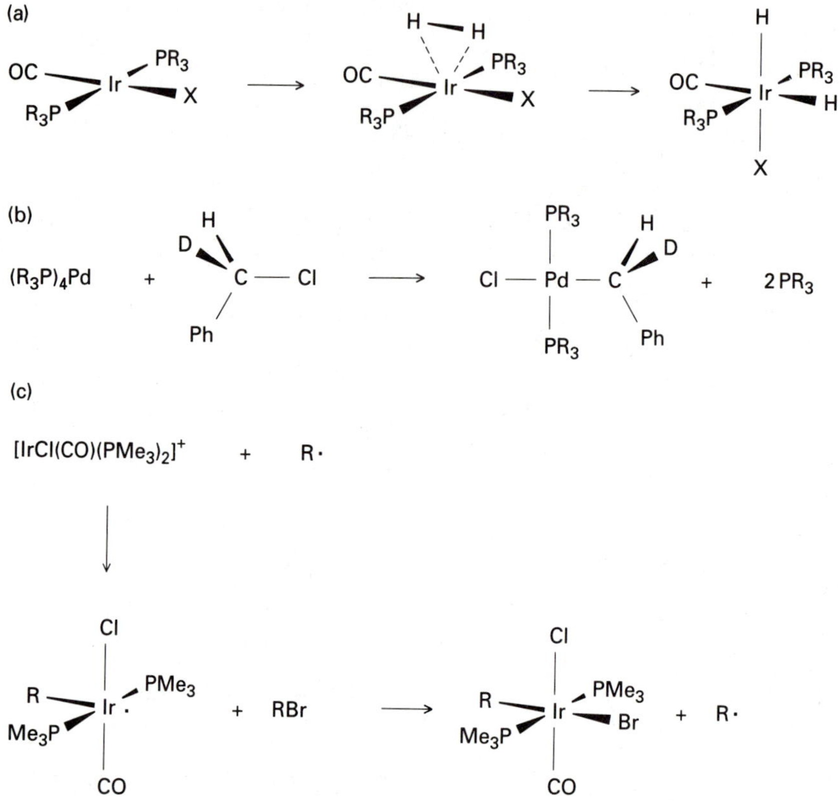

(a)

(b)

(c)

[IrCl(CO)(PMe$_3$)$_2$]$^+$ +   R·

15.12 Three mechanisms of oxidative addition. (a) Concerted reaction, which is characteristic of (but not limited to) H$_2$, (b) nucleophilic attack by the metal on the C atom, (c) a radical reaction (in this case, by a chain mechanism).

bond enthalpy of H$_2$ (436 kJ mol^{-1}), suggesting that little bond breaking has occurred in the formation of the activated complex. This inference is confirmed by replacing H by D. The fact that dihydrogen (as distinct from dihydrido) complexes have been prepared in which H$_2$ is bonded side-on to a metal atom suggests that the activated complex for oxidative addition also is a side-on M—H$_2$ arrangement that subsequently collapses to the *cis* dihydrido product.

The oxidative addition of haloalkanes often proceeds by nucleophilic attack of the metal atom on the halogen-bearing C atom. As shown in Fig. 15.12(b), attack by Pd displaces Cl$^-$ in the rate-determining step; the subsequent addition of Cl$^-$ is then rapid.[9] The fact that these reactions are not very stereoselective at the metal atom is consistent with this mechanism. In contrast, inversion does occur at the C atom, which demonstrates that nucleophilic displacement is indeed occurring. The reactivity order for haloalkanes indicates that steric hindrance does inhibit attack, for reaction rates lie in the order

$$CH_3X > CH_3CH_2X > CHR_2X > \text{cyclohexyl-X} \qquad X = \text{halogen}$$

Haloalkenes, haloarenes, and α-haloesters show evidence of the radical pathway depicted in Fig. 15.12c in reactions with [IrCl(CO)(PMe$_2$)$_2$].[10]

The reverse of oxidative addition is **reductive elimination**. As we

[9] J. Halpern, *Acc. Chem. Res.*, **3**, 386 (1970).
[10] J. A. Labinger, J. A. Osborne, and N. J. Coville, *Inorg. Chem.*, **19**, 3236 (1980).

shall see in Chapter 17, the combination of oxidative addition and reductive elimination can be used to modify organic compounds in ways that are useful in the construction of catalytic cycles. A simple example is the oxidative addition

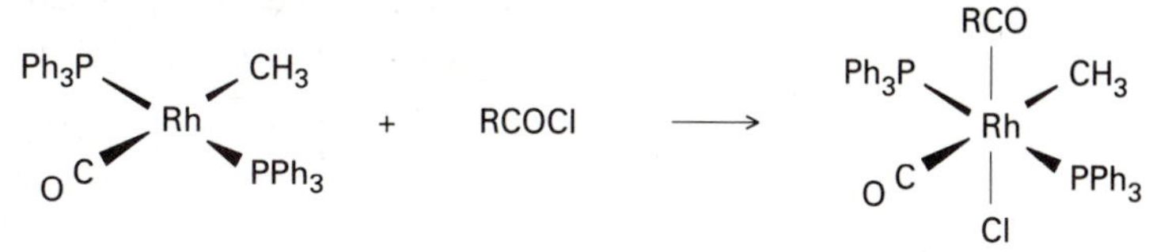

followed by the reductive elimination

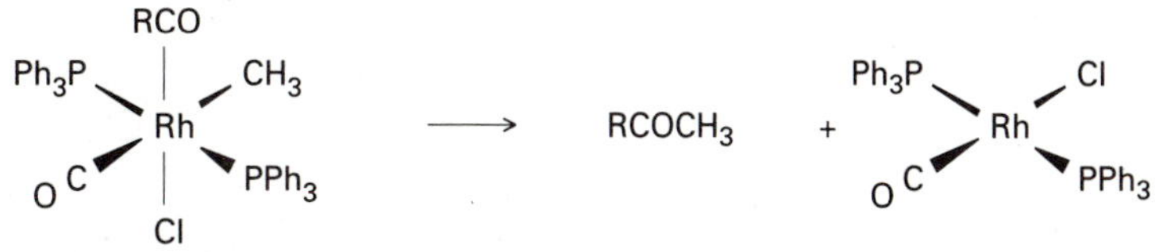

The overall reaction is the conversion of an acyl chloride to a ketone.

PHOTOCHEMICAL REACTIONS

The absorption of a photon increases the energy of a complex, typically by between 170 and 600 kJ mol^{-1}. Because these energies are larger than typical activation energies it should not be surprising that new reaction channels are opened. However, when the high energy of a photon is used to provide the energy of the primary forward reaction, the back reaction is almost always very favorable, and much of the design of efficient photochemical systems lies in trying to avoid it.

15.14 Prompt and delayed reactions

In some cases, the excited states dissociate promptly. Examples include formation of the pentacarbonyl intermediates that initiate ligand substitution in carbonyls

$$Cr(CO)_6 \xrightarrow{h\nu} Cr(CO)_5 + CO$$

and the scission of Co—Cl bonds

$$[Co^{III}Cl(NH_3)_5]^{2+} \xrightarrow{h\nu\,(\lambda<350\,nm)} [Co^{II}(NH_3)_5]^{2+}, \cdot Cl^-$$

Both these processes occur in less than 50 fs and hence are called **prompt reactions**.

In the second reaction, the **quantum yield**, the amount of reaction per mole of photons absorbed, increases as the wavelength is decreased and the photon energy increases. Energy in excess of the bond energy is available to the newly formed fragments and contributes to the probability that they will escape from each other through the solution before they have an opportunity to recombine.

Some excited states have long lifetimes. They may be regarded as energetic isomers of the ground state that can participate in **delayed**

reactions. The excited state of $[Ru^{II}(bipy)_3]^{2+}$ created by photon absorption in the metal-to-ligand charge transfer band (Section 15.3) may be regarded as a Ru(III) complexed to a radical anion of the ligand. Its redox reactions can be explained by adding the excitation energy (expressed as a potential using $-FE = \Delta G$ and equating ΔG to the molar excitation energy) to the ground state reduction potential:[11]

$$[Ru(bipy)_3]^{3+} + e^- \rightarrow [Ru(bipy)_3]^{2+}$$

$$E^\ominus = +1.26 \text{ V } (-122 \text{ kJ mol}^{-1})$$

$$[Ru(bipy)_3]^{2+} + h\nu(590 \text{ nm}) \rightarrow [{}^\ddagger Ru(bipy)_3]^{2+}$$

$$\Delta G = +202 \text{kJ mol}^{-1}$$

$$[Ru(bipy)_3]^{3+} + e^- \rightarrow [{}^\ddagger Ru(bipy)_3]^{2+}$$

$$E^{\ominus\ddagger} = -0.84 \text{ V } (+80 \text{ kJ mol}^{-1})$$

15.15 *d–d* and charge-transfer reactions

We have already seen (Sections 14.3 and 14.4) that there are two main types of spectroscopically observable electron promotion in metal complexes, namely *d–d* transitions and charge-transfer transitions. A *d–d* transition corresponds to the essentially angular redistribution of electrons within a *d* shell. In octahedral complexes, this redistribution often corresponds to the occupation of M–L antibonding e_g orbitals. An example is the ${}^4T_{1g} \leftarrow {}^4A_{2g}$ ($t_{2g}^2 e_g^1 \leftarrow t_{2g}^3$) transition in $[Cr(NH_3)_6]^{3+}$. The occupation of the antibonding e_g orbital results in a quantum yield close to 1 (specifically 0.6) for the photosubstitution

$$[Cr(NH_3)_6]^{3+} + H_2O \xrightarrow{h\nu} [Cr(NH_3)_5(OH_2)]^{3+} + NH_3$$

This is a prompt reaction, for it occurs in less than 5 ps.

Charge-transfer transitions correspond to the *radial* redistribution of electron density. They involve to the promotion of electrons into more predominantly ligand orbitals if the transition is metal-to-ligand or into orbitals of greater metal character if the transition is ligand-to-metal. The former process corresponds to metal oxidation and the latter to metal reduction. These excitations commonly initiate photoredox reactions of the kind already mentioned in connection with Co(III) and Ru(II).

Although it is a very useful first approximation to associate photosubstitution and photoisomerization with *d–d* transitions and photoredox with charge transfer transitions, the rule is not absolute. For example, it is not uncommon for a charge transfer band to result in photosubstitution by an indirect path:

$$[Co^{III}Cl(NH_3)_5]^{2+} + H_2O \rightarrow [Co^{II}(NH_3)_5(OH_2)]^{2+} + \cdot Cl$$

$$[Co^{III}(NH_3)_5(OH_2)]^{2+} + \cdot Cl \rightarrow [Co^{III}(NH_3)_5(OH_2)]^{3+} + Cl^-$$

---

[11] For a discussion, see T. J. Meyer, Excited state electron transfer, *Prog. Inorg. Chem.*, **30**, 389 (1983).

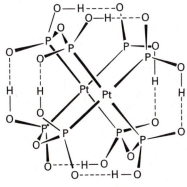

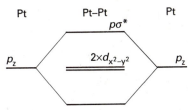

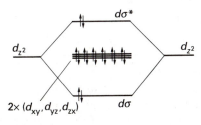

15.13 The dinuclear complex [Pt$_2$(μ-P$_2$O$_5$H$_2$)$_4$]$^{4-}$ called informally 'PtPOP'. The structure consists of two face-to-face square-planar complexes held together by a bridging pyrophosphito ligand. The metal p_z–p_z and d_{z^2}–d_{z^2} orbitals interact along the Pt—Pt axis. The other p and d orbitals are considered to be nonbonding.

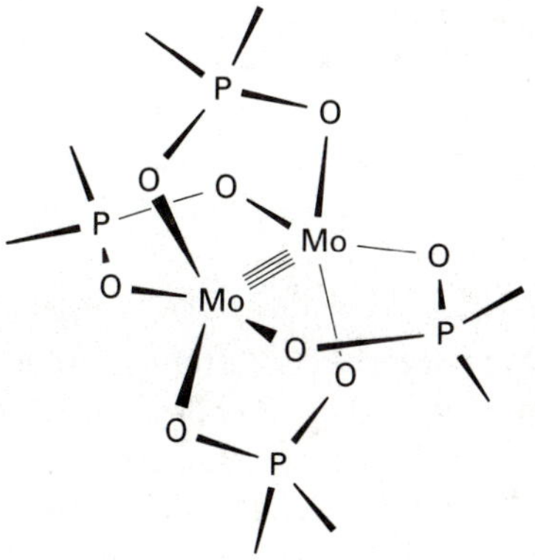

In this case, the aqua complex formed after the homolytic fission of the Co—Cl bond is reoxidized by the Cl atom. The net result leaves the Co oxidized and substituted. Conversely, the long lived ^{2}E state of [Cr(bipy)$_3$]$^{3+}$ is a pure d–d state which lacks substitutional reactivity in acidic solution. Its lifetime of several microseconds allows the excess energy to enhance its redox reactions. The standard potential (+1.3 V) calculated by adding the excitation energy to the ground state value accounts for its function as a good oxidizing agent, in which it undergoes reduction to [Cr(bipy)$_3$]$^{2+}$.

15.16 Transitions in metal–metal bonded systems

We encountered the $\delta^* \leftarrow \delta$ transition of multiple metal–metal bonds in Chapter 14, and it should not be surprising that population of an antibonding orbital of the metal–metal system can sometimes initiate photodissociation. It is more interesting that such excited states have been shown to initiate multielectron redox photochemistry.

One of the best characterized systems is the dinuclear platinum complex [Pt$_2$(μ-P$_2$O$_5$H$_2$)$_4$]$^{4-}$ called informally 'PtPOP' (Fig. 15.13).[12] There is no metal–metal bonding in the ground state of this Pt(II)-Pt(II) d^8–d^8 species. The HOMO–LUMO pattern shown in Fig. 15.13 indicates that excitation populates a bonding orbital between the metals. The lowest lying excited state has a lifetime of 9 μs. It is a powerful reducing agent, and reacts by both electron and halogen atom transfer. The most interesting oxidation products are Pt(III)—Pt(III) that contain X$^-$ ligands (where X is halogen or pseudohalogen) at both ends and a metal-metal single bond. Irradiation in the presence of (Bu)$_3$SnH gives a dihydrido product which can eliminate H$_2$.

Irradiation of the quadruply bonded dinuclear cluster Mo$_2$(O$_2$P(OC$_6$H$_5$)$_2$)$_4$ (**16**) at 500 nm in the presence of ClCH$_2$CH$_2$Cl results in production of ethene and the addition of two Cl atoms to the two Mo atoms, with a two-electron oxidation.[13] The reaction proceeds in steps of one electron, and requires a complex with the metals blocked by sterically crowding ligands. If smaller ligands are present, the reaction that occurs instead is a photochemical oxidative addition of the organic molecule.

FURTHER READING

Recent treatments are:

R. B. Jordan, *Reaction mechanisms of inorganic and organometallic systems*. Oxford University Press, New York (1991).

R. G. Wilkins, *Kinetics and mechanism of reactions of transition metal complexes*. VCH, Weinheim (1991).

J. D. Atwood, *Inorganic and organometallic reaction mechanisms*. Brooks-Cole (1985).

[12] The chemistry of this interesting complex is reviewed in D. M. Roundhill, H. B. Gray, and C.-M. Chi, *Acc. Chem. Res.*, **22**, 55 (1989).
[13] I. J. Chang and D. C. Nocera, *Inorg. Chem.*, **24**, 4309 (1989).

16

Important older books include:

F. Basolo and R. G. Pearson, *Mechanisms of inorganic reactions*. Wiley, New York (1967). This book is the classic in the field.

M. L. Tobe, *Inorganic reaction mechanisms*. Nelson, London (1972). This is a very readable treatment.

C. H. Langford and H. B. Gray, *Ligand substitution processes*. W. A. Benjamin, New York (1966). This book introduced the associative/dissociative notation.

A series of reviews summarizing current work is:

M. V. Twigg (ed.), *Mechanisms of inorganic and organometallic reactions*, Vols 1–6. Plenum Press, New York (1983–89).

KEY POINTS

1. Ligand substitution reactions
Ligand substitution reactions involve the replacement of a coordinated ligand by an incoming ligand from solution.

2. Types of mechanism
Stoichiometric mechanisms are designated D (dissociative) if the first step is dissociation (loss) of a coordinated ligand, A (associative) if the first step is coordination of an incoming ligand, and I (interchange) if the reaction involves addition of the incoming ligand with an expansion of the coordination shell in the activated complex (an energy maximum). Interchanges are further differentiated as I_a if the activated complex is primarily favored by bond making with the incoming ligand, or I_d if the energy maximum is determined primarily by breaking the bond with the outgoing ligand.

3. Identification of mechanism
Mechanisms are inferred from the concentration, temperature, and pressure dependence of rates, from trends in rates as a function of incoming and outgoing ligand, and from the stereochemistry of reactants and properties.

4. The trans effect
The structure of the products may be controlled by the *trans* effect in substitutions on square planar complexes.

5. Isomerization reactions
Isomerization reactions may occur by substitution pathways or by intermolecular rearrangements. Two of the latter for octahedral complexes are the Bailar twist and the Ray–Dutt twist. The Berry pseudorotation is common in five-coordinate complexes.

6. Redox reaction mechanisms
Redox reactions involve changes in oxidation state of the reacting complexes. Inter-sphere redox reactions involve a transient bridging ligand, which carries electron density from one metal atom to the other. Outer-sphere redox reactions involve electron transfer from one reactant species to the other.

7. Photochemical processes
Photochemical reactions involve the absorption of light to produce a highly reactive electronic state.

8. Prompt reactions
Prompt photochemical reactions often involve the rapid dissociation of a ligand upon photoexcitation. These photosubstitution reactions often result from *d–d* transitions.

9. Delayed reactions
Delayed photochemical reactions often involve electron transfer, either ligand-to-metal or metal-to-ligand. Long-lived excited states associated with electronic transitions in metal–metal bonded compounds may also lead to delayed photochemical reactions.

EXERCISES

15.1 Classify (a) NH_3, (b) Cl^-, (c) Ag^+, (d) S^{2-}, (e) Al^{3+} as nucleophiles or electrophiles.

15.2 The reactions of $Ni(CO)_4$ in which phosphanes or phosphites replace CO to give the family $Ni(CO)_3L$ occur at the same rate for different phosphanes or phosphites. Is the reaction *d* or *a*?

15.3 Figure 15.1 shows how the Gibbs free energy varies with reaction coordinate for the *D* process in which the rate determining step is the breaking of the bond to the leaving group X. Draw the corresponding diagram for a *D* reaction in which the rate determining step is the addition of the entering group Y to the intermediate.

15.4 In experiment (a), $^{36}Cl^-$ is added to the solution during the study of the *cis*-to-*trans* isomerism of $[CoCl_2(en)_2]^+$ to determine whether isomerization is accompanied by substitution. In experiment (b), D replaces H in the ammine ligands of $[Cr(NCS)_4(NH_3)_2]^-$ and reduces the rate of replacement of NCS^- by water. Which experiment studies the stoichiometric mechanism and which studies the intimate mechanism?

15.5 Write the rate law for formation of $[MnX(OH_2)_5]^+$ from the aqua ion and X^-. How would you undertake to determine whether the reaction is *d* or *a*?

15.6 Octahedral complexes of metal centers with high oxidation numbers or of *d* metals of the second and third series are less labile than those of the low oxidation number and *d* metals of the first series of the block. Account for this on the basis of dissociative activation.

15.7 A Pt(II) complex of tetraethyldiethylenetriamine is attacked by Cl^- 10^5 times less rapidly than the diethylenetriamine analog. Explain this observation in terms of associative activation.

15.8 The rate of loss of chlorobenzene, PhCl, from $[W(CO)_4L(PhCl)]$ increases with increase in the cone angle of L. What does this observation suggest about the mechanism? When PhCl is replaced by a phosphane ligand, the rate of reaction depends upon the concentration of the phosphane in experiments conducted at low phosphane concentration. Is this observation in conflict with the mechanistic conclusion drawn from cone angle effects?

15.9 Does the fact that $[Ni(CN)_5]^{3-}$ can be isolated help to explain why substitution reactions of $[Ni(CN)_4]^{2-}$ are very rapid?

15.10 Design two-step syntheses of *cis*- and *trans*-$[PtCl_2(NO_2)(NH_3)]^-$ starting from $[PtCl_4]^{2-}$.

15.11 How does each of the following modifications affect the rate of a square planar complex substitution reaction? (a) Changing a *trans* ligand from H to Cl. (b) Changing the leaving group from Cl to I. (c) Adding a bulky substituent to a *cis* ligand. (d) Increasing the positive charge on the complex.

15.12 The rate of attack on Co(III) by an entering group Y is nearly independent of Y with a spectacular exception of the rapid reaction with OH^-. Explain the anomaly. What is the implication of your explanation for the behavior of a complex lacking Brønsted acidity on the ligands?

15.13 Predict the products of the following reactions:
(a) $[Pt(PR_3)_4]^{2+} + 2Cl^-$
(b) $[PtCl_4]^{2-} + 2PR_3$
(c) *cis*-$[Pt(NH_3)_2(py)_2]^{2+} + 2Cl^-$

15.14 Put in order of increasing rate of substitution by H_2O the complexes (a) $[Co(NH_3)_6]^{3+}$, (b) $[Rh(NH_3)_6]^{3+}$, (c) $[Ir(NH_3)_6]^{3+}$, (d) $[Mn(OH_2)_6]^{2+}$, (e) $[Ni(OH_2)_6]^{2+}$.

15.15 State the effect on the rate of dissociatively activated reactions of Rh(III) complexes of (a) an increase in the overall charge on the complex; (b) changing the leaving group from NO_3^- to Cl^-; (c) changing the entering group from Cl^- to I^-; (d) changing the *cis* ligands from NH_3 to H_2O; (e) changing an ethylenediamine ligand to propylenediamine when the leaving ligand is Cl^-.

15.16 The mechanism of CO insertion is thought to be

$$RMn(CO)_5 \rightleftharpoons (RCO)Mn(CO)_4$$

$$(RCO)Mn(CO)_4 + L \rightleftharpoons (RCO)MnL(CO)_4$$

In the first step, the acyl intermediate is formed, leaving a vacant coordination position. In the second step, a ligand enters. At high ligand concentrations, the rate is independent of [L]. At this limit, what rate constant can be extracted from rate data?

15.17 The pressure dependence of the replacement of chlorobenzene (PhCl) by piperidine (pip) in the complex $[W(CO)_4(PPh_3)(PhCl)]$ has been studied. The volume of activation is found to be $+11.3$ cm^3 mol^{-1}. What does this value suggest about the mechanism?

15.18 Write out the inner- and outer-sphere pathways for reduction of azidopentaamminecobalt(III) ion with $V^{2+}(aq)$. What experimental data might be used to distinguish between the two pathways?

15.19 The intermediate $[Fe(SCN)(OH_2)_5]^{2+}$ can be detected in the reaction of $[Co(NCS)(NH_3)_5]^{2+}$ with $Fe^{2+}(aq)$ to give $Fe^{3+}(aq)$ and $Co^{2+}(aq)$. What does this observation suggest about the mechanism?

15.20 The photochemical substitution of $[W(CO)_5(py)]$ (py = pyridine) with triphenylphosphane gives $W(CO)_5(P(C_6H_5)_3)$. In the presence of excess phosphane, the quantum yield is approximately 0.4. A flash photolysis study reveals a spectrum that can be assigned to the intermediate $W(CO)_5$. What product and quantum yield do you predict for substitution of $[W(CO)_5(py)]$ in the presence of excess triethylamine? Is this reaction expected to be initiated from the ligand field or MLCT excited state of the complex?

15.21 From the spectrum of $[CrCl(NH_3)_5]^{2+}$ shown in Fig. 14.7, propose a wavelength for transient photoreduc-

tion of Cr(III) to Cr(II) accompanied by oxidation of a ligand.

15.22 Reactions of $[Pt(Ph)_2(SMe_2)_2]$ with the bidentate ligand, 1,10-phenanthroline (phen) give $[Pt(Ph)_2phen]$. There is a kinetic pathway with activation parameters $\Delta H^{\ddagger} = +101 \text{ kJ mol}^{-1}$ and $\Delta S^{\ddagger} = +42 \text{ J K}^{-1} \text{ mol}^{-1}$. Propose a mechanism.

PROBLEMS

15.1 Solutions of $[PtH_2(P(CH_3)_3)_2]$ exist as a mixture of *cis* and *trans* isomers. Addition of excess $P(CH_3)_3$ led to formation of $[PtH_2(P(CH_3)_3)_3]$ at a concentration that could be detected using NMR. This complex exchanged phosphane ligands rapidly with the *trans* isomer but not the *cis*. Propose a pathway. What are the implications for the *trans* effect of H versus $P(CH_3)_3$? (See D. L. Packett and W. G. Trogler, *Inorg. Chem.*, **27**, 1768 (1988).)

15.2 Given the following mechanism for the formation of a chelate complex

$$[Ni(OH_2)_6]^{2+} + L—L \rightleftarrows [Ni(OH_2)_6]^{2+}, L—L$$

$$K_E, \text{ rapid}$$

$$[Ni(OH_2)_6]^{2+}, L—L \rightleftarrows [Ni(OH_2)_5L—L]^{2+} + H_2O$$

$$k_a, k'_a$$

$$[Ni(OH_2)_5L—L]^{2+} \rightleftarrows [Ni(OH_2)_4LL]^{2+} + H_2O$$

$$k_b, k'_b$$

Derive the rate law for the formation of the chelate. Discuss the step that is different from that for two monodentate ligands. It is found that the formation of chelates with strongly bound ligands occurs at the rate of formation of the analogous monodentate complex, but that the formation of chelates of weakly bound ligands is often significantly slower. Assuming an I_d mechanism, explain this observation. (See R. G. Wilkins, *Acc. Chem. Res.*, **3**, 408 (1970).)

15.3 The complex $[PtH(PEt_3)_3]^+$ was studied in deuterated acetone in the presence of excess PEt_3, where Et is CH_3CH_2. In the absence of excess ligand the ^{1}H-NMR spectrum in the hydride region exhibits two triplets. As excess PEt_3 ligand is added the triplets begin to collapse, the line shape depending on the ligand concentration. Give a mechanism to account for the effects of excess PEt_3. The coupling to ^{195}Pt and the protons of the ethyl groups have been eliminated from the spectrum illustrated.

15.4 The substitution reactions of the bridged dinuclear Rh(II) complex $[Rh_2(\mu\text{-}O_2CCH_3)_4XY]$ (17) have been studied by M. A. S. Aquino and D. H. Macartney (*Inorg. Chem.*, **26**, 2696 (1987)). Reaction rates show little dependence on the choice of the entering group. The table below shows the dependence on the leaving group, X, and the

ligand on the opposite Rh, (*trans*) Y at 298 K. What conclusions can you draw concerning the mechanism?

X	Y	$k/\text{L mol}^{-1}\text{ s}^{-1}$
H_2O	H_2O	10^5–10^7
CH_3OH	CH_3OH	2×10^6
CH_3CN	CH_3CN	1.1×10^5
PPh_3	PPh_3	1.5×10^5
CH_3CN	PR_3	10^7–10^9
PR_3	CH_3CN	10^{-1}–10^2
N-donor	H_2O	10^2–10^3

Note that the complex has a d^7 configuration at each Rh and a single Rh—Rh bond.

15.5 Figure 15.14 (which is from J. B. Goddard and F. Basolo, *Inorg. Chem.*, **7**, 936 (1968)) shows the observed first-order rate constants for the reaction of $[PdBrL]^+$ with various Y^- to give $[PdYL]^+$ where L is $Et_2NCH_2CH_2NHCH_2CH_2NH_2$. Notice the large slope for $S_2O_3^{2-}$ and zero slopes for $X = N_3^-$, I^-, NO_2^-, and SCN^-. Propose a mechanism.

15.6 The activation enthalpy for the reduction of *cis*-$[CoCl_2(en)_2]^+$ by $Cr^{2+}(aq)$ is -24 kJ mol^{-1}. Explain the negative value. (See R. C. Patel, R. E. Ball, J. F. Endicott, and R. G. Hughes, *Inorg. Chem.*, **9**, 23 (1970).)

17

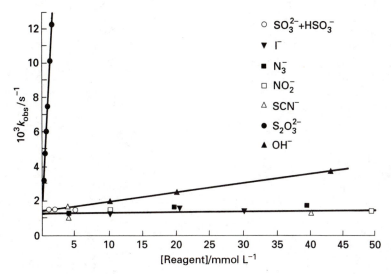

15.14 The data required for Problem 15.5.

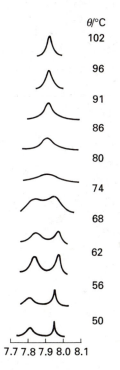

15.15 The ^{1}H-NMR spectra required for Problem 15.8.

15.7 The rate of reduction of $[Co(NH_3)_5(OH_2)]^{3+}$ by Cr(II) is seven orders of magnitude slower than reduction of its conjugate base, $[Co(NH_3)_5(OH)]^{2+}$ by Cr(II). For the corresponding reductions with $[Ru(NH_3)_6]^{2+}$ the two differ by less than a factor of 10. What do these observations suggest about mechanisms? Comment on H_2O and OH^- as bridging ligands.

15.8 Using the NMR spectra in Fig. 15.15, estimate the rate of exchange between free and coordinated ammonia for the alkyl Co(III) complex represented. (See R. J. Guschl, R. Stewart, and T. L. Brown, *Inorg. Chem.*, **13**, 959 (1974).)

15.9 Calculate the rate constants for outer sphere reactions from the following data. Compare your results to the measured values in the last column.

Reaction	k_1/(L mol^{-1} s^{-1})	k_2/(L mol^{-1} s^{-1})	$E^{\ominus}$/V	k_{obs}/(L mol^{-1} s^{-1})
$Cr^{2+} + Fe^{2+}$	2×10^{-5}	4.0	$+1.18$	2.3×10^3
$[W(CN)_8]^{4-}$ $+ Ce(IV)$	$>4 \times 10^4$	4.4	$+0.90$	$>10^8$
$[Fe(CN)_6]^{4-}$ $+ MnO_4^-$	7.4×10^2	3×10^3	$+0.20$	1.7×10^5
$[Fe(phen)_3]^{2+}$ $+ Ce(IV)$	$>3 \times 10^7$	4.4	$+0.36$	1.4×10^5

15.10 Oxidative addition of CH_3I to the Rh(I) complex (**18**) has the rate law:

$$k_{obs} = k_S + k_2[CH_3I]$$

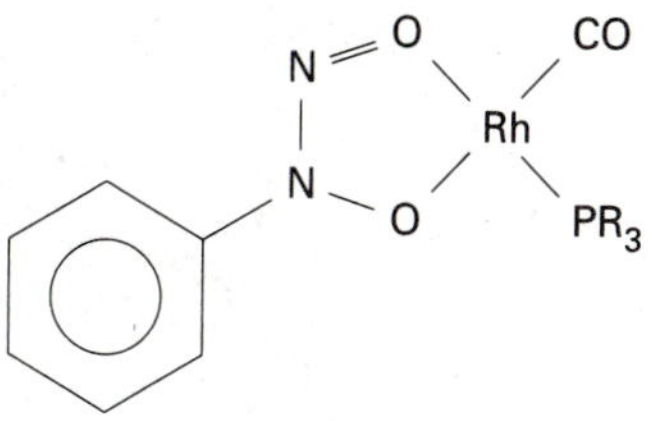

18

The term k_S is sensitive to solvent nucleophilicity. It is suggested (S. J. Basson, J. G. Leipoldt, A. Roodt, and J. A. Venter, *Inorg. Chim. Acta*, **128**, 31 (1987)) that the rate law implies competing pathways for the initial step; one involving electrophilic attack by the CH_3 group to give a five coordinate species and a second in which the solvent forms the five coordinate species prior to CH_3I attack. Write pathways for the oxidative addition and identify the electrophilic and nucleophilic steps.

15.11 Figure 15.16 (which is from W. R. Muir and C. H. Langford, *Inorg. Chem.*, **7**, 1032 (1968)) shows plots of observed first-order rate constants against $[X^-]$ for the reactions

$$[Co(NO_2)(dmso)(en)_2]^{2+} + X^-$$

$$\rightarrow [Co(NO_2)X(en)_2]^{2+} + dmso$$

Assume that all three reactions have the same mechanism. (a) If the mechanisms were D, to what would the limiting rate constants correspond? (b) If the mechanisms were I_d, to what would the limiting rate correspond? (c) What is the significance of the fact that the dmso exchange rate constant is larger than the limits for anion attack? (d) The limiting

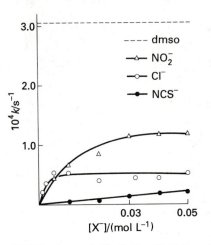

15.16 The data required for Problem 15.11. Rate of reaction of *cis*-$[Co(en)_2NO_2dmso]^{2+}$ ion as a function of the concentration of the entering anion X^-: $\triangle$, $X^- = NO_2^-$; $\bigcirc$, $X^- = Cl^-$; $\bullet$, $X^- = SCN^-$. The broken line shows the rate of the dmso-exchange reaction.

rate constants are 5×10^{-4} s^{-1}, 1.2×10^{-4} s^{-1}, and about 1×10^{-4} s^{-1} for Cl^-, NO_2^-, and NCS^-, respectively. Does the evidence favor a D or an I_d mechanism?

16

d- and *f*-block organometallic compounds

Much of the basic organometallic chemistry of the p-block metals was understood by the early part of the twentieth century, but that of the d and f blocks has been developed much more recently. Since the mid 1950s the latter field has grown into a thriving area that spans interesting new types of reactions, unusual structures, and practical applications in organic synthesis and industrial catalysis. We begin by considering the structures, bonding, and reactions of metal carbonyls, which form the foundation of much of d-block organometallic chemistry. Then we consider hydrocarbon ligands, and their variety of structures, bonding modes, and reactions. Finally, we deal with the structures and reactions of metal cluster compounds, which has been one of the most recent areas of the subject to develop. Chapter 17 will take the story further by describing how d-block organometallic compounds are used in catalysis.

A few *d*-block organometallic compounds were synthesized and partially characterized in the nineteenth century. The first of them (**1**), an ethene complex of platinum(II), was prepared by W. C. Zeise in 1827. The next major discovery was the metal carbonyl tetracarbonylnickel (**2**), which was synthesized by Ludwig Mond, Carl Langer, and Friedrich Quinke in 1890. Beginning in the 1930s, Walter Hieber in Munich synthesized a wide variety of metal carbonyl cluster compounds, many of which are anionic, including $[Fe_4(CO)_{13}]^{2-}$ (**3**). It was clear from this work that metal carbonyl chemistry was potentially a very rich field. However, as the structures of these and other *d*- and *f*-block organometallic compounds are difficult or impossible to deduce by chemical means alone, fundamental advances had to await the development of X-ray diffraction for precise structural data on solid samples and of IR and NMR spectroscopy for structural information about organometallic complexes in solution. The discovery of the

1 $[PtCl_3(C_2H_4)]^-$ **2** $Ni(CO)_4$

3 $[Fe_4(CO)_{13}]^{2-}$

4 $Fe(C_5H_5)_2$

5

remarkably stable organometallic compound ferrocene, $Fe(C_5H_5)_2$, occurred at a time (in 1951) when these techniques were becoming widely available. The 'sandwich' structure of ferrocene (**4**) was soon correctly inferred from its IR spectrum and then determined in detail by X-ray crystallography.

The stability, structure, and bonding of ferrocene defied the classical Lewis description and therefore captured the imagination of chemists. This puzzle in turn set off a train of synthesizing, characterizing, and theorizing that led to rapid development of *d*-block organometallic chemistry. Two highly productive research workers in the formative stage of the subject, Ernst Fischer in Munich and Geoffrey Wilkinson in London, were awarded the Nobel Prize in 1973 for their contributions.[1] Similarly, *f*-block organometallic chemistry blossomed soon after the discovery in the late 1970s that the pentamethylcyclopentadienyl ligand, C_5Me_5, forms stable *f*-block compounds (**5**).

We shall adhere to the convention that an **organometallic compound** contains at least one metal–carbon (M—C) bond. Thus, compounds (**1**) to (**5**) clearly qualify as organometallic, whereas a complex such as $[Co(en)_3]^{3+}$, which contains carbon but has no C—M bonds, does not. Cyano complexes, such as hexacyanoferrate(II) ions, do have M—C bonds, but their properties are more akin to those of Werner complexes and they are generally not considered to be organometallic. In contrast, complexes of the isoelectronic ligand CO are considered to be organometallic. The justification for this somewhat arbitrary distinction is that many metal carbonyls are significantly different from Werner complexes both chemically and physically.

Following the recommended convention, we shall use the same nomenclature as for Werner complexes (Section 6.2). Ligands are listed in alphabetical order followed by the name of the metal, all of which is written as one word. The name may be followed by the oxidation number of the metal in parentheses. The nomenclature employed in research journals, however, does not always obey these rules strictly, and it is common to find the name of the metal buried in the middle of the name of the compound and the oxidation state of the metal omitted. For example, (**6**) is sometimes referred to as benzenemolybdenumtricarbonyl, rather than the preferred name benzene(tricarbonyl)molybdenum(0). All ligands that are radicals when neutral have the suffix -yl, as in methyl, cyclopentadienyl, and allyl.

The IUPAC recommendation for formulas is to write them in the same form as for Werner complexes: metal first followed by formally anionic ligands, listed in alphabetical order. The neutral ligands are then listed in alphabetical order based on their chemical symbol. We shall follow these conventions unless a different order of ligands helps to clarify a particular point.

A further notational point is that organic ligands possess a special kind of versatility: a single ligand may attach to a central metal atom using several of its atoms simultaneously (ferrocene is an example).

---

[1] Wilkinson's personal account captures the excitement of these developments: *J. Organomet. Chem.*, **100**, 273 (1975).

The **hapticity**, η, of a ligand is the number of its atoms that are within bonding distance of the metal atom.[2] The definition is independent of a model of the bonding itself. For example, a CH_3 group attached by a single M—C bond is **monohapto**, η^1. The ethene ligand is **dihapto**, η^2, if its two C atoms are both within bonding distance of the metal (**1**). The bonding of C_5H_5 to the Fe atom in ferrocene (**4**) is **pentahapto**, η^5. The C_5H_5 ligand can also be monohapto (**7**) and trihapto (**8**).

BONDING

In organometallic chemistry, as in most areas of chemistry, the concepts of closed electronic shells and (to a lesser extent) oxidation numbers help to rationalize the structures and reactivity of the compounds.

16.1 Valence electron count

In the 1920s, the British chemist N. V. Sidgwick recognized that the metal atom in a simple metal carbonyl, such as $Ni(CO)_4$, has the same valence electron count (18) as the noble gas that terminates the long period to which the metal belongs. Sidgwick coined the term 'inert gas rule' for this indication of stability, but it is now usually referred to as the **18-electron rule**. It should be noted, however, that the 18-electron rule is not as uniformly obeyed for *d*-block organometallic compounds as the octet rule is obeyed for compounds of Period 2 elements. The 18-electron rule does not apply to *f*-block complexes. We shall see that some of the exceptions to the rule can be rationalized.

When electrons are counted, each metal atom and ligand is treated as neutral. If the complex is charged, then we simply add or subtract the appropriate number of electrons to the total. We must include in the count all valence electrons of the metal atom and all the electrons donated by the ligands. For example, $Fe(CO)_5$ acquires 18 electrons from the eight valence electrons on the Fe atom and the ten electrons donated by the five CO ligands. Table 16.1 lists the maximum number of electrons available for donation to a metal for several common ligands.

The 18-electron rule and formulas of metal carbonyls

The 18-electron rule systematizes the formulas of metal carbonyls. As shown in Table 16.2, the carbonyls of the Period 4 elements of Groups 6 to 10 have alternately one and two metal atoms and a decreasing number of CO ligands. The two-metal carbonyls are formed by elements of the odd-numbered groups, which have an odd number of valence electrons and therefore dimerize by forming metal–metal (M—M) bonds. The decrease in the number of CO ligands per metal atom from left to right across a period matches the need for fewer CO ligands to achieve 18 valence electrons.

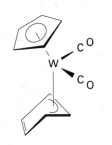

6 $Mo(\eta^6\text{–}C_6H_6)(CO)_3$

7 $Ti(\eta^1\text{–}C_5H_5)_2(\eta^5\text{–}C_5H_5)_2$

8 $W(\eta^3\text{–}C_5H_5)(\eta^5\text{–}C_5H_5)(CO)_2$

[2] This terminology is based on the Greek word haptein (ἅπτειν), which means to fasten. It was introduced by F.A. Cotton, *J. Am. Chem. Soc.*, **90**, 6230 (1968).

Table 16.1 Some organic ligands

Available electrons*	Hapticity	Ligand	Metal–ligand structure
1	η^1	Methyl, alkyl $\cdot CH_3, \cdot CH_2R$	M — CH_3
2	η^1	Alkylidene (carbene)	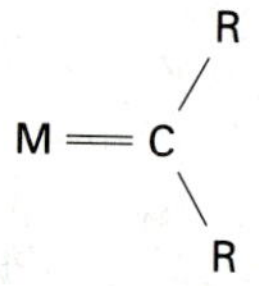
2	η^2	Alkene $H_2C=CH_2$	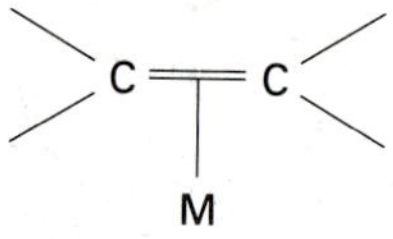
3	η^3	π-Allyl C_3H_5	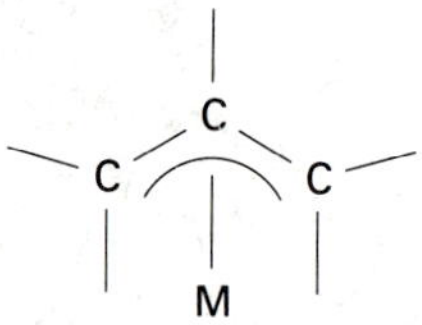
3	η^1	Alkylidyne (carbyne) $C-R$	$M\equiv C$ — R
4	η^4	1,3-Butadiene C_4H_6	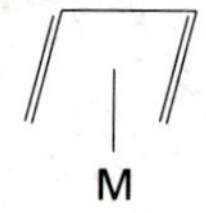
4	η^4	Cyclobutadiene C_4H_4	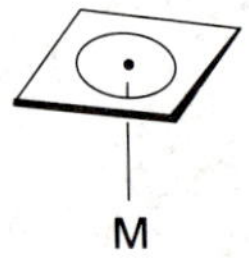
5 (3) (1)	η^5 η^3 η^1	Cyclopentadienyl C_5H_5 (Cp)	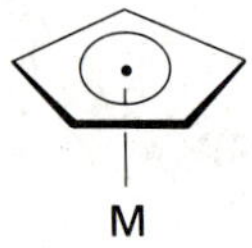
6	η^6	Benzene C_6H_6	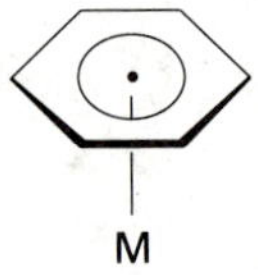
6	η^7	Tropylium $C_7H_7^+$	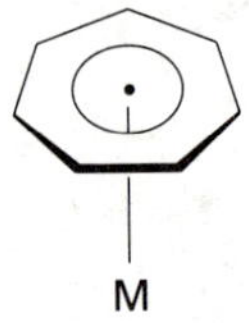
6	η^6	Cycloheptatriene C_7H_8	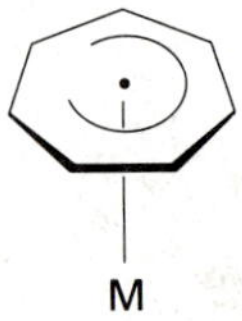

Table 16.1 (*continued*)

Available electrons*	Hapticity	Ligand	Metal–ligand structure
8†	η^8	Cyclooctatetraene	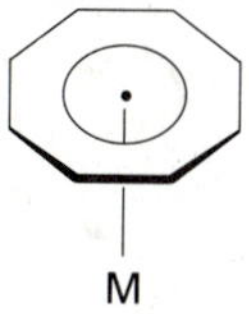
(6)	η^6	C_8H_8 (cot)	
(4)	η^4		

*For neutral ligands.

†As with several other polyene ligands in this list, the cot ligand may bond to a metal through only some of the available electron pairs. Variable hapticity is common with cot.

Table 16.2 Formulas and electron count for some Period 4 carbonyls

Group	Formula	Valence electrons		Structure
6	$Cr(CO)_6$	Cr	6	
		6(CO)	12	
			$\overline{18}$	
7	$Mn_2(CO)_{10}$	Mn	7	
		5(CO)	10	
		M—M	1	
			$\overline{18}$	
8	$Fe(CO)_5$	Fe	8	
		5(CO)	10	
			$\overline{18}$	
9	$Co_2(CO)_8$	Co	9	
		4(CO)	8	
		M—M	1	
			$\overline{18}$	
10	$Ni(CO)_4$	Ni	10	
		4(CO)	8	
			$\overline{18}$	

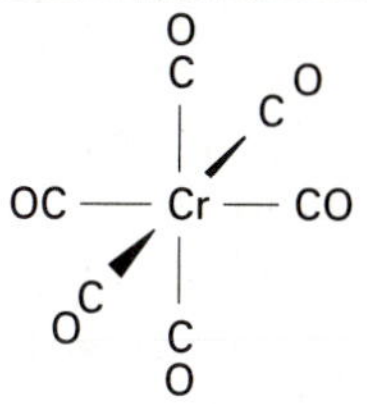

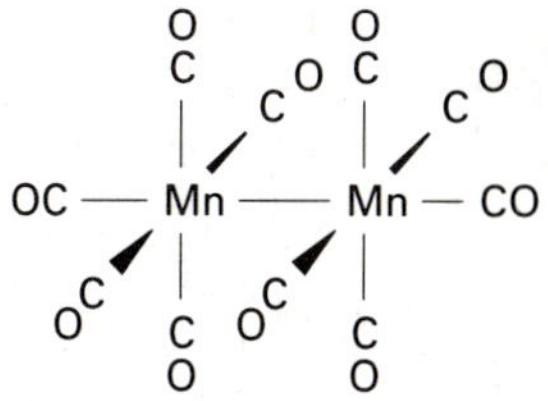

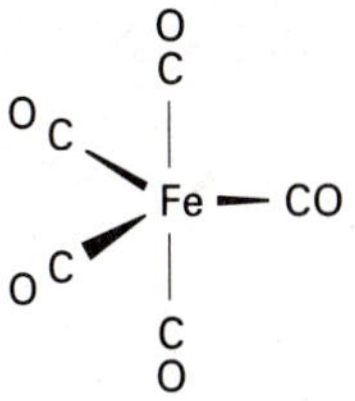

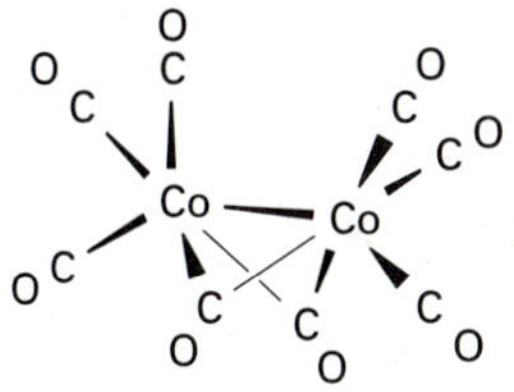

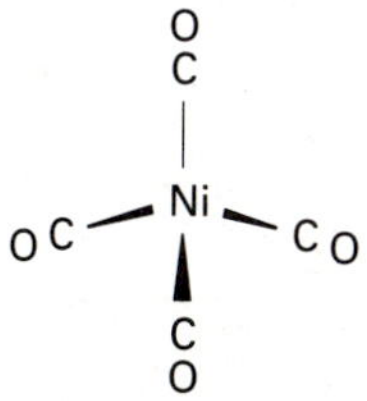

Table 16.3 Scope of the 16/18-electron rule for *d*-block organometallic compounds

Usually less than 18			Usually 18 electrons			16 or 18 electrons	
Sc	Ti	V	Cr	Mn	Fe	Co	Ni
Y	Zr	Nb	Mo	Tc	Ru	Rh	Pd
La	Hf	Ta	W	Re	Os	Ir	Pt

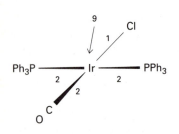

9 $Fe_2(CO)_4(C_5H_5)_2$

10 *trans*–[IrCl(CO)(PPh$_3$)$_2$]

The same principles apply when other soft ligands occupy the metal coordination shell. Two examples are benzene(tricarbonyl)molybdenum(0) (**6**) and bis(cyclopentadienyl)(tetracarbonyl)diiron(0) (**9**), both of which obey the 18-electron rule. The 18-electron rule can be extended to simple polynuclear carbonyls by adding one electron to the valence-electron count of a specific metal atom in the cluster for each M—M bond to that metal atom. However, we shall see later that the 18-electron rule does not apply to M_6 and larger clusters.

Sixteen-electron complexes

Organometallic complexes with 16 valence electrons are common on the right of the *d* block, particularly in Groups 9 and 10 (Table 16.3). Examples of such complexes (which are generally square planar) include [IrCl(CO)(PPh$_3$)$_2$] (**10**) and the anion of Zeise's salt, [PtCl$_3$(C$_2$H$_4$)]$^-$ (**1**). Square-planar 16-electron complexes are particularly common for the d^8 metals of the heavier elements in Groups 9 and 10, especially for Rh(I), Ir(I), Pd(II), and Pt(II). It should be recalled from Section 6.5 that the ligand-field stabilization energy of d^8 complexes favors a low-spin square planar configuration when Δ is large, as is typical of Period 5 and 6 *d*-metal atoms and ions. The $d_{x^2-y^2}$ orbital is then empty and d_{z^2}, which is doubly occupied, is stabilized.

Example 16.1: *Counting the valence electrons on a metal atom*
Do (a) IrBr$_2$(CH$_3$)(CO)(PPh$_3$) and (b) Cr(η^5-C$_5$H$_5$)(η^6-C$_6$H$_6$) obey the 18-electron rule?

Answer. (a) An Ir atom (Group 9) has nine valence electrons; the two Br atoms and the CH$_3$ radical are one-electron donors; CO and PPh$_3$ are two-electron donors. Thus the number of valence electrons on the metal atom is $9 + 3 \times 1 + 2 \times 2 = 16$, in accord with the common occurrence of either 16 or 18 electron complexes of Group 9 metals. (b) A Cr atom (Group 6) has six valence electrons, the η^5-C$_5$H$_5$ ligand donates five electrons, and the η^6-C$_6$H$_6$ ligand donates six; so the number of metal valence electrons is $6 + 5 + 6 = 17$. This complex does not obey the 18 electron rule and is not stable. A related but stable 18-electron compound is Cr(η^6-C$_6$H$_6$)$_2$.

Exercise E16.1. Is Mo(CO)$_7$ likely to exist?

Exceptions to the 16/18 electron rule[3]

Exceptions to the 16/18 electron rule are common on the left of the *d* block. Here steric and electronic factors are in competition, and it is not possible to crowd enough ligands around the metal either to satisfy the rule or to permit dimerization. For example, the simplest carbonyl in Group 5, which is $V(CO)_6$, is a 17-electron complex. Other examples of the deviation include $W(CH_3)_6$, which has 12 electrons, and $Cr(CO)_2(\eta^5\text{-}C_5H_5)(PPh_3)$, with 17 electrons. The latter compound provides a good example of the role of steric crowding. When the compact CO ligand is present in place of bulky triphenylphosphane, a dimeric compound with a long but definite $Cr—Cr$ bond is observed in the solid state and in solution. The formation of the $Cr—Cr$ bond in $[Cr(CO)_3(\eta^5\text{-}C_5H_5)]_2$ raises the electron count on each metal to 18.

Deviations from the 16/18-electron rule are common in neutral bis(cyclopentadienyl) complexes, which are known for most of the *d*-block metals. Since two η^5-cyclopentadienyl ligands jointly contribute ten valence electrons, the 18-electron rule can be satisfied only for neutral compounds of the Group 8 metals. The complexes of this kind that do obey the 18-electron rule (such as ferrocene itself) are the most stable. Their stability is suggested by their bond lengths and their redox reactions. For example, the 19-electron complex $[Co(C_5H_5)_2]$ is readily oxidized to the 18-electron cation $[Co(\eta^5\text{-}C_5H_5)_2]^+$. The bonding and electronic structures of the biscyclopentadienyl complexes are described in greater detail in Section 16.8. A broad-guide is that the 18-electron rule is reliable only when the central *d*-metal atom has a low charge, so that the 3*d*, 4*s*, and 4*p* orbitals have similar energies.

16.2 Oxidation numbers and formal ligand charges

Oxidation numbers are of less importance in the organometallic chemistry of the *d*-block metals than in many other areas of inorganic chemistry (such as the chemistry of ammine and halo complexes of the *d*-block metals or the oxo-chemistry of the halogens). However, oxidation numbers do help to systematize reactions such as oxidative addition (Section 15.13), which we shall consider further in Section 16.7. They also bring out analogies between the chemical properties of organometallic complexes and Werner complexes. Thus, the oxidation states U(IV), Fe(II), and Fe(III) are known in both types of complex. The principal difference between organometallic complexes and Werner complexes with hard ligands is that the soft organic ligands in the former help to stabilize metals in low oxidation states. In organometallic complexes, as in halides and oxides, there is a decreasing ability for the metal to achieve high oxidation numbers on moving from Group 6 to the right of the *d* block.

The rules for calculating the oxidation number of an element in an organometallic complex are the same as for conventional compounds.

---

[3] For a series of articles on 17- and 19-electron organometallic compounds see. W. C. Trogler (ed.), *Organometallic radical processes*, Elsevier, Amsterdam (1990); D. R. Tyler, *Acc. Chem. Res.*, **24**, 325 (1991).

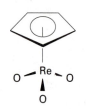

11 Re(η^5–C_5H_5)O_3

Accordingly, nonmetal ligands such as —H, —CH_3, and —C_5H_5 are formally considered to take an electron from the metal atom, and hence are assigned oxidation number -1. Thus, ferrocene can be considered to be a compound of Fe(II) and the ferrocenium ion, $[Fe(\eta^5\text{-}C_5H_5)_2]^+$, a compound of Fe(III). It follows that the formal name of ferrocene is bis(η^5-cyclopentadienyl)iron(II). The oxidation number of the metal atom is often omitted when no ambiguity results. When planar, the η^8-cyclooctatetraene ligand is considered to extract two electrons and to become (formally) the aromatic dinegative ligand $[C_8H_8]^{2-}$. With two such dinegative ligands in uranocene $U(\eta^8\text{-}C_8H_8)_2$, the U atom must be assigned an oxidation number of $+4$, and the formal name of uranocene is bis(η^8-cyclooctatetraenyl)uranium(IV).

In addition to the high oxidation state of uranium in uranocene, another example of a compound with a metal in high oxidation state is the unstable complex $W(CH_3)_6$, which contains formal W(VI). Another example is Re(VII) in Re(η^5-C_5H_5)O_3 (**11**).

Example 16.2: *Assigning oxidation states*
Determine the oxidation states of iridium and chlorine in $IrCl(CO)(PPh_3)_2$.

Answer. The ligands CO and PPh_3 are neutral two-electron donors and thus do not influence the oxidation state of iridium. A Cl atom is a one-electron ligand, and in combination with metals is assigned oxidation number -1. Because the complex as a whole is electrically neutral (implying that the oxidation numbers sum to zero), the oxidation state of the Ir atom is $+1$.

Exercise E16.2. Assign the oxidation state of cobalt in $Co(\eta^5\text{-}C_5H_5)(CO)_2$.

As remarked in Section 16.1, the valence electron count on a metal in an organometallic compound can be determined without assigning oxidation states. Therefore, it is sensible to ignore oxidation states when counting the number of valence shell electrons on a metal atom in an organometallic compound. To establish the electron count, it is necessary to consider only the number of electrons on the neutral metal atom, the electrons donated by neutral ligands, and the overall charge of the compound.

d-BLOCK CARBONYLS

A **homoleptic carbonyl** is a complex containing only carbonyl ligands. Simple homoleptic metal carbonyls can be prepared for most of the *d* metals, but those of palladium and platinum are so unstable that they exist only at low temperatures, and no simple neutral metal carbonyls are known for copper, silver, and gold or for the members of Group 3.[4] The metal carbonyls are useful synthetic precursors for

---

[4] The silver carbonyl cation $[OCAgCO]^+$ is described by P. K. Hurlburt, J. J. Rack, S. F. Dec, O. P. Anderson, and S. H. Strauss, *Inorg. Chem.*, **32**, 373 (1993).

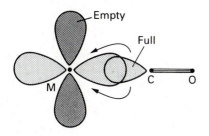

12

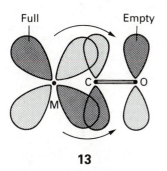

13

Table 16.4 The influence of coordination and charge on CO stretching wavenumbers

Compound	$\bar{\nu}/cm^{-1}$
CO(g)	2143
$[Mn(CO)_6]^+$	2090
$Cr(CO)_6$	2000
$[V(CO)_6]^-$	1860
$[Ti(CO)_6]^{2-}$	1750

Source: K. Nakamoto, *Infrared and Raman spectra of inorganic and coordination compounds*, Wiley, New York (1986); data for $[Ti(CO)_6]^{2-}$ are from S. R. Frerichs, B. K. Stein, and J. E. Ellis, *J. Am. Chem. Soc.*, **109**, 5558 (1987).

other organometallics and are used in organic syntheses and as industrial catalysts.

16.3 Carbon monoxide as a ligand

Carbon monoxide is the most common π-acceptor ligand in organometallic chemistry. Its primary mode of attachment to metal atoms is through the C atom.

The molecular orbitals of CO

The molecular orbital scheme for CO (Fig. 16.1) shows that the HOMO has σ symmetry and is essentially a lobe that projects away from the C atom. When CO acts as a ligand, this σ orbital serves as a very weak donor to a metal atom, and forms a σ bond with the central metal atom (**12**). The LUMOs of CO are the π* orbitals. These two orbitals play a crucial role because they can overlap metal *d* orbitals that have local π symmetry (such as the t_{2g} orbitals in an O_h complex (**13**)). The π interaction leads to the delocalization of electrons from filled *d* orbitals on the metal into the empty π* orbitals on the CO ligands, so the ligand also acts as a π acceptor. The ability of CO to delocalize electron density from the metal accounts for the prevalence of *d*-metal carbonyls in zero or negative oxidation states.

Trends in CO bond lengths and stretching frequencies obtained from IR spectra are in general agreement with the bonding model just described. Thus, both strong σ donor ligands attached to a mixed ligand metal carbonyl or a formal negative charge on a metal carbonyl anion results in slightly greater CO bond lengths and significantly lower CO stretching frequencies. This is consistent with the bonding model because an increase in the electron density on the metal atom is delocalized over the CO ligands, by populating the carbonyl π* orbital (**13**) thus weakening the CO bond. A good illustration of the effect is the decrease in stretching frequency with increasing negative charge shown in Table 16.4 (in terms of wavenumbers).

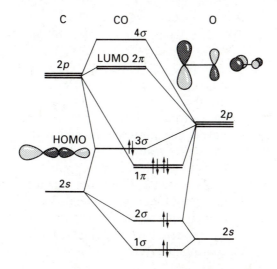

16.1 The molecular orbital energy level diagram for CO. The filled 3σ and vacant 2π orbitals are important in metal complex formation.

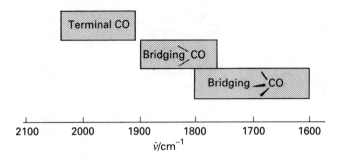

16.2 Approximate ranges for the CO stretching wavenumber in neutral metal carbonyls. Note that high wavenumbers (and hence high frequencies) are on the left, in keeping with the way in which infrared spectrometers generally plot spectra.

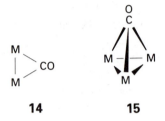

14   **15**

Carbon monoxide is versatile as a ligand because it can bridge two (**14**) or three (**15**) metal atoms. Although the description of the bonding is now more complicated, the concepts of σ-donor and π-acceptor ligands remain useful. The CO stretching frequencies generally follow the order $MCO > M_2CO > M_3CO$ (Fig. 16.2), which suggests an increasing occupation of the π^* orbital as the CO molecule bonds to more metal atoms.

Related π-acceptor ligands

Many other π-acceptor ligands are found in organometallic complexes, but not all of them contain carbon. Some of them (most importantly, those isoelectronic with CO) can be arranged according to increasing π-acceptor strength:

$$C\equiv N^- < N\equiv N < C\equiv NR < C\equiv O < C\equiv S < N\equiv O^+$$

Carbonyl stretching frequencies are often used to determine the order of π-acceptor strengths for the other ligands present in a complex. The basis of the approach is that the CO stretching frequency is decreased when it serves as a π acceptor. However, as other π acceptors in the same complex complete for the metal's d electrons, they cause the CO frequency to increase (Fig. 16.3). This behavior is opposite to that observed with donor ligands, which cause the CO stretching frequency to decrease as they supply electrons to the metal and hence, indirectly, to the CO π^* orbital.

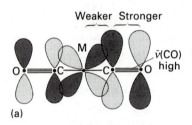

(a)

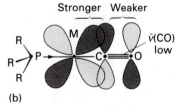

(b)

16.3 The influence of metal–ligand π bonding on CO bonds in metal carbonyls. (a) An electron-poor metal center that has d orbitals lowered in energy because of the positive charge on the metal atom or competing π acceptor ligands. In this case, the CO bond length is reduced and the CO stretching frequency is raised. (b) An electron-rich center, which occurs with strong σ-donor ligands or negative charge. Now the CO π antibonding orbital is more highly populated, so the CO bond is weakened. As a result, the CO bond lengthens slightly and the CO stretching frequency decreases.

The ligands at the ends of the π-acceptor series, CN^- and NO^+, differ significantly from CO and are not always good analogs. For example, CN^-, which is electron-rich, is a good σ donor and a weak π acceptor and therefore forms many classical Werner complexes with metal atoms in high oxidation states. At the other extreme, NO^+ is very strongly electron withdrawing. It forms a linearly linked M—NO ligand in complexes in which it is considered to be NO^+ (**16**), but when NO takes on an additional electron to form NO^-, the M—NO link is bent (**17**). For example, in $[IrCl(CO)(NO)(PPh_3)_2]^+$ the IrNO angle is $124°$. The other ligands in the series (N_2 through CS) are generally found with metals having low oxidation numbers, and their compounds resemble each other quite closely.

Another useful perspective is obtained by regarding linearly-coordinated NO as a neutral three-electron ligand. Thus two NO

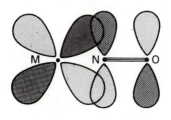

16 M—NO

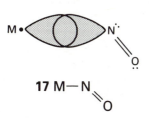

17 M—N
$\diagdown$O

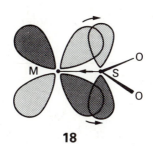

18

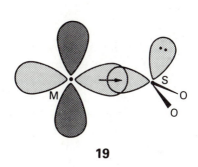

19

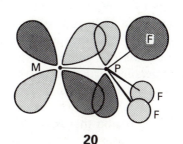

20

ligands can replace three CO ligands and preserve the same 18-electron count on the metal. The tetrahedral nitrosyl compound $Cr(NO)_4$, for instance, has the same electron count as the octahedral carbonyl compound $Cr(CO)_6$.

Two other important π-acceptor ligands are SO_2 and PF_3. The former can serve as a fairly strong π-acceptor ligand. Like NO, it binds in various ways depending on the number of electrons available. One common case is the coplanar M—SO_2 fragment, in which the bonding is σ donation of an electron pair from the S atom to the metal atom in conjunction with π backbonding from the metal to the SO_2 (**18**). In another extreme, the plane of the SO_2 ligand is tilted with respect to the M—S link (**19**). This type of bonding is generally explained in terms of σ donation to the SO_2 ligand of a lone pair from an electron-rich metal center. In the latter case, SO_2 acts as a simple Lewis acid in its interaction with the metal.[5]

As judged from spectroscopic data, the π-acceptor character of PF_3 is comparable to that of CO. This was originally explained in terms of the vacant P3d orbitals acting as electron acceptors. Accurate molecular orbital calculations on PF_3, however, indicate that the acceptor orbital is primarily the P—F antibonding orbital formed from P3p orbitals (**20**) and that the P3d orbitals play only a minor role. Experimental data on complexes and calculations both indicate that P(OR)$_3$ ligands are somewhat weaker π acceptors than PF_3. When the attached groups are less electronegative, as in PH_3 and alkylphosphanes, the P atom becomes a weaker π acceptor but a stronger σ donor.[6]

16.4 Synthesis

The two principal methods for the synthesis of binary metal carbonyls are direct combination of carbon monoxide with a finely divided metal and the reduction of a metal salt in the presence of carbon monoxide under pressure. Many polymetallic carbonyls are synthesized from monometallic carbonyls.

Direct combination

Mond, Langer, and Quinke used the direct combination of nickel and carbon monoxide in their synthesis of the first metal carbonyl, $Ni(CO)_4$:[7]

$$Ni(s) + 4CO(g) \xrightarrow[\text{1 atm}]{30\,°C} Ni(CO)_4(l)$$

[5] For a review of the reactions of SO_2 with metal coordination compounds, including organometallics see: G. J. Kubas and R. R. Ryan, *Polyhedron*, **1/2**, 473 (1986).

[6] A theoretical interpretation of phosphane–metal bonding is given by G. Pacchioni and P. S. Bagus, *Inorg. Chem.*, **31**, 4391 (1992) and P. Fantucci, *Comments Inorg. Chem.*, **31**, 241 (1992).

[7] Mond, Langer, and Quinke discovered $Ni(CO)_4$ in the course of studying the corrosion of nickel valves in process gas containing CO. They quickly applied their discovery to develop a new industrial process for the separation of nickel from cobalt. The centennial of this important discovery is celebrated in a special volume of *J. Organomet. Chem.*, **383**, (1990), which is devoted to metal carbonyl chemistry.

Tetracarbonylnickel(0) is in fact the metal carbonyl that is most readily synthesized in this way.

Other metal carbonyls, such as $Fe(CO)_5$, are formed more slowly. They are therefore synthesized at high pressures and temperatures (Fig. 16.4):

$$Fe(s) + 5CO(g) \xrightarrow[\text{200 atm}]{\text{200 °C}} Fe(CO)_5 \text{ (l at room temperature)}$$

$$2Co(s) + 8CO(g) \xrightarrow[\text{35 atm}]{\text{150 °C}} Co_2(CO)_8 \text{(s at room temperature)}$$

Reductive carbonylation

Direct reaction is impractical for most of the remaining *d* metals and **reductive carbonylation**, the reduction of a salt or metal complex in the presence of CO, is normally employed instead. Reducing agents vary from active metals such as aluminum and sodium, to alkylaluminum compounds, H_2, and CO itself:

$$CrCl_3(s) + Al(s) + 6CO(g) \xrightarrow[\text{Benzene}]{\text{AlCl}_3} AlCl_3(\text{soln}) + Cr(CO)_6(\text{soln})$$

$$3Ru(acac)_3(\text{soln}) + H_2(g) + 12CO(g)$$

$$\xrightarrow[\text{200 atm, CH}_3\text{OH}]{\text{150 °C}} Ru_3(CO)_{12}(\text{soln}) + \ldots$$

$$Re_2O_7(s) + 17CO(g) \xrightarrow[\text{350 atm}]{\text{250 °C}} Re_2(CO)_{10} + 7CO_2(g)$$

16.5 Structure

Metal carbonyl molecules generally have well defined, simple, symmetrical shapes that correspond to the CO ligands taking up the most-distant locations, like electron pairs in VSEPR theory. Thus, the Group 6 hexacarbonyls are octahedral (**21**), pentacarbonyliron(0) is trigonal bipyramidal (**22**), and tetracarbonylnickel(0) is tetrahedral (**2**). Decacarbonyldimanganese(0) consists of two square-pyramidal $Mn(CO)_5$ groups joined by a metal–metal bond. In one isomer of octacarbonyldicobalt(0) the metal–metal bond is bridged by CO.

Infrared and ^{13}C-NMR spectroscopy are widely used to determine the arrangement of atoms in metal carbonyl compounds as separate ^{13}C-NMR signals are observed for inequivalent CO ligands when the molecule is not fluxional (on an NMR timescale), the terminal ligand ^{13}C nucleus being the more shielded. When the molecule is not fluxional, NMR spectra generally contain more detailed structural information than IR spectra. However, IR spectra are often simpler to obtain, and are particularly useful for following reactions. Most CO stretching bands occur in the range 2100 to 1700 cm^{-1}, a region that is generally uncluttered by bands from organic groups. The range of CO stretching

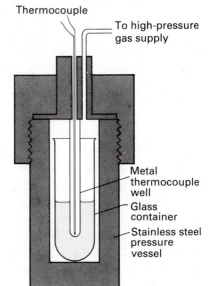

Thermocouple

To high-pressure gas supply

Metal thermocouple well

Glass container

Stainless steel pressure vessel

16.4 A high-pressure metal reaction vessel. The reaction mixture is in the glass container.

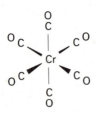

21 $Cr(CO)_6$, O_h

22 $Fe(CO)_5$, D_{3h}

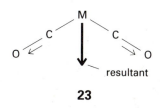

23

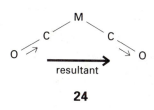

24

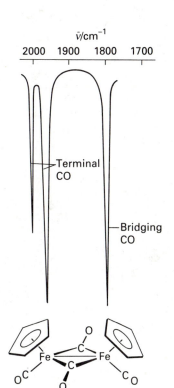

16.5 The infrared spectrum of $(C_5H_5)_2Fe_2(CO)_4$. Note the two high-frequency terminal CO stretches and the lower frequency absorption of the bridging CO ligands. Although two bridging CO bands would be expected on account of the low symmetry of the complex, a single band is observed because the two bridging CO groups are nearly collinear.

frequencies (Fig. 16.2) and the number of CO bands (Table 16.5) are both important for making structural inferences.

If the CO ligands are not related by a center of inversion or a threefold or higher axis of symmetry, the molecule should have one CO stretching absorption for each CO ligand. Thus a bent OC—M—CO group (with only a twofold symmetry axis) will have two infrared absorptions because both the symmetric (**23**) and antisymmetric (**24**) stretches cause the electric dipole moment to change and are IR active. Highly symmetric molecules have fewer bands than CO ligands. Thus, in a linear OC—M—CO group, only one IR band is observed in the CO stretching region because the symmetric stretch leaves the overall electric dipole moment unchanged. As shown in Fig. 16.5, the positions of CO ligands in a metal carbonyl compound may be more symmetrical than the point group of the whole compound suggests, and then fewer bands will be observed than are predicted on the basis of the overall point group.

Infrared spectroscopy is also useful for distinguishing terminal CO (MCO) from two-metal bridging CO (μ_2-CO) and face-bridging CO (μ_3-CO). As explained earlier, the CO stretches occur at lower frequency in the more highly bridging structures (Figs. 16.2 and 16.5).

Example 16.3: *Determining the structure of a carbonyl from IR data*

The complex $[Cr(CO)_4(PPh_3)_2]$ has one very strong IR absorption band at 1889 cm^{-1} and two other very weak bands in the CO stretching region. What is the probable structure of this compound? (The CO stretching wavenumbers are lower than in the corresponding hexacarbonyl because the phosphane ligands are better σ-donors and poorer π-acceptors than CO.)

Answer. A disubstituted hexacarbonyl may exist with either *cis* or *trans* configurations. In the *cis* isomer the four CO ligands are in low symmetry (C_{2v}) and therefore four IR bands should be observed, as indicated in Table 16.5. The *trans* isomer has a square planar array of four CO ligands (D_{4h}), for which only one band in the CO stretching region is expected (Table 16.5). The *trans* CO arrangement is indicated by the data, since it is reasonable to assume that the weak bands reflect a small departure from D_{4h} symmetry imposed by the PPh$_3$ ligands.

Exercise E16.3. The IR spectrum of $Ni_2(\eta^5\text{-}C_5H_5)_2(CO)_2$ has a pair of CO stretching bands at 1857 cm^{-1} (strong) and 1897 cm^{-1} (weak). Does this complex contain bridging or terminal CO ligands, or both? (Substitution of η^5-C$_5$H$_5$ ligands for CO ligands leads to small shifts in the CO stretching frequency for a terminal CO.)

It was seen in Section 3.2 that a motionally averaged NMR signal is observed when a molecule undergoes changes in structure more rapidly than the technique can resolve. Although this phenomenon is quite common in the NMR spectra of organometallic compounds, it is not usually observed in their IR or Raman spectra. An example of

Table 16.5 Relation of the number of CO stretching bands in the IR spectrum to structure

Complex	Isomer	Structure*	Point group	Number of bands†
$M(CO)_6$			O_h	1
$M(CO)_5L$		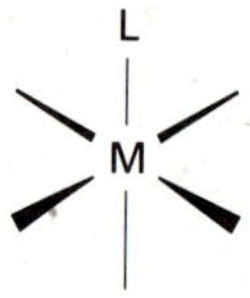	C_{4v}	3‡
$M(CO)_4L_2$	*trans*	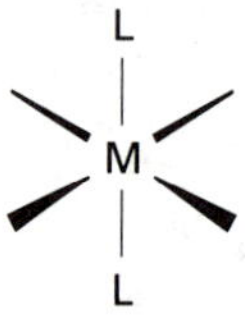	D_{4h}	1
	cis	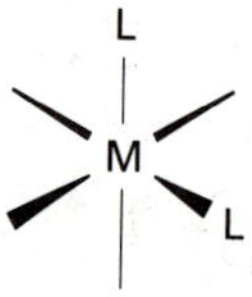	C_{2v}	4‡‡
$M(CO)_3L_3$	*mer*	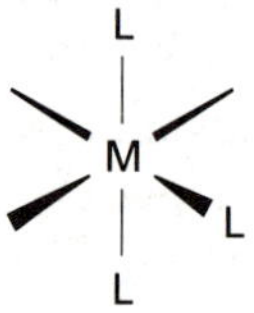	C_{2v}	3‡‡
	fac	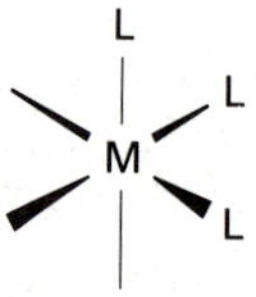	C_{3v}	2
$M(CO)_5$		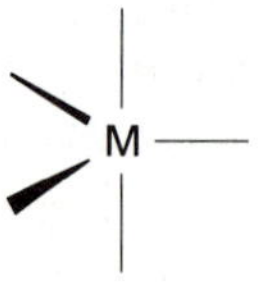	D_{3h}	2
$M(CO)_4L$	*ax*	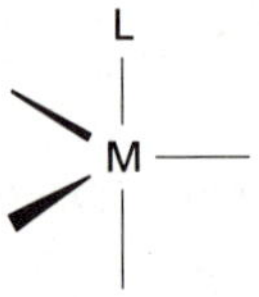	C_{3v}	3**
	eq	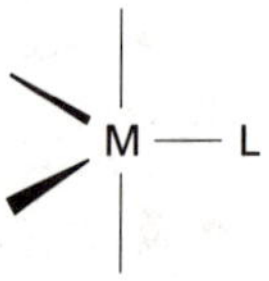	C_{2v}	4

Table 16.5 (*continued*)

Complex	Isomer	Structure*	Point group	Number of bands†
$M(CO)_3L_2$		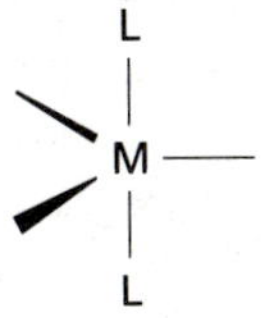	D_{3h}	1
		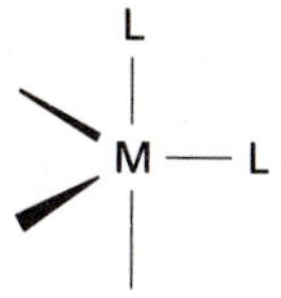	C_s	3
$M(CO)_4$		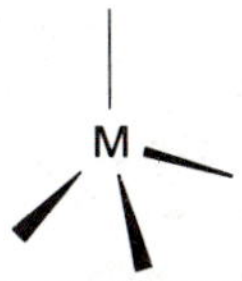	T_d	1

*Each bonding line that is not terminated by an L is a CO.

†The number of IR bands expected in the CO stretching region is based on formal selection rules. In some cases, fewer bands are observed, as explained in the footnotes.

‡If the fourfold array of CO ligands lies in the same plane as the metal atom, two bands will be observed.

‡‡If the *trans* CO ligands are nearly collinear, one fewer band will be observed.

**If the threefold array of CO ligands is nearly planar, only two bands will be observed.

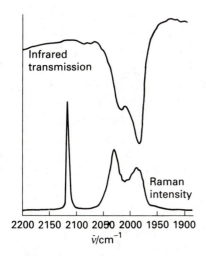

16.6 Infrared and Raman vibrational spectra of liquid $Fe(CO)_5$ in the CO stretching region. Note that, in accord with Table 16.5, only two CO stretching bands are observed for this D_{3h} complex. The totally symmetric band at 2115 cm^{-1} is absent in the infrared spectrum but strong in the Raman spectrum because the selection rules are different in the two cases.

this difference is $Fe(CO)_5$, for which the ^{13}C-NMR signal shows a single line at $\delta \approx 210$, whereas IR and Raman spectra are consistent with a trigonal-bipyramidal structure (Fig. 16.6). We shall meet more NMR evidence for fluxional behavior when we discuss cyclic polyene ligands and metal carbonyl clusters.

16.6 Properties and reactions

Iron and nickel carbonyls are liquids at room temperature and pressure, but all other common carbonyls are solids. All the mononuclear carbonyls are volatile; their vapor pressures at room temperature range from approximately 350 Torr for tetracarbonylnickel(0) to approximately 0.1 Torr for hexacarbonyltungsten(0). The high volatility of $Ni(CO)_4$ coupled with its extremely high toxicity requires unusual care in handling it. Although the other carbonyls appear to be less toxic, they too must not be inhaled or allowed to touch the skin.

Since they are nonpolar, all the mononuclear and many of the polynuclear carbonyls are soluble in hydrocarbon solvents. The most striking exception among the common carbonyls is enneacarbonyldiiron(0), $Fe_2(CO)_9$, which has a very low vapor pressure and is insoluble in solvents with which it does not react. In contrast, $Mn_2(CO)_{10}$ and $Co_2(CO)_8$ are soluble in hydrocarbon solvents and readily sublime.

Most of the mononuclear carbonyls are colorless or lightly colored. Polynuclear carbonyls are colored, the intensity of the color increasing with the number of metal atoms. For example, pentacarbonyliron(0) is a light straw-colored liquid, enneacarbonyldiiron(0) forms golden-yellow flakes, and dodecacarbonyltriiron(0) is a deep green compound that looks black in the solid state. The colors of polynuclear carbonyls arise from electronic transitions between orbitals that are largely localized on the metal framework.

The principal reactions of the metal center of simple metal carbonyls are substitution, condensation into clusters, reduction, and oxidation. In certain cases, the CO ligand itself is also subject to attack by nucleophiles or electrophiles.

Substitution

Substitution reactions provide a route to organometallic compounds containing a variety of other ligands. The simplest examples involve the replacement of CO by another electron pair donor, such as a phosphane. Studies of the rates at which alkylphosphanes and other ligands replace CO in $Ni(CO)_4$, $Fe(CO)_5$, and the hexacarbonyls of the chromium group indicate that a dissociative mechanism (D or I_d, Section 15.1) is in operation. In the D mechanism, an intermediate of reduced coordination number, or more probably a solvated intermediate such as $Cr(CO)_5(THF)$, is produced. This intermediate then combines with the entering group in a bimolecular process:

$$Cr(CO)_6 + Sol \rightleftharpoons Cr(CO)_5(Sol) + CO$$

$$Cr(CO)_5(Sol) + PR_3 \rightarrow Cr(CO)_5PR_3 + Sol$$

(Sol denotes a solvent molecule.) In keeping with this interpretation, solvated metal carbonyls have been observed by IR spectroscopy as the photolysis products of metal carbonyls in polar solvents such as THF. The evidence for an I_d mechanism is that the rate law for reactions of strong nucleophiles at high concentrations with Group 6 carbonyls have a second-order term. Thus the proposed reaction pathways are

$$Cr(CO)_6 \overset{-CO}{\underset{+Sol}{\nearrow}} [Cr(CO)_5Sol] \xrightarrow[-Sol]{+PMe_3} [Cr(CO)_5(PMe_3)] \qquad D$$

$$\underset{+PMe_3}{\searrow} [Cr(CO)_6(PMe_3)]^{\ddagger} \xrightarrow{-CO} [Cr(CO)_5(PMe_3)] \qquad I_d$$

Loss of the first CO group from $Ni(CO)_4$ occurs easily, and substitution is fast at room temperature. The CO ligands are much more tightly bound in the Group 6 carbonyls, and loss of CO often needs to be promoted thermally or photochemically. For example, the substitution of CH_3CN for CO is carried out in refluxing acetonitrile, using a stream of nitrogen to sweep away the carbon monoxide and thus drive the reaction to completion. To achieve photolysis, mononuclear carbonyls (which do not absorb strongly in the visible) are exposed to near-UV radiation in an apparatus like that shown in Fig. 16.7. As with the thermal process, there is strong evidence that the photo-

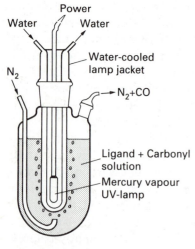

16.7 Apparatus for photochemical ligand substitution of metal carbonyls.

assisted substitution reaction leads to the formation of a labile intermediate complex with the solvent (Sol), which is then displaced by the entering group:

$$Cr(CO)_6 \xrightarrow[-CO, +Sol]{hv} [Cr(CO)_5(Sol)] \xrightarrow[-Sol]{L} Cr(CO)_5L$$

Influence of steric repulsion

As in the reactions of conventional complexes, we can expect steric crowding between ligands to accelerate dissociative processes and to decrease associative processes (Section 15.1). The extent to which various ligands crowd each other is approximated by a cone with an angle determined from a space-filling model (Fig. 15.6 and Table 15.8). Ligands with large cone angles have considerable steric repulsion with each other when packed around a metal center.

We can see how the cone angle of a ligand influences the equilibrium constant for ligand binding by examining the dissociation constant of $Ni(PR_3)_4$ complexes (Table 16.6). These complexes are slightly dissociated in solution if the phosphane ligands are compact, such as PMe_3, with a cone angle 118°. However, a complex such as $Ni[(P(t-Bu)_3]_4$, where the cone angle is large (182°), is highly dissociated. The ligand $P(t-Bu)_3$ is so bulky that the 14-electron complex $[Pt(P(t-Bu)_3)_2]$ can be isolated.

Table 16.6 Cone angle θ and dissociation constant K_d for some Ni complexes*

L	$\theta/°$	$K_d/(mol\ L^{-1})$
$P(CH_3)_3$	118	$<10^{-9}$
$P(C_2H_5)_3$	137	1.2×10^{-5}
$P(CH_3)Ph_2$	136	5.0×10^{-2}
PPh_3	145	Large
$P(t-Bu)_3$	182	Large

*Data are for $NiL_4 \rightleftharpoons NiL_3 + L$ (in benzene, as 25 °C); from C. A. Tolman, *Chem. Rev.*, **77**, 313 (1977).

Example 16.4: *Preparing substituted metal carbonyls*

Starting with MoO_3 as a source of Mo, and CO and PPh_3 as the ligand sources, plus other reagents of your choice, give equations and conditions for the synthesis of $Mo(CO)_5PPh_3$.

Answer. A typical procedure is to synthesize $Mo(CO)_6$ first and then carry out a ligand substitution. Reductive carbonylation of MoO_3 might be performed using $Al(CH_2CH_3)_3$ as a reducing agent in the presence of carbon monoxide under pressure. The temperature and pressure required for this reaction is less than those for the direct combination of molybdenum and carbon monoxide.

$$MoO_3 + Al(CH_2CH_3)_3 + 6CO \xrightarrow[150°C, heptane]{50\ atm}$$
$$Mo(CO)_6 + \text{oxidation products of } Al(CH_2CH_3)_3$$

The subsequent substitution could be carried out photochemically using the apparatus illustrated in Fig. 16.7.

$$Mo(CO)_6 + PPh_3 \xrightarrow[hv]{THF} Mo(CO)_5PPh_3 + CO$$

The progress of the reaction can be followed by IR spectroscopy in the CO stretching region using small samples that are periodically removed from the reaction vessel.

Exercise E16.4. If the highly substituted complex $Mo(CO)_3L_3$ is desired, which of the ligands $P(CH_3)_3$ or $P(t-Bu)_3$ would be preferred? Give reasons for your choice.

Influence of electronic structure on CO substitution

Extensive studies of CO substitution reactions of simple carbonyl complexes have revealed systematic trends in mechanisms and rates.[8]

The 18-electron metal carbonyls generally undergo dissociatively activated substitution reactions. As we saw in Section 15.2, this class of reaction is characterized by insensitivity of the rate to the identity of the entering group. Dissociative activation of an 18-electron complex is usually favored because the alternative, associative activation, would require the formation of an energetically unfavorable 20-electron activated complex.

The rates of substitution of 16-electron complexes are observed to be sensitive to the identity and concentration of the entering group, which indicates associative activation (Section 15.1). For example, the reactions of $IrCl(CO)(PPh_3)_2$ with triethylphosphane are associatively activated. As with Werner coordination compounds, 16-electron organometallic complexes appear to undergo associatively activated substitution reactions because the 18-electron activated complex is energetically more favorable than the 14-electron activated complex that would occur in dissociative activation.

Although these generalizations apply to a wide range of reactions, some exceptions are observed, especially if cyclopentadienyl or nitrosyl ligands are present. In these cases it is common to find evidence of associatively activated substitution even for 18-electron complexes. The common explanation is that NO may switch from being a three-electron donor when M—NO is linear (as in **16**), to being a one-electron donor when it is angular, since two of the electrons can then be regarded as localized on NO (as in **17**). Similarly, the η^5-Cp five-electron donor can slip relative to the metal and become an η^3-Cp three-electron donor. In this case the Cp ligand is regarded as having a three-carbon interaction with the metal while the remaining two electrons form a simple C=C bond that is not engaged with the metal (**25**), and the electron-deficient central metal is then susceptible to substitution.

M

25

One final general observation is that the rate of CO substitution in six-coordinate metal carbonyls often decreases as more strongly basic ligands replace CO, and two or three alkylphosphane ligands often represent the limit of substitution. With bulky phosphane ligands, further substitution may be thermodynamically unfavorable on account of ligand crowding. Increased electron density on the metal center, which arises when a π-acceptor ligand is replaced by a net donor ligand, appears to bind the remaining CO ligands more tightly and

---

[8] For an informative article on the development of the major principles see F. Basolo, *Inorg. Chim. Acta*, **50**, 65 (1981).

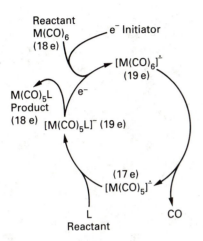

16.8 Schematic diagram of an electron transfer catalyzed CO substitution. After addition of a small amount of a reducing initiator, the cycle continues until the limiting reagent $M(CO)_6$ or L has been consumed.

therefore reduce the rate of CO dissociative substitution. The donor character of P(III) ligands as judged by ease of protonation and other data lie in the order[9]

$$PF_3 \ll P(OAr)_3 < P(OR)_3 < PAr_3 < PR_3$$

where Ar is aryl and R is alkyl.

The explanation of the influence of σ-donor ligands on CO bonding is that the increased electron density contributed by the phosphane leads to stronger back π-bonding to the remaining CO ligands and therefore strengthens the M—CO bond. This stronger M—C bond decreases the tendency of CO to leave the metal and therefore decreases the rate of dissociative substitution. In some instances these bonding effects may lead to greater thermodynamic stability for the metal carbonyl than for its substitution product.

It has been found that for some metal carbonyls, CO substitution can be catalyzed by electron transfer processes that create anion or cation radicals. A typical process of this type is illustrated in Fig. 16.8. As can be seen, the key features are the lability of CO in the 19-electron anion radical and the smaller reduction potential of the phosphane-substituted product than for the metal carbonyl starting material. Similarly, the rare-metal compounds with 19 or 17 electrons are labile with respect to substitution.

Reduction to form metal carbonyl anions

Most metal carbonyls can be reduced to metal carbonyl anions. In monometallic carbonyls, two-electron reduction is generally accompanied by loss of the two-electron donor CO ligand:

$$Fe(CO)_5 \xrightarrow{Na,THF} [Fe(CO)_4]^{2-} + CO$$

The metal carbonyl anion contains Fe with oxidation number -2, and it is rapidly oxidized by air. That much of the negative charge is delocalized over the CO ligands is confirmed by the observation of a low CO stretching frequency in the IR spectrum, corresponding to about 1730 cm^{-1}.

Polynuclear carbonyls containing M—M bonds are generally cleaved by strong reducing agents. Once again the 18-electron rule is obeyed and a mononegative mononuclear carbonyl results:

$$2Na + (OC)_5Mn-Mn(CO)_5 \xrightarrow{THF} 2Na^+[Mn(CO)_5]^-$$

Some metal carbonyls disproportionate in the presence of a strongly basic ligand, producing the ligated cation and a carbonylate anion. Much of the driving force for this reaction is the stability of the metal cation when it is surrounded by strongly basic ligands. Octacarbonyl-dicobalt(0) is highly susceptible to this type of reaction when exposed

---

[9] For a discussion of various measures of phosphane basicity, see R. C. Bush and R. J. Angelici, *Inorg. Chem.*, **27**, 681 (1988).

to a good Lewis base such as pyridine (py):

$$3Co_2(CO)_8 + 12py \rightarrow 2[Co(py)_6][Co(CO)_4]_2 + 8CO$$

Oxidation number:　0　　　　　　　　　　+2　　　−1

It is also possible for the CO ligand to be oxidized in the presence of the strongly basic ligand OH^-, the net outcome being the reduction of a metal center:

$$3Fe(CO)_5 + 4OH^- \rightarrow [Fe_3(CO)_{11}]^{2-} + CO_3^{2-} + 2H_2O + 3CO$$

　0　+2　　　　　　　　　　$-\frac{2}{3}$　　　+4

Metal basicity

Many organometallic compounds can be protonated at the metal center. Metal carbonyl anions provide many examples of this metal basicity:

$$[Mn(CO)_5]^-(aq) + H^+(aq) \rightarrow HMn(CO)_5(s)$$

The affinity of metal carbonyl anions for the proton varies widely (Table 16.7). It is observed that the more negative the anion, the higher its Brønsted basicity and hence the lower the acidity of its conjugate acid (the metal carbonyl hydride).

The *d*-block M—H complexes are commonly referred to as hydrides. This name reflects the assignment of oxidation number −1 to an H atom attached to a metal atom. Nevertheless, most of the hydrides of metals to the right of the *d* block are Brønsted acids. This odd situation simply reflects the formal nature of oxidation numbers. In striking contrast to *p*-block hydrogen compounds, the Brønsted acidity of *d*-block M—H compounds decreases on descending a group.

Neutral metal carbonyls (such as pentacarbonyliron) can be protonated in air-free concentrated acid. Compounds having metal–metal bonds are even more easily protonated. The Brønsted basicity of a metal atom with oxidation number 0 is associated with the presence of nonbonding *d* electrons. Similarly, the tendency to protonate metal–metal bonds can be explained by the transformation of a 2c,2e metal–metal bond into a 3c,2e M—H—M interaction (**26**). A characteristic feature of the Groups 6 to 10 organometallic hydrides is the appearance of a highly shielded ^{1}H-NMR signal ($\delta \approx -8$ to -60); the hydrogen signal is significantly more shielded than one attached to most elements (see Further Information 3).

Metal hydrides play an important role in inorganic chemistry and not all of them are generated by protonation. For example, in Section 9.3 we encountered complexes in which an η^2-H_2 is coordinated to a metal center. Even when H_2 adds to a metal center giving conventional H—M—H complexes, it is likely that a transitory η^2-H_2 complex is formed. We will discuss the addition of H_2 to metal complexes in Section 16.7 and the role of hydrogen complexes in homogeneous catalysis in Section 17.4.

Table 16.7 Acidity constant of *d*-metal hydrides in acetonitrile at 25 °C

Hydride	pK_a (MH)
$HCo(CO)_4$	8.3
$HCo(CO)_3P(OPh)_3$	11.3
$H_2Fe(CO)_4$	11.4
$CpCr(CO)_3H$	13.3
$CpMo(CO)_3H$	13.9
$HMo(CO)_5$	15.1
$HCo(CO)_3PPh_3$	15.4
$CpW(CO)_3H$	16.1
$Cp^*Mo(CO)_3H$	17.1
$H_2Ru(CO)_4$	18.7
$CpFe(CO)_2H$	19.4
$CpRu(CO)_2H$	20.2
$H_2Os(CO)_4$	20.8
$HRe(CO)_5$	21.1
$Cp^*Fe(CO)_2H$	26.3
$CpW(CO)_2(PMe_3)H$	26.6

Source: E. J. Moore, J. M. Sullivan, and J. R. Norton, *J. Am. Chem. Soc.*, **108**, 2257 (1986). Cp* is the pentamethylcyclopentadienyl ligand, C_5Me_5.

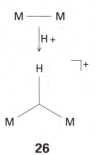

26

Metal basicity is turned to good use in the synthesis of a wide variety of organometallic compounds. For example, alkyl and acyl groups can be attached to metal atoms by the reaction of an alkyl or acyl halide with an anionic metal carbonyl:

$$[Mn(CO)_5]^- + CH_3I \rightarrow (H_3C)Mn(CO)_5 + I^-$$

$$[Co(CO)_4]^- + CH_3COI \rightarrow Co(CO)_4(COCH_3) + I^-$$

A similar reaction with organometallic halides may be used to form M—M bonds:

$$[Mn(CO)_5]^- + ReBr(CO)_5 \rightarrow (OC)_5Mn—Re(CO)_5 + Br^-$$

Oxidation to form metal carbonyl halides

Metal carbonyls are susceptible to oxidation by air above room temperature. Although uncontrolled oxidation produces the metal oxide and CO or CO_2, of more interest in organometallic chemistry are the controlled reactions that give rise to organometallic halides. One of the simplest of these is the oxidative cleavage of an M—M bond:

$$(OC)_5Mn—Mn(CO)_5 + Br_2 \rightarrow 2MnBr(CO)_5$$

	↑	↑	↑
Oxidation number:	0	0	+1

In keeping with the loss of electron density from the metal when a halogen atom is attached, the CO stretching frequencies of the product are significantly higher than those of $Mn_2(CO)_{10}$.

Reactions of the CO ligand

The C atom of CO is susceptible to attack by nucleophiles if it is attached to a metal atom that is not electron rich. Thus, terminal carbonyls with high CO stretching frequencies are liable to attack by nucleophiles. The d electrons in these neutral or cationic metal carbonyls are not extensively delocalized on to the carbonyl C atom, so that atom can be attacked by electron-rich reagents. For example, strong nucleophiles (such as methyllithium, Section 10.6) attack the CO in many neutral metal carbonyl compounds:

$$\tfrac{1}{4}Li_4(CH_3)_4 + Mo(CO)_6 \rightarrow Li[Mo(COCH_3)(CO)_5]$$

The resulting acyl compound reacts with carbocation reagents to produce a stable and easily handled neutral product:

$$\overset{\displaystyle O}{\underset{\displaystyle \parallel}{}}$$

$$Li[Mo(\overset{O}{\overset{\parallel}{C}}Me)(CO)_5] + [Et_3O][BF_4] \rightarrow$$

$$Mo(=\overset{OEt}{\underset{|}{C}}Me)(CO)_5 + LiBF_4 + Et_2O$$

Compounds of this type, with a direct $M{=}C$ bond, are often called **Fischer carbenes** because they were discovered in E. O. Fischer's laboratory. The attack of a nucleophile on the C atom is also important for the mechanism of the hydroxide-induced disproportionation of metal carbonyls:

$$(OC)_nMCO + OH^- \rightarrow (OC)_nMC{\overset{\overset{\displaystyle OH}{|}}{=}}O\, \Big]^- \xrightarrow{3OH^-} [M(CO)_n]^{2-} + CO_3^{2-} + 2H_2O$$

In electron-rich metal carbonyls considerable electron density is delocalized onto the CO ligand and in some cases the O atom of a CO ligand is susceptible to attack by electrophiles. Once again, IR data give an indication of when this should be expected, for a low CO stretching frequency indicates significant back-donation to the CO ligand, and hence appreciable electron density on the O atom. Thus a bridging carbonyl is particularly susceptible to attack at the O atom:

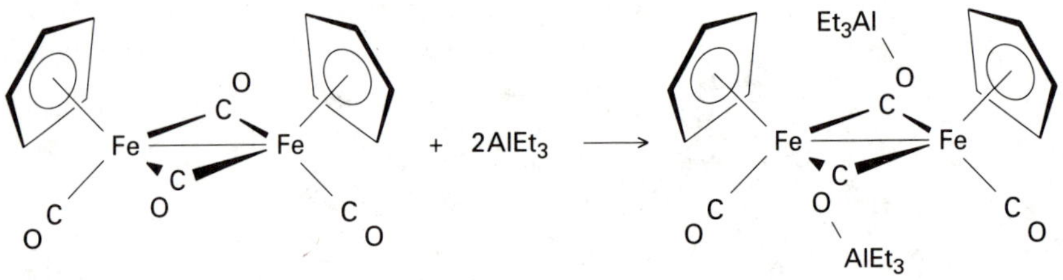

27

The attachment of an electrophile to the oxygen of a CO ligand promotes migratory insertion reactions (Section 15.10) and C—O cleavage reactions (Section 16.12).[10]

Some alkyl-substituted metal carbonyls undergo a **migratory insertion** reaction, which is the conversion of an alkyl ligand and a CO ligand to an acyl ligand, $-(CO)R$. As discussed in Section 15.10, there is good evidence that the reaction of $Mn(CH_3)(CO)_5$ occurs by initial migration of the CH_3 group to CO, producing a low concentration of the solvent-coordinated acyl compound. That intermediate then takes on a ligand to produce the stable acyl product:

$$\underset{\underset{\displaystyle Me}{|}}{L_nM}{-}CO \underset{-Sol}{\overset{+Sol}{\rightleftharpoons}} \underset{\underset{\displaystyle Sol}{|}}{L_nM}{-}\overset{\overset{\displaystyle O}{\|}}{C}{-}Me \xrightarrow{+L} L_{n+1}M{-}\overset{\overset{\displaystyle O}{\|}}{C}{-}Me + Sol$$

We shall see the importance of this reaction in catalytic reactions in Section 17.2.

---

[10] These and other reactions of C- and O-bonded metal carbonyls are described by C. P. Horwitz and D. F. Shriver, *Adv. Organomet. Chem.*, **23**, 219 (1984).

Example 16.5: *Converting CO to carbene and acyl ligands*

Propose a set of reactions for the formation of $W(CO)_5(C(OCH_3)Ph)$, starting with hexacarbonyltungsten(0) and other reagents of your choice.

Answer. As with its neighbor in Group 6, the CO ligands in hexacarbonyltungsten(0) are susceptible to attack by nucleophiles, and the reaction with phenyllithium should give a C-phenyl intermediate:

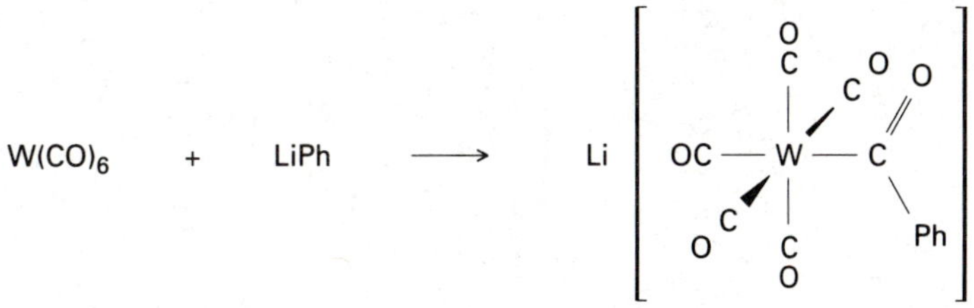

The anion can then be alkylated on the O atom of the PhCO ligand by an electrophile:

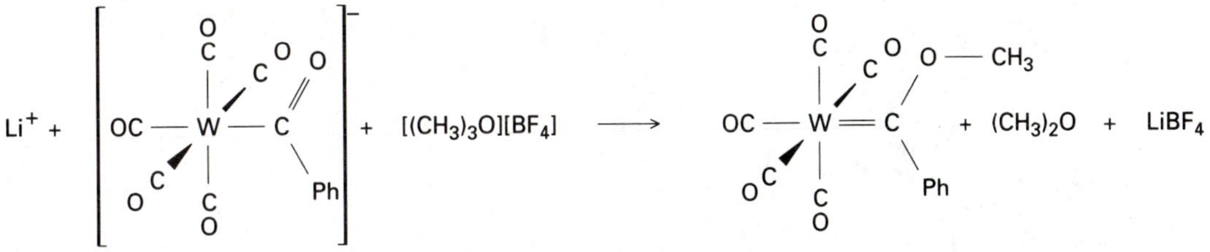

Exercise E16.5. Propose a synthesis for $Mn(CO)_4(PPh_3)(COCH_3)$ starting with $Mn_2(CO)_{10}$, PPh_3, Na, and CH_3I.

OTHER ORGANOMETALLIC COMPOUNDS

A large number of *d*-block complexes containing hydrocarbon ligands have been discovered and investigated from the mid 1950s to the present. We shall discuss the basic structures, bonding, and reactions here and continue the discussion of reactions in Chapter 17 when we describe catalysis. As we work through the following examples it will be useful to keep an eye on Table 16.1, which summarizes the number of electrons that an organic ligand may donate to a metal.

16.7 Hydrogen and open-chain hydrocarbon ligands

Hydrogen

The one-electron ligand H is intimately involved in organometallic chemistry. We have already seen how an M—H bond can be produced by protonation of neutral and anionic metal carbonyls (Section 16.5); in some cases a similar reaction gives a stable product with other organometallic compounds. For example, ferrocene can be protonated in strong acid to produce an Fe—H bond. As a consequence of the way in which negative oxidation states are assigned to nonmetal ligands, such as H or halides, this protonation reaction leads to a change in oxidation state on the metal:

$$Cp_2Fe + HBF_4 \rightarrow [Cp_2Fe-H]^+[BF_4]^-$$

Oxidation number: +2   +4

One of the most interesting and important methods for introducing hydrogen into a metal complex is by oxidative addition (Section 15.13). In this reaction an XY molecule adds to a 16-electron complex ML_4 with cleavage of the X—Y bond to give $M(X)(Y)L_4$ and an increase in the oxidation state of the metal. Thus the addition of either H_2 or HX to a 16-electron complex may produce an 18-electron hydrido complex:

$$IrCl(CO)(PPh_3)_2 + H_2 \rightarrow IrCl(H)_2(CO)(PPh_3)_2$$

16e Ir(I)   18e Ir(III)

$$IrCl(CO)(PPh_3)_2 + HCl \rightarrow IrCl_2(H)(CO)(PPh_3)_2$$

The upper reaction may proceed through a dihydrogen complex (Section 9.3) in which η^2-H_2 forms transiently.

Alkyl ligands

An alkyl ligand forms an M—C single bond, and in doing so the alkyl group acts as a one-electron monohapto ligand. Many such compounds are known, but they are less common in the *d* block than in the *s* and *p* blocks. This may in part be a result of the modest M—C bond strengths (Table 16.8). The main reasons for the decomposition of *d*-block alkyl compounds are the kinetically facile reactions, such as *β*-hydrogen elimination, CO insertion, and reductive elimination, which lead to the transformation of the alkyl ligand into other groups.

In a *β*-hydrogen elimination reaction, an H atom on the *β*-C atom of an alkyl group is transferred to the metal atom and an alkene is eliminated

$$L_nM-CH_2CH_3 \rightarrow L_nM-H + H_2C=CH_2$$

We discuss the reverse of this reaction, alkene insertion into the M–H bond, in Chapter 17. Both reactions are thought to proceed through a cyclic intermediate involving a 3c,2e M—H—C bond, which is called an **agostic interaction**:[11]

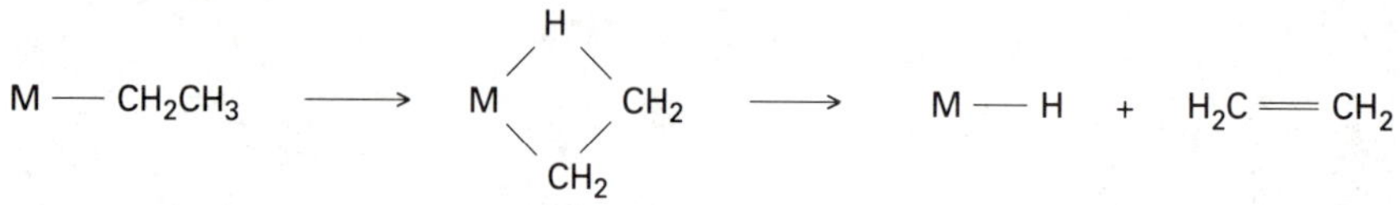

Table 16.8 Bond dissociation enthalpies

	B/(kJ mol^{-1})
$(OC)_5Mn-CH_2Ph$	87
$(OC)_5Mn-Mn(CO)_5$	94
$(OC)_5Mn-CH_3$	153
$(OC)_5Mn-H$	213

[11] These agostic interactions have been identified spectroscopically and by diffraction, and they are often invoked in organometallic mechanisms, see: M. Brookhart and M. L. H. Green, *J. Organomet. Chem.*, **205**, 395 (1983); A. J. Schulz *et al.*, *Science*, **220**, 197 (1983).

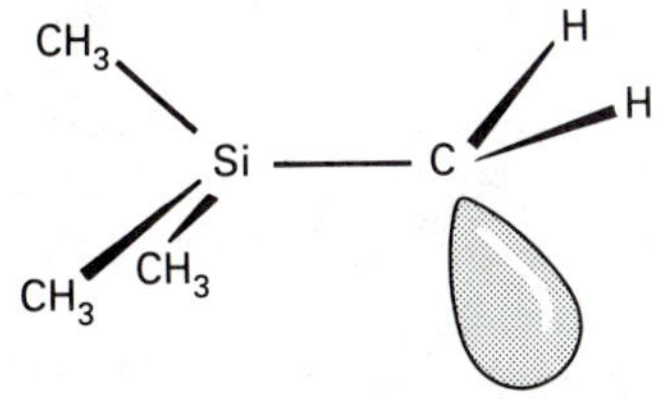

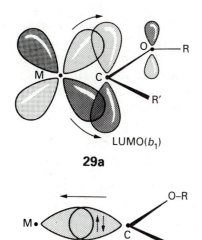

28

Because β-hydrogen elimination is blocked if there are no H atoms on the β-C atom, benzyl $-CH_2(C_6H_5)$ and trimethylsilylmethyl $-CH_2Si(CH_3)_3$ (**28**) ligands are much more robust than ethyl ligands. Similarly, the lack of β-H atoms on the methyl group accounts for the greater stability of complexes containing methyl ligands rather than ethyl ligands. The β-hydrogen elimination reaction is also seen with some *p*-block organometallic compounds, most notably alkylaluminum compounds (Section 10.10).

Alkylidene ligands

The methylidene groups, CH_2, CHR, or CR_2, are monohapto, two-electron ligands that form the M=C *d-p* double bonds, such as the Fischer carbenes (**29a**, **29b**, Section 16.6). Fischer carbenes are attacked at the C atom by nucleophiles. For example, the attack of an amine on the electrophilic C atom of a Fischer carbene results in the displacement of the $-OR$ group to yield a new carbene ligand:[12]

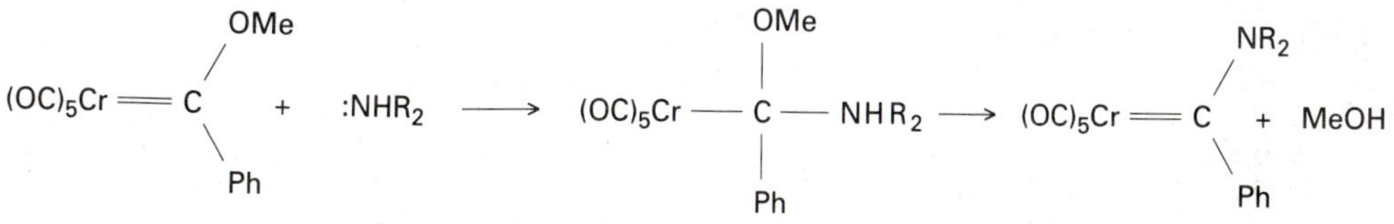

$$(OC)_5Cr = C \underset{Ph}{\overset{OMe}{<}} \; + \; :NHR_2 \longrightarrow (OC)_5Cr - \underset{Ph}{\overset{OMe}{\underset{|}{\overset{|}{C}}}} - NHR_2 \longrightarrow (OC)_5Cr = C \underset{Ph}{\overset{NR_2}{<}} \; + \; MeOH$$

This type of reaction is synthetically useful for the preparation of a wide range of Fischer carbenes. The Fischer carbenes are formed by middle to late *d*-block metals. The electrophilicity of the metal-bound carbon atom in the Fischer carbenes is attributed to the significant electronegativity of the middle to late *d*-block metals. The corresponding molecular orbital explanation, which is illustrated in Fig. 16.9, is that for the late *d*-block metals the metal *dπ* orbitals are at lower energy than the carbon *p* orbitals, so the lower energy and therefore doubly occupied π orbital has an electron density build up on the metal. Conversely the vacant π orbital is mainly located on carbon; thus the C atom is susceptible to attack by nucleophiles.

The reactions of Fischer carbene complexes with alkynes have considerable utility in organic synthesis.[13] For example, naphthyl compounds can be synthesized by the reaction of methoxy phenyl Fischer carbenes with an alkyne:

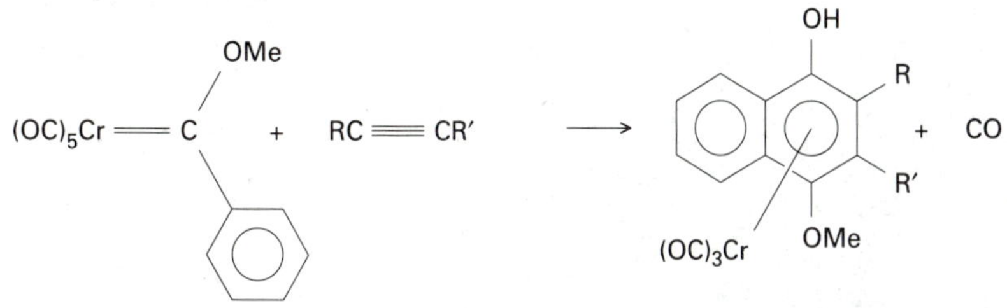

LUMO(b_1)

29a

HOMO(a_1)

29b

$$(OC)_5Cr = C \overset{OMe}{<} \; + \; RC \equiv CR' \longrightarrow$$

The ten C atoms in the naphthalene rings are contributed by a CO ligand (1 atom), the carbene ligand (7 atoms), and the acetylene reagent

[12] E. O. Fischer, *Adv. Organomet. Chem.*, **14**, 1 (1976).
[13] K. H. Dotz, *Angew. Chem., Int. Ed. Engl.*, **23**, 587 (1984).

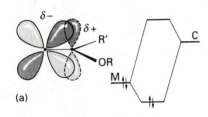

(a)

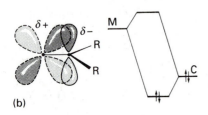

(b)

16.9 In the Fischer carbene (a), the vacant p_z orbital of CR'(OR) is higher in energy than the metal $d\pi$ orbitals, so the π-electron density is largely concentrated on the metal atom and the carbene C is electrophilic. For the Schrock carbene (b), the p_z orbital of CR_2 is lower in energy than the $d\pi$ orbitals, so π-electron density is concentrated in the carbon and the carbene C is nucleophilic.

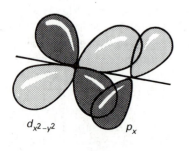

30a

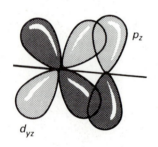

30b

(2 atoms). The chromium can be removed from the organic product by mild oxidation.

The metals on the left of the *d* block form another series of methylidene compounds called **Schrock carbenes** after their discoverer.[14] The chemistry and spectroscopy of the Schrock carbenes indicate that the M=C bond is polarized so as to put negative charge on the metal-bound C atom (Fig. 16.9b). This polarization is consistent with the low electronegativity of the metal atoms on the left of the *d*-block. The chemical evidence for the higher electron density on that atom is the reaction of the metal-bound C atom with electrophiles such as Me_3SiBr:

$$Cp_2Ti(CH_2)Me + Me_3SiBr \rightarrow [Cp_2Ti(CH_2SiMe_3)Me]^+ + Br^-$$

An interesting reaction of the Schrock carbenes is the **alkene metathesis reaction**:

As indicated in the chemical equation, this reaction appears to proceed through a four-membered metallocycle.

Alkylidyne ligands

The alkylidyne ligands have the general formula CH or CR.[15] They are monohapto three-electron ligands and are bound to the metal by an M≡C triple bond involving one σ bond and two (*d*, *p*)-overlap π bonds (**30**). The simplest member of this series is methylidyne, CH; the next simplest is ethylidyne, CCH_3. There are several routes to these compounds, and we have already described their preparation by the cleavage of M≡M bonds (Section 8.9). The first synthesis of a metal alkylidyne involved the abstraction of an alkoxide group from a Fischer carbene by BBr_3:

This reaction with BBr_3 was originally carried out in an attempt to

[14] Alkylidene complexes of niobium and tantalum; R. R. Schrock, *Acc. Chem. Res.*, **12**, 98 (1984).
[15] H. Fischer, P. Hoffmann, F. R. Kreissl, R. R. Schrock, U. Schubert, and K. Weiss, *Carbyne complexes*. VCH, Weinheim (1988).

convert the carbene ligand to its bromo analog, M=CBrPh. As with many exploratory synthetic projects, the outcome was different from that intended, and in this case much more interesting.

Alkene ligands

A C=C π bond can act as a source of electron pairs in complex formation, as in Zeise's salt (**1**). This salt is prepared by bubbling ethylene through an aqueous solution of tetrachloroplatinate(II) ions in the presence of Sn(II), which aids the removal of Cl$^-$ ions from the Pt(II) coordination spheres (see **1**):

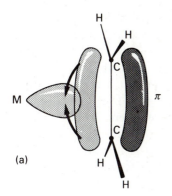

$$K_2[PtCl_4] + H_2C{=}CH_2 \xrightarrow{SnCl_2} K[PtCl_3(\eta^2\text{-}C_2H_4)] + KCl$$

Another route to alkene complexes is the removal of H$^-$ from a coordinated alkyl ligand:

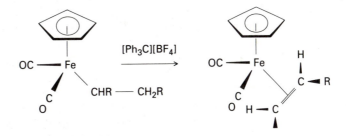

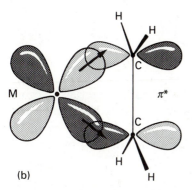

16.10 Interaction of ethylene with a metal atom. (a) Donation of electron density from the filled π molecular orbital of ethylene to a vacant metal σ orbital. (b) Acceptance of electron density from a filled $d\pi$ orbital into the vacant π^* orbital of ethylene.

Simple alkene ligands are dihapto two-electron donors because the filled π orbital projects toward the metal and donates electrons to suitably oriented metal orbitals (the s, p_z, d_{z^2}, and $d_{x^2-y^2}$ orbitals in Fig. 16.10). In addition, the π^* orbital of the ligand can accept electron density from filled metal orbitals of the appropriate symmetry (the p_x and d_{xz} orbitals in Fig. 16.10). This description of metal–alkene bonding in terms of electron donation from the filled bonding π orbital of the ligand and simultaneous electron acceptance into the empty antibonding π^* orbital is called the **Dewar–Chatt model**.

Electron donor and acceptor character appear to be fairly evenly balanced in most ethene complexes of the d metals, but the degree of donation and back-donation can be altered by substituents. An extreme example is tetracyanoethene (**31**), which is an abnormally strong electron acceptor ligand on account of its electron-withdrawing cyano groups. Tetracyanoethene qualifies as a π-acceptor ligand because its role as an acceptor dominates its role as a donor. When the degree of donation of electron density from the metal atom to the alkene ligand is small, substituents on the ligand are bent only slightly away from the metal, and the C—C bond length is only slightly greater than in the free alkene. With electron-rich metals or electron-withdrawing substituents on the alkene, the back-donation is greater. Substituents on the alkene are then bent away from the metal, and the C—C bond length approaches that of a single C—C bond. These differences lead

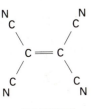

31 TCNE

32

33

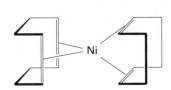

34 Ni(cod)$_2$

35 Fe(η^1–C$_3$H$_5$)(CO)$_2$(η^5–Cp)

some chemists to depict the first group as simple π complexes (**32**) and the second as metallocycles (**33**) with M—C single bonds.

Diene and polyene ligands

Diene (–C=C—C=C–) and polyene ligands present the possibility of polyhapto bonding. As with the chelate effect in Werner complexes, the resulting polyene complexes are usually more stable than the equivalent complex with individual ligands. This is because the entropy of dissociation of the complex is much smaller than when the liberated ligands can move independently (the chelate effect, Section 6.7). For example, bis(η^4-cycloocta-1,5-diene)nickel(0) (**34**) is more stable than the corresponding complex containing four ethylene ligands. Cyclo-octa-1,5-diene is a fairly common ligand in organometallic chemistry, where it is referred to engagingly as cod, and is introduced into the metal coordination sphere by simple ligand displacement reactions. An example is:

Metal complexes of cod are often used as starting materials because they often have intermediate stability. Many of them are sufficiently stable to be isolated and handled, but cod can be displaced by many stronger ligands. For example, if the highly toxic Ni(CO)$_4$ molecule is needed in a reaction, then it may be generated from Ni(cod)$_2$ directly in the reaction flask:

$$\text{Ni(cod)}_2(\text{soln}) + 4\text{CO(g)} \rightarrow \text{Ni(CO)}_4(\text{soln}) + 2\text{cod(soln)}$$

The π-allyl ligand

The allyl ligand can bind to a metal atom in either of two configurations. As an η^1-ligand (**35**) it is a one-electron donor; as an η^3-ligand (**36**) it is a three-electron donor. In the latter case the terminal substituents are bent slightly out of the plane of the three-carbon backbone and are either *exo* (**37**) or *endo* (**38**) relative to the metal. It is common to observe *exo* and *endo* group exchange, which in some cases is fast on an NMR timescale. A mechanism that involves the transformation $\eta^3 \rightarrow \eta^1 \rightarrow \eta^3$ is often invoked to explain this exchange. Because of this flexibility in type of bonding, η^3-allyl complexes are often highly reactive.

There are many routes to allyl complexes. One is the nucleophilic attack of an allyl Grignard reagent with a metal halide:

$$2\text{C}_3\text{H}_5\text{MgBr} + \text{NiCl}_2 \xrightarrow{\text{Ether}} \text{Ni}(\eta^3\text{-C}_3\text{H}_5)_2 + 2\text{MgBrCl}$$

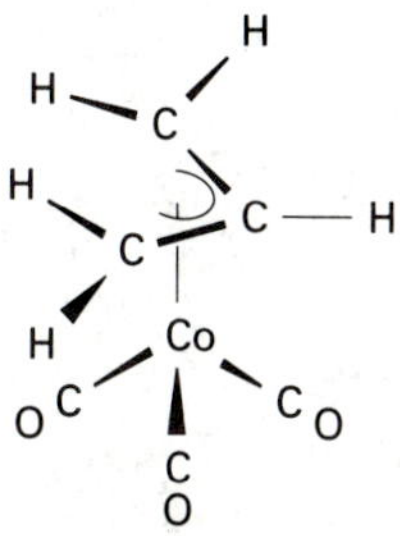

36 Co(η³–C₃H₅)(CO)₃

37 *exo*

38 *endo*

39 Co₂(PhC₂Ph)(CO)₆

Conversely, electrophilic attack of a haloalkane on a low-oxidation state metal center yields allyl complexes:

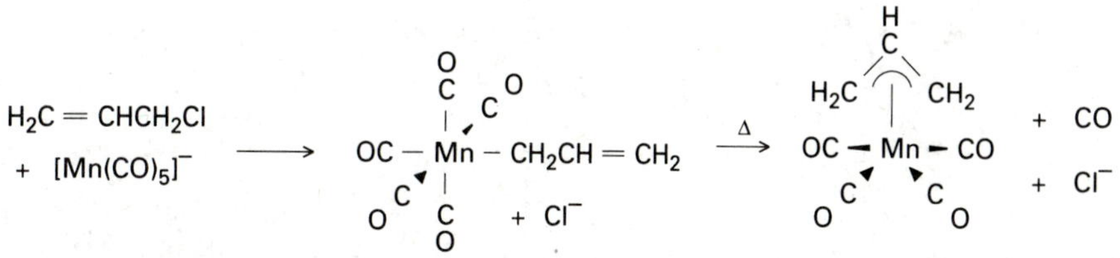

$$H_2C = CHCH_2Cl + [Mn(CO)_5]^-$$

If the metal center is not basic, the protonation of a butadiene complex leads to a π-allyl complex:

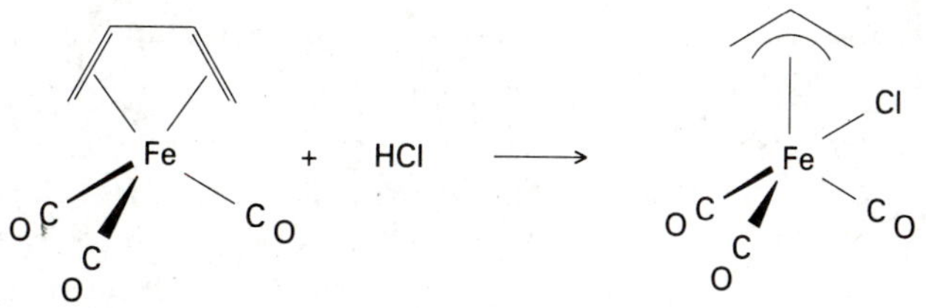

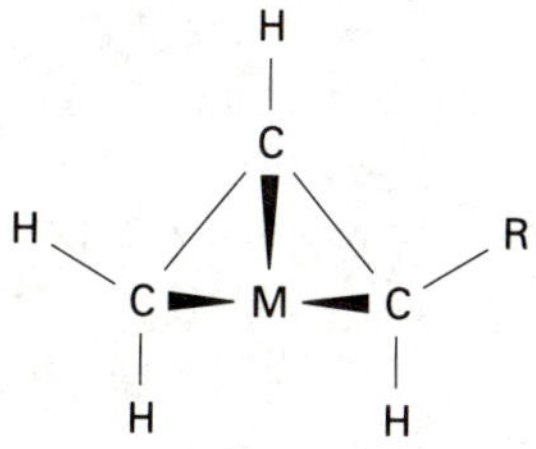

$$+ \ HCl \longrightarrow$$

Alkyne ligands

Acetylene (ethyne) has two π bonds and hence is a potential four-electron donor. However, it is not always clear from experimental information that it acts in this way. Substituted acetylenes form very stable polymetallic complexes in which the acetylene can be regarded as a four-electron donor. An example is η^2-diphenylethyne(hexacarbonyl)dicobalt(0) (**39**) where we can view one π bond as donating to one of the Co atoms and the second π bond as overlapping with the other Co atom (**40**). As in this example, the alkyl or aryl groups present on the acetylene impart stability. They do so by reducing the tendency toward secondary reactions of the coordinated acetylene, such as loss of the slightly acidic acetylenic H atom to the metal.

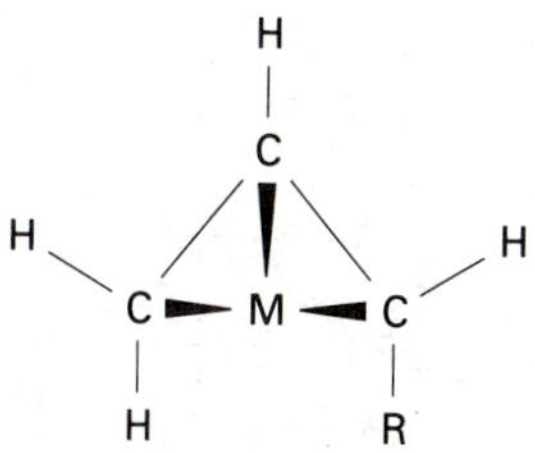

16.8 Cyclic polyene complexes

Cyclic polyene ligands are among the most important in organometallic chemistry. In this section we shall consider cyclic polyene ligands

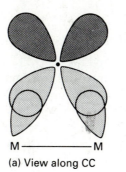
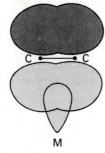

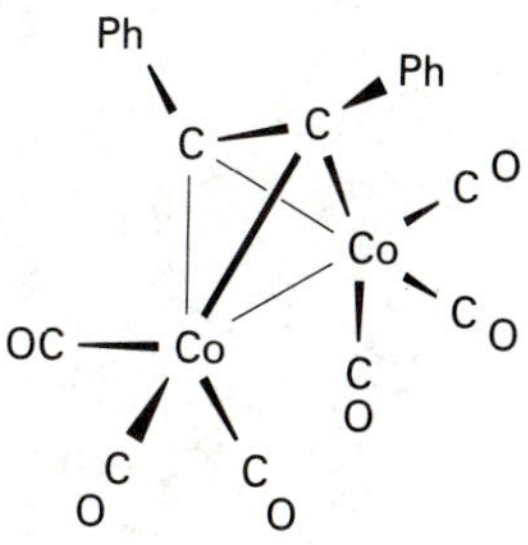

(a) View along CC   (b) View along MM

40

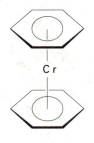

41 Cr(η^6–C$_6$H$_6$)$_2$

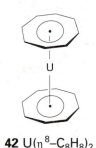

42 U(η^8–C$_8$H$_8$)$_2$

ranging from cyclobutadiene to cyclooctatetraene.[16] These ligands are capable of forming the metallocenes ferrocene (**4**), bis(benzene)chromium(0) (**41**), and uranocene (**42**). A **metallocene** consists of a metal atom between two planar polyhapto rings (as in ferrocene) and they are informally called 'sandwich compounds'. Cyclic polyenes are known to form complexes in which they are bound to a metal atom through some but not all of their C atoms. In these cases, the ring is nonplanar. Beside these homoleptic cyclopolyene complexes there are many other organometallic compounds in which other ligands share the coordination sphere of a metal with a cyclic polyene.

Metallo cyclobutadiene compounds

The simplest common ligand of the group, cyclobutadiene, is a four-electron donor. It is unstable as the free molecule, but stable complexes are known, including Ru(CO)$_3$(η^4-C$_4$H$_4$) (**43**). This species is one of many in which coordination to a metal atom stabilizes an otherwise unstable molecule. Because of the instability of cyclobutadiene, the ligand must be generated in the presence of the metal to which it is coordinated. This synthesis can be accomplished in a variety of ways, one of which being the dehalogenation of a halogenated cyclobutene:

Another synthesis is the dimerization of a substituted acetylene:

43 Ru(C$_4$H$_4$)(CO)$_3$

Metallo cyclopentadienyl compounds

The cyclopentadienyl ligand C$_5$H$_5$ (abbreviated Cp) has played a major role in the development of organometallic chemistry and continues to be the archetype of cyclic polyene ligands. A huge number of metal cyclopentadienyl compounds are known. Some have Cp as a mono-hapto ligand, in which case it contributes one electron to a σ bond to the metal (**7**), others contain Cp as a trihapto ligand, when it donates three electrons. Usually, though, Cp is present as a pentahapto ligand

---

[16] The cyclopropyl cation is the simplest member in the series and some η^3-cyclopropenyl complexes are known, but are not common.

contributing five electrons. Structure (**8**) shows a compound that contains both η^3 and η^5 ligands.

Sodium cyclopentadienide is a common starting material for the preparation of cyclopentadienyl compounds. It can be prepared by the action of Na on cyclopentadiene in tetrahydrofuran solution:

$$2Na + 2C_5H_6 \xrightarrow{\text{THF}} 2Na[C_5H_5] + H_2$$

Sodium cyclopentadienide reacts with *d*-metal halides to produce metallocenes:

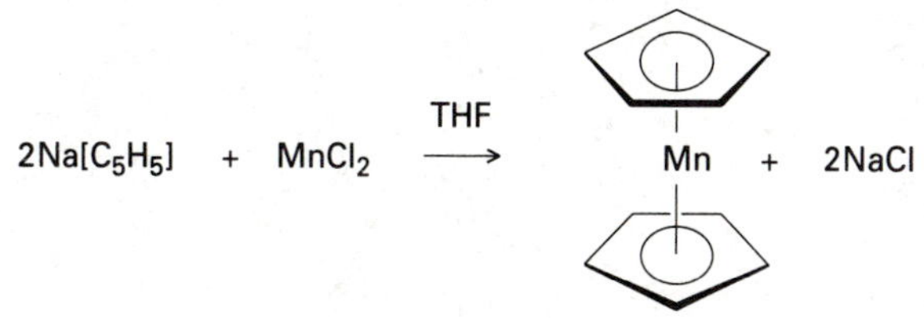

$$2Na[C_5H_5] \quad + \quad MnCl_2 \quad \xrightarrow{\text{THF}} \quad \text{Mn} \quad + \quad 2NaCl$$

The overall result is simple ligand displacement, but the mechanism may be more complicated than this suggests. The bis(cyclopentadienyl) complexes of iron, cobalt, and nickel are also readily prepared by this method.

Reactions

Because of their great stability, the 18-electron Group 8 compounds ferrocene, ruthenocene, and osmocene maintain their ligand–metal bonds under rather harsh conditions, and it is possible to carry out a variety of transformations on the cyclopentadienyl ligands. For example, they undergo reactions similar to those of simple aromatic hydrocarbons, such as Friedel–Crafts substitution:

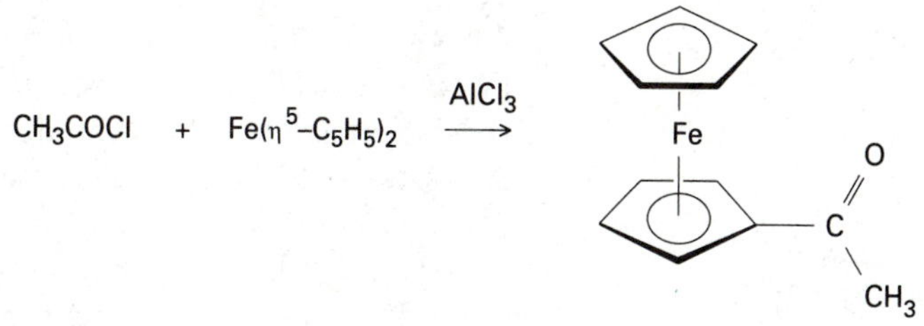

$$CH_3COCl \quad + \quad Fe(\eta^5\text{–}C_5H_5)_2 \quad \xrightarrow{\text{AlCl}_3} \quad$$

It also is possible to replace H on a Cp ring by Li:

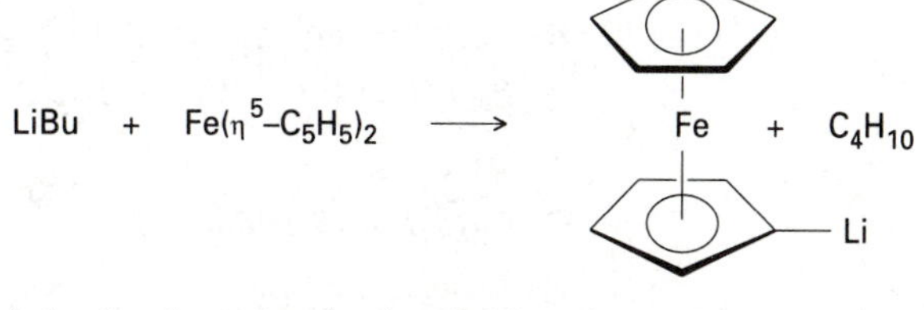

$$LiBu \quad + \quad Fe(\eta^5\text{–}C_5H_5)_2 \quad \longrightarrow \quad \text{Fe} \quad + \quad C_4H_{10}$$

As might be imagined, the lithiated product is an excellent starting material for the synthesis of a wide variety of ring-substituted products

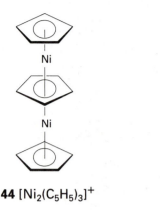

44 [Ni₂(C₅H₅)₃]⁺

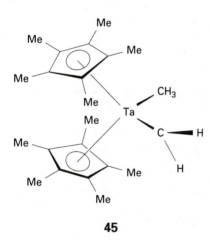

45

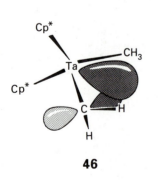

46

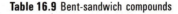

Table 16.9 Bent-sandwich compounds

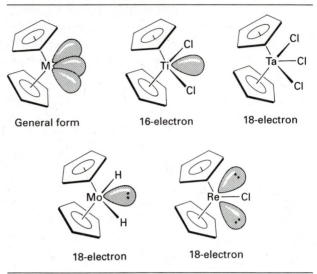

and in this respect resembles simple organolithium compounds (Section 10.6).

There are many interesting related structures in addition to the simple bis(cyclopentadienyl) and bis(arene) complexes. In the jargon of this area, these species are referred to as 'bent sandwich compounds' (Table 16.9), 'half-sandwich' or 'piano stool' compounds (**6**), and, inevitably, 'triple deckers' (**44**). We have already met triple deckers and multidecker sandwich compounds that incorporate $[B_3C_2H_5]^{4-}$, which is isoelectronic with $[C_5H_5]^-$ but has a greater tendency to form stacked sandwich compounds than the cyclopentadienyl ligand (Section 11.3).

Bent sandwich compounds play a major role in the organometallic chemistry of the early and middle *d*-block elements, and we have already encountered one example, the Schrock carbene $[Ta(\eta^5\text{-}Cp^*)_2(CH_3)(CH_2)]$, (**45**). Other examples include $[Re(\eta^5\text{-}Cp)_2(Cl)]$, $[W(\eta^5\text{-}Cp)_2(H)_2]$, and $[Nb(\eta^5\text{-}Cp)_2(Cl)_3]$. As shown in Table 16.9, these bent sandwich compounds occur with a variety of electron counts and stereochemistries. Their structures can be systematized in terms of a model in which three metal atom orbitals project toward the open face of the bent Cp_2M moiety. According to this model, the metal atom often satisfies its electron deficiency when the electron count is less than 18 by interaction with lone pairs or C—H groups on the ligands. For example, this model explains why the sterically disfavored orientation of CH_2 exists in the Schrock carbene, because by adopting this conformation the filled $p\pi$ orbital on CH_2 donates electron density to the central orbital on the Ta atom (**46**).

Bonding in ferrocene

Although details of the bonding in ferrocene are not settled, the molecular orbital energy level diagram shown in Fig. 16.11 accounts for a number of experimental observations. This diagram refers to the

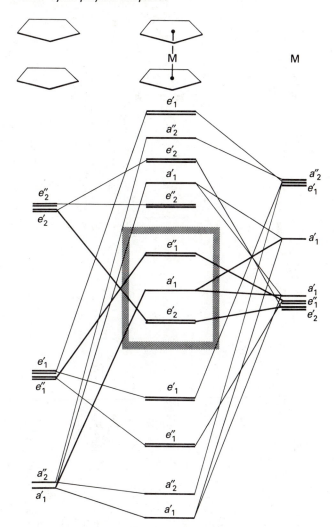

16.11 Molecular orbital energy diagram of a metallocene. The energies of the symmetry-adapted π orbitals of the C_5H_5 ligands are shown on the left, relevant d orbitals of the metal are on the right, and the resulting molecular orbital energies are in the center. Eighteen electrons can be accommodated by filling the molecular orbitals up to and including the a'_1 frontier orbital in the box. Three frontier orbitals are indicated in the box rather than the usual two because the wide range of electron counts for metallocenes involve the filling of these sets of orbitals.

eclipsed (D_{5h}) form of the complex, which in the gas phase is about 4 kJ mol^{-1} more stable than the staggered conformation. We shall focus our attention on the frontier orbitals. As shown in Fig. 16.12, the e''_1 symmetry-adapted linear combinations of ligand orbitals have the same symmetry as the d_{zx} and d_{yz} orbitals of the metal atom. The lower energy frontier orbital (a'_1) is composed of d_{z^2} and the corresponding symmetry-adapted combination of ligand orbitals (Fig. 16.12). However, there is little interaction between the ligands and the metal orbitals because the ligand p orbitals happen to lie—by accident—in the conical nodal surface of the metal's d_{z^2} orbital. In ferrocene and the other 18-electron bis(cyclopentadienyl) complexes the a'_1 frontier orbital and all lower orbitals are full but the e''_1 frontier orbital and all higher orbitals are empty.

The frontier orbitals are neither strongly bonding nor strongly antibonding. This characteristic permits the possibility of the existence of metallocenes that diverge from the 18-electron rule, such as the 17-electron complex [FeCp$_2$]$^+$ and the 20-electron complex NiCp$_2$. Deviations from the 18-electron rule, however, do lead to significant changes

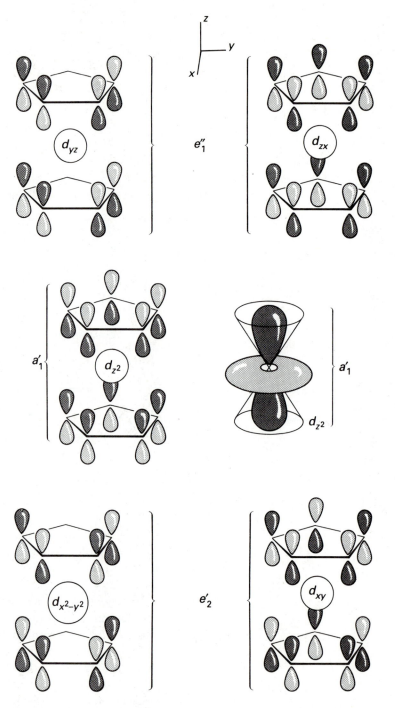

16.12 Symmetry-adapted orbital combinations giving rise to the three metallocene frontier orbitals. The accidental coincidence of the nodal surface of d_{z^2} and the C_5H_5 π orbitals results in negligible interaction despite its being symmetry-allowed.

in M—C bond lengths that correlate quite well with the molecular orbital scheme (Table 16.10). Similarly, the redox properties of the complexes can be understood in terms of the electronic structure. Thus, we have already mentioned that ferrocene is fairly readily oxidized to the ferrocinium ion, $[Fe(C_5H_5)_2]^+$. From an orbital viewpoint, this oxidation corresponds to the removal of an electron from the non-bonding a_1' orbital. The 19-electron complex $[Co(\eta^5\text{-}C_5H_5)_2]$ is much

Table 16.10 Electronic configuration and M—C bond length in $M(\eta^5\text{-}C_5H_5)_2$ complexes

Complex	Valence electrons	Electron configuration	$R(M-C)/\text{Å}$
$V(C_5H_5)_2$	15	$e_2'^2 a_1'^1$	2.28
$Cr(C_5H_5)_2$	16	$e_2'^3 a_1'^1$	2.17
$Mn(C_5H_4CH_3)_2^*$	17	$e_2'^3 a_1'^2$	2.11
$Fe(C_5H_5)_2$	18	$e_2'^4 a_1'^2$	2.06
$Co(C_5H_5)_2$	19	$e_2'^4 e_1''^1 a_1'^2$	2.12
$Ni(C_5H_5)_2$	20	$e_2'^4 e_1''^2 a_1'^2$	2.20

*Data are quoted for this complex because $Mn(C_5H_5)_2$ has a high-spin configuration $a_1'^2$, $e_1'^2$, $e_1''^2$ and hence an anomalously long M—C bond length (2.38 Å).

more readily oxidized than ferrocene because the electron is lost from the antibonding e_1'' orbital to give the 18-electron $[Co(\eta^5\text{-}C_5H_5)_2]^+$ ion.

Another useful comparison can be made with octahedral complexes. The e_1'' frontier orbital of a metallocene is the analog of the e_g orbital in an octahedral complex, and the a_1' orbital plus the e_2' pair of orbitals are analogous to the t_{2g} orbitals of an octahedral complex. This formal similarity extends to the existence of high- and low-spin bis(cyclo-pentadienyl) complexes (Table 16.10). The redox chemistry of metallocenes is permeated by similarities to that of simple octahedral complexes.

Example 16.6: *Identifying metallocene electronic structure and stability*
By referring to Fig. 16.11, discuss the occupancy and nature of the HOMO in $[Co(\eta^5\text{-}C_5H_5)_2]^+$ and the change in metal-ligand bonding relative to neutral cobaltocene.

Answer. The $[Co(\eta^5\text{-}C_5H_5)_2]^+$ ion contains 18 valence electrons (9 from Co, 10 from the two Cp ligands, less one for the single positive charge). Assuming that the molecular orbital energy level diagram for ferrocene applies, the 18-electron count leads to double occupancy of the orbitals up through a_1'. The 19-electron cobaltocene molecule has an additional electron in the e_1'' orbital, which is antibonding with respect to the metal and ligand. Therefore the metal-ligand bond should be stronger and shorter in $[Co(\eta^5\text{-}C_5H_5)_2]^+$ than in $Co(\eta^5\text{-}C_5H_5)_2$. This is borne out by structural data.

Exercise E16.6. Using the same molecular orbital diagram, comment on whether the removal of an electron from $Fe(\eta^5\text{-}C_5H_5)_2$ to produce $[Fe(\eta^5\text{-}C_5H_5)_2]^+$ should produce a substantial change in M—C bond length relative to neutral ferrocene.

Metallo-arene compounds

Benzene and its derivatives are most often hexahapto six-electron donors (as in **41**) in which all six π electrons are shared with the metal. Photochemical or thermal activation of metal carbonyls is often an effective route to introduce neutral cyclic polyene ligands into the metal coordination sphere. For example, hexacarbonylchromium(0)

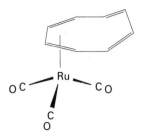

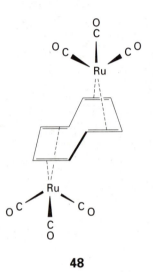

47 Ru(η^4–C$_8$H$_8$)(CO)$_3$

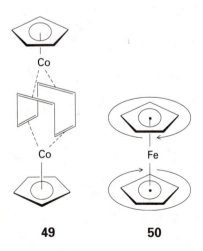

48

49   **50**

16.13 (a) The fluxional process in Ge(η^1-C$_5$H$_5$)(CH$_3$)$_3$ occurs by a series of 1,2-shifts. (b) The fluxionality of Ru(η^4-C$_8$H$_8$)(CO)$_3$ can be described similarly. We need to imagine that the Ru atom is out of the plane of the page; the CO ligands have been omitted.

can be refluxed with an arene to produce the arene tricarbonyl:

$$Cr(CO)_6 + C_6H_6 \xrightarrow{\Delta} Cr(\eta^6\text{-}C_6H_6)(CO)_3 + 3CO$$

Metallo-cyclooctatetraene compounds

Cyclooctatetraene is a large ligand that is found in a wide variety of bonding arrangements. It uses some of its π electrons for bonding to metals when it is an η^2, an η^4 (**47**), or an η^6 ligand, and in these cases the ring is puckered. Once again, photochemistry provides a good route to cyclooctatetraene carbonyl complexes:

$$Fe(CO)_5 + C_8H_8 \xrightarrow{h\nu} Fe(\eta^4\text{-}C_8H_8)(CO)_3 + 2CO$$

The bonding versatility of cyclooctratetraene is also shown by its ability to act as a bridge between two metal atoms (**48** and **49**). As we have already mentioned, when C$_8$H$_8$ is a planar η^8 ligand (**5**), it is best regarded as the planar aromatic C$_8$H$_8^{2-}$ group.

Fluxional cyclic polyene complexes

One of the most remarkable aspects of many cyclic polyene complexes is their stereochemical nonrigidity. For example, at room temperature the two rings in ferrocene rotate rapidly relative to each other (**50**). This type of fluxional process is called **internal rotation**, and is similar to the process in ethane.

Of greater interest is the stereochemical nonrigidity that is often seen when a conjugated cyclic polyene is attached to a metal atom through some but not all of its C atoms. In such complexes the metal–ligand bonding may hop around the ring, a property called **fluxionality** and, in the informal jargon of organometallic chemists, 'ring whizzing'. A simple example is found in Ge(η^1-C$_5$H$_5$)(CH$_3$)$_3$, where the single site of attachment of the Ge atom to the cyclopentadiene ring hops around the ring in a series of 1,2-shifts, or motion in which a C—M bond is replaced by a C—M bond to the next C atom around the ring (Fig. 16.13a). The great majority of fluxional

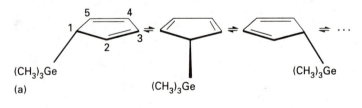

(b)

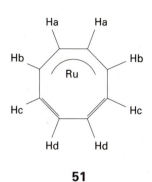

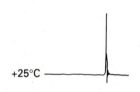

51

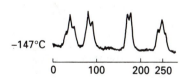

16.14 Observed ^{1}H-NMR spectra for (η^4-C_8H_8)Ru(CO)$_3$ at various temperatures. (From: F. A. Cotton, in *Dynamic nuclear magnetic resonance spectroscopy* (ed. L. M. Jackman and F. A. Cotton), Chapter 10, Academic Press, New York (1975).)

conjugated polyene complexes that have been investigated migrate by **1,2-shifts**, but it is not known whether these shifts are controlled by a principle of least motion or by some aspect of orbital symmetry.

NMR provides the primary evidence for the existence and mechanism of these fluxional processes, for they occur on a time scale of 10^{-2} to 10^{-4} s, and can be studied by ^{1}H- and ^{13}C-NMR. The compound Ru(η^4-C_8H_8)(CO)$_3$ (**47**) is an illustration of the approach. At room temperature, its ^{1}H-NMR spectrum consists of a single, sharp line (Fig. 16.14) that could be interpreted as arising from a symmetrical η^8-C_8H_8 ligand. However, X-ray diffraction studies of single crystals show unambiguously that the ligand is tetrahapto. This conflict is resolved by ^{1}H-NMR spectra at lower temperatures, for as the sample is cooled the signal broadens and then separates into four peaks (Fig. 16.14). These peaks are expected for the four pairs of protons of a η^4-C_8H_8 ligand (**51**). The interpretation is that at room temperature the ring is whizzing around the metal atom rapidly (Fig. 16.13b) compared with the timescale of the NMR experiment, so an averaged signal is observed (Fig. 16.14). At lower temperatures the motion of the ring is slower, and the distinct conformations exist long enough to be resolved. A detailed analysis of the line shape of the NMR spectra can be used to measure the activation energy of the migration.[17]

16.9 Reactivity of early *d*-metal and *f*-metal organometallic compounds[18]

There are strong similarities between the chemical properties of the early *d*-block organometallic compounds (those of Groups 3 to 5) and those of the *f*-block:

> Early *d*-block organometallic compounds and those of the *f*-block are oxophilic and halophilic, highly reactive toward C—H bond cleavage, and have little tendency to form M—M bonds without bridging ligands.

Under normal laboratory conditions, binary neutral metal carbonyl compounds are unknown for the *f*-block elements and are unknown for Groups 3 and 4 in the early *d*-block. We have already encountered some of these factors in the discussion of Schrock carbenes. Many of these properties can be rationalized in terms of the hard, electropositive character, low electron count, and limited number of accessible oxidation states that characterize these *d*- and *f*-block elements.

Oxophilicity

That the early *d*- and *f*-block elements are hard is shown by their affinity for hard ligands such as O and Cl. The affinity for O is evident

17 L. M. Jackman and F. A. Cotton (ed.), *Dynamic nuclear magnetic resonance spectroscopy*. Academic Press, New York (1975).

18 For a review of organolanthanide chemistry see: W. J. Evans, *Adv. Organomet. Chem.*, **24**, 131 (1985). For organoactinides see: T. J. Marks and A. Streitwieser p. 1547, and T. J. Marks p. 1588, in G. Seaborg, J. Katz, and L. R. Morss, *Chemistry of the actinide elements*. Vol. 2, Chapman and Hall, London (1986).

in the properties of the carbonyls of these elements. For example, the reaction of carbon monoxide with $ZrCp_2(CH_3)_2$ gives a C- and O-bonded η^2-acetyl (**52**) instead of a simple C-bonded η^1-acetyl typical of late *d*-metals:

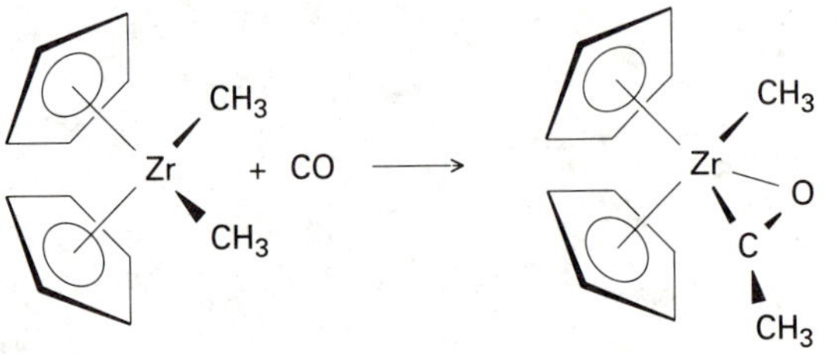

In this reaction, the vacant central orbital on the starting material $ZrCp_2(CH_3)_2$ is utilized in the product by the η^2-acetyl ligand. Similarly, the reaction between a zirconium dihydride complex and CO in the presence of H_2 results in reduction and migration of the oxygen into the coordination sphere of the Zr atom:

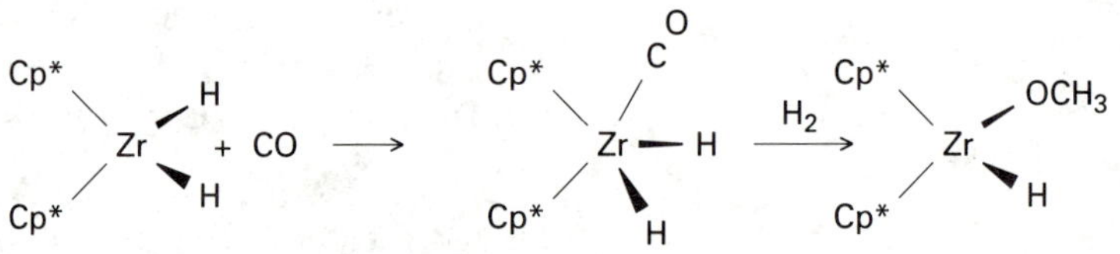

(Cp* is pentamethylcyclopentadienyl.) As we explain below, pentamethylcyclopentadiene confers greater stability than the unsubstituted cyclopentadiene ligand on early *d*-block and *f*-block organometallic compounds.

C–H cleavage

The early *d* metals have a marked tendency to activate C—H bonds. This tendency is so pronounced, for instance, that attempts to prepare $TiCp_2$ in fact produce the dimer (**53**) in which C—H bonds have oxidatively added to the Ti atom. To suppress this and other unwanted reactions, the pentamethylcyclopentadienyl ligand (Cp*) is often used. This ligand has no H atoms on the ring carbons, so C—H cleavage is avoided. It is also more electron-rich than Cp and therefore forms somewhat stronger metal–ligand bonds. Moreover, the bulk of the Cp* ligand blocks access to the metal and thus hinders associative reactions.

A striking example of C—H activation, the activation of C—H bonds in methane by a lanthanide organometallic compound, was discovered by Patricia Watson (Dupont). The discovery was based on the observation that $^{13}CH_4$ exchanges ^{13}C with the CH_3 group attached to Lu:

$$LuCp^*{}_2(CH_3) + {}^{13}CH_4 \rightarrow LuCp^*{}_2({}^{13}CH_3) + CH_4$$

This reaction can be carried out in deuterated cyclohexane with no evidence for activation of the cyclohexane C—D bond, presumably

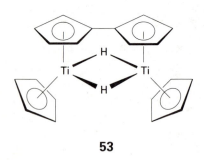

53

because cyclohexane is too bulky to gain access to the metal center. Methane activation has also been observed with organoactinides and late d-block complexes. Electrophilic metal centers promote this reaction, and a four-center intermediate has been proposed:

$$M - R \quad + \quad R' - H \quad \longrightarrow \quad \left[\begin{array}{cc} R' --- H \\ | \quad\quad | \\ M --- R \end{array} \right]^{\ddagger} \quad \longrightarrow \quad \begin{array}{cc} R' \quad\quad H \\ | \quad\quad | \\ M \quad + \quad R \end{array}$$

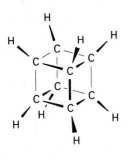

54 C_8H_8

METAL–METAL BONDING AND METAL CLUSTERS

Organic chemists must go to heroic efforts to synthesize their cluster molecules, such as cubane (**54**). In contrast, one of the distinctive characteristics of inorganic chemistry is the large number of closed polyhedral molecules, such as the tetrahedral P_4 molecule (Section 12.1), the octahedral halide-bridged early d-block clusters (Section 8.9), the polyhedral carboranes (Section 11.3), and the organometallic cluster compounds that we discuss here. We shall include some nonorganometallic compounds in this section so as to permit a fuller discussion of metal–metal multiple bonds.

16.10 Structure

We define **metal clusters** to be molecular complexes with metal–metal bonds that form triangular or larger closed structures. This definition excludes linear M—M—M compounds. It also excludes **cage compounds**, in which several metal atoms are held together exclusively by ligand bridges.[19] The presence of bridging ligands in a cluster (such as **55**), raises the possibility that the atoms are held together by M—L—M interactions rather than M—M bonds. Bond lengths are of some help in resolving this issue. If the M—M distance is much greater than twice the metallic radius, it is reasonable to conclude that the M—M bond is either very weak or absent. However, if the metal atoms are within a reasonable bonding distance, the proportion of the bonding that is attributable to direct M—M interaction is ambiguous. For example, there has been debate in the literature over the extent of Fe—Fe bonding in $Fe_2(CO)_9$ (**56**).

Metal–metal bond strengths in metal complexes cannot be determined with great precision, but a variety of evidence—such as the stability of compounds and M—M force constants—indicates that there is an increase in M—M bond strengths down a group in the d block. This trend contrasts with that in the p block where element-element bonds are usually weaker for the heavier members of a group. As a consequence of this trend, metal–metal bonded systems are most numerous for the Period 4 and 5 d metals.

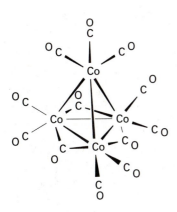

55 $Co_4(CO)_{12}$

56 $Fe_2(CO)_9$

[19] This definition is sometimes relaxed. Occasionally a dinuclear M—M bonded system is referred to as a cluster, and quite often chemists do not make the distinction between cage and cluster compounds.

Electron count and structure of clusters[20]

Organometallic cluster compounds are rare for the early d metals and unknown for the f metals, but a large number of metal carbonyl clusters exist for the elements of Groups 6 to 10. The bonding in the smaller clusters can be readily explained in terms of local M—M and M—L electron pair bonding and the 18-electron rule, but octahedral M_6 and larger clusters do not conform to this pattern.

A semiempirical correlation of electron count and structure of the larger organometallic clusters was introduced by K. Wade and refined by D. M. P. Mingos and J. Lauher. These **Wade–Mingos–Lauher rules** are summarized in Table 16.11; they apply most reliably to metal clusters in Groups 6 to 9. In general (and as in the boron hydrides where similar considerations apply), more open structures (which have fewer metal-metal bonds) occur when there is a higher cluster valence electron (CVE) count.

As an illustration of the influence of correlation between electron count and structure, consider $Rh_4(CO)_{12}$ and $[Re_4(CO)_{16}]^{2-}$:

$Rh_4(CO)_{12}$ tetrahedral	$[Re_4(CO)_{16}]^{2-}$ butterfly
4Rh: $4 \times 9 = 36$	4Re: $4 \times 7 = 28$
12CO: $12 \times 2 = 24$	16CO: $16 \times 2 = 32$
	charge: 2
$\overline{60}$	$\overline{62}$

Example 16.7: *Correlating spectroscopic data, cluster valence electron count, and structure*

The reaction of chloroform with $Co_2(CO)_8$ yields a compound of formula $Co_3(CH)(CO)_9$. NMR and IR data indicate the presence of only terminal CO ligands and the presence of a CH group. Propose a structure consistent with the spectra and the correlation of CVE with structure.

Answer. Electrons available for the cluster are: 27 for three Co atoms, 18 for nine CO ligands, and 3 for CH (assuming the latter is simply C-bonded: one C electron is used for the C—H bond, so three are available for bonding in the cluster). The resulting total CVE of 48 indicates a triangular cluster. A structure consistent with this conclusion and the presence of only terminal CO ligands and a capping CH ligand is illustrated in (**57**).

Exercise E16.7. The compound $Fe_4Cp_4(CO)_4$ is a dark green solid. Its IR spectrum shows a single CO stretch at 1640 cm^{-1}. The ^{1}H-NMR is a single line even at low temperatures. From this spectroscopic data and the CVE propose a structure for $Fe_4Cp_4(CO)_4$.

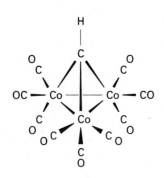

57 $Co_3(CH)(CO)_9$

---

[20] D. M. P. Mingos and D. J. Wales, *Introduction to cluster chemistry*. Prentice Hall, Englewood Cliffs (1990). J. W. Lauher, *J. Am. Chem. Soc.*, **100**, 5305 (1978).

Table 16.11 Correlation of cluster valence electron count and structure*

Number of metal atoms	Structure of metal framework	Cluster valence electron count	Example
1 Single metal		18	$Ni(CO)_4$ (**2**)
2 Linear	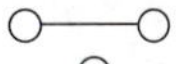	34	$Mn_2(CO)_{10}$†
3 Closed triangle	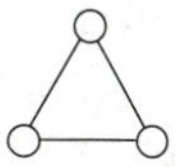	48	$Co_3(CO)_9CH$ (**56**)
4 Tetrahedron	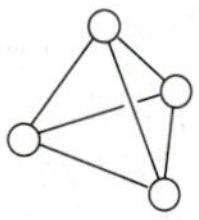	60	$Co_4(CO)_{12}$ (**54**)
Butterfly	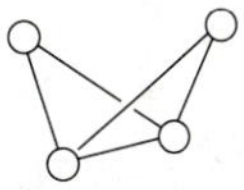	62	$[Fe_4(CO)_{12}C]^{2-}$ (**58**)
Square	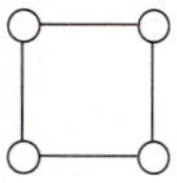	64	$Pt_4(O_2CCH_3)_8$
5 Trigonal bipyramid	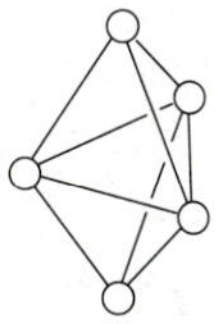	72	$Os_5(CO)_{16}$
Square pyramid	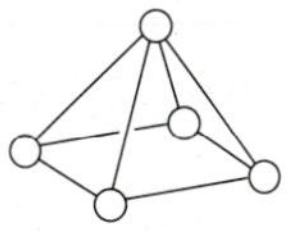	74	$Fe_5C(CO)_{15}$
6 Octahedron	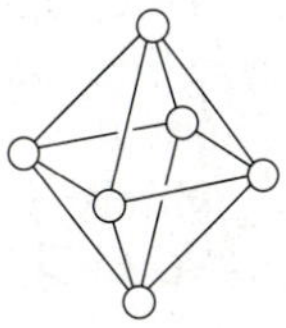	86	$Ru_6C(CO)_{17}$
Trigonal prism	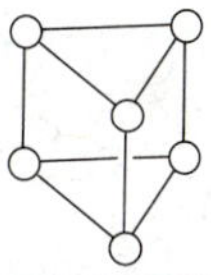	90	$[Rh_6C(CO)_{15}]^{2-}$

*For a more extensive table, see J. W. Lauher, *J. Am. Chem. Soc.*, **100**, 5305 (1978).
†Table 16.2.

Structure and isolobal analogies[21]

A good way to picture the incorporation of hetero atoms into a metal cluster is to use isolobal analogies. These let us draw a parallel between $Co_3(CO)_9(CH)$ (**57**) and $Co_4(CO)_{12}$ (**55**), both of which can be regarded as triangular $Co_3(CO)_9$ fragments capped on one side either by $Co(CO)_3$ or by CH. The CH and $Co(CO)_3$ groups are isolobal because each one has three orbitals and three electrons available for participa-

---

[21] The application of isolobal analogies to metal cluster compounds are described in the Nobel Prize lecture by R. Hoffmann, *Angew. Chem., Int. Ed. Engl.*, **21**, 711 (1982). Structures of organometallics and clusters are also nicely systematized by isoelectronic relations: J. Ellis, *J. Chem. Educ.*, **53**, 2 (1976).

tion in framework bonding. A minor complication in this comparison is the occurrence of $Co(CO)_2$ groups together with bridging CO ligands in $Co_4(CO)_{12}$ because bridging and terminal ligands often have similar energies.

Isolobal analogies allow us to draw many parallels between metal–metal bonded systems containing only d metals and mixed systems containing d-block and p-block metals. Additional examples are given in Table 16.12, where it will be noted that a P atom is isolobal with CH; accordingly, a cluster similar to (**57**) is known, but with a capping P atom. Similarly, the ligands CR_2 and $Fe(CO)_4$ are both capable of bonding to two metal atoms in a cluster; CH_3 and $Mn(CO)_5$ can bond to one metal atom.

16.11 Syntheses[22]

One of the oldest methods for the synthesis of metal clusters is the thermal expulsion of CO from a metal carbonyl. The pyrolytic formation of metal cluster compounds can be viewed from the standpoint of electron count: a decrease in valence electrons around the metal resulting from loss of CO is compensated by the formation of M—M bonds. One example is the synthesis of $Co_4(CO)_{12}$ by heating $Co_2(CO)_8$:

$$2Co_2(CO)_8 \xrightarrow{\Delta} Co_4(CO)_{12} + 4CO$$

This reaction proceeds slowly at room temperature, so samples of octacarbonyldicobalt are usually contaminated with dodecacarbonyltetracobalt.[23]

A widely employed and more controllable reaction is based on the condensation of a carbonyl anion and a neutral organometallic complex:

$$[Ni_5(CO)_{12}]^{2-} + Ni(CO)_4 \rightarrow [Ni_6(CO)_{12}]^{2-} + 4CO$$

$$76 \text{ CVE} \qquad\qquad\qquad 86 \text{ CVE}$$

The descriptive name **redox condensation** is often given to reactions of this type, which are very useful for the preparation of anionic metal carbonyl clusters. In this example, a trigonal bipyramidal cluster containing Ni with oxidation number $-\frac{2}{5}$ and $Ni(CO)_4$ containing Ni(0) is converted into an octahedral cluster having Ni with oxidation number $-\frac{1}{3}$. The $[Ni_5(CO)_{12}]^{2-}$ cluster, which has four electrons in excess of the 72 expected for a trigonal bipyramid, illustrates a fairly common tendency for the Group 10 metal clusters to have an electron count in excess of that expected from the Wade–Mingos–Lauher rules.

A third method, pioneered by F. G. A. Stone,[24] is based on the

[22] R. D. Adams, Chapter 3 in *The chemistry of metal cluster complexes*, (ed. D. F. Shriver, H. D. Kaesz, and R. D. Adams). VCH, Weinheim (1990).
[23] A wide range of osmium and ruthenium cluster compounds with interesting structures have been prepared by the pyrolysis method: J. Lewis and B. F. G. Johnson, *Pure Appl. Chem.*, **54**, 97 (1982).
[24] F. G. A. Stone, *Pure Appl. Chem.*, **23**, 89 (1984).

Table 16.12 Some isolobal *p*-block and *d*-block fragments*

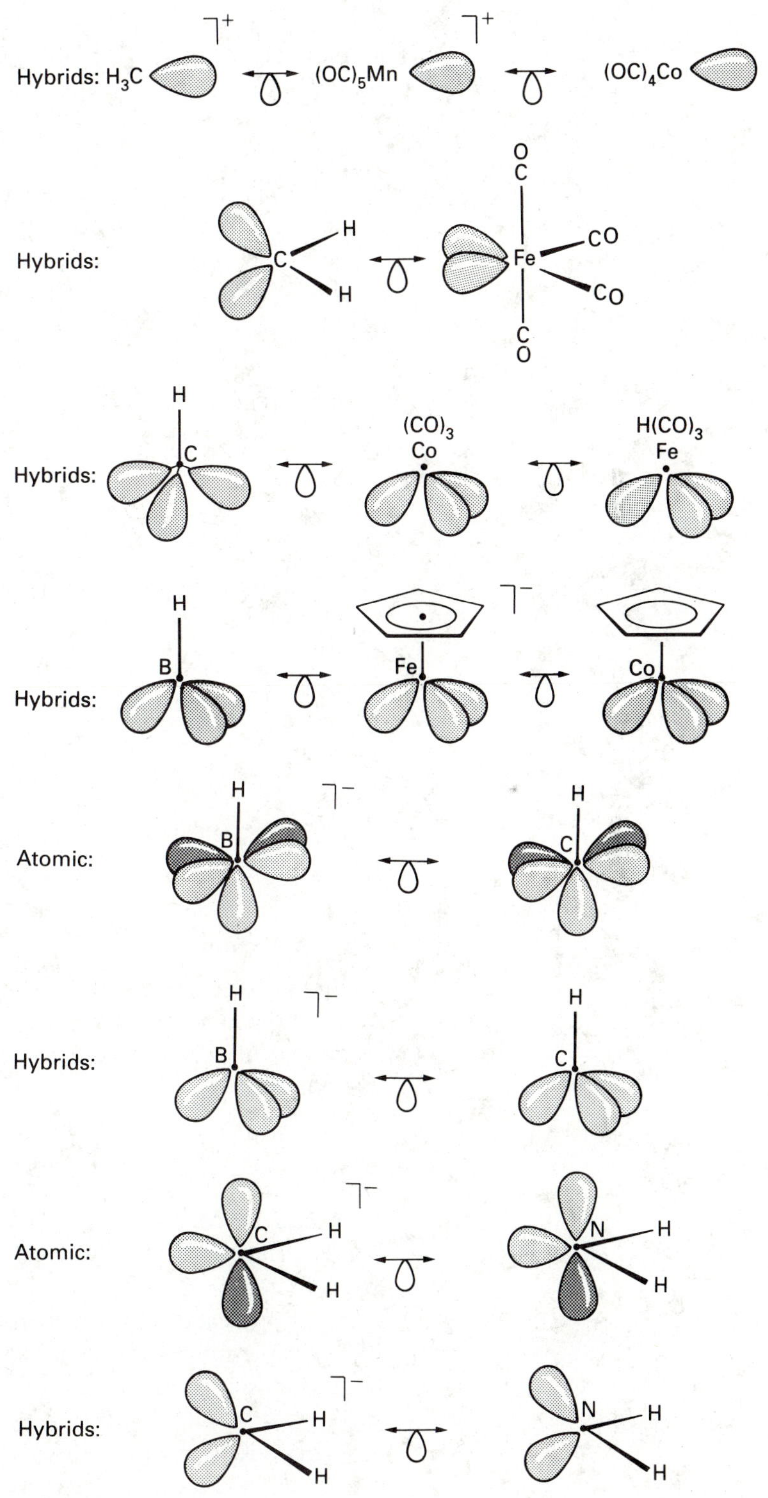

*Showing both atomic orbital and hybrid orbital representations.

condensation of an organometallic complex containing displaceable ligands with an unsaturated organometallic compound. The unsaturated complex may be a metal alkylidene, $L_nM=CR_2$, a metal alkylidyne, $L_nM\equiv CR$, or a compound with multiple metal–metal bonds:

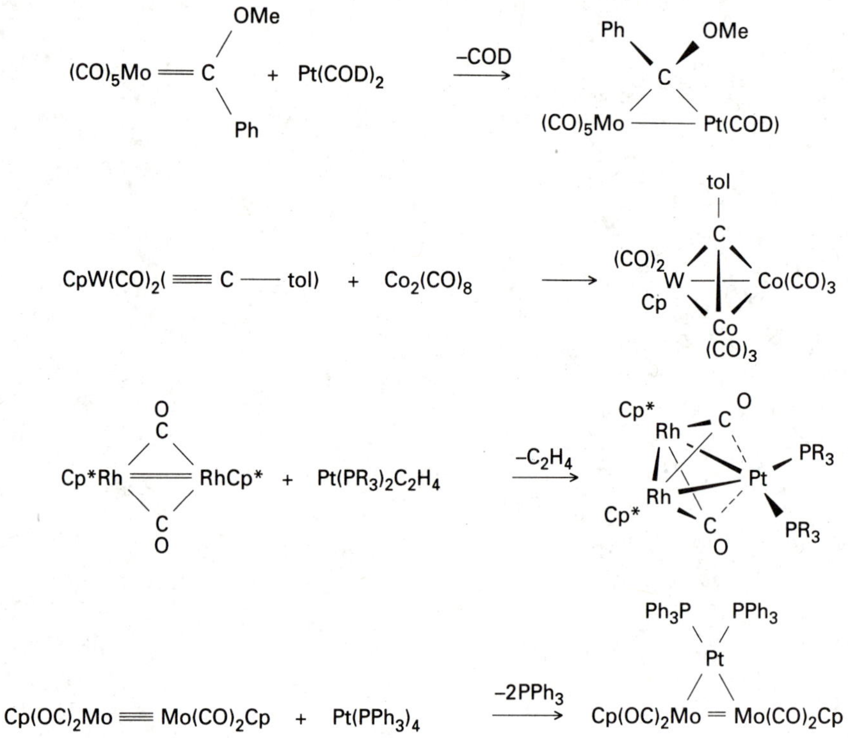

16.12 Reactions

We have already seen some examples of the reactivity of metal clusters in the above discussion of metal cluster synthesis. Some other common reactions of clusters are ligand substitution, fragmentation, and protonation.

Substitution versus fragmentation

Because M—M bonds are generally comparable in strength to M—L bonds, there is often a delicate balance between ligand substitution and cluster fragmentation. For example, dodecacarbonyltriiron(0), $Fe_3(CO)_{12}$, reacts with triphenylphosphane under mild conditions to yield simple mono- and disubstituted products as well as some cluster fragmentation products

$$Fe_3(CO)_{12} + PPh_3 \rightarrow Fe_3(CO)_{11}(PPh_3)$$
$$+ Fe_3(CO)_{10}(PPh_3)_2 + Fe(CO)_5 + Fe(CO)_4(PPh_3)$$
$$+ Fe(CO)_3(PPh_3)_2 + CO$$

However, for somewhat longer reaction times or elevated temperatures, only monoiron cleavage products are obtained. Since the strength of M—M bonds increases down a group, substitution products of the heavier clusters such as $Ru_3(CO)_{10}(PPh_3)_2$ or $Os_3(CO)_{10}(PPh_3)_2$ can be prepared without significant fragmentation into mononuclear complexes.

Protonation

The tendency of protons to attach to metal atoms in clusters is even more pronounced than that of mononuclear carbonyls. This Brønsted basicity of clusters is associated with the ready protonation of M—M bonds to produce a formal 3c,2e bond like that in diborane:

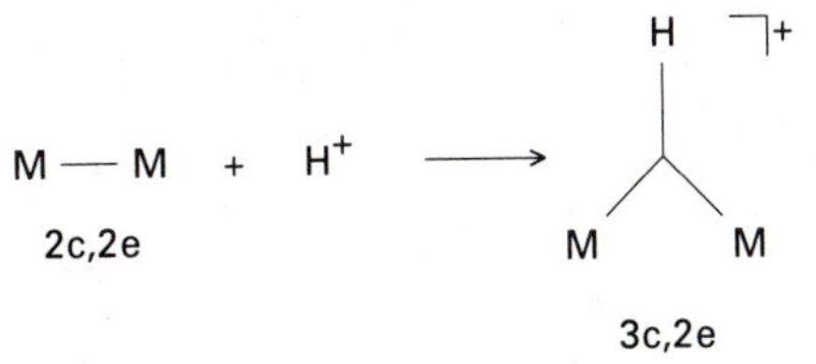

For example:

$$[Fe_3(CO)_{11}]^{2-} + H^+ \rightarrow [Fe_3H(CO)_{11}]^-$$

As expected, metal cluster anions are much stronger bases than their neutral analogs. The M—H—M bridge is by far the most common bonding mode of hydrogen in clusters, but hydrogen also is known to bridge a triangular three-atom face and to reside inside a metal polyhedron.

Cluster-assisted ligand transformations[25]

Transformations of ligands on metal clusters sometimes mimic those on single metal centers, but a cluster of metal atoms often facilitates ligand reactions because the presence of several metal atoms near a ligand provides additional opportunities for interactions. Two examples will be given, one involving the CH_3 ligand and the other CO.

The osmium cluster $[Os_3(CO)_{10}(CH_3)(H)]$ has been shown to undergo a transfer of H atoms to the metal-atom framework. The resulting equilibrium mixture of products can lose a CO ligand. This loss is accompanied by the transfer of two H atoms to the Os atom and the formation of a μ_3-CH ligand; the latter caps the three metal atom array (**58**).[26] A similar set of transformations may occur on catalytic metal surfaces. However, the chemistry of metal surfaces cannot be studied in the detail that was possible in this study of a metal cluster compound. In solution, NMR proved to be very useful,

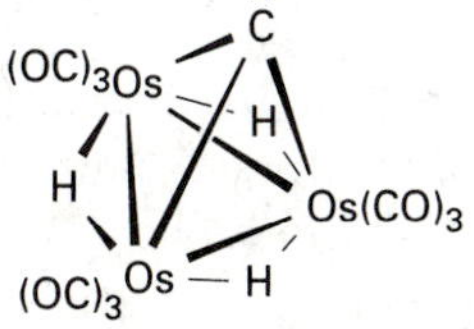

58

[25] Cluster-Assisted Ligand Transformations: G. Lavigne in *The chemistry of metal cluster complexes* (ed. D. F. Shriver, H. D. Kaesz, and R. D. Adams), p. 201. VCH, Weinheim (1990).
[26] See J. R. Shapley and coworkers, *J. Am. Chem. Soc.*, **99**, 5225 (1977); **100**, 6240 and 7726 (1978).

and in the solid state, X-ray and neutron diffraction measurements provided precise structural information.

Both CO migratory insertion and the attack on CO by nucleophiles have also been observed in clusters. Of greater interest, however, are reactions in which the metal atoms work in concert to bring about ligand transformations. One example is the greater ease of cleaving the CO ligand in a cluster:

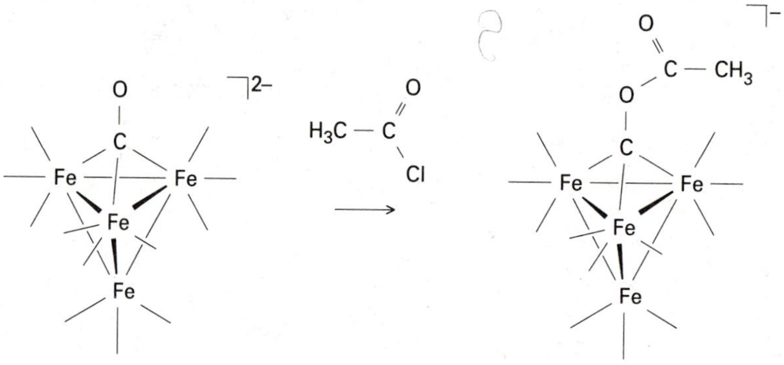

$$[Fe_4(CO)_{12}(\mu_3\text{--CO})]^{2-} \quad\quad [Fe_4(CO)_{12}(\mu_3\text{--COAc})]^-$$

where Ac denotes CH_3CO. This acylation of the O atom of CO has been observed only in clusters. The acetate group so produced is a good leaving group and it can be removed by electron transfer to the cluster, which maintains the electron count on the cluster when the acetate has departed. This process yields a metal carbide cluster with the carbon in a four-coordinate environment, but one that is far from tetrahedral:

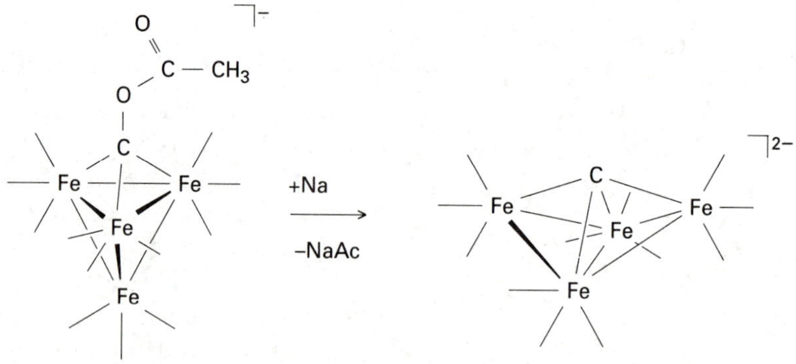

$$[Fe_4(CO)_{12}(\mu_3\text{--COAc})]^- \quad\quad [Fe_4(CO)_{12}(\mu_4\text{--C})]^{2-}$$

58

In penta- and hexa-iron clusters, the C atom is found in square pyramidal and octahedral coordination environments. From these examples and other areas of inorganic chemistry—such as carboranes (Section 11.3) and solid metal carbides (Section 11.7)—we see that carbon participates in a wide range of bonding arrangements that are unprecedented in organic chemistry.

FURTHER READING

Some excellent introductory texts are available:

Ch. Elschenbroich and A. Salzer, *Organometallics*. VCH, Weinheim (1992).

P. Powell, *Principles of organometallic chemistry*. Chapman and Hall, London (1988).

R. H. Crabtree, *The organometallic chemistry of the transition metals*. Wiley–Interscience, New York (1988).

A. Yamamoto, *Organotransition metal chemistry*. Wiley–Interscience, New York (1986).

C. M. Lukehart, *Fundamental transition metal organometallic chemistry*. Brooks Cole, Belmont (1985).

More detailed coverage is available in:

J. P. Collman, L. S. Hegedus, J. R. Norton, and R. G. Finke, *Principles and applications of organotransition metal chemistry*. University Science Books, Mill Valley (1987).

G. Wilkinson, F. G. A. Stone, and E. W. Abel (ed.), *Comprehensive organometallic chemistry*. Pergamon Press, Oxford (1982). This multivolume set provides excellent introductions and details for all areas of organometallic chemistry. More recent developments will be the topic of a forthcoming second edition.

D. F. Shriver, H. D. Kaesz, and R. D. Adams (ed.), *The chemistry of metal cluster complexes*. VCH, New York (1990).

KEY POINTS

1. Organometallic compounds
Organometallic compounds contain at least one metal–carbon bond.

2. Hapticity
The hapticity η of a compound designates the number of points of attachment of an organic ligand to a metal atom.

3. The 16/18 electron rule
d-Block organometallic compounds of metals in Groups 6 to 8 generally have 18 valence electrons around the central metal atom. For Groups 9 and 10, the norm is 16 or 18 electrons.

4. CO–metal interaction
The ligand CO is counted as a two-electron σ-donor. Additionally, it is a net acceptor of electrons via π interaction from the coordinated metal atom. It usually bonds through C to one, two, or three metal atoms.

5. Metal carbonyl syntheses
Metal carbonyls are usually synthesized by direct reaction of CO with the metal or reaction between CO and a metal salt in the presence of a reducing agent.

6. Metal carbonyl IR spectra
The number of observed IR bands correlates with the structures of substituted mononuclear carbonyls: CO stretching frequencies are lowered by metal bridging ($-CO > \overset{\diagdown}{\underset{\diagup}{C}}O > \overset{\diagdown}{-}CO$) and by electron density on the metal (neutral > mononegative > dinegative complexes).

7. Substitution of CO
Substitution of CO by other ligands generally occurs by initial CO dissociation, and is promoted by heat or photolysis.

8. β-hydrogen elimination
β-hydrogen elimination from an ethyl or larger hydrocarbon ligand produces an M—H bond and free alkene.

9. Alkylidene ligands
Alkylidene ligands have a formal M=C double bond. Alkylidenes designated Fischer carbenes are

formed with mid to late *d*-block metals and are electrophilic at the metal-bound C atom. Schrock carbenes are formed with early *d*-block metals and the C atom is nucleophilic.

10. Alkene and polyene ligands

For electron counting purposes, alkene and polyene ligands donate two-electrons for each metal-bound C—C group.

11. Early d- and f-block organometallic compounds

The organometallic compounds of early *d*- and *f*-block elements are highly oxophilic, and are aggressive C—H bond cleavers.

12. Metal clusters

Metal clusters involve direct M—M bonds and generally obey the 18-electron rule for M_5 and smaller clusters. The structures of clusters are correlated with electron count by the Wade–Mingos–Lauher rules.

13. Metal cluster syntheses

Metal clusters may be synthesized by thermal ligand dissociation followed by M—M bond formation between the coordinatively unsaturated fragments, redox condensation, or addition of electron rich metal compounds to unsaturated MM or MC bonds.

14. Multi-metal interactions

Such interactions may promote reactivity of alkyl or CO ligands.

EXERCISES

16.1 Name the species and draw the structures of (a) $Fe(CO)_5$, (b) $Ni(CO)_4$, (c) $Mo(CO)_6$, (d) $Mn_2(CO)_{10}$, (e) $V(CO)_6$, (f) $[PtCl_3(C_2H_4)]^-$.

16.2 (a) Sketch an η^2 interaction of 1,3-butadiene with a metal atom, and (b) do the same for an η^4 interaction.

16.3 Assign oxidation numbers to the metal atom in (a) $[Fe(\eta^5\text{-}C_5H_5)_2][BF_4]$, (b) $Fe(CO)_5$, (c) $[Fe(CO)_4]^{2-}$, and (d) $Co_2(CO)_8$.

16.4 Count the number of valence electrons for each metal atom in the complexes listed in Exercises 16.1 and 16.3. Do any of them deviate from the 18-electron rule? If so, is this reflected in their structure or chemical properties?

16.5 State the two common methods for the preparation of simple metal carbonyls and illustrate your answer with chemical equations. Is the selection of method based on thermodynamic or kinetic considerations?

16.6 Suggest a sequence of reactions for the preparation of $Fe(diphos)(CO)_3$, given iron metal, CO, diphos ($Ph_2PCH_2CH_2PPh_2$), and other reagents of your choice.

16.7 Suppose that you are given a series of metal tricarbonyl compounds having the respective symmetries C_{2v}, D_{3h}, and C_s. Without consulting reference material, which of these should display the greatest number of CO stretching bands in the IR spectrum? Check your answer and give the number of expected bands for each by consulting Table 16.5.

16.8 The compound $Ni_3(C_5H_5)_3(CO)_2$ has a single CO stretching absorption at 1761 cm^{-1}. The IR data indicate that all C_5H_5 ligands are pentahapto and probably in identical environments. (a) On the basis of these data, propose a structure. (b) Does the electron count for each metal in your structure agree with the 18-electron rule? If not, is nickel in a region of the periodic table where deviations from the 18-electron rule are common?

16.9 Decide which of the two complexes (a) $W(CO)_6$ or (b) $IrCl(PPh_2)_2(CO)$ should undergo the fastest exchange with ^{13}CO. Justify your answer.

16.10 Which metal carbonyl in each of (a) $[Fe(CO)_4]^{2-}$ or $[Co(CO)_4]^-$, (b) $[Mn(CO)_5]^-$ or $[Re(CO)_5]^-$ should be the most basic toward a proton? What are the trends upon which your answer is based?

16.11 For each of the following mixtures of reactants, give (i) a plausible chemical equation and (ii) structure for the organometallic product, and (iii) general reason for the course of the reaction: (a) methyllithium and $W(CO)_6$, (b) $Co_2(CO)_8$ and $AlBr_3$.

16.12 What hapticities are possible for the interaction of each of the following ligands with a single *d*-block metal atom such as cobalt? (a) C_2H_4, (b) cyclopentadienyl, (c) C_6H_6, (d) butadiene, (e) cyclooctatetraene.

16.13 Draw plausible structures and give the electron count of (a) $Ni(\eta^3\text{-}C_3H_5)_2$, (b) η^4-cyclobutadiene-η^5-cyclopentadienylcobalt, (c) $(\eta^3\text{-}C_3H_5)Co(CO)_2$. If the electron count deviates from 18, is the deviation explicable in terms of periodic trends?

16.14 Write out the *d* block of the periodic table. (By this stage, you should not have to consult a periodic table.) Indicate on this table (a) the elements that form 18-electron neutral Cp_2M compounds, (b) the Period 4 elements for which the simplest carbonyls are dimeric, (c) the Period 4 elements that form neutral carbonyls with six, five, and four carbonyl ligands, and (d) the elements that most commonly obey the 18-electron rule.

16.15 Using the 18-electron rule as a guide, indicate the probable number of carbonyl ligands in (a) $W(\eta^6\text{-}C_6H_6)(CO)_n$, (b) $Rh(\eta^5\text{-}C_5H_5)(CO)_n$, and (c) $Ru_3(CO)_n$.

16.16 Give plausible equations, different from those in the text, to demonstrate the utility of metal carbonyl anions in the synthesis of M—C, M—H, and M—M' bonds.

16.17 Propose a synthesis for $MnH(CO)_5$, starting with $Mn_2(CO)_{10}$ as the source of Mn and other reagents of your choice.

16.18 Give the probable structure of the product obtained when $Mo(CO)_6$ is allowed to react first with LiPh and then with a strong carbocation reagent, $CH_3OSO_2CF_3$.

16.19 Provide plausible reasons for the differences in IR wavenumbers between each of the following pairs: (a) $Mo(PF_3)_3(CO)_3$ 2040, 1991 cm^{-1}, versus $Mo(PMe_3)_3(CO)_3$ 1945, 1851 cm^{-1}, (b) $MnCp(CO)_3$ 2023, 1939 cm^{-1}, versus $MnCp^*(CO)_3$ 2017, 1928 cm^{-1}.

16.20 Which compound would you expect to be more stable, $Rh(\eta^5\text{-}C_5H_5)_2$ or $Ru(\eta^5\text{-}C_5H_5)_2$? Give a plausible explanation for the difference in terms of simple bonding concepts.

16.21 Give the equation for a workable reaction that will convert $Fe(\eta^5\text{-}C_5H_5)_2$ into $Fe(\eta^5\text{-}C_5H_5)(\eta^5\text{-}C_5H_4COCH_3)$.

16.22 Sketch the a_1' symmetry-adapted orbitals for the two eclipsed C_5H_5 ligands stacked together with D_{5h} symmetry. Identify the s, p, and d orbitals of a metal atom lying between the rings that may have nonzero overlap, and state how many a_1' molecular orbitals may be formed.

16.23 The compound $Ni(\eta^5\text{-}C_5H_5)_2$ readily reacts with HF to yield $[Ni(\eta^5\text{-}C_5H_5)(\eta^4\text{-}C_5H_6)]^+$ whereas $Fe(\eta^5\text{-}C_5H_5)_2$ reacts with strong acid to yield $[Fe(\eta^5\text{-}C_5H_5)_2H]^+$. In the latter compound the H atom is attached to the Fe atom. Provide a reasonable explanation for this difference.

16.24 Write a plausible mechanism, giving your reasoning, for the reactions (a) $[Mn(CO)_5(CF_2)]^+ + H_2O \rightarrow [Mn(CO)_6]^+ + 2HF$ and (b) $Rh(CO)(C_2H_5)(PR_3)_2 \rightarrow RhH(CO)(PR_3)_2 + C_2H_4$

16.25 Contrast the general chemical characteristics of the organometallic complexes of d-block elements in Groups 6 to 8 with those in Groups 3 and 4 with respect to (a) stability of the $\eta^5\text{-}C_5H_5$ ligand, (b) hydridic or protonic character of M—H bonds, and (c) adherence to the 18-electron rule.

16.26 (a) What cluster valence electron (CVE) count is characteristic of octahedral and trigonal prismatic complexes? (b) Can these CVE values be derived from the 18-electron rule? (c) Determine the probable geometry (octahedral or trigonal prismatic) of $[Fe_6(C)(CO)_{16}]^{2-}$ and $[Co_6(C)(CO)_{16}]^{2-}$. (The C atom in both cases resides in the center of the cluster and can be considered to be a four-electron donor.)

16.27 Based on isolobal analogies, choose the groups that might replace the group in boldface in

(a) $Co_2(CO)_9$**CH** OCH_3, $N(CH_2)_2$, or $SiCH_3$

(b) $(OC)_5Mn$**Mn(CO)**$_5$ I, CH_2, or CCH_3

16.28 Ligand substitution reactions on metal clusters are often found to occur by associative mechanisms, and it is postulated that these occur by initial breaking of a M—M bond, thereby providing an open coordination site for the incoming ligand. If the proposed mechanism is applicable, which would you expect to undergo the fastest exchange with added ^{13}CO? $Co_4(CO)_{12}$ or $Ir_4(CO)_{12}$? Suggest an explanation.

PROBLEMS

16.1 Propose the structure of the product obtained by the reaction of $[Re(CO)(\eta^5\text{-}C_5H_5)(PPh_3)(NO)]^+$ with Li[HBEt$_3$]. The latter contains a strongly nucleophilic hydride. (For full details see: W. Tam, G. Y. Lin, W. K. Wong, W. A. Kiel, V. Wong, and J. A. Gladysz, *J. Am. Chem. Soc.*, **104**, 141 (1982).

16.2 Develop a qualitative molecular orbital diagram for $Cr(CO)_6$ based on the symmetry-adapted linear combinations in Appendix 4. First consider the influence of σ bonding and then the effects of π bonding. (For help see T. A. Albright, J. K. Burdett, and M. H. Whangbo, *Orbital interactions in chemistry*, Wiley–Interscience, New York (1985).)

16.3 When several CO ligands are present in a metal carbonyl, an indication of the individual bond strengths can be determined by means of force constants derived from the experimental IR frequencies. In $Cr(CO)_5(PPh_3)$ the *cis*-CO ligands have the higher force constants, whereas in $Ph_3SnCo(CO)_4$ the force constants are higher for the *trans*-CO. Explain which carbonyl C atoms should be susceptible to nucleophiles to attack in these two cases. (For details see D. J. Darensbourg and M. Y. Darensbourg, *Inorg. Chem.*, **9**, 1691 (1970).)

16.4 The complex $Ru_2Cp_2(CO)_4$ is known to exist as three isomers in hydrocarbon solution, one of which contains no CO bridges and presumably has nearly free rotation about the Ru—Ru bond. The two other isomers both have two CO bridges; one has a *cis* disposition of the Cp and terminal CO ligands and the other has a *trans* disposition of these ligands. The IR spectrum of this equilibrium mixture of isomers in solution contains CO stretching bands at 2030, 2024, 2021*, 1993, 1988, 1967*, 1945*, 1834 and 1795 cm^{-1}. The bands marked with * have been assigned to the nonbridged isomer. The variation of the IR spectra with $AlEt_3$ concentration indicate that at an intermediate concentration, a species is produced with the simpler spectrum 2030, 2024, 1993, 1988, 1834, and 1679 cm^{-1}. When the concentration of $AlEt_3$ is increased further the IR spectrum is even simpler: 2046, 2011, 2006, and 1679 cm^{-1}. Propose a chemical basis for the simplification of the IR spectrum at high $AlEt_3$ concentrations. Accompany this answer with equations for the probable reactions and an interpretation of the IR data for the two reaction products. (For futher details see A. A. Alich, N. J. Nelson, D. Strope, and D. F. Shriver, *Inorg. Chem.*, **11**, 2976 (1972).)

16.5 The dinitrogen complex $Zr_2(\eta^5\text{-Cp*})_4(N_2)_3$ has been isolated and its structure determined by single-crystal X-ray diffraction. Each Zr atom is bonded to two Cp* and one terminal N_2. The third N_2 bridges between the Zr atoms in a nearly linear ZrN=NZr array. Before consulting the reference, write a plausible structure for this compound that accounts for the ^{1}H-NMR spectrum obtained on a sample held at $-7\,°C$. This spectrum is a doublet indicating

that the Cp* rings are in two different environments. At somewhat above room temperature these rings become equivalent on the NMR time scale and ^{15}N-NMR indicates that N_2 exchange between the terminal ligands and dissolved N_2 is correlated with the process that interconverts the Cp* ligand sites. Propose a way in which this equilibration could interconvert the sites of the Cp* ligands. (For further details see J. M. Manriquez, D. R. McAlister, E. Rosenberg, H. M. Shiller, K. L. Willamson, S. I. Chan, and J. E. Bercaw, *J. Am. Chem. Soc.* **108**, 3078 (1978).)

16.6 Devise a qualitative molecular orbital scheme for $Ni(\eta^5\text{-}C_5H_5)_2$ assuming D_{5h} symmetry and making use of the symmetry adapted orbitals in Appendix 4.

16.7 If any of the following statements are incorrect, provide a corrected counterpart and an example that bears our your correction. (a) A high-field NMR signal is characteristic of the hydride ligand in Groups 8 and 9 organometallic complexes. (b) Fischer carbenes are synthesized from carbonyl compounds of the early *d*-block metals. (c) The appearance of a single feature in the ^{1}H-NMR spectrum of a cyclooctatetraene metal complex is proof that the ligand is bound in a symmetric η^8 manner. (d) Square planar 16-electron compounds generally undergo ligand substitution by a dissociative mechanism. (e) Metal–metal bonds in cluster compounds generally become weaker going down a group in the *d* block.

17

Catalysis

In this chapter we apply the concepts of organometallic chemistry and coordination chemistry to catalysis. We emphasize general principles, such as the nature of catalytic cycles, in which a catalytic species is regenerated in a reaction, and the delicate balance of reactions required for a successful cycle. We shall see that there are four requirements for a successful catalytic process: the reaction being catalyzed must be thermodynamically favorable and fast enough when catalyzed; the catalyst must have an appropriate selectivity toward the desired product and a lifetime long enough to be economical. We then survey the five reaction types that are commonly encountered in the homogeneous catalysis of hydrocarbon interconversion and show how these reactions are invoked in proposals about mechanisms. The final part of the chapter develops a similar theme in heterogeneous catalysis, and we shall see many parallels between homogeneous and hetero-geneous catalysis that lurk beneath differences in terminology. In neither type of catalysis are the mechanisms finally settled, and there is still considerable scope for making new discoveries.

A **catalyst** is a substance that increases the rate of a reaction but is not itself consumed. Catalysts are widely used in nature, in industry, and in the laboratory, and it is estimated that they contribute to one-sixth of the value of all manufactured goods in industrialized countries. As shown in Table 17.1, 13 of the top 20 synthetic chemicals are produced directly or indirectly by catalysis. For example, a key step in the production of the dominant industrial chemical, sulfuric acid, is the catalytic oxidation of SO_2 to SO_3. Ammonia, another chemical essential for industry and agriculture, is produced by the catalytic reduction of N_2 by H_2. Inorganic catalysts are also used for the production of the major organic chemicals and petroleum products, such as fuels, petrochemicals, and polyalkene plastics. Catalysts play a

Table 17.1 The top twenty synthetic chemicals

Synthetic chemical	Rank†	Catalytic process
Sulfuric acid	1	SO_2 oxidation, heterogeneous
Ethylene	2	*a*
Ammonia	3	$N_2 + H_2$; heterogeneous
Calcium hydroxide	4	Not catalytic
Phosphoric acid	5	Not catalytic
Sodium hydroxide	6	Not catalytic
Propylene	7	*a*
Chlorine	8	Electrocatalysis, heterogeneous
Sodium carbonate	9	Not catalytic
Urea	10	*b*
Nitric acid	11	$NH_3 + O_2$; heterogeneous
1,2-Dichloroethane	12	$C_2H_4 + Cl_2$; homogeneous
Ammonium nitrate	13	*b*
Vinyl chloride	14	Chlorination of C_2H_4; heterogeneous
Benzene	15	Petroleum refining; heterogeneous
Ethylbenzene	16	Alkylation of benzene; homogeneous
Carbon dioxide	17	Not catalytic
Methyl tert-butyl ether	18	Heterogeneous
Styrene	19	Dehydrogenation of ethylbenzene; heterogeneous
Methanol	20	$CO + H_2$, heterogeneous

† Based on mass, from *Chemical and Engineering News* survey of US industrial chemicals April 12, (1993).
a Primarily used in a catalytic polymerization.
b Synthesis based on starting materials produced by a catalytic process.

steadily increasing role in achieving a cleaner environment, both through the destruction of pollutants (as with the automotive catalytic exhaust converters) and through the development of cleaner industrial processes with less abundant byproducts. Enzymes, a class of elaborate biochemical catalysts, are discussed in Chapter 19.

In addition to their economic importance and contribution to the quality of life, catalysts are interesting for the subtlety with which they go about their business. The understanding of the mechanisms of catalytic reactions has improved greatly in recent years because of the availability of isotopically labeled molecules, improved methods for determining reaction rates, and improved spectroscopic and diffraction techniques.

GENERAL PRINCIPLES

Catalysts are classified as *homogeneous* if they are present in the same phase as the reagents; this normally means that they are present as solutes in a liquid reaction mixture. Catalysts are *heterogeneous* if they are present in a different phase from that of the reactants. Both types of catalysis are discussed in this chapter, and it will be seen that they are fundamentally similar. Of the two, heterogeneous catalysis has a much greater economic impact.

The term 'negative catalyst' is sometimes applied to substances that retard reactions. We shall not use the term because these substances are best considered to be **catalyst poisons** that block one or more elementary steps in a catalytic reaction.

17.1 Description of catalysts

A catalyzed reaction is faster (or, in some cases, more specific) than an uncatalyzed version of the same reaction because the catalyst provides a different reaction pathway with a lower activation energy. To describe these characteristics we develop in this section some of the terminology used to express the speed of a catalytic reaction and its mechanism.

Catalytic efficiency

The **turnover frequency** N (formerly 'turnover number') is often used to express the efficiency of a catalyst. For the conversion of A to B catalyzed by Q and with a rate v,

$$A \xrightarrow{\;Q\;} B \quad v = \frac{d[B]}{dt}$$

the turnover frequency is given by

$$N = \frac{v}{[Q]}$$

if the rate of the uncatalyzed reaction is negligible. A highly active catalyst, one that results in a fast reaction even in low concentrations, has a high turnover frequency.

In heterogeneous catalysis, the reaction rate is expressed in terms of the rate of change in the amount of product (in place of concentration) and the concentration of catalyst is replaced by the amount present. The determination of the number of active sites in a heterogeneous catalyst is particularly challenging, and often the denominator [Q] is simply replaced by the surface area of the catalyst.

Catalytic cycles

The essence of catalysis is a cycle of reactions that consumes the reactants, forms products, and regenerates the catalytic species. In the example shown in Fig. 17.1, an equilibrium generates a weakly coordinated Rh(I) complex (a). The reactants, hydrogen and an alkene, enter the cycle by reaction with the complex to produce in succession a hydrido complex (b) and then an alkene complex (c). In the final step, from (d) to (a), the hydrogenated product leaves the loop with the regeneration of the coordinatively unsaturated Rh(I) complex, and the cycle can continue.

The rhodium in Fig. 17.1 transforms through a series of complexes, each of which facilitates a step in the overall reaction. This multiple role is common in homogeneous catalysis, and strictly speaking there is no single catalyst: a catalytic cycle may contain several complexes, each one of which participates in the overall process. The complex

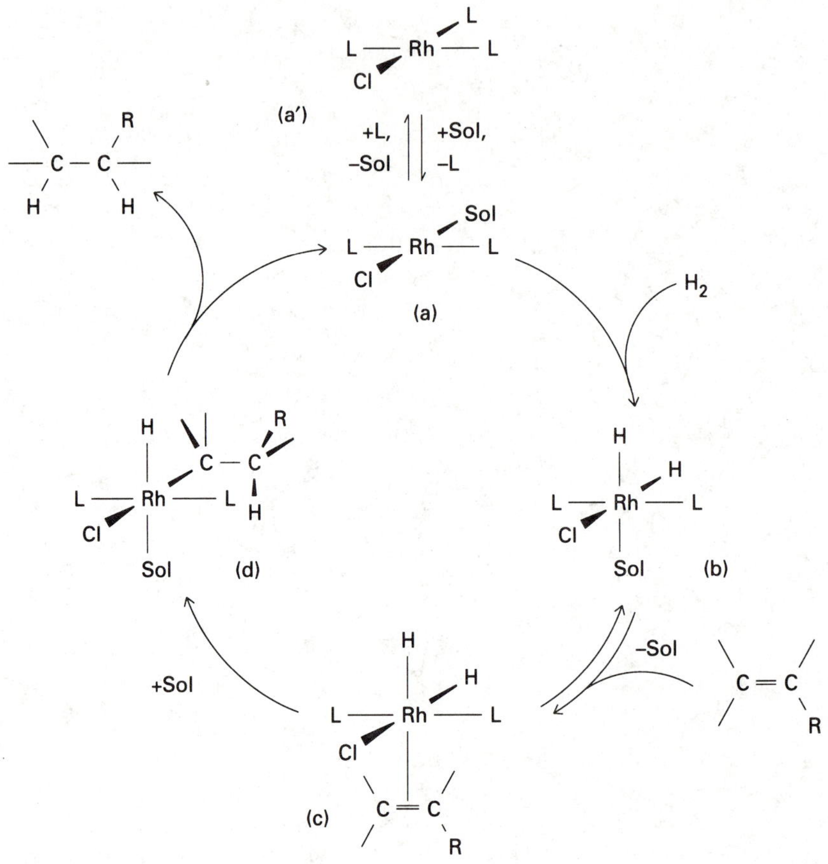

17.1 The main catalytic cycle in the homogeneous hydrogenation of an alkene by rhodium phosphane complexes, L = PPh_3.

introduced into the system, in this case $RhCl(PPh_3)_3$ (a′), is often referred to as 'the catalyst', but in examples like that illustrated in Fig. 17.1 it is more precise to call it a **catalyst precursor** because it is not within the catalytic cycle.

As with all mechanisms, the cycle has been proposed on the basis of a range of information like that summarized in Table 17.2. Many of the components shown in the table were encountered in Chapter 14 in connection with the determination of mechanisms of substitution reactions. However, the elucidation of catalytic mechanisms is complicated by the occurrence of several delicately balanced reactions, which often cannot be studied in isolation.

Two stringent tests of any proposed mechanism are the determination of rate laws and the elucidation of stereochemistry. If intermediates are postulated, then their detection by NMR and IR also provides support. If specific atom transfer steps are proposed, then isotope tracer studies may serve as a test. The influence of different ligands and different substrates are also sometimes informative. Although rate data and the corresponding laws have been determined for many *overall* catalytic cycles, it is also necessary to determine rate laws for the individual steps in order to have reasonable confidence in the mechanism. However, because of experimental complications, it is rare that catalytic cycles are studied in this detail.

Table 17.2 Determination of catalytic mechanisms

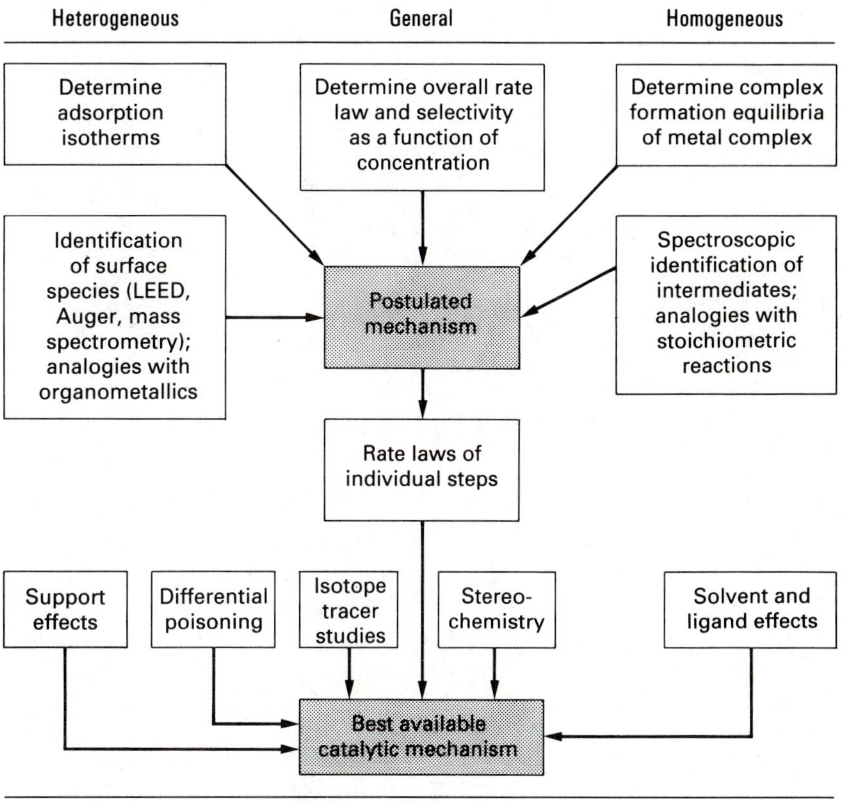

Heterogeneous	General	Homogeneous

Determine adsorption isotherms

Determine overall rate law and selectivity as a function of concentration

Determine complex formation equilibria of metal complex

Identification of surface species (LEED, Auger, mass spectrometry); analogies with organometallics

Postulated mechanism

Spectroscopic identification of intermediates; analogies with stoichiometric reactions

Rate laws of individual steps

Support effects	Differential poisoning	Isotope tracer studies	Stereo-chemistry	Solvent and ligand effects

Best available catalytic mechanism

Energetics

A catalyst increases the rates of processes by introducing new pathways (mechanisms) with lower Gibbs free energies of activation ($\Delta G^{\ddagger}$). It is important to focus on the free energy profile, of a catalytic reaction, not just the enthalpy or energy profile, because the new elementary steps that occur in the catalyzed process are likely to have quite different entropies of activation.[1] A catalyst does not affect the Gibbs free energy of the overall reaction ($\Delta G^{\ominus}$) because G is a state function (that is, G depends only on the current state of the system and not on the path that led to the state). The difference is illustrated in Fig. 17.2, where the overall reaction Gibbs free energy is the same in both energy profiles. Thus reactions that are thermodynamically unfavorable cannot be made favorable by a catalyst.

Figure 17.2 also shows that the free energy profile of a catalyzed reaction contains no high peaks and no deep troughs. The new pathway introduced by the catalyst changes the mechanism of the reaction. As a result, the free energy profile is quite different from that of the uncatalyzed reaction, and the free energy maxima are lower. However, an equally important point is that stable catalytic intermediates do not occur in the cycle. Similarly, the product must be released in a thermo-

[1] For further discussion of this point see, A. Haim, Catalysis: new reaction pathways, not just lowering of the activation energy, *J. Chem. Educ.*, **66**, 731 (1989).

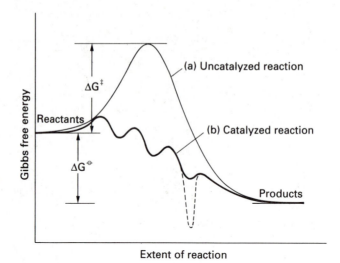

Gibbs free energy

(a) Uncatalyzed reaction

$\Delta G^{\ddagger}$

Reactants

(b) Catalyzed reaction

$\Delta G^{\ominus}$

Products

Extent of reaction

17.2 Schematic representation of the energetics in a catalytic cycle. The uncatalyzed reaction (a) has a higher Gibbs energy of activation $\Delta G^{\ddagger}$ than any step in the catalyzed reaction (b). The Gibbs free energy of reaction, $\Delta G^{\ominus}$ for the overall reaction is unchanged from (a) to (b).

dynamically favorable step. If, as shown by the dotted line in Fig. 17.2, a stable complex is formed with the catalyst, it would turn out to be the product of the reaction and the cycle would not continue. Similarly, impurities may suppress catalysis by coordinating strongly to catalytically active sites. Such impurities act as catalyst poisons.

17.2 Properties of catalysts

Although the increased rate of a catalytic reaction is important, it is not the only criterion. Of similar importance are a minimum of side products and a long catalyst lifetime.

Selectivity

A **selective catalyst** yields a high proportion of the desired product with minimum amounts of side products. In industry, there is considerable economic incentive to develop selective catalysts. For example, when metallic silver is used to catalyze the oxidation of ethylene with oxygen to produce ethylene oxide (**1**), the reaction is accompanied by the more thermodynamically favored but undesirable formation of CO_2 and H_2O. This lack of selectivity increases the consumption of ethylene, so chemists are constantly trying to devise a more selective catalyst for ethylene oxide synthesis. Selectivity can be ignored in only a very few simple inorganic reactions, where there is essentially only one thermodynamically favorable product, as in the formation of NH_3 from H_2 and N_2.

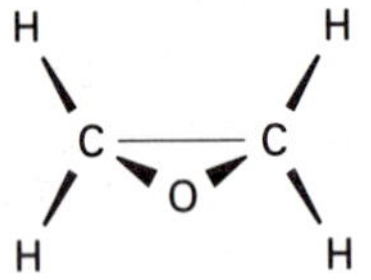

1 Ethylene oxide

Lifetime

A small amount of catalyst must survive through a large number of cycles if it is to be economically viable. However, a catalyst may be destroyed by side reactions to the main catalytic cycle or by the presence of small amounts of impurities in the starting materials (the feedstock). For example, many alkene polymerization catalysts are destroyed by O_2, so in the synthesis of polyethylene and polypropylene the concentration of O_2 in the ethylene or propylene feedstock should be no more than a few parts per billion.

Some catalysts can be regenerated quite readily. For example, the supported metal catalysts used in the reforming reactions that convert hydrocarbons to high octane gasoline become covered with carbon because the catalytic reaction is accompanied by a small a amount of dehydrogenation. These supported metal particles can be cleaned by periodically interrupting the catalytic process and then burning off the accumulated carbon.

HOMOGENEOUS CATALYSIS

In this section we describe some important homogeneous catalytic reactions and describe their currently favored mechanisms. As with nearly all mechanistic chemistry, catalytic mechanisms are inferred from a range of indirect evidence, and so a mechanism is subject to refinement or change as more detailed experimental information becomes available. It should also be borne in mind that the mechanism of homogeneous catalysis is more accessible to detailed investigation than heterogeneous catalysis because the interpretation of rate data is frequently easier. Moreover, species in solution are often easier to characterize than those on a surface.

From a practical standpoint, homogeneous catalysis is attractive because it is often highly selective toward the formation of a desired product. In large-scale industrial processes homogeneous catalysts may be preferred for exothermic reactions because it is easier to dissipate heat from a solution than from the solid bed of a heterogeneous catalyst.

17.3 Catalytic steps

In this section we review five types of reaction (and in some cases their reverse) which, in combination, account for most of the homogeneous catalytic cycles that have been proposed for hydrocarbon transformations. This list will undoubtedly alter and grow as our knowledge of catalysis deepens through further research.

Ligand coordination and dissociation

The catalysis of molecular transformations generally requires facile coordination of reactants to metal ions and equally facile loss of products from the coordination sphere. Both processes must occur with low activation free energy, so highly labile metal complexes are required. These labile complexes are **coordinatively unsaturated** in the sense that they contain an open coordination site or, at most, a site that is only weakly coordinated.

Square-planar 16-electron complexes are coordinatively unsaturated and are often employed to catalyze the reactions of organic molecules. In Sections 6.8 and 15.5 we met a number of examples of associative reactions between square-planar complexes and entering groups. These facile reactions are quite common for catalytic systems involving ML_4 complexes of Pd(II), Pt(II), and Rh(I), such as the hydrogenation catalyst $[RhCl(PPh_3)_3]$ featured in Fig. 17.1.

Insertion and elimination

The migration of alkyl and hydride ligands to unsaturated ligands, as in the reaction

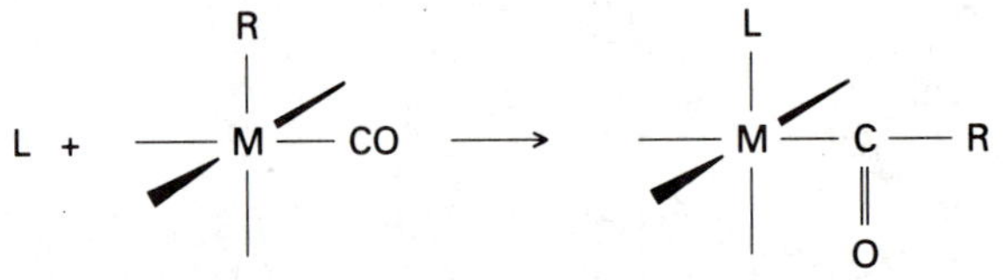

is an example of a migratory insertion reaction (or simply 'insertion reaction') discussed in Section 16.5. Another example of the same kind is the migration of an H ligand to a coordinated alkene to produce a coordinated alkyl ligand:

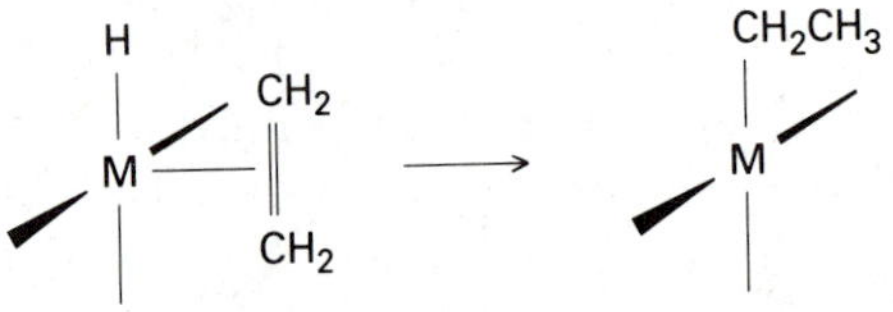

The conversion of (c) to (d) in Fig. 17.1 is a hydrogen migration of this type.

The reverse of insertion is elimination. Elimination reactions include the important β-hydrogen elimination (Section 16.7):

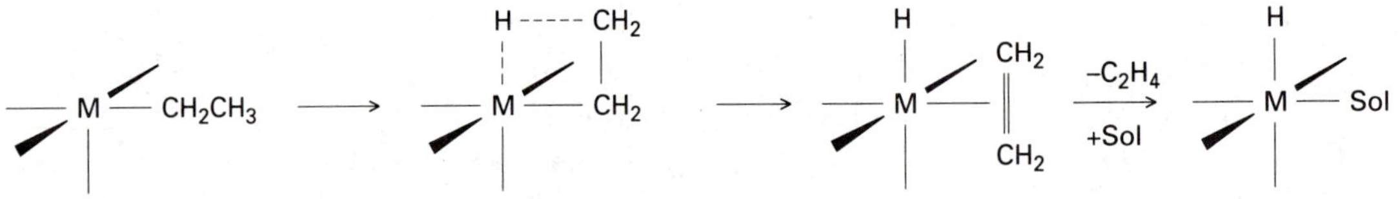

Nucleophilic attack on coordinated ligands

The coordination of ligands such as CO and alkenes to metal ions in positive oxidation states results in the activation of the coordinated C atoms toward attack by nucleophiles. Such reactions are useful in catalysis as well as in organometallic chemistry (Chapter 16).

The hydration of ethylene bound to Pd(II) is a good example of catalysis by nucleophilic activation. The stereochemical evidence indicates that the reaction occurs by direct attack on the most highly substituted C atom of the coordinated alkene:

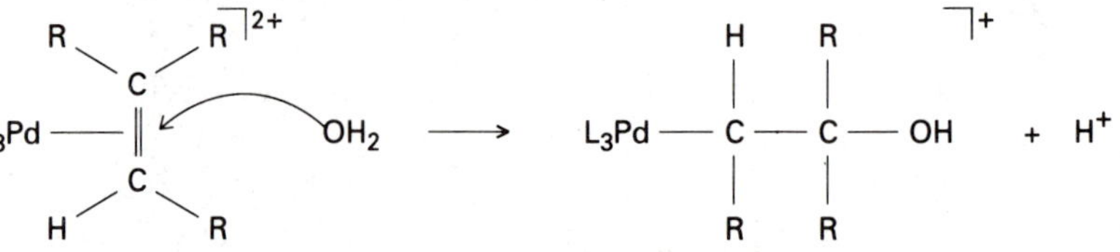

It also is possible for the hydroxylation of a coordinated alkene to take place by the prior coordination of H_2O to a metal complex, followed by an insertion reaction:

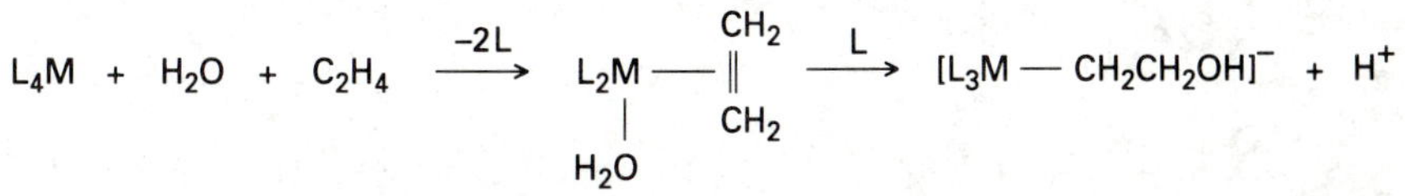

$$L_4M + H_2O + C_2H_4 \xrightarrow{-2L} \underset{\underset{H_2O}{|}}{L_2M}\underset{}{-}\overset{CH_2}{\underset{CH_2}{\|}} \xrightarrow{L} [L_3M-CH_2CH_2OH]^- + H^+$$

Similarly, a coordinated CO ligand is attacked by an OH^- ion at the C atom, forming a $-CO(OH)$ ligand, which subsequently loses CO_2:

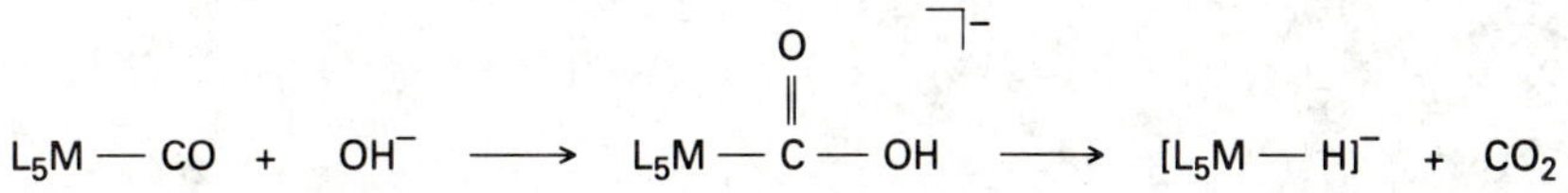

$$L_5M-CO + OH^- \longrightarrow \left[L_5M-\overset{\overset{\displaystyle O}{\|}}{C}-OH \right]^- \longrightarrow [L_5M-H]^- + CO_2$$

We discussed this reaction in Section 16.6 in connection with the formation of metal carbonyl anions. It is thought to be a critical step in the water-gas shift reaction catalyzed by metal carbonyl complexes or metal ions on solid surfaces:

$$CO + H_2O \xrightarrow{catalyst} CO_2 + H_2$$

Oxidation and reduction [2]

Metal complexes are often used for the catalytic oxidation of organic substrates, and in the catalytic cycle the metal atom alternates between two oxidation states. Some common catalytic one-electron couples are Cu^{2+}/Cu^+, Co^{3+}/Co^{2+}, and Mn^{3+}/Mn^{2+}; a few examples of catalytic two-electron processes are also known, such as with the couple Pd^{2+}/Pd.

Catalysts containing metal ions are used in large scale processes for the oxidation of hydrocarbons, as in the oxidation of *p*-xylene to terephthalic acid (1,4-benzenedicarboxylic acid). The metal ion can play various roles in these radical oxidations, as we can see by considering a mechanism of the following general form:

Initiation (In· = radical initiator):

$$In· + R-H \to In-H + R·$$

Propagation:

$$R· + O_2 \to R-O-O·$$

$$R-O-O· + R-H \to R-O-O-H + R·$$

Termination:

$$2R· \to R_2$$

$$R· + R-O-O· \to R-O-O-R$$

$$2R-O-O· \to R-O-O-O-O-R \to O_2 + \text{nonradicals}$$

[2] Good general references are: R. S. Sheldon and J. K. Kochi, *Metal-catalyzed oxidations of organic compounds*. Academic Press, Boca Raton (1981); P. M. Henry, *Palladium-catalyzed oxidation of hydrocarbons*. Reidel, Dordrecht (1980).

The metal ions control the reaction by contributing to the formation of the R—O—O· radicals:

$$Co(II) + R—O—O—H \rightarrow Co(ROOH) \rightarrow Co(III)OH + R—O·$$

$$Co(III) + R—O—O—H \rightarrow Co(II) + R—O—O· + H^+$$

The metal atom shuttles back and forth between oxidation states in this pair of reactions. A metal ion can also act as an initiator. For example, when arenes are involved it is believed that initiation occurs by a simple redox process:

$$ArCH_3 + Co(III) \rightarrow ArCH_3^+ + Co(II)$$

Oxidative addition and reductive elimination

We saw in Sections 15.13 and 16.6 that the oxidative addition of a molecule AX to a complex brings about dissociation of the A—X bond and coordination of the two fragments:

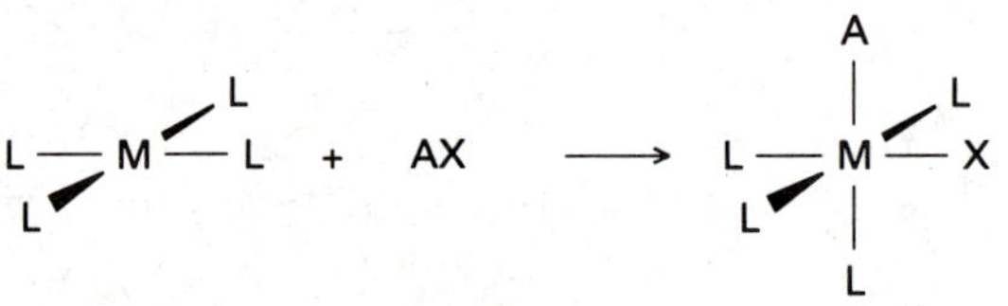

Reductive elimination is the reverse of oxidative addition, and often follows it in a catalytic cycle.

The mechanisms of oxidative addition reactions vary. Depending upon reaction conditions and the nature of the reactants, there is evidence for oxidative addition by simple concerted reaction, heterolytic (ionic) addition of A^+ and X^-, or radical addition of A· and X·. Despite this diversity of mechanism it is found that the rates of oxidative addition of alkyl halides generally follow the orders

primary alkyl < secondary alkyl < tertiary alkyl

$F \ll Cl < Br < I$

Example 17.1: *Interpreting the reactions in a catalytic cycle*

Figure 17.1 contains two examples of simple association reactions and one of a simple dissociation reaction. Identify them and write balanced chemical equations.

Answer. The catalyst precursor undergoes a ligand dissociation:

$$\underset{\text{(a')}}{RhCl(PPh_3)_3} + Sol \rightarrow PPh_3 + \underset{\text{(a)}}{RhCl(PPh_3)_2(Sol)}$$

The product is $RhCl(PPh_3)_2(Sol)$ if a solvent Sol is weakly coordinated. One of the association reactions is the reverse reaction:

$$PPh_3 + RhCl(PPh_3)_2(Sol) \rightarrow RhCl(PPh_3)_3 + Sol$$

Another association reaction involves ethylene:

$$RhClH_2(PPh_3)_2(Sol) + H_2C=CHR$$
$$(b)$$

$$\rightarrow RhClH_2(PPh_3)_2(H_2C=CHR) + Sol$$
$$(c)$$

Exercise E17.1. There is one oxidative addition reaction and one reductive elimination reaction in Fig. 17.1. Give balanced chemical equations for them both and assign oxidation numbers to all the rhodium complexes in the equations.

17.4 Examples

We shall now see how several individual elementary reactions combine forces and contribute to catalytic cycles. The following examples illustrate our current understanding of the mechanisms of some important types of catalytic reactions and give some insight into the manner in which catalytic activity and selectivity might be altered. As we have already remarked, there is usually even more uncertainty associated with catalytic mechanisms than with mechanisms of simpler kinds of reactions. Unlike simple reactions, a catalytic process frequently contains many steps over which the experimentalist has little control. Moreover, highly reactive intermediates are often present in concentrations too low to be detected spectroscopically. The best attitude to adopt toward these catalytic mechanisms is to learn the pattern of transformations and appreciate their implications, but to be prepared to accept new mechanisms that might be indicated by future work.

The scope of homogeneous catalysis can be appreciated from the examples given in Tables 17.1 and 17.3. The reactions cited there include hydrogenation, oxidation, and a host of other processes. Often the complexes of all metal atoms in a group will exhibit catalytic activity in a particular reaction, but the 4*d*-metal complexes are often superior as catalysts to their lighter and heavier congeners. In some cases the difference may be associated with the greater substitutional lability of 4*d* organometallic compounds in comparison with their 3*d* and 5*d* analogs. It is often the case that the complexes of costly metals must be used on account of their superior performance compared with the complexes of cheaper metals.

Hydrogenation of alkenes

The addition of hydrogen to an alkene to form an alkane is highly favored thermodynamically ($\Delta G^\ominus = -101$ kJ mol^{-1} for the conversion of ethylene to ethane). However, the reaction rate is negligible at ordinary conditions in the absence of a catalyst. Efficient homogeneous and heterogeneous catalysts are known for the hydrogenation of alkenes and are used in such diverse areas as the manufacture of margarine, pharmaceuticals, and petrochemicals.

Table 17.3 Some homogeneous catalytic processes

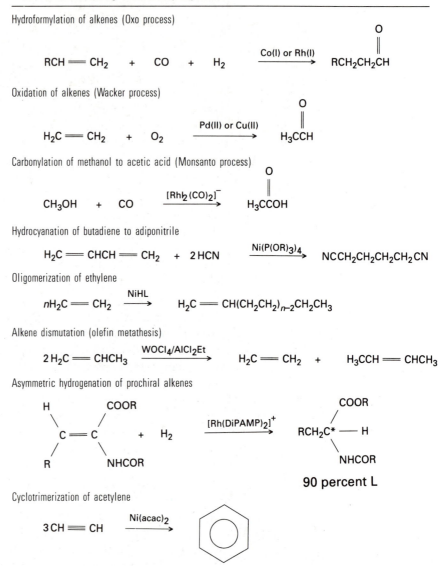

Hydroformylation of alkenes (Oxo process)

$$RCH = CH_2 \quad + \quad CO \quad + \quad H_2 \quad \xrightarrow{\text{Co(I) or Rh(I)}} \quad RCH_2CH_2\overset{\displaystyle O}{\overset{\displaystyle \|}{C}}H$$

Oxidation of alkenes (Wacker process)

$$H_2C = CH_2 \quad + \quad O_2 \quad \xrightarrow{\text{Pd(II) or Cu(II)}} \quad H_3C\overset{\displaystyle O}{\overset{\displaystyle \|}{C}}H$$

Carbonylation of methanol to acetic acid (Monsanto process)

$$CH_3OH \quad + \quad CO \quad \xrightarrow{[Rh_2(CO)_2]^-} \quad H_3C\overset{\displaystyle O}{\overset{\displaystyle \|}{C}}OH$$

Hydrocyanation of butadiene to adiponitrile

$$H_2C = CHCH = CH_2 \quad + \quad 2\,HCN \quad \xrightarrow{\text{Ni(P(OR)}_3)_4} \quad NCCH_2CH_2CH_2CH_2CN$$

Oligomerization of ethylene

$$n H_2C = CH_2 \quad \xrightarrow{\text{NiHL}} \quad H_2C = CH(CH_2CH_2)_{n-2}CH_2CH_3$$

Alkene dismutation (olefin metathesis)

$$2\,H_2C = CHCH_3 \quad \xrightarrow{\text{WOCl}_4/\text{AlCl}_2\text{Et}} \quad H_2C = CH_2 \quad + \quad H_3CCH = CHCH_3$$

Asymmetric hydrogenation of prochiral alkenes

$$\underset{R}{\overset{H}{\diagdown}}C = C\underset{NHCOR}{\overset{COOR}{\diagup}} \quad + \quad H_2 \quad \xrightarrow{[Rh(DiPAMP)_2]^+} \quad RCH_2C^* \underset{NHCOR}{\overset{COOR}{\diagup}} H$$

90 percent L

Cyclotrimerization of acetylene

$$3\,CH = CH \quad \xrightarrow{\text{Ni(acac)}_2} \quad \bigcirc$$

Adapted from J. Halpern, *Inorg. Chim. Acta*, **50**, 11 (1981).

One of the most studied catalytic systems is the Rh(I) complex [RhCl(PPh$_3$)$_3$] (a' in Fig. 17.1), which is often referred to as 'Wilkinson's catalyst'. This useful catalyst hydrogenates a wide variety of alkenes at pressures of hydrogen close to 1 atm or less. The cycle shown in Fig. 17.1 is close to the original mechanism proposed by Wilkinson and his coworkers but some details have been revised, primarily through the careful mechanistic studies undertaken in Jack Halpern's laboratory at the University of Chicago. This is one of the few catalytic cycles in which each of the postulated steps has been checked kinetically, so we can go through the mechanism in some detail.

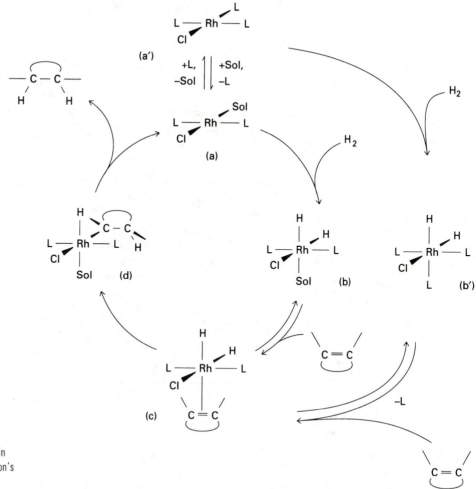

17.3 The catalytic cycles implicated in cyclohexene hydrogenation by Wilkinson's catalyst, [RhCl(PPh$_3$)$_3$].

The dominant cycle for simple alkenes such as cyclohexene (Fig. 17.3) appears to involve an early step in which Rh(I) oxidatively adds H$_2$. Kinetic data indicate that in the primary catalytic cycle H$_2$ adds not to the precursor complex (a′) in Fig. 17.3 but instead to (a), the complex generated by dissociation of the substituted phosphane from the 16-electron precursor. There is also kinetic evidence for an alternative slower but more direct route, which is also shown in Fig. 17.3. In this route, H$_2$ adds directly to the precursor complex [RhCl(PPh$_3$)$_3$] to yield (b′). The involvement of complex (a), or of its solvent-free version, has been a point of considerable controversy because it violates opinions about the instability of 14-electron complexes and also because (a) has not yet been detected spectroscopically in the catalytic reaction mixture.[3] The kinetic evidence indicates that (a) reacts with H$_2$ about 10^7 times faster than (a′) reacts.

The reaction of the Rh(III) dihydrido complex (b) with a large excess of cyclohexene (effectively making the concentration of cyclo-

[3] However, flash photolysis of [RhCl(CO)(PPh$_3$)$_2$] in benzene is known to yield the transient species [RhCl(PPh$_3$)$_2$] (D. Wink and P. C. Ford, *J. Am. Chem. Soc.*, **107**, 1794 (1985)).

hexene constant) yields a rate law which is pseudofirst order in the dihydrido complex:

$$-\frac{d[RhClH_2L_2]}{dt} = k[RhClH_2L_2]$$

More detailed analysis with varying concentrations of the alkene yield a more complex rate expression consistent with a pre-equilibrium reaction between (b) and the 18-electron dihydrido alkene complex (c). This pre-equilibrium is followed by a slower (rate-determining) migratory insertion reaction to produce an alkyl hydrido complex (d). The subsequent reductive elimination of alkane from (d) yields (a) and the cycle is set to repeat. Kinetic data with H_2 and D_2 indicate a faster reaction with H_2, which is consistent with the hydrogen migratory insertion reaction being rate determining.

Experiments with the catalytic addition of D_2 to cyclic alkenes indicate that the *cis* addition of hydrogen to cyclic alkenes is highly stereospecific. Spectroscopic investigation of the reaction mixture shows the presence of several rhodium complexes, but none of the active complexes (a) through (d) has been detected in the reaction mixture. For example, NMR demonstrates the presence of a halogen bridged dimer (2) of the unobserved complex (a) in Fig. 17.3. The early controversy about this catalytic mechanism stemmed from the inclination of some investigators to include spectroscopically observed complexes in the proposed mechanism without kinetic evidence for their direct involvement.

Wilkinson's catalyst is highly sensitive to the nature of the phosphane ligand and the alkene substrate. Analogous complexes with alkylphosphane ligands are inactive, presumably because they are more strongly bound to the metal and do not readily dissociate. Similarly, the alkene must be just the right size: highly hindered alkenes or the sterically unencumbered ethylene are not hydrogenated by the catalyst. It is presumed that the sterically crowded alkenes do not coordinate and that ethylene forms a strong complex which does not react further. These observations emphasize the point made earlier that a catalytic cycle is usually a delicately poised sequence of reactions, and anything that upsets their flow may block catalysis or alter the mechanism.

Wilkinson's catalyst is used in laboratory-scale organic synthesis and in the production of fine chemicals. Related Rh(I) phosphane catalysts that contain a chiral phosphane ligand have been developed to synthesize optically active products in **enantioselective** reactions. The alkene to be hydrogenated must be **prochiral**, which means that it must have a structure that leads to R or S chirality when complexed to the metal (Fig. 17.4).[4] The resulting complex will thus have two diastereomeric forms depending on the mode of coordination of the

2 [Rh$_2$Cl$_2$L$_4$]

---

[4] The designations *R* and *S* for a chiral center are determined as follows. With the element with the lowest atomic number (*Z*) away from the viewer, the center is *R* if the sequence of *Z* from highest to lowest for the remaining three atoms decreases in a clockwise manner. The center is *S* if the decrease in *Z* occurs in a counterclockwise manner. (There are additional rules in the event of atoms having identical atomic numbers.)

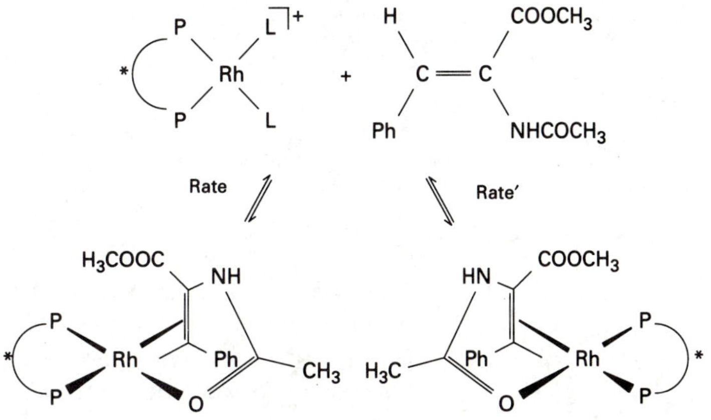

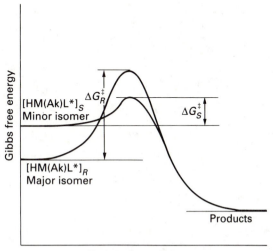

17.4 The diastereomeric complexes that may form from a complex with chiral phosphane ligands (*) and a prochiral alkene.

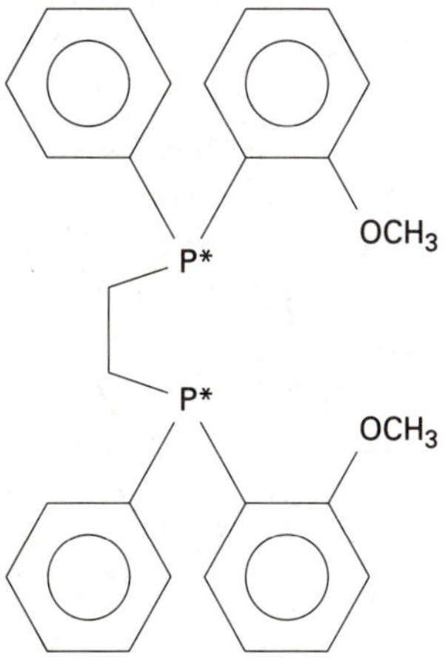

alkene. In general, diastereomers have different stabilities and labilities, and in favorable cases one or the other of these effects leads to product enantioselectivity.

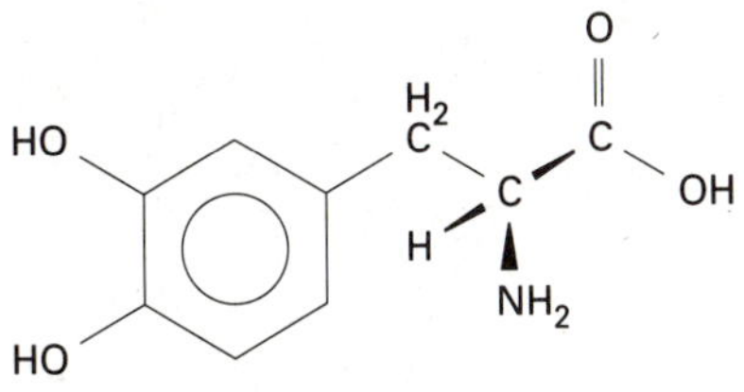

An enantioselective hydrogenation catalyst containing a chiral phosphane ligand referred to as DiPAMP (**3**) is used by Monsanto to synthesize L-dopa (**4**).[5] An interesting detail of the process is that the minor diastereomer in solution leads to the major product. The explanation of the greater turnover frequency of the minor isomer lies in the difference in activation Gibbs free energies (Fig. 17.5). Spurred by clever ligand design, this field is growing rapidly and providing clinically useful compounds.[6]

3 DiPAMP

4 L-Dopa

17.5 Kinetically controlled stereoselectivity. L* is a chiral ligand, Ak is the alkene ligand undergoing hydrogenation. Note that $\Delta G^{\ddagger}_S < \Delta G^{\ddagger}_R$, so the minor isomer reacts faster than the major isomer.

[5] L-Dopa is a chiral amino acid used to treat Parkinson's disease. The development of this catalyst and the origin of stereoselectivity in the catalytic synthesis are described by W. S. Knowles, *Acc. Chem. Res.*, **16**, 106 (1983) and by J. Halpern, *Pure Appl. Chem.*, **55**, 99 (1983), respectively.

[6] W. A. Nugent, T. V. RajanBabu, and M. J. Burk, Beyond nature's chiral pool: enantioselective catalysis in industry. *Science*, **259**, 479 (1993).

Hydroformylation

In a **hydroformylation reaction** an alkene, CO, and H_2 react to form an aldehyde containing one more C atom than in the original alkene:

$$R-CH=CH_2 \;+\; CO \;+\; H_2 \;\longrightarrow\; R-CH_2-CH_2-\overset{\displaystyle O}{\overset{\|}{C}}-H$$

Both cobalt and rhodium complexes are employed as catalysts. Aldehydes produced by hydroformylation are normally reduced to alcohols that are used as solvents, plasticizers, and in the synthesis of detergents. The scale of production is enormous, amounting to billions of kilograms per year.

The term 'hydroformylation' derived from the idea that the product resulted from the addition of formaldehyde to the alkene, and the name has stuck even though experimental data indicate a different mechanism. A less common but more appropriate name is **hydrocarbonylation**. The general mechanism of cobalt carbonyl catalyzed hydroformylation (Fig. 17.6) was proposed in 1961 by Heck and Breslow (of the Hercules Powder Company) by analogy with reactions familiar from organometallic chemistry. Their general mechanism is still invoked, but has proved difficult to verify in detail.

In the proposed mechanism, a pre-equilibrium is established in which octacarbonyldicobalt combines with hydrogen at high pressure to yield the known tetracarbonylhydridocobalt complex:

$$Co_2(CO)_8 + H_2 \;\rightleftharpoons\; 2[CoH(CO)_4]$$

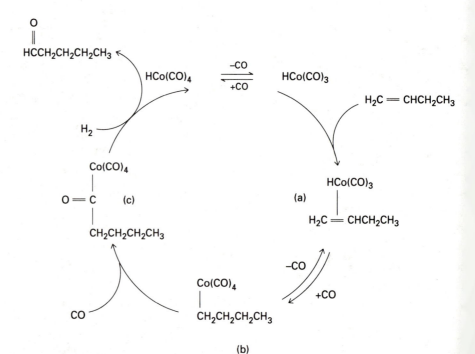

17.6 Proposed mechanism for the hydroformylation of 1-butene to pentanal, catalyzed by cobalt carbonyl complexes.

This complex, it is proposed, loses CO to produce the coordinatively unsaturated complex $[CoH(CO)_3]$:

$$[CoH(CO)_4] \rightleftharpoons [CoH(CO)_3] + CO$$

It is then thought that $[CoH(CO)_3]$ coordinates to an alkene, producing (a) in Fig 17.6, which undergoes an insertion reaction with the coordinated hydrido ligand. The product at this stage is a normal alkane complex (b). In the presence of CO at high pressure, (b) undergoes migratory insertion yielding the acyl complex (c) which has been observed by IR spectroscopy under catalytic reaction conditions. The formation of the aldehyde product is thought to occur by attack of either H_2 (as depicted in Fig. 17.6) or by the strongly acidic complex $[CoH(CO)_4]$ to yield an aldehyde and regenerate the coordinatively unsaturated $[CoH(CO)_3]$.

A significant portion of branched aldehyde is also formed in the cobalt-catalyzed hydroformylation. This product may result from a 2-alkylcobalt intermediate formed when isomerization of (a) is followed by insertion of CO:

$$\begin{array}{ccc}
\text{Co(CO)}_4 & & \text{Co(CO)}_4 \\
| & \xrightarrow{\text{CO}} & | \\
\text{H}_3\text{CCHCH}_2\text{CH}_3 & & \text{O}=\text{C} \\
& & | \\
& & \text{H}_3\text{CCHCH}_2\text{CH}_3
\end{array}$$

Hydrogenation then yields a branched aldehyde:

$$\begin{array}{ccccc}
\text{Co(CO)}_4 & & & \text{O}=\text{C}-\text{H} & \\
| & & & | & \\
\text{O}=\text{C} & + \text{ H}_2 & \longrightarrow & \text{H}_3\text{CCHCH}_2\text{CH}_3 & + \text{ HCo(CO)}_4 \\
| & & & & \\
\text{H}_3\text{CCHCH}_2\text{CH}_3 & & & &
\end{array}$$

The linear aldehyde is preferred for some applications, such as for the synthesis of biodegradable detergents, and then it is desirable to suppress the isomerization. It is found that addition of an alkylphosphane to the reaction mixture gives much higher selectivity for the linear product. One plausible explanation is that the replacement of CO by a bulky ligand disfavors the formation of complexes of sterically crowded 2-alkenes:

$$\begin{array}{ccc}
\text{Co(CO)}_2\text{PBu}_3 & & \text{Co(CO)}_2\text{PBu}_3 \\
| & \longrightarrow & | \qquad K \ll 1 \\
\text{CH}_2\text{CH}_2\text{CH}_2\text{CH}_3 & & \text{H}_3\text{CCHCH}_2\text{CH}_3
\end{array}$$

Here again we see an example of the powerful influence of ancillary ligands on catalysis.

In keeping with our comments about the high catalytic activity of 4*d*-metal complexes, rhodium phosphane complexes are even more active hydroformylation catalysts than cobalt complexes. One effective catalyst precursor, $[RhH(CO)(PPh_3)_3]$ (5), loses a phosphane ligand to form the coordinatively unsaturated 16-electron complex

$$\begin{array}{c}
\text{H} \\
| \\
\text{Ph}_3\text{P} \diagdown \quad \diagup \text{PPh}_3 \\
\text{Rh} \\
\text{Ph}_3\text{P} \diagup \quad | \\
\text{C} \\
\text{O}
\end{array}$$

5 $[RhH(CO)(PPh_3)_3]$

[RhH(CO)(PPh$_3$)$_2$], which promotes hydroformylation at moderate temperatures and 1 atm. This behavior contrasts with the cobalt carbonyl catalyst, which typically requires 150 °C and 250 atm. Because of its effectiveness under convenient conditions, the rhodium complex is useful in the laboratory. Because it favors linear aldehyde products, it competes with the phosphane-modified cobalt catalyst in industry.

Example 17.2: *Interpreting the influence of chemical variables on a catalytic cycle*
An increase in CO partial pressure above a certain threshold decreases the rate of the cobalt-catalyzed hydroformylation of 1-pentene. Suggest an interpretation of this observation.

Answer. The decrease in rate with increasing partial pressure suggests that CO suppresses the concentration of one of the catalytic species. An increase in CO pressure will lower the concentration of CoH(CO)$_3$ in the equilibrium

$$[CoH(CO)_4] \rightleftharpoons [CoH(CO)_3] + CO$$

This type of evidence was used as the basis for postulating the existence of [CoH(CO)$_3$], which is not detected spectroscopically in the reaction mixture.

Exercise E17.2. Predict the influence of added triphenylphosphane on the rate of hydroformylation catalyzed by [RhH(CO)(PPh$_3$)$_3$].

Monsanto acetic acid synthesis

The time-honored method for synthesizing acetic acid is by aerobic bacterial action on dilute aqueous ethanol, which produces vinegar. However, this process is uneconomical as a source of concentrated acetic acid for industry. A highly successful commercial process is based on the rhodium-catalyzed carbonylation of methanol:

$$CH_3OH + CO \xrightarrow{[RhI_2(CO)_2]^-} CH_3COOH$$

The reaction is catalyzed by all three members of Group 9 (cobalt, rhodium, and iridium), but complexes of the 4d metal rhodium are the most active. Originally a cobalt complex was employed, but the rhodium catalyst developed at Monsanto greatly reduced the cost of the process by allowing lower pressures to be used. As a result, the **Monsanto process** is licensed for use throughout the world.

   The principal catalytic cycle in the Monsanto process is illustrated in Fig. 17.7. Under normal operating conditions, the rate determining step is oxidative addition of iodomethane to the four-coordinate, 16-electron complex [RhI$_2$(CO)$_2$]$^-$ (a), producing the six-coordinate 18-electron complex [(H$_3$C)RhI$_3$(CO)$_2$]$^-$ (b). This step is followed by CO migratory insertion, yielding a 16-electron acyl complex (c). Coordination of CO restores an 18-electron complex (d), which is then set to undergo reductive elimination of acetyl iodide with the regeneration

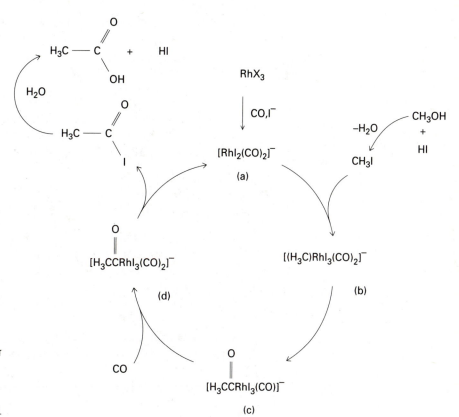

17.7 One cycle in the proposed mechanism for the conversion of methanol and CO to acetic acid, catalyzed by rhodium phosphane complexes. The acetyl iodide produced by this cycle undergoes hydrolysis to yield acetic acid.

of $[RhI_2(CO)_2]^-$. Water then hydrolyzes the acetyl iodide to acetic acid and regenerates HI:

$$CH_3COI + H_2O \rightarrow CH_3COOH + HI$$

No other anion works as well as iodide in this catalytic system, and its special ability arises from several factors. Among them is the greater rate of oxidative addition of iodomethane relative to the other haloalkanes in the rate determining step. In addition, the soft I^- ion, which is a good ligand for the soft Rh(I), appears to form a five-coordinate complex, $[RhI_3(CO)_2]^{2-}$, which undergoes oxidative addition with iodomethane more rapidly than $[RhI_2(CO)_2]^-$. The strong acid HI is also effective in halogenating methanol:

$$CH_3OH + HI \rightarrow CH_3I + H_2O$$

Wacker oxidation of alkenes

The **Wacker process** is primarily used to produce acetaldehyde from ethylene and oxygen:

$$C_2H_4(g) + \tfrac{1}{2}O_2(g) \rightarrow CH_3CHO(g) \quad \Delta G^{\ominus} = -197 \text{ kJ mol}^{-1}$$

Its invention at the Wacker Consortium für Elektrochemische Industrie (West Germany) in the late 1950s marked the beginning of an era of production of chemicals from petroleum feedstock.[7] Although this is no longer a major industrial process, it has some interesting new mechanistic features that are worth noting. An initial step is thought to be

$$C_2H_4 + PdCl_2 + H_2O \rightarrow CH_3CHO + Pd(0) + 2HCl$$

The exact nature of the Pd(0) is unknown, but it is probably present as a complex. The slow oxidation of Pd(0) back to Pd(II) by oxygen is catalyzed by the addition of Cu(II), which shuttles back and forth to Cu(I):

$$Pd(0) + 2[CuCl_4]^{2-} \rightarrow Pd^{2+} + 2[CuCl_2]^- + 4Cl^-$$

$$2[CuCl_2]^- + \tfrac{1}{2}O_2 + 2H^+ + 4Cl^- \rightarrow 2[CuCl_4]^{2-} + H_2O$$

The overall catalytic cycle is shown in Fig. 17.8. Detailed stereochemical studies on related systems indicate that the hydration of the alkene–Pd(II) complex (a) occurs by the attack of H_2O from the solution on the coordinated ethylene rather than the insertion of coordinated OH.[8] Hydration, to form (b), is followed by two steps that isomerize the coordinated alcohol. First, β-hydrogen elimination occurs with the formation of (c), and then migratory insertion results in the formation of (d). Elimination of the acetaldehyde and a hydrogen ion then leaves Pd(0). The Pd(0) is converted back to Pd(II) by the auxiliary copper(II) catalyzed air oxidation cycle mentioned above.

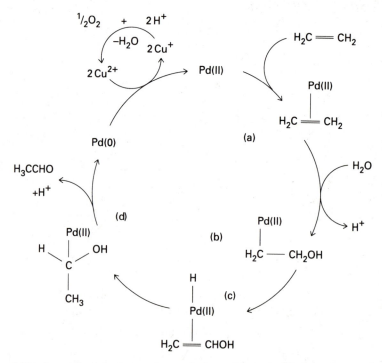

17.8 Proposed mechanism for the oxidation of ethylene to acetaldehyde in the Wacker process. Chloride ligands have been omitted. The oxidation number of palladium is +2 at all stages of this cycle except the upper left where reductive elimination of acetaldehyde gives Pd(0), which is oxidized by Cu(II).

[7] Ethylene and other alkenes, the starting materials in the Wacker process and many other petrochemical syntheses are produced by the thermal cracking of saturated hydrocarbons.

[8] This and related details of the mechanism are still being debated, see J.-E. Bäkvall, B. Åkermark, and S. O. Ljunggren, *J. Am. Chem. Soc.*, **101**, 2411 (1979), and references therein.

Alkene ligands coordinated to Pt(II) are also susceptible to nucleophilic attack, but only palladium leads to a successful catalytic system. The principal reason appears to be the greater lability of the $4d$ Pd(II) complexes in comparison with their $5d$ Pt(II) counterparts. Furthermore, the potential for the oxidation of Pd(0) to Pd(II) is more favorable than for the corresponding platinum couple.

Alkene polymerization

Polyalkenes, which are among the most common and useful class of synthetic polymers, are most often prepared by use of organometallic catalysts, either in solution or supported on a solid surface. The latter are heterogeneous catalysts but we shall mention them here for convenience.

In the 1950s Karl Ziegler, working in Germany, developed a catalyst for ethene polymerization based on a catalyst formed from $TiCl_4$ and $Al(C_2H_5)_3$, and soon thereafter G. Natta in Italy utilized this type of catalyst for the stereospecific polymerization of propene. These developments ushered in a revolution in packaging materials, fabrics, and materials of construction. A **Ziegler–Natta catalyst** and another widely-used chromium polymerization catalyst are solid particles on which the polymerization takes place. Largely on account of their heterogeneous character, the mechanism of action of these catalysts has defied precise determination. Related homogeneous catalysts, however, and analogies with known organometallic reactions provide some hints on the course of the heterogeneous reaction. One homogeneous model is the tipped ring complex $[(\eta^5\text{-}C_5H_5)_2Zr(CH_3)L]^+$ (**6**), which catalyzes alkene polymerization by successive insertion steps that are thought to involve prior coordination of the alkene to the electrophilic Zr center.[9] This type of homogeneous catalyst is now employed commercially for the synthesis of specialized polymers. With a chiral zirconium catalyst (**7**), it is possible to produce optically active polypropylene.

Cycloalkenes other than cyclohexene are susceptible to ring-opening polymerization (ROMP) which is catalyzed by certain organometallic complexes:[10]

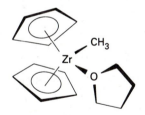

6 $[Zr(Cp)_2(CH_3)(THF)]^+$

7

W(CH–t–Bu) + x ⟶ W=⟨...⟩= CH(t–Bu)$_x$

+PhCHO ↓ –[W(O)(NAr)(O–t–Bu)$_2$]

PhHC =⟨...⟩= CH(t–Bu)$_x$

In this polymerization reaction an alkene inserts into the metal carbene, giving rise to a new carbene complex with a longer side chain on the

[9] R. F. Jordan, *J. Chem. Educ.*, **65**, 285 (1988).
[10] R. H. Grubbs and W. Tumas, *Science*, **243**, 907 (1989); R. H. Schrock, *Acc. Chem. Res.*, **23**, 158 (1990).

carbene. Successive insertions lead to longer chains. The polymerization reaction is thermodynamically favored by the relief of ring strain when the cyclic compound breaks open. A wide variety of metal carbenes are known to catalyze this reaction and many of these result in a 'living polymerization'.

One characteristic of a living polymerization is that the polymerization continues until all of the monomer is consumed and then can be restarted by simply adding more monomer. The polymer isolated from a living polymerization is often highly monodisperse (that is, has a very narrow range of molar masses). These characteristics result from a minimum of side reactions, so each catalytic metal center has a chain of similar length attached to it during the polymerization process. The polymerization is terminated by hydrolysis.

HETEROGENEOUS CATALYSIS

Heterogeneous catalysts are used very extensively in industry. One attractive feature is that many of these solid catalysts are robust at high temperatures and therefore make available a wide range of operating conditions. Another reason for their widespread use is that extra steps are not needed to separate the product from the catalyst. Typically, gaseous or liquid reactants enter a tubular reactor at one end, pass over a bed of the catalyst, and products are collected at the other end. This same simplicity of design applies to the catalytic converter employed to oxidize CO and hydrocarbons and reduce nitrogen oxides in automobile exhausts (Fig. 17.9).

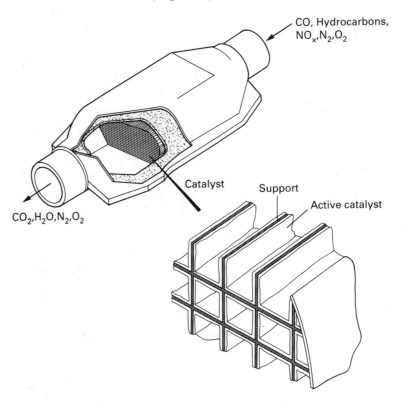

CO, Hydrocarbons, NO_x, N_2, O_2

CO_2, H_2O, N_2, O_2

Catalyst

Support

Active catalyst

17.9 A reactor for heterogeneous catalysis. This automobile catalytic converter oxidizes CO and hydrocarbons and reduces nitrogen oxides. The metal catalyst is supported on a ceramic honeycomb, which is more robust in this application than a bed of loose particles.

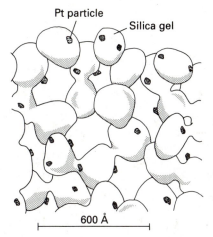

17.10 Schematic diagram of metal particles supported on silica gel.

17.5 The nature of heterogeneous catalysts

Information on the structure of a heterogeneous catalyst can sometimes be determined by spectroscopy and related studies of reactive molecules adsorbed on single crystal surfaces. However, these studies are generally carried out in ultra-high vacuum conditions, which are very different from those used in practice. Practical heterogeneous catalysts are high surface area materials that may contain several different phases and which operate at 1 atm and higher pressures. In some cases the bulk of a high-surface-area material serves as the catalyst, and such a material is called a **uniform catalyst**. One example is the catalytic zeolite ZSM-5, which contains channels through which reacting molecules diffuse. More often **multiphasic catalysts** are used. These consist of a high surface area material which serves as a support on to which an active catalyst is deposited (Fig. 17.10).[11]

Surface area and porosity

In homogeneous catalysis the interaction of reactants and catalysts occurs readily, but special measures must be taken in heterogeneous catalysis to ensure that the reactive molecules achieve contact with catalytic sites. For many applications, an ordinary dense solid is unsuitable as a catalyst because its surface area is quite low. Thus α-alumina, which is a dense material with a low surface area, is used much less as a catalyst support than the microcrystalline solid γ-alumina, which can be prepared with small particle size, and therefore a high surface area. The high surface area results from the many small but connected particles like those shown in Fig. 17.10, and a gram or so of a typical catalyst support has an internal surface area equal to that of a tennis court. Similarly, quartz is not used as a catalyst support but the high surface area versions of SiO_2, the silica gels, are widely employed.

Both γ-alumina and silica gel are metastable materials, but under ordinary conditions they do not convert to their more stable phases (α-alumina and quartz, respectively). The preparation of γ-alumina involves the dehydration of an aluminum oxide hydroxide:

$$2AlO(OH) \xrightarrow{\Delta} \gamma\text{-}Al_2O_3 + H_2O$$

Similarly, silica gel is prepared from the acidification of silicates to produce $Si(OH)_4$, which rapidly forms a hydrated silica gel from which much of the adsorbed water can be removed by gentle heating (Chapter 5). When viewed with an electron microscope, the texture of silica gel or alumina appears to be that of a rough gravel bed with irregularly shaped voids between the interconnecting particles (as in Fig. 17.10).

Zeolites (Section 11.11) are examples of uniform catalysts.[12] They are prepared as very fine crystals that contain large regular channels

---

[11] The distinctions between uniform and multiphasic catalysts are developed by J. M. Thomas, *Angew. Chem., Int. Ed. Engl.*, **12**, 1673, (1988).

[12] An introductory discussion of zeolite structure and catalysis is given by A. Dyer, *An introduction to zeolite molecular sieves*. Wiley, Chichester (1988).

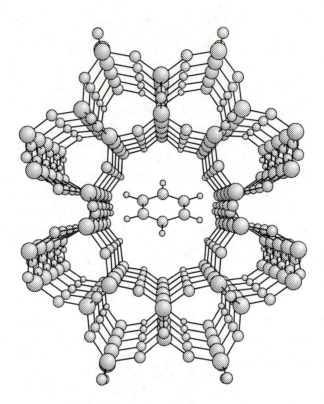

17.11 A view through the channels in Theta-1 zeolite with an adsorbed benzene molecule in one of the channels. (Reproduced by permission from A. Dyer, *An introduction to molecular sieves*, p. 51, Wiley, Chichester (1988).)

and cages defined by the crystal structure (Fig. 17.11). The openings in these channels vary from one crystalline form of the zeolite to the next, but are typically 3 to 10 Å in their smallest diameter. The zeolite absorbs molecules small enough to enter the channels and excludes larger molecules. This selectivity, in combination with catalytic sites inside the cages, provides a degree of control over catalytic reactions that is unattainable with silica gel or γ-alumina. The synthesis of new zeolites and similar shape-selective solids and the introduction of catalytic sites into them is a vigorous area of research (see Section 11.11).

Surface acidic and basic sites

When exposed to atmospheric moisture, the surface of γ-alumina is covered with adsorbed water molecules. Dehydration at 100 to 150 °C leads to the desorption of water, but surface OH groups remain and act as weak Brønsted acids:

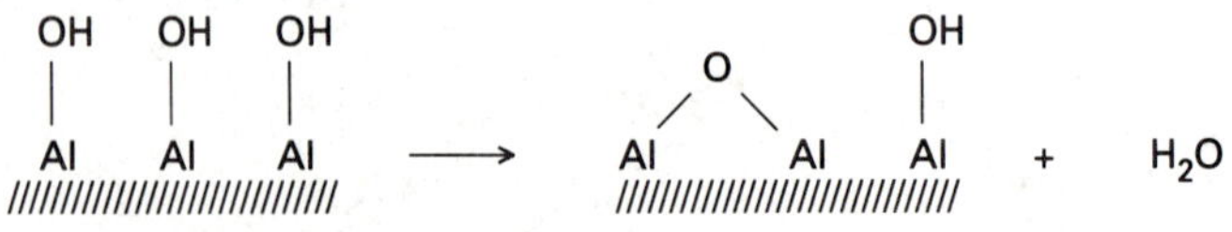

$$
\begin{array}{ccc}
\text{OH} & \text{OH} & \text{OH} \\
| & | & | \\
\text{Al} & \text{Al} & \text{Al}
\end{array}
\longrightarrow
\begin{array}{ccccc}
 & \text{O} & & & \text{OH} \\
 & \diagup\diagdown & & & | \\
\text{Al} & & \text{Al} & & \text{Al}
\end{array}
+ \; \text{H}_2\text{O}
$$

At even higher temperatures adjacent OH groups condense to liberate more H_2O and generate exposed Al^{3+} Lewis acid sites as well as O^{2-} Lewis base sites (**8**). The rigidity of the surface permits the coexistence of these strong Lewis acid and base sites, which would otherwise immediately combine to form Lewis acid–base complexes. Surface

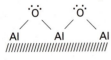

8

acids and bases are highly active for catalytic reactions such as the dehydration of alcohols and isomerization of alkenes. Similar Brønsted and Lewis acid sites exist on the interior of certain zeolites.

Example 17.3: *Using IR spectra to probe molecular interaction with surfaces*

Infrared spectra of hydrogen bonded and Lewis acid complexes of pyridine (py), such as $Cl_3Al(py)$, show that bands near 1540 cm^{-1} can be ascribed to hydrogen bonded pyridine and that bands near 1465 cm^{-1} are due to Al—py Lewis acid–base interactions. A sample of γ-alumina that has been pre-treated by heating to 200 °C and then cooled and exposed to pyridine vapor has absorption bands near 1540 cm^{-1} and none near 1465 cm^{-1}. Another sample that was heated to 500 °C, cooled, and then exposed to pyridine had bands near 1540 cm^{-1} and 1465 cm^{-1}. Correlate these results with the statements made in the text concerning the effect of heating γ-alumina. (Much of the evidence for the chemical nature of the γ-alumina surface comes from experiments like these.)

Answer. At about 200 °C, surface H_2O is lost but OH$^-$ bound to Al^{3+} remains. These groups appear to be mildly acidic as judged by color indicators and bands around 1540 cm^{-1} that indicate the presence of hydrogen bonded pyridine generated in the reaction:

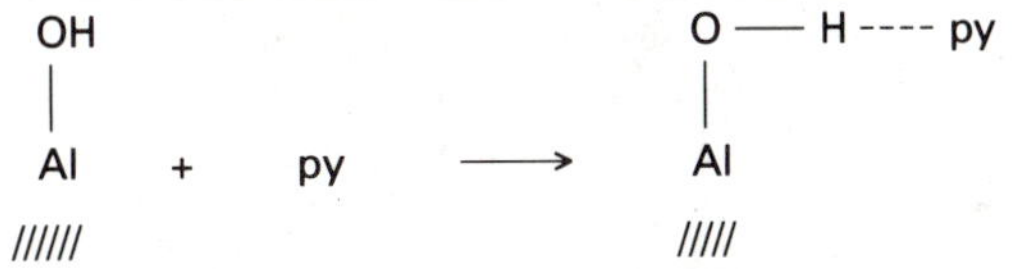

When heated to 500 °C, much of but not all the OH is lost as H_2O, leaving behind O^{2-} and exposed Al^{3+}. The evidence is the appearance of the 1465 cm^{-1} bands, which are indicative of Al^{3+}—NC_5H_5, as well as the 1540 cm^{-1} band.

Exercise E17.3. If the statements in this section about the effect of heating γ-alumina are correct, what will be the changes in the intensities of the diagnostic IR bands if the γ-alumina sample is pretreated at 900 °C, cooled, exposed to pyridine vapor, and a spectrum determined?

Surface metal sites

Metal particles are often deposited on supports to provide a catalyst. For example, finely divided platinum–rhenium alloys distributed through the pore space of γ-alumina are used to interconvert hydrocarbons, and finely divided platinum–rhodium alloy particles supported on γ-alumina are employed in automobile catalytic converters to promote the combination of O_2 with CO and hydrocarbons to CO_2 and reduction of nitrogen oxides to nitrogen. A supported metal particle about 25 Å in diameter has about 40 percent of its atoms on the surface, and the particles are protected from fusing together into bulk metal by their separation. The high proportion of exposed atoms is a great

Table 17.4 Chemisorbed ligands on surfaces

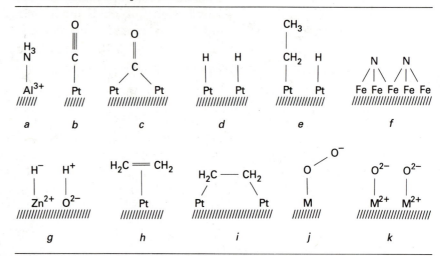

a Ammonia adsorbed on the Lewis and Al^{3+} sites of γ-alumina.

b,c CO coordinated to platinum metal.

d Hydrogen dissociatively chemisorbed on platinum metal.

e Ethane dissociatively chemisorbed on platinum.

f Nitrogen dissociatively chemisorbed on iron.

g H_2 dissociatively chemisorbed on ZnO.

h Ethylene η^2 coordinated to a Pt atom.

i Ethylene bonded to two Pt atoms.

j O_2 bound as a superoxide to a metal surface.

k O_2 dissociatively chemisorbed on a metal surface.

Adapted from: R. L. Burwell, Jr, Heterogeneous catalysis. In *Survey of Progress in Chemistry*, **8**, 2 (1977).

advantage for these small supported particles, particularly for metals such as platinum and the even more expensive rhodium.

The metal atoms on the surface of metal clusters are capable of forming bonds such as M—CO, M—CH_2R, M—H, and M—O (Table 17.4). Often the nature of surface ligands is inferred by comparison of IR spectra with those of organometallic or inorganic complexes. Thus, both terminal and bridging CO can be identified on surfaces by IR spectroscopy, and the IR spectra of many hydrocarbons ligands on surfaces are similar to those of discrete organometallic complexes. The case of the N_2 ligand is an interesting contrast, because coordinated N_2 was identified by IR spectroscopy on metal surfaces before dinitrogen complexes were prepared and characterized by inorganic chemists.

The development of new techniques for studying single crystal surfaces has greatly expanded our knowledge of the surface species that may be present in catalysis. For example, the desorption of molecules from surfaces (thermally, or by ion or atom impact) combined with mass spectrometric analysis of the desorbed substance, provides insight into the chemical identity of surface species. Similarly, Auger and X-ray photoelectron spectroscopy (XPS) provide information on the elemental composition of surfaces.

Low-energy electron diffraction (LEED) provides information about the structure of single crystal surfaces and, when adsorbate molecules are present, their arrangement on the surface. One important finding from LEED is that the adsorption of small molecules may bring about a structural modification of a surface. This surface reconstruction is often observed to reverse when desorption occurs. We are just beginning to see the wide use of scanning tunneling microscopy for the location of adsorbates on surfaces. This striking technique provides a contour map of single crystal surfaces at or close to atomic resolution.[13]

Although most of these modern surface techniques cannot be applied to the study of supported multiphasic catalysts, they are very helpful for revealing the range of probable surface species and circumscribing the structures that may plausibly be invoked in a mechanism of heterogeneous catalysis. The application of these techniques to heterogeneous catalysis is similar to the use of X-ray diffraction and spectroscopy to the characterization of organometallic homogeneous catalyst precursors and model compounds.

17.6 Catalytic steps[14]

There are many parallels between the individual reaction steps encountered in heterogeneous and homogeneous catalysis.

Chemisorption and desorption

The adsorption of molecules on surfaces often activates molecules just as coordination activates molecules in complexes. The desorption of product molecules that is necessary to refresh the active sites in heterogeneous catalysis is analogous to the dissociation of a complex in homogeneous catalysis.

Before a heterogeneous catalyst is used it is usually **activated**. Activation is a catch-all term; in some instances it refers to the desorption of adsorbed molecules such as water from the surface, as in the dehydration of γ-alumina. In other cases it refers to the preparation of the active site by a chemical reaction, such as by reduction of metal oxide particles to produce active metal particles.

An activated surface can be characterized by the adsorption of various inert and reactive gases. The adsorption may be either **physisorption**, when no new chemical bond is formed, or **chemisorption**, when surface–adsorbate bonds are formed (Fig. 17.12). Low-temperature physisorption of a gas like nitrogen is useful for the determination of the total surface area of a solid, whereas chemisorption is used to determine the number of exposed reactive sites. For example, the dissociative chemisorption of H_2 on supported platinum particles reveals the number of exposed surface Pt atoms.

There are many parallels between the interaction of small molecules with metal surfaces and with low oxidation state metal complexes.

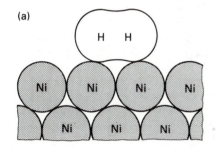

(a)

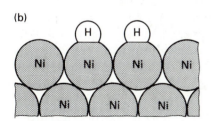

(b)

17.12 Schematic representation of (a) physisorption and (b) chemisorption of H_2 on a nickel metal surface.

[13] These and other modern surface techniques and their help in the elucidation of catalysis are summarized by G. A. Samorjai in *Chemistry in two dimensions*. Cornell University Press (1981).

[14] For an entertaining and informative account of the development of mechanistic concepts in heterogeneous catalysis see R. L. Burwell, Jr, in *Chemtech*, **17**, 586 (1987).

Table 17.5 The tendency of metals to chemisorb simple molecules

	Gases						
	O_2	C_2H_2	C_2H_2	CO	H_2	CO_2	N_2
Ti, Zr, Hf, V, Nb, Ta, Cr, Mo, W, Fe, Ru, Os	+	+	+	+	+	+	+
Ni, Co	+	+	+	+	+	+	+
Rh, Pd, Pt, Ir	+	+	+	+	+	+	+
Mn, Cu	+	+	+	+	±	+	+
Al, Au	+	+	+	+	−	−	−
Na, K	+	+	−	−	−	−	−
Ag, Zn, Cd, In, Si, Ge, Sn, Pb, As, Sb, Bi	+	−	−	−	−	−	−

+ Strong chemisorption, ± weak, − unobservable.
Adapted from G. C. Bond, *Heterogeneous catalysis*, p. 29, Oxford University Press (1987).

Table 17.5 shows that a wide range of metals chemisorb CO, and that many fewer are capable of chemisorbing N_2, just as there is a much wider variety of metals that form carbonyls than form N_2 complexes. Furthermore, just as with metal carbonyl complexes, both bridging and terminal CO surface species have been identified by infrared spectroscopy. The dissociative chemisorption of H_2 (Fig. 17.12) is analogous to the oxidative addition of H_2 to metal complexes.

Even though adsorption is essential for heterogeneous catalysis to occur, it must not be so strong as to block the catalytic sites and prevent further reaction. This factor is in part responsible for the limited number of metals that are effective catalysts. The catalytic decomposition of formic acid on metal surfaces

$$HCOOH \xrightarrow{\text{M}} CO + H_2O$$

provides a good example of this balance between adsorption and catalytic activity. It is observed that the catalysis is fastest on metals for which the metal formate is of intermediate stability (Fig. 17.13). The plot in Fig. 17.13 is an example of a **volcano diagram**, and is typical of many catalytic reactions. The implication is that the earlier *d*-block metals form very stable surface compounds whereas the later noble metals such as silver and gold form very weak surface compounds, both of which are detrimental to a catalytic process. Between these extremes the metals in Groups 8 to 10 have high catalytic activity, and this is especially true of the platinum metals. In Section 17.2 we saw a similar high activity of platinum metal complexes in the homogeneous catalysis of hydrocarbon transformations.

The active sites of heterogeneous catalysts are not uniform and many diverse sites are exposed on the surface of a poorly crystalline solid such as *γ*-alumina or a noncrystalline solid such as silica gel. However, even highly crystalline metal particles are not uniform. A crystalline solid has typically more than one type of exposed plane, each with its

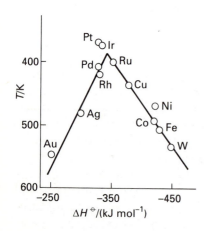

17.13 A volcano diagram. In this case the reaction temperature for a set rate of formic acid decomposition is plotted against the stability of the corresponding metal formate as judged by heat of formation. (W. J. M. Rootsaert and W. M. H. Sachtler, *Z. Physik. Chem.*, **26**, 16 (1960).)

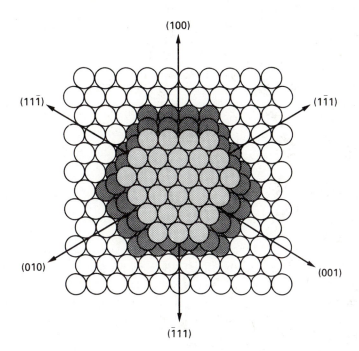

17.14 A collection of metal crystal planes that might be exposed to reactive gases (111), ($\bar{1}\bar{1}\bar{1}$), etc. are close-packed hexagonal planes. The planes represented by (100), (010), etc. have square arrays of atoms.

characteristic pattern of surface atoms (Fig. 17.14). In addition, single crystal metal surfaces have irregularities such as steps that expose metal atoms with low coordination numbers (Fig. 17.15). These highly exposed, coordinatively unsaturated sites appear to be particularly reactive. As a result, the different sites on the surface may serve different functions in catalytic reactions. The variety of sites also accounts for the lower selectivity of many heterogeneous catalysts in comparison with homogeneous analogs.

Surface migration

We saw in Chapter 16 that ligands in organometallic clusters are often fluxional. The surface analog of fluxionality in clusters is diffusion, and there is abundant evidence for the diffusion of chemisorbed molecules or atoms on metal surfaces. For example, adsorbed H atoms and adsorbed CO molecules are known to move over the surface of a metal particle. This mobility is important in catalytic reactions for it allows atoms or molecules to approach one another. Surface diffusion rates are difficult to measure and reliable data have been obtained only recently.

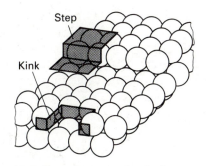

17.15 Schematic representation of surface irregularities, steps, and kinks.

17.7 Examples

A huge range of industrial processes are facilitated by heterogeneous catalysis. We give a few examples here that illustrate some of this range. Despite some superficial similarities the first two examples, hydrogenation of alkenes and ammonia synthesis (the hydrogenation

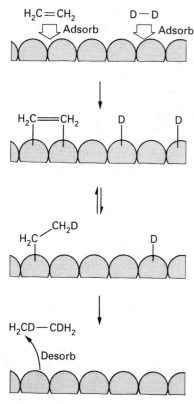

17.16 Schematic diagram of the hydrogenation of ethylene by D_2 on a metal surface.

of nitrogen) employ quite different catalysts and conditions. The example of the isomerization of aromatics on zeolites illustrates acid catalysis. Finally, we illustrate catalysis at the electrode of an electrochemical cell.

Hydrogenation of alkenes

A milestone in heterogeneous catalysis was Paul Sabatier's observation that nickel catalyzes the hydrogenation of alkenes (1900). He was actually attempting to synthesize $Ni(C_2H_4)_4$ in response to Mond, Langer, and Quinke's synthesis of $Ni(CO)_4$ (Section 16.4). However, when he passed ethylene over heated nickel he detected ethane. His curiosity was sparked, so he included hydrogen with the ethylene, whereupon he observed a good yield of ethane. Major industrial applications soon followed.[15]

The hydrogenation of alkenes on supported metal particles is thought to proceed in a manner very similar to that in metal complexes. As pictured in Fig. 17.16, H_2, which is dissociatively chemisorbed on the surface, is thought to migrate to an adsorbed ethylene molecule, giving first a surface alkyl and then the saturated hydrocarbon. When ethylene is hydrogenated with D_2 over platinum, the simple mechanism depicted in Fig. 17.16 indicates that CH_2D—CH_2D should be the product. In fact, a complete range of $C_2H_nD_{6-n}$ ethane molecules is observed. It is for this reason that Step 3 is written as reversible; the rate of the reverse reaction must be greater than the rate at which the ethane molecule is formed and desorbed in the final step.

Ammonia synthesis

The synthesis of ammonia

$$N_2 + 3H_2 \rightarrow 2NH_3$$

has already been discussed from several different viewpoints (Sections 9.14 and 12.2). Here we concentrate on details of the catalytic steps. Ammonia is an exoergic and exothermic compound at 25 °C, the relevant thermodynamic data being $\Delta G_f^\ominus = -16.5$ kJ mol^{-1}, $\Delta H_f^\ominus = -46.1$ kJ mol^{-1}, and $\Delta S_f^\ominus = -99.4$ J K^{-1} mol^{-1}. The negative entropy of formation reflects the fact that two molecules of gas replace four molecules of reactant gases.

The great inertness of N_2 (and to a lesser extent H_2) requires that a catalyst be used to effect the reaction. Iron metal together with small quantities of alumina and potassium salts and other promoters is used as the catalyst. Extensive studies on the mechanism of ammonia synthesis indicate that the rate determining step under normal operating conditions is the dissociation of N_2 coordinated to the catalyst surface. The other reactant, H_2, undergoes much more facile dissociation on

---

[15] See P. Sabatier, *Bull. Soc. Chim. Fr.*, **6**, 1261, (1930).

the metal surface and a series of insertion reactions between adsorbed species leads to the production of NH_3:

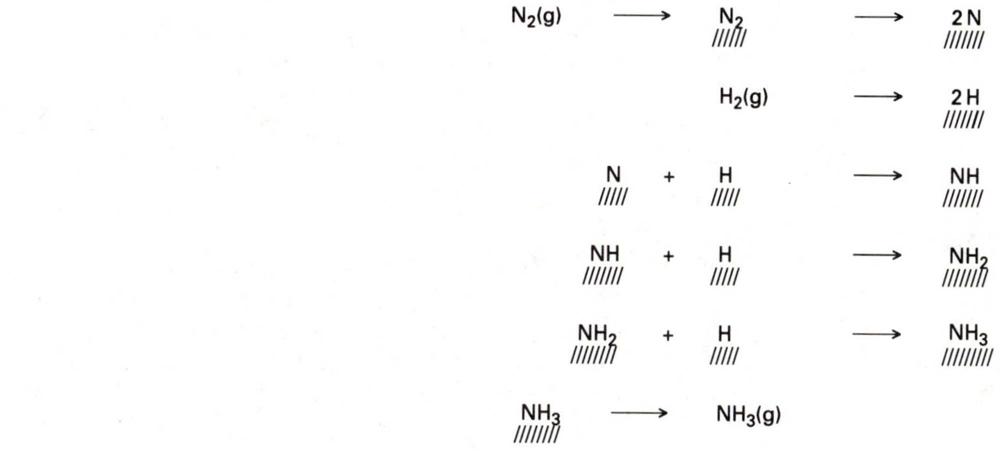

$$N_2(g) \longrightarrow N_2 \text{(adsorbed)} \longrightarrow 2N \text{(adsorbed)}$$

$$H_2(g) \longrightarrow 2H \text{(adsorbed)}$$

$$N \text{(ads)} + H \text{(ads)} \longrightarrow NH \text{(ads)}$$

$$NH \text{(ads)} + H \text{(ads)} \longrightarrow NH_2 \text{(ads)}$$

$$NH_2 \text{(ads)} + H \text{(ads)} \longrightarrow NH_3 \text{(ads)}$$

$$NH_3 \text{(ads)} \longrightarrow NH_3(g)$$

Because of the slowness of the N_2 dissociation, it is necessary to run the ammonia synthesis at high temperatures, typically 400 °C. However, since the reaction is exothermic, high temperature greatly reduces the equilibrium constant of the reaction, for from the van't Hoff equation

$$\frac{\mathrm{d} \ln K}{\mathrm{d} T} = \frac{\Delta H^{\ominus}}{R T^2} < 0$$

so K decreases as T is increased. To recover some of this reduced yield, pressures on the order of 100 bar are employed. A catalyst that would operate at room temperature would give good equilibrium yields of NH_3, but such catalysts have not yet been discovered (Section 12.2).

In the course of developing the original ammonia synthesis process, Haber and his coworkers investigated the catalytic activity of most of the metals in the periodic table, and found that the best are iron, ruthenium, and uranium. Cost and toxicity considerations led to the choice of iron as the basis of the commercial catalyst. Other metals, such as lithium, are more active for the critical $N\equiv N$ cleavage step, but in these cases the succeeding steps are not favorable on account of the great stability of the metal nitride. This observation once again emphasizes the detrimental nature of a highly stable intermediate in a catalytic cycle.

SO$_2$ oxidation

The oxidation of SO_2 to SO_3 is a key step in the production of sulfuric acid (Section 12.8). The reaction of sulfur with oxygen to produce SO_3 gas is exoergic ($\Delta G^{\ominus} = -371$ kJ mol^{-1}) but very slow, and the principal product of the combustion of sulfur is SO_2:

$$S(s) + O_2(g) \rightarrow SO_2(g)$$

The combustion is followed by the catalytic oxidation of SO_2:

$$SO_2(g) + \tfrac{1}{2}O_2(g) \rightarrow SO_3(g)$$

This step is also exothermic and, as with ammonia synthesis, has a less favorable equilibrium constant at elevated temperatures. The process is therefore generally run in stages. In the first stage, the combustion of sulfur raises the temperature to about 600 °C, but by cooling before the catalytic stage the equilibrium is driven to the right and high conversion of SO_2 to SO_3 is achieved.

Several quite different catalytic systems have been used to catalyze the combination of SO_2 with O_2. The most widely used catalyst at present is potassium vanadate supported on a high-surface-area silica (diatomaceous earth). One interesting aspect of this catalyst is that the vanadate is a molten salt under operating conditions. The current view of the mechanism of the reaction is that the rate determining step is the oxidation of V^{4+} to V^{5+} by O_2. Schematically, the cycle is

$$\tfrac{1}{2}O_2 + 2V^{4+} \rightarrow O^{2-} + 2V^{5+}$$

$$SO_2 + 2V^{5+} + O^{2-} \rightarrow 2V^{4+} + SO_3$$

In the melt the vanadium and oxide ions are part of a polyvanadate complex (Section 5.8), but little is known about the evolution of the oxo species.

Interconversion of aromatics by zeolites

The zeolite-based heterogeneous catalysts play an important role in the interconversion of hydrocarbons and the alkylation of aromatics as well as in oxidation and reduction.

The synthetic zeolite ZSM-5 (which was developed in the research laboratories of Mobil Oil; the initials stand for Zeolite Socony-Mobil) is widely used by the petroleum industry for a variety of hydrocarbon interconversions. ZSM-5 is an aluminosilicate zeolite with a higher silica content and a much lower aluminum content than the zeolites discussed in Section 11.11. Its channels consist of a three-dimensional maze of intersecting tunnels (Fig. 17.17). As with other aluminosilicate catalysts, the aluminum sites are strongly acidic. The charge imbalance of Al^{3+} in place of tetrahedrally coordinated Si^{4+} requires the presence of an added positive ion. When this ion is H^+ (Fig. 17.18) the Brønsted acidity of the aluminosilicate can be as great as that of concentrated H_2SO_4 and the turnover frequency for hydrocarbon reactions at these sites can be exceedingly high.[16]

Reactions such as xylene isomerization and toluene disproportionation are illustrative of the selectivity that can be achieved with acidic zeolite catalysis. The shape selectivity of these catalysts has been attributed to faster diffusion of product molecules that have dimensions compatible with the channels. According to this idea, molecules that do not fit the channels diffuse slowly, and because of their long resid-

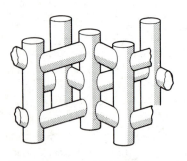

17.17 The structure of ZSM-5 has intersecting channels defined by the crystal structure. The tubes in this diagram represent the channels.

[16] W. O. Haag, R. M. Lago, and P. B. Weisz, *Nature*, **309**, 589 (1984).

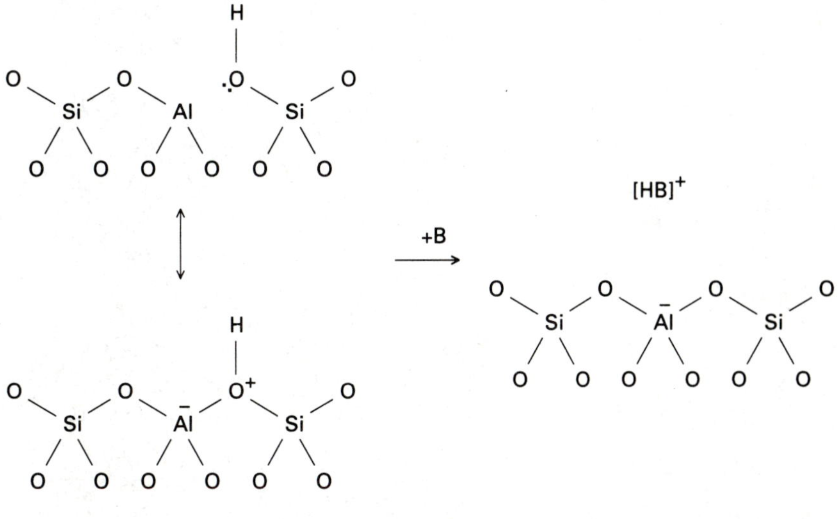

17.18 The Brønsted acid site in HZSM-5 and its interaction with a base. (From W. O. Haag, R. M. Lago, and P. B. Weisz, *Nature*, **309**, 589 (1984).)

ence in the zeolite, have ample opportunity to be converted to the more mobile isomers that can escape rapidly. A currently more favored view, however, is that the orientation of reactive intermediates within the zeolite channels favors specific products, such as *p*-dialkylbenzenes.

Aside from their important shape selectivity, acidic zeolite catalysts appear to promote reactions by standard carbonium ion mechanisms. For example, the isomerization of *m*-xylene to *p*-xylene,

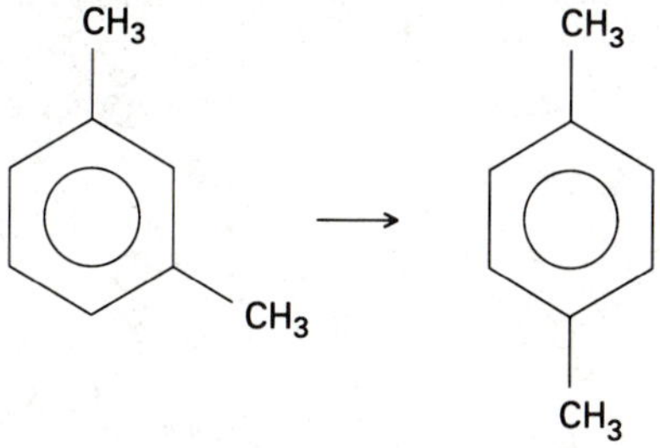

may occur by the following steps:

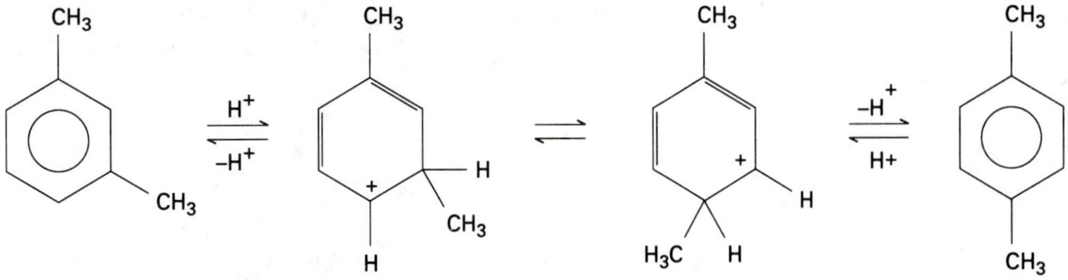

Another common reaction in zeolites is the alkylation of aromatics with alkenes.

Example 17.4: *Proposing a plausible mechanism for the alkylation of benzene in ZSM-5*

In its protonated form, ZSM-5 catalyzes the reaction of ethylene with benzene to produce ethylbenzene. Write a plausible mechanism for the reaction.

Answer. The acidic form of ZSM-5 is strong enough to generate carbo-cations:

$$CH_2{=}CH_2 + H^+ \rightarrow [CH_2{-}CH_3]^+$$

As we saw in Section 5.10, the carbocation can attack benzene. Subsequent deprotonation of the product yields ethylbenzene:

$$[CH_2{-}CH_3]^+ + C_6H_6 \rightarrow C_6H_5{-}CH_2CH_3 + H^+$$

Exercise E17.4. A pure silica analog of ZSM-5 can be prepared. Would you expect this compound to be an active catalyst for benzene alkylation? Explain your reasoning.

Electrocatalysis

The kinetic barrier to electrochemical reactions involving H_2 and O_2 was discussed in Chapter 7. Kinetic barriers are quite common for electrochemical reactions at the interface between a solution and an electrode and, as we saw in Section 7.4, it is common to express these barriers as overpotentials.

The overpotential η of an electrolytic cell is the potential in addition to the zero-current cell potential that must be applied in order to bring about a nonspontaneous reaction within the cell. The overpotential is empirically related to the current density j (the current per unit area of electrode) that passes through the cell by[17]

$$\eta = a + b \log j$$

or equivalently

$$j = c e^{\eta d}$$

where a, b, c, and d are empirical constants. The constant c, the **exchange current density**, is a measure of the rates of the forward and reverse electrode reactions at dynamic equilibrium in the absence of an overpotential. For systems obeying these relations, the reaction rate (as measured by the current density) increases rapidly with increasing applied potential when $\eta d > 1$. If the exchange current density is high, an appreciable reaction rate may be achieved for only a small overpotential. If the exchange current density is low, a high overpotential is necessary. There is therefore of considerable interest in controlling the exchange current density. In an industrial process an overpotential in a synthetic step is very costly because it represents wasted energy. However, a low exchange current density may also be beneficial for a side reaction, for then it blocks undesirable pathways.

A catalytic electrode surface can increase the exchange current den-

---

[17] The exponential relation between the current and the overpotential is accounted for by the 'Butler–Volmer equation', which is derived by applying activated complex theory to dynamical processes at electrodes. See Chapter 29 of P. W. Atkins, *Physical Chemistry*. Oxford University Press and W. H. Freeman & Co. (1994).

sity and hence dramatically decrease the overpotential required for sluggish electrochemical reactions, such as H_2, O_2, or Cl_2 evolution and consumption. For example, 'platinum black', a finely divided form of platinum, is very effective for increasing the exchange current density and hence decreasing the overpotential of reactions involving the consumption or evolution of H_2. The role of platinum is to dissociate the strong H—H bond and thereby to reduce the large barrier that this strength imposes on reactions involving H_2. Palladium metal also has a high exchange current density (and hence requires a relatively low overpotential) for H_2 evolution or consumption.

The effectiveness of metals can be judged from Fig. 17.19, which also gives insight into the process. The volcano-like plot of exchange current density against M—H bond enthalpy suggests that M—H bond formation and cleavage are both important in the catalytic process. It appears that an intermediate M—H bond energy leads to the proper balance for the existence of a catalytic cycle and the most effective metals for electrocatalysis are clustered around Group 10.

Ruthenium dioxide is an effective catalyst for both O_2 and Cl_2 evolution and it also is a good electrical conductor. It turns out that at high current densities RuO_2 is more effective for the catalysis of Cl_2 evolution than for O_2 evolution. Therefore RuO_2 is extensively used as an electrode material in commercial chlorine production. The electrode processes that contribute to this subtle catalytic effect do not appear to be well understood.

There is great interest in devising new catalytic electrodes, particularly ones that decrease the O_2 overpotential on surfaces such as graphite. Thus, tetrakis(4-*N*-methylpyridyl)tetraphenylporphyriniron(II), [Fe(TMPyP)] (**9**), has been deposited on the exposed edges of graphite electrodes (on which O_2 reduction requires a high overpotential) and the resulting electrode surface was found to catalyze the electrochemical reduction of O_2. A plausible explanation for this catalysis is that

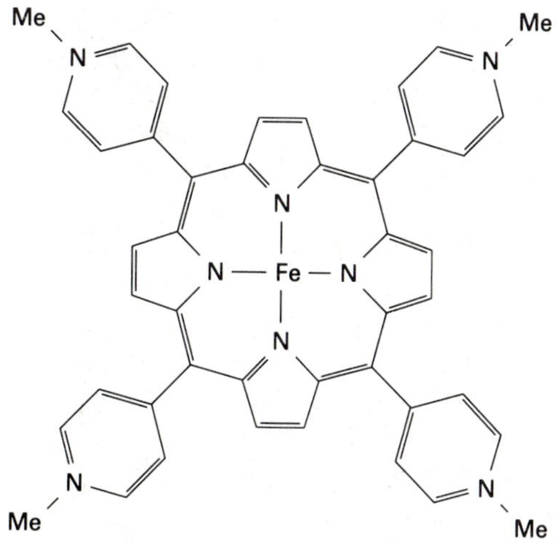

9

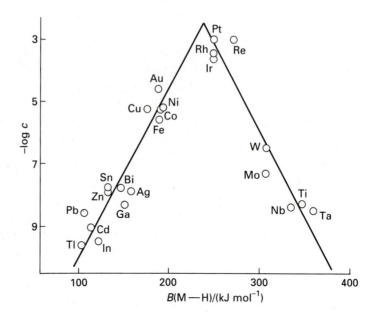

17.19 Rate of H_2 evolution expressed as the logarithm of the exchange current density versus M—H bond energies.

[$Fe^{III}(TMPyP)$] attached to the electrode (indicated below by *) is first reduced electrochemically:

$$[Fe^{III}(TMPyP)]^* + e^- \rightarrow [Fe^{II}(TMPyP)]^*$$

The resulting [$Fe^{II}(TMPyP)$] forms an O_2 complex:

$$[Fe^{II}(TMPyP)]^* + O_2 \rightarrow [Fe^{II}(O_2)(TMPyP)]^*$$

This iron(II) porphyrin oxygen complex is then susceptible to reduction to both water and hydrogen peroxide:

$$[Fe^{II}(O_2)(TMPyP)]^* + ne^- + mH^+ \rightarrow$$

$$[Fe^{II}(TMPyP)]^* + H_2O_2, H_2O$$

Although details of the mechanism are still elusive the general set of reactions given above is in harmony with electrochemical measurements and the known properties of iron porphyrins. The investigation of porphyriniron(III) complex as a catalyst was undoubtedly motivated by nature's use of metalloporphyrins for oxygen activation, a subject we explore in Chapter 19.

FURTHER READING
Homogeneous catalysis

G. W. Parshall and S.D. Ittel, *Homogeneous catalysis*. Wiley, New York, (1992). This extensive revision of the highly successful first edition, documents the chemistry of homogeneous catalysts that are of importance in industrial and laboratory-scale organic synthesis.

C. Masters, *Homogeneous transition-metal catalysis: a gentle art*. Chapman and Hall, London (1981).

B. R. James, *Homogeneous hydrogenation*. Wiley, New York (1973).

R. A. Shelton and J. K. Kochi, *Metal catalyzed oxidations of organic compounds.* Academic Press, New York (1981).

J. P. Collman, L. S. Hegedus, J. R. Norton, and R. G. Finke, *Principles and applications of organotransition metal chemistry.* University Science Books, Mill Valley (1987). This important book covers both stoichiometric and catalytic organometallic chemistry.

Heterogeneous catalysis

R. L. Burwell, Jr, Heterogeneous catalysis. In *Survey of progress in chemistry*, **8**, 2 (1977). A brief and instructive account.

G. C. Bond, *Heterogeneous catalysis.* Oxford University Press (1987). Another brief and general introduction to the field.

R. D. Srivatava, *Heterogeneous catalytic science.* CRC Press, Boca Raton (1988). A survey of experimental methods and several major catalytic processes, including the oxidation of hydrocarbons, the hydrogenation of CO, and hydrocarbon reforming.

J. M. Thomas and K. I. Zamaraev (ed.), *Perspectives in catalysis.* (1992). An edited volume with a modern perspective.

B. C. Gates, *Catalytic chemistry.* Wiley-Interscience (1991). This text discusses both homogeneous and heterogeneous catalysis.

H. H. Kung, Transition metal oxides. In *Surface chemistry and catalysis*,. Elsevier, Amsterdam (1989). A modern and chemically interesting discussion of oxide catalysts.

KEY POINTS

1. Definition of catalysis
Catalysis may be viewed as a cyclic process in which a species, the catalyst, increases the rate of conversion of reactants to products and is constantly regenerated.

2. Action of catalysts
The catalyzed reaction follows a different path from the uncatalyzed reaction, and has a lower activation free energy. A catalyst does not influence the position of equilibrium of a reaction.

3. Homogeneous and heterogeneous catalysis
In homogeneous catalysis the catalyst and reactants are in the same phase, usually liquid but sometimes gas. Heterogeneous catalysts are present in a different phase from the reactants. Most heterogeneous catalysts are solids, which are easily separated from liquid or gaseous products.

4. Metal complexes as homogeneous catalysts
Some common steps in the catalysis of hydrocarbon reactions by metal complexes are coordination of a reactant to the metal, insertion of a group into a metal–carbon bond (or its reverse, elimination), oxidative addition (or its reverse, reductive elimination), and dissociation of a product from the metal center.

5. Catalytic cycles
Since the catalyst is not consumed, it is convenient to depict catalytic processes as a cycle in which the metal complex interacts with reactants and eventually produces products with regeneration of the catalyst.

6. Important homogeneous catalytic reactions
Examples include alkene hydrogenation, hydroformylation (combination of an alkene with CO and hydrogen), acetic acid synthesis from methanol and carbon monoxide, the formation of acetaldehyde from ethene and carbon monoxide, and alkene polymerization.

7. Commercial advantages of heterogeneous catalysis

Solid catalysts (often consisting of metal particles on a solid or acid sites on a solid) are often preferred in commercial processes because the separation of catalyst from products is simple and therefore economical.

8. Uniform and multiphasic heterogeneous catalysts

Uniform solid catalysts consist of a single phase material such as a reactive zeolite. Multiphasic catalysts often consist in large proportion of a more or less inert high-surface-area-support (often silica or γ-alumina) which holds fine particles of a catalytic species, such as platinum or rhodium metal.

9. Mechanism of heterogeneous catalysis

Chemisorption brings a reactive molecule on to a surface and modifies its character. Products are released from the surface by desorption. Surface migration and combination are other steps characteristic of many heterogeneous catalytic reactions.

10. Important heterogeneous catalytic reactions

Some of the many important heterogeneous catalytic processes are oxidation of SO_2 to SO_3 in sulfuric acid production, hydrogenation of alkenes, conversion of hydrogen and nitrogen to ammonia, polymerization of alkenes, catalytic total oxidation of hydrocarbon pollutants (automobile catalytic converters), surface acid catalyzed hydrocarbon isomerization, and electrochemical catalysis as in chlorine production.

EXERCISES

17.1 Which of the following constitute genuine examples of catalysis and which do not? Present your reasoning. (a) The addition of H_2 to C_2H_4 when the mixture is brought into contact with finely divided platinum. (b) The reaction of an H_2/O_2 gas mixture when an electric arc is struck. (c) The combination of N_2 gas with lithium metal to produce Li_3N, which then reacts with H_2O to produce NH_3 and LiOH.

17.2 Define the terms (a) turnover frequency, (b) selectivity, (c) catalyst, (d) catalytic cycle, and (e) catalyst support.

17.3 Classify the following as homogeneous or heterogeneous catalysis and present your reasoning. (a) The increased rate in the presence of $NO(g)$ of $SO_2(g)$ oxidation by $O_2(g)$ to $SO_3(g)$. (b) The hydrogenation of liquid vegetable oil using a finely divided nickel catalyst. (c) The conversion of an aqueous solution of D-glucose to a D,L mixture catalyzed by $HCl(aq)$.

17.4 You are approached by an industrialist with the proposition that you develop catalysts for the following processes at $80\,°C$ with no input of electrical energy or electromagnetic radiation.

    (a) The splitting of water into H_2 and O_2.
    (b) The decomposition of CO_2 into C and O_2.
    (c) The combination of N_2 with H_2 to produce NH_3.
    (d) The hydrogenation of the double bonds in vegetable oil.

The industrialist's company will build the plant to carry out the process and the two of you will share equally in the profits. Which of these would be easy to do, which are plausible candidates for investigation, and which are

unreasonable? Describe the chemical basis for the decision in each case.

17.5 Addition of PPh_3 to a solution of Wilkinson's catalyst, $[RhCl(PPh_3)_3]$, reduces the turnover frequency for the hydrogenation of propylene. Give a plausible mechanistic explanation of this observation.

17.6 The rates of H_2 gas absorption (in $L\,mol^{-1}\,s^{-1}$) by alkenes catalyzed by $[RhCl(PPh_3)_3]$ in benzene at $25\,°C$ are: hexene, 2910; *cis*-4-methyl-2-pentene, 990; cyclohexene, 3160; 1-methylcyclohexene, 60. Suggest the origin of the trends and identify the affected reaction step in the proposed mechanism (Fig. 17.3).

17.7 Infrared spectroscopic investigation of a mixture of CO, H_2, and 1-butene under conditions that bring about hydroformylation indicate the presence of compound (c) in Fig. 17.6 in the reacting mixture. The same reacting mixture in the presence of added tributylphosphane was studied by infrared spectroscopy and neither (c) nor an analogous phosphane-substituted complex was observed. What does the first observation suggest as the rate limiting reaction in the absence of phosphane? Assuming the sequence of reactions remains unchanged, what are the possible rate limiting reactions in the presence of tributylphosphane?

17.8 (a) Starting with the alkene complex shown in Fig. 17.8(a) with *trans*-DHC=CHD in place of C_2H_4, assume dissolved OH^- attacks from the side opposite the metal. Give a stereochemical drawing of the resulting compound of the type shown in Fig. 17.8(b). (b) Assume attack on the coordinated *trans*-DHC=CHD by an OH^- ligand coordinated to Pd, and draw the stereochemistry of the resulting compound. (c) Does the stereochemistry differentiate these proposed steps in the Wacker process?

17.9 Formic acid is thermodynamically unstable with respect to CO_2 and H_2. Give a plausible mechanism for the catalysis of formic acid decomposition by $[IrCl(H)_2(PPh_3)_3]$.

17.10 At elevated temperatures, finely divided titanium readily reacts with $N_2(g)$ to form a stable nitride. The rate determining step in ammonia synthesis over iron is $N \equiv N$ bond breaking. Why then is titanium ineffective whereas iron is effective as a catalyst for ammonia synthesis?

17.11 Aluminosilicate surfaces were described in the text as strong Brønsted acids, whereas silica gel is a very weak acid. (a) Give an explanation for the enhancement of acidity by the presence of Al^{3+} in a silica lattice. (b) Name three other ions that might enhance the acidity of silica gel.

17.12 Indicate the difference between each of the following: (a) The Lewis acidity of γ-alumina that has been heated to $900\,°C$ versus γ-alumina that has been heated to $100\,°C$.

(b) The Brønsted acidity of silica gel versus γ-alumina. (c) The structural regularity of the channels and voids in silica gel versus that for ZSM-5.

17.13 Why is the platinum–rhodium in automobile catalytic converters dispersed on the surface of a ceramic rather than used in the form of thin foil?

17.14 Describe the concept of shape-selective catalysts and how this shape selectivity is attained.

17.15 Alkanes are observed to exchange hydrogen atoms with deuterium gas over some platinum metal catalysts. When 3,3-dimethylpentane in the presence of D_2 is exposed to a platinum catalyst and the gases are observed before the reaction has proceeded very far the main product is $CH_3CH_2C(CH_3)_2CD_2CD_3$ plus unreacted 3,3-dimethylpentane. Devise a plausible mechanism to explain this observation.

17.16 The effectiveness of platinum in catalyzing the reaction $2H^+(aq) + 2e^- \rightarrow H_2(g)$ is greatly decreased in the presence of CO. Suggest an explanation.

PROBLEMS

17.1 Carbon monoxide is known to undergo dissociative chemisorption on nickel at elevated temperatures, which results in surface carbide and oxide. With this as an initial step, propose a series of plausible reactions for the nickel catalyzed conversion of CO and H_2 to CH_4 and H_2O.

17.2 Consider the truth of each of the following statements and provide corrections where required. (a) A catalyst introduces a new reaction pathway with lower enthalpy of activation. (b) Since the Gibbs free energy is more favorable for a catalytic reaction, yields of the product are increased by catalysis. (c) An example of a homogeneous catalyst is the Ziegler–Natta catalyst made from $TiCl_4(l)$ and $Al(C_2H_5)_3(l)$. (d) Highly favorable free energies for the attachment of reactants and products to a homogeneous or heterogeneous catalyst are the key to high catalytic activity.

17.3 When direct evidence for a mechanism is not available, chemists frequently invoke analogies with similar sys-

tems. Describe how J. E. Bäckvall, B. Åkermark, and S. O. Ljunggren (*J. Am. Chem. Soc.*, **101**, 2411 (1979)) inferred the attack of uncoordinated water on η^2-C_2H_4 in the Wacker process.

17.4 The removal of the sulfur atom (as H_2S) from organosulfur compounds in crude oil (hydrodesulfurization) is an important process in petroleum refining. Metal sulfides are used as catalysts in hydrodesulfurization and both the $4d$ and $5d$ metal sulfides display catalytic activity that is greatest in Group 8 (Ru and Os). The $3d$ metal sulfides have relatively low catalytic activity and do not display maximum activity for the Group 8 element, iron. After reviewing the paper by S. Harris and R. R. Chianelli (*J. Catal.*, **86**, 400 (1984)) describe the feature of the electronic structures of these sulfides that correlates with the catalytic activity and how that parameter might be relevant to the catalysis.

Structures and properties of solids

The physical and chemical properties of solids are a pervasive aspect of inorganic chemistry, and have been discussed throughout the text. As part of the current enthusiasm for 'materials chemistry' there has been a great increase in the synthesis and study of new inorganic solids. In this chapter we will touch on some of the areas of current investigation. Some basic concepts of solids including the structures of some simple prototypical solids, lattice enthalpies and both ionic and covalent bonding models have been discussed in Chapter 4. In previous chapters we have only touched on defects, and this important feature will now be developed more thoroughly. The manner in which interruptions in the uniformity of a crystal structure influence properties such as the migration of ions through it is developed. In addition, we extend the discussion of intercalation compounds, and see how their formation, structures, and properties correlate with the crystal structure and the electronic band structure of the parent compound. We shall see in particular that the appropriate choice of a localized or a delocalized description of the electronic structure of a solid is of great help in correlating their various properties.

Solid state inorganic chemistry is concerned with the synthesis, structures, and properties of solids. New solid compounds often exhibit new phenomena or possess desirable properties, such as high-temperature superconductivity or ferromagnetism. It is a vigorous and exciting area of inorganic research,[1] partly on account of the technological applications of these properties but also because the structures and properties are challenging to understand. We shall draw on some of the concepts developed in earlier chapters for discussing the solid

---

[1] The vigor of this area can be judged by the rapid growth of publications. See for example: L. V. Interrante, Chemistry of materials: entering our fifth year of publication, *Chem. Mater.*, **5**, 1 (1993).

state, such as lattice enthalpies and band structure, and introduce some additional concepts that are needed if we are to discuss the dynamics of events that occur in the interior of solids. We shall also need to go beyond the view that solids have a well defined stoichiometry, for interesting phenomena arise from nonstoichiometry and atom mobility.

SOME GENERAL PRINCIPLES

In Chapter 4 we described the structures of some typical inorganic solids in terms of infinitely repeating regular arrays of atoms. It is now time to modify this idealized picture.

18.1 Defects

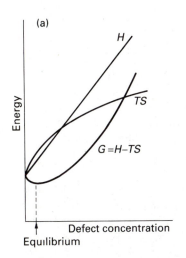

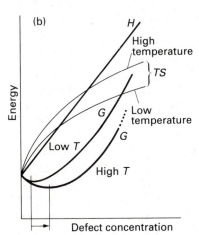

18.1 (a) The variation of the enthalpy and entropy of a crystal as the number of defects increases. The resulting Gibbs free energy $G = H - TS$ has a minimum at a nonzero concentration, and hence defect formation is spontaneous. (b) As the temperature is increased, the minimum in the Gibbs free energy moves to higher defect concentrations, and so more defects are present at equilibrium at higher temperatures than at low.

All solids contain **defects**, or imperfections of structure or composition. Defects are important because they influence properties such as mechanical strength, electrical conductivity, corrosion, and chemical reactivity. We need to consider both **intrinsic defects**, which are present for thermodynamic reasons, and **extrinsic defects**, which are not required by thermodynamics and can often be controlled by purification or by changing the synthetic conditions. It is also common to distinguish **point defects**, which occur at single sites, from **extended defects**, which are ordered in one, two, and three dimensions. Point defects are random errors in a periodic lattice, such as the absence of an atom at its usual site (a vacancy) or the presence of an atom at a site that is not ordinarily occupied (an interstitial). Extended defects involve, for example, various irregularities in the stacking of planes.

Why crystals have defects

All solids have a thermodynamic tendency to acquire point defects, because defects introduce disorder into an otherwise perfect structure and hence increase its entropy. The Gibbs free energy, $G = H - TS$, of a solid with defects has contributions from the enthalpy and the entropy of the sample. Because entropy is a measure of the disorder of a system, any solid in which some of the atoms are not in their usual sites has a higher entropy than in a perfect crystal. It follows that defects contribute a negative term to the Gibbs free energy of the solid. The formation of defects is usually endothermic (so H is higher in the presence of defects), but so long as $T > 0$, the Gibbs free energy will have a minimum at a non-zero concentration of defects (Fig. 18.1(a)), and their formation will be spontaneous. Moreover, as the temperature is raised, the minimum in G shifts to higher concentrations of defects (Fig. 18.1(b)).

Intrinsic point defects

Point defects are difficult to detect directly. X-Ray diffraction, for example, samples the periodic structure of a solid over thousands of

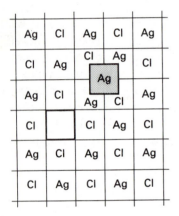

1 Shottky defect

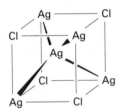

2 Frenkel defect

3 Interstitial Ag^+

angstroms, so small random deviations from periodicity often go unnoticed. Occasionally spectroscopic measurements indicate the presence of an ion in an unusual position or an electron trapped at a particular site in the lattice. The presence of an appreciable number of vacancies or excess atoms may sometimes be inferred from the difference between the measured and calculated densities of a sample (as we illustrate in Example 18.1). The electrical conductivity of a sample, another bulk property, has also been used to infer the presence of defects. Modern electron microscopy has improved our ability to detect atomic defects because its resolution is so high that defects can sometimes be observed directly.

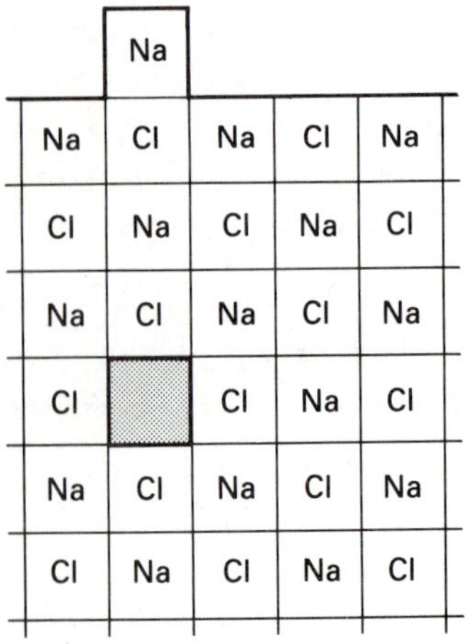

In the 1930s, two solid state physicists—Schottky in Germany and Frenkel in Russia—used conductivity and density data to identify specific types of point defects in solids. A **Schottky defect** (1) is a vacancy in an otherwise perfect lattice. That is, it is a point defect in which an atom or ion is missing from its normal site in the lattice. The overall stoichiometry of a solid is not usually affected by the presence of Schottky defects because there are equal numbers of vacancies at M and X sites so as to preserve charge balance. A **Frenkel defect** (2) is a point defect in which an atom or ion has been displaced into an interstitial site. For example, in silver chloride, which has the rock-salt structure, a small number of Ag^+ ions reside in normally unoccupied tetrahedral sites (3). The stoichiometry of the compound is unchanged when a Frenkel defect forms. A generalization is:

> Frenkel defects are most often encountered in the more open structures such as wurtzite and sphalerite, where the coordination numbers are low and the open structure provides sites that can accommodate interstitial atoms.

The concentration of Schottky defects varies considerably from one type of compound to the next. The concentration of vacancies is very low in the alkali metal halides, being of the order of 10^{-12} mol L^{-1} at 130 °C. That concentration corresponds to about one defect per 10^{14} formula units. On the other hand, some *d*-metal oxides, sulfides, and hydrides have very high concentrations of vacancies. An extreme example is the high-temperature form of TiO, which has vacancies on both the cation and anion sites at a concentration of about 12 mol L^{-1}, corresponding to about one defect per 10 formula units. A useful generalization is that:

> Schottky defects are commonly encountered when the metal ions can readily assume more than one oxidation state.

Example 18.1: *Inferring the type of defect from density measurements*

Titanium monoxide has a rock-salt structure (Fig. 4.11). X-Ray diffraction data show that the length of one edge of the cubic unit cell for TiO with a 1:1 ratio of Ti to O is 4.18 Å, and the density determined by volume and mass measurements is 4.92 g cm^{-3}. Do the data indicate that defects are present? If so, are they vacancy or interstitial defects?

Answer. The presence of vacancies (Schottky defects) on the Ti and O sites should be reflected in a lower measured density than that calculated from the size of the unit cell and the assumption that every Ti and O site is occupied. Interstitial (Frenkel) defects would give little if any difference between the measured and theoretical densities. There are four TiO formula units per unit cell (Fig. 4.17). The mass of one formula unit is 63.88 u, so the corresponding theoretical mass in one unit cell is 4×63.88 u $= 255.52$ u, which corresponds to 4.24×10^{-22} g. The corresponding theoretical density is

$$\rho = \frac{4.24 \times 10^{-22} \text{ g}}{(4.18 \times 10^{-8} \text{ cm})^3} = 5.81 \text{ g cm}^{-3}$$

which is significantly greater than the measured density. Therefore, the crystal must contain numerous vacancies. Because the overall composition of the solid is TiO, there must be equal numbers of vacancies on the cation and anion sites.

Exercise E18.1. The measured density of VO (1:1 stoichiometry) is 5.92 g cm^{-3} and the theoretical density is 6.49 g cm^{-3}. Do the data indicate the presence of vacancies or interstitials?

Schottky and Frenkel defects are only two of the many possible types of defect. Another example is an **atom interchange defect,** which consists of an interchanged pair of M and X atoms or ions. Atom interchange defects are also sometimes called **anti-site defects.** This type of defect is common in metal alloys. For example, a copper–gold alloy of overall composition CuAu has extensive disorder at high temperatures and a significant fraction of Cu and Au atoms are interchanged (4).

Extrinsic point defects

Extrinsic defects are inevitable because perfect purity is unattainable in crystals of any significant size. For example, highly purified silica glass for optical fibers contains about 1 part in 10^9 of metal impurities. In Section 2.10 we discussed impurities that had been introduced intentionally, where a pure substance was deliberately doped with another, such as when As atoms are substituted for Si atoms in a single crystal of silicon in order to modify its semiconducting properties. We can begin to see how the electronic structure of the solid influences the type of defect it is likely to form. Thus, when As replaces Si the added electron from each As atom finds its way into the conduction band. In the more ionic substance ZrO$_2$, the introduction of Ca^{2+} impurities in place of Zr^{4+} ions is accompanied by the formation of an O^{2-} ion vacancy to maintain charge neutrality (Fig. 18.2). In some cases a change in oxidation state (if one is possible) may be induced by an impurity. For example the introduction of Li$_2$O into NiO places Li$^+$ in Ni^{2+} sites, and the system balances charge by oxidation of one

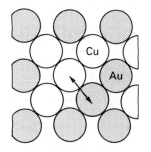

4 Atom interchange

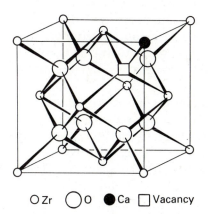

O Zr   ◯ O   ● Ca   ☐ Vacancy

18.2 Introduction of a calcium ion into the ZrO$_2$ lattice produces a vacancy on the oxide sublattice and helps to stabilize the fluorite structure.

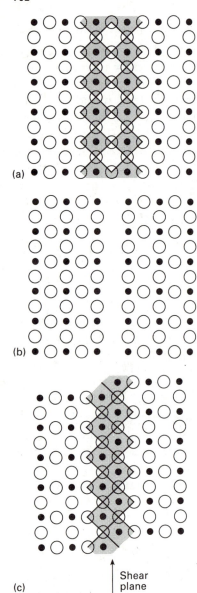

(a)

(b)

(c)

↑ Shear plane

18.3 The concept of a crystallographic shear plane illustrated by the [100] plane of the ReO_3 structure. (a) A plane of metal, ●, and oxygen, ○, atoms. The octahedron around each metal atom is completed by a plane of O atoms above and below the plane illustrated here. Some of the octahedra are shaded to clarify the processes that follow. (b) Oxygen atoms in the plane perpendicular to the page are removed, leaving two planes of metal atoms that lack their sixth oxygen ligand. (c) The octahedral coordination of the two planes of metal atoms is restored by translating the right slab as shown. This creates a plane (labeled the shear plane) vertical to the paper in which the MO_6 octahedra share edges.

Ni^{2+} ion to Ni^{3+} for each Li^+ ion present. As with the doping of silicon by boron, this process introduces holes in the valence band of NiO and the conductivity increases greatly.

Another example of an extrinsic point defect is a **color center**, a generic term for defects responsible for modifications to the infrared, visible, and ultraviolet absorption characteristics of solids that have been irradiated or exposed to chemical treatment. One type of color center is produced by heating an alkali halide crystal in the vapor of the alkali metal. The process results in an alkali metal cation at a normal cation site, but the electron the atom brings into the crystal occupies a halide ion vacancy. A color center consisting of an electron in a halide ion vacancy (5), is called an **F-center** (from the German word for color center, *Farbenzenter*). The color results from the excitation of the electron in the localized environment of its surrounding ions, and its quantized energy levels resemble those of an electron in a spherical box (which are like those for particles in rectangular boxes, Section 1.4).

Extended defects

All the point defects discussed so far entail a significant local distortion of the lattice and in some instances localized charge imbalances too. Therefore it should not be surprising that defects may cluster together and sometimes form lines and planes. These extended defects control the mechanical properties of metals and ceramics. They may also influence their electrical and magnetic properties.

Tungsten oxide illustrates the formation of a plane of defects. As illustrated in Fig. 18.3(a), the idealized structure of WO_3 (which is usually referred to as the 'ReO$_3$ structure'; see below) consists of WO_6 octahedra sharing all vertices. To picture the formation of the defect plane we should imagine the removal of shared O atoms along a diagonal. Then adjacent slabs slip past each other in a motion that results in the completion of the vacant coordination sites around each W atom (Fig. 18.3(b)). This shearing motion creates edge-shared octahedra along a diagonal direction. The resulting structure was named a **crystallographic shear plane** by the Australian crystallographer A. D. Wadsley, who first devised this way of describing extended planar defects.

Crystallographic shear planes randomly distributed in the solid are called **Wadsley defects**. Such defects lead to a continuous range of compositions, as in tungsten oxide, which ranges from WO_3 to $WO_{2.93}$. If, however, the crystallographic shear planes are distributed in a nonrandom, periodic manner, we should regard the material as a new stoichiometric phase. Thus, when even more O^{2-} ions are removed from tungsten oxide, a series of discrete phases having ordered crystallographic shear planes and compositions W_nO_{3n-2} ($n = 20, 24, 25,$ and 40) are observed. Closely spaced shear plane phases are known for oxides of tungsten, molybdenum, titanium, and vanadium. They can be identified by X-ray diffraction, but electron microscopy provides a

Na	Cl	Na	Cl	Na
Cl	Na	Cl	Na	Cl
Na	e⁻	Na	Cl	Na
Cl	Na	Cl	Na	Cl
Na	Cl	Na	Cl	Na
Cl	Na	Cl	Na	Cl

5 F-center

more general and direct method (Fig. 18.4) because it reveals both ordered and random arrays of shear planes.

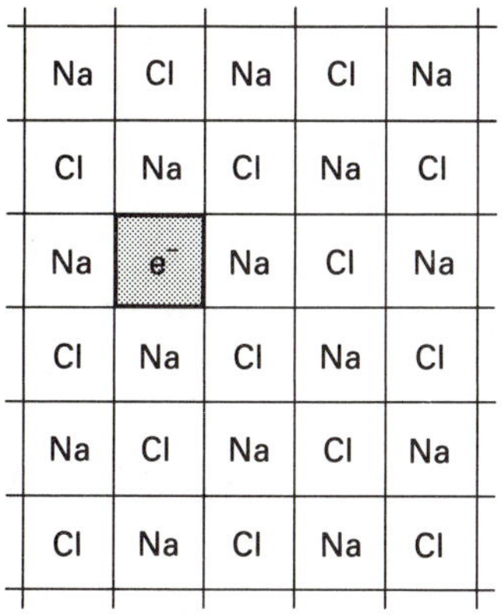

18.2 Nonstoichiometric compounds

A **nonstoichiometric compound** is a substance with variable composition but which retains essentially the same basic structure. For example, at $1000\,°C$ the composition of wüstite (which is nominally FeO) can vary from $Fe_{0.89}O$ to $Fe_{0.96}O$, and throughout this composition range the major peaks in the X-ray diffraction pattern indicate that the main features of the rock-salt structure are retained. There are gradual changes in the unit cell size as the composition is varied, and some added broad and weak X-ray reflections are observed, but the essential structure remains the same. Some representative nonstoichiometric hydrides, oxides, and sulfides are listed in Table 18.1. A helpful generalization is as follows:

> Deviations from stoichiometry are common with d-, f-, and some p-block metals in combination with soft anions, such as S^{2-} and H^- ions, as well as with the harder O^{2-} anion. In contrast, hard anions such as fluorides, chlorides, sulfates, and nitrates form fewer nonstoichiometric compounds.

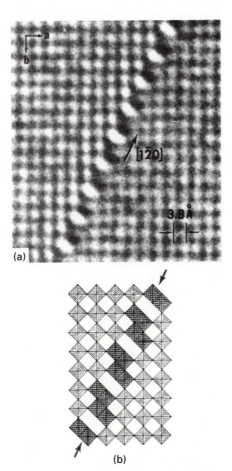

18.4 (a) High resolution electron micrograph lattice image of a crystallographic shear plane (120) in WO_{3-x}. (b) Drawing of the oxygen octahedral polyhedron that surround the tungsten atoms imaged in the electron micrograph. Note the edge-shared octahedra along the crystallographic shear plane. (Reproduced by permission from S. Iijima, *J. Solid State Chem.*, **14**, 52 (1975).)

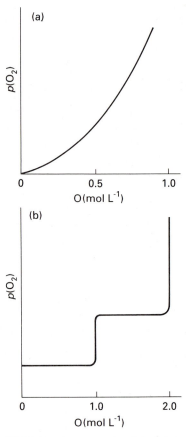

18.5 Schematic representation of the variation of the partial pressure of oxygen with composition at constant temperature for (a) a nonstoichiometric oxide; (b) a stoichiometric pair of metal oxides MO and MO_2.

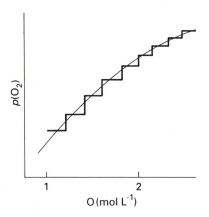

18.6 The stepped line indicates the variation in pressure of O_2 in equilibrium with a series of closely spaced discrete phases at constant temperature. Two solid phases are present in the horizontal regions. The smooth curve indicates the observed pressure when the reaction is so slow that equilibrium is not established.

Table 18.1 Representative composition ranges† of nonstoichiometric binary hydrides, oxides, and sulfides

d block			*f* block		
Hydrides					
TiH_x	1 to 2			Fluorite type	Hexagonal
ZrH_x	1.5 to 1.6		GdH_x	1.8–2.3	2.85–3.0
HfH_x	1.7 to 1.8		ErH_x	1.95–2.31	2.82–3.0
NbH_x	0.64 to 1.0		LuH_x	1.85–2.23	1.74–3.0
Oxides					
	Rock-salt type	Rutile type			
TiO_x	0.7–1.25	1.9-2.0			
VO_x	0.9–1.20	1.8–2.0			
NbO_x	0.9–1.04				
Sulfides					
ZrS_x	0.9–1.0				
YS_x	0.9–1.0				

†Expressed as the range of values that *x* may take.

Nonstoichiometric compounds can be regarded as solid solutions in the sense that, as with liquid solutions, the chemical potentials of the components vary continuously as the composition changes. For example, as the partial pressure of oxygen is varied while equilibrium is maintained in the presence of a metal oxide, the composition of the oxide varies continuously (Fig. 18.5).

It may be very difficult to distinguish a nonstoichiometric compound from a series of discrete phases if the Gibbs free energies of the individual phases are similar. For example, the partial pressure of oxygen over a series of closely spaced phases may appear to vary continuously rather than in steps (Fig. 18.6), and the presence of separate phases would not be apparent. Discrete stoichiometric phases can also be detected by X-ray diffraction and often by optical or electron microscopy.

18.3 Atom and ion diffusion

Diffusion in solids is generally much slower than diffusion in gases and liquids. However, there are some striking exceptions to this generalization as we shall soon see. Diffusion in solids is in fact very important in many areas of solid state technology, such as semiconductor manufacture, the synthesis of new solids, metallurgy, and heterogeneous catalysis.

General principles of diffusion

The diffusion of particles through any medium is governed by the **diffusion equation**

$$\frac{\partial c}{\partial t} = D \frac{\partial^2 c}{\partial x^2}$$

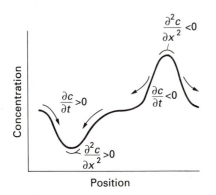

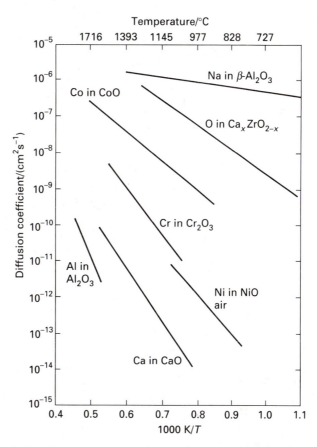

where c is the concentration of the diffusing species and D is the **diffusion coefficient**. The diffusion equation summarizes the fact that the net rate of change of the concentration of particles in a region is proportional to the spatial inhomogeneity of the concentration there (Fig. 18.7). Where there are more particles than average (in the sense that $\partial^2 c/\partial x^2 < 0$), the particles will tend to disperse and lower the concentration locally ($\partial c/\partial t < 0$). If there is a dip in the concentration locally (and $\partial^2 c/\partial x^2 > 0$), particles will tend to diffuse into the region ($\partial c/\partial t > 0$).[2]

The rate of approach to a uniform distribution of particles is high when D is large. In a solid, D may depend more strongly than in a liquid on the concentration of particles, for the presence of foreign particles affects the local structure of the solid and has a marked effect on the ability of atoms to migrate. The diffusion coefficient increases with temperature, because particle migration is an activated process, and we can write the Arrhenius-like expression

$$D = D_0 e^{-E_a/RT}$$

where E_a is the activation energy for diffusion. Figure 18.8 shows the

18.7 A schematic representation of the content of the diffusion equation. Where there is a trough in the concentration, particles tend to accumulate to remove it; where there is a peak, particles tend to spread away.

18.8 Arrhenius plot of the logarithm of the diffusion coefficient of the mobile ion in a series of solids versus $1/T$.

[2] The diffusion equation can be regarded as a mathematical form of the remark that 'Nature abhors a wrinkle' (see P. W. Atkins, *Physical chemistry*. Oxford University Press and W. H. Freeman and Co. (1994)). Peaks in distributions tend to disperse, and troughs tend to fill.

temperature dependence of the diffusion coefficients for some representative solids at high temperature. According to the equation above, the slopes of the lines are proportional to the activation energy for atom or ion transport. Thus Na^+ is highly mobile and has a low activation energy for motion in β-alumina, whereas Ca^{2+} in CaO is much less mobile and has a much higher activation energy for diffusion.

Mechanisms of diffusion

The diffusion coefficients of ions can often be understood in terms of the mechanism for their migration and the activation barriers the ions encounter as they move. In particular, diffusion coefficients in solids are markedly dependent on the presence of defects. The role defects play is summarized in Fig. 18.9, which shows some commonly postulated mechanisms for atom or ion motion in solids. In general, the diffusion rate is proportional to the defect density of the controlling defect.

As with the mechanisms of chemical reactions, the evidence for the operation of a particular diffusion process is circumstantial. The individual events are never directly observed but are inferred from the influence of experimental conditions on atom or ion diffusion rates. In addition, a detailed analysis of the thermal motion of ions in crystals based on X-ray and neutron diffraction provides strong hints about the ability of ions to move through the crystal, including their most likely paths. Theoretical models, which are often elaborations of the ionic model (Section 4.7), give very useful guidance to the feasibility of these migration mechanisms.

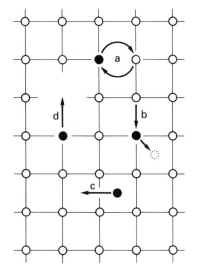

18.9 Some diffusion mechanisms for ions or atoms in a solid: (a) exchange, (b) interstitialcy, (c) interstitial, and (d) vacancy.

Solid electrolytes

Any electrochemical cell such as a battery, fuel cell, electrochromic display, or electrochemical sensor, requires an electrolyte. Recently there has been considerable interest in the use of solid electrolytes in these applications, and a commercial device that utilizes a solid oxide-ion conductor to measure O_2 pressure is described in Section 18.5. Silver tetraiodomercurate(II), Ag_2HgI_4, and β-alumina, a sodium-doped version of alumina of composition $Na_{1+x}Al_{11}O_{17+x/2}$, provide examples of two quite different types of solids in which cations are highly mobile. The former is a soft material with a lattice of low rigidity and the latter is a hard refractory solid. In general, the diffusion rate is proportional to the defect density of the controlling defect.

Below 50 °C, Ag_2HgI_4 has an ordered crystal structure in which Ag^+ and Hg^{2+} ions are tetrahedrally coordinated by I^- ions (Fig. 18.10(a)) and there are unoccupied tetrahedral holes. At this temperature its electronic and ionic conductivity is low. Above 50 °C, the Ag^+ and Hg^{2+} cations of Ag_2HgI_4 are distributed in a disorderly manner over all the tetrahedral sites (Fig. 18.10(b)). At this temperature the material is a good electrical conductor largely on account of the mobility of the Ag^+ ions. The weak and easily deformed close-packed array of I^- ions gives rise to a low activation energy for ion

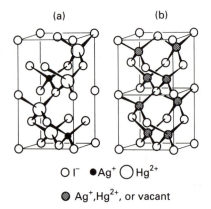

(a)   (b)

○ I^- ● Ag^+ ○ Hg^{2+}

◉ Ag^+, Hg^{2+}, or vacant

18.10 (a) Low temperature ordered structure of Ag_2HgI_4. (b) High-temperature disordered structure. Ag_2HgI_4 is an Ag^+ ion conductor in the high temperature form.

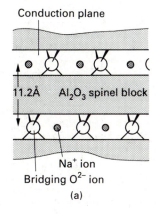

Conduction plane

11.2Å   Al$_2$O$_3$ spinel block

Na$^+$ ion
Bridging O^{2-} ion

(a)

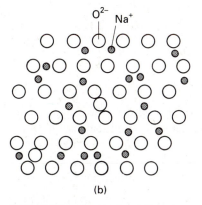

O^{2-} Na$^+$

(b)

18.11 (a) Schematic side view of β-alumina showing the Na$_2$O conduction planes between Al$_2$O$_3$ slabs. The O atoms in this plane bridge the two slabs. (b) A view of the conduction plane. Note the abundance of mobile ions and vacancies in which they can move.

migration from one lattice site to the next. Solid inorganic electrolytes generally have a low-temperature phase in which the ions are ordered on a subset of lattice sites and at higher temperatures the ions become disordered over the sites and the ionic conductivity increases. There are many similar solid electrolytes having soft lattices, such as AgI and RbAg$_4$I$_5$, both of which are Ag$^+$ conductors. The mobility of ions in solids can be quite high: the conductivity of RbAg$_4$I$_5$ at room temperature is greater than that of an aqueous sodium chloride solution.

Sodium β-alumina is an example of a mechanically hard material that is a good ionic conductor. In this case the rigid and dense Al$_2$O$_3$ slabs are bridged by a sparse array of O^{2-} ions (Fig. 18.11). The plane containing these bridging ions also contains Na$^+$ ions which can move from site to site because there are no major bottlenecks to hinder their motion. Many similar rigid materials having planes or channels through which ions can move are known, and are called **framework electrolytes**. Another closely related material, sodium β''-alumina, has even less restricted motion of ions than in β-alumina and it has been found possible to substitute dipositive cations such as Mg^{2+} or Ni^{2+} for Na$^+$. Even the large lanthanide cation Eu^{2+} can be introduced into β''-alumina, although the diffusion of such ions is slower than their smaller counterparts. These examples illustrate a fairly common strategy for the formation of new solids by diffusion processes at much lower temperatures than the melting point of the host material. Such **low temperature syntheses** provide routes to a wide variety of solids that might not be thermodynamically stable and therefore cannot be prepared from the melt. The rare earth β''-aluminas prepared in this manner are of interest for possible applications in lasers.

Example 18.2: *Correlating conductivity and ion size in a framework electrolyte*
Conductivity data on β-alumina containing monopositive ions of various radii show that Ag$^+$ and Na$^+$ ions, both of which have $r \approx 1.0$ Å, have activation energies for conductivity close to 17 kJ mol^{-1} whereas that for Tl$^+$ ($r \approx 1.4$ Å) is about 35 kJ mol^{-1}. Suggest an explanation of the difference.

Answer. In sodium β-alumina and related β-aluminas a fairly rigid framework provides a two-dimensional network of passages that permit ion migration. Judging from the experimental results, the bottlenecks for ion motion appear to be large enough to allow Na$^+$ or Ag$^+$ to pass quite readily (with a low activation energy) but too small to let the larger Tl$^+$ pass through as readily.

Exercise E18.2. Why does increased pressure reduce the conductivity of K$^+$ in β-alumina more than Na$^+$ in β-alumina?

The common characteristics of solid electrolytes are high concentrations of mobile ions and vacancies and the existence of pathways between the vacancies that provide routes for ion motion with low activation energy. In both Ag$_2$HgI$_4$ and sodium β-alumina there are

Table 18.2 Monoxides of the Period 4 metals

Compound	Structure	Stoichiometry MO_x	Electrical properties
CaO_x	Rock salt	1	Insulator
TiO_x	Rock salt	0.65-1.25	Metallic
$VO_x{}^a$	Rock salt	0.79–1.29	Metallic
MnO_x	Rock salt	1–1.15	Semiconductor
FeO_x	Rock salt	1.04–1.17	Semiconductor
CoO_x	Rock salt	1–1.01	Semiconductor
NiO_x	Rock salt	1–1.001	Semiconductor
CuO_x	PtS	1	Semiconductor
ZnO_x	Wurtzite	slight Zn excess	Wide gap n-type semiconductor

[a] At low temperatures, VO adopts a less symmetrical structure.

high concentrations of carrier ions (Ag^+ and Na^+ respectively) as well as of vacant sites into which these carriers may diffuse.

PROTOTYPICAL OXIDES AND FLUORIDES

In this section we explore compounds of the hard oxide and fluoride anions with metals. These compounds provide chemical insight into defects, nonstoichiometry, and ion diffusion, and into the influence of these characteristics on physical properties. This section also describes the structures of solids that are of interest in the synthesis of magnetic and superconducting materials.

18.4 Monoxides of the 3*d* metals

The monoxides of most of the 3*d* metals have the rock-salt structure, so it might at first appear that there would be little more to say about them and that their properties should be simple. In fact, the structures and properties of these oxides (Table 18.2) are so interesting that they have been repeatedly investigated. In particular, the compounds provide examples of how mixed oxidation states and defects lead to non-stoichiometry, for TiO, VO, FeO, CoO, and NiO can all be prepared with significant deviations from the stoichiometry MO. Their range of chemical character, from mild to aggressive reducing agents, is reflected in the various ways in which they are prepared.

We can see from the phase diagram for iron and oxygen (Fig. 18.12) that the preparation of FeO must be carried out above 570 °C. The partial pressure of oxygen must also be carefully controlled. This control is achieved by using an atmosphere of CO and CO_2, for at the high temperature used for the preparation the equilibrium

$$2CO_2(g) \rightleftharpoons 2CO(g) + O_2(g) \qquad K = 1.4 \times 10^{-9} \text{ at } 830 °C$$

is rapidly established, and the partial pressure of O_2 is governed by the relative proportions of the two carbon oxides:

$$p(O_2)/\text{bar} = K \times \frac{p(CO)^2}{p(CO_2)^2}$$

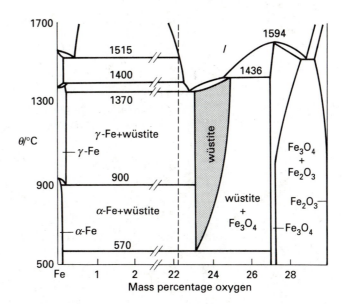

18.12 A part of the iron-oxygen phase diagram. The vertical dashed line corresponds to the stoichiometry FeO.

Thus a mixture in which $p(CO_2)$ and $p(CO)$ are equal has an equilibrium partial pressure of O_2 of 1.4 nbar at 830 °C. The two carbon oxides form a buffer to maintain the pressure at this constant value: CO_2 can release oxygen if the latter's partial pressure falls, and CO can absorb it if the partial pressure rises.

Defects and nonstoichiometry

The origin of the nonstoichiometry in FeO has been studied in more detail than that in most other compounds. The consensus is that it arises from the creation of vacancies on the Fe^{2+} octahedral sites, and that each vacancy is charge-compensated by the conversion of two Fe^{2+} ions to two Fe^{3+} ions. The relative ease of oxidizing Fe(II) to Fe(III) accounts for the rather broad range of compositions of FeO_x. The Fe^{3+} ions that are formed move to tetrahedral sites of the fcc oxide lattice but the symmetry of the rock-salt structure is not broken because the Fe^{3+} ions are not distributed in an orderly way. At high temperatures the interstitial Fe^{3+} ions associate with the Fe^{2+} vacancies to form clusters distributed throughout the structure (Fig. 18.13). A similar clustering of defects appears to occur with all other 3d monoxides, with the possible exception of NiO in which the range of composition is the smallest for all 3d compounds. Over this small range, the conductivity and the rate of atom diffusion vary with O_2 partial pressure in a manner that suggests the presence of isolated point defects.

Both CoO and NiO occur in a metal-deficient state, although their range of stoichiometry is not as broad as that of FeO. As indicated by standard potentials in aqueous solution (Section 7.3), Fe(II) is more easily oxidized to Fe(III) than is either Co^{2+} or Ni^{2+}, and this solution redox chemistry correlates well with the much smaller range of oxygen deficiency in NiO and CoO. Chromium(II) oxide does not exist because

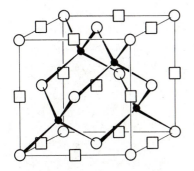

☐ Vacant octahedral sites

● Fe^{3+} ions in tetrahedral sites

○ ccp O^{2-} ions

18.13 Defect sites proposed for FeO. Note that the tetrahedral Fe^{3+} interstitials and octahedral Fe^{2+} vacancies tend to cluster together.

it is unstable with respect to disproportionation:

$$3CrO(s) \rightarrow Cr_2O_3(s) + Cr(s)$$

However, it is possible to stabilize CrO as a solid solution in CuO. Significant deviations from the stoichiometry MO is out of the question for Group 2 metal oxides such as CaO because for these metals M^{3+} ions are chemically inaccessible.

Electrical conductivity

The $3d$ monoxides MnO, FeO, CoO, and NiO have low electrical conductivities that increase with temperature; hence they are semiconductors. The electron or hole migration in these oxide semiconductors is attributed to a hopping mechanism. In this model the electron or hole hops from one localized metal site to the next. When it lands on a new site it causes the surrounding ions to adjust their locations and the electron or hole is trapped temporarily in the potential well this atomic polarization produces. The electron resides at its new site until it is thermally activated to migrate into another nearby site. Another aspect of this charge hopping mechanism is that the electron or hole tends to associate with local defects so the activation energy for charge transport may also include the energy of freeing the hole from its position next to a defect. The conduction process in MnO, FeO, CoO, and probably NiO occurs by such a hopping mechanism, but it is generally not clear whether site-to-site hopping or dissociation from a defect is rate limiting.

Hopping contrasts with the band model for semiconductivity discussed in Section 2.9, where the conduction and valence electrons occupy orbitals that spread through the whole crystal. The difference stems from the less diffuse d orbitals in the monoxides of the mid to late $3d$ metals, which are too compact to form the broad bands necessary for metallic conduction. We have already mentioned that when NiO is doped with Li_2O in an O_2 atmosphere a solid solution of composition $Li_xNi^{2+}_{1-x}Ni^{3+}_xO$ is obtained, which has greatly increased conductivity for reasons similar to the increase in conductivity of silicon when doped with boron (Section 2.10). The characteristic exponential increase in electrical conductivity with increasing temperature of metal oxide semiconductors is used in 'thermistors' to measure temperature.

In contrast to the semiconductivity of the monoxides in the center and right of the $3d$ series, TiO and VO have high electric conductivities that decrease with increasing temperature. This metallic conductivity persists over a broad composition range from highly oxygen-rich to metal-rich. In these compounds a conduction band is formed by the overlap of t_{2g} orbitals of metal ions in neighboring octahedral sites that are oriented toward each other (Fig 18.14). The radial extension of the d orbitals of these early d-block elements is greater than for elements later in the period, and a band results from their overlap (Fig. 18.15); this band is in general only partly filled. The widely varying compositions of these monoxides also appear to be associated with the electronic

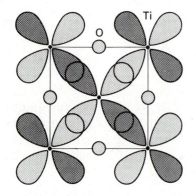

18.14 Overlap of d_{zx} orbitals in TiO.

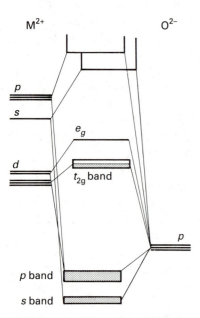

18.15 Energy level diagram for early d-block metal monoxides. The t_{2g} band is only partly filled and metallic conductivity results.

delocalization: the conduction band serves as a rapidly accessible source and sink of electrons that can readily compensate for the formation of vacancies.

18.5 Higher oxides

MO₂ and fluorite

The fluorite structure (Fig. 4.14) contains cations that are surrounded by eight anions. As with the cesium-chloride structure, this high coordination number is favored by relatively large cations. The structure is found for fluorides such as PbF_2 and oxides such as ZrO_2.

Michael Faraday reported in 1834 that red-hot solid PbF_2 is a good conductor of electricity. Much later it was recognized that the conductivity arises from the mobility of F^- ions through the solid. The property of anion conductivity is shared by other crystals having the fluorite structure. Ion transport in these solids is thought to be by an interstitial mechanism in which an F^- ion first migrates from its normal position into an interstitial site and then moves to a vacant F^- site.

Certain oxides that have the fluorite structure conduct by O^{2-} ion migration. The best known of these is ZrO_2 with added CaO to stabilize the fluorite phase (which in its absence is stable only at high temperature). For every Ca^{2+} substituted for Zr^{4+}, a vacancy on an O^{2-} site is created to maintain electroneutrality, and the presence of these vacancies improves the O^{2-} mobility.

A CaO-doped ZrO_2 electrolyte is used in a solid state electrochemical sensor for the oxygen partial pressure in automobile exhaust systems (Fig. 18.16).[3] The platinum electrodes in this cell adsorb O atoms, and if the partial pressure of oxygen is different between the sample and reference side there is a thermodynamic tendency for oxygen to migrate through the electrolyte as the O^{2-} ion. The thermodynamically favored processes are:

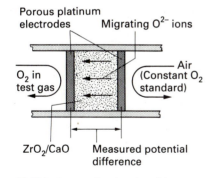

Porous platinum electrodes   Migrating O^{2-} ions

O_2 in test gas

Air (Constant O_2 standard)

ZrO_2/CaO   Measured potential difference

18.16 An O_2 sensor based on the solid electrolyte ZrO_2/CaO.

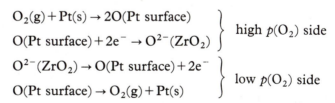

$$O_2(g) + Pt(s) \rightarrow 2O(Pt \text{ surface})$$
$$O(Pt \text{ surface}) + 2e^- \rightarrow O^{2-}(ZrO_2)$$
high $p(O_2)$ side

$$O^{2-}(ZrO_2) \rightarrow O(Pt \text{ surface}) + 2e^-$$
$$O(Pt \text{ surface}) \rightarrow O_2(g) + Pt(s)$$
low $p(O_2)$ side

The cell potential is related to the oxygen partial pressures through the Nernst equation (Section 7.3), so a simple measurement of the potential provides a measure of the O_2 partial pressure in the exhaust.

M₂O₃ corundum structure

α-Alumina (the mineral corundum) adopts a structure that can be described as a hexagonal close packed array of O^{2-} ions with the cations in two-thirds of the octahedral holes. The details of the structure are complex and we need not go into them. The corundum

---

[3] The signal from this sensor is used to adjust the air/fuel ratio and thereby the composition of the exhaust gas being fed to the catalytic converter (Fig. 17.9).

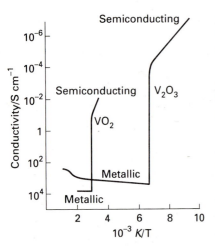

18.17 Temperature dependence of the conductivity of VO_2 and V_2O_3 showing the metal-to-semiconductor transition.

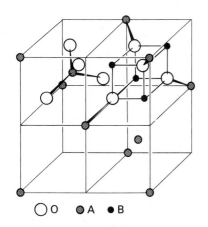

○ O   ◉ A   ● B

18.18 A segment of the spinel (AB_2O_4) unit cell showing the tetrahedral environment of A ions and the octahedral environment of B ions.

structure is adopted by the oxides of titanium, vanadium, chromium, iron, and gallium in their $+3$ oxidation states. Two of these oxides, Ti_2O_3 and V_2O_3, exhibit metallic-to-semiconducting transitions below 410 and 150 K respectively (Fig. 18.17). In V_2O_3 the transition is accompanied by antiferromagnetic ordering of the spins. The two insulators Cr_2O_3 and Fe_2O_3 also display antiferromagnetic ordering.

Another interesting aspect of the M_2O_3 compounds is the formation of solid solutions of dark green Cr_2O_3 and colorless Al_2O_3 to form brilliant red ruby. As we remarked in Section 14.5, this shift in the ligand field transitions of Cr^{3+} stems from the compression of the O^{2-} ions around Cr^{3+} in the Al_2O_3 host lattice. (In Al_2O_3 $a = 4.75$ Å and $c = 13.00$ Å; in Cr_2O_3 the lattice constants are 4.93 Å and 13.56 Å, respectively.) The compression shifts the absorption toward the blue and the solid appears red in white light. The responsiveness of the absorption (and fluorescence) spectrum of Cr^{3+} ions to compression is sometimes used to measure pressure in high-pressure spectroscopic experiments. In this application, a tiny crystal of ruby in one segment of the sample can be interrogated by visible light and the shift in its fluorescence spectrum provides an indication of the pressure inside the cell.

Spinels

The simple d-block oxides Fe_3O_4, Co_3O_4, and Mn_3O_4 and many related mixed-metal compounds have interesting properties. They have structures related to spinel, $MgAl_2O_4$, and may be given the general formula AB_2O_4. The **spinel structure**, which is rather complex (Fig. 18.18), consists of an fcc array of O^{2-} ions in which the A ions reside in one-eighth of the tetrahedral holes and the B ions inhabit half the octahedral holes. It is sometimes useful to emphasize this arrangement of cations by enclosing the octahedral ions in square brackets, and writing $A[B_2]O_4$.

Some compounds with the composition AB_2O_4 possess an **inverse spinel structure**, in which the cation distribution is $B[AB]O_4$. Lattice enthalpy calculations based on a simple ionic model indicate that for A^{2+} and B^{3+} the normal spinel structure, $A[B_2]O_4$, should be the most stable. The observation that many d-metal spinels do not conform to this expectation has been traced to the effect of ligand field stabilization energies on the site preferences of the ions.

The **occupation factor** λ of a spinel is the fraction of B atoms in the tetrahedral sites. Thus $\lambda = 0$ for a normal spinel and 0.5 for an inverse spinel, $B[AB]O_4$. The distribution of cations in (A^{2+}, B^{3+}) spinels (Table 18.3) illustrates that for d^0 A and B ions the normal structure is preferred ($\lambda = 0$) as predicted by electrostatic considerations. Table 18.3 shows that when the A^{2+} is a d^6, d^7, d^8, or d^9 ion and B^{3+} is Fe^{3+}, the inverse structure is generally favored. This preference can be traced to the lack of ligand field stabilization of the high-spin d^5 Fe^{3+} ion in either the octahedral or the tetrahedral site and the ligand field stabilization of the other d^n ions in the octahedral

Table 18.3 Cation site preference, λ, in some spinels[a]

B	A	Mg^{2+} d^0	Mn^{2+} d^5	Fe^{2+} d^6	Co^{2+} d^7	Ni^{2+} d^8	Cu^{2+} d^9	Zn^{2+} d^{10}
Al^{3+}	d^0	0	0	0	0	0.38	0	
Cr^{3+}	d^3	0	0	0	0	0	0	0
Fe^{3+}	d^5	0.45	0.1	0.5	0.5	0.5	0.5	0
Mn^{3+}	d^4	0						0
Co^{3+}	d^6					0		0

[a] $\lambda = 0$ corresponds to a normal spinel; $\lambda = 0.5$ corresponds to an inverse spinel.

site. It is important to note that simple ligand field stabilization appears to work over this limited range of cations, but more detailed analysis is necessary when cations of different radius are introduced. Moreover, λ is often found to depend on the temperature.

The inverse spinels of formula AFe_2O_4 are sometimes called **ferrites**. When $kT > J$, where J is the energy of interaction of the spins on different ions, ferrites are paramagnetic. However, when $kT < J$, a ferrite may be either **ferromagnetic** or **antiferromagnetic** (see Box 18.1). The antiparallel alignment of spins characteristic of antiferromagnetism is illustrated by $ZnFe_2O_4$, which has the cation distribution $Fe[ZnFe]O_4$. In this compound the Fe^{3+} ions (with $S = \frac{5}{2}$) in the tetrahedral and octahedral sites are antiferromagnetically coupled below 9.5 K to give nearly zero net magnetic moment to the solid as a whole.

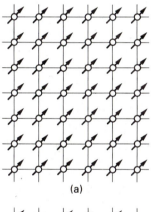

(a)

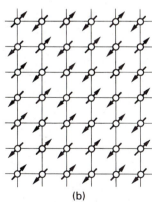

(b)

B18.1 (a) The parallel alignment of individual magnetic moments in a ferromagnetic material and (b) the antiparallel arrangement of spins in an antiferromagnetic material.

Box 18.1: Cooperative magnetism

The diamagnetic and paramagnetic properties of compounds were covered in Section 6.4; they are characteristic of individual atoms or complexes. In contrast, properties such as ferromagnetism and antiferromagnetism depend on interactions between electron spins on many atoms and arise from the cooperative behavior of many unit cells in a crystal.

In a **ferromagnetic** substance the spins on different metal centers are coupled into a parallel alignment (Fig. B18.1(a)) that is sustained over thousands of atoms in a magnetic **domain**. The net magnetic moment may be very large because the magnetic moments of individual spins augment each other. Moreover, once established and the temperature maintained below the **Curie temperature** (T_C), the magnetization persists because the spins are locked together. Ferromagnetism is exhibited by materials containing unpaired electrons in d or f orbitals that couple with unpaired electrons in similar orbitals on surrounding atoms. The key feature is that this interaction is strong enough to align spins but not so strong as to form covalent bonds, where the electrons are paired.

The magnetization M of a ferromagnet is not linearly propor-

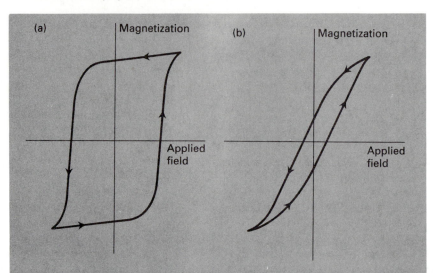

B18.2 Magnetization curves for ferromagnetic materials. A hysteresis loop results because the magnetization of the sample with increasing field (→ —) is not retraced as the field in decreased (— ←). (a) A hard ferromagnet, (b) a soft ferromagnet.

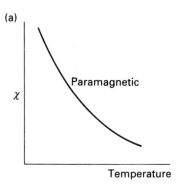

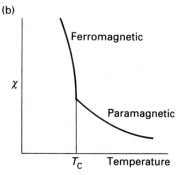

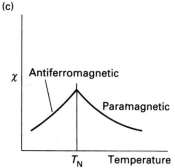

B18.3 Comparison of the tempreature dependence of paramagnetic, ferromagnetic, and antiferromagnetic substances.

tional to the applied field H. Instead, a hysteresis loop (Fig. B18.2) is observed. For **hard ferromagnets** the loop is broad (Fig. B18.2(a)) and M is large when the applied field has been reduced to zero. Hard ferromagnets are used for permanent magnets. A **soft ferromagnet** has a narrower hysteresis loop and is therefore much more responsive to the applied field (Fig. B18.2(b)). Soft ferromagnets are used in transformers where they must respond to a rapidly oscillating field.

In an **antiferromagnetic** substance, neighboring spins are locked into an antiparallel alignment (Fig. B18.1(b)) and the sample has a low magnetic moment. Antiferromagnetism is often observed when a paramagnetic material is cooled to a low temperature and is signaled by a decrease in magnetic susceptibility with decreasing temperature (Fig. B18.3). The critical temperature for the onset of antiferromagnetism is called the **Néel temperature** (T_N). The spin coupling responsible for antiferromagnetism generally occurs through intervening ligands by a mechanism called **superexchange**. As indicated in Fig. B18.4, the spin on one metal atom induces a small spin polarization on an occupied orbital of a ligand and this results in alignment of spin on adjacent metal atom that is antiparallel to the first. Many compounds exhibit antiferromagnetic behavior in the appropriate temperature range; for example MnO is antiferromagnetic below $-151\,°C$, and Cr_2O_3 is antiferromagnetic below $37\,°C$. Coupling of spins through intervening ligands is frequently observed in molecular complexes containing two ligand-bridged metal ions.

In a third type of collective magnetic interaction, **ferrimagnetism**, net magnetic ordering is observed below the Curie temperature. However, ferrimagnets differ from ferromagnets because ions with different local moments are present. These ions order with opposed spins, as in antiferromagnetism, but because of the different magnitude of the individual spin moments there is incomplete cancellation and the sample has a net overall moment.

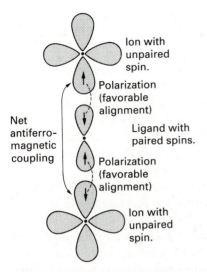

B18.4 Antiferromagnetic coupling between two metal centers created by spin polarization of a bridging ligand.

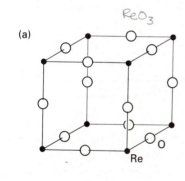

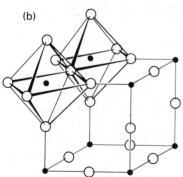

18.19 (a) The ReO_3 unit cell; (b) the ReO_3 unit cell plus some oxygen atoms from adjacent unit cells to show the corner shared octahedra.

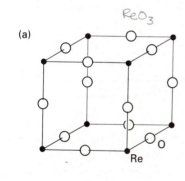

ReO₃

(a)

(b)

As with antiferromagnetism, these interactions are generally transmitted by the ligand.

[References: R. L. Carlin, *Magnetochemistry*, Springer-Verlag, Berlin, (1986); C. Kittel, *Introduction to solid state physics*, Wiley, New York (1986); J.B. Goodenough, *Magnetism and the chemical bond*, Interscience–Wiley, New York (1963).]

Example 18.3: *Predicting the structures of spinel compounds*

Is $MnCr_2O_4$ likely to have a normal or inverse spinel structure?

Answer. A normal spinel structure is expected because Cr^{3+} will have a ligand field stabilization energy in the octahedral site whereas the d^5 Mn^{2+} ion will not. Table 18.3 shows that this prediction is verified experimentally.

Exercise E18.3. Table 18.3 indicates that $FeCr_2O_4$ is a normal spinel. Rationalize this observation.

Perovskites and related phases

To introduce the structure of perovskites we first examine the related but simpler ReO_3 structure, which consists of corner-shared octahedra. Figure 18.19(a) shows that the unit cell of the ReO_3 structure contains M atoms at the corners of a cube and O atoms on the edges. The presence of the corner-shared octahedra is clarified by extending the structure to more than one unit cell (Fig. 18.19(b)). Note in particular that the ReO_3 structure is very open and that at the center it has a large hole with C.N. = 12.

The perovskites have the general formula ABX_3, in which the 12-coordinate hole of ReO_3 is occupied by a large A ion (Fig. 18.20; a different view of this structure was given in Fig. 4.18). The X ion is generally O^{2-} or F^-; perovskite itself is $CaTiO_3$ and an example of a fluoride perovskite is $NaFeF_3$. The structure is often observed to be distorted in such a manner that the unit cell is no longer centrosymmetric and the crystal has an overall permanent electric polarization. Some polar crystals are ferroelectric, for they resemble ferromagnets, but instead of a large number of spins being aligned over a region the electric polarizations of many unit cells are aligned. As a result the relative permittivities of a ferroelectric material often exceed 1×10^3 and can be as high as 15×10^3; for comparison the relative permittivity of liquid water is about 80 at room temperature.[4] Another characteristic of many crystals that lack a center of symmetry is **piezoelectricity**, the ability to generate an electrical field when the crystal is under stress or to change dimensions when an electrical field is applied to the crystal. Piezoelectric materials are used for a variety of applications such as pressure transducers, ultramicromanipulators, and sound

---

[4] For a general discussion of ferroelectrics see J. M. Herbert, *Chem. Br.*, **19**, 728 (1983).

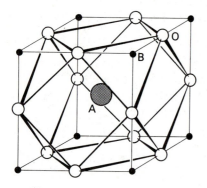

18.20 A view of the perovskite (ABO_3) structure emphasizing the dodecahedral coordination of the large A ion. The octahedral environment of the B ion has been shown in Fig. 18.19.

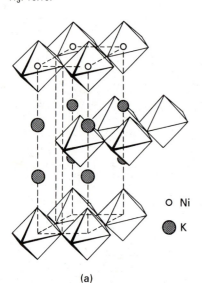

O Ni

○ K

(a)

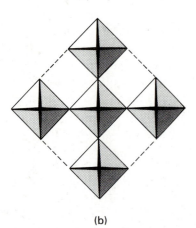

(b)

18.21 The K_2NiF_4 structure. (a) The displaced layers of NiF_6 octahedra and (b) a view of the NiF_6 octahedra showing the corner sharing of F.

detectors. Some important examples are $BaTiO_3$, $NaNbO_3$, $NaTaO_3$, and $KTaO_3$. Although a noncentrosymmetric structure is required for both ferroelectric and piezoelectric behavior, whereas all ferroelectric materials are piezoelectric, the opposite is not true. For example, quartz is piezoelectric but not ferroelectric. Quartz is widely used to set the clock rate of microprocessors and watches because a thin sliver oscillates at a specific frequency to produce a small oscillating electrical field.

Another prototypical structure, that of potassium tetrafluoronickelate(II) (K_2NiF_4, Fig. 18.21) is related to perovskite and has a bearing on the mechanism of high-temperature superconduction. The compound can be thought of as containing layers from the perovskite structure that share four F atom vertices within the layer and have terminal vertices above and below the layer. These layers are offset, so the terminal vertex in one layer is capped by a K atom in the next layer (Fig. 18.21). Compounds with the K_2NiF_4 structure have come under renewed investigation because some high-temperature superconductors, such as $La_{1.85}Sr_{0.15}CuO_4$, crystallize with this structure.[5] Aside from their importance in superconductivity, compounds with the K_2NiF_4 structure are also of interest because they provide an opportunity to investigate two-dimensional domains of magnetic interactions.

Superconductors

The versatility of the perovskites extends to superconductivity, because some of the high-temperature superconductors first reported in 1986 can be viewed as variants of the perovskite structure. Superconductors have two striking characteristics. Below the critical temperature (T_c) at which the superconducting state is formed, they have zero electrical resistance. They also exhibit the **Meissner effect**, the exclusion of a magnetic field (Fig. 18.22).

Following the discovery in 1911 that mercury is a superconductor below 4.2 K, physicists and chemists made slow but steady progress in the discovery of superconductors with higher values of T_c, at a rate of about 3 K per decade. After 75 years, T_c had been edged up to 23 K. Then, in 1986, **high-temperature superconductors** were discovered.[6] Several materials are now known with T_c well above 77 K, the boiling point of the inexpensive refrigerant liquid nitrogen, and in a few years T_c had increased by more than a factor of five.

Two types of superconductors are known.[7] Those classed as **Type I** exclude a magnetic field completely until a **critical field** H_c is reached, and they then lose their superconductivity abruptly. **Type II** superconductors, which include high-temperature materials, show a gradual magnetic field penetration above a lower critical field H_{c1} and a gradual loss of superconductivity. This superconductivity is lost completely

[5] Structural similarities among a wide range of oxygen-deficient perovskites are described by M. T. Anderson, J. T. Vaughey, and K. R. Poeppelmeier, *Chem. Mater.*, **5**, 151 (1993).

[6] See an article based on the Nobel Prize address by A. Müller and J. G. Bednorz, The discovery of a class of high-temperature superconductors, *Science*, **217**, 1133 (1987).

[7] For a good survey of the range of superconducting materials see J. Etourneau, in *Solid state chemistry, compounds* (ed. A. K. Cheetham and P. Day), Chapter 3, pp. 60–111, Oxford University Press (1992).

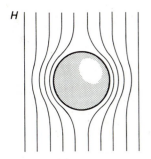

18.22 The exclusion of a magnetic field by a superconducting material below T_c (the Meissner effect).

above an upper critical field H_{c2}. Because type II materials withstand higher magnetic fields, superconducting magnets for NMR are usually made from materials of this type (such as Nb/Ti). The Chevrel phases (Section 18.8) have the highest observed values of H_c. Figure 18.23 shows that there is a degree of periodicity in the elements that exhibit superconductivity. Note in particular that the ferromagnetic metals iron, cobalt, and nickel do not display superconductivity; nor do the alkali metals or the coinage metals copper, silver, and gold. For simple metals, ferromagnetism and superconductivity never coexist, but in some of the oxide superconductors ferromagnetism and superconductivity appear to coexist on different sublattices of the solid.

One of the most widely studied oxide superconductor materials $YBa_2Cu_3O_7$ (informally called '123', from the proportions of metal atoms in the compound) has a structure similar to perovskite but with missing O atoms. In terms of the structure shown in Fig. 18.24, the A atoms of the original perovskite are yttrium and barium in $YBa_2Cu_3O_7$ and the B atoms are copper. The square-planar CuO_4 units are arranged in chains. Sheets and chains of CuO_4 or CuO_5 units can also be detected in other high-temperature superconductors, and it is thought that they are an important component of the mechanism of superconduction.

There is as yet no settled explanation of high-temperature superconductivity. It is believed that the Cooper pairs that are responsible for conventional superconductivity (Section 2.11) are important in the high-temperature materials but the mechanism for pairing is hotly debated.[8]

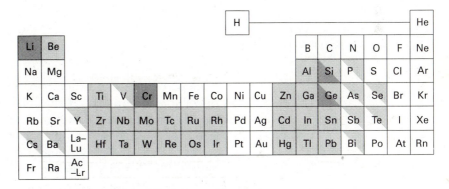

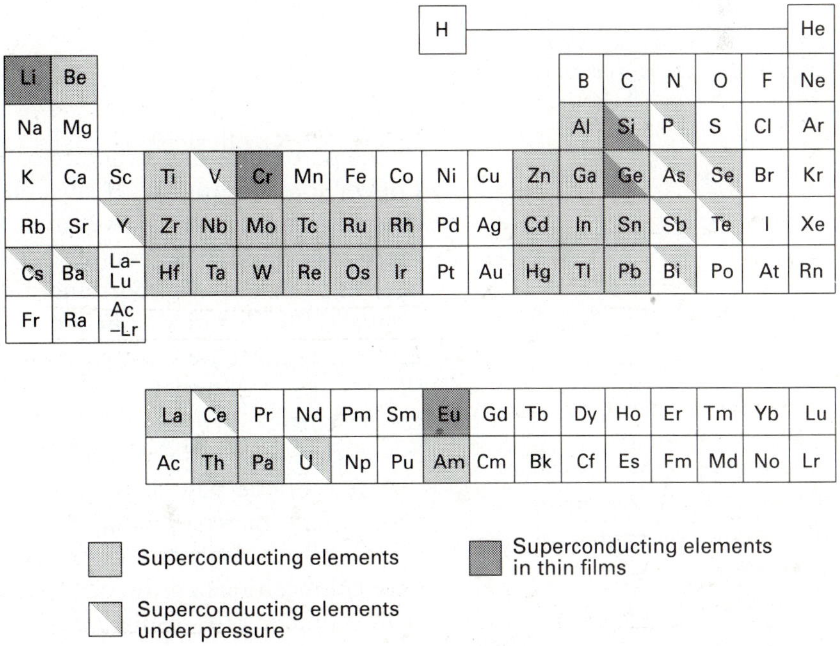

18.23 Superconducting elements (from J. Etourneau, Superconducting materials, Chapter 3 in *Solid state chemistry: compounds* (ed. A. K. Cheetham and P. Day), Oxford University Press (1992)).

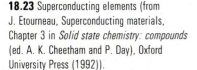

Superconducting elements

Superconducting elements in thin films

Superconducting elements under pressure

[8] For an entertaining account of the people investigating superconductivity see: R. M. Hazen, *The Breakthrough. The race for the superconductor*. Summit Books, New York (1988).

Table 18.4 Some superconductors

Elements	T_c/K	Compounds	T_c/K
Zn	0.88	Nb_3Ge	23.2
Cd	0.56	Nb_3Sn	18.0
Hg	4.15	$LiTiO_4$	13
Pb	7.19	$K_{0.4}Ba_{0.6}BiO_3$	29.8
Nb	9.50	$YBa_2Cu_3O_7$	95
		$Tl_2Ba_2Ca_2Cu_3O_{10}$	122

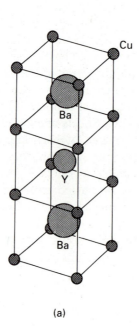

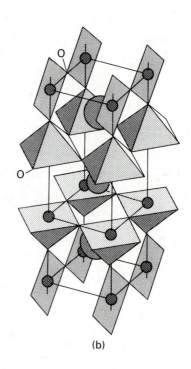

18.24 Structure of the $YBa_2Cu_3O_7$ ('123') superconductor. (a) Metal atom positions. (b) Oxygen polyhedra around the metal atoms, unlike the octahedral environment in perovskite metal ions in 123 are in square-planar and square-pyramidal coordination environments rather than the octahedra in perovskite.

(a)   (b)

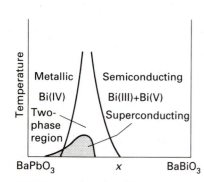

18.25 Schematic correlation of phase instability and superconductivity.

The perovskite-like structure of $YBa_2Cu_3O_7$ is shown in Fig. 18.24. Copper ions occupy the B sites and Y or Ba ions occupy the A sites of the layered lattice. However, unlike in a true perovskite structure, the B sites are not surrounded by an octahedron of O atoms. Some have five O atom neighbors and others have only four (Fig. 18.24 (b)). Similarly, Ba^{2+} and Y^{3+} in the A site have less than 12-coordination. Another point to note is that if we assign the usual oxidation numbers (Y $+3$, Ba $+2$, and O -2), then the average oxidation number of copper turns out to be $+2.33$, so it is inferred that $YBa_2Cu_3O_7$ is a mixed oxidation state material that contains Cu^{2+} and Cu^{3+}.

A few of the new superconductors are listed in Table 18.4 along with other superconducting materials. As illustrated in Fig. 18.25 for $BaPb_{1-x}Bi_xO_3$, superconductivity is often observed for materials that have compositions on the borderline between metallic conducting and semiconducting (or insulating) phases. A possible interpretation is that the types of processes of instability that lead to changes in the more ordinary conduction process may at low temperatures give rise to superconductivity.

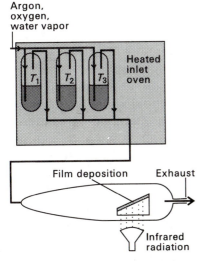

Argon, oxygen, water vapor

Heated inlet oven

T_1 T_2 T_3

Film deposition   Exhaust

Infrared radiation

18.26 Schematic diagram of the chemical vapor deposition of superconductor thin films. A reactive carrier gas mixture (Ar, O_2, and H_2O) is passed through traps of the volatile metal precursors held at temperatures T_1, T_2 and T_3 which provide the desired vapor pressure of each reactant. Deposition occurs on the wedge-shaped block, which is heated to a high temperature by an IR lamp. After the deposition, the resulting film is annealed at high temperature to improve the crystallinity of the film.

Me₃C

C — O

HC

C — O

F₇C₃

6 f od

The synthesis of high-temperature superconductors has been guided by a variety of qualitative considerations, such as the demonstrated success of the layered structures and of mixed oxidation state copper in combination with heavy *p*-block elements. Additional considerations are the radii of ions and their preference for certain coordination geometries. Some of these materials are prepared simply by heating a mixture of the metal oxides to 800–900 °C in an open alumina crucible or a sealed gold tube.

Thin films are needed for the use of superconductors in electronic devices. Their preparation is an active area of research and a promising method is **chemical vapor deposition**.[9] The general strategy is to form the thin film by decomposing a thermally unstable compound on a hot solid substrate material (Fig. 18.26). Fluorinated acetylacetonato complexes of the metals (using ligands such as fod, **6**) are sometimes used because they are more volatile than simple acetylacetonato complexes. Complexes such as $Cu(acac)_2$, $Y(dmpm)_3$, and $Ba(fod)_2$ are swept into the reaction chamber by slightly moist oxygen gas. When conditions are properly controlled this gaseous mixture reacts on the hot substrate to produce the desired $YBa_2Cu_3O_{7-\delta}$ film. The film may be amorphous and require subsequent heating to form a crystalline product.

18.6 Glasses

The term **ceramic** is often applied to all inorganic nonmetallic, non-molecular materials, including both amorphous and crystalline materials. The term **glass** is used in a variety of contexts, but for present purposes it implies a noncrystalline ceramic with a viscosity so high that it can be considered rigid. A substance in its glassy form is said to be in its **vitreous state**. Although ceramics and glasses have been utilized since antiquity, their development is currently an area of rapid scientific and technological progress. This enthusiasm stems from interest in the scientific basis of their properties and the development of novel synthetic routes to new high performance materials. We confine our attention here to glasses. The most familiar glasses are alkali metal or alkaline earth metal silicates or borosilicates.

A glass is prepared by cooling the melt more quickly than its rate of crystallization. Thus, cooling molten silica (quartz) gives vitreous silica. Under these conditions the solid mass has no long-range periodicity as judged by the lack of X-ray diffraction peaks but spectroscopic and other data indicate that each Si atom is surrounded by a tetrahedral array of O atoms (as in crystalline quartz). The lack of long-range order results from variations of the Si—O—Si angles. Figure 18.27 illustrates in two dimensions how a local coordination environment can be preserved but long-range order lost by variation of the bond angles around oxygen. Silicon dioxide readily forms a glass because the three-dimensional network of strong covalent Si—O bonds in the

[9] L. M. Tonge, D. S. Richeson, T. J. Marks, J. Zhao, J. Shang, B. W. Wessels, H. O. Marcy, and C. R. Kannewurf, *Advances in Chemistry Series (ACS)*, **226**, 351 (1990).

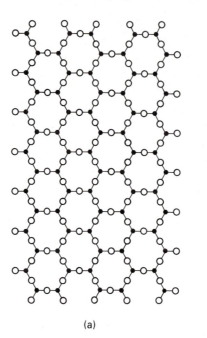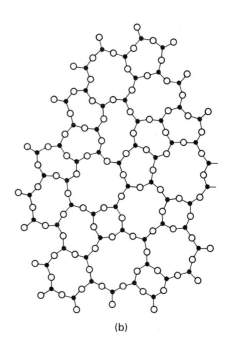

(a)   (b)

18.27 Schematic representation of (a) two-dimensional crystal and (b) two-dimensional glass. (Reproduced with permission from W. H. Zachariasen, *J. Am. Chem. Soc.*, **54**, 3841, (1932).)

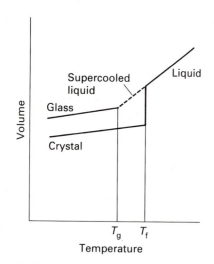

18.28 Comparison of the volume change for supercooled liquids and glasses with that for a crystalline material.

melt does not readily break and reform upon cooling. The lack of strong directional bonds in metals and simple ionic substances makes it much more difficult to form glasses from these materials. Recently, however, techniques have been developed for ultrafast cooling, and as a result a wide variety of metals and simple inorganic materials can now be frozen into a vitreous state.

The concept that the local coordination sphere of the glass-forming element is preserved but that bond angles around O are variable was originally proposed by the American crystallographer W. H. Zacharisen in 1932. He reasoned that these conditions would lead to similar free energies and molar volumes for the glass and its crystalline counterpart. Zacharisen also proposed that the vitreous state is favored by corner-shared O atoms, rather than edge or face shared, which would enforce greater order. These and other **Zacharisen rules** hold for common glass-forming oxides, but exceptions are known.

An instructive comparison between vitreous and crystalline materials is seen in their change in volume with temperature (Fig. 18.28). When a molten material crystallizes an abrupt change in volume (usually a decrease) occurs. In contrast, a glass-forming material which is cooled sufficiently rapidly persists in the liquid state to form a metastable supercooled liquid. When cooled below the **glass transition temperature**, T_g, the supercooled liquid becomes rigid, and this change is accompanied by only an inflection in the cooling curve.

Although vitreous silica is a strong glass that can withstand rapid cooling or heating without cracking, it has a high glass transition temperature and therefore must be worked at inconveniently high temperatures. Therefore a **modifier**, such as a basic oxide Na_2O or CaO, is generally added to SiO_2. These modifiers transfer O^{2-} ions to the Lewis acid Si(IV). This transfer disrupts some of the Si—O—Si

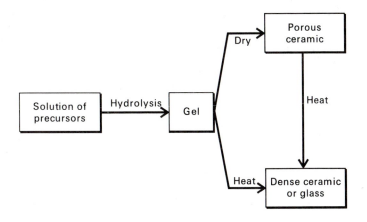

7

linkages and replace them by terminal $Si—O^-$ links that associate with the cation (**7**). The consequent partial disruption of the $Si—O$ network leads to glasses that have lower softening points. The common glass used in bottles and windows is called 'sodalime glass' and contains Na_2O and CaO as modifiers. When B_2O_3 is used as a modifier, the resulting 'borosilicate glasses' have lower thermal expansion coefficients than sodalime glass and thus are less likely to crack when heated. Borosilicate glass is therefore widely used for ovenware and laboratory glassware.

Glass formation is a property of many oxides, and practical glasses have been made from sulfides, fluorides, and other anionic constituents. Some of the best glass formers are the oxides of elements near silicon in the periodic table (B_2O_3, GeO_2, and P_2O_5), but the solubility in water of most borate and phosphate glasses and the high cost of germanium limit their usefulness.

A current technological revolution is the replacement of electrical signals for voice and data transmission by optical signals. Similarly, optical circuit elements are being developed that may eventually replace electrical integrated circuits. These developments are stimulating considerable research in transparent crystalline and vitreous materials for light transmission and processing. For example, optical fibers for light transmission are currently being produced with a composition gradient from the interior to the surface. This composition gradient modifies the index of refraction and thereby decreases light loss. Fluoride glasses are also being investigated as possible substitutes for oxide glasses, because oxide glasses contain small numbers of OH groups, which absorb near-infrared radiation.

Most glasses are made simply by melting the component oxides and cooling faster than the rate of crystallization, which is very slow for many complex metal silicates, phosphates, and borates. Another route to glasses is the **sol–gel process**, which is described schematically in Fig. 18.29.[10] As shown in Fig. 18.29, the sol–gel process is also used to produce crystalline ceramic materials and high surface area com-

18.29 Schematic diagram of the sol–gel process. When the gel is dried at high temperatures dense ceramics or glasses are formed. Drying at low temperatures above the critical pressure of water produces microporous solids known as xerogels or aerogels.

[10] Present status and future potential of the sol–gel process, R. C. Mehrotra, *Structure and bonding*, **77**, 1 (1992); R. Roy, *Science*, **238**, 1665 (1987).

pounds such as silica gel. A typical process involves the addition of a metal alkoxide precursor to an alcohol (the *sol*ution part of the process), followed by the addition of water to hydrolyze the reactants. This hydrolysis leads to a thick gel which can be dehydrated and **sintered** (heated below its melting point to produce a compact solid). For example, ceramics containing TiO_2 and Al_2O_3 can be prepared in this way at a much lower temperatures than required to produce the ceramic from the simple oxides. Often a special shape can be fashioned at the gel stage. Thus, the gel may be shaped into a fiber and then heated to expel water and produce a glass or ceramic fiber at much lower temperatures than would be necessary if the fiber were made from the melt of the components.

PROTOTYPICAL SULFIDES AND RELATED COMPOUNDS

The soft chalcogens sulfur, selenium, and tellurium form binary compounds with metals that often have quite different structures from the corresponding oxides. As we saw in Section 4.6, this difference is consistent with the greater covalency of the compounds of sulfur and its heavier congeners. For example, we noted there that MO compounds generally adopt the rock-salt structure whereas ZnS and CdS adopt the sphalerite or the wurtzite structures. Similarly, the *d*-block monosulfides generally adopt the more characteristically covalent nickel arsenide structure rather than the rock-salt structure of alkaline earth oxides such as MgO. Even more striking are the layered MS_2 compounds of many *d*-block elements in contrast to the fluorite or rutile structures of many *d*-block dioxides (Section 12.9).

Before we discuss these compounds, we should note that there are many metal-rich compounds that disobey simple valence rules and do not conform to an ionic model. Some examples are Ti_2S, Pd_4S, V_2O, and Fe_3N and even the alkali metal suboxides, such as Cs_3O (Section 12.9).[11] The occurrence of metal-rich phases is generally associated with the formation of M—M bonds (Section 8.9). Many other intermetallic compounds display stoichiometries that cannot be understood in terms of conventional valence rules. Included among these are the important permanent magnet materials $Nd_2Fe_{17}B$ and $SmCo_5$.

18.7 Layered MS_2 compounds and intercalation

We introduced the layered sulfides and their intercalation compounds in Section 12.10. Here we develop a broader picture of their structures and properties.

Synthesis and crystal growth

The chalcogens react with most metals at or above room temperature, and a wide range of compounds are known. The preparation of crystalline disulfides suitable for chemical and structural studies is often

---

[11] A discussion of the structure and bonding in metal-rich compounds is given by H. F. Franzen, *Prog. Solid State Chem.*, **12**, 1 (1978).

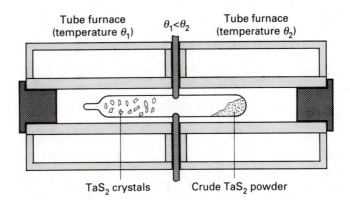

18.30 Vapor transport crystal growth and purification of TaS$_2$. A small quantity of I$_2$ is present to serve as a transport agent.

Tube furnace (temperature θ_1)

$\theta_1 < \theta_2$

Tube furnace (temperature θ_2)

TaS$_2$ crystals

Crude TaS$_2$ powder

performed by **chemical vapor transport**, which is a generally useful technique in solid state chemistry. It is possible sometimes simply to sublime a compound, but the vapor transport technique can also be applied to a wide variety of nonvolatile compounds.

Typically, the crude material is loaded into one end of a borosilicate or fused quartz tube. After evacuation, a small pressure of a chemical transport agent is introduced, the tube is sealed, and placed in a furnace with a temperature gradient. The polycrystalline and possibly impure metal chalcogenide is vaporized at one end and redeposited as pure crystals at the other (Fig. 18.30). The technique is called **chemical vapor transport** rather than sublimation because the chemical transport agent, which may be a halogen, produces an intermediate volatile species such as a metal halide. Generally only a small amount of transport agent is needed because upon crystal formation it is released and diffuses back to pick up more reactant. For example, TaS$_2$ can be transported with I$_2$ in a thermal gradient. At 850 °C the reaction with I$_2$ to produce gaseous products is endothermic:

$$TaS_2(s) + 2I_2(g) \rightarrow TaI_4(g) + S_2(g)$$

Because the reaction is endothermic, its equilibrium constant increases with increasing temperature. Therefore, the partial pressure of volatile products is greater at 850 °C than at 750 °C, so TaS$_2$ is deposited at the lower temperature. If, as occasionally is the case, the transport reaction is exothermic, the solid is carried from the cooler to the hotter end of the tube.

Structure

As we saw in Section 12.10, the *d*-block disulfides fall into two classes. There are the layered materials formed by metals on the left of the *d*-block, and in the middle and toward the right of the block there are compounds containing formal S$_2^{2-}$ ions, such as pyrite (FeS$_2$). We shall concentrate on the layered materials.

In TaS$_2$ and many other layered disulfides the *d*-metal ion is located in octahedral holes between close packed AB layers (Fig. 18.31(a)).[12]

---

[12] Recall the notation ABC for stacks of close-packed layers in Section 4.2.

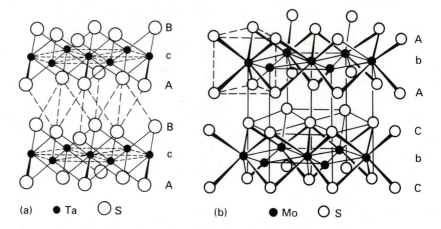

(a) ● Ta ○ S   (b) ● Mo ○ S

18.31 (a) The structure of TaS_2 (CdI_2-type). The Ta atoms reside in octahedral, c, sites between the AB layers of sulfur. (b) The MoS_2 structure; the Mo atoms reside in trigonal prismatic sites between AA or CC sulfide layers.

The Ta ions form a close packed-type layer denoted c, and so the metal and adjoining sulfide layers can be portrayed as an AcB sandwich. These AcB sandwich-like slabs form a three dimensional crystal by stacking in sequences such as AcB AcB AcB etc., where the strongly bound AcB slabs are held to their neighbors by weak dispersion forces. The Mo atoms in MoS_2 reside in the trigonal prismatic holes between sulfide layers that are in register with one another (AA, Fig. 18.31(b)). The Mo atoms, which are strongly bonded to the adjacent sulfide layers, form a close-packed array, and so we can represent each slab as AbA or CbC. These slabs form a three-dimensional crystal by stacking in a pattern AbA CbC AbA CbC etc. Weak dispersion forces are also responsible for holding these AbA and CbC slabs together.

Polytypes (Section 4.2, structures that differ only in the stacking arrangement along a direction perpendicular to the plane of the slabs) also occur. Thus MoS_2 forms several polytypes including one with the sequence AbA CbC and another with the sequence AbA BcB CaC. In addition to variations in stacking sequences, the structures of many *d*-block dichalcogenides show subtle periodic distortions within the individual slabs. These distortions are called **charge density waves** and they will be described after we have established some background information on the band model of metal dichalcogenides.

In the discussion above it was convenient to describe the layered structures in terms of cations and anions. Now we shall use a band model, for then it is easier to account for their significant covalency, electrical conductivities, and chemical properties. Some approximate band structures derived primarily from molecular orbital calculations and photoelectron spectra are shown in Fig. 18.32. They show that the dichalcogenides with octahedral and trigonal prismatic metal sites have low lying bands primarily composed of chalcogen *s* and *p* orbitals, higher energy bands derived primarily from metal *d* orbitals, and above these metal and chalcogen bands.

When the metal atom resides in an octahedral site, its *d* orbitals are split into a lower t_{2g} set and a higher e_g set, just as in localized complexes. These atomic orbitals combine to give a t_{2g} band and an e_g band. The broad t_{2g} band may accommodate up to six electrons per

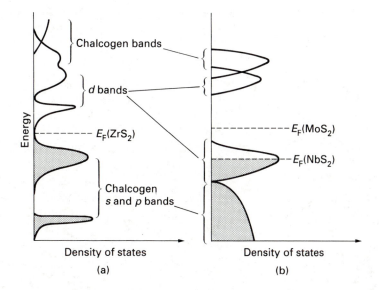

18.32 Approximate band structures of dichalcogenides. (a) Octahedral MS$_2$ compounds; for ZrS$_2$ (which is d^0 and is illustrated), only the sulfur s and p bands are full and the compound is a semiconductor. TaS$_2$ (d^1) has a partially filled d-band and is metallic. (b) Trigonal prismatic MS$_2$ compounds. NbS$_2$ (d^1, as illustrated) is metallic whereas MoS$_2$ (d^2) is a semiconductor.

metal atom. Thus TaS$_2$, with one d electron per metal, has an only partly filled t_{2g} band, and is a metallic conductor.

A trigonal prismatic ligand field leads to a low energy a'_1 level and two doubly degenerate higher energy orbitals designated e' and e'' respectively. Now the lowest band needs only two electrons per atom to fill it. Thus MoS$_2$, with trigonal prismatic Mo sites and two d electrons, has a filled a'_1 band and is a semiconductor.

The formation of charge density waves in some metal dichalcogenides is similar to the onset of the Peierls distortion of one-dimensional structures (Section 2.9). For a partly filled conduction band in one- and two-dimensional systems, a distortion of the lattice will split the band so that the resulting lower component may be filled with electrons and the upper component emptied. Since the energy advantage of this distortion is not large, a static modulation of the structure is usually only evident at low temperatures. Charge density waves are observed at low temperatures in the Group 5 dichalcogenides VS$_2$, NbS$_2$, and TaS$_2$. In these compounds four of the five electrons from each metal are required to fill the s and p band formed from the sulfur orbitals, and the remaining electron occupies but does not fill the d-band. At sufficiently low temperatures the lattice distorts and the d-band splits.

Intercalation and insertion

We have already introduced the idea that alkali metal ions may insert between graphite sheets (Section 11.6) and metal disulfide slabs (Section 12.11) to form intercalation compounds. For a reaction to qualify as an intercalation, or as an **insertion reaction**, the basic structure of the host should not be altered when it occurs.[13] Insertion

---

[13] Reactions in which the structure of one of the solid starting materials is not radically altered are called 'topotactic' reactions. They are not limited to the type of insertion chemistry we are discussing here. For example, hydration, dehydration, and ion exchange reactions may be topotactic.

Table 18.5 Some alkali metal intercalation compounds

Compound	$\Delta d/\text{Å}$†
$KZrS_2$	1.60
$NaTiS_2$	1.17
$KTiS_2$	1.92
$RbSnS_2$	2.24
$KSnS_2$	2.67
$Na_{0.6}MoS_2$ (2H)	1.35
$K_{0.4}MoS_2$ (2H)	2.14
$Rb_{0.3}MoS_2$ (2H)	2.45
$Cs_{0.3}MoS_2$ (2H)	3.66

†Change in interlayer spacing in comparison with the parent MS_2. (2H signifies a particular layer sequence.)

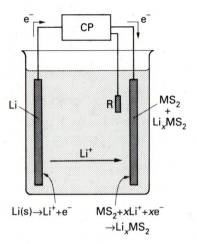

18.33 Experimental arrangement for electrointercalation. A polar organic solvent (e.g. propylene carbonate) containing a lithium salt is used as the electrolyte. R is a reference electrode and CP is a coulometer (to measure the charge passed) and a potential controller.

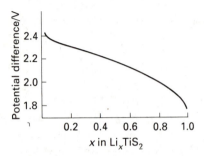

18.34 Potential versus composition diagram for the electrointercalation of lithium into titanium disulfide. The composition, x in Li_xTiS_2 is inferred from the charge passed in the course of electrointercalation.

compound formation is the basis of a rechargeable battery technology. These 'high energy cells', consist of a lithium anode, a chalcogenide cathode, and an electrolyte dissolved in an inert polar organic solvent. Such a cell can produce well over 2 V and store over twice the energy per unit volume of a Ni/Cd rechargeable cell. The recharging of such cells is fast because no solid phase transition is involved.

The π conduction and valence bands of graphite are contiguous in energy (we have seen in fact that graphite is formally a semimetal, Section 2.9) and the favorable free energy for intercalation arises from the transfer of an electron from the alkali metal atom to the graphite conduction band. The insertion of an alkali metal atom into a dichalcogenide involves a similar process. Now, though, the electron is accepted into the d band. In both graphite and TiS_2 the charge-compensating alkali metal ion diffuses to positions between the slabs. Some representative alkali metal insertion compounds are listed in Table 18.5.

The insertion of alkali metal ions into host lattices can be achieved by direct combination of the alkali metal and the disulfide:

$$TaS_2 + x Na \xrightarrow{800\,°C} Na_xTaS_2 \qquad x = 0.4 \text{ to } 0.7$$

or by utilizing a highly reducing alkali metal compound or the electrochemical technique of **electrointercalation** (Fig. 18.33). One advantage of electrointercalation is that it is possible to measure the amount of alkali metal incorporated by monitoring the amount of electrons (It/F) passed during the synthesis. It also is possible to distinguish solid solution formation from discrete phase formation. As illustrated in Fig. 18.34, the formation of a solid solution is characterized by a gradual change in potential as intercalation proceeds, whereas the formation of a new discrete phase yields a steady potential over the range in which one solid phase is being converted into the other, followed by an abrupt change in potential when that reaction is complete.

Insertion compounds are examples of mixed ionic and electronic conductors. In general the insertion process can be reversed either chemically or electrochemically. This makes it possible to recharge a lithium cell by removal of lithium from the compound. In a clever synthetic application of these concepts, the previously unknown layered disulfide VS_2 was prepared by first making the known layered compound $LiVS_2$ in a high temperature process. The lithium was then removed by reaction with I_2 at lower temperatures to produce the metastable layered VS_2, which was found to have the TiS_2 structure:

$$2LiVS_2 + I_2 \rightarrow 2LiI + 2VS_2$$

Insertion compounds also can be formed with molecular guests. Perhaps the most interesting are the metallocenes $Co(\eta^5\text{-}C_5H_5)_2$ and $Cr(\eta^5\text{-}C_5H_5)_2$ which are incorporated into a variety of hosts, such as TiS_2, $TiSe_2$, and TaS_2, to the extent of about 0.25 $M(\eta^5\text{-}C_5H_5)_2$ per MS_2 or MSe_2. This limit appears to correspond to the space available for forming a complete layer of $M(\eta^5\text{-}C_5H_5)_2^+$. The organometallic

Table 18.6 Some three-dimensional intercalation compounds

Phase	Composition (x)
$Li_x[Mo_6S_8]$	2.4 to 0.6
$Na_x[Mo_6S_8]$	3.6
$Ni_x[Mo_6Se_8]$	1.8
H_xWO_3	0.6
Li_xWO_3	0.6
Li_xNiPS_3	0 to 1.5

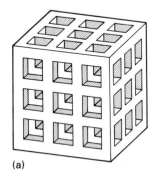

(a)

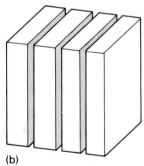

(b)

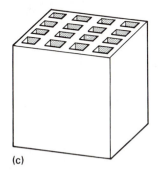

(c)

18.35 Schematic representation of host materials for intercalation reactions. (a) A three-dimensional host with intersecting channels, (b) a two-dimensional layered compound, and (c) one-dimensional channels.

compound appears to undergo oxidation upon intercalation, so the favorable free energy in these reactions arises in the same way as in alkali metal intercalation. In agreement with this interpretation, $Fe(\eta^5-C_5H_5)_2$, which is harder to oxidize than its chromium or cobalt analogs, does not intercalate.

We can imagine the insertion of ions into one-dimensional channels, two-dimensional planes of the type we have been discussing, or channels that intersect to form three dimensional networks (Fig. 18.35). Aside from the availability of a site for a guest to enter, the host must provide a conduction band of suitable energy to take up electrons reversibly (or, in some cases, be able to donate electrons to the host). Table 18.6 illustrates that a wide variety of hosts are possible, including metal oxides and various ternary and quaternary compounds. We see that intercalation chemistry is by no means limited to graphite and layered disulfides.

18.8 Chevrel phases[14]

We close this chapter with a discussion of an interesting class of ternary compounds first reported by the French solid state inorganic chemist R. Chevrel and his collaborators in 1971. These compounds, which illustrate three-dimensional intercalation, have formulas such as Mo_6S_8 and $M_xMo_6S_8$; Se or Te may take the place of S and the intercalated M atom may be a variety of metals such as Li, Mn, Fe, Cd, and Pb. The parent compounds Mo_6Se_8 and Mo_6Te_8 are prepared by heating the elements in sealed silica glass ampoules at about 1000 °C. A structural unit common to this series is M_6S_8, which may be viewed as an octahedron of M atoms face bridged by S atoms, or alternatively as an octahedron of M atoms in a cube of S atoms (Fig. 18.36). This type of cluster is also observed for some halides of the Period 4 and 5 early d-block elements, such as the $[M_6X_8]^{4+}$ cluster found in molybdenum and tungsten dichlorides, bromides, and iodides.

Figure 18.37 shows that in the three-dimensional solid the Mo_6S_8 clusters are canted relative to each other and relative to the sites occupied by intercalated ions. The Mo_6S_8 appear to be canted to allow secondary donor acceptor interaction between vacant Mo d_{z^2} orbitals (which project outward from the faces of the Mo_6S_8 cube) and a filled donor orbital on S atoms on of adjacent clusters (Fig. 18.38).

A simple molecular orbital treatment of the Mo_6S_8 cluster indicates a capacity for 20 metal cluster valence electrons to fill the bonding d-orbitals and another four to fill a nonbonding e_g level. Six Mo atoms have a total of 36 valence electrons (six electrons per Mo atom) and the formation of eight S^{2-} ions will consume 16 of them; hence compounds of composition Mo_6S_8 have 20 cluster electrons, just filling the bonding orbitals. Metal atom guests provide additional electrons that can be accommodated in the nonbonding e_g orbital. In the infinite solid the interaction between these clusters leads to band formation

[14] Ø. Fisher, *Appl. Phys. A*, **16**, 1 (1978); T. Hughbanks and R. Hoffman, *J. Am. Chem. Soc.*, **105**, 1150 (1983); R. Schöllhorn, *Angew. Chem., Int. Ed. Engl.*, **19**, 983 (1980).

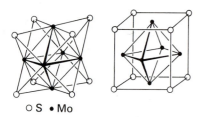

○ S • Mo

18.36 Two views of the Mo_6S_8 unit in a Chevrel compound.

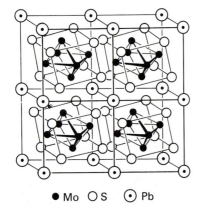

● Mo ○ S   ⊙ Pb

18.37 Structure of a Chevrel compound showing the canted Mo_6S_8 unit in the slightly distorted cube of Pb atoms.

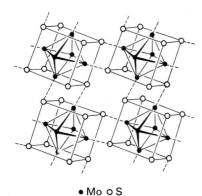

• Mo ○ S

18.38 Explanation for the tilt of M_6S_8 units in a Chevrel phase in terms of donor-acceptor interaction between M in one cube with S in the next.

18.39 Potential versus amount of Li intercalated into Mo_6S_8 as determined by electrointercalation.

from the orbitals we have just discussed, but the general picture in so far as electron counting is concerned does not change.

The electrointercalation of lithium into Mo_6S_8 (Fig. 18.39) nicely illustrates that three phases occur as the lithium content is changed. The two plateaus indicate the formation of discrete compounds, $Li_{0.8}Mo_6S_6$ and $Li_{2.4}Mo_6S_8$, and the broad sloping region indicates a nonstoichiometric phase, $Li_xMo_6S_8$ ($2.4 < x < 3.6$). The explanation of the formation of these individual phases is beyond the scope of simple theory but the upper limit of intercalation ($x = 3.6$) agrees with theory in that it does not exceed the capacity of four electrons in the nonbonding e_g band.

Example 18.4: *Determining the nature of a mixed metal Chevrel phase*

Chevrel and his coworkers prepared a compound $Mo_4Ru_2Se_8$. Assuming the frontier orbitals are similar to those in Mo_6S_8, determine the capacity of this mixed-metal Chevrel phase to intercalate sodium.

Answer. We may compare the number of metal valence electrons available in the mixed metal system with the maximum of 24 for the Mo_6 Chevrel compounds. Four Mo atoms give 24 electrons and two Ru atoms give 16. Removing 16 for the formation of eight S^{2-} ions yields 24 electrons for the cluster overall, and we predict that it should not intercalate sodium. Thus, if we were simply trying to make another intercalated Chevrel phase, $Mo_4Ru_2Se_8$ would not appear to be a good choice. On the other hand, if we are curious about the outcome, we might try the intercalation reaction because the uncertainty of approximate electronic structure arguments combined with the subtle balance of energies in many reactions leaves the door open for new discoveries. (In fact, the experiment has been tried, and electrointercalation of Li has been achieved. Presumably the electrons populate antibonding orbitals.)

Exercise E18.4. Should $Mo_4Ru_2Se_8$ be expected to exhibit metallic conduction or semiconduction?

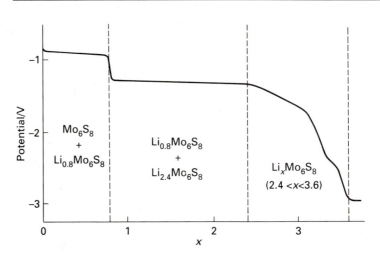

One of the physical properties that has drawn attention to the Chevrel phases is their superconductivity. Superconductivity persists up to 14 K in $PbMo_6S_8$ and it also persists to very high magnetic fields, which is of considerable practical interest because many applications involve high fields. In this respect the Chevrel phases may be superior to the newer oxide-based high-temperature superconductors.

FURTHER READING

The first three books provide good general overviews of inorganic solid state chemistry:

A. R. West, *Solid state chemistry and its applications*. Wiley, New York (1984).

C. N. R. Rao (ed.), *Solid state chemistry*. Dekker, New York (1974).

C. N. R. Rao and J. Gopalakrishnan, *New directions in solid state chemistry*. Cambridge University Press (1986).

A. K. Cheetham and P. Day (ed.), *Solid state chemistry techniques*. Oxford University Press (1987). This book includes an informative chapter on the preparation of solids; otherwise it is devoted to techniques for the characterization of solids.

A. K. Cheetham and P. Day (ed.), *Solid state chemistry compounds*. Oxford University Press (1992). This book provides very readable discussions of electronic structures of solids, chain compounds, superconductors, metal-rich compounds, heterogeneous catalysts, intercalation, zeolites, and ferroelectrics.

A. Wold and K. Dwight, *Solid state chemistry*. Chapman and Hall, London (1993). A good introduction to the structures, synthesis, and characterization of some important oxides and sulfides.

P. A. Cox, *The electronic structure and chemistry of solids*. Oxford University Press (1987).

J. K. Burdett, *Prog. Solid State Chem.*, **15**, 173 (1984). This very accessible review article discusses the relationships between electronic structures and properties of solids.

A. F. Wells, *Structural inorganic chemistry*. Oxford University Press (1984). This large book is an excellent resource for structural information and structural correlations.

B. G. Hyde and S. Andersson, *Inorganic crystal structures*. Wiley, New York (1989). Hyde and Andersson provides a system for understanding complex inorganic crystal structures and apply it to many important structures.

D. M. Adams, *Inorganic solids*. Wiley–Interscience, New York (1984). Adams's more concise book provides a good introduction to the structures of prototypical solids.

U. Müller, *Inorganic structure chemistry*. Wiley, Chichester (1993). A compact and well illustrated text mainly devoted to solids.

The next two books provide good coverage of the topics indicated by their titles:

R. J. D. Tilley, *Defect crystal chemistry and its applications*. Blackie, London and Chapman and Hall, New York (1987).

D. B. Brown (ed.), *Mixed-valence compounds*. Riedel, New York (1981).

R. E. Newnham, *Structure–property relations*, Springer-Verlag, Berlin (1975). This is an accessible volume stressing correlations between structures and properties such as electron transport, ion transport, and ferroelectric and magnetic behavior.

P. W. Atkins, *Physical chemistry*. Oxford University Press and W. H. Freeman & Co., New York (1994). See Chapter 8 for a discussion of the phase rule.

P. G. Bergeron and S. H. Risbud, *Introduction to phase equilibria in ceramics*. The American Ceramics Society, Columbus (1984). A large and very useful collection of phase diagrams is given in *Phase diagrams for ceramists*, which is a continuing series published by The American Ceramics Society.

KEY POINTS

Defects
Point defects in structure are found in solids and may influence properties such as ionic migration and color. Two common point defects are the Schottky defect, which is a *vacancy* created by the displacement of an atom from its normal site to the surface of the crystal and the Frenkel defect, which is the displacement of an atom from its normal site to an interstitial site in the crystal.

Nonstoichiometric compounds
Nonstoichiometric compounds deviate from simple whole-number ratios for the constituent atoms in the compound. They are common for the hydrogen compounds of early *d*-block metals and for oxides of metals that can readily adopt more than one oxidation state (such as iron).

Solid electrolytes
In solid electrolytes one or more types of ion diffuses at an appreciable rate. Examples are Ag_2HgI_4 above 50 °C and sodium β-alumina where Ag^+ and Na^+, respectively, are mobile.

Cooperative magnetism
Ferromagnetism and antiferromagnetism involve the long range order of unpaired electrons on the individual ions.

Superconductivity
Superconductivity is observed for many metals and some compounds of nonmetals. High temperature superconductors are metal oxides that have structures related to that of perovskite.

Glasses
The vitreous phase of materials usually contain local order (e.g. four O atoms around a Si atom) but no long range order.

Metal sulfides
The sulfides of some metals exist in layered structures and some of them (TaS_2, for instance) can accommodate ions or molecules between the layers by the process of intercalation or insertion. Intercalation can be performed chemically or electrochemically.

EXERCISES

18.1 Describe the nature of Frenkel and Schottky defects, and some experimental tests for their occurrence. Do either of these defects by themselves give rise to nonstoichiometry?

18.2 From each of the pairs of compounds (a) NaCl and NiO, (b) CaF_2 and PbF_2, and (c) Al_2O_3 and Fe_2O_3, pick the one most likely to have a high concentration of defects. Describe the possible defects and give your reasoning.

18.3 Distinguish intrinsic from extrinsic defects and give an example of each in actual compounds.

18.4 Draw one unit cell in the ReO_3 structure showing the M and O atoms. Does this structure appear to be sufficiently open to undergo Na^+ ion intercalation? If so, where might the Na^+ ions reside?

18.5 Is a crystallographic shear plane a defect, a way of describing a new structure, or both? Explain your answer.

18.6 Sketch the rock-salt crystal structure and locate in the sketch the interstitial sites to which a cation might migrate. What is the nature of the bottleneck between the normal and interstitial sites?

18.7 Contrast the nature of the defects in TiO with those in FeO.

18.8 How might you distinguish experimentally the existence of a solid solution from a series of crystallographic shear plane structures for a material which appears to have variable composition?

18.9 Label the regions in the iron–oxygen phase diagram

(Fig. 18.12) in the region 27 to 30 percent oxygen, and state the general principles that you used in assigning the labels.

18.10 Bearing in mind that the redox chemistry of metal oxides often can be correlated with the redox behavior of the ions in solution, explain which of TiO, MnO, and NiO is (a) the most difficult to oxidize with air, (b) the easiest to reduce.

18.11 Describe the electrical conductivity of TiO and NiO and describe the interpretation in terms of the trends in electronic structure across the $3d$ series.

18.12 Above $146\,°C$, AgI is an Ag^+ ion conductor. Suppose a pellet of this compound is sandwiched between two silver electrodes at $165\,°C$ and a potential difference of 0.1 V is imposed between the electrodes. Sketch the apparatus and indicate the changes that occur with time at the two electrodes.

18.13 What will be the quantitative changes in the mass of the cathode, the AgI, and the anode after 10^{-3} mol of electrons have been passed through the cell described in Exercise 18.12.

18.14 Contrast the mechanism of interaction between unpaired spins in a ferromagnetic material with that for an antiferromagnet.

18.15 Magnetic measurements on the ferrite $CoFe_2O_4$ indicate 3.4 spins per formula unit. Suggest a distribution of cations between octahedral and tetrahedral sites that would satisfy this observation.

18.16 The superconducting compound $YBa_2Cu_3O_7$ is described as having a perovskite-like structure. Ignoring problems of charge on the ions, describe the difference between this structure and that of perovskite.

PROBLEMS

18.1 Given the ligand-field Δ_O and Δ_T values that follow, determine the site preference for $A = Ni^{2+}$ and $B = Fe^{3+}$ in normal compared with inverse spinel, assuming that ligand field stabilization is dominant. Δ_O: Fe^{3+} ($1400\ cm^{-1}$), Ni^{2+} ($860\ cm^{-1}$); Δ_T: Fe^{3+} ($620\ cm^{-1}$), Ni^{2+} ($380\ cm^{-1}$)

18.2 Magnetite, Fe_3O_4, has the spinel structure and is thus a member of the ferrite group of compounds. In a normal spinel unit cell eight tetrahedral holes would be filled with Fe^{2+} and 16 octahedral holes would contain Fe^{3+}. At room temperature magnetite has high conductivity (about $2 \times 10^2\ S\ cm^{-1}$). Are the conductivity data consistent with the normal structure? If not, propose one possible variant of the spinel structure that would lead to high conductivity.

19

Bioinorganic chemistry

In contrast to the simpler and more static mineral world, living matter is based on much more intricate structures (such as cells) that persist in states that are far from equilibrium by maintaining a steady flow of nutrients and energy. The biological role of inorganic compounds and ions is often associated with the management of these flows and with the creation of structure. To put bioinorganic chemistry in its broad context, we start by comparing the distribution of the elements in the lithosphere and biosphere. Then we explore the variety of roles metal ions may play in biology. In most cases, a detailed understanding of their role has been achieved in only a few key steps, and intelligent speculation plays a more prominent role here than in other areas of inorganic chemistry. Most of this discussion hinges on the function of metal ions in specific steps. In this discussion we encounter interesting coordination environments for metal ions and common inorganic reaction types introduced earlier in the text, including Brønsted and Lewis acid–base reactions, redox reactions that depend on electron and group transfer, and photochemically induced electron transfer. The main focus of the discussion will be on biochemical catalysts, the enzymes, many of which employ a metal atom as a key component of the active site. In each case we meet a biochemical step that is reasonably well understood and some inorganic model systems that behave in ways that shed light on the biochemical step.

The exchange of ideas between inorganic chemistry and biochemistry runs in both directions. Just as we can bring our knowledge of inorganic reactions to bear on biochemical processes, the structures that Nature has devised have stimulated the synthesis and characterization of new inorganic compounds, some of which will be described in this chapter.

Nature does not employ the elements in living systems in the same order as their abundance in the Earth's crust. For example, Table 19.1

Table 19.1 Descending mass abundance of the elements

Earth		Humans
Crust	Oceans	
O	O	H
Si	H	O
Al	Cl	C
Fe	Na	N
Ca	Mg	Na
Mg	Ca	K
Na	K	Ca
K	C	Mg
Ti	Br	P
H	B	S
B	Sr	Cl

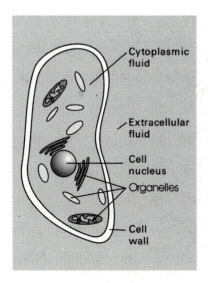

19.1 Schematic diagram of a cell immersed in its extracellular fluid.

shows a high crustal abundance for aluminum and silicon, but these elements are not abundant in humans, or in mammals in general. Some elements occur in trace quantities terrestrially but play a central role in organisms' fight to avoid equilibrium. Copper and selenium are two examples. Although some elements of low terrestrial abundance, such as beryllium, thallium, and uranium, are not known to be used in biochemical functions, new discoveries are constantly being made in Nature's utilization of the elements and one day they too may be found to be important.

It is helpful to classify the functions of the elements in relation to their biological environment (Fig. 19.1):

1. In the extracellular fluids and on the outer face of cell membranes.
2. In cell walls, the membranes that define the boundaries of cells and control the entry and exit of material.
3. Inside cells, starting from the inner surface of the membrane in the aqueous phase known as the cytoplasmic fluid.
4. In the organelles inside cells, including nuclei and mitochondria (the localized structures in a cell commonly identified with its source of power).

The identification of the location and the host environment of the elements gives the first clue to the role they play in biological systems. Some examples are given in Table 19.2, where it will be noted that cells typically incorporate potassium, magnesium, and iron, but discriminate against sodium and calcium.

Table 19.3 is a survey of biomolecules that utilize metal ions; many of these molecules are proteins. Indeed, about 30 percent of enzymes (biological catalysts) are **metalloenzymes**, or enzymes that have a metal atom at the active site. Metalloenzymes are involved in acid-catalyzed hydrolysis (hydrolases), redox reactions (oxidases and oxygenases), processes that rearrange carbon–carbon bonds (synthases and isomerases). A major theme of this chapter will be the typical mechanisms of metal-containing hydrolases and oxidases. It will also be seen from Table 19.3 that metal-ion containing molecules in general (not just enzymes) serve important functions as electron carriers, metal storage sites, and O_2 binding and storage sites, and participate in signal transduction. Metal ions are also involved in sites of photoredox reactions, as in the energy-harvesting molecule chlorophyll. Although some of these functions are also shared by nonmetal species, it is clear that

Table 19.2 The distribution of elements in the biological cell

Extracellular	Organelles	Cytoplasm
Na^+, Ca^{2+}	K^+, Mg^{2+}	K^+, Mg^{2+}
Cu^{2+} (Mo)	Fe, Co	Co
	Zn, Ni, Mn	Zn
Cl, Si	P (S)	P (S)
Al	Se	Se

Table 19.3 The classification of some biomolecules containing metal ions

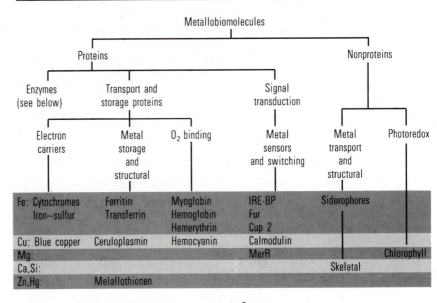

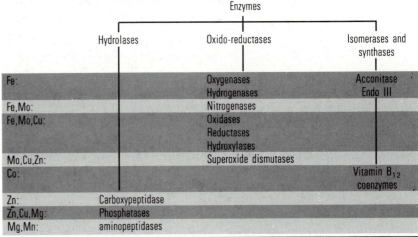

Augmented version of J. A. Ibers and R. H. Holm, *Science*, **209**, 223 (1980).

metal ions play a major role in biochemical processes. Their roles can be summarized as follows:

> Metal ions exert an inductive effect by coordination to the site of reaction and serve as redox sites that function by either electron or atom transfer. Selectivity is achieved in Nature by the deployment of metal ions having the appropriate size, stereochemical preference, hard–soft character, or reduction potential.

Enzymes are particularly important and, as we shall have much to say about them, we need to know the broad outline of their structures. Small complexes are insufficiently specific as catalysts to maintain the subtle dynamic balance of a living system, so most enzymes are macromolecules, and specifically are polypeptides. Because of its intricate

Table 19.4 The classification of amino acids $NH_2CHRCOOH$

Type	Name	Abbreviation	R—
Hydrophobic *R*	Glycine	gly	H—
	Alanine	ala	CH_3—
	Valine	val	$(CH_3)_2CH$—
	Leucine	leu	$(CH_3)_2CHCH_2$—
	Isoleucine	ile	$CH_3CH_2CH(CH_3)$—
	Phenylalanine	phe	⟨benzene ring⟩—CH_2—
Inert heteroatom in *R*	Tryptophan	trp	⟨indole ring⟩—CH_2—
Hydroxylic *R*	Serine	ser	$HOCH_2$—
	Threonine	thr	$HOCH(CH_3)$—
	Tyrosine	tyr	HO—⟨benzene ring⟩—CH_2—
Carboxylic *R*	Aspartic acid	asp	$HOOCCH_2$—
	Glutamic acid	glu	$HOOCCH_2CH_2$—
Amine *R*	Lysine	lys	$H_2NCH_2CH_2CH_2CH_2$—
	Arginine	arg	$H_2N-\underset{\underset{NH}{\|\|}}{C}-NHCH_2CH_2CH_2$—
Amide *R*	Aspargine	asn	$H_2N\underset{\underset{O}{\|\|}}{C}CH_2$—
	Glutamine	gln	$H_2N\underset{\underset{O}{\|\|}}{C}CH_2CH_2$—
Imidazole *R*	Histidine	his	⟨imidazole ring⟩
Sulfur-containing *R*	Cysteine	cys	$HSCH_2$—
	Methionine	met	$CH_3SCH_2CH_2$—
Others	Proline	pro	⟨pyrrolidine ring structure⟩

structure, a macromolecule achieves precise organization of the environment of the reactive site and manages the flow of reactants.

Polypeptides are macromolecules formed from α-amino acids linked by peptide bonds (**1**). We shall sometimes use the term 'peptide residue', which denotes the component of an amino acid that remains in the chain once the peptide link has formed by elimination of H_2O. Twenty different α-amino acids occur naturally in proteins, and the different side chains that occur along the polypeptide chain possess a range of functional groups, such as alkyl, carboxyl (—COOH), amino (—NH_2), hydroxyl (—OH), and thiol (—SH) groups (Table 19.4). These functional groups confer hydrophobic and hydrophilic character, Brønsted acidity and basicity, and the Lewis basicity necessary for complexation with metal ions. One very important role for the functional groups of the peptide residues is to modify the immediate envir-

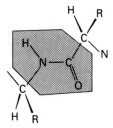

1 Peptide link

Table 19.5 Minerals in biological structural materials

Mineral	Formula	Organism	Location
Calcite	$CaCO_3$	Birds	Eggshells
Aragonite	$CaCO_3$	Mollusks	Shell
Hydroxyapatite	$Ca_5(PO_4)_3(OH)$	Vertebrates	Bone
		Mammals	
Silicon dioxide	$SiO_2 \cdot nH_2O$	Diatoms	Cell wall
		Limpets	Teeth
		Plants	Leaves

onment of metal ions in metalloenzymes, and we shall concentrate on this subtle interplay of inorganic and organic chemistry.

Aside from their role in the dynamics of biological processes and the stabilization of the conformations of large biomolecules, metal ions in the form of crystalline minerals or amorphous compounds are important as structural materials in many organisms (Table 19.5).[1] In Section 12.1 we saw that the principal commercial source of phosphates and phosphorus is the mineral hydroxyapatite. Hydroxyapatite is widespread in the animal kingdom as a principal constituent of bones, teeth, and shells. Usually the mineral materials exist along with other phases; for example, in bones and teeth hydroxyapatite coexists with the protein collagen that controls the morphology of hydroxyapatite growth and influences its mechanical properties. Hydrated silicon dioxide occurs in tiny diatoms that live in the sea and deposits of their dead cell walls form diatomaceous earth, which is a useful filtering agent in the laboratory and in industry.

PUMPS AND TRANSPORT PROTEINS

As we observed in Chapter 5, the phosphate polymer ATP (adenosine triphosphate)[2] stores energy for cells in the form of the free energy

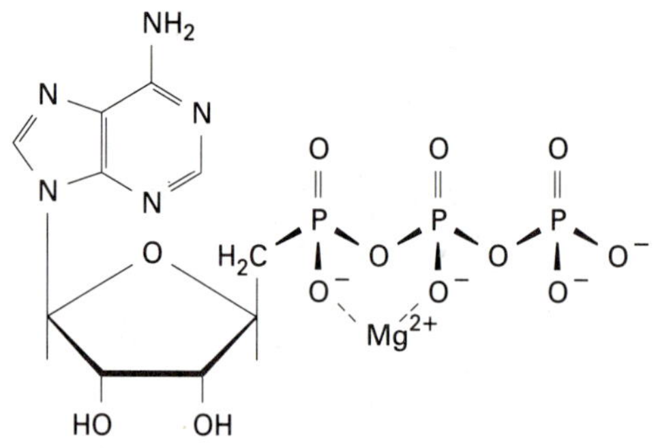

2 MgATP

---

[1] For a useful and brief review see: S. Mann, Biomineralization: the hard part of bioinorganic chemistry! *J. Chem. Soc., Dalton Trans.*, 1 (1993).

[2] ATP is a simplified notation. In the cell, the molecule is present as a Mg^{2+} complex of the polyphosphate ligand. One possible structure is (**2**).

that can be released on hydrolysis of the $-P-O-P-$ polyphosphate link. Because this method of energy storage is very general, the means by which ATP is synthesized are of central importance to an understanding of the global biochemistry of an organism.

The key features of the conditions favoring condensation of two $-P-OH$ units to $-P-O-P-$ with loss of water are control of pH and electric charge. The pH and charge gradients required are created across a cell membrane by reactions that exploit the electron acceptor properties of the O_2 molecule, and so oxygen is ultimately responsible for driving the build-up of the energy carrier of the cell. This example hints at the intricacy of the network of transport processes that interact to sustain an organism, and we shall begin our discussion of bioinorganic chemistry by considering the operation of one of the pumps that creates ion concentration gradients.

19.1 Ion pumps

We deal here with the pump that contributes to **active transport**, the transport of ions through membranes against their natural tendency to diffuse toward lower concentration. The charge distribution established across the membrane by active transport results in an electric potential difference as well as a difference in chemical composition of the fluids inside and outside the cell.

The sodium/potassium pump[3]

Most animal cells have a higher concentration of K^+ ions inside the cell membrane than outside, and a higher concentration of Na^+ ions outside the cell than inside. The immediate source of the energy to sustain this disequilibrium is ATP, and a large amount of ATP is consumed to maintain the concentration gradient.

One clue to the operation of the ion pump came in 1957, when Jens Skou discovered an enzyme of molar mass about 110 kDa (110 kg mol^{-1}) that hydrolyzes ATP only if Na^+ and K^+ are present in addition to the Mg^{2+} required by all ATPases (the enzymes that catalyze hydrolysis of ATP). The activity of this enzyme correlates quantitatively with the extent of ion transport. Another clue came from the observation that this ATPase is phosphorylated (that is, undergoes formation of an enzyme–phosphate bond) at an aspartate site (**3**) only in the presence of Na^+ and Mg^{2+} ions. Moreover, the phosphorylated product is hydrolyzed if K^+ ions are present. These observations are summarized by the catalytic cycle shown in Fig. 19.2. It is also observed that the enzyme undergoes a conformational change when it is phosphorylated.

From these clues, the mechanism outlined in Fig. 19.3 was proposed. First, ATP and three Na^+ ions bind to the inside of the membrane and the enzyme is phosphorylated. The product of the reaction undergoes a conformational change called **eversion**, like motion

3 β-Aspartyl phosphate

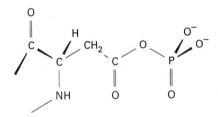

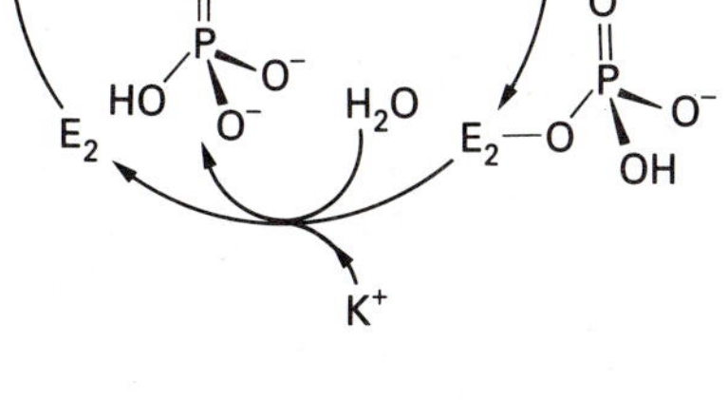

19.2 The cycle of enzyme reactions that accomplish sodium pumping. E_1 and E_2 denote the two enzyme conformations.

[3] L. Stryer, *Biochemistry*, pp. 950–4. W. H. Freeman and Co., New York (1988).

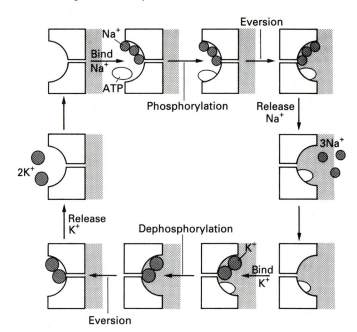

19.3 A schematic diagram of the mechanism of sodium pumping showing eversion, which transfers ions between the inside and outside of the cell. The shading denotes the extracellular fluid.

through a revolving door, that brings the bound Na^+ ions to the outside of the cell membrane. There the three Na^+ ions are replaced by two K^+ ions. The attachment of the K^+ ions induces dephosphorylation, and the hydrolysis of ATP to ADP creates a conformational change that carries the two K^+ ions into the interior of the cell, where they are released. The process builds up a charge gradient across the membrane because three Na^+ ions are expelled for two K^+ ions incorporated, and the outer surface becomes relatively positively charged.

An observation that supports the mechanism is that minute (10^{-9} M) concentrations of vanadate ions (VO_4^{3-}) can inhibit the operation of ATPase. The difference between phosphate and vanadate ions is that the five-coordinate phosphate intermediate formed when water attacks is unstable and readily breaks down, so completing the hydrolysis. However, the five-coordinate species that vanadate can form does not break down, the conformational change does not take place, and the bound K^+ ions remain on the outside of the cell membrane.

The selectivity of the process

The novel feature of active transport from an inorganic standpoint is the selectivity of complexation between K^+ and Na^+, which is uncommon in simple systems. We need to examine how an enzyme can be selective first to Na^+ and then to K^+ in its different conformations. We have seen that Group 1 cations are usually not strongly complexed, so selectivity in simple systems is mainly dependent upon coulombic forces and differences in ionic radii.

The ion channels that permit passage of ions through membranes are typically constructed from protein polymers in helical conformations. These helices form the walls of the ion channels, and their precise conformations produce specific geometrical arrangements of a number of potential ligand–metal binding sites. The differences between channels selective toward either Na^+ and K^+ are remarkable:

1. Only Tl^+ approaches K^+ in permeability of the K^+ channels. No other ion, larger or smaller, passes well.
2. The K^+ channels exclude NH_4^+ more rigorously than do the Na^+ channels despite the fact that the NH_4^+ and K^+ ions have more similar radii.
3. Large hydrogen bonding spherical molecules and ions (e.g. guanidinium) pass the Na^+ channels. No methylated compounds (e.g. $CH_3NH_3^+$) pass either channel.

These facts can be explained by the assumption that the Na^+ channel is larger and hydrated. The more strongly hydrated Na^+ ion moves into the sites in the channel while exchanging H_2O ligands to bind to the H_2O molecules of the channel interior. The larger and less strongly hydrated K^+ ion passes through a dehydrated channel where it complexes to protein side-chain anions. (Note that the solubilities of Na^+ salts of poorly hydrated anions, such as ClO_4^-, are much greater than those of the corresponding K^+ salts.)

Example 19.1: *Interpreting the properties of membrane ion channels*
What is the significance of the fact that channels for Na^+ transport are quite permeable to H^+ whereas K^+ channels are not?

Answer. It was suggested that the Na^+ channels are hydrated and accept ions by coordination to the water molecules on the interior of the channel. These water molecules can readily transport protons by the Grotthus mechanism (Section 5.1). The K^+ channels are formed from ligand sites of the proteins that may resemble the crown ether (**4**).

Exercise E19.1. Why do hydrogen bonding molecules of large radius pass the Na^+ channels more readily than the K^+ channels?

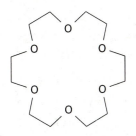

4 Crown ether

An impression of the role of steric factors is given by one simple biochemically important ligand. It is known that crown ethers that have antibiotic activity increase the permeability of cell membranes to cations. The chelating ligand shown in (**4**), for instance, binds both Na^+ and K^+ ions but strongly favors K^+ by a factor of nearly 100. The origin of the selectivity is stereochemical, for a K^+ ion fits neatly into the cavity in the hollow crown. Steric selectivity in complexation equilibria was discussed in Section 6.7.

Calcium biochemistry

Calcium is the major cation in the structural materials bone and shell, where its role stems from the insolubility of its carbonates and phosphates. However, Ca^{2+} ions also have diverse non-structural roles, for they function as messengers for hormonal action, in the triggering of muscle contraction and nerve signals, in the initiation of blood clotting, and in the stabilization of protein structures.

The diversity of the roles of the Ca^{2+} ion is surprising because it is not known for forming robust complexes with simple ligands in solution. In this regard, the differences between Ca^{2+} and its congener Mg^{2+} are illuminating and arise from three factors. First, as a consequence of its low selectivity, Ca^{2+} can bind neutral oxygen donor ligands (carbonyls and alcohols) in competition with water; it also binds N donors. Second, Ca^{2+} resembles Na^+ and K^+ (and lanthanide cations) in favoring high coordination numbers and irregular coordination geometry. With its charge number of $+2$ it can bind anions that alkali metal ions do not. Third, the rates of binding to and dissociation from Ca^{2+} are high: the rate of attachment is diffusion limited and the rate of dissociation is proportional to stability. In contrast, the water exchange rate at Mg^{2+} is $10^5 \ s^{-1}$, nearly five orders of magnitude slower than at Ca^{2+}. These features make Ca^{2+} an ideal bridging species, for it is rapidly exchangeable and responsive to changing coordination geometries.

Calcium binding proteins are typically rich in aspartate (asp) and glutamate (glu), both of which have carboxylate groups as side chains and hence can act as hard anionic ligands. As an illustration of the kind of structure that occurs, one Ca^{2+} ion binding site that has been characterized by X-ray diffraction consists of four carboxylate O atoms in a distorted tetrahedral array (**5**). The Ca^{2+} ion frequently functions as a bridge between different protein segments, binding to anionic sidegroups of different amino acids or even carbonyl groups. The utilization of this property for the control of chain folding is illustrated schematically in (**5**), which shows how the protein chain folds to allow four $-CO_2^-$ groups to coordinate to Ca^{2+}. The outcome of this movement is a shift in the locations of the side groups on one helical protein region relative to the side groups on the second helical region.

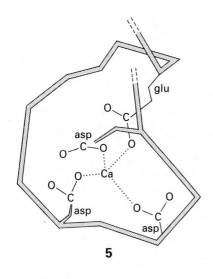

5

19.2 Oxygen transport

Oxygen is a recurring theme in this chapter because its role in respiration and its production in photosynthesis provide examples of a range of biologically important redox reactions, electron transfers, atom transfers, and photochemical processes.

The distribution of oxygen carriers

First, we consider molecules that transport oxygen to the cell. All oxygen carriers that contain iron are found inside cells, whereas carriers that contain copper are also found in extracellular fluids. A part of the reason for the distinction is that the cell interior is a reducing medium that can sustain iron as Fe(II). In addition, the cell supplies a porphyrin ring (**6**) as a ligand that can form a non-labile Fe(II) complex. Because the porphyrin ring is susceptible to oxidative attack it also benefits from the protection of the reductases in the cell. On the other hand, both Cu(I) and Cu(II) can form strong complexes with the imidazole

6 Metalloporphyrin

side chains of polypeptide residues, and the copper-based carriers may survive free in the bloodstream. The iron and copper carriers can play distinct roles in the same organism since they adopt different strategies for managing oxygen concentration gradients in different compartments.

The most common oxygen carriers are all based on helical proteins. In hemoglobin these coiled proteins act like springs that can respond to the strain generated when oxygen binds at one site and can transmit that strain to other sites.[4] Thus, the chemistry of O_2 complexation can be a combination of effects of complexation at the metal site and of alteration of the protein environment. The oxygen carriers are not very selective to the identity of the incoming π-acceptor ligand, except in respect of its size. Thus, hemoglobin and myoglobin bind NO, CO, CN^-, RNC, N_3^-, and SCN^- as well as O_2.

Structure of the O_2 binding site at Fe(II)

We now examine the oxygen carrier proteins hemoglobin (Hb) and myoglobin (Mb), both of which are built round Fe(II).[5] The oxidized forms, metmyoglobin and methemoglobin, contain Fe(III) and do not bind O_2.

Naked 'heme', the Fe–porphyrin complex without the accompanying polypeptide, is oxidized irreversibly to Fe(III) by molecular O_2 to give a μ-O_2^{2-} bridged dimeric intermediate and ultimately a stable μ-O^{2-} product (7) which cannot function as an O_2 carrier. This would seem to be a fatal flaw for its biological function. However, this unproductive reaction is prevented by the protein environment in which the heme is imbedded in Hb or Mb.

The structure of deoxy Mb, the oxygen-free form of Mb, is shown in outline in Fig. 19.4. The porphyrin active site is entwined in the polypeptide chain, which is represented by the tubes in the illustration. The protein pocket in which the heme is carried is formed from amino acid residues with nonpolar side groups, and is hydrophobic. The same groups block access of larger molecules to the Fe atom neighborhood and so prevent the formation of the Fe—O_2—Fe bridged species involved in the oxidative reaction of protein-free heme. The hydrophobic groups also deny solvation to ions produced in the oxidation of the heme complex. The result is that the Fe(II) complex can survive long enough to bind and release O_2. This is a typical example of the delicate control of reaction environments that proteins can exercise.

Deoxy Mb is a five-coordinate high-spin Fe(II) complex with four of the coordination positions occupied by the porphyrin ring N atoms. The fifth position is occupied by the N atom of an imidazole ligand of a histidine residue (Fig. 19.5) which couples the heme to the protein. Such five-coordinate heme complexes of Fe(II) are always high spin. The diameter of a high-spin $t_{2g}^4 e_g^2$ Fe(II) ion is larger than that of the

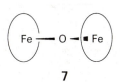

7

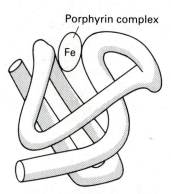

Porphyrin complex

Fe

19.4 An outline of the structure of myoglobin. The tubular structure represents the polypeptide chain, and straight sections indicate helical regions.

[4] R. J. P. Williams, *Chimica Scripta*, **28A**, 5 (1988). The argument just quoted is developed in this paper, which surveys the biological reactions of oxygen.
[5] R. E. Dickerson and I. Geis, *Hemoglobin: structure, function, evolution and pathology*. Benjamin-Cummings, Menlo Park (1983).

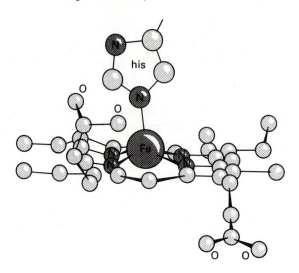

19.5 The coordination environment of iron in myoglobin.

central hole in the porphyrin ring, and hence the Fe(II) lies about 0.40 Å above the plane of the ring (**8**). High-spin Fe(II) porphyrin complexes, including the active site in Hb and Mb, involve puckering and twisting of the porphyrin macrocycle, whereas a low-spin Fe(II) ion has a smaller diameter and can fit comfortably into the ring. When O_2 completes the six-coordination environment, the complex converts to low spin (corresponding to the t_{2g}^6 configuration of Fe(II)), the Fe ion shrinks a little, and moves into the plane (**9**). This structural change is verified by X-ray diffraction.

Models of O_2 binding

Several investigations of the interaction of O_2 with metal complexes have been motivated by the hope of understanding the function of oxygen carriers. The results of the work have contributed not only to unraveling the biological problem but also to an understanding of the pathways by which molecular O_2 functions as an oxidizing agent. Among the *d*-block metals that bind O_2 and might provide helpful model systems it is probably cobalt that has provided the best general picture of the oxidative reactions of oxygen.

Like simple Fe(II) complexes, Co(II) complexes can react with O_2 with electron transfer:

$$LCo^{2+} + O_2 \rightarrow [LCo^{3+}O_2^-]$$

The product is formally a Co(III) complex of the superoxide ion, O_2^-. It reacts very readily with a second Co(II) complex to give a bridged complex of the peroxide ion O_2^{2-}:

$$[LCo^{3+}O_2^-] + LCo^{2+} \rightarrow LCo^{3+} - (O_2^{2-}) - Co^{3+}L$$

The structure of the product $[(NH_3)_5Co-O_2-Co(NH_3)_5]^+$ has been determined (**10**), and the O—O bond distance of 1.47 Å is satisfactorily close to the typical peroxide bond length of 1.49 Å.

The Co(II) complexes of ligands such as salen (the Schiff base

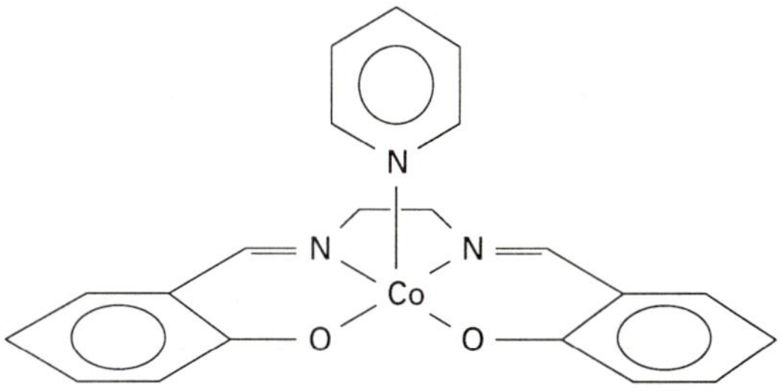

11 [Co(salen)(py)]

chelating ligand in **11**) and a base such as pyridine react with O_2 in a rapid, reversible reaction which is suggestive of the reactions of Mb or Hb:

$$[Co(salen)(py)] + O_2 \rightleftharpoons [Co(O_2)(py)(salen)]$$

An X-ray study of the complex reveals an O—O length of 1.26 Å, which is between the values for O_2 (1.21 Å) and O_2^- (1.34 Å).

EPR spectra are consistent with a single unpaired electron interacting with the ^{59}Co nucleus. The single unpaired electron cannot be on a Co(II) atom because a high-spin d^7 complex has three unpaired electrons, and studies of complexes with comparable ligands show that the ligand field is not strong enough to force the complex to have the low-spin configuration (which would have one unpaired electron). A low spin d^7 formulation is also not supported by an interpretation of the EPR hyperfine coupling constants. This observation suggests that an O_2^- ligand is coordinated to a low-spin d^6 Co(III) center.

An alternative proposal for O_2 bonding to Co accounts for the observed long O—O bond length in terms of potential σ-donor and π-acceptor character of O_2. The suggestion is that O_2 donates a lone pair of electrons in a σ fashion and that the π^* orbitals of O_2 accept electrons from the d_{zx} or d_{yz} orbitals of Co. Such back-donation weakens and lengthens the O—O bond. The π-acceptor character of the ligand *trans* to O_2 should exert a substantial influence by competing for or supplying electrons to the π system.

Example 19.2: *Diagnosing the extent of electron transfer to O_2*

Figure 19.6 shows that the logarithms of the equilibrium constant for O_2 binding by cobalt Schiff base complexes, [Co(Schiff base)(B)], with various axial ligands, B, correlate linearly with standard potentials for the Co(III)/Co(II) complex couples. What can you conclude from this relation?

Answer. Because log K is proportional to the $\Delta G^{\ominus}$ of the reaction, and the standard potential is also proportional to a $\Delta G^{\ominus}$, the correlation is a linear free energy relation (Section 15.5). Because the free energies of reduction and O_2 binding increase together, it is reasonable to suggest that electron transfer from Co (and reduction of O_2) is involved

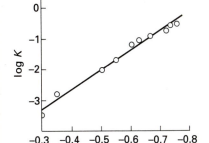

19.6 The linear correlation of log K and E for a series of Schiff-base complexes.

in O_2 binding. However, an increase of 0.4 V (40 kJ mol^{-1}) in $E^{\ominus}$ leads to an increase in log K of only about 2.1 (12 kJ mol^{-1}). The difference suggests that electron transfer to O_2 is incomplete and that $Co(II)—O_2$ and $Co(III)—O_2^-$ are idealized extremes.

Exercise E19.2. The $O—O$ bond lengths in O_2, KO_2, and BaO_2, are 1.21, 1.34, and 1.49 Å, respectively. These values provide reference data on the relation between bond length and oxidation state. For the complexes $[Co(CN)_5O_2]^{3-}$, $[Co(bzacen)(py)(O_2)]$ (where bzacen is an acyclic tetradentate ligand suggestive of salen with two N and two O donor atoms), $[(NH_3)_5Co(O_2)Co(NH_3)_5]^{4+}$, and $[(NH_3)_5Co(O_2)Co(NH_3)_5]^{5+}$, the $O—O$ bond lengths are 1.24, 1.26, 1.47, and 1.30 Å, respectively. Comment on the extent of Co to O_2 electron transfer in each complex.

Some Fe porphyrin models of reversible O_2 coordination have been synthesized. As we have remarked, the challenge is to prevent binuclear bridged complex formation, and three approaches have been successful. One is the use of steric inhibition in the porphyrin ligand to prevent the approach of a second complex or an additional ligand (as in the biological system itself).[6] A second is the use of low temperatures to slow the dimerization reaction, and the third is to anchor the Fe complex to a surface (e.g. silica gel) so that dimerization is prevented. The first is the one most closely related to the behavior of Mb and Hb.

Steric hindrance of dimer formation is achieved with so-called **picket-fence porphyrins**, where the 'pickets' are a group of blocking substituents projecting from one side of the planar ring. Coordination of a bulky ligand, such as an N-alkylimidazole, can occur only on the unhindered side. An imidazole (Im) is an effective σ-donor that favors coordination of a π-acceptor lying *trans* to itself. It gives the complex an affinity for O_2 similar to that of Mb, and the blocking substituents create a pocket for O_2 and prevent formation of $[Fe(porph)(Im)_2]$. The pickets also prevent reaction with a second Fe center which would give the inactive μ-O_2 species.

One such picket-fence complex is shown in Fig. 19.7. It binds O_2 to give a structure very similar to the structure shown as (**8**); the Fe—O—O angle is $136°$ and the $O—O$ bond length is 1.25 Å. This model system provides the best guidance available on the structure of Mb and Hb oxygen complexes. That the complex is diamagnetic (low spin), might be regarded as evidence of a low-spin d^6 Fe(II) complex of singlet O_2, which would mean that Fe oxidation and O_2 reduction is less important than in the Co complexes. However, we must be cautious with this interpretation because other evidence points in a different direction.

The $O—O$ stretch lies at 1107 cm^{-1}, which is closer to the O_2^- value of 1145 cm^{-1} than the O_2 value of 1550 cm^{-1}. This suggests the

---

[6] Various strategies for blocking one face of the porphyrin ring are displayed in K. S. Suslick and T. J. Reinert, *J. Chem. Educ.*, **62**, 974 (1985).

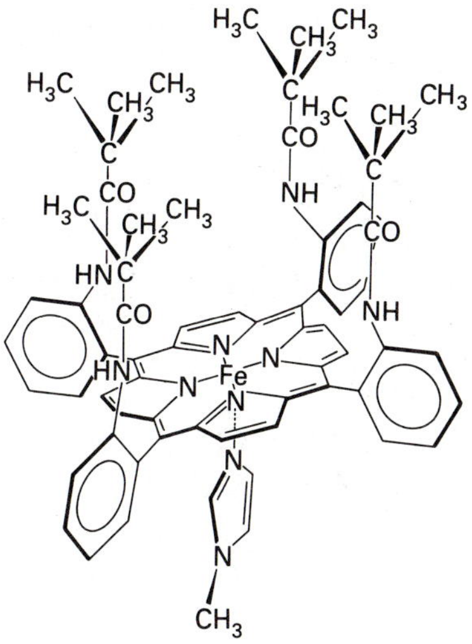

19.7 Structure of a picket-fence porphyrin. (J. Collman, J. I. Brauman, E. Rose, and K. S. Suslick, *Proc. Natl. Acad. Sci.*, **75**, 1053 (1978).)

formulation O_2^-, which is a spin-$\frac{1}{2}$ ion, in combination with low-spin Fe^{3+}, which is also a spin-$\frac{1}{2}$ ion. The low-spin character observed for the complex could arise from spin pairing of Fe(III) and O_2^-, in which case the complex would be similar to the cobalt model compounds. That such a possibility is very real is underlined by the observation that a d^3 Cr(III) porphyrin complex of oxygen has been synthesized which has only two unpaired electrons. Because d^3 Cr(III) has three unpaired electrons, the only explanation for the spin observed is spin-pairing with an electron from O_2^-. These conflicting indications demonstrate once again that assigning precise charges to the central metal ion and to the ligands is an oversimplification. It is employed only because it often helps in the interpretation of reactions and structures, but must never be taken too literally.

Hemoglobin and myoglobin functions

The function of hemoglobin (Hb) is to bind O_2 at the high oxygen partial pressures found in lung tissue, to carry it without loss through the blood, and then to release it to myoglobin (Mb). The sequence requires Mb to have a greater oxygen affinity than Hb at low partial pressures. This is shown to be the case in Fig. 19.8 where the degree of saturation is always less for Hb than Mb.

The shape of the Mb curve is easily explained in terms of the equilibrium

$$Mb + O_2 \rightleftharpoons MbO_2 \qquad\qquad K = \frac{[MbO_2]}{[Mb]p}$$

where p is the partial pressure of oxygen. The **fractional oxygen saturation** α is the ratio of the concentration of the Mb present as MbO_2 to the total concentration of Mb:

$$\alpha = \frac{[MbO_2]}{[Mb] + [MbO_2]}$$

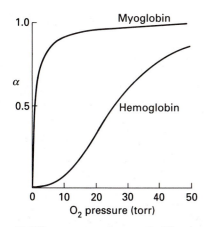

19.8 The oxygen saturation curves for Mb and Hb showing the fractional oxygen saturation α as a function of the oxygen partial pressure at pH = 7.2.

It follows from the equilibrium expression that

$$\alpha = \frac{Kp}{1 + Kp}$$

The Mb curve in the illustration conforms to this equation but the Hb curve does not.

The Hb curve can be reproduced if the dependence of α on the partial pressure of oxygen is changed to p^n, with n between 2 and 3. Another feature of the binding site is that O_2 binding by Hb is pH dependent despite the absence of an acid group at the heme site, and it is observed that O_2 is released more readily at lower pH. Thus, O_2 is released more readily in cells where metabolism is active and that have a high concentration of CO_2, which lowers the pH. A final detail to accommodate is that the binding of organic phosphates to the protein of Hb far from the heme site also affects O_2 binding.

The structural difference between Mb and Hb is that whereas the former has a single heme group, the latter is essentially a tetramer of Mb, with four heme groups (Fig. 19.9). The difference is crucial, because it allows the four heme units of Hb to bind O_2 *cooperatively*: once one O_2 is bound to Hb, the affinity for subsequent O_2 molecules is greater. The origin of this cooperativity was suggested by M. F. Perutz, who determined the crystal structure of Hb, and argued that oxygenation of one Fe atom in Hb leads to structural changes in its partners.[7] When the high-spin Fe atom that lies 0.4 Å above the porphyrin plane coordinates O_2 and changes to low-spin and moves into the plane, it pulls the histidine residue of the protein along with it. As a result, the shape of the protein is adjusted and the binding characteristics of the other sites are modified. The pH and phosphate dependence are ascribed to conformational influences stemming from points moderately distant from the metal atom. The latter change underlines yet again the sensitivity of structures to influences at distant sites.

19.9 A schematic structure of Hb showing the relationship among the four subunits of the tetramer.

ENZYMES EXPLOITING ACID CATALYSIS

A metalloenzyme can act as an acid catalyst either on account of the Lewis acidity of the metal ion itself or the Brønsted acidity of a ligand being enhanced by the metal ion. The metal ion in a metalloenzyme may also be able to adopt roles denied to it as a part of a small complex in solution. For instance, the active site of the enzyme may be protected from the bulk aqueous environment in such a way that solvent leveling (Section 5.3) or the effects of solution permittivity are modified. Some of these features have already been seen in the discussion of oxygen transport by Hb.

The Zn^{2+} ion is a very common Lewis acid in biochemical systems, and it is of some interest to explore why zinc occurs rather than Mg^{2+}

[7] M. F. Perutz, G. Fermi, B. Luisi, B. Shaanan, and R. C. Liddington give an up to date summary of the link between stereochemistry and mechanism in *Acc. Chem. Res.*, **20**, 309 (1987); M. F. Perutz, *Mechanism of cooperativity and allosteric regulation in proteins.* Cambridge University Press (1990).

Table 19.6 Special characteristics of zinc

1.	More available than Ni, Cd, Fe, Cu
2.	More strongly complexed than Mn(II), Fe(II)
3.	Faster ligand exchange than Ni(II), Mg(II)
4.	Not redox active compared with Cu(II), Fe(II), Mn(II)
5.	More flexible coordination geometry than Ni(II), Mg(II)
6.	Good Lewis acid; among M^{2+} ions, only Cu(II) is stronger

Adapted from J. J. R. Fraústo da Silva and R. J. P. Williams, *The biological chemistry of the elements*, p. 300. Oxford University Press (1991).

and Cu^{2+}, which are also small dipositive ions. A feature that distinguishes zinc from magnesium is the electron affinity of Zn^{2+} (which is equal to the sum of the first and second ionization energies of the neutral Zn atom). Its value, 27.4 eV, which is greater than for Mg^{2+} (22.6 eV), means that Zn^{2+} is a stronger Lewis acid. The electron affinity of Cu^{2+} (28.0 eV) is similar, so another feature must account for zinc's difference from copper. Zinc differs from copper in that it does not have a variable oxidation state, so redox reactions are unlikely to complicate its behavior and there is little risk of generating radicals.

A further point is that catalysis in general requires reorganization of atoms. It follows that fast reactions will be more characteristic of a metal ion that, while affecting ligands structurally, can take up and release ligands rapidly. In fact, Zn^{2+} complexes are much more labile than the corresponding complexes of Mg^{2+} and Ni^{2+}. Moreover, geometrical rearrangement (an aspect of fluxional behavior) is usually facile for zinc complexes but not for the corresponding complexes of Mg^{2+} and Cu^{2+}. The only ion closely similar to Zn^{2+}, and which might be a competitor for use by biochemical systems, is Cd^{2+}; but cadmium is a much rarer element than zinc, and so is less available for incorporation. The unique features of zinc are summarized in Table 19.6.

19.3 Oxaloacetate decarboxylase

We turn now to enzymes and model catalysts. One of the simplest and most important reactions subject to acid catalysis is CO_2 loss from oxaloacetic acid (**12**) with the formation of pyruvic acid (**13**):

$$HO_2CCOCH_2CO_2H(aq) \rightarrow HO_2CCOCH_3(aq) + CO_2(g)$$

This reaction is a step in the widely distributed **citric acid cycle**, or **Krebs cycle**, of carbohydrate metabolism, which is the final step in the oxidation of carbohydrates, amino acids, and fatty acids. A comparison of the enzyme and nonbiological catalysts provides a good indication of the influence of proteins on the catalytic activity of metal ions.

The conversion of oxaloacetic acid to pyruvic acid and CO_2 is catalyzed by many aqua metal cations. It is accomplished biochemically by the enzyme oxaloacetate decarboxylase, which has a metal ion (Mn(II), Zn(II)) at its active site. The rate constant for decarboxylation of the simple complex ZnA (where A = oxaloacetate) is 0.007 s^{-1} at

COOH
|
CH$_2$
|
C $=$ O
|
COOH

12 Oxaloacetic acid

CH$_3$
|
C $=$ O
|
COOH

13 (keto)–Pyruvic acid

25 °C and that for CuA is 0.17 s^{-1}. The uncatalyzed decarboxylation of A^{2-} has a rate constant of only 2×10^{-5} s^{-1}. Notice that the stronger Lewis acid, Cu^{2+}, is only 25 times better as a catalyst than Zn^{2+}. The enzyme catalyzed pathway, however, is 10^4 times faster than the Cu^{2+} catalyzed pathway. This difference provides the first glimpse of the striking contribution of protein complexity.

The metal-ion catalyzed reaction is believed to follow the mechanism shown in Fig. 19.10. A ^{13}C kinetic isotope effect of about 6 percent is observed for the metal-ion catalyzed reaction, which indicates that the cleavage of the C—C bond is rate determining. In contrast, the enzyme catalyzed reaction shows no ^{13}C isotope effect, but is slower in D$_2$O than in H$_2$O. It is clear that C—C bond breaking is no longer the rate determining step in the enzyme catalyzed reaction, but that proton transfer is involved.

These observations do not prove that the enzyme catalyzed reaction follows a different pathway through different intermediates, for if the enzyme can enhance the rate of the C—C bond cleavage sufficiently, then a subsequent step may become rate limiting. The solvent isotope effect does in fact suggest that a proton transfer step has become rate determining. The conversion of the enol form of pyruvate (**14**) to the keto form (**13**), in which the CH$_2$= group becomes CH$_3$—, is such a step. The rate is determined by the transformation (d) → (e) in Fig. 19.10. The rate determining step of the reaction catalyzed by a

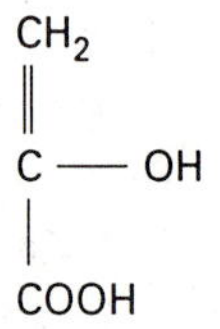

CH$_2$
‖
C — OH
|
COOH

14 (enol)–Pyruvic acid

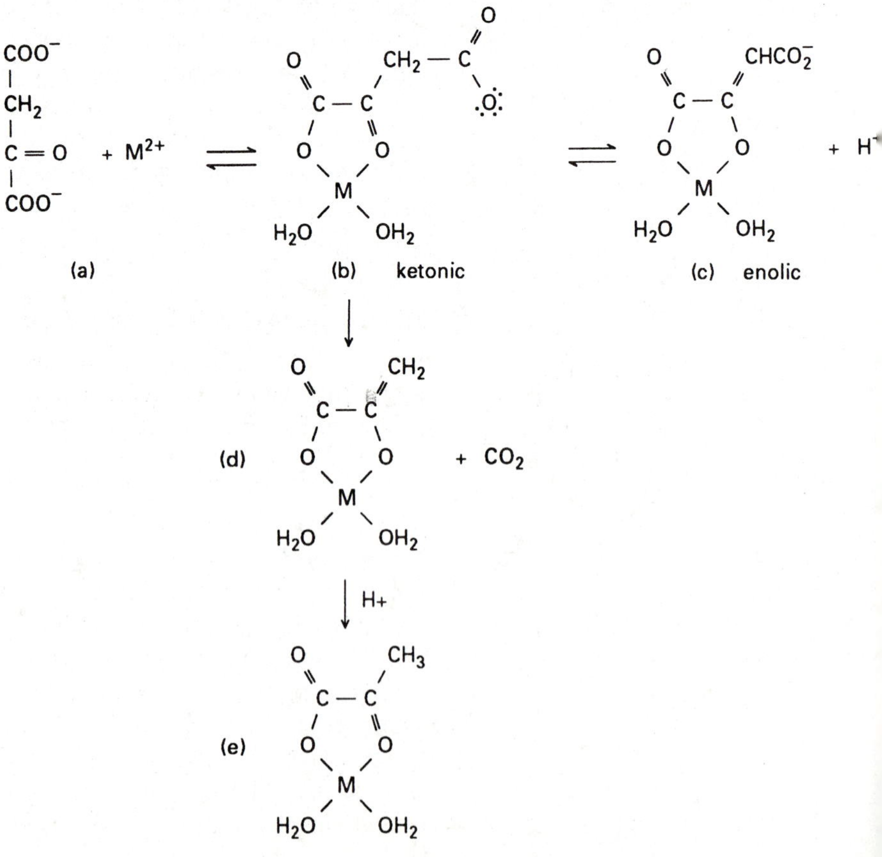

19.10 The mechanism of the conversion of oxaloacetate to pyruvate catalyzed by an M^{2+} Lewis acid.

simple metal ion is probably (b) → (d) as a result of the stabilization of (d) by the metal ion.

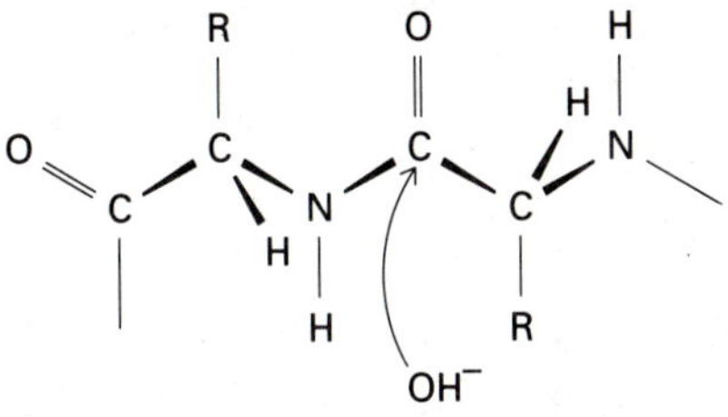

19.4 Carboxypeptidases[8]

We now consider a hydrolytic enzyme. The carboxypeptidases are enzymes containing Zn(II) that catalyze the hydrolysis of peptide bonds in peptides and proteins:

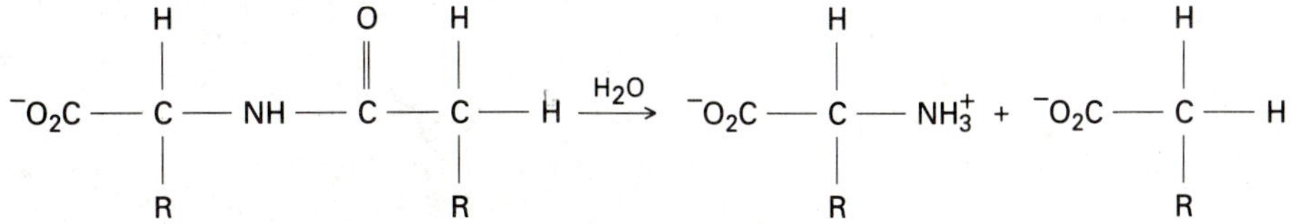

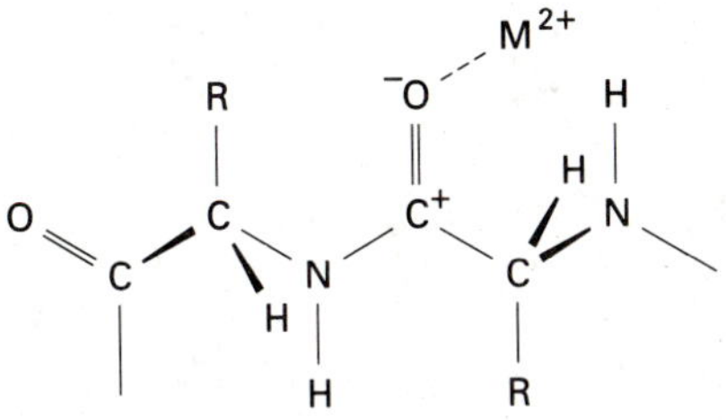

Bovine carboxypeptidase A is an enzyme composed of 307 residues in a single polypeptide chain which binds one Zn^{2+} ion per molecule. The protein has been crystallized, and the hydrogen bonding between groups, the coordination environment of the Zn^{2+} ion, and the folding of the protein chain that defines the secondary and tertiary structures have all been identified from X-ray studies. Moreover, comparison of crystals with and without simple substrates show some of the details of the probable reaction precursor.

The hydrolysis reaction requires the promotion of nucleophilic attack of an O atom on the carbonyl group of the peptide bond (**15**). One way in which a metal ion might serve is as a Lewis acid that promotes proton loss from the H_2O molecule, so making it into the more nucleophilic OH^- species. Alternatively, the metal ion may act as a Lewis acid catalyst by coordination to the peptide carbonyl group and thereby reduce the electron density at its C atom (**16**). In general, either the hydroxo or the Lewis acid mechanisms or both can operate, but to learn the most probable path requires study of the enzyme structure. We shall set the stage for the presentation of the reaction mechanism by considering some model systems.

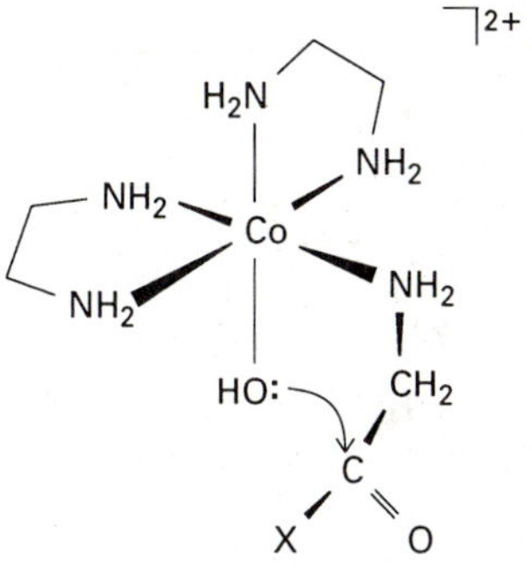

Model studies

The hydrolysis of peptides and the analogous reaction of ester hydrolysis have been modeled with octahedral complexes of Co(III), in which the hydroxo mechanism is believed to operate. The hydroxo mechanism is illustrated in (**17 a** → **b**), where the aqua ligand *cis* to the peptide loses a proton to become a hydroxo ligand (OH^-) that attacks the carbonyl group. The Lewis acid mechanism is illustrated in (**18 a** → **b**), where the carbonyl group is coordinated to Co(III), and the latter is a source of Lewis acid catalysis for the attack of a water molecule or an OH^- ion from the solution. The cobalt(III) model system has been developed into a useful laboratory method for both hydrolytic breakdown and analysis of peptides and the reverse, their synthesis.[9]

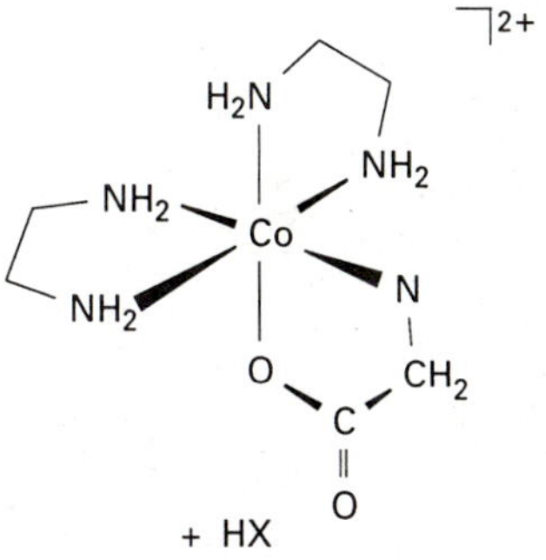

---

[8]   M. W. Makinen, G. B. Wells, and S. O. Kang, *Adv. Inorg. Biochem.*, **6**, 1 (1984).
[9]   The Co(III) peptide hydrolysis story is well told by P. A. Sutton and D. A. Buckingham in *Acc. Chem. Res.*, **20**, 357 (1987).

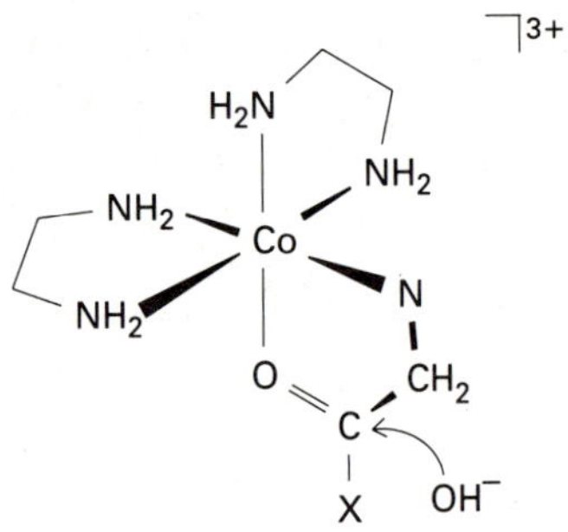

18a

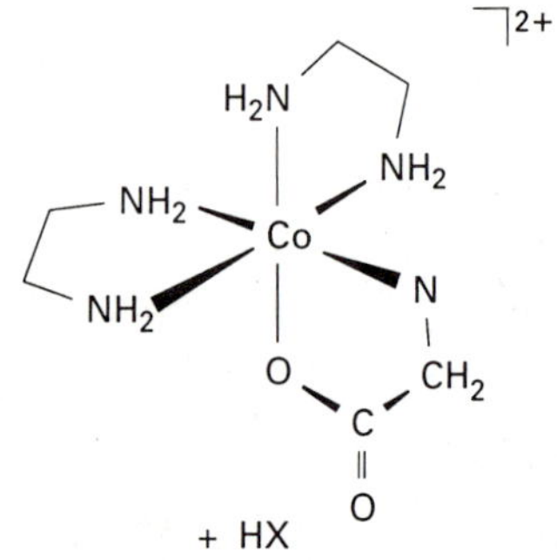

18b

Bovine carboxypeptidase

Figure 19.11 is a representation of the structure of the bovine carboxypeptidase A molecule. The Zn^{2+} ion is situated in a pocket near its center. The four donor atoms of the ligands come from two histidine residues from the protein (69 and 196 in the chain) which are N donors, a glutamic acid residue (which coordinates through an O atom), and an H_2O molecule. The illustration also shows the distorted tetrahedral coordination environment of the Zn^{2+} ion. A second H_2O molecule occupies the enzyme pocket at about 3.50 Å from the Zn^{2+} ion.

As with most enzyme mechanisms, the identification of the correct path is tricky and depends on clever inferences; more than one possibility must be evaluated. First, we consider the evidence that favors the Lewis acid mechanism of the enzyme's action. Four features are observed when the model peptide substrate glycyltyrosine (**19**) is bound to the active site and mimics a substrate in a structure which may be structurally characterized by X-ray diffraction (Fig. 19.12). One is that the side chain ($HO-C_6H_4-CH_2-$ in this case) of the terminal amino acid residue enters a hydrophobic region of the enzyme pocket. Secondly, the terminal carboxylate group interacts coulombically with the positive $-NH_3^+$ group belonging to an arginine residue at site 145. Thirdly, the carbonyl O atom of the peptide, which is to be severed, displaces the H_2O molecule from the coordination sphere of the Zn^{2+} ion and coordinates in its place. Fourthly, the phenolic OH group of the tyrosine residue at location 248 migrates through about 12 Å to form a hydrogen bond with the NH group of the peptide to be cleaved. The resulting coordination environment is inconsistent with the hydroxo ligand mechanism because no water remains directly coordinated to the Zn^{2+} ion. Therefore, it was proposed that Lewis acid catalysis by Zn^{2+} facilitates nucleophilic attack at the carbonyl C atom.

The Lewis acid mechanism just described is based on a study of the

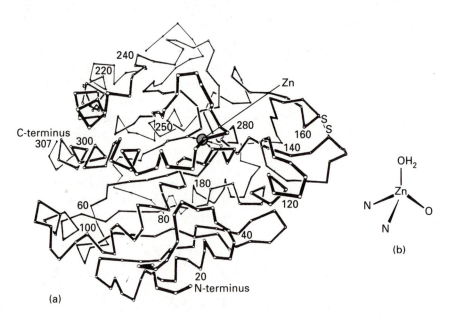

19.11 (a) An outline of the structure of carboxypeptidase and (b) the distorted tetrahedral environment of the Zn^{2+} ion. The numbers identify the specific amino acid residues, with the numbering starting at the N-terminus.

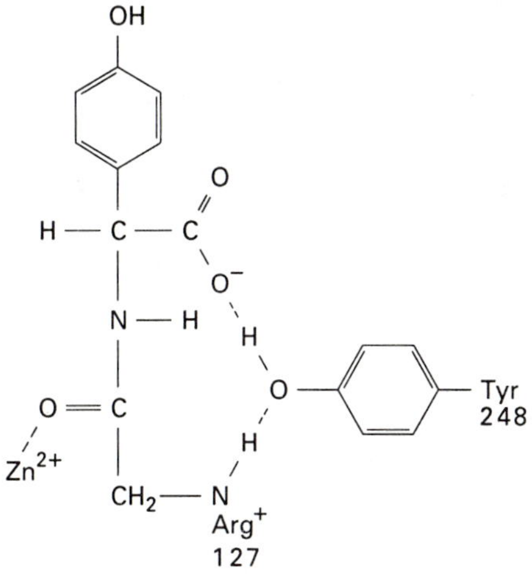

19

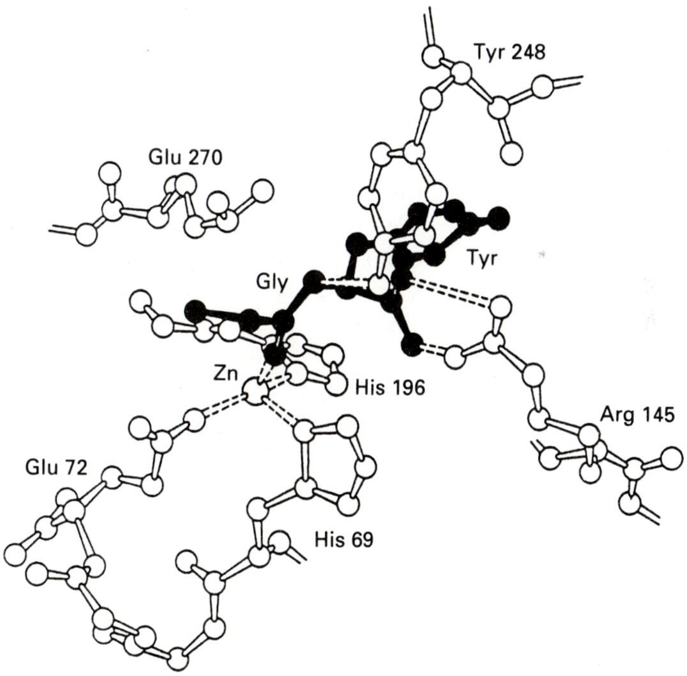

19.12 X-Ray structure of glycyltyrosine bound to the active site in carboxypeptidase A. (From D. M. Blow and T. A. Steitz, *Ann. Rev. Biochem.*, **39**, 79 (1970).)

substrates. It has been criticized on the grounds that the only ligands that can replace an H_2O ligand by an O atom of a carbonyl group are those that form chelates with Zn^{2+}, and such ligands are not actually substrates susceptible to hydrolysis.

A clue to what may be the true mechanism is found in the considerable movement of tyrosine residue 248 toward the active site during substrate binding, which suggests that it is very likely to be involved in the mechanism. Its role has been established by using the technique of site-specific mutagenesis[10] to alter the genetic coding for the enzyme from tyrosine at location 248 to phenylalanine. The mutant enzyme is found to have almost the same catalytic activity as in the original enzyme, which suggests that the role of tyrosine is in binding the substrate and not in the cleavage step.

X-Ray studies of a model substrate have led to the development of an alternative hydroxo mechanism, and the hitherto conflicting aspects of the mechanism are now falling into place.[11] In this picture, the carbonyl group to be attacked is hydrogen bonded to arginine residue 127; an H_2O molecule remains in the coordination sphere of the Zn^{2+} ion and attacks the substrate by the hydroxo mechanism. The primary intermediate in the reaction is the diol structure shown in Fig. 19.13a. This proposal is consistent with the kinetics of the reaction and with observations on the binding of analogs of the substrate. Figure 19.13b shows a model compound that has been structurally characterized. Notice that the mechanism implies the existence of transient five-coordination at the Zn^{2+} ion.

The Zn^{2+} ion of the native enzyme can be replaced by a variety of other M^{2+} ions. The Co^{2+} ion is particularly important because its ligand field spectrum in solution indicates that the ion is in a distorted tetrahedral environment, and hence that the structure determined from the crystal does represent the functioning enzyme in solution.

Mechanisms of acid catalysis

Carboxypeptidase teaches us that the hydroxo or Lewis acid mechanisms available with enzymes are much more elaborate than the mechanisms for the reactions of simple metal complexes in solution. For instance, the substrate is bound into a favorable place by hydrophobic

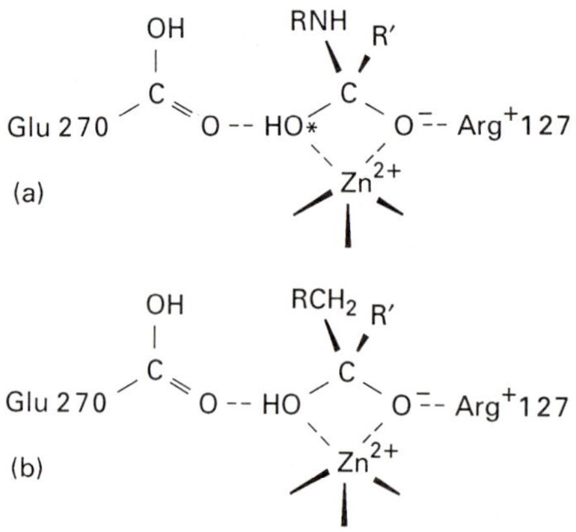

19.13 (a) The diol primary product of the reaction of a substrate catalyzed by carboxypeptidase according to the hydroxo mechanism. The starred O atom marks the hydroxo group from the attacking Zn^{2+} ion. (b) An analog of the reaction intermediate that has been studied by X-ray diffraction.

---

[10] The technique of site specific mutagenesis is an extraordinarily powerful tool which has become a central element of the strategy for the study of enzyme mechanisms with the emergence of molecular genetics techniques. This technique changes the genetic coding for protein synthesis to achieve the replacement of a particular amino acid residue in a protein so that its function can be tested.

[11] D. W. Christianson and W. N. Lipscomb, *Acc. Chem. Res.*, **22**, 62 (1989) summarize a series of studies.

and ionic interactions. If the attacking nucleophile H_2O is not activated by the metal it may still be activated by a carboxylate group from the protein. The environment in the protein fold may balance aqueous and hydrophobic zones to give the most favorable solvent effect, and all these features can operate together.

Example 19.3: *Proposing an enzyme-catalyzed reaction mechanism*

Carbonic anhydrase[12] catalyzes the interconversion of CO_2 to HCO_3^-. The reaction occurs on a millisecond timescale when it is uncatalyzed and a thousand times faster when catalyzed. Despite the simplicity of CO_2 hydration, the enzyme has many features in common with peptide hydrolysis enzymes. It has a Zn^{2+} ion coordinated to three histidine N atoms and an H_2O molecule at the active site. Propose a Lewis acid mechanism for the attack of water on CO_2 when the latter is bound to zinc at the active site.

Answer. A Lewis acid mechanism would require coordination of CO_2 to the Zn(II) center, possibly as $Zn-O=C=O$. This structure would then be attacked by an H_2O molecule at the C atom to convert it to a coordinated HCO_3^- ion, accompanied by proton transfer to the enzyme.

Exercise E19.3. The X-ray data on crystallized carbonic anhydrase tend to argue against a Lewis acid mechanism. The active site includes a number of H_2O molecules in a very well ordered array, which suggests that CO_2 is not coordinated, but that the coordinated H_2O is activated by the Zn for nucleophilic attack on a C atom in a hydroxo mechanism. Propose a mechanism for the carbonic anhydrase catalyzed attack of water on CO_2 that is related to the hydroxo mechanism discussed above for carboxypeptidase A.

REDOX CATALYSIS

Because photosynthesis and respiration, the two primary energy conversion processes, involve redox reactions, it is clear that a major metabolic role must be played by the enzymes that catalyze oxidation and reduction. As we saw in Chapters 7 and 15, such reactions may occur by electron transfer, atom transfer, and group transfer, and metalloenzymes can function in all three of these ways. Here we shall concentrate on iron, the one metallic element that is common to all forms of life.

Reduction and oxidation in a living cell do not occur in a single step but usually involve a series of compounds called **mediators**, which act like a series of locks on a canal, allowing oxidation to occur in stages. Once again, control is of paramount importance, for uncontrolled oxidation by oxygen is combustion.

---

[12] Carbonic anhydrase is reviewed by D. N. Silverman and S. Lindskog in *Acc. Chem. Res.*, **21**, 30 (1988). They discuss a hydroxo mechanism, and alternatives where proton transfer is rate determining.

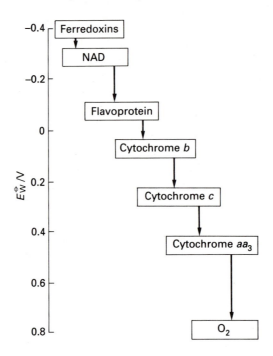

19.14 Reduction potentials of some important electron transfer mediators in biological cells at pH = 7.

Figure 19.14 illustrates the range of standard potentials of some mediators found in mitochondria (the organelles in which oxygen utilization is accomplished). They include the heme-containing cytochromes and the flavoproteins, which also mediate the reactions that insert oxygen into organic molecules. Nicotinamide adenine dinucleotide (**20**, NAD) is a widely distributed substance that participates in reductions by achieving the equivalent of H$^-$ ion transfer. The most strongly reducing member of the chain in Fig. 19.14 is ferredoxin, a member of a very interesting group of iron–sulfur proteins.

19.5 Iron–sulfur proteins and non-heme iron

The porphyrin ligand environment of iron which was seen in Hb and Mb is also important in redox enzymes. Thus the large class of biochemically important heme proteins have Fe coordinated to a porphyrin ligand. All other iron proteins are defined as non-heme. They constitute a large group, among which those containing iron in a tetrahedral environment of four S atoms are very important. Iron–sul-

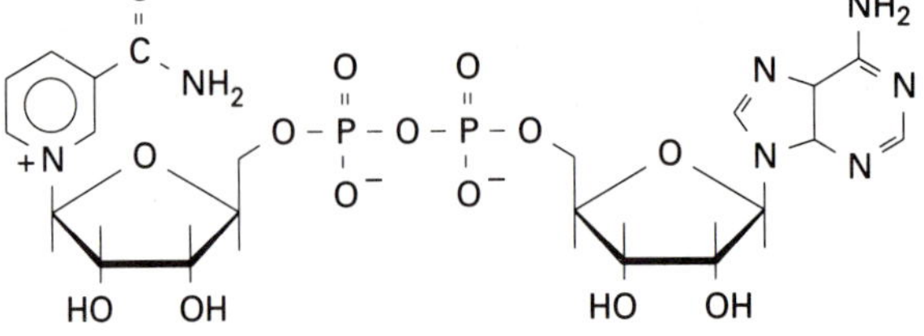

20 NAD

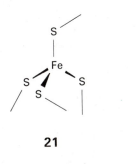

21

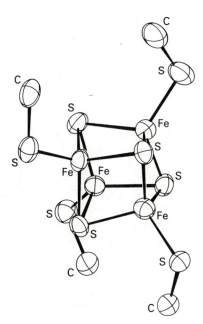

22

23 $[Fe_4S_4(SCH_2Ph)_4]^{2-}$

fur clusters were not familiar to inorganic chemists prior to the recognition of their biochemical importance. They display reduction potentials from 0.0 to -0.5 V and are involved in reducing NAD to NADH and N_2 to NH_3; they can generate H_2 from acidic solutions.

One or more Fe atoms may participate at the active site in iron–sulfur proteins. Examples are the isolated Fe atom in a tetrahedral environment of RS^- from cystine (**21**), two Fe atom clusters formed with S^{2-} bridging atoms (**22**), which are called two-iron ferredoxins, and the 4Fe,4S ferredoxin shown in the model structure (**23**).

The fact that the Fe—S cluster is always held in a fold of the protein enables control to be exercised over access by substrates. The Fe—S cluster shows little change in bond length on gain of an electron despite the resulting conversion of high spin Fe(III) to high spin Fe(II). As a result, there is little barrier to electron transfer arising from reorganization, and the common function of these proteins is one-electron transfer. The iron–sulfur proteins are usually found in membranes or cytoplasm. Various side groups can render the whole protein water-soluble or lipid-soluble to adapt to the cytoplasm or membrane, respectively.

Modeling these Fe—S clusters has stimulated a significant area of synthetic and structural inorganic chemistry.[13] For instance, it has been found that the reaction of $FeCl_3$, $NaOCH_3$, NaHS, and benzyl-thiol in methanol gives $[Fe_4S_4(SCH_2Ph)_4]^{2-}$ (**23**). The magnetic susceptibility, electronic spectrum, redox properties, and ^{57}Fe Mössbauer spectrum are analogous to those of ferredoxins. Oxidation number rules suggest two Fe(II) atoms and two Fe(III) atoms, but all spectroscopic methods, including X-ray photoelectron spectroscopy, suggest that all four Fe atoms are equivalent. Consequently, a delocalized electron description is most appropriate, and the electron transfer reactions involve orbitals that are delocalized over the cluster with the S and Fe atoms both sharing in the change of oxidation state.

Example 19.4: *Writing redox reactions that involve thiolate ligands used in cluster synthesis*

One of the problems that has plagued synthetic chemists in their attempts to prepare model compounds for cysteine-complexed metal ions in metalloproteins is the easy oxidation of thiolate anions (RS^-) to RS—SR. Simple complexes with Cu^{2+}—SR and Fe^{3+}—SR bonds which might serve as models for cytochrome P-450 and the ferredoxins (and, in addition, blue copper proteins) are unstable because of this reaction. Write balanced equations for the decomposition of $[Cu(II)L_n(SR)]$ and $[Fe(III)L_n(SR)]$.

Answer. If thiolate ligands are oxidized to disulfides, the Cu(II) or Fe(III) must be reduced to Cu(I) or Fe(II), respectively. The balanced equations, ignoring the possible charges on the complexes, are

$$2[Cu(II)L_n(SR)] \rightarrow 2[Cu(I)L_n] + RSSR$$

$$[Fe(III)L_n(SR) \rightarrow 2[Fe(II)L_n] + RSSR$$

[13] R. H. Holm, *Adv. Inorg. Chem.*, **38**, 1 (1992).

Exercise E19.4. Early attempts to prepare models of the 2Fe-2S ferredoxins with simple thiolate ligands did not lead to $[Fe_2S_2(RS)_4]^{2-}$ clusters but instead to $[Fe_4S_4(SR)_4]^{2-}$. By considering the average oxidation state of Fe in these clusters, suggest a reason for the difficulty in their preparation.

19.6 Cytochromes of the electron transport chain

Oxygen is a powerful and potentially dangerous oxidizing agent. If the cell is to use it in metabolism, then it must keep a safe distance between the sites of oxidation by O_2 and the variety of metabolic reactions being fueled. This isolation is accomplished by the series of proteins of decreasing standard potential which pass electrons toward oxygen in the mitochondria of the cell. A schematic version of the mitochondrial electron transport chain is shown in Fig. 19.15.[14] The most prominent participants are the **cytochromes**, a group of heme proteins with Fe in a porphyrin type ligand environment.

There are large Gibbs free energy changes at two stages in Fig. 19.15, one at the reaction between the pair of cytochromes denoted cyt b/cyt c_1 and the second where cytochrome a reacts with oxygen (denoted cyt a/O_2). The energy released in these stages can be stored for use elsewhere in the cell by coupling the reaction to the formation of ATP from ADP and HPO_4^{2-}. The chain might seem unnecessarily complex, but it breaks the reaction of O_2 with NADH into discrete steps. This allows coupling to the synthesis of several ATP molecules as well as organizing the process spatially from one side of the membrane (where the cytochromes are located) to the other.

Outer-sphere character

Cytochromes function by shuttling iron between Fe(II) and Fe(III) at the active site and hence are one-electron transfer reagents. The Fe atoms are in porphyrin ring coordination environments buried in the middle of the protein. By using a computer to simulate steric and electrostatic interactions, it is possible to generate plausible models of the complex that is formed when two of these proteins meet to exchange

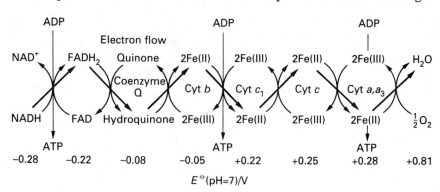

19.15 The sequence of reactions in the electron transport chain in the mitochondria of cells where oxygen is utilized.

[14] G. R. Moore and G. W. Pettigrew, *Cytochrome c: structural and physicochemical aspects.* Springer Verlag, Berlin (1990).

Table 19.7 Some rates of electron transfer reactions among redox proteins[†]

Couple	$E^{\ominus}/V$	$R/\text{Å}$[‡]	k/s^{-1}
$Fe^{II}cytb_5/Fe^{III}cytc$	0.2	8	1.5×10^3
$Znapocytc^*/Fe^{II}b_5$	0.8	8	3×10^5
$H_2porfc^*/Fe^{III}b_5$	0.4	8	1×10^4
$Zncytb_5/Fe^{III}cytc$	1.1	8	1×10^3
$RuHis33/cytc$	0.15	11	40
$RuHis/azurin$	0.2	10	2.5
$RuHis/Mb$	0.05	13	0.02
$Fe^{II}ccp/Fe^{III}cytc$	0.4	16	0.025
$Fe^{III}ccp/porfcytc$	1.0	16	180

[†]Most of the arcane abbreviations are designations of proteins. They are elaborated in biochemical texts. For our purposes we do not need to know the details. All that the different abbreviations mean here is that different members of the family of electron transfer proteins are associated with the different values of $E^{\ominus}$ and RuHis refers to the group $(NH_3)_5Ru$ attached to a histidine (number indicated if necessary) on the enzyme. Distances can be varied by changing the histidine unit to which the Ru is bound. The notation * indicates a reaction using a photoexcited state to make the reaction more exoergic.
[‡]Distance between electron donor and acceptor metal centers in Å.

an electron. Figure 19.16 shows a computer model of the encounter between two cytochromes, and it is easy to believe that because the Fe atoms of the two encountering complexes always remain far apart, the reaction cannot be a case of inner-sphere electron transfer. The reaction must take place by long-distance outer-sphere electron transfer.[15]

Table 19.7 shows some rate constants, standard potentials, and distances of closest approach of the edge of the porphyrin ring containing the Fe atom and its electron transfer partner. It is instructive to compare these protein–protein parameters with similar parameters for reactions of a protein with simple metal complexes, and some of the latter are included in the table too. Of particular interest is the variation of the reaction rate with distance and energy.

Distance dependence and tunneling

The distance dependence of the rate constant is usually interpreted in terms of quantum mechanical tunneling, in which a particle may escape

19.16 One stage in a computer simulation of the encounter between two electron transfer proteins. The shaded areas are the iron porphyrins.

[15] An interesting question is whether the pathway is along specific directions of the peptide chain or through space. Evidence for the former is discussed in D. N. Beratan, J. N. Onuchic, J. N. Betts, B. E. Bowler, and H. B. Gray, *J. Am. Chem. Soc.*, **112**, 7915 (1990).

through a barrier despite having insufficient energy to surmount it. In simple models of tunneling, the probability of finding the particle outside the barrier declines exponentially with the width of the barrier. The data in Table 19.7 confirm that the distance between the metal centers has the expected effect if comparisons are limited to similar structures.

The energy dependence and Marcus theory

The variation of reaction rate with energy can be considered in the light of Marcus theory (Section 15.12). In particular, we can consider whether the theories developed to explain outer-sphere electron transfer between small molecules also work for large molecules that make contact only at large distances.

We saw in Section 15.12 that the Marcus theory allows us to express the rate constant k_r of any outer sphere electron transfer as

$$k_r^2 = k_1 k_2 K f \qquad f \approx 1$$

and equivalently

$$\ln k_r = \ln (k_1 k_2)^{1/2} - \frac{\Delta G^\ominus}{2RT}$$

The product $k_1 k_2$ of self-exchange rate constants reflects the intrinsic barrier to electron transfer of the two complexes. The equilibrium constant K is a measure of the overall reaction free energy $\Delta G^\ominus$. Unfortunately, the intrinsic barrier is not easily estimated for many proteins because it is not feasible to study self-exchange between two oxidation states of a protein.

The best approach seems to be to consider the outer-sphere reaction of a protein and a small molecule. In such cases the reaction free energy and the self-exchange rate of the small molecule couple can be known. The difficulty lies only in finding the self-exchange rate constant for the protein couple. If we suppose that the Marcus theory does indeed apply to the reaction, then we can calculate a self-exchange rate constant for the protein couple using the known values of k_r and k_1. Then we can use the value of k_2 so obtained to discuss other reactions of the protein. In general, this approach is quite successful and supports the view that the Marcus relation is valid.

The variation of k_r with the overall reaction free energy (and, equivalently, with the standard potentials of the proteins through $\Delta G^\ominus = -nFE^\ominus$ with $n = 1$) is also as predicted by Marcus theory for proteins with similar structures. However, the structure also plays a role in determining k_r, and different types of protein structures have rate constants that may differ by orders of magnitude. For instance, the first three entries in Table 19.7 form a series with the dependence on $E^\ominus$ predicted by the Marcus equation. Similarly, the last two entries may also represent a structurally similar pair. The role of the enzyme structure in determining the intrinsic barrier is clearly indicated by comparing the fourth entry to the first three. The potential is more

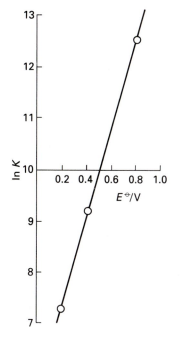

19.17 The plot of the data in Example 19.5.

favorable than any of the first three, yet the rate constant is smaller.[16] The general conclusion appears to be that the cytochromes and several other redox proteins are simple outer-sphere one-electron transfer reagents.

Example 19.5: *Testing the Marcus theory*

Test the assertion that the first three points in Table 19.7 show the dependence on Gibbs free energy expected from Marcus theory.

Answer. To test the theory, we plot the logarithm of the rate constants against the overall free energy change for the electron transfer. Because $E^{\ominus}$ is proportional to the reaction free energy, we may use the standard potentials. The logarithms of the first three rate constants are 7.31, 12.61, and 9.21. The corresponding potentials are 0.2, 0.8, and 0.4 V, respectively. The straight line (Fig. 19.17) confirms the approximately linear relationship predicted by the theory.

Exercise E19.5. Compare the first and the sixth, the third and the eighth, and the fourth and the ninth entries in Table 19.7. The $E^{\ominus}$ values are nearly the same but rates and distances differ. Compare the change in distance to change in $\ln k_r$. Do these results confirm exponential distance dependence of the intrinsic barriers to electron transfer or must another factor be invoked?

19.7 Cytochrome P-450 enzymes

'Cytochrome P-450' is the designation of a family of enzymes with iron porphyrin active sites that catalyze the addition of oxygen to a substrate.[17] The designation 'P-450' is taken from the position of the characteristic blue to near-ultraviolet absorption band of the porphyrin; the band is called the 'Soret band' and is red shifted to 450 nm in carbonyl complexes of these molecules. The most important representative of this class of reactions is the insertion reaction:

$$R-H + \tfrac{1}{2}O_2 \rightarrow R-OH$$

The insertion of O into an R—H bond—which is just the sort of redox reaction likely to occur by an atom transfer mechanism—is a part of the body's defense against hydrophobic compounds such as drugs, steroid precursors, and pesticides. The hydroxylation of RH to ROH renders the target compounds more water soluble and thereby aids their elimination.

---

[16] Physical factors controlling long range electron transfer are analyzed in G. L. Closs and J. R Miller in *Science*, **240**, 440 (1988). Two excellent surveys of long range electron transfer in enzymes are given by G. McLendon in *Acc. Chem. Res.*, **21**, 160 (1988), and C. C. Moser, J. M. Keske, K. Worneka, R. S. Farid, and P. L. Dutton, *Nature*, **355**, 776 (1992).

[17] G. R. Moore and G. W. Pettigrew *Cytochrome c: structural and physicochemical aspects.* Springer Verlag, Berlin (1990); D. Ostovic and T. C. Bruce, Mechanism of alkene epoxidation by iron, chromium, and manganese higher valent oxo-metalloporphyrins. *Acc. Chem. Res.*, **25**, 314 (1992).

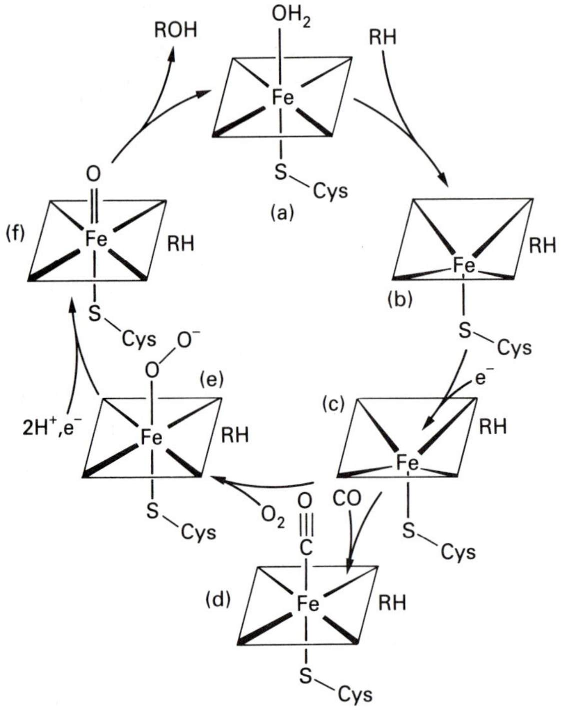

19.18 The cycle of reactions of cytochrome P-450. The resulting state of the enzyme is (a) and the important Fe(IV) oxo species is (f).

The proposed catalytic cycle for P-450 is shown in Fig. 19.18. The sequence begins at (a) in the figure with the enzyme in a resting state with iron present as Fe(III). The hydrocarbon substrate then binds (b), and one electron is transferred (c). The resulting Fe(II) complex with bound substrate proceeds to bind O_2 (e). (At this point in the cycle, a competing reaction with CO to give (d) leads to a species which is easily identified and is responsible for the absorption at 450 nm which gives the family its name.) A key reaction is the reduction of the porphyrin ring of the oxygen complex (e) by a second electron, which produces the ring radical anion. Uptake of two H^+ ions then leads to the formation of the Fe(IV) complex (f) which attacks the substrate to insert oxygen. Loss of ROH and uptake of an H_2O molecule at the vacated coordination position brings the cycle back to the resting state. The key to the cycle is the formation of the Fe(IV) oxo complex.

Structure of the active site

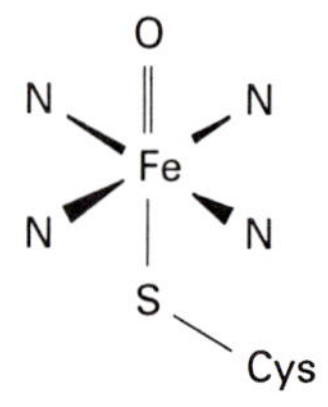

The amino acid sequence of the protein creates a folded structure around the heme site and its Fe atom, so forming an environment shielded from the solution and with a low permittivity. The hydrocarbon substrate is bound nearby, and the C—H bond under attack may be as little as 5 Å from the Fe site. The ligand on the Fe atom lying *trans* to the site of oxygen binding is a thiolate side chain of a cysteine residue (**24**). When O_2 coordinates, the outer end of the O_2 molecule projects into the solution, but the substrate is bound to the enzyme within the hydrophobic pocket.

24

Oxygenation mechanism

The precise mechanism of oxygenation remains a subject of intense research. There are thought to be two possibilities, one involving the generation of an oxygen radical species which can attack the C—H bond, and the other the transfer of an O atom to the C—H bond. There are no good inorganic precedents for oxidation of an organic compound by $Fe^{(IV)}O$, and the current debate about the mechanism of the P-450 reaction will almost certainly enlarge the range of possible oxidation mechanisms known to inorganic chemists.[18] The proposal of a radical mechanism entails some unconventional features. Radical oxidations are usually relatively unselective and not stereospecific; P-450 oxidations are quite selective and preserve the optical activity of chiral substrates.

The proposal that oxygenation is accomplished by electrophilic attack of a positive oxygen center on the C—H bond is just one example of a novel mechanism proposed to explain an enzyme reaction. It is supported by evidence of retention of stereochemical configuration and the reduction of reactivity for cases in which the substrate has substituents that block the approach of the O atom to the C—H bond. The transfer of what is effectively a neutral O unit leaves the iron porphyrin reduced from Fe(IV) to Fe(III) while the organic substrate is oxidized.

The role of the S atom in this pathway can be expressed in terms of molecular orbitals by focusing attention on the atomic orbitals of the S—Fe—O fragment. If we limit our attention to the sulfur $3p_z$, the iron $3d_{z^2}$, $3d_{zx}$, $3d_{yz}$, and oxygen $2p$ orbitals, then we obtain the orbital scheme shown in Fig. 19.19. The 13 electrons available in the fragment occupy the bonding and weakly antibonding levels, and the last electron occupies the strongly antibonding 3σ orbital. The high formal oxidation number of the iron in the complex can therefore be rationalized in terms of its stabilization by considerable charge donation from the readily polarized S atom into the 2σ and 2π orbitals of the fragment. The special role of the *trans* S atom is to control the charge on the Fe atom so that an O atom can be transferred.

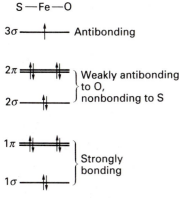

19.19 The molecular orbital scheme of the linear S—Fe—O fragment in a cytochrome P-450 enzyme.

Example 19.6: *Judging the electron configuration of a metalloenzyme*

The Fe=O oxidizing center in P-450 is characterized as an Fe(IV) complex with the porphyrin also oxidized by one electron. Such a picture receives strong support from the ^{57}Fe Mössbauer spectrum[19] of the species. The complex has a magnetic susceptibility that indicates the presence of three unpaired electrons. To what orbitals are they assigned?

---

[18] A. E. Martell and D. T. Sawyer (ed.), *Oxygen complexes and oxygen activation by transition metals.* Plenum, New York (1988).

[19] Mössbauer spectroscopy measures the resonant absorption of a γ-ray by a nucleus held rigidly in a crystal framework so that the emission and absorption of γ-ray photons is recoil-free. The chemical environment of the nucleus affects the absorption frequency, and so the spectrum is a probe of the electron distribution. The elements most commonly studied are Fe, Sn, and I. The technique may be used to identify the oxidation state of iron.

25

Answer. The Fe(IV) oxidation state corresponds to a d^4 configuration. Assuming it to be approximately octahedral, in a strong field environment it should be t_{2g}^4 with two unpaired electrons (**25**). The metal ion is in a triplet state ($S = 1$). The oxidation of the porphyrin ring removes one electron from the π HOMO, leaving the third unpaired electron there. This configuration could be considered to be a doublet state of the oxidized free ligand.

Exercise E19.6. In the resting state of P-450, the complex is treated as five-coordinate Fe(III) complexed to four porphyrin N atoms and one cysteine S atom, with overall C_{4v} symmetry. Assign electrons to the iron d orbitals.

19.8 Nitrogen fixation

In Section 17.7 we discussed the high pressure and high temperature catalysis of the reaction of H_2 with N_2 to produce NH_3. Without the questionable benefit of naturally occurring high temperatures and pressures in the biosphere, Nature has taken an entirely different and much more elaborate route to NH_3. The reducing agent used by Nature is ATP and the reduction half-reaction can be represented as

$$N_2 + 16MgATP + 8e^- + 8H^+ \rightarrow 2NH_3 + 16MgADP + 16P_i + H_2$$

where P_i is inorganic phosphate. This process is probably much less efficient than the industrial Haber process because energy is put into producing hydrogen and protecting the highly reducing biological system from atmospheric oxygen. The appealing feature, however, is that the process occurs at room temperature and pressure in the *Rhizobium* organisms which live in the root nodules of various legumes (clover, alfalfa, beans, peas, etc.) as well as several bacteria and blue–green algae. The enzyme nitrogenase carries out the reaction under essentially anaerobic conditions. The problem the enzyme has solved is how to overcome the great inertness of the N≡N molecule.

The mechanistic details of the fixation of N_2 are unclear, but it is known that nitrogenase employs an iron–sulfur protein and a molybdenum–iron–sulfur protein, as indicated in the sequence in Fig. 19.20.[20]

Biochemists have isolated the metal-containing cofactors involved in this catalysis, and bioinorganic chemists have prepared many potential structural models of the active site. A major breakthrough in this area was the successful crystallization and X-ray structure determination of the $MoFe_7S_8$ cofactor (Fig. 19.21(a)) and an associated 'P' cluster (Fig. 19.21(b)) which contains two 4Fe,4S clusters joined by sulfur bridges.[21] As discussed in Section 19.5, the latter type of cluster is common in electron transfer systems and remains intact during sequential electron transfer reactions.

The X-ray structure of the Mo—Fe—S cluster at which N_2 reduc-

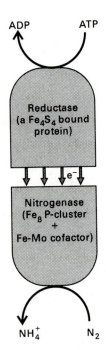

19.20 The coupling between the dephosphorylation of ATP to ADP and the reduction of dinitrogen to NH_4^+.

[20] W. H. Orme-Johnson, *Science*, **257**, 1639 (1992).
[21] M. K. Chan, J. Kim, and D. C. Rees, *Science*, **260**, 792 (1993).

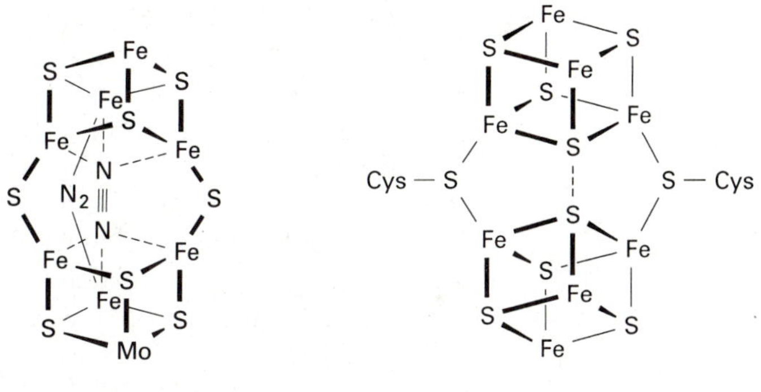

19.21 (a) The MoFe$_7$S$_8$ cofactor in nitrogenase. The nitrogen was not located in the X-ray diffraction experiment, but it was positioned in this drawing to fill an apparent void in the structure. (b) The Fe$_8$ 'P cluster' associated with (a). See M. K. Chan, J. Kim, and D. C. Rees, *Science*, **260**, 792 (1993).

tion appears to take place has an open site. It is speculated that in a more reduced form of this cluster the N$_2$ might bind as illustrated in Fig. 19.21(a) and undergo reduction along with the addition of protons. We are far from knowing the details of this process or indeed whether the N$_2$ is bound in this unprecedented manner, but the new structural data provide a more definite model for the active site than had existed previously.

Dinitrogen complexes are well known in inorganic chemistry (Section 12.2), but they are quite different from the proposed biological structure. In the presence of strong acid, the N$_2$ in some of the inorganic complexes undergoes a proton-induced reduction to ammonium ion:[22]

$$[\text{Mo}(\text{PR}_3)(\text{N}_2)_2] + 8\text{H}^+ \rightarrow 2\text{NH}_4^+ + \text{N}_2 + \text{Mo(VI)} + \dots$$

The idea that multiple metal binding sites might facilitate the reduction of N$_2$ is echoed in organometallic chemistry, where it has been demonstrated that metal clusters facilitate the proton induced reduction of the isoelectronic molecule CO.[23]

19.9 Photosynthesis

One of the most remarkable redox reactions is the thermodynamically 'uphill' conversion of water and carbon dioxide into carbohydrates and oxygen using solar radiation as an energy source. Formally, the production of carbohydrates involves the reduction of CO$_2$ and the oxidation of 2H$_2$O to O$_2$. There are two photochemical reaction centers, **photosystem I** and **photosystem II** (PS I and PS II), and the photosynthetic system is organized in the green leaf organelles called **chloroplasts**.[24] A general point to bear in mind in the following is that the primary steps following photoexcitation are outer-sphere electron transfers which have been directly observed by picosecond spectroscopy.

[22] G. J. Leigh, *Acc. Chem. Res.*, **25**, 177 (1992).
[23] M. A. Drezdzon, K. H. Whitmire, A. A. Battacharryya, W. L. Hsu, C. C. Nagel, S. G. Shore, and D. F. Shriver, *J. Am. Chem. Soc.*, **104**, 6247 (1982).
[24] PS II has been reviewed by J. B. Vincent and G. Cristou, *Adv. Inorg. Chem.*, **33**, 244 (1989) and G. W. Brudvig, H. H. Thorp, and R. W. Crabtree, *Acc. Chem. Res.*, **24**, 311 (1991).

Photosystem I is based on chlorophyll a_1, a magnesium dihydroporphyrin complex (**26**). After PS I has been photoexcited, it can act as a reducing agent for an iron–sulfur complex, and the electrons it transfers to the complex are ultimately used to reduce CO_2. The oxidized form of PS I that remains after the electron transfer step is not strong enough to oxidize water. Its return to the reduced form is used to drive the conversion of two ADP molecules to two ATP molecules through a sequence of mediating species that include several iron-based redox couples and a quinone called plastoquinone.

The oxidized form of PS II is a strong enough oxidizing agent to oxidize water. However, the reaction involves the net transfer of four electrons

$$2H_2O(l) \rightarrow 4H^+(aq) + O_2(g) + 4e^-$$

and is achieved only by a complex series of reactions that utilize the redox reactions of a manganese based enzyme which transfers electrons to the photochemically active center. Experiments with successive light flashes show that the enzyme system is oxidized in four one-electron steps before accomplishing the four-electron oxidation of water to O_2. If the redox reservoir involves the reduction of Mn(IV) to Mn(II), the enzyme must contain at least two Mn atoms to accommodate the four electrons.

A specific proposal for the cycle of events that take place in the system is shown in Fig. 19.22. The resting state S_0 is oxidized to S_1 by the loss of one electron and one proton. In the second step, S_2 is formed by transfer of a second electron. In the conversion of S_2 to S_3, the third level of oxidation, the electron loss is accompanied by uptake of the two H_2O molecules which supply the oxygen, and a proton is lost. After one more electron has been transferred to give S_4, the critical reaction occurs in which O_2 is released and the Mn enzyme is reduced by the equivalent of four electrons. A proposal for the active site in the S_3 site, based largely on spectroscopic data, is illustrated in Fig. 19.23.[25] Additional information indicates that this site is a mixed oxidation state cluster that can be represented as {2Mn(II), 2Mn(IV)}, or possibly as 4Mn(III). This is an active and important area of research where it is not yet possible to provide definitive information on the nature of the active site of the enzyme through all four steps.

26 Chlorophyll a_1

19.22 The four steps of oxidation which precede the release of O_2 in the oxygen evolving system of PS II.

Chlorophyll

The chromophore in the photosystem, chlorophyll a_1 (**26**), has the absorption spectrum shown in Fig. 19.24. The strong absorption band in the red (called the Q band) and the band in the blue to near ultraviolet (the Soret band) are characteristic of porphyrins. Both arise from promotion of electrons from the porphyrin π HOMO to the π^* LUMO. The non-absorbing region between these two bands accounts

[25] V. K. Yachandra, V. J. De Rose, M. J. Latimer, I. Makerji, K. Sauer, and M. P. Klein, Where plants make oxygen: A structural model for the photosynthetic oxygen-evolving manganese cluster. *Science*, **260**, 675 (1993).

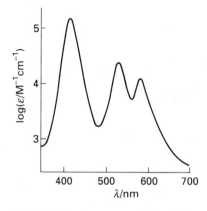

19.23 Proposed structure of the S_1 state complex in photosystem II, which appears to contain a 2Mn(II), 2Mn(IV) mixed oxidation state cluster. (From ref. 25).

19.24 The absorption spectrum of chlorophyll a_1 in the visible region.

27

for the characteristic green color of vegetation. Other pigments—called 'antenna' pigments—are also present in leaves to harvest some light in the absorption gap and transfer the energy to the reaction center.

The initial absorption is from the singlet ground state to a singlet excited state of the chlorophyll Q band (^{1}Q). Isolated chlorophyll molecules in solution undergo rapid intersystem crossing into the lower lying spin triplet (^{3}Q), and this state is responsible for the photochemical electron transfer reactions of the molecule in solution. In contrast, the PS reaction center shortcircuits the intersystem crossing by immediately transferring an electron from ^{1}Q to electron acceptors that are adjacent to the chlorophyll in the chloroplast. We see yet again that spatial organization is a key aspect of biochemistry. A simplified model of such a reaction is obtained using the porphyrin–amide–quinone molecule (**27**) which exhibits fluorescence properties, suggesting that charge transfer from the singlet state of the porphyrin to the quinone is 10^4 times faster than relaxation to the triplet.

Reaction center organization

The spatial organization in the molecule allows the rapid transfer (within 10 ps) of the energy of the singlet excited state. The singlet state is at higher energy than the triplet and its use results in increased energy efficiency: intersystem crossing into a triplet squanders energy as heat. At this stage, spatial organization plays another essential role. The products of reactions that store energy could revert to reactants, so wasting energy, but the system is so organized that electrons are transferred along the reaction sequence faster than the reverse reactions can take place.

A reaction center of a photosynthetic bacterium has been crystallized and the structure determined by X-ray diffraction.[26] A 'special pair' of bacteriochlorophylls, which has long been recognized to be essential to the rapid initial photoelectron transfer, lies at the site of light absorption. Around this pair is organized a double chain of the acceptor molecules. One puzzle is that the series of acceptors seems to be duplicated, but the chemical evidence suggests that one component of the structure is dormant and that only one of the pair is active in electron transport.

SUMMARY

We have seen a variety of important roles for metal ions in biological systems, including energy storage and release, oxygen transport and storage, hydrolytic enzyme action, electron transfer, selective oxidation of C—H bonds, nitrogen fixation, and photosynthesis. Other interesting issues beyond the scope of this chapter are how *d*-metal ions are transported to the sites where they are needed and how metal ion concentrations are sensed and controlled. Indeed, so complex and

[26] H. Michel, O. Epp, and J. Deisenhofer, *Eur. Mol. Bio. Org.*, **5**, 2445 (1986); J. Deisenhofer, O. Epp, K. Michi, R. Huber, and H. Michel, *J. Mol. Biol.*, **180**, 385 (1984).

intricate are the inorganic aspects of organisms, that there can be no doubt that they will occupy inorganic chemists for years to come and be a constant and continuing source of stimulation and inspiration.

FURTHER READING

L. Stryer, *Biochemistry*. W. H. Freeman & Co., New York (1988). A general textbook of biochemistry which presents mechanisms involving inorganic species in their biochemical context.

S. J. Lippard and J. M. Berg, *Principles of bioinorganic chemistry*. University Science Books, Mill Valley (1993). This book develops the general concepts of the field including: the properties of biomolecules such as proteins, RNA, and DNA; the binding of metals in biomolecules.

J. J. R. Fraústo da Silva and R. J. P. Williams, *The biological chemistry of the elements: The inorganic chemistry of life*. Oxford University Press (1991). This is another bioinorganic text with a broad view of the subject. The emphasis is on the way the chemical properties of the elements determine their role in biology.

R. J. P. Williams, *Coord. Chem. Rev.*, **100**, 758 (1990). A view of the implication of considering biological systems as dynamic structures and the role of the elements.

R. W. Hay, *Bioinorganic chemistry*. Ellis Horwood, Chichester and New York (1984). A good brief general introduction.

S. Mann, J. Webb, and R. J. P. Williams, *Biomineralization: chemical and biochemical perspectives*. VCH, Weinheim (1989). This more specialized book stresses skeletal materials. The 6 August 1993 issue of *Science*, **216**, 699–730, contains a series of articles devoted to bioinorganic chemistry. It begins with a survey by S. J. Lippard, and includes discussions of metalloenzymes (K. D. Karlin), ribozymes (A. M. Pyle), metals in gene expression (T. V. O'Halloran), and medicinal application of metal compounds (M. J. Abrams amd B. A. Murrer).

KEY POINTS

1. Inorganic components of organisms

The inorganic components of organisms maintain structure, participate in many dynamic processes such as ion, energy, electron, and oxygen transport, and are central to the operation of enzymes.

2. Biologically important elements

The light and abundant elements C, H, N, O, and P are the most abundant in living systems and occur in proteins, carbohydrates, and the structural materials of plants and animals. At a lower but still quite significant level, two alkali metal ions Na^+ and K^+, two alkaline earth metal ions Mg^{2+} and Ca^{2+}, and the nonmetals S, Cl, and I occur in biological electrolytes and structural materials. A variety of other elements, such as many of the common *d* metals, are present in low abundance but play essential roles in enzymes, transport and storage proteins, and photosynthesis.

3. Sodium/potassium pump

The distribution of Na^+ ions primarily outside the cells and of K^+ ions primarily inside the cells of organisms is maintained by the conversion of ATP to ADP. Selectivity in the transport of these two ions is achieved by hydrated channels which transport Na^+ and the larger water-free channels for K^+.

4. Calcium

Calcium plays diverse roles in animals, ranging from the structural minerals apatite and calcite to muscle proteins. It is typically bound by carboxylate groups on the proteins, where it may induce conformational changes.

5. Iron porphyrins

Iron porphyrins play a major role in O_2 binding in the oxygen-carrying protein hemoglobin and the oxygen-storage protein myoglobin. In both cases the iron is bound by a porphyrin ring. Synthetic porphyrins (picket-fence porphyrins) that can mimic this O_2 uptake provide a pocket that shields the coordinated O_2 from coordination to a second iron center.

6. Zinc

Zinc is in the active site of the enzyme oxaloacetate decarboxylase, where it provides an acid site that catalyzes a step in the oxidation of a C—C bond in the Krebs cycle. It also is present in carboxypeptidase A, which catalyzes hydrolysis of peptide bonds in polypeptides and proteins.

7. Iron–sulfur proteins

Iron–sulfur proteins, such as the ferredoxins, are found in the reducing environment of cell membranes or cytoplasmic fluid, where they participate in outer-sphere electron transfer. One common active site contains the four-iron, four-sulfur cubane structure.

8. Cytochrome P-450

Cytochrome P-450 is a class of iron–porphyrin enzymes that catalyze oxidation reactions, such as the selective insertion of O into certain C—H bonds.

9. Nitrogen fixation

The fixation of dinitrogen is catalyzed by the enzyme nitrogenase, which contains a 4Fe,4S protein plus a molybdenum–iron cluster protein. The first protein provides electrons and the second catalyzes the dinitrogen reduction.

10. Photosynthesis

Photosynthesis in plants involves two reaction centers (photosystems I and II). PS I harvests light with chlorophyll a_1, a magnesium dihydroporphyrin, and through a series of electron transfer steps the oxidizing power is transferred to PS II which oxidizes water to O_2. The latter reaction is a four-electron process that involves a series of reactions with a manganese-containing enzyme.

EXERCISES

19.1 Consider the elements O, N, K, and Ca. Identify where they are concentrated in animals and describe a major function for each one.

19.2 Give the chemical equation and the associated standard reaction free energy for ATP hydrolysis to ADP. Speculate on why this process is adopted in Nature to supply and store energy in so many biochemical processes.

19.3 Describe the origin of the selectivity of some ion channels for Na^+ and others for the similar ion K^+.

19.4 Describe the characteristics of the ligands that are adopted for binding Ca^{2+} to proteins and those used to bind Fe^{2+} in the oxygen-carrying protein hemoglobin. Suggest some reasons for the differences.

19.5 In the discussion of O_2 coordination, O_2, O_2^-, and O_2^{2-} were considered as limiting forms. Consider the molecular orbital energy level diagram for O_2 from Chapter 2. What is the implication for bond length and net spin of the choice of each of these O_2 species as a model of the ligand?

19.6 (a) Does the equilibrium position of the following reaction lie to the left or the right?

$$Hb + Hb(O_2)_4 \rightleftharpoons 2Hb(O_2)_2$$

Use Fig. 19.7 to explain your reasoning. (b) Does the equilibrium position of the reaction

$$Hb(O_2)_4 + 4Mb \rightleftharpoons Hb + 4Mb(O_2)$$

lie to the left or the right? Does the equilibrium depend on the partial pressure of O_2?

19.7 Oxygen is a σ-donor and a π-acceptor. Carbon monoxide is also an excellent example of this type of ligand. Can you use these facts to propose a mechanism for CO poisoning?

19.8 The use of Co(III) centers for synthesis of peptides depends on the reaction

In this scheme, A stands for a peptide of arbitrary length. What is the nucleophile and what center is subject to catalyzed nucleophilic attack?

19.9 The substitution of Co(II) for Zn(II) gives a 'spectral probe'. What Co(II) spectral features are exploited? Why does Zn(II) lack them? If Co(II) is to serve as a useful probe for the zinc site, what assumption must be satisfied?

19.10 The diameter of a high-spin Fe(II) ion is larger than that of the 'hole' at the center of the porphyrin ring, whereas a low-spin Fe(II) ion is smaller than the hole. (a) Give the electron configurations for the two spin states in an octahedral environment. Why is the high-spin ion larger? (b) Give examples of ligands that might lead to six-coordinate high- and low-spin [Fe(porph)L_2] complexes.

19.11 Why are d metals such as manganese, iron, cobalt, and copper used in redox enzymes in preference to zinc, gallium, and calcium?

19.12 By considering the specific character of enzyme substrate binding, can you suggest why it is difficult to measure the self-exchange rate constants of redox enzymes?

19.13 Suggest two ways in which the protein parts of the enzymes contribute to making an outer-sphere electron transfer highly specific to one oxidant with one reductant partner.

19.14 Identify one significant role in biological processes for the elements iron, manganese, molybdenum, copper, and zinc.

19.15 What prevents simple iron porphyrins from functioning as O_2 carriers?

19.16 Sketch the steps illustrating a metal complex functioning as (a) a Brønsted acid and (b) a Lewis acid in an enzyme catalyzed reaction.

19.17 What is the average change in oxidation number of each of the four Mn atoms in the proposed reaction producing O_2 in the S_4 to S_0 step in the PS II cycle? See Fig. 19.22.

19.18 Describe the differences between the source of an electron in the reaction of the excited state of chlorophyll and its source in the Fe(II) state of a cytochrome. Could magnesium be used as the metal ion in a cytochrome?

PROBLEMS

19.1 Correct any of the following statements that are erroneous. Calcium carbonate is the main mineral constituent of bones. The sodium/potassium pump maintains a higher concentration of Na^+ than of K^+ in the cytoplasm. The attachment of one O_2 to hemoglobin increases the affinity for the next O_2 because of simple entropy considerations. Carboxypeptidase A utilizes a Zn^{2+} site to bring about the hydrolysis of peptides and two different mechanisms may apply.

19.2 A topic of current interest, the simulation of selective ion transport across membranes, is summarized by H. Tsukube in *J. Coord. Chem. B*, **16**, 101 (1987). (a) Describe the experimental arrangement for these measurements and contrast it with the nature of a cell immersed in extracellular fluids. (b) The 'armed' crown ethers described in the review display interesting selectivity. Describe the probable origin of that selectivity.

19.3 The attachment of Fe(TPP), where TPP is tetraphenylporphyrin, to a rigid silica gel support containing a 3-imidazoylpropyl group as a nitrogen donor ligand produces a good reversible oxygen carrier. How does this system prevent irreversible oxidation without a 'picket fence'? (See *Acc. Chem. Res.*, **8**, 384 (1975).)

19.4 The structures of Hb may be classified as 'relaxed' (R) or 'tense' (T) as alternative terms to oxygenated and deoxygenated. The R and T structures differ in both the relation among the four subunits (the quaternary structure) and the conformation within a subunit (the tertiary structure). Explain how these structural differences relate to the difference in the oxygen binding curve of Hb as compared to Mb.

19.5 Discuss the probable difference in the pockets present in carboxypeptidase and carbonic anhydrase.

19.6 Consider the four features of the binding of glycyltyrosine to carboxypeptidase listed in Section 19.4. Which are related to the spatial organization of the binding of the substrate to the enzyme and which promote the cleavage of the peptide bond? Explain your reasoning.

19.7 Chelating ligands that are specific for a particular metal ion are useful for the removal of toxic concentrations of metal ions from the body. See: K. N. Raymond and W. L. Smith, *Structure and Bonding*, **43**, 159 (1981) or F. L. Weith and K. N. Raymond, *J. Am. Chem. Soc.*, **102**, 2289 (1980). After reviewing either paper, summarize the way in which coordination number, charge, and hard–soft character can be utilized to remove the highly toxic Pu(IV) from mammals.

19.8 The substituted carboxypeptidase enzymes show a reactivity order Mn < Zn < Ni < Co. To what extent does this order follow the order of LFSE for tetrahedral complexes and thus support the notion that the determining factor is the bond of the metal to the carbonyl oxygen? What order would you predict if the coordination geometry were octahedral?

19.9 Consider the Frost diagram for oxygen (Fig. 7.10). Using a table of standard potentials, add a point for superoxide. Discuss the biological significance of these potentials. Keep in mind that a biological system is a reduced system that is threatened by oxidizing agents.

19.10 Consider the energy available from a photon at a wavelength of 700 nm. If the photochemical reaction at PS I generates a potential difference of 1 V, what is the efficiency of harvest of energy?

19.11 When photosynthesis uses the combination of PS I and PS II to accomplish the difficult process of water oxidation, is the arrangement the analog of connecting batteries in parallel or in series? When the system is organized by stacking chlorophyll in grana, which are in turn stacked in the chloroplasts, is this increasing efficiency by the analog of parallel or series connection?

Further information

FURTHER INFORMATION 1:
Nomenclature

We present in this section a concise set of rules and examples for naming inorganic compounds according to IUPAC conventions. With minor exceptions these rules are followed in this text. For information on detailed structural descriptors and on alternative schemes, consult the IUPAC official nomenclature book, G. J. Leigh (ed.), *Nomenclature of inorganic chemistry*. Blackwell Scientific Publications, Oxford (1990); CRC Press, Boca Raton (1990). This publication is widely referred to as the *Red book*. Another useful presentation is given by B. P. Block, W. H. Powell, and W. C. Fernelius, *Inorganic chemical nomenclature: Principles and practice*. American Chemical Society, Washington DC (1990). In the following sections, examples are given in parentheses after a rule is introduced.

Chemical formulas

Simple ionic compounds

For ionic compounds, the cation (more electropositive element) should always be first (KCl, Na_2S). If several cations are present, then they are listed in alphabetical order, followed by the anions in alphabetical order ($KMgClF_2$). An exception is the proton, which is listed last in the sequence of cations ($RbHF_2$).

Sequence of atoms in polyatomic ions and molecules

There is significant lattitude in writing the formulas of many compounds. Examples of acceptable formulas for acids, simple salts, and coordination compounds, are given in Tables F1.1 through F1.4, respectively. For neutral molecules or polyatomic ions with a central

Table F1.1 Systematic and traditional names for some chain and ring compounds

Formula	Systematic name	Traditional name
O_2	dioxygen	oxygen
O_3	catena-trioxygen	ozone
S_8	cyclo-octasulfur	sulfur
P_4	tetraphosphorus	white phosphorus
Hg_2^{2+}	dimercury($2+$)	mercurous
O_2^{2-}	dioxide($2-$)	peroxide
O_2^-	dioxide($1-$)	superoxide
C_2^{2-}	dicarbide($2-$)	acetylide*
N_3^-	trinitride($1-$)	azide
I_3^-	triodide($1-$)	triodide

*Confusingly, this traditional name also applies to HC_2^-.

atom, the central atom is generally listed first followed by the attached atoms or attached polyatomic groups in alphabetical order (SO_4^{2-}, $NClH_2$, PCl_3O, SO_3, CF_3). In the last example, CF_3 is a polyatomic attached group; its location in a formula is determined by the letter C. This scheme is similar to that used for coordination compounds. The chemical symbols are sometimes listed in an order that represents the structure, particularly for linear species (SCN^-, OCN^-, and CNO^-).

Coordination compounds

The ligand–metal array is written with the central metal atom given first followed by anionic ligands in alphabetical order, and then neutral ligands in alphabetical order of the first symbol of their formulas. Ligand abreviations may be used in place of complete formulas (en, for $H_2NC_2H_4NH_2$). A list of ligand abbreviations is given inside the back cover. The formula for a metal–ligand entity, which in this text we call a complex (a usage frowned on by IUPAC), is enclosed in brackets whether it is charged or uncharged ($[Co(Cl)_3(NH_3)_3]$). When a compound consists of charged complexes, the cation is listed first ($K_2[Ni(CN)_4]$, $[CoCl_2(NH_3)_4]Cl$).

Descriptors that indicate the spatial arrangement of ligands (*cis-*, *trans-*, *mer-*, *fac-*) may be included as prefixes (*cis*-$[Co(Cl)_2(NH_3)_4]^+$). Further information on stereochemical descriptors and examples are given in Section 6.2 and more detailed descriptors are given in the *Red book*.

Chemical names

Homoatomic species

The prefix *catena* is used for chains, and *cyclo* for rings (Table F1.1).

Heteroatomic species

Acids. The IUPAC recommendation is to use trivial names for only the most common acids and systematic names for the rest (Table F1.2).

Table F1.2 Traditional and systematic names for acids

Formula	Traditional name	Hydrogen nomenclature	Acid nomenclature
H_3BO_3	boric acid	trihydrogen trioxoborate	trioxoboric acid
H_4SiO_4	orthosilicic acid	tetrahydrogen tetraoxosilicate	tetraoxosilicic acid
H_2CO_3	carbonic acid	dihydrogen trioxocarbonate	trioxocarbonic acid
HNO_3	nitric acid*	hydrogen trioxonitrate(1 −)	trioxonitric acid
HNO_2	nitrous acid*	hydrogen dioxonitrate(1 −)	dioxonitric acid
HPH_2O_2	phosphinic acid	hydrogen dihydridodioxophosphate(1 −)	dihydridodioxophosphoric acid
H_3PO_3	phosphorous acid*	trihydrogen trioxophosphate(3 −)	trihydridotrioxophosphoric (2 −) acid
H_2PHO_3	phosphonic acid	dihydrogen hydridotrioxophosphate(2 −)	hydridotrioxophosphoric (2 −) acid
H_3PO_4	phosphoric acid* orthophosphoric acid	trihydrogen tetraoxophosphate(3 −)	tetraoxophosphoric acid
$H_4P_2O_7$	diphosphoric acid	tetrahydrogen μ-oxo-hexaoxodiphosphate	μ-oxo-hexaoxo-phosphoric acid
$(HPO_3)_n$	metaphosphoric acid	poly[hydrogen trioxophosphate(1 −)]	polytrioxophosphoric acid
H_3AsO_4	arsenic acid*	trihydrogen tetraoxoarsenate	tetraoxoarsenic acid
H_3AsO_3	arsenous acid*	trihydrogen trioxoarsenate(3 −)	trioxoarsenic acid
H_2SO_4	sulfuric acid*	dihydrogen tetraoxosulfate	tetraoxosulfuric acid
$H_2S_2O_3$	thiosulfuric acid	dihydrogen trioxothiosulfate	trioxothiosulfuric acid
$H_2S_2O_6$	dithionic acid	dihydrogen hexaoxodisulfate	hexaoxodisulfuric acid
$H_2S_2O_4$	dithionous acid	dihydrogen tetraoxodisulfate	tetraoxodisulfuric acid
H_2SO_3	sulfurous acid*	dihydrogen trioxosulfate	trioxosulfuric acid
H_2CrO_4	chromic acid*		
$H_2Cr_2O_7$	dichromic acid*		
$HClO_4$	perchloric acid*	hydrogen tetraoxochlorate	tetraoxochloric acid
$HClO_3$	chloric acid	hydrogen trioxochlorate	trioxochloric acid
$HClO_2$	chlorous acid	hydrogen dioxochlorate	dioxochloric acid
$HClO$	hypochlorous acid*	hydrogen monoxochlorate	monooxochloric acid
HIO_3	iodic acid	hydrogen trioxoiodate	trioxoiodic acid
HIO_4	periodic acid	hydrogen tetraoxoiodate	tetraoxoiodic acid
H_5IO_6	orthoperiodic acid	pentahydrogen hexaoxoiodate(5 −)	hexaoxoiodic acid
$HMnO_4$	permanganic acid		tetraoxomanganic(1 −) acid
H_2MnO_4	manganic acid		tetraoxomanganic(2 −) acid

*Common oxoacids for which traditional names are widely understood.

We indicate in the table the acids that can be considered common. A point worth noting is the generally accepted distinction between hydrogen halides and hydrohalic acids. For example, molecular HCl, as in the gas phase or hydrocarbon solvent, should be named hydrogen chloride and not hydrochloric acid. The latter name is reserved for the aqueous solution.

The traditional names of oxoacids and their systematic IUPAC names based on the scheme or the *acid nomenclature* scheme are displayed in Table F1.2. Note that for neutral molecular acids, hydrogen is left as a separate word (dihydrogen trioxocarbonate); in anions derived from an oxoacid, the attached hydrogen is named as one word (HCO_3^-, hydrogencarbonate ion).

Salts. The principles of nomenclature for salts should be clear from the representative names for simple salts given in Table F1.3. Note that cations are named first. When multiple cations are present they are listed in alphabetical order, disregarding prefixes such as di-, tri-, etc. The same rule holds for anions.

Mononuclear coordination compounds. Ligands are named in alphabetical order followed by the name of the central metal atom. Note that, unlike in formulas, the names of charged and neutral ligands are intermingled as dictated by alphabetization. When two or more simple ligands are present, their number is indicated by a prefix (di-, tri-, tetra-, penta-, hexa-, etc.). However, when the ligand name itself contains one of these prefixes (as in ethylenediamine), then the multiplicative prefix is bis-, tris-, tetrakis-, pentakis-, hexakis-, etc. ($[Co(NH_3)_6]^{3+}$, hexaamminecobalt(III) or hexa(ammine)cobalt(III); $[Co(en)_3]^{3+}$, tris(ethylenediamine)cobalt(III)). Note that when a multiplicative prefix is present, the ligand may be enclosed in parenthesis to improve the ease of reading.

The oxidation state of the metal atom is indicated by Roman numerals in parentheses following the name of the metal. It is also permissible to indicate the oxidation state by giving the overall charge on the complex ion by Arabic numerals and sign, enclosed in parentheses following the name of the complex. As a third alternative the number of counterions present may be given (see Table F1.4). The

Table F1.3 Illustrative names for some salts

Formula	Name
$KMgF_3$	magnesium potassium fluoride
$NaTl(NO_3)_2$	sodium thallium(I) nitrate, or sodium thallium dinitrate
$MgNH_4PO_4 \cdot 6H_2O$	ammonium magnesium phosphate hexahydrate
$NaHCO_3$	sodium hydrogencarbonate
LiH_2PO_4	lithium dihydrogenphosphate
$CsHSO_4$	cesium hydrogensulfate, or cesium hydrogentetraoxosulfate(VI), or cesium hydrogentetraoxosulfate(1 −)
$NaCl \cdot NaF \cdot 2Na_2SO_4$, $Na_6ClF(SO_4)_2$	hexasodium chloride fluoride sulfate
$Ca_5F(PO_4)_3$	pentacalcium fluoride tris(phosphate)

Table F1.4 Illustrative systematic names of coordination compounds

Formula	Name
$K_4[Fe(CN)_6]$	potassium hexacyanoferrate(II)*
	potassium hexacyanoferrate(4 −)
	tetrapotassium hexacyanoferrate
$[Pt(Cl)_2(C_5H_5N)(NH_3)]$	amminedi(chloro)pyridineplatinum(II)

*Preferred in this text.

suffix -ate is appended to the metal name in all anionic complexes. Note that the Latin name of the metal is sometimes used in conjunction with the -ate suffix (ferrate, argentate, aurate). The Latin names are all suggested by the symbols for the element (Fe, Ag, Au). Mercury is an exception, presumably because hydragyrate is a tongue twister.

As in some of the above examples, anionic ligands end in -o rather than -e of the free anion, and NH_3 as a ligand is designated ammine, with a double m. A few common ligand names are illustrated in Table F1.5.

Stereochemical descriptors, such as *cis* or *trans*, are given as prefixes in italics (when writing, underlines are used to indicate italics) and connected to the name of the complex by a hyphen (*cis*-diamminedichloroplatinum(II)). The Δ and Λ descriptors for chirality (Section 6.3) may be similarly appended (Δ-tris(ethylenediamine)cobalt(III)).

When there is ambiguity, the site of attachment of a ligand may be indicated by the element symbol surrounded by hyphens and placed after the ligand name ($[RhNO_2(NH_3)_5]^{2+}$ may exist in either of two isomers: penta(ammine)nitro-O-rhodium(III) cation, containing a Rh—O—N—O link, or penta(ammine)nitro-N-rhodium(III) cation, containing Rh—NO_2).

Polynuclear coordination compounds. Bridging ligands are indicated by means of the prefix μ (as in $[\{Cr(NH_3)_5\}_2(\mu\text{-}OH)]Cl_5$ μ-hydroxo-bis(penta(ammine)chromium)(III) pentachloride).

Organometallic compounds. The nomenclature rules follow those given for metal complexes. The hapticity, the number of sites, n, of attachment, of a ligand is commonly specified by η^n (as in bis(η^5-cyclopentadienyl)iron). Examples of η^n organic ligands are given in Table 16.1 and many examples may be seen throughout that chapter.

Hydrogen compounds and their derivatives. The common names for p-block hydrogen compounds (which IUPAC continues to refer to as *hydrides* despite the inappropriateness of the name in many cases) have been given in Table 9.3. The systematic names given in Table F1.6 form the basis for a substitutive system of nomenclature in which

Table F1.5 Illustrative ligand names

Formula	Name in a complex
CN^-	cyano
H^-	hydrido
$CH_3CO_2^-$	acetato
$(CH_3)_2N^-$	dimethylamido
O^{2-}	oxo
$(O_2)^{2-}$	peroxo
NH_3	ammine
$NH_2C_2H_4NH_2$	ethylenediamine

Table F1.6 Illustrative substitutive names

Formula	Substitutive		Other
	Systematic	Traditional	
PH_2CH_3	methylphosphane	methylphosphine	
$B(C_2H_5)_3$	triethylborane		borontriethyl
$S(C_6H_5)_2$	diphenylsulfane		diphenyl sulfide
$Sb(C_2H_4)_3$	trivinylstibane	trivinylstibine	

the groups replacing hydrogen are indicated. This substitutive nomenclature is generally confined to boron aluminum, gallium, and the elements from Groups 14/IV through 16/VI.

FURTHER INFORMATION 2:
Nuclear magnetic resonance

Nuclear magnetic resonance (NMR) is the most powerful and widely used spectroscopic method for the determination of molecular structures in solution, pure liquids, and gases. In many cases, it provides information about shape and symmetry with greater certainty than is possible with other spectroscopic techniques, such as infrared and Raman spectroscopy. However, unlike X-ray diffraction (Box 3.1), NMR studies of molecules in solution generally do not provide detailed bond distance and angle information. NMR also provides information about the rate and nature of the interchange of ligands in fluxional molecules (Section 16.8).

NMR can be observed only for compounds containing elements with magnetic nuclei (those with nonzero nuclear spin). The sensitivity is dependent on several parameters, including the abundance of the isotope and the size of its nuclear magnetic moment. For example, ^1H with 99.98 percent natural abundance and a large magnetic moment, is easier to observe than ^{13}C, which has a smaller magnetic moment and only 1.1 percent natural abundance. With modern multinuclear NMR techniques it is easy to observe spectra for approximately 20 different nuclei, including many elements that are important in inorganic chemistry, such as ^1H, ^7Li, ^{11}B, ^{13}C, ^{15}N, ^{19}F, ^{23}Na, ^{27}Al, ^{29}Si, ^{31}P, ^{195}Pt, and ^{199}Hg. With more effort, useful spectra can also be obtained using many other nuclei.

Measurement

A nucleus with spin I can takes up $2I + 1$ distinct orientations relative to the direction of an applied magnetic field. Each orientation has a different energy, with the lowest level (marginally) the most highly populated. The energy of the transition between these nuclear spin states is measured by exciting nuclei in the sample with a radiofrequency pulse or pulse sequence and then observing the return of the nuclear magnetization back to equilibrium. After data processing (Fourier transformation), the data are displayed as an absorption spectrum (Fig. F2.1), with peaks at frequencies corresponding to transitions between the different nuclear energy levels.

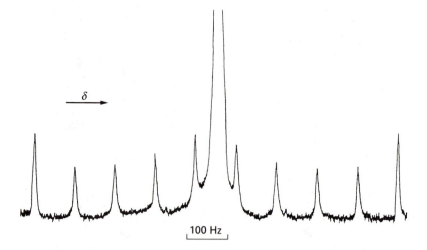

F2.1 The ^{1}H-NMR spectrum of GeH$_4$. (Source: E. A. V. Ebsworth, D. W. H. Rankin, and S. Cradock, *Structural methods in inorganic chemistry.* Blackwell, Oxford (1991).)

Chemical shifts

The frequency of an NMR transition depends on the *local* magnetic field, B_{loc}, the nucleus experiences. This local field is the sum of the field of the magnet, B_{appl}, and an additional local field arising from the effect of the applied field on the molecule, which is written $-\sigma B_{appl}$:

$$B_{loc} = (1 - \sigma)B_{appl}$$

The constant σ, which expresses the role of the local chemical environment, is called the **shielding constant**. In practice, the position of an NMR signal is expressed as the **chemical shift** δ. This is defined in terms of the difference between the resonance frequency of nuclei in the sample and that of a reference compound:

$$\delta = \frac{\nu_{sample} - \nu_{ref}}{\nu_{ref}} \times 10^6$$

A common standard for ^{1}H, ^{13}C, or ^{29}Si NMR spectra is tetramethylsilane, Si(CH$_3$)$_4$ (TMS).

When δ for a signal is negative, the nucleus involved is said to be **shielded** relative to the standard. Conversely, a positive δ corresponds to nucleus that is **deshielded** with respect to the reference. An H atom bound to a closed-shell, low-oxidation-state, *d*-block element from Groups 6 through 10 (such as HCo(CO)$_4$) is generally found to be highly shielded whereas in an oxoacid, such as H$_2$SO$_4$, it is deshielded (in each case, relative to tetramethylsilane). From these examples it might be supposed that the higher the electron density around a nucleus the greater its shielding. However, as several factors contribute to the shielding, a simple physical interpretation of chemical shifts in terms of electron density is generally not possible.

The chemical shifts of ^1H and other nuclei in various chemical environments are tabulated, and so empirical correlations can often be used to identify compounds or the element to which the resonant nucleus is bound. For example, the ^1H chemical shift in CH_4 is only 0.1 because the H nuclei are in an environment similar to that in tetramethylsilane, but the ^1H chemical shift is $\delta = 3.1$ for H bonded to Ge in GeH_4. Chemical shifts are different for the same element in inequivalent positions within a molecule. Thus in ClF_3 the chemical shift of the equatorial ^{19}F nucleus is separated by $\delta = 120$ from that of the axial F nuclei.

Spin-spin coupling

Structural assignment is often helped by the observation of the **spin–spin coupling** of nuclei, which gives rise to a multiplet of lines in the spectrum. The strength of spin–spin coupling, which is reported as the **spin–spin coupling constant** J (in hertz, Hz), decreases rapidly with distance through chemical bonds, and in many cases is greatest when the two atoms are directly bonded to each other. For simple so-called **first-order spectra**, which are being considered here, the coupling constant is equal to the separation of adjacent lines in a multiplet. As can be seen in Fig. F2.1, $J(^1H-^{73}Ge) \approx 100\,Hz$. The chemical shift is measured at the center of the multiplet.

The allowed transitions contributing to a multiplet all occur at the same frequency when the nuclei are related by symmetry. Thus, a single ^1H signal is observed for the H_3Cl molecule because the three H nuclei are related to each other by a threefold axis. Similarly, in the spectrum of GeH_4 the single central line arises from the four equivalent H atoms in GeH_4 molecules that contain germanium isotopes of zero nuclear spin. This strong central line is flanked by 10 evenly spaced but less intense lines that arise from a small fraction of GeH_4 that contains ^{73}Ge, for which $I = \frac{9}{2}$. The properties of spin–spin coupling are such that a multiplet of $2I + 1$ lines result in the spectrum of a spin-$\frac{1}{2}$ nucleus when that nucleus (or a set of symmetry related spin-$\frac{1}{2}$ nuclei) is coupled to a nucleus of spin I. In the present case, the ^1H nuclei are coupled to the ^{73}Ge nucleus to yield a $2 \times \frac{9}{2} + 1 = 10$ line multiplet.

The coupling of the nuclear spins of different elements is called **heteronuclear coupling**: the Ge—H coupling discussed above is an example. **Homonuclear coupling** between nuclei of the same element is detectable when the nuclei are unrelated by the symmetry operations of the molecule, as in the ^{19}F-NMR spectrum of ClF_3 (Fig. F2.2). The signal ascribed to the two axial F nuclei is split into a doublet by the single equatorial F nucleus, and the latter is split into a triplet by the two axial F nuclei. Thus the pattern of ^{19}F resonances readily distinguishes this unsymmetrical structure from two more symmetric possibilities, trigonal planar and trigonal pyramidal, both of

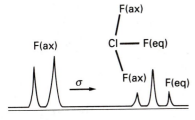

F2.2 The ^{19}F-NMR spectrum of ClF_3. (Source: R. S. Drago, *Physical methods in chemistry.* Saunders, Philadelphia (1992).)

which would have equivalent F nuclei and hence a single ^{19}F resonance.

The sizes of coupling constants are often related to the geometry of a molecule by noting empirical trends. In square planar Pt(II) complexes, $J(Pt—P)$ is sensitive to the group *trans* to a phosphane ligand and the value of $J(Pt—P)$ increases in the following order of *trans* ligands:

$$PR_3 < H^- < R^- < NH_3 < Br^- \le Cl^-$$

For example, *cis*-[PtCl$_2$(PEt$_3$)$_2$], where Cl$^-$ is trans to P, has $J(Pt—P) = 3.5$ kHz, whereas *trans*-[PtCl$_2$(PEt$_3$)$_2$], with P *trans* to P, has $J(Pt—P) = 2.4$ kHz. These systematics permit us to differentiate *cis* and *trans* isomers quite readily.

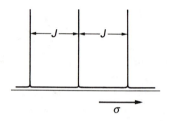

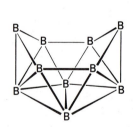

Intensities

Two aspects of line intensities are useful in the analysis of simple NMR spectra. The first is the intensity pattern within a multiplet. The second is the relative integrated intensity for signals of magnetically inequivalent nuclei.

The general rule for a multiplet that arises from coupling to spin-$\frac{1}{2}$ nuclei is that the ratios of the line intensities are given by Pascal's triangle:

Intensity ratio	Coupling to
1	0 other nuclei
1 1	1 spin-$\frac{1}{2}$ nucleus
1 2 1	2 spin-$\frac{1}{2}$ nuclei
1 3 3 1	3 spin-$\frac{1}{2}$ nuclei
1 4 6 4 1	4 spin-$\frac{1}{2}$ nuclei

This pattern is different from those generated by nuclei with higher spin moments. For example the ^{1}H-NMR spectrum of HD is a triplet as a result of coupling with the ^{2}H nucleus ($I = 1$, so $2I + 1 = 3$), but the intensities of the components in this triplet are equal (Fig. F2.3).

The integrated intensity under a signal arising from a set of symmetry-related nuclei is proportional to the number of nuclei in the set. As an example we return to ClF$_3$ (Fig. F2.2) where the relative integrated intensities are 2 (doublet) to 1 (triplet). This pattern is consistent with the structure and the splitting pattern, for it indicates the presence of two symmetry-related F nuclei and one magnetically inequivalent F nucleus.

Sometimes is it advantageous to eliminate spin–spin coupling by special electronic techniques. Figure F2.4 shows the ^{11}B-NMR spectrum of B$_{10}$H$_{14}$ in solution, collected under conditions in which proton spin coupling has been eliminated. Each set of symmetry-related B nuclei gives rise to a signal that has an intensity approximately proportional to the number of B atoms it contains.

F2.3 Schematic ^{1}H-NMR spectrum of HD.

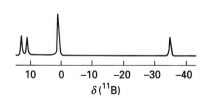

F2.4 The ^{11}B-NMR spectrum of B$_{10}$H$_{14}$. ^{11}B—^{1}H coupling has been suppressed and no B—B coupling is observed. The most intense peak belongs to the four equivalent B nuclei; the rest correspond to signals for each of the three equivalent pairs of B nuclei.

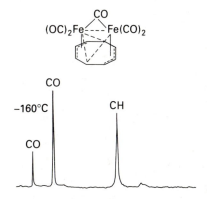

F2.5 The CPMAS spectrum of solid $Fe_2(C_8H_8)(CO)_5$. From C. A. Fyfe, *Solid state NMR for chemists*. CFC Press, Guelph, Ontario (1983).

Solid-state NMR

Recent developments of far reaching importance in inorganic chemistry have made possible the observation of high resolution NMR spectra on solids. One of these high resolution techniques, referred to as CPMAS-NMR, involves high speed sample spinning at a 'magic angle' with respect to the field axis; the technique is a combination of magic angle spinning (MAS) with cross polarization (CP). Compounds containing ^{13}C, ^{31}P, and ^{29}Si and many other nuclei have been studied in the solid state using these techniques. An example is the use of ^{29}Si MAS-NMR spectra to infer the positions of Si atoms in natural and synthetic aluminosilicates. It is also used to study molecular compounds in the solid state. For example the ^{13}C-CPMAS spectrum of $Fe_2(C_8H_8)(CO)_5$, at $-160\,°C$ (Fig. F2.5) indicates that all C atoms in the C_8 ring are equivalent on the timescale of the experiment. The interpretation of this observation is that the molecule is fluxional in the solid state.

FURTHER READING

The first three books provide introductory presentations with an inorganic perspective:

E. A. V. Ebsworth, D. W. H. Rankin, and S. Cradock, *Structural methods in inorganic chemistry*, Chapter 2. Blackwell Scientific, Oxford (1991).

R. S. Drago, *Physical methods for chemists*, Chapter 7. Harcourt Brace Javanovich, Philadelphia (1992).

R. V. Parrish, *NMR, NQR, EPR and Mössbauer spectroscopy in inorganic chemistry*. Ellis Horwood, Chichester (1990).

J. M. Sanders and B. K. Hunter, *Modern NMR spectroscopy: A guide for chemists*. Oxford University Press (1993). This book provides a good introduction to some of the modern NMR techniques.

C. A. Fyfe, *Solid state NMR for chemists*. CFC Press, Guelph, Ontario (1983).

FURTHER INFORMATION 3:
Group theory

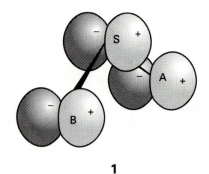

1

The origin of character tables is the representation of the effects of symmetry operations by matrices. As an illustration, consider the C_{2v} molecule SO_2 and the valence p_x orbitals on each atom (**1**) which we shall denote p_S, p_A, and p_B. Under σ_v, the change

$$(p_S, p_A, p_B) \to (p_S, p_B, p_A)$$

takes place. We can express this transformation under a reflection by using matrix multiplication:

$$(p_S, p_A, p_B)\begin{pmatrix} 1 & 0 & 0 \\ 0 & 0 & 1 \\ 0 & 1 & 0 \end{pmatrix} = (p_S, p_B, p_A)$$

This relation can be expressed more succinctly as

$$(p_S, p_A, p_B)\boldsymbol{D}(\sigma_v) = (p_S, p_B, p_A) \quad \text{where } \boldsymbol{D}(\sigma_v) = \begin{pmatrix} 1 & 0 & 0 \\ 0 & 0 & 1 \\ 0 & 1 & 0 \end{pmatrix}$$

The matrix $\boldsymbol{D}(\sigma_v)$ is called a **representative** of the operation σ_v. Representatives take different forms according to the basis (the set of orbitals) that has been adopted.

We can use the same technique to find matrices that reproduce the other symmetry operations. For instance, C_2 has the effect

$$(p_S, p_A, p_B) \to (-p_S, -p_B, -p_A)$$

and its representative is

$$D(C_2) = \begin{pmatrix} -1 & 0 & 0 \\ 0 & 0 & -1 \\ 0 & -1 & 0 \end{pmatrix}$$

The effect of σ_v' is

$$(p_S, p_A, p_B) \rightarrow (-p_S, -p_A, -p_B)$$

and its representative is

$$D(\sigma_v') = \begin{pmatrix} -1 & 0 & 0 \\ 0 & -1 & 0 \\ 0 & 0 & -1 \end{pmatrix}$$

The identity operation has no effect on the basis and so its representative is the unit matrix:

$$D(E) = \begin{pmatrix} 1 & 0 & 0 \\ 0 & 1 & 0 \\ 0 & 0 & 1 \end{pmatrix}$$

The set of matrices that represents all the operations of the group is called a **matrix representation** of the group for the particular basis we have chosen. We denote this three-dimensional representation by the symbol $\Gamma^{(3)}$. The discovery of a matrix representation of the group means that we have found a link between the symbolic manipulations of the operations and algebraic manipulations involving numbers. It may readily be verified that the matrices, when multiplied together, reproduce the group multiplication table.

The **character**, χ, of an operation in a particular matrix representation is the sum of the diagonal elements of each representative. Thus, in the basis we are illustrating, the characters of the representatives are

$D(E)$	$D(C_2)$	$D(\sigma_v)$	$D(\sigma_v')$
3	-1	1	-3

The character of an operation depends on the basis.

The representatives in the basis we have chosen are three dimensional (i.e. they are 3×3 matrices), but inspection shows that they are all of the form

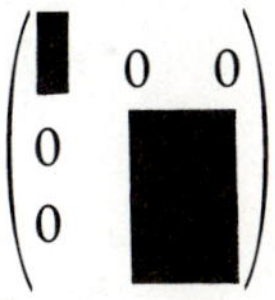

$$\begin{pmatrix} \blacksquare & 0 & 0 \\ 0 & \blacksquare & \\ 0 & & \end{pmatrix}$$

and that the symmetry operations never mix p_S with the other two functions. This suggests that the basis can be cut into two parts, one consisting of p_S alone and the other of (p_A, p_B). It is readily verified that the p_S orbital itself is a basis for the one-dimensional representation

$$\boldsymbol{D}(E) = 1 \quad \boldsymbol{D}(C_2) = -1 \quad \boldsymbol{D}(\sigma_v) = 1 \quad \boldsymbol{D}(\sigma_v') = -1$$

which we shall call $\Gamma^{(1)}$. The remaining two basis functions are a basis for the two-dimensional representation $\Gamma^{(2)}$:

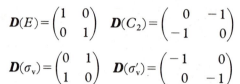

$$\boldsymbol{D}(E) = \begin{pmatrix} 1 & 0 \\ 0 & 1 \end{pmatrix} \quad \boldsymbol{D}(C_2) = \begin{pmatrix} 0 & -1 \\ -1 & 0 \end{pmatrix}$$

$$\boldsymbol{D}(\sigma_v) = \begin{pmatrix} 0 & 1 \\ 1 & 0 \end{pmatrix} \quad \boldsymbol{D}(\sigma_v') = \begin{pmatrix} -1 & 0 \\ 0 & -1 \end{pmatrix}$$

These matrices are the same as those of the original three-dimensional representation, except for the loss of the first row and column. We say that the original three-dimensional representation has been **reduced** to the **direct sum** of a one-dimensional representation **spanned** by p_S and a two-dimensional representation spanned by (p_A, p_B). This reduction is consistent with the common sense view that the central orbital plays a role different from the other two. The reduction is denoted symbolically by

$$\Gamma^{(3)} = \Gamma^{(1)} + \Gamma^{(2)}$$

The one-dimensional representation cannot be reduced any further, and is called an **irreducible representation** of the group. We can demonstrate that the two-dimensional representation is reducible (for this basis) by switching attention to the linear combinations $p_1 = p_A + p_B$ and $p_2 = p_A - p_B$. These combinations are sketched in **2**.

The representatives in the new basis can be constructed from the old. For example, because under σ_v

$$(p_A, p_B) \to (p_B, p_A)$$

it follows (by applying these transformations to the linear combinations) that

$$(p_1, p_2) \to (p_1, -p_2)$$

The transformation is achieved by writing

$$(p_1, p_2)\boldsymbol{D}(\sigma_v) = (p_1, -p_2) \quad \text{with } \boldsymbol{D}(\sigma_v) = \begin{pmatrix} 1 & 0 \\ 0 & -1 \end{pmatrix}$$

which gives us the representative $\boldsymbol{D}(\sigma_v)$ in the new basis. The remaining three representatives may be found similarly, and the complete representation is

$$\boldsymbol{D}(E) = \begin{pmatrix} 1 & 0 \\ 0 & 1 \end{pmatrix} \quad \boldsymbol{D}(C_2) = \begin{pmatrix} -1 & 0 \\ 0 & 1 \end{pmatrix}$$

$$\boldsymbol{D}(\sigma_v) = \begin{pmatrix} 1 & 0 \\ 0 & -1 \end{pmatrix} \quad \boldsymbol{D}(\sigma_v') = \begin{pmatrix} -1 & 0 \\ 0 & -1 \end{pmatrix}$$

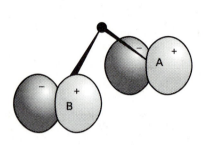

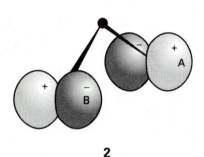

2

The new representatives are all in block diagonal form

and the two combinations are not mixed with one another by any operation of the group. We have therefore achieved the reduction of $\Gamma^{(2)}$ to the direct sum of two one-dimensional representations. Thus, p_1 spans

$$\boldsymbol{D}(E)=1 \quad \boldsymbol{D}(C_2)=-1 \quad \boldsymbol{D}(\sigma_v)=1 \quad \boldsymbol{D}(\sigma_v')=-1$$

which is the same one-dimensional representation as that spanned by p_s, and p_2 spans

$$\boldsymbol{D}(E)=1 \quad \boldsymbol{D}(C_2)=1 \quad \boldsymbol{D}(\sigma_v)=-1 \quad \boldsymbol{D}(\sigma_v')=-1$$

which is a different one-dimensional representation; we shall denote it $\Gamma^{(1)'}$. It is easy to check that either set of 1×1 matrices is a representation by multiplying pairs together and seeing that they reproduce the original group multiplication table.

Now we can make the final link to the material in the text. The **character table** of a group is the list of the characters of all its irreducible representations. At this point we have found two irreducible representations of the group C_{2v}. Their characters are

	E	C_2	σ_v	σ_v'
$\Gamma^{(1)}$	1	-1	1	-1
$\Gamma^{(1)'}$	1	1	-1	-1

The two irreducible representations are normally labeled B_1 and A_2, respectively. An A or a B is used to denote a one-dimensional representation; A is used if the character under the principal rotation is $+1$ and B is used if the character is -1. (The letter E denotes a two-dimensional irreducible representation and T denotes a three-dimensional irreducible representation: all the irreducible representations of C_{2v} are one-dimensional.) There are in fact only two more species of irreducible representations of this group, for it is a surprising theorem of group theory that

Number of symmetry species = Number of classes

In C_{2v} there are four classes (four columns in the character table), and so there are only four species of irreducible representation.

The most important applications in chemistry of group theory, the construction of symmetry adapted orbitals and the analysis of selection rules, are based on the **little orthogonality theorem**:

$$\sum_C g(C)\chi^{(\Gamma)}(C)\chi^{(\Gamma')}(C)=0$$

The sum is over the classes of operation (the columns in the character table), g is the number of operations in each class (such as the 2 in $2C_3$), and Γ and Γ' are two *different* irreducible representations. If Γ

and Γ' are the same irreducible representations, then

$$\sum_C g(C)\chi^{(\Gamma)}(C)\chi^{(\Gamma')}(C) = h$$

where h is the order of the group.

   To find out whether a reducible representation contains a given irreducible representation, we use an expression derived from the little orthogonality theorem. Thus, for a given operation the character of a reducible representation is a linear combination of the characters of the irreducible representations of the group:

$$\chi(C) = \sum_\Gamma c_\Gamma \chi^{(\Gamma)}(C)$$

To find the coefficients for a given irreducible representation Γ', we multiply both sides by $g(C)\chi^{(\Gamma')}$ and sum over all the classes of operation C:

$$\sum_C g(C)\chi^{(\Gamma')}(C)\chi(C) = \sum_C \sum_\Gamma c_\Gamma g(C)\chi^{(\Gamma')}(C)c_\Gamma \chi^{(\Gamma)}(C)$$

When the right hand side is summed over C, the little orthogonality theorem gives 0 for all terms for which Γ is not equal to Γ'. However, because we are summing over all Γ, a term with $\Gamma = \Gamma'$ is guaranteed to be present. Only that term contributes, and the right hand side is then equal to $hc_{\Gamma'}$. It then follows that

$$c_{\Gamma'} = \frac{1}{h}\sum_\Gamma g(\Gamma)\chi^{(\Gamma')}(C)\chi(C)$$

This formula is exceptionally important for finding the decomposition of a reducible representation, because Γ' can be set equal to each irreducible representation in turn, and the coefficients c determined. It takes an even easier form if we only want to know if the totally symmetric irreducible representation is present, because all the characters of that representation are 1, so

$$c_{A_1} = \frac{1}{h}\sum_C g(C)\chi(C)$$

   To form an orbital of symmetry species Γ we form $P\psi$, where

$$P\psi = \sum_R \chi^{(\Gamma)}(R)R\psi$$

R is an operation of the group. Note that the actual operations occur in the formula, not the classes as in the earlier expressions. The quantity P is called a **projection operator**. As an example of its form, to project out a B_1 symmetry-adapted linear combination in the group C_{2v} we would use

$$P = \chi^{(B_1)}(E)E + \chi^{(B_1)}(C_2)C_2 + \chi^{(B_1)}(\sigma_v)\sigma_v + \chi^{(B_1)}(\sigma_v')\sigma_v'$$

$$= E - C_2 + \sigma_v - \sigma_v'$$

FURTHER INFORMATION 4:
The addition of angular momenta

The total angular momentum quantum numbers L and S are found by working out the various ways that the individual quantized spin and orbital angular momenta of the individual electrons may combine to give a quantized resultant. However, in practice a short cut is provided by the **Clebsch–Gordan series**. This series states that the total orbital angular momentum quantum number of an atom with two electrons with individual orbital angular momentum quantum numbers l_1 and l_2 may be one of the values

$$L = l_1 + l_2, \; l_1 + l_2 - 1, \; \dots \; |l_1 - l_2|$$

The value with $L = l_1 + l_2$ corresponds to a state in which the two electrons have orbital angular momentum in the same direction (like the planets) and hence combine to give a large resultant angular momentum. The smallest value, $|l_1 - l_2|$, arises when the orbital momenta are opposed and their combined momentum is small. For two p electrons, $l_1 = 1$ and $l_2 = 1$, so the possible values of L are

$$L = 2, 1, 0$$

If there are three electrons present, the total orbital angular momentum is obtained by combining l_3 with each value of L obtained by coupling l_1 and l_2. Thus, for each value of L we obtain

$$L' = L + l_3, \; L + l_3 - 1, \; \dots \; |L - l_3|$$

and so on for any more electrons that may be present.

As an example, consider the configuration $p^2 d^1$. First, we couple the two p electrons together, with $l_1 = l_2 = 1$:

$$L = 2, 1, 0$$

Then we couple the single d electron ($l_3 = 2$) to each of these values of L:

From $L = 2$, $L' = 4, 3, 2, 1, 0$

From $L = 1$, $L' = 3, 2, 1$

From $L = 0$, $L' = 2$

Therefore, the terms that arise are one G, two F, three D, two P, and one S.

A similar Clebsch–Gordan series applies to the spin. When there are two electrons present, the total spin quantum number may be one of the values

$$S = s_1 + s_2, \, s_1 + s_2 - 1, \, \dots \, |s_1 - s_2|$$

Since s_1 and s_2 are both equal to $\frac{1}{2}$ for electrons, the total spin quantum number S may be 1 (spins parallel) or 0 (spins opposed). If there are three electrons, each value of S is coupled to the spin of the third electron, giving

$$S' = S + s_3, \, S + s_3 - 1, \, \dots \, |S - s_3|$$

APPENDIX 1:
Electronic properties of the elements

Ground-state electron configurations of atoms are determined experimentally from spectroscopic and magnetic measurements. The results of these determinations are listed below. They can be rationalized in terms of the building-up principle, in which electrons are added to the available orbitals in a specific order in accord with the Pauli exclusion principle. Some variation in order is encountered in the d- and f-block elements to accommodate the effects of electron–electron interaction more faithfully. The closed shell configuration $1s^2$ characteristic of helium is denoted [He] and likewise for the other noble-gas element configurations. The ground state electron configurations and term symbols listed below have been taken from S. Fraga, J. Karwowski, and K. M. S. Saxena, *Handbook of atomic data.* Elsevier, Amsterdam (1976).

The first three ionization energies of an element E are the energies required for the following processes:

$$I_1: \qquad E(g) \rightarrow E^+(g) + e^-(g)$$

$$I_2: \qquad E^+(g) \rightarrow E^{2+}(g) + e^-(g)$$

$$I_3: \qquad E^{2+}(g) \rightarrow E^{3+}(g) + e^-(g)$$

The electron affinity E_{ea} is the energy *released* when an electron attaches to a gas-phase atom:

$$E_{ea}: \quad E(g) + e^-(g) \rightarrow E^-(g)$$

The values given here are taken from various sources, particularly C. E. Moore, *Atomic energy levels*, NBS Circular 467, Washington (1970) and W. C. Martin, L. Hagan, J. Reader, and J. Sugar, *J. Phys. Chem. Ref. Data*, **3**, 771 (1974). Values for the actinides are taken from

J. J. Katz, G. T. Seaborg, and L. R. Morss (ed.), *The chemistry of the actinide elements.* Chapman and Hall, London (1986). Electron affinities are from H. Hotop and W. C. Lineberger, *J. Phys. Chem. Ref. Data,* **14,** 731 (1985).

For conversions to kilojoules per mole and reciprocal centimeters, use:

$$1 \text{ eV} = 96.485 \text{ kJ mol}^{-1} \qquad 1 \text{ eV} = 80.655 \text{ cm}^{-1}$$

Atom	Ionization energy/eV			Electron affinity, E_a/eV
	I_1	I_2	I_3	
1 H $1s^1$	13.60			+ 0.754
2 He $1s^2$	24.59	54.51		− 0.5
3 Li [He]$2s^1$	5.320	75.63	122.4	+ 0.618
4 Be [He]$2s^2$	9.321	18.21	153.9	− 0.5
5 B [He]$2s^2 2p^1$	8.297	25.15	37.93	+ 0.277
6 C [He]$2s^2 2p^2$	11.257	24.38	47.88	+ 1.263
7 N [He]$2s^2 2p^3$	14.53	29.60	47.44	− 0.07
8 O [He]$2s^2 2p^4$	13.62	35.11	54.93	+ 1.461
9 F [He]$2s^2 2p^5$	17.42	34.97	62.70	+ 3.399
10 Ne [He]$2s^2 2p^6$	21.56	40.96	63.45	− 1.2
11 Na [Ne]$3s^1$	5.138	47.28	71.63	+ 0.548
12 Mg [Ne]$3s^2$	7.642	15.03	80.14	− 0.4
13 Al [Ne]$3s^2 3p^1$	5.984	18.83	28.44	+ 0.441
14 Si [Ne]$3s^2 3p^2$	8.151	16.34	33.49	+ 1.385
15 P [Ne]$3s^2 3p^3$	10.485	19.72	30.18	+ 0.747
16 S [Ne]$3s^2 3p^4$	10.360	23.33	34.83	+ 2.077
17 Cl [Ne]$3s^2 3p^5$	12.966	23.80	39.65	+ 3.617
18 Ar [Ne]$3s^2 3p^6$	15.76	27.62	40.71	− 1.0
19 K [Ar]$4s^1$	4.340	31.62	45.71	+ 0.502
20 Ca [Ar]$4s^2$	6.111	11.87	50.89	− 0.3
21 Sc [Ar]$3d^1 4s^2$	6.54	12.80	24.76	
22 Ti [Ar]$3d^2 4s^2$	6.82	13.58	27.48	
23 V [Ar]$3d^3 4s^2$	6.74	14.65	29.31	
24 Cr [Ar]$3d^5 4s^1$	6.764	16.50	30.96	
25 Mn [Ar]$3d^5 4s^2$	7.435	15.64	33.67	
26 Fe [Ar]$3d^6 4s^2$	7.869	16.18	30.65	
27 Co [Ar]$3d^7 4s^2$	7.876	17.06	33.50	
28 Ni [Ar]$3d^8 4s^2$	7.635	18.17	35.16	
29 Cu [Ar]$3d^{10} 4s^1$	7.725	20.29	36.84	
30 Zn [Ar]$3d^{10} 4s^2$	9.393	17.96	39.72	
31 Ga [Ar]$3d^{10} 4s^2 4p^1$	5.998	20.51	30.71	+ 0.30
32 Ge [Ar]$3d^{10} 4s^2 4p^2$	7.898	15.93	34.22	+ 1.2
33 As [Ar]$3d^{10} 4s^2 4p^3$	9.814	18.63	28.34	+ 0.81
34 Se [Ar]$3d^{10} 4s^2 4p^4$	9.751	21.18	30.82	+ 2.021
35 Br [Ar]$3d^{10} 4s^2 4p^5$	11.814	21.80	36.27	+ 3.365
36 Kr [Ar]$3d^{10} 4s^2 4p^6$	13.998	24.35	36.95	− 1.0
37 Rb [Kr]$5s^1$	4.177	27.28	40.42	+ 0.486
38 Sr [Kr]$5s^2$	5.695	11.03	43.63	− 0.3
39 Y [Kr]$4d^1 5s^2$	6.38	12.24	20.52	
40 Zr [Kr]$4d^2 5s^2$	6.84	13.13	22.99	
41 Nb [Kr]$4d^4 5s^1$	6.88	14.32	25.04	
42 Mo [Kr]$4d^5 5s^1$	7.099	16.15	27.16	
43 Tc [Kr]$4d^5 5s^2$	7.28	15.25	29.54	
44 Ru [Kr]$4d^7 5s^1$	7.37	16.76	28.47	
45 Rh [Kr]$4d^8 5s^1$	7.46	18.07	31.06	
46 Pd [Kr]$4d^{10}$	8.34	19.43	32.92	
47 Ag [Kr]$4d^{10} 5s^1$	7.576	21.48	34.83	

48 Cd [Kr]$4d^{10}5s^2$	8.992	16.90	37.47	
49 In [Kr]$4d^{10}5s^25p^1$	5.786	18.87	28.02	+ 0.3
50 Sn [Kr]$4d^{10}5^25p^2$	7.344	14.63	30.50	+ 1.2
51 Sb [Kr]$4d^{10}5s^25p^3$	8.640	18.59	25.32	+ 1.07
52 Te [Kr]$4d^{10}5s^25p^4$	9.008	18.60	27.96	+ 1.971
53 I [Kr]$4d^{10}5s^25p^5$	10.45	19.13	33.16	+ 3.059
54 Xe [Kr]$4d^{10}5s^25p^6$	12.130	21.20	32.10	− 0.8
55 Cs [Xe]$6s^1$	3.894	25.08	35.24	
56 Ba [Xe]$6s^2$	5.211	10.00	37.51	
57 La [Xe]$5d^16s^2$	5.577	11.06	19.17	
58 Ce [Xe]$4f^15d^16s^2$	5.466	10.85	20.20	
59 Pr [Xe]$4f^36s^2$	5.421	10.55	21.62	
60 Nd [Xe]$4f^46s^2$	5.489	10.73	20.07	
61 Pm [Xe]$4f^56s^2$	5.554	10.90	22.28	
62 Sm [Xe]$4f^66s^2$	5.631	11.07	23.42	
63 Eu [Xe]$4f^76s^2$	5.666	11.24	24.91	
64 Gd [Xe]$4f^75d^16s^2$	6.140	12.09	20.62	
65 Tb [Xe]$4f^96s^2$	5.851	11.52	21.91	
66 Dy [Xe]$4f^{10}6s^2$	5.927	11.67	22.80	
67 Ho [Xe]$4f^{11}6s^2$	6.018	11.80	22.84	
68 Er [Xe]$4f^{12}6s^2$	6.101	11.93	22.74	
69 Tm [Xe]$4f^{13}6s^2$	6.184	12.05	23.68	
70 Yb [Xe]$4f^{14}6s^2$	6.254	12.19	25.03	
71 Lu [Xe]$4f^{14}5d^16s^2$	5.425	13.89	20.96	
72 Hf [Xe]$4f^{14}5d^26s^2$	6.65	14.92	23.32	
73 Ta [Xe]$4f^{14}5d^36s^2$	7.89	15.55	21.76	
74 W [Xe]$4f^{14}5d^46s^2$	7.89	17.62	23.84	
75 Re [Xe]$4f^{14}5d^56s^2$	7.88	13.06	26.01	
76 Os [Xe]$4f^{14}5d^66s^2$	8.71	16.58	24.87	
77 Ir [Xe]$4f^{14}5d^76s^2$	9.12	17.41	26.95	
78 Pt [Xe]$4f^{14}5d^96s^1$	9.02	18.56	29.02	
79 Au [Xe]$4f^{14}5d^{10}6s^1$	9.22	20.52	30.05	
80 Hg [Xe]$4f^{14}5d^{10}6s^2$	10.44	18.76	34.20	
81 Tl [Xe]$4f^{14}5d^{10}6s^26p^1$	6.107	20.43	29.83	
82 Pb [Xe]$4f^{14}5d^{10}6s^26p^2$	7.415	15.03	31.94	
83 Bi [Xe]$4f^{14}5d^{10}6s^26p^3$	7.289	16.69	25.56	
84 Po [Xe]$4f^{14}5d^{10}6s^26p^4$	8.42	18.66	27.98	
85 At [Xe]$4f^{14}5d^{10}6s^2\,6p^5$	9.64	16.58	30.06	
86 Rn [Xe]$4f^{14}5d^{10}6s^26p^6$	10.75			
87 Fr [Rn]$7s^1$	4.15	21.76	32.13	
88 Ra [Rn]$7s^2$	5.278	10.15	34.20	
89 Ac [Rn]$6d^17s^2$	5.17	11.87	19.69	
90 Th [Rn]$6d^27s^2$	6.08	11.89	20.50	
91 Pa [Rn]$5f^26d^17s^2$	5.89	11.7	18.8	
92 U [Rn]$5f^36d^17s^2$	6.19	14.9	19.1	
93 Np [Rn]$5f^46d^17s^2$	6.27	11.7	19.4	
94 Pu [Rn]$5f^67s^2$	6.06	11.7	21.8	
95 Am [Rn]$5f^77s^2$	5.99	12.0	22.4	
96 Cm [Rn]$5f^76d^17s^2$	6.02	12.4	21.2	
97 Bk [Rn]$5f^97s^2$	6.23	12.3	22.3	
98 Cf [Rn]$5f^{10}7s^2$	6.30	12.5	23.6	
99 Es [Rn]$5f^{11}7s^2$	6.42	12.6	24.1	
100 Fm [Rn]$5f^{12}7s^2$	6.50	12.7	24.4	
101 Md [Rn]$5f^{13}7s^2$	6.58	12.8	25.4	
102 No [Rn]$5f^{14}7s^2$	6.65	13.0	27.0	
103 Lr [Rn]$5f^{14}6d^17s^2$	4.6	14.8	23.0	

APPENDIX 2:
Standard potentials

The standard potentials quoted here are presented in the form of Latimer diagrams (Section 7.8) and are arranged according to the blocks of the periodic table in the order s, p, d, f. Data and species in parentheses are uncertain. Most of the data, together with occasional corrections, come from A. J. Bard, R. Parsons, and J. Jordan (ed.), *Standard potentials in aqueous solution*. Marcel Dekker, New York (1985). Data for the actinides are from L. R. Morss, *The chemistry of the actinide elements*, Vol. 2 (ed. J. J. Katz, G. T. Seaborg, and L. R. Morss. Chapman and Hall, London (1986). The value for $[Ru(bipy)_3]^{3+/2+}$ is from B. Durham, J. L. Walsh, C. L. Carter, and T. J. Meyer, *Inorg. Chem.*, **19**, 860 (1980). Potentials for carbon species and some d-block elements are taken from S. G. Bratsch, *J. Phys. Chem. Ref. Data*, **18**, 1 (1989). For further information on standard potentials of unstable radical species see D. M. Stanbury in *Adv. Inorg. Chem.*, **33**, 69 (1989). Potentials are occasionally reported relative to the standard calomel electrode (SCE) and may be converted to the H^+/H_2 scale by adding 0.2412 V. For a detailed discussion of other reference electrodes, see D. J. G. Ives and G. J. Janz, *Reference electrodes*. Academic Press, New York (1961).

s Block · Group 1/I

Acidic solution

+1   0

H^+ ——————0——————→ H_2

Li^+ ——————-3.040——————→ Li

Na^+ ——————-2.713——————→ Na

K^+ ——————-2.925——————→ K

Rb^+ ——————-2.924——————→ Rb

Cs^+ ——————-2.923——————→ Cs

Basic solution

+1   0

H_2O ——————-0.828——————→ H_2

s Block · Group 2/II

Acidic solution

+2   0

Be^{2+} ——————-1.97——————→ Be

Mg^{2+} ——————-2.356——————→ Mg

Ca^{2+} ——————-2.84——————→ Ca

Sr^{2+} ——————-2.89——————→ Sr

Ba^{2+} ——————-2.92——————→ Ba

Ra^{2+} ——————-2.916——————→ Ra

Basic solution

+2   0

$Mg(OH)_2$ ——————-2.687——————→ Mg

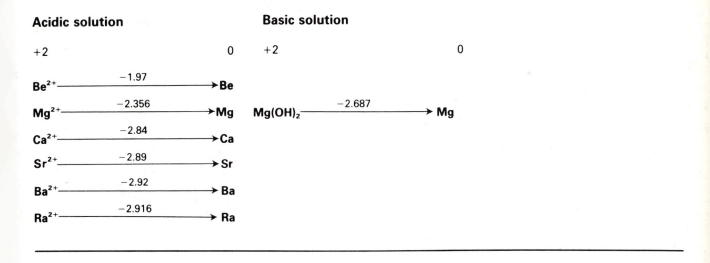

p Block · Group 13/III

Acidic solution	Basic solution

Acidic solution

Basic solution

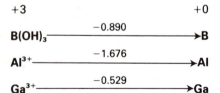

+3   +0

$B(OH)_3 \xrightarrow{-0.890} B$

$Al^{3+} \xrightarrow{-1.676} Al$

$Ga^{3+} \xrightarrow{-0.529} Ga$

+3

$B(OH)_4^- \xrightarrow{-1.24} BH_4^-$

$Al(OH)_4^- \xrightarrow{-2.310} Al$

$GaO(OH)_2^- \xrightarrow{-1.22} Ga$

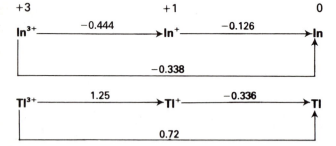

+3   +1   0

$In^{3+} \xrightarrow{-0.444} In^+ \xrightarrow{-0.126} In$

$\xrightarrow{-0.338}$

$Tl^{3+} \xrightarrow{1.25} Tl^+ \xrightarrow{-0.336} Tl$

$\xrightarrow{0.72}$

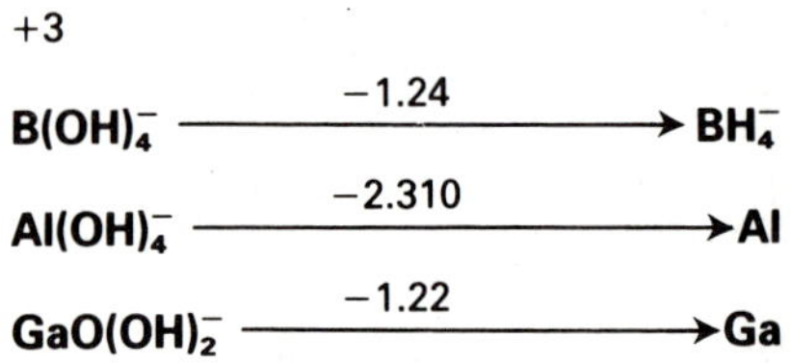

p Block · Group 14/IV

Acidic solution

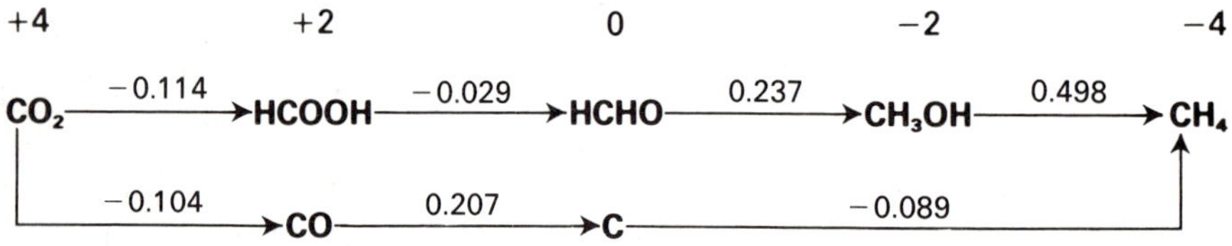

+4   +2   0   −2   −4

$CO_2 \xrightarrow{-0.114} HCOOH \xrightarrow{-0.029} HCHO \xrightarrow{0.237} CH_3OH \xrightarrow{0.498} CH_4$

$CO_2 \xrightarrow{-0.104} CO \xrightarrow{0.207} C \xrightarrow{-0.089} CH_4$

Basic solution

$CO_3^{2-} \xrightarrow{-0.930} HCO_2^- \xrightarrow{-1.160} HCHO \xrightarrow{-0.591} CH_3OH \xrightarrow{-0.245} CH_4$

$C \xrightarrow{-1.148} CH_3OH$

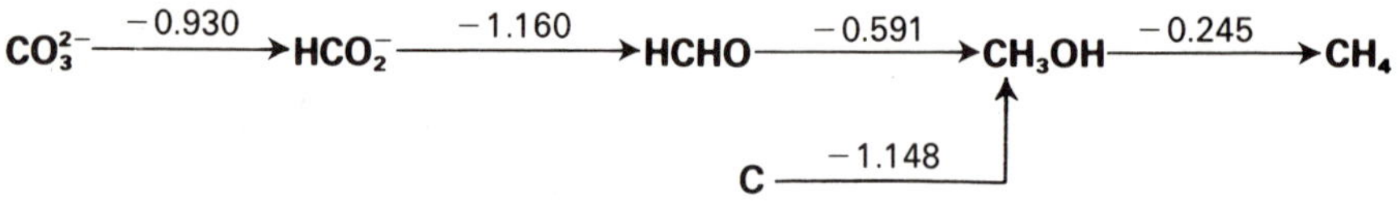

p Block · Group 14/IV (continued)

Acidic solution

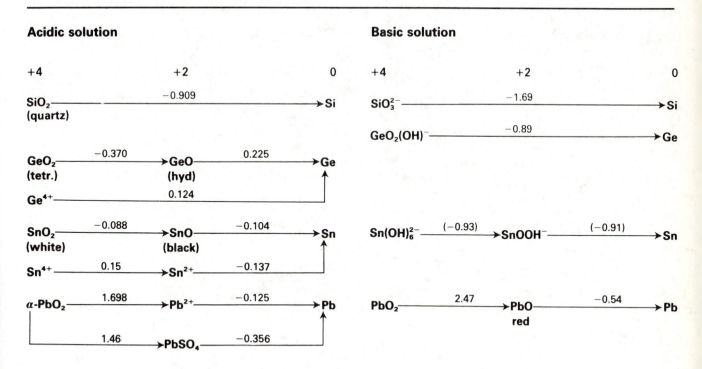

+4   +2   0

SiO_2
(quartz) ————————— -0.909 —————————→ Si

GeO_2 —— -0.370 →→ GeO —— 0.225 ——→ Ge
(tetr.)   (hyd)
Ge^{4+} ———— 0.124 ————

SnO_2 —— -0.088 →→ SnO —— -0.104 ——→ Sn
(white)   (black)
Sn^{4+} —— 0.15 →→ Sn^{2+} —— -0.137 ——

α-PbO_2 —— 1.698 →→ Pb^{2+} —— -0.125 ——→ Pb
     1.46 →→ $PbSO_4$ —— -0.356 ——

Basic solution

+4   +2   0

SiO_3^{2-} ————— -1.69 —————→ Si

$GeO_2(OH)^-$ ——— -0.89 ———→ Ge

$Sn(OH)_6^{2-}$ —— (-0.93) →→ $SnOOH^-$ —— (-0.91) ——→ Sn

PbO_2 —— 2.47 →→ PbO —— -0.54 ——→ Pb
             red

---

p Block · Group 15/V

Acidic solution

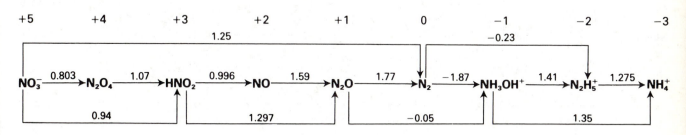

+5   +4   +3   +2   +1   0   −1   −2   −3

NO_3^- — 0.803 → N_2O_4 — 1.07 → HNO_2 — 0.996 → NO — 1.59 → N_2O — 1.77 → N_2 — -1.87 → NH_3OH^+ — 1.41 → $N_2H_5^+$ — 1.275 → NH_4^+

(upper: 1.25 from NO_3^- to HNO_2; -0.23 from N_2 to $N_2H_5^+$)

(lower: 0.94; 1.297; -0.05; 1.35)

Basic solution

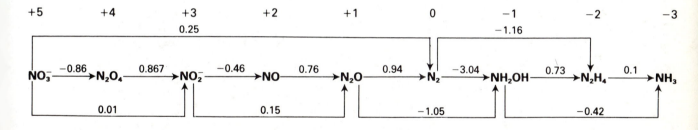

+5   +4   +3   +2   +1   0   −1   −2   −3

NO_3^- — -0.86 → N_2O_4 — 0.867 → NO_2^- — -0.46 → NO — 0.76 → N_2O — 0.94 → N_2 — -3.04 → NH_2OH — 0.73 → N_2H_4 — 0.1 → NH_3

(upper: 0.25; -1.16)

(lower: 0.01; 0.15; -1.05; -0.42)

p Block · Group 15/V (continued)

Acidic solution

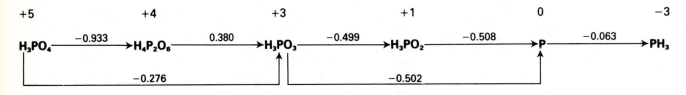

+5   +4   +3   +1   0   −3

H_3PO_4 →(−0.933)→ $H_4P_2O_6$ →(0.380)→ H_3PO_3 →(−0.499)→ H_3PO_2 →(−0.508)→ P →(−0.063)→ PH_3

H_3PO_4 →(−0.276)→ H_3PO_3

H_3PO_3 →(−0.502)→ P

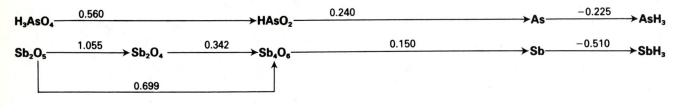

H_3AsO_4 →(0.560)→ $HAsO_2$ →(0.240)→ As →(−0.225)→ AsH_3

Sb_2O_5 →(1.055)→ Sb_2O_4 →(0.342)→ Sb_4O_6 →(0.150)→ Sb →(−0.510)→ SbH_3

Sb_2O_5 →(0.699)→ Sb_4O_6

(Bi^{+5}) →(2)→ (Bi^{3+}) →(0.317)→ Bi

Basic solution

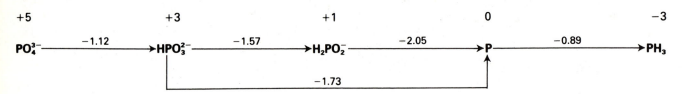

+5   +3   +1   0   −3

PO_4^{3-} →(−1.12)→ HPO_3^{2-} →(−1.57)→ $H_2PO_2^-$ →(−2.05)→ P →(−0.89)→ PH_3

HPO_3^{2-} →(−1.73)→ P

AsO_4^{3-} →(−0.67)→ AsO_2^- →(−0.68)→ As →(−1.37)→ AsH_3

$Sb(OH)_6^-$ →(−0.465)→ $Sb(OH)_4^-$ →(−0.639)→ Sb →(−1.338)→ SbH_3

Bi_2O_3 →(−0.452)→ Bi

p Block · Group 16/VI

Acidic solution

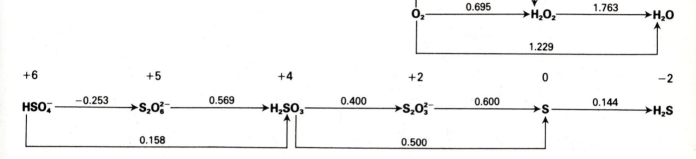

```
0                    -1                    -2

         -0.125   HO₂    1.51
        ┌──────→      ─────────┐
        │                      ↓
O₂ ──────────→ H₂O₂ ─────────→ H₂O
       0.695          1.763
        └─────────────────────↑
              1.229
```

```
+6                +5                 +4                  +2                 0                  -2

HSO₄⁻ ──────→ S₂O₆²⁻ ──────→ H₂SO₃ ──────→ S₂O₃²⁻ ──────→ S ──────→ H₂S
      -0.253         0.569          0.400          0.600        0.144
        │                            ↑      │                ↑
        └──────────────────────────┘      └────────────────┘
             0.158                              0.500
```

$$\left[\text{S}_2\text{O}_8^{2-} \xrightarrow{\ 1.96\ } \text{SO}_4^{2-}\right]$$

```
+6                +4                          0                  -1                  -2

SeO₄²⁻ ──────→ H₂SeO₃ ──────────────────→ Se ──────→ H₂Se
      1.15               0.74                    -0.11

H₂TeO₄ ──────→ (Te⁴⁺) ──────→ Te ──────→ Te₂²⁻ ──────→ H₂Te
      0.93            0.57         -0.74          -0.64
        │                          ↑
        └──────→ TeO₂ ────────────┘
          1.00           0.53
```

Basic solution

```
0                    -1                    -2

         -0.33   O₂⁻    0.20
        ┌──────→     ─────────┐
        │                      ↓
O₂ ──────────→ HO₂⁻ ─────────→ OH⁻
       -0.0649        0.867
        └─────────────────────↑
              0.401
```

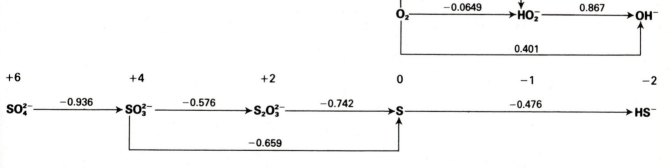

```
+6                +4                 +2                  0                  -1                  -2

SO₄²⁻ ──────→ SO₃²⁻ ──────→ S₂O₃²⁻ ──────→ S ──────→ HS⁻
      -0.936        -0.576         -0.742        -0.476
        │                            ↑
        └───────────────────────────┘
              -0.659

SeO₄²⁻ ──────→ SeO₃²⁻ ──────────────────→ Se ──────→ Se²⁻
       0.03               -0.36                 -0.67

TeO₄²⁻ ──────→ TeO₃²⁻ ──────────────────→ Te ──────→ Te₂²⁻ ──────→ Te²⁻
       0.07               -0.42                -0.84          -1.445
        │                                       │               ↑
        └───────────────────────────────────────┘
                       -1.143
```

p Block · Group 17/VII

Acidic solution

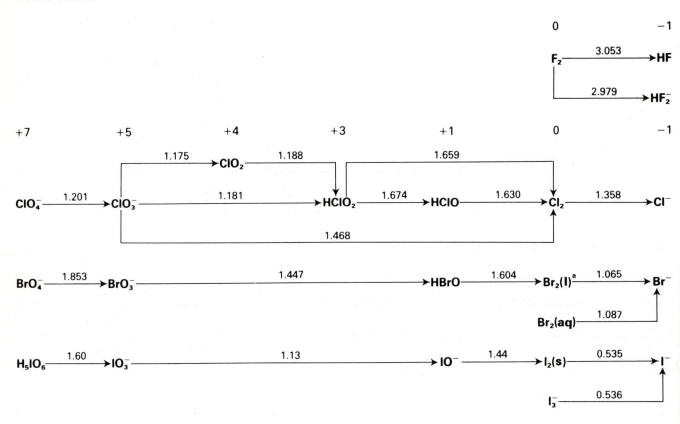

[a]Bromine is not sufficiently soluble in water at room temperature to achieve unit activity. Therefore the value for a saturated solution in contact with $Br_2(l)$ should be used in all practical calculations.

p Block · Group 17/VII (continued)

Basic solution

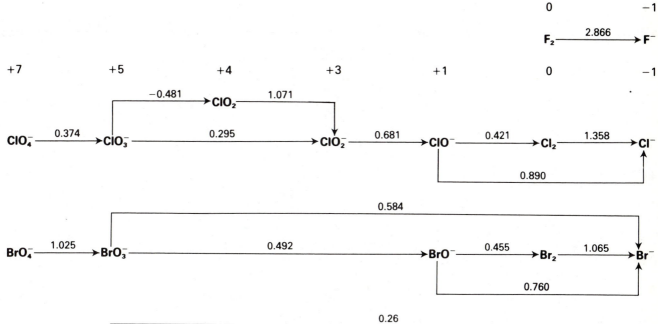

0   −1

F_2 —— 2.866 —→ F^-

+7   +5   +4   +3   +1   0   −1

ClO_4^- —0.374→ ClO_3^- —−0.481→ ClO_2 —1.071→ ClO_2^- ; ClO_3^- —0.295→ ClO_2^- —0.681→ ClO^- —0.421→ Cl_2 —1.358→ Cl^- ; ClO^- —0.890→ Cl^-

BrO_4^- —1.025→ BrO_3^- —0.584→ Br^- ; BrO_3^- —0.492→ BrO^- —0.455→ Br_2 —1.065→ Br^- ; BrO^- —0.760→ Br^-

$H_3IO_6^{2-}$ —0.65→ IO_3^- —0.26→ I^- ; IO_3^- —0.15→ IO^- —0.42→ I_2 —0.535→ I^- ; IO^- —0.48→ I^-

p Block · Group 18/VIII

Acidic solution

+8   +6   0

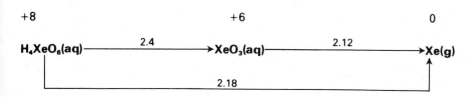

$H_4XeO_6(aq)$ —2.4→ $XeO_3(aq)$ —2.12→ $Xe(g)$; $H_4XeO_6(aq)$ —2.18→ $Xe(g)$

Basic solution

$HXeO_6^{3-}$ —0.99→ $HXeO_4^-$ —1.24→ $Xe(g)$

d Block · Group 3

Acidic solution   **Basic solution**

+3   0   +3   0

Sc^{3+} —————-2.03————→ Sc ScF^{2+} ———-2.16———→ Sc $Sc(OH)_3$ ———-2.60———→ Sc

                                           ScF_3 ———-2.37———→ Sc
                                           (aq)

Y^{3+} —————-2.37————→ Y

La^{3+} —————-2.38————→ La

d Block · Group 4

Acidic solution   **Basic solution**

+4   +3   +2   0   +4   +3   +2   0

                    -0.86

TiO^{2+} —0.1→ Ti^{3+} —-0.37→ Ti^{2+} —-1.63→ Ti TiO_2 —-1.38→ Ti_2O_3 —-1.95→ TiO —-2.13→ Ti

                           -1.21

TiO_2 —-0.56→ Ti_2O_3 —-1.23→ TiO —-1.31→ Ti

                           +4   0

Zr^{4+} —————-1.55————→ Zr

Hf^{4+} —————-1.70————→ Hf

d Block · Group 5

Acidic solution

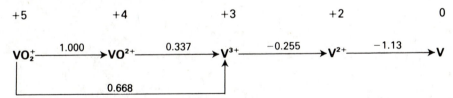

+5   +4   +3   +2   0

$$VO_2^+ \xrightarrow{1.000} VO^{2+} \xrightarrow{0.337} V^{3+} \xrightarrow{-0.255} V^{2+} \xrightarrow{-1.13} V$$

0.668

Weakly acidic solution, pH about 3.0–3.5

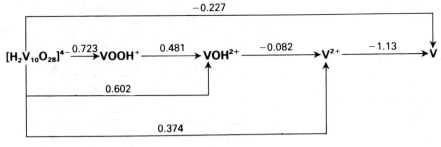

−0.227

$$[H_2V_{10}O_{28}]^{4-} \xrightarrow{0.723} VOOH^+ \xrightarrow{0.481} VOH^{2+} \xrightarrow{-0.082} V^{2+} \xrightarrow{-1.13} V$$

0.602

0.374

Basic solution

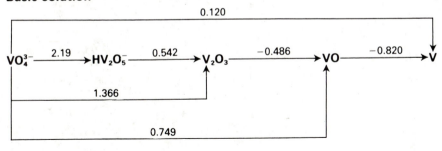

0.120

$$VO_4^{3-} \xrightarrow{2.19} HV_2O_5^- \xrightarrow{0.542} V_2O_3 \xrightarrow{-0.486} VO \xrightarrow{-0.820} V$$

1.366

0.749

Acidic solution

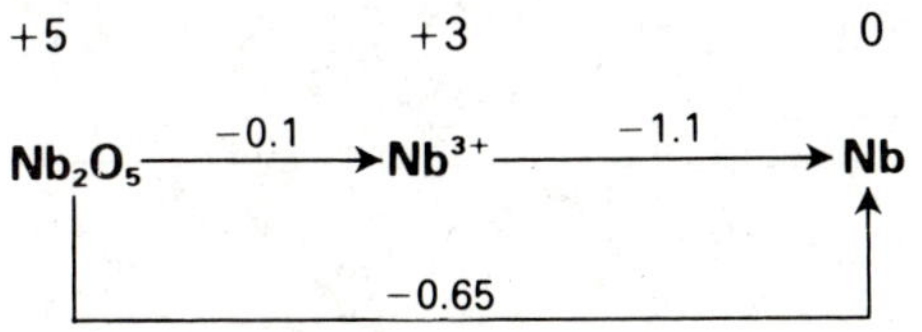

+5   +3   0   +5   0

$$Nb_2O_5 \xrightarrow{-0.1} Nb^{3+} \xrightarrow{-1.1} Nb$$ $$Ta_2O_5 \xrightarrow{-0.81} Ta$$

−0.65

$$TaF_7^{2-} \xrightarrow{-0.45} Ta$$

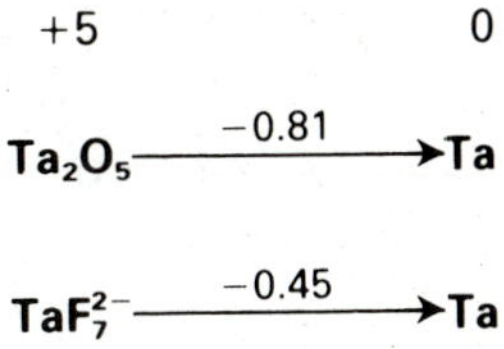

d Block · Group 6

Acidic solution

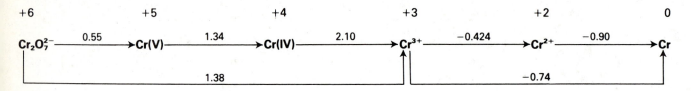

+6	+5	+4	+3	+2	0

$$Cr_2O_7^{2-} \xrightarrow{0.55} Cr(V) \xrightarrow{1.34} Cr(IV) \xrightarrow{2.10} Cr^{3+} \xrightarrow{-0.424} Cr^{2+} \xrightarrow{-0.90} Cr$$

1.38   −0.74

Neutral solution

+3   +2

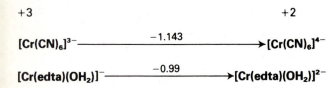

$$[Cr(CN)_6]^{3-} \xrightarrow{-1.143} [Cr(CN)_6]^{4-}$$

$$[Cr(edta)(OH_2)]^- \xrightarrow{-0.99} [Cr(edta)(OH_2)]^{2-}$$

Basic solution

+6   +3   0

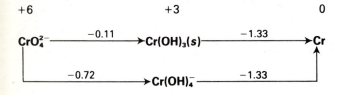

$$CrO_4^{2-} \xrightarrow{-0.11} Cr(OH)_3(s) \xrightarrow{-1.33} Cr$$

$$\xrightarrow{-0.72} Cr(OH)_4^- \xrightarrow{-1.33}$$

Acidic solution

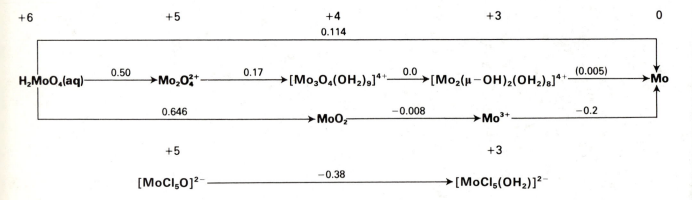

+6	+5	+4	+3	0

0.114

$$H_2MoO_4(aq) \xrightarrow{0.50} Mo_2O_4^{2+} \xrightarrow{0.17} [Mo_3O_4(OH_2)_9]^{4+} \xrightarrow{0.0} [Mo_2(\mu-OH)_2(OH_2)_8]^{4+} \xrightarrow{(0.005)} Mo$$

$$\xrightarrow{0.646} MoO_2 \xrightarrow{-0.008} Mo^{3+} \xrightarrow{-0.2}$$

+5   +3

$$[MoCl_5O]^{2-} \xrightarrow{-0.38} [MoCl_5(OH_2)]^{2-}$$

Neutral solution

+5   +4

$$[Mo(CN)_8]^{3-} \xrightarrow{0.725} [Mo(CN)_8]^{4-}$$

d Block · **Group 6** (continued)

Basic solution

+6   +4   0

$$MoO_4^{2-} \xrightarrow{-0.780} MoO_2 \xrightarrow{-0.980} Mo$$

$$\xrightarrow{-0.913}$$

Acidic solution

+6   +5   +4   0

$$WO_3 \xrightarrow{-0.029} W_2O_5 \xrightarrow{-0.031} WO_2^* \xrightarrow{-0.119} W$$

$$\xrightarrow{-0.090}$$

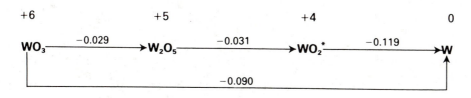

$$[CoW_{12}O_{40}]^{6-} \xrightarrow{-0.046} [H_2CoW_{12}O_{40}]^{6-}$$

$$[PW_{12}O_{40}]^{3-} \xrightarrow{0.22} [PW_{12}O_{40}]^{4-}$$

Neutral solution

+5   +4

$$W(CN)_8^{3-} \xrightarrow{0.457} W(CN)_8^{4-}$$

Basic solution

+6   +4   0

$$WO_4^{2-} \xrightarrow{-1.259} WO_2 \xrightarrow{-0.982} W$$

$$\xrightarrow{-1.074}$$

$$[W(CN)_4(OH)_4]^{2-} \xrightarrow{-0.702} [W(CN)_4(OH)_4]^{4-}$$

*Probably $[W_3(\mu_3-O)(\mu-O)_3(OH_2)_9]^{4+}$ See S.P. Gosh and E.S. Gould, *Inorg. Chem.*, **30**, 3662 (1991).

d Block · Group 7

Acidic solution

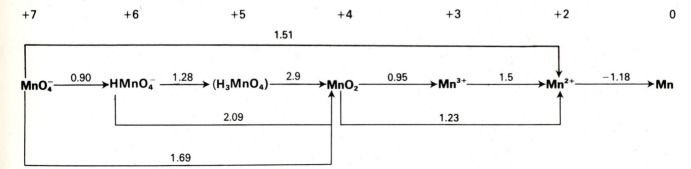

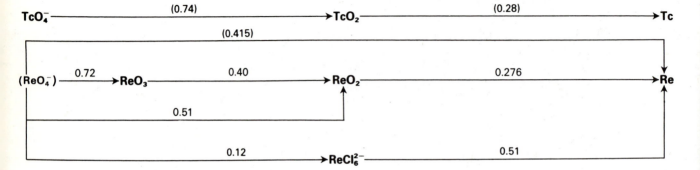

Basic solution

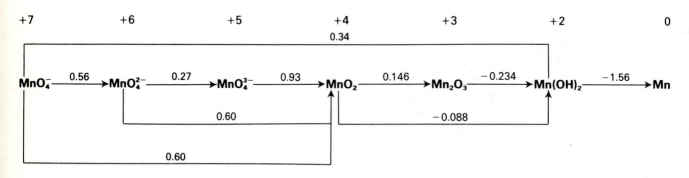

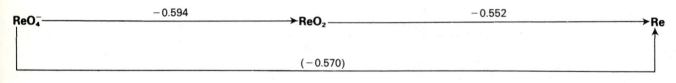

d Block · Group 8

Acidic solution

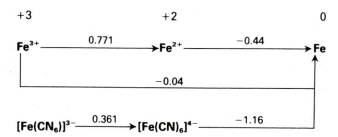

+3   +2   0

Fe^{3+} ———0.771——→ Fe^{2+} ———−0.44——→ Fe

                    −0.04

$[Fe(CN_6)]^{3-}$ —0.361→ $[Fe(CN)_6]^{4-}$ ———−1.16———

Basic solution

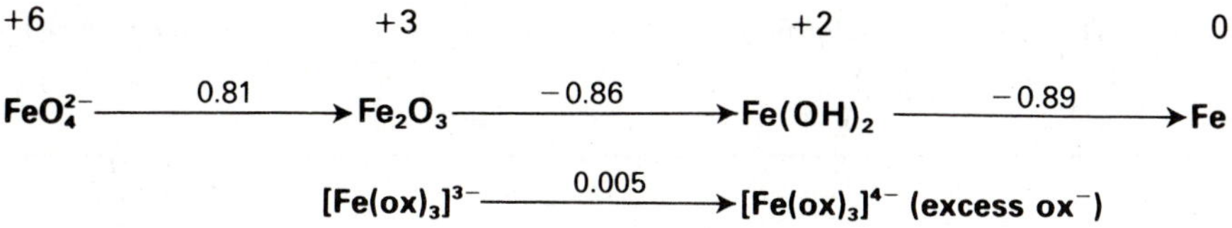

+6   +3   +2   0

FeO_4^{2-} ——0.81——→ Fe_2O_3 ——−0.86——→ $Fe(OH)_2$ ——−0.89——→ Fe

$[Fe(ox)_3]^{3-}$ ——0.005——→ $[Fe(ox)_3]^{4-}$ (excess ox^-)

Acidic solution

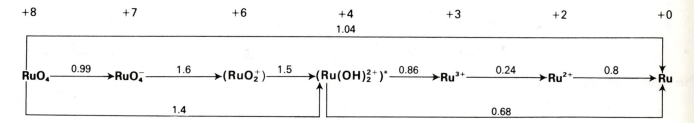

+8   +7   +6   +4   +3   +2   +0
                                    1.04

RuO_4 —0.99→ RuO_4^- —1.6→ (RuO_2^+) —1.5→ $(Ru(OH)_2^{2+})^*$ —0.86→ Ru^{3+} —0.24→ Ru^{2+} —0.8→ Ru

          1.4   0.68

Neutral solution

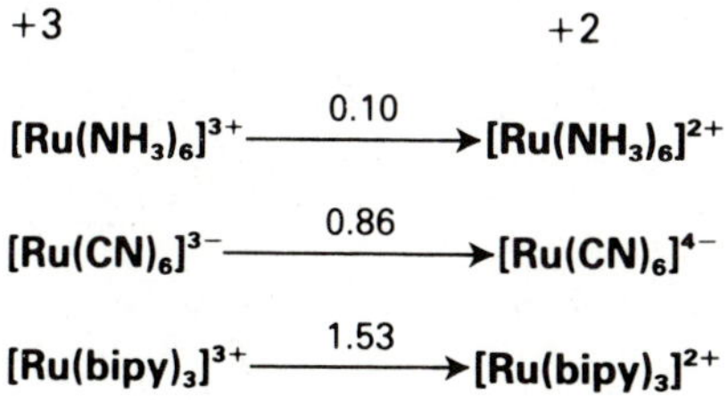

+3   +2

$[Ru(NH_3)_6]^{3+}$ ——0.10——→ $[Ru(NH_3)_6]^{2+}$

$[Ru(CN)_6]^{3-}$ ——0.86——→ $[Ru(CN)_6]^{4-}$

$[Ru(bipy)_3]^{3+}$ ——1.53——→ $[Ru(bipy)_3]^{2+}$

*Likely to be $H_n[Ru_4O_6(OH_2)_{12}]^{(4+n)+}$. See A. Patel and D.T. Richen, *Inorg. Chem.* **30**, 3792 (1991).

d Block · **Group 8** (continued)

Acidic solution

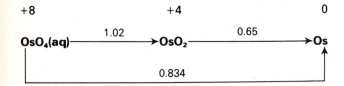

+8   +4   0

OsO$_4$(aq) $\xrightarrow{1.02}$ OsO$_2$ $\xrightarrow{0.65}$ Os

$\xrightarrow{0.834}$

+4   +3   +3   +2

[OsCl$_6$]$^{2-}$ $\xrightarrow{0.45}$ [OsCl$_6$]$^{3-}$ [Os(CN)$_6$]$^{3-}$ $\xrightarrow{0.634}$ [Os(CN)$_6$]$^{4-}$

[OsBr$_6$]$^{2-}$ $\xrightarrow{0.45}$ [OsBr$_6$]$^{3-}$ [Os(bipy)$_3$]$^{3+}$ $\xrightarrow{0.885}$ [Os(bipy)$_3$]$^{2+}$

d Block · **Group 9**

Acidic solution

+4   +3   +2   0

CoO$_2$ $\xrightarrow{1.4}$ Co^{3+} $\xrightarrow{1.92}$ Co^{2+} $\xrightarrow{-0.282}$ Co

Basic solution

+4   +3   +2   0

CoO$_2$ $\xrightarrow{0.7}$ Co(OH)$_3$ $\xrightarrow{0.42}$ Co(OH)$_2$ $\xrightarrow{-0.733}$ Co

Neutral solution

+3   +2

[Co(NH$_3$)$_6$]$^{3+}$ $\xrightarrow{0.058}$ [Co(NH$_3$)$_6$]$^{2+}$

[Co(phen)$_3$]$^{3+}$ $\xrightarrow{0.33}$ [Co(phen)$_3$]$^{2+}$

[Co(ox)$_3$]$^{3-}$ $\xrightarrow{0.57}$ [Co(ox)$_3$]$^{4-}$

Acidic solution

+3   0

Rh^{3+} $\xrightarrow{0.76}$ Rh

[RhCl$_6$]$^{3+}$ $\xrightarrow{0.5}$ Rh

Neutral solution

+3   +2

[Rh(CN)$_6$]$^{3-}$ $\xrightarrow{0.9}$ [Rh(CN)$_6$]$^{4-}$

d Block · **Group 9** (continued)

Acidic solution

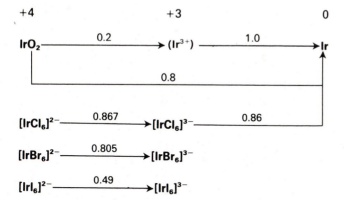

+4   +3   0

IrO_2 —— 0.2 ——→ (Ir^{3+}) —— 1.0 ——→ Ir

—— 0.8 ——

$[IrCl_6]^{2-}$ —— 0.867 ——→ $[IrCl_6]^{3-}$ —— 0.86 ——

$[IrBr_6]^{2-}$ —— 0.805 ——→ $[IrBr_6]^{3-}$

$[IrI_6]^{2-}$ —— 0.49 ——→ $[IrI_6]^{3-}$

d Block · **Group 10**

Acidic solution

+4   +3   +2   0

NiO_2 —————— 1.5 —————→ Ni^{2+} —— −0.257 ——→ Ni

Basic solution

NiO_2 —— 0.7 ——→ NiOOH —— 0.52 ——→ $Ni(OH)_2$ —— −0.72 ——→ Ni

Neutral solution

$[Ni(NH_3)_6]^{2+}$ —— −0.49 ——→ Ni

Acidic solution

+4   +2   0

PdO_2 —— 1.194 ——→ Pd^{2+} —— 0.915 ——→ Pd

$[PdCl_6]^{2-}$ —— 1.47 ——→ $[PdCl_4]^{2-}$ —— 0.60 ——→ Pd

$[PdBr_4]^{2-}$ —— 0.49 ——→ Pd

---

d Block · **Group 10** (continued)

Basic solution

+4   +2   0

PdO$_2$ $\xrightarrow{\quad 1.47 \quad}$ PdO $\xrightarrow{\quad 0.897 \quad}$ Pd

Acidic solution

+4   +2   0

PtO$_2$(s) $\xrightarrow{\quad 1.01 \quad}$ PtO(s) $\xrightarrow{\quad 0.98 \quad}$ Pt

[PtCl$_6$]$^{2-}$ $\xrightarrow{\quad 0.726 \quad}$ [PtCl$_4$]$^{2-}$ $\xrightarrow{\quad 0.758 \quad}$ Pt

[PtBr$_6$]$^{2-}$ $\xrightarrow{\quad 0.631 \quad}$ [PtBr$_4$]$^{2-}$ $\xrightarrow{\quad 0.698 \quad}$ Pt

[PtI$_6$]$^{2-}$ $\xrightarrow{\quad 0.329 \quad}$ [PtI$_4$]$^{2-}$ $\xrightarrow{\quad 0.40 \quad}$ Pt

---

d Block · **Group 11**

Acidic solution

+2   +1   0

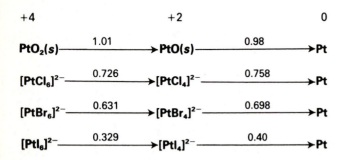

Cu^{2+} $\xrightarrow{\quad 0.159 \quad}$ Cu$^+$ $\xrightarrow{\quad 0.520 \quad}$ Cu

0.340

+2   +1   0

[Cu(NH$_3$)$_4$]$^{2+}$ $\xrightarrow{\quad 0.10 \quad}$ [Cu(NH$_3$)$_2$]$^+$ $\xrightarrow{\quad -0.10 \quad}$ Cu

Cu^{2+} $\xrightarrow{\quad 1.12 \quad}$ [Cu(CN)$_2$]$^-$ $\xrightarrow{\quad -0.44 \quad}$ Cu

d Block · **Group 11** (continued)

Acidic solution

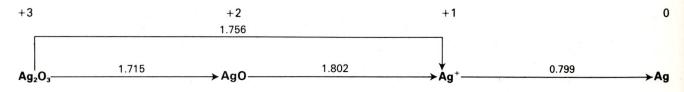

+3   +2   +1   0

Ag_2O_3 —1.715→ AgO —1.802→ Ag^+ —0.799→ **Ag**

(1.756: Ag_2O_3 → Ag^+)

Basic solution

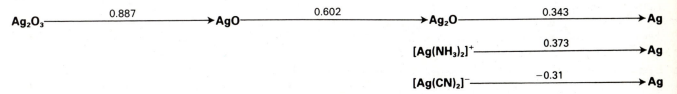

Ag_2O_3 —0.887→ AgO —0.602→ Ag_2O —0.343→ **Ag**

$[Ag(NH_3)_2]^+$ —0.373→ **Ag**

$[Ag(CN)_2]^-$ —−0.31→ **Ag**

Acidic solution

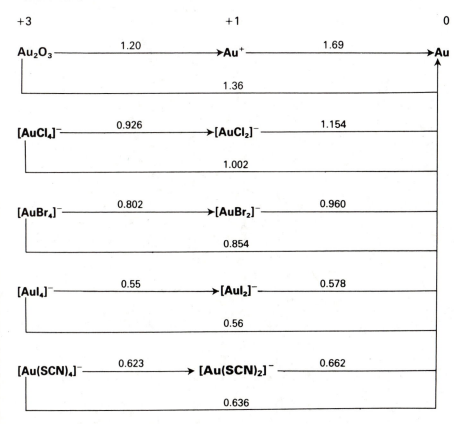

+3   +1   0

Au_2O_3 —1.20→ Au^+ —1.69→ **Au** (1.36)

$[AuCl_4]^-$ —0.926→ $[AuCl_2]^-$ —1.154→ **Au** (1.002)

$[AuBr_4]^-$ —0.802→ $[AuBr_2]^-$ —0.960→ **Au** (0.854)

$[AuI_4]^-$ —0.55→ $[AuI_2]^-$ —0.578→ **Au** (0.56)

$[Au(SCN)_4]^-$ —0.623→ $[Au(SCN)_2]^-$ —0.662→ **Au** (0.636)

d Block · Group 12

Acidic solution

+2   0

Zn^{2+} ——————-0.762—————→ **Zn**

Basic solution

$[Zn(OH)_4]^{2-}$ ———-1.199———→ **Zn**

$Zn(OH)_2$ ———-1.246———→ **Zn**

Acidic solution

Cd^{2+} ————-0.402————→ **Cd**

Basic solution

$Cd(OH)_2(s)$ ——-0.824——→ **Cd**

Acidic solution

+2   +1   0

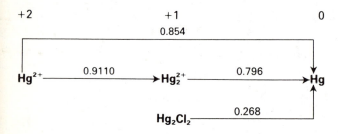

Basic solution

HgO ——————0.0977——————→ **Hg**

f Block · **Lanthanides**

Acidic solution

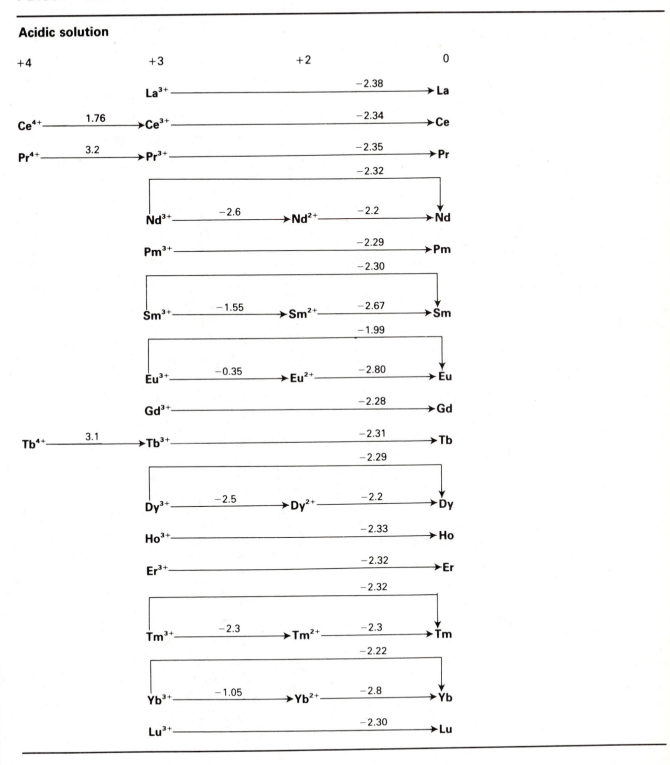

+4   +3   +2   0

La³⁺ ——————— −2.38 ——————→ La

Ce⁴⁺ —— 1.76 —→ Ce³⁺ ——————— −2.34 ——————→ Ce

Pr⁴⁺ —— 3.2 —→ Pr³⁺ ——————— −2.35 ——————→ Pr

Nd³⁺ ⎡ −2.32 ⎤ → Nd
Nd³⁺ —— −2.6 —→ Nd²⁺ —— −2.2 —→ Nd

Pm³⁺ ——————— −2.29 ——————→ Pm

Sm³⁺ ⎡ −2.30 ⎤ → Sm
Sm³⁺ —— −1.55 —→ Sm²⁺ —— −2.67 —→ Sm

Eu³⁺ ⎡ −1.99 ⎤ → Eu
Eu³⁺ —— −0.35 —→ Eu²⁺ —— −2.80 —→ Eu

Gd³⁺ ——————— −2.28 ——————→ Gd

Tb⁴⁺ —— 3.1 —→ Tb³⁺ ——————— −2.31 ——————→ Tb

Dy³⁺ ⎡ −2.29 ⎤ → Dy
Dy³⁺ —— −2.5 —→ Dy²⁺ —— −2.2 —→ Dy

Ho³⁺ ——————— −2.33 ——————→ Ho

Er³⁺ ——————— −2.32 ——————→ Er

Tm³⁺ ⎡ −2.32 ⎤ → Tm
Tm³⁺ —— −2.3 —→ Tm²⁺ —— −2.3 —→ Tm

Yb³⁺ ⎡ −2.22 ⎤ → Yb
Yb³⁺ —— −1.05 —→ Yb²⁺ —— −2.8 —→ Yb

Lu³⁺ ——————— −2.30 ——————→ Lu

f Block · Actinides

Acidic solution

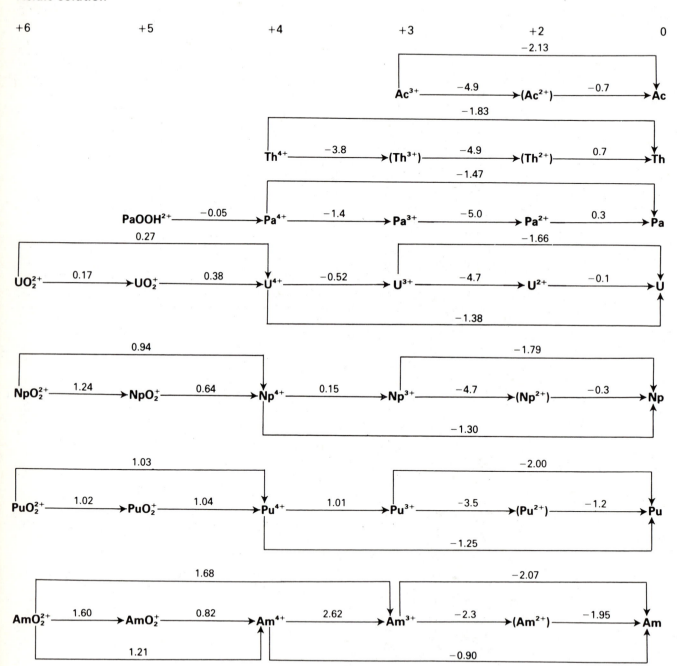

+6	+5	+4	+3	+2	0

f Block · **Actinides** (continued)

Acidic solution

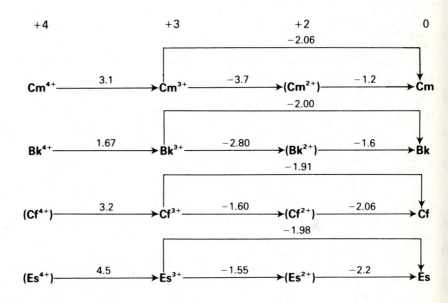

APPENDIX 3:
Character tables

The character tables that follow are for the most common point groups encountered in inorganic chemistry. Each one is labeled with the symbol adopted in the *Schoenflies system* of nomenclature (such as C_{3v}). Point groups that qualify as crystallographic point groups (because they are also applicable to unit cells) are also labeled with the symbol adopted in the *International system* (or the Hermann–Mauguin system, such as $2/m$). In the latter system, a number n represents an n-fold axis and a letter m represents a mirror plane. A diagonal line indicates that a mirror plane lies perpendicular to the symmetry axis and a bar over the number indicates that the rotation is combined with an inversion.

The symmetry species of the p and d orbitals are shown on the right of the tables. Thus, in C_{2v}, a p_x orbital (which is proportional to x) has B_1 symmetry. The functions x, y, and z also show the transformation properties of translations and of the electric dipole moment. The set of functions that span a degenerate representation (such as x and y, which jointly span E in C_{3v}) are enclosed in parentheses. The transformation properties of rotation are shown by the letters R on the right of the tables.

The groups C_1, C_s, C_i

C_1 (1)	E	$h=1$
A	1	

$C_s = C_h$ (m)	E	σ_h		$h=2$
A′	1	1	x, y, R_z	x^2, y^2, z^2, xy
A″	1	−1	z, R_x, R_y	yz, xz

$C_i = S_2$ ($\bar{1}$)	E	i		$h=2$
A_g	1	1	R_x, R_y, R_z	$x^2, y^2, z^2, xy, xz, yz$
A_u	1	−1	x, y, z	

The groups C_n

C_2 (2)	E	C_2		$h=2$
A	1	1	z, R_z	x^2, y^2, z^2, xy
B	1	−1	x, y, R_x, R_y	yz, xz

C_3 (3)	E	C_3	C_3^2	$\varepsilon = \exp(2\pi i/3)$	$h=3$
A	1	1	1	z, R_z	$x^2 + y^2, z^2$
E	$\begin{Bmatrix}1 & \varepsilon & \varepsilon^* \\ 1 & \varepsilon^* & \varepsilon\end{Bmatrix}$			$(x, y)(R_x, R_y)$	$(xy^2 - y^2, xy)(yz, xz)$

C_3 (4)	E	C_4	C_2	C_4^3		$h=4$
A	1	1	1	1	z, R_z	$x^2 + y^2, z^2$
B	1	−1	1	−1		$x^2 - y^2, xy$
E	$\begin{Bmatrix}1 & i & -1 & -i \\ 1 & -i & -1 & i\end{Bmatrix}$				$(x, y)(R_x, R_y)$	(yz, xz)

The groups C_{nv}

C_{2v}	E	C_2	$\sigma_v(xz)$	$\sigma_v'(yz)$	$h = 4$	
$(2mm)$						
A_1	1	1	1	1	z	x^2, y^2, z^2
A_2	1	1	-1	-1	R_z	xy
B_1	1	-1	1	-1	x, R_y	xz
B_2	1	-1	-1	1	y, R_x	yz

C_{3v}	E	$2C_3$	$3\sigma_v$	$h = 6$	
$(3m)$					
A_1	1	1	1	z	$x^2 + y^2, z^2$
A_2	1	1	-1	R_z	
E	2	-1	0	$(x, y)(R_x, R_y)$	$(x^2 - y^2, xy)(xz, yz)$

C_{4v}	E	$2C_4$	C_2	$2\sigma_v$	$2\sigma_d$	$h = 8$	
$(4mm)$							
A_1	1	1	1	1	1	z	$x^2 + y^2, z^2$
A_2	1	1	1	-1	-1	R_z	
B_1	1	-1	1	1	-1		$x^2 - y^2$
B_2	1	-1	1	-1	1		xy
E	2	0	-2	0	0	$(x, y)(R_x, R_y)$	(xz, yz)

C_{5v}	E	$2C_5$	$2C_5^2$	$5\sigma_v$	$h = 10, \alpha = 72°$	
A_1	1	1	1	1	z	$x^2 + y^2, z^2$
A_2	1	1	1	-1	R_z	
E_1	2	$2\cos\alpha$	$2\cos 2\alpha$	0	$(x, y)(R_x, R_y)$	(xz, yz)
E_2	2	$2\cos 2\alpha$	$2\cos\alpha$	0		$(x^2 - y^2, xy)$

C_{6v}	E	$2C_6$	$2C_3$	C_2	$3\sigma_v$	$3\sigma_d$	$h = 12$	
$(6mm)$								
A_1	1	1	1	1	1	1	z	$x^2 + y^2, z^2$
A_2	1	1	1	1	-1	-1	R_z	
B_1	1	-1	1	-1	1	-1		
B_2	1	-1	1	-1	-1	1		
E_1	2	1	-1	-2	0	0	$(x, y)(R_x, R_y)$	(xz, yz)
E_2	2	-1	-1	2	0	0		$(x^2 - y^2, xy)$

$C_{\infty v}$	E	C_2	$2C_\phi$	$\infty\sigma_v$		$h = \infty$
A_1 (Σ^+)	1	1	1	1	z	$z^2, x^2 + y^2$
A_2 (Σ^-)	1	1	1	-1	R_z	
E_1 (Π)	2	-2	$2\cos\phi$	0	$(x, y)(R_x, R_y)$	(xz, yz)
E_2 (Δ)	2	2	$2\cos 2\phi$	0		$(xy, x^2 - y^2)$
$\vdots$	$\vdots$	$\vdots$	$\vdots$	$\vdots$		

The groups D_n

D_2 (222)	E	$C_2(z)$	$C_2(y)$	$C_2(x)$		$h = 4$
A	1	1	1	1		x^2, y^2, z^2
B_1	1	1	-1	-1	z, R_z	xy
B_2	1	-1	1	-1	y, R_y	xz
B_3	1	-1	-1	1	x, R_x	yz

D_3 (32)	E	$2C_3$	$3C_2$		$h = 6$
A_1	1	1	1		$x^2 + y^2, z^2$
A_2	1	1	-1	z, R_z	
E	2	-1	0	$(x, y)(R_x, R_y)$	$(x^2 - y^2, xy)(xz, yz)$

The groups D_{nh}

D_{2h} (*mmm*)	E	$C_2(z)$	$C_2(y)$	$C_2(x)$	i	$\sigma(xy)$	$\sigma(xz)$	$\sigma(yz)$		$h = 4$
A_g	1	1	1	1	1	1	1	1		x^2, y^2, z^2
B_{1g}	1	1	-1	-1	1	1	-1	-1	R_z	xy
B_{2g}	1	-1	1	-1	1	-1	1	-1	R_y	xz
B_{3g}	1	-1	-1	1	1	-1	-1	1	R_x	yz
A_u	1	1	1	1	-1	-1	-1	-1		
B_{1u}	1	1	-1	-1	-1	-1	1	1	z	
B_{2u}	1	-1	1	-1	-1	1	-1	1	y	
B_{3u}	1	-1	-1	1	-1	1	1	-1	x	

D_{3h} ($\bar{6}m2$)	E	$2C_3$	$3C_2$	σ_h	$3S_3$	$3\sigma_v$		$h=4$
A_1'	1	1	1	1	1	1		$x^2+y^2+z^2$
A_2'	1	1	-1	1	1	-1	R_z	
E'	2	-1	0	2	-1	0	(x, y)	(x^2-y^2, xy)
A_1''	1	1	1	-1	-1	-1		
A_2''	1	1	-1	-1	-1	1	z	
E''	2	-1	0	-2	1	0	(R_x, R_y)	(xz, yz)

The groups D_{nh} (continued)

D_{4h} ($4/mmm$	E	$2C_4$	C_2	$2C_2'$	$2C_2''$	i	$2S_4$	σ_h	$2\sigma_v$	$2\sigma_d$		$h=16$
A_{1g}	1	1	1	1	1	1	1	1	1	1		x^2+y^2, z^2
A_{2g}	1	1	1	-1	-1	1	1	1	-1	-1	R_z	
B_{1g}	1	-1	1	1	-1	1	-1	1	1	-1		x^2-y^2
B_{2g}	1	-1	1	-1	1	1	-1	1	-1	1		xy
E_g	2	0	-2	0	0	2	0	-2	0	0	(R_x, R_y)	(xy, yz)
A_{1u}	1	1	1	1	1	-1	-1	-1	-1	-1		
A_{2u}	1	1	1	-1	-1	-1	-1	-1	1	1	z	
B_{1u}	1	-1	1	1	-1	-1	1	-1	-1	1		
B_{2u}	1	-1	1	-1	1	-1	1	-1	1	-1		
E_u	2	0	-2	0	0	-2	0	2	0	0	(x, y)	

D_{5h}	E	$2C_5$	$2C_5^2$	$5C_2$	σ_h	$2S_5$	$2S_5^3$	$5\sigma_v$		$h=20,\ \alpha=72°$
A_1'	1	1	1	1	1	1	1	1		x^2+y^2, z^2
A_2'	1	1	1	-1	1	1	1	-1	R_z	
E_1'	2	$2\cos\alpha$	$2\cos 2\alpha$	0	2	$2\cos\alpha$	$2\cos 2\alpha$	0	(x, y)	
E_2'	2	$2\cos 2\alpha$	$2\cos\alpha$	0	2	$2\cos 2\alpha$	$2\cos\alpha$	0		(x^2-y^2, xy)
A_1''	1	1	1	1	-1	-1	-1	-1		
A_2''	1	1	1	-1	-1	-1	-1	1	z	
E_1''	2	$2\cos\alpha$	$2\cos 2\alpha$	0	-2	$-2\cos\alpha$	$-2\cos 2\alpha$	0	(R_x, R_y)	(xy, yz)
E_2''	2	$2\cos 2\alpha$	$2\cos\alpha$	0	-2	$-2\cos 2\alpha$	$-2\cos\alpha$	0		

D_{6h} (6/mmm)	E	$2C_6$	$2C_3$	C_2	$3C_2'$	$3C_2''$	i	$2S_3$	$2S_6$	σ_h	$3\sigma_d$	$3\sigma_v$	$h=24$	
A_{1g}	1	1	1	1	1	1	1	1	1	1	1	1		$x^2+y^2,\ z^2$
A_{2g}	1	1	1	1	-1	-1	1	1	1	1	-1	-1	R_z	
B_{1g}	1	-1	1	-1	1	-1	1	-1	1	-1	1	-1		
B_{2g}	1	-1	1	-1	-1	1	1	-1	1	-1	-1	1		
E_{1g}	2	1	-1	-2	0	0	2	1	-1	-2	0	0	(R_x, R_y)	$(xy,\ yz)$
E_{2g}	2	-1	-1	2	0	0	2	-1	-1	2	0	0		$(x^2-y^2,\ xy)$
A_{1u}	1	1	1	1	1	1	-1	-1	-1	-1	-1	-1		
A_{2u}	1	1	1	1	-1	-1	-1	-1	-1	-1	1	1	z	
B_{1u}	1	-1	1	-1	1	-1	-1	1	-1	1	-1	1		
B_{2u}	1	-1	1	-1	-1	1	-1	1	-1	1	1	-1		
E_{1u}	2	1	-1	-2	0	0	-2	-1	1	2	0	0	(x, y)	
E_{2u}	2	-1	-1	2	0	0	-2	1	1	-2	0	0		

$D_{\infty h}$	E	$\infty C_2'$	$2C_\phi$	i	$\infty\sigma_v$	$2S_\phi$	$h=\infty$		
$A_{1g}\ (\Sigma_g^+)$	1	1	1	1	1	1		$z^2,\ x^2+y^2$	
$A_{1u}\ (\Sigma_u^+)$	1	-1	1	-1	1	-1	z		
$A_{2g}\ (\Sigma_g^-)$	1	-1	1	1	-1	1	R_z		
$A_{2u}\ (\Sigma_u^-)$	1	1	1	-1	-1	-1			
$E_{1g}\ (\Pi_g)$	2	0	$2\cos\phi$	2	0	$-2\cos\phi$	(R_x, R_y)	$(xz,\ yz)$	
$E_{1u}\ (\Pi_u)$	2	0	$2\cos\phi$	-2	0	$2\cos\phi$	(x, y)	$(xy,\ x^2-y^2)$	
$E_{2g}\ (\Delta_g)$	2	0	$2\cos 2\phi$	2	0	$2\cos 2\phi$			
$E_{2u}\ (\Delta_u)$	2	0	$2\cos 2\phi$	-2	0	$-2\cos 2\phi$			
$\vdots$	$\vdots$	$\vdots$	$\vdots$	$\vdots$	$\vdots$	$\vdots$			

The groups D_{nd}

$D_{2d}=V_d$ ($\bar{4}2\,m$)	E	$2S_4$	C_2	$2C_2'$	$2\sigma_d$	$h=8$	
A_1	1	1	1	1	1		$x^2+y^2,\ z^2$
A_2	1	1	1	-1	-1	R_z	
B_1	1	-1	1	1	-1		x^2-y^2
B_2	1	-1	1	-1	1	z	xy
E	2	0	-2	0	0	(x, y) (R_x, R_y)	$(xz,\ yz)$

D_{3d} ($\bar{3}m$)	E	$2C_3$	$3C_2$	i	$2S_6$	$3\sigma_d$		$h = 12$
A_{1g}	1	1	1	1	1	1		$x^2 + y^2$, z^2
A_{2g}	1	1	-1	1	1	-1	R_z	
E_g	2	-1	1	2	-1	0	(R_x, R_y)	$(x^2 - y^2, xy)$
								(xz, yz)
A_{1u}	1	1	1	-1	-1	-1		
A_{2u}	1	1	-1	-1	-1	1	z	
E_u	2	-1	1	-2	1	0	(x, y)	

D_{4d}	E	$2S_8$	$2C_4$	$2S_8^3$	C_2	$4C_2'$	$4\sigma_d$		$h = 16$
A_1	1	1	1	1	1	1	1		$x^2 + y^2$, z^2
A_2	1	1	1	1	1	-1	-1	R_z	
B_1	1	-1	1	-1	1	1	-1		
B_2	1	-1	1	-1	1	-1	1	z	
E_1	2	$\sqrt{2}$	0	$-\sqrt{2}$	-2	0	0	(x, y)	
E_2	2	0	-2	0	2	0	0		$(x^2 - y^2, xy)$
E_3	2	$-\sqrt{2}$	0	$-\sqrt{2}$	-2	0	0	(R_x, R_y)	(xz, yz)

The cubic groups

T_d ($\bar{4}3m$)	E	$8C_3$	$3C_2$	$6S_4$	$6\sigma_d$		$h = 24$
A_1	1	1	1	1	1		$x^2 + y^2 + z^2$
A_2	1	1	1	-1	-1		
E	2	-1	2	0	0		$(2z^2 - x^2 - y^2,$ $x^2 - y^2)$
T_1	3	0	-1	1	-1	(R_x, R_y, R_z)	
T_2	3	0	-1	-1	1	(x, y, z)	(xy, xz, yz)

The cubic groups (continued)

O_h ($m3m$)	E	$8C_3$	$6C_2$	$6C_4$	$3C_2$ ($=C_4^2$)	i	$6S_4$	$8S_6$	$3\sigma_h$	$6\sigma_d$	$h=48$	
A_{1g}	1	1	1	1	1	1	1	1	1	1		$x^2+y^2+z^2$
A_{2g}	1	1	−1	−1	1	1	−1	1	1	−1		
E_g	2	−1	0	0	2	2	0	−1	2	0		$(2z^2-x^2-y^2,$ $x^2-y^2)$
T_{1g}	3	0	−1	1	−1	3	1	0	−1	−1	(R_x, R_y, R_z)	
T_{2g}	3	0	1	−1	−1	3	−1	0	−1	1		(xz, yz, xy)
A_{1u}	1	1	1	1	1	−1	−1	−1	−1	−1		
A_{2u}	1	1	−1	−1	1	−1	1	−1	−1	1		
E_u	2	−1	0	0	2	−2	0	1	−2	0		
T_{1u}	3	0	−1	1	−1	−3	−1	0	1	1	(x, y, z)	
T_{2u}	3	0	1	−1	−1	−3	1	0	1	−1		

The icosahedral group

I	E	$12C_5$	$12C_5^2$	$20C_3$	$15C_2$	$h=60$	
A	1	1	1	1	1		$x^2+y^2+z^2$
T_1	3	$\frac{1}{2}(1+\sqrt{5})$	$\frac{1}{2}(1+\sqrt{5})$	0	−1	(x, y, z) (R_x, R_y, R_z)	
T_2	3	$\frac{1}{2}(1+\sqrt{5})$	$\frac{1}{2}(1+\sqrt{5})$	0	−1		
G	4	−1	−1	1	0		
H	5	0	0	−1	1		$(2z^2-x^2-y^2,$ $x^2-y^2, xy, yz, zx)$

Further information: P. W. Atkins, M. S. Child, and C. S. G. Phillips, *Tables for group theory*. Oxford University Press (1970).

APPENDIX 4:
Symmetry adapted orbitals

Table A4.1 gives the symmetry classes of the s, p, and d orbitals of the central atom of an AB_n molecule of the specified point group. In most cases, the z-axis is the principal axis of the molecule; in C_{2v} the x-axis lies perpendicular to the molecular plane.

The orbital diagrams that follow show the linear combinations of atomic orbitals on the peripheral atoms of AB_n molecules of the specified point groups. Where a view from above is shown, the dot representing the central atom is either in the plane of the paper (for the D groups) or above the plane (for the corresponding C groups). Different phases of the atomic orbitals ($+$ or $-$ amplitudes) are shown by tinting. Where there is a large difference in the magnitudes of the orbital coefficients in a particular combination, the atomic orbitals have been drawn large or small to represent their relative contributions to the linear combination. In the case of degenerate linear combinations (those labeled E or T), any linearly independent combination of the degenerate pair is also of suitable symmetry. In practice, these different linear combinations look like the ones shown here, but their nodes are rotated by an arbitrary axis around the z-axis.

Molecular orbitals are formed by combining an orbital of the central atom (as in Table A4.1) with a linear combination of the same symmetry.

Table A4.1

	$D_{\infty h}$	C_{2v}	D_{3h}	C_{3v}	D_{4h}	C_{4v}	D_{5h}	C_{5v}	D_{6h}	C_{6v}	T_d	O_h
s	σ	A_1	A_1'	A_1	A_{1g}	A_1	A_1'	A_1	A_{1g}	A_1	A_1	A_{1g}
p_x	π	B_1	E'	E	E_u	E	E_1'	E_1	E_{1u}	E_1	T_2	T_{1u}
p_y	π	B_2	E'	E	E_u	E	E_1'	E_1	E_{1u}	E_1	T_2	T_{1u}
p_z	σ	A_1	A_2''	A_1	A_{2u}	A_1	A_2''	A_1	A_{2u}	A_1	T_2	T_{1u}
d_{z^2}	σ	A_1	A_1'	A_1	A_{1g}	A_1	A_1'	A_1	A_{1g}	A_1	E	E_g
$d_{x^2-y^2}$	δ	A_1	E'	E	B_{1g}	B_1	E_2'	E_2	E_{2g}	E_2	E	E_g
d_{xy}	δ	A_2	E'	E	B_{2g}	B_2	E_2'	E_2	E_{2g}	E_2	T_2	T_{2g}
d_{yz}	π	B_2	E''	E	E_g	E	E_1''	E_1	E_{1g}	E_1	T_2	T_{2g}
d_{zx}	π	B_1	E''	E	E_g	E	E_1''	E_1	E_{1g}	E_1	T_2	T_{2g}

$D_{\infty h}$	C_{2v}	
σ_g	A_1	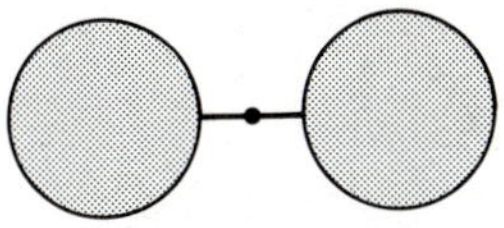
π_g	A_2	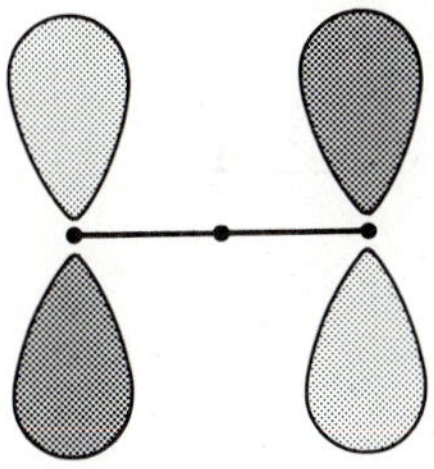
π_u	B_1	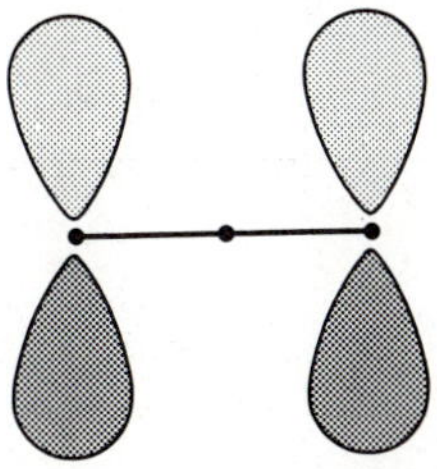
σ_u	B_2	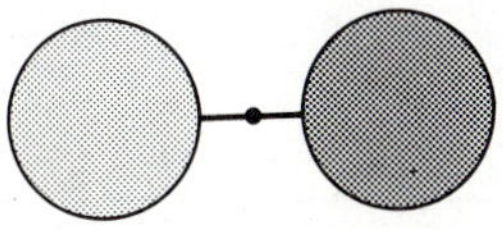
	A_1	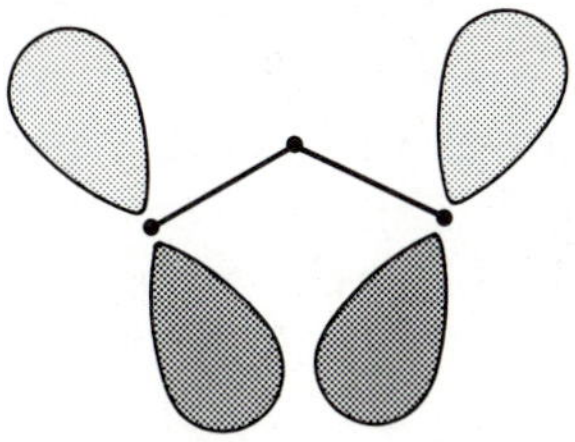
	B_2	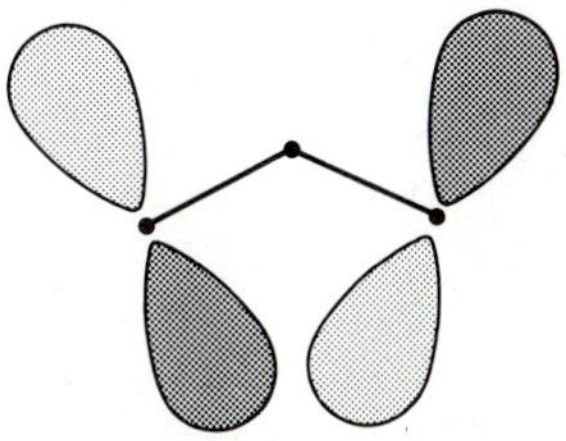

D_{3h}	C_{3v}	
A_1'	A_1	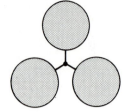
A_2'	A_2	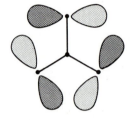
E'	E	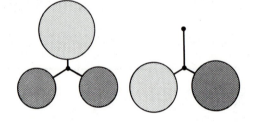
E'	E	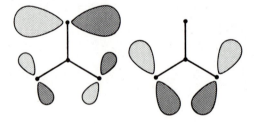
A_1''	A_1	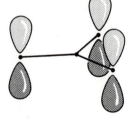

D_{3h}	C_{3v}	(cont.)

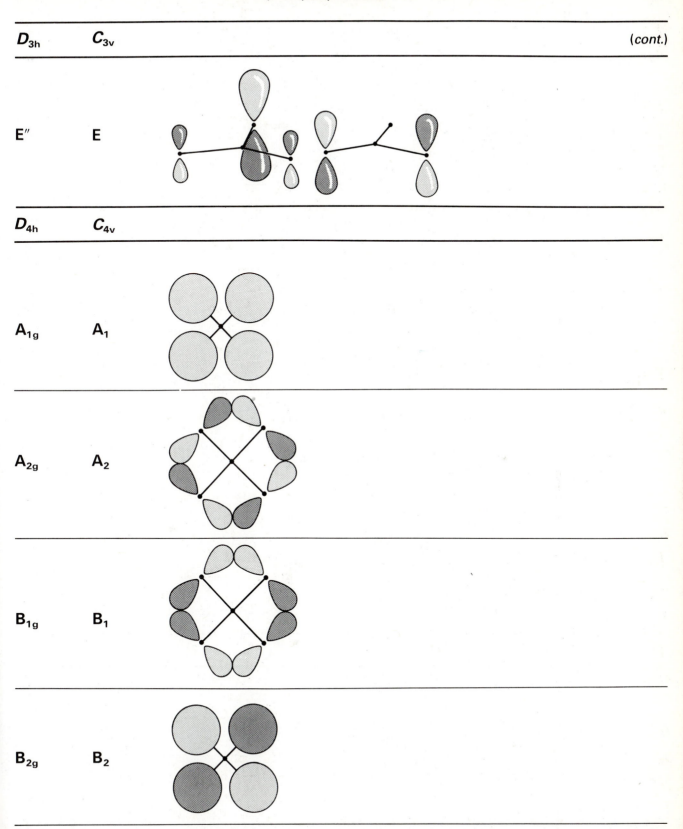

| E'' | E |

D_{4h}	C_{4v}

| A_{1g} | A_1 |

| A_{2g} | A_2 |

| B_{1g} | B_1 |

| B_{2g} | B_2 |

D_{4h}	C_{4v}	(cont.)
E_u	E	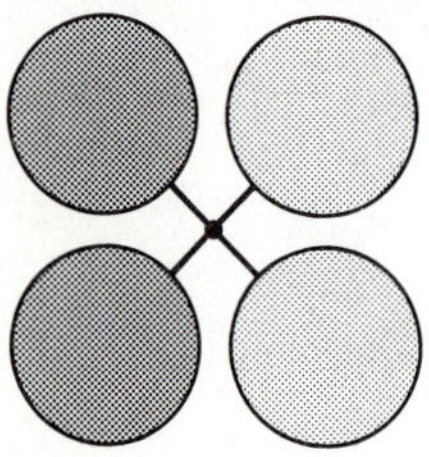 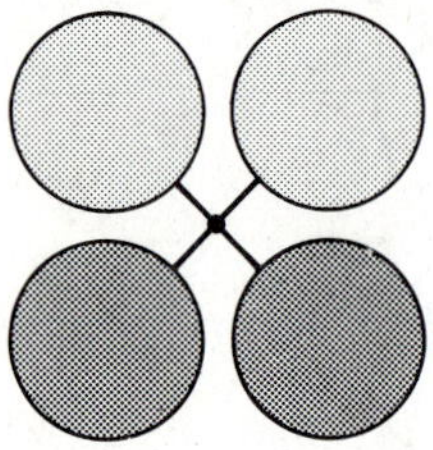
A_{2u}	A_1	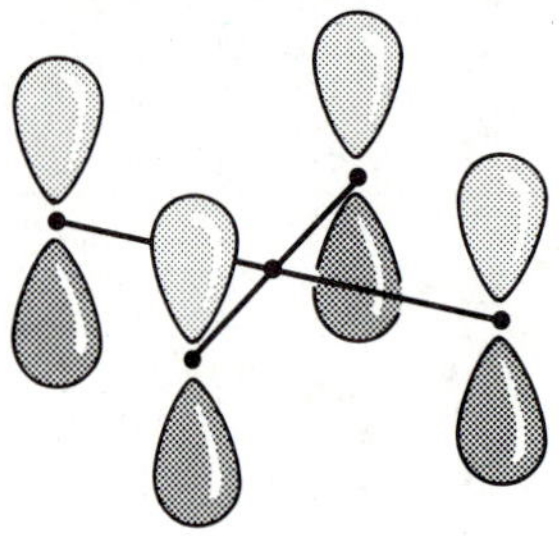
E_g	E	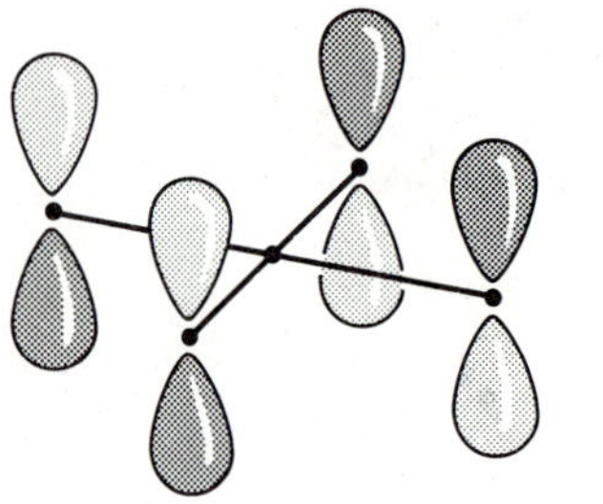 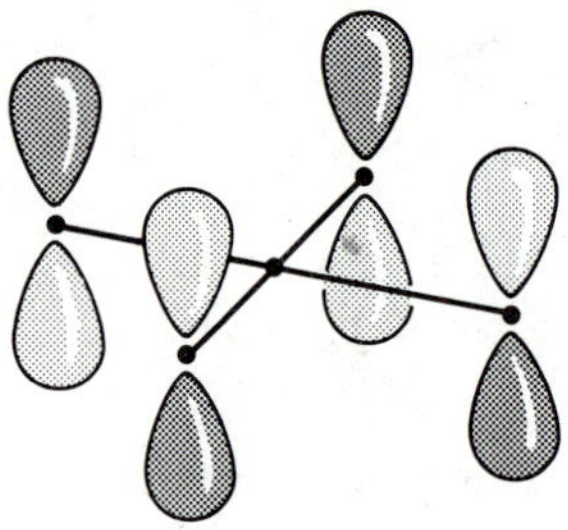
B_{2u}	B_2	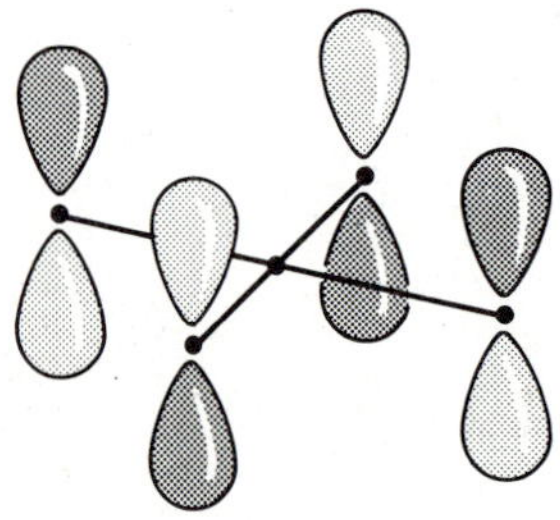
D_{5h}	C_{5v}	
A_1'	A_1	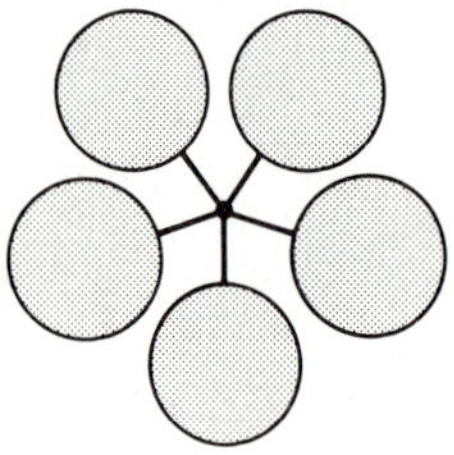

D_{5h}	C_{5v}			(cont.)
A_2'	A_2			
E_1'	E_1			
E_2'	E_2			
A_2''	A_1			
E_1''	E_1			
E_2''	E_2			

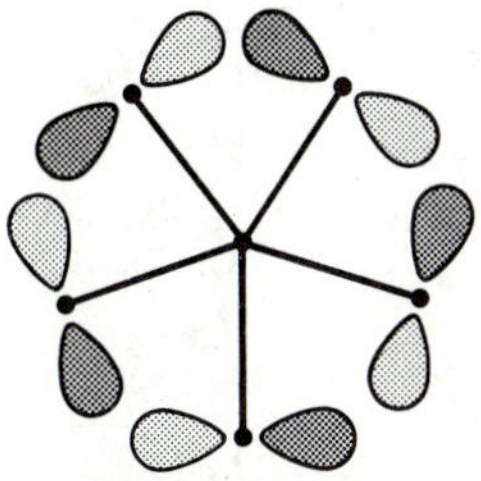

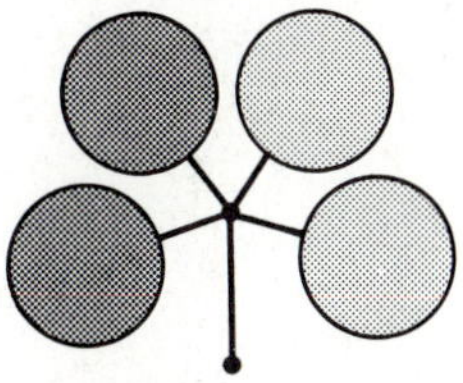

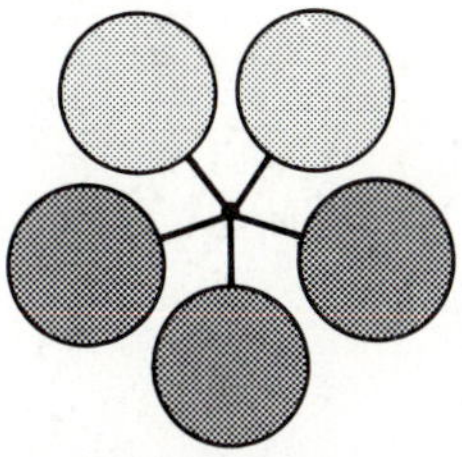

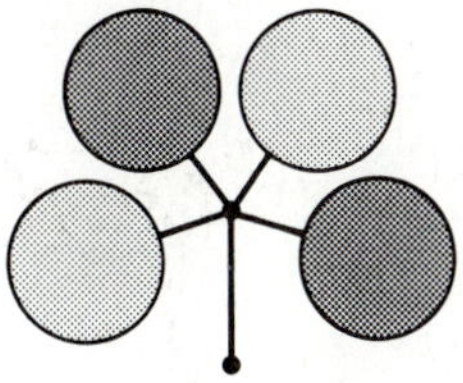

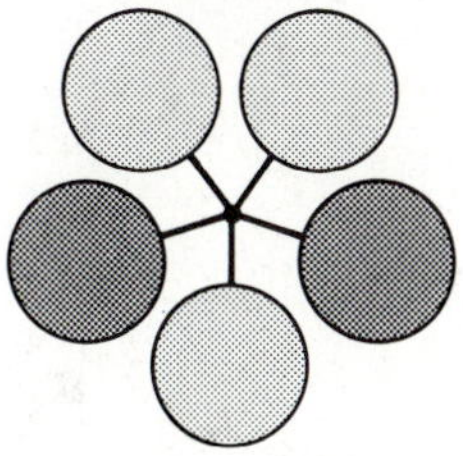

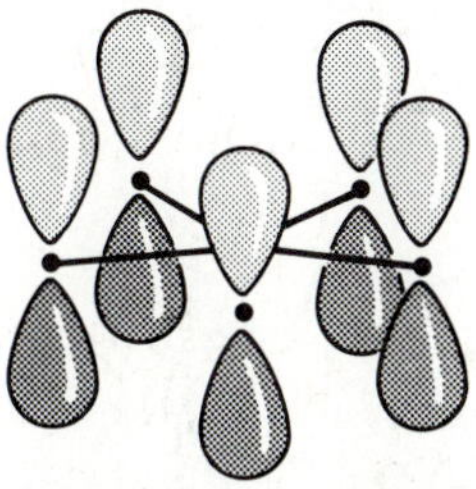

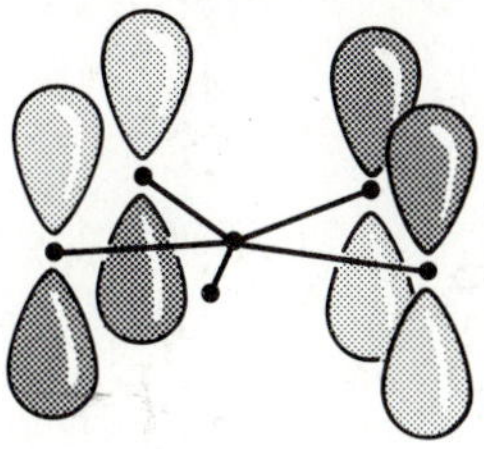

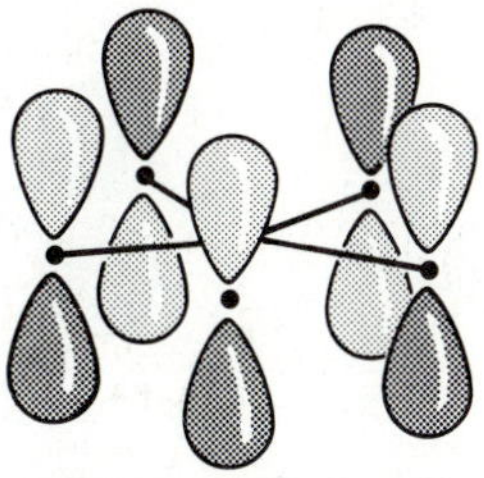

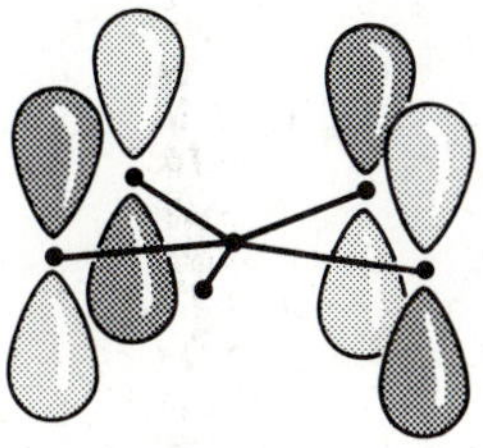

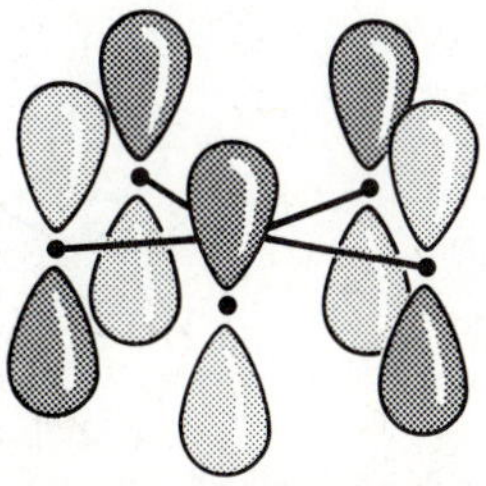

D_{6h}	C_{6v}		
A_{1g}	A_1	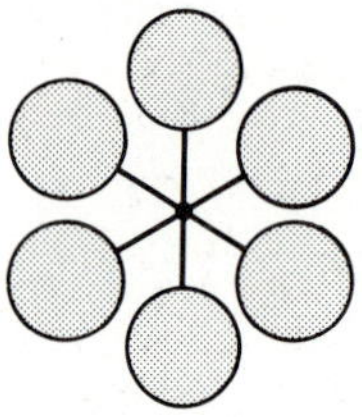	
A_{2g}	A_2	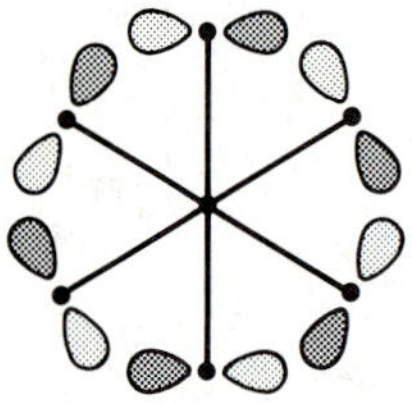	
B_{1u}	B_1	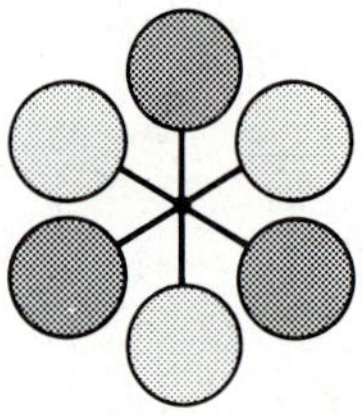	
B_{2u}	B_2	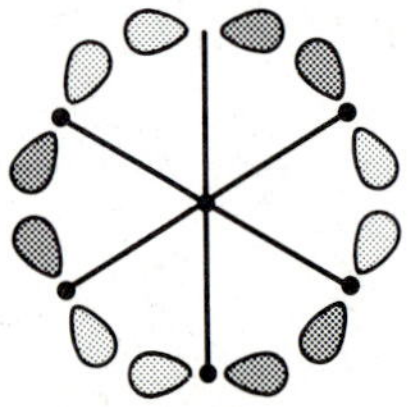	
E_{1u}	E_1	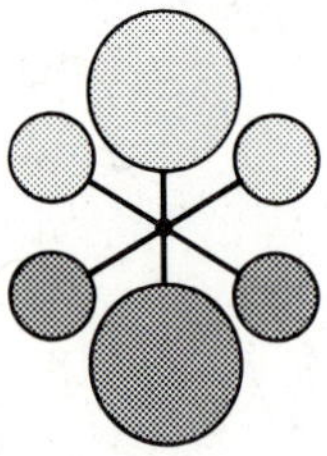	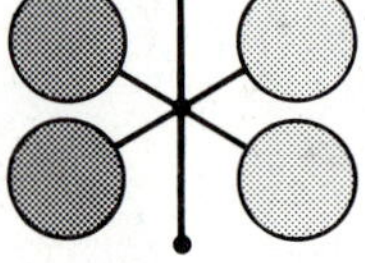
E_{2g}	E_2	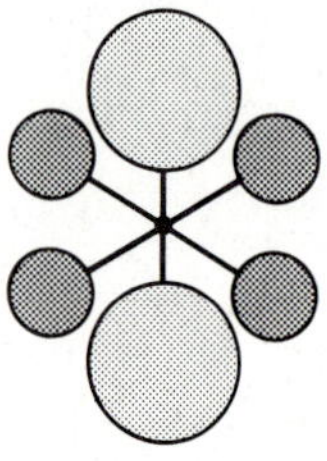	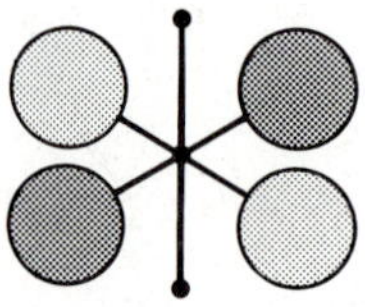

D_{6h}	C_{6v}		(cont.)

A_{2u}	A_1	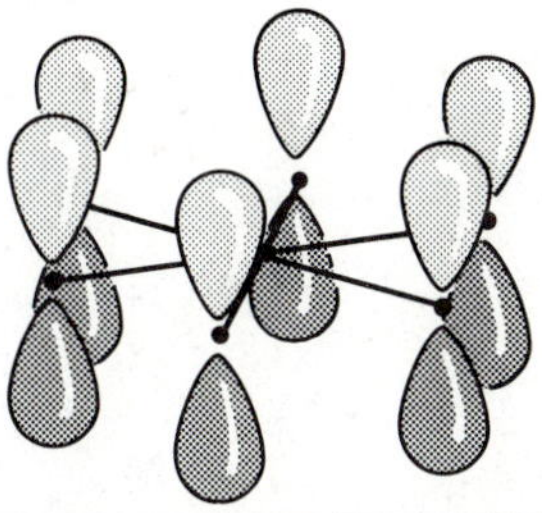	
B_{2g}	B_1	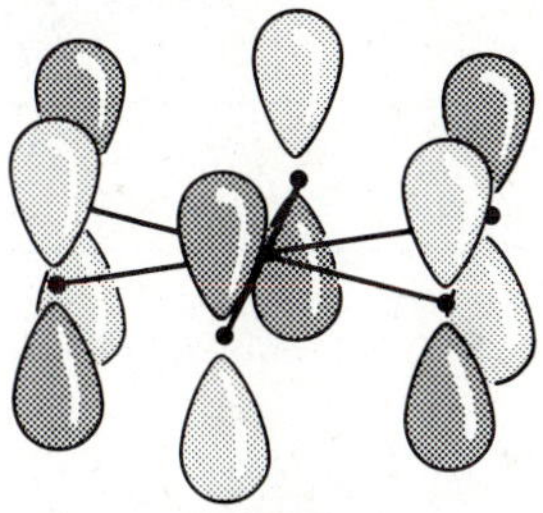	
E_{1g}	E_1	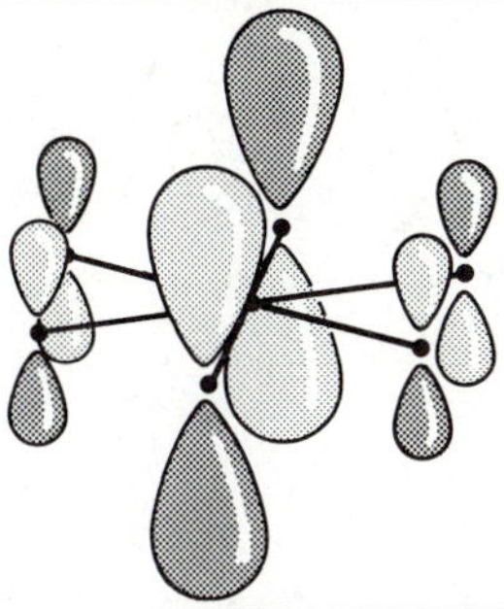	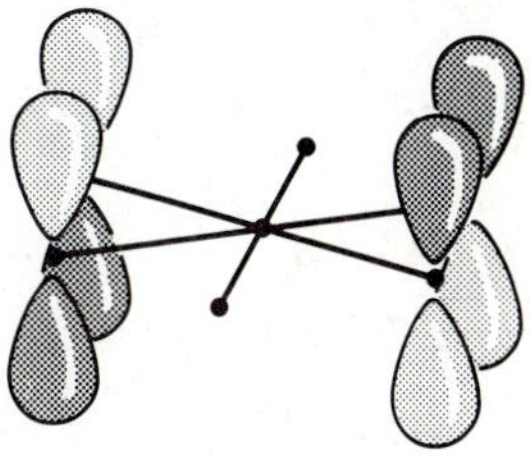
E_{2u}	E_2	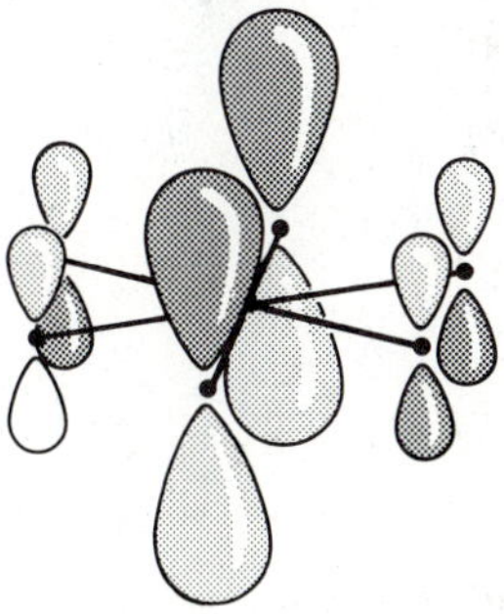	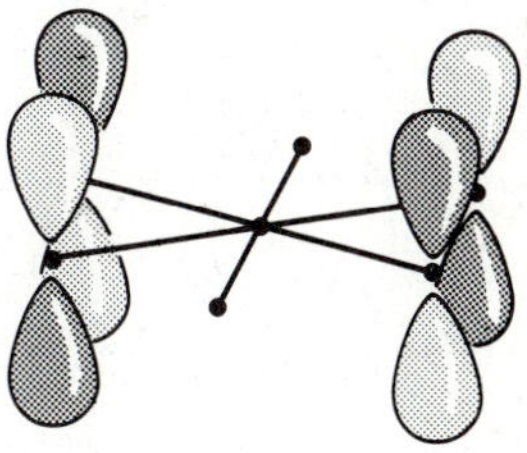

T_d			
A_1			

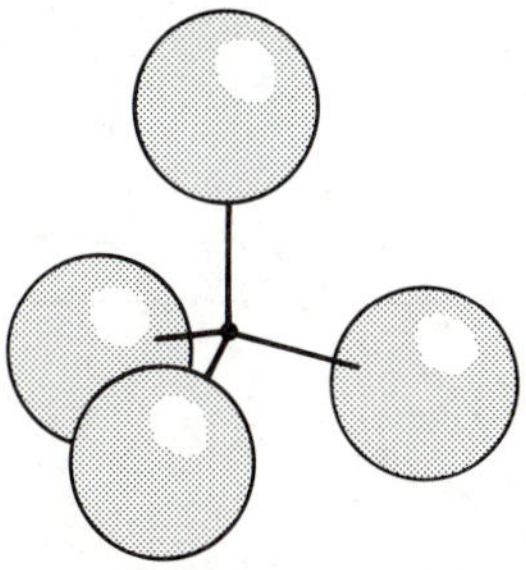

T_d   *(cont.)*

T_2

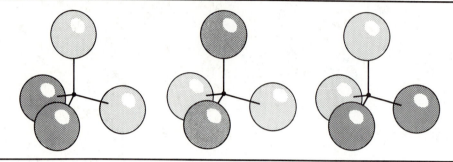

$O_h(\sigma)$

A_{1g}

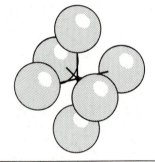

E_g

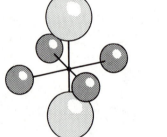

T_{1u}

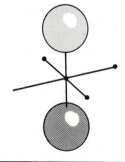 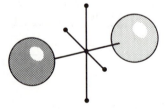 

$O_h(\pi)$

T$_{1u}$

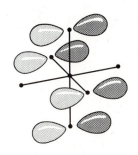
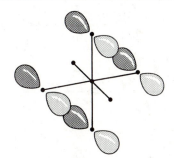
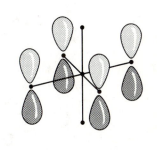

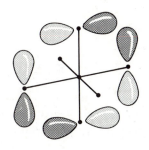
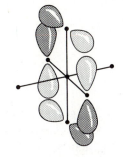
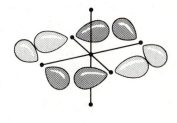

T$_{2g}$

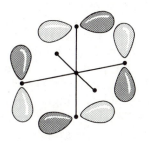
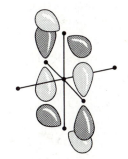
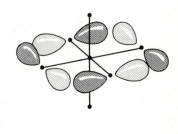

T$_{1g}$

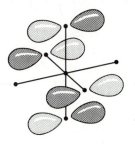
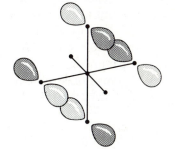
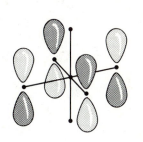

T$_{2u}$

APPENDIX 5:
Tanabe-Sugano diagrams

This appendix collects together the Tanabe–Sugano diagrams for octahedral complexes with electron configurations d^2 to d^9. The diagrams, which were introduced in Section 14.3, show the dependence of the term energies on ligand field strength. The term energies E are expressed as the ratio E/B, where B is a Racah parameter, and the ligand field splitting Δ_O is expressed as Δ_O/B. Terms of different multiplicity are included in the same diagram by making specific, plausible choices about the value of the Racah parameter C, and these choices are given for each diagram. The term energy is always measured from the lowest energy term, and so there are discontinuities of slope where a low-spin term displaces a high-spin term at sufficiently high ligand-field strengths for d^4 to d^8 configurations. Moreover, the non-crossing rule requires terms of the same symmetry to mix rather than to cross, and this mixing accounts for the curved rather than the straight lines in a number of cases. The term labels are those of the point group O_h.

The diagrams were first introduced by Y. Tanabe and S. Sugano, *J. Phys. Soc. Japan*, **9**, 753 (1954). They may be used to find the parameters Δ_O and B by fitting the ratios of the energies of observed transitions to the lines. Alternatively, if the ligand field parameters are known, then the ligand field spectra may be predicted.

1. d^2 with $C = 4.42B$

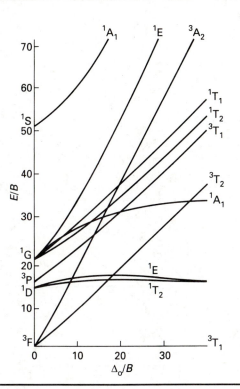

3. d^4 with $C = 4.61B$

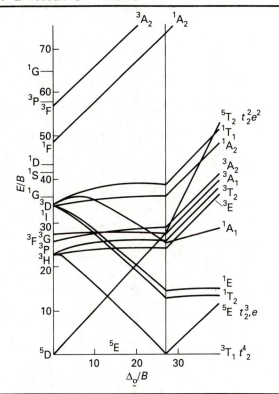

2. d^3 with $C = 4.5B$

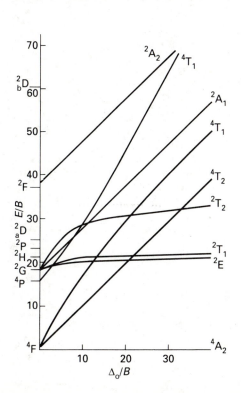

4. d^5 with $C = 4.477B$

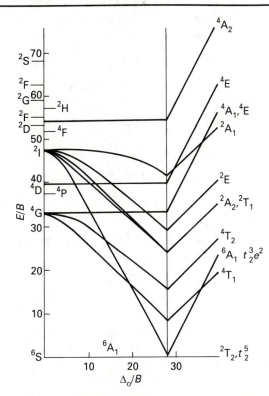

5. d^6 with $C = 4.8B$

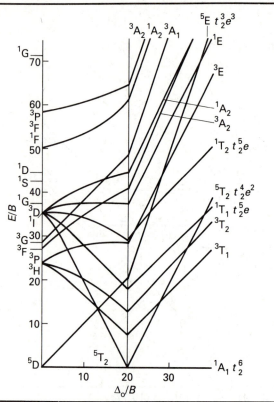

7. d^8 with $C = 4.709B$

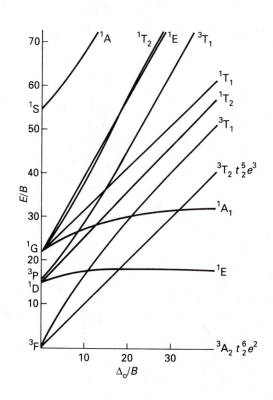

6. d^7 with $C = 4.633B$

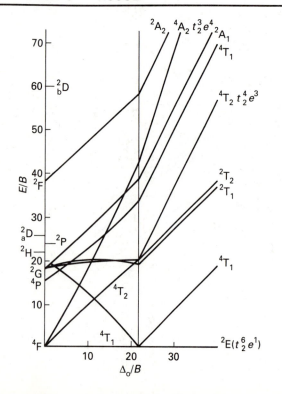

Answers to Exercises

Chapter 1

In-chapter exercises

1.1 $^{80}_{35}\mathrm{Br} + \mathrm{n} \rightarrow {}^{81}_{35}\mathrm{Br} + \gamma$

1.2 3.87×10^{-19} J.

1.3 $3p$.

1.4 Added p electron is in a different (p) orbital so it is less shielded.

1.5 Ni: $[\mathrm{Ar}]3d^8 4s^2$, $[\mathrm{Ar}]3d^8$

1.6 Going down a group the atomic radius increases and the first ionization energy generally decreases.

1.7 For C the p orbitals are all singly occupied. On going to C the added electron must go into a half filled orbital with added electron-electron repulsion.

End-of-chapter exercises

1.1 (a) $^{14}_{7}\mathrm{N} + {}^{4}_{2}\mathrm{He} \rightarrow {}^{17}_{8}\mathrm{O} + {}^{1}_{1}\mathrm{p} + \gamma$
(b) $^{12}_{6}\mathrm{C} + {}^{1}_{1}\mathrm{p} \rightarrow {}^{13}_{7}\mathrm{N} + \gamma$
(c) $^{14}_{7}\mathrm{N} + {}^{0}_{1}\mathrm{n} \rightarrow {}^{12}_{6}\mathrm{C} + {}^{3}_{1}\mathrm{H}$

1.2 $^{22}_{10}\mathrm{Ne} + {}^{4}_{2}\alpha \rightarrow {}^{25}_{12}\mathrm{Mg} + \mathrm{n}$

1.3 Check inside front cover

1.4 0.25.

1.5 -13.2 eV.

1.6 0 up to $n-1$.

1.7 n^2.

1.8 See Figs. 1.14 through 1.18.

1.9 See Table 1.7 and discussion on p40.

1.10 Table 1.7 shows Sr > Ba < Ra. Ra is anomalous because of higher Z_{eff} due to lanthanide contraction.

1.11 Anomalously high value for Cr is associated with the stability of 1/2-filled d shell.

1.12 (a) $[\mathrm{He}]2s^2 2p^2$; (b) $[\mathrm{He}]2s^2 2p^5$; (c) $[\mathrm{Ar}]4s^2$; (d) $[\mathrm{Ar}]3d^{10}$; (e) $[\mathrm{Xe}]4f^{14}5d^{10}6s^2 3p^3$; (f) $[\mathrm{Xe}]4f^{14}5d^{10}6s^2$.

1.13 (a) $[\mathrm{Ar}]3d^{10}s^2$; (b) $[\mathrm{Ar}]3d^2$; (c) $[\mathrm{Ar}]3d^5$; (d) $[\mathrm{Ar}]3d^4$; (e) $[\mathrm{Ar}]3d^6$; (f) $[\mathrm{Ar}]$.

1.14 (a) $[\mathrm{Xe}]4f^{14}5d^4 6s^2$; (b) $[\mathrm{Kr}]4d^6$; (c) $[\mathrm{Xe}]4f^6$; (d) $[\mathrm{Xe}]4f^7$; (e) $[\mathrm{Ar}]$; (f) $[\mathrm{Kr}]4d^2$.

1.15 (a) I_i increases across group except for a dip at S; (b) A_e tends to increase except for Mg (filled subshell), P (half filled subshell) and Ar (filled shell).

1.16 Radii of period 4 and 5 d-block metals are similar because of lanthanide contraction.

1.17 χ increases steadily in most scales. For χ_{M} anomalies in I are generally offset by those in A.

1.18 $2s^2$ and $2p^0$.

1.19 With some exceptions generally associated with half-full or full subshells I and χ increase and r decreases across a period.

Chapter 2

In-chapter exercises

2.1
:Cl: P :Cl:
   :Cl:

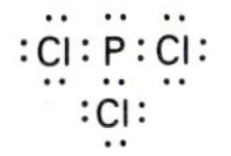

2.2 $+1/2$, $+5$

2.3
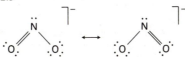

2.4 24 kJ mol^{-1}.

2.5 S_2^{2-}: $1\sigma_g^2\, 2\sigma_u^2\, 3\sigma_g^2\, 1\pi_u^4\, 2\pi_g^4$; Cl_2^-: $[\mathrm{S}_2^{2-}]\, 4\sigma_u^1$.

2.6 Similar to Fig. 2.17 with Cl $3s$ and $3p$ on left and O $2s$ and $2p$ on right.

2.7 Lowest energy: totally symmetric (no nodal planes); doubly degenerate (MO pair with $\perp$ nodal planes); doubly degenerate (each has two nodal planes); doubly degenerate (each has three nodal planes); highest energy; nodes between each nucleus.

2.8 0.7065.

End-of-Chapter exercises

2.1 (a)
:Cl:
:Ge: Cl:
:Cl:

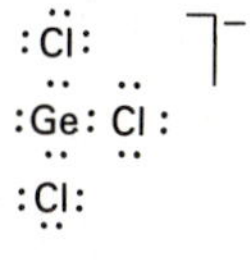

(b)
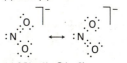

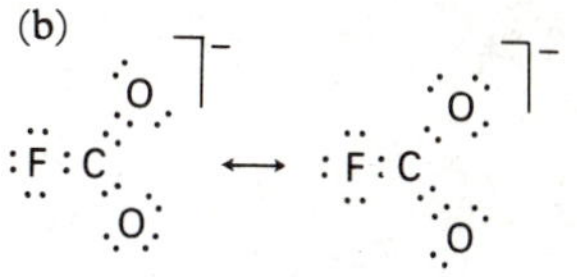

(c)
[:O:C:O:]$^{2-}$ $\longleftrightarrow$ [:O:C:O:]$^{2-}$ $\longleftrightarrow$ [:O::C:O:]$^{2-}$

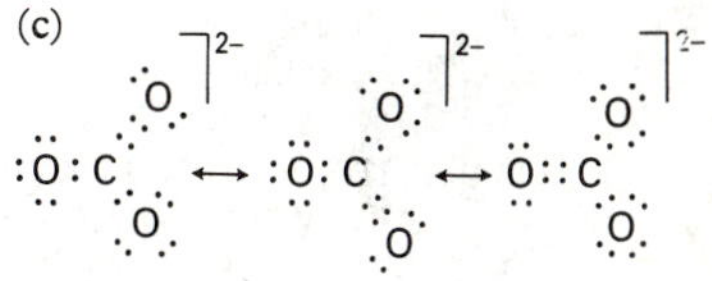

(d)
[:Cl: Al :Cl: with :Cl: above and :Cl: below]$^-$

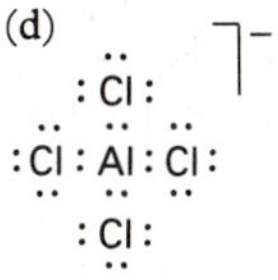

(e)
N, :F:, :O:

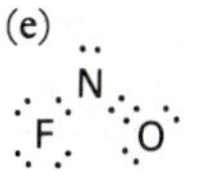

2.2
[:O:N:::C:]$^-$ $\longleftrightarrow$ [:O::N::C:]$^-$

2.3 (a) and (b):
[:N:O: with :O: above]$^-$ $\longleftrightarrow$ [:N:O: with :O: above]$^-$

(c) N($+3$), O(-2)
(i) formal charge, (ii) oxidation number, (iii) neither

2.4 (a) Xe with 4 F (XeF$_4$) (b) P with F, F, F, F, F
(c) BrF (d) Te with 4 Cl
(e) [:Cl: I :Cl:]$^-$

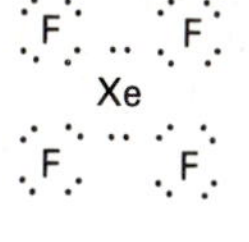

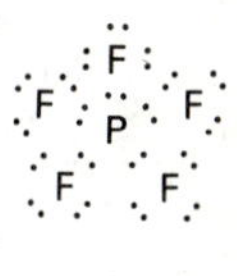

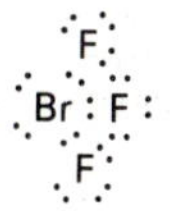

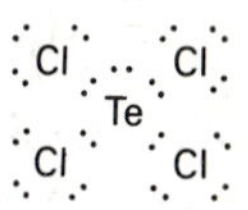

2.5   (a) 1.76 Å, (b) 2.17 Å, (c) 2.21 Å.

2.6 $2(Si—O) = 932$ kJ $> Si=O = 640$ kJ, therefore two $Si—O$ are preferred and SiO_2 should (and does) have four single $Si—O$ bonds.

2.7   Bond enthalpy data (Table 2.5) indicate $2N\equiv N \rightarrow N_4$ $\Delta H^\ominus \approx +912$ kJ $2P_2 \rightarrow P_4$ $\Delta H^\ominus \approx -238$ kJ. Multiple bonds are much stronger for 2nd period elements than heavier elements.

2.8   483 kJ difference is smaller than expected because bond energies are not accurate.

2.9   (a) 0, (b) 205 kJ mol^{-1}.

2.10 (a) One; (b) one; (c) none; (d) one.

2.11 (a) $1\sigma_g^2 2\sigma_u^2$, (b) $1\sigma_g^2 2\sigma_u^2 1\pi_u^2$, (c) $1\sigma_g^2 2\sigma_u^2 1\pi_u^4 3\sigma_g^1$, (d) $1\sigma_g^2 2\sigma_u^2 3\sigma_g^2 1\pi_u^4 2\pi_g^3$.

2.12 (a) 2; (b) 1; (c) 2.

2.13 (a) $+0.5$; (b) -0.5; (c) $+0.5$.

2.14 (a) 4; (b) see E_4 in Fig. 2.24 where the atomic orbitals are all $1s$; (c) as with E_4 in Fig. 2.24 where bottom orbital is lowest in energy.

2.15 (a) See E_3 in Fig. 2.24; (b) as with E_3 Fig. 2.24 with bottom orbital lowest in energy; (b) possibly stable in isolation (only bonding and non bonding orbitals are filled); not stable in solution because solvents would have higher proton affinity than He.

2.16 1.

2.17 HOMO exclusively F; LUMO mainly S.

2.18 (a) Electron deficient; (b) electron precise.

2.19 (a) NH_3; (b) BH_3 (c) NH_3.

2.20 (a) Compare Fig. 2.43 for a metal with Fig. 2.50 for a semiconductor; (b) for metal σ (conductivity) decreases moderately with increasing T; for semiconductor σ increases strongly with increasing T; (c) similar qualitatively but σ for insulator is lower.

2.21 (a) n; (b) p; (c) neither.

2.22 3.55 eV.

2.23 n.

2.24 n.

2.25 518 nm.

2.26 665.

Chapter 3

In-chapter exercises

3.1   Linear.

3.2 2.89×10^{-4} s.

3.3 MgH_2 in gas phase.

3.4   Four such axes, one along each of the NH bonds.

3.5   (a) D_{3h}; (b) T_d.

3.6   No.

3.7   Yes (but the two chiral forms rapidly interconvert).

3.8   Triple.

3.9 B_{2g}.

3.10 A_1.

3.11 Yes.

3.12 T_{1u}.

End-of-chapter exercises

3.1   (a) Planar (in gas phase); (b) planar; (c) square pyramidal.

3.2   Tetrahedral and octahedral, respectively.

3.3   X-ray p.e.s. (least), vis-uv, IR, NMR (most).

3.4   (a) 5.3×10^{-13} s; (b) 6.4×10^{-4} s.

3.5   (a) 1.13×10^3 s^{-1}.

3.6   (N), C_3 through N and the center of the H_3 triangle, three σ_v through N and one of each of the H atoms; C_4 through Pt and perpendicular to plane of the complex; σ_h is in the plane of the complex.

3.7   (1) CO_2, C_2H_2; (2) SO_4^{2-}.

3.8   (a) C_s; (b) D_{3h}; (c) T_d; (d) $C_{\infty v}$; (e) C_1; (f) T_d.

3.9   (a) ∞ C_n axes; ∞ σ; i, (b) C_∞, ∞ σ; (c) 3σ, $3C_2$; (d) σ, C_∞, ∞C_2.

3.10 (a) $\sigma \perp$ to C_n axis, i, multiple non-collinear C_n axes; (b) a, d, and e may be plar.

3.11 (a) Lack of S_n symmetry element (s); (b) only SiFClBrI.

3.12 (a) C_{3v}; (b) double degeneracy; (c) p_x and p_y (where C_3 is along z axis of SO_3^{2-}).

3.13 (a) D_{3h}; (b) doubly degenerate; (c) p_x and p_y (where C_3 is along z axis of PF_5).

3.14 (a) (A_{1g}); (E_u); (B_{2g}); (b) D_{4h}, symmetry labels given in part a. (c) d_{z^2} (A_{1g}), and d_{xy} (B_{2g}).

3.15 (a) 5; (b) 1.

3.16 (a) none; (b) E'.

3.17 A_2, B_1, B_2.

Chapter 4

In-chapter exercises

4.1   2.

4.2   12.

4.3   6.

4.4   2421 kJ mol^{-1}.

4.5   2.63 MJ mol^{-1}.

4.6 T (decomp): $MgSO_4 < CaSO_4 < SrSO_4 < BaSO_4$.

4.7 $NaClO_4$.

End-of-chapter exercises

4.1   Closest packed: (a), (b), (d).

4.2   Use Fig. 4.3 as a guide.

4.3   Let diagonal of a square array of anions, with $r = 1$ be d. Then $d = 2\sqrt{2} = 2.828$; this minus the radii of the two anions $= 0.828$. Radius ratio: $(0.828)/2 = 0.414$.

4.4   (a) see p147 and 149; (b) polymorphs: diamond and graphite, polytypes often occur in layered compounds, such as TaS_2.

4.5   Closest packed W with C in octahedral holes.

4.6   (a) rock salt 6:6, cesium chloride 8:8; (b) cesium chloride.

4.7   (a) Re: C.N. $= 6$, O: C.N. $= 2$; (b) perovskite.

4.8   Cations: Groups 1 and 2 (except Be), anions: Group 7 and oxygen.

4.9   (a) 6:6; (b) 12; (c) use Figs. 4.4 and 4.6 as a guide.

4.10 (a) 8:8; (b) 6.

4.11 (a) One of each; (b) four of each.

4.12 Use Fig. 4.17 taking into account the sharing of peripheral atoms with the surrounding unit cells.

4.13 Use Fig. 4.18 and the suggestion in the above answer.

4.14 Fluorite (CaF_2) structure.

4.15 $r(Ca^{2+}) = 1.08$ Å, $r(Sr^{2+}) = 1.19$, Å, $r(Ba^{2+}) = 1.38$ Å.

4.16 (a) 6:6; (b) 6:6; (c) 6:6, (d) 6:6 or 4:4. The small Be may lead to covalency which favors 4:4 coordination.

4.17 (a) 2.4 MJ mol^{-1}; (b) large second ionization energy for K.

4.18 (b) $<$ (c) $<$ (a), based on size and charge.

4.19 (a) $MgCO_3$; (b) CsI_3.

4.20 (a) $MgSO_4$; (b) $NaBF_4$.

Chapter 5

In-chapter exercises

5.1   Acid/conjugate base: (a) HNO_3/NO_3^-; (b) H_2O/OH^-; H_2S/HS^-.

5.2 $[Na(OH_2)_6]^+ < [Mn(OH_2)_6]^{2+} < [Ni(OH_2)_6]^{2+} < [Sc(OH_2)_6]^{3+}$.

5.3 pK_a estimated by Pauling's rules: (a) 3; (b) 8; (c) 13.

5.4   Aq. NH_3 precipitates TiO_2, which dissolves in NaOH.

5.5   Acid/base: (a) $FeCl_3/Cl^-$; (b) I_2/I^-; (c) $Mn(CO)_5/SnCl_3^-$.

5.6 $(H_3Si)_3N$, trigonal planar; $(H_3C)_3N$, pyramidal.

5.7   Aluminosilicates: Rb, Cr, Sr; sulfides: Cd, Pb, Pd.

5.8   B coordinated to O; pyramidal C_2OB array.

End-of-chapter exercises

5.1 Check element symbols inside cover. Check acidity and basicity in Fig. 5.5.

5.2 $[Co(NH_3)_5(OH)]^{4+}$; SO_4^{2-}; CH_3O^-; HPO_4^{2-}; $[Si(OH)_3O]^-$; S^{2-}.

5.3 $C_5H_5NH^+$; $H_2PO_4^-$; H_2O; $CH_3CO_2H_2^+$; $HCo(CO)_4^-$; HCN.

5.4 $I^- < F^- < HS^- < NH_2^-$.

5.5 (a) In water too strong; O^{2-}; too weak: ClO_4^-, NO_3^-; measurable; (b) (in H_2SO_4) too strong: ClO_4^-; measurable: NO_3^-, HSO_4^-.

5.6 CN is electron withdrawing.

5.7 From Pauling's rules, $pK_a = 13$.

5.8 $HSiO_4^{3-} < HPO_4^{2-} < HSO_4^- < HClO_4$.

5.9 (a) $[Fe(OH_2)_6]^{3+}$, higher ξ; (b) $[Al(OH_2)_6]^{3+}$, higher ξ; (c) $Si(OH)_4$, higher ξ; (d) $HClO_4$, (higher ox. no. or Paulings rules); (e) $HMnO_4$ (higher ox. no.); (f) H_2SO_4 (higher ox. no. or Pauling's rules).

5.10 $Cl_2O_7 < SO_3 < CO_2 < B_2O_3 < Al_2O_3 < BaO$.

5.11 $NH_3 < CH_3GeH_3 < H_4SiO_4 < HSO_4^- < H_3O^+ < HSO_3F$.

5.12 Ag^+.

5.13 Polyanions: As, B, Si, Mo; polyoxocations: Al, Ti, Cu.

5.14 Reduction of one.

5.15 $4PO_4^{3-} + 8H^+ \rightarrow P_4O_{12}^{4-} + 4H_2O$, $2[Fe(OH_2)_6]^{3+} \rightarrow [(H_2O)_4Fe(\mu\text{-}OH)_2Fe(OH_2)_4]^{4+} + 2H_3O^+$.

5.16 (a) $H_3PO_4 + HPO_4^{2-} \rightarrow 2H_2PO_4^-$; (b) $CO_2 + CaCO_3 + H_2O \rightarrow Ca^{2+} + 2HCO_3^-$.

5.17 Check symbols inside cover; Lewis acids: BF_3, $AlCl_3$, M(I) or M(II) halides of Ga, In, Td; Cl_2; M(IV) halides of Si, Ge, Sn; PCl_5; M(V) halides of P, Sb, As; $BiCl_3$; dioxides of S, Se, Te; Br_2, I_2, IF_7.

5.18 Acids: (a) SO_3 Brønsted; Hg^{2+}; $SnCl_2$; SbF_5; hydrogen bond.

5.19 (a) BCl_3, BCl_3, $B(n\text{-}Bu)_3$; (b) Me_3N, $4\text{-}CH_3C_5H_4N$.

5.20 $K > 1$: (b), (d).

5.21 BH_3 binds to P, BF_3 binds to N.

5.22 Steric repulsion.

5.23 (a) DMSO stronger base to hard and soft acid. (b) SMe_2 stronger base toward soft acids than DMSO.

5.24 $SO_2 + 4HF \rightarrow H_2O + SiF_4$.

5.25 $Al_2S_3 + 3H_2O \rightarrow Al_2O_3 + 3H_2S$.

5.26 (a) and (b) Hard hydrogen bonding solvent; (c) hard donor eg. H_2O; (d) softer base than Cl^-, possibly PR_3.

5.27 I_2^+ and Se_8^{2+} disproportionate in or coordinate to bases; S_4^{2-} and Pb_9^{4-} disproportionate in or coordinate to acids.

5.28 Al_2O_3 acid sites coordinate Cl^-.

5.29 HgS soft − soft, ZnO hard − hard.

5.30 (a) $EtOH + HF \rightarrow EtOH_2^+ + F^-$; (b) $NH_3 + HF \rightarrow NH_4^+ + F^-$; (c) $PhCOOH + HF \rightleftharpoons PhCOO^- + H_2F^+$.

5.31 Both.

5.32 Hard.

5.33 SiO_2 stronger.

5.34 Base O^{2-}, acid $S_2O_7^{2-}$.

5.35 AsF_5 trigonal bipyramidal (D_{3h}); AsF_6^{2-} octahedral (O_h); C_{3v} and D_{2h}.

Chapter 6

In-chapter exercises

6.1 A is *trans*, B is *cis*.

6.2 (a) *cis*-$[PtCl_2(H_2O)_2]$; (b) $[Cr(NCS)_4(NH_3)_2]^-$; (c) $[Rh(en)_3]^{3+}$.

6.3 Replace Co for Cr in **42a** and **42b**.

6.4 Only (a).

6.5 $t_{2g}^3 e_g^2$.

6.6 See discussion of Fig. 6.11.

6.7 Increased LFSE for high to low spin state.

End-of-chapter exercises

6.1 Check element positions inside front cover, tetrahedral halide complexes: Fe^{2+}, Co^{2+}, Ni^{2+}, Cu^{2+} and Zn^{2+}.

6.2 (a) Rh^+, Ir^+, Ni^{2+}, Pd^{2+}, Au^{3+}; (b) $PtCl_4^{2-}$, $Pt(NH_3)_4^{2+}$, $Ni(CN)_4^{2-}$.

6.3 (a) See **12** and **13**; (b) the trigonal prism; (c) $[Cr(OH_2)_6]^{3+}$, $[Co(NH_3)_6]^{3+}$, $[Fe(CN)_6]^{4-}$.

6.4 (a) Tetracarbonylnickel(0), tetrahedral; (b) tetracyanonickelate(2−), square planar; (c) tetrachlorocobaltate(2−); (d) hexaamminenickel(2+).

6.5 (a) See **37**; (b) see ox^{2-} formula and description in Table 6.2 and Fig. 6.7, (c), (d) and (e) see ligands in Table 6.2.

6.6 (a) Fig. 6.1, **14**; (c) $[NCAgCN]^-$.

6.7 (a) $[CoCl(NH_3)_5]^{2+}$; (b) $[Fe(OH_2)_6]NO_3$; (c) $[cis\text{-}RuCl_2(en)_2]$; (d) $[Cr_2(\mu\text{-}OH)(NH_3)_5]Cl_5$.

6.8 (a) *cis*-tetraammine-dichlorochromium(+1), *trans*-tetraisocyanatodiamminechromium(−1); (c) oxalato*bis*-(ethylenediamine)cobalt(+1). [In each cases oxidation state rather than charge.]

6.9 (a) *cis* and *trans*; (b) square planar with *trans* phosphanes; (c) one isomer; (d) two isomers; (e) three isomers.

6.10 *trans*-carbonylchloro*bis*-(phosphane)iridium(I); dicarbonylchloro*bis*(triphenylphosphane) iridium(I); dicarbonylhydrido*bis*(triphenylphosphane)iridium(I).

6.11 Chiral: (a) (d) (e).

6.12 Pink $[Co(NH_3)_5(OH_2)]Cl_3$; pentaammineaquacobalt(III) chloride; purple $[CoCl(NH_3)_5]Cl_2$, pentaamminechlorocobalt(III) chloride.

6.13 Violet: $[Cr(OH_2)_6]Cl_3$; green $[Cr(OH_2)_5Cl]Cl_2$; both complexes are octahedral.

6.14 *trans*-diamminediaquaplatinum(II), square planar.

6.15 $[Pt(NH_3)_4][PtCl_4]$ tetraammineplatinum(II) tetrachloroplatinate(II); $[Pt(NH_3)_4][NO_3]_2$, tetraammineplatinum(II) nitrate; $Ag_2[PtCl_4]$, silver(I) tetrachloroplatinate(II).

6.16 Both are square planar; the β isomer is *trans* and the other is *cis*.

6.17 The PX_3 ligands are equivalent within each isomer, but the chemical shifts will differ between the isomers. (^{31}P coupling will be quite different between the isomers).

6.18 (a) For ^{31}P NMR see 6.17; The ^{13}C NMR will show equivalent CO ligands for the *cis* isomer and two inequivalent sets for the *trans* isomer. (b) Only one P chem. shift and one ^{13}C shift for the axial phosphane isomer; a single chemical shift for ^{31}P in the phosphane isomer but two separate ^{13}C shifts in a 1:2 intensity ratio.

6.19 The bonding orbital consists of the two $P\sigma$ orbitals in phase with Ptd_{z^2}, and in the antibonding orbital the $P\sigma$ orbitals are out of phase with d_{z^2}.

6.20 (a) and (e) t_{2g}^6, $2.4\Delta_O$; (b) $t_{2g}^4 e_g^2$, $0.4\Delta_O$; (c) t_{2g}^5, $2\Delta_O$; (d) t_{2g}^3, $1.2\Delta_O$; (e) t_{2g}^6;(f) $e^2 t^3$, 0; (g) $e^2 t^4$, 0.

6.21 No; H^-, strong σ donor no π interaction; PPh_3 σ donor + π acceptor.

6.22 (a) 0; (b) 4.9, (c) 1.7, (d) 3.9, (e) 0; (f) 4.9, (g) 0.

6.23 Yellow, pink, and blue, respectively.

6.24 See Fig. 6.11 and associated discussion.

6.25 Perchlorate: square planar $(d_{xz}, d_{yz})^4 (d_{xy})^2 (d_{z^2})^2$ diamagnetic; thiocyanate approximately octahedral $t_{2g}^6 e_g^2$ with unpaired electron in e_g.

6.26   Distorted octahedral.
6.27   Jahn-Teller distortion in excited state.
6.28 D mechanism.
6.29   Rate of associative process depends on identity entering $\times$ and therefore, it isn't an inherent property of $[M(OH_2)_6]^{n+}$.

Chapter 7

In-chapter exercises

7.1   Above $c.$ 1750 °C.
7.2   Yes; $E^\ominus$ for reaction with Cl^- only slightly favorable (reaction is very slow).
7.3   0.85 V.
7.4   No.
7.5   No.
7.6   Oxidation by O_2 plus H_2O to H_2SO_4.
7.7   1.43 V.
7.8 Tl^+ at a minimum ($nE^\ominus = -0.34$ V), Tl^{+3} the highest point ($nE^\ominus = +2.19$ V).
7.9 Mn^{2+}
7.10   Much stronger in acid solution.
7.11   Fe(II) favored.
7.12 $Ni(en)_3^{2+}/Ni$ potential more negative (less favored).

End-of-chapter exercises

7.1   Above 1400 °C.
7.2   Higher overpotential for O_2 formation because of a more demanding mechanism.
7.3   Thermodynamically suitable in acid solution: (a) $HClO$, α-PbO_2 etc.; (b) Fe, Zn, etc. (c) Al, Fe, Zn etc.; (d) same as c.
7.4   (a) $4Cr^{2+} + O_2 + 4H^+ \rightarrow 4Cr^{3+} + 2H_2O$; (b) $4Fe^{2+} + O_2 + 4H^+ \rightarrow 4Fe^{3+} + 2H_2O$; (c) NR; (d) NR; (e) $2Zn + O_2 + 4H^+ \rightarrow 2Zn^{2+} + 2H_2O$.
7.5   (a) $4Fe^{2+} + O_2 + 4H^+ \rightarrow 4Fe^{3+} + 2H_2O$; (b) $4Ru^{2+} + O_2 + 4H^+ \rightarrow 4Ru^{3+} + 2H_2O$ and/or $3Ru^{2+} \rightarrow Ru(s) + 2Ru^{3+}$; (c) $3HClO_2 + H_2O \rightarrow ClO_3^- + 2HClO + H^+$ (followed by ClO_3^- disproportionation) $HClO_2 + O_2 \rightarrow ClO_3^-$ (slow due to O_2 overpotential). (d) NR
7.6   (a) $E = E^\ominus - (0.059/4)$ [log $p_{O_2}^{-1} + 4$ pH] (b) $E = E^\ominus - (0.059/6)$ [13.8 pH]
7.7   For $Cr_2O_4^{2-}$: $\Delta G^\ominus = +31.8$ kJ mol^{-1}, $K = 2.7 \times 10^{-6}$ For $[Cu(NH_3)_2]^+$: $\Delta G^\ominus = +9.65$ kJ mol^{-1}, $K = 2.0 \times 10^{-2}$ Despite similar $E^\ominus$, $\Delta G^\ominus$ and K differ because of differing n.

7.8   (a) Disproportionation: $Cl_2 + 2OH^- \rightarrow Cl^- + ClO^- + H_2O$; (b) very little disproportionation; (c) In acid kinetic; in base thermodynamic.
7.9   (a) $5N_2O + 2OH^- \rightarrow 2NO_3^- + 4N_2 + H_2O$; (b) $Zn + I_3^- \rightarrow Zn^{2+} + 3I^-$; (c) $3I_2 + 5HClO + 3H_2O \rightarrow 6IO_3^- + 5Cl^- + H^+$.
7.10   Added acid (lower pH) (a) disfavored; (b) favored; (c) disfavored; (d) no effect.
7.11   (a) Atom transfer; (b) outer sphere electron transfer; (c) multistep.
7.12   + 1.39 V; $2ClO_4^- + 16H^+ + 14e^- \rightarrow Cl_2 + 8H_2O$.
7.13 3.7×10^{38}
7.14   (a) Fe_2O_3; (b) Mn_2O_3; (c) HSO_4^-.
7.15 $c.$ -0.1 V.
7.16   Potential less positive.
7.17 $Fe^{2+}/Fe(OH)_2$; $Fe/Fe(OH)_3$.
7.18   If potential for reduction shifts upon addition of ligand, complex formation is indicated.
7.19 CO_2 lowers pH and favors oxidation.

Chapter 8

In-chapter exercises

8.1   (a) 2.2.2 crypt, $Rb^+ > Na^+ \approx Cs^+ > Li$; (b) EDTA, $Fe^{3+} > Cu^{2+}$.
8.2   V(V), VO_2^+.
8.3 $Re_3Cl_9(PPh_3)_3$ Re_3 triangle with bridging and terminal Cl and one PPh_3 on each Re.
8.4   Layered structure with weak van der Waals interaction between layers.
8.5   (a) $(Me)_2SAlCl_3 + GaBr_3 \rightarrow Me_2SGaBr_3 + AlCl_3$ (Ga$-$S soft/soft), (b) $2TlI_3 + CH_2O + H_2O \rightarrow 2TlCl + CO_2 + 4H^+ + 4Cl^-$.
8.6 UO_2^{2+} which contains U(VI).

End-of-chapter exercises

8.1   Check position inside cover; (a) decrease down a group; (b) increase down a group; (c) increase up a group.
8.2   (a) Mg^{2+} (larger q/r); (b) Sr; (c) K^+.
8.3   (a) CaF_2 8 coord. (cubic), MoS_2 6 coord. (trigonal prism); (b) CdI_2 layered, $MoCl_2$ contains Mo_6 clusters; (c) BeO 4:4 coord., CaO 6:6 coord.; (d) acetate-bridged Mo_2 dimer (27); central O surrounded by 4 Be^{2+} (5).
8.4   See inside cover, **C**: Sr^{3+}, Ti^{4+}, V^{5+}; **O**; Cr, Mn; **N**: Fe, Co, Ni, Cu, Zn.

8.5 d block: higher ox. states more stable, p block higher ox. states less stable; see Appendix 2.
8.6   (a) $Cr^{2+} + Fe^{3+} \rightarrow Cr^{3+} + Fe^{2+}$ ($+2$ ox. state more stable to the right of a d-block series); (b) $2CrO_4^{2-} + 2MoO_2 + 2OH^- \rightarrow Cr_2O_3 + 2$ $MoO_4^{2-} + H_2O$ (stability of highest oxidation state increases down a group); (c) $6MnO_4^- + 10Cr^{3+} + 11H_2O \rightarrow 6Mn^{2+} + 5Cr_2O_7^{2-} + 22H^+$ (maximum ox. state less stable from right to left in 3d series).
8.7   (a) Ni^{2+}; (b) for the same ox. no. metal ions become softer going toward the right of a d series; (c) $Ni^{2+} + H_2S \rightarrow NiS + 2H^+$.
8.8   (a) Check inside front cover; (b) TiF_2 through NiF_2; (c) ScY and lanthanides, Zr and HF, Nb and Ta, Mo and W, Ti and Re; one example is Mo_6Cl_{12}.
8.9 cis-$[Ru^{II}LCl(OH_2)]^+ + OH^- \rightarrow cis$-$[Ru^{III}LCl(OH)]^+ + H_2O + e^-$ (e^- might be transferred to an electrode which maintains the potential on the system). Higher ox. state promotes acidity thus $[Fe(OH_2)_6]^{3+}$ is more acidic than $[Fe(OH_2)_6]^{2+}$.
8.10   (a) NR; (b) $6MoO_4^{2-} + 10H^+ \rightarrow [Mo_6O_{19}]^{2-} + 5H_2O$; (c) $3ReCl_5 + 2MnO_4^- + 8H_2O \rightarrow 3ReO_4^- + 2MnO_2 + 16H^+$; (d) $6MoCl_2 + 2HCl + 2H_2O \rightarrow [H_3O]_2[Mo_6Cl_{14}]$.
8.11 $trans$-octahedral; cis-octahedral (d^0), $trans$-octahedral (d^2), sq. pyramid.
8.12   Ionic or borderline: NiI_2, ionic: $NbCl_4$, FeF_2; more covalent PtS; $M-M$ bonded: WCl_2.
8.13   (a) $2TiO + 6HCl \rightarrow 2Ti^{3+} + H_2 + 2H_2O + 6Cl^-$; (b) $Ce^{4+} + Fe^{2+} \rightarrow Fe^{3+} + Ce^{3+}$; (c) $2Rb_9O_2 + 14H_2O \rightarrow 18Rb^+ + 18OH^- + 5H_2$; (d) $2Na + 2CH_3OH \rightarrow 2Na(OCH_3) + H_2$.
8.14   (a) $\sigma^2\pi^4\delta^2$, B.O. $= 4$; (b) $\sigma^2\pi^4\delta^2$, B.O. $= 4$; (c) $\sigma^2\pi^4\delta^2\delta^{*2}\pi^{*2}\sigma^{*2}$, B.O. $= 0$.
8.15   (a) $Zn(NH_2)_2 + 2NH_4^+ \rightarrow Zn(NH_3)_6^{2+}$; (b) $Zn(NH_2)_2 + 2KNH_2 \rightarrow Zn(NH_2)_4 + 2K^+$.
8.16   (a) $Hg^{2+} + Cd \rightarrow Hg + Cd^{2+}$; (b) $Tl^{3+} + Ga \rightarrow Tl + Ga^{3+}$; (c) N.R.
8.17   (a) Group 13 max ox. state $3+$; ignoring element$-$element bonded systems such as $B_{10}H_{12}^{2-}$ $+3$ most stable for light elements B; Al. The $+1$ state is progressively more stable from Ga to Tl. Similarly in Group 14 ignoring element$-$element

bonded system $+4$ is most stable for lighter elements and $+2$ is most stable for Pb.

(b) (i) $Sn^{2+} + PbO_2 + 4H^+ \rightarrow$ $Sn^{4+} + Pb^{2+} + 2H_2O$ (fits); (ii) $Tl^{3+} + Al \rightarrow Al^{3+} + Tl$; (iii) $3In^+ \rightarrow In^{3+} + 3In$; (iv) $2Sn^{2+} + O_2 + 4H^+ \rightarrow 2Sn^{4+} + 2H_2O$; (v) N.R.

8.18 (i) 1.32V; (ii) 2.40V; (iii) 0.32V, (iv) 1.08V; (v) -0.02V.

8.19 (a) $2Ln(s) + 6H^+ \rightarrow 2Ln^{3+} + 3H_2$. (b) The $+3$ state is by far the most stable. (c) Cerium $(+4)$ and europium $(+2)$.

8.20 $Ce(IO_3)_4$ and $EuSO_4$ have low solubility compared with the Ln^{3+} salts of these anions.

8.21 See Fig. 8.34; ^{90}Sr and ^{144}Ce have high yields and present serious hazards.

8.22 The $4f$ orbitals in lanthanides do not have significant probability density at the ionic radii. The $5f$ orbitals are more exposed to ligands.

Chapter 9

In-chapter exercises

9.1 1301 cm^{-1}.

9.2 It may terminate the radical chain reaction.

9.3 Group 2: MgH_2, saline; Group 11: CuH, intermediate; Group 12 H_2O molecular. Of these only H_2O forms hydrogen bonds.

9.4 $Et_3SnH + CH_3Br \rightarrow Et_3SnCH_3 + HBr$.

9.5 $B_2H_6 + 2THF \rightarrow 2H_3B \cdot THF$ $H_3B \cdot THF + CH_3CH = CH_2 \rightarrow CH_3CH_2BH_2 \cdot THF$.

End-of-chapter exercises

9.1 (a) H$(+1)$, S(-2); (b) K$(+1)$, H(-1); (c) Re$(+7)$, H(1); (d) H$(+1)$, S$(+6)$, O(-2), (e) H$(+1)$, P$(+1)$, O(-2).

9.2 Industry at high temperature: $CH_4 + H_2O \rightarrow CO + 3H_2$ $CO + H_2O \rightarrow CO_2 + H_2$ $C + 2H_2O \rightarrow CO_2 + 4H_2$ Laboratory at ambient temperature: $2HCl + Zn \rightarrow ZnCl_2 + H_2$ $NaH + CH_3OH \rightarrow NaOCH_3 + H_2$.

9.3 (a) Check with Fig. 9.3; (b) check with Table 9.5; (c) Group 13, Group 14, Groups 15 through 17.

9.4 (a) Barium hydride, saline; (b) silane, electron precise; (c) ammonia electron rich, molecular; (d) arsane, same as NH_3; (e) palladium hydride, metallic hydrogen iodide, electron rich.

9.5 (a) $BaH_2 + 2H_2O \rightarrow 2H_2 + Ba(OH)_2$; (b) $HI + NH_3 \rightarrow NH_4I$; (c) $PdH_x \overset{\Delta}{\rightarrow} PdH_{x-\delta} + (\delta/2)H_2$; (d) $NH_3 + BMe_3 \rightarrow H_3NBMe_3$.

9.6 Solids: BaH_2, $PdH_{0.9}$; liquids none, gases: SiH_4, NH_3, AsH_3, HI; PdH_x electrical conductor.

9.7 H_2Se, bent, C_{2v}; P_2H_4, pyramidal around each P, C_2; H_3O^+, pyramidal, C_{3v}.

9.8 (b)

9.9 Me_3SnH, weak Sn—H bond.

9.10 (a) $H_2O < H_2S < H_2Se$; (b) $H_2Se < H_2S < H_2O$.

9.11 (i) Direct combination: $Pd + x/2$ $H_2 \rightarrow PdH_x$; (ii) protonation of a salt: $NaCl + H_2SO_4 \rightarrow HCl + NaHSO_4$; (iii) metathesis: $4BCl_3 + 3LiAlH_4 \rightarrow 2B_2H_6 + LiAlCl_4$.

9.12 (a) $CaSe + HCl(aq) \rightarrow CaCl_2 + H_2Se$; (b) $SiCl_4 + LiAlD_4 \rightarrow SiD_4 + LiAlCl_4$; (c) $2GeMe_2Cl_2 + LiAlH_4 \rightarrow 2GeMe_2H_2 + LiAlCl_4$; (d) $Si + 2HCl \rightarrow SiH_2Cl_2$ (+ other chlorosilanes), $2SiH_2Cl_2 \rightleftharpoons SiH_4 + SiCl_4$.

9.13 (i) No, $B_2H_6 + 3O_2 \rightarrow B_2O_3 + 3H_2O$, (ii) See Box 9.1.

9.14 $BH_4^- \ll AlH_4^- > GaH_4^-$; AlH_4^- strongest reducing agent; $GaH_4^- + 4HCl \rightarrow GaCl_4^- + 4H_2$.

9.15 (a) $2NaBH_4 + 2H_3PO_4 \rightarrow 2NaHPO_4 + B_2H_6$ $B_2H_6 + THF \rightarrow 2H_3BTHF$ $H_3BTHF + 3C_2H_4 \rightarrow 3BEt_3 + THF$ (b) BH_3THF (from above) $+ Et_3N \rightarrow Et_3NBH_3 + THF$ or $NaBH_4 + NH_4Br$ $\overset{THF}{\longrightarrow} H_3 + H_2 + NaBr$.

9.16 $Si + 2Cl_2 \rightarrow SiCl_2$ (distill); $SiCl_4 + 2H_2 \rightarrow Si + 4HCl$.

9.17 Period 2 hydrides are volatile except for BeH_2; Period 3 CaH_2 and AlH_3 are not volatile, the rest are volatile. Period 2 hydrides generally more thermodynamically stable (B_2H_6 vs AlH_3 is an exception).

9.18 Clathrate hydrates in which the krypton is contained in a hydrogen-bonded $(H_2O)_n$ cage.

9.19 See Fig. 9.11.

Chapter 10

In-chapter exercises

10.1 $Mg + HgMe_2 \rightarrow MgMe_2 + Hg$.

10.2 Homolytic Hg—CH_3 bond cleavage.

10.3 Bridging hydrogens and terminal Bu groups.

10.4 Lower force constant for Si—O—Si bend.

10.5 $Al_2Me_6 + 6$ MeOH $\rightarrow$ $2Al(OMe)_3 + 6CH_4$; No reaction of $SiMe_4 + MeOH$; $SiCl_4 + 4MeOH \rightarrow Si(OMe)_4 + 4HCl$.

10.6 GeH_4, GeR_4, AsH_3, AsR_3. Hydrogen and alkyl compounds are similar for each element. This may reflect similar H and C electronegativity.

End-of-chapter exercises

10.1 Organometallic: (a), (c), (d) and (g) because each contains a metal (or metalloid) − carbon bond.

10.2 Check answer with Fig. 10.1 and Table 10.2.

10.3 (a) $BiMe_3$, trimethylbismuthane; (b) $SiPh_4$, tetraphenylsilicon (IV), (c) $[AsPh_4][Br]$, tetraphenylarsenic$(1+)$ bromide; (d) $K[BPh_4]$, potassium tetraphenylboron$(1-)$.

10.4 (a) Triethylsilane (tetrahedral monomer, electron precise); (b) diphenylchloroborane (trigonal monomer, electron deficient); (c) tetraphenyldichloroaluminum (two Al − Cl − Al bridges, in this structural form it is electron precise); (d) ethyllithium or, more precisely, tetraethyltetralithium (tetrahedral Li_4 array with a phenyl carbon bridging each face, electron deficient); methylrubidium, salt-like.

10.5 (a) Li tetrahedron with each face capped by CH_3 (structure **2**); (b) planar triangular array of B and C; (c) four terminal CH_3 and two CH_3 bridges in a diborane-like structure; (d) tetrahedral; (e) pyramidal; (f) pseudotetrahedral.

10.6 The small B atom bonds to three CH_3 groups. The larger Al is a dimer with each Al bonded to four carbons. Although similar in size to Al, $GaMe_3$ does not associate into dimers. $GeMe_4$ is a simple tetrahedral molecule, $InMe_3$ is a monomer in solution.

10.7 (1): (a), (b), (c), (e); (2): (a), (c), (d), (e); (3): $AsMe_3$; (4) (a), (b), (c), (e).

10.8 (a) $2AsCl_3 + 3ZnMe_2 \rightarrow 2AsMe_3 + 3ZnCl_2$; $SiPh_2Cl_2 + ZnMe_2 \rightarrow ZnCl_2$; (b) $NH_3 + BMe_3 \rightarrow H_3NBMe_3$; (c) $AsMe_3 + [HgMe][BF_4] \rightarrow [Me_3AsHgMe][BF_4]$.

10.9 (a) $Mg + MeBr \rightarrow MeMgBr$, active metal (Me); (b) $3HgMe_2 + 2Al \rightarrow 3Hg + Al_2Me_6$, organometallic of less active metal (Me) plus a more active metal; (c) $3Li_4Me_4 + 4BCl_3 \rightarrow 4BMe + 12$ $LiCl$, organometallic of more active metal with a metal halide.

10.10 (a) $Na[C_{10}H_8]$, less delocalized (stabilized) anion; (b) $Na_2[C_{10}H_8]$, the anion is more highly reduced.

10.11 (a) $Ca + HgMe_2 \rightarrow CaMe_2 + Hg$, Ca more electropositive than Hg; (b) NR, converse of (a); (c) $Li_4Me_4 + 4SiPh_3Cl \rightarrow 4LiCl + 4SiPh_3Me$, the Me^- prefers the more electronegative element; (d) NR, converse of (c); (e) $SiMe_3H + C_2H_4 \rightarrow SiMeEt$; hydrosilation.

10.12 (a) $Si + 2MeCl \rightarrow SiMe_2Cl_2$ (elevated temp and Cu catalyst); (b) $2SiCl_2Me_2 \rightleftharpoons SiClMe_3 + SiCl_3Me$ (heat).

10.13 Reaction (a) above followed by hydrolysis: $4SiMe_2Cl + 4H_2O \rightarrow (SiMe_2O)_4 + 8HCl$, $n(SiMe_2O)_4 \rightarrow 4(-Si-O-)_n$, H_2SO_4 catalyzed.

10.14 Group 13: $+3$ predominates, some Tl(I) organometallics; Group 14: $+4$ predominates, some Sn(II) and Pb(II); Group 15: $+3$ predominates some As(II) and Sb(V).

10.15 (a) $M-C$ enthalpy decreases down a group; (b) Lewis acidity decreases from AlR_3 to TlR_3; (c) Basicity of central metal only in group 15, $:ER_3$, compounds, order of basicity: $As > Sb > Bi$.

10.16 (a) $PbMe_4 > SiMe_4$; (b) $BMe_3 > Li_4Me_4 > SiMeCl_3 >> SiMe_4$; (c) $AsMe_3 >>> SiMe_4$; $Li_4Me_4 >> HgMe_2$.

10.17 (a) Schlenk ware; (b) BMe_3 is highly volatile, vacuum line or solutions in Schlenk ware; (c) Schlenk ware; (d) open vessels; (e) open vessels.

10.18 Delocalized σ antibonding orbitals.

10.19 $3SiR_2Cl_2 + 6Na \rightarrow Si_3R_6 + 6NaCl$, $Si_3R_6 \xrightarrow{hv} R_2Si=SiR_2$, R must be bulky.

Chapter 11

In-chapter exercises

11.1 (a) $BCl_3 + 3EtOH \rightarrow B(OEt)_3 + 3HCl$, protolysis; (b) $BCl_3 + NC_5H_5 \rightarrow Cl_3BNC_5H_5$, Lewis acid−base; (c) $BBr_3 + F_3BNMe_3 \rightarrow Br_3BNMe_3 + BF_3$, Lewis acidity $BF_3 < BBr_3$.

11.2 $3NMeH_2 + 3BCl_3 \rightarrow Me_3N_3B_3Cl_3 + 6HCl$. $Me_3N_3B_3Cl_3 + 3MgMeBr \rightarrow Me_3N_3B_3Me_3 + 3MgBrCl$.

11.3 8 skeletal electron pairs.

11.4 7 skeletal framework electron pairs; consult Table 11.4 for structure.

11.5 $2[B_{10}H_{13}]^- + Al_2Me_6 \rightarrow 2[B_{10}H_{11}AlMe]^- + 4CH_4$.

11.6 $1,2\text{-}B_{10}C_2H_2 + 2LiBu \rightarrow 1,2\text{-}B_{10}C_2H_{10}Li_2 + 2C_4H_{10}$. $1,2\text{-}B_{10}C_2H_{10}Li_2 + 2SiMe_2Cl_2 \rightarrow 1,2\text{-}B_{10}C_2H_{10}(SiMe_2Cl)_2 + 2LiCl$.

11.7 (a) Electrons are added to the π bond; (b) electrons are withdrawn from the π bond.

11.8 $^{13}CO + 2MnO_2 \rightarrow {}^{13}CO_2 + Mn_2O_3$. $^{13}CO_2 + LiBMe_3H \rightarrow Li[^{13}CHO_2] + BMe_3$.

11.9 Cyclic $Si_4O_{12}{}^{n-}$ (with one bridging O between each Si and two terminal O on each Si) $4Si^{4+} \rightarrow 16+$, $12O^{2-} \rightarrow 24-$ therefore $n=8$.

11.10 Fig. 11.32 (a) has 24 vertices, therefore the total number of Si and Al is 24.

End-of-chapter exercises

11.1 Check inside front cover; (a) Group 13 metals Al through Tl, Group 14 metals Sn and Pb; (b) C through Sn in Group 14; (c) highly oxophilic: B, Al, C, Si.

11.2 Work from Fig. 11.2, two B atoms are closest to viewer in this orientation.

11.3 $B_2O_3 + 3Mg \rightarrow 2B + 3MgO$; $SiO_2 + 2C \rightarrow Si + 2CO$; $GeO_2 + H_2 \rightarrow Ge + 2H_2O$; Even though C is the cheapest reducing agent a high temperature arc furnace is used to make Si, and the Ge recovery is probably the most efficient.

11.4 $SiF_4 < BF_3 < BCl_3 < AlCl_3$; (i) $F_4SiNMe_3 + BF_3 \rightarrow F_3BNMe_3 + SiF_4$; (ii) $F_3BNMe_3 + BCl_3 \rightarrow Cl_3BNMe_3 + BF_3$; (iii) NR (reaction is controlled by hard−soft).

11.5 $BCl_3 + Hg \rightarrow B_2Cl_4 + HgCl_2$ (electric discharge); $B_2Cl_4 + C_2H_4 \rightarrow Cl_2BCH_2CH_2BCl_2$; $3Cl_2BCH_2CH_2BCl_2 + 4AsF_3 \rightarrow 3F_2BCH_2CH_2BF_2 + 4AsCl_3$.

11.6 B: C.N. = 3 or 4, C: C.N. = 3; Si: C.N. = 4; partly determined by size of the central atom: $C < B << Si$, also multiple bonding in CO and BO bonds.

11.7 B_6H_{10} *nido* boranes more stable than *arachno* (B_6H_{12}).

11.8 (a) $2B_5Hg(l) + 12O_2(g) \rightarrow 5B_2O_3(s) + 9H_2O(g)$; (b) the solid B_2O_3 would deposit inside the engine.

11.9 (a) $B_{10}H_{14}$ is *nido*; (b) 12 skeletal (framework) electron pairs.

11.10 Check against *closo*-$B_{10}C_2H_{12}$ synthesis, p. 477, followed by base cleavage and reaction with $FeCl_2$, p. 478.

11.11 (a) Hexagonal array of atoms in stacked sheets; different stacking, B over N in BN, C offset from C in graphite; (b) graphite reacts with Na and with Br_2, BN does not; (c) large HOMO−LUMO gap in BN does not provide source and sink for e^-.

11.12 (a) $3PhNH_2 + 3BCl_3 \rightarrow Ph_3N_3B_3Cl_3 + 6HCl$; (b) $Ph_3N_3B_3Cl_3 + 3NaBH_4 \rightarrow Ph_3N_3B_3H_3 + 3NaBH_3Cl$.

11.13 Check against Table 11.4, and structure **23**.

11.14 $B_2H_6 < B_5H_9 < B_{10}H_{14}$.

11.15 (a) E_g: $C > Si > Sn \approx 0$; $BN > AlP > GaAs$; (b) increase; (c) $\sigma = \sigma^\circ \exp[E_g/2kT]$, so σ for AlP with a larger E_g is more temperature sensitive.

11.16 Check against Fig. 11.28.

11.17 (a) Graphite + K(am) $\rightarrow KC_8$; (b) C heated with $Cu \rightarrow CuC_2$; (c) K(vapor) + $C_{60} \rightarrow K_8C_{60}$.

11.18 Carbon black (paint pigment); activated carbon (adsorbent), silica gel (adsorbent), amorphous silicon (photovoltaic cells).

11.19 (a) (i) B hard semiconductor, Al malleable metal; (ii) diamond harder and larger E_g than Si; (b) CO_2 linear molecule, gas at room T + P; SiO_2 hard 3-dimensional solid; (c) CX_4 are not simple Lewis acids; $SiCl_4$ mild Lewis acid; (d) BX_3 molecular for X = F, Cl, Br, I; AlX_3 various extended structures in solid, Al_2Cl_6 and Al_2Br_6 in non-coordinating solvents.

11.20 $K_2CO_3 + 2HCl \rightarrow 2KCl + H_2O + CO_2$. $Na_2SiO_4 + 2H^+ \rightarrow 2Na^+ + SiO_2 \cdot H_2O$ (silica gel).

11.21 Jadeite linear chains of $(SiO_3^{2-})_n$; kaolinite layered see Fig. 11.30. Si is 4-coordinate in both materials.

11.22 (a) 48 bridging O; (b) see Fig. 11.32.

11.23 See Fig. 11.31, 2:1 layers consisting of SiO tetrahedra, AlO_6 octahedra, SiO tetrahedra; pyrophyllite has neutral layers; muscovite has charged layers held together by K^+ ions and is more brittle.

Chapter 12

In-chapter exercises

12.1 According to the VSEPR interpretation the 3-coordinate Bi may have a lone pair on Bi.

12.2 Sulfur (Group 16)/VI should be a stronger oxidant than P (group 15) since electronegativity increases with group number.

12.3 Nucleophilic displacement reactions, see equations on p. 521.

12.4 Equation on p. 524, substituting $[Na^+][N(CH_3)_2^-]$ for $C_2F_5O^-$.

12.5 Both Br^- and Cl^- are potential catalysts.

12.6 (a) SO_3: see resonance forms for isoelectronic NO_3^- on p. 57, D_{3h}; (b) SO_3F^-: see isoelectronic XO_4^{n-} Table 2.1, p. 53; C_{3v}.

12.7 38 e^-, (19 pairs), excluding 9 radial lone pairs gives 10 skeletal e^- pairs, *closo*, consistent with **38**.

End-of-chapter exercises

12.1 Check inside front cover; (a), (b), (c), (d): Bi is a metalloid the rest are nonmetals; all but oxygen can achieve max ox. no.; Bi strong inert pair effect.

12.2 (a) Middle equation p. 508, followed by $P_4 + 5O_2 \rightarrow P_4O_{10}$, then $P_4O_{10} + 6H_2O \rightarrow 4H_3PO_4$; (b) fertilizer grade top equation p. 508; (c) high energy cost for P_4 preparation.

12.3 (a) $6Li + N_2 \rightarrow 2Li_3N$, $Li_3N + 3H_2O \rightarrow 3LiOH + NH_3$; (b) $3H_2 + N_2 \rightarrow NH_3$; (c) H_2 cheaper than Li to produce.

12.4 NCl_3 highly unstable, no NCl_5; PCl_3 and PCl_5 stable.

12.5 (a) PCl_4^+ tetrahedral; (b) PCl_4^- see-saw; (c) $AsCl_5$ trigonal bipyramid.

12.6 (a), (b) see 12.2(a); (c) $2H_3PO_4 + 3CaCl_2 \rightarrow Ca_3(PO_4)_2 + 6HCl$.

12.7 10e^-: N_2, NO^+, (CN^-); 11e^-: O_2^+, (NO); 11e^-: O_2^+, (NO); 14e^-: (O_2^{2-}); $N_2H_5^+$. Strongest oxidant is italic; strongest base in parenthesis.

12.8 (a) CO_3^{2-}; (b) CO_2^{2-}; (c) $C_2O_4^{2-}$; (d) C_2^{2-}; (e) CH_3^-.

12.9 (a) $4NH_3 + 7O_2 \rightarrow 6H_2O + 2NO_2$; $NO_2 + H_2O \rightarrow 2HNO_3 + NO$; (b)$2NO_2 + 2OH^- \rightarrow NO_2^- + NO_3^- + H_2O$; (c) $NO_3^- + 2HSO_3^- + H_2O \rightarrow NH_3OH^+ + SO_4^{2-}$, $NH_3OH^+ + OH^- \rightarrow NH_2OH + H_2O$; (d) $NaNH_2 + N_2O \rightarrow NaN_3 + NaOH + NH_3$.

12.10 $P_4(s) + 5O_2(g) \rightarrow P_4O_{10}(s)$; P_4 is not the most stable allotrope.

12.11 Check against Fig. 12.5; Bi(V) is readily reduced; the Bi(III),

P(III) and P(V) oxo species are much less prone to redox.

12.12 Faster: the good electrophile NO^+ is formed and the mechanisms generally proceed through attack on NO^+ by nucleophiles.

12.13 The rate law contains a $p(NO)^3$ term. Thus at low partial pressure NO is not rapidly oxidized.

12.14 (a) $PCl_5 + H_2O \rightarrow POCl_3 + 2HCl$; (b) $PCl_5 + 8H_2O \rightarrow 2H_3PO_4 + 10HCl$; (c) $PCl_5 + AlCl_3 \rightarrow [PCl_4][AlCl_4]$; (d) $3PCl_5 + 3NH_4Cl \rightarrow N_3P_3Cl_6 + 12HCl$.

12.15 $H_3PO_2 + Cu^{2+} + H_2O \rightarrow H_3PO_4 + Cu + 3H^+$, $E^\ominus = +0.839V$, H_3PO_2 is a useful reducing agent.

12.16 (a) $+1.068$ V; (b) Cr^{2+} is not a good catalyst for H_2O_2 disproportionation.

12.17 (a) en; (b) en; (c) possibly SO_2; basic solvents are compatible with highly reducing anions. Cd_2^{2+} would disproportionate with en to form a Ca^{2+} complex with en and Cd.

12.18 most reducing SO_3^{2-}, SO_4^{2-}, $S_2O_8^{2-}$ most oxidizing.

12.19 (a) $Te(OH)_6$; SO_4^{2-}, (b) The large Te(VII) accommodates more ligands and is less acidic; (c) $PO(OH)_3$ and $Sb(OH)_5$ show the same trend in Group 15.

12.20 (a) Square; (b) possibly P_4S_3 structure, **41**

Chapter 13

In-chapter exercises

13.1 Both SO_2 and Sn^{2+} are thermodynamically suitable, but the oxidation products of the former are easier to separate from the product and S is much cheaper than Sn.

13.2 Pyramidal, C_s.

13.3 I_3^-, IBr_2^-, Br_3^-, central atom hypervalent.

13.4 In acid solution: F_2 and Cl_2. F_2 reaction is fast.

13.5 $2XeO_3^{2-} \rightarrow XeO_6^{4-} + Xe + O_2$, $2XeO_3^{2-} + 2H_2O \rightarrow 2Xe + 4OH^- + 3O_2$, two independent reactions (disproportionation of xenate and oxidation of water); their ratio will depend on conditions.

End-of-chapter exercises

13.1 Check with Table 13.1, F_2 light green-yellow, Cl_2 green-yellow, Br_2 brown, I_2 purple in gas or

dilute hydrocarbon solution; Noble gases are colorless.

13.2 Br_2 or I_2 by oxidation of halide solution with Cl_2; Cl_2 aqueous electrolysis, F_2 nonaqueous electrolysis. See p. 544–546.

13.3 Check with Fig. 13.3; $Cl_2 + OH^- \rightarrow HClO + Cl^-$.

13.4 $\bigcirc\!\!\times\!\!\bigcirc\ \ \bigcirc\!\!\times\!\!\bigcirc$, vacant lobes on each end accept electrons from donor molecules.

13.5 (a) NH_3 is hydrogen bonded (b) high electronegativity of F decreases basicity of N.

13.6 (a) $NCCN + 2OH^- \rightarrow NCO^- + CN^- + H_2O$; (b) $2SCN^- + MnO_2 + H^+ \rightarrow Mn^{2+} + (SCN)_2 + 2H_2O$; (c) C_{3v}.

13.7 (a) Octahedral, pentagonal bipyramid; (b) $IF_7 + SbF_5 \rightarrow [IF_5][SbF_6]$.

13.8 (a) Stronger, (b) no change, (c) weaker.

13.9 (a) No, Sb in maximum oxidation state; (b) explosive CH_3OH is a reducing agent, BrF_5 a strong oxidizing agent; (c) BrF_3 can be oxidized, but probably not explosively; (d) explosive, S_2Cl_2 easily oxidized.

13.10 $Br_3^- + I_2 \rightarrow 2IBr + Br^-$.

13.11 Large cations stabilize large unstable anions.

13.12 ClO_2 angular radical; I_2O_5 has oxo bridge between two IO_2.

13.13 $HBrO_4$ is more acidic but less stable than H_5IO_6.

13.14 (a) Reduction potential increases with decreasing pH; (b) at $E^\ominus$ (pH 7) = 0.787V vs. $E^\ominus$ (pH = 0) = $+1.201$V.

13.15 Reduction of the oxoanion is more favorable because protons are consumed in that half reaction.

13.16 Periodic I—O bonds are more labile.

13.17 (a) ClO_2^-; (b) ClO^-.

13.18 (a) and (d) because they have reducible cations associated with a strongly oxidizing anion, ClO_4^-.

13.19 He has sufficient speed to escape from the atmosphere.

13.20 (a) He; (b) Xe; (c) Ar.

13.21 (a) $Xe + F_2 \rightarrow XeF_2$ sunlight; (b) $Xe + 3F_2 \rightarrow XeF_6$ high pressure of F_2, 300 °C; (c) $XeF_6 + 3H_2O \rightarrow XeO_3 + 6HF$.

13.22 (a) XeF_4 (square planar); (b) XeF_2 linear; (c) XeO_3 pyramidal; (d) IF^+.

13.23 (a) All three share a single electron pair and have an octet around each atom. Formal charges: Cl(0) O(-1); Br(0); Xe($+1$), F(0); (b) They are isolobal; (c) All three display some electrophilicity in the order $ClO^- < Br_2 << XeF^+$. Positive formal charge on XeF^+ correlates with its Lewis acidity and electrophilicity.

13.24 (a) Eight electron pairs around Xe one is a lone pair. (b) The configuration of these pairs may lead to a square antiprism (**19** p. 235) or dodecahedron (**20** p235) with one unoccupied vertex. If the lone pair is not stereochemically active typical 7-coordinate structures may result: a pentagonal bipyramid (**15** p. 235) or capped octahedron (**16** p. 235). The latter two structures seem most probable.

Chapter 14

In-chapter exercises

14.1 ^3D or ^1D.
14.2 (a) ^3P; (b) ^2D.
14.3 $^3T_{1g}$, $^3T_{2g}$, $^3A_{2g}$ and $^1T_{2g}$, ^1E_g.
14.4 18,400 cm^{-1}, 27,500 cm^{-1}.
14.5 $^4A_{2g}$ to ^2E_g, $^4T_{1g}$, $^4T_{2g}$, ligand.
14.6 An extra band.
14.7 Class 1.

End-of-chapter exercises

14.1 (a) ^6S; (b) ^4F; (c) ^2D; (d) ^3P.
14.2 (a) ^3F; (b) ^5D; (c) ^6S.
14.3 (a) ^2S, (b) ^3P (ground), ^1D, ^1S.
14.4 $3p^5$: F; $4d^{10}$: Cd^{2+}.
14.5 B^+: ^2P; Na: ^2S; Tl^{2+}: ^2S; Ag^+: ^1S.
14.6 $B = 861$ cm^{-1}, $C = 3168$ cm^{-1}.
14.7 d^6, $^1A_{1g}$; d^1, $^2T_{2g}$; d^5, $^6A_{1g}$.
14.8 From $T-S$ diagram: (a) $B \approx 770$ cm^{-1}, $\Delta_0 \approx 8500$ cm^{-1}; (b) $B \approx 720$ cm^{-1}, $\Delta_0 \approx 10750$ cm^{-1}.
14.9 $^5T_{2g}$, charge transfer, LMCT.
14.10 Weak: spin forbidden LF; medium: spin allowed LF transitions; strong: CT transition.
14.11 $[FeF_3]^{3-}$ has only one spin-forbidden LF band; $[CoF_6]^{3-}$ has $^5E_g \leftarrow {}^5T_{2g}$ which is spin allowed.
14.12 CN^- is more covalent.
14.13 A spin-forbidden LF band, two spin-allowed LF bands, and a LMCT band respectively.
14.14 High spin d^5 has only spin-forbidden LF bands.
14.15 $[Cr(OH_2)_6]^{3+}$: LF bands which are weak; CrO_4^{2-}: allowed LMCT bands.
14.16 a_{1g}; (a) d_{xz}, d_{yz}; (b) d_{z^2}; (c) lowest xy; (xz, yz), z^2, x^2-y^2 highest energy.
14.17 Mn(VII) is d^0 and has no LF bands.

14.18 The two LMCT bands are to t_{2g} and e_g. The difference is Δ_T.
14.19 Metal−metal CT.

Chapter 15

In-chapter exercises

15.1 5.1 L mol^{-1} s^{-1}.
15.2
$$[PtCl_4]^{2-} \xrightarrow{PH_3} [PtCl_3PPH_3]^-$$
$$\xrightarrow{NH_3} trans\text{-}[PtCl_2NH_3PPH_3];$$
$$[PtCl_4]^{2-} \xrightarrow{NH_3} [PtCl_3NH_3]^-$$
$$\xrightarrow{PPh_3} cis\text{-}[PtCl_2NH_3PPh_3].$$
15.3 $K_E \approx 1$, $k = 1.2 \times 10^2$ s^{-1}.
15.4 2 *cis* to 1 *trans*.
15.5 $K = 3 \times 10^{18}$, $k = 3.6 \times 10^{10}$ L mol^{-1} s^{-1}.

End-of-chapter exercises

15.1 Nucleophiles; NH_3, Cl^-, S^{2-}; electrophiles; Ag^+, Al^{3+}.
15.2 d.
15.3 Maximum after intermediate.
15.4 (a) Stoichiometric; (b) intimate.
15.5 Rate $= k[Mn(OH_2)_6^{2+}]$ $[X^-]/(1 + K_a[X^-])$, vary X^-.
15.6 Stronger covalent bonds to be broken for higher ox. states and heavier metals.
15.7 Steric hindrance to nucleophilic attack.
15.8 The rate controlling step may be dissociative. The influence of phosphane is consistent with this conclusion.
15.9 Yes, a favorable A intermediate.
15.10
$$[PtCl_4]^{2-} \xrightarrow{NH_3} [PtCl_3NH_3]^-$$
$$\xrightarrow{NO_2^-} cis\text{-}[PtCl_2(NO_2)(NH_3)],$$
$$[PtCl_4]^{2-} \xrightarrow{NO_2^-} [PtCl_3NO_2]^- \xrightarrow{NH_3}$$
$$trans\text{-}[PtCl_2(NO_2)(NH_3)].$$
15.11 (a) Decrease rate; (b decrease rate; (c) decrease rate; (d) increase rate.
15.12 Conjugate base path; Brønsted acidity is required.
15.13 (a) cis-$[PtCl_2(PR_3)_2]$; (b) $trans$-$[PtCl_2(PR_3)_2]$; (c) $trans$-$[PtCl_2(py)_2]$.
15.14 $[Ir(NH_3)_6]^{3+} < [Rh(NH_3)_6]^{3+} < [Co(NH_3)_6]^{3+} < [Ni(OH_2)_6]^{2+} < [Mn(OH_2)_6]^{2+}$.
15.15 (a) Decreases; (b) decreases; (c) little effect; (d) decreases; (e) increases.
15.16 The rate constant for the initial CO insertion step.
15.17 Dissociative.
15.18 Inner sphere goes through N_3^- bridged complex. Outer sphere has no ligand bridge.

15.19 Inner-sphere mechanism.
15.20 $[W(CO)_5(NEt_3)]$ 0.4; ligand field.
15.21 250 nm.
15.22 Rate determining step is SMe_2 dissociation.

Chapter 16

In-chapter exercises

16.1 No (20 e^-).
16.2 $+1$.
16.3 Bridging only.
16.4 $P(CH_3)_3$, less steric repulsion.
16.5 Prepare: $[Mn(CO)_5]^-$ with Na (p. 677 top), then $Mn(CO)_5CH_3$ with CH_3I (p. 679), migratory insertion with L = PPh_3 (p. 677 bottom).
16.6 Small change.
16.7 CVE = 60; tetrahedral Fe_4 with Cp on each vertex, CO on each face.

End-of-chapter exercises

16.1 (a) Pentacarbonyliron(0);
      (b) tetracarbonylnickel(0);
      (c) hexacarbonylmolybdenum(0);
      (d) decacarbonyldimanganese(0);
      (e) hexcarbonylvanadium(0);
      (f) trichloroethyleneplatinate(II).
16.2 (a)

      (b)

16.3 (a) $+3$; (b) 0; (c) -2; (d) 0.
16.4 (a) to (d) 18e^- in both 16.1 and 16.3; $V(CO)_6$ 17e^-, no influence on structure, easily reduced; $[PtCl_3(C_2H_4)]^-$ 16e^-, square planar; $[(\eta^5\text{-}C_5H_5)_2Fe][BF_4]$, 17$e^-$ slightly longer Fe−C distances; easily reduced.
16.5 Direct combination, reductive carbonylation.
16.6 $$Fe + 5CO \xrightarrow{\text{high pressure}} Fe(CO)_5,$$
      $diphos + Fe(CO)_5 \rightarrow Fe(CO)_3$-$(diphos) + 2CO$.
16.7 C_s (3); C_{3v} (2); D_{3h} (1).
16.8 Ni_3 triangle each Ni capped by Cp, CO above and below nickel triangle and collinear; 18 1/3 e^- per Ni atom, deviations from $18-e^-$ rule are common in Group 10.
16.9 (b) Reacts by an associative process.
16.10 (a) $[Fe(CO)_4]^{2-}$ more negative, complex is more basic; (b) $[Re(CO)_5]^-$ heavier metal, more basic.

16.11

Products: (a) $Li[W(CO)_5(-\overset{\overset{\textstyle O}{\|}}{\underset{\underset{\textstyle CH_3}{\backslash}}{C}})]$,

(b) $AlBr_3$ coordinated to bridging CO ligands.

16.12 (a) η^2; (b) η^1, η^3, η^5; (c) η^2, η^4, η^6; in practice the last is observed; (d) η^2, η^4; (e) $\eta^2, \eta^4, \eta^6, \eta^8$.

16.13 (a) 16e sandwich; (b) 18e; (c) 18e.

16.14 (a) Fe, Ru, Os; (b Mn, Co; (c) Cr, Fe, Ni; (d) Table 16.3.

16.15 (a) 3; (b) 2; (c) 12.

16.16 See p. 682, 683, 687, and 688.

16.17 Prepare $Mn(CO)_5^-$ (p. 677), protonate with HCl.

16.18

$Mo(CO)_5C\overset{\overset{\textstyle OCH_3}{/}}{\underset{\underset{\textstyle Ph}{\backslash}}{}}$

16.19 (a) PF_3 π-acceptor ligand (increases $v(CO)$), $P(CH_3)_3$ σ donor ligand (decreases $v(CO)$); (b) Cp* stronger donor, and therefore lower $v(CO)$.

16.20 $RuCp_2$ (18e) more stable than $RhCp_2$ (19e); $(\eta^5\text{-}C_5H_5)_2Fe + CH_3C(O)Cl \xrightarrow{AlCl_3} (\eta^5\text{-}C_5H_5)(\eta^5\text{-}C_5H_4COCH_3)Fe + HCl$.

16.21 $(\eta^5\text{-}C_5H_5)_2Fe + LiBu \rightarrow (\eta^5\text{-}C_5H_5)(\eta^5\text{-}C_5H_4Li)Fe + BuH$, $(\eta^5\text{-}C_5H_5)\text{-}(\eta^5\text{-}C_5H_4Li)Fe + ClC(O)CH_3 \rightarrow (\eta^5\text{-}C_5H_5)(\eta^5\text{-}C_5H_4COOH)Fe + LiCl$.

16.22 See Appendix 6.

16.23 C_5H_6 is a tetrahapto four-electron ligand. Its formation reduces the metal electron count to 18 on Ni; protonation of Fe in $FeCp_2$ leaves the 18e count unchanged.

16.24 (a) Attack of H_2O on the CF_2 carbon followed by elimination of HF; (b) β-H elimination.

16.25 (a) Groups $6-9$; stable Cp compounds, Groups 3 and 4; C—H insertion; (b) Groups 6 to 8; MH more acidic than in Groups 3 and 4. (c) 18e rule violation for many Group 3 and 4 and some Group 9 metals.

16.26 (a) Octahedral (86), trigonal prismatic, CVE = 90; (b) no; (c) $[Fe_6C(CO)_{16}]^{2-}$, CVE = 86, octahedral; $[Co_6C(CO)_{16}]^{2-}$, CVE = 92, trigonal prismatic.

16.27 (a) $Si(CH_3)$; (b) I.

16.28 $Co_4(CO)_{12}$ has weaker MM bonds.

Chapter 17

In-chapter exercises

17.1 Oxidative addition: $Rh^{(I)}ClL_2(Sol) + H_2 \rightarrow Rh^{(III)}ClH_2L_2(Sol)$; reductive elimination: $Rh^{(III)}ClH(C_2H_4R)L_2(Sol) \rightarrow Rh^{(I)}ClL_2(Sol) + C_2H_5R$.

17.2 Decrease rate.

17.3 Single $1465\ cm^{-1}$ band expected.

17.4 No, it lacks acidic sites.

End-of-chapter exercises

17.1 (a) Catalytic; (b) noncatalytic; (c) noncatalytic.

17.2 See (a) p. 711; (b) p. 714; (c) p. 709; (d) p. 711; (e) 731.

17.3 (a) Homogeneous; (b) heterogeneous; (c) homogeneous.

17.4 Accomplished with existing technology: (d); thermodynamically prohibited: (a) and (b); need research: (c).

17.5 Added PPh_3 suppresses equilibrium concentration of $RhCiL_2(Sol)$.

17.6 Steric effects reduce equilibrium formation of (c) in Fig. 17.3.

17.7 Formation of (a) or (c) may be rate-limiting with added phosphane.

17.8 Attack of coordinated OH^- gives one enantiomer, attack of solution OH^- gives the other.

17.9 Coordination of formic acid with loss of L, reductive elimination of H_2, ox. addn. HCO_2H, and CO_2 elim.

17.10 Titanium nitride is too stable.

17.11 (a) An extra proton is required to charge-compensate Al^{3+} substituted for Si^{4+}. (b) Ga(III), Sc(III), and Fe(III).

17.12 (a) 900 °C treatment Lewis acid; (b) silica gel more Brønsted acidic; (c) ZSM-5 is highly regular, silica gel is not.

17.13 To achieve high surface area.

17.14 See p.740.

17.15 Initial chemisorption as $C^*H_2C^*HC(CH_3)_2CH_2CH_3$, exchange of H for D at surface attached carbon atoms (*), dissociation (via reductive elimination).

17.16 CO strongly chemisorbs and blocks sites for H_2.

Chapter 18

In-chapter exercises

18.1 Vacancies on both anion and cation sites.

18.2 Reduction in thickness of conduction plane impedes the larger cation more.

18.3 $Cr^{(III)}$ in octahedral holes provides maximum LFSE.

18.4 Semiconduction.

End-of-chapter exercises

18.1 Frenkel, interstitial, Schottky, holes; no.

18.2 (a) NiO; (b) PbF_2; (c) Fe_2O_3.

18.3 Intrinsic in pure substance, extrinsic (due to impurities).

18.4 Fig. 18.19; Na^+ in center of cell.

18.5 Both. It is a defect when disordered, and describes a new phase when ordered.

18.6 See Fig. 18.13 for interstitial sites; triangular oxygen arrays.

18.7 TiO vacancies on both Ti and O sites; FeO; Fe^{3+} interstitials with vacancies on Fe^{2+} sites.

18.8 Electron microscopy, electron diffraction.

18.9 For example, from 27 to 30 percent O at 1300 °C; Fe_3O_4 ss (solid solution); Fe_3O_4 ss plus Fe_2O ss; Fe_2O_3 ss.

18.10 (a) NiO (MnO is much more easily oxidized than the aqueous potentials suggest); (b) NiO.

18.11 TiO metallic, NiO semiconductor.

18.12 Positive Ag electrode: $Ag(s) \rightarrow Ag^+$ (in AgI) $+ e^-$ (external circuit); AgI electrolyte: Ag^+ to $-$electrode; negative Ag electrode: Ag^+(in AgI) $+ e^- \rightarrow Ag(s)$.

18.13 Positive electrode mass loss 108 mg, AgI no change, negative electrode mass gain 108 mg.

18.14 Ferromagnet: domains of parallel spins on atoms; antiferromagnet: domains of antiparallel spins on atoms.

18.15 In the inverse spinel structure, Fe[CoFe] with Co^{2+} and Fe^{3+}; three unpaired electrons are expected from the d^7 octahedral Co^{2+}. The structure appears to be close to the inverse spinel formulation.

18.16 See p.768.

Chapter 19

In-chapter exercises

19.1 Hydrogen bonding molecules will partition into the H_2O molecules in the Na^+ channels.

19.2 1.24 Å is just slightly above O_2, suggesting little electron transfer; 1.26 Å is just less than O_2^- suggesting nearly one electron transferred. 1.47 Å is near O_2^- suggesting transfer of mainly one electron from each Co, and 1.30 Å is just over O_2^- suggesting transfer of just over one electron between the two Co centers.

19.3

19.4 The average oxidation number of Fe in the latter is 2.5. The former is 3. Reduction and disulfide formation can convert the Fe_2 clusters to Fe_4 clusters.

19.5 First and sixth entries have same $E^\ominus$ but change of R by 2 Å correlates with greater than 10^2 rate change. The comparison of third and eighth entries is similar. Fourth and ninth show doubling R increases rate by a factor of 10. All imply qualitative agreement with Marcus theory.

19.6 C_{4v} has the order of d orbitals $xz, yz < xy < z^2 < x^2 - y^2$. The configuration is $(xz)^1(yz)^1(xy)^1(z^2)^1(x^2-y^2)^1$.

End-of-chapter exercises

19.1 Neutral O_2 with net double bond, triplet; O_2^- longer; 1 1/2 order bond, doublet; O_2^{2-} longer single bond, singlet.

19.2 Left for HbO_2; right for Mb at low $p(O_2)$, left at high $p(O_2)$.

19.3 Binding to hemoglobin blocking O_2.

19.4 Nucleophile is N end of peptide, site of attack is coordinated carbonyl.

19.5 LF spectrum, Zn is d^{10}. Assume same geometry.

19.6 (a) High spin $t_{2g}^4 e_g^2$, low spin t_{2g}^6; (b) high H_2O, low $(C_6H_5)_3P$:.

19.7 Variable oxidation number.

19.8 The two oxidation states are not configured to conform to each other. They bind substrates.

19.9 Specificity in binding to allow close approach, control of inner sphere reorganization energies.

19.10 Consult text.

19.11 $Fe-O-O-Fe$ bridge formation.

19.12 (a) $ZnOH_2$ transfers a proton to promote $-OH$ attack; (b) $Zn-O=C$ promotes nucleophilic attack on C.

19.13 $+2.5$.

19.14 O_2 evolution requires $4e^-$, over four Mn atoms, Mn charge changes by -1.

19.15 Chlorophyll$-$porphyrin π^*, cytochrome Fe(II). Mg could not be used; Mg(III) inaccessible.

Copyright acknowledgements

Figure B3.1 from A. K. Cheetham and P. Day (eds.) *Solid state chemistry techniques*, Oxford University Press (1987); Figure B3.2 from J. M. Manoli, C. Potash, J. M. Begeaut, and W. P. Griffith. *J. Chem. Soc., Dalton*, 192 (1980), The Royal Society of Chemistry; Figure 10.7 from M. J. Fink, M. J. Michalczyk, K. J. Haller, R. West, and J. Michl, *Organometallics*, **3**, 793 (1984), American Chemical Society; Figure 8.18 and Figure 8.19 from J. D. Corbett, *Acc. Chem. Res.*, **14**, 239 (1981), American Chemical Society; Figure 17.14 and Figure 17.15 from G. A. Somorjai, *Chemistry in two dimensions*, Cornell University Press, Ithaca, N. Y.; Figure 18.4 from S. Iijima, *J. Solid State Chem.*, **14**, 52 (1975), Academic Press; Figure 18.37 and Figure 18.38 from T. Hubanks and R. Hoffman, *J. Am. Chem. Soc.*, **105**, 1152 (1983), American Chemical Society; Figure 3.33 from R. Layton, D. W. Sink, and J. R. Durig, *J. Inorg. Nucl. Chem.*, **28**, 1965 (1966); Figure 6.20 from B. R. Higgenson, D. R. Loyd, P. Burroughs, D. M. Gibson, and A. F. Orchard, *J. Chem. Soc. Faraday Trans. II*, **69**, 1659 (1973); Figure 8.9 from S. B. Dawes, D. L. Ward, R. H. Huang, and J. L. Dye, *J. Am. Chem. Soc.*, **108**, 3534 (1986); Figure 8.30 from R. E. Sievers (ed.) *Nuclear magnetic resonance shift reagents*, Academic Press, N. Y. (1973); Figure 8.32 from J. J. Katz, G. T. Seaborg, and L. Moss, *Chemistry of actinide elements*, Chapman and Hall (1986); Figure 9.10 from N. B. Colthup, L. H. Daly, and S. E. Wiberley, *Introduction to infrared and Raman spectroscopy*, Academic Press (1975); Figure B10.1 Lab Conco, Kansas City, MO; Figure 10.7 from Fink, M. J. Michalczyk, K. J. Haller, R. West, and J. Michl, *Organometallics*, **3**, 793 (1984), American Chemical Society; Figure 10.8 from R. West, *Angew. Chemie Intl. Ed. Engl.*, **26**, 1201 (1983); Figure 13.7 from J. Wiley, *The structures of the elements*, N. Y. 1974; Figure 14.15 from A. H. Maki and B. R. McGarvey, *J. Chem. Phys.*, **29**, 35 (1958); Figure 18.23 from J. Etourneau, *Superconducting materials*, Chapter 3 in *Solid state chemistry: compounds*, (ed. A. K. Cheetham and P. Day), Oxford University Press (1992)).

Formula Index

D: Discussion (e.g. properties, reactions, spectra), E: electronic structure, L: Lewis structure, P: preparation, S: Structure.

Subject Index

T signifies tabular matter. Bold type signifies key entries.

Useful relations

At 298.15 K, $RT = 2.4790$ kJ mol^{-1} and $RT/F = 25.693$ mV

1 atm = 101.325 kPa = 760 Torr (exactly)

1 bar = 10^5 Pa

1 eV = $1.602\ 18 \times 10^{-19}$ J = 96.485 kJ mol^{-1} = 8065.5 cm^{-1}

1 cm^{-1} = 1.986×10^{-23} J = 11.96 J mol^{-1} = 0.1240 meV

1 cal = 4.184 J (exactly)

1 D (debye) = $3.335\ 64 \times 10^{-30}$ C m

1 T = 10^4 G

1 Å (angstrom) = 100 pm

1 M = 1 mol dm^{-3}

General data and fundamental constants

Quantity	Symbol	Value
Speed of light	c	$2.997\ 925 \times 10^8$ m s^{-1}
Elementary charge	e	$1.602\ 177 \times 10^{-19}$ C
Faraday constant	$F = eN_A$	9.6485×10^4 C mol^{-1}
Boltzmann constant	k	$1.380\ 66 \times 10^{-23}$ J K^{-1}
		8.6174×10^{-5} eV K^{-1}
Gas constant	$R = kN_A$	$8.314\ 51$ J K^{-1} mol^{-1}
		$8.205\ 78 \times 10^{-2}$ dm^3 atm K^{-1} mol^{-1}
Planck constant	h	$6.626\ 08 \times 10^{-34}$ J s
	$\hbar = h/2\pi$	$1.054\ 57 \times 10^{-34}$ J s
Avogadro constant	N_A	$6.022\ 14 \times 10^{23}$ mol^{-1}
Atomic mass unit	u	$1.660\ 54 \times 10^{-27}$ kg
Mass of electron	m_e	$9.109\ 39 \times 10^{-31}$ kg
Vacuum permittivity	ε_0	$8.854\ 19 \times 10^{-12}$ J^{-1} C^2 m^{-1}
	$4\pi\varepsilon_0$	$1.112\ 65 \times 10^{-10}$ J^{-1} C^2 m^{-1}
Bohr magneton	$\mu_B = eh/2m_e$	$9.274\ 02 \times 10^{-24}$ J T^{-1}
Bohr radius	$a_0 = 4\pi\varepsilon_0 h^2/m_e e^2$	$5.291\ 77 \times 10^{-11}$ m
Rydberg constant	$R_\infty = m_e e^4/8h^3 c\varepsilon_0^2$	$1.097\ 37 \times 10^5$ cm^{-1}

Prefixes

f	p	n	μ	m	c	d	k	M	G
femto	pico	nano	micro	milli	centi	deci	kilo	mega	giga
10^{-15}	10^{-12}	10^{-9}	10^{-6}	10^{-3}	10^{-2}	10^{-1}	10^3	10^6	10^9